Lecture Notes in Computer Science 2461
Edited by G. Goos, J. Hartmanis, and J. van Leeuwen

Springer
Berlin
Heidelberg
New York
Hong Kong
London
Milan
Paris
Tokyo

Rolf Möhring   Rajeev Raman (Eds.)

Algorithms – ESA 2002

10th Annual European Symposium
Rome, Italy, September 17-21, 2002
Proceedings

 Springer

Series Editors

Gerhard Goos, Karlsruhe University, Germany
Juris Hartmanis, Cornell University, NY, USA
Jan van Leeuwen, Utrecht University, The Netherlands

Volume Editors

Rolf Möhring
Sekr. MA 6-1, Institut für Mathematik
Fakultät II: Mathematik und Naturwissenschaften
Technische Universität Berlin
Strasse des 17. Juni 136, 10623 Berlin, Germany
E-mail: moehring@math.tu-berlin.de

Rajeev Raman
Department of Mathematics and Computer Science
University of Leicester, University Road, Leicester LE1 7RH, UK
E-mail: r.raman@mcs.le.ac.uk

Cataloging-in-Publication Data applied for

Die Deutsche Bibliothek - CIP-Einheitsaufnahme

Algorithms : 10th annual European symposium ; proceedings / ESA 2002,
Rome, Italy, September 17 - 21, 2002. Rolf Möhring ; Rajeev Raman (ed.). -
Berlin ; Heidelberg ; New York ; Hong Kong ; London ; Milan ; Paris ;
Tokyo : Springer, 2002
 (Lecture notes in computer science ; Vol. 2461)
 ISBN 3-540-44180-8

CR Subject Classification (1998): F.2, G.1-2, E.1, F.1.3, I.3.5, C.2.4, E.5

ISSN 0302-9743
ISBN 3-540-44180-8 Springer-Verlag Berlin Heidelberg New York

This work is subject to copyright. All rights are reserved, whether the whole or part of the material is concerned, specifically the rights of translation, reprinting, re-use of illustrations, recitation, broadcasting, reproduction on microfilms or in any other way, and storage in data banks. Duplication of this publication or parts thereof is permitted only under the provisions of the German Copyright Law of September 9, 1965, in its current version, and permission for use must always be obtained from Springer-Verlag. Violations are liable for prosecution under the German Copyright Law.

Springer-Verlag Berlin Heidelberg New York
a member of BertelsmannSpringer Science+Business Media GmbH

http://www.springer.de

© Springer-Verlag Berlin Heidelberg 2002
Printed in Germany

Typesetting: Camera-ready by author, data conversion by DA-TeX Gerd Blumenstein
Printed on acid-free paper   SPIN: 10871217   06/3142   5 4 3 2 1 0

Preface

This volume contains the 74 contributed papers and abstracts of 4 of the 5 invited talks presented at the 10th Annual European Symposium on Algorithms (ESA 2002), held at the University of Rome "La Sapienza", Rome, Italy, 17-21 September, 2002.

For the first time, ESA had two tracks, with separate program committees, which dealt respectively with:

- the design and mathematical analysis of algorithms (the "Design and Analysis" track);
- real-world applications, engineering and experimental analysis of algorithms (the "Engineering and Applications" track).

Previous ESAs were held in Bad Honnef, Germany (1993); Utrecht, The Netherlands (1994); Corfu, Greece (1995); Barcelona, Spain (1996); Graz, Austria (1997); Venice, Italy (1998); Prague, Czech Republic (1999); Saarbrücken, Germany (2000), and Århus, Denmark (2001). The predecessor to the Engineering and Applications track of ESA was the Annual Workshop on Algorithm Engineering (WAE). Previous WAEs were held in Venice, Italy (1997), Saarbrücken, Germany (1998), London, UK (1999), Saarbrücken, Germany (2000), and Århus, Denmark (2001).

The proceedings of the previous ESAs were published as Springer LNCS volumes 726, 855, 979, 1284, 1461, 1643, 1879, and 2161. The proceedings of WAEs from 1999 onwards were published as Springer LNCS volumes 1668, 1982, and 2161.

Papers were solicited in all areas of algorithmic research, including but not limited to: Computational Biology, Computational Finance, Computational Geometry, Databases and Information Retrieval, External-Memory Algorithms, Graph and Network Algorithms, Graph Drawing, Machine Learning, Network Design, On-line Algorithms, Parallel and Distributed Computing, Pattern Matching and Data Compression, Quantum Computing, Randomized Algorithms, and Symbolic Computation. The algorithms could be sequential, distributed, or parallel. Submissions were strongly encouraged in the areas of mathematical programming and operations research, including: Approximation Algorithms, Branch-and-Cut Algorithms, Combinatorial Optimization, Integer Programming, Network Optimization, Polyhedral Combinatorics, and Semidefinite Programming.

Each extended abstract was submitted to exactly one of the two tracks, and some abstracts were switched from one track to the other at the discretion of the program chairs, before the reviewing process. The extended abstracts were read by at least three referees each, and evaluated on their quality, originality, and relevance to the symposium. The program committees of both tracks met at TU Berlin on May 24 and 25. Of 149 abstracts submitted for the Design and

Analysis track, 50 were selected for presentation. Of 52 abstracts submitted for the Engineering and Applications track, 24 were selected for presentation.

The members of the program committees of the two tracks were:

Design and Analysis Track		Engineering and Applications Track	
Susanne Albers	(Freiburg)	Karen Aardal	(Utrecht)
Stephen Alstrup	(ITU, København)	Camil Demetrescu	(Roma)
János Csirik	(Szeged)	Olivier Devillers	(INRIA)
Thomas Erlebach	(ETH, Zürich)	Thomas Liebling	(EPF, Lausanne)
Sándor Fekete	(Braunschweig)	Michael Mitzenmacher	(Harvard)
Lisa Fleischer	(Carnegie-Mellon)	David Mount	(Maryland)
Kazuo Iwama	(Kyoto)	Matthias Müller-Hannemann	(Bonn)
Alberto Marchetti-Spaccamela	(Roma)	S. Muthukrishnan	(Rutgers)
Rolf Möhring	(TU Berlin, chair)	Petra Mutzel	(TU Wien)
Günter Rote	(FU Berlin)	Rajeev Raman	(Leicester, chair)
Andreas Schulz	(MIT)	Peter Sanders	(MPI Informatik)
Jiří Sgall	(CAS, Praha)		
Christos Zaroliagis	(Patras)		
Uri Zwick	(Tel Aviv)		

ESA 2002 was held along with the 5th International Workshop on Approximation Algorithms for Combinatorial Optimization (APPROX 2002), the 2nd Workshop on Algorithms in Bioinformatics (WABI 2002), and the 3rd Workshop on Approximation and Randomization Algorithms in Communication NEtworks (ARACNE 2002) in the context of the combined conference ALGO 2002.

The organizing committee of ALGO 2002 consisted of: Giorgio Ausiello, Fabrizio d'Amore, Camil Demetrescu, Silvana Di Vincenzo, Paolo Giulio Franciosa, Daniele Frigioni, Stefano Leonardi, Alberto Marchetti-Spaccamela, and Alessandro Panconesi.

ESA 2002 was sponsored by EATCS (the European Association for Theoretical Computer Science). The EATCS sponsorship included an award of EUR 500 for the best student paper at ESA 2002. This award was shared by Mayur Datar for his paper *Multibutterflies and Peer-to-Peer Networks* and by Marcin Peczarski for his paper *Sorting 16 Elements Requires 34 Comparisons*.

July 2002

Rolf Möhring and Rajeev Raman
Program Chairs, ESA 2002

Referees

Dimitris Achlioptas
Oswin Aichholzer
Tatsuya Akutsu
Jochen Alber
Itai Alon
Helmut Alt
Ernst Althaus
Christoph Ambuehl
Sai Anand
Lars Arge
Estie Arkin
Franz Aurenhammer
Giorgio Ausiello
Adi Avidor
Ricardo Baeza-Yates
Georg Baier
Brenda Baker
Nikhil Bansal
Gill Barequet
Luca Becchetti
Rene Beier
Alexander Below
Amir Ben-Amram
Michael Bender
Therese Biedl
Philip Bille
Avrim Blum
Jens-Peter Bode
Hans Bodlaender
Joan Boyar
Peter Braß
Gerth Brodal
Britta Broser
Dan Brown
Adam Buchsbaum
Ioannis Caragiannis
Frederic Cazals
Marco Cesati
Chandra Chekuri
Marek Chrobak
William Cook
Derek Corneil
Artur Czumaj
Fabrizio d'Amore

Mark de Berg
Brian Dean
Romuald Debruyne
Erik Demaine
Camil Demetrescu
Luc Devroye
Giuseppe Di Battista
Vida Dujmovic
Sebastian Egner
Stephan Eidenbenz
Andreas Eisenblätter
Ioannis Emiris
David Eppstein
Leah Epstein
Patricia Evans
Guy Even
Rolf Fagerberg
Amr Farahat
Lene Monrad Favrholdt
Stefan Felsner
Paolo Ferragina
Jiri Fiala
Amos Fiat
Dimitris Fotakis
Alan Frieze
Stefan Funke
Hal Gabow
Naveen Garg
Cyril Gavoille
Joachim Giesen
Inge Li Gørtz
Catherine Greenhill
Roberto Grossi
Dimitrios Gunopulos
Anupam Gupta
Jens Gustedt
Torben Hagerup
Peter Hajnal
Alexander Hall
Toru Hasunuma
Laura Heinrich-Litan
John Hershberger
Winfried Hochstaettler
Frank Hoffmann

Thomas Hofmeister
Takashi Horiyama
Keiko Imai
Michiko Inoue
Toshimasa Ishii
Giuseppe F. Italiano
Hiro Ito
Chuzo Iwamoto
Riko Jacob
Klaus Jansen
Berit Johannes
Tibor Jordan
Vanessa Kääb
Lars Kaderali
Volker Kaibel
Haim Kaplan
Sanjiv Kapoor
Menelaos Karavelas
Juha Kärkkäinen
Jyrki Katajainen
Michael Kaufmann
Makino Kazuhisa
Claire Kenyon
Richard Kenyon
Lutz Kettner
Sanjeev Khanna
Ralf Klasing
Gunnar Klau
Sandi Klavžar
Rolf Klein
Jon Kleinberg
Bettina Klinz
Christian Knauer
Gabriele Kodydek
Ekkehard Köhler
Stavros Kolliopoulos
Petr Kolman
Jochen Könemann
Ulrich Kortenkamp
Guy Kortsarz
Arie Koster
Daniel Kral
Georg Kraml
Klaus Kriegel

Referees

Sven Krumke
Sebastian Leipert
Stefano Leonardi
Adam Letchford
Andrzej Lingas
Ivana Ljubić
Marco Lübbecke
Frank Lutz
Christos Makris
Alberto Marchetti-Spaccamela
Chip Martel
Conrado Martinez
Toshimitsu Masuzawa
Madhav Marathe
Yossi Matias
Ross McConnell
Colin McDiarmid
Terry McKee
Henk Meijer
Adam Meyerson
Joseph Mitchell
Hiroyoshi Miwa
Shuichi Miyazaki
Géraldine Morin
Walter Morris
Christian W. Mortensen
Nicolas Stier Moses
Rudolf Müller
David Mount
S. Muthukrishnan
Stefan Näher
Hiroshi Nagamochi
Shin-ichi Nakano
Giri Narasimhan
Gonzalo Navarro
Nathan Netanyahu
Rolf Niedermeier
Mads Nielsen
Sotiris Nikoletseas
Koji Obokata

Jesper Holm Olsen
Joe O'Rourke
Anna Östlin
Rasmus Pagh
Christian N. S. Pedersen
Itsik Pe'er
Marco Pellegrini
Shietung Peng
Seth Pettie
Marc Pfetsch
Athanasios Poulakidas
Kirk Pruhs
Tomasz Radzik
Naila Rahman
Jörg Rambau
Edgar Ramos
Theis Rauhe
R. Ravi
Eric Remila
Franz Rendl
Ares Ribó
Liam Roditty
Amir Ronen
Peter Rossmanith
Ingo Schiermeyer
Stefan Schirra
Elmar Schömer
Martin Schönhacker
Steve Seiden
Maria Jose Serna
Jay Sethuraman
Micha Sharir
Bruce Shepherd
Tetsuo Shibuya
Akiyoshi Shioura
Amitabh Sinha
Spyros Sioutas
Rene Sitters
Steve Skiena
Martin Skutella
Michiel Smid

Jack Snoeyink
Roberto Solis-Oba
Ewald Speckenmeyer
Bettina Speckmann
Jeremy Spinrad
Paul Spirakis
S. Srinivasa Rao
Yiannis Stamatiou
Stamatis Stefanakos
Cliff Stein
Leen Stougie
Subhash Suri
Mario Szegedy
Takashi Takabatake
Ayellet Tal
Roberto Tamassia
Arie Tamir
Vanessa Teague
Sven Thiel
Mikkel Thorup
Takeshi Tokuyama
Kostas Tsichlas
Tatsuie Tsukiji
Marc Uetz
Takeaki Uno
Hein van der Holst
Rob van Stee
Rakesh Vohra
Danica Vukadinovic
Uli Wagner
Dorothea Wagner
Arnold Wassmer
René Weiskircher
Carola Wenk
Peter Widmayer
Gerhard Woeginger
Alexander Wolff
Günter Ziegler
Martin Zinkevich

Table of Contents

Invited Lectures

Solving Traveling Salesman Problems 1
William Cook

Computing Shapes from Point Cloud Data 2
Tamal K. Dey

Mechanism Design for Fun and Profit 3
Anna R. Karlin

On Distance Oracles and Routing in Graphs 4
Mikkel Thorup

Contributed Papers

Kinetic Medians and kd-Trees 5
Pankaj K. Agarwal, Jie Gao, and Leonidas J. Guibas

Range Searching in Categorical Data: Colored Range Searching on Grid 17
Pankaj K. Agarwal, Sathish Govindarajan, and S. Muthukrishnan

Near-Linear Time Approximation Algorithms for Curve Simplification 29
Pankaj K. Agarwal, Sariel Har-Peled, Nabil H. Mustafa, and Yusu Wang

Translating a Planar Object to Maximize Point Containment 42
*Pankaj K. Agarwal, Torben Hagerup, Rahul Ray, Micha Sharir,
Michiel Smid, and Emo Welzl*

Approximation Algorithms for k-Line Center 54
Pankaj K. Agarwal, Cecilia M. Procopiuc, and Kasturi R. Varadarajan

New Heuristics and Lower Bounds
for the Min-Max k-Chinese Postman Problem 66 64
Dino Ahr and Gerhard Reinelt

SCIL – Symbolic Constraints in Integer Linear Programming 75
*Ernst Althaus, Alexander Bockmayr, Matthias Elf, Michael Jünger,
Thomas Kasper, and Kurt Mehlhorn*

Implementing I/O-efficient Data Structures Using TPIE 88
Lars Arge, Octavian Procopiuc, and Jeffrey Scott Vitter

On the k-Splittable Flow Problem 101
Georg Baier, Ekkehard Köhler, and Martin Skutella

Partial Alphabetic Trees 114
Arye Barkan and Haim Kaplan

Classical and Contemporary Shortest Path Problems in Road Networks:
Implementation and Experimental Analysis of the TRANSIMS Router 126
*Chris Barrett, Keith Bisset, Riko Jacob, Goran Konjevod,
and Madhav Marathe*

Scanning and Traversing: Maintaining Data for Traversals
in a Memory Hierarchy 139
*Michael A. Bender, Richard Cole, Erik D. Demaine,
and Martin Farach-Colton*

Two Simplified Algorithms for Maintaining Order in a List 152
*Michael A. Bender, Richard Cole, Erik D. Demaine,
Martin Farach-Colton, and Jack Zito*

Efficient Tree Layout in a Multilevel Memory Hierarchy 165
Michael A. Bender, Erik D. Demaine, and Martin Farach-Colton

A Computational Basis for Conic Arcs and Boolean Operations
on Conic Polygons 174
*Eric Berberich, Arno Eigenwillig, Michael Hemmer, Susan Hert,
Kurt Mehlhorn, and Elmar Schömer*

TSP with Neighborhoods of Varying Size 187
*Mark de Berg, Joachim Gudmundsson, Matthew J. Katz,
Christos Levcopoulos, Mark H. Overmars, and A. Frank van der Stappen*

1.375-Approximation Algorithm for Sorting by Reversals 200
Piotr Berman, Sridhar Hannenhalli, and Marek Karpinski

Radio Labeling with Pre-assigned Frequencies 211
*Hans L. Bodlaender, Hajo Broersma, Fedor V. Fomin,
Artem V. Pyatkin, and Gerhard J. Woeginger*

Branch-and-Bound Algorithms for the Test Cover Problem 223
*Koen M.J. De Bontridder, B.J. Lageweg, Jan K. Lenstra,
James B. Orlin, and Leen Stougie*

Constructing Plane Spanners of Bounded Degree and Low Weight 234
Prosenjit Bose, Joachim Gudmundsson, and Michiel Smid

Eager st-Ordering 247
Ulrik Brandes

Three-Dimensional Layers of Maxima 257
Adam L. Buchsbaum and Michael T. Goodrich

Optimal Terrain Construction Problems and Applications
in Intensity-Modulated Radiation Therapy 270
*Danny Z. Chen, Xiaobo S. Hu, Shuang Luan, Xiaodong Wu,
and Cedric X. Yu*

Geometric Algorithms for Density-Based Data Clustering 284
Danny Z. Chen, Michiel Smid, and Bin Xu

Balanced-Replication Algorithms for Distribution Trees 297
Edith Cohen and Haim Kaplan

Butterflies and Peer-to-Peer Networks 310
Mayur Datar

Estimating Rarity and Similarity over Data Stream Windows 323
Mayur Datar and S. Muthukrishnan

Efficient Constructions of Generalized Superimposed Codes
with Applications to Group Testing and Conflict Resolution
in Multiple Access Channels 335
Annalisa De Bonis and Ugo Vaccaro

Frequency Estimation of Internet Packet Streams with Limited Space 348
Erik D. Demaine, Alejandro López-Ortiz, and J. Ian Munro

Truthful and Competitive Double Auctions 361
*Kaustubh Deshmukh, Andrew V. Goldberg, Jason D. Hartline,
and Anna R. Karlin*

Optimal Graph Exploration without Good Maps 374
Anders Dessmark and Andrzej Pelc

Approximating the Medial Axis from the Voronoi Diagram
with a Convergence Guarantee 387
Tamal K. Dey and Wulue Zhao

Non-independent Randomized Rounding and an Application
to Digital Halftoning 399
Benjamin Doerr and Henning Schnieder

Computing Homotopic Shortest Paths Efficiently 411
Alon Efrat, Stephen G. Kobourov, and Anna Lubiw

An Algorithm for Dualization in Products of Lattices and
Its Applications 424
Khaled M. Elbassioni

Determining Similarity of Conformational Polymorphs 436
Angela Enosh, Klara Kedem, and Joel Bernstein

Minimizing the Maximum Starting Time On-line 449
Leah Epstein and Rob van Stee

Vector Assignment Problems: A General Framework 461
Leah Epstein and Tamir Tassa

Speeding Up the Incremental Construction of the Union
of Geometric Objects in Practice 473
Eti Ezra, Dan Halperin, and Micha Sharir

Simple and Fast: Improving a Branch-And-Bound Algorithm
for Maximum Clique 485
Torsten Fahle

Online Companion Caching 499
Amos Fiat, Manor Mendel, and Steven S. Seiden

Deterministic Communication in Radio Networks with Large Labels 512
Leszek Gąsieniec, Aris Pagourtzis, and Igor Potapov

A Primal Approach to the Stable Set Problem 525
Claudio Gentile, Utz-Uwe Haus, Matthias Köppe, Giovanni Rinaldi, and Robert Weismantel

Wide-Sense Nonblocking WDM Cross-Connects 538
Penny Haxell, April Rasala, Gordon Wilfong, and Peter Winkler

Efficient Implementation of a Minimal Triangulation Algorithm 550
Pinar Heggernes and Yngve Villanger

Scheduling Malleable Parallel Tasks:
An Asymptotic Fully Polynomial-Time Approximation Scheme 562
Klaus Jansen

The Probabilistic Analysis of a Greedy Satisfiability Algorithm 574
Alexis C. Kaporis, Lefteris M. Kirousis, and Efthimios G. Lalas

Dynamic Additively Weighted Voronoi Diagrams in 2D 586
Menelaos I. Karavelas and Mariette Yvinec

Time-Expanded Graphs for Flow-Dependent Transit Times 599
Ekkehard Köhler, Katharina Langkau, and Martin Skutella

Partially-Ordered Knapsack and Applications to Scheduling 612
Stavros G. Kolliopoulos and George Steiner

A Software Library for Elliptic Curve Cryptography 625
Elisavet Konstantinou, Yiannis Stamatiou, and Christos Zaroliagis

Real-Time Dispatching of Guided
and Unguided Automobile Service Units with Soft Time Windows 637
Sven O. Krumke, Jörg Rambau, and Luis M. Torres

Randomized Approximation Algorithms
for Query Optimization Problems on Two Processors 649
Eduardo Laber, Ojas Parekh, and R. Ravi

Covering Things with Things 662
Stefan Langerman and Pat Morin

On-Line Dial-a-Ride Problems under a Restricted Information Model 674
Maarten Lipmann, X. Lu, Willem E. de Paepe, Rene A. Sitters, and Leen Stougie

Approximation Algorithm
for the Maximum Leaf Spanning Tree Problem for Cubic Graphs 686
Krzysztof Loryś and Grażyna Zwoźniak

Engineering a Lightweight Suffix Array Construction Algorithm 698
Giovanni Manzini and Paolo Ferragina

Complexity of Compatible Decompositions of Eulerian Graphs
and Their Transformations 711
Jana Maxová and Jaroslav Nešetřil

External-Memory Breadth-First Search with Sublinear I/O 723
Kurt Mehlhorn and Ulrich Meyer

Frequency Channel Assignment on Planar Networks 736
Michael Molloy and Mohammad R. Salavatipour

Design and Implementation of Efficient Data Types for Static Graphs 748
Stefan Näher and Oliver Zlotowski

An Exact Algorithm for the Uniformly-Oriented Steiner Tree Problem 760
Benny K. Nielsen, Pawel Winter, and Martin Zachariasen

A Fast, Accurate and Simple Method for Pricing European-Asian
and Saving-Asian Options 772
*Kenichiro Ohta, Kunihiko Sadakane, Akiyoshi Shioura,
and Takeshi Tokuyama*

Sorting 13 Elements Requires 34 Comparisons 785
Marcin Peczarski

Extending Reduction Techniques for the Steiner Tree Problem 795
Tobias Polzin and Siavash Vahdati Daneshmand

A Comparison of Multicast Pull Models 808
Kirk Pruhs and Patchrawat Uthaisombut

Online Scheduling for Sorting Buffers 820
Harald Räcke, Christian Sohler, and Matthias Westermann

Finding the Sink Takes Some Time: An Almost Quadratic Lower Bound
for Finding the Sink of Unique Sink Oriented Cubes 833
Ingo Schurr and Tibor Szabó

Lagrangian Cardinality Cuts and Variable Fixing
for Capacitated Network Design 845
Meinolf Sellmann, Georg Kliewer, and Achim Koberstein

Minimizing Makespan and Preemption Costs on a System
of Uniform Machines 859
Hadas Shachnai, Tami Tamir, and Gerhard J. Woeginger

Minimizing the Total Completion Time On-line on a Single Machine,
Using Restarts 872
Rob van Stee and Han La Poutré

High-Level Filtering for Arrangements of Conic Arcs 884
Ron Wein

An Approximation Scheme for Cake Division
with a Linear Number of Cuts 896
Gerhard J. Woeginger

A Simple Linear Time Algorithm for Finding Even Triangulations
of 2-Connected Bipartite Plane Graphs 902
Huaming Zhang and Xin He

Author Index 915

Solving Traveling Salesman Problems
— Invited Lecture —

William Cook

School of Industrial and Systems Engineering, Georgia Tech.
wcook@isy.gatech.edu

Abstract Given the cost of travel between each pair of a finite number of cities, the *traveling salesman problem* (TSP) is to find the cheapest tour passing through all of the cities and returning to the point of departure. We will present a survey of recent progress in algorithms for large-scale TSP instances, including the solution of a million city instance to within 0.09% of optimality and the exact solution a 15,112-city instance. We will also discuss extensions of TSP techniques to other path-routing problems, and describe the solution of the WhizzKids'96 problem in vehicle routing. This talk is based on joint work with David Applegate, Robert Bixby, Vasek Chvátal, Sanjeeb Dash and Andre Rohe.

Computing Shapes from Point Cloud Data
— Invited Lecture —

Tamal K. Dey

The Ohio State University
Columbus OH 43210, USA
tamaldey@cis.ohio-state.edu
http://cis.ohio-state.edu/~tamaldey

Abstract. The problem of modeling a shape from point cloud data arises in many areas of science and engineering. Recent advances in scanning technology and scientific simulations can generate sample points from a geometric domain with ease. Perhaps the most usual sampled domain is the boundary of a 3D object, which is a surface in 3D. Consequently, the problem of surface reconstruction has been a topic of intense research in recent years. In this talk we will present a surface reconstruction algorithm popularly known as Cocone that simplified and improved its predecessor called Crust. We will discuss its mathematically provable guarantees and issues related to its successful implementation.

The concepts used in Cocone have been further extended to other problems in sample based modeling. We show how these concepts can be used to decimate a point cloud dataset while preserving the shape features. We extend the Cocone theory and methods to detect the dimension of a manifold from its point samples, where the manifold is embedded in an Euclidean space. In a recent work we used the Cocone concepts to approximate the medial axis of a smooth surface from the Voronoi diagram of its point samples. We will discuss these extensions and present our theoretical and experimental results.

Mechanism Design for Fun and Profit
— Invited Lecture —

Anna R. Karlin

Computer Science Department, University of Washington
karlin@cs.washington.edu

Abstract The emergence of the Internet as one of the most important arenas for resource sharing between parties with diverse and selfish interests has led to a number of fascinating and new algorithmic problems. In these problems, one must solicit the inputs to each computation from participants (or agents) whose goal is to manipulate the computation to their own advantage. Until fairly recently, failure models in computer science have not dealt the notion of selfish participants who "play by the rules" only when it fits them. To deal with this, algorithms must be designed so as to provide motivation to the participants to "play along". Recent work in this area, has drawn on ideas from game theory and microeconomics, and specifically from the field of *mechanism design*. The goal is to design protocols so that rational agents will be motivated to adhere to the protocol. A specific focus has been on *truthful mechanisms* in which selfish agents are motivated to reveal their true inputs.

In the first part of the talk, we survey recent work in the area of *algorithm mechanism design*. In the second part of the talk, we focus on mechanism design specifically geared at maximizing the profit of the mechanism designer. In particular, we consider a class of dynamic pricing problems motivated by the same computational and economic trends. We describe a class of generalized auction problems as well as a competitive framework that can be used to evaluate solutions to these problems. We present a number of results on the design of profit-maximizing truthful generalized auctions. This is joint work with Amos Fiat, Andrew Goldberg and Jason Hartline.

On Distance Oracles and Routing in Graphs
— Invited Lecture —

Mikkel Thorup

AT&T Labs—Research, Shannon Laboratory
180 Park Avenue, Florham Park, NJ 07932, USA
mthorup@research.att.com
http://www.research.att.com/~mthorup

Abstract. We review two basic problems for graphs:
- Constructing a small space distance oracle that, for any pair (v,w) of nodes, provides a quick and good estimate of the distance from v to w.
- Distributing the distance oracle in a labeling scheme that can be used for efficient routing.

For general graphs, near-optimal trade-offs between space and precision are discussed. Better results for planar and bounded tree-width graphs are also discussed.

Kinetic Medians and *kd*-Trees

Pankaj K. Agarwal[1], Jie Gao[2], and Leonidas J. Guibas[2]

[1] Department of Computer Science, Duke University
Durham, NC 27708, USA
pankaj@cs.duke.edu
[2] Department of Computer Science, Stanford University
Stanford, CA 94305, USA
{jgao,guibas}@cs.stanford.edu

Abstract. We propose algorithms for maintaining two variants of *kd*-trees of a set of moving points in the plane. A pseudo *kd*-tree allows the number of points stored in the two children to differ by a constant factor. An overlapping *kd*-tree allows the bounding boxes of two children to overlap. We show that both of them support range search operations in $O(n^{1/2+\epsilon})$ time, where ϵ only depends on the approximation precision. As the points move, we use event-based kinetic data structures to update the tree when necessary. Both trees undergo only a quadratic number of events, which is optimal, and the update cost for each event is only polylogarithmic. To maintain the pseudo *kd*-tree, we develop algorithms for computing an approximate median level of a line arrangement, which itself is of great interest. We show that the computation of the approximate median level of a set of lines or line segments can be done in an online fashion smoothly, i.e., there are no expensive updates for any events. For practical consideration, we study the case in which there are speed-limit restrictions or smooth trajectory requirements. The maintenance of the pseudo *kd*-tree, as a consequence of the approximate median algorithm, can also adapt to those restrictions.

1 Introduction

Motion is ubiquitous in the physical world. Several areas such as digital battlefields, air-traffic control, mobile communication, navigation system, geographic information systems, call for storing moving objects into a data structure so that various queries on them can be answered efficiently; see [22, 24] and the references therein. The queries might relate either to the current configuration of objects or to a configuration in the future — in the latter case, we are asking to predict the behavior based on the current information. In the last few years there has been a flurry of activity on extending the capabilities of existing database systems to represent moving-object databases (MOD) and on indexing moving objects; see, e.g., [14, 21, 22]. The known data structures for answering range queries on moving points either do not guarantee a worst-case bound on the query time [25, 24, 19] or are too complicated [1, 15].

In this paper we develop kinetic data structures for kd-trees, a widely used data structure for answering various proximity queries in practice [10], which can efficiently answer range queries on moving points. The *kinetic data structure* framework, originally proposed by Basch et al. [5], has led to efficient algorithms for several geometric problems involving moving objects; see [13] and references therein. The main idea in the kinetic framework is that even though the objects move continuously, the relevant combinatorial structure of the data structure changes only at certain discrete times. Therefore one does not have to update the data structure continuously. The *kinetic updates* are performed on the data structure only when certain *kinetic events* occur; see [5, 13] for details.

Recall that a kd-tree on a set of points is a binary tree, each node v of which is associated with a subset S_v of points. The points in S_v are partitioned into two halves by a vertical or horizontal line at v and each half is associated with a child of v. The orientation of the partition line alternates as we follow a path down the tree. In order to develop a kinetic data structure for kd-trees, the first step is to develop a kinetic data structure for maintaining the median of a set of points moving on a line. In fact, this subroutine is needed for several other data structures.

Related work. The problem of maintaining the point of rank k in a set S of points moving on the x-axis is basically the same as computing the k-level in the arrangement of curves; the *k-level* in an arrangement of x-monotone curves is the set of edges of the arrangement that lie above exactly k curves [4]. If the points in S are moving with fixed speed, i.e., their trajectories are lines in the xt-plane, then the result by Dey [11] implies that the point of rank k changes $O(nk^{1/3})$ times; the best-known lower bound is $ne^{\Omega(\sqrt{\log k})}$ [23]. Only much weaker bounds are known if the trajectories of points are polynomial curves. Edelsbrunner and Welzl [12] developed an algorithm for computing a level in an arrangements of lines. By combining their idea with kinetic tournaments proposed by Basch et al. [6] for maintaining the maximum of a set of moving points on a line, one can construct an efficient kinetic data structure for maintaining the median of a set of moving points.

Since no near-linear bound is known on the complexity of a level, work has been done on computing an approximate median level. A δ-approximate median-level is defined as a x-monotone curve that lies between $(1/2-\delta)n$- and $(1/2+\delta)n$-levels. Edelsbrunner and Welzl [12] showed that a δ-approximate median level with at most $\lceil \lambda(n)/(\delta n) \rceil$ edges, where $\lambda(n)$ is the number of vertices on the median level of the arrangement, can be computed for line arrangements. Later Matoušek [16] proposed an algorithm to obtain a δ-approximate median level with constant complexity for an arrangement of n lines. However no efficient kinetic data structure is known for maintaining an approximate median of a set of moving points.

Because of their simplicity, kd-trees are widely used in practice and several variants of them have been proposed [3, 7, 17]. It is well known that a kd-tree can answer a two-dimensional orthogonal range query in $O(\sqrt{n} + k)$ time, where k is the number of points reported. Variants of kd-trees that support insertions and deletions are studied in [18, 9]. No kinetic data structures are known for kd-trees.

Agarwal et al. [1] were the first to develop kinetic data structures for answering range-searching queries on moving points. They developed a kinetic data structure that answers a two-dimensional range query in $O(\log n + k)$ time using $O(n \log n/(\log \log n))$ space, where k is the output size. The amortized cost of a kinetic event is $O(\log^2 n)$, and the total number of events is $O(n^2)$.[1] They also showed how to modify the structure in order to obtain a tradeoff between the query bound and the number of kinetic events. Kollios et al. [15] proposed a data structure for range searching among points moving on the x-axis, based on partition trees [3]. The structure uses $O(n)$ space and answers queries in $O(n^{1/2+\epsilon} + k)$ time, for an arbitrarily small constant $\epsilon > 0$. Agarwal et al. [1] extended the result to 2D. Unlike kinetic data structures, these data structures are time oblivious, in the sense that they do not evolve over time. However all these data structures are too complex to be used in practice.

In the database community, a number of practical methods have been proposed for accessing and searching moving objects (see [25, 24, 19] and the references therein). Many of these data structures index the trajectories of points either directly or by mapping to higher dimensions. These approaches are not efficient since trajectories do not cluster well. To alleviate this problem, one can parametrize a structure such as the R-tree, which partitions the points but allows the bounding boxes associated with the children of a node to overlap. To expect good query efficiency, the areas of overlap and the areas of the bounding boxes must be small. Although R-tree works correctly even when the overlap areas are too large, the query performance deteriorates in this case. Kinetic data structures based on R-tree were proposed in [24, 20] to handle range queries over moving points. Unfortunately, these structures also do not guarantee sublinear query time in the worst case.

Our results. Let $S = \{p_1, \ldots, p_n\}$ be a set of n points in $\mathbb{R}^1$, each moving independently. The position of a point p_i at time t is given by $p_i(t)$. We use $\bar{p}_i = \bigcup_t (p_i(t), t)$ to denote the *graph* of the trajectory of p_i in the xt-space. The user is allowed to change the trajectory of a point at any time. For a given parameter $\delta > 0$, we call a point x, not necessarily a point of S, a δ-*approximate median* if its rank is in the range $[(1/2 - \delta)n, (1/2 + \delta)n]$. We first describe an off-line algorithm that can maintain a δ-approximate median of S in a total time of $O((\mu/(n^2\delta^2)) + 1/\delta)n \log n)$, where μ is the number of times two points swap position. We then show how a δ-median can be maintained on-line as the points of S move. Our algorithm maintains a point x^* on the x-axis, not necessarily one of the input points, whose rank is in the range $(1/2 \pm \delta)n$. As the input points move, x^* also moves, and its trajectory depends on the motion of input points. We show that the speed of this point is not larger than the fastest moving

---

[1] Agarwal et al. [1] actually describe their data structure in the standard two level I/O model, in which the goal is to minimize the memory access time. Here we have described their performance in the standard pointer-machine model.

point, and that our algorithm can be adapted so that the trajectory of x^* is C^k-continuous for any $k \geq 0$.

Next, let S be a set of n points in $\mathbb{R}^2$, each moving independently. We wish to maintain the kd-tree of S, so that *range searching* queries, i.e., given a query rectangle R at time t, report $|S(t) \cap R|$, can be answered efficiently, Even if the points in S are moving with fixed velocity, the kd-tree on S can change $\Theta(n^2)$ times, and there are point sets on which many of these events can cause a dramatic change of the tree, i.e., each of them requires $\Omega(n)$ time to update the tree. We therefore propose two variants of kd-trees, each of which answers a range query in time $O(n^{1/2+\varepsilon} + k)$, for any constant $\varepsilon > 0$, (k is the number of points reported), processes quadratic number of events, and spends polylogarithmic (amortized) time at each event. The first variant is called the δ-*pseudo kd-tree*, in which if a node has m points stored in the subtree, then each of its children has at most $(1/2 + \delta)m$ points in its subtree. In the second variant, called δ-*overlapping kd-tree*, the bounding boxes of the points stored in the subtrees of two children of a node can overlap. However, if the subtree at a node contains m points, then the overlapping region at that node contains at most δm points.

As the points move, both of the trees are maintained in the standard kinetic data structure framework [5]. The correctness of the tree structure is certified by a set of conditions, called *certificates*, whose failure time is precomputed and inserted into an event queue. At each certificate failure, denoted as an *event*, the KDS certification repair mechanism is invoked to repair the certificate set and possibly the tree structure as well. In the analysis of a KDS, we assume the points follow *pseudo-algebraic motions*. By this we mean that all certificates used by the KDS can switch from TRUE to FALSE at most a constant number of times. In some occasions, we may make stronger assumptions, like the usual scenario of linear, constant velocity motions. For both pseudo and overlapping kd-trees, we show that the set of certificates has linear size, that the total number of events is quadratic, and that each event has polylogarithmic amortized update cost. Dynamic insertion and deletion are also supported with the same update bound as for an event.

The pseudo kd-tree data structure uses our δ-approximate median algorithms as a subroutine. Unlike pseudo kd-trees, the children of a node of an overlapping kd-tree store equal number of points and the minimum bounding boxes of the children can overlap. However, maintaining overlapping kd-trees is somewhat simpler, and the analysis extends to pseudo-algebraic motion of points.

2 Maintaining the Median

2.1 Off-Line Maintenance

Matoušek [16] used a level shortcutting idea to compute a polygonal line, with complexity $O(1/\delta^2)$, which is a δ-approximate median. Here we extend his idea to a set S of line segments in the plane. Let μ denote the number of intersection points between the segments.

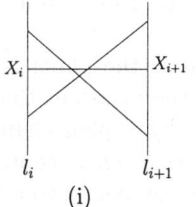 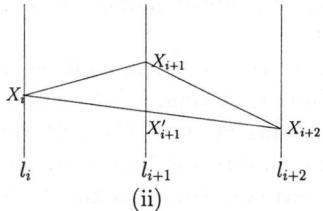

Fig. 1. (i) Approximate median level of n lines; (ii)When points change motion plan

We divide the plane by vertical lines l_i's into strips such that there are at most $\delta^2 n^2/16$ intersections and at most $\delta n/4$ endpoints inside each strip. In addition, we can assume that each strip either has exactly $\delta^2 n^2/16$ intersections or exactly $\delta n/4$ endpoints, otherwise we can always enlarge the strip. The number of strips is $O(\mu/(n^2\delta^2) + 1/\delta)$. Along each l_i, we compute the median X_i of the intersections between S and l_i. We connect adjacent medians. We claim that the resulting polygonal line is a δ-approximate median level. Indeed, consider a strip bounded by l_i and l_{i+1}. We first delete the segments of S that have at least one endpoint inside the strip. Let $U \subseteq S$ be the set of line segments that intersect the segment $X_i X_{i+1}$ and that lie above X_i at l_i. Similarly let $V \subseteq S$ be the set of line segments that intersect $X_i X_{i+1}$ and that lie below X_i at l_i. $|U| = u$, $|V| = v$. Since X_i and X_{i+1} are on the exact median level and we have deleted at most $\delta n/4$ segments, $|u - v| \leq \delta n/4$. Suppose $u \leq v \leq u + \delta n/4$. The intersection of a segment in U and a segment in V must be inside the strip (see Figure 1 (i)), therefore $uv \leq \delta^2 n^2/16$, thereby implying that $u \leq \delta n/4$ and $v \leq \delta n/2$. Consequently, $X_i X_{i+1}$ is intersected at most δn times, and thus it is a δ-approximation of the median level between l_i and l_{i+1}. Computing the partition line l_i involves computing the kth leftmost intersection in an arrangement. This can be done in $O(n \log n)$ time, using the slope-selection algorithm [8]. Computing a median along a cut line costs $O(n)$. So the total computational cost is $O((\mu/(n^2\delta^2) + 1/\delta)n \log n)$.

Theorem 1. *Let S be a set of n segments in the plane with a total of μ intersection points. A δ-approximate median level of S with at most $O(\mu/(n^2\delta^2) + 1/\delta)$ vertices can be computed in $O((\mu/(n\delta^2) + n/\delta) \log n)$ time.*

2.2 On-Line Maintenance

In previous subsection we assume that we know the motion plan of the points beforehand and compute the approximate median ahead of time. However, when the points change their motion plan, we want to maintain the approximate median online and distribute the amount of work more uniformly over time. Assume that the total number of motion plan changes over time is only linear in the number of points. We will show that the approximate median can be maintained smoothly, with linear number of events and polylogarithmic time update cost per event.

Consider the time-space plane. As in Theorem 1, the plane is divided by vertical lines l_i at time t_i into strips. We maintain the invariant that each strip either contains at most $\delta^2 n^2/36$ vertices or $\delta n/3$ flight plan changes. Also if the strip contains less than $\delta n/3$ flight plan changes, then it contains at least $\delta^2 n^2/72$ vertices. We approximate the median level by computing a polygonal line $X_1 X_2 ... X_m$ in which X_i lies on l_i. We can see that the total number of strips is bounded by $O(1/\delta^2)$ since there are only linear number of flight plan changes in total.

However, instead of computing the $O(1/\delta^2)$ cut lines all at one time, we compute them one by one. We begin with computing the median on the two leftmost cut lines l_1 and l_2 (l_1 corresponds to the starting time). There are $\delta^2 n^2/72$ vertices between l_1 and l_2. The approximate median then moves along line segment $X_1 X_2$. This preprocessing costs $O(n \log n)$. Then we compute the position of the cut line l_{i+2} and the median along l_{i+2} gradually, as we are moving along $X_i X_{i+1}$. The position of l_{i+2} is computed such that there are $\delta^2 n^2/72$ vertices between l_{i+1} and l_{i+2}. So X_{i+2} is our prediction of the median on l_{i+2}, assuming there is no flight plan change between t_{i+1} and t_{i+2}. The cost of computing l_{i+2} and X_{i+2}, which is a slope-selection problem, is amortized over the time interval $[t_i, t_{i+1}]$, using the technique proposed by Agarwal et al. [2]. In an online version, we need to accommodate the flight plan changes in the future. If there are $\delta n/6$ motion plans changes before we reach l_{i+1}, we start another session to compute X_{i+2} based on the current (changed) trajectories. If we reach X_{i+1} before another $\delta n/6$ motion updates, we switch to the segment $X_{i+1} X_{i+2}$ computed from the first construction. Otherwise, if we see another $\delta n/6$ motion plan changes (i.e., a total of $\delta n/3$ changes since t_i) before we reach l_{i+1}, then we just stop and move l_{i+1} to the current position and X_{i+1} to the current position of the approximate median. The next line segment that the approximate median needs to follow is $X_{i+1} X_{i+2}$, where X_{i+2} is the value returned by the second computation (the one initiated after $\delta n/6$ motion-plan changes). Thus inside one strip, the number of intersections is at most $\delta^2 n^2/36$, and the number of flight plan changes is at most $\delta n/3$. Next we prove that the polygonal line we compute is a δ-approximation of the median level.

The only problem is that if some points change their flight plans between time t_i and t_{i+1}, then X_{i+1}, which is computed before t_i, is no longer the real median X'_{i+1}. See Figure 1 (ii). However, only the points between X_{i+1} and X'_{i+1} can cross $X_{i+1} X_{i+2}$ without crossing $X'_{i+1} X_{i+2}$. Since there are only at most $\delta n/3$ flight plan changes, there are at most $\delta n/3$ points between X_{i+1} and X'_{i+1}. Since $X'_{i+1} X_{i+2}$ is crossed by at most $2\delta n/3$ lines, $X_{i+1} X_{i+2}$ is still a δ-approximation.

Theorem 2. *Let S be a set of n points, each moving with a constant velocity. Suppose the flight paths of S change a total of m times. Then a δ-approximate median of S can be maintained in an online fashion so that the median moves continuously with at most $O(1/\delta^2)$ flight plan changes. The total maintenance* cost is $O((n+m)\log n/\delta^2)$. Each flight plan change costs polylogarithmic update time.

Remark. Finding the next cut line in Theorem 2 is based on complex algorithms and data structures (e.g., slope selection). We can use a considerably simpler randomized algorithm, based on the random-sampling technique, for computing the next cut line. We omit the details from this version of the paper.

2.3 Approximate Median with Speed Limit

In many applications like mobile networks and spatial databases, the maximum velocity of the points is restricted. We'll show an approximate median that moves no faster than the fastest point.

Lemma 1. *The approximate median computed above cannot move faster than the maximum speed of the n points, if the points do not change flight plans.*

Proof. Observe that the approximate median is composed of line segments connecting two median points X and Y at different time steps. So if there is a line l crossing XY and going up, there must be another line l' crossing XY and going down. So the slope of XY is bounded by the slope of l and l'. If XY is not crossed by any lines, X and Y lie on the same line as they are both medians. Hence, the approximate median moves along one of the input points between X and Y. This proves the lemma.

If the points can change their flight plan, our scheme in Theorem 2 does not guarantee that the approximate median moves within the limit restriction. But we can adapt it as follows. Let X_i and X'_i denote the precomputed and exact median, respectively. We let the approximate median to move toward the precomputed median. If we can reach there without violating the speed constraint, then everything is fine. Otherwise, we just move toward it with the maximum speed. Figure 2 (i) shows the second case. W.l.o.g., we assume X_{i+1} is above X'_{i+1}. Assume the approximate median gets to the point X''_{i+2} at time t_{i+2}. Since the approximate median moves with the maximum speed, the points above X_{i+1} will stay above X''_{i+2}. So at most $\delta n/3$ points, the ones that lie between X_{i+1}

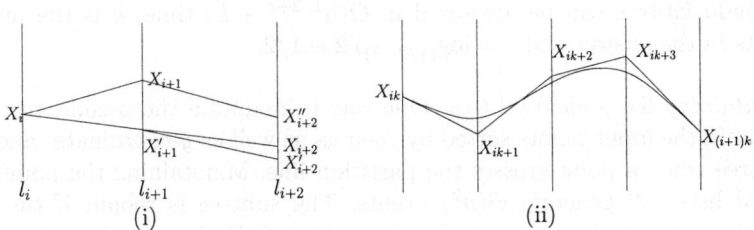

Fig. 2. (i) Speed restricted median; (ii) Smooth medians

and X'_{i+1}, can cross $X_{i+1}X''_{i+2}$ without crossing $X'_{i+1}X_{i+2}$. Following the argument in Theorem 2, $X'_{i+1}X_{i+2}$ is crossed by at most $2\delta n/3$ lines. So $X_{i+1}X''_{i+2}$ is a δ-approximate median level. There are $n/2$ points below X'_{i+1} and X'_{i+2}, respectively, so the number of points between X''_{i+2} and X'_{i+2} can only be less than the number of points between X_{i+1} and X'_{i+1}. This imples that we can continue this process for the next time interval and we obtain the following result.

Theorem 3. *Let S be a set of n points, each moving with fixed velocity. We can maintain a δ-approximate median of S that preserves all the properties of Theorem 2 and that cannot move faster than the maximum speed of the input points.*

2.4 Smooth Medians

For a set of moving points, the approximate median we computed above, follows a polygonal line composed of line segments. Although the approximate median moves continuously, it may have to make sharp turns. In practice, a smooth trajectory is preferred. A curve is said to have C^k continuity if its k-th derivative is continuous. The idea is to use Bézier interpolation on the medians along the vertical lines, see Figure 2 (ii). We will give the theorem whose proof is omitted.

Theorem 4. *Given an arrangement of n lines in the plane, we can compute a curve with C^k continuity which is a δ-approximate median level. The curve is determined by $O(k^2/\delta^2)$ control points and is computed in $O(k^2 n \log n/\delta^2)$ time.*

3 Pseudo kd-Tree

Overmars [18] proposed the pseudo kd-tree as a dynamic data structure for range searching that admits efficient insertion and deletion of points.

Definition. A δ-*pseudo kd-tree* is defined to be a binary tree created by alternately partitioning the points with vertical and horizontal lines, so that for each node v with m points in the subtree, there are at most $(1/2 + \delta)m$ points in the subtrees rooted at the children of v. A δ-pseudo kd-tree is an almost balanced tree with depth $\log_{2/(1+2\delta)} n$. We denote by $d(u)$ the depth of a node u. The same analysis as for the standard kd-tree implies that a range query in a δ-pseudo kd-tree can be answered in $O(n^{1/2+\varepsilon} + k)$ time, k is the number of points in the answer and $\varepsilon = \log_{(1/2+\delta)^2} 2 - 1/2$.

Maintaining the pseudo kd-tree. One way to maintain the pseudo kd-tree is to maintain the input points sorted by their x- as well as y-coordinates and update the tree when a point crosses the partition line. Maintaining the points in two sorted lists will generate $\Theta(n^2)$ events. The subtree is rebuilt if the number of points in one child exceeds fraction $1/2 + \delta$. Each event has an amortized update cost of $O(\log n)$. Another way of maintaining the pseudo kd-tree is to use the dynamic data structure proposed by Overmars [18]. However, it maintains a forest of kd-trees instead of a single tree. Here, we will show how to maintain a single kd-tree with only polylogarithmic update cost per event without any rebuilding ,assume the points move with constant velocity.

To maintain a pseudo kd-tree, we need to maintain a hierarchical partition of the points, which always supports a δ-pseudo kd-tree. Since points are inserted into or deleted from a subtree, the trajectories of the points stored in a node are actually line segments. Maintaining the partition line of a node involves finding the approximate median of a set of line segments in an online fashion. The idea for online maintenance in Section 2.2 works here as well. The difference is, points may come in and out of the range of a node. So inside one strip the number of intersections, endpoints of the line segments and flight plan changes should be bounded all at the same time. The endpoints of the line segments can be treated in the same way as the events of motion plan changes we described before. We omit the details here and prove the following lemma and theorem in an off-line setting. We define a δ-*approximate partition line* of depth k to be a δ-approximate median level in the arrangement of the trajectories of points stored in the subtree of a node at depth k. Let V^k denote the set of nodes at depth k.

Lemma 2. *There exists a set of δ-approximate partition lines of V^k with total complexity $O(k/(\delta\alpha^k)^2)$, and it can be computed in $O((kn\log n)/(\delta\alpha^k)^2)$ time, where $\alpha = 1/2 - \delta$.*

Proof. A key observation is that the combination of the trajectories of all the points in V_k is the arrangement of n lines. So the total number of intersections of the segments in V_k is bounded by $O(n^2)$. Assume a node of depth k has n_k points associated with it. Since the child of a node with m points has at least $(1/2-\delta)m$ points, we have $n_k \geq n\alpha^k$ where $\alpha = 1/2 - \delta$. Let s_k be the total number of line segments in V^k, and let c_k be the total complexity of the δ-approximate partition lines for V^k. By Theorem 1, we have that $c_k = O((n/(\delta n_k))^2 + s_k/(\delta n_k))$. When a point crosses the partition line at a node of depth k, it ends the trajectory in the old child and begins a trajectory in the new child. Therefore we have $s_{k+1} = 2c_k \times \delta n_k = O(n^2/(\delta n_k)) + 2s_k$. By induction, we get $s_k = O(kn/(\delta\alpha^k))$, $c_k = O(k/(\delta^2\alpha^{2k}))$. The running time is bounded in the same way as in Theorem 1.

An event happens when a point crosses a partition line. Updating the pseudo tree involves moving a point from one subtree to the other, which costs $\log_{2/(1+2\delta)} n$. The number of events of depth k is bounded by $s_{k+1}/2$. So the total number of events is $O((n^2/\delta) \log_{2/(1+2\delta)} n)$. In summary, we have

Theorem 5. *For a set of n moving points on the plane, each moving with a constant velocity, we can find a moving hierarchical partition of the plane which supports a δ-pseudo kd-tree at any time. The number of events in the kinetic data structure is $O((n^2/\delta)\log_{2/(1+2\delta)} n)$ in total. Update cost for each event is $O(\log_{2/(1+2\delta)} n)$.*

4 Overlapping kd-Tree

In the second variant of the kd-tree, the bounding boxes associated with the children of a node can overlap.

Definition. Assume $S(v)$ is the set of points stored in the subtree of a node v, $B(v)$ is the minimum bounding box of $S(v)$, and $d(v)$ is the depth of v. Define $\sigma(w) = B(u) \cap B(v)$ to be the *overlapping region* of two children u and v of node w. A δ-overlapping kd-tree is a perfectly balanced tree so that for each node w with m points in the subtree, there can be at most δm points in the overlapping region $\sigma(w)$, $0 < \delta < 1$. The overlapping kd-tree has linear size and depth $O(\log n)$. We call a node x-*partitioned* (y-*partitioned*) if it is partitioned by a vertical (resp. horizontal) line. An orthogonal range query can be answered in the same way as in the standard kd-tree.

Lemma 3. *Let ℓ be a vertical (or horizontal) line, and let v, w be two x-partitioned (y-partitioned) nodes of a δ-overlapping kd-tree such that w is a descendant of v. If ℓ intersects both $\sigma(v)$ and $\sigma(w)$, then $d(w) \geq d(v) + \log_2 \gamma$, where $\gamma = (1 - 2\delta)/(2\delta)$.*

Theorem 6. *A range query in a δ-overlapping kd-tree can be answered in $O(n^{1/2+\varepsilon} + k)$ time, where $\varepsilon = \log_\gamma 2$, $\gamma = (1 - 2\delta)/(2\delta)$, k is the number of points reported.*

Proof. As for the standard kd-tree, it suffices to bound the number of nodes whose bounding boxes intersect a vertical line ℓ. Let v be a x-partitioned node at depth k, and let n_v be the points stored in the subtree rooted at v. The query procedure visits both children and thus all four grandchildren of v only if ℓ intersects $\sigma(v)$. Otherwise, ℓ intersects at most two grandchildren of v. The granchildren of v are also x-partitioned. By Lemma 3, ℓ does not intersect $\sigma(w)$ for any descendent w of v at depth less than $k + \log_2 \gamma$. An easy calculation shows that ℓ intersects at most $8\sqrt{\gamma}$ nodes of depth at most $k + \log_2 \gamma$ and at most $2\sqrt{\gamma}$ nodes of depth $k + \log_2 \gamma$. On the other hand, any descendent of v at depth $k + \log_2 \gamma$ has at most n_v/γ points. Let $C(n_v)$ denote the number of nodes intersected by ℓ in the subtree rooted at a x-partitioned node that contains at most n_v points. Then we obtain the following recurrence

$$C(n_v) \leq 2\sqrt{\gamma} C(n_v/\gamma) + 8\sqrt{\gamma}$$

whose solution is $C(n_v) = O(n_v^{1/2+\varepsilon})$, where $\varepsilon = \log_\gamma 2$.

Maintaining the overlapping kd-tree. Maintaining the δ-overlapping kd-tree mainly involves tracking the number of points in the overlapping region. Consider the root of the kd-tree, assume n points are divided into $n/2$ red points and $n/2$ blue points by a vertical line at the beginning. W.l.o.g, assume red points have bigger coordinates. Consider the time-space plane, the goal is to track the depth of the lower(upper) envelope of red(blue) curves in the arrangement of blue(red) curves. An event happens when those two numbers add up to δn. So at least $\delta n/2$ points in the overlapping region have the same color, suppose they are red. Then the highest blue point must be above at least $\delta n/2$ red points. We define a red-blue inversion to be the intersection of a red curve and a blue curve. So there must be at least $\Omega(\delta n)$ red-blue inversions since the last recoloring. The number of recoloring events in the root of the kd-tree is bounded by $O(n/\delta)$, if the points follow pseudo-algebraic motion.

Since the combination of the trajectories of all the points in nodes of depth k becomes the arrangement of the trajectories of n points, the total number of recoloring events in depth k is bounded by $O(n2^k/\delta)$, using the same argument as above. When an event happens at depth k, we rebuild the subtree, which costs $O(n \log n / 2^k)$. So the total update cost sumed up over all events is $O(n^2 \log n/\delta)$. The amortized cost for one event is therefore $O(\log n)$. Putting everything together, we obtain the following.

Theorem 7. *Let S be a set of n points moving in the plane, and let $\delta > 0$ be a constant. We can maintain a δ-overlapping kd-tree by spending $O(\log n)$ amortized time at each event. Under pseudo-algebraic motion of S, the number of events is $O(n^2/\delta)$.*

Acknowledgements

The authors wish to thank John Hershberger and An Zhu for numerous discussions and for their contribution to the arguments in this paper. Research by all three authors is partially supported by NSF under grant CCR-00-86013. Research by P. Agarwal is also supported by NSF grants EIA-98-70724, EIA-01-31905, and CCR-97-32787, and by a grant from the U.S.–Israel Binational Science Foundation. Research by J. Gao and L. Guibas is also supported by NSF grant CCR-9910633 and grants from the Stanford Networking Research Center, the Okawa Foundation, and the Honda Corporation.

References

[1] P. K. Agarwal, L. Arge, and J. Erickson. Indexing moving points. In *Proc. Annu. ACM Sympos. Principles Database Syst.*, 2000. 175–186.
[2] P. K. Agarwal, L. Arge, and J. Vahrenhold. Time responsive external data structures for moving points. In *Workshop on Algorithms and Data Structures*, pages 50–61, 2001.
[3] P. K. Agarwal and J. Erickson. Geometric range searching and its relatives. In B. Chazelle, J. E. Goodman, and R. Pollack, editors, *Advances in Discrete and Computational Geometry*, volume 223 of *Contemporary Mathematics*, pages 1–56. American Mathematical Society, Providence, RI, 1999.
[4] P. K. Agarwal and M. Sharir. Arrangements and their applications. In J.-R. Sack and J. Urrutia, editors, *Handbook of Computational Geometry*, pages 49–119. Elsevier Science Publishers B.V. North-Holland, Amsterdam, 2000.
[5] J. Basch, L. Guibas, and J. Hershberger. Data structures for mobile data. *J. Alg.*, 31(1):1–28, 1999.

[6] J. Basch, L. J. Guibas, and G. D. Ramkumar. Reporting red-blue intersections between two sets of connected line segments. In *Proc. 4th Annu. European Sympos. Algorithms*, volume 1136 of *Lecture Notes Comput. Sci.*, pages 302–319. Springer-Verlag, 1996.

[7] J. L. Bentley. Multidimensional binary search trees used for associative searching. *Communications of the ACM*, 18(9):509–517, 1975.

[8] R. Cole, J. Salowe, W. Steiger, and E. Szemerédi. An optimal-time algorithm for slope selection. *SIAM J. Comput.*, 18(4):792–810, 1989.

[9] W. Cunto, G. Lau, and P. Flajolet. Analysis of kdt-trees: kd-trees improved by local reorganisations. *Workshop on Algorithms and Data Structures (WADS'89)*, 382:24–38, 1989.

[10] M. de Berg, M. van Kreveld, M. Overmars, and O. Schwarzkopf. *Computational Geometry: Algorithms and Applications*. Springer-Verlag, Berlin, 1997.

[11] T. K. Dey. Improved bounds on planar k-sets and k-levels. In *IEEE Symposium on Foundations of Computer Science*, pages 165–161, 1997.

[12] H. Edelsbrunner and E. Welzl. Constructing belts in two-dimensional arrangements with applications. *SIAM J. Comput.*, 15:271–284, 1986.

[13] L. J. Guibas. Kinetic data structures — a state of the art report. In P. K. Agarwal, L. E. Kavraki, and M. Mason, editors, *Proc. Workshop Algorithmic Found. Robot.*, pages 191–209. A. K. Peters, Wellesley, MA, 1998.

[14] R. H. Güting, M. H. Böhlen, M. Erwig, C. S. Jensen, N. A. Lorentzos, M. Schneider, and M. Vazirgiannis. A foundation for representing and querying moving objects. *ACM Trans. Database Systems*, 25(1):1–42, 2000.

[15] G. Kollios, D. Gunopulos, and V. J. Tsotras. On indexing mobile objects. In *Proc. Annu. ACM Sympos. Principles Database Syst.*, pages 261–272, 1999.

[16] J. Matoušek. Construction of ϵ-nets. *Discrete Comput. Geom.*, 5:427–448, 1990.

[17] J. Nievergelt and P. Widmayer. Spatial data structures: Concepts and design choices. In J.-R. Sack and J. Urrutia, editors, *Handbook of Computational Geometry*, pages 725–764. Elsevier Science Publishers B.V. North-Holland, Amsterdam, 2000.

[18] M. H. Overmars. *The Design of Dynamic Data Structures*, volume 156 of *Lecture Notes Comput. Sci.* Springer-Verlag, Heidelberg, West Germany, 1983.

[19] D. Pfoser, C. J. Jensen, and Y. Theodoridis. Novel approaches to the indexing of moving object trajectories. In *Proc. 26th Intl. Conf. Very Large Databases*, pages 395–406, 2000.

[20] C. M. Procopiuc, P. K. Agarwal, and S. Har-Peled. Star-tree: An efficent self-adjusting index for moving points. In *Proc. 4th Workshop on Algorithm Engineering and Experiments*, 2002.

[21] A. P. Sistla and O. Wolfson. Temporal conditions and integrity constraints in active database systems. In *Proceedings of the 1995 ACM SIGMOD International Conference on Management of Data*, pages 269–280, 1995.

[22] A. P. Sistla, O. Wolfson, S. Chamberlain, and S. Dao. Modeling and querying moving objects. In *Proc. Intl Conf. Data Engineering*, pages 422–432, 1997.

[23] G. Tóth. Point sets with many k-sets. *Discrete & Computational Geometry*, 26(2):187–194, 2001.

[24] S. Šaltenis, C. S. Jensen, S. T. Leutenegger, and M. A. Lopez. Indexing the positions of continuously moving objects. In *Proc. ACM SIGMOD International Conference on Management of Data*, pages 331–342, 2000.

[25] O. Wolfson, A. P. Sistla, S. Chamberlain, and Y. Yesha. Updating and querying databases that track mobile units. *Distributed and Parallel Databases*, pages 257–287, 1999.

Range Searching in Categorical Data: Colored Range Searching on Grid*

Pankaj K. Agarwal[1], Sathish Govindarajan[1], and S. Muthukrishnan[2]

[1] Department of Computer Science, Duke University
Durham, NC 27708
{pankaj,gsat}@cs.duke.edu
[2] AT&T Labs, Shannon Laboratory
180 Park Ave., Florham Park, NJ 07932
muthu@research.att.com

Abstract. Range searching, a fundamental problem in numerous applications areas, has been widely studied in computational geometry and spatial databases. Given a set of geometric objects, a typical range query asks for reporting all the objects that intersect a query object. However in many applications, including databases and network routing, input objects are partitioned into categories and a query asks for reporting the set of categories of objects that intersect a query object. Moreover in many such applications, objects lie on a grid. We abstract the category of an object by associating a color with each object. In this paper, we present efficient data structures for solving the colored range-searching and colored point-enclosure problem on $U \times U$ grid. Our data structures use near- linear space and answer a query in $O(\log \log U + k)$ time, where k is the output size. As far as we know, this is the first result on colored range-searching for objects lying on a grid.

1 Introduction

We are given a set of geometric objects – points, lines, polygons – to preprocess. Given a query object, the *range searching* problem is to return the intersection of the query with the given set of objects. In the past few decades, range searching has been extensively studied. See [1, 16] for recent surveys. The fascination with range searching is because it has myriad applications in areas of database retrieval, computer aided design/manufacturing, graphics, geographic information systems, etc. Specifically, range querying is exceedingly common in database systems (Who are the students with GPAs greater than 3.8? Name the customers whose age is in [20 − 40] and whose income is in greater than 100k?). Every commercial database system in the market has optimized data structures

---

* Research by the first two authors is supported by NSF under grants CCR-00-86013 EIA-98-70724, EIA-01-31905, and CCR-97-32787, and by a grant from the U.S.– Israel Binational Science Foundation. Part of this work was done when the second author was visiting the third author at DIMACS.

for solving various range searching problems. Thus, range searching data structures remain one of the few data structures that actually get used in practice at commercial scale.

In this paper, we study an extension to the basic range searching problem. Our motivation arose from database applications where we observed that range searching amid objects are common, but what was highly prevalent was range searching in which the objects have categories and the query often calls for determining the (distinct) list of categories on the objects that intersected the query object. Here are two examples:

(i) A canonical example is an analyst who is interested in "What are the sectors (telecom, biotech, automobile, oil, energy, ..) that had $5-10\%$ increase in their stock value?". Here each stock has a category that is the industry sector it belongs to, and we consider a range of percentage increase in the stock value. We are required to report all the distinct sectors that have had one or more of their stocks in desired range of growth, and not the specific stocks themselves. This is one-dimensional range searching in categorical data.[1]

(ii) As another example, consider a database of IP packets sent through an Internet router over time. IP packets have addresses, and organizations get assigned a contiguous range of IP addresses that are called subnets (eg., all hosts within Rutgers Univ, all hosts within uunet, are distinct examples of subnets, and all IP addressed within a subnet share a long common prefix). An analyst of the IP traffic may ask "Between time 10 AM and 11 AM today, which are the subnets to which traffic went from uunet through the given router?". Here the uunet specifies a range of IP addresses and time specifies an orthogonal range, and each IP packet that falls within the cross-product of these two ranges falls into a category based on the subnet of the destination of the packet. This is an example of a two-dimensional range searching in categorical data.

Thus range searching (in all dimensions) in categorical data is highly prevalent in database queries. Range searching in categorical data also arises in document retrieval problems [15] and in indexing multidimensional strings [11].

The range-searching problem in categorical data, as they arise in database applications, have the following common characteristics. First, the points (and endpoints of objects) are on a grid (eg., IP addresses are 32 bit integers and time, stock prices etc. are typically rounded). Second, while it is important to optimize the preprocessing time and space, query time is very critical. In data analysis applications, response time to a query must be highly optimized. Third, often the number of categories is very large (for example, in the router traffic example above, the number of subnets is in thousands). Finally, the dataset may be clustered along one of the dimensions or in a category, and hence, an efficient algorithm cannot afford to retrieve all points in the given range and

---

[1] Flip Korn of AT&T Research clarified that in classical database language, this may be thought of as GROUP BY based on the category of the stock following a range SELECT. Therefore, this is of fundamental interest in databases.

search through them to find the distinct categories they belong to; the output set may be far smaller than the set of all points in the range. In other words, the naive algorithm of doing a classical range query first and followed by selecting distinct categories form the answer set will be inefficient.

In this paper, we study two very basic problems in range searching in categorical data, keeping the above characteristics in mind. The points (and end points of rectangles) come from an integer grid. We study the range-searching and point-enclosure problems. We present highly efficient algorithms for both problems, in particular, the query time is $O(\log \log U)$ where U is the grid size.

1.1 Problems

In abstraction, the problems of our interest are as follows. The notion of the category associated with an object is abstracted as the *color* associated with it.

1. *Colored range searching.* We are given P, a set of n colored points in $[0, U]^2$, for preprocessing. Given a query rectangle q whose endpoints lie on the grid $[0, U]^2$, the problem is to output the set of *distinct* colors of points contained in q.
2. *Colored point enclosure problem.* We are given P, a set of n colored rectangles whose endpoints lie on the grid $[0, U]^2$, for preprocessing. Given a query point $q = [q_1, q_2]$, where $q_1, q_2 \in [0, U]$, the problem is to output the set of *distinct* colors of rectangles that contain q.

The problems above are stated in two dimensions, and for most part of the paper, we focus on the two dimensional case. However, our approach gives best-known bounds for higher dimensional cases as well, and they are detailed in Section 4.

Typically, the data structure results on the grid are interpreted under the assumption that U is polynomial in n. In our problem, U is indeed polynomial in n since the input is static and we can renumber the points and corners into an integer set of size $1 \ldots O(n)$; any query point will first get mapped into the $O(n)$ range using a predecessor query (this is a standard trick, taking time $O(\log \log U)$) and then the query processing begins. As a result $U = O(n)$, provided we are willing to have $O(\log \log U)$ additive term in the preprocessing of the query. We refer to this technique as *input mapping* in the rest of the discussion.

1.2 Our Results

We present data structures for both problems that are highly optimized for query processing.

Colored range searching. We construct an $O(n \log^2 U)$ sized data structure that answers colored range searching queries in $O(\log \log U + k)$ time, where k is the output size.

Previously known result for this problem did not use the grid properties and produced a data structure that uses $O(n \log^2 n)$ size and answers a query in $O(\log n + k)$ time [14, 13]. Previous algorithms that work on the grid only studied the classical range searching problem and produced a $O(n \log^\epsilon n)$ size data structure with $O(\log \log U + k)$ query time [3]; the query time of such a data structure for colored range searching can be arbitrarily bad.

Colored point enclosure. For any integer parameter ℓ, we can construct a $O(n\ell^2 \log_\ell^2 U)$-size data structure that can answer a colored point enclosure queries in $O(k \log_\ell U)$ time, where k is the output size.

We can obtain a space-time tradeoff for our data structure by choosing appropriate values for ℓ. For example, setting $\ell = U^\epsilon$ and using the input mapping, we get a data structure of $O(n^{1+\epsilon})$-size that can answer colored point enclosure queries in $O(k + \log \log U)$ time where k is the output size. Most of the previous results make an additional assumption that $U = O(n^c)$ for some constant c. If we make this assumption, the query time reduces to $O(k)$. Note that the $\log \log U$ term in the query is due to input mapping, which is not required if $U = O(n^c)$. This result, surprisingly, matches the best known bound for what appears to be a simpler problem of point enclosure problem on the grid without considering the colors, i.e., the standard point enclosure problem [10]. Other previous results for the colored point enclosure problem do not assume an underlying grid and produce a data structure that uses $O(n \log n)$ space and answers queries in $O(\log n + k)$ time [13].

1.3 Technical Overview of Our Results

We use contrasting techniques to obtain the two results.

For the range searching problem, best-known results typically use recursive divide and conquer on one dimension after another [4, 5, 6]. We start with a one dimensional result. There are at least three known solutions [14, 13, 15] for the colored range searching problem in one dimension. We use their intuition but present yet another solution which, in contrast to the others, we are able to make dynamic as well as persistent. Then we use a sweep-line approach to obtain our result in $2D$ (rather than the recursive divide and conquer).

For the point-enclosure problem, the previously known approach [10] relies on building a one dimensional data structure using exponential trees and then extending it to two dimensions by dynamizing and making it persistent. We build the desired one dimensional data structure efficiently for colored point enclosure (this may be of independent interest), but then generalize it to two dimensions directly to obtain two-dimensional exponential trees. This gives us our bound for the colored point enclosure problem.

In Section 2, we present our results for colored range searching in two dimensions. In Section 3, we present our results for colored point enclosure in two dimensions. In Section 4, we present extensions to higher dimensions.

2 Colored Range Searching in 2D

In this section, we describe a data structure to solve the colored range-searching problem on $U \times U$ grid. In Section 2.1, we present a structure to answer the 1-dimensional colored range query. In Section 2.2 we extend the data structure to answer two dimensional colored range queries. The main idea of the extension is as follows. We make the 1-dimensional structure partially persistent and use the sweep-line paradigm to answer three-sided queries. We then extend this structure to answer four-sided range queries.

2.1 Colored Range Searching in 1D

Let P be a set of n colored points in $[0, U]$. Let C denote the set of distinct colors in the point set P. We first solve the colored range-searching problem for the special case of semi-infinite query q i.e, $q = [q_1, \infty]$. For each color $c \in C$, we pick the point $p_c \in P$ with color c having the maximum value. Let P^{max} denote the set of all such points. Let L be the linked list of points in P^{max}, sorted in non-increasing order. We can answer the colored range query by walking along the linked list L until we reach the value q_1 and report the colors of the points encountered. It can be shown that there exists a point $l \in P$ of color c in interval q if and only if there exists a unique point $m \in P^{max}$ of color c in interval q. We can solve the other case, i.e, $q = [-\infty, q_2]$ in a similar manner.

We now build a data structure to answer a general colored range query $q = [q_1, q_2]$. The data structure is a trie T [2] built on the values of points $p \in P$. For each node $v \in T$, let P_v denote the set of points contained in the subtree of T rooted at v. At each internal node v, we store a secondary structure, which consists of two semi-infinite query data structures L_v and R_v corresponding to the queries $[q, \infty]$ and $[-\infty, q]$. Note that the structures as described above are sorted linked lists on max/min coordinates of each color in P_v^{max} and P_v^{min}. For each non-root node $v \in T$, let $B(v) = 0$ if v is a left child of its parent and $B(v) = 1$ if v is a right child of its parent. To efficiently search in the trie T, we adopt the hash table approach of Overmars [17]. We assign an index I_v, called node index, for each non-root node $v \in T$. I_v is an integer whose bit representation corresponds to the concatenation of $B(w)$'s, where w is in the path from root to v in T. Define the level of a node v as the length of the path from the root to v in T. We build a static hash table H_i on the indices I_v of all nodes v at level i, $1 \leq i \leq \log U$ [12]. We store the pointer to node $v \in T$ along with I_v in the hash table. The hash tables H_i uses linear space and provides $O(1)$ worst case lookup. The number of nodes in the trie T is $O(n \log U)$. Since each point $p \in P$ might be stored at most once at each level in the lists R_v, L_v, and the height of the trie T is $O(\log U)$, the total size of the secondary structure is $O(n \log U)$. Thus the size of the entire data structure is $O(n \log U)$.

The trie T can be efficiently constructed level by level in a top down fashion. Initially, we sort the point set P to get the sorted list of P_{root}. Let us suppose we have constructed the secondary structures at level $i - 1$. Let z be a node at level i and let v and w be the children of z in T. We partition the sorted list of points in P_z into sorted list of points in P_v and P_w. We then construct in $O(|P_v|)$ time the lists L_v and R_v at node v by scanning P_v once. We then construct the hash table H_i on indices I_v for all nodes v in level i. The total time spent in level i is $O(n)$. Hence the overall time for constructing the data structure is $O(n \log n + n \log U) = O(n \log U)$.

Let $[q_1, q_2]$ be the query interval. Let z_1(resp. z_2) be the leaf of T to which the search path for q_1(resp. q_2) leads. We compute the least common ancestor v of z_1 and z_2 in $O(\log \log U)$ time by doing a binary search on the height of the trie [17]. Let w and z be the left and right child of v. All the points $p \in [q_1, q_2]$ are contained in P_w or P_z. Hence we perform the query $[q_1, q_2]$ on P_w and P_z. Since all points in P_w have value $\leq q_2$ the queries $[q_1, q_2]$ and $[q_1, \infty]$ on P_w gives the same answer. Similarly, queries $[q_1, q_2]$ and $[-\infty, q_2]$ on P_z gives the same answer. Thus we perform two semi-infinite queries $[q_1, \infty]$ and $[-\infty, q_2]$ on P_w and P_z respectively, and the final output is the union of the output of the two semi-infinite queries. Each color in the output list is reported at most twice.

Theorem 1. *Let P be a set of n colored points in $[0, U]$. We can construct in $O(n \log U)$ time, a data structure of size $O(n \log U)$ so that a colored range searching query can be answered in $O(\log \log U + k)$ time, where k is the output size.*

Remark. The previous best-known result for 1-dimensional colored range query uses $O(n)$ space and $O(\log \log U + k)$ query time [15]. Their solution relies on Cartesian trees and Least Common Ancestors(LCA) structures. While Least Common Ancestor structures can be dynamized [7], it remains a challenge to make the Cartesian trees dynamic and persistent. In contrast, our 1-dimensional structure above can be made dynamic and persistent and hence can be extended to 2-dimensions using the sweep-line paradigm, as we show next.

2.2 Colored Range Searching in 2D

We first consider the case when the query is 3-sided i.e., $q = [x_1, x_2] \times [-\infty, y_2]$. Let M denote the 1D data structure described in the previous section. Our approach is to make M dynamic and partially persistent. Then we use the standard sweep-line approach to construct a partially persistent extension of M so that a 3-sided range query can be answered efficiently.

First we describe how to dynamize M. Note that M consists of the trie T with secondary structure L_v, R_v at each node v of T and static hash tables H_i on node indices I_v for all nodes v on a given level i. We use the dynamic hash table of Dietzfelbinger et al [8] instead of the static hash table H_i used in the 1D structure. We maintain for each node v, a static hash table $\mathcal{H}_v$ on the colors of points in P_v. We also maintain a balanced binary search tree TL_v (resp. TR_v) on L_v (resp. R_v).

To insert a point p with color c, we insert p in the trie T. We also may have to insert p in the secondary structures L_v, R_v for all nodes v in the root to leaf search path of p in T. We insert p in L_v(resp. R_v) if P_v^{max} (resp. P_v^{min}) has no points with color c, or if the value of p is greater (resp. less) than the value of point with color c in P_v^{max} (resp. P_v^{min}). We can check the above condition using $\mathcal{H}_v$. If the condition is satisfied, we insert p into L_v(resp. R_v) in $O(\log n)$ time using the binary search tree TL_v (resp. TR_v). Finally if a new trie node v is created while inserting p into T, we insert the index I_v into the appropriate hash table H_i. Since we might insert p in all the nodes in the root-leaf path of p in T, the insert operation costs $O(\log n \log U)$ time. We do not require to delete points from M since the sweep-line paradigm would only insert points.

We now describe how to make the above structure partially persistent:

(i) The trie structure is made persistent by using the node copying technique [9] (creating copies of nodes along the search path of p). We number the copies of node v by integers 1 through n. Let c_v denote this copy number of node v.
(ii) We make the lists L_v and R_v persistent across various versions of v using [9].
(iii) We make the hash table H_i, corresponding to each level i of the trie T, persistent by indexing each node v using the static index I_v and the copy number c_v of node v, i.e., new index I'_v is bit concatenation of I_v and c_v.

The color hash table $\mathcal{H}_v$ and the binary search trees TL_v, TR_v need not be made persistent since they are not used by the query procedure.

We perform a sweep-line in the $(+y)$-direction. Initially the sweep-line is at $y = -\infty$ and our data structure D is empty. Whenever the sweep line encounters a point in P, we insert p in D using the persistent scheme described earlier. When the sweep line reaches $y = \infty$, we have the required persistent structure D. Note that D consists of $D_i, 1 \leq i \leq n$, the different versions of the dynamic 1D structure got by inserting points in the sweep-line process. We also build a Van Embe Boas tree $\mathcal{T}$ [18] on the y-coordinates of P.

The node-copying technique [9], used to make the trie T and the lists L_v, R_v persistent, only introduces a constant space and query time overhead. Thus, the space used by our persistent data structure D is still $O(n \log U)$ and the query time to perform a 1D colored range-searching query on any given version D_i of D is $O(k)$ where k is the output size.

Let $[x_1, x_2] \times [-\infty, y]$ be a three sided query. We perform a predecessor query on y using the Van Embe Boas tree $\mathcal{T}$ to locate the correct version of the persistent structure. This version contains only those points of P whose y-coordinates are at most y, so we perform a 1-dimensional query on this version with the interval $[x_1, x_2]$, as described in Section 2.1. We locate the splitting node v of x_1, x_2 using the hash tables H_i and then walk along the correct version of the secondary structure, i.e. lists L_v, R_v. The predecessor query can be done in $O(\log \log U)$ time and we can perform the 1D query in $O(\log \log U + k)$ time where k is the number of distinct colors of points contained in the query.

Theorem 2. *Let P be a set of n colored points in $[0, U]^2$. We can construct a data structure of size $O(n \log U)$ in $O(n \log n \log U)$ time so that we can answer a 3-sided colored range-searching query in $O(\log \log U + k)$ time, where k is the output size.*

We now extend the 3-sided query structure to answer a general 4-sided query. The approach is exactly the same as the one in Section 2.1 where we extended a semi-infinite query structure to a 1D query structure. We build a trie T on y-coordinates of P. At each node v, we store two secondary structures U_v and W_v, which are the 3-sided query structures for point set P_v, built as described above. U_v corresponds to the 3-sided query $q = [x_1, x_2] \times [y, \infty]$ and W_v corresponds to the 3-sided query $q = [x_1, x_2] \times [-\infty, y]$. We also build a hash table H_i for all nodes v of a given level i of the trie T.

The query process is also similar to the 1-dimensional query. We locate the splitting node v in the search path of y_1 and y_2 using the hash table H_i in constant time. Let w and z be the left and right son of v respectively. We then perform two 3-sided queries $q = [x_1, x_2] \times [y_1, \infty]$ and $q = [x_1, x_2] \times [-\infty, y_2]$ on secondary structures U_w and W_z respectively. The final output is the union of the output of the semi-infinite queries. Each color is reported at most four times, at most twice in each 3-sided query.

Theorem 3. *Let P be a set of n colored points in $[0, U]^2$. We can construct a data structure of size $O(n \log^2 U)$ in $O(n \log n \log^2 U)$ time so that we can answer a 4-sided colored range-searching query in $O(\log \log U + k)$ time, where k is the output size.*

3   Colored Point Enclosure in 2D

In this section, we describe a data structure to solve the colored point-enclosure problem in $2D$. We first describe a data structure to solve the problem in $1D$. Then we extend this data structure to answer two dimensional colored point-enclosure queries.

The 1D structure. Let P be a set of n colored intervals whose endpoints lie on $[0, U]$. Fix a parameter $\ell \geq 0$. Starting with the interval $I = [0, U]$, we recursively partition I into ℓ equal sized intervals. The recursion stops when an interval I contains no endpoints of intervals in P. Let T be the tree that corresponds to this recursive divide and let I_v denote the interval corresponding to a node v in T. Let $p(v)$ denote the parent of v in T. There are at most $2n\ell$ leaves in T.

We call an interval $p \in P$ *short* at a node v if (at least) one of the endpoints of p lies inside I_v. We call p *long* at v if $I_v \subseteq p$. At each node $v \in T$, we store C_v, the set of distinct colors of intervals of P that are long at v but short at $p(v)$. Since any interval $p \in P$ is short for at most two nodes (nodes v where interval I_v contains the endpoints of p) of any given level, the color of p is stored in at most 2ℓ nodes of any given level. Thus,

$$\sum_{v \in T} |C_v| \leq 2n\ell \log_\ell U.$$

Hence, T requires $O(n\ell \log_\ell U)$ space. We can adapt the above recursive partitioning scheme so that we can compute short and long intervals and the set C_v in $O(n\ell)$ time at each level. Hence, T can be constructed in $O(n\ell \log_\ell U)$ time.

Let q be a query point. The query procedure searches for q and visits all nodes v in T in the root-leaf search path of q in T. At each node, we report all the colors in C_v. Each color can be reported once in each level. Hence, the query time is $O(k \log_\ell U)$, where k is the output size.

We improve the query time to $O(\log_\ell U + k)$ using a simple technique. After we build the data structure as described above, we traverse every root-leaf path in T starting from the root and perform the following operation. Let $v_0, v_1, \ldots v_m$ be any root to leaf path in T. Starting with $i = 1$, we recursively calculate $C'_{v_i} = C_{v_i} \setminus C'_{v_{i-1}}$ for $1 \leq i \leq m$. We store C'_v at v. By definition, $C'_{v_i} \cap C'_{v_j} = \emptyset, 1 \leq i, j \leq m, i \neq j$. This implies that each color is present in at most one list C'_v in any root to leaf path of T. The above operation can be efficiently performed by traversing T in a top-down fashion level by level. The query time thus reduces to $O(\log_\ell U + k)$, where k is the output size.

Theorem 4. *Let P be a set of n colored intervals whose endpoints lie in $[0, U]$. We can construct a $O(n\ell \log_\ell U)$ sized data structure in $O(n\ell \log_\ell U)$ time so that a colored point-enclosure query can be answered in $O(\log_\ell U + k)$ time where k is the output size.*

The 2D structure. We now extend the 1D structure to 2D. Let P be a set of n colored rectangles whose endpoints lie on a $U \times U$ grid. The data structure is similar to a multi-resolution structure like quad-tree. Starting with the square $[0, U]^2$, we recursively divide the current square into ℓ^2 equal-sized squares, where ℓ is a parameter. The recursion stops when a square does not contain any endpoints of the input rectangles. Let T be the tree denoting this recursive partition, and let the square S_v correspond to the node v in T. Some of the notation used in the description of the $1D$ structure will be redefined below.

We call a rectangle r *long* at v if $S_v \subseteq r$; we call r *short* if at least one vertex of r lies in S_v; we say that r *straddles* v if an edge of r intersects S_v but r is not short at v. Figure 1 illustrates the above cases. If r straddles v, then either a horizontal edge or a vertical edge of r intersects S_v, but not both.

For each node v, let C_v denote the set of distinct colors of rectangles that are long at v but short at $p(v)$. Let Σ_v denote the set of rectangles that straddle v and are short at $p(v)$. Set $\chi_v = |C_v|$ and $m_v = |\Sigma_v|$. If a rectangle r is short at v, its color could be stored in all children of v, but it straddles at most 4ℓ children of v. Since a rectangle is short for at most four nodes of any given level,

$$\sum_{v \in T} \chi_v = O(n\ell^2 \log_\ell U) \quad \text{and} \quad \sum_{v \in T} m_v = O(n\ell \log_\ell U).$$

We store C_v at v. We also store at v, two 1D point-enclosure data structures $T_v^=, T_v^\|$ as its secondary structures. Let $\Sigma_v^=$ (resp. $\Sigma_v^\|$) denote the set of rectangles in Σ_v whose horizontal (resp. vertical) edges intersect S_v. For a rectangle $r \in S_v$, let $r^=$ (resp. $r^\|$) denote the x-projection (resp. y-projection) of $r \cap S_v$. See Figure 1 (iii). Let $\mathcal{I}_v^= = \{r^= \mid r \in \Sigma_v^=\}$ and $\mathcal{I}_v^\| = \{r^\| \mid r \in \Sigma_v^\|\}$. $T_v^=$ (resp. $T_v^\|$) is the 1D point-enclosure data structure on the set of intervals $\mathcal{I}_v^=$

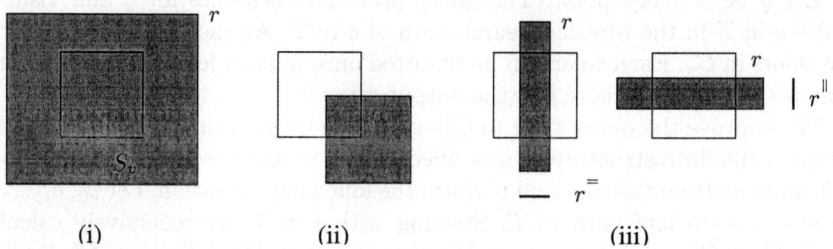

Fig. 1. (i) r long at v; (ii) r short at v; (iii) r straddles v

(resp. $\mathcal{I}_v^{\|}$). By Theorem 4, $T_v^{=}, T_v^{\|}$ require $O(m_v \ell \log_\ell U)$ space. Hence, T requires $O(n\ell^2 \log_\ell^2 U)$ space. T can be constructed in $O(n\ell^2 \log_\ell^2 U)$ time in a top-down manner.

Let $q = (q_x, q_y) \in [0, U]^2$ be a query point. The query procedure searches for q in T and visits all nodes v in T in the root-leaf search path of q in T. At each node, we report all the colors in the list C_v. For a rectangle $r \in \Sigma_v^{=}$, $q \in r$ if and only if $q_x \in r^{=}$. Similarly for $r \in \Sigma_v^{\|}$, $q \in r$ if and only if $q_y \in r^{\|}$. Therefore we query the 1D structures $T_v^{=}$ and $T_v^{\|}$ with q_x and q_y respectively. The query time for the 1D structure is $O(\log_\ell U + k)$, where k is the number of distinct colors of intervals containing the query point p. Since we make exactly two 1D queries at each level and each color can be duplicated once in each level of T, the total query time is $O(k \log_\ell U)$, where k is the number of distinct colors of rectangles containing the query point q.

Theorem 5. *Let P be a set of n colored rectangles whose endpoints lie in $[0, U]^2$. We can construct a $O(n\ell^2 \log_\ell^2 U)$ sized data structure in $O(n\ell^2 \log_\ell^2 U)$ time so that a colored point-enclosure query can be answered in $O(k \log_\ell U)$ time where k is the output size.*

Setting $\ell = U^\epsilon$ and using input mapping, we get

Theorem 6. *Let P be a set of n colored rectangles whose endpoints lie on the grid $[0, U]^2$. We can construct a $O(n^{1+\epsilon})$ sized data structure in $O(n^{1+\epsilon})$ time so that a colored point-enclosure query can be answered in $O(\log \log U + k)$ time where k is the output size.*

Note that if we make the assumption that $U = O(n^c)$, we can remove the input mapping and thus the above query time gets reduced to $O(k)$, where k is the output size.

4 Extensions

We show how to extend our solutions to higher dimensions. We will first sketch how to extend our two dimensional colored point-enclosure data structure to higher dimensions. Using this, we construct a data structure to solve the colored range searching problem in higher dimensions as well.

Colored point enclosure in higher dimensions. We show how to extend the above data structure to answer point-enclosure in d-dimensions. Let P be a set of n colored hyper-rectangles whose endpoints lie on the grid $[0, U]^d$. We construct a tree T using a recursive partitioning scheme. At each node v of T, we define long and short hyper-rectangles and hyper-rectangles that straddle at v. For each node v, let C_v denote the set of distinct colors of rectangles that are long at v but short at $p(v)$. At v, we store C_v and a family of $(d-1)$-dimensional point-enclosure structures as secondary structure. The query process is similar to the 2D case. We omit the details of the structure for lack of space. Using the same analysis as in Section 3, we can show that the space used by the data structure is $O(n\ell^d \log_\ell^d U)$ and the query time is $O(k \log_\ell^{d-1} U)$, where k is the output size. This gives us an analogous result as in Theorem 5, for any dimension d. In particular, for a fixed d, setting $\ell = U^\epsilon$ and using input mapping, we obtain the following.

Theorem 7. *Let P be a set of n colored hyper-rectangles whose endpoints lie on the grid $[0, U]^d$, for a fixed d. We can construct a $O(n^{1+\epsilon})$ size data structure in $O(n^{1+\epsilon})$ time so that a colored point-enclosure query can be answered in $O(\log \log U + k)$ time, where k is the the output size.*

Note that we can reduce the above query time to $O(k)$ by removing the input mapping, if we make an assumption that $U = O(n^c)$ for some constant c.

Colored range searching in higher dimensions. Our approach is to first solve the more general colored rectangle intersection problem in higher dimensions, which is defined as follows. We wish to preprocess a set P of n hyper-rectangles, whose endpoints lie in $[0, U]^d$, into a data structure so that we can report the list of distinct colors of hyper-rectangles in P that intersect a query hyper-rectangle r. We present a simple reduction between the rectangle-intersection problem in d dimensions and the point-enclosure problem in $(2d)$- dimensions.

Let the hyper-rectangle r be represented as $[x_{11}, x_{12}] \times \cdots \times [x_{d1}, x_{d2}]$. We map r to a hyper-rectangle r' in $(2d)$- dimensions given by $[-\infty, x_{12}] \times [x_{11}, \infty] \times \cdots \times [-\infty, x_{d2}] \times [x_{d1}, \infty]$. The query hyper-rectangle $q = [y_{11}, y_{12}] \times \cdots \times [y_{d1}, y_{d2}]$ is mapped to a $(2d)$- dimensional point $q' = (y_{11}, y_{12}, \ldots, y_{d1}, y_{d2})$. It is easy to see that a hyper-rectangle r intersects a hyper-rectangle q iff the hyper-rectangle r' contains the point q'. This observation gives us a reduction between rectangle-intersection problem and point-enclosure problem. Applying the reduction and using Theorem 7, we obtain the following.

Theorem 8. *Let P be a set of n colored hyper-rectangles whose endpoints lie on the grid $[0, U]^d$. We can construct a $O(n^{1+\epsilon})$ sized data structure in $O(n^{1+\epsilon})$ time so that a colored rectangle-intersection query can be answered in $O(\log \log U + k)$ time, where k is the output size*

The colored rectangle-intersection problem reduces to a colored range-searching problem when the input hyper-rectangles are points. Hence we get analogous results as Theorem 8 for colored range-searching in higher dimensions too. Note that in 2D, we can answer a colored range query in $O(\log \log U + k)$ time using a $O(n \log^2 U)$ size data structure, as described in Section 2.

References

[1] P. K. Agarwal and J. Erickson. Geometric range searching and its relatives. In B. Chazelle, J. E. Goodman, and R. Pollack, editors, *Advances in Discrete and Computational Geometry*, volume 223 of *Contemporary Mathematics*, pages 1–56. American Mathematical Society, Providence, RI, 1999.

[2] A. V. Aho, J. E. Hopcroft, and J. D. Ullman. *Data Structures and Algorithms*. Addison Wesley Press, 1983.

[3] S. Alstrup, G. Brodal, and T. Rauhe. New data structures for ortogonal range searching. In *Proc. 41th Annual IEEE Symp. Foundations of Comp. Sci.*, pages 198–207, 2000.

[4] J. L. Bentley. Multidimensional divide-and-conquer. *Commun. ACM*, 23(4):214–229, 1980.

[5] B. Chazelle. Filtering search: A new approach to query-answering. *SIAM J. Comput.*, 15(3):703–724, 1986.

[6] B. Chazelle. A functional approach to data structures and its use in multidimensional searching. *SIAM J. Comput.*, 17(3):427–462, June 1988.

[7] R. Cole and R. Hariharan. Dynamic LCA queries. In *Proc. 10th Annual Symposium on Discrete Algorithms*, pages 235–244, 1999.

[8] M. Dietzfelbinger, A. Karlin, K. Mehlhorn, F. M. auf der Heide, H. Rohnert, and R. E. Tarjan. Dynamic perfect hashing: Upper and lower bounds. *SIAM J. Comput.*, 23:738–761, 1994.

[9] J. R. Driscoll, N. Sarnak, D. D. Sleator, and R. E. Tarjan. Making data structures persistent. *Journal of Computer and System Sciences*, 38:86–124, 1989.

[10] D. Eppstein and S. Muthukrishnan. Internet packet filter management and rectangle geometry. In *Proc. 12th Annual Symp. on Discrete Algorithms*, pages 827–835, 2001.

[11] P. Ferragina, N. Koudas, D. Srivastava, and S. Muthukrishnan. Two-dimensional substring indexing. In *Proc. of Intl Conf. on Principles of Database Systems*, pages 282–288, 2001.

[12] M. L. Fredman, J. Komlos, and E. Szemeredi. Storing a sparse table with o(1) worst case access time. *J. Assoc. Comput. Mach.*, 31:538–544, 1984.

[13] J. Gupta, R. Janardan, and M. Smid. Further results on generalized intersection searching problems: Counting,reporting and dynamization. In *Proc. 3rd Workshop on Algorithms and Data structures*, pages 237–245, 1993.

[14] R. Janardan and M. Lopez. Generalized intersection searching problems. *J. of Comp. Geom. and Appl.*, 3:39–70, 1993.

[15] S. Muthukrishnan. Efficient algorithms for document retrieval problems. In *Proc. 13th Annual Symposium on Discrete Algorithms*, 2002.

[16] J. Nievergelt and P. Widmayer. Spatial data structures: Concepts and design choices. In J.-R. Sack and J. Urrutia, editors, *Handbook of Computational Geometry*, pages 725–764. Elsevier Science Publishers B.V. North-Holland, Amsterdam, 2000.

[17] M. Overmars. Efficient data structures for range searching on a grid. *Journal of Algorithms*, 9:254–275, 1988.

[18] P. van Emde Boas. Preserving order in a forest in less than logarithmic time and linear space. *Information Processing Letters*, 6:80–82, 1977.

Near-Linear Time Approximation Algorithms for Curve Simplification*

Pankaj K. Agarwal[1], Sariel Har-Peled[2], Nabil H. Mustafa[1], and Yusu Wang[1]

[1] Department of Computer Science, Duke University
Durham, NC 27708-0129, USA
{pankaj,nabil,wys}@cs.duke.edu

[2] Department of Computer Science, DCL 2111, University of Illinois
1304 West Springfield Ave., Urbana, IL 61801, USA
sariel@cs.uiuc.edu

Abstract. We consider the problem of approximating a polygonal curve P under a given error criterion by another polygonal curve P' whose vertices are a subset of the vertices of P. The goal is to minimize the number of vertices of P' while ensuring that the error between P' and P is below a certain threshold. We consider two fundamentally different error measures — Hausdorff and Fréchet error measures. For both error criteria, we present near-linear time approximation algorithms that, given a parameter $\varepsilon > 0$, compute a simplified polygonal curve P' whose error is less than ε and size at most the size of an optimal simplified polygonal curve with error $\varepsilon/2$. We consider monotone curves in the case of Hausdorff error measure and arbitrary curves for the Fréchet error measure. We present experimental results demonstrating that our algorithms are simple and fast, and produce close to optimal simplifications in practice.

1 Introduction

Given a polygonal curve, the curve simplification problem is to compute another polygonal curve that approximates the original curve, according to some predefined error criterion, and whose complexity is as small as possible. Curve simplification has useful applications in various fields, including geographic information systems (GIS), computer vision, graphics, image processing, and data compression. The massive amounts of data available from various sources make efficient processing of this data a challenging task. One of the major applications of this data is for cartographic purposes, where the information has to be visualized and presented as a simple and easily readable map. Since the information is too dense, the maps are usually simplified. To this end, curve simplification is used to simplify the representation of rivers, roads, coastlines, and other features when a map at large scale is produced. There are many advantages of the

---

* Research by the first, the third and the fourth authors is supported by NSF grants ITR-333-1050, EIA-98-70724, EIA-01-31905, CCR-97-32787, and CCR-00-86013. Research by second author is partially supported by a NSF CAREER award CCR-0132901.

simplification process, such as removing unnecessary cluttering due to excessive detail, saving disk and memory space, and reducing the rendering time.

1.1 Problem Definition

Let $P = \langle p_1, \ldots, p_n \rangle$ denote a polygonal curve in $\mathbb{R}^2$ or $\mathbb{R}^3$, where n is the size of P. A curve P in $\mathbb{R}^2$ is x-monotone if the x-coordinates of p_i are increasing. A curve $P \in \mathbb{R}^3$ is xy-monotone if both the x-coordinates and y-coordinates of p_i are increasing. A curve is *monotone* if there exists a coordinate system for which it is x-monotone (or xy-monotone). A polygonal curve $P' = \langle p_{i_1}, \ldots, p_{i_k} \rangle \subseteq P$ *simplifies* P if $1 = i_1 < \cdots < i_k = n$.

Let $d(\cdot, \cdot)$ denote a distance function between points. In this paper we use L_1, L_2, L_∞, and uniform metrics to measure the distance between two points. Uniform metric is defined in $\mathbb{R}^2$ as follows: For two points $a = (a_x, a_y), b = (b_x, b_y)$ in $\mathbb{R}^2$, $d(a,b)$ is $|a_y - b_y|$ if $a_x = b_x$ and ∞ otherwise. The distance between a point p and a segment e is defined as $d(p,e) = \min_{q \in e} d(p,q)$.

Let $\delta_M(p_i p_j, P)$ denote the error of a segment $p_i p_j$ under error measure M. M can be either Hausdorff (H) or Fréchet (F) error measure and will be defined formally in Section 1.3. The error of simplification $P' = \langle p_{i_1}, \ldots, p_{i_k} \rangle$ of P is defined as

$$\delta_M(P', P) = \max_{1 \le j < k} \delta_M(p_{i_j} p_{i_{j+1}}, P).$$

Call P' an ε-*simplification* of P if $\delta_M(P', P) \le \varepsilon$. The *curve-simplification problem* is to compute the smallest size ε-simplification of P, with its size denoted as $\kappa_M(\varepsilon, P)$. $\delta_M(\varepsilon, P)$ and $\kappa_M(\varepsilon, P)$ will be denoted as $\delta_M(\varepsilon)$ and $\kappa_M(\varepsilon)$ respectively when P is clear from the context.

If we remove the constraint that the vertices of P' are a subset of the vertices of P, then P' is called a *weak ε-simplification* of P.

1.2 Previous Results

The problem of approximating a polygonal curve has been studied extensively during the last two decades, both for computing an ε-simplification and a weak ε-simplification (see [Wei97] for a survey). Imai and Iri [II88] formulated this simplification problem as computing a shortest path between two nodes in a directed acyclic graph. Under the Hausdorff measure with uniform metric, their algorithm runs in $O(n^2 \log n)$ time. Chin and Chan [CC92], and Melkman and O'Rourke [MO88] improve the running time of their algorithm to quadratic or near quadratic. Agarwal and Varadarajan [AV00] improve the running time to $O(n^{4/3+\delta})$ for L_1 and uniform metric, for $\delta > 0$, by implicitly representing the underlying graph.

Curve simplification using the Fréchet error measure was first proposed by Godau [God91], who showed that $\kappa_F(\varepsilon)$ is smaller than the size of the optimal weak $\varepsilon/7$-simplification. Alt and Godau [AG95] also proposed the first algorithm to compute Fréchet distance between two polygonal curves in $\mathbb{R}^d$ in time $O(mn)$, where m and n are the complexity of the two curves.

Since the problem of developing a near-linear time algorithm for computing an optimal ε-simplification remains elusive, several heuristics have been proposed over the years. The most widely used heuristic is the Douglas-Peucker method [DP73] (together with its variants), originally proposed for simplifying curves under the Hausdorff error measure. Its worst case running time is $O(n^2)$ in $\mathbb{R}^d$. In $\mathbb{R}^2$, the running time was improved to $O(n \log n)$ by Snoeyink et al. [HS94]. However, the Douglas-Peucker heuristic does not offer any guarantee on the size of the simplified curve.

The second class of simplification algorithms compute a weak ε-simplification of the polygonal curve P. Imai and Iri [II86] give an optimal $O(n)$ time algorithm for finding an optimal weak ε-simplification of a given monotone curve under Hausdorff error measure. For weak ε-simplification of curves in $\mathbb{R}^2$ under Fréchet distance, Guibas et al. [GHMS93] proposed a factor 2 approximation algorithm with $O(n \log n)$ running time, and an $O(n^2)$ exact algorithm using dynamic programming.

1.3 Our Results

In this paper, we study the curve-simplification problem under both the Fréchet and Hausdorff error measures. We present simple near-linear time algorithms for computing an ε-simplification of size at most $\kappa(\varepsilon/c)$, where $c \geq 1$ is a constant. In particular, our contributions are:

Hausdorff Error Measure. Define the *Hausdorff* error of a line segment $p_i p_j$ w.r.t. P, where $p_i, p_j \in P$, $i < j$, to be

$$\delta_H(p_i p_j, P) = \max_{i \leq k \leq j} d(p_k, p_i p_j)$$

We prove the following theorem in Section 2.

Theorem 1. *Given a monotone polygonal curve P and a parameter $\varepsilon > 0$, one can compute an ε-simplification with size at most $\kappa_H(\varepsilon/2, P)$ in:*

(i) $O(n)$ time and space, under the L_1, L_2, L_∞ or uniform metrics in $\mathbb{R}^2$;
(ii) $O(n \log n)$ time and $O(n)$ space, under L_1 or L_∞ metrics in $\mathbb{R}^3$.

We have implemented the algorithm in $\mathbb{R}^2$ and present experimental results.

Fréchet Error Measure. Given two curves $f : [a, a'] \to \mathbb{R}^d$, and $g : [b, b'] \to \mathbb{R}^d$, the Fréchet distance $\mathcal{F}_D(f, g)$ between them is defined as:,

$$\mathcal{F}_D(f,g) = \inf_{\substack{\alpha : [0,1] \to [a,a'] \\ \beta : [0,1] \to [b,b']}} \max_{t \in [0,1]} d(f(\alpha(t)), g(\beta(t)))$$

where α and β range over continuous and increasing functions with $\alpha(0) = a, \alpha(1) = a', \beta(0) = b$ and $\beta(1) = b'$. The *Fréchet* error of a line segment $p_i p_j$ where $p_i, p_j \in P$, $i < j$, is defined to be

$$\delta_F(p_i p_j, P) = \mathcal{F}_D(\pi(p_i, p_j), p_i p_j),$$

where $\pi(p,q)$ denotes the subcurve of P from p to q. We prove the following result in Section 3:

Theorem 2. *Given a polygonal curve P in $\mathbb{R}^d$ and a parameter $\varepsilon \geq 0$, an ε-simplification of P with size at most $\kappa_F(\varepsilon/2, P)$ can be constructed in $O(n \log n)$ time and $O(n)$ space.*

The algorithm is independent of any monotonicity properties. To our knowledge, it is the first efficient, simple approximation algorithm for curve simplification in dimension higher than two under the Fréchet error measure. We provide experimental results for polygonal chains in $\mathbb{R}^3$ to demonstrate the efficiency and quality of our approximation algorithm.

Relations between simplifications. We further analyze the relations between simplification under Hausdorff and Fréchet error measures, and Fréchet and weak Fréchet ε-simplification in Section 4.

2 Hausdorff Simplification

Let $P = \langle p_1, \ldots, p_n \rangle$ be a monotone polygonal curve in $\mathbb{R}^2$ or $\mathbb{R}^3$. For a given distance function $d(\cdot, \cdot)$, let $D(p, r) = \{q \mid (p, q) \leq r\}$ be the disk of radius r centered at p. Let D_i denote $D(p_i, \varepsilon)$. Then $p_i p_j$ is a valid segment, i.e. $\delta_H(p_i p_j) \leq \varepsilon$, if and only if $p_i p_j$ intersects $D_{i+1}, \ldots, D_{j-1}$ in order. We now define a general problem, and use it to compute ε-simplification of polygonal curves under different distance metrics.

2.1 Segment Cover

Let $E = \langle e_1, e_2, \ldots, e_m \rangle$ be a sequence of segments. E is called an ε-*segment cover* of P if there exists a subsequence of vertices $p_{i_1} = p_1, p_{i_2}, \ldots, p_{i_{m+1}} = p_n$, $i_1 < i_2 < \ldots < i_{m+1}$, such that e_j intersects $D_{i_j}, D_{i_j+1}, \ldots, D_{i_{j+1}}$ in order, and the endpoints of e_j lie in D_{i_j} and $D_{i_{j+1}}$ respectively. An ε-segment cover is optimal if its size is minimum among all ε-segment covers. The following lemma is straightforward.

Lemma 1. *Let $\mu_\varepsilon(P)$ denote the size of an optimal ε-segment cover. For a monotone curve P, $\mu_\varepsilon(P) \leq \kappa_H(\varepsilon, P)$.*

Lemma 2. *Let $E = \langle e_1, e_2, \ldots, e_m \rangle$ be an $\varepsilon/2$-segment cover of size m of a monotone polygonal curve P. Then an ε-simplification of size at most m can be computed in $O(m)$ time.*

PROOF. By the definition of an $\varepsilon/2$-segment cover, there exists a subsequence of vertices $p_{i_1} = p_1, p_{i_2}, \ldots, p_{i_{m+1}} = p_n$ such that e_j intersects $D(p_{i_j}, \varepsilon/2)$, $D(p_{i_j+1}, \varepsilon/2)$, ..., $D(p_{i_{j+1}}, \varepsilon/2)$, and the endpoints of e_j lie in $D(p_{i_j}, \varepsilon/2)$ and $D(p_{i_{j+1}}, \varepsilon/2)$. See Figure 1 for an example of an optimal ε-segment cover under uniform metric. Define the polygonal curve $P' = \langle p_{i_1} = p_1, \ldots, p_{i_m}, p_{i_{m+1}} = p_n \rangle$.

Using the triangle inequality one can easily verify that the segment $p_{i_j}p_{i_{j+1}}$ intersects all the disks $D(p_{i_j},\varepsilon), D(p_{i_j+1},\varepsilon),\ldots,D(p_{i_{j+1}},\varepsilon)$ in order. Hence $p_{i_j}p_{i_{j+1}}$ is a valid segment. Therefore, the polygonal curve P' is an ε-simplification of P, and it can be constructed in $O(m)$ time. □

2.2 An Approximation Algorithm

In this section, we present near-linear approximation algorithms for computing an ε-simplification of a monotone polygonal curve P in $\mathbb{R}^2$ and $\mathbb{R}^3$ under the Hausdorff error measure.

Algorithm. The approximation algorithm to compute an ε-simplification for a monotone curve P, denoted HausdorffSimp, in fact computes an optimal $\varepsilon/2$-segment cover of P. It then follows from Lemma 2 that the vertex set of an optimal $\varepsilon/2$-segment cover also forms an ε-simplification P' of P. The size of P' is at most $\kappa_H(\varepsilon/2, P)$ by Lemma 1.

An ordered set of disks $\mathcal{D}$ has a line transversal if there exists a line intersecting all the disks $D_i \in \mathcal{D}$ in order. We use the greedy method of Guibas et al. [GHMS93] to compute an optimal ε-segment cover: start with the set $\mathcal{D} = \langle D_1 \rangle$. Now iteratively add each disk D_k, $k = 2, 3, \ldots$ to $\mathcal{D}$. If there does not exist a line transversal for $\mathcal{D}$ after adding disk D_k, then add the vertex $p_{i_1} = p_k$ to our segment-cover, set $\mathcal{D} = \emptyset$, and continue. Let $S = \langle p_{i_1},\ldots,p_{i_m}\rangle$ be the polygonal curve computed by algorithm HausdorffSimp. Clearly, the segments $\mathcal{C}(S) = \langle e_j = p_{i_j}p_{i_{j+1}}, j = 1, \ldots, (m-1) \rangle$ form a $\varepsilon/2$-segment cover. It can be shown that the resulting set $\mathcal{C}(S)$ computed is an optimal $\varepsilon/2$-cover.

Analysis. Given a set of i disks $\mathcal{D} = \langle D_1, \ldots, D_i \rangle$, it takes linear time to compute a line that stabs $\mathcal{D}$ in order in $\mathbb{R}^2$ under L_1, L_2, $L\infty$, and uniform metrics [Ame92, GHMS93]. The algorithm is incremental, and we can use a data structure so that it only takes constant time to update the data structure while adding a new disk. Thus our greedy approach uses $O(n)$ time and space overall in $\mathbb{R}^2$. In $\mathbb{R}^3$, the line transversal under L_1 and L_∞ metrics can be computed in $O(i)$ time using linear programming, and one needs to update it efficiently when a new disk is added. (Of course, we can use techniques for dynamic linear programming [Ram00], but we describe a faster and simpler approach.) Therefore, we use an exponential-binary search method using the linear programming algorithm as a black-box, and obtain an $O(n \log n)$ running time. The exponential-binary

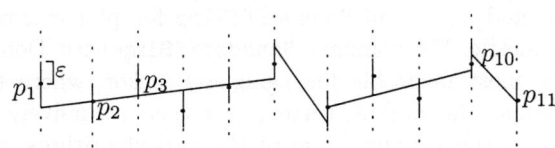

Fig. 1. Covering the vertical segments of length 2ε with maximal set of stabbing lines

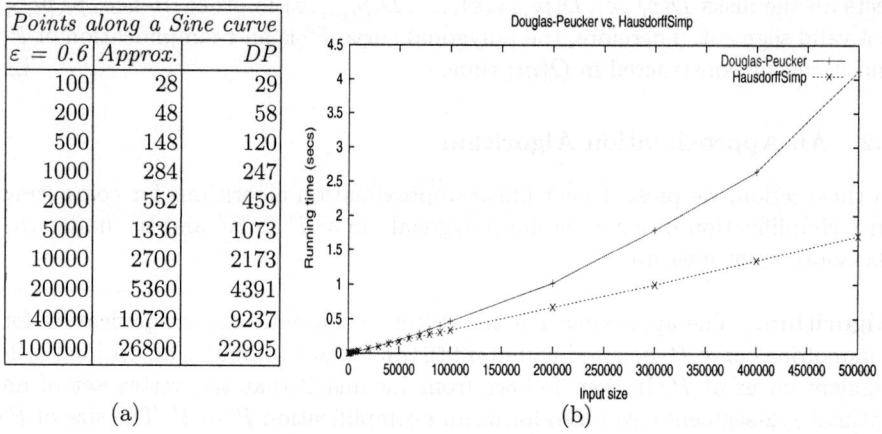

Fig. 2. (a) Sizes of ε-simplifications computed by `HausdorffSimp` and Douglas-Peucker, (b) comparing running times of `HausdorffSimp` and Douglas-Peucker for $\varepsilon = 0.6$

search method will be described in more detail in the next section. Putting everything together proves Theorem 1.

Remark. It can be shown that for monotone curves, the size of the Hausdorff simplification is in fact equal to the size of the Fréchet simplification. In the next section, we extend this approach (with an extra logarithmic overhead) to work for simplifying arbitrary curves under the Fréchet error measure.

2.3 Experiments

In this section, we compare the results of our approximation algorithm for curve simplification under the Hausdorff error measure with the Douglas-Peucker heuristic. For our experiments, there are two input parameters — the size of the input polygonal curve, and the error threshold ε for simplification. Similarly, there are two output parameters — the size of the simplified curve for a particular ε, and the time for the simplification algorithm. All our experiments were run on a Sun Blade-100 machine running SunOS 5.8 with 256MB RAM.

We implemented algorithm `HausdorffSimp` for planar x-monotone curves under uniform metric. We compare `HausdorffSimp` with Douglas-Peucker on inputs that are most favorable for Douglas-Peucker, where the curve is always partitioned at the middle vertex, and then recursively simplified. Figure 2(b) compares the running time of the two algorithms, where the curve consists of point sets of varying sizes sampled from a sinusoidal curve. As expected, `HausdorffSimp` exhibits empirically linear running time, outperforming the Douglas-Peucker heuristic. Figure 2(a) shows the sizes of ε-simplifications produced when $\varepsilon = 0.6$, and the curves again consist of points sampled from a sinusoidal curve.

3 Fréchet Simplification

We now present algorithms for simplification under the Fréchet error measure. It is easy to verify that $\delta_F(\varepsilon)$ can be computed exactly in $O(n^3)$ time following the approach of Imai and Iri [II88]. Therefore we focus on approximation algorithms below.

Lemma 3. *Given two directed segments uv and xy in $\mathbb{R}^d$,*

$$\mathcal{F}_D(uv, xy) = \max\{d(u,x), d(v,y)\},$$

where $d(\cdot, \cdot)$ represents the L_1, L_2 or L_∞ norm.

PROOF. Let δ denote the maximum of $d(u,x)$ and $d(v,y)$. Note that $\mathcal{F}_D(uv, xy) \geq \delta$, since u (resp. v) has to be matched to x (resp. y). Assume the natural parameterization for segment uv, $A(t) : [0,1] \to uv$, such that $A(t) = u + t(v - u)$. Similarly, define $B(t) : [0,1] \to xy$ for segment xy, such that $B(t) = x + t(y - x)$. For any two matched points $A(t)$ and $B(t)$, let $C(t) = A(t) - B(t) = (1-t)(u-x) + t(v-y)$. Since $C(t)$ is a convex function, $\|C(t)\| \leq \delta$ for any $t \in [0,1]$. Therefore $\mathcal{F}_D(uv, xy) \leq \delta$. □

Lemma 4. *Given a polygonal curve P and two directed line segments uv and xy,*

$$|\mathcal{F}_D(P, uv) - \mathcal{F}_D(P, xy)| \leq \mathcal{F}_D(uv, xy).$$

Lemma 5. *Let $P = \langle p_1, p_2, \ldots, p_n \rangle$ be a polygonal curve. For $l \leq i \leq j \leq m$, $\delta_F(p_i p_j, P) \leq 2 \cdot \delta_F(p_l p_m, P)$.*

PROOF. Let $\delta^* = \delta_F(p_l p_m)$. Suppose under the optimal matching between $\pi(p_l, p_m)$ and $p_l p_m$, p_i and p_j are matched to $\hat{p}_i$ and $\hat{p}_j \in p_l p_m$ respectively (see Figure 3 for an illustration). Then obviously, $\mathcal{F}_D(\pi(p_i, p_j), \hat{p}_i\hat{p}_j) \leq \delta^*$. In particular, we have that $d(p_i, \hat{p}_i) \leq \delta^*$, and $d(p_j, \hat{p}_j) \leq \delta^*$. Now by Lemma 3, $\mathcal{F}_D(p_i p_j, \hat{p}_i\hat{p}_j) \leq \delta^*$. It then follows from Lemma 4 that

$$\delta_F(p_i p_j) = \mathcal{F}_D(\pi(p_i, p_j), p_i p_j) \leq \mathcal{F}_D(\pi(p_i, p_j), \hat{p}_i\hat{p}_j) + \delta^* \leq 2\delta^*.$$

□

3.1 An $O(n \log n)$ Algorithm

The algorithm (denoted FrechetSimp) will compute an ε-simplification P' of P in a greedy manner: set the initial simplification as $P' = \langle p_{i_1} = p_1 \rangle$, and iteratively add vertices to P' as follows. Assume $P' = \langle p_{i_1}, \ldots, p_{i_j} \rangle$. The algorithm finds an index $k > i_j$ such that (i) $\delta_F(p_{i_j}, p_k) \leq \varepsilon$ and (ii) $\delta_F(p_{i_j}, p_{k+1}) > \varepsilon$. Set $i_{j+1} = k$, and repeat the above procedure till the last vertex of P.

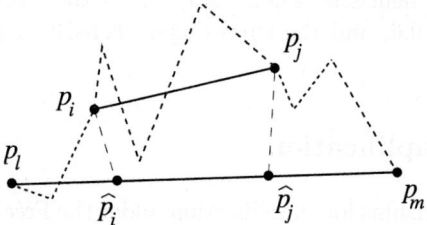

Fig. 3. Bold dashed curve is $\pi(p_l, p_m)$, p_i and p_j are matched to $\hat{p}_i$ and $\hat{p}_j$ respectively

Lemma 6. FrechetSimp *computes P' such that $\delta_F(P', P) \leq \varepsilon$, and $|P'| \leq \kappa_F(\varepsilon/2)$.*

PROOF. It is clear that the algorithm computes a curve that is an ε-simplification of P. It remains to show that the size of the curve P' is bounded by $\kappa_F(\varepsilon/2)$.

Let $Q = \langle p_{j_1} = p_1, \ldots, p_{j_l} = p_n \rangle$ be the optimal $\varepsilon/2$-simplification of P, where $1 \leq j_m \leq n$ for $1 \leq m \leq l$. Let $P' = \langle p_{i_1} = p_1, \ldots, p_{i_k} = p_n \rangle$, where $1 \leq i_m \leq n$ for $1 \leq m \leq k$.

The proof proceeds by induction. The following invariant will always be true: $i_m \geq j_m$, for all m. This implies that $k \leq l$, and therefore $k \leq \kappa_F(\varepsilon/2)$.

Assume $i_{m-1} \geq j_{m-1}$. Let $i' = i_m + 1$. Then note that $\delta_F(p_{i_{m-1}}, p_{i'}) > \varepsilon$, and $\delta_F(p_{i_{m-1}}, p_{i'-1}) \leq \varepsilon$ By the inductive step, $i' > j_{m-1}$. If $i' > j_m$, we are done. So assume $i' \leq j_m$. Since Q is an $\varepsilon/2$-simplification, $\delta_F(p_{j_{m-1}} p_{j_m}) \leq \varepsilon/2$. Lemma 5 implies that for all $j_{m-1} \leq i_{m-1} \leq j' \leq j_m$, $\delta_F(p_{i_{m-1}} p_{j'}) \leq \varepsilon$. But since $\delta_F(p_{i_{m-1}}, p_{i'}) > \varepsilon$, $i' > j_m$ and hence $i_m \geq j_m$. □

After computing vertex $p_{i_m} \in P'$, find the next vertex $p_{i_{m+1}}$ as follows: let b_ρ be a bit that is one if $\delta_F(p_{i_m} p_{i_m+\rho}) > \varepsilon$ and zero otherwise. b_ρ can be computed in $O(\rho)$ time by the algorithm proposed in [AG95]. Recall our goal: finding two consecutive bits b_Δ and $b_{\Delta+1}$ such that $b_\Delta = 0$ and $b_{\Delta+1} = 1$. Clearly, then the index of the next vertex is $i_{m+1} = i + \Delta$. Δ can be computed by performing an exponential search, followed by a binary search. First find the smallest j such that $b_{2^j} = 1$ by computing the bits $b_{2^{j'}}$, $j' \leq j$. The total time can be shown to be $O(i_{m+1} - i_m)$. Next, use binary search to find two consecutive bits in the range $b_{2^{j-1}}, \ldots, b_{2^j}$. Note that this is not strictly a binary search, as the bits which are ones are not necessarily consecutive. Nevertheless, it is easy to verify that the same divide and conquer approach works. This requires computing $O(j-1) = O(\log(i_{m+1} - i_m))$ bits in this range, and it takes $O(i_{m+1} - i_m)$ time to compute each of them. Therefore computing $p_{i_{m+1}}$ takes $O((i_{m+1} - i_m) \log(i_{m+1} - i_m))$ time. Summing over all i_j's yields the running time of $O(n \log n)$, proving Theorem 2.

Table 1. Comparing the size of simplifications produced by `FrechetSimp` with the optimal algorithm

	Curve 1		Curve 2		Curve 3	
Size:	327		1998		9777	
ε	Aprx.	Exact	Aprx.	Exact	Aprx.	Exact
0.05	327	327	201	201	6786	6431
0.08	327	327	168	168	4277	3197
0.12	327	327	134	134	1537	651
1.20	254	249	42	42	178	168
1.60	220	214	36	36	140	132
2.00	134	124	32	32	115	88

3.2 Experiments

We now present experiments comparing our $O(n \log n)$ algorithm `FrechetSimp` with (i) the optimal $O(n^3)$ time Fréchet simplification algorithm for quality; and (ii) with the Douglas-Peucker algorithm under Hausdorff error measure (with L_2 metric) to demonstrate its efficiency. Our experiments were run a Sun Blade-100 machine with 256 RAM.

Recall that the $O(n^3)$ running time of the optimal algorithm is independent of the input curve, or the error measure ε — it is always $\Omega(n^3)$. Therefore, it is *orders* of magnitude slower than the approximation algorithm, and so we omit its empirical running time. We focus on comparing the quality (size) of simplifications produced by `FrechetSimp` and the optimal Fréchet simplification algorithm. The results are presented in Table 3.2. *Curve 1* is a protein backbone, *Curve 2* is a set of points forming a circle, and *Curve 3* is a protein backbone with some artificial noise. As seen from Table 3.2, the size of the simplifications produced by our approximation algorithm is always close to the optimal sized simplification.

We compare the efficiency of `FrechetSimp` with the Douglas-Peucker heuristic. Figure 4 compares the running time of our approximation algorithm with Douglas-Peucker for a protein backbone (with artificial noise added) with 49633 vertices. One can make an interesting observation: as ε decreases, Douglas-Peucker's performance decreases. However, as ε decreases, the performance of our approximation algorithms increases or remains nearly the same. This is due to the fact that Douglas-Peucker tries to find a line segment that simplifies a curve, and recurses into subproblems only if that fails. Thus, as ε decreases, it needs to make more recursive calls. Our approximation algorithm, however, proceeds in a linear fashion from the first vertex to the last vertex, and hence it is more stable towards changes in ε.

4 Comparisons

In this section, we compare the output under two different error measures, and we relate two different Fréchet simplifications.

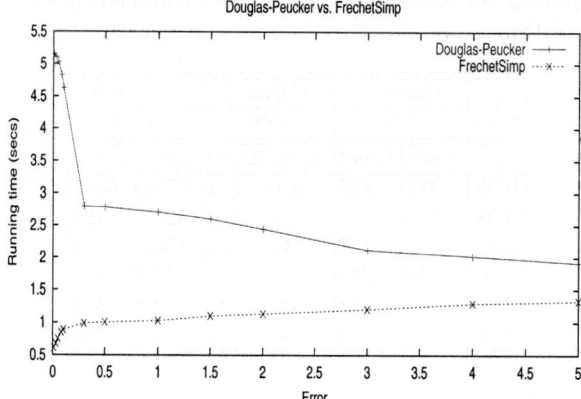

Fig. 4. Comparing running times of `FrechetSimp` and Douglas-Peucker for varying ε for a curve with 49633 vertices

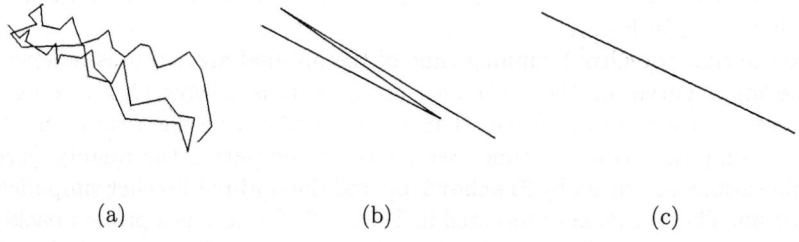

          (a)   (b)   (c)

Fig. 5. (a) Polygonal chain composed of three alpha-helices, (b) its Fréchet ε-simplification and (c) its Douglas-Peucker Hausdorff ε-simplification

Hausdorff vs. Fréchet . One natural question is to compare the quality of simplifications produced under the Hausdorff and the Fréchet error measures. Given a curve $P = \langle p_1, \ldots, p_n \rangle$, it is not too hard to show that $\delta_H(p_ip_j) \leq \delta_F(p_ip_j)$. The converse however does not hold.

The Fréchet error measure takes the order along the curve into account, and hence is more useful in some cases especially when the order of the curve is important (such as curves derived from protein backbones). Figure 5 illustrates a substructure of a protein backbone, where ε-simplifying under Fréchet error measure preserves the overall structure, while ε-simplifying under Hausdorff error measure is unable to preserve it.

Weak Fréchet vs. Fréchet . In the previous section we described a fast approximation algorithm for computing an ε-simplification of P under Fréchet error measure, where we used the Fréchet measure in a local manner: we restrict the curve $\langle p_i, \ldots, p_j \rangle$ to match to the line segment p_ip_j. We can remove this restriction to make the measure more global by instead looking at the *weak*

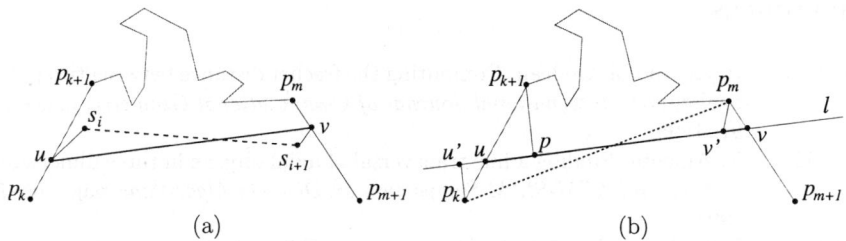

Fig. 6. In (a), u and v are the points that s_i and s_{i+1} are mapped to in the optimal matching between P and S. In (b), $j_i = k$ and $j_{i+1} = m$

Fréchet ε-simplification. More precisely, given P and $S = \langle s_1, s_2, \ldots, s_m \rangle$, where it is not necessary that $s_i \in P$, S is a weak ε-simplification under Fréchet error measure if $\mathcal{F}_D(P, S) \le \varepsilon$. The following lemma shows that the size of the optimal Fréchet simplification can be bounded by the size of the optimal weak Fréchet simplification:

Lemma 7. *Given a polygonal curve P, let $\hat{\kappa}_F(\varepsilon)$ denote the size of the minimum weak ε-simplification of P. Then*

$$\kappa_F(\varepsilon) \le \hat{\kappa}_F(\varepsilon/4)$$

PROOF. Assume $S = \langle s_1, \ldots, s_t \rangle$ is an optimal weak $\varepsilon/4$-simplification of P, i.e., $\mathcal{F}_D(S, P) \le \varepsilon/4$. For any edge $s_i s_{i+1}$, let $u \in p_k p_{k+1}$ and $v \in p_m p_{m+1}$ denote the points on P that s_i and s_{i+1} are mapped to respectively in the optimal matching between P and S. See Figure 6 (a) for an illustration. Let p_{j_i} (resp. $p_{j_{i+1}}$) denotes the endpoint of $p_k p_{k+1}$ (resp. $p_m p_{m+1}$) that is closer to u (resp. v). In other words, $\|p_{j_i} - u\| \le 1/2 \|p_{k+1} - p_k\|$, and $\|p_{j_{i+1}} - v\| \le 1/2 \|p_{m+1} - p_m\|$. Set $P' = \langle p_{j_1} = p_1, \ldots, p_{j_t} = p_n \rangle$. It is easy to verify that $j_i \le j_r$ for any $1 \le i < r \le t$.

It remains to show that P' as constructed above is indeed an ε-simplification of P. $\mathcal{F}_D(\pi(u,v), uv) \le \varepsilon/2$ follows from $\mathcal{F}_D(\pi(u,v), s_i s_{i+1}) \le \varepsilon/4$ and Lemma 4. Let l denote the line containing segment uv. We construct a segment $u'v' \subset l$ such that $\mathcal{F}_D(\pi(p_{j_i}, p_{j_{i+1}}), u'v') \le \varepsilon/2$. We describe how to compute u', and v' can be computed similarly. Let $p \in uv$ denote the point that p_{k+1} is mapped to in the optimal matching between $\pi(u,v)$ and uv. If $j_i = k+1$, i.e. p_{j_i} is the right endpoint of edge $p_k p_{k+1}$, then set $u' = p$. Otherwise, u' is the point on l such that $p_k u'$ is parallel to $p_{k+1} p$. See Figure 6 (b) for an illustration. Note that in both cases, $\|p_{j_i} - u'\| \le \varepsilon/2$, (resp. $\|p_{j_{i+1}} - v'\| \le \varepsilon/2$) which together with Lemma 3 implies that $\mathcal{F}_D(u'v', p_{j_i} p_{j_{i+1}}) \le \varepsilon/2$. On the other hand, the original optimal matching between uv and $\pi(u,v)$ can be modified into a matching between $u'v'$ and $\pi(p_{j_i}, p_{j_{i+1}})$ such that $\mathcal{F}_D(u'v', \pi(p_{j_i}, p_{j_{i+1}})) \le \varepsilon/2$ (proof omitted due to lack of space). It then follows from Lemma 4 that $\mathcal{F}_D(\pi(p_{j_i}, p_{j_{i+1}}), p_{j_i} p_{j_{i+1}}) \le \varepsilon$ for $i = 1 \ldots t$, implying that $\delta_F(P', P) \le \varepsilon$. □

References

[AG95]   H. Alt and M. Godeau. Computing the frechet distance between two polygonal curves. *International Journal of Computational Geometry*, pages 75–91, 1995.

[Ame92]   N. Amenta. Finding a line transversal of axial objects in three dimensions. In *Proc. 3rd ACM-SIAM Symposium on Discrete Algorithms*, pages 66–71, 1992.

[AV00]   P. K. Agarwal and K. R. Varadarajan. Efficient algorithms for approximating polygonal chains. *Discrete Comput. Geom.*, 23:273–291, 2000.

[CC92]   W. S. Chan and F. Chin. Approximation of polygonal curves with minimum number of line segments. In *Proc. 3rd Annual International Symposium on Algorithms and Computation*, pages 378–387, 1992.

[DP73]   D. H. Douglas and T. K. Peucker. Algorithms for the reduction of the number of points required to represent a digitized line or its caricature. *Canadian Cartographer*, 10(2):112–122, 1973.

[GHMS93] L. J. Guibas, J. E. Hershberger, J. B. Mitchell, and J. S. Snoeyink. Approximating polygons and subdivisions with minimum link paths. *International Journal of Computational Geometry and Applications*, 3(4):383–415, 1993.

[God91]   M. Godau. A natural metric for curves: Computing the distance for polygonal chains and approximation algorithms. In *Proc. of the 8th Annual Symposium on Theoretical Aspects of Computer Science*, pages 127–136, 1991.

[HS94]   J. Hershberger and J. Snoeyink. An $O(n \log n)$ implementation of the Douglas-Peucker algorithm for line simplification. In *Proc. 10th Annual ACM Symposium on Computational Geometry*, pages 383–384, 1994.

[II86]   H. Imai and M. Iri. An optimal algorithm for approximating a piecewise linear function. *Information Processing Letters*, 9(3):159–162, 1986.

[II88]   H. Imai and M. Iri. Polygonal approximations of a curve-formulations and algorithms. In G. T. Toussaint, editor, *Computational Morphology*, pages 71–86. North-Holland, Amsterdam, Netherlands, 1988.

[MO88]   A. Melkman and J. O'Rourke. On polygonal chain approximation. In G. T. Toussaint, editor, *Computational Morphology*, pages 87–95. North-Holland, Amsterdam, Netherlands, 1988.

[Ram00]   E. A. Ramos. Linear optimization queries revisited. In *Proc. 16th Annual ACM Symposium on Computational Geometry*, pages 176–181, 2000.

[Wei97]   Robert Weibel. Generalization of spatial data: principles and selected algorithms. In Marc van Kreveld, Jürg Nievergelt, Thomas Roos, and Peter Widmayer, editors, *Algorithmic Foundations of Geographic Information System*. Springer-Verlag Berlin Heidelberg New York, 1997.

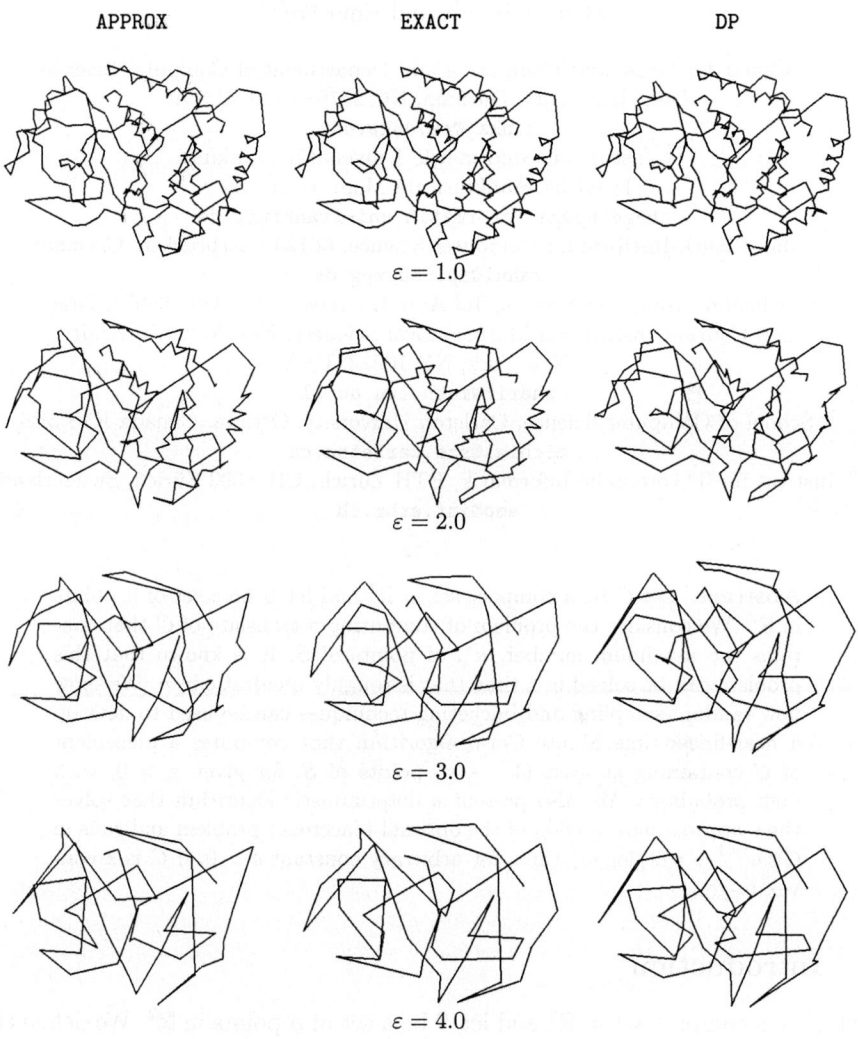

Fig. 7. Simplifications of a protein (1cja) backbone

Translating a Planar Object to Maximize Point Containment*

Pankaj K. Agarwal[1], Torben Hagerup[2], Rahul Ray[3], Micha Sharir[4],
Michiel Smid[5], and Emo Welzl[6]

[1] Center for Geometric Computing and Department of Computer Science
Duke University, Durham, NC 27708-0129, U.S.A.
pankaj@cs.duke.edu
[2] Institut für Informatik, Universität Frankfurt
D–60054 Frankfurt am Main, Germany
hagerup@ka.informatik.uni-frankfurt.de
[3] Max-Planck-Institute for Computer Science, 66123 Saarbrücken, Germany
rahul@mpi-sb.mpg.de
[4] School of Computer Science, Tel Aviv University, Tel Aviv 69978, Israel
and Courant Institute of Mathematical Sciences, New York University
New York, NY 10012, USA
sharir@math.tau.ac.il
[5] School of Computer Science, Carleton University, Ottawa, Canada K1S 5B6
michiel@scs.carleton.ca
[6] Institut für Theoretische Informatik, ETH Zürich, CH–8092 Zürich, Switzerland
emo@inf.ethz.ch

Abstract. Let C be a compact set in $\mathbb{R}^2$ and let S be a set of n points in $\mathbb{R}^2$. We consider the problem of computing a translate of C that contains the maximum number, κ^*, of points of S. It is known that this problem can be solved in a time that is roughly quadratic in n. We show how random-sampling and bucketing techniques can be used to develop a near-linear-time Monte Carlo algorithm that computes a placement of C containing at least $(1-\varepsilon)\kappa^*$ points of S, for given $\varepsilon > 0$, with high probability. We also present a deterministic algorithm that solves the ε-approximate version of the optimal-placement problem and runs in $O((n^{1+\delta} + n/\varepsilon) \log m)$ time, for arbitrary constant $\delta > 0$, if C is a convex m-gon.

1 Introduction

Let C be a compact set in $\mathbb{R}^2$ and let S be a set of n points in $\mathbb{R}^2$. We define the *optimal-placement* problem to be that of computing a point $t \in \mathbb{R}^2$ for which

---

* Agarwal was supported by NSF grants ITR–333–1050, EIA–9870724, EIA–9972879, CCR-00-86013, and CCR–9732787, and by a grant from the U.S.-Israeli Binational Science Foundation. Sharir was supported by NSF Grants CCR-97-32101 and CCR-00-98246, by a grant from the Israel Science Fund (for a Center of Excellence in Geometric Computing), by the Hermann Minkowski–MINERVA Center for Geometry at Tel Aviv University, and by a grant from the U.S.-Israeli Binational Science Foundation. Smid was supported by NSERC.

the translate $C + t$ of C contains the maximum number of points of S. Set

$$\kappa^*(C, S) = \max_{t \in \mathbb{R}^2} |S \cap (C + t)|.$$

Motivated by applications in clustering and pattern recognition (see [14, 16]), the optimal-placement problem has received much attention over the last two decades. Chazelle and Lee [7] presented an $O(n^2)$-time algorithm for the case in which C is a circular disk, and Eppstein and Erickson [11] proposed an $O(n \log n)$-time algorithm for rectangles. Efrat et al. [10] developed an algorithm for convex m-gons with a running time of $O(n\kappa^* \log n \log m + m)$, where $\kappa^* = \kappa^*(C, S)$, which was subsequently improved by Barequet et al. [3] to $O(n \log n + n\kappa^* \log(m\kappa^*) + m)$.

All the algorithms above, except the one for rectangles, require at least quadratic time in the worst case, which raises the question of whether a near-linear approximation algorithm exists for the optimal-placement problem. In this paper we answer the question in the affirmative by presenting Monte-Carlo and deterministic approximation algorithms.

We call a translate $C + t$ of C an ε-*approximate placement* if $|S \cap (C + t)| \geq (1 - \varepsilon)\kappa^*(C, S)$, and an algorithm that produces an ε-approximate placement is called an ε-*approximation* algorithm. We define the ε-*optimal-placement* problem to be the one that asks for an ε-approximate placement.

We make the following assumptions about C and S:

(A1) The boundary of C, denoted by ∂C, is connected and consists of m edges, each of which is described by a polynomial of bounded degree. The endpoints of these edges are called the vertices of C. We will refer to C as a *disk*.

(A2) For all distinct $t, t' \in S$, the boundaries of the two translates $C+t$ and $C+t'$ of C intersect in at most s points, and the intersections can be computed in $I(m)$ time. By computing an intersection, we here mean determining the two edges, one of each translate, that are involved in the intersection. Moreover, for every point $p \in \mathbb{R}^2$, we can decide in $Q(m)$ time whether $p \in C$.

(A3) C is sandwiched between two axes-parallel rectangles whose widths and heights differ by factors of at most α and β, respectively, for $\alpha, \beta \geq 1$.

We call C *fat* if α and β in assumption (A3) are constants. Schwarzkopf et al. [19] showed that after a suitable affine transformation, a compact convex set is fat with $\alpha = \beta = 2$.

We assume a model of computation in which the roots of a bounded-degree polynomial can be computed in $O(1)$ time. This implies that primitive operations on the edges of C, e.g., computing the intersection points between two edges of two translates $C + t$ and $C + t'$, with $t, t' \in S$, can be performed in $O(1)$ time.

In Section 2, we present two algorithms for the optimal-placement problem and show how bucketing can be used to expedite the running time, especially if C is fat. In particular, let $T(n)$ be the running time of an algorithm for the optimal-placement problem on a set of n points, and let $T_g(n)$ denote the time

Table 1. Worst-case running times (constant factors omitted) and error probabilities of our ε-approximation algorithms if C is a circular disk; δ and ε are arbitrarily small positive constants and γ is a nonnegative real number, possibly depending on n

Time	Error probability	Reference
$n + (\gamma n)^2$	$ne^{-\gamma \kappa^*}$	Section 3.1
$n \log n$	$e^{-\sqrt{\kappa^* \log n}}$	Section 3.2
n	$ne^{-\sqrt{\kappa^*}}$	Section 3.2
$n^{1+\delta}$	0	Section 4

required to partition n points into the cells of an integer grid. Then the bucketing algorithm can compute an optimal placement in time $O(T_g(n) + nT(\kappa^*)/\kappa^*)$, where $\kappa^* = \kappa^*(C, S)$. Besides being interesting in its own right, this will be crucial for the approximation algorithms.

In Section 3, we show that using random sampling and/or bucketing, we can transform any deterministic algorithm for the optimal-placement problem to a Monte-Carlo algorithm for the ε-optimal-placement problem. Given a parameter $\gamma \geq 0$, the first algorithm in Section 3, based on a random-sampling technique, computes an ε-approximate placement in $O(n + T(\gamma n))$ time with error probability at most $sne^{-\varepsilon^2 \gamma \kappa^*}$. The second algorithm combines the random-sampling technique with the bucketing technique and computes an ε-approximate placement in $O(T_g(n) + nQ(m) + nT(\alpha\beta\gamma\kappa^*)/\kappa^*)$ time with error probability at most $s\alpha\beta\kappa^* e^{-\varepsilon^2 \gamma \kappa^*}$. For circular disks and constant ε, the running time becomes $O(n \log n)$ and the error probability is at most $e^{-\sqrt{\kappa^* \log n}}$; see Table 1. If C is fat and $m = O(1)$, by combining two levels of random sampling with the bucketing technique, we can compute an ε-approximate placement in $O(n)$ time for constant ε with error probability at most $ne^{-\sqrt{\kappa^*}}$.

Finally, in Section 4, we present a deterministic algorithm, based on the cutting algorithm of Chazelle [6], that computes an ε-approximate placement for sets C that satisfy (A1)–(A3). If C has $O(1)$ edges, the algorithm runs in $O(n^{1+\delta} + n/\varepsilon)$ time, for any given constant $\delta > 0$ (where the constant of proportionality depends on δ). If C is a convex m-gon, the running time is $O((n^{1+\delta} + n/\varepsilon) \log m)$.

2 Preliminaries and Exact Algorithms

Let C be a disk satisfying assumptions (A1)–(A3) and let S be a set of n points in $\mathbb{R}^2$. For simplicity, we assume that the origin lies inside C. For a point $p \in \mathbb{R}^2$, define $C_p = \{p - c \mid c \in C\}$. Let $\mathcal{C} = \{C_p \mid p \in S\}$. For a point set R and a point

$x \in \mathbb{R}^2$, the *depth* of x with respect to R, denoted by $d_R(x)$, is the number of points $p \in R$ for which C_p contains x. This is the same as $|R \cap (C + x)|$, so that $\kappa^*(C, S) = \max_{x \in \mathbb{R}^2} d_S(x)$. Hence, the problem of computing an optimal placement reduces to computing a point of maximum depth with respect to S.

Consider the arrangement $\mathcal{A}(\mathcal{C})$ defined by the boundaries of the sets C_p, where $p \in S$. Each vertex in $\mathcal{A}(\mathcal{C})$ is either a vertex of C_p, for some point $p \in S$ (a *type 1* vertex), or an intersection point of C_p and C_q, for two points $p, q \in S$ (a *type 2* vertex). The maximum depth of a point is realized by a type 2 vertex of $\mathcal{A}(\mathcal{C})$ (unless the maximum depth is 1). Using this observation, $\kappa^*(C, S)$ can be computed as follows.

Simple algorithms for computing κ^*. As in [10], we repeat the following step for each point $p \in S$. Compute the intersections between ∂C_p and all other boundaries ∂C_q, where $q \in S$, and sort them along ∂C_p. Compute the depth of one intersection with respect to S by brute force and step through the remaining ones in sorted order while maintaining the depth. Finally report a point of the maximum depth encountered for any $p \in S$. The total time spent is $O(n^2(I(m) + Q(m)) + sn^2 \log(sn))$.

Alternatively, we can compute $\mathcal{A}(\mathcal{C})$ with the algorithm of Amato et al. [2] and use a standard graph-traversal algorithm, such as depth-first search, to compute a type 2 vertex of maximum depth (with respect to S). Since $\mathcal{A}(\mathcal{C})$ has $O(mn + sn^2)$ vertices, this algorithm takes $O(mn \log(mn) + nQ(m) + sn^2)$ time. Hence, we obtain the following.

Theorem 1. *Let S be a set of n points in the plane and let C be a disk satisfying assumptions (A1)–(A3). The value of $\kappa^*(C, S)$ can be computed in time $O\left(n^2(I(m) + Q(m)) + sn^2 \log(sn)\right)$ or in time $O\left(mn \log(mn) + nQ(m) + sn^2\right)$.*

If $m = O(1)$, then $s, Q(m) = O(1)$, and if C is convex, then $s = 2$ [15] and $I(m), Q(m) = O(\log m)$ [18]. (For the upper bounds on $I(m)$ and $Q(m)$, an $O(m)$-time preprocessing step is needed.) Therefore Theorem 1 implies the following.

Corollary 1. *Let S be a set of n points in the plane and let C be a disk satisfying assumptions (A1)–(A3). Then $\kappa^*(C, S)$ can be computed in $O(n^2)$ time if C has $O(1)$ edges, and in $O(n^2 \log(mn) + m)$ or $O(mn \log(mn) + n^2)$ time if C is a convex m-gon.*

Bucketing and estimating $\kappa^*(C, S)$. For any two positive real numbers r and r', we denote by $\mathcal{G}_{r,r'}$ the two-dimensional grid through the origin whose cells have horizontal and vertical sides of lengths r and r', respectively. Hence, each cell of $\mathcal{G}_{r,r'}$ is of the form $B_{ij} = [ir, (i+1)r) \times [jr', (j+1)r')$ for some integers i and j. We call the pair (i, j) the *index* of B_{ij}.

We need an algorithm that groups the points of S according to the cells of some grid $\mathcal{G}_{r,r'}$, i.e., stores S in a list such that for each grid cell B, all points of S in B occur together in a contiguous sublist. This operation is similar to a sorting of the elements of S by their associated grid cells, but does not require the full power of sorting. Let $T_g(n)$ denote the time needed to perform such a grouping of n points according to some grid and assume that T_g is nondecreasing and smooth in the sense that $T_g(O(n)) = O(T_g(n))$ (informally, a smooth function grows polynomially). The following lemma is straightforward.

Lemma 1. *Let S be a set of n points in $\mathbb{R}^2$ and let C be a disk satisfying assumptions (A1)–(A3). Let $a, b > 0$ be such that $R_0 \subseteq C \subseteq R_1$ for axes-parallel rectangles R_0 of width a and height b and R_1 of width αa and height βb. Let M be the maximum number of points of S contained in any cell of the grid $\mathcal{G}_{a,b}$. Then $M \leq \kappa^*(C, S) \leq (\alpha + 1)(\beta + 1)M$.*

Lemma 1 shows that an approximation M to $\kappa^* = \kappa^*(C, S)$ with $M \leq \kappa^* \leq (\alpha + 1)(\beta + 1)M$ can be computed in $O(T_g(n))$ time. Let us see how the grouping of S can be implemented. It is clear that once each point p of S has been mapped to the index (i, j) of the cell containing p of a grid under consideration, S can be grouped with respect to the grid in $O(n \log n)$ time by sorting the pairs (i, j) lexicographically. The mapping of points to grid indices uses the nonalgebraic floor function. To avoid this, we can replace the grid by the *degraded grid* introduced in [8, 17], which can be constructed in $O(n \log n)$ time without using the floor function, and for which Lemma 1 also holds. Given any point $p \in \mathbb{R}^2$, the cell of the degraded grid that contains p can be found in $O(\log n)$ time, so that the grouping can be completed in $O(n \log n)$ time.

In a more powerful model of computation, after mapping S to grid indices, we can carry out the grouping by means of hashing. Combining the universal class of Dietzfelbinger et al. [9] with a hashing scheme of Bast and Hagerup [4], we obtain a Las Vegas grouping algorithm that finishes in $O(n)$ time with probability at least $1 - 2^{-n^\mu}$ for some fixed $\mu > 0$. We omit the details.

A bucketing algorithm. We can use Lemma 1 and Theorem 1 to obtain a faster algorithm for computing $\kappa^* = \kappa^*(C, S)$ in some cases. Suppose we have an algorithm $\mathcal{A}$ that computes $\kappa^*(C, S)$ in time $T(n)$. Using Lemma 1, we first compute M and, for each pair $(i, j) \in \{0, 1, 2\}^2$, consider the grid $\mathcal{G}^{ij}$ obtained by shifting $\mathcal{G}_{3\alpha a, 3\beta b}$ by the vector $(i\alpha a, j\beta b)$. For each cell B of $\mathcal{G}^{ij}$ that contains at least M points of S, we run $\mathcal{A}$ on the set $S \cap B$ to obtain a point y_B of maximum depth k_B with respect to $S \cap B$. Finally we return a point y_B for which k_B is maximum over all runs of $\mathcal{A}$.

To see the correctness of the algorithm, let $x \in \mathbb{R}^2$ satisfy $d_S(x) = \kappa^*$. Observe that for some $0 \leq i, j \leq 2$, x lies in the middle ninth of some cell B of $\mathcal{G}^{ij}$, in which case $C + x \subseteq B$. It is now clear that $|S \cap B| \geq d_S(x) = \kappa^* \geq M$ and $k_B = \kappa^*$.

Let us analyze the running time. For each of $0 \leq i, j \leq 2$, the algorithm spends $T_g(n)$ time to partition S among the grid cells. Since at most n/M cells of $\mathcal{G}^{ij}$ contain at least M points of S and no cell contains more than $9\alpha\beta M$ points, the total running time is $O(T_g(n) + nT(\alpha\beta M)/M) = O(T_g(n) + nT(\alpha\beta\kappa^*)/\kappa^*)$, where in the last step we used the relation $M \leq \kappa^*$ and the assumption that $T(n)/n$ is nondecreasing. We thus obtain the following result.

Theorem 2. *Let S be a set of n points in the plane, let C be a disk satisfying assumptions (A1)–(A3), and let $\mathcal{A}$ be an algorithm that computes $\kappa^* = \kappa^*(C,S)$ in time $T(n)$. Then κ^* can be computed in $O(T_g(n) + nT(\alpha\beta\kappa^*)/\kappa^*)$ time.*

Corollary 2. *Let $C, S,$ and $T(\cdot)$ be as in Theorem 2. The value of $\kappa^* = \kappa^*(C,S)$ can be computed in $O(T_g(n) + n\kappa^*)$ time if C is fat and has $O(1)$ edges, and in $O(T_g(n) + n\kappa^* \log(m\kappa^*) + m)$ or $O(T_g(n) + mn\log(m\kappa^*) + n\kappa^*)$ time if C is a convex m-gon.*

3 Monte-Carlo Algorithms

In this section we present Monte-Carlo algorithms for the ε-optimal-placement problem. These algorithms use one of the deterministic algorithms described in Theorem 1 as a subroutine. We will refer to this algorithm as $\mathcal{A}$ and to its running time as $T(n)$. We assume that $T(n)$ and $T(n)/n$ are nondecreasing and that T is smooth.

3.1 A Random-Sampling Approach

We first present an algorithm based on the *random-sampling* technique. We carry out the probabilistic analysis using the following variant of the well-known Chernoff bounds (see, e.g., Hagerup and Rüb [13]).

Lemma 2. *Let Y be a binomially distributed random variable and let $0 \leq \lambda \leq 1$.*

1. *For every $t \leq E(Y)$, $\Pr[Y \leq (1 - \lambda)t] \leq e^{-\lambda^2 t/2}$.*
2. *For every $t \geq E(Y)$, $\Pr[Y \geq (1 + \lambda)t] \leq e^{-\lambda^2 t/3}$.*

Theorem 3. *Let S be a set of n points in the plane and let C be a disk satisfying assumptions (A1)–(A3). For arbitrary $\varepsilon, \gamma > 0$, an ε-approximate solution to the optimal-placement problem can be computed in $O(n + T(\gamma n))$ time with error probability at most $sne^{-\varepsilon^2 \gamma \kappa^*}$, where $\kappa^* = \kappa^*(C,S)$.*

Proof. Let $\bar{\varepsilon} = \min\{\varepsilon, 1/2\}$ and $\bar{\gamma} = \min\{288\gamma, 1\}$. The algorithm first draws a $\bar{\gamma}$-sample S' of S, i.e., includes every point of S in S' with probability $\bar{\gamma}$ and independently of all other points. If $|S'| > 2\bar{\gamma}n$ (the sampling *fails*), the algorithm returns an arbitrary point. Otherwise it uses $\mathcal{A}$ to return a point y of maximum depth with respect to S'.

Since $\bar{\gamma} = O(\gamma)$ and T is smooth, it is clear that the algorithm can be executed in $O(n + T(\gamma n))$ time. By Lemma 2, the sampling fails with probability at most $e^{-\bar{\gamma}n/3}$. If $\varepsilon \geq 1$ or $\bar{\gamma} = 1$, the output is obviously correct. Assume that this is not the case and that the sampling succeeds.

Let us write d for d_S and d' for $d_{S'}$ and let $x \in \mathbb{R}^2$ be a point with $d(x) = \kappa^*$. Informally, our proof proceeds as follows. Let $Z = \{z \in \mathbb{R}^2 \mid d(z) < (1 - \varepsilon)\kappa^*\}$ be the set of "bad" points. The error probability is equal to $\Pr[y \in Z]$. We first show that $d'(y)$ is likely to be large, where "large" means at least $(1 - \bar{\varepsilon}/2)\bar{\gamma}\kappa^*$.

Subsequently we show that for every $z \in Z$, $d'(z)$ is not likely to be large. Combining the two assertions shows that except with small probability, $y \notin Z$.

The first part is easy: Since $E(d'(x)) = \bar{\gamma}\kappa^*$ and $d'(y) \geq d'(x)$, Lemma 2 implies that
$$\Pr[d'(y) < (1 - \bar{\varepsilon}/2)\bar{\gamma}\kappa^*] \leq e^{-\bar{\varepsilon}^2\bar{\gamma}\kappa^*/8}.$$
Now fix $z \in Z$. Since $1 - \bar{\varepsilon}/2 \geq (1 + \bar{\varepsilon}/2)(1 - \bar{\varepsilon})$ and $E(d'(z)) = \bar{\gamma}d(z) < (1-\varepsilon)\bar{\gamma}\kappa^* \leq (1-\bar{\varepsilon})\bar{\gamma}\kappa^*$, we have
$$\Pr[d'(z) \geq (1 - \bar{\varepsilon}/2)\bar{\gamma}\kappa^*] \leq \Pr[d'(z) \geq (1 + \bar{\varepsilon}/2)(1 - \bar{\varepsilon})\bar{\gamma}\kappa^*] \leq e^{-\bar{\varepsilon}^2(1-\bar{\varepsilon})\bar{\gamma}\kappa^*/12}.$$

The preceding argument applies to a fixed $z \in Z$. A priori, we have to deal with an infinite number of candidate points $z \in Z$. However, using the fact that the arrangement defined by the boundaries of the sets C_p, with $p \in S$, has $O(sn^2)$ vertices of type 2, it is not difficult to see that there is a set X with $|X| = O(sn^2)$ such that for every $z \in Z$, there is a $\hat{z} \in X \cap Z$ with $d'(\hat{z}) = d'(z)$. Therefore the probability that $d'(z) \geq (1 - \bar{\varepsilon}/2)\bar{\gamma}\kappa^*$ for some $z \in Z$ is $O(sn^2 e^{-\bar{\varepsilon}^2(1-\bar{\varepsilon})\bar{\gamma}\kappa^*/12})$. The other failure probabilities identified above are no larger. Now, $\bar{\varepsilon} \geq \varepsilon/2$, $1 - \bar{\varepsilon} \geq 1/2$, and $\bar{\gamma} = 3 \cdot 8 \cdot 12\gamma$. Moreover, we can assume that $e^{\varepsilon^2 \gamma \kappa^*} \geq sn$, since otherwise the theorem claims nothing. But then $sn^2 e^{-\bar{\varepsilon}^2(1-\bar{\varepsilon})\bar{\gamma}\kappa^*/12} \leq sn^2 e^{-3\varepsilon^2 \gamma \kappa^*} \leq (1/s)e^{-\varepsilon^2 \gamma \kappa^*}$. Therefore, except if n is bounded by some constant (in which case we can use a deterministic algorithm), the failure probability is at most $sne^{-\varepsilon^2 \gamma \kappa^*}$. □

Combining Theorem 3 with Corollary 1, we obtain the following result.

Corollary 3. *Let S be a set of n points in the plane and let C be a disk satisfying assumptions (A1)–(A3). For arbitrary $\varepsilon, \gamma > 0$, an ε-approximate placement can be computed with error probability at most $ne^{-\varepsilon^2 \gamma \kappa^*}$ in $O(n+(\gamma n)^2)$ time if C has $O(1)$ edges, and in $O(n+(\gamma n)^2 \log(\gamma mn)+m)$ or $O(n+\gamma mn \log(\gamma mn)+(\gamma n)^2)$ time if C is a convex m-gon.*

Our random-sampling approach can also be used to solve related problems. As an example, consider the problem of computing a point of maximum depth in a set H of n halfplanes. If we denote this maximum depth by κ^*, then $\kappa^* \geq n/2$. By computing and traversing the arrangement defined by the bounding lines of the halfplanes, one can compute κ^* in $O(n^2)$ time. Since a corresponding decision problem is 3SUM-hard (see Gajentaan and Overmars [12]), it is unlikely that it can be solved in subquadratic time. If we apply our random-sampling transformation with $\gamma = c/\sqrt{n}$ for a suitable constant $c > 0$, we obtain the following result.

Theorem 4. *Let H be a set of n halfplanes. For arbitrary constant $\varepsilon > 0$, in $O(n)$ time we can compute a point in $\mathbb{R}^2$ whose depth in H is at least $(1-\varepsilon)\kappa^*$, except with probability at most $e^{-\sqrt{n}}$.*

3.2 Bucketing and Sampling Combined

We now present a Monte Carlo algorithm that combines Theorem 3 with the bucketing algorithm described in Section 2.

First compute M, as defined in Lemma 1. Next, for each $(i,j) \in \{0,1,2\}^2$, consider the grid $\mathcal{G}^{ij}$ as in Section 2. Fix a parameter $\gamma \geq 0$. For each cell B of $\mathcal{G}^{ij}$ with $|S \cap B| \geq M$, run the algorithm described in Section 3.1 on the set $S \cap B$ to obtain a point y_B and compute the value $k_B = d_{S \cap B}(y_B)$. Finally return a point y_B for which k_B is maximum.

Theorem 5. *Let S be a set of n points in the plane and let C be a disk satisfying assumptions (A1)–(A3). For arbitrary $\varepsilon, \gamma > 0$, an ε-approximate placement of C can be computed in $O(T_g(n) + nQ(m) + nT(\alpha\beta\gamma\kappa^*)/\kappa^*)$ time by a Monte Carlo algorithm with error probability at most $s\alpha\beta\kappa^* e^{-\varepsilon^2 \gamma \kappa^*}$.*

Corollary 4. *Let S be a set of n points in the plane and let C be a disk satisfying assumptions (A1)–(A3). For arbitrary constant $\varepsilon > 0$, an ε-approximate placement of C can be computed in $O(n \log n)$ time with probability of error at most $e^{-\sqrt{\kappa^* \log n}}$ if C is fat and has $O(1)$ edges, and in $O(n\log(mn) + m)$ time with probability of error at most $e^{-\sqrt{\kappa^* \log(mn)/\log(m\kappa^*)}}$ if C is a convex m-gon.*

Proof. To prove the first claim, let $\mathcal{A}$ be the first algorithm of Corollary 1. Then $T(n) = O(n^2)$. Applying Theorem 5 to $\mathcal{A}$, we obtain an ε-approximation algorithm for the optimal-placement problem with running time $O(n \log n + \gamma^2 n\kappa^*)$ and error probability $O(\kappa^* e^{-\varepsilon^2 \gamma \kappa^*})$. If we choose $\gamma = (2/\varepsilon^2)\sqrt{(\log n)/M}$, where M is as in Lemma 1, the running time and the error probability are as claimed for n larger than some constant.

For the proof of the second claim, we take $\mathcal{A}$ to be the second algorithm of Corollary 1. Then $T(n) = O(n^2 \log(mn) + m)$. If we apply Theorem 5 to $\mathcal{A}$, we obtain an ε-approximation algorithm for the optimal-placement problem with running time $O(n\log(mn) + n\gamma^2\kappa^* \log(m\gamma\kappa^*) + m)$ and error probability $O(\kappa^* e^{-\varepsilon^2 \gamma \kappa^*})$. We choose $\gamma = c\sqrt{\log(mn)/(M\log(mM))}$, where M is as in Lemma 1 and c is a sufficiently large constant. Except if $\gamma > 1$, in which case we can use the second algorithm of Corollary 2, this gives the running time and the error probability claimed for n sufficiently large. □

Theorem 6. *Let S be a set of n points in the plane and let C be a fat disk satisfying assumptions (A1)–(A3) and having $O(1)$ edges. For arbitrary constant $\varepsilon > 0$, an ε-approximate placement of C can be computed in $O(n)$ time with error probability at most $ne^{-\sqrt{\kappa^*}}$.*

Proof. We use the following algorithm, which might be described as sampling followed by bucketing followed by sampling: Draw a random γ-sample S' of S, where $\gamma = \min\{L/\log n, 1\}$ for a constant $L > 0$ to be chosen below. If $|S'| > 2\gamma n$, return an arbitrary point. Otherwise apply the first algorithm of Corollary 4 to S', but with approximation parameter $\varepsilon/4$, rather than ε, and return the point returned by that algorithm. The overall running time is clearly $O(n)$.

As in the proof of Theorem 3, we write d for d_S and d' for $d_{S'}$. Assume that $\varepsilon \leq 1$ and that $|S'| \leq 2\gamma n$ and let κ' be the maximum value of $d'(x)$ over all points x in the plane. By Lemma 2, $\Pr[\kappa' < (1-\varepsilon/4)\gamma\kappa^*] \leq e^{-c_1\gamma\kappa^*}$ for some constant $c_1 > 0$. Moreover, the analysis in the proof of Theorem 3 shows the probability that $d'(z) \geq (1-\varepsilon/2)\gamma\kappa^*$ for some $z \in \mathbb{R}^2$ with $d(z) < (1-\varepsilon)\kappa^*$ to be $O(ne^{-c_2\gamma\kappa^*})$ for some other constant $c_2 > 0$. Finally, the probability that the algorithm of Corollary 4 returns a point y with $d'(y) < (1-\varepsilon/4)\kappa'$ is at most $e^{-\sqrt{\kappa'\log n}}$. Observe that $\kappa' \geq (1-\varepsilon/4)\gamma\kappa^*$ and $d'(y) \geq (1-\varepsilon/4)\kappa'$ imply $d'(y) \geq (1-\varepsilon/2)\gamma\kappa^*$. Therefore, for some constant $c > 0$, the answer is correct, except with probability $O(ne^{-c\gamma\kappa^*} + e^{-c\sqrt{\gamma\kappa^*\log n}})$. If $\gamma < 1$, the probability under consideration is $O(ne^{-cL\kappa^*/\log n} + e^{-c\sqrt{L\kappa^*}})$. We can assume that $\kappa^* \geq (\log n)^2$, since otherwise there is nothing to prove. But then it is clear that for L chosen sufficiently large and for n larger than some constant, the error probability is at most $ne^{-\sqrt{\kappa^*}}$. □

4 A Deterministic Approximation Algorithm

In this section we present a deterministic algorithm for the ε-optimal-placement problem. For simplicity, we assume that C is convex and has $O(1)$ edges. The algorithm can be extended to more general cases. For example, at the cost of an extra $O(\log m)$-factor in the running time, the algorithm can be extended to arbitrary convex m-gons. Throughout this section, S denotes a set of n points in the plane, $\mathcal{C}$ denotes the set of n translates C_p, with $p \in S$, $\mathcal{A}(\mathcal{C})$ denotes the arrangement defined by the boundaries of the elements of $\mathcal{C}$, and κ^* denotes $\kappa^*(C, S)$.

4.1 Cuttings

We will refer to a simply connected region with at most four edges (left and right edges being vertical segments and top and bottom edges being portions of the boundaries of translates of $-C$) as a *pseudo-trapezoid*. For technical reasons, we will also regard vertical segments and portions of the boundaries of translates of $-C$ as 1-dimensional pseudo-trapezoids. For a pseudo-trapezoid Δ and a set $\mathcal{F}$ of translates of $-C$, we will use $\mathcal{F}_\Delta \subseteq \mathcal{F}$ to denote the set of all elements of $\mathcal{F}$ whose boundaries intersect Δ. The vertical decomposition $\mathcal{A}^{\|}(\mathcal{C})$ of $\mathcal{A}(\mathcal{C})$ partitions the plane into pseudo-trapezoids. Let $\mathcal{F}$ be a subset of $\mathcal{C}$ and let Δ be a pseudo-trapezoid. We will denote by $\chi(\mathcal{F}, \Delta)$ the number of pairs of elements of $\mathcal{F}$ whose boundaries intersect inside Δ. If Δ is the entire plane, we use the notation $\chi(\mathcal{F})$ to denote $\chi(\mathcal{F}, \Delta)$, for brevity. Given a parameter $r \geq 1$, a partition Ξ of Δ into a collection of pairwise openly-disjoint pseudo-trapezoids is called a $(1/r)$-*cutting* of $(\mathcal{F}, \Delta)$ if $|\mathcal{F}_\tau| \leq |\mathcal{F}|/r$ for every pseudo-trapezoid $\tau \in \Xi(\mathcal{F}, \Delta)$. The *conflict list* of a pseudo-trapezoid τ in Ξ is the set $\mathcal{F}_\tau$.

We state the following technical result in full generality, since it is of independent interest and may find additional applications. We apply it here only with Δ being the whole plane.

Theorem 7. *Let Δ be a pseudo-trapezoid, let $r \geq 1$, and let $\delta > 0$ be an arbitrarily small constant. A $(1/r)$-cutting of $(\mathcal{C}, \Delta)$ of size $O(r^{1+\delta} + \chi(\mathcal{C}, \Delta)r^2/n^2)$, along with the conflict lists of its pseudo-trapezoids, can be computed in $O(nr^\delta + \chi(\mathcal{C}, \Delta)r/n)$ time, where the constants of proportionality depend on δ.*

We prove this theorem by adapting Chazelle's cutting algorithm [6] to our setting. We call a subset $\mathcal{F}$ of $\mathcal{C}$ a $(1/r)$-*approximation* if, for every pseudo-trapezoid Δ,

$$\left| \frac{|\mathcal{C}_\Delta|}{|\mathcal{C}|} - \frac{|\mathcal{F}_\Delta|}{|\mathcal{F}|} \right| < \frac{1}{r}.$$

A result by Brönnimann et al. [5] implies that a $(1/r)$-approximation of $\mathcal{C}$ of size $O(r^2 \log r)$ can be computed in time $nr^{O(1)}$. Moreover, an argument similar to the one in [6] implies the following:

Lemma 3. *Let $\mathcal{F}$ be a $(1/r)$-approximation of $\mathcal{C}$. For every pseudo-trapezoid Δ,*

$$\left| \frac{\chi(\mathcal{C}, \Delta)}{|\mathcal{C}|^2} - \frac{\chi(\mathcal{F}, \Delta)}{|\mathcal{F}|^2} \right| < \frac{1}{r}.$$

Next, we call a subset $\mathcal{H}$ of $\mathcal{C}$ a $(1/r)$-*net* of $\mathcal{C}$ if, for any pseudo-trapezoid Δ, $|\mathcal{C}_\Delta| > |\mathcal{C}|/r$ implies that $\mathcal{H}_\Delta \neq \emptyset$. A $(1/r)$-net $\mathcal{H}$ is called *sparse* for Δ if

$$\frac{\chi(\mathcal{H}, \Delta)}{\chi(\mathcal{C}, \Delta)} \leq 4 \left(\frac{|\mathcal{H}|}{n} \right)^2.$$

As in [6], we can prove the following.

Lemma 4. *Given a pseudo-trapezoid Δ and a parameter $r \geq 1$, we can compute, in $n^{O(1)}$ time, a $(1/r)$-net of $\mathcal{C}$ having size $O(r \log n)$ and that is sparse for Δ.*

Proof of Theorem 7. Using Lemmas 3 and 4, we compute a $(1/r)$-cutting of $(\mathcal{C}, \Delta)$ as follows. Let c be a sufficiently large constant whose value will be chosen later. For every $0 \leq j \leq \lceil \log_c r \rceil$, we compute a $(1/c^j)$-cutting Ξ_j of $(\mathcal{C}, \Delta)$. The final cutting is a $(1/r)$-cutting of $(\mathcal{C}, \Delta)$. While computing Ξ_j, we also compute the conflict lists of the pseudo-trapezoids in Ξ_j.

Ξ_0 is Δ itself, and $\mathcal{C}$ is the conflict list of Δ. We compute Ξ_j from Ξ_{j-1} as follows. For each pseudo-trapezoid $\tau \in \Xi_{j-1}$, if $|\mathcal{C}_\tau| \leq n/c^j$, then we do nothing in τ. Otherwise, we first compute a $(1/2c)$-approximation $\mathcal{F}_\tau$ of $\mathcal{C}_\tau$ of size $O(c^2 \log c)$ and then a $(1/2c)$-net $\mathcal{H}_\tau$ of $\mathcal{F}_\tau$ of size $O(c \log c)$ that is sparse for τ. Note that $\mathcal{H}_\tau$ is a $(1/c)$-net of $\mathcal{C}_\tau$. We then compute the vertical decomposition $\mathcal{A}^\|(\mathcal{H}_\tau)$ of $\mathcal{A}(\mathcal{H}_\tau)$ within τ. $\mathcal{A}^\|(\mathcal{H}_\tau)$ consists of $O(|\mathcal{H}_\tau| + \chi(\mathcal{H}_\tau, \tau))$ cells. We replace τ with the pseudo-trapezoids of $\mathcal{A}^\|(\mathcal{H}_\tau)$. Repeating this for all $\tau \in \Xi_{j-1}$, we form Ξ_j from Ξ_{j-1}. It is easy to see that Ξ_j is a $(1/c^j)$-cutting of $(\mathcal{C}, \Delta)$.

By an analysis similar to the one in [6], it can be shown that by choosing the constant c sufficiently large (but depending on δ), the size of the final cutting is $O(r^{1+\delta} + \chi(\mathcal{C}, \Delta)r^2/n^2)$ and that the running time of the algorithm is $O(nr^\delta + \chi(\mathcal{C}, \Delta)r/n)$. This completes the proof of Theorem 7. □

By a result of Sharir [20], $\chi(\mathcal{C}, \Delta) = O(n\kappa^*)$, so Theorem 7 implies the following.

Corollary 5. *Let $r \geq 1$ and let $\delta > 0$ be an arbitrarily small constant. A $(1/r)$-cutting $\Xi(\mathcal{C})$ of size $O(r^{1+\delta} + \kappa^* r^2/n)$, along with the conflict lists, can be computed in $O(nr^\delta + \kappa^* r)$ time.*

4.2 The Approximation Algorithm

Let $\delta, \varepsilon > 0$ be real numbers. We now present a deterministic ε-approximation algorithm. We will need the following lemma.

Lemma 5. *Let $r \geq 1$ and let Ξ be a $(1/r)$-cutting of $\mathcal{C}$. Then we can compute a point of depth (with respect to S) at least $\kappa^* - n/r$ in $O(|\Xi|n/r)$ time.*

Proof. Let Δ be a pseudo-trapezoid of Ξ. The maximum depth (with respect to S) of any point inside Δ is at most n/r plus the depth (with respect to S) of any vertex of Δ. It thus suffices to return a vertex of (a pseudo-trapezoid of) Ξ of maximum depth (with respect to S). We can compute the depths of all vertices of Ξ by following an Eulerian tour on the dual graph of the planar subdivision induced by Ξ; see, e.g., [1]. As shown in [1], the time spent in this step is proportional to the total size of all the conflict lists in Ξ. Since the size of each conflict list is at most n/r, the claim follows. □

Our algorithm works in two stages. In the first stage, we estimate the value of κ^* to within a factor of 9, and then we use Lemma 5 to compute an ε-approximation of κ^*.

I. Using the bucketing algorithm of Section 2, we obtain a coarse estimate k_0 of κ^* that satisfies $k_0/9 \leq \kappa^* \leq k_0$. (Since we assume that C is convex, we have $\alpha = \beta = 2$, as follows from [19], which leads to the constant 9 in the estimate above.)

II. We set $r = \min\left\{\frac{9n}{\varepsilon k_0}, n\right\}$, compute a $(1/r)$-cutting Ξ of $\mathcal{C}$, and return a point of maximum depth (with respect to S) among the vertices of Ξ. Denote this maximum depth by k.

By Lemma 5, and assuming that $r = \frac{9n}{\varepsilon k_0}$,

$$k \geq \kappa^* - \frac{n}{r} = \kappa^* - \frac{n}{9n/(\varepsilon k_0)} = \kappa^* - \frac{\varepsilon k_0}{9} \geq (1-\varepsilon)\kappa^*.$$

If $r = n$, then $k \geq \kappa^* - n/r = \kappa^* - 1$, which is at least $(1-\varepsilon)\kappa^*$, since we may assume that $\varepsilon \geq 1/\kappa^*$.

As shown in Section 2, Step I can be implemented in $O(n \log n)$ time. Using Theorem 7 and Lemma 5, Step II takes time

$$O\left(nr^\delta + \frac{\kappa^* n}{\varepsilon k_0}\right) = O\left(n^{1+\delta} + \frac{n}{\varepsilon}\right).$$

Hence, we conclude the following.

Theorem 8. *Let S be a set of n points in the plane and let C be a convex disk with $O(1)$ edges. Assume that assumptions $(A1)$–$(A3)$ are satisfied and let $\varepsilon > 0$ be a real number. An ε-approximate placement of C can be computed in $O(n^{1+\delta} + n/\varepsilon)$ time, for any given constant $\delta > 0$.*

References

[1] P. K. Agarwal, B. Aronov, M. Sharir, and S. Suri. Selecting distances in the plane. *Algorithmica*, 9:495–514, 1993.

[2] N. M. Amato, M. T. Goodrich, and E. A. Ramos. Computing the arrangement of curve segments: Divide-and-conquer algorithms via sampling. In *Proc. 11th ACM-SIAM Sympos. Discrete Algorithms*, pages 705–706, 2000.

[3] G. Barequet, M. Dickerson, and P. Pau. Translating a convex polygon to contain a maximum number of points. *Comput. Geom. Theory Appl.*, 8:167–179, 1997.

[4] H. Bast and T. Hagerup. Fast and reliable parallel hashing. In *Proc. 3rd Annual ACM Symposium on Parallel Algorithms and Architectures*, pages 50–61, 1991.

[5] H. Brönnimann, B. Chazelle, and J. Matoušek. Product range spaces, sensitive sampling, and derandomization. *SIAM J. Comput.*, 28:1552–1575, 1999.

[6] B. Chazelle. Cutting hyperplanes for divide-and-conquer. *Discrete Comput. Geom.*, 9:145–158, 1993.

[7] B. Chazelle and D. T. Lee. On a circle placement problem. *Computing*, 36:1–16, 1986.

[8] A. Datta, H.-P. Lenhof, C. Schwarz, and M. Smid. Static and dynamic algorithms for k-point clustering problems. *J. Algorithms*, 19:474–503, 1995.

[9] M. Dietzfelbinger, T. Hagerup, J. Katajainen, and M. Penttonen. A reliable randomized algorithm for the closest-pair problem. *J. Algorithms*, 25:19–51, 1997.

[10] A. Efrat, M. Sharir, and A. Ziv. Computing the smallest k-enclosing circle and related problems. *Comput. Geom. Theory Appl.*, 4:119–136, 1994.

[11] D. Eppstein and J. Erickson. Iterated nearest neighbors and finding minimal polytopes. *Discrete Comput. Geom.*, 11:321–350, 1994.

[12] A. Gajentaan and M. H. Overmars. On a class of $O(n^2)$ problems in computational geometry. *Comput. Geom. Theory Appl.*, 5:165–185, 1995.

[13] T. Hagerup and C. Rüb. A guided tour of Chernoff bounds. *Inform. Process. Lett.*, 33:305–308, 1990.

[14] D. P. Huttenlocher and S. Ullman. Object recognition using alignment. In *Proc. 1st Internat. Conf. Comput. Vision*, pages 102–111, 1987.

[15] K. Kedem, R. Livne, J. Pach, and M. Sharir. On the union of Jordan regions and collision-free translational motion amidst polygonal obstacles. *Discrete Comput. Geom.*, 1:59–71, 1986.

[16] Y. Lamdan, J. T. Schwartz, and H. J. Wolfson. Object recognition by affine invariant matching. In *Proc. IEEE Internat. Conf. Comput. Vision Pattern. Recogn.*, pages 335–344, 1988.

[17] H. P. Lenhof and M. Smid. Sequential and parallel algorithms for the k closest pairs problem. *Internat. J. Comput. Geom. Appl.*, 5:273–288, 1995.

[18] F. P. Preparata and M. I. Shamos. *Computational Geometry: An Introduction.* Springer-Verlag, New York, NY, 1985.

[19] O. Schwarzkopf, U. Fuchs, G. Rote, and E. Welzl. Approximation of convex figures by pairs of rectangles. *Comput. Geom. Theory Appl.*, 10:77–87, 1998.

[20] M. Sharir. On k-sets in arrangements of curves and surfaces. *Discrete Comput. Geom.*, 6:593–613, 1991.

Approximation Algorithms for k-Line Center*

Pankaj K. Agarwal, Cecilia M. Procopiuc, and Kasturi R. Varadarajan

[1] Duke University, Durham, NC 27708-0129
pankaj@cs.duke.edu
[2] AT&T Research Lab, Florham Park, NJ 07932
magda@research.att.com
[3] University of Iowa, Iowa City, IA 52242-1419
kvaradar@cs.uiowa.edu

Abstract. Given a set P of n points in $\mathbb{R}^d$ and an integer $k \geq 1$, let w^* denote the minimum value so that P can be covered by k cylinders of radius at most w^*. We describe an algorithm that, given P and an $\varepsilon > 0$, computes k cylinders of radius at most $(1 + \varepsilon)w^*$ that cover P. The running time of the algorithm is $O(n \log n)$, with the constant of proportionality depending on k, d, and ε. We first show that there exists a small "certificate" $Q \subseteq P$, whose size does not depend on n, such that for any k-cylinders that cover Q, an expansion of these cylinders by a factor of $(1 + \varepsilon)$ covers P. We then use a well-known scheme based on sampling and iterated re-weighting for computing the cylinders.

1 Introduction

Problem statement and motivation. Given a set P of n points in $\mathbb{R}^d$, an integer $k \geq 1$, and a real $\varepsilon > 0$, we want to compute k cylinders of radius at most $(1+\varepsilon)w^*$ that cover P (that is, the union of the cylinders contains P), where w^* denotes the minimum value so that P can be covered by k cylinders of radius at most w^*.

This problem is a special instance of projective clustering. In a more general formulation, a *projective clustering* problem can be defined as follows. Given a set S of n points in $\mathbb{R}^d$ and two integers $k < n$ and $q \leq d$, find k q-dimensional flats $h_1, \ldots, h_k$ and partition S into k subsets $S_1, \ldots, S_k$ so that $\max_{1 \leq i \leq k} \max_{p \in S_i} d(p, h_i)$ is minimized. That is, we partition S into k clusters, and each cluster S_i is projected onto a q-dimensional linear subspace so that the maximum distance between a point p and its projection p^* is minimized. In this paper we study the special case in which $q = 1$, i.e., we wish to cover S by k congruent cylinders of smallest minimum radius.

Clustering is a widely used technique for data mining, indexing, and classification [24]. Most of the methods — both theoretical and practical — proposed in the last few years [2, 14, 24] are "full-dimensional," in the sense that they give

---

* Research by the first author is supported by NSF under grants CCR-00-86013 ITR–333–1050, EIA-98-70724, EIA-01-31905, and CCR-97-32787, and by a grant from the U.S.–Israel Binational Science Foundation.

equal importance to all the dimensions in computing the distance between two points. While such approaches have been successful for low-dimensional datasets, their accuracy and efficiency decrease significantly in higher dimensional spaces (see [21] for an excellent analysis and discussion). The reason for this performance deterioration is the so-called dimensionality curse. A full-dimensional distance for moderate-to-high dimensional spaces is often irrelevant. Methods such as principal component analysis, singular value decomposition, and randomized projection reduce the dimensionality of the data by projecting all points on a subspace so that the information loss is minimized. A full-dimensional clustering method is then used in this subspace. However, these methods do not handle well those situations in which different subsets of the points lie on different lower-dimensional subspaces. Motivated by the need for increased flexibility in reducing the data dimensionality, recently a number of methods have been proposed for *projective clustering*, in which points that are closely correlated in some subspace are clustered together [5, 6, 10, 26]. Instead of projecting the entire dataset on a single subspace, these methods project each cluster on its associated subspace, which is generally different from the subspace associated with another cluster.

Previous results. Meggido and Tamir [27] showed that it is NP-complete to decide whether a set of n points in the plane can be covered by k lines. This immediately implies that projective clustering is NP-Complete even in the planar case. In fact, it also implies that approximating the minimum width within a constant factor is NP-Complete. Approximation algorithms for hitting compact sets by minimum number of lines are presented in [19]. Fitting a $(d-1)$-hyperplane through S is the classical *width problem*. The width of a point set can be computed in $\Theta(n \log n)$ time for $d = 2$ [22], and in $O(n^{3/2+\varepsilon})$ expected time for $d = 3$ [1]. Duncan et al. gave an algorithm for computing the width approximately in higher dimensions [13]. Several algorithms with near-quadratic running time are known for covering a set of n points in the plane by two strips of minimum width; see [25] and references therein. Har-Peled and Varadarajan [18] have recently given a polynomial-time approximation scheme for the projective clustering problem in high dimensions for any fixed k and q.

Besides these results, very little is known about the projective clustering problem, even in the plane. A few Monte Carlo algorithms have been developed for projecting S onto a single subspace [23]. An exact solution to the projective clustering problem can be solved in $n^{O(dk)}$ time. We can also use the greedy algorithm [11] to cover points by congruent q-dimensional hyper-cylinders. More precisely, if S can be covered by k hyper-cylinders of radius r, then the greedy algorithm covers S by $O(k \log n)$ hyper-cylinders of radius r in time $n^{O(d)}$. The approximation factor can be improved to $O(k \log k)$ using the technique by Brönimann and Goodrich [8]. For example, this approach computes a cover of $S \subseteq \mathbb{R}^d$ by $O(k \log k)$ hyper-cylinders of a given radius r in time $O(n^{O(d)} k \log k)$, assuming that S can be covered by k hyper-cylinders of radius r each. Agarwal and Procopiuc [3] give a significantly faster scheme to cover S with $O(dk \log k)$ hyper-cylinders of radius at most r in $O(dnk^2 \log^2 n)$ time. Combining this result with parametric search, they also give an algorithm that computes, in time $O(dnk^3 \log^4 n)$, a cover of S by $O(dk \log k)$ hyper-cylinders of radius at most $8w^*$, where w^* is the minimum radius of a cover by k hyper-cylinders.

The problem that we consider can also be thought of as an instance of a shape fitting problem, where the quality of the fit is determined by the maximum distance of a point from the shape. In these problems one generally wants to fit a shape, for example a line, hyperplane, sphere, or cylinder, through a given set of points P. Approximation algorithms with near-linear dependence on the size of P are now known for most of these shapes [7, 9, 17]. Trying to generalize these techniques to more complicated shapes seems to be the next natural step. The problem that we consider, that of trying to fit a point set with $k > 1$ lines, is an important step in this direction. The only previous result giving a $(1 + \varepsilon)$-approximation in near-linear time in n, for $k > 1$, is the algorithm of Agarwal et al. [4] for $k = 2$. However, their techniques do not generalize to higher dimensions or to the case of $k > 2$.

Our results and techniques We present an $(1+\varepsilon)$-approximation algorithm, with $O(n \log n)$ running time, for the k-line center problem in any fixed dimension; the constant of proportionality depends on k, ε, and d. We believe that the techniques used in showing this result are quite useful in themselves. We first show that there exists a small "certificate" $Q \subseteq P$, whose size does not depend on n, such that for any k-cylinders that cover Q, an expansion of these cylinders by a factor of $(1 + \varepsilon)$ covers P. The proof of this result is non-constructive in the sense that it does not give us an efficient way of constructing a certificate. The ideas used in this proof offer some hope for simplifying some known results as well as for proving the existence of a small certificate for other, more difficult problems.

We then observe that a well-known scheme based on sampling and iterated re-weighting [12] gives us an efficient algorithm for solving the problem. Only the existence of a small certificate is used to establish the correctness of the algorithm. This technique is quite general and can be used in other contexts as well. Thus it allows us to focus our attention on trying to prove the existence of small certificates, which seems to be the right approach for more complex shapes.

We present a few definitions in Section 2, a proof of the existence of small certificates in Section 3, and our algorithm in Section 4.

2 Preliminaries

A cylinder in $\mathbb{R}^d$, defined by specifying a line ℓ in $\mathbb{R}^d$ and a non-negative real number $r \geq 0$, is the set of all points within a distance of r from the line ℓ. We refer to ℓ as the *axis* and r as the *radius* of the cylinder. For $\varepsilon \geq 0$, an *ε-expansion* of a cylinder with axis ℓ and radius r is the cylinder with axis ℓ and radius $(1 + \varepsilon)r$. We define an *ε-expansion* of a finite set of cylinders to be the set obtained by replacing each cylinder with its ε-expansion.

Definition 2.1. Let P be a set of points in $\mathbb{R}^d$, $\varepsilon > 0$, and $k \geq 1$. We say that a subset $Q \subseteq P$ is an (ε, k)-*certificate* for P if for any set Σ of k congruent cylinders that covers Q, an ε-expansion of Σ results in a cover of P. We stress that the cylinders in Σ must have equal radius.

Let $I = [a, b]$ be an interval. For $\varepsilon \geq 0$, we define its ε-*expansion* to be the interval $I = [a - \varepsilon(b - a), b + \varepsilon(b - a)]$. We define an ε-expansion of a set of intervals on the real line analogously.

Definition 2.2. Let P be a set of points in $\mathbb{R}^1$, $\varepsilon > 0$ a real number, and $k \geq 1$ an integer. We say that a subset $Q \subseteq P$ is an (ε, k)-*certificate* for P if for any set $\mathcal{I}$ of k intervals that covers Q, an ε-expansion of $\mathcal{I}$ results in a cover for P. We emphasize that the intervals in $\mathcal{I}$ are allowed to have different lengths.

Though we are using the term (ε, k)-certificate to mean two different notions, the context will clarify which one we are referring to.

3 Existence of Small Certificates

We show in this section that for any point set P in $\mathbb{R}^d$, $\varepsilon > 0$, and integer $k \geq 1$, there is an (ε, k)-certificate for P whose size is independent of the size of P. In order to do this, we first have to show the existence of such certificates for points in $\mathbb{R}^1$.

Lemma 3.1. *Let P be any set of points in $\mathbb{R}^1$, $\varepsilon > 0$ be a real number, and $k \geq 1$ be an integer. There is an (ε, k)-certificate $Q \subseteq P$ for P with $g(\varepsilon, k)$ points, where $g(\varepsilon, k) = (k/\varepsilon)^{O(k)}$.*

Proof: The proof is by induction on k. If $k = 1$, we let Q be the two extreme points of P.

If $k > 1$, Q is picked as follows. Let Δ be the length of the interval I spanned by P. We divide I into k intervals of length Δ/k. Let $a_0 < a_1 \cdots < a_k$ denote the endpoints of these intervals (thus, $I = [a_0, a_k]$). For each a_i, $1 \leq i \leq k-1$, and $1 \leq j \leq k-1$, we compute an (ε, j)-certificate for $P \cap [a_0, a_i]$ and an $(\varepsilon, k-j)$-certificate for $P \cap [a_i, a_k]$. Let Q_1 denote the union of all these certificates. Obviously,

$$|Q_1| \leq \sum_{i=1}^{k-1}\sum_{j=1}^{k-1} g(\varepsilon, j) + g(\varepsilon, k-j) \leq 2k^2 g(\varepsilon, k-1).$$

Next, we divide I into k/ε intervals of length $\varepsilon\Delta/k$; we call these *basic* intervals. Let us call an interval of the real line *canonical* if it is a union of basic intervals and has length at least Δ/k. There are $\Theta(k^2/\varepsilon^2)$ canonical intervals. For each canonical interval I', we compute an $(\varepsilon, k-1)$-certificate for the points in P lying *outside* I'. Let Q_2 denote the union of these certificates. $|Q_2| \leq ck^2/\varepsilon^2 g(\varepsilon, k-1)$, where c is a constant.

We let $Q = Q_1 \cup Q_2$. We obtain the following recurrence for $g(\varepsilon, k)$:

$$g(\varepsilon, k) \leq ck^2/\varepsilon^2 g(\varepsilon, k-1) + 2k^2 g(\varepsilon, k-1).$$

The solution to the above recurrence is $g(\varepsilon, k) = (k/\varepsilon)^{O(k)}$, as claimed.

We now argue that Q is an (ε, k)-certificate. Let $\Sigma = \{s_1, s_2, \ldots, s_k\}$ be k intervals that covers Q. We first consider the case where all these segments have length smaller than Δ/k. In this case there exists an a_i, $1 \leq i \leq k-1$, that is not contained in any of these segments. Indeed, each interval in Σ can cover at most one a_i, and a_0, a_k are covered by Σ. Without loss of generality, we may assume that a_ξ, for some $\xi < k$, is not covered by Σ and that the segments $s_1, \ldots, s_j$ lie to the left of a_ξ and the segments $s_{j+1}, \ldots, s_k$ lie to the right of a_ξ, for some $1 \leq j \leq k-1$. Since Q includes an (ε, j)-certificate for the points $P \cap [a_0, a_\xi]$, and $s_1, \ldots, s_j$ cover this certificate, we conclude that an ε-expansion of $s_1, \ldots, s_j$ contains $P \cap [a_0, a_\xi]$. By a symmetric argument, we conclude that an ε-expansion of $s_{j+1}, \ldots, s_k$ covers $P \cap [a_\xi, a_n]$, and we are done.

We now consider the case when one of the segments, say s_1, has length at least Δ/k. Let I' denote the smallest canonical interval containing s_1; note that I' is covered by an ε-expansion of s_1. Since Q includes an $(\varepsilon, k-1)$-certificate for the points in P outside I', and $s_2, \ldots, s_k$ cover this certificate, we conclude that an ε-expansion of $s_2, \ldots, s_k$ covers the points in P outside I'. Since an ε-expansion of s_1 covers $P \cap I'$, the result follows. □

We use Lemma 3.1 to prove the existence of a small certificate for any set of points in $\mathbb{R}^d$.

Lemma 3.2. *Let P be any set of points in $\mathbb{R}^d$, $\varepsilon > 0$, and $k \geq 1$. There exists an (ε, k)-certificate for P with $f(\varepsilon, k)$ points, where $f(\varepsilon, k) = k^{O(k)}/\varepsilon^{O(d+k)}$.*

Proof: Consider a cover of P by k congruent cylinders of minimum radius, and let this radius be denoted by w^*. Let $\delta = c\varepsilon$, where $c > 0$ is a sufficiently small constant. Clearly, there is a set L of $O(k/\varepsilon^{d-1})$ lines such that for any point $p \in P$, there is a line $\ell \in L$ with $d(p, \ell) \leq \delta w^*$. Indeed, for every cylinder C in the cover, draw a grid of size ε in the $(d-1)$-dimensional ball forming the base of C and draw a line parallel to the axis of C from each of the grid points.

Let P' be the point set obtained by projecting each point in P to the nearest line in L. Let $P'(\ell) \subseteq P'$ denote the points on the line ℓ. Let $Q'(\ell)$ be a (δ, k)-certificate, computed according to Lemma 3.1, for the points $P'(\ell)$ (by treating ℓ as the real line). Let $Q' = \bigcup_{\ell \in L} Q'(\ell)$, and let Q be the original points in P corresponding to Q'.

The bound on the size of Q is easily verified. Consider k cylinders of radius r that cover Q. Expanding each cylinder by an *additive* factor of δw^*, we cover Q'. For each $\ell \in L$, the segments formed by intersecting the cylinders with ℓ cover $Q'(\ell)$. Therefore, the δ-expansion of the segments results in a cover of $P'(\ell)$. It follows that a δ-expansion of the cylinders results in a cover for P' (this step needs a geometric claim that is easily verified). Expanding further by an additive

factor of δw^*, we get a cover of P. Since w^* is the radius of the optimal cover of P, we have
$$(r + \delta w^*)(1 + \delta) + \delta w^* \geq w^*.$$
Assuming that $\delta < 1/10$, this yields $w^* \leq 2r$. Thus the radius of the cylinders that cover P is at most
$$(r + \delta w^*)(1 + \delta) + \delta w^* \leq r((1 + 2\delta)(1 + \delta) + 2\delta) \leq (1 + \varepsilon)r.$$

□

The ideas used above in proving the existence of a small certificate may prove useful in other contexts also. To illustrate this point, we use these ideas to establish a known result whose earlier proofs relied on powerful tools like the Lowner-John ellipsoid [16].

Theorem 3.1. *Let P be a set of points in $\mathbb{R}^d$, and let $\varepsilon > 0$. There exists a subset $Q \subseteq P$ with $O(1/\varepsilon^{d-1})$ points such that for any slab that contains Q an ε-expansion of the slab contains P.*

Proof: If $d = 1$, Q consists of the two extreme points. If $d > 1$, consider the minimum width slab that covers P, and let w^* be its width. We find $1/\varepsilon$ slabs such that each point in P is within εw^* of the nearest slab. We move each point in P to the nearest slab. We now have $1/\varepsilon$ $(d-1)$-dimensional point sets. We recursively compute a $(d-1)$-dimensional certificate for each of these point sets, let Q' be their union, and Q the corresponding points in P. We argue as in Lemma 3.2 that Q is a certificate for P. □

4 The Algorithm

Let $P \in \mathbb{R}^d$ be a set of n points in $\mathbb{R}^d$, and let $\varepsilon > 0$ be a given parameter. Let $w^* \geq 0$ denote the smallest number such that there are k cylinders of radius w^* that cover P. We describe an efficient algorithm to compute a cover of P using k cylinders of radius at most $(1 + \varepsilon)w^*$. The algorithm relies on the fact P has a small (ε, k)- certificate. Let $\mu \leq f(\varepsilon, k)$ denote the size of this certificate. The algorithm is an adaptation of the iterated re-weighting algorithm of Clarkson [12] for linear programming.

The algorithm maintains an integer weight $\mathrm{Wt}(p)$ for each point $p \in P$, which is initialized to 1. For any subset $Q \subseteq P$, let $\mathrm{Wt}(Q)$ denote the sum of the weights $\sum_{q \in Q} \mathrm{Wt}(q)$. Our algorithm proceeds as follows.

I. We repeat the following sampling process until we meet with success.
   (a) Let $r = (2\log_2 e)\mu$ and R be a random sample from P of size $c(d,k)r \log r$, where $c(d,k)$ is a sufficiently large constant. R is constructed by $|R|$ independent repetitions of the sampling process in which a point $p \in P$ is picked with probability $\mathrm{Wt}(p)/\mathrm{Wt}(P)$.

(b) We compute $\mathcal{C}(R)$, the optimal k-cylinder cover for R, using the brute-force exact algorithm in $O(|R|^{2dk})$ time. Let S be the set of points in P not covered by $\mathcal{C}(R)$. If $\text{Wt}(S) > \text{Wt}(P)/r$, then we return to Step (I.a). Otherwise, we proceed to Step (II).

II. We check whether the ε-expansion of $\mathcal{C}(R)$ covers P. If it does, our algorithm returns this ε-expansion as the approximate k-cylinder cover of P and halts. Otherwise, we double the weight $\text{Wt}(s)$ of each point $s \in S$ and return to Step (I).

It is clear that if the algorithm does halt, it returns an ε-approximate k-cylinder cover of P because $\mathcal{C}(R)$ is an *optimal* k-cylinder cover of R and an ε-expansion of $\mathcal{C}(R)$ covers P. We now argue that the algorithm does indeed halt.

Lemma 4.1. *Step (II) of the algorithm is executed at most $2\mu \log n$ times.*

Proof: Let Q be an (ε, k)-certificate for P with size μ. Let us assume that Step (II) is executed ℓ times, and each time the ε-expansion of $\mathcal{C}(R)$ does not cover P. This means that $\mathcal{C}(R)$ did not cover some point from Q in each iteration. Let $q \in Q$ be the point that was not covered the most number of times. Then q was not covered at least ℓ/μ times, and so its weight after ℓ executions of Step (II) is at least $2^{\ell/\mu}$.

Note that whenever Step (II) is executed, we have $\text{Wt}(S) \leq \text{Wt}(P)/r$. Since we double the weights of only the points in S, it follows that $\text{Wt}(P)$ increases by a factor of at most $(1 + 1/r)$ each time Step (II) is executed. After ℓ executions of Step (II), we have that $\text{Wt}(P) \leq n(1 + 1/r)^\ell$. Since we must have $\text{Wt}(q) \leq \text{Wt}(P)$, we have
$$2^{\ell/\mu} \leq n(1 + 1/r)^\ell.$$
Taking logarithms, and rearranging the terms, we get
$$\ell \leq \frac{\log_2 n}{(1/\mu - \log_2(1 + 1/r))}.$$
Using the fact that $\log_2(1 + a) \leq (\log_2 e)a$ for any $a \geq 0$, and substituting the value of r, it follows that $\ell \leq 2\mu \log n$. □

Lemma 4.2. *Let R be a random sample of P as constructed in Step (I.a) of the algorithm, and let S be the set of points in P not covered by $\mathcal{C}(R)$. Then the probability that $\text{Wt}(S) > \text{Wt}(P)/r$ is at most $1/2$ if the constant $c(d, k)$ is chosen large enough.*

Proof: This follows from the theory of ε-nets [20]. Let $\mathcal{C}$ be the set of all cylinders in $\mathbb{R}^d$, and let $\mathcal{C}^k$ be the family of k-tuples in $\mathcal{C}$. It can be shown that VC-dimension of the range space $(\mathbb{R}^d, \mathcal{C}^k)$ is finite and depends only on k and d. Assuming that the constant $c(d, k)$ is larger than the VC-dimension of the range space, the lemma follows from a result by Haussler and Welzl [20]. □

Lemma 4.2 implies that the expected number of times we have to iterate Steps (I.a) and (II.b) before we find a sample R for which $\text{Wt}(S) \le \text{Wt}(P)/r$ is at most 2. Combining this with Lemma 4.1, we see that the expected running time of the algorithm is $O(\ell n + \ell(\mu \log \mu)^{2dk})$, where $\ell = \mu \log n$. We have thus obtained the main result of this paper.

Theorem 4.3. *Let P be a set of n points in $\mathbb{R}^d$, $w^* \ge 0$ denote the smallest number such that there are k cylinders of radius w^* that cover P, and $\varepsilon > 0$ be a parameter. We can compute k cylinders of radius at most $(1 + \varepsilon)w^*$ that cover P in $O(n \log n)$ time, with the constant of proportionality depending on k, ε, and d.*

5 Conclusion

We presented an ε-approximation algorithm for computing a k-line-center of a set of points in $\mathbb{R}^d$ whose running time is $O(n \log n)$; the constant of proportionality depends on d, k, ε. We showed the existence of a small certificate for P whose size does not depend on n and used this result to prove the correctness of the algorithm.

It is easy to see that the algorithm is fairly general and would work in related contexts, provided we can demonstrate the existence of a small certificate. One disadvantage is the large dependence of the running time on d and k. We have not tried to optimize this dependence. Some simple techniques, like computing a constant factor approximation first and then refining it to a factor of $(1 + \varepsilon)$, may help improve the dependence on some of the parameters.

Another interesting open question is whether our approach can be extended to general projective clustering.

References

[1] P. K. Agarwal and M. Sharir. Efficient randomized algorithms for some geometric optimization problems. *Discrete Comput. Geom.*, 16 (1996), 317–337.
[2] P. K. Agarwal and M. Sharir. Efficient algorithms for geometric optimization. *ACM Comput. Surv.*, 30 (1998), 412–458.
[3] P. K. Agarwal and C. M. Procopiuc. Approximation algorithms for projective clustering. In *Proc. 11th ACM-SIAM Sympos. Discrete Algorithms*, 2000, pp. 538–547.
[4] P. K. Agarwal, C. M. Procopiuc, and K. R. Varadarajan. A $(1+\varepsilon)$-approximation algorithm for 2-line-center. Submitted for publication.
[5] C. C. Aggarwal, C. M. Procopiuc, J. L. Wolf, P. S. Yu, and J. S. Park. Fast algorithms for projected clustering. In *Proc. of ACM SIGMOD Intl. Conf. Management of Data*, 1999, pp. 61–72.
[6] C. C. Aggarwal and P. S. Yu. Finding generalized projected clusters in high dimensional spaces. In *Proc. of ACM SIGMOD Intl. Conf. Management of Data*, 2000, pp. 70–81.

[7] G. Barequet and S. Har-Peled. Efficiently approximating the minimum-volume bounding box of a point set in three dimensions. In *Proc. 10th ACM-SIAM Sympos. Discrete Algorithms*, 1999, pp. 82–91.

[8] H. Brönnimann and M. T. Goodrich. Almost optimal set covers in finite VC-dimension. *Discrete Comput. Geom.*, 14 (1995), 263–279.

[9] T. M. Chan. Approximating the diameter, width, smallest enclosing cylinder and minimum-width annulus. In *Proc. 16th Annu. ACM Sympos. Comput. Geom.*, 2000, pp. 300–309.

[10] K. Chakrabarti and S. Mehrotra. Local dimensionality reduction: A new approach to indexing high dimensional spaces. In *Proc. 26th Intl. Conf. Very Large Data Bases*, 2000, pp. 89–100.

[11] V. Chvátal. A greedy heuristic for the set-covering problem. *Math. Oper. Res.*, 4 (1979), 233–235.

[12] K. L. Clarkson. Las Vegas algorithms for linear and integer programming. *J. ACM*, 42:488–499, 1995.

[13] C. A. Duncan, M. T. Goodrich, and E. A. Ramos. Efficient approximation and optimization algorithms for computational metrology, In *Proc. 8th ACM-SIAM Sympos. Discrete Algorithms*, 1997, pp. 121–130.

[14] M. Ester, H.-P. Kriegel, J. Sander, and X. Xu. Density-connected sets and their application for trend detection in spatial databases. In *Proc. 3rd Intl. Conf. Knowledge Discovery and Data Mining*, 1997.

[15] C. Faloutsos and K.-I. Lin, FastMap: A fast algorithm for indexing, data-mining and visualization of traditional and multimedia databases. In *Proc. ACM Intl SIGMOD Conf. Management of Data*, 1995, pp. 163–173.

[16] P. M. Gruber. Aspects of approximation of convex bodies. *Handbook of Convex Geometry*, volume A, (P. M. Gruber and J. M. Wills, eds.), North-Holland, Amsterdam, 1993, pp. 319–345.

[17] S. Har-Peled and K. Varadarajan. Approximate Shape Fitting via Linearization. In *Proc. IEEE Symp. Foundations of Comp. Sci.*, 2001, pp. 66–73.

[18] S. Har-Peled and K. Varadarajan. Projective clustering in high dimensions using core-sets. To appear in *ACM Symp. Comput. Geom.*, 2002.

[19] R. Hassin and N. Megiddo. Approximation algorithms for hitting objects by straight lines. *Discrete Appl. Math.*, 30 (1991), 29–42.

[20] D. Haussler and E. Welzl. Epsilon-nets and simplex range queries. *Discrete Comput. Geom.*, 2 (1987), 127–151.

[21] A. Hinneburg and D. A. Keim. Optimal grid-clustering: Towards breaking the curse of dimensionality in high-dimensional clustering. In *Proc. 25th Intl. Conf. Very Large Data Bases*, 1999, pp. 506–517.

[22] M. E. Houle and G. T. Toussaint. Computing the width of a set. *IEEE Trans. Pattern Anal. Mach. Intell.*, PAMI-10 (1988), 761–765.

[23] P. Indyk, R. Motwani, P. Raghavan, and S. Vempala. Locality-preserving hashing in multidimensional spaces. In *Proc. 29th Annu. ACM Sympos. Theory Comput.*, 1997, pp. 618–625.

[24] A. K. Jain, M. N. Murty, and P. J. Flynn. Data clustering: A review. *ACM Comput. Surv.*, 31 (1999), 264–323.

[25] J. W. Jaromczyk and M. Kowaluk. The two-line center problem from a polar view: A new algorithm and data structure. *Proc. 4th Workshop Algorithms Data Structures, Lecture Notes Comput. Sci.*, Vol. 955, Springer-Verlag, 1995, pp. 13–25.

[26] C. M. Procopiuc and M. Jones and P. K. Agarwal and T. M. Murali. A Monte Carlo algorithm for fast projective clustering. In *Proc. ACM SIGMOD Intl. Conf. Management of Data*, 2002, pp. 418–427.

[27] N. Megiddo and A. Tamir. On the complexity of locating linear facilities in the plane. *Oper. Res. Lett.*, 1 (1982), 194–197.

New Heuristics and Lower Bounds for the Min-Max k-Chinese Postman Problem

Dino Ahr and Gerhard Reinelt

Institute for Computer Science, University of Heidelberg
Im Neuenheimer Feld 368, 69120 Heidelberg, Germany
{dino.ahr,gerhard.reinelt}@informatik.uni-heidelberg.de

Abstract. Given an undirected edge-weighted graph and a depot node, postman problems are generally concerned with traversing the edges of the graph (starting and ending at the depot node) while minimizing the distance traveled. For the Min-Max k-Chinese Postman Problem (MM k-CPP) we have $k > 1$ postmen and want to minimize the longest of the k tours. We present two new heuristics and improvement procedures for the MM k-CPP. Furthermore, we give three new lower bounds in order to assess the quality of the heuristics. Extensive computational results show that our algorithms outperform the heuristic of Frederickson et al. [12].

Keywords: Arc Routing, Chinese Postman Problem, Min-Max Optimization, Heuristics, Lower Bounds.

1 Introduction

1.1 Problem Definition and Contributions

We consider arc routing problems with multiple postmen. Given an undirected graph $G = (V, E)$, weights $w : E \to \mathbb{R}^+$ for each edge (which we usually interprete as distances), a distinguished depot node $v_1 \in V$ and a fixed number $k > 1$ of postmen, we want to find k tours where each tour starts and ends at the depot node and each edge $e \in E$ is covered by at least one tour. In contrast to the usual objective to minimize the total distance traveled by the k postmen, for the *Min-Max k-Chinese Postman Problem* (MM k-CPP) we want to minimize the length of the longest of the k tours.

This kind of objective function is preferable when the aim is to serve each customer as early as possible. Furthermore, tours will be enforced to be more balanced resulting in a "fair" scheduling of tours.

The MM k-CPP was introduced in [12] and shown to be $\mathcal{NP}$-hard by a reduction from the k-partition problem. Furthermore, in [12] a $(2 - \frac{1}{k})$-approximation algorithm for the MM k-CPP is proposed, which we call *FHK-heuristic* in the following. To the best of our knowledge these are the only results for the MM-k-CPP in the literature.

In this paper we present two new constructive heuristics for the MM k-CPP as well as improvement procedures which are used as postprocessing steps for the constructive heuristics. We implemented our new heuristics and the improvement procedures as well as the FHK-heuristic and compared them on a variety of test instances from the literature as well as on random instances. We were able to improve the results obtained by the FHK-heuristic for most instances, sometimes considerably. In order to assess the solution quality of the heuristics in general, we also derived three new lower bounds for the MM k-CPP.

1.2 Related Work

Arc routing problems of a great variety are encountered in many practical situations such as road or street maintenance, garbage collection, mail delivery, school bus routing, etc. For an excellent survey on arc routing we refer the reader to [8], [2], [10] and [11].

The *k-Chinese Postman Problem* (k-CPP) has the same set of feasible solutions as the MM k-CPP except that the objective is to minimize the total distance traveled by the k postmen. In [22] it is shown that the k-CPP is polynomially solvable by applying a lower bound algorithm for the Capacitated Arc Routing Problem from [3].

Because of its great practical relevance much attention has been paid to the *Capacitated Chinese Postman Problem* (CCPP) introduced in [4] and the *Capacitated Arc Routing Problem* (CARP) introduced in [14]. Both problems have the same input data as the MM k-CPP but in addition there are edge demands $(q : E \to \mathbb{R}^+)$, which can be interpreted as the amount of load to collect or deposit along the edge, and a fixed vehicle capacity Q. A feasible set of k tours $C_1, \ldots, C_k$, each tour starting and ending at the depot node, has to cover all edges e with positive demand $(q(e) > 0)$ while obeying the restricted capacity of each vehicle $(\sum_{e \in C_i} q(e) \leq Q$ for $i = 1, \ldots, k)$. The objective is to find a feasible set of k tours which minimizes the total distance traveled. The CCPP is a special case of the CARP, where all edges have a positive demand. Both problems are hard to solve, in fact in [14] it is shown that even 1.5-approximation of the CCPP is $\mathcal{NP}$-hard.

The CARP is strongly related to the MM k-CPP since again tours are forced to be balanced because of the capacity constraints.

Great effort has been spent to devise heuristics for the CARP. We will mention some of them, inspiring our heuristics, in the subsequent sections. For an extensive survey of CARP heuristics we refer the reader to [19].

Recently the postoptimization tools developed in [18] for the Undirected Rural Postman Problem have been embedded into a tabu search algorithm for the CARP [17] outperforming all other heuristics. Motivated by these results, we developed improvement procedures suitable for the MM k-CPP with comparable success.

In [1] a special Min-Max Vehicle Routing Problem is considered, where the objective is also to minimize the longest tour. The difference to the MM k-CPP is that nodes have to be traversed instead of edges. In particular, the paper describes the solution process of a special instance with 4 vehicles and 120 customer nodes, issued in the scope of a mathematical contest in 1996. The fact that this instance could only be solved with a sophisticated distributed branch-and-cut based implementation, taking 10 days on a distributed network of 188 processors, impressively demonstrates the inherent difficulty of problems subjected to a min-max objective.

The paper is organized as follows. In Sect. 2 we describe the heuristics and improvement procedures we have implemented for the MM k-CPP. Sect. 3 deals with the lower bounds developed for the MM k-CPP. In the subsequent section we report computational results which show that our new algorithms perform well. Finally, we give conclusions and directions for future work.

2 Algorithms

2.1 Terminology

For the MM k-CPP we have the following input data: a connected undirected graph $G = (V, E)$, weights for each edge $w : E \to \mathbb{R}^+$, a distinguished depot node $v_1 \in V$ and a fixed number $k > 1$ of postmen. A feasible solution, called *k-postman tour*, is a set $\mathcal{C}$ of k closed walks, $\mathcal{C} = \{C_1, \ldots, C_k\}$, such that each tour C_i contains the depot node v_1, all edges $e \in E$ are covered by at least one tour C_i and each postman is involved.

We extend the weight function w to edge sets $F \subseteq E$ by defining $w(F) = \sum_{e \in F} w(e)$. Now, for a k-postman tour $\mathcal{C}$, let us denote the maximum weight attained by a single tour C_i as $w_{\max}(\mathcal{C})$, i.e.

$$w_{\max}(\mathcal{C}) = \max_{i=1,\ldots,k} w(C_i).$$

The objective of the MM k-CPP is to find a k-postman tour $\mathcal{C}^*$ which minimizes $w_{\max}$ among all feasible k-postman tours, i.e.

$$w_{\max}(\mathcal{C}^*) = \min\{w_{\max}(\mathcal{C}) \mid \mathcal{C} \text{ is a } k\text{-postman tour}\}.$$

We denote by $SP(v_i, v_j)$ the set of edges on the shortest path between nodes $v_i, v_j \in V$. The distance of the shortest path between v_i and v_j is given by $w(SP(v_i, v_j))$.

For each node set W let $E(W) = \{e \in E \mid e = \{v_i, v_j\} \text{ and } v_i, v_j \in W\}$ be the set of edges having both endnodes in W.

2.2 General Remarks

Heuristics for multiple vehicle arc routing problems can be broadly classified into three categories: *simple constructive methods*, *two-phase constructive methods* and adaptions of *meta-heuristics* resp. *improvement procedures*. The class of two-phase constructive heuristics can be further subdivided into *route first - cluster second* and *cluster first - route second* approaches.

In the next section we describe the FHK-heuristic which follows the route first - cluster second paradigm. In Sect. 2.4 and 2.5 we present our new algorithms *Augment-Merge* and *Cluster* which are a simple constructive heuristic and a cluster first - route second heuristic, respectively. In the last subsection we explain our improvement procedures used as postprocessing steps for the augment-merge and cluster heuristics.

2.3 The Frederickson-Hecht-Kim Algorithm

The strategy of the FHK-heuristic [12] is to compute a single tour (a 1-postman tour) covering all edges $e \in E$ in a first step and then to partition the tour into k parts.

Computing a 1-postman tour is the well-known *Chinese Postman Problem* (CPP), first stated by Guan [16], which is polynomially solvable by the algorithm of Edmonds and Johnson [9].

The 1-postman tour R computed in the first step will be subdivided in the following way: First, $k-1$ so called *splitting nodes* on R are determined in such a way that they mark tour segments of R approximately having the same length. Then k tours are constructed by connecting these tour segments with shortest paths to the depot node. For details see [12].

The time complexity of the algorithm is dominated by the computation of a 1-postman tour which can be accomplished in $\mathcal{O}(|V|^3)$ [9]. In [12] it is shown that this algorithm yields a $(2 - \frac{1}{k})$-approximation for the MM k-CPP.

2.4 The Augment-Merge Algorithm

This new algorithm is based on the augment-merge algorithm for the CARP given in [14]. The idea of the algorithm is roughly as follows. We start with a closed walk C_e for each edge $e = \{v_i, v_j\} \in E$, which consists of the edges on the shortest path between the depot node v_1 and v_i, the edge e itself, and the edges on the shortest path between v_j and v_1, i.e. $C_e = (SP(v_1, v_i), e, SP(v_j, v_1))$. Then we successively merge two closed walks - trying to keep the tour weights low and balanced - until we arrive at k tours. In detail:

Algorithm: AUGMENT-MERGE

(1) Sort the edges e in decreasing order according to their weight $w(C_e)$.
(2) In decreasing order according to $w(C_e)$, for each $e = \{v_i, v_j\} \in E$, create the closed walk $C_e = (SP(v_1, v_i), e, SP(v_j, v_1))$, if e is not already covered by an existing tour. Let $\mathcal{C} = (C_1, \ldots, C_m)$ be the resulting set of tours. Note that the tours are sorted according to their length, i.e. $w(C_1) \geq w(C_2) \geq \ldots \geq w(C_m)$.

If $m \leq k$ we are done and have computed an optimal k-postman tour, since no tour is longer than the shortest path tour lower bound (see Sect. 3.1). If $m < k$ we add $k - m$ "dummy" tours to $\mathcal{C}$, each consisting of twice the cheapest edge incident to the depot node.
(3) While $|\mathcal{C}| > k$ we merge tour C_{k+1} with a tour from $C_1, \ldots, C_k$ such that the weight of the merged tour is minimized.

The time complexity of the algorithm is $\mathcal{O}(|E|^2)$ since in the third step we have to merge $\mathcal{O}(|E|)$ times two tours. Merging of two tours can be accomplished in linear time in the size of the input tours which is $\mathcal{O}(|E|)$.

2.5 The Cluster Algorithm

Our second new algorithm follows the cluster first - route second approach. In the first step we divide the edge set E into k clusters and in the second step we compute a tour for each cluster.

Going back to the second step of the augment-merge algorithm, we observe that for some edges e we explicitly have to create a tour C_e and for other edges we do not, since they are contained in an already existing tour. Let us denote the first kind of edges as *critical edges* since we have to take care that they are served, whereas serving edges classified as non-critical is "for free", since they are on some shortest path connecting a critical edge and the depot node.

Motivated by this observation the cluster step of our algorithm first performs a k-clustering of the critical edges into edge sets $F_1, \ldots, F_k$. After that, each cluster F_i is supplemented by shortest path edges connecting the contained critical edges to the depot. The routing step consists of computing the optimum 1-postman tour for each subgraph induced by F_i.

The *k-clustering step* is based on the *farthest-point clustering* algorithm of [15] and works as follows. Let F be the set of critical edges. First, we determine k *representative* edges $f_1, \ldots, f_k \in F$. Using a distance function $d : E \times E \to \mathbb{R}^+$, let $f_1 \in F$ be the edge having the maximum distance from the depot and $f_2 \in F$ the edge having maximum distance from f_1. Then the representatives $f_i \in F, i = 3, \ldots, k$ are successively determined by maximizing the minimum distance to the already existing representatives $f_1, \ldots, f_{i-1}$. The remaining critical edges g of F will be assigned to the edge set F_i which minimizes the distance between its representative f_i and g.

Algorithm: CLUSTER

(1) Let $\phi_{SP}(e) = 0$ for all $e \in E$. For each edge $e = \{v_i, v_j\} \in E$ increment the frequency $\phi_{SP}(g)$ by one for edges g on the shortest paths $SP(v_1, v_i)$ and $SP(v_j, v_1)$. Now the set of critical edges is $F = \{e \in E \mid \phi_{SP}(e) \leq 1\}$.

(2) Define the distance d between two edges $e = \{u, v\}, f = \{w, x\} \in F$ as

$$d(e, f) = \max\{w(SP(u, w)), w(SP(u, x)), w(SP(v, w)), w(SP(v, x))\}.$$

(3) Compute the k-clustering $F_1, \ldots, F_k$ according to the distance function d as described above.

(4) Extend edge sets F_i by adding all edges on shortest paths between the endnodes of edges contained in F_i and the depot node.

(5) Compute an optimum 1-postman tour on $G[F_i], i = 1, \ldots, k$, with the algorithm of Edmonds and Johnson [9].

The expensive parts of the cluster algorithm are the computation of an all pairs shortest path ($\mathcal{O}(|V|^3)$) needed for the distance function d, determining the frequency $\phi_{SP}(e)$ of each edge e with time complexity $\mathcal{O}(|E|^2)$ and the computation of the k optimum 1-postman tours which costs $\mathcal{O}(|V|^3)$.

2.6 Improvement Procedures

Having implemented the augment-merge and cluster heuristics and having tested them on a variety of instances, we observed that the solutions produced still leave potential for further improvements. We devised two general deterministic improvement strategies which try to reduce the weight of the longest tour in a greedy manner. These can be applied to an arbitrary k-postman tour $\mathcal{C}$ resulting in a k-postman tour $\tilde{\mathcal{C}}$ with $w_{\max}(\tilde{\mathcal{C}}) \leq w_{\max}(\mathcal{C})$.

Thus, we can use the algorithms augment-merge and cluster to produce good initial solutions of different structure and then apply the improvement procedures as a postoptimization step.

The basic idea of our first improvement procedure is to eliminate edges from the longest tour C_i if these edges are already covered by other tours. In order to easily maintain the closed walk property of C_i, we focus on those edges occurring at least twice in C_i.

Algorithm: ELIMINATE REDUNDANT EDGES

(1) Let $\phi_i(e)$ be the frequency of e occurring in C_i and $\phi(e) = \sum_{i=1}^{k} \phi_i(e)$ the frequency of e occurring in all tours. An edge e is called *multi-redundant* for C_i if $\phi_i(e) \geq 2$ and $\phi(e) - \phi_i(e) > 0$.

(2) As long as there exists a tour which contains multi-redundant edges, choose the longest tour C_i among them, and find the multi-redundant edge e of C_i, which leads to the maximum reduction of the tour length when removed. In detail:
- If $\phi_i(e)$ is even and removal of $\phi_i(e)$ times edge e *keeps* connectivity of C_i, set $n = \phi_i(e)$.
- If $\phi_i(e)$ is even and removal of $\phi_i(e)$ times edge e *destroys* connectivity of C_i, set $n = \phi_i(e) - 2$, i.e. we leave two edges for keeping connectivity.
- If $\phi_i(e)$ is odd, set $n = \phi_i(e) - 1$, i.e. we leave one edge for connectivity.

Now we choose the edge e which maximizes $n \cdot w(e)$. Remove n times edge e from C_i and update the frequencies $\phi_i(e)$ and $\phi(e)$.

In contrast to the first improvement procedure which tries to reduce the tour lengths by eliminating redundant edges totally, we now want to improve the longest tour by removing edges from it and integrate them into a shorter tour.

Algorithm: TWO EDGE EXCHANGE

(1) As long as the longest tour has been improved do the following:
  (1.1) Let C_i be the tour with maximum length.

(1.2) Traverse C_i and find consecutive edges $\tilde{e}$ and $\tilde{f}$ which will achieve the maximum reduction of the length of C_i when deleted and replaced by the shortest path between appropriate endnodes of $\tilde{e}$ and $\tilde{f}$.

(1.3) Consider tours C_j with $j \neq i$ and find the tour C_j^* for which the resulting tour after adding $\tilde{e}$ and $\tilde{f}$ (and other edges required to keep the closed walk property) has minimum length.

(1.4) If the tour length of C_j^* is shorter than the original tour length of C_i then replace $\tilde{e}$ and $\tilde{f}$ by the shortest path in C_i and integrate them into C_j^*. If the tour length of C_j^* is not shorter than the original tour length of C_i then goto Step 1.2, find the next better edges $\hat{e}$ and $\hat{f}$ and proceed.

3 Lower Bounds

3.1 Shortest Path Tour Lower Bound (SPT-LB)

This lower bound is based on the observation that in an optimal solution $\mathcal{C}^*$ the length of the longest tour must have at least the length of the shortest path tour $C_{\tilde{e}} = (SP(v_1, v_i), \tilde{e}, SP(v_j, v_1))$ traversing the edge $\tilde{e} = \{v_i, v_j\} \in E$ farthest away from the depot. Since the number of postmen is not taken into account this bound will only produce good results for instances where the number of postmen is suitable for the size of the graph.

3.2 CPP Tour Lower Bound (CPP/k-LB)

We simply compute the optimum 1-postman tour and divide its weight by k. It is clear that we get a valid lower bound since the 1-postman tour includes a minimum augmentation to make the given graph G Eulerian.

3.3 LP Relaxation Lower Bound (LP-LB)

We consider the following integer programming formulation of the MM k-CPP.

$$\min T$$

s.t.

$$\sum_{e \in E} w(e)x^i(e) + w(e)y^i(e) \leq T \quad i = 1, \ldots, k$$

$$\sum_{i=1}^{k} x^i(e) = 1 \qquad \text{for all } e \in E$$

$$\sum_{e \in \delta(v)} x^i(e) + y^i(e) \equiv 0 \pmod{2} \text{ for all } v \in V, i = 1, \ldots, k$$

$$x^i(\delta(S)) + y^i(\delta(S)) \geq 2x^i(e) \qquad \text{for all } S \subseteq V \setminus \{v_1\}, e \in E(S), i = 1, \ldots, k$$

$$T \in \mathbb{R}^+$$

$$x^i(e) \in \{0,1\}, y^i(e) \in \mathbb{N}_0 \qquad \text{for all } e \in E, i = 1, \ldots, k$$

The binary variables $x^i(e)$ indicate if edge e is serviced by tour C_i or not. Integer variables $y^i(e)$ count how often edge e is traversed by tour C_i without servicing it. The variable T models the length of the longest tour which is expressed with the first constraint. The second set of equations ensures that each edge is serviced exactly once. Each single tour of a valid k-postman tours must be a closed walk containing the depot. These requirements are enforced by the third and fourth set of constraints, respectively. For our lower bound we computed the LP relaxation of the above integer program, omitting the fourth set of constraints.

4 Computational Results

We benchmarked the heuristics and lower bounds for $k = 2, \ldots, 10$ on the following instances.

- Ten random generated CARP instances from [3].
- 23 CARP instances from [13].
- Two CARP instances from [20] and [21].
- 24 random generated *Rural Postman Problem (RPP)* instances from [5, 7, 6].
- Three RPP instances from [7, 6].
- Seven random generated instances.

Due to space restrictions we give explicit results only for seven selected instances and summarize the other results.

Table 1 shows the results for the three RPP instances from [7], Table 2 for the two CARP instances from [21] and Table 3 for four selected random instances.

Table 1. RPP instances from [7]

Inst.	$|V|$	$|E|$	Algorithm	\multicolumn{9}{c|}{$w_{\max}$ for $k =$}								
				2	3	4	5	6	7	8	9	10
A1	116	174	FHK	8681	7003	5825	5265	4929	4529	4278	4085	4218
			AM+	9379	7609	6119	5384	5199	4528	4180	4094	3936
			C+	8736	**6754**	**5540**	**4780**	**4396**	**4084**	**4058**	**3648**	**3620**
			SPT-LB	3476	3476	3476	3476	3476	3476	3476	3476	3476
			CPP/k-LB	7703	5136	3852	3082	2568	2201	1926	1712	1541
			LP-LB	6738	4451	3370	2722	2290	1981	1749	1569	1425
A2	102	160	FHK	8968	6858	5680	5252	4900	4596	4428	4236	4148
			AM+	9238	7160	5952	5427	4963	4458	4257	3989	3798
			C+	**8639**	**6597**	**5412**	**4850**	**4290**	**3812**	**3764**	**3764**	**3532**
			SPT-LB	3436	3436	3436	3436	3436	3436	3436	3436	3436
			CPP/k-LB	7706	5137	3853	3083	2569	2202	1927	1713	1542
			LP-LB	6816	4591	3477	2832	2401	2094	1863	1684	1540
A3	90	144	FHK	8673	6307	5306	4655	4246	3917	4050	3625	3556
			AM+	8073	6411	5395	**4457**	4047	3866	3452	**3422**	3248
			C+	**7892**	**5920**	**5000**	4470	**3748**	**3650**	**3316**	4196	**3124**
			SPT-LB	3124	3124	3124	3124	3124	3124	3124	3124	3124
			CPP/k-LB	7199	4800	3600	2880	2400	2057	1800	1600	1440
			LP-LB	6298	4163	3126	2519	2114	1824	1607	1439	1304

Table 2. CARP instances from [21]

Inst.	$|V|$	$|E|$	Algorithm	\multicolumn{9}{c}{$w_{\max}$ for $k =$}								
				2	3	4	5	6	7	8	9	10
e1	77	98	FHK	2093	1559	1270	1180	1180	1126	1126	1014	1014
			AM+	**1915**	**1551**	**1265**	**1133**	1070	1038	1023	983	924
			C+	1930	1593	1291	1156	**1020**	890	874	874	872
			SPT-LB	820	820	820	820	820	820	820	820	820
			CPP/k-LB	1685	1124	843	674	562	482	422	375	337
			LP-LB	1471	991	760	621	528	462	412	373	343
e2	140	190	FHK	3010	**2171**	**1813**	**1550**	**1431**	1421	1387	1341	**1209**
			AM+	3040	2353	2111	1920	1780	1667	1623	1545	1550
			C+	**2824**	2240	1817	1620	1477	**1374**	**1338**	**1251**	**1209**
			SPT-LB	1027	1027	1027	1027	1027	1027	1027	1027	1027
			CPP/k-LB	2607	1738	1304	1043	869	745	652	580	522
			LP-LB	2324	1560	1179	951	798	690	608	544	494

The first four columns of the tables contain the instance name, the number of nodes $|V|$, the number of edges $|E|$ and the name of the algorithm or lower bound. The remaining columns contain the value $w_{\max}$ computed for $k = 2, \ldots, 10$. The best result is printed in boldface. We use the following abbreviations: FHK for the FHK-heuristic, AM+ and C+ for the augment-merge heuristic and cluster heuristic followed by both improvement procedures, SPT-LB for the shortest path cycle lower bound, CPP/k-LB for the CPP tour lower bound and LP-LB for the LP relaxation lower bound.

For instances from [3] we were able to improve the FHK-heuristic in 91% of all cases with an average improvement of 10%. For instances from [13] we improved the results in 50% of all cases with an average improvement of 7%. For instances from [5] we improved the results in 66% of all cases with an average improvement of 12%.

5 Conclusions

We have presented new heuristics, improvement procedures and lower bounds for the MM k-CPP. In computational experiments we compared our algorithms for a variety of test instances with the only heuristic existing for the MM k-CPP to date, the FHK-heuristic. The results showed that we were able to improve the results obtained by the FHK-heuristic considerably in nearly all cases.

Our future work will comprise further enhancements of the heuristics and improvement procedures as well as the development of stronger lower bounds. The goal is to embed these results in a branch-and-cut framework in order to achieve optimal solutions for the MM k-CPP.

Table 3. Random instances

Inst.	$	V	$	$	E	$	Algorithm	\multicolumn{9}{c}{$w_{\max}$ for $k =$}
				2	3	4	5	6	7	8	9	10
r1	20	33	FHK	5394	4381	4057	3690	3191	2928	2888	2869	2637
			AM+	5809	5012	4721	**3173**	**2991**	**2687**	**2687**	2687	**2545**
			C+	**4520**	**4100**	**3655**	3306	3061	2802	2802	**2676**	2615
			SPT-LB	2545	2545	2545	2545	2545	2545	2545	2545	2545
			CPP/k-LB	4375	2917	2188	1750	1459	1250	1094	973	875
			LP-LB	4086	2704	2116	1732	1510	1352	1234	1141	1068
r2	40	70	FHK	**7758**	6100	5183	4284	4019	3702	3702	3536	3184
			AM+	8624	6017	5104	4324	4032	3971	3394	3525	3026
			C+	8112	**5676**	**4697**	**3910**	**3813**	**3484**	**3295**	**2939**	**2973**
			SPT-LB	2211	2211	2211	2211	2211	2211	2211	2211	2211
			CPP/k-LB	7311	4874	3656	2925	2437	2089	1828	1625	1463
			LP-LB	6542	4390	3368	2755	2346	2054	1835	1664	1528
r3	100	199	FHK	12398	**8826**	7588	6628	5726	5267	4803	4475	4437
			AM+	12266	9500	7868	7038	6350	6420	6106	5082	4788
			C+	**11983**	9207	**7152**	**6001**	**5310**	**4732**	**4490**	**4256**	**4102**
			SPT-LB	2900	2900	2900	2900	2900	2900	2900	2900	2900
			CPP/k-LB	11155	7437	5578	4462	3719	3188	2789	2479	2231
			LP-LB	9826	6464	4859	3897	3255	2797	2453	2186	1972
r4	100	200	FHK	**11804**	8426	7134	5990	5681	4887	4505	4059	4034
			AM+	11979	9296	7572	6806	5924	5941	5116	4873	4722
			C+	11834	**8252**	**6755**	**5805**	**5074**	**4526**	**4074**	**3833**	**3647**
			SPT-LB	2637	2637	2637	2637	2637	2637	2637	2637	2637
			CPP/k-LB	10877	7251	5439	4351	3626	3108	2720	2417	2176
			LP-LB	9906	6460	4871	3918	3282	2828	2488	2223	2011

References

[1] D. Applegate, W. Cook, S. Dash, and A. Rohe. Solution of a Min-Max Vehicle Routing Problem. *INFORMS Journal on Computing*, 14(2):132–143, 2002.
[2] A. A. Assad and B. L. Golden. Arc Routing Methods and Applications. In M. G. Ball, T. L. Magnanti, C. L. Monma, and G. L. Nemhauser, editors, *Network Routing*, volume 8 of *Handbooks in Operations Research and Management Science*, chapter 5, pages 375–483. Elsevier, 1995.
[3] E. Benavent, V. Campos, A. Corberán, and E. Mota. The Capacitated Arc Routing Problem: Lower Bounds. *Networks*, 22:669–690, 1992.
[4] N. Christofides. The Optimum Traversal of a Graph. *Omega*, 1:719–732, 1973.
[5] N. Christofides, V. Campos, A. Corberán, and E. Mota. An Algorithm for the Rural Postman Problem. Technical Report I. C. O. R. 81.5, Imperial College, 1981.
[6] A. Corberán, A. Letchford, and J. M. Sanchis. A Cutting Plane Algorithm for the General Routing Problem. *Mathematical Programming Series A*, 90(2):291–316, 2001.
[7] A. Corberán and J. M. Sanchis. A polyhedral approach to the rural postman problem. *European Journal of Operational Research*, 79:95–114, 1994.

[8] M. Dror. *Arc Routing: Theory, Solutions and Applications*. Kluwer Academic Publishers, 2000.

[9] J. Edmonds and E. L. Johnson. Matching, Euler Tours and the Chinese Postman. *Mathematical Programming*, 5:88–124, 1973.

[10] H. A. Eiselt, M. Gendreau, and G. Laporte. Arc routing problems I: The Chinese Postman Problem. *Operations Research*, 43(2):231–242, 1995.

[11] H. A. Eiselt, M. Gendreau, and G. Laporte. Arc routing problems II: The Rural Postman Problem. *Operations Research*, 43(3):399–414, 1995.

[12] G. N. Frederickson, M. S. Hecht, and C. E. Kim. Approximation Algorithms for some routing problems. *SIAM Journal on Computing*, 7(2):178–193, May 1978.

[13] B. L. Golden, J. S. DeArmon, and E. K. Baker. Computational experiments with algorithms for a class of routing problems. *Computers & Operations Research*, 10(1):47–59, 1983.

[14] B. L. Golden and R. T. Wong. Capacitated arc routing problems. *Networks*, 11:305–315, 1981.

[15] T. Gonzalez. Clustering to minimize the maximum intercluster distance. *Theoretical Computer Science*, 38:293–306, 1985.

[16] M. Guan. Graphic Programming using odd and even points. *Chinese Mathematics*, 1:273–277, 1962.

[17] A. Hertz, G. Laporte, and M. Mittaz. A Tabu Search Heuristic for the Capacitated Arc Routing Problem. *Operations Research*, 48(1):129–135, 2000.

[18] A. Hertz, G. Laporte, and P. Nanchen Hugo. Improvement Procedures for the Undirected Rural Postman Problem. *INFORMS Journal on Computing*, 11(1):53–62, 1999.

[19] A. Hertz and M. Mittaz. Heuristic Algorithms. In M. Dror, editor, *Arc Routing: Theory, Solutions and Applications*, chapter 9, pages 327–386. Kluwer Academic Publishers, 2000.

[20] L. Y. O. Li. *Vehicle Routeing for Winter Gritting*. PhD thesis, Department of Management Science, Lancaster University, 1992.

[21] L. Y. O. Li and R. W. Eglese. An Interactive Algorithm for Vehicle Routeing for Winter-Gritting. *Journal of the Operational Research Society*, 47:217–228, 1996.

[22] W. L. Pearn. Solvable cases of the k-person Chinese postman problem. *Operations Research Letters*, 16(4):241–244, 1994.

SCIL – Symbolic Constraints in Integer Linear Programming*

Ernst Althaus[1], Alexander Bockmayr[2], Matthias Elf[3], Michael Jünger[3], Thomas Kasper[4], and Kurt Mehlhorn[5]

[1] International Computer Science Institute
1947 Center St., Suite 600, Berkeley, CA 94704-1198, USA
althaus@icsi.berkeley.edu
[2] Université Henri Poincaré, LORIA
B.P. 239, F-54506 Vandœuvre-lès-Nancy, France
bockmayr@loria.fr
[3] Universität zu Köln, Institut für Informatik
Pohligstraße 1, 50969 Köln, Germany
{elf,mjuenger}@informatik.uni-koeln.de
[4] SAP AG, GBU Supply Chain Management
Neurottstraße 16, 69189 Walldorf, Germany
thomas.kasper@sap.com
[5] Max-Planck-Institute für Informatik
Stuhlsatzenhausweg 85, 66123 Saarbrücken, Germany
mehlhorn@mpi-sb.mpg.de

Abstract. We describe a new software system SCIL that introduces symbolic constraints into branch-and-cut-and-price algorithms for integer linear programs. Symbolic constraints are known from constraint programming and contribute significantly to the expressive power, ease of use, and efficiency of constraint programming systems.

1 Introduction

Many combinatorial optimization problems are naturally formulated through constraints. Consider the *traveling salesman problem* (TSP). It asks for the minimum cost Hamiltonian cycle[1] in an undirected graph $G = (V, E)$ with edge weights $(w_e)_{e \in E}$. Formulated as an optimization task:

Find a subset T of the edges of G such that "T is a Hamiltonian cycle" and $\sum_{e \in T} w_e$ is minimum.

---

* Partially supported by the Future and Emerging Technologies programme of the EU under contract number IST-1999-14186 (ALCOM-FT). Die Arbeit wurde mit der Unterstützung eines Stipendiums im Rahmen des Hochschulsonderprogramms III von Bund und Ländern über den DAAD ermöglicht.
[1] A Hamiltonian cycle ("tour") in a graph is a cycle passing exactly once through every node of the graph.

Our vision is that the sentence above (written in some suitable language) suffices to obtain an efficient algorithm for the TSP. Efficiency is meant in a double sense. We want short development time (= efficient use of human resources) and runtime efficiency (= efficient use of computer resources). The software system SCIL is our first step towards realizing this vision. Section 3 shows a SCIL program for the traveling salesman problem.

We propose a programming system for *(mixed) integer linear programming* (ILP) based on a *branch-and-cut-and-price* framework (BCP) that features *symbolic constraints*.

Integer linear programming and more specifically the branch-and-cut-and-price paradigm for solving ILPs is one of the most effective approaches to find exact or provably good solutions of hard combinatorial optimization problems [NW88, EGJR01]. It has been used for a wide range of problems including the TSP [ABCC99, Nad02], maximum-cut-problems [SDJ+96], cutting-stock-problems [VBJN94], or crew-scheduling [BJN+98]. For more applications we refer to the overview [CF97]. The implementation of a BCP algorithm requires significant expert knowledge, a fact that has hindered the spread of the method outside the scientific world. It consists of various involved components, each with wide influence on the algorithm's overall performance. Almost all parts of a BCP algorithm considerably rely on *linear programming* (LP) methods or properties of the linear programming relaxation of the underlying ILP. Many components are problem independent and can be provided by existing software packages (see [JT00]). But there is still a major problem dependent part: an ILP formulation has to be found, appropriate linear programs have to be derived, the various methods for exploiting LP solutions to find feasible solutions or speedup the computation have to be designed and implemented. To our knowledge there is no BCP system for combinatorial optimization that covers the problem dependent part in an appropriate way.

SCIL closes this gap by introducing *symbolic constraints*, one of the key achievements constraint programming [vHS96] into integer linear programming. It simplifies the implementation of BCP-algorithms by supporting high-level specifications of combinatorial optimization problems with linear objective functions. It provides a library of symbolic constraints, which allow one to convert high-level specifications into efficient BCP-algorithms. A user may extend SCIL by defining new symbolic constraints or by changing the standard behavior of the algorithms. We have used SCIL already in several projects like curve reconstruction [AM01] and computing protein dockings [AKLM]. A documentation of SCIL, a list of available symbolic constraints and more examples can be found on the SCIL home page (http://www.mpi-sb.mpg.de/SCIL).

The rest of the paper is organized as follows. We give a short introduction on BCP algorithms for combinatorial optimization in Section 1.1, we relate SCIL to extant work in Section 2, give a first SCIL-program in Section 3, formalize the BCP-paradigm and define the role of symbolic constraints in Section 4, and discuss the implementation of symbolic constraints in Section 5.

SCIL also allows one to solve problems using the Lagrangean relaxation technique. Furthermore we developed a numerical exact BCP algorithm for SCIL. For details of these issues we refer to a forthcoming paper.

1.1 ILP Formulations and BCP Algorithms

Again we use the TSP to explain how combinatorial optimization problems are modeled as ILPs and solved by BCP. First, one has to find an ILP formulation. For the TSP (as for many other problems) one starts from a first high-level description, e.g., the description we gave in the introduction. Then one tries to characterize the problem in more detail by identifying problem defining structures. This may yield another high-level description like (1b) which is suitable to be formulated as an ILP. The resulting well-known ILP formulation (*subtour elimination formulation*) is given in (1a).

$$
(P) \begin{vmatrix} \min \sum_{e \in E} w_e x_e & & & \min_{T \subseteq E} \sum_{e \in T} w_e \\ \text{s.t.} \quad \sum_{j \in V} x_{ij} = 2 & \text{for all } i \in V & \text{s.t. each node in T has degree 2} \\ \sum_{i,j \in S} x_{ij} \leq |S| - 1 & \begin{array}{l} \text{for all } S \subsetneq V, \\ 3 \leq |S| \leq \lfloor \frac{|V|}{2} \rfloor \end{array} & T \text{ contains no subtour} \\ 0 \leq x_e \leq 1 & \text{for all } e \in E \\ x_e \text{ integral} & \text{for all } e \in E \end{vmatrix} \quad (1)
$$

a) ILP formulation   b) high-level description

In this formulation there is a connection between edges e and binary variables x_e. The value of x_e is 1 if the edge belongs to the tour T and 0 otherwise. Furthermore, nodes $i \in V$ correspond to *degree equations* and subsets S correspond to *subtour elimination constraints*. Such correspondences between combinatorial objects and components of the ILP formulation are inevitable and turn out to be very useful in the modeling environment SCIL. Beside the two classes of constraints used in (1a) there are many other classes of inequalities used in state of the art TSP solvers [ABCC99, Nad02].

In the following we sketch a BCP algorithm for the formulation (1a) which also applies to every other ILP problem. A BCP-algorithm for the minimization problem P is basically a branch-and-bound algorithm in which the bounds are solutions of LP-relaxations. These relaxations are naturally derived from the linear description of our problem by dropping the integrality constraints. However, our formulation contains an exponential number of constraints for an instance with n nodes (there are $\Theta(2^n)$ subsets). This leads to the idea to solve the relaxation with a subset of all constraints, check whether the optimum solution violates any other of the constraints, and if so, append them to the relaxation. Identifying violated constraints is called *separation* and added constraints are called *cutting planes*. The procedure is a *cutting plane algorithm*.

Every subproblem in the enumeration of the branch-and-bound algorithm has its own LP relaxation which is iteratively strengthened by the cutting plane procedure. This results in increasing *local lower bounds* on the optimum solution for a subproblem. Subproblems can be discarded if the local lower bound is not smaller than the objective function value g of the best feasible solution found so far. The value g is a *global upper bound* on the optimum solution. Branching, i.e., splitting a subproblem into two or more, is performed by introducing appropriate linear constraints to the resulting subproblems. The LP relaxation of a successor node in the enumeration-tree is created by adding appropriate "branching constraints" to the relaxation of the predecessor.

In some applications where the LP relaxations of the subproblems contain too many variables in order to solve the LPs efficiently the LP relaxations are solved by *column generation*. This method solves the LP on a small subset of the variables. All missing variables are assumed to have value 0. In order to prove that the solution of the restricted LP is optimal for the entire set of variables one has to check the reduced costs of the ignored variables. This step is called *pricing*.

2 Comparison to Extant Work

The cutting plane procedure was invented in 1954 by Dantzig et. al. [DFJ54] for the traveling salesman problem. The branch-and-cut-paradigm was first formulated for and successfully applied to the linear ordering problem by Grötschel et. al. in 1984 [GJR84]. The first state-of-the-art BCP algorithm was developed in 1991 by Padberg and Rinaldi [PR91]. Since then, the method was applied to a huge number of problems. The LP-solver was treated as an independent black-box almost from the beginning. The software system ABACUS [JT00] is an object oriented software framework for BCP-algorithms. It provides the problem independent part, e.g., the enumeration tree, the interface with the LP-solver and various standard strategies, e.g. for branching or exploiting LP properties. It has concepts like subproblem, cut-generation, and pricing. It has no modeling language and does not provide a library of separation and pricing routines.

Another direction are general ILP-solvers using the BC-paradigm. These solvers are mostly extensions of existing linear programming solvers, e.g. CPLEX [CPL] or XPRESS [Das00]. They solve arbitrary ILPs using general cuts, e.g., Gomory cuts. Modeling languages, such as AMPL [FGK92] or GAMS [Cor02], can be used to specify the ILP. However, modeling is on the level of ILPs, all constraints must be given explicitly, and the support of problem specific cuts and variable generation is rudimentary. Modeling systems for general ILP solvers translate a high-level problem description into a matrix. During this process most of the structural knowledge contained in the model gets lost. The solver only sees a matrix.

Symbolic constraints are one of the main achievements of *constraint programming* (CP). They were first introduced under the name *global constraints* in the CP-system CHIP [BC94, DvHS+88] and later used in further CP-systems like ILOG OPL Studio [ILO] and OZ [Smo96]. On the one hand, symbolic constraints allow for high-level and intuitive problem formulations. On the other hand, they improve efficiency by integrating problem-specific algorithms into a general solver.

Bockmayr and Kasper [BK98, Kas98] suggested to extend ILP by symbolic constraints. SCIL realizes this suggestion. An optimization problem is specified by a set of variables (ranging over the rational numbers) a linear objective function and a set of constraints. A constraint may be a *linear constraint* (a linear inequality or equation) or a *symbolic constraint*. In principle, a symbolic constraint can be every subset of the set of all assignments of values to variables. Note, however, that not all subsets of assignments are realizable in our framework. The system of linear constraints can be solved efficiently by an LP-solver whereas the symbolic constraints typically make the problem hard.

Symbolic constraints in integer programming allow the modeler to include large families of linear constraints, variables, and branching rules into the model, without writing them down explicitly. An important symbolic constraint is the integrality constraint. It is the only symbolic constraint which is available in all BCP-systems. In its simplest form the integrality constraint for a variable x includes only the branching rules $x \leq c \vee x \geq c+1$ for all integers c into the system. Another example, is the `tour` constraint mentioned earlier in the paper. It includes the degree and the subtour elimination constraints into the model. Declaratively, this constraint is equivalent to exponentially many linear inequalities. Operationally, however, only some of these inequalities will be added to the model at runtime (as cutting planes). Concerning efficiency, symbolic constraints allow one to integrate specialized cutting plane algorithms based on polyhedral combinatorics into a general branch-and-cut solver. Symbolic constraints give the modeler the possibility to identify some specific structure in the problem, which later can be exploited when the model is solved. For example, when we solve a model containing the symbolic constraint `tour`, we can enhance our general branch-and-cut solver by computing specialized cutting planes for the traveling salesman problem instead of using general cutting planes for arbitrary linear 0-1 programs.

All available systems for ILP have in common that the variables are only an indexed set. But as mentioned above we often have a correspondence between combinatorial objects and variables and constraints. This turns out to be a key to elegant modeling. SCIL takes this into account by introducing variable maps and constraint maps. See section 3 for details.

3 Specification of Optimization Problems

As our implementation is based on C++, it is natural that problems, subproblems, symbolic constraints, linear constraint and variables are C++ objects. Since variables and constraints are C++ objects, they can be easily created or extended by inheritance. They can be added to a specific problem by appropriate member functions.

As mentioned in section 1.1 many ILP formulations for combinatorial problems use structured collections of variables and linear constraints, e.g., we may have a variable for each edge of a graph and a linear constraint for each vertex. In SCIL we provide variable maps (similarly constraint maps) that allow one to establish a correspondence between variables and objects of an arbitrary type. A `var_map<T>` where T is an arbitrary C++ class stores a mapping between objects of type T and ILP variables. Read- and write-access to variable maps is via the subscript-operator.

The following SCIL-program solves the TSP-problem on a graph G with edge weights w. It returns the tour as a list of edges. SCIL imports combinatorial objects from LEDA [MN99, LED] and hence its syntax is remeniscient of LEDA, e.g., `edge_array<double>` is an array of double values indexed by the edges of a graph. Even for a non-experienced C++ programmer the following program should be clear.

```
void compute_tsp_tour(graph& G, edge_array<double>& w, list<edge>& tour)
{
    // define a minimization problem P
    ILP_problem P(optsense_min);

    // add a 0-1 variable for each edge e of G and make w[e] it coefficient
    // in the objective function. The var_map EV establishes the connection
    // between edges and variables
    var_map<edge> EV(nil);
    forall_edges(e,G) EV[e] = P.add_binary_variable(w[e]);

    // add the tour constraint for the edges of G to P
    P.add_constraint(new TOUR(G, EV));

    // solve P
    P.solve();

    // extract the solution
    forall_edges(e,G) if ( P.get_value(EV[e]) == 1 ) tour.add(e);
}
```

The tour constraint is a built-in constraint of SCIL and makes the program above particularly short. The concept of variable maps makes the program very readable.

In many applications there are side constraints to a central symbolic constraint. Side constraints for the traveling salesman tour arise for example, if the salesman must respect a partial ordering of the cities, e.g., if he picks additional items from some cities for other cities.

In contrast to the previous problem, we use the directed version of the tour constraint to model the Hamiltonian cycle. The partial ordering can be modeled as follows. Let s be the start node of the tour and u and v be two nodes so that u must be visited before v. Even if one ignores all edges adjacent to s there must be a path between u and v. To model this we can use path constraints. The *path constraint* `path(G,u,v,EV)` enforces that the set of edges, whose corresponding variables have value one, describe a superset of a path between u and v.

The polyhedral structure of the traveling salesman problem with precedence constraints was investigated in [BFP95] and a specialized branch-and-cut algorithm was presented in [AJR00].

4 The BCP-Framework with Symbolic Constraints

We define the BCP-paradigm by a set of transition rules and then explain the role played by symbolic constraints. The description of an algorithm by transition rules is typical for CP-systems, but uncommon for BCP-systems. Nevertheless, we found this description useful to explore the exact requirements for the symbolic constraints to guarantee the correctness and termination of the algorithm. We define those requirements for symbolic constraints and prove that the BCP-method terminates when the requirements are met.

4.1 Transition Rules for BCP

We consider a minimization problem M formulated as an ILP over a set V of variables and use $\mathcal{F}$ to denote its set of feasible solutions. For a set $\mathcal{F}$ of solutions we use $C(\mathcal{F})$ to denote the set of all linear constraints that are valid for all assignments in $\mathcal{F}$. For a finite set P of linear constraints on the variables in V and a subset $A \subseteq V$ of variables, $LP(P, A)$ denotes the optimal objective function value of the linear program (P, A) with M's objective function where all variables in $V \setminus A$ are set to 0. If (P, A) is infeasible, $LP(P, A) = \infty$. For a set P of linear constrains, $\mathcal{F}_P = \{f \in \mathcal{F}; f$ satisfies all $c \in P\}$ denotes the feasible solutions of M selected by the constraints in P. For any set P of linear constraints feasible for M we have $LP(P, V) \leq LP(C(\mathcal{F}_P), V)$ and the value on the right side is the optimal objective value of any feasible solution in $\mathcal{F}_P$ since our objective function is linear. We also maintain a global upper bound g, the minimal objective value of any feasible solution encountered so far. Initially, $g = \infty$ and P and A are arbitrary subsets of $C(\mathcal{F})$ and V respectively.

The BCP-paradigm uses the following transition rules to transform problems (P, A, g). In each case, the old problem is given above the horizontal bar, the new problem is given below, and the condition for applicability is given to the right of the bar. The name of the rule is indicated on the left.

(Cut) $\dfrac{P \quad A \quad g}{P \cup \{c\} \quad A \quad g}$ c is linear and valid for $\mathcal{F}$

(Price) $\dfrac{P \quad A \quad g}{P \quad A \cup \{v\} \quad g}$ $v \in V \setminus A$

(Branch) $\dfrac{P \quad A \quad g}{P \cup \{c_1\} \; A \; g \mid P \cup \{c_2\} \; A \; g}$ c_1 and c_2 are linear and $c_1 \vee c_2$ is valid for $\mathcal{F}$.

(Improved Bound) $\dfrac{P\ A\ g}{P\ A\ g'}$ g' is the objective value of a feasible solution and $g' < g$

(Fathom) $\dfrac{P\ A\ g}{\_}$ $LP(P,V) \geq g$

If the objective function is known to be integral, the condition of the fathom rule can be relaxed.

(Integral Fathom) $\dfrac{P\ A\ g}{\_}$ $LP(P,V) > g - 1$

Notice that the fathom rule requires that the LP value for the whole set of variables has to be greater or equal to g. Otherwise the LP value might not be a lower bound on the objective value of the best feasible solution.

4.2 The Role of Symbolic Constraints

We explain the role of symbolic constraints in rule application[2]. We are working on a subproblem (P, A, g) with LP-objective value $g' = LP(P, A)$. If $g' < \infty$, we also have an LP-solution s; s is an assignment to the variables in A. We can view it as an assignment to all variables by setting $s(v) = 0$ for $v \in V \setminus A$. We ask each symbolic constraint SC for its opinion about s; whether s is a feasible solution for SC or not and in the latter case, whether the constraint suggests a variable in $V \setminus A$ for the price rule, a cut c for the cut rule, or a disjunction $c_1 \vee c_2$ for the branching rule.

If s is feasible for all symbolic constraints we have found a new feasible solution to our minimization problem. If $g' < g$, we can improve the upper bound. The user may supply a heuristic that attempts to construct a feasible solution from an infeasible s. If s is feasible, we also try to apply the fathom rule or the price rule.

A variable v is *local* to a symbolic constraint SC if the variable is used in no constraint generated outside SC. This includes especially branching constraints. We also say that SC owns v. The nice feature of local variables is that the symbolic constraint owning a variable can compute its reduced cost. Non-local variables are called *shared*. We may apply the fathom rule if A contains all shared variables and the local variables of all symbolic constraints have non-negative reduced cost. We initialize A with the set of shared variables and hence it suffices to ensure that the local variables have non-negative reduced cost. Therefore, the symbolic constraints are asked to suggest variables for the price rule. In symbolic constraints that contain too many variables to compute the reduced costs directly the pricing problem is solved in an algorithmic way.

We require:

Requirement 1: If a symbolic constraint has local variables of negative reduced cost, it suggests one; if $g' = \infty$, all local variables are considered to have negative reduced cost.

---

[2] We will later discuss requirements on symbolic constraints and rule application which guarantee termination.

So assume that $g' < g$ and s is infeasible for M. We ask the symbolic constraints to supply cuts and/or disjunctions for the branching rule. A symbolic constraint may supply any cut or disjunction that is feasible for its feasible solutions (or for its feasible solutions selected by P). The built-in integrality constraints always suggest the obvious branching rules for fractional variables. We require:

Requirement 2: If s satisfies all integrality constraints, but does not satisfy a particular symbolic constraint, the constraint must suggest a cut c or a disjunction $c_1 \vee c_2$ which cuts off s, i.e., in the case of a cut, s does not satisfy $P \cup c$, and in the case of a disjunction, s satisfies neither $P \cup c_1$ nor $P \cup c_2$.

4.3 Termination

Having defined the BCP-framework with symbolic constraints, we now argue termination under the following assumptions:

1. The minimization problem is either infeasible or has a finite optimum and we start with a set of constraints that guarantees that none of the LPs generated is unbounded[3].
2. The integral variables are bounded, i.e., for each integral variable v there are constants l_v and u_v and constraints $l_v \leq v \leq u_v$.
3. Every symbolic constraint satisfies requirements 1 and 2.
4. **Requirement 3:** Every symbolic constraint SC has the strong termination property, i.e., the method terminates if it is run just with constraint SC and if an adversary is allowed to modify subproblems (P, A, g) by adding an arbitrarily chosen set P' of additional constraints; the LP-solver is then required to return an optimal solution for the problem $(P \cup P', A, g)$.

The proofs of the following results will be given in a long version of this paper.

Lemma 1. *In every step of the algorithm, at least one rule is applicable.*

Theorem 1. *If assumptions (1) to (4) are satisfied, the method terminates.*

Lemma 2. *The integrality constraint satisfies Requirements 1, 2 and 3.*

4.4 Order of Rule Application

The efficiency of the BCP-method depends on the order of rule application. In most applications preferring cutting to branching turns out to be advantageous. Pricing should be performed before cutting.

For each of the rules there are many strategies to deal with the situation that we can choose from several possibilities. Take for example the situation that more

---

[3] We make this assumption because an optimal solution to an unbounded LP contains a ray and, as stated, symbolic constraints only know how to exploit finite solutions.

than one violated constraint is detected. Which of them should be added to the system? A popular strategy is to add constraints according to their violation and disregard those with small violation. Another strategy is to rank cutting planes with respect to the Euclidean distance of the previously found LP solution to the hyperplanes defined by the violated constraints. ABACUS implements both strategies and makes them available to SCIL. The user of SCIL may implement his or her own strategy.

5 Implementation of Symbolic Constraints

We first discuss how LP matrices are specified, then we discuss the implementation of the symbolic constraint for Hamiltonian cycles.

5.1 The LP-Matrix

Every pair of variable and linear constraint has an entry in the LP-matrix. The default value is 0. There are several ways to create coefficients different from 0. These methods are very similar to the methods provided by typical LP solvers.

The basic method defines a single entry of the matrix by the call of an appropriate member function of either the main problem or a subproblem.

Frequently, we want to generate rows or columns in the LP-matrix which follow a common pattern. Consider, for example, the degree equations for the TSP. There are as many degree equations as nodes in the graph and all of them generate a 1 in the LP-matrix for all edges which are incident to the node. SCIL provides *constraint-schemas* and *variable-schemas* to implement patterns. In order to implement a constraint-schema, one derives a class from the system class `cons_obj`. The non-zero entries for the constraint are created in a specific member function.

5.2 The Implementation of the Tour Constraint

We sketch the implementation of the tour constraint. The tour constraint has no local variables and hence it must provide only three functions beside the constructor, a function `init` that adds an initial set of constraints, a function `is_feasible` that determines whether the current LP-solution satisfies the constraint, and a function `separate` that generates cuts. We give more details. In the constructor the graph and the variables that correspond to the edges are given. The precondition is that all given variables are binary. We check the precondition and save references to the graph and to the variable-map.

The init function simply adds degree equations for all nodes of g. We add the degree equation permanently to the LP, i.e., we do not allow the system to remove them.

An LP-solution is feasible if it is integral and if the subgraph induced by the variables with value 1 forms a tour. Since we know that the degree equations are added permanently to the LPs, they are always satisfied. Hence, an integer solution is a collection of cycles and a tour if and only if it contains no subtour, i.e., none of the subtour constraints may be violated. We check this using the separation routine for subtour constraints described next.

The member function that needs most work is the **separate** function. In a first implementation we might look for violated subtour elimination constraints. In the presence of degree constraints, the subtour elimination constraints are equivalent to the fact that for each subset $S \subset V$ of the nodes with $\emptyset \neq S \neq V$ the induced cut has value at least two. The value of a cut is the sum of the values of the LP-variables associated with edges cross the cut. A cut of value less than two corresponts to a violated subtour elimination constraint. It can be found efficiently by a minimum capacity cut computation.

We need to argue that our implementation of the tour constraint satisfies requirements 1 and 2 and has the strong termination property. It has property 1 since it has no local variables, it has property 2 since any integral solution of the degree constraints consists of a set of cycles and hence is either a tour or violates a subtour elimination constraint. The strong termination property holds since there are only finitely many subtour elimination constraints and since all variables are binary.

There are many methods known to improve the performance of the TSP solvers implemented above. All of them are easily added to the implementation of our symbolic constraint. A possible modification may be a branching rule different from the standard rule of branching on fractional integer variables. Some TSP codes reduce the size of the enumeration tree by branching on a hyperplane.

6 Conclusion

We gave the linear framework how to enrich integer programming solvers by symbolic constraints. The success of the system strongly depends on the quality of the symbolic constraints in the library. On the one hand they must be general enough to have a broad area of applications, on the other hand, they must be specific enough to have an efficient implementation.

References

[ABCC99] D. Applegate, R. Bixby, V. Chvátal, and W. Cook. Tsp-solver "concorde". http://www.keck.caam.rice.edu/concorde.html, 1999.

[AJR00] N. Ascheuer, M. Jünger, and G. Reinelt. A branch & cut algorithm for the asymmetric traveling salesman problem with precedence constraints. *Computational Optimization and Applications*, 17(1):61–84, 2000.

[AKLM] E. Althaus, O. Kohlbacher, H.-P. Lenhof, and P. Müller. A combinatorial approach to protein docking with flexible side-chains. In *Proceedings of the 4th Annual International Conference on Computational Molecular Biology (RECOMB-00)*.

[AM01] E. Althaus and K. Mehlhorn. Traveling salesman-based curve reconstruction in polynomial time. *SIAM Journal on Computing*, 31(1):27–66, 2001.

[BC94] N. Beldiceanu and E. Contejean. Introducing global constraints in CHIP. *Mathl. Comput. Modelling*, 20(12):97 – 123, 1994.
[BFP95] E. Balas, M. Fischetti, and W. R. Pulleyblank. The precedence-constrained asymmetric traveling salesman polytope. *Mathematical Programming*, 68:241–265, 1995.
[BJN+98] C. Barnhart, E. Johnson, G. Nemhauser, M. Savelsbergh, and P. Vance. Branch-and-price: column generation for solving huge integer programs. *Operations Research*, 46:316–329, 1998.
[BK98] A. Bockmayr and T. Kasper. Branch and infer: a unifying framework for integer and finite domain constraint programming. *INFORMS Journal on Computing*, 10:287–300, 1998.
[CF97] A. Caprara and M. Fischetti. *Annotated bibliographies in combinatorial optimization*, chapter Branch-and-cut algorithms, pages 45–64. Wiley, 1997.
[Cor02] GAMS Development Corporation. Gams: General algebraic modeling system, 2002.
[CPL] CPLEX. www.cplex.com.
[Das00] Dash Associates. *XPRESS 12 Reference Manual: XPRESS-MP Optimizer Subroutine Library XOSL*, 2000.
[DFJ54] G. B. Dantzig, D. R. Fulkerson, and S. M. Johnson. Solution of a large scale traveling salesman problem. *Operations Research*, 2:393–410, 1954.
[DvHS+88] M. Dincbas, P. van Hentenryck, H. Simonis, A. Aggoun, and T. Graf. The constraint logic programming language CHIP. In *Fifth Generation Computer Systems, Tokyo, 1988*. Springer, 1988.
[EGJR01] M. Elf, C. Gutwenger, M. Jünger, and G. Rinaldi. *Computational Combinatorial Optimization*, volume 2241 of *Lecture Notes in Computer Science*, chapter Branch-and-cut algorithms for combinatorial optimization and their implementation in ABACUS, pages 157–222. Springer, 2001.
[FGK92] R. Fourer, D. M. Gay, and B. W. Kernighan. *AMPL: A modeling language for Mathematical Programming*. Duxbury Press/Wadsworth Publishing, 1992.
[GJR84] M. Grötschel, M. Jünger, and G. Reinelt. A cutting plane algorithm for the linear ordering problem. *Operations Research*, 32:1195–1220, 1984.
[ILO] Ilog. www.ilog.com.
[JT00] M. Jünger and S. Thienel. The abacus system for branch and cut and price algorithms in integer programming and combinatorial optimization. *Software Practice and Experience*, 30:1325–1352, 2000.
[Kas98] T. Kasper. *A Unifying Logical Framework for Integer Linear Programming and Finite Domain Constraint Programming*. PhD thesis, Fachbereich Informatik, Universität des Saarlandes, 1998.
[LED] LEDA (Library of Efficient Data Types and Algorithms). www.algorithmic-solutions.com.
[MN99] K. Mehlhorn and S. Näher. *The LEDA Platform for Combinatorial and Geometric Computing*. Cambridge University Press, 1999. 1018 pages.
[Nad02] D. Naddef. *The Traveling Salesman Problem and its Variations*, chapter Polyhedral theory, branch and cut algorithms for the symmetric traveling salesman problem. Kluwer Academic Publishing, 2002.
[NW88] G. L. Nemhauser and L. A. Wolsey. *Integer and Combinatorial Optimization*. John Wiley & Sons, 1988.

[PR91]   M. Padberg and G. Rinaldi. A branch and cut algorithm for the resolution of large scale symmetric traveling salesman problems. *SIAM Review*, 33(60–10), 1991.

[SDJ+96]   C. Simone, M. Diehl, M. Jünger, P. Mutzel, and G. Rinaldi. Exact ground states of two-dimensional $\pm J$ ising spin glasses. *Journal of Statistical Physics*, 84:1363–1371, 1996.

[Smo96]   G. Smolka. Constraints in OZ. *ACM Computing Surveys*, 28(4es):75, December 1996.

[VBJN94]   P. H. Vance, C. Barnhart, E. L. Johnson, and G. Nemhauser. Solving binary cutting stock problems by column generation and branch-and-bound. *Computational Optimization and Applications*, 3:111—130, 1994.

[vHS96]   P. van Hentenryck and V. Saraswat. Strategic directions in constraint programming. *ACM Computing Surveys*, 28(4):701 – 726, 1996.

Implementing I/O-efficient Data Structures Using TPIE

Lars Arge*, Octavian Procopiuc**, and Jeffrey Scott Vitter***

Center for Geometric and Biological Computing, Department of Computer Science,
Duke University, Durham, NC 27708, USA
{large,tavi,jsv}@cs.duke.edu

Abstract. In recent years, many theoretically I/O-efficient algorithms and data structures have been developed. The TPIE project at Duke University was started to investigate the practical importance of these theoretical results. The goal of this ongoing project is to provide a *portable*, *extensible*, *flexible*, and *easy to use* C++ programming environment for *efficiently* implementing I/O-algorithms and data structures. The TPIE library has been developed in two phases. The first phase focused on supporting algorithms with a *sequential* I/O pattern, while the recently developed second phase has focused on supporting on-line I/O-efficient data structures, which exhibit a more *random* I/O pattern. This paper describes the design and implementation of the second phase of TPIE.

1 Introduction

In many modern massive dataset applications I/O-communication between fast internal memory and slow disks, rather than actual internal computation time, is the bottleneck in the computation. Examples of such applications can be found in a wide range of domains such as scientific computing, geographic information systems, computer graphics, and database systems. As a result, much attention has been focused on the development of I/O-efficient algorithms and data structures (see e.g. [4, 20]). While a lot of practical and often heuristic I/O-efficient algorithms and data structures in ad-hoc models have been developed in the database community, most theoretical work on I/O-efficiency in the algorithms community has been done in the Parallel Disk Model of Vitter and Shriver [21]. To investigate the practical viability of the theoretical work, the TPIE[1] project was started at Duke University. The goal of this ongoing project is to provide a

---

* Supported in part by the National Science Foundation through ESS grant EIA–9870734, RI grant EIA–9972879, CAREER grant CCR–9984099, ITR grant EIA–0112849, and U.S.-Germany Cooperative Research Program grant INT–0129182.
** Supported in part by the National Science Foundation through ESS grant EIA–9870734 and RI grant EIA–9972879.
*** Supported in part by the National Science Foundation through research grants CCR–9877133 and EIA–9870734 and by the Army Research Office through MURI grant DAAH04–96–1–0013.
[1] TPIE: Transparent Parallel I/O Environment. Pronunciation: 'tE-'pI (like tea-pie)

portable, extensible, flexible, and *easy to use* programming environment for *efficiently* implementing algorithms and data structures developed for the Parallel Disk Model. A project with similar goals, called LEDA-SM [13] (an extension of the LEDA library [15]), has also been conducted at the Max-Planck Institut in Saarbrucken.

TPIE is a templated C++ library[2] consisting of a kernel and a set of I/O-efficient algorithms and data structures implemented on top of it. The kernel is responsible for abstracting away the details of the transfers between disk and memory, managing data on disk and in main memory, and providing a unified programming interface appearing as the Parallel Disk Model. Each of these tasks is performed by a separate module inside the kernel, resulting in a highly extensible and portable system. The TPIE library has been developed in two phases. The first phase focused on supporting algorithms with a *sequential* I/O pattern, that is, algorithms using primitives such as scanning, sorting, merging, permuting, and distributing [19]. However, in recent years, a growing number of on-line I/O-efficient data structures have been developed, and the sequential framework is ill-equipped to handle the more *random* I/O patterns exhibited in these structures. Therefore a second phase of the TPIE development has focused on providing support for random access I/O patterns. Furthermore, just like a large number of batched algorithms were implemented in the first phase of the project, a large number of data structures have been implemented in the second phase, including B-trees [12], persistent B-trees [9], R-trees [10], CRB-trees [1], K-D-B-trees [18] and Bkd-trees [17]. The two parts of TPIE are highly integrated, allowing seamless implementation of algorithms and data structures that make use of both random and sequential access patterns.[3]

While the first part of TPIE has been described in a previous paper [19], this paper describes the design and implementation details of the second phase of the system. In Section 2 the Parallel Disk Model, as well as disk and operation technology that influenced the TPIE design choices are first described. Section 3 then describes the architecture of the TPIE kernel and its design goals. Section 4 presents a brief description of the data structures implemented using the random access framework, and finally Section 5 contains a case study on the implementation of the K-D-B-tree [18] along with some experimental results.

2 The I/O Model of Computation

In this section we discuss the physical disk technology motivating the Parallel Disk Model, as well as some operating system issues that influenced important TPIE design choices.

---

[2] The latest TPIE release can be downloaded from http://www.cs.duke.edu/TPIE/
[3] The LEDA-SM library takes a slightly different approach, optimized for random access patterns. The algorithms and data structures implemented using LEDA-SM are somewhat complementary to the ones implemented in TPIE. The kernels of the two libraries are compatible, and as a result algorithms and structures implemented for one of the systems can easily be ported to the other.

Magnetic Disks. The most common external memory storage device is the magnetic disk. A magnetic disk drive consists of one or more rotating platters with one read/write head per platter. Data are stored on the platter surface in concentric circles called tracks. To read or write data at a certain position on the platter, the read/write head must *seek* to the correct track and then *wait* for the desired position on the track to pass by. Because mechanical movement is involved, the typical read or write time is on the order of milliseconds. By comparison, the typical transfer time of main memory is a few nanoseconds—a factor of 10^6 faster! Since the seek and wait time is much larger than the time needed to read a unit of data, magnetic disks transfer a large *block* of contiguous data items at a time. Accessing a block involves only one seek and wait, so the amortized cost per unit of data is much smaller than the cost of accessing a single unit.

Parallel Disk Model. The Parallel Disk Model (PDM) was introduced by Vitter and Shriver [21] (see also [3]) in order to more accurately model a two-level main memory-disk system with block transfers. PDM has become the standard theoretical model for designing and analyzing I/O-efficient algorithms. The model abstracts a computer as a three-component system: a processor, a fixed amount of main memory, and one or more independent disk drives. Data is transferred between disks and main memory in fixed-size *blocks* and one such transfer is called an *I/O operation* (or simply an *I/O*). The primary measures of algorithm performance in the model are the number of I/Os performed, the amount of disk space used, and the internal computation time. To be able to quantify these measures, N is normally used to denote the number of items in the problem instance, M the maximum number of items that fit in main memory, B the number of items per block, and D the number of independent disks. In this paper we only consider the one-disk model.

Random versus sequential I/O. Using the Parallel Disk Model, a plethora of theoretically I/O-efficient algorithms and data structures have been developed— refer to [4, 20] for surveys. Experimental studies (many using the TPIE system) have shown the model's accuracy at predicting the relative performance of algorithms—refer to the above mentioned surveys for references. However, they have also revealed the limits of the model, primarily its inability to distinguish between the complexities of sequential and random I/O patterns. Intuitively, accessing data sequentially is more efficient than accessing blocks in a random way on disk, since the first pattern leads to less seeks and waits than the latter. Furthermore, since studies have shown that the typical use of a disk file is to open it, read its entire contents sequentially, and close it, most operating systems are optimized for such sequential access.

UNIX I/O primitives. In the UNIX operating system, optimization for sequential access is implemented using a *buffer cache*. It consists of a portion of the main memory reserved for caching data blocks from disk. More specifically, when a user requests a data block from disk using the `read()` system call, the block is looked up in the buffer cache, and if not there, it is fetched from disk into the cache. From there, the block is copied into a user-space memory location. To optimize for sequential access a prefetching strategy is also implemented, such that blocks following a recently accessed block are loaded into the buffer cache while computation is performed. Similarly, when a block is written using a `write()` system call, the block is first copied to the buffer cache, and if necessary, a block from the buffer cache is written to disk.

When the I/O-pattern is random rather than sequential, the buffer cache is mostly useless and can actually have a detrimental effect. Not only are resources wasted on caching and prefetching, but the cache also incurs an extra copy of each data block. Therefore most UNIX-based operating systems offer alternative I/O routines, called `mmap()` and `munmap()`, which avoid using the buffer cache. When the user requests a disk block using `mmap()`, the block is mapped directly in user-space memory.[4] The mapping is released when `munmap()` is called, allowing the block to be written back to disk. If properly implemented, these routines achieve a zero-copy I/O transfer, resulting in more efficient I/O than the `read()`/`write()` functions in applications that exhibit a random I/O access pattern. Another important difference between the two sets of functions is that in order to achieve zero-copy transfer, the `mmap()`/`munmap()` functions control which user-space location a block is mapped into. In the `read()`/`write()` case, where a copy is incurred anyway, it is the application that controls the placement of blocks in user-space. As described in the next section, all the above issues have influenced the design of the random access part of TPIE.

3 The TPIE Kernel

In this section we describe in some detail the architecture of the TPIE kernel and the main goals we tried to achieve when designing it. The kernel, as well as the rest of the TPIE library, are written in C++. We assume the reader is familiar with object-oriented and C++ terminology, like classes, templates, constructors and destructors.

3.1 Overview

As mentioned in the introduction, the TPIE library has been built in two phases. The first phase was initially developed for algorithms based on sequential scanning, like sorting, permuting, merging, and distributing. In these algorithms, the computation can be viewed as a continuous process in which data is fed in *streams* from an outside source and streams of results are written behind. This stream-based view of I/O computation is not utilizing the full power of the parallel disk model, but it provides a layer of abstraction that frees the developer from worrying about details like managing disk blocks and scheduling I/Os.

The TPIE kernel designed in this first phase consists of three modules: the *Stream-based Block Transfer Engine (BTE)*, responsible for packaging data into

---

[4] Some systems use the buffer cache to implement `mmap()`, mapping pages from the buffer cache into user-space memory.

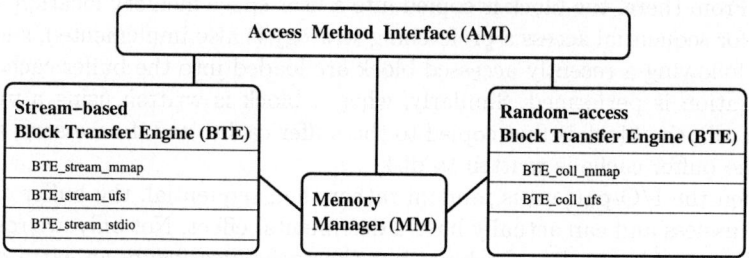

Fig. 1. The structure of the TPIE kernel

blocks and performing I/O transfers, the *Memory Manager (MM)*, responsible for managing main memory resources, and the *Application Method Interface (AMI)*, which provides the public interface and various tools. The BTE implements a stream using a UNIX file and its functionality, like reading, writing, or creating a block, is implemented using UNIX I/O calls. Since the performance of these calls is paramount to the performance of the entire application, different BTE implementations are provided, using different UNIX I/O calls (see Figure 1): BTE_stream_mmap uses mmap() and munmap(), BTE_stream_ufs uses the read() and write() system calls, and BTE_stream_stdio uses fread() and fwrite(). Other implementations can easily be added without affecting other parts of the kernel. The Memory Manager (MM) module maintains a pool of main memory of given size M and insures that this size is not exceeded. When either a TPIE library component or the application makes a memory allocation request, the Memory Manager reserves the amount requested and decreases a global counter keeping track of the available memory. An error is returned if no more main memory is available. Finally, the AMI provides a high-level interface to the stream functionality provided by the BTE, as well as various tools, including templated functions for scanning, merging, and sorting streams. In addition, the AMI provides tools for testing and benchmarking: a tool for logging errors and debugging messages and a mechanism for reporting statistics.

As mentioned, the stream-based view of I/O provided by these modules is ill-equipped for implementing on-line I/O-efficient data structures. The second phase of TPIE provides support for implementing these structures by using the full power of the disk model. Maintaining the design framework presented above, the new functionality is implemented using a new module, the Random-access BTE, as well as a new set of AMI tools. Figure 1 depicts the interactions between the various components of the TPIE kernel. The rest of this section describes the implementation of the Random-access BTE module and the new AMI tools.

3.2 The Random Access Block Transfer Engine (BTE)

The Random-access BTE implements the functionality of a *block collection*, which is a set of fixed-size blocks. A block can be viewed as being in one of two states: on disk or in memory. To change the state of a block, the block collection should support two operations: *read*, which loads a block from disk to memory, and *write*, which stores an in-memory block to disk. In addition, the block collection should be able to *create* a new block and *delete* an existing block. In order to support these operations, a unique *block ID* is assigned to each block in a block collection. When requesting a new block using the create operation, the collection returns a new block ID, which can then be used to read, write, or delete the block.

In our implementation, a block collection is organized as a linear array of blocks stored in a single UNIX file. A block from a collection is uniquely determined by its index in this array—thus this index is used as the block ID. When performing a read or write operation we start by computing the offset of the block in the file using the block ID. This offset is then used to seek in the file and transfer the requested block. Furthermore, the write operation uses a *dirty flag* to avoid writing blocks that have not been modified since they were read. This per-block dirty flag should be set by the user-level application whenever the contents of the in-memory block are modified. Unlike read and write, the create and delete operations modify the size of the block collection. To implement these operations, we employ a stack storing the IDs of the blocks previously deleted from the collection. When a block is deleted, its ID is simply pushed onto this stack. When a new block is requested by the create procedure, the stack is first checked, and if it's not empty, the top ID is popped and returned; if the stack is empty, the block ID corresponding to a new block at the end of the file is returned. The use of the stack avoids costly reorganization of the collection during each delete operation. However, it brings up a number of issues that need to be addressed. First, the stack has to reside on disk and has to be carried along with the file storing the blocks. In other words, a collection consists of two files: one containing data blocks and one containing block IDs. The second issue concerns space overhead. When multiple create and delete operations are performed on a collection, the number of blocks stored in the data blocks file can be much larger than the number of blocks in the collection. To eliminate this space overhead, the collection can be reorganized. However, such a reorganization would change the IDs of some blocks in the collection and therefore it cannot be performed by the BTE, which has no knowledge of the contents of the blocks: If a block contains IDs of other blocks, then the contents of that block would need to be updated as well. A reorganization procedure, if needed, should therefore be implemented on the application level.

We decided to implement the block collection on top of the UNIX file system and not, e.g., on the raw disk, to obtain portability and ease of use. Using raw disk I/O would involve creating a separate disk partition dedicated to TPIE data files, a non-trivial process that usually requires administrator privileges. Some operating systems offer user-space raw I/O, but this mechanism is not standardized and is offered only by a few operating systems (such as Solaris and Linux). On the other hand, storing a block collection in a file allows the use of existing file utilities to copy and archive collections.

A BTE block collection is implemented as a C++ class that provides a standard interface to the AMI module. The interface consists of a constructor for opening a collection, a destructor for closing the collection, the four block-handling routines described above, as well as other methods for reporting the size of the collection, error handling, etc. Except for the four block-handling routines, all methods are implemented in a base class. We implemented two BTE collections as extensions to this base class: BTE_coll_mmap and BTE_coll_ufs. As the names suggest, BTE_coll_mmap uses mmap() and munmap() to perform I/O transfers, while BTE_coll_ufs uses the read() and write() system calls. The implementation of choice for most systems is BTE_coll_mmap since, as mentioned in Section 2, the mmap()/munmap() functions are more suited for the random I/O pattern exhibited by online algorithms. We implemented BTE_coll_ufs to compare its performance with BTE_coll_mmap and to account for some systems where mmap() and munmap() are very slow.

3.3 The Access Method Interface (AMI)

The AMI tools needed to provide the random access I/O functionality consist of a front-end to the BTE block collection and a typed view of a disk block. In the BTE, we viewed the disk block as a fixed-size sequence of "raw" bytes. However, when implementing external memory data structures, disk blocks often have a well-defined internal structure. For example, a disk block storing an internal node of a B$^+$-tree contains the following: an array of b pointers to child nodes, an array of $b-1$ keys, and a few extra bytes for storing the value of b. Therefore, the AMI contains a templated C++ class called AMI_block<E,I>. The contents of a block are partitioned into three fields: an array of zero or more links to other blocks (i.e., block IDs), an array of zero or more elements of type E (given as a template parameter), and an *info* field of type I (also given as a template parameter). Each link is of type AMI_bid and represents the ID of another block in the same collection. This way the structure of a block is uniquely determined by three parameters: the types E and I and the number of links. Easy access to elements and links is provided by simple array operators. For example, the ith element of a block b is referenced by b.el[i], and the jth link is referenced by b.lk[j].

The AMI_block<E,I> class is more than just a structuring mechanism for the contents of a block. It represents the in-memory image of a block. To this end, constructing an AMI_block<E,I> object loads the block in memory, and destroying the object unloads it from memory. The most general form of the constructor is as follows.

```
AMI_block<E,I>(AMI_collection* c, size_t links, AMI_bid bid=0)
```

When a non-zero block ID is passed to the constructor, the block with that ID is loaded from the given collection. When the block ID is zero, a new block is created in the collection. When deleting an AMI_block<E,I> object, the block is deleted from the collection or written back to disk, depending on a *persistence flag*. By default, a block is *persistent*, meaning that it is kept in the collection after the in-memory image has been destroyed. The persistence flag can be changed for individual blocks.

As its name suggests, the AMI_collection type that appears in the above constructor represents the AMI interface to a block collection. The functionality of the AMI_collection class is minimal, since the block operations are handled by the AMI_block<E,I> class. The main operations are open and close, and are performed by the constructor and destructor. An instance of type AMI_collection is constructed by providing a file name, an access type, and a block size.

```
AMI_collection(char* fn, AMI_collection_type ct, size_t bl_sz)
```

This constructor either opens an existing collection or creates and opens a new one. The destructor closes the collection and, if so instructed, deletes the collection from disk. The exact behavior is again determined by a *persistence flag*, which can be set before calling the destructor.

3.4 Design Goals

This subsection summarizes the main goals we had in mind when designing the TPIE kernel and the methods we used to achieve these goals.

Ease of use. The usability of the kernel relies on its intuitive and powerful interface. We started from the parallel disk model and built the Random-access BTE module to simulate it. Multiple BTE implementations exist and they are easily interchangeable, allowing an application to use alternative low-level I/O routines. The user interface, provided within the AMI, consists of a typed view of a block—the AMI_block<E,I> class—and a front-end to the block collection—the AMI_collection class. The design of these two classes provides an easy to understand, yet powerful application interface. In addition, the AMI_block<E,I> class provides structure to the contents of the block in order to facilitate the implementation of external memory data structure.

Flexibility. As illustrated in Figure 1, the TPIE kernel is composed of four modules with a well-defined interface. Each of the modules has at least a default implementation, but alternative implementations can be provided. The best candidates for alternative implementations are the two BTE modules, since they allow the use of different I/O mechanisms. The Stream-based BTE has three implementations, using different system calls for performing the I/O in stream operations. The Random-access BTE has two implementations, which use different low-level system calls to perform the I/O in block collection operations.

Efficiency. In order to obtain a fast library, we paid close attention to *optimizing disk access, minimizing CPU operations*, and *avoiding unnecessary in-memory data movement*. To optimize disk access, we used a per-block dirty flag that indicates whether the block needs to be written back or not. To minimize CPU operations, we used templated classes with no virtual functions; because of their inherently dynamic nature, virtual functions are not typically inlined by C++ compilers and have a relatively high function call overhead. Finally, to avoid in-memory copying, we used the mmap()/munmap() I/O system calls; they typically transfer blocks of data directly between disk and user-space memory.

Portability. Being able to easily port the TPIE kernel on various platforms was one of our main goals. As discussed in Section 3.1, the default methods used for performing the disk I/O are those provided by the UNIX-based operating systems and were chosen for maximum portability. Alternative methods can be added, but the existing implementations insure that the library works on all UNIX-based platforms.

4 Data Structures

As mentioned in the introduction, there are various external memory data structures implemented using the TPIE kernel. They are all part of the extended TPIE library. In this section, we briefly survey these data structures.

B-tree. The B-tree [12] is the classical external memory data structure for online searching. In TPIE we implemented the more general (a,b)-tree [14], supporting insertion, deletion, point query, range query, and bulk loading.[5] All these operations are encapsulated in a templated C++ class. The template parameters allow the user to choose the type of data items to be indexed, the key type, and the key comparison function. A full description of the (a,b)-tree implementation will be given in the full version of this paper.

Persistent B-tree. The persistent B-tree [9] is a generalization of the B-tree that records all changes to the initial structure over a series of updates, allowing queries to be answered not only on the current structure, but on any of the previous ones as well. The persistent B-tree can be used to answer 3-sided range queries and vertical ray shooting queries on segments in $\mathbb{R}^2$. More details on the implementation can be found in [5].

R-tree. The R-tree and its variants are widely used indexing data structures for spatial data. The TPIE implementation uses the insertion heuristics proposed by Beckmann et al. [10] (their variant is called the R*-tree) and various bulk loading procedures. More details are given in [6, 7].

Logarithmic method. The logarithmic method [16] is a generic dynamization method. Given a static index with certain properties, it produces a dynamic structure consisting of a set of smaller static indexes of geometrically increasing sizes. We implemented the external memory versions of this method, as proposed by Arge and Vahrenhold [8] and Agarwal et al. [2]. More details on the implementation can be found in [17].

K-D-B-tree. The K-D-B-tree [18] combines the properties of the kd-tree [11] and the B-tree to handle multidimensional points in an external memory setting. Our implementation supports insertion, deletion, point query, range query and bulk loading. More details on the implementation can be found in [17].

---
[5] Bulk loading is a term used in the database literature to refer to constructing an index from a given data set from scratch.

Bkd-tree. The Bkd-tree [17] is a data structure for indexing multidimensional points. It uses the kd-tree [11] and the logarithmic method to provide good worst-case guarantees for the update and query operations. More details can be found in [17].

5 Case Study: Implementing the K-D-B-Tree

We conclude this paper with some details of the K-D-B-tree implementation in order to illustrate how to implement a data structure using TPIE. We chose the K-D-B-tree because it is a relatively simple yet typical example of a tree-based structure implementation.

The K-D-B-tree is a data structure for indexing multidimensional points that attempts to combine the query performance of the kd-tree with the update performance of the B-tree. More precisely, a K-D-B-tree is a multi-way tree with all leaves on the same level. In two dimensions, each internal node v corresponds to a rectangular region r and the children of v define a disjoint partition of r obtained by recursively splitting r using axis-parallel lines (similar to the kd-tree [11] partitioning scheme). The points are stored in the leaves of the tree, and each leaf or internal node is stored in one disk block.

The implementation of the K-D-B-tree is parameterized on the type c used for the point coordinates and on the dimension of the space d.

```
template<class c, size_t d> class Kdbtree;
```

The K-D-B-tree is stored in two block collections: one for the (internal) nodes, and one for the leaves. Using two collections to store the K-D-B-tree allows us to choose the block size of nodes and that of leaves *independently*; it also allows us to have the nodes clustered on disk, for improved performance.

By the flexible design of the AMI_block class, we can simply extend it and use the appropriate template parameters in order to provide the required structure for nodes and leaves.

```
template<class c, size_t d>
class Kdbtree_node: AMI_block<box<c, d>, kdbtree_node_info>;
template<class c, size_t d>
class Kdbtree_leaf: AMI_block<point<c, d>, kdbtree_leaf_info>;
```

In other words, a Kdbtree_node<c,d> object consists of an array of d-dimensional boxes of type box<c,d>, an array of links pointing to the children of the node, and an info element of type kdbtree_node_info. The info element stores the actual fanout of the node (which is equal to the number of boxes stored), the weight of the node (i.e., the number of points stored in the subtree rooted at that node), and the splitting dimension (a parameter used by the insertion procedure, as described in [18]). The maximum fanout of a node is computed (by the AMI_block class) using the size of the box<c,d> class and the size of the block, which is a parameter of the nodes block collection. A Kdbtree_leaf<c,d> object consists of an array of d-dimensional points of type point<c,d>, no links, and an info element of type `kdbtree_leaf_info` storing the number of points, a pointer to another leaf (for threading the leaves), and the splitting dimension.

As already mentioned, the operations supported by this implementation of the K-D-B-tree are insertion, deletion, point query, window query, and bulk loading. It has been shown that batched algorithms for bulk loading can be much faster than using repeated insertions [6]. For the K-D-B-tree, we implemented two different bulk loading algorithms, as described in [17]. Both algorithms start by sorting the input points and then proceed to build the tree level by level, in a top down manner. The implementation of these algorithms shows the seamless integration between the stream-handling AMI tools and the block handling AMI tools: The initial sorting is done by the built-in AMI sort function, and the actual building is done by scanning the sorted streams and producing blocks representing nodes and leaves of the K-D-B-tree.

The update operations (insertion and deletion) are implemented by closely following the ideas from [18]. The query operations are performed as in the kd-tree [11]. Figure 2 shows the implementation of the simple point query procedure. Starting from the root, the procedure traverses the path to the leaf that might contain the query point. The traversal is done by iteratively fetching a node using its block ID (line 7), finding the child node containing the query point (line 8), and releasing the node (line 10). When the child node is a leaf, that leaf is fetched (line 12), its contents are searched for the query point (line 13), and then the leaf is released (line 14). These pairings of fetch and release calls are typical examples of how applications use the TPIE kernel to perform I/O. Intuitively, `fetch_node` reads a node from disk and `release_node` writes it back. The point query procedure is oblivious to how the I/O is performed or whether any I/O was performed at all. Indeed, the fetch and release functions employ a *cache manager* to improve I/O performance. By using application-level caching (instead of fixed, kernel-level caching) we allow the application developer to choose the most appropriate caching algorithm. A few caching algorithms are

```
1   bool find(point_t& p) {
2     bool ans;  size_t i;
3     Kdbtree_node<c,d>* bn;
4     region_t<c,d> r;
5     kdb_item_t<c,d> ki(r, header_.root_bid, header_.root_type);
6     while (ki.type != BLOCK_LEAF) {
7       bn = fetch_node(ki.bid);
8       i = bn->find(p);
9       ki = bn->el[i];
10      release_node(bn);
11    }
12    Kdbtree_leaf<c,d>* bl = fetch_leaf(ki.bid);
13    ans = (bl->find(p) < bl->size());
14    release_leaf(bl);
15    return ans;
16  }
```

Fig. 2. Implementation of the point query procedure

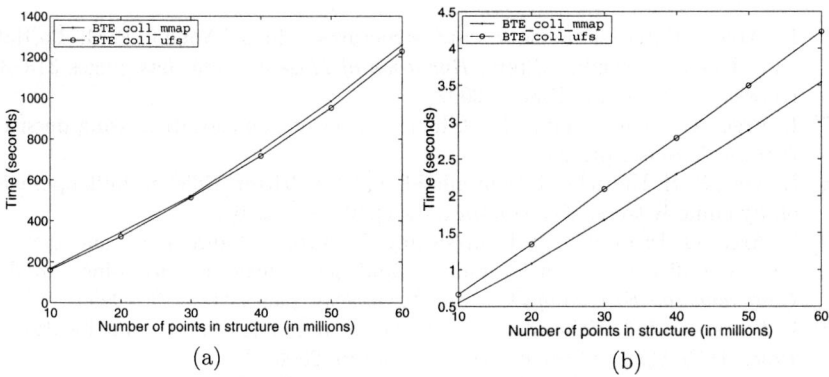

Fig. 3. (a) Performance of K-D-B-tree bulk loading (b) Performance of a range query (averaged over 10 queries, each returning 1% of the points in the structure)

already provided in TPIE, and more can be easily added by extending the cache manager base class.

Experiments. Using the K-D-B-tree implementation, we performed experiments to show how the choice of I/O system calls affects performance. We bulk loaded and performed range queries on K-D-B-trees of various sizes.[6] The data sets consisted of uniformly distributed points in a squared-shaped region. The graph in Figure 3(a) shows the running times of bulk loading, while the graph in Figure 3(b) shows the running time of one range query, averaged over 10 similar-size queries. Each experiment was performed using the two existing Random-access BTE implementations: BTE_coll_mmap and BTE_coll_ufs. As expected, the running time of the bulk loading procedure—a highly sequential process—is not affected by the choice of Random-access BTE. On the other hand, the performance of a range query is affected significantly by this choice: Using the ufs-based Random-access BTE results in higher running times. This validates our analysis from Section 2 and confirms that BTE_coll_mmap is the implementation of choice for the Random-access BTE.

References

[1] P. K. Agarwal, L. Arge, and S. Govindarajan. CRB-tree: An optimal indexing scheme for 2d aggregate queries. Manuscript, 2002.
[2] P. K. Agarwal, L. Arge, O. Procopiuc, and J. S. Vitter. A framework for index bulk loading and dynamization. In *Proc. 28th Intl. Colloq. Automata, Languages and Programming (ICALP)*, 2001.
[3] A. Aggarwal and J. S. Vitter. The Input/Output complexity of sorting and related problems. *Communications of the ACM*, 31(9):1116–1127, 1988.

---

[6] All experiments were performed on a dedicated Pentium III/500MHz computer running FreeBSD 4.4, with 128MB of main memory and an IBM Ultrastar 36LZX SCSI disk.

[4] L. Arge. External memory data structures. In J. Abello, P. M. Pardalos, and M. G. C. Resende, editors, *Handbook of Massive Data Sets*, pages 313–358. Kluwer Academic Publishers, 2002.

[5] L. Arge, A. Danner, and S.-M. Teh. I/O-efficient point location using persistent B-trees. Manuscript, 2002.

[6] L. Arge, K. H. Hinrichs, J. Vahrenhold, and J. S. Vitter. Efficient bulk operations on dynamic R-trees. *Algorithmica*, 33(1):104–128, 2002.

[7] L. Arge, O. Procopiuc, S. Ramaswamy, T. Suel, J. Vahrenhold, and J. S. Vitter. A unified approach for indexed and non-indexed spatial joins. In *Proc. Conference on Extending Database Technology*, pages 413–429, 1999.

[8] L. A. Arge and J. Vahrenhold. I/O-efficient dynamic planar point location. In *Proc. ACM Symp. Computational Geometry*, 2000.

[9] B. Becker, S. Gschwind, T. Ohler, B. Seeger, and P. Widmayer. An asymptotically optimal multiversion B-tree. *VLDB Journal*, 5(4):264–275, 1996.

[10] N. Beckmann, H.-P. Kriegel, R. Schneider, and B. Seeger. The R*-tree: An efficient and robust access method for points and rectangles. In *Proc. SIGMOD Intl. Conf. on Management of Data*, pages 322–331, 1990.

[11] J. L. Bentley. Multidimensional binary search trees used for associative searching. *Commun. ACM*, 18(9):509–517, Sept. 1975.

[12] D. Comer. The ubiquitous B-tree. *ACM Comput. Surv.*, 11:121–137, 1979.

[13] A. Crauser and K. Mehlhorn. LEDA-SM: Extending LEDA to secondary memory. In *Proc. Workshop on Algorithm Engineering*, 1999.

[14] S. Huddleston and K. Mehlhorn. A new data structure for representing sorted lists. *Acta Informatica*, 17:157–184, 1982.

[15] K. Mehlhorn and S. Näher. *LEDA: A Platform for Combinatorial and Geometric Computing*. Cambridge University Press, Cambridge, UK, 2000.

[16] M. H. Overmars. *The Design of Dynamic Data Structures*, volume 156 of *Lecture Notes Comput. Sci.* Springer-Verlag, Heidelberg, West Germany, 1983.

[17] O. Procopiuc, P. K. Agarwal, L. Arge, and J. S. Vitter. Bkd-tree: A dynamic scalable kd-tree. Manuscript, 2002.

[18] J. T. Robinson. The K-D-B-tree: A search structure for large multidimensional dynamic indexes. In *Proc. SIGMOD Intl. Conf. on Management of Data*, pages 10–18, 1981.

[19] D. E. Vengroff and J. S. Vitter. Supporting I/O-efficient scientific computation in TPIE. In *Proc. IEEE Symp. on Parallel and Distributed Computing*, pages 74–77, 1995.

[20] J. S. Vitter. External memory algorithms and data structures: Dealing with MASSIVE data. *ACM Computing Surveys*, 33(2):209–271, 2001.

[21] J. S. Vitter and E. A. M. Shriver. Algorithms for parallel memory, I: Two-level memories. *Algorithmica*, 12(2–3):110–147, 1994.

On the k-Splittable Flow Problem[*]

Georg Baier, Ekkehard Köhler, and Martin Skutella

Technische Universität Berlin
Fakultät II — Mathematik und Naturwissenschaften, Institut für Mathematik
Sekr. MA 6–1, Straße des 17. Juni 136, D–10623 Berlin, Germany
{baier,ekoehler,skutella}@math.tu-berlin.de

Abstract. In traditional multi-commodity flow theory, the task is to send a certain amount of each commodity from its start to its target node, subject to capacity constraints on the edges. However, no restriction is imposed on the number of paths used for delivering each commodity; it is thus feasible to spread the flow over a large number of different paths. Motivated by routing problems arising in real-life applications, such as, e. g., telecommunication, unsplittable flows have moved into the focus of research. Here, the demand of each commodity may not be split but has to be sent along a single path.
In this paper, a generalization of this problem is studied. In the considered flow model, a commodity can be split into a bounded number of chunks which can then be routed on different paths. In contrast to classical (splittable) flows and unsplittable flows, already the single-commodity case of this problem is NP-hard and even hard to approximate. We present approximation algorithms for the single- and multi-commodity case and point out strong connections to unsplittable flows. Moreover, results on the hardness of approximation are presented. It particular, we show that some of our approximation results are in fact best possible, unless P=NP.

1 Introduction

The k-splittable flow problem is a multi-commodity flow problem in which each commodity may be shipped only on a restricted number of different paths. The number of possible paths can be the same for all commodities or it may depend on the particular commodity. Problems of this kind occur, for instance, in transportation logistics: A number of different commodities has to be delivered; for each commodity, only a limited number of transportation devices is available. Thus, we have to bound the number of paths used for delivering a commodity by the number of available transportation devices for this commodity. Similar problems appear in communication networks: Customers request connections of given capacities between certain pairs of terminals in the network. If these capacities

---

[*] This work was supported in part by the EU Thematic Network APPOL I+II, Approximation and Online Algorithms, IST-1999-14084, IST-2001-30012, and by the Bundesministerium für Bildung und Forschung (bmb+f), grant no. 03-MOM4B1.

are large, it might be impossible for the network administrator to realize them unsplittably, that is, on single paths without increasing net capacity. On the other hand, the customer might not want to handle many connections of small capacity. Thus, one has to find a flow fulfilling the different customer demands and restrictions on connecting lines while respecting all capacity constraints.

Problem definition and notation. An instance of the *k-splittable flow problem* is defined on a directed or undirected graph $G = (V, E)$ with edge capacities $u_e \in \mathbb{Z}_{>0}$, $e \in E$. For arbitrary source and target nodes $s, t \in V$, let $\mathcal{P}_{s,t}$ denote the set of simple s-t-paths in G. Then, a *k-splittable s-t-flow* F is specified by k pairs $(P_1, f_1), \ldots, (P_k, f_k) \in \mathcal{P}_{s,t} \times \mathbb{R}_{\geq 0}$ of s-t-paths P_i and flow values f_i. We do not require that the paths $P_1, \ldots, P_k$ are distinct. In particular, any k-splittable flow is also k'-splittable for all $k' \geq k$. The sum $\sum_{i=1}^{k} f_i$ is called the *s-t-flow value of F*. The flow F is *feasible* if it respects edge capacities, that is, for each edge $e \in E$ the sum of flow values on paths containing this edge must be bounded by its capacity u_e. The *maximal k-splittable s-t-flow problem* asks for a feasible k-splittable s-t-flow of maximal value.

Obviously, k-splittable s-t-flows form a special class of s-t-flows in graphs. In fact, it is a well-known result from classical network flow theory that any s-t-flow can be decomposed into the sum of at most $|E|$ flows on s-t-paths and a circulation; apart from the circulation, it is thus an $|E|$-splittable s-t-flow in our terminology. In particular, for $k \geq |E|$ the maximal k-splittable s-t-flow problem can be solved efficiently by standard network flow techniques; see, e.g., [1]. On the other hand, we show that the problem for directed graphs is NP-hard for $k = 2$. All our results except for hardness hold for directed and undirected graphs but for simpler notation we concentrate on the directed case.

We also study k-splittable s-t-flows with additional restrictions. A k-splittable s-t-flow in which all paths with non-zero flow value carry identical amounts of flow is called a *uniform k-splittable s-t-flow*. Moreover, a uniform k-splittable s-t-flow which sends flow on exactly k paths is called *uniform exactly-k-splittable s-t-flow*. For our algorithms it is essential that we do *not* require k distinct paths. If there are also edge costs $c_e \in \mathbb{R}_{\geq 0}$, $e \in E$, we can consider the problem with additional budget constraint: Find a maximal k-splittable s-t-flow whose cost does not exceed a given budget $B \geq 0$. In this setting, let $c_P = \sum_{e \in P} c_e$ denote the cost of path P. Then, the cost of a k-splittable s-t-flow $(P_1, f_1), \ldots, (P_k, f_k)$ can be written as $\sum_{i=1}^{k} f_i c_{P_i}$, which is equal to the sum over all edges of edge cost times flow value on the edge.

In the multi-commodity variant of the k-splittable flow problem, there are ℓ terminal pairs $(s_1, t_1), \ldots, (s_l, t_l)$ of source and target nodes, and a bound k_i on the number of paths allowed for each terminal pair (s_i, t_i), $i = 1, \ldots, \ell$. A *k-splittable multi-commodity flow* (or simply *k-splittable flow*) F is the sum of k_i-splittable s_i-t_i-flows, for $i = 1, \ldots, \ell$. If all k_i-splittable s_i-t_i-flows are uniform exactly-k_i-splittable s_i-t_i-flows, then F is called *uniform exactly-k-splittable multi-commodity flow*. Notice that this definition allows different flow values per path for different commodities. For given demand values $D_1, \ldots, D_\ell > 0$, the *k-splittable multi-commodity flow problem* (or simply *k-splittable flow problem*) is to find a feasible k-splittable flow F with s_i-t_i-flow value D_i, for $i = 1, \ldots, \ell$. Here, the flow sent from s_i to t_i is also referred to as *commodity i*.

We use the following terminology: A flow whose flow value on any edge is a multiple of a given value $\alpha > 0$ is called *α-integral*. If we restrict to α-integral flows, we can round down edge capacities to the nearest multiple of α without abandoning any solution. Thus, the problem of finding α-integral flows is equivalent to finding integral flows in graphs with integral capacities; it can therefore be solved efficiently in the single-commodity case by standard flow techniques.

Related results from the literature. To the best of our knowledge, the k-splittable flow problem has not been considered in the literature before. However, it contains the well-known unsplittable flow problem as a special case: Setting $k_i = 1$ models the requirement that commodity i must be routed unsplittably, i. e., on a single s_i-t_i-path. The unsplittable flow problem was introduced by Kleinberg [2, 3] as a generalization of the disjoint path problem. Kleinberg shows that it comprises several NP-complete problems from areas such as packing, partitioning, scheduling, load balancing, and virtual-circuit routing.

Kleinberg introduced several optimization versions of the unsplittable flow problem. In the "minimum congestion" version, the task is to find the smallest value $\lambda \geq 1$ such that there exists an unsplittable flow that violates the capacity of any edge at most by a factor λ. An equivalent problem is the "concurrent flow" problem where the aim is to maximize the routable fraction of the demand, i. e., find the maximal factor by which the given demands can be multiplied such that there still exists a feasible unsplittable flow satisfying the resulting demands. The "minimum number of rounds" version asks for a partition of the set of commodities into a minimal number of subsets (rounds) and a feasible unsplittable flow for each subset. Finally, the "maximum routable demand" problem is to find a single round of maximum total demand.

Since, apart from some special cases, all of these problems are NP-hard, much effort has been spent in obtaining approximation results. A *ρ-approximation algorithm* is an algorithm running in polynomial time and always finding a feasible solution whose value is at most a factor of ρ away from the optimum. The value ρ is called the *performance ratio* or *performance guarantee* of the algorithm. In the following we use the convention that the performance ratio for a maximization problem is less than 1, while for a minimization problem it is greater than 1.

A summary on results of all mentioned unsplittable flow problems can be found in [4]. Due to space limitations we only give some references [2, 3, 5, 6, 7, 8, 9, 10, 11]. The unsplittable flow problem is much easier if all commodities share a common single source; nevertheless, the resulting *single source unsplittable flow problem* remains NP-hard. Again we only give the references [2, 3, 6, 7, 12, 13]

In a recent work Bagchi, Chaudhary, Scheideler, and Kolman [14] consider fault tolerant routings in networks. They independently define problems similar to our uniform exactly k-splittable flow problem. To ensure connection for each commodity for up to k edge failure in the network, they require edge disjoint flow-paths per commodity. The question of bounded splittablily is also considered in the scheduling context in a recent paper by Krysta, Sanders, and Vöcking [15].

Contribution of this paper. In Section 2, approximation results for the maximal k-splittable s-t-flow problem are derived. We start with the special cases $k = 2$ and $k = 3$ in Section 2.1 and develop fast approximation algorithms with performance ratio $2/3$. Surprisingly, this approximation result is best possible, unless P=NP. In order to derive approximation algorithms for higher values of k, it is first shown in Section 2.2 that a maximal uniform k-splittable s-t-flow can be computed in polynomial time. It is interesting to mention that there exists a natural cut definition such that the maximal-flow minimum-cut relation holds for uniform k-splittable s-t-flows. Due to space limitations we omit further details in this extended abstract. In Section 2.3 it is shown that the value of a maximal uniform exactly-k-splittable s-t-flow approximates the value of a maximal k-splittable s-t-flow within a factor of $1/2$. This result can be extended to the problem with costs. In Section 3, approximation results for the maximal concurrent flow version of the k-splittable multi-commodity flow problem are presented. Finally, in Section 4 a result on the hardness of approximation is presented.

Due to space limitations, we omit some proofs in this extended abstract. More details can be found in [4].

2 The Single-Commodity Case

In this section we consider the *maximal k-splittable s-t-flow problem*, that is, the case of only one commodity whose demand must be routed on at most k paths from source s to target t. Already this problem is strongly NP-hard since it contains the single-source unsplittable flow problem as a special case: Assume we are given an instance of the single-source unsplittable flow problem, that is, a graph G with source s, sinks $t_1, \ldots, t_k$, and corresponding demands $D_1, \ldots, D_k$. We add a new node t together with edges $t_i t$ of capacity D_i, for $i = 1, \ldots, k$. Computing a k-splittable flow from source s to target t of flow value $\sum_{i=1}^{k} D_i$ is equivalent to solving the given single-source unsplittable flow problem. Thus, the k-splittable s-t-flow problem is NP-hard, see, e. g., [2].

As mentioned above, the k-splittable s-t-flow problem can be solved efficiently by classical flow techniques if $k \geq |E|$. In the remainder of this section we therefore assume that $k < |E|$. We start by discussing the special cases $k = 2$ and $k = 3$.

2.1 Approximating Maximal 2-Splittable and 3-Splittable s-t-Flows

We start with the maximal 2-splittable flow problem and give a simple combinatorial 2/3-approximation algorithm. As we will show in Section 4, this result is best possible in the sense that no approximation algorithm with strictly better performance ratio exists for this problem, unless P=NP.

Our algorithm (described in the proof of Theorem 1) uses a special variant of the classical augmenting path algorithm for maximal flows which chooses an augmenting s-t-path of maximal residual capacity in each iteration. Notice that such a *maximum capacity path* can be found in $\mathcal{O}(|E|\log|E|)$ time by a modified Dijkstra labeling algorithm. This variant is known as the *maximum capacity augmenting path algorithm* (see, e.g., [1, Chapter 7.3]).

Lemma 1. *After two iterations of the augmenting path algorithm, the resulting flow can be decomposed into the sum of a 3-splittable flow and a circulation. Moreover, such a decomposition can be computed in linear time.* □

Lemma 2. *After two iterations of the maximum capacity augmenting path algorithm, the value of the resulting flow is an upper bound on the value of a maximal 2-splittable flow.*

Proof. Let F^* be a maximal 2-splittable flow solution which sends flow on paths P_1^* and P_2^* with flow values f_1^* and f_2^*, respectively. Moreover, let P_1 and P_2 denote the two augmenting paths found by the maximum capacity augmenting path algorithm with corresponding flow augmentations f_1 and f_2. We have to show that $f_1 + f_2 \geq f_1^* + f_2^*$.

Consider the subgraph G' of G induced by all edges in paths P_1, P_1^*, and P_2^*. Set the capacity of each edge in G' to the smallest value such that both the optimal solution $\{(P_1^*, f_1^*), (P_2^*, f_2^*)\}$ and the flow of value f_1 on path P_1 are feasible. Consider now the residual graph of G' with respect to the flow of value f_1 on P_1. We will show that there is an s-t-path of capacity at least $f_1^* + f_2^* - f_1$ in this residual graph. This yields $f_2 \geq f_1^* + f_2^* - f_1$ since the augmenting path algorithm could have chosen this path.

If $f_1^* + f_2^* - f_1 \leq 0$, there is nothing to be shown. We can thus assume that $f_1 < f_1^* + f_2^*$. Moreover, we know by choice of P_1 that $f_1 \geq \max\{f_1^*, f_2^*\}$. Let P be a thickest s-t-path in the residual graph with bottleneck edge e. The residual capacity of edge e is strictly positive since the value of a maximal s-t-flow in G' is at least $f_1^* + f_2^* > f_1$. We distinguish the following three cases.

Case 1: edge e is a forward edge of P_1. Since e has positive residual capacity, both P_1^* and P_2^* must contain e such that the residual capacity of e is $f_1^* + f_2^* - f_1$.

Case 2: edge e is a backward edge of P_1. In this case, the residual capacity of e is $f_1 \geq f_1^* + f_2^* - f_1$.

Case 3: edge e is not contained in P_1. In this case, the residual capacity of e is at least $\min\{f_1^*, f_2^*\} \geq f_1^* + f_2^* - f_1$. □

Lemma 1 and Lemma 2 together yield the following approximation result for the maximal 2-splittable flow problem.

Theorem 1. *There exists a 2/3-approximation algorithm for the maximal 2-splittable s-t-flow problem with running time $\mathcal{O}(|E|\log|E|)$.*

Proof. Run two iterations of the maximum capacity augmenting path algorithm. The resulting flow can be decomposed into three paths and a circulation by

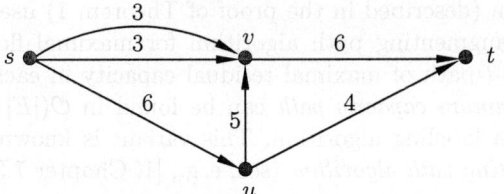

Fig. 1. Simple counterexample showing that Lemma 2 cannot be generalized to values of k greater than 2. A 3-splittable s-t-flow can use the paths s-v-t, s-v-t, and s-u-t with flow values 3, 3, and 4, respectively. Thus, a maximal 3-splittable flow has value 10. However, in its first two iterations, the maximum capacity augmenting path algorithm chooses the paths s-u-v-t and s-v-u-t of bottleneck capacity 5 and 3. Now both edges incident to t have residual capacity 1. Thus, the maximal capacity augmenting path algorithm yields after three augmentations a flow of value 9

Lemma 1. Deleting the circulation and the path with smallest flow value yields a 2-splittable flow. The performance guarantee of 2/3 is an immediate consequence of Lemma 2. Finally, the running time is dominated by the modified Dijkstra labeling algorithm which is $\mathcal{O}(|E| \log |E|)$. □

In Section 4 we show that this result is in fact tight, that is, there does not exist an approximation algorithm for the 2-splittable flow problem with performance guarantee $2/3 + \varepsilon$ for any $\varepsilon > 0$, unless P=NP. The result for $k = 2$ from Theorem 1 can easily be carried over to the maximal 3-splittable flow problem.

Corollary 1. *Two iterations of the maximum capacity augmenting path algorithm yield a $2/k$-approximation algorithm for the maximal k-splittable s-t-flow problem with running time $\mathcal{O}(|E| \log |E|)$. In particular there is a $2/3$-approximation algorithm for $k = 3$.* □

Unfortunately, the straightforward approach which led to the approximation result in Theorem 1 cannot be extended directly to a larger number of iterations. Firstly, k iterations of the maximum capacity augmenting path algorithm do not necessarily yield an upper bound on the value of a maximal k-splittable flow; see Figure 1. Secondly, for arbitrary k, it is difficult to bound the number of flow-carrying paths needed in a path decomposition of the flow resulting from k iterations of the maximum capacity augmenting path algorithm. In Figure 2, we show that one augmentation may even double this number.

2.2 Computing Maximal Uniform k-Splittable s-t-Flows

Before we turn to the general k-splittable s-t-flow problem, we first discuss the problem of determining a maximal k-splittable s-t-flow where all paths with positive flow value carry the same amount of flow. We refer to this problem as *uniform* k-splittable s-t-flow problem. In contrast to the problem with arbitrary

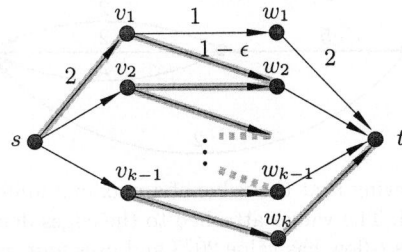

Fig. 2. A single augmentation can double the number of necessary paths in a path decomposition. Assume we have a flow of value 1 on each of the $k-1$ horizontal s-t-paths via v_i and w_i. As augmenting path we use the gray marked s-t-path via $v_1, w_2, v_2, w_3, \ldots, w_{k-1}, v_{k-1}, w_k$ with residual capacity $1-\epsilon$. Increasing the flow along this path requires $k-1$ additional paths

flow values, which is strongly NP-hard, the uniform k-splittable s-t-flow problem turns out to be solvable in polynomial time. This result will be used in Section 2.3 as an important building block for the construction of approximate solutions to the general problem.

Consider first a further specialization of the k-splittable s-t-flow problem where flows of equal value must be sent on *exactly* k paths. We refer to this problem as *uniform exactly-k*-splittable s-t-flow problem. Assume that we know the maximal flow value D. Then, our problem is equivalent to finding a D/k-integral s-t-flow of value D; notice that such a flow can easily be decomposed into a sum of k path-flows of value D/k plus a circulation (iteratively, take any flow-carrying s-t-path and reduce the flow on it by D/k). As mentioned above, this problem is efficiently solvable by standard flow techniques. Thus, we can find an optimal flow if we know the optimal flow value D.

Unfortunately, neither the flow value per path D/k nor the total flow value D have to be integral in general. Consider a graph consisting of two parallel edges from s to t, each with capacity 1. For odd $k = 2q + 1 > 2$, there is an optimal solution that uses q times the first edge and $q+1$ times the second edge as paths. Thus, the maximal flow value per path is $1/(q+1)$ which yields a non-integral uniform exactly-k-splittable flow of value $2 - 1/(q+1)$. Even if we do not require exactly k paths, the optimal value may be non-integral; an example is given in Figure 3.

How 'fractional' is an optimal solution in the 'worst' case? Consider an arbitrary optimal solution. Since the flow value is maximal, there must exist at least one edge with a tight capacity constraint. Thus, the flow value on any path is equal to the capacity of this edge divided by the number of paths using this edge in the optimal solution under consideration. Hence, possible flow values are of the form u_e/i, for some edge $e \in E$ and some number $i \in \{1, \ldots, k\}$. That is, there are at most $|E|k$ possible values which can be enumerated in polynomial time.

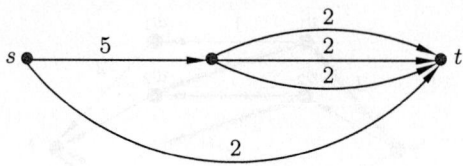

Fig. 3. An instance showing that the value of a maximal uniform k-splittable s-t-flow is in general not integral. The values attached to the edges denote the edge capacities. A maximal 4-splittable s-t-flow has value $20/3$ and uses four paths with flow value $5/3$

Observation 1. *The maximal uniform exactly-k-splittable s-t-flow problem can be solved in polynomial time.*

If we add costs and consider the budget-constrained maximal uniform exactly-k-splittable s-t-flow problem, the situation gets more complicated. In this case, the flow value per path may be forced by the budget constraint instead of a capacity constraint. Nevertheless, we can prove the following result.

Theorem 2. *A maximal budget-constrained uniform exactly-k-splittable s-t-flow can be computed in time $\mathcal{O}(k^2|E|^2 \log |E|)$.*

Proof. Consider an arbitrary optimal solution to the problem under consideration and let $\alpha \geq 1$ be the maximal factor by which the solution can be scaled without violating the capacity of an edge. Clearly, the scaled flow is a uniform exactly-k-splittable s-t-flow of flow value $k\,u_e/i$ for some tight edge $e \in E$ and some number $i \in \{1, \ldots, k\}$. Moreover, assuming that the budget constraint is tight, its cost is αB where B is the given budget. Therefore, a minimum-cost u_e/i-integral flow with flow value $k\,u_e/i$ has cost βB with $\beta \leq \alpha$. Thus, scaling it by a factor $\min\{1/\beta, 1\}$ yields an optimal budget-constrained uniform exactly-k-splittable s-t-flow.

Using this insight, we can now state an algorithm which solves the maximal budget-constrained uniform exactly-k-splittable s-t-flow in polynomial time. For all $e \in E$ and $i \in \{1, \ldots, k\}$, compute a minimum-cost u_e/i-integral flow with flow value $k\,u_e/i$ and scale it down until it obeys the cost-constraint. From the resulting $k\,|E|$ budget-constrained uniform exactly-k-splittable s-t-flows take one with maximal flow value. As shown above, this yields a maximal budget-constrained uniform exactly-k-splittable s-t-flow.

The running time of this algorithm is dominated by the $k\,|E|$ minimum-cost flow computations. Using the successive shortest path algorithm each minimum-cost flow can be computed in time $\mathcal{O}(k\,|E|\log |E|)$. □

In the following we present an algorithm with better running time for the problem without costs. The algorithm is a variant of the augmenting path algorithm with exactly k iterations. As augmenting paths we use maximum capacity paths in a special residual graph which is defined as follows.

Let $P_1, \ldots, P_i$ be the paths found in iterations 1 to i, and let f_i denote their common flow value after iteration i. The resulting s-t-flow of value $i f_i$ is denoted by F_i. We construct the following residual graph $\overline{G}_i$. For each edge e of the original graph, let q_e^i be the number of paths in $\{P_1, \ldots, P_i\}$ containing e. The residual capacity of edge e in $\overline{G}_i$ is set to $u_e/(1+q_e^i)$. Moreover, if $q_e^i > 0$, we additionally insert a backward edge of capacity f_i into $\overline{G}_i$.

In $\overline{G}_i$ we compute a maximum capacity s-t-path P with bottleneck capacity $f_{i+1} := c_P$. Then, a new flow F_{i+1} of value $(i+1)f_{i+1}$ is defined as follows. Send f_{i+1} units of flow on each path in $\{P, P_1, \ldots, P_i\}$. By definition of the residual graph $\overline{G}_i$, the resulting flow F_{i+1} is feasible in G. However, since the path P may contain backward edges, we still have to argue that F_{i+1} is a uniform exactly-$(i+1)$-splittable flow. However, this follows immediately since F_{i+1} is f_{i+1}-integral and has flow value $(i+1)f_{i+1}$.

Theorem 3. *The above algorithm finds a maximal uniform exactly-k-splittable s-t-flow in time $\mathcal{O}(k|E|\log|E|)$.*

We prove Theorem 3 by induction on k. Therefore we first mention two observations on the behavior of the flow value with increasing k. The maximal s-t-flow value of a uniform exactly-k-splittable flow is not monotone in k. Consider the graph with two parallel unit-capacity s-t edges discussed above. For k even, the maximal flow value is 2. However, for k odd, the maximal flow value is $2 - \frac{2}{k+1}$. On the other hand, although the total flow value is not monotone, the flow value per path is monotonically non-increasing for increasing k. Otherwise we would get a contradiction to optimality.

Proof (of Theorem 3). We use induction on k. The theorem is certainly true for $k=1$. Assume we have an optimal solution F_{k+1}^* for $k+1$ paths with a flow value of f_{k+1}^* per path. We have to show that the bottleneck capacity f_{k+1} of the path found in iteration $k+1$ of the algorithm is at least f_{k+1}^*.

Consider the graph G^* which is obtained by rounding down the edge capacities of the original graph G to the nearest multiples of f_{k+1}^*. The flow F_k can be scaled down to a feasible flow F_k' in the rounded graph G^* by rounding down the flow value on each of the k paths in F_k to f_{k+1}^*. Since the total flow value of F_{k+1}^* exceeds the total flow value of F_k' by f_{k+1}^*, there exists an s-t-path P in the residual graph of G^* with respect to F_k'. Path P also exists in $\overline{G}_k$ and we show that its capacity in $\overline{G}_k$ is at least f_{k+1}^*.

Consider an arbitrary edge e of P. If e is a backward edge, its capacity is f_k. By induction and monotonicity we get $f_k = f_k^* \geq f_{k+1}^*$. Thus, assume in the following that e is a forward edge. Then, the capacity of e in $\overline{G}_k$ is $u_e/(1+q_e^k)$. It remains to show that $u_e \geq (1+q_e^k)f_{k+1}^*$. This follows since the rounded capacity of e in G^* suffices to hold $1+q_e^k$ flow paths of value f_{k+1}^*. (Notice that F_k' and the edge capacities in G^* are f_{k+1}^*-integral.)

The running time of the algorithm is bounded by k times the time required to find a maximum capacity s-t-path. The latter problem can be solved by a modified Dijkstra algorithm in time $\mathcal{O}(|E|\log|E|)$. □

As mentioned above, the flow value of a maximal uniform exactly-k-splittable flow is not necessarily increasing with k. However, since our algorithm computes a maximal uniform exactly-i-splittable flow for all $i = 1, \ldots, k$, we can easily determine the number of paths maximizing the flow value.

Theorem 4. *The maximal uniform k-splittable s-t-flow problem can be solved efficiently in time $\mathcal{O}(k|E|\log|E|)$.* □

2.3 Approximating Maximal k-Splittable s-t-Flows

We show that computing a maximal (budget-constrained) uniform k-splittable s-t-flow yields a constant factor approximation for the maximal (budget-constrained) k-splittable s-t-flow problem.

Theorem 5. *Computing a maximal (budget-constrained) uniform exactly-k-splittable s-t-flow is an approximation algorithm with performance ratio $1/2$ for the maximal (budget-constrained) k-splittable s-t-flow problem.*

Proof. Consider an optimal solution F of value OPT to the general (budget-constrained) k-splittable s-t-flow problem. We show that there exists a (budget-constrained) uniform exactly-k-splittable s-t-flow which sends $D := \text{OPT}/(2k)$ units of flow on each path. In particular, the value of this flow is exactly $\text{OPT}/2$.

In the given optimal solution F, any flow-carrying path P with flow value $f_P > 0$ can be replaced by $\lfloor f_P/D \rfloor$ copies of P, each carrying D units of flow. It suffices to show that the resulting solution consists of at least k paths, each with flow value D. Using the fact that the number of paths used by F is bounded by k, we get

$$\sum_{P:f_P>0} \lfloor f_P/D \rfloor \geq \frac{\text{OPT}}{D} - k = k \ .$$

Finally, observe that the given conversion never increases flow on an edge and therefore never increases cost. This concludes the proof. □

On the other hand, we can show that the result in Theorem 5 is tight, that is, there exist instances with an asymptotic gap of 2 between the value of a maximal k-splittable s-t-flow and the value of a maximal uniform k-splittable s-t-flow. Consider a graph with two nodes s and t, and k parallel s-t edges. One of them has capacity $k-1$, the others have unit capacity. A maximal uniform k-splittable s-t-flow in this graph has value k; the maximal k-splittable s-t-flow has value $2(k-1)$.

3 The Multi-commodity Case

In this section we will extend some techniques and results from the last section to the general situation of multiple commodities. We consider the (budget-constrained) concurrent flow problem where the aim is to maximize the routable fraction, i.e., maximize the factor by which we can multiply all demands and still get a realizable instance respecting the given capacities (and the budget constraint). As in Section 2.3, we use uniform exactly-k-splittable flows in order to find approximate solutions.

Theorem 6. *For the multi-commodity case, the value of a maximal concurrent (budget-constrained) uniform exactly-k-splittable flow approximates the optimal value of the corresponding general concurrent (budget-constrained) k-splittable flow problem within a factor of $1/2$.*

Proof. As in the proof of Theorem 5, we turn an arbitrary k-splittable flow into a uniform exactly-k-splittable flow which satisfies at least one half of the original demands. Notice that the technique described in the proof of Theorem 5 can be applied to the commodities one after another. Since flow on an edge is never increased during this conversion, the resulting uniform exactly-k-splittable flow is feasible (and obeys the budget-constraint). □

However, in contrast to the single-commodity case, the multi-commodity case of the maximal concurrent uniform exactly-k-splittable flow problem is NP-hard, since it contains the unsplittable flow problem as a special case (set all k_i to 1, for $i = 1, \ldots, \ell$). Thus, we can only expect approximate solutions.

Assume that we want to route concurrently a fraction μ of the demand of each commodity. Then, each path of commodity i has to carry exactly $\mu D_i / k_i$ units of flow. In particular, there is no choice for the number of paths and for the amount of flow if μ is known. Thus, the problem is equivalent to the following unsplittable flow problem on the same graph: For each terminal pair (s_i, t_i) of our k-splittable flow problem, $i = 1, \ldots, \ell$, we introduce k_i terminal pairs (s_i, t_i) and assign a demand value of D_i / k_i to all of them. A concurrent unsplittable flow for the new problem instance with value μ is also a concurrent flow solution for the uniform exactly-k-splittable flow problem of the same value, and vice versa. Combining this observation with Theorem 6 yields the following result.

Theorem 7. *Any ρ-approximation algorithm for the maximal concurrent (budget-constrained) unsplittable flow problem yields an approximation algorithm with performance guarantee 2ρ for the maximal concurrent (budget-constrained) k-splittable flow problem.*

There are various approximation algorithms known for the concurrent unsplittable flow problem, see [2, 16, 3, 6, 12, 13].

4 Complexity and Non-approximability Results

In this section we mention NP-hardness results for k-splittable s-t-flow problem in directed graphs. Due to space limitations we omit the proofs.

Theorem 8. *It is NP-hard to approximate instances of the maximal 2-splittable s-t-flow problem with an approximation ratio strictly better than $2/3$.*

The proof is similar to and inspired by the proof of [13, Theorem 9 and Corollary 8]. It uses a reduction from the NP-complete SAT problem.

In Section 2.1 we have seen that the maximal uniform k-splittable s-t-flow problem can be solved in polynomial time. However, as soon as we allow small differences in the flow values on the paths, the problem becomes NP-hard again.

Corollary 2. *For arbitrary $\epsilon > 0$, it is NP-hard to compute a maximal 2-splittable s-t-flow, already if the flow values on the two paths may differ by a factor of no more than $1 + \epsilon$.*

On the other hand, the problem described in Corollary 2 is easier to approximate since a maximal uniform k-splittable s-t-flow is obviously a $(1 + \epsilon)$-approximate solution.

References

[1] Ahuja, R. K., Magnanti, T. L., Orlin, J. B.: Network Flows. Prentice-Hall (1993)
[2] Kleinberg, J.: Approximation Algorithms for Disjoint Paths Problems. PhD thesis, MIT (1996)
[3] Kleinberg, J.: Single-source unsplittable flow. In: Proceedings of the 37th Annual Symposium on Foundations of Computer Science. (1996) 68–77
[4] Baier, G., Köhler, E., Skutella, M.: On the k-splittable flow problem. Technical Report 739, Technische Universität Berlin (2002)
[5] Kolliopoulos, S. G., Stein, C.: Approximating disjoint-path problems using greedy algorithms and packing integer programs. In: Proceedings of the 6th Conference on Integer Programming and Combinatorial Optimization. (1998)
[6] Kolliopoulos, S. G., Stein, C.: Improved approximation algorithms for unsplittable flow problems. In: Procceedings of the 38th Annual Symposium on Foundations of Computer Science. (1997) 426–435
[7] Kolliopoulos, S. G.: Exact and Approximation Algorithms for Network Flow and Disjoint-Path Problems. PhD thesis, Dartmouth College (1998)
[8] Baveja, A., Srinivasan, A.: Approximation algorithms for disjoint paths and related routing and packing problems. Mathematics of Operations Research **25** (2000) 255–280
[9] Guruswami, V., Khanna, S., Rajaraman, R., Shepherd, B., Yannakakis, M.: Near-optimal hardness results and approximation algorithms for edge-disjoint paths and related problems. In: Proceedings of the thirty-first Annual ACM Symposium on Theory of Computing. (1999) 19–28
[10] Kolman, P., Scheideler, C.: Improved bounds for the unsplittable flow problem. In: 13th Annual ACM-SIAM Symposium on Discrete Algorithms. (2002) 184–193
[11] Azar, Y., Regev, O.: Strongly polynomial algorithms for the unsplittable flow problem. In: Proceedings of the 8th Conference on Integer Programming and Combinatorial Optimization. (2001) 15–29
[12] Dinitz, Y., Garg, N., Goemans, M. X.: On the single-source unsplittable flow problem. Combinatorica **19** (1999) 17–41
[13] Skutella, M.: Approximating the single source unsplittable min-cost flow problem. Mathematical Programming **91** (2002) 493–514

[14] Bagchi, A., Chaudhary, A., Scheideler, C., Kolman, P.: Algorithms for fault-tolerant routing in circuit switched networks. In: Fourteenth ACM Symposium on Parallel Algorithms and Architectures. (2002) To appear
[15] Krysta, P., Sanders, P., Vöcking, B.: Scheduling and traffic allocation for tasks with bounded splittability (2002) Manuscript
[16] Raghavan, P., Thompson, C.: Randomized rounding: A technique for provably good algorithms and algorithmic proofs. Combinatorica **7** (1987) 365–374

Partial Alphabetic Trees

Arye Barkan and Haim Kaplan

School of Computer Science, Faculty of Exact Sciences, Tel-Aviv University
Tel Aviv 69978, Israel
arye@oblicore.com
haimk@post.tau.ac.il

Abstract. In the partial alphabetic tree problem we are given a multiset of nonnegative weights $W = \{w_1, \ldots, w_n\}$, partitioned into $k \leq n$ blocks $B_1, \ldots, B_k$. We want to find a binary tree T where the elements of W resides in its leaves such that if we traverse the leaves from left to right then all leaves of B_i precede all leaves of B_j for every $i < j$. Furthermore among all such trees, T has to minimize $\sum_{i=1}^{n} w_i d(w_i)$, where $d(w_i)$ is the depth of w_i in T. The partial alphabetic tree problem generalizes the problem of finding a Huffman tree over W (there is only one block) and the problem of finding a minimum cost alphabetic tree over W (each block consists of a single item). This fundamental problem arises when we want to find an optimal search tree over a set of items which may have equal keys and when we want to find an optimal binary code for a set of items with known frequencies, such that we have a lexicographic restriction for some of the codewords.
Our main result is a pseudo-polynomial time algorithm that finds the optimal tree. Our algorithm runs in $O\left(\left(\frac{W_{sum}}{W_{min}}\right)^{2\alpha} \log(\frac{W_{sum}}{W_{min}})n^2\right)$ time where $W_{sum} = \sum_{i=1}^{n} w_i$, $W_{min} = \min_i w_i$, and $\alpha = \frac{1}{\log \phi} \approx 1.44$ [1]. In particular the running time is polynomial in case the weights are bounded by a polynomial of n. To bound the running time of our algorithm we prove an upper bound of $\lfloor \alpha \log(W_{sum}/W_{min}) + 1 \rfloor$ on the depth of the optimal tree.
Our algorithm relies on a solution to what we call the layered Huffman forest problem which is of independent interest. In the layered Huffman forest problem we are given an unordered multiset of weights $W = \{w_1, \ldots, w_n\}$, and a multiset of integers $D = \{d_1, \ldots, d_m\}$. We look for a forest F with m trees, $T_1, \ldots, T_m$, where the weights in W correspond to the leaves of F, that minimizes $\sum_{i=1}^{n} w_i d_F(w_i)$ where $d_F(w_i)$ is the depth of w_i in its tree plus d_j if $w_i \in T_j$. Our algorithm for this problem runs in $O(kn^2)$ time.

1 Introduction

Let T be a binary tree with n leaves each associated with a weight[2] $w_i \geq 0$. We define the *cost* of T to be $cost(T) = \sum_{i=1}^{n} w_i d_T(w_i)$ where $d_T(w_i)$ is the length

---

[1] $\phi = \frac{\sqrt{5}+1}{2} \approx 1.618$ is the golden ratio
[2] One can think of the leaves as containing items $\{v_1, \ldots, v_n\}$ where $w(v_i) = w_i$. To simplify the notation we identify the item v_i with its weight w_i.

of the path (number of edges) from the root to the leaf w_i. We call $d_T(w_i)$ the *depth* of w_i.

Given a sequence of n weights $w_1,\ldots,w_n$ the well known Huffman coding problem [9] looks for a binary tree of minimum cost whose leaves are associated with $w_1,\ldots,w_n$. A related and also well studied problem is the minimum cost alphabetic tree problem [6, 8]. In this problem we also look for a binary tree of minimum cost whose leaves are associated with $w_1,\ldots,w_n$ but this time we also require that w_i appears before w_j in an inorder traversal of the tree for every $i < j$.

In this paper we study what we call the *Partial Alphabetic Tree* (PAT) problem which is a natural generalization of both the Huffman problem and the minimal cost alphabetic tree problem. In the PAT problem we are given a set of weights $w_1,\ldots,w_n$ partitioned into k blocks $B_1,\ldots,B_k$, where each block can have a different size. We want to find a minimum cost binary tree whose leaves are associated with $w_1,\ldots,w_n$ such that w_i precedes w_j in an inorder traversal of the tree if $w_i \in B_\ell$ and $w_j \in B_{\ell'}$ for $\ell < \ell'$. In less technical terms, we want to find a minimum cost binary tree such that all items in B_i are "to the left" of all items in B_j for every $i < j$. The order of the items within each block is arbitrary. Note that the partition of weights into blocks defines a partial order, and the order of leaves in the resulting tree is consistent with that partial order.

The Huffman problem is a special case of the PAT problem where there is only one block. The minimal cost alphabetic tree problem is also a special case of the PAT problem when there are $k = n$ blocks, and block i contains the single element w_i.

For the optimal alphabetic tree problem there is a straightforward dynamic programming algorithm proposed by Gilbert and Moore [6]. This algorithm solves $O(n^2)$ subproblems where each such problem is of the form "find the optimal alphabetic tree for $w_i,\ldots,w_j$". In contrast, for the PAT problem there is no such straightforward approach even if one is willing to settle for a worse but polynomial running time. Since the items are not totally ordered the number of subproblems in the straightforward approach of Gilbert and Moore potentially becomes exponential. We overcome this difficulty by using a dynamic programming scheme in which subproblems encapsulate structural conditions on the boundaries of the target trees.

Our main result presented in this paper is an algorithm that solves the partial alphabetic tree problem in $O\left(\left(\frac{W_{sum}}{W_{min}}\right)^{2\alpha} \log\left(\frac{W_{sum}}{W_{min}}\right) n^2\right)$ time where $W_{sum} = \sum_{i=1}^n w_i$, and $W_{min} = \min_{i \in \{1,\ldots,n\}} w_i$, and $\alpha = \frac{1}{\log \phi} \approx 1.44$. In particular, our algorithm runs in polynomial time if the weights are bounded by a polynomial function of n. Our technique is general and can be used to find optimal partial alphabetic trees for several objective functions other than the average path length. The problem of whether there exists an algorithm for the partial alphabetic tree problem that runs in polynomial time for any set of weights remains an intriguing open question.

Our algorithm for the PAT problem relies on a solution to what we call the *layered Huffman forest* problem which is of independent interest. In the *lay-*

ered *Huffman forest* problem we are given a sequence of weights $w_1, \ldots, w_n$ and a sequence of *depths* $d_1, \ldots, d_m$. The goal is to find a forest F of m binary trees $T_1, ldots, T_m$ that minimizes $\sum_{i=1}^{n} w_i d_F(w_i)$ where $d_F(w_i) = d_{T_j}(w_i) + d_j$ if $w_i \in T_j$. We describe an algorithm for this problem that runs in $O(mn^2)$ time.

To bound the running time of our algorithm we have to bound the number of subproblems that we solve. Recall that our subproblems are defined based upon structural conditions on the boundaries of the target tree. Therefore to bound the number of such problems we prove an upper bound of $\lfloor \alpha \log(W_{sum}/W_{min}) + 1 \rfloor$ on the depth of the minimal-cost alphabetic tree for $w_1, \ldots, w_n$. We also show that this bound is tight.

Applications: Our immediate interest in the PAT problem stems from recent algorithms to label trees for ancestor and least common ancestor queries [1, 4, 10, 3, 14]. The need for such algorithms arises in XML search engines and internet routing. Recent algorithms for these problems first balance the tree by partitioning it into heavy-paths. Then they label the edges outgoing of the vertices of each heavy path such that labels of edges outgoing from deeper vertices get lexicographic larger labels. There is no restriction, however, on the relative order of labels of edges outgoing from the same vertex. Therefore, we get exactly an instance of the PAT problem for each such heavy path where block B_i contains the edges outgoing from the ith vertex on the path. The weights in these instances of the PAT problem are all bounded by n, where n is the number of nodes in the XML file.

Partial alphabetic trees are also useful when we want to code a set of items with known frequencies subject to an alphabetic restriction on some of the codewords.

An application of the layered Huffman forest problem, is finding an optimum variable length prefix encoding for some alphabet, where the codewords for some of the symbols in the alphabet are fixed in advance. Consider for example a scenario where you have different alphabets for different languages (German, Japanese, Russian, etc...). Each such alphabet contains symbols specific to the particular language and symbols common to all alphabets such as English letters, digits, and various punctuation symbols. To allow easy conversions of streams between alphabets and common applications that can work on the common parts of all alphabets one may want to encode the character sets such that common symbols have the same encodings across all alphabets. To encode a new alphabet (say Hebrew), we want to find codewords to the special symbols of the new set which are consistent with the (already assigned) codewords of the common symbols. When the frequencies of the special symbols are known in advance we can find an optimal encoding to the new characters by solving the layered Huffman forest problem.

Related work: The Huffman tree problem and the optimal alphabetic tree problem are two classical problems, which were defined about 50 years ago [9, 6], and have been heavily studied [2] since then. In 1952, Huffman [9] defined the well-known Huffman codes and suggested an algorithm for the Huffman problem that runs in $O(n \log n)$ time. About 7 years later, Gilbert and Moore [6] presented the minimum cost alphabetic tree problem, and suggested a dynamic programming algorithm to solve it in $O(n^3)$ time and $O(n^2)$ space. Later Knuth [13] improved the dynamic programming scheme of Gilbert and Moore and reduced its running time to $O(n^2)$. Hu and Tucker [8] gave a different algorithm, somewhat greedy in nature, that can be implemented to run in $O(n \log n)$ time. Their algorithm, though simple to describe, has a complicated correctness proof. Garcia and Wachs [5] describe another algorithm that solves the minimal cost alphabetic tree problem in $O(n \log n)$ time. The algorithm of Garcia and Wachs and the algorithm of Hu and Tucker are related [11].

Hu et al. [7] extended the approach of the Hu-Tucker algorithm to find optimal search trees for a wider range of objective functions different from the average access time. Also notice that the dynamic programming algorithms of Gilbert and Moore and Knuth extend to the related problem where items reside in internal nodes rather than the leaves. It is not clear however how to extend the faster algorithms of Hu and Tucker and of Garcia and Wachs to such a scenario. Our dynamic programming scheme does extend to other objective functions as well as the variant in which items reside at internal nodes.

The structure of the paper is as follows. Section 2 describes our main result, which is an algorithm that solves the PAT problem in polynomial time when the weights are bounded by a polynomial of n. Section 3 gives an upper bound, in terms of W_{sum}/W_{min}, on the depth of the minimal-cost alphabetic tree. Section 4 describes an algorithm to solve the layered Huffman forest problem. Section 5 defines some closely related problems and suggests directions for further research.

2 Finding an Optimal Partial Alphabetic Tree

Our algorithm for the PAT problem is based on dynamic programming. In order to describe its high level structure we need some notation regarding binary trees. We call the path from the root to the leftmost leaf in a binary tree T the *leftmost path* of T and the path from the root to the rightmost leaf the *rightmost path*. We shall consider binary trees in which every internal node has a left child and a right child except possibly nodes on the leftmost and rightmost paths. A node on the leftmost path may be missing its left child and a node on the rightmost path may be missing its right child. We call such a binary tree a *cut-tree*. We call a cut-tree in which no node on the rightmost path is missing a right child a *left cut-tree* and we call a cut-tree in which no node on the leftmost path is missing a left child a *right-cut-tree*.

The notion of a cut-tree is useful to us since we can build larger cut-trees by *concatenating* together smaller cut-trees. Specifically, let T_1 and T_2 be two cut-trees, and let $p_1 = <v_1, \ldots, v_\ell>$ be the path from the root of T_1 to the last node on its rightmost path that is missing a right child. Similarly let $p_2 = <u_1, \ldots, u_{\ell'}>$ be the path from the root of T_2 to the last node on its leftmost path that is missing a left child.[3] We can concatenate T_1 and T_2 to form a larger

---

[3] Path p_1 is empty if T_1 is a left cut-tree and path p_2 is empty if T_2 is a right cut-tree.

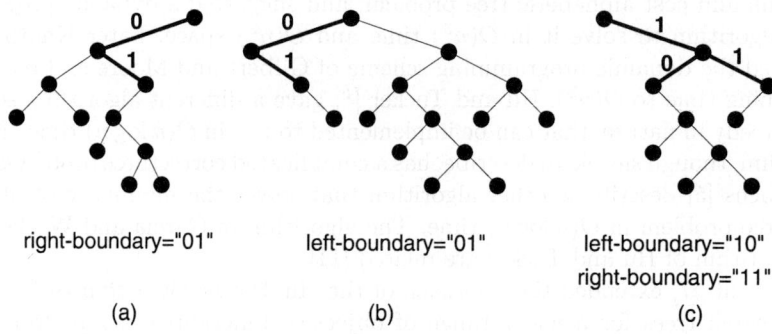

Fig. 1. (a) Only right-boundary (Right-cut-tree). (b) Only left-boundary (Left-cut-tree). (c) Both right and left boundaries (Cut-tree). Bold edges represent the boundaries

cut-tree T if $\ell = \ell' \geq 1$ and there exists $1 \leq j < \ell$ such that a) if $1 \leq i \leq j-1$ then v_i and u_i have only a right child or v_i and u_i have only a left child, b) if $j \leq i < \ell$ then v_i is missing a right child if and only if u_i has two children, c) v_ℓ has only a left child and u_ℓ has only a right child.

Let T be a cut-tree. Let p_1 be the path from the root of T to the last node missing a left child on the leftmost path. similarly, let p_2 be the path from the root to the last node missing a right child on the rightmost path. We will encode p_1 and p_2 by binary strings which we call the right boundary and the left boundary of T, respectively.[4] We denote these binary strings by $lb(T)$ and $rb(T)$, respectively. The string $lb(T)$ contain 0 at the ith position if the ith edge on p_1 is from a node to its left child and 1 otherwise. Similarly the string $rb(T)$ contain 0 at the ith position if the ith edge on p_2 is from a node to its left child and 1 otherwise. If p_1 is empty we say that $lb(T)$ does not exist, and similarly if p_2 is empty we say that $rb(T)$ does not exist. Note however that $lb(T)$ and $rb(T)$ can be empty strings when p_1 and p_2 contains only one node.

In particular note that we can *concatenate* T_1 and T_2 as defined above if $rb(T_1)$ and $lb(T_2)$ exists, and $rb(T_1) = lb(T_2)$. Note that when we concatenate a right cut-tree T_1 to a cut-tree T_2 we obtain a right cut-tree. When we concatenate a right cut-tree with a left cut-tree we obtain a full binary tree in which every node has two children. Also note that when we concatenate T_1 and T_2 then $rb(T_1)$ encodes some path from the root to some node in the resulting tree T. See figure 1.

We say that a cut tree T is a *partial alphabetic cut-tree* over $B_1, \ldots, B_i$ if its leaves corresponds to the items in $B_1 \cup B_2 \cup \cdots \cup B_i$ and they are ordered consistently with the partial order defined by $B_1, \ldots, B_i$. Our dynamic programming

---

[4] We will actually use the terms right boundary and left boundary to refer both to the paths and to their encodings, no confusion will arise.

algorithm is based on the following straightforward two lemma whose proof is omitted.

Lemma 1. *Let T_1 be a right cut-tree over $B_1, \ldots, B_{i-1}$ and let T_2 be a cut-tree over B_i, such that $rb(T_1) = lb(T_2)$. Then the concatenation of T_1 to T_2 is a right cut-tree T over $B_1, \ldots, B_i$ and $rb(T) = rb(T_2)$.*

We also have the following converse.

Lemma 2. *Let T be an optimal partial alphabetic right cut-tree over $B_1, \ldots, B_i$. There exists*

1. *A boundary b,*
2. *An optimal right cut-tree T_1 over $B_1, \ldots, B_{i-1}$ such that $rb(T_1) = b$,*
3. *An optimal cut-tree T_2 over B_k such that $lb(T_2) = b$ and $rb(T_2) = rb(T)$,*

and T is the concatenation of T_1 and T_2.

In particular we obtain as a corollary to Lemma 2 that if T is an optimal alphabetic tree over $B_1, \ldots, B_k$ then there exists a boundary b, an optimal right cut-tree T_1 over $B_1, \ldots, B_{k-1}$ with $rb(T_1) = b$, and an optimal left cut-tree T_2 over B_k with $lb(T_2) = b$, such that T is a concatenation of T_1 and T_2.

At a very high level our dynamic programming algorithm builds the optimal partial alphabetic tree by concatenating optimal cut-trees with particular boundaries over each of the blocks $B_1, \ldots, B_k$. Since the boundaries of these cut trees end up to be paths in the optimal alphabetic tree it suffices to consider cut-trees whose boundaries are shorter than the maximum depth of the optimal tree. We denote by h the maximum depth of an optimal tree, and we define a boundary to be *feasible* if its length is shorter than h.

Equipped with this notation we can now describe our dynamic programming scheme. For every prefix of the sequence of the blocks $B_1, \ldots, B_i$, $i \leq k - 1$ and every feasible boundary b we calculate an optimal partial alphabetic right cut-tree T_1 over $B_1, \ldots, B_i$ such that $rb(T_1) = b$. For every feasible boundary b we also calculate an optimal left cut-tree T_2 over B_k such that $lb(T_2) = b$. Finally, we find the optimal partial alphabetic tree over the whole sequence $B_1, \ldots, B_k$ by finding a boundary b where the cost of an optimal right cut-tree T_1 over $B_1, \ldots, B_{k-1}$ with $rb(T_1) = b$ plus the cost of the optimal left cut-tree T_2 over B_k with $lb(T_2) = b$ is minimized. By Lemma 1 and Lemma 2 the optimal tree T is the concatenation of T_1 and T_2.

We find the optimal right cut-tree over $B_1, \ldots, B_i$ with right boundary b by increasing values of i. We start by calculating the optimal right cut-tree over B_1 with right boundary b for every feasible right boundary. Assume that for every boundary b' we have already calculated the optimal right cut-tree $T_1^{b'}$ over $B_1, \ldots, B_{i-1}$ such that $rb(T_1^{b'}) = b'$. For every such b' we also calculate the optimal cut-tree $T_2^{b'}$ over B_i with $lb(T_2^{b'}) = b'$ and $rb(T_2^{b'}) = b$. Then we locate the boundary b' for which the sum of the costs of $T_1^{b'}$ and $T_2^{b'}$ is minimized. The concatenation of $T_1^{b'}$ and $T_2^{b'}$ is the optimal right cut-tree over $B_1, \ldots, B_i$

with right boundary b. The correctness of this computation also follows from Lemma 1 and Lemma 2.

The following easy lemma gives a characterization for which pairs of binary strings b_1 and b_2 there exists a cut-tree T with $lb(T) = b_1$ and $rb(T) = b_2$. We make use of it to limit the choices of possible boundaries b' for which a cut-tree T over B_i exists where $b' = lb(T)$ and $b = rb(T)$. To state the lemma we need the following definition. Let lb and rb be two binary strings. We say that lb *precedes* rb, and we write $lb <_p rb$ if one of the following three conditions holds: 1) lb is a prefix of rb and the first digit of rb which is not in lb is "1". 2) rb is a prefix of lb and the first digit of lb which is not in rb is "0". 3) The strings lb and rb are not prefix of one another, and $lb < rb$ lexicographically. The following lemma easily follows from our definitions.

Lemma 3. *(1) For every cut-tree T, which is not a right-cut-tree, and not a left-cut-tree, $lb(T) <_p rb(T)$ (2) For every two binary strings $lb <_p rb$, there exists a cut-tree T such that $lb(T) = lb$ and $rb(T) = rb$.*

Here is a high level pseudo-code for our algorithm.

Procedure PAT($B_1, \ldots, B_k$)
1. (basic subproblems)

   For every boundary b such that $|b| \leq h$ find an optimal right cut-tree over B_1 whose right boundary is b. Denote the cost of this tree by $RT(b, 1)$.

   For every pair of boundaries $lb <_p rb$ such that $|lb|, |rb| \leq h$, and $2 \leq i \leq k-1$ find an optimal cut-tree over B_i whose left boundary is lb and whose right boundary is rb. Denote the cost of this tree by $CT(lb, rb, i)$.

   For every boundary b such that $|b| \leq h$ find an optimal left cut-tree over B_k whose left boundary is b. Denote the cost of this tree by $LT(b, k)$.

2. Loop for every i from 2 to $k-1$
3.   for every binary string rb, $|rb| \leq h$
4.   Set $RT(rb, i) = \min_{c, |c| \leq h} \{RT(c, i-1) + CT(c, rb, i))\}$.
5.   The cost of the optimal tree is $\min_{c, |c| \leq h} \{RT(c, k-1) + LT(c, k)\}$.

It is easy to extend this algorithm so that it maintain an optimal tree for each subproblem in addition to its cost.

To complete the description of the algorithm we still have to show how to solve each of the basic problems of finding an optimal cut-tree with particular boundaries over a single block (Step 1). We do that in Section 4. The time bound of the algorithm that we describe there is $O(n^2 h)$.

To analyze the performance of the algorithm notice that the number of possible boundaries of length at most h is 2^{h+1}. Therefore we have $O(2^{2h}k)$ basic subproblems. If $n_i = |B_i|$ for $1 \leq i \leq k$ then the time it takes to obtain solution to all the basic subproblems is $O(2^{h+1}hn_1^2 + 2^{h+1}hn_k^2 + \sum_{i=2}^{k-1} 2^{h+1}2^{h+1}hn_i^2) = O(2^{2h}hn^2)$

The number of nonbasic subproblems over a prefix $B_1, \ldots, B_i$, where $2 \leq i \leq k-1$ is $O(2^h k)$. To solve each of these problems we need to take the minimum of a set of size $O(2^h)$ so it takes $O(2^{2h} k)$ time to solve all these subproblems.

To summarize we obtain that the running time of the algorithm is dominated by the time needed to solve the basic subproblems, that is $O(2^{2h}hn^2)$. Since at any one time it suffices to store the answers to the set of basic subproblems corresponding to a single block, and the set of answers to the nonbasic subproblems corresponding to a single prefix of blocks, it is possible to implement the algorithm using $O(2^{2h})$ space.

For this to be efficient we had to find an upper bound on h, which would be as small as possible. We prove in the next section a tight upper bound of $\lfloor \alpha \log(W_{sum}/W_{min}) \rfloor + 1$ on h, where $W_{sum} = \sum_{w \in W} w$, $W_{min} = \min_{w \in W} w$, and $\alpha = \frac{1}{\log \phi} \approx 1.44$. Plugging this upper bound into our algorithm we obtain the following theorem.

Theorem 1. *The algorithm described above solves the* PAT *problem in* $O\left(\left(\frac{W_{sum}}{W_{min}}\right)^{2\alpha} \log(\frac{W_{sum}}{W_{min}}) n^2\right)$ *time.*

3 Maximal Depth of Alphabetic Tree

In this section we show a bound on the depth of a minimal-cost alphabetic tree in terms of $s = \frac{W_{sum}}{W_{min}}$ where $W_{sum} = \sum_{w \in W} w$ and $W_{min} = \min_{w \in W} w$. Katona and Nemetz [12] showed a bound on the depth of a Huffman tree in terms of s. We show that the same bound holds for minimal-cost alphabetic trees. Since the bound of Katona and Nemetz [12] is tight for Huffman trees, it is also tight for alphabetic trees. The bound we show is: $\left\lfloor \frac{1}{\log \phi} \log s + 1 \right\rfloor \approx \lfloor 1.44 \log s + 1 \rfloor$ where $\phi = \frac{1+\sqrt{5}}{2} \approx 1.618$ is the golden ratio.

Let T be a binary tree, and let v be a node in T. We denote by $p(v)$ the parent of v, and by $uncle(v)$ the sibling of $p(v)$. The *weight* of an internal node v in T is the sum of the weights of the leaves in the subtree rooted by v. We denote the weight of v by $w(v)$. The proof of Katona and Nemetz [12], relies on the fact that for every internal node v in the Huffman tree, we have $w(v) \leq w(uncle(v))$. We show that the same is true for optimal alphabetic tree, and thus the bound of Katona and Nemetz [12] holds for optimal alphabetic trees as well.

To show that for every internal node v in the optimal alphabetic tree T we have $w(v) \leq w(uncle(v))$, we show a transformation of T that preserves the order of the leaves, and which decreases by one the depth of every node in the subtree rooted by v, and increases by one the depth of $uncle(v)$. Figure 2 shows this transformation for the case when v is to the left of $uncle(v)$. The case when v is to the right of $uncle(v)$ is symmetric. Now, if we suppose by contradiction that for some node v in T, $w(v) > w(uncle(v))$, then we can perform the transformation described above, and get another alphabetic tree with better cost, contradicting the optimality of T.

4 Layered Huffman Forests

In this section we show how to find an optimal cut-tree with boundaries lb and rb over a block $B_i = \{w_1, \ldots, w_s\}$. We use this algorithm to solve the basic

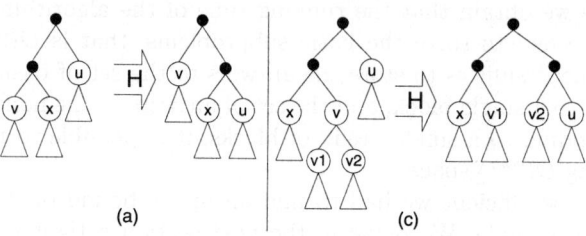

Fig. 2. The order preserving transformation

subproblems of the dynamic programming scheme in Section 2. We use the following mapping from cut-trees to forests of Huffman trees.

Let $F = \{T_1, \ldots, T_m\}$ be a forest of Huffman trees. We say that F is *over* $\{w_1, \ldots, w_s\}$ if there is a partition $S_1, \ldots, S_m$ of $\{w_1, \ldots, w_s\}$ such that T_i is a Huffman tree of S_i. We say that F is a *layered Huffman forest* if each tree T_i has a depth $d(T_i)$ associated with it. Let T be a cut-tree such that $lb(T) = lb$ and $rb(T) = rb$. The tree T defines a layered Huffman forest F of full binary trees $T_1, \ldots, T_m$ over $\{w_1, \ldots, w_s\}$ which we obtain by removing the nodes on the right boundary and the nodes on the left boundary of T. We define *the depth of* T_i to be the depth of the root of T_i in T. It is not hard to see that the number of trees in F and their depths is a function only of the left boundary and the right boundary of T. We denote the multiset of depths defined by lb and rb by $D(lb, rb)$. We define the cost of a binary layered Huffman forest over $\{w_1, \ldots, w_s\}$ to be $cost(F) = \sum_{i=1}^{s} d_F(w_i)w_i$, where $d_F(v) = d_{T_i}(v) + d(T_i)$, and $d_{T_i}(v)$ is the depth of v in T_i. With this definition the cost of a cut-tree T and the cost of the layered Huffman forest that it defines is the same.

This mapping from a cut-tree T with $lb(T) = lb$ and $rb(T) = rb$ to a layered Huffman forest over $\{w_1, \ldots, w_s\}$ with depths $D(lb, rb)$ of the same cost is reversible. Given a layered Huffman forest over $\{w_1, \ldots, w_s\}$ with depths $D(lb, rb)$ we can assemble a cut-tree with $lb(T) = lb$ and $rb(T) = rb$ as follows. We first construct the right boundary and the left boundary. Then we hang a tree of depth d over every boundary node of depth $d - 1$ that has two children. If the last node on the right boundary is of depth $d - 1$ and it does not belong to the left boundary then we make a subtree of rank d its left child. Similarly if the last node on the left boundary is of depth $d - 1$ and it does not belong to the right boundary then we make a subtree of rank d its right child. Since the forest is of depths $D(lb, rb)$ we have exactly the subtrees of the appropriate depths to hang. It is easy to see that the cost of the resulting cut-tree equals to the cost of the layered Huffman forest we started out with.

Using this cost preserving mapping from cut-trees to layered Huffman forests and vice versa we in fact reduced the problem of finding an optimal cut-tree T with $lb(T) = lb$ and $rb(T) = rb$ over $\{w_1, \ldots, w_s\}$ to the problem of finding an optimal layered Huffman forest with depths $D(lb, rb)$ over $\{w_1, \ldots, w_s\}$. Figure 3 illustrate this reduction.

We next show how to find an optimal layered Huffman forest with depths $D = \{d_1, \ldots, d_m\}$ over weights $W = \{w_1, \ldots, w_s\}$. We assume without loss of generality that $w_1 \leq w_2 \leq \cdots \leq w_s$, and that $d_1 \geq d_2 \geq \cdots \geq d_m$, and T_i is the tree of depth d_i. For every layered Huffman forest $F = \{T_1, \ldots, T_m\}$ over W, we define a total order $<_F$ on W as follows. For every two leaves $w_i, w_j \in W$, we say that $w_i <_F w_j$ if 1) w_i and w_j are in the same tree T_a, and w_i appears before w_j in an inorder traversal of T_a, or 2) w_i and w_j are in different trees T_a and T_b respectively, and $a < b$. The following lemma is the basis for a dynamic programming algorithm for finding an optimal layered Huffman forest. (The proof is easy and we omit it from this extended abstract.)

Lemma 4. *There exists a minimal cost layered Huffman forest F over $W = \{w_1, \ldots, w_s\}$ with depths $D = \{d_1, \ldots, d_m\}$, such that w_i is on the left of w_j only if $w_i \leq w_j$ (In other words $w_1 <_F w_2 <_F \cdots <_F w_s$).*

Lemma 4 indicates that we can find an optimal layered Huffman forest via straightforward dynamic programming approach. It follows from Lemma 4 that if F is an optimal forest over $w_1, \ldots, w_s$ with depths $d_1, \ldots, d_m$ then there exists a $m - 1 < s' \leq s$ such that T_m is a Huffman tree over $\{w_{s'}, \ldots w_s\}$ and $T_1, \ldots, T_{m-1}$ is an optimal layered Huffman forest over $w_1, \ldots, w_{s'-1}$ with depths $d_1, \ldots, d_{m-1}$. For every $\ell \leq s$ and $p \leq m$ our algorithm would find the cost of an optimal layered Huffman forest over $w_1, \ldots, w_\ell$ with depths $d_1, \ldots, d_p$. We denote this cost by $S(p, \ell)$. We compute $S(p, \ell)$ by increasing values of p using the following formula.

$$S(p, \ell) = \min_{\ell' < \ell} \{S(p-1, \ell') + Huffman(w_{\ell'+1}, \ldots, w_\ell) + d_p \sum_{i=\ell'+1}^{\ell} w_i$$

where $Huffman(w_{\ell'+1}, \ldots, w_\ell)$ is the cost of a Huffman tree over $w_{\ell'+1}, \ldots, w_\ell$. To initialize the algorithm, we set $S(1, \ell) = Huffman(w_1, \ldots, w_\ell) + d_1 \sum_{i=1}^{\ell} w_i$ for every $1 \leq \ell \leq s$.

We next analyze the worst-case complexity of this simple dynamic programming scheme. We have $m * s$ subproblems to solve. Solving each subproblem requires calculating the best of the s possibilities to split the items between the last tree and the other trees. To calculate the score of each split we need to solve a Huffman problem on a subsequence of the weights. Once we have the scores, the best split is the one of minimal score. In total we have $O(s^2)$ Huffman problems. If we presort the weights, we can solve each subproblem in $O(s)$ time. However, we can use a technique of Knuth [13] to solve all the $O(s^2)$ subproblems in $O(s^2)$ time.[5] Calculating the best (minimum score) split for each subproblem takes $O(s)$ time. So the total running time is dominated by the time to calculate these minima which is $O(ms^2)$ for all subproblems. It is not hard to see that the algorithm requires $O(s^2)$ space.

---

[5] The technique as presented by Knuth is for alphabetic trees. But since there is an Huffman tree in which the leaves are monotonically decreasing we can apply it in our context.

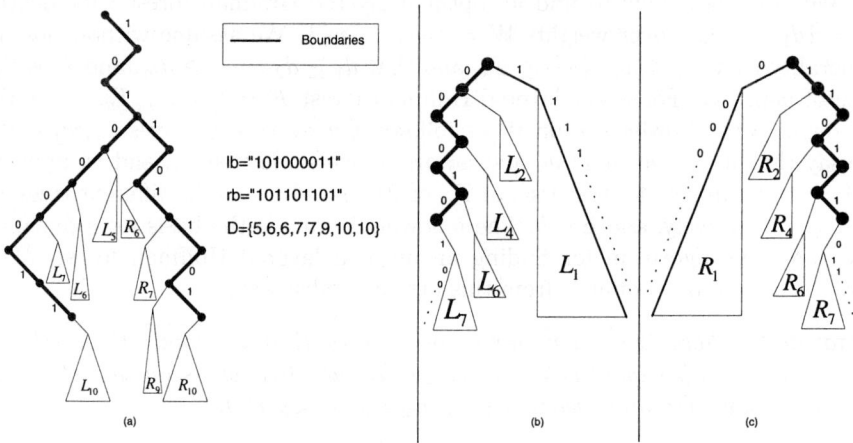

Fig. 3. Illustration of the reduction from optimal cut-tree problem to the optimal layered Huffman forest problem. (a) The case of a general cut-tree. (b) The case of a left-cut-tree. (c) The case of a right-cut-tree

Finally we get back to our optimal cut-tree problem with left boundary lb and right boundary rb. In every basic subproblem which we need to solve in the algorithm presented in Section 2 the set of items corresponds to a block so its cardinality is at most n. Furthermore, the lengths of lb and rb is at most h (the depth of an optimal partial alphabetic tree) and therefore the number of subtrees in the corresponding layered Huffman forest problem is $O(h)$. To conclude we obtain that the time it takes to solve each of the basic subproblems in the dynamic programming scheme of Section 2 is $O(n^2 h)$.

5 Concluding Remarks

A partial alphabetic tree over a set of weights W that is partitioned into block $B_1, \ldots, B_k$ is a binary tree with items in W residing in its leaves such that for every $w \in B_k$ and $w' \in B_{k'}$, such that $k < k'$, w is in a leaf which is to the left of the leaf of w'.

We describe a pseudo polynomial algorithm that finds an optimal partial alphabetic tree for a given set of weights partitioned into blocks. The tree is optimal in the sense that it minimizes the average path length over all partial alphabetic trees with weights W. Our technique is general and applies to several other objective functions as well.

We know of several special cases of the PAT for which a polynomial time algorithm exists. Two such special cases are the classical Huffman problem and the classical optimal alphabetic tree problem. Two other special cases that can be solved in polynomial time are the case when the size of every block is bounded by some constant c, and the case when only the leftmost and rightmost blocks contain more than one item. Perhaps the most intriguing problem left open by our work is whether a strongly polynomial time algorithm exists for the partial alphabetic tree problem.

The partial alphabetic tree problem generalizes nicely to arbitrary partial orders as follows. Let $W = \{w_1, \ldots, w_n\}$ be a multiset of nonnegative weights, and let $\prec$ be a partial order defined on W. Let $\mathcal{T}$ be the set of binary trees each having the weights in W as leaves such that if $w_i \prec w_j$ then w_i precedes w_j in an inorder traversal of the tree. The problem is to find a tree $T \in \mathcal{T}$ that minimizes $\sum_{i=1}^{n} w_i d(w_i)$, where $d(w_i)$ is the depth of w_i in T. The complexity status of this problem is also an open problem.

References

[1] S. Abiteboul, H. Kaplan, and T. Milo. Compact labeling schemes for ancestor queries. In 12th Annual ACM Symposium on Discrete Algorithms (SODA), pages 547–556, 2001.

[2] J. Abrahams. Code and parse trees for lossless source encoding, 1997.

[3] S. Alstrup, C. Gavoille, H. Kaplan, and T. Rauhe. Identifying nearest common ancestors in a distributed environment. unpublished manuscript, july 2001.

[4] S. Alstrup and T. Rauhe. Improved labeling scheme for ancestor queries. In *Proc. ACM-SIAM Symposium on Discrete Algorithms (SODA)*, (to appear) January 2002.

[5] A. M. Garsia and M. L. Wachs. A new algorithm for minimum cost binary trees. *SIAM J. Comput*, 6(4):622–642, Dec 1977.

[6] E. N. Gilbert and E. F. Moore. Variable-length binary encodings. *Bell System Technical Journal*, 38:933–967, July 1959.

[7] T. C. Hu, D. J. Kleitman, and J. K. Tamaki. Binary trees optimum under various criteria. *SIAM J. Appl. Math.*, 37(2):514–532, October 1979.

[8] T. C. Hu and C. Tacker. Optimum computer search trees. *SIAM J. Appl. Math.*, 21:514–532, 1971.

[9] D. A. Huffman. A method for the construction of minimum-redundancy codes. In *Proc IRE, 40*, pages 1098–1101, 1952.

[10] H. Kaplan, T. Milo, and R. Shabo. A comparison of labeling schemes for ancestor queries. In *Proc. ACM-SIAM Symposium on Discrete Algorithms (SODA)*, (to appear) January 2002.

[11] Marek Karpinski, Lawrence L. Larmore, and Wojciech Rytter. Correctness of constructing optimal alphabetic trees revisited. *Theoretical Computer Science*, 180(1-2):309–324, June 1997.

[12] G. O. H. Katona and T. O. H. Nemetz. Huffman codes and self information. *IEEE Trans. on Information Theory*, IT-22(3):337–340, May 1976.

[13] D. E. Knuth. Optimum binary search trees. *Acta Infomatica*, 1:14–25, 1971.

[14] M. Thorup and U. Zwick. Compact routing schemes. In 13th *ACM Symposium on Parallel Algorithms and Architectures (SPAA)*, 2001.

Classical and Contemporary Shortest Path Problems in Road Networks: Implementation and Experimental Analysis of the TRANSIMS Router

Chris Barrett[1], Keith Bisset[1], Riko Jacob[2],
Goran Konjevod[3], and Madhav Marathe[1]

[1] Los Alamos National Laboratory, P.O. Box 1663, MS M997, Los Alamos, NM
{barrett,bisset,marathe}@lanl.gov
[2] BRICS, Department of Computer Science, University of Aarhus, Denmark
rjacob@brics.dk
[3] Department of Computer Science, Arizona State University, Tempe, AZ
goran@asu.edu

Abstract. We describe and analyze empirically an implementation of some generalizations of Dijkstra's algorithm for shortest paths in graphs. The implementation formed a part of the TRANSIMS project at the Los Alamos National Laboratory. Besides offering the first implementation of the shortest path algorithm with regular language constraints, our code also solves problems with time-dependent edge delays in a quite general first-in-first-out model.

We describe some details of our implementation and then analyze the behavior of the algorithm on real but extremely large transportation networks. Even though the questions we consider in our experiments are fundamental and natural, it appears that they have not been carefully examined before. A methodological contribution of the present work is the use of formal statistical methods to analyze the behaviour of our algorithms. Although the statistical methods employed are simple, they provide a possibly novel approach to the experimental analysis of algorithms.

Our results provide evidence for our claims of efficiency of the algorithms described in a very practical setting.

1 Introduction

TRANSIMS is a multi-year project at the Los Alamos National Laboratory funded by the Department of Transportation and by the Environmental Protection Agency. Its purpose is to develop models and methods to answer planning questions, such as the economic and social impact of building new roads in a large metropolitan area. We refer the reader to [TR+95a] and the website http://transims.tsasa.lanl.gov for more extensive descriptions of the TRANSIMS project.

The basic purpose of the TRANSIMS module we describe (the *route planner*) is to use activity information generated earlier from demographic data to determine the optimal mode choices and travel routes for each individual traveler. The routes need to be computed for a large number of travelers (in the Portland case study 5–10 million trips). After planning, the routes are executed by a microsimulation that places the travelers (in vehicles and on foot) in the network and simulates their behavior. In order to remove the forward causality artificially introduced by this design, and with the goal of bringing the system to a "relaxed" state, TRANSIMS uses a feedback mechanism: the link delays observed in the microsimulation are used by the route planner to repeatedly re-plan a fraction of the travelers.

Clearly, this requires high computational throughput. The high level of detail in planning and the efficiency demand are both important design goals; methods to achieve reasonable performance are well known if only one of the goals needs to be satisfied. Here, we propose a framework that uses two independent extensions of the basic shortest path problem to simultaneously cope with both.

1.1 Shortest Paths

In the first part of this paper, we describe our implementation of a generalized Dijkstra's shortest path algorithm. The general problem we solve is that of finding regular-language-constrained shortest paths [BJM98] in graphs with time-dependent edge delays with a first-in-first-out assumption. Several authors have studied special cases (such as traffic-light networks or special cases of the regular language constraint), but these studies seem to be isolated and largely independent of a larger real-life system or application. As far as we are aware, ours is the first implementation scalable to problems with millions of vertices and edges that can find shortest-path problems in the presence of **both** language constraints and time dependence.

The various models and practical settings, especially in the context of multi-modal urban transportation systems, are discussed in a companion paper [BB+02].

1.2 Formal Language Constraints

Consider a pedestrian bridge across a river or a highway. Suppose we are asked to find a shortest path between some two points in the network. The bridge can only be used by pedestrians, so we must take care not to route cars across it. Similarly, we should not use highways as parts of the routes planned for pedestrians or bicyclists. In order not to have to update the network for every single routing question, we annotate the network with information needed to deal with these problems. More precisely to each edge and/or vertex of the network, we assign a label $\ell \in \Sigma$. We also refer to the finite set Σ as the *alphabet*. We call such labels *modes* and say that a labeled network is *multimodal*.

By concatenation, the edge and/or vertex labeling extends to walks. The resulting string of labels is called the *label* of the walk. This walk-label determines whether or not the walk is acceptable as a particular traveler's itinerary. We usually refer to walks in the network as paths; in other words, we allow our paths to repeat edges and/or vertices and instead use the term *simple path* to denote paths that do not repeat edges or vertices.

The problem of finding a shortest path subject to a formal-language constraint can be stated as follows: given a finite set (*language*) $L \subseteq \Sigma^*$ over the alphabet Σ, a source node s and a destination node d, find a shortest path p from s to d whose label belongs to L. Our results in [BJM98] prove that this problem can be solved in polynomial time.

Regular languages as models for shortest-path problems were also suggested independently by Romeuf [Rom88], Yannakakis [Ya90] and Mendelzon and Wood [MW95]. For more details on the theoretical background, refer to [BJM98].

1.3 Time Dependence

Finding optimal paths in time-dependent networks is an important problem [Ch97a, ZM95]. Unless restrictive assumptions are made, time-dependence of delays implies NP-hardness [OR90]. A natural assumption is that the traffic on each link has the first-in-first-out property. Our model, using *piecewise-linear delay functions*, is a natural implementation of this assumption. It has been rediscovered independently at least once more —by Sung et al. [SB+00]—but the full power of this model for various problems arising in transportation science is not evident from their paper.

We argue that this model is (1) adequate for the rapidly changing conditions on roadways and (2) flexible enough to describe more complicated scenarios such as scheduled transit and time-window constraints but also (3) allows computationally efficient algorithms. Several general versions of this problem can be solved efficiently in our framework (more details in the companion paper on models for transportation problems [BB+02]).

1.4 Experimental Analysis

In the second part of this paper we describe experimental results. We ran our algorithm on a set of shortest path instances similar to trips that typical urban travelers take each day. Origin-destination pairs were placed in the street network of Portland, Oregon, and a total of 280000 shortest paths found for several categories of travelers and trip types. This allowed us to analyze the behavior of the algorithm on a typical set of problem instances generated in the TRANSIMS framework.

The TRANSIMS network for Portland has been divided into approximately 1200 traffic analysis zones (TAZs). From these a distance matrix was created from the Euclidean distance between each pair of TAZs. Source and destination TAZ pairs were selected from this matrix so that the distance between the source and destination ranged from 1000 to 50000 meters (±10%) in increments of 500 meters, with 50 trips of each size selected. For each TAZ pair, starting and ending points were randomly selected from within the given TAZ, producing 5000 trips.

Through the experiments, we attempt to infer the scalability of our methods and empirical improvements obtained by augmenting the basic algorithm with heuristic methods.

For example, we show the power and limits of the Sedgewick-Vitter heuristic when applied to a not strictly Euclidean graph. We see that the heuristic is only effective up to a certain point, and study its running time in comparison with ordinary Dijkstra's algorithm. We find that the running time of the heuristic is statistically linear in the number of edges of the solution path (thus, a constant fraction of vertices examined ends up in the solution path). The exact algorithm, on the other hand, seems to follow a logistic response function. We believe this to be just an artifact of our experiment design (the correct answer having the form $k \log k$) and we will need to do more experiments to complete this study.

1.5 Discussion

Even though our questions are fundamental, it appears that they have not been examined before. In order to experimentally analyze the behavior of the algorithms in realistic settings we employ simple statistical techniques (e.g. ANOVA), that have to our knowledge not been used earlier in the field of experimental algorithmics. In our opinion, formal statistical techniques such as ANOVA and experimental design may provide an excellent tool for empirical analysis of algorithms. We have recently used similar tools in analyses of certain flow algorithms and interaction of communication protocols [MM+02, BCF+01].

This research should be viewed as experimental analysis of a well-known algorithm and its generalizations and variants in a realistic setting. No references we have been able to locate cover more than a small part of our theoretical framework, and the situation is similar with actual implementations. To the best of our knowledge TRANSIMS contains the first unified implementation of these results and it is both reasonably complete from a theoretician's point of view and useful in practice. Nevertheless, we argue that our algorithm and conclusions on the implementation are not TRANSIMS-specific, but applicable to a number of other realistic transportation problems.

2 Implementation of the TRANSIMS Router

2.1 Algorithm for Linear Regular Expressions

First, some (standard) notation: w^+ denotes one or more repetitions of a word (string) w, $x + y$ denotes either x or y, Σ typically denotes the alphabet, that is the set of all available symbols.

TRANSIMS currently supports *linear* (or *simple-path*) regular expressions. These are the expressions of the form $x_1^+ x_2^+ \cdots x_k^+$, where $x_i \in \Sigma \cup (\Sigma + \Sigma)$.

Algorithm 1 *Input:* A linear regular expression R (as the string $R[0 \ldots |R|-1])$, a directed edge-labeled weighted graph G, vertices s and $d \in V(G)$. *Output:* A minimum-weight path p^* in G from s to d such that $l(p^*) \in R$.

Conceptually, the algorithm consists of running Dijkstra's algorithm on the direct product of G and the finite automaton $M(R)$ representing R.

For efficiency, we do not explicitly construct $G \times M(R)$, but concatenate the identifier of each vertex of G with the identifier of the appropriate vertex in $M(R)$. In other words, we run Dijkstra's shortest-path algorithm on G with the following changes: each vertex is referred to by the pair consisting of its index in G and an integer $0 \leq a \leq |R| - 1$ denoting the location within R.

In the first step, $a = 0$ and the only "explored" vertex is $(s, 0)$. In each subsequent exploration step of Dijkstra's algorithm, consider only the edges e leaving the current vertex (v, a) with $l(e) = R[a]$ or $l(e) = R[a + 1]$. If an edge $e = vw$ with $l(e) = R[a+1]$ is explored, then the vertex reached will be $(w, a+1)$. Otherwise the vertex reached is (w, a). The algorithm halts when it reaches the vertex $(d, |R| - 1)$. □

Theorem 2. *Algorithm 1 computes the shortest R-constrained path in G (with nonnegative edge-weights) in time $O(T(|R||G|))$, where $T(n)$ denotes the running time of a shortest-path algorithm on a graph with n nodes.*

2.2 Time-Dependent Delays

To get a class of functions that is flexible enough to model various applications, but computationally feasible, we use **monotonic piecewise-linear (MPL)** functions. Among other properties, they allow fast lookup of values and this is important, being a part of the innermost loop of the algorithm.

We represent MPL functions using a sorted set of pairs that can be searched for both x and (linearly interpolated) y values. For functions that do not need to be modified frequently (for example, a real TRANSIMS situation in which the traversal functions are only modified after planning a substantial number of travelers), an implementation using arrays and binary search performs very well.

For details, see our modeling paper [BB+02].

2.3 Data Structures and Network Representation

The network used by the route planner is substantially different from the TRANSIMS network used by the microsimulation and some other modules of the system. The reason for extra work is the improved efficiency achieved by streamlining the description and removing the features ignored by the route planner and reorganizing some others. A schematic drawing of the network used by the planner is given in Figure 1.

Priority Queue. Dijkstra's algorithm requires the implementation of a priority queue in order to maintain the fringe vertices amongh which it chooses the next vertex to explore. We used the simple binary heap. Due to a relatively regular structure of graphs involved, the size of the heap never became too large to allow significant improvements by using a more complex data structure.

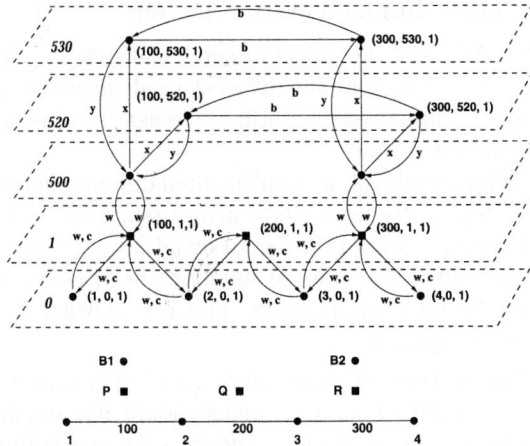

Fig. 1. The network as seen by the route planner. The underlying street network is shown in the bottom of the figure, with nodes 1, 2, 3 and 4 and street segments connecting them. Nodes P, Q and R represent parking locations (present on most links). Associated with the links joining 1 with 2 and 3 with four are also bus stops B1 and B2. The internal network constructed by the planner is shown in the upper portion of the figure, with "layers" separated for clarity. Layers 0 and 1 describe the street network with links that can be traversed by drivers (mode c) and pedestrians (mode w). Nodes in layer 0 are intersections, those in layer 1 parking locations. Bus stops are located in layer 500 and each bus stop is associated with a parking location (and only accessible from it). To actually board a bus, however, a traveler must walk from a bus stop to a node associated with a specific bus line (layer $500 + x$ for bus line x). This forces walking to transfer between buses and helps avoid some modeling artifacts

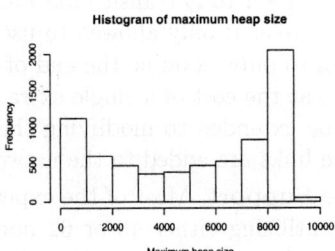

Fig. 2. The maximum heap size over 30000 car trips planned. The heap size never exceeds 10000 vertices, so the standard binary heap is still quite efficient

2.4 Some Low-Level Implementation Details

Software Design. We used the object oriented features as well as the template mechanism of C++ to easily combine different implementations. As we did not want to introduce any unnecessary runtime overhead, we avoided for example the concept of virtual inheritance.

For example, using templates, we implemented different versions of Dijkstra's algorithm for some frequently used mode strings (such as "wcw" for simple car plans and "t" for transit plans) in order to avoid all overhead associated with checking whether certain links may be used by the traveler and most of the overhead of the regular-language Dijkstra. This is somewhat ironic, considering our claim of a unified algorithm.

Compile-Time Optimization. A simpler algorithm suffices for some types of plans. For example, all-car or all-walk plans should not require the overhead of examining the finite automaton that specifies the constraint language, and may be planned more efficiently if only a part of the network is examined. In addition to the network modification trick to be described, for such cases we use a separately optimized and compiled procedure implemented using the C++ template mechanism.

Thus in fact we are able to implement several variants of the algorithm and transparently vary the underlying network independently so that the appropriate algorithm and network can be chosen and run on a traveler-by-traveler basis with no overhead, we just disguise them as one using C++ templates. In the case where a simple mode is used by a large proportion of travelers, the implementation and program size overhead is small compared to the savings.

Implicit vs. Explicit Network Modification. As described earlier, the network consists of *layers* representing *car*, *walk* and *transit* links, with walk links also crossing between car and transit layers. It is possible to order the creation of edges in the network so that edges with a fixed label form a consecutive interval in the adjacency list of each vertex. Thus for example, edges numbered 0 to i_1 will be car links, those from $i_1 + 1$ to i_2 transit links and those from $i_2 + 1$ to i_3 transit links. Then if the traveler is only allowed to use walk and transit links, we ask Dijkstra's algorithm to only examine the end of each adjacency list and ignore car links completely, at the cost of a single extra table lookup per vertex examined. This trick can be extended to modifying the adjacency list in more general ways as long as the links are added to the network in a sensible order.

Hardware and Software Support. Most of the experiments were performed on an MPP Linux cluster utilizing either 46 or 62 nodes with Dual 500 Mhz Pentium II processors in each node. Each node had 1 Gb of main memory. Many of the experiments were done by executing independent shortest paths runs at each node. For this we had to create copies of the network. Fortunately the network fits in just under 1 Gb of memory and thus did not cause problems. On each node, the route planner was run using 2 routing threads and 1 output thread.

Parallelization. The implementation may use multiple threads running in parallel and it may also be distributed across multiple machines using MPI. Threads enable the parallel execution of several copies of the path-finding algorithm on a shared-memory machine. Each thread uses the same copy of the network. Because separate threads are used for reading, writing and planning, improvements in the running time may be observed even with a single-processor machine.

3 The Sedgewick-Vitter Heuristic

One of the additional optimizations we've used for all-car plans is the Sedgewick-Vitter [SV86] heuristic for Euclidean shortest paths that biases the search in the direction of the source-destination vector.

We can modify the basic Dijkstra's algorithm by giving an appropriate weight to the distance from x to t. By choosing an appropriate multiplicative factor, we can increase the contribution of the second component in calculating the label of a vertex. From a intuitive standpoint this corresponds to giving the destination a high potential, in effect biasing the search towards the destination. This modification will in general **not** yield shortest paths, nevertheless our experimental results suggest that the errors produced can be kept reasonably small. This multiplicative factor is called the *overdo* parameter.

4 A First Look at the Data

As mentioned, the experiments were carried out on a multi-modal transportation network spanning the city of Portland. The network representation is very detailed and contains *all* the streets in Portland. In fact, the data also specifies the lanes, grade, pocket/turn lanes, etc. Much of this was not required in the route planner module, but most of it was used by at least some parts of TRANSIMS (usually the microsimulation or the emissions analyzer).

Types	Street	Parking	Activity	Bus+Rail	Route
Vertices	100511	121503	243423	9771+56	30874
Edges	249222	722745	2285594	55676	30249

In the basic TRANSIMS network, there are a total of 475 264 external nodes and 650 994 external links. The internal network thus grows to over three million edges (see Section 2.3).

Measured Quantities. We base our results on measurements and counts of the following quantities: **cpu**: running time used for finding the shortest path (no i/o), **nodes**: number of nodes on the path found by the algorithm, **hadd**: number of nodes added to the heap during the execution, **max**: maximum size of the heap during the execution, **touched**: total number of nodes touched (a node may be counted multiple times here), **unique**: number of unique nodes touched, **edist**: Euclidean (straight line) distance between the origin and destination, **time**: time to traverse the path found by the algorithm,

In addition, each observation can be categorized according to its *mode* (walk, auto, transit, light rail, park-and-ride, bus), *overdo factor* (strength of bias when/if using the Sedgewick-Vitter heuristic—0 (none), 0.15, 0.25, 0.5), *delay*

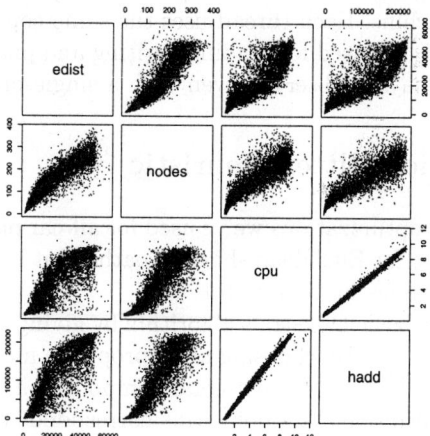

Fig. 3. Plots of Euclidean origin-destination distance, trip duration, running time and total number of nodes added to the heap. The linear relation between the running time and the number of nodes added to the heap during the execution is obvious from the plot and also very clear from the algorithm statement (as long as the time for individual heap operations does not vary too much). We do not consider the Euclidean distance further in this extended abstract

(for car trips— free-speed link delays, or those produced by feedback from the microsimulation after 7 or 24 iterations).

Figure 3 illustrates the relationships between some parameters that are easy to understand. In the later sections, we focus on a few that are not completely obvious. Due to lack of space, our discussion focuses on car trips and on free-speed link delays. We will have more to say (most importantly about the distinctions between various modes and about the adjustments to link delays produced by the microsimulation feedback) in the full version of the paper.

4.1 Varying the Sedgewick-Vitter Bias

We now take a look at the results obtained by setting different values of the bias parameter (overdo) in the Sedgewick-Vitter heuristic. To summarize briefly, it appears that a value of more than 0.15 is not very useful, as it gives only a marginal improvement in the running time, whereas the path quality continues to decrease. However, as we increase the overdo parameter from 0 to 0.15 the running time improves quite substantially.

	Estimate	Std. Error	t value	$P[>|t|]$
Intercept	0.8247292	0.0043564	189.32	$< 2 \cdot 10^{-16}$
Slope	0.0002389	0.0000197	12.13	$< 2 \cdot 10^{-16}$

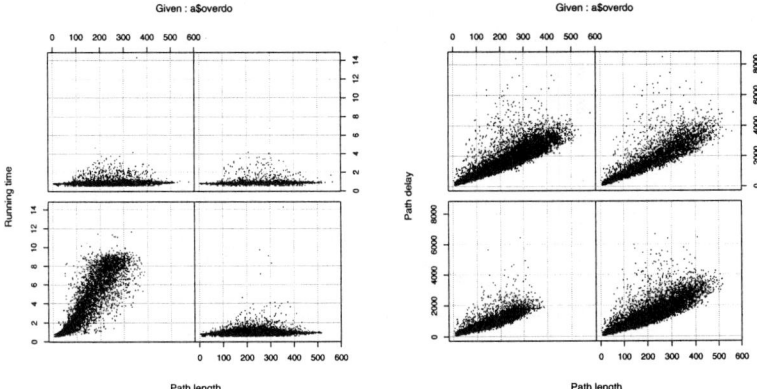

Fig. 4. Running time and path delay against the number of nodes on the path for four different values of overdo (0 0.15,0.25,0.5). The figures correspond to increasing overdo parameters and are to be read row wise and bottom up. Note the steep decrease in running time as we go from overdo = 0 to overdo = .15; the improvement is small after that. We will take a closer look at the horn-shape with overdo= 0, but note that the running time appears linear in the other three plots. In fact, it seems *constant*, but a linear fit shows a very low slope, as is to be expected. However, the quality of the paths output continues to decay with increased overdo

The quality of paths found continues to worsen as the bias parameter is increased, without a corresponding improvement in the running time. Thus we must conclude that it is not useful to push overdo beyond 0.15. We still need to investigate the values between 0 and 0.15 to find the best tradeoff.

The reason why the running time can be expected to be linear in the length of the path produced when running the Sedgewick-Vitter heuristic is precisely because of the bias: instead of performing the depth-first-search and expanding equally in all directions (where the number of nodes examined for example in a grid would be proportional to the square of the path length), the search expands primarily in the single direction towards the destination. Note that this is even a stronger claim than the theoretical result applicable to graphs with Euclidean distance functions, which says that the running time is linear in the size of the graph. However, there are some caveats we should be aware of. By studying the plot for overdo= 0.15 (bottom right in Figure 4), we see that only the lower envelope of the data set is a straight line. The upper envelope is not. These points correspond to the cases where the bias led the algorithm astray, for example where the direct geometric route led to the river bank, hoping to get across but not finding a bridge in the vicinity.

An interesting phenomenon is the similarity of running times with the overdo set to 0.15, 0.25 and 0.5. The average running times are practically equal, and

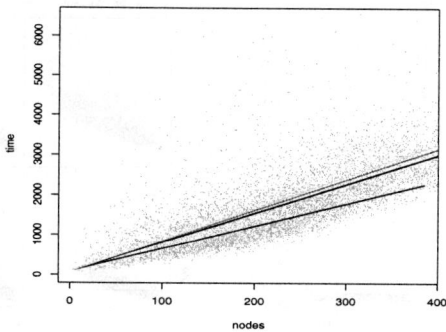

Fig. 5. Delay (quality) of paths produced for various values of the bias parameter. The color scheme for the points and the fitted lines is as follows: red/pink: 0, blue/light blue 0.15, green/light green 0.25, yellow, light yellow 0.5

a formal analysis of variance (ANOVA) test shows that we should not reject the hypothesis of equality of the running times with the bias at 0.15 and 0.25. (Admittedly, it is not at all clear that all the assumptions necessary to validate the test are satisfied.)

4.2 Nonlinear Dependence of Running Time on Path Length

Finally, let us take a closer look at the dependence of running time on the path length in the case of exact Dijkstra's algorithm (no Sedgewick-Vitter heuristic). The best fit for the data appears to be a logistic response curve of the form $y = \frac{a}{1+e^{\frac{b-x}{c}}}$.

How to explain this behavior? Consider a regular two-dimensional grid (a reasonable approximation to the dense street network of a city). If the path length from the origin to the destination is k, then there are roughly k^2 nodes to be examined before the destination is reached. Thus, we may expect that the running time grows as a square of the path length. However, in a finite grid, if the explored area reaches a boundary, the number of nodes examined must grow slower simply because there are no more nodes to be examined. Thus, after a certain path length is achieved without finding the destination, the growth of the running time should slow down. Indeed, the point of inflection on the logistic curve is just under half the maximum path length among our planned trips, and intuitively, this is where the curve tapers off—when the algorithm runs out of new useless nodes to explore.

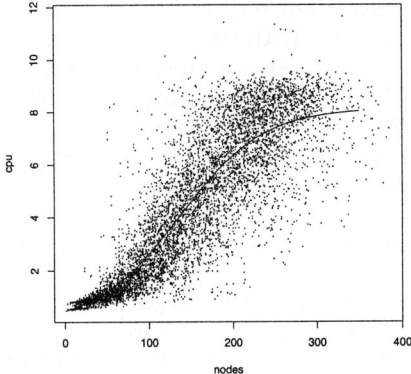

Fig. 6. Running time plotted against the number of nodes on the path produced by the algorithm for overdo= 0. The parameters were estimated at $a = 8.14297$, $b = 137.083$ and $c = 49.629$. Notice that the point of inflection is a little under half the maximum path length.

Acknowledgements

The research was performed under contract from Department of Transportation and the Environmental Protection Agency and also supported by Department of Energy under Contract W-7405-ENG-36. Riko Jacob and Goran Konjevod worked on this project while visiting Los Alamos National Laboratory. We thank the members of the TRANSIMS team in particular, Richard Beckman, Kathy Birkbigler, Brian Bush, Stephan Eubank, Deborah Kubicek, Kristie Henson, Phil Romero, Jim Smith, Ron Smith, Paula Stretz, for providing the software infrastructure, pointers to related literature and numerous discussions on topics related to the subject. This paper would not have been possible without their support. We also thank David Bernstein, Joseph Cheriyan, Terrence Kelly, Ana Maria Kupresanin, Alberto Mendelzon, Kai Nagel, S.S. Ravi, Daniel Rosenkrantz, Prabhakar Ragde, R. Ravi and Aravind Srinivasan for constructive comments and pointers to related literature.

References

[BB+02] C. Barrett, K. Bisset, R. Jacob, G. Konjevod and M. Marathe *Algorithms and models for routing and transportation problems in in time-dependent and labeled networks*, in preparation, 2002.

[TR+95a] C. Barrett, K. Birkbigler, L. Smith, V. Loose, R. Beckman, J. Davis, D. Roberts and M. Williams, *An Operational Description of TRANSIMS*, Technical Report, LA-UR-95-2393, Los Alamos National Laboratory, 1995.

[BCF+01] C. Barrett, D. Cook, V. Faber, G. Hicks, A. Marathe, M. Marathe, A. Srinivasan, Y. J. Sussmann and H. Thornquist, *Experimental analysis of algorithms for bilateral-contract clearing mechanisms arising in deregulated power industry*, Proc. WAE 2001, pp. 172–184.

[BJM98] C. Barrett, R. Jacob, M. Marathe, *Formal Language Constrained Path Problems* in SIAM J. Computing, 30(3), pp. 809-837, June 2001.

[Ch97a] I. Chabini, *Discrete Dynamic Shortest Path Problems in Transportation Applications: Complexity and Algorithms with Optimal Run Time,* Presented at 1997 Transportation Research Board Meeting.

[CGR96] B. Cherkassky, A. Goldberg and T. Radzik, *Shortest Path algorithms: Theory and Experimental Evaluation*, Mathematical Programming, Vol. 73, 1996, pp. 129–174.

[JMN99] R. Jacob, M. Marathe and K. Nagel, *A Computational Study of Routing Algorithms for Realistic Transportation Networks,* invited paper appears in ACM J. Experimental Algorithmics, 4, Article 6, 1999. http://www.jea.acm.org/1999/JacobRouting/ Preliminary version appeared in Proc. 2nd Workshop on Algorithmic Engineering, Saarbrucken, Germany, August 1998.

[MM+02] C. L. Barrett, M. Drozda, A. Marathe, and M. Marathe, *Characterizing the interaction between routing and MAC protocols in ad-hoc networks*, to appear in Proc. ACM MobiHoc 2002.

[MW95] A. Mendelzon and P. Wood, *Finding Regular Simple Paths in Graph Databases*, SIAM J. Computing, vol. 24, No. 6, 1995, pp. 1235-1258.

[OR90] A. Orda and R. Rom, *Shortest Path and Minimum Delay Algorithms in Networks with Time Dependent Edge Lengths*, J. ACM, Vol. 37, No. 3, 1990, pp. 607-625.

[Rom88] J. F. Romeuf, *Shortest Path under Rational Constraint* Information Processing Letters 28 (1988), pp. 245-248.

[SV86] R. Sedgewick and J. Vitter *Shortest Paths in Euclidean Graphs*, Algorithmica, 1986, Vol. 1, No. 1, pp. 31-48.

[SB+00] K. Sung and M. G. H. Bell and M. Seong and S. Park, *Shortest paths in a network with time-dependent flow speeds*, European Journal of Operational Research 121 (1) (2000), pp. 32–39.

[Ya90] M. Yannakakis "Graph Theoretic Methods in Data Base Theory," invited talk, *Proc. 9th ACM SIGACT-SIGMOD-SIGART Symposium on Database Systems (ACM-PODS)*, Nashville TN, 1990, pp. 230-242.

[ZM95] A. Ziliaskopoulos and H. Mahmassani, *Minimum Path Algorithms for Networks with General Time Dependent Arc Costs*, Technical Report, December 1997.

Scanning and Traversing: Maintaining Data for Traversals in a Memory Hierarchy

Michael A. Bender[1]*, Richard Cole[2]**,
Erik D. Demaine[3]***, and Martin Farach-Colton[4]†

[1] Department of Computer Science, State University of New York at Stony Brook
Stony Brook, NY 11794-4400, USA
bender@cs.sunysb.edu.
[2] Courant Institute, New York University
251 Mercer Street, New York, NY 10012, USA
cole@cs.nyu.edu.
[3] MIT Laboratory for Computer Science, 200 Technology Square
Cambridge, MA 02139, USA
edemaine@mit.edu.
[4] Google Inc., 2400 Bayshore Parkway, Mountain View, CA 94043, USA
and Department of Computer Science, Rutgers University
Piscataway, NJ 08855, USA
farach@cs.rutgers.edu

Abstract. We study the problem of maintaining a dynamic ordered set subject to insertions, deletions, and traversals of k consecutive elements. This problem is trivially solved on a RAM and on a simple two-level memory hierarchy. We explore this traversal problem on more realistic memory models: the cache-oblivious model, which applies to unknown and multi-level memory hierarchies, and sequential-access models, where sequential block transfers are less expensive than random block transfers.

1 Introduction

A basic computational task is to maintain a dynamic ordered set of elements subject to insertions, deletions, and *logical traversals*. By a logical traversal we mean an in-order access of the k elements following an element x, for a given k and x. These three operations are performed by nearly any computer program that uses even the most common data structures, such as linked lists or search trees.

At first glance it does not seem that there is much to study. On a RAM, the problem is trivially solved by the lowly linked-list. Even on a two-level memory

---

* Supported in part by HRL Laboratories, NSF Grant EIA-0112849, and Sandia National Laboratories.
** Supported in part by NSF Grants CCR-9800085 and CCR-0105678.
*** Supported in part by NSF Grant EIA-0112849.
† Supported in part by NSF Grant CCR-9820879.

hierarchy, as described by the Disk Access Model (DAM model) [1], we can guarantee optimal results: For blocks of size B, we obtain $O(\lceil k/B \rceil)$ memory accesses per traversal and $O(1)$ accesses per insertion and deletion by maintaining $\Theta(B)$ contiguous elements in each block.

The meat of the problem lies in exploring more realistic memory models. For example, the DAM model makes no distinction between sequential and random-access block transfers. Furthermore, the solution proposed above is sensitive to the parameter B. Can we design a data structure that works well on more than two levels or for all values of B? The main contribution of this paper is a systematic study of the memory-traversal problem under a variety of tradeoffs and assumptions about memory, e.g.: cache aware versus cache oblivious, minimizing accesses versus minimizing seeks, and amortized versus worst case.

Some special cases of the traversal problem have been explored previously. By taking a systematic approach to memory issues—e.g., when does it pay to keep elements in sorted order—we improve bounds for almost all versions of the problem. By proving separations between the versions, we demonstrate the impact of making particular assumptions on the memory hierarchy.

1.1 Hierarchical Memories

The *Disk Access Model* (DAM model), introduced by Aggarwal and Vitter [1], is the standard two-level memory model. In this model, the memory hierarchy consists of an internal memory of size M and an arbitrarily large external memory partitioned into blocks of size B. Memory transfers between the two levels are performed in blocks.

The *cache-oblivious model* enables us to reason about a simple two-level memory model, but prove results about an unknown, multilevel memory model. The idea is to avoid any memory-specific parameterization, that is, to design algorithms that avoid using information about memory-access times or about cache-line and disk-block sizes. Such algorithms are optimal at all levels of the memory hierarchy, because the analysis is not tailored to any particular level.

The cache-oblivious model was introduced by Frigo, Leiserson, Prokop, and Ramachandran [14, 22]. They show how several basic problems—namely matrix multiplication, matrix transpose, Fast Fourier Transform, and sorting—have optimal algorithms that are cache-oblivious. More recently, cache-oblivious data structures matching the bounds of B-trees, priority queues, and tries have been developed [2, 4, 5, 6, 7].

Most memory models do not distinguish between random block transfers and sequential block transfers. The difference in access times is caused by the seek times and latencies on disk and by prefetching in disk and main memory. The current difference in speed between random access and sequential access on disk is around a factor of 10 [15], and this factor appears to increase by 5 every two years [23, 21]. We could incorporate these issues into the models by specifying the balance between physical scans and random block access, but this is another unwieldy parameter at each level of the memory hierarchy. Instead, we make the restriction that all operations in the traversal problem involve only a constant

Table 1. Number of memory transfers for a cluster of k updates in a structure supporting fast traversals, amortized except where noted as worst-case. All results are new, except those marked with a reference. Uncaptured by this table are the precise traversal bounds: they are all $O(\lceil k/B \rceil)$ except for the polyloglog cache-oblivious result; in this case, the bound is off by a small factor in the worst case, but holds in the amortized sense

	DAM	Cache-Oblivious
Unsorted, unrestricted block transfers	$O(\lceil k/B \rceil)$ [well-known]	$O\left(\left\lceil k \frac{(\log \log N)^{2+\varepsilon}}{B} \right\rceil\right)$
Sorted, unrestricted block transfers	$O\left(\left\lceil k \frac{\log^2(N/(k+B))}{B} \right\rceil\right)$ amortized [4]	$O\left(\left\lceil k \frac{\log^2(N/k)}{B} \right\rceil\right)$ amortized [4]
	$O\left(\left\lceil k \frac{\log^3(N/(k+B))}{B} \right\rceil\right)$ worst-case	$O\left(\left\lceil k \frac{\log^3(N/k)}{B} \right\rceil\right)$ worst-case
Sorted, $O(1)$ physical scans	$O\left(\left\lceil \frac{\log^2(N/B)}{B} \right\rceil\right)$	$O\left(\left\lceil \frac{\log^2 N}{B} \right\rceil\right)$

number of scans. This restriction can lead to better performance in both the DAM model and the cache-oblivious model. The idea of minimizing scans was also considered by Farach-Colton, Ferragina, and Muthukrishnan [12].

1.2 Results

We present the first positive progress on the traversal problem in almost two decades (cf. Section 1.3). Results are summarized in Table 1. We consider a variety of settings, depending on whether the memory hierarchy has two levels with a known block size (DAM) or multiple levels with unknown parameters (cache oblivious); whether the block accesses must be in sequential order or may be in arbitrary order; and whether the elements must be kept sorted[1]. One encouraging aspect of the traversal problem is that the update cost is surprisingly low (polylogarithmic or better) in all contexts, especially compared to the naïve update cost of $\Theta(N)$ when the elements are stored in sorted order without gaps.

When the elements must be sorted, the best general lower bound is $\Omega((\log N)/B)$ memory transfers, which follows directly from [9], and the best upper bound was $O((\log^2 N)/B)$ amortized memory transfers for unrestricted block transfers [4, 17, 31]. We make two main contributions in this setting. In Section 4 we extend the upper-bound result to hold even when updates can make only a constant number of physical scans. This restriction is important because it causes block accesses to be mainly sequential. Finally, we strengthen the result in a different direction, achieving even worst-case bounds. This algorithm follows

---

[1] By keeping elements in sorted order, we mean that their logical order is monotone with respect to their memory addresses. This notion is valuable for algorithms that use memory addresses as proxies for order. See, for example, [6].

a different, de-amortized strategy, and is more complex. We defer the description of this algorithm to the full paper.

One of our main contributions is to remove the implicit assumption in the literature that elements have to be sorted. Allowing elements to be stored out-of-order leads to update bounds exponentially below the lower bound of $\Omega((\log N)/B)$ for the sorted case. In particular, in Section 2 we develop a cache-oblivious structure using $O(\lceil (\log \log N)^{2+\varepsilon}/B^1 \rceil)$ unrestricted block transfers, for any $\varepsilon > 0$. This is our only result for which the traversal bound is (slightly) worse than the optimal $\Theta(\lceil k/B \rceil)$; if k is at least $B^{1-\varepsilon}$, then there is an additive B^ε term. However, in Section 3, we show how to modify scans to be self-adjusting (following splay trees [26]) and achieve the optimal $O(\lceil k/B \rceil)$ bound in an amortized sense, while maintaining the near-optimal worst case.

This polyloglog structure uses a weak version of the tall-cache assumption: $M = \Omega(B^\tau)$ for some $\tau > 1$. This assumption is commonly made in other external-memory and cache-efficient algorithms and usually holds in practice.

Our result is an exponential improvement in update cost over the best known solutions for sequential block accesses and for when elements must be in sorted order. An intriguing open problem is to prove a nonconstant lower bound.

A natural question is whether the per-operation performance of a traversal structure can be improved when the updates are batched and possibly clustered around a common element. Although some improvement is possible for large clusters, as shown in Table 1, it seems that this improvement is marginal.

Our results form a body of tools for manipulating dynamic data in unknown and multilevel memory hierarchies. In particular, they can be used to improve cache-oblivious B-trees. The only cache-oblivious B-tree structures that support traversals optimally [4, 6, 7] require $O(\log_B N + \frac{\log^2 N}{B})$ amortized memory transfers per update. By applying the structure in Section 4, we obtain the following improvement:

Corollary 1. *There is a cache-oblivious data structure that maintains an ordered set subject to searches in $O(\log_B N)$ memory transfers, insertions and deletions in $O(\log_B N)$ amortized memory transfers, and traversals of k consecutive elements in $O(\lceil k/B \rceil + [B^\varepsilon$ if k is at least $B^{1-\varepsilon}])$ memory transfers in the worst-case and $O(\lceil k/B \rceil)$ memory transfers in the amortized sense. Furthermore, scanning all N element requires $O(\lceil N/B \rceil)$ worst case memory transfers.*

1.3 Related Work

The main body of related work maintains a set of N elements ordered in $O(N)$ locations of memory, subject to insertions and deletions at specific locations in the order. This *ordered-file maintenance* problem is one version of the traversal problem, in which the elements must be sorted. Most of this work analyzes running time, but these results often adapt to multilevel memory hierarchies.

Itai, Konheim, and Rodeh [17] give a structure using $O(\log^2 N)$ amortized time per update. Similar results were also obtained by Melville and Gries [20] and Willard [29]. Willard [29, 30, 31] gives a complicated structure using $O(\log^2 N)$

worst-case time per update. A modification to the structure of Itai et al. results in a cache-oblivious traversal algorithm running in $O((\log^2 N)/B)$ amortized memory transfers per update.

The best lower bounds for this problem are $\Omega(\log N)$ time in general [9] and $\Omega(\log^2 N)$ time for "smooth relabeling strategies" [9, 10, 32]. The problem has also been studied in the context of average-case analysis [13, 16, 17].

Raman [23] gives a scheme for maintaining N elements in order using polynomial space $(\Theta(N^{1+\varepsilon}))$ and $O(\log^2 N)$ time per update. The update bound can be improved to $O(\log N)$ by tuning existing algorithms [3, 8, 11, 27], and this bound is optimal [9]. However, such a blowup in space is disastrous for data locality, so this work does not apply to the traversal problem.

2 Cache-Oblivious Traversal in $O((\log \log N)^{2+\varepsilon})$

We first consider the cache-oblivious traversal problem without restriction on the number of scans. In this context, we can store the data out-of-order, but because the cache-oblivious algorithm does not know the block size, we do not know how out-of-order the data can be. To resolve this conflict, our data layout keeps the data "locally ordered" to obtain the following result:

Theorem 1. *There is a cache-oblivious data structure supporting traversals of k elements in $O(\lceil k/B \rceil + [B^\varepsilon \text{ if } B^{1-\varepsilon} \leq k \leq B])$ memory transfers, and insertions and deletions in $O(\lceil (\log \log N)^{2+\varepsilon}/B \rceil)$ amortized memory transfers, for any $\varepsilon > 0$. Each update moves $O((\log \log N)^{2+\varepsilon})$ amortized elements. More generally, a block of k consecutive elements can be inserted or deleted in $O(\lceil k(\log \log N)^{2+\varepsilon}/B \rceil)$ memory transfers, for any $\varepsilon > 0$. If $k = N$, then the number of memory transfers is $O(\lceil N/B \rceil)$. These bounds assume that $M = \Omega(B^\tau)$ for some $\tau > 1$.*

For intuition, we first describe the structure top-down, but for formality, we define the structure bottom-up in the next paragraph. Consider the entire memory region storing the N elements as a *widget* W_N. This widget is recursively laid out into *subwidgets* as follows. Each widget W_ℓ consists of $(1 + 1/(\log \log \ell)^{1+\varepsilon})\ell^{1-\alpha}$ subwidgets of type W_{ℓ^α} for some $\varepsilon > 0$ and $0 < \alpha < 1$. The constant α is chosen near 1, depending both on ε (see Lemma 2) and on the weak tall-cache assumption. Specifically, if $M = \Omega(B^\tau)$, then $\alpha \geq 1/\tau$. For example, under the standard tall-cache assumption in cache-oblivious and external-memory algorithms, $\tau = 2$ and $\alpha \geq 1/2$. The closer τ and α are to 1, the weaker the assumption on the memory hierarchy.

This top-down definition of widgets suffers from rounding problems. Instead, we define the finest widget, $W_{(\log \log N)^{2+\varepsilon}}$, to consist of $\Theta((\log \log N)^{2+\varepsilon})$ elements. From finer widgets of type W_ℓ we build widgets of type $W_{\ell^{1/\alpha}}$, from which we build widgets of type $W_{\ell^{(1/\alpha)^2}}$, etc. The coarsest widget type is between W_N and $W_{N^{1/\alpha}}$, specifically, $W_{\tilde{N}}$ where $\tilde{N} = (1/\alpha)^{\lceil \log_{1/\alpha} N \rceil}$. Thus, conceptually, we keep the top widget underfull to avoid rounding at every level. In fact, this approach wastes too much space, and we partition the structure into widgets of various sizes, as described at the end of this section.

At each recursive level of detail, each widget W_ℓ is either *empty*, which means it stores no elements, or *active*, which means it stores at least ℓ elements. We recursively impose the restriction that if a widget W_ℓ is active, at least $\ell^{1-\alpha}$ of its subwidgets are active.

Lemma 1. *Widget W_ℓ occupies $\Theta(\ell)$ space.*

Proof. Let $S(\ell)$ denote the space occupied by widget W_ℓ. Then $S(\ell) = (1 + 1/(\log \log \ell)^{1+\varepsilon}) \ell^{1-\alpha} S(\ell^\alpha)$, which has the solution $S(\ell) = \Theta(\ell)$ for any constant $\varepsilon > 0$ and $0 < \alpha < 1$. It is for this space bound that we need $\varepsilon > 0$. □

Within a widget W_ℓ, the subwidgets of type W_{ℓ^α} are stored in a consecutive segment of memory (for fast traversals), but out-of-order (for fast updates). This "unordered divide-and-conquer" recursive layout is powerful, allowing us to break through the polylogarithmic barrier.

Lemma 2. *Traversing k elements in a widget W_ℓ requires $O(\lceil k/B \rceil + [B^\varepsilon \text{ if } B^\varepsilon \leq k \leq B])$ memory transfers.*

Proof. Let j be the smallest integer such that widget W_j has size greater than B. Thus subwidget W_{j^α} fits into a block. There are three cases.

First, if $k = O(j^\alpha) \leq O(B)$, then by Lemma 1 the elements to be traversed fit in a constant number of subwidgets of type W_{j^α}. Each subwidget W_{j^α} occupies a contiguous region of memory, and hence occupies at most two memory blocks. Thus, the traversal takes $O(1)$ memory transfers, which is $O(\lceil k/B \rceil)$ as desired.

Second, if $k = \Omega(j) > \Omega(B)$, then the number of elements to be traversed is $\Omega(1)$ times the size of a widget W_j. Because $j^\alpha \leq O(B)$, the weak tall-cache assumption implies that $j = (j^\alpha)^{1/\alpha} \leq O(B^{1/\alpha}) \leq O(B^\tau) \leq O(M)$. Hence, each widget W_j fits in memory. Thus we spend at most $O(\lceil j/B \rceil)$ memory transfers to read an entire widget or part of a widget. Hence, the total number of memory transfers is $O(\lceil k/j \rceil \lceil j/B \rceil)$, which is $O((k/j)(j/B)) = O(k/B)$ because $k = \Omega(j)$ and $j = \Omega(B)$.

Third, if k is $\Omega(j^\alpha)$ and $O(j)$, then k is also $\Omega(B^\alpha)$ and $O(B)$. Accessing each subwidget of type W_{j^α} only costs $O(1)$ block transfers because $j^\alpha = O(B)$, but the number of subwidgets of type W_{j^α} may be as much as $j^{1-\alpha} = O(B^{1/\alpha})^{1-\alpha} = O(B^{1/\alpha - 1})$. Setting $\varepsilon = 1/\alpha - 1$, ε approaches 0 as α approaches 1, and the desired bound holds. □

We now outline the behavior of this layout under insertions. At the top level, when $W_{\tilde{N}}$ is active, we have at least $\tilde{N}^{1-\alpha}$ active subwidgets, so at most $\tilde{N}^{1-\alpha}/(\log \log \tilde{N})^{1+\varepsilon}$ empty subwidgets. When all of a widget's subwidgets are active, we say that the widget is *hyperactive*. Insertions are handled locally if possible. Insertion into a hyperactive widget may find room in a subwidget, or it may "overflow." When a widget overflows, it passes its logically last subwidget to its parent widget for handing. At this point, the widget is no longer hyperactive, because it has a (single) empty subwidget. The remaining issues are how the parent widget "handles" a given subwidget, and when that causes an overflow.

More precisely, the following steps define the possible outcomes of an insertion into a W_ℓ widget. We will not move individual elements (except at the finest level of detail); rather, we will manipulate entire subsubwidgets of type $W_{\ell^{\alpha^2}}$, inserting them into subwidgets of type W_{ℓ^α}.

1. Conceptually, we recursively insert the element into the appropriate subwidget S of type W_{ℓ^α}, and if that widget does not overflow, we are finished. In fact, the insertion algorithm is not recursive; it begins at the finest level and works its way up to coarser levels.
2. If the subwidget S overflows (in particular, it was hyperactive), then S gave us one of its subsubwidgets T of type $W_{\ell^{\alpha^2}}$ to handle. We insert this subsubwidget T into the logical successor S' of S. If S' was not hyperactive, it has room for T and we are finished. If S' was hyperactive, we push T into S' anyway, and out pops the last subsubwidget of S', T'. We pass T' onto the successor of S', and *cascade* in this fashion for $(\log \log \ell)^{1+\varepsilon}$ steps, or until we find a nonhyperactive subwidget. (In fact, if a cascade causes the last subsubwidget of S to pop out of subwidget S after fewer than $(\log \log \ell)^{1+\varepsilon}$ steps, then we reverse-cascade; below for simplicity we only consider cascading forward.)
3. If we finish the cascade without finding a nonhyperactive subwidget, we enter the *activation stage*. Consider the last subwidget L of the cascade. Take its logically last $\ell^{(1-\alpha)^2}$ subsubwidgets of type $W_{\ell^{\alpha^2}}$, and form a new subwidget of type W_{ℓ^α} from these. This creates a new active subwidget (to be placed later), but leaves L nearly empty with exactly $\ell^{(1-\alpha)^2}/(\log \log \tilde{N})^{1+\varepsilon}$ subsubwidgets (because L was hyperactive), which is not enough for L to be active. Next take enough subsubwidgets from L's logical predecessor to make L active. Repeat this process back through the cascade chain until the beginning. Because all cascade subwidgets are hyperactive, we have enough subsubwidgets to activate a new subwidget, while leaving every cascade subwidget active.
4. If the W_ℓ widget was not hyperactive, we store the new active subwidget from the previous step into one of the empty subwidgets. Otherwise, the W_ℓ widget *overflows* to its parent, passing up the logically last subwidget.

Theorem 2. *The amortized cost of an insertion into this layout is $O((\log \log N)^{2+\varepsilon})$ time and $O(\lceil (\log \log N)^{2+\varepsilon}/B \rceil + [B^\varepsilon \text{ if } B^\varepsilon \leq k \leq B])$ memory transfers.*

Proof. Within a widget of type W_ℓ, activation of a subwidget of type W_{ℓ^α} requires a logically contiguous segment of $(\log \log \ell)^{1+\varepsilon}$ subwidgets of type W_{ℓ^α} all to be hyperactive. After such an activation, these subwidgets are all nonhyperactive, each having an empty fraction of $1/(\log \log \ell^\alpha)^{1+\varepsilon}$. Thus, each of these subwidgets will not participate in another activation until these

$\ell^{\alpha(1-\alpha)}/(\log\log \ell^\alpha)^{1+\varepsilon}$ subsubwidgets of type $W_{\ell^{\alpha^2}}$ are refilled. These refills are caused precisely by activations of subsubwidgets of type $W_{\ell^{\alpha^2}}$ within a subwidget of type W_{ℓ^α}. Thus, the activation of a subwidget of type W_{ℓ^α} requires the activation of $(\log\log \ell)^{1+\varepsilon}\ell^{\alpha(1-\alpha)}/(\log\log \ell^\alpha)^{1+\varepsilon} = \Theta(\ell^{\alpha(1-\alpha)})$ subsubwidgets of type $W_{\ell^{\alpha^2}}$. Working our way down to the bottom level, where widgets of type W_1 are created by insertions, the activation of a subwidget of type W_{ℓ^α} requires $\Theta(\ell^{\alpha(1-\alpha)}\ell^{\alpha^2(1-\alpha)}\cdots 1)$ insertions. The exponents form a geometric series, and we obtain that $\Theta(\ell^{(\alpha+\alpha^2+\cdots)(1-\alpha)}) \leq \Theta(\ell^{(\alpha/(1-\alpha))(1-\alpha)}) = \Theta(\ell^\alpha)$ insertions are needed for an activation of a subwidget of type W_{ℓ^α}.

An activation of a subwidget of type W_{ℓ^α} involves the movement of $O((\log\log \ell)^{1+\varepsilon})$ subwidgets, which costs $O((\log\log \ell)^{1+\varepsilon}\ell^\alpha)$ time and $O((\log\log \ell)^{1+\varepsilon}\lceil \ell^\alpha/B \rceil)$ memory transfers. We charge this cost to the $\Theta(\ell^\alpha)$ insertions that caused the activation. As a result, each insertion is charged at most $\log\log N$ times, once for each level. The amortized time for the activation is therefore $O((\log\log \ell)^{2+\varepsilon})$.

To compute the amortized number of memory transfers, there are three cases. If $\ell^\alpha \geq B$ so $\ell \geq B$, we have $O(\lceil (\log\log N)^{2+\varepsilon}/B \rceil)$ amortized memory transfers. If $\ell \leq B$ so $\ell^\alpha \leq B$, then the number of memory transfers is $O(1)$. If $\ell^\alpha < B$ and $\ell > B$, then at this particular level, there is an additional amortized cost of $(\log\log \ell)^{1+\varepsilon}\lceil \ell^\alpha/B \rceil/\ell^\alpha = (\log\log \ell)^{1+\varepsilon}/\ell^\alpha = o(1)$.

Cascades are more frequent than activations, because several cascades occur at a particular level between two activations, but a cascade can only be triggered by an activation at the next lower level. Also, cascades are cheaper than activations, because they only touch one subsubwidget per subwidget involved. Therefore, the cost of a cascade within widget W_ℓ is $O((\log\log \ell)^{1+\varepsilon}\ell^{\alpha^2})$, which can be amortized over the $\Omega(\ell^{\alpha^2})$ insertions that must take place to cause a subsubwidget to activate. The amortized number of memory transfers is bounded as above. □

The overall structure consists of widgets of several types, because one widget is insufficient by itself to support a wide range of values of N. In the ideal situation, the structure consists of some number of active widgets of type $W_{(1/\alpha)^{\lceil \log_{1/\alpha} N \rceil}}$, followed by an equal number of empty widgets of that type, followed by some number of active widgets of the next smaller type, followed by an equal number of empty widgets of that type, etc. The number of widgets of each type can be viewed roughly as a base-$(1/\alpha)$ representation of N. Whenever all widgets of a particular type become active and another is needed, the entire structure to the right of those widgets is moved to make room for an equal number of empty widgets of that type.

A slight modification of our data structure supports deletions, by allowing each widget W_ℓ to consist of between $(1 - 1/(\log\log \ell)^{1+\varepsilon})\ell^{1-\alpha}$ and $(1 + 1/(\log\log \ell)^{1+\varepsilon})\ell^{1-\alpha}$ subwidgets of type W_{ℓ^α}. We cannot use the global rebuilding technique, in which we mark deleted nodes as ghost nodes and rebuild whenever the structure doubles in size, because many ghost nodes in a region of elements decreases the density, causing small traversals within that region to become expensive.

3 Cache-Oblivious Traversal with Self-Adjusting Scans

In this section we modify the data structure in the previous section so that, in addition to achieving near-optimal worst-case traversal bounds, the structure achieves optimal traversal bounds in an amortized setting. The idea is to allow expensive traversals to adjust the data structure in order to improve the cost of future traversals. Specifically, out-of-order subwidgets make traversal expensive, and this jumbling is caused by insertions. Thus, we augment traversals to sort the subwidgets traversed. The main difficulty is that there is little extra space in any widget, and thus little room to re-order subwidgets.

We partition each widget W_ℓ into two parts: the "main" left part, which is the initial set of subwidgets, and the "extra" right part, which is a $\Theta(1/\log \log \ell)$ fraction of the size. We never move the widgets in the main part, so the subwidgets within the main part remain sorted with respect to each other at all times. The subwidgets in the extra part serve as little additions in between adjacent subwidgets in the main part, but they are stored off to the right side.

We enforce the constraint that only a third (or any constant fraction $\leq 1/3$) of the extra part is actually occupied by subwidgets; the remaining two thirds is (a small amount of) extra empty space that we use for memory management.

Consider a traversal in widget W_ℓ. The critical case is when ℓ is asymptotically larger than B, but the size of a subwidget W_{ℓ^α} is asymptotically smaller than B, so each random access to a subwidget causes an additional not-totally-used memory transfer.

We consider the traversal as having two working sets. On the one hand, it may visit subwidgets in the main part of the widget. In this context, no additional memory transfers are incurred, because the subwidgets in the main part are in sorted order and thus are visited consecutively.

On the other hand, the traversal may visit subwidgets in the extra part of the widget. Sometimes these subwidgets are stored in sorted order, and sometimes they are out-of-order. We count the number r of consecutive *runs* of subwidgets stored in order that the traversal visits. Each distinct run incurs $O(1)$ extra memory accesses. We ignore the cost of the first run, but charge the remaining $O(r-1)$ memory accesses to the previous insertions that caused those runs to split in the first place.

Now the traversal concatenates these $r-1$ runs, so that future traversals do not also charge to the same insertions. We proceed by sorting the r runs, i.e., extracting the runs and writing them in the correct order at the end of the extra part of the widget, and we erase the original copies of the runs.

Ideally, there is enough unused space at the end of the extra part of the widget to fit the new concatenated run. If there is not enough space, at least the first two thirds of the extra part must be occupied (with holes), which means that we have already concatenated several runs, an amount equal to a third of the entire size of the extra part. We charge to this prior concatenation the cost of *recompactification*: shifting items in the extra part to the left so as to fill all holes and maintain the relative order of the items. Now only the first third of the extra part is occupied, so there is room for the new run at the right end.

The total cost, therefore, is $O(\lceil k/B \rceil)$ amortized memory transfers for traversing k elements. Each insertion is only charged at most $O(1)$ extra memory transfers, so the cost of insertions does not change.

Theorem 3. *There is a data structure achieving all bounds claimed in Theorem 1, with the additional property that traversing k elements uses only $O(\lceil k/B \rceil)$ amortized memory transfers.*

4 Constant Number of Physical Scans

We now impose the restriction that every operation uses only $O(1)$ physical scans of memory. In the cache-oblivious model, this restriction requires us to keep the elements close to their sorted order. Thus, we base our data structure on the ordered-file structures mentioned in Section 1.3. Specifically, our structure is motivated by the "packed-memory structure" of [4]. Unfortunately, the packed-memory structure performs $\Theta(\log N)$ physical scans per update in the worst case.

In this section, we show how to reduce the number of physical scans to $O(1)$ per operation in the worst case. Our structure does not use an implicit binary-tree structure on top of the array as in [4, 17]. Instead, we always rebalance in intervals to the right of the updated element. This algorithm requires a different analysis because we can no longer charge costs to internal nodes in the binary tree. Nonetheless, we show that the same performance bounds are achieved, thereby proving the following theorem:

Theorem 4. *There is a cache-oblivious data structure supporting traversals of k elements in $O(\lceil k/B \rceil)$ memory transfers and $O(1)$ physical scans, and insertions and deletions in $O(\lceil (\log^2 N)/B \rceil)$ amortized memory transfers and $O(1)$ physical scans. Each update moves $O(\log^2 N)$ amortized elements. More generally, a block of k consecutive elements can be inserted or deleted in $O(\lceil (\log^2(N/k))/B \rceil)$.*

The algorithm works as follows. We consider rebalancing intervals in the circular array, where the smallest interval has size $\log N$ and the intervals increase as powers of two up to $\Theta(N)$. The *density* of an interval is the fraction of space occupied by elements. Let $h = \log N - \log \log N$ be the number of different interval sizes. Associated with each interval are two *density thresholds* which are guidelines for the acceptable densities of the interval. (The density of an interval may deviate beyond the thresholds, but as soon as the deviation is "discovered," the densities of the intervals are adjusted to be within the thresholds.)

The density thresholds are determined by the size of the interval. We denote the upper and lower density thresholds of an interval of size $(\log N)2^j$ by τ_j and ρ_j, respectively. These thresholds satisfy $\rho_h < \rho_{h-1} < \cdots < \rho_1 < \rho_0 < \tau_0 < \tau_1 < \cdots < \tau_h$. The values of the densities are determined according to an arithmetic progression. Specifically, let $\tau_0 = \alpha < 1$ and let $\tau_h = 1$. Let $\delta = (\tau_h - \tau_0)/h$. Then define density threshold τ_k to be $\tau_k = \tau_0 + k \cdot \delta$. Similarly, let $\rho_0 = \beta < \alpha$

and $\rho_h = \gamma < \beta$. Let $\delta' = (\rho_0 - \rho_h)/h$. Then define density threshold ρ to be $\rho = \rho - k \cdot \delta'$.

The insertion and deletion algorithms work as follows. We begin with the interval of size $\log N$ starting at the updated element. If the density of the interval is outside its density thresholds, we grow the interval rightward to double its size. Then we evenly space all of the elements within the interval.

5 Traversals in the DAM Model

The standard traversal structure on a DAM model can be applied recursively to support a multilevel memory hierarchy:

Theorem 5. *For any constant $c > 1$, there is a data structure on an ℓ-level memory hierarchy supporting traversals of k elements in $O(\lceil c^\ell k / B_i \rceil)$ memory transfers at level i, and insertions and deletions in $O(c^\ell + 1/(c-1)^\ell)$ memory transfers.*

An important consequence of this theorem is that this approach does not apply when the memory hierarchy has $\omega(1)$ levels, because of the exponential blowup in space and consequently the exponential decrease in density.

To support operations in $O(1)$ physical scans, we modify the approach from Section 4. We divide the data into blocks of size $\Theta(B)$, and store them in a circular array of blocks. Within each block, the elements are unsorted, but there is a total order among the blocks. Thus, the rank of each element in the circular array is within $O(B)$ of its actual rank. When we traverse k elements, the block access pattern is sequential, even though the element access pattern is not. Thus traversals use $O(1)$ physical scans.

An insertion or deletion may cause the block to become too full or too empty, that is, cause a split and/or merge. Splitting or merging blocks means that we must insert or delete a block into the circular array. We maintain the order among the $\Theta(N/B)$ blocks by employing the algorithm in Section 4. The same bound as before applies, except that we now manipulate $\Theta(N/B)$ blocks of size $\Theta(B)$ instead of N individual elements. Each insertion or deletion of a block moves $O(B \log^2(N/B))$ elements for a cost of $O(\log^2(N/B))$ memory transfers; but such a block operation only happens every $\Omega(B)$ updates. Thus, we obtain the following theorem:

Theorem 6. *There is a data structure supporting traversals of k elements in $O(\lceil k/B \rceil)$ memory transfers and $O(1)$ physical scans, and insertions and deletions in $O(\lceil (\log^2(N/B))/B \rceil)$ amortized memory transfers and $O(1)$ physical scans. Each update moves $O(\log^2(N/B))$ amortized elements. More generally, a block of k consecutive elements can be inserted or deleted in $O(\lceil (k \log^2(N/\max\{B, k\}))/B \rceil)$.*

References

[1] A. Aggarwal and J. S. Vitter. The input/output complexity of sorting and related problems. *CACM*, 31(9):1116–1127, Sept. 1988.
[2] L. Arge, M. A. Bender, E. D. Demaine, B. Holland-Minkley, and J. I. Munro. Cache-oblivious priority queue and graph algorithm applications. In *STOC*, 2002.
[3] M. A. Bender, R. Cole, E. D. Demaine, and M. Farach-Colton. Two simplified algorithms for maintaining order in a list. In *ESA*, 2002.
[4] M. A. Bender, E. Demaine, and M. Farach-Colton. Cache-oblivious search trees. In *FOCS*, 2000.
[5] M. A. Bender, E. D. Demaine, and M. Farach-Colton. Efficient tree layout in a multilevel memory hierarchy. In *ESA*, 2002.
[6] M. A. Bender, Z. Duan, J. Iacono, and J. Wu. A locality-preserving cache-oblivious dynamic dictionary. In *SODA*, 2002.
[7] G. S. Brodal, R. Fagerberg, and R. Jacob. Cache oblivious search trees via binary trees of small height (extended abstract). In *SODA*, 2002.
[8] P. Dietz. Maintaining order in a linked list. In *STOC*, 1982.
[9] P. Dietz, J. I. Seiferas, and J. Zhang. A tight lower bound for on-line monotonic list labeling. In *SWAT*, 1994.
[10] P. Dietz and J. Zhang. Lower bounds for monotonic list labeling. In *SWAT*, 1990.
[11] P. F. Dietz and D. D. Sleator. Two algorithms for maintaining order in a list. In *STOC*, 1987.
[12] M. Farach-Colton, P. Ferragina, and S. Muthukrishnan. Overcoming the memory bottleneck in suffix tree construction. In *FOCS*, 1998.
[13] W. R. Franklin. Padded lists: Set operations in expected $O(\log \log N)$ time. *IPL*, 9(4):161–166, 1979.
[14] M. Frigo, C. E. Leiserson, H. Prokop, and S. Ramachandran. Cache-oblivious algorithms. In *FOCS*, 1999.
[15] J. Gray and G. Graefe. The five minute rule ten years later. *SIGMOD Record*, 26(4), 1997.
[16] M. Hofri and A. G. Konheim. Padded lists revisited. *SICOMP*, 16:1073, 1987.
[17] A. Itai, A. G. Konheim, and M. Rodeh. A sparse table implementation of priority queues. In S. Even and O. Kariv, editors, *ICALP*, 1981.
[18] R. Ladner, J. Fix, and A. LaMarca. Cache performance analysis of algorithms. In *SODA*, 1999.
[19] A. LaMarca and R. E. Ladner. The influence of caches on the performance of sorting. *Journal of Algorithms*, 31:66–104, 1999.
[20] R. Melville and D. Gries. Controlled density sorting. *IPL*, 10:169–172, 1980.
[21] D. Patterson and K. Keeton. Hardware technology trends and database opportunities. In *SIGMOD*, 1998. Keynote address.
[22] H. Prokop. Cache-oblivious algorithms. Master's thesis, MIT, 1999.
[23] V. Raman. Locality preserving dictionaries: theory and application to clustering in databases. In *PODS*, 1999.
[24] S. Sen and S. Chatterjee. Towards a theory of cache-efficient algorithms. In *SODA*, 2000.
[25] D. D. Sleator and R. E. Tarjan. Amortized efficiency of list update and paging rules. *CACM*, 28(2):202–208, 1985.

[26] D. D. Sleator and R. E. Tarjan. Self-adjusting binary search trees. *Journal of the ACM*, 32(3):652–686, July 1985.
[27] A. Tsakalidis. Maintaining order in a generalized linked list. *Acta Informatica*, 21(1):101–112, May 1984.
[28] J. S. Vitter. External memory algorithms. *LNCS*, 1461, 1998.
[29] D. E. Willard. Maintaining dense sequential files in a dynamic environment. In *STOC*, 1982.
[30] D. E. Willard. Good worst-case algorithms for inserting and deleting records in dense sequential files. In *SIGMOD*, 1986.
[31] D. E. Willard. A density control algorithm for doing insertions and deletions in a sequentially ordered file in good worst-case time. *Information and Computation*, 97(2):150–204, Apr. 1992.
[32] J. Zhang. Density control and on-line labeling problems. Technical Report TR481, University of Rochester, Computer Science Department, Dec. 1993.

Two Simplified Algorithms for Maintaining Order in a List

Michael A. Bender[1]*, Richard Cole[2]**, Erik D. Demaine[3]***,
Martin Farach-Colton[4]†, and Jack Zito[1]

[1] Dept of Computer Science
SUNY Stony Brook, Stony Brook, NY 11794, USA
bender@cs.sunysb.edu
JackZito@JackZito.com
[2] Courant Institute, New York University
251 Mercer Street, New York, NY 10012, USA
cole@cs.nyu.edu
[3] MIT Laboratory for Computer Science
200 Technology Square, Cambridge, MA 02139, USA
edemaine@mit.edu
[4] Google Inc., 2400 Bayshore Parkway, Mountain View, CA 94043, USA
and Department of Computer Science, Rutgers University
Piscataway, NJ 08855, USA
farach@cs.rutgers.edu

Abstract. In the *Order-Maintenance Problem*, the objective is to maintain a total order subject to insertions, deletions, and precedence queries. Known optimal solutions, due to Dietz and Sleator, are complicated. We present new algorithms that match the bounds of Dietz and Sleator. Our solutions are simple, and we present experimental evidence that suggests that they are superior in practice.

1 Introduction

The *Order-Maintenance Problem* is to maintain a total order subject to the following operations:

1. Insert (X, Y): Insert a new element Y immediately after element X in the total order.
2. Delete (X): Remove an element X from the total order.
3. Order (X, Y): Determine whether X precedes Y in the total order.

We call any such data structure an *order data structure*.

---

* Supported in part by HRL Laboratories, NSF Grant EIA-0112849, and Sandia National Laboratories.
** Supported in part by NSF Grants CCR-9800085 and CCR-0105678.
*** Supported in part by NSF Grant EIA-0112849.
† Supported in part by NSF Grant CCR-9820879.

1.1 History

The first order data structure was published by Dietz [4]. It supports insertions and deletions in $O(\log n)$ amortized time and queries in $O(1)$ time. Tsakalidis improved the update bound to $O(\log^* n)$ and then to $O(1)$ amortized time [10].

The fastest known order data structures was presented in a classic paper by Dietz and Sleator [3]. They proposed two data structures: one supports insertions and deletions in $O(1)$ amortized time and queries in $O(1)$ worst-case time; the other is a complicated data structure that supports all operations in $O(1)$ worst-case time.

Special cases of the order-maintenance problem include the *online list labeling problem* [1, 5, 6, 7] also called the *file maintenance problem* [11, 12, 13, 14]. In online list labeling, we maintain a mapping from a dynamic set of n elements to the integers in the range from 1 to u (*tags*), such that the order of the elements matches the order of the corresponding tags. Any solution to the online list labeling problem yields an order data structure. However, the reverse is not true: there is an $\Omega(\log n)$ lower bound on the list labeling problem [5, 6]. In file maintenance, we require that $u = O(n)$, since it is the problem of maintaining a dynamic file stored sequentially in memory with gaps for insertions.

1.2 Results

$O(1)$ Amortized Order Data Structure: The drawback of Dietz and Sleator's amortized $O(1)$ solution is its proof. It relies on a complicated potential-function argument, which is not just unintuitive, but counterintuitive. This data structure assigns tags to elements à la online list labeling[1]. If we use more bits for the tags (say, $3 \log n$ rather than $2 \log n$ bits per tag), the amortized number of relabels *should* decrease. However, the proof of the Dietz and Sleator algorithm gives a paradoxical *increasing* upper bound on the amortized number of relabels. This was cited in their paper as a shortcoming of their analysis.

We present a similar list labeling-based algorithm that has several advantages. First, the proof is straightforward. Second, it yields a sensible tradeoff between tag size and relabeling. We verify experimentally that this tradeoff is essentially tight for our algorithm.

Furthermore, our analysis yields a counter-intuitive insight. Intuitively, it seems advantageous to assign the tag of a new element to be the average of its neighbors. Otherwise, an adversary could insert in the smaller half, thus inducing more relabelings. Indeed, Dietz and Sleator's proof relies on this average-tag choice. Our proof is independent of the tag-choice strategy; thus, the bound holds no matter how the tag is chosen. Perhaps our bounds are simply tighter for bad choices of tags. However, we confirm the insensitivity to tag selection experimentally.

In summary, we present a simple amortized solution along with a simple proof. This proof elucidates the tradeoffs and quirks of the algorithm, and we

---

[1] They use a standard trick of indirection to overcome the list labeling lower bound. We use the same trick.

verify these experimentally. Furthermore, we show experimentally that our solution dominates Dietz and Sleator's in terms of the bit/relabel tradeoff.

$O(1)$ **Worst-Case Order Data Structure:** The Dietz and Sleator worst-case constant-time solution relies on an algorithm by Willard [11, 12, 13, 14] for the *file maintenance problem*. Instead, we give a direct deamortization of the simple amortized algorithm above. Due to space considerations, we defer the discussion of this algorithm to the full paper.

$O(\log^2 n)$ **Worst-Case File Maintenance:** This result matches the bounds of the afore-mentioned Willard result. However, a complete description of our result takes 5 pages (See Section 5), rather than Willard's 54.

2 Preliminaries

2.1 Indirection: Shaving a Log

Below, we give a solution to the order-maintenance problem that takes $O(\log n)$ amortized time to update and $O(1)$ worst-case time for queries. The standard technique of having a two-level data structure, where the bottom level has $\Theta(\log n)$ elements and the top level has $\Theta(n/\log n)$ elements can be applied to any data structure with such a complexity to improve the update time to $O(1)$ amortized time, while keeping $O(1)$ worst case time queries.

We omit the details, which can be found, e.g., in [3].

2.2 Connection to Weight-Balanced Trees

Our order maintenance algorithms rely on a close connection between list labeling and weight-balanced trees. Suppose we have a list where each element has been assigned an integer tag from $[0, u)$. The bits of the tags implicitly form a binary tree, which is the *trie* of the bit strings. We call this trie the *virtual tree*.

The leaves of the virtual tree correspond to the tags, with each root-to-leaf path specifying the tag's binary representation. The depth of the virtual tree is $\log u$. As we insert new elements into the list, we assign them new tags, effectively inserting new leaves into the virtual tree. When we run out of tags, we redistribute tags within some subrange of the list, and this redistribution corresponds to a rebalancing operation in the implicit tree.

The next two lemmas give the algorithmic connection between this tree viewpoint and the list-labeling viewpoint:

Lemma 1. *Any strategy for maintaining n tags in the universe $[0, u)$, $u = n^c$ for some constant c, with amortized relabel cost $f(n)$ yields a strategy for maintaining balanced binary trees of height $\log u$ with amortized insertion time of $O(f(n))$.*

If the converse were true, we would immediately have an $O(1)$ amortized solution to the order maintenance problem (e.g., by using red-black trees plus the indirection noted above). However, updates in balanced tree structures might change many root-to-leaf paths. For example, a root rotation in red-black tree changes the path label of every leaf. Therefore, the cost of modifying (i.e., splitting/merging/rotating) a node v in the virtual tree is proportional to the *number of descendants* of v, which we refer to as the *weight* of v.

The *weight cost* of an operation on a tree is the running time of the operation, plus the sum of the weights of all modified nodes. For example, the red-black tree has $O(\log n)$ weight-cost searches and $O(n)$ weight-cost insertions and deletions. This definition extends naturally to *amortized weight cost*, and we note that red-black trees have $O(n)$ amortized weight-cost insertion and deletions as well.

Theorem 1. *Any balanced-tree structure with (amortized) weight cost $f(n)$ for insertions, maximum degree d, and depth h yields a strategy for list labeling with (amortized) cost $O(f(n))$ and tags from universe $[0, d^h)$.*

We require a balanced-tree structure where the *weight costs* of all operations are small. Fortunately, a few such trees exist, such as BB[α] trees [8], skip lists [9], and weight-balanced B-trees [2]. All these trees are *weight balanced*, in the sense that, for every non-leaf node, the weights of its children are within constant factors of each other.

These tree, when combined with indirection, give simple constant-time amortized solutions to the order-maintenance problem. The advantage of our solution below, as well as Dietz and Sleator's, is that it does not maintain the tree explicitly.

2.3 Roadmap

In Section 3, we show an amortized order-maintenance data structure that does not store the weight-balanced tree explicitly. The algorithm is similar to Dietz and Sleator's original algorithm, but our analysis is simple enough to teach to undergraduates, and it elucidates the performance issues. These performance issues are further explored and experimentally verified in Section 4. In Section 5, we give an $O(\log^2 n)$ algorithm for file-maintenance.

3 The Amortized Order-Maintenance Algorithm

Our algorithm does not use an explicit tree, but our analysis does. Let u, the tag universe size, be a power of two, and consider the complete binary tree in which the leaves represent the u tags. At any state of the algorithm, n of these leaves are occupied, namely, the leaves that correspond to the tags used to label list elements.

When we insert an element f between elements e and g, we are free to choose any tag between the tags of e and g. If e and g occupy adjacent leaves (their tags differ by 1), then we relabel a sublist of the linked list to make room for the new element.

Each internal node of the tree corresponds to a (possibly empty) sublist of the linked list, namely those linked elements that occupy leaves below the node. Each leaf has $\log u$ such *enclosing tag ranges*. The *density* of a node is the fraction of its descendant leaves that are occupied.

When we insert f between e and g, if we need to relabel, we examine the enclosing tag ranges of e until we find the smallest enclosing tag range with low-enough density. We calculate the density by traversing the elements within a certain tag range (instead of walking up the implicit tree).

It remains to define when an enclosing region is considered "sparse enough." We define the *overflow threshold* as follows. Let T be a constant between 1 and 2. For a range of size 2^0, that is, a leaf of the tree, the overflow threshold τ_0 is 1. Otherwise, for ranges of size 2^i, $\tau_i = \tau_{i-1}/T = T^{-i}$. We say that a range is *in overflow* if its density is above the overflow threshold.

We summarize the algorithm as follow: relabel the smallest enclosing range that is not overflowing. Note that larger enclosing ranges may be in overflow. We lazily stop at the smallest non-overflowing range.

This algorithm is similar to Dietz and Sleator's. Whereas we scan the list in geometrically increasing tag ranges, they scan the list in geometrically increasing number of list elements. Whereas we have a stopping condition that is a function of the density and size of our subrange, their stopping condition is that the density drops by some constant between sublists. We therefore call our algorithm the *Tag-range Relabeling Algorithm*, and Dietz and Sleator's the *List-range Relabeling Algorithm*.

We now show that the labels assigned to list elements are logarithmic in length, and that the amortized cost of an insertion is $O(\log n)$.

Lemma 2. *The Tag-range Relabeling Algorithm uses $O(\log n)$ bits per tag.*

Proof. Note that u is a function of n, but n may change wildly. Thus, we pick a u that works for any value from $n/2$ to $2n$. If at any point we have too few or too many list elements, we rebuild the data structure for the new value of n. This rebuilding introduces only a constant amortized overhead.

The root range must never be allowed to overflow. It is the subrange of last recourse. That is, we should always be able to relabel at the root. The density at the root is at most $2n/u$, because we rebuild if we get more than $2n$ elements. So we set $\tau_{\log u} = T^{-\log u} = u^{-\log T} = 2n/u$. Thus, $u = (2n)^{1/1-\log c}$, so we use $\log n/(1 - \log T)$ bits. □

We conclude that the worst-case query complexity is $O(1)$ since tags fit in a constant number of machine words.

Lemma 3. *The amortized cost of insertion in the Tag-range Relabeling Algorithm is $O(\log n)$.*

Proof. When we relabel a range of tags of size 2^i, the resultant density throughout the range becomes uniformly no more than τ_i, because we only relabel ranges that are not overflowing. The time complexity for this relabeling is $O(2^i \tau_i) = O((2/T)^i)$.

Before we relabel this range again, one of the two child subranges must overflow. After the relabeling, the density of the child subranges is at most τ_i and when they next overflow their density will be at least τ_{i-1}. This requires at least $(2 - T/2)(2/T)^i$ insertions, thus yielding an amortized insertion cost of $2 - T/2 = O(1)$.

Finally, by the previous lemma, each insertion has $O(\log n)$ enclosing subranges. Because the amortized cost *per enclosing subrange* is $O(1)$, the total amortized cost is $O(\log n)$. □

Using one level of indirection, we obtain

Theorem 2. *The order-maintenance problem can be solved in $O(1)$ amortized insertion and deletion time and $O(1)$ worst-case query time.*

4 Experiments

4.1 Choosing Tags

Surprisingly, the method for picking the label of the new element does not play a rôle in the proof. Consider two cases. The natural choice for a label is the average of the neighbors, as in Dietz and Sleator. A seemingly worse choice for the label is one more than the predecessor's label, which is vulnerable to adversarial relabeling. We refer to these strategies as *Average Inserts* and *Consecutive Inserts*, respectively.

Our proof suggests that the two schemes have equal performance. But it may be that our bound is simply tighter for the Consecutive Inserts case, and that Average Inserts is in fact a constant factor faster. To determine the constant factors, we performed the following experiment. We inserted n elements, each at the front. Thus, the Consecutive Inserts algorithm generates a relabeling at every other step.

It turns out that the total number of relabelings is almost the same for both schemes, with only an additive-constant overhead. See Figure 1. What appears to be happening is the following: Consecutive inserts trigger relabels after every insert, but the number of elements relabeled is small. On the other hand, average inserts have fewer triggers, and the number of elements relabeled is larger. With Consecutive Inserts, the average number of relabels per element remains only a small additive amount higher.

4.2 Space/Time Tradeoff

Note that the number of bits needed for an order maintenance structure that hold n elements is $\log n / (1 - \log T)$, where T is between 1 and 2. On a real

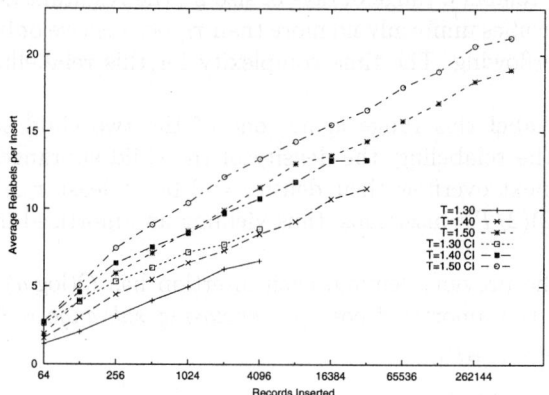

Fig. 1. 32-bit tags with consecutive vs. average inserts

machine, the number of bits is fixed, and thus there is a tradeoff between n and T. Recall that the amortized cost of insertions is $(2-T/2)(\log n)/(1-\log T)$. Thus, as we require our machine word to serve for more and more elements, we require a larger T, the cost of an insertion increases. While this is not surprising, Dietz and Sleator's proof had a paradoxical, opposite tradeoff in the bounds.

In our proof, we concentrate on the duration of time that the root range does not overflow, and when it does, we increase our word size and rebuild the structure. This, of course, does not work on a real computer. So for each T, the order data structure should have some maximum capacity, after which the root overflows, and at some point after which we start thrashing with relabels. In the next experiment, we measured how close to theoretical overflow we get thrashing, for a variety of choices of T, and for Dietz and Sleator's List-range algorithm.

Figure 2 shows relabels per insertion as we insert more and more elements into a list with a variety of choices of T, as well as for Dietz and Sleator (labeled DNS in the figure). The theoretical capacity given by each choice of T is indicated with an arrow.

In Figure 2, the y-axis represents the average number of relabels per inserted element. The x-axis represents the number of elements inserted. For this graph, lower points are better, representing fewer relabels per element inserted. Note that for any given threshold T, the algorithm thrashes soon after the "rated" maximum limit for that T has been exceeded. It is interesting to note that a threshold of $T = 1.5$ is comparable to Dietz and Sleator's algorithm for "worst case" inserts. While Dietz and Sleator's analysis suggests that the capacity of their structure is approximately 64,000, our structure with $T = 1.5$ is rated to approximately 430,000 elements. Thus Dietz and Sleator's analysis has substantial slack.

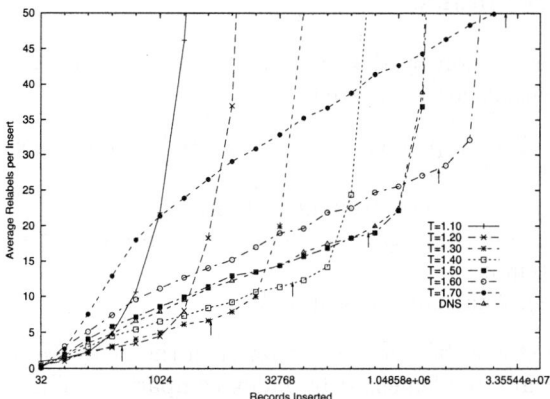

Fig. 2. 32-bit tags with constant thresholds

4.3 Variable T

Finally, note that although standard RAM models assume $O(\log n)$ word size, computers have fixed word sizes. The choice of T depends on n, the number of inserted elements. For any particular n, we would like T to be as small as possible, to reduce the cost of insertions. One final note about our algorithm is that there is nothing in our analysis that requires T to be fixed. At any step, we can calculate the smallest T that keep the root range from overflowing, and we use that T to compute the τ's. We call this algorithm the *Variable Threshold* Algorithm. Note that this algorithm does not produce theoretical improvements over the fixed-threshold algorithm, because they assume incomparable machine models. However, it seems to perform well, as Figure 3 shows.

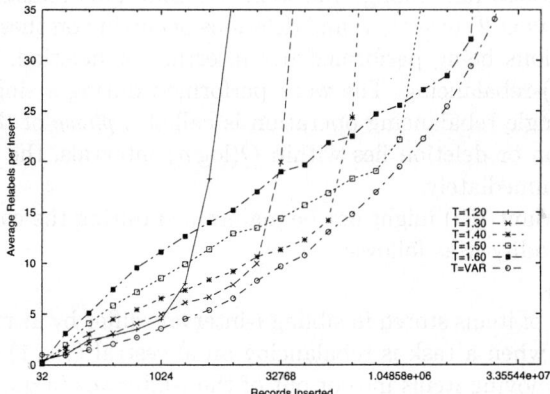

Fig. 3. 32-bit tags with fixed versus variable thresholds

5 File Maintenance

We begin our description by showing how to store n items in an array of size $O(n)$. Four operations are supported:

- Insert(e, f): Given a pointer to already stored item e, insert f immediately after e.
- Delete(e): Given a pointer to item e, delete it.
- Scan-right(k, e): Given a pointer to item e, scan the next k items in the list starting at item e.
- Scan-left(k, e): Analogous to scan-right.

Our solution stores the items in list order in the array. Note that the order-operation is trivially implemented by index comparison. The insert and delete operations require $O(\log^2 n)$ worst case time and the scan requires $O(k)$ time. Willard gave a data structure with this performance. Unfortunately, his solution is quite involved.' (The journal version requires some 54 pages).

As in Willard's data structure, the array is implicitly recursively partitioned into an $O(\log n)$ level balanced binary tree. The leaves correspond to length $a \log n$ intervals of the array for a suitable constant a; we call such intervals *cells*. As is standard, the nodes in the tree are assigned levels, incrementing from the leaf level and with the leaves having level zero. The subinterval corresponding to a node at level i comprises 2^i cells, and is called an i-interval. It is convenient to extend tree terminology (e.g., parent, child, sibling) to the i-intervals.

The key invariant below ensures there is always room to insert an item.

Invariant 1.
(i) An i-interval stores at most $(a \log n - 2i) \cdot 2^i$ items.
(ii) The number of items stored in sibling i-intervals differ by at most $2 \cdot 2^i$ items.

The basic update operation is the following: when the number of items stored by two sibling intervals differ by $2 \cdot 2^i$ items, 2^i items are redistributed from the fuller to the less full sibling, one item per cell. This redistribution is performed over the next 2^i insertions and deletions occurring on these two intervals, $O(\log n)$ operations being performed per insertion or deletion. The process is called an $(i+1)$-rebalancing. The work performed during a single insertion or deletion on a single rebalancing operation is called a *phase* of the rebalancing. As each insertion or deletion lies within $O(\log n)$ intervals, the $O(\log^2 n)$ time bound follows immediately.

In fact, Invariant 1(ii) might not be maintained during the course of a rebalancing, so we modify it as follows:

Invariant 1(ii).
(ii) The number of items stored in sibling i-intervals differ by at most $2 \cdot 2^i$ items, except possibly when a task is rebalancing an ancestral $(j+1)$-interval, $j \geq i$, and the task is moving items into or out of the i-intervals in question.

Since only 2^i insertions or deletions are performed during the course of a rebalancing, clearly Invariant 1(ii) holds after the rebalancing, given that it held at the start. Justification of Invariant 1(i) is deferred. In order to detect when the condition for rebalancing holds, with each node of the tree, we record its imbalance, namely how many more items its left subinterval holds than its right subinterval. The imbalance is stored in the cell immediately to the left of the middle of the interval.

The main part of the implementation lies in the rebalancing procedure itself. The challenge is to show that the rebalancings at different levels of the tree do not interfere.

WLOG, we consider a rebalancing of an $(i+1)$-interval whose left half is fuller. The rebalancing scans the $(i+1)$-interval from right to left, shifting items to the right, with the net effect of adding one item to each cell in the right half and removing one item from each cell in the left half. Ignoring the effect of further insertions and deletions, it is clear that the balance of any subinterval of the $(i+1)$-interval is unchanged.

In order to control the interaction of the various rebalancings, we need to store $O(i)$ auxiliary data with an i-rebalancing and $O(1)$ data with each cell.

To understand the data stored with a rebalancing process, it is helpful to consider the arrangement of data in an $(i+1)$-interval that is being rebalanced: at the right end, there will be data that has already been copied over to the new density, in the middle there will be a gap with empty cells, and at the left end will be the data that has yet to be moved. Finally, there may be two partially filled cells, one at either end of the gap. The algorithm is slightly simplified by requiring that at the end of its phase, the cell at the front (left end) of the gap be empty (in the case of a gap less than one cell in length, we consider the cell contents to be to the rear of the gap). We need to record how the data is to be distributed in the gap, and what density of data is to be moved from the left end (this is non-trivial owing to the interaction with rebalancing procedures associated with subintervals of the $(i+1)$-interval, as will become clear).

More specifically, the left end, comprising the as-yet-not-moved data, is partitioned into up to $i+2$ maximal h-intervals, $h \leq i$. For each such interval, we record an integral number of items to be copied to or from each cell in the interval (it is the same number for every cell). This data is stored as a linked list, called the *increment list*. At the start of a rebalancing there will be two i-intervals, one with count 1 and the other with count -1.

Similarly, the gap is partitioned into up to $2i$ maximal h-intervals, $h < i$. For each interval, we record how many items are to be moved into it. The partially filled cell at the rear of the gap, if any, forms a one-cell interval within the gap. This data is also kept as a linked list, called the *gap list*.

Invariant 2. The number of items to be moved into the gap, as recorded in the list of gap intervals, equals the change to the number of items in the "left end", as recorded in the list of intervals for the left end.

Both lists are kept in interval order, to facilitate combining of intervals as needed. When the contents of a cell are moved the natural changes are made to the lists. One detail remains to be specified: as the right end of the gap is filled, the interval at the right end of the gap will be split in two: the associated count is simply divided as evenly as possible. This has the effect of somewhat smoothing the local fluctuations in density within the $(i+1)$-interval.

In order to avoid unnecessary rebalancings at $(h+1)$-subintervals due to the gap created in the $(i+1)$-interval by its rebalancing, we do not alter the imbalance value for the $(h+1)$-subinterval due to the data movements of the $(i+1)$-rebalancing. Of course, if the gap of the rebalancing includes the whole $(h+1)$-subinterval, as the subinterval begins to be filled, its imbalance is reset to $-1, 0$, or 1, as appropriate.

Thus we understand Invariant 1(ii) as applying to the imbalances apparently recording the relative number of items in a pair of subintervals as opposed to the actual number which may differ due to a rebalancing of a larger interval containing these two subintervals.

The only difficulty with multiple rebalancing operations occurs when two gaps meet. There are just two possibilities: either they cross or they combine, as we detail. The task rebalancing the larger subinterval is named *Big* and the task for the smaller subinterval, contained in *Big*'s subinterval, is named *Small*.

WLOG, suppose that *Big* is moving items from left to right.

Case 1. *Small* moves items from left to right.

Case 1a. The gap for *Small* is to the left of the gap for *Big* when they meet (due to *Big*'s progress).

Then the rebalancing for *Small* is terminated, and *Big*'s process takes over Small's task by combining *Small*'s list with its own, as follows. Basically, the collections of lists are merged so as to record the desired incremental changes to intervals that have not had items moved, and to record the desired size for intervals in the gap. This may entail the splitting or joining of some intervals in order to combine information of equal sized intervals and to then produce maximal intervals once more.

Case 1b. The gap for *Big* overlaps the right end of *Small*'s interval or is just to the right of the right end of *Small*'s interval when *Small* is initiated, so in the latter case *Small*'s gap meets *Big*'s gap as *Small* is initiated.

Small is terminated, and its lists combined with those for *Big* as in Case 1a.

Case 1c. The gap for *Big* is to the left of the gap for *Small* when they meet (due to *Small*'s progress). This is handled as in Case 1a.

Each of the above cases uses $O(\log n)$ time.

Case 2. *Small* moves items from right to left. Details left for the full paper.

To ensure an $(i+1)$-rebalancing completes within 2^i phases, in each phase the contents of one cell are moved from the front end to the rear end of the rebalancing's gap. To ensure a smooth interaction with the processing of Cases 1 and 2 above, operations are ordered as follows. In a phase, first, adjacent gaps moving in opposite directions, if any, are merged; then the contents of a cell are moved across the gap; finally, adjacent gaps moving in the same direction, if any, are merged.

Lemma 4. *Invariant 2 is unaffected by the processing of Cases 1 and 2 above and by insertions and deletions.*

Lemma 5. *Invariant 1(ii) holds.*

Proof. A rebalancing of an $(i + 1)$-interval is completed within 2^i insertions and/or deletions to the interval, and thus the difference between the number of items in the two halves remains bounded by $2 \cdot 2^i$. The only reason it would not finish its task is through its being terminated and its work taken over by a $(j+1)$-rebalancing, for some $j > i$. But whenever the $(i+1)$-rebalancing would have performed a phase, the $(j+1)$ rebalancing would also perform a phase, and hence the $(j + 1)$-rebalancing will finish the processing of the $(i + 1)$-interval at least as soon as the $(i+1)$-rebalancing would have. Consequently Invariant 1(ii) is maintained.

Lemma 6. *Invariant 1(i) holds.*

Proof. Left to the full version.

References

[1] A. Andersson and O. Petersson. Approximate indexed lists. *Journal of Algorithms*, 29(2):256–276, 1998.
[2] L. Arge and J. Vitter. Optimal dynamic interval management in external memory. In *FOCS*, 1996.
[3] P. Dietz and D. Sleator. Two algorithms for maintaining order in a list. In *STOC*, 1987.
[4] P. F. Dietz. Maintaining order in a linked list. In *STOC*, 1982.
[5] P. F. Dietz, J. I. Seiferas, and J. Zhang. A tight lower bound for on-line monotonic list labeling. In *SWAT*, 1994.
[6] P. F. Dietz and J. Zhang. Lower bounds for monotonic list labeling. In *SWAT*, 1990.
[7] A. Itai, A. Konheim, and M. Rodeh. A sparse table implementation of priority queues. In *ICALP*, 1981.
[8] J. Nievergelt and E. M. Reingold. Binary search trees of bounded balance. *SIComp*, 2:33–43, 1973.
[9] W. Pugh. Skip lists: a probabilistic alternative to balanced trees. In *WADS*, 1989.
[10] A. K. Tsakalidis. Maintaining order in a generalized linked list. *Acta Informatica*, 21(1):101–112, May 1984.
[11] D. Willard. Inserting and deleting records in blocked sequential files. Technical Report TM81-45193-5, Bell Laboratories, 1981.
[12] D. Willard. Maintaining dense sequential files in a dynamic environment. In *STOC*, 1982.
[13] D. Willard. Good worst-case algorithms for inserting and deleting records in dense sequential files. In *SIGMOD*, 1986.
[14] D. Willard. A density control algorithm for doing insertions and deletions in a sequentially ordered file in good worst-case time. *Information and Computation*, 97(2):150–204, 1992.

A Final Thought

Dietz and Sleator is quite influential
With its tags and its proofs by potential
But to teach it in class
Is a pain in the —
So our new result is preferential.

Efficient Tree Layout in a Multilevel Memory Hierarchy

Michael A. Bender[1*], Erik D. Demaine[2**], and Martin Farach-Colton[3***]

[1] Department of Computer Science, State University of New York at Stony Brook
Stony Brook, NY 11794-4400, USA
bender@cs.sunysb.edu
[2] MIT Laboratory for Computer Science
200 Technology Square, Cambridge, MA 02139, USA
edemaine@mit.edu
[3] Google Inc., 2400 Bayshore Parkway, Mountain View, CA 94043, USA
and Department of Computer Science, Rutgers University
Piscataway, NJ 08855, USA
farach@cs.rutgers.edu

Abstract. We consider the problem of laying out a tree or trie in a hierarchical memory, where the tree/trie has a fixed parent/child structure. The goal is to minimize the expected number of block transfers performed during a search operation, subject to a given probability distribution on the leaves. This problem was previously considered by Gil and Itai, who show optimal but high-complexity algorithms when the block-transfer size is known. We propose a simple greedy algorithm that is within an additive constant strictly less than 1 of optimal. We also present a relaxed greedy algorithm that permits more flexibility in the layout while decreasing performance (increasing the expected number of block transfers) by only a constant factor. Finally, we extend this latter algorithm to the *cache-oblivious* setting in which the block-transfer size is unknown to the algorithm; in particular this extension solves the problem for a multilevel memory hierarchy. The query performance of the cache-oblivious layout is within a constant factor of the query performance of the optimal layout with known block size.

1 Introduction

The B-tree [4] is the classic optimal search tree for external memory, but it is only optimal when accesses are uniformly distributed. In practice, however, most distributions are nonuniform, e.g., distributions with heavy tails arise almost universally throughout computer science. Examples of nonuniform distributions include access distributions in file systems [3, 16, 19, 22].

---

[*] Supported in part by HRL Laboratories, NSF Grant EIA-0112849, and Sandia National Laboratories.
[**] Supported in part by NSF Grant EIA-0112849.
[***] Supported in part by NSF Grant CCR-9820879.

Consequently, there is a large body of work on optimizing search trees for nonuniform distributions in a variety of contexts:

1. *Known distribution on a RAM* — optimal binary search trees [1, 15] and variations [12], and Huffman codes [13].
2. *Unknown distribution on a RAM* — splay trees [14, 20].
3. *Known distribution in external memory* — optimal binary search trees in the HMM model [21].
4. *Unknown distribution in external memory* — alternatives to splay trees [14].[1]

Fixed Tree Topology. Search trees frequently encode decision trees that cannot be rebalanced because the operations lack associativity. Such trees naturally arise in the context of string or geometric data, where each node represents a character in the string or a geometric predicate on the data. Examples of such structures include tries, suffix trees, Cartesian trees, k-d trees and other BSP trees, quadtrees, etc. These data structures are among the most practical in computer science. Almost always their contents are not uniformly distributed, and often these search trees are unbalanced.

How can we optimize these fixed-topology trees when the access distribution is known? On a RAM there is nothing to optimize because there is nothing to vary. In external memory, however, we can choose the layout of the tree structure in memory, that is, which nodes of the tree are stored in which blocks in memory. This problem was proposed by Gil and Itai [10, 11] at ESA'95. Among other results described below, they presented a dynamic-programming algorithm for optimizing the partition of the N nodes into blocks of size B, given the probability distribution on the leaves. The algorithm runs in $O(NB^2 \log \Delta)$ time, where Δ is the maximum degree of a node, and uses $O(B \log N)$ space.

This problem is so crucial because when trees are unbalanced or distributions are skewed, there is even more advantage to a good layout. Whereas uniform distributions lead to B-trees, which save a factor of only $\lg B$ over the standard $\lg N$, the savings grow with nonuniformity in the tree. In the extreme case of a linear-height tree or a very skewed distribution we obtain a dramatic factor of B savings over a naïve memory layout.

Recently, there has been a surge of interest in data structures for multilevel memory hierarchies. In particular, Frigo, Leiserson, Prokop, and Ramachandran [9, 17] introduced the notion of *cache-oblivious algorithms*, which have asymptotically optimal memory performance for all possible values of the memory-hierarchy parameters (block size and memory-level size). As a consequence, such algorithms tune automatically to arbitrary memory hierarchies and exploit arbitrarily many memory levels. Examples of cache-oblivious *data structures* include cache-oblivious B-trees [6] and its simplifications [7, 8, 18], cache-oblivious persistent trees [5], and cache-oblivious priority queues [2]. However, all of these data structures assume a uniform distribution on operations.

---

[1] Although [14] does not explicitly state its results in the external-memory model, its approach easily applies to this scenario.

Our Results. In this paper, we design simple greedy algorithms for laying out a fixed tree structure in a hierarchical memory. The objective is to minimize the expected number of blocks visited on a root-to-leaf path, for a given probability distribution on the leaves. Our results are as follows:

1. We give a greedy algorithm whose running time is $O(N \log B)$ and whose query performance is within an additive constant strictly less than 1 of optimal.
2. We show that this algorithm is robust even when the greedy choices are within a constant factor of the locally best choices.
3. Using this greedy approach, we develop a simple cache-oblivious layout whose query performance is within a small constant factor of optimal at every level of the memory hierarchy.

Related Work. In addition to the result mentioned above, Gil and Itai [10, 11] consider other algorithmic questions on tree layouts. They prove that minimizing the number of distinct blocks visited by each query is equivalent to minimizing the number of block transfers over several queries; in other words, caching blocks over multiple queries does not change the optimization criteria. Gil and Itai also consider the situation in which the total number of blocks must be minimized (the external store is expensive) and prove that optimizing the tree layout subject to this constraint is NP-hard. In contrast, with the same constraint, it is possible optimize the expected query cost within an additive $1/2$ in $O(N \log N)$ time and $O(B)$ space. This algorithm is obtained by taking their polynomial-time dynamic program for the unconstrained problem and tuning it.

2 Tree Layout with Known Block Size

Define the *probability of an internal node* to be the probability that the node is on a root-to-leaf path for a randomly chosen leaf, that is, the probability of an internal node is the sum of the probabilities of the leaves in its subtree. These probabilities can be computed and stored at the nodes in linear time.

All of our algorithms are based on the following structural lemma:

Lemma 1. *There exists an optimal layout of a tree T such that within the block containing the root of T (the* root block*) the nodes form a connected subtree.*

Proof. The proof follows an exchange argument. To obtain a contradiction, suppose that in all optimal layouts the block containing the root does not form a connected subtree. Let r be the root of the tree T. Define the *root block* of a given layout to be the block containing r. Define the *root-block tree* to be the maximal-size (connected) tree rooted at r and entirely contained within the root block. Consider the optimal layout $\mathcal{L}^*$ such that the root-block tree is largest. We will exhibit a layout $\hat{\mathcal{L}}$ whose search cost is no larger and where the root-block tree contains at least one additional node. Thus, we will obtain a contradiction.

Consider nodes $u, v, w \in T$, where node u is the parent of node v, and node w is a descendant of node v; furthermore, node u is in the root-block tree (and therefore is stored in the root block), node v is stored in a different block, and node w is stored in the root block (but by definition is not in the root-block tree). Such nodes u, v, and w must exist by the inductive hypothesis.

To obtain the new layout $\hat{\mathcal{L}}$ from $\mathcal{L}^*$, we exchange the positions of v and w in memory. This exchange increases the number of nodes in the root-block tree by at least one. We show that the search cost does not increase; there are three cases. (1) The search visits neither v nor w. The expected search cost is unchanged in this case. (2) The search visits both v and w. Again the expected search cost is unchanged in this case. (3) The search visits v but not w. The search cost in this case stays the same or decreases because the block that originally housed v may no longer need to be transferred. Therefore, the search cost does not increase, and we obtain a contradiction. □

2.1 Greedy Algorithm

The *greedy algorithm* chooses the root block that maximizes the sum of the probabilities of the nodes within the block. To compute this root block, the algorithm starts with the root node, and progressively adds maximum-probability nodes adjacent to nodes already in the root block. Then the algorithm conceptually removes the root block and recurses on the remaining subtrees. The base case is reached when a subtree has at most B nodes; these nodes are all stored in a single block.

Equivalently, the greedy algorithm maximizes the expected depth of the leaves in the root block. Let p_i denote the probability of a leaf node in the root block, and let d_i denote the depth of that leaf. If we write out the sum of the probabilities of every node in the root block, the term p_i will occur for each ancestor of the corresponding leaf, that is, d_i times. Thus, we are choosing the root block to maximize $\sum_{i=1}^{\ell} p_i \cdot d_i$, which is the expected depth of a leaf.

Because the expected depth of the leaves in the root block plus the expected depth of the leaves in the remaining subtrees equals the total expected depth of the leaves in the tree, the greedy algorithm is equivalent to minimizing the expected depth of leaves in the remaining subtrees.

The greedy algorithm has the following performance:

Theorem 1. *The expected block cost of the greedy algorithm is at most the optimal expected block cost plus $(B-1)/B < 1$.*

Proof. Consider a *smooth* cost model for evaluation, in which the cost of accessing a block containing j elements is j/B instead of 1. By the properties of the greedy algorithm, the only blocks containing $j < B$ elements are *leaf blocks* (those containing all descendents of all contained nodes). This cost model only decreases the cost, so that the optimal expected smooth cost is a lower bound on the optimal expected block cost. We claim that the greedy algorithm produces a layout having the optimal expected smooth cost. Hence, the expected smooth cost of the greedy algorithm is at most the optimal expected block cost. Therefore, the expected block cost of the greedy algorithm is at most $(B-1)/B$ plus the optimal expected block cost.

Now we prove the claim. Consider an optimal layout according to the expected smooth cost. Assume by induction on the number of nodes in the tree that only the root block differs from the greedy layout. Pick for removal or *demotion* a minimum-probability leaf d in the root block, and pick for addition or *promotion* a maximum-probability child p of another leaf in the root block. Assuming the root block differs from the greedy layout, the probability of d is less than the probability of p. We claim that an "exchange," consisting of demoting d and promoting p (as described below) strictly decreases the expected smooth cost.

First consider demoting d from the root block. We push d into one of its child blocks. (Recall that d is a leaf of the root block, so its children are in different blocks.) This push causes a child block to overflow, and we recursively remove the smallest-probability leaf in that block, which has smaller probability than d. In the end, we either push a node d' into a nonempty leaf block, in which case we simply add d' to that block, or we reach a full leaf block, so that an excess node d' cannot be pushed into a child block because there are no child blocks. In the latter case, we create a new child block that just stores d'. In either case, the increase in the expected smooth cost is $\Pr[d']/B \leq \Pr[d]/B$. Our re-arrangement could only be worse than the optimal, which is equal to greedy on these subproblems by induction.

Second consider promoting p to the root block. This corresponds to removing p from its previous block b. We claim that this will decrease the expected smooth cost of accessing a leaf by at least $\Pr[p]/B$. If a child of p is not in block b, then accessing every leaf below that child now costs one block access less than before, because they no longer have to route through block b (and because by the greedy strategy, block b is full). Thus, we only need to consider children of p that are in block b.

If k of p's children are in block b, then the promotion of p up out of block b effectively partitions the block into k sub-blocks. In any case, any child previously in block b ends up in a block that stores strictly less than B nodes. Thus, we can recursively add to that block (at least) the maximum-probability child of a leaf in the block. In this way, the additions propagate to all leaf blocks below p. In the end, we move the maximum-probability leaf node from every leaf block into its parent block. This move decreases the smooth cost of that leaf by at least $1/B$, because it no longer has to access the leaf block which had size at least 1. The move also decreases the smooth cost of every leaf remaining in that leaf block by at least $1/B$, because the leaf block got smaller by one element.

Thus, the total decrease in expected smooth cost from promoting p to the root block is at least $\Pr[p]/B$. Again our re-arrangement could only be worse than the optimal, which is equal to greedy on these subproblems by induction.

In total the expected smooth cost decreases by at least $(\Pr[p] - \Pr[d])/B > 0$, contradicting the assumption that we started with an optimal layout according to expected smooth cost. Hence the optimal smooth-cost layout is in fact greedy. □

The greedy algorithm can be implemented in $O(N \log B)$ time on a RAM as follows. Initialize each root-block-selection phase by creating a priority queue containing only the root node. Until the root block has size B, remove the maximum-probability node from the priority queue and "add" that node. To add a node, place it into the root block, compute an order statistic to find the B maximum-probability children of the node, and add those children into the priority queue. We can get away with keeping track of only the top B children because we will select only B nodes total. Any remaining children become the roots of separate blocks. Hence, the priority queue has size at most B^2 at any moment, so accessing it costs $O(\log B)$ time. We compute order statistics only once for each node, so the total cost for computing order statistics is $O(N)$.

An interesting open problem is how to implement the greedy algorithm efficiently in a hierarchical memory given a tree in some prescribed form (e.g., a graph with pointers stored in a contiguous segment of $O(N)$ memory cells). The bound of $\Theta(N)$ follows trivially by implementing the priority queue of size B^2 as a B-tree of height 2. Can we reduce the number of memory transfers below $\Theta(N)$?

2.2 Relaxed Greedy Algorithm

For a constant $0 < \varepsilon < 1$, the *relaxed greedy algorithm* chooses nodes for the root block in a less greedy manner: at each step, it picks any node whose probability is at least ε times the highest-probability node that could be picked. Once the root block is chosen, the remaining subtrees are laid out recursively as before.

Theorem 2. *The relaxed greedy algorithm has an expected block cost of at most $1/\varepsilon$ times optimal, plus $(B-1)/B < 1$.*

Proof. In our analysis we start with a greedy layout and gradually transform it to a relaxed greedy layout. As we perform the transformation, we bound the increase in cost. In the greedy layout, we work up from the bottom of the tree, and change each block into the relaxed-greedy choice for the block. The expected smooth cost of leaf nodes is the same as before in Theorem 1.

The proof of the claim follows the exchange argument in the proof of Theorem 1. By induction suppose that only the root block of the tree has not yet been converted into the relaxed-greedy layout. Consider exchanging nodes, one at a time, from the greedy layout to the relaxed-greedy layout. Each exchange causes one highest-probability node d to be demoted from the root block, and causes one node p to be promoted, where p has probability at least ε times the highest probability.

By the argument in the proof of Theorem 1, the demotion of d causes the expected smooth cost to increase by at most $1/B$ times the probability of d;

and the promotion of p causes the expected smooth cost to decrease by at least $1/B$ times the probability of p. In particular, the increase in expected smooth cost at most $\Pr[d]/B - \Pr[p]/B$. Because $\varepsilon \Pr[d] < \Pr[p]$, this quantity is at most $\left(\frac{1}{\varepsilon} - 1\right) \frac{\Pr[p]}{B} \leq \left(\frac{1}{\varepsilon} - 1\right) \frac{1}{B}$. Because there are at most B exchanges in the root block, the expected smooth cost increases by at most $\left(\frac{1}{\varepsilon} - 1\right)$. Since the original smooth cost of accessing the (entirely filled) root block was 1, there is a $1/\varepsilon$ factor increase in the smooth cost in order to change the root block from the greedy layout to the relaxed greedy layout. This proves the claim. □

3 Cache-Oblivious Tree Layout

We now present the cache-oblivious layout. We follow the greedy algorithm with unspecified block size B, repeatedly adding the highest-probability node to the root block, until the number of nodes in the root block is roughly equal to the expected number of nodes in the child subtrees (rounding so that the root block may be slightly larger). Then we recursively lay out the root block and each of the child subtrees and concatenate the resulting recursive layouts (in any order). At any level of detail, we distinguish a node as either a *root block* or a *leaf block*, depending on the rôle it played during the last split.

Theorem 3. *The cache-oblivious layout has an expected block cost of at most 4 times optimal, plus 4.*

Proof. Consider the level of detail at which every block has more than B nodes and such that, at the next finer level of detail, a further refinement of each block induces a *refined root block* with at most B nodes. Thus, the *refined child block* that is visited by a random search has an expected number of nodes that is at most B. That is, if we multiply the size of each refined child block within a block by the probability of entering that refined child block (the probability of its root), then the aggregate (expectation) is at most B. Therefore, the expected number of memory transfers within each block is at most 4: at most 2 for the refined root block (depending on the alignment with block boundaries) and at most 2 for a randomly visited refined child block.

Thus, each block has size more than B and has expected visiting cost of at most 4 memory transfers. The partition into blocks can be considered an execution of the greedy algorithm in which blocks have varying maximum size that is always more than B. The expected number of blocks visited can thus only be better than the expected block cost of greedy in which every block has size exactly B, which by Theorem 1 is at most optimal plus 1. Hence, the total expected number of memory transfers is at most 4 times optimal plus 4. □

Acknowledgment

We thank Ian Munro for many helpful discussions.

References

[1] A. Aho, J. Hopcroft, and J. Ullman. *The Design and Analysis of Computer Algorithms*. Addison-Wesley, 1974.

[2] L. Arge, M. A. Bender, E. D. Demaine, B. Holland-Minkley, and J. I. Munro. Cache-oblivious priority queue and graph algorithm applications. In *Proc. 34th Annual ACM Symposium on Theory of Computing*, pages 268–276, 2002.

[3] M. Baker, J. Hartman, M. Kupfer, K. Shirriff, and J. Ousterhout. Measurements of a distributed file system. In *Proc. 13th Symposium on Operating Systems Principles*, pages 198–212, 1991.

[4] R. Bayer and E. M. McCreight. Organization and maintenance of large ordered indexes. *Acta Informatica*, 1(3):173–189, February 1972.

[5] M. A. Bender, R. Cole, and R. Raman. Exponential structures for efficient cache-oblivious algorithms. In *Proc. 29th International Colloquium on Automata, Languages, and Programming*, 2002. To appear.

[6] M. A. Bender, E. Demaine, and M. Farach-Colton. Cache-oblivious B-trees. In *Proc. 41st Annual Symposium on Foundations of Computer Science*, pages 399–409, 2000.

[7] M. A. Bender, Z. Duan, J. Iacono, and J. Wu. A locality-preserving cache-oblivious dynamic dictionary. In *Proc. 13th Annual ACM-SIAM Symposium on Discrete Algorithms*, pages 29–38, 2002.

[8] G. S. Brodal, R. Fagerberg, and R. Jacob. Cache oblivious search trees via binary trees of small height (extended abstract). In *Proc. 18th Annual ACM-SIAM Symposium on Discrete Algorithms*, 2002.

[9] M. Frigo, C. E. Leiserson, H. Prokop, and S. Ramachandran. Cache-oblivious algorithms. In *Proc. 40th Annual Symposium on Foundations of Computer Science*, pages 285–297, New York, October 1999.

[10] J. Gil and A. Itai. Packing trees. In *Proc. 3rd Annual European Symposium on Algorithms)*, pages 113–127, 1995.

[11] J. Gil and A. Itai. How to pack trees. *Journal of Algorithms*, 32(2):108–132, 1999.

[12] T. C. Hu and P. A. Tucker. Optimal alphabetic trees for binary search. *Information Processing Letters*, 67(3):137–140, 1998.

[13] D. A. Huffman. A method for the construction of minimum-redundancy codes. *Proc. IRE*, 40(9):1098–1101, 1952.

[14] J. Iacono. Alternatives to splay trees with $O(\log n)$ worst-case access times. In *Proc. 11th Symposium on Discrete Algorithms*, pages 516–522, 2001.

[15] D. E. Knuth. *The Art of Computer Programming, V. 3: Sorting and Searching*. Addison-Wesley, Reading, 1973.

[16] J. K. Ousterhout, H. D. Costa, D. Harrison, J. A. Kunze, M. Kupfer, and J. G. Thompson. A trace-driven analysis of the UNIX 4.2 BSD File System. In *Proc. 10th Symposium on Operating Systems Principles*, pages 15–24, 1985.

[17] H. Prokop. Cache-oblivious algorithms. Master's thesis, Massachusetts Institute of Technology, Cambridge, MA, June 1999.

[18] N. Rahman, R. Cole, and R. Raman. Optimized predecessor data structures for internal memory. In *Proc. 5th Workshop on Algorithms Engineering*, pages 67–78, 2001.

[19] D. Roselli, J. Lorch, and T. Anderson. A comparison of file system workloads. In *Proc. 2000 USENIX Annual Technical Conference*, pages 41–54, 2000.

[20] D. D. Sleator and R. E. Tarjan. Self-adjusting binary search trees. *Journal of the ACM*, 32(3):652–686, July 1985.
[21] S. Thite. Optimum binary search trees on the hierarchical memory model. Master's thesis, Department of Computer Science, University of Illinios at Urbana-Champaign, 2001.
[22] W. Vogels. File system usage in Windows NT 4.0. In *Proc. 17th ACM Symposium on Operating Systems Principles*, pages 93–109, 1999.

A Computational Basis for Conic Arcs and Boolean Operations on Conic Polygons*

Eric Berberich, Arno Eigenwillig, Michael Hemmer, Susan Hert, Kurt Mehlhorn, and Elmar Schömer

Max-Planck-Institut für Informatik
Stuhlsatzenhausweg 85, 66123 Saarbrücken, Germany
{eric,arno,hemmer,hert,mehlhorn,schoemer}@mpi-sb.mpg.de

Abstract. We give an exact geometry kernel for conic arcs, algorithms for exact computation with low-degree algebraic numbers, and an algorithm for computing the arrangement of conic arcs that immediately leads to a realization of regularized boolean operations on conic polygons. A conic polygon, or polygon for short, is anything that can be obtained from linear or conic halfspaces (= the set of points where a linear or quadratic function is non-negative) by regularized boolean operations. The algorithm and its implementation are complete (they can handle all cases), exact (they give the mathematically correct result), and efficient (they can handle inputs with several hundred primitives).

1 Introduction

We give an exact geometry kernel for conic arcs, algorithms for exact computation with low-degree algebraic numbers, and a sweep-line algorithm for computing arrangements of curved arcs that immediately leads to a realization of regularized boolean operations on conic polygons. A conic polygon, or polygon for short, is anything that can be obtained from linear or conic halfspaces (= the set of points where a linear or quadratic function is non-negative) by regularized boolean operations (Figure 1). A regularized boolean operation is a standard boolean operation (union, intersection, complement) followed by regularization. Regularization replaces a set by the closure of its interior and eliminates dangling low-dimensional features.

Our algorithm and implementation are *complete* and *exact*. They are complete in the sense that they can handle all inputs including arbitrary degeneracies. They are exact in that they always deliver the mathematically correct result. Complete and exact implementations for the linear case are available, *e.g.*, in the generalized polygon class [21, Section 10.8] of LEDA and in the planar map class [15] of CGAL. However, existing implementations for conic polygons are either incomplete or inexact, except for the very recent work by Wein [25].

---

* Partially supported by the IST Programme of the EU as a Shared-cost RTD (FET Open) Project under Contract No IST-2000-26473 (ECG - Effective Computational Geometry for Curves and Surfaces).

Fig. 1. We compute the union of two curved polygons (left panel). The first input polygon is created from a regular 10-gon by replacing each straight edge with a half-circle with that edge as diameter. The second polygon is created from the first by rotating it around the origin by $\alpha = \pi/20$ radians. On the right is the union of these two polygons. We can compute the correct union of such rotated n-gons for any n and any α. For $n = 1000$ and $\alpha = \pi/2000$, the time required to compute the correct union with 2702 edges is less than 20 minutes on an 846 MHz Pentium III processor

There are three main parts to our work: (1) a sweep-line algorithm for computing arrangements of curved arcs, (2) predicates and functions for conic arcs, and (3) algorithms for the exact computation with low-degree algebraic numbers. For part (1), the sweep-line algorithm extends the Bentley-Ottmann sweep-line algorithm for segments [1]. The handling of many curves passing through the same point is considerably more involved than in the straight-line case. For (2), we give algorithms for basic predicates and functions on conics and use them to realize the functionality required in the sweep algorithm. For (3), we have integrated the representation of algebraic numbers as roots of polynomials and the representation as explicit expressions involving square roots.

Our implementation consists of a basic layer providing polynomials, roots of polynomials, low-degree algebraic numbers, conics, and predicates and functions on conics and conic arcs and an algorithmic layer that provides arrangements of conics, the sweep-line algorithm for curves, and boolean operations on conic polygons and on polygons with circular and straight-line arcs. We have tested our implementations on inputs of various sizes and with various degeneracies; see for example Figure 2. We provide evidence of the efficiency of the approach presented here by comparing the results to the implementations in LEDA for polygons with line-segment edges. Our implementation can handle scenes with several hundred conic segments.

The rest of the paper is organized as follows. We first summarize related work (Section 2) and then review the Bentley-Ottmann sweep-line algorithm. We extend it to curved arcs and derive the required set of predicates and functions (Section 3). In Section 4 we discuss conics, functions and predicates on conics, and computation with low-degree algebraic numbers. In Section 5 we give more details of our implementations and provide the results of our experiments. Section 6 offers some conclusions.

2 Related Work

The work of three communities is relevant for our work: computational geometry, solid modeling, and computer algebra. The solid modeling community has always dealt with curved objects, and CAD systems dealing with curved objects in two and three dimensions have been available since the 60's. None of these systems is complete or exact, not even for straight-line objects. The question of complete and exact implementations has been addressed only recently. MAPC [20] provides a set of classes for manipulating algebraically defined points and curves in the plane, which includes an implementation of the naïve $O(n^2)$ algorithm for computing an arrangement of n curves in the plane. The algorithms are not complete; they handle some but not all degeneracies. Also the use of Sturm sequences to handle degenerate cases, such as tangential curves or degenerate segments, results in unnecessarily slow computations in these cases. ESOLID [19] performs accurate boundary evaluation of low-degree curved solids. It is explicitly stated that degeneracies are not treated.

Surprisingly little work in computational geometry deals with curved objects; some examples are [24, 12, 11]. The sweep-line algorithm of Bentley-Ottmann [1] is known to work for x-monotone curves, at least in the absence of degeneracies. Degeneracies have been discussed for straight-line segments only. Several papers [3, 4, 6] have looked into the question of using restricted predicates to report or compute segment intersections, the rationale being that lower-degree predicates are simpler to evaluate. All papers have to exclude at least some degenerate cases.

The exact and efficient implementation of the predicates required in our algorithms is non-trivial, since they involve algebraic numbers. One of the predicates used in our algorithms, the lexicographical comparison of vertices in an arrangement of circles and lines, has been considered by Devillers *et al.* [10]. A very efficient realization is given. The more complex predicates also needed in our algorithms are not discussed in their work and it is not clear whether the technique generalizes.

CGAL's planar map class also supports the computation of arrangements of circular arcs and line segments. Very recently, the implementation was extended to conic arcs by Wein [25]. The implementation is in some respects similar to ours. However, the sweep method is not yet available and the the computation of boolean operations on polygons is implemented only indirectly.

The papers [13, 17] show how to compute arrangements of quadrics in three-space. The algorithms are complete, but (as of now) can handle only a small number of quadrics. Root isolation of univariate polynomials, resultant computation, and exact treatment of algebraic numbers are well studied problems in computer algebra. We use the standard techniques.

3 Conic Polygons and the Sweep-Line Algorithm

The Bentley-Ottmann Sweep-line Algorithm: The Bentley-Ottmann sweep-line algorithm for computing an arrangement of segments in the plane [1] was originally formulated for sets of segments, no three of which pass through a common point and no two of which overlap each other. A vertical line is swept across the plane and the ordered sequence of intersections between the sweep line and the segments is maintained (= Y-structure). The status of the sweep line changes when a segment starts, when a segment ends, and when two segments cross. Bentley and Ottmann observed that the algorithm can actually handle any set of x-monotone curves. Of course, when two such curves meet they may either cross or touch, which requires a minor modification to the algorithm. The events are maintained in a priority queue, referred to as the X-structure.

It was later observed, see for example [9, 21], that in the case of straight-line segments the algorithm can also handle arbitrary degeneracies. A number of small extensions are required. For example, when the algorithm sweeps across a point in which several segments meet, the y-order of the segments meeting at this point is reversed.

We next argue that the algorithm can also handle degenerate situations in the case of curves. The main problem is sweeping across a point where many curves meet and we restrict our attention to this problem here.

Consider a point p and assume that arcs C_1 to C_k pass through p. We assume that the curves are numbered according to their y-order just left of p. Let s_i be the multiplicity of intersection of the curves C_i and C_{i+1} at p; see [2, Chapters I and IV] for a formal definition. Intuitively, the multiplicity is one if the curves meet at p and have different slopes, the multiplicity is two if the curves have identical tangent but different radii of curvature, the multiplicity is three if the curves have same tangent and identical radii of curvature but different .... Two curves meeting at p cross at p if the multiplicity of the intersection is odd, and they touch, but do not cross, if the multiplicity is even. The multiplicity of intersection between C_i and C_j for $i < j$ is $\min\{s_i, \ldots, s_{j-1}\}$ because the multiplicity of intersection is the number of identical initial coefficients in the local Taylor series expansion. For distinct conics, the multiplicity of intersection at any point is at most 4. For example, the multiplicity of intersection at the origin between $y(1-x) = x^2$ and $y = x^2 + y^2$ is three.

The following algorithm determines the y-order of our curves C_1 to C_k just to the right of p in time $O(k)$. Make four passes over the sequence of curves passing through p. In the j-th pass, $4 \geq j \geq 1$, form maximal subsequences of curves, where two curves belong to the same subsequence if they are not separated by a multiplicity less than j, and reverse the order of each subsequence.

Lemma 1. *The algorithm above correctly computes the y-order of the segment passing through a common point p immediately to the right of p from the order immediately to the left of p.*

Proof. Consider two arbitrary curves C_h and C_i with $h < i$. Their y-order right of p differs from their y-order left of p iff $s = \min\{s_h, \ldots, s_{i-1}\}$ is odd. Next observe that s is also exactly the number of times C_h and C_i belong to the same subsequence, i.e., the number of times their order is reversed. We conclude the order of C_i and C_h is reversed iff C_i and C_h cross at p.

Of course, the algorithm just outlined will also work if the maximal multiplicity M of intersection is arbitrary. However, its running time will be M times k, the number of curves passing through the point. Lutz Kettner (personal communication) has shown that the permutation can always be computed in time linear in the length of the subsequence.

Conic Polygons: Regularized boolean operations on straight-line or conic polygons can be built on top of the sweep-line algorithm for segment intersection, see [21, Section 10.8]. The corresponding data structure in LEDA is called generalized polygons. We reused it with only one small change. When curved edges are used, a polygonal chain with only two edges is possible. We also made the implementation more general by parameterizing it with the type of the polygon used to represent the boundaries (Section 5).

Required Predicates and Functions: We are now ready to summarize the predicates and functions that must be defined for the points and segments used in the sweep-line algorithm and in the computation of generalized polygons.

compare_xy(p, q) – compares points p and q lexicographically. This predicate is used to maintain the order of the events in the X-structure.

seg.y_order(p) – determines if a point p in the x-range of segment seg is vertically above, below, or on seg. This predicate is used to insert starting segments into the Y-structure.

seg.compare_right_of_common_point(seg2, p) – compares seg and $seg2$ to determine their y-ordering just right of their common point p. This predicate is also used to insert starting segments into the Y-structure. The order of segments starting at the same point is determined by this predicate.

seg.common_point_multiplicity(seg2, p) – returns the multiplicity of intersection of seg and $seg2$ at their common point p. This predicate is used to handle curves passing through a common point as described above.

seg.has_on(p) – determines if p lies on seg. This predicate is not needed in the sweep line algorithm, but in the algorithm for boolean operations. It is used to determine the containment of one boundary inside another.

intersect(seg1, seg2, result) – determines if $seg1$ and $seg2$ intersect or not and, if so, inserts their intersection points in lexicographical order into $result$

4 Conics and Computations with Conics

We discuss predicates and constructions on conics and their algorithmic realization in this section.

Every conic is defined as the zero-set in $\mathbb{R}^2$ of a quadratic implicit equation P in the variables (x, y) i.e.,

$$\alpha_1 x^2 + \alpha_2 y^2 + 2\alpha_3 xy + 2\alpha_4 x + 2\alpha_5 y + \alpha_6 = 0$$

with $\neg(\alpha_1 = \alpha_2 = \alpha_3 = 0)$. We restrict attention to the non-degenerate conics in this paper as lines bring nothing new algorithmically. For every value x there are at most two values of y satisfying the equation. We can obtain an x-monotone parameterization by rewriting the equation as a quadratic equation for y and solving for y. We obtain $Q = a(x)y^2 + b(x)y + c(x) = 0$. Solving for y yields

$$y = \begin{cases} \frac{-b(x) \pm \sqrt{b(x)^2 - 4a(x)c(x)}}{2a(x)} & \text{if } \alpha_2 \neq 0 \\ \frac{-c(x)}{b(x)} & \text{if } \alpha_2 = 0 \text{ and } b(x) \neq 0 \end{cases}$$

Degenerate conics are easily recognized [18].

A *conic arc* is an x-monotone curve that corresponds to one choice of sign in the equation above. A conic decomposes into either one or two conic arcs. In the latter case we refer to the two arcs as the lower and the upper arc. For a conic C we use C_0 and (if it exists) C_1 to denote the arcs of C. We view the arcs as functions of x, i.e., $C_1(x)$ is the y-value of the upper arc of C at x.

Intersection of Two Conics: Consider two conics

$$Q_1 = a_1 y^2 + b_1 y + c_1 = 0 \qquad Q_2 = a_2 y^2 + b_2 y + c_2 = 0.$$

Then there is a polynomial R in x of degree at most four such that the x-coordinates of the intersections of Q_1 and Q_2 are roots of R.

$$R = a_2 b_1^2 c_2 + a_2^2 c_1^2 - a_2 c_1 b_2 b_1 - 2 a_2 a_1 c_1 c_2 - b_2 a_1 b_1 c_2 + c_2^2 a_1^2 + b_2^2 a_1 c_1 = 0$$

is called the *resultant* of Q_1 and Q_2; see [8] for a discussion of resultants. A root x of R does not necessarily correspond to an intersection of Q_1 and Q_2 in $\mathbb{R}^2$; the corresponding y-values may have non-zero imaginary parts.

Low-Degree Algebraic Numbers: The x-coordinates (and similarly the y-coordinates) of intersections points are roots of polynomials of degree at most four. We call an algebraic number a *one-root-number* if it is of the form $\alpha + \beta \sqrt{\gamma}$ with $\alpha, \beta, \gamma \in \mathbb{Q}$. The following well-known lemma is useful.

Lemma 2. *A degree-four polynomial p either has four simple roots or all roots of p are one-root-numbers. The two cases are easily distinguished and the one-root-numbers can be determined in the latter case.*

Proof. (Sketch) Follows from a case distinction on the degree of $p/gcd(p, p')$.

We represent algebraic numbers x in one of two ways. Either as one-root-numbers or as triples (P, l, r) where P is a polynomial with only simple roots, l and r are rational numbers, P has exactly one real root in the open interval (l, r) and $P(l) \neq 0 \neq P(r)$. Such an interval is called an *isolating interval* for the root. In our implementation, l and r have the additional property that their denominators are powers of two. We determine isolating intervals by means of Uspensky's algorithm [7, 23]. Isolating intervals are easily *refined* by considering the point $m = (l + r)/2$. If $P(m) = 0$, we have a one-root-number for x. Otherwise, we replace the isolating interval by either (l, m) or (m, r) depending on the sign of $P(m)$.

One-root-numbers are represented as objects of the number type leda-real, cf. [5] or [21, Section 4.4]. Integers are leda-reals, and if x and y are leda-reals, so are $x \pm y$, $x * y$, x/y, and $\sqrt[k]{x}$ for arbitrary integer k. Leda-reals have exact comparison operators $\leq$, $<$ and $=$. In particular, if x is a leda-real and P is a polynomial with integer coefficients, we can determine the sign of $P(x)$.

We next describe how to compare two algebraic numbers x and y. If both of them are given as leda-real, we use the comparison operator of leda-real. If $x = (P, l, r)$ and y is a leda-real, we proceed as follows: If $y \leq l$ or $y \geq r$, the outcome of the comparison is clear. So assume $l < y < r$. If $P(y) = 0$, $x = y$. So assume that $P(y) \neq 0$. Then(!!!) $x \neq y$. We now refine the isolating interval for x as described in the preceding paragraph until $y \notin (l, r)$.

In order to compare two algebraic numbers $x = (P, l_x, r_x)$ and $y = (Q, l_y, r_y)$ we have to work slightly harder. If the isolating intervals are disjoint, we are done. Otherwise, let $I = (l, r)$ be the intersection of the isolating intervals. We have $x = y$ iff P and Q have a common root in I. We first refine the isolating intervals of x and y using the endpoints of I. Then it is either the case that both intervals are I or the intervals are disjoint. If they are disjoint, then we are done. Otherwise, we know that P and Q both have exactly one simple root in I. These roots are equal if $g = gcd(P, Q)$ has a root z in I, in which case z must be a simple root. Thus the degree of g can be used to decide quickly about equality or inequality in certain cases. For example, if $deg(g) = 0$, we know that x and y are not equal, and similarly, if $deg(g) = 4$, we know they are equal. Otherwise we use the fact that g has only simple roots. So $x = y = z$ iff $sign(g(l)) \neq sign(g(r))$. Furthermore, depending on the degree of g, a rational or one-root representation for x and y may be obtained if they are zeros of g, P/g, or Q/g and if the respective degree is ≤ 2.

The *critical values* of a conic are the roots of $h(x)$ if $\alpha_2 \neq 0$ and is the root of $b(x)$ if $\alpha_2 = 0$. At the critical values the conic either has a vertical tangent or a pole. Critical values are one-root-numbers.

Intersection of Two Conics, Continued: Let R be the resultant of conics C and D and let x be a root of R. Do arcs C_i and D_j intersect at x?

If x is given as a leda-real, we simply compare $C_i(x)$ and $D_j(x)$ using leda-reals. Otherwise, by Lemma 2, $x = (R, l, r)$ is a simple root of R and hence arcs C_i and D_j cross at x if they intersect at all. Let (R, l, r) be the representation of x. We refine the representation of x until the isolating interval contains no critical value of either C or D. Then C_i and D_j are defined on the entire interval $[l, r]$ and they intersect at most once in $[l, r]$. Thus C_i and D_j intersect at x iff the signs of $C_i(r) - D_j(r)$ and $C_i(l) - D_j(l)$ differ. We compute the signs using leda-reals.

Conic Points, Conic Segments, and Comparisons: We specify conic points by an x-coordinate (= an algebraic number) and a conic arc C_i. The point has coordinates $(x, C_i(x))$. If x is a leda-real, we can compute the y-coordinate as a leda-real. A conic segment is the part of a conic arc between two conic points.

The x-compare of two conic points is tantamount to the comparison of two algebraic numbers, which was discussed above. We turn to the xy-compare. Let (x_1, C_i) and (x_2, D_j) be two conic points. If $x_1 \neq x_2$, we are done. If $x_1 = x_2$ and we know a one-root-number for x_1, we simply compare $C_i(x_1)$ and $D_j(x_2)$. So assume otherwise. We compute the resultant R of C and D and compare x_1 to the roots of R. Three cases arise: x_1 is either not a root of R, a multiple root of R (yielding a one-root-number for it), or a simple root of R. In the latter two cases, we proceed as described in the paragraph on intersection of two conics. In the first case, we refine the isolating interval of x_1 until is contains no critical values of C and D and then use the fact that the y-order of the two arcs at x_1 is the same as the y-order at the right end of the isolating interval of x_1. The strategy just described also resolves the *has_on* predicate as well as *y_order*.

Multiplicity of Intersection: Let $p = (x, y)$ be an intersection point of arcs C_i and D_j. Then x is a zero of the resultant of C and D with multiplicity $m > 0$. If $m = 1$, the multiplicity of the intersection is 1. So assume that $m > 1$. Then we know a one-root-number for x. We have to deal with two difficulties. The first difficulty is that there might be intersections with non-real y values, but this cannot happen since the y values would come in conjugate pairs. Thus the conic would have three points on the complex line through x parallel to the y axis and hence the conic would be degenerate.

The second difficulty arises when both conics have two arcs and hence hence there might be intersections between C_{1-i} and D_{1-j} at x. Let d be the (currently unknown) multiplicity of the intersection at p and let e be the multiplicity of the intersection of C_{1-i} with D_{1-j} at x. Then $d + e = m$. We test $e \geq 1$ by testing the equality $C_{1-i}(x) = D_{1-j}(x)$ using leda-reals. If $e = 0$, we are done. Otherwise, let q be the corresponding intersection point. We next test $d \geq 2$, by checking whether the normal vectors $(C_x(p), C_y(p))$ and $(D_x(p), D_y(p))$ are parallel. Here C_x and C_y are the partial derivatives of C. The check is again a computation using leda-reals since $p = (x, C_i(x))$. If $d < 2$ then $d = 1$ and we are done. So assume $d \geq 2$. If $m = 3$ then $d = 2$ and we are done. Assume otherwise, i.e., $m = 4$. We check whether $e \geq 2$. If $e = 1$ then $d = m - 1 = 3$. If $e \geq 2$ then $d = e = 2$.

Order Immediately to the Right of a Common Point: Consider arcs C_i and D_j containing $p = (x, y)$ and extending to the right. Assume that we know a leda-real u such that $x < u$, arcs C_i and D_j are defined on $[x, u]$ and do not intersect in the interval $(x, u]$. Then the y-order just right of x is the same as the y-order at u; the latter can be computed using leda-reals. How can we obtain u? If x is a simple root of the resultant of C and D, we refine the isolating interval of x until it contains no critical values of C and D and take u as the right endpoint of the isolating interval. If x is a multiple root of the resultant, we have one-root-numbers for all roots and we simply take a rational point to the right of x and within the x-range of the two arcs. This can be, for example, the midpoint of x and the next larger root of the resultant if there is one.

5 Implementation and Empirical Results

We describe our implementations and report about experiments. All our implementations are in C++. They use components of LEDA, CGAL, and the standard template library.

Algebraic Numbers and Conics: We have implemented four classes: *alg_number*, *conic*, *arrangement2*, and *X_mono_conic_segment*. The first realizes algebraic numbers as discussed in Section 4. The main ingredients are polynomials, gcd of polynomials, resultant computations, Uspensky's algorithm for root isolation, and leda-reals. The main functionality is exact comparison between algebraic numbers. The class *conic* realizes conics and the predicates and constructions discussed in Section 4. The class *arrangement2* uses the functionality of the two other classes to construct the arrangement of two conics. The class *X_mono_conic_segment* realizes an implementation of a segment type as required by the sweep-line algorithm and generalized polygon class.

The Naive Algorithm for Computing Arrangements of Conics: This algorithm takes a set of conics and computes the subdivision of the plane defined by them. It first computes the arrangement of any pair of conics and then merges the pairwise arrangements into a single arrangement. This implementation provides a testbed for classes *alg_number*, *conic*, and *arrangement2*, and serves as a reference implementation for the sweep-line algorithm.

The first phase of the naive algorithm produces a list of all intersections (x, C_i, D_j). We generate two conic points (x, C_i) and (x, D_j) for each intersection and sort them using *compare_xy*. Then it is easy to remove duplicates, and the vertices of the arrangement A are known. We create additional vertices for the points of vertical tangency and also create a dictionary that maps conic points to the vertices of A. Then we sort our conic points a second time. For the second sorting, the conic arc is the main key and the second key is the x-coordinate. We obtain the sorted list of intersection points on each arc. Using the map from conic points to vertices of A, we create the edges of A. Finally, we determine the cyclic order of the edges incident to each vertex using the predicate *compare_right_of_common_point*. The final result is a planar map.

We checked the correctness of the implementation both manually (for small examples) and by checking Euler's equation for planar maps, which is a good heuristic test. We ran the algorithm on: (1) conics in general position, (2) conics with carefully designed degeneracies, and (3) conics with perturbed degeneracies (*i.e.*, almost degenerate points). Figure 2 shows two examples of the second kind. Both feature high-degree intersection points with various multiplicities of intersections. The example on the left also contains intersection points with equal x-coordinate. Table 1 shows that, as one would expect, running time increases when degeneracies are present and also when the input precision is increased to represent nearly degenerate cases. However, the increase is not unreasonable. With the naive algorithm, we are able to compute, an arrangement of 200 conics with approximately 55000 intersection points in around 28 minutes.

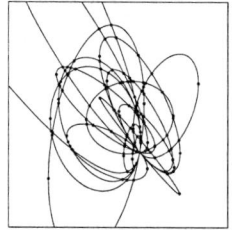 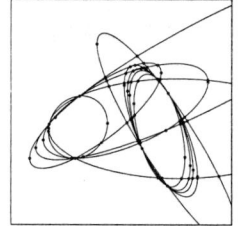

Fig. 2. On the left, a set of 15 ellipses all of which intersect at one point in the lower right quadrant of the picture with varying multiplicities. On the right is a set of 10 ellipses, also in highly degenerate positions

A Generic Implementation of the Sweep Algorithm: Our implementation of the algorithms, classes and predicates described in Sections 3 and 4 is based on the CGAL geometry kernel and the original implementations in LEDA of the sweep-line algorithm and the generalized polygons. To easily accommodate different segment types, point types and predicate implementations, we have followed the generic programming paradigm [22] and used the concept of geometric traits classes introduced with CGAL [14]. By supplying different traits classes, the same algorithm can be applied to different kinds of objects or using different predicate implementations.

Such a generic implementation made light work of producing the empirical results presented below that compare different segment types, different predicate implementations and different underlying kernels.

Sweeping Circular Arcs and Straight Line Segments: When only circular arcs and line segments are used, the implementation of the predicates and functions become easier since the coordinates of all intersection points are one-root-numbers. This means in particular that when the circles and lines supporting the segments are specified through rational points, much of the computation can be carried out using rational numbers. Only when it is time to compute the coordinates of the points using the *sqrt* function will the leda-real number type be used. This is generally a big efficiency win. Furthermore, we are able to exploit the fact that the intersection point between any two rational circles can be described via the

Table 1. Average computation time required by the naive algorithm per pair of conics

	processing time per pair of conics	input precision
general position	30 ms	50 bits
degenerate position	48 ms	50 bits
perturbed degenerate position	48 ms	100 bits

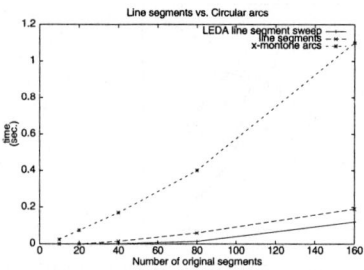 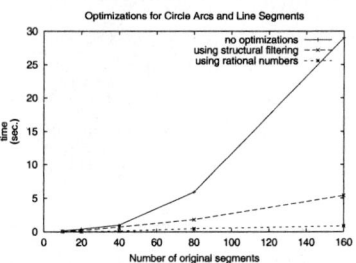

Fig. 3. On the left, a comparison of running times for the optimized version of the LEDA line segment sweep algorithm, our generic implementation of the sweep-line algorithm using line segments and the same implementation using circular arcs. On the right, a comparison of running times using various optimizations for circular arcs and line segments. Running times were recorded on an 846 MHz Pentium III processor

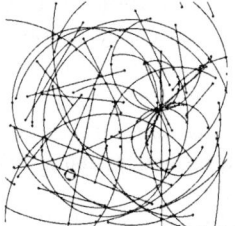 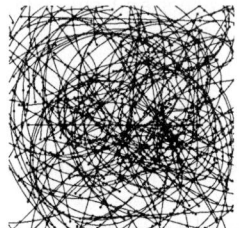

Fig. 4. On the left the planar graph that results from the sweep of a set of 50 circular arcs and line segments with many degeneracies. The graph was produced in less than 1 second on an 846 MHz Pentium III processor. On the right is an example with 241 circular arcs and line segments, which required approximately 2 seconds to compute

intersection of a rational line and a rational circle [10] in our implementation of the *compare_xy* predicate for points that lie on circular arcs or line segments. That is, before comparing the actual coordinates of the points p and q, we first determine if they were constructed in the same way. If so, they are equal and we are done. This technique is called *structural filtering* [16, 25]. Notice that, in contrast to usual numeric filters used in the exact computation paradigm, which filter out the easy cases that can be dealt with quickly by imprecise number types (*i.e.*, when the points' coordinates are vastly different from each other), this filtering technique works by filtering out the most difficult case (when the points' coordinates are identical) and thus assures that the cases left are usually relatively easy. The graph in figure 3 illustrates the advantages of using these optimizations.

Sweeping Conic Segments: When constructing arrangements of general conic segments such as the ones shown in Figure 2, running times are naturally higher.

Examples similar to the ones shown in Figure 2 with 30 and 60 conic segments require, respecitvely, 19 seconds and 49 seconds using our current implementation. Table 1 indicates that, as one would expect, running time degrades when degeneracies are introduced. However, with the application of appropriate filtering techniques in our predicate implementations we are confident that the running times for conic segments will come in line with those for circular arcs.

6 Conclusions

We described an exact kernel for conic arcs, algorithms for dealing with low-degree algebraic numbers, a sweep-line algorithm for curved segments, and algorithms for boolean operations on conic polygons. Our algorithms and their implementations are complete, exact, and efficient. We feel that it was crucial for our work that we had all three goals in mind right from the beginning of our work.

References

[1] J. Bentley and T. Ottmann. Algorithms for reporting and counting geometric intersections. *IEEE Transaction on Computers C 28*, pages 643–647, 1979.
[2] R. Bix. *Conics and Cubics: A Concrete Introduction to Algebraic Curves*. Springer Verlag, 1998.
[3] J.-D. Boissonnat and F. P. Preparata. Robust plane sweep for intersecting segments. Research Report 3270, INRIA, Sophia Antipolis, Sept. 1997.
[4] J.-D. Boissonnat and J. Snoeyink. Efficient algorithms for line and curve segment intersection using restricted predicates. In *Proc. 15th Annu. ACM Sympos. Comput. Geom.*, pages 370–379, 1999.
[5] C. Burnikel, S. Funke, K. Mehlhorn, S. Schirra, and S. Schmitt. A separation bound for real algebraic expressions. In *ESA 2001*, volume 2161 of *LNCS*, pages 254–265, 2001.
[6] T. M. Chan. Reporting curve segment intersection using restricted predicates. *Computational Geometry*, 16(4):245–256, 2000.
[7] G. E. Collins and A.-G. Akritas. Polynomial real root isolation using Descartes' rule of sign. In *SYMSAC*, pages 272–275, Portland, OR, 1976.
[8] D. Cox, J. Little, and D. O'Shea. *Ideals, Varieties, and Algorithms*. Springer-Verlag New York, Inc., 2nd edition, 1997.
[9] M. de Berg, M. Kreveld, M. Overmars, and O. Schwarzkopf. *Computational Geometry: Algorithms and Applications*. Springer, 1997.
[10] O. Devillers, A. Fronville, B. Mourrain, and M. Teillaud. Exact predicates for circle arcs arrangements. In *Proc. 16th Annu. ACM Sympos. Comput. Geom.*, 2000.
[11] D. P. Dobkin and D. L. Souvaine. Computational geometry in a curved world. *Algorithmica*, 5:421–457, 1990.
[12] D. P. Dobkin, D. L. Souvaine, and C. J. Van Wyk. Decomposition and intersection of simple splinegons. *Algorithmica*, 3:473–486, 1988.
[13] L. Dupont, D. Lazard, S. Lazard, and S. Petitjean. A new algorithm for the robuts intersection of two general quadrics. submitted to Solid Modelling 2002.

[14] A. Fabri, G.-J. Giezeman, L. Kettner, S. Schirra, and S. Schönherr. On the design of CGAL, the computational geometry algorithms library. *Software – Practice and Experience*, 30:1167–1202, 2000.

[15] E. Flato, D. Halperin, I. Hanniel, and O. Nechushtan. The design and implementation of planar maps in CGAL. In *Proceedings of the 3rd Workshop on Algorithm Engineering*, volume 1668 of *Lecture Notes in Computer Science*, pages 154–168. Springer, 1999.

[16] S. Funke and K. Mehlhorn. Look - a lazy object-oriented kernel for geometric computation. In *Proceedings of the 16th Annual Symposium on Computational Geometry (SCG-00)*, pages 156–165, Hong Kong, China, June 2000. Association of Computing Machinery (ACM), ACM Press.

[17] N. Geismann, M. Hemmer, and E. Schömer. Computing a 3-dimensional cell in an arrangement of quadrics: Exactly and actually. In *ACM Conference on Computational Geometry*, 2001.

[18] M. Hemmer. Reliable computation of planar and spatial arrangements of quadrics. Master's thesis, Max-Planck-Institut für Informatik, 2002.

[19] J. Keyser, T. Culver, M. Foskey, S. Krishnan, and D. Manocha. Esolid - a system for exact boundary evaluation. submitted to Solid Modelling 2002.

[20] J. Keyser, T. Culver, D. Manocha, and S. Krishnan. MAPC: A library for efficient and exact manipulation of algebraic points and curves. Technical Report TR98-038, University of N. Carolina, Chapel Hill, 1998.

[21] K. Mehlhorn and S. Näher. *The LEDA Platform for Combinatorial and Geometric Computing*. Cambridge University Press, 1999. 1018 pages.

[22] D. R. Musser and A. A. Stepanov. Generic programming. In *1st Intl. Joint Conf. of ISSAC-88 and AAEC-6*, pages 13–25. Springer LNCS 358, 1989.

[23] F. Rouillier and P. Zimmermann. Efficient isolation of polynomial real roots. Technical Report 4113, INRIA, 2001.

[24] A. A. Schäffer and C. J. Van Wyk. Convex hulls of piecewise-smooth Jordan curves. *J. Algorithms*, 8:66–94, 1987.

[25] R. Wein. High-level filtering for arrangements of conic arcs. In *Proceedings of ESA 2002*, 2002.

TSP with Neighborhoods of Varying Size

Mark de Berg[1], Joachim Gudmundsson[1]*, Matthew J. Katz[2]**,
Christos Levcopoulos[3], Mark H. Overmars[1], and A. Frank van der Stappen[1]

[1] Institute of Information and Computing Sciences, Utrecht University
P.O.Box 80.089, 3508 TB Utrecht, the Netherlands
{markdb,joachim,markov,frankst}@cs.uu.nl
[2] Department of Computer Science, Ben-Gurion University
Beer-Sheva 84105, Israel
matya@cs.bgu.ac.il
[3] Department of Computer Science, Lund University
Box 118, 221 00 Lund, Sweden
christos@cs.lth.se

Abstract. In TSP with neighborhoods we are given a set of objects in the plane, called *neighborhoods*, and we seek the shortest tour that visits all neighborhoods. Until now constant-factor approximation algorithms have been known only for cases where the objects are of approximately the same size. We present the first polynomial-time constant-factor approximation algorithm for disjoint convex fat objects of arbitrary size. We also show that the problem is APX-hard and cannot be approximated within a factor of 391/390 in polynomial time, unless $P = NP$.

1 Introduction

The Traveling Salesman Problem (TSP) is one of the best known and most widely studied optimization problems. In the Euclidean setting, one is given a set of n points in the plane (or a higher-dimensional Euclidean space), and one wants to find the tour—that is, closed curve—of shortest length that visits all points from the set. A natural generalization of the Euclidean TSP, first studied by Arkin and Hassin [3], is the TSP-with-neighborhoods problem (TSPN): given a set of objects in the plane, called *neighborhoods*, find a shortest tour that visits all neighborhoods—see Figure 1 for an example. (Stated in the traditional TSP setting, one can phrase the problem as follows. A salesman wants to meet some potential buyers, each of whom specifies a region in the plane where he is willing to meet the salesman. The salesman now wants to find a shortest tour such that he can meet with all potential buyers.) Since TSPN is a generalization of the Euclidean TSP, it is NP-hard [8, 16].

One can think of TSPN as a generalization of the Errand Scheduling Problem [20], sometimes also called the One-of-a-Set TSP [14]. In the One-of-a-Set

---

* Supported by The Swedish Foundation for International Cooperation in Research and Higher Education.
** Supported by grant no. 2000160 from the U.S.-Israel Binational Science Foundation.

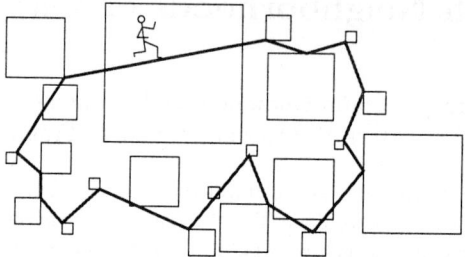

Fig. 1. An example of TSPN when the input objects are axis-parallel squares

TSP one is given a collection of sets of points and the aim is to visit at least one point from each set. This problem has applications in communication-network design [10], VLSI routing [17], and in the manufacturing industry [19]. The difference between TSPN and One-of-a-Set TSP is that in the latter problem each set to be visited has a finite number of points, unlike the neighborhoods in TSPN.

Mata and Mitchell [12] provided a general framework that gives an $O(\log k)$-approximation algorithm for TSPN with time complexity $O(n^5)$, where k is the number of neighborhoods and n is the total complexity of the input. The time complexity was later improved by Gudmundsson and Levcopoulos [9] to $O(n^2 \log n)$. Although no algorithms with a better approximation factor are known for the general TSPN problem, they are known for various special cases. For example, Arkin and Hassin [3] gave $O(1)$-approximation algorithms for the following cases: parallel unit-length segments, translates of a convex polygonal object and, more generally, for objects with diameter segments that are parallel to a common direction and a ratio between the longest and the shortest diameter that is bounded by a constant. Later, Dumitrescu and Mitchell [7] extended and improved these results to include neighborhoods with comparable diameter, unit disks, and infinite lines. Gudmundsson et al. [9] gave a polynomial-time approximation scheme (PTAS) for the special case when the tour is short compared to the size of the neighborhoods.

The results quoted above leave the main question in this area still open: is there a constant-factor approximation, or perhaps even a PTAS, for TSPN? In this paper we present two results, which give a partial answer to this question.

First of all, in Section 2 we use a reduction from the Vertex-Cover Problem to show that TSPN is APX-hard and cannot be approximated within a factor of 391/390 in polynomial time, unless P=NP.[1] Thus, a PTAS does not exist for TSPN, unless P=NP. This is in contrast to the standard Euclidean TSP, for which there is a PTAS allowing one to compute, for any given $\varepsilon > 0$, a $(1+\varepsilon)$-approximation in $O(n^{O(1/\varepsilon)})$ time [4, 15].

---

[1] This was recently improved by Schwartz and Safra [18] who showed that TSPN cannot be approximated within a factor of $(2-\varepsilon)$, unless $NP \subseteq TIME(n^{O(\log \log n)})$.

Second, in Section 3 we give a new constant-approximation algorithm for TSPN. Although we are not able to solve the general case, our algorithm is able to handle cases that could not be handled before. In particular, all special cases that were successfully solved until now have one common property: it is required that the input objects have roughly the same size. The main contribution of our paper is a polynomial-time constant-factor approximation algorithm for a class of arbitrary-size input objects, namely for input sets S consisting of n disjoint convex fat objects.

Before we proceed we introduce some notation. We use $|\tau|$ to denote the length of a curve τ. In particular, for two points p and q, we use $|pq|$ to denote the length of the line segment pq. Furthermore, $\text{dist}(\sigma, \sigma')$ denotes that distance between objects σ and σ', that is, $\text{dist}(\sigma, \sigma') := \min_{p \in \sigma, q \in \sigma'} |pq|$. Finally, for a set S of objects we use $T_{opt}(S)$ to denote an optimal (that is, shortest) tour visiting each object in S.

2 TSPN is APX-Hard

In this section we show that TSPN is APX-hard. We use a reduction from VERTEXCOVER, which is defined as follows. Let $\mathcal{G} = (V, A)$ be an undirected graph. A vertex cover for $\mathcal{G}$ is a subset $V' \subset V$ of vertices such that for any arc $(u, v) \in A$ we have $\{u, v\} \cap V' \neq \emptyset$—that is, any arc has at least one incident vertex in V'. VERTEXCOVER asks for a vertex cover of minimum cardinality.

Lemma 1. *(Berman and Karpinski [5])* VERTEXCOVER *is APX-hard, and cannot be approximated within a factor of* $79/78$ *in polynomial time unless P=NP, even when the degree of each vertex is bounded by four.*

Let $\mathcal{G} = (V, A)$ be an undirected graph with n vertices of maximum degree at most four. We can restrict ourselves to consider connected graphs with at least n arcs since the cases when this is not true easily can be handled separately: if the graph is not connected then we can consider each connected component separately, and in the case when the graph is a tree an optimal vertex cover can be found by dynamic programming.

We will construct a set $S(\mathcal{G})$ of $n + |A|$ objects such that an approximation of $T_{opt}(S(\mathcal{G}))$ can be transformed into an approximation of a minimum vertex cover for $\mathcal{G}$. The construction uses two types of objects: points and objects consisting of a line segment followed by a circular arc followed by another line segment. (The circular arc can easily be replaced by a few line segments.) The second type of objects will correspond to the arcs of $\mathcal{G}$, hence they are called *arc objects*. The points are called *helper objects*.

> **The helper objects.** The helper objects are the vertices of a regular n-gon of unit edge length.
> **The arc objects.** To define the arc objects, we first assign a location to each node of $\mathcal{G}$. Note that these locations are not objects in $S(\mathcal{G})$. The n node locations are defined as follows. For each edge pq of the n-gon defining the

helper objects, the point x outside the n-gon such that $|px| = |qx| = 1$ is a node location. We assign each node in V a unique node location in an arbitrary manner. Let C be the circle with the same center as the n-gon and at distance $2n$ from the node locations. We call the line segment of length $2n$ connecting a node location to C the *spoke* of the node. Now the arc object for an arc $(u,v) \in A$ consists of the spokes of u and v and one of the circular arcs (it doesn't matter which one) along C connecting the spokes.

Figure 2 illustrates the construction. The set of helper objects is denoted by $\mathcal{S}_H$ and the set of arc objects by $\mathcal{S}_A$. We let $\mathcal{S}(\mathcal{G}) := \mathcal{S}_H \cup \mathcal{S}_A$. In the remainder of this section we prove that if we can approximate $\mathcal{T}_{opt}(\mathcal{S}(\mathcal{G}))$, then we can approximate a minimum vertex cover for $\mathcal{G}$. From this it will follow that TSPN is APX-hard.

We start by proving a bound on the length of any tour visiting a given subset of the spokes.

Lemma 2. *Let S be an arbitrary subset of the spokes in the construction above. Then $|\mathcal{T}_{opt}(\mathcal{S}_H \cup S)| = n + |S|$.*

Proof. An optimal tour for $\mathcal{S}_H \cup S$ has $n + |S|$ links, each connecting a pair of objects in $\mathcal{S}_H \cup S$. The distance between any two objects is at least 1, so $|\mathcal{T}_{opt}(\mathcal{S}_H \cup S)| \geq n + |S|$. Obviously, a tour of this length also exists: connect (the node location of) each spoke to the two closest helper points and complete the tour by adding the appropriate subset of edges of the n-gon.

Next we show that an approximation for $\mathcal{T}_{opt}(\mathcal{S}(\mathcal{G}))$ induces an approximation of a minimum vertex cover for $\mathcal{G}$.

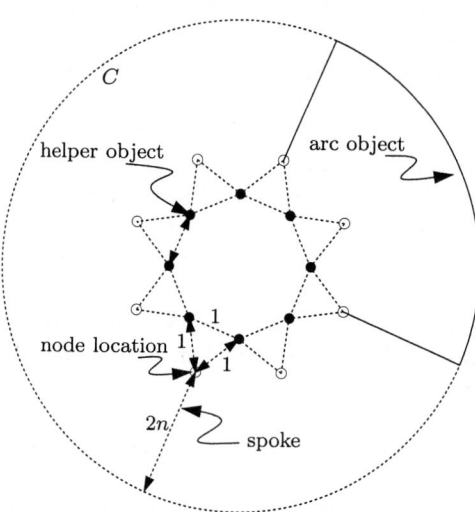

Fig. 2. The construction used in the reduction from VERTEXCOVER to TSPN

Lemma 3. *Let $\mathcal{T}$ be a tour visiting each object in $\mathcal{S}(\mathcal{G})$, such that $|\mathcal{T}| = c \cdot |\mathcal{T}_{opt}(\mathcal{S}(\mathcal{G}))|$ for some constant $1 \leq c < 2$. Then we can construct in polynomial time a vertex cover for $\mathcal{G}$ whose cardinality is at most $5c - 4$ times the minimum cardinality of a vertex cover for $\mathcal{G}$.*

Proof. Let m be the minimum cardinality of a vertex cover for $\mathcal{G}$. Note that $m \geq n/4$, since the degree of the nodes in $\mathcal{G}$ is bounded by four and since the number of arcs is at least n. By Lemma 2 this implies that there is a tour for $\mathcal{S}(\mathcal{G})$ of length $n + m$. Since a tour of length $n + m$ or less can never reach the circular-arc parts of the arc objects, we even have that $|\mathcal{T}_{opt}(\mathcal{S}(\mathcal{G}))| = n + m$. Using Lemma 2 again, and the fact that $\mathcal{T}$ cannot visit the circular-arc parts of the arc objects (because $c < 2$), we see that $\mathcal{T}$ can visit at most

$$c(n+m) - n = ((c-1)n/m + c)m \leq (5c-4)m$$

spokes. The nodes corresponding to these spokes form a vertex cover, which can trivially be found in polynomial time.

The main result of this section now follows easily.

Theorem 1. *TSP with neighborhoods is APX-hard, and cannot be approximated within a factor of $391/390$ in polynomial time unless P=NP.*

Proof. Lemma 3 implies that if we can approximate TSPN within a factor $391/390$, then we can approximate VERTEXCOVER (for graphs of maximum degree four) within a factor $5 \cdot (391/390) - 4 = 79/78$. Lemma 1, however, states that this is impossible in polynomial time unless P=NP.

3 A Constant-Approximation Algorithm for Fat Objects

We now turn our attention to computing an approximation of the optimal tour through a given set of α-fat disjoint bounded convex objects. The definition of fatness we use was introduced by Van der Stappen [21]. For convex objects it is basically equivalent to other definitions [1, 2, 11, 13].

Definition 1. *[21] Let α be a parameter with $\alpha \geq 1$. An object σ is α-fat if for any disk D whose center lies in σ and whose boundary intersects σ, we have $\text{area}(D) \leq \alpha \cdot \text{area}(\sigma \cap D)$.*

In our setting of bounded planar objects it turns out that $\alpha \geq 4$ [21].

We shall need the following property of convex fat objects, which was proved by Chew et al. [6].

Lemma 4. *[6] Let σ be a convex α-fat object, and let p, q be two points on $\partial \sigma$. Then the length of the shorter path from p to q along $\partial \sigma$ is at most $(1 + 4\alpha/\pi) \cdot |pq|$.*

We also need the following lemma. Its proof, and the proofs of Lemma 6 and 7 stated later, are omitted in this extended abstract.

Lemma 5. *Let ℓ and ℓ' be two parallel lines intersecting a convex α-fat object σ. Let p and q be two points between ℓ and ℓ' and on opposite sides of σ. Then $\min\{\text{dist}(q, \ell), \text{dist}(q, \ell')\} \leq (1 + 4\alpha/\pi) \cdot |pq|$.*

3.1 The Algorithm

Let $\mathcal{S}$ be a set of n disjoint bounded convex α-fat objects, for some constant α (≥ 4). The overall approach we take is quite simple: for each object $\sigma \in \mathcal{S}$ we compute a collection $\mathcal{R}_\sigma$ of representative points, and then we compute a short tour that visits (that is, passes through at least one point of) each set $\mathcal{R}_\sigma$. The latter step is done using an algorithm by Slavik [20]. His algorithm computes, for a collection of n planar point sets, in polynomial time a tour that visits each set and whose length is at most $3k/2$ times the optimal length of such a tour, where k is the maximum cardinality of the sets. Our goal is thus to construct the sets $\mathcal{R}_s$ in such a way that (i) their maximum cardinality is bounded by a constant, and (ii) the optimal tour visiting each set $\mathcal{R}_\sigma$ is not much longer than the optimal tour visiting each object σ.

The Representative Points. Fix an object $\sigma \in \mathcal{S}$. The set $\mathcal{R}_\sigma$ of representative points of σ contains at most three types of points: several *corners*, at most one *anchor*, and possibly several *guards*. Next we define these points, and we prove some basic properties of them.

The corners. Let γ be a small constant such that $\pi/2 \geq \gamma > 0$ and $t = 2\pi/\gamma$ is an integer; later we shall see that we need to choose γ so that $\tan(\gamma/2) \leq \pi/(21\alpha)$. For some i with $0 \leq i < t$, we let p_i be a point on $\partial\sigma$ that has a tangent making an angle $i\gamma$ with the positive x-axis. Here (and in the sequel) the tangent is a directed line, whose direction is consistent with a counterclockwise orientation of $\partial\sigma$. For convenience, we assume that the points p_i are unique; since the objects have constant complexity—in particular, each object has only a constant number of straight segments on its boundary—we can easily enforce this by a suitable rotation of the coordinate axes. We call the points p_i the *corners* of σ. The portions of $\partial\sigma$ connecting consecutive corners are called *sides* of σ. The side connecting corners p_i and p_{i+1} is denoted by e_i. It can happen that $p_i = p_{i+1}$, in which case the side e_i can be ignored. When necessary, we shall write $p_i(\sigma)$, $e_i(\sigma)$, and so on, to indicate to which object the corner or side belongs.

Before we proceed, we need to introduce one more concept. Consider a side e_i of σ, connecting the corners p_i and p_{i+1}. For a point p and an angle ϕ, let $\ell(p, \phi)$ be the line through p that makes an angle ϕ with the positive x-axis. Define $\phi_i := (i + 1/2)\gamma - \pi/2$. The *slab of* e_i, denoted $\mathrm{slab}(e_i)$, is the slab bounded by the lines $\ell(p_i, \phi_i)$ and $\ell(p_{i+1}, \phi_i)$. The part of this slab bounded by e_i, $\ell(p_i, \phi_i)$, and $\ell(p_{i+1}, \phi_i)$, and not containing σ, is called the *semi-slab of* e_i; it is denoted $\mathrm{slab}(e_i)^+$. Fig. 3 illustrates these definitions.

Next we prove two properties of the partitioning into sides. The first lemma is basically a consequence of the fact that a side is 'almost straight'—the tangent does not vary too much along a side—and that the slab boundaries are 'almost perpendicular' to the side.

Lemma 6. *Let σ be a convex object, and e_i be one of its sides. Let ℓ be any line parallel to and inside* $\mathrm{slab}(e_i)$, *and let $p := \ell \cap e_i$. Then for any point $q \in e_i$ we have* $|qp| \leq (1/\cos(\gamma/2)) \cdot \mathrm{dist}(q, \ell)$.

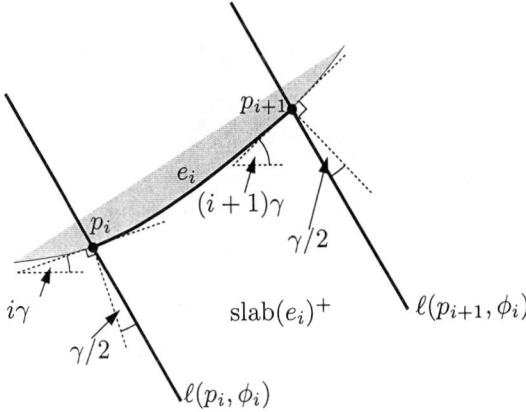

Fig. 3. Corners, sides and (semi-)slabs

The next lemma uses the fact that the sides are almost straight, as well as the fact that we deal with fat objects.

Lemma 7. *Let σ and σ' be two convex α-fat objects such that $e_i(\sigma')$ lies at least partially inside $\mathrm{slab}(e_i(\sigma))^+$. Let p be any point on $e_i(\sigma)$ such that the line through p parallel to $\mathrm{slab}(e_i(\sigma))$ intersects σ', and let q be any point in $\sigma' \cap \mathrm{slab}(e_i(\sigma))^+$. If γ is such that $\tan(\gamma/2) \leq \pi/(21\alpha)$, then $|pq| \leq c \cdot \mathrm{dist}(e_i(\sigma), e_i(\sigma'))$, for $c = 2 + 10\alpha/\pi$.*

The anchor. Consider a side $e_i(\sigma)$. We say that $e_i(\sigma)$ *dominates* a side $e_{i'}(\sigma')$ of another object σ' if $i = i'$ and $\sigma' \subset \mathrm{slab}(e_i(\sigma))^+$. We define an anchor for σ if σ has a side that dominates a side of another object, as follows. Of all the sides of other objects that are dominated by any of the sides of σ, take the one whose distance to its dominating side is minimal. Suppose this 'closest dominated side' is $e_j(\sigma^*)$. Let $\ell := \ell(p_j(\sigma^*), \phi_j)$. Note that ℓ is parallel to $\mathrm{slab}(e_j(\sigma))$. Then anchor($\sigma$), the anchor of σ, is the point $\ell \cap e_j(\sigma)$—see Figure 4.

The guards. Consider a side e_i of σ. We call e_i *extreme* if all other objects σ' intersect the semi-slab $\mathrm{slab}(e_i)^+$. For each extreme side of σ, we add one or two more representative points, called guards. They are defined as follows.

Suppose there are lines parallel to and inside $\mathrm{slab}(e_i)$ that stab all the objects in $\mathcal{S}$. The collection of all such stabbing lines forms a subslab of $\mathrm{slab}(e_i)$. We add two guards on e_i to $\mathcal{R}_\sigma$, namely the intersection points of the lines bounding this subslab and e_i—see Figure 5(a).

If there are no stabbing lines, then there must be a line ℓ parallel to and inside $\mathrm{slab}(e_i)$ such that both half-planes defined by ℓ fully contain at least one object of $\mathcal{S}$. (Thus the line ℓ separates at least two objects.) In this case we add one guard to $\mathcal{R}_\sigma$, namely the intersection point of ℓ and e_i—see Figure 5(b).

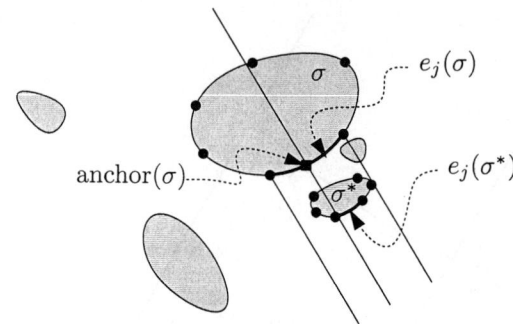

Fig. 4. The anchor of an object σ

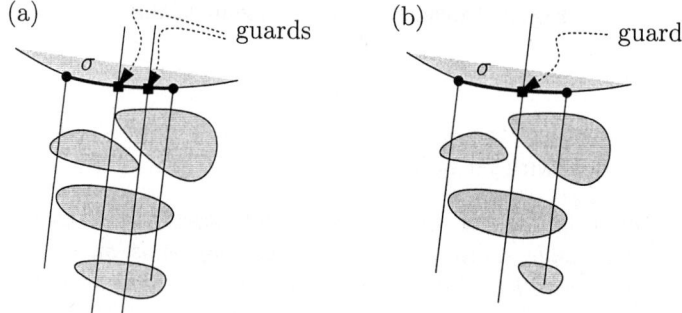

Fig. 5. The guards of an object σ

3.2 The Analysis

To show that our algorithm achieves a good approximation factor, we have to show that the optimal tour visiting each set $\mathcal{R}_\sigma$ is not much longer than the optimal tour visiting each object σ.

Theorem 2. *Let $\mathcal{S}$ be a collection of n disjoint convex α-fat objects in the plane, and let $\mathcal{R}(\mathcal{S}) := \{\mathcal{R}_\sigma : \sigma \in \mathcal{S}\}$ be the collection of sets of representative points, as defined above. Let $\mathcal{T}_{opt}(\mathcal{R}(\mathcal{S}))$ be an optimal tour visiting each set $\mathcal{R}_\sigma \in \mathcal{R}(\mathcal{S})$, and let $\mathcal{T}_{opt}(\mathcal{S})$ be an optimal tour visiting each object $\sigma \in \mathcal{S}$. Then*

$$|\mathcal{T}_{opt}(\mathcal{R}(\mathcal{S}))| \leq (1+c) \cdot |\mathcal{T}_{opt}(\mathcal{S})|,$$

for $c = (4 + 20\alpha/\pi)t$ (where $t = 2\pi/\gamma$ depends on α).

Proof. The strategy of our proof is to add detours to $\mathcal{T}_{opt} := \mathcal{T}_{opt}(\mathcal{S})$, of total length $c \cdot |\mathcal{T}_{opt}|$, such that the extended tour visits each set $\mathcal{R}_\sigma$.

Let p_{start} be an arbitrary point on $\mathcal{T}_{opt}$ outside all the objects. Imagine traversing the tour (in some arbitrary direction), starting at p_{start}. We define the *entry* of an object $\sigma \in \mathcal{S}$ to be the point on σ where the tour first meets σ during the traversal. If the entry of σ happens to be one of its corners, then the set $\mathcal{R}_\sigma$ is already visited by $\mathcal{T}_{opt}$ and no detour has to be added for σ. The objects for which this is not the case have their entry in the interior of one of their sides. We partition the collection of these objects into t subsets, $\mathcal{S}_0, \ldots, \mathcal{S}_{t-1}$, such that $\mathcal{S}_i$ contains the objects σ whose entry lies in the interior of $e_i(\sigma)$.

Fix a non-empty subset $\mathcal{S}_i$. We claim that we can add detours of total length $(4 + 20\alpha/\pi)|\mathcal{T}_{opt}|$ to $\mathcal{T}_{opt}$, such that each set $\mathcal{R}_\sigma$ with $\sigma \in \mathcal{S}_i$ is visited. Proving this claim will establish the theorem. We distinguish two cases.

Case (i): $|\mathcal{S}_i| > 1$.

Number the objects in $\mathcal{S}_i$ as $\sigma_1, \sigma_2, \ldots$ in the order as they are encountered during the traversal of $\mathcal{T}_{opt}$. Consider an object $\sigma_j \in \mathcal{S}_i$, and let $p_i := p_i(\sigma_j)$ and $p_{i+1} := p_{i+1}(\sigma_j)$. Denote the entry of σ_j by q_j, and let $\mathcal{T}_{opt}(q_j, q_{j+1})$ be the portion of $\mathcal{T}_{opt}$ from q_j to q_{j+1}, with indices taken modulo $(|\mathcal{S}_i|+1)$. We have three subcases.

- If $q_{j+1} \notin \mathrm{slab}(e_i(\sigma_j))^+$, then we add the shorter of the two detours $q_j p_i q_j$ and $q_j p_{i+1} q_j$ to $\mathcal{T}_{opt}$. If $q_{j+1} \notin \mathrm{slab}(e_i(\sigma_j))$ then $\mathcal{T}_{opt}(q_j, q_{j+1})$ crosses one of the bounding lines of $\mathrm{slab}(e_i(\sigma_j))$ and we can use Lemma 6 to bound the length of the detour by $2 \cdot (1/\cos(\gamma/2)) \cdot |\mathcal{T}_{opt}(q_j, q_{j+1})|$. Otherwise q_{j+1} lies inside $\mathrm{slab}(e_i(\sigma_j))$ but on the opposite side of σ_j as seen from q_j. In this case we can use Lemma 4 to bound the length of the detour as follows. Let $q' \neq q_j$ be the intersection point of $q_j q_{j+1}$ with $\partial \sigma_j$.

$$\min\{|q_j p_i q_j|, |q_j p_{i+1} q_j|\} \leq 2 \cdot |\text{shortest path along } \partial \sigma_j \text{ from } q_j \text{ to } q'|$$
$$\leq 2 \cdot (1 + 4\alpha/\pi) \cdot |q_j q'|$$
$$\leq 2 \cdot (1 + 4\alpha/\pi) \cdot |q_j q_{j+1}|$$
$$\leq 2 \cdot (1 + 4\alpha/\pi) \cdot |\mathcal{T}_{opt}(q_j, q_{j+1})|.$$

- If $q_{j+1} \in \mathrm{slab}(e_i(\sigma_j))^+$ and σ_{j+1} lies only partly in $\mathrm{slab}(e_i(\sigma_j))^+$, then we add one of the two detours $q_{j+1} p_i q_{j+1}$ and $q_{j+1} p_{i+1} q_{j+1}$ to $\mathcal{T}_{opt}$ as follows. Let $p \in \{p_i, p_{i+1}\}$ be a corner of $e_i(\sigma_j)$ such that the bounding line of $\mathrm{slab}(e_i(\sigma_j))$ that passes through p intersects σ_{j+1}. We add the detour $q_{j+1} p q_{j+1}$, and use Lemma 7 to bound the length of the detour by $2 \cdot (2 + 10\alpha/\pi) \cdot |\mathcal{T}_{opt}(q_j, q_{j+1})|$.
- If $q_{j+1} \in \mathrm{slab}(e_i(\sigma_j))^+$ and σ_{j+1} lies entirely in $\mathrm{slab}(e_i(\sigma_j))^+$, then we add a detour to $\mathrm{anchor}(\sigma_j)$. (Note that $\mathrm{anchor}(\sigma_j)$ must exist in this case.) More precisely, if σ^* denotes the object that determines $\mathrm{anchor}(\sigma_j)$, then we add a detour from the entry of σ^* to $\mathrm{anchor}(\sigma_j)$ and back. (That σ^* need not be in $\mathcal{S}_i$ does not matter.)

Let $e_k(\sigma_j)$ be the side containing anchor(σ_j). Then, using Lemma 7 and the definition of the anchor, we have

$$\begin{aligned}|\text{length of detour}| &\leq 2 \cdot (2 + 10\alpha/\pi) \cdot \text{dist}(e_k(\sigma_j), e_k(\sigma^*)) \\ &\leq 2 \cdot (2 + 10\alpha/\pi) \cdot \text{dist}(e_i(\sigma_j), e_i(\sigma_{j+1})) \\ &\leq 2 \cdot (2 + 10\alpha/\pi) \cdot |\mathcal{T}_{opt}(q_j, q_{j+1})|.\end{aligned}$$

The total length of the detours we add in Case (i) is bounded by

$$\sum_{\sigma_j \in \mathcal{S}_i} 2 \cdot \max\{\tfrac{1}{\cos(\gamma/2)}, (2 + \tfrac{10\alpha}{\pi})\} \cdot |\mathcal{T}_{opt}(q_j, q_{j+1})| \leq \max\{\tfrac{2}{\cos(\gamma/2)}, (4 + \tfrac{20\alpha}{\pi})\} \cdot |\mathcal{T}_{opt}|.$$

Case (ii): $|\mathcal{S}_i| = 1$.

Let σ be the object in $\mathcal{S}_i$ and let q be its entry. There are two subcases.
- If $\mathcal{T}_{opt}$ does not stay inside slab$(e_i(\sigma))^+$, then we can follow the proof of the first subcase of Case (i) to show that we can add a detour for σ of length at most $\max\{1/\cos(\gamma/2), (1 + 4\alpha/\pi)\} \cdot |\mathcal{T}_{opt}|$.
- Otherwise $\mathcal{T}_{opt}$ lies completely inside slab$(e_i(\sigma))^+$, so $e_i(\sigma)$ is extreme. In this case we add the shortest detour from q to a guard on $e_i(\sigma)$ and back.
  To bound the length of the detour, first suppose that there is no stabbing line for $\mathcal{S}$ parallel to and inside slab$(e_i(\sigma))$. In this case we added a single guard on $e_i(\sigma)$, which lies on a line ℓ that has objects on both sides of it. But then $\mathcal{T}_{opt}$ must cross ℓ, so we can use Lemma 6 to bound the length of the detour by $(1/\cos(\gamma/2)) \cdot |\mathcal{T}_{opt}|$.
  Next suppose that there is a collection of lines inside and parallel to slab$(e_i(\sigma))$ that stab all objects. Let ℓ and ℓ' be the two lines bounding the subslab of all such stabbing lines. The guards we added in this case were the points $\ell \cap e_i(\sigma)$ and $\ell' \cap e_i(\sigma)$. If $\mathcal{T}_{opt}$ does not lie entirely inside the subslab, then we can use Lemma 6 again to bound the length of the detour by $(1/\cos(\gamma/2)) \cdot |\mathcal{T}_{opt}|$. If, on the other hand, $\mathcal{T}_{opt}$ stays inside the subslab, then we can bound the length of the detour as follows. Let σ' be the object that lies directly below σ inside the subslab. Clearly $\mathcal{T}_{opt}$ has to cross σ' to reach a point p below σ' (assuming there are at least three objects). Lemma 5 tells us that $\min\{\text{dist}(q, \ell), \text{dist}(q, \ell')\} \leq (1 + 4\alpha/\pi) \cdot |pq|$. If we combine this with Lemma 6 and use the fact that $\mathcal{T}_{opt} \geq 2|pq|$, then we see that we can bound the length of the detour by $((1 + 4\alpha/\pi)/\cos(\gamma/2)) \cdot |\mathcal{T}_{opt}|$.

We conclude that in Case (ii) we can bound the length of the detour by $((1 + 4\alpha/\pi)/\cos(\gamma/2)) \cdot |\mathcal{T}_{opt}|$.

The condition $\tan(\gamma/2) \leq \pi/(21\alpha)$ needed for Lemma 7 implies that $0.98 < \cos(\gamma/2)$, so the length of the detour in both cases can be bounded by $(4 + 20\alpha/\pi)|\mathcal{T}_{opt}|$. Since there are at most t non-empty subsets $\mathcal{S}_i$, we obtain that $c = (4 + 20\alpha/\pi)t$. Notice that $t = 2\pi/\gamma$ depends on α because of the condition on γ above.

Our algorithm computes an approximation of the optimal tour visiting each set $\mathcal{R}_\sigma$, using Slavik's algorithm. The latter algorithm has an approximation factor $3k/2$, where k is the maximum cardinality of any of the sets. In our case, the maximum size of $\mathcal{R}_\sigma$ is $3t+1$ (t corners, $2t$ guards and 1 anchor), where $t = t(\alpha)$ is the number of sides of an object in $\mathcal{S}$. Combining this with the theorem above, we get the following result.

Corollary 1. *Let $\mathcal{S}$ be a collection of n disjoint convex α-fat objects in the plane. We can compute in polynomial time a tour visiting each object in σ, whose length is at most $3(3t+1)/2 \cdot (1+c)$ times the length of an optimal tour, where $t = \pi/\arctan(\pi/21\alpha)$ is the number of sides of an object in $\mathcal{S}$ and $c = (4 + 20\alpha/\pi)t$.*

Remark 1. A very rough estimate of the approximation factor would be $12000\alpha^3$. A more careful analysis would probably give a better value, although still large.

To improve the approximation factor one might be tempted to increase the value of t, i.e. the number of representative points, but due to the limitations of our proof technique this will only increase the value of the approximation factor.

4 Conclusions and Open Problems

In this paper we have made two steps towards a better understanding of the approximability of TSPN. We have shown that in the most general setting—intersecting and non-convex neighborhoods—TSPN is APX-hard, and cannot be approximated within a factor 391/390, unless P=NP. We also presented a constant-factor approximation algorithm for disjoint neighborhoods that are convex and fat. This is the first constant-factor approximation algorithm that does not require the neighborhoods to have roughly the same size.

Despite this progress, the problem is far from resolved. Probably the most interesting case to focus on is that of disjoint convex objects. Here the best known approximation factor is $O(\log k)$ [9, 12], where k is the number of neighborhoods. In fact, even for horizontal line segments this is the best known bound. We have tried to apply our techniques to obtain a constant-factor approximation for this case, but without success. There might even be a PTAS for disjoint convex neighborhoods; our lower bound result uses intersecting non-convex neighborhoods, and it seems hard to get around this.

Recently Schwartz and Safra [18] showed that TSPN cannot be efficiently approximated to within a factor of $(2-\varepsilon)$, unless $NP \subseteq TIME(n^{O(\log \log n)})$, by using a similar construction as in Section 2. In the same paper the authors also show that TSPN cannot be approximated within a factor of $\Omega(\log \log^{1/6} n)$ in 3D and $\Omega(\log \log^{\frac{d-2}{3(d-1)}} n)$ in d dimensions in polynomial time.

Acknowledgement

We thank V. Kann for suggesting to use the Vertex-Cover problem in the reduction proving the APX-hardness of TSPN. Many thanks also go to the anonymous referees who gave many helpful comments and pointed out some errors in a preliminary version.

References

[1] P. Agarwal, M. Katz and M. Sharir. Computing depth orders for fat objects and related problems. *Computational Geometry - Theory & Appl.*, 5:187–206, 1995.

[2] H. Alt, R. Fleischer, M. Kaufmann, K. Mehlhorn, S. Näher, S. Schirra and C. Uhrig. Approximate motion planning and the complexity of the boundary of the union of simple geometric figures. *Algorithmica*, 8:391–406, 1992.

[3] E. M. Arkin and R. Hassin. Approximation algorithms for the geometric covering salesman problem. *Discrete Applied Mathematics*, 55:197–218, 1994.

[4] S. Arora. Polynomial time approximation schemes for Euclidean traveling salesman and other geometric problems. *Journal of the ACM*, 45(5):753–782, 1998.

[5] P. Berman and M. Karpinski. On some tighter inapproximability results. Technical Report 98-029, ECCC, 1998.

[6] L. P. Chew, H. David, M. J. Katz and K. Kedem. Walking around fat obstacles. *Information Processing Letters*, to appear.

[7] A. Dumitrescu and J. S. B. Mitchell. Approximation algorithms for TSP with neighborhoods in the plane. In *Proc. 12th Annual ACM-SIAM Symposium on Discrete Algorithms*, 2001.

[8] M. R. Garey, R. L. Graham and D. S. Johnson. Some NP-complete geometric problems. In *Proc. 8th Annual ACM Symposium on Theory of Computing*, 1976.

[9] J. Gudmundsson and C. Levcopoulos. A fast approximation algorithm for TSP with neighborhoods. *Nordic Journal of Computing*, 6:469–488, 1999.

[10] J. Hershberger and Subhash Suri. A pedestrian approach to ray shooting: Shoot a ray, take a walk. *Journal of Algorithms*, 18:403–431, 1995.

[11] M. van Kreveld. On fat partitioning, fat covering, and the union size of polygons. In *Proc. 3rd Workshop on Algorithms and Data Structures*, volume 709 of *Lecture Notes in Computer Science*, pages 452–463. Springer-Verlag, 1993.

[12] C. Mata and J. S. B. Mitchell. Approximation algorithms for geometric tour and network design problems. In *Proc. 11th Annual ACM Symposium on Computational Geometry*, pages 360–369, 1995.

[13] J. Matoušek, N. Miller, J. Pach, M. Sharir, S. Sifrony and E. Welzl. Fat triangles determine linearly many holes. In *Proc. 32nd Annual Annual IEEE Symposium on Foundations of Computer Science*, pages 49–58, 1991.

[14] J. S. B. Mitchell. Geometric shortest paths and network optimization. In J.-R. Sack and J. Urrutia (eds.), *Handbook of Computational Geometry*, Elsevier Science.

[15] J. S. B. Mitchell. Guillotine subdivisions approximate polygonal subdivisions: A simple polynomial-time approximation scheme for geometric TSP, k-MST, and related problems. *SIAM Journal of Computing*, 28(4):1298–1309, 1999.

[16] C. H. Papadimitriou. The Euclidean traveling salesman problem is NP-complete. *Theoretical Computer Science*, 4:237–244, 1977.

[17] G. Reich and P. Widmayer. Beyond Steiner's problem: A VLSI oriented generalization. In *Proc. 15th Internat. Workshop Graph-Theoret. Concepts Comput. Sci.*, volume 411 of Lecture Notes in Comp. Sci., pages 196–210, Springer-Verlag, 1989.

[18] O. Schwartz and S. Safra. On the complexity of approximating TSP with neighborhoods and related problems. Manuscript, 2002.
[19] B. Shaleooi. Algorithms for cutting sheets of metal. LUNDFD6/NFCS-5189/1–44/2001, Master thesis, Department of Computer Science, Lund University, 2001.
[20] P. Slavík. The Errand Scheduling Problem. Technical report 97-02, Department of Computer Science and Engineering, SUNY Buffalo, 1997.
[21] A. F. van der Stappen. Motion Planning amidst Fat Obstacles, Ph.D. Dissertation, Department of Computer Science, Utrecht University, Utrecht, Netherlands, 1994.

1.375-Approximation Algorithm for Sorting by Reversals*

Piotr Berman[1] **, Sridhar Hannenhalli[2], and Marek Karpinski[3] ***

[1] Dept. of Computer Science and Engineering, The Pennsylvania State University
berman@cse.psu.edu
[2] Celera Genomics Co.
45 W. Gude Dr., Rockville, MD 20350
Sridhar.Hannenhalli@celera.com
[3] Dept. of Computer Science, University of Bonn
marek@cs.uni-bonn.de

Abstract. Analysis of genomes evolving by inversions leads to a general combinatorial problem of *Sorting by Reversals*, MIN-SBR, the problem of sorting a permutation by a minimum number of reversals. Following a series of preliminary results, Hannenhalli and Pevzner developed the first exact polynomial time algorithm for the problem of sorting signed permutations by reversals, and a polynomial time algorithm for a special case of unsigned permutations. The best known approximation algorithm for MIN-SBR, due to Christie, gives a performance ratio of 1.5. In this paper, by exploiting the polynomial time algorithm for sorting signed permutations and by developing a new approximation algorithm for maximum cycle decomposition of breakpoint graphs, we design a new 1.375-algorithm for the MIN-SBR problem.

1 Introduction

A *reversal* $\rho = \rho(i,j)$ applied to a permutation $\pi = \pi_1 \ldots \pi_{i-1} \pi_i \ldots \pi_j \pi_{j+1} \ldots \pi_n$ reverses the order of elements $\pi_i \ldots \pi_j$ and transforms π into permutation $\pi \cdot \rho = \pi_1 \ldots \pi_{i-1} \pi_j \ldots \pi_i \pi_{j+1} \ldots \pi_n$. Reversal distance $d(\pi, \sigma)$ is defined as the minimum number of reversals $\rho_1, \ldots, \rho_t$ needed to transform π into the permutation σ, i.e., $\pi \cdot \rho_1 \cdots \rho_t = \sigma$. Let $id = 12..n$ be the identity permutation, then $d(\pi, \sigma) = d(\pi \cdot \sigma^{-1}, id)$. The problem of computing the reversal distance for

---
* Most proofs are omitted in this extended abstract. The full version with complete proofs is available at ftp://ftp.eccc.uni-trier.de/pub/eccc/reports/2001/TR01-047.
** Research done in part while visiting Dept. of Computer Science, University of Bonn. Work partially supported by NSF grant CCR-9700053, NIH grant 9R01HG02238-12 and DFG grant Bo 56/157-1.
*** Research done in part while vising Dept. of Computer Science, Princeton University. Work partially supported by DFG grants, DIMACS, and IST grant 14036 (RAND-APX).

given two permutation is equivalent to the problem of *Sorting by reversal*, MIN-SRB, where for a given π we compute $d(\pi, id)$. This problem received a lot of attention because it models *global genome rearrangements*. The importance of computational methods to analyze genome rearrangements was first recognized by Sankoff et al. [SCA90]. See Pevzner [P00] for the most complete discussion of these problems. Similar combinatorial problems were investigated by Gates and Papadimitriou [GP79], Amato *et al.* [ABIR89] and Cohen and Blum [CB95].

Biologists derive gene orders either by sequencing entire genomes or by comparative physical mapping. Sequencing provides information about directions of genes and allows one to represent a genome by a *signed* permutation (Kececioglu and Sankoff [KS93]). Most of currently available experimental data on gene orders are based on comparative physical maps. Physical maps usually do not provide information about directions of genes and, therefore lead to representation of a genome as an *unsigned* permutation π.

Kececioglu and Sankoff [KS93] have found a 2-approximation algorithm and conjectured that the problem is NP-hard. They were first to exploit the link between the reversal distance and the number of *breakpoints* in a permutation. Bafna and Pevzner [BP93] improved the performance ratio to 1.75 for unsigned permutations and 1.5 for signed permutations. Hannenhalli and Pevzner [HP95] found however an exact polynomial algorithm for sorting *signed* permutations by reversals, a problem which also was believed to be NP-hard (see [BH96] and [KST97] for faster algorithms). However, MIN-SBR, the problem of sorting an unsigned permutation, was shown to be NP-hard by Caprara [C97] thus proving the conjecture. Later, this problem was also shown to be MAX-SNP hard by Berman and Karpinski [BK99], while Christie [Ch98] improved the performance ratio for MIN-SBR to 1.5.

In this paper, by exploiting a polynomial time algorithm for sorting a signed permutation by reversals, and by developing a new approximation algorithm for maximum cycle decomposition of breakpoint graphs, we design a 1.375-approximation algorithm for sorting by reversals. This improvement over 1.5 ratio of Christie is obtained here by a different method and a substantially more complicated algorithm.

Bafna and Pevzner [BP93] revealed important links between the breakpoint graph of a permutation and the reversal distance. In particular, they showed a strong correspondence between the maximum cycle decomposition of the breakpoint graph of the permutation and its reversal distance. Moreover, for all known biological instances, it was observed that the maximum cycle decomposition is sufficient to estimate the reversal distance exactly. Although, in general, the maximum cycle decomposition does not suffice to compute the reversal distance precisely, it does suffice to compute the reversal distance approximately with a guaranteed performance.

2 From MIN-SBR to Alternating Cycles

Solving MIN-SBR problem directly does not seem feasible, because as yet it is not known how to evaluate individual reversals, and sequences of reversals form exponentially large searching space. Fortunately, Hannenhalli and Pevzner reduced this problem to the one of finding an optimal decomposition of a certain graph with two edge colors. We will have two goals: finding a maximally large family of edge-disjoint cycles, while in the same time minimizing the number of so-called hurdles that this family of cycles defines.

Because we only want to approximate the optimal solution, we will be able to simplify the task dramatically. To begin with, we will be searching for cycles that consists of at most 8 edges. Moreover, we will be able to eliminate the explicit counting of the hurdles altogether.

We start from precise definitions and then proceed with an amortized analysis that will reveal the relative importance of various cycles and hurdles from the point of view of approximation. Importantly, we will show that we can neglect the existence of certain classes of objects, and eliminate another class of objects by applying certain kinds of greedy choices. We will conclude this Section with an algorithm for approximating MIN-SBR that uses as a subroutine the algorithm for certain simpler problem.

2.1 Definitions and Graph-Theoretic Background

Bafna and Pevzner [BP93], Hannenhalli and Pevzner [HP95, HP96] have described how to reduce the MIN-SBR to a purely graph-theoretic problem, Maximum Decomposition into Alternating Cycles, or MDAC in short. In this section, we will paraphrase the results in [HP96], where they describe an exact algorithm for MIN-SBR problem that is polynomial in certain biologically important cases.

We use $[i, j]$ to denote the set of integers $\{i, i+1, \ldots, j\}$. A permutation π is a 1-1 mapping of $[1, n]$ onto itself, and π_i is the value or π for an argument i. We extend π to one extra argument by setting $\pi_0 = 0$. To avoid modulo notation, we will assume $\pi_i = \pi_{i+n+1}$ for every integer i.

A *breakpoint graph of* π, G_π, has a node set $[0, n]$ and two sets of edges:

$$breaks = \{\{\pi_i, \pi_{i+1}\} : i \in [0, n]\};$$
$$chords = \{\{i, i+1\} : i \in [0, n]\}.$$

If a chord happens to be a break, we count it as a separate object, and say that this is a *short chord*. For that reason, our edge sets can be actually multisets. An *alternating cycle*, AC for short, is a connected set of edges C with the following property: if a node belongs to i breaks of C, than it also belongs to i chords of C. A *decomposition into alternating cycles*, DAC for short, is a partition of the edges of G_π into ACs.

A DAC $\mathcal{C}$ of G_π can be represented by the following *consecutive* relation: edges e and e' are consecutive edges on a cycle; here cycle is identified with its traversal. In turn, this relation uniquely determines a *spin* of π, a signed permutation π' such that $\pi'_i = \pm \pi_i$ (see [BP93, HP96]); a sequence of reversals that sorts π' obviously sorts π as well, and any sequence of reversals that sorts π sorts one of its spins. Because we can find an optimum reversal sequence for a spin of π in polynomial time [HP95], the search for an optimal reversal sequence for π is equivalent to the search for an optimal spin of π, and, in turn, the search for an optimal DAC of G_π.

A given cycle decomposition $\mathcal{C}$ defines a set of *hurdles* (defined later). The following theorem of Hannenhalli and Pevzner [HP96] is crucial:

Theorem 1 *Given a cycle decomposition of the breakpoint graph of π, there exists a polynomial time algorithm that finds a sequence of $n - c + h + f$ reversals that sorts permutation π, where c is the number of cycles in the decomposition, h is the number of hurdles and $f \in [0, 1]$.*

Because we are interested in an approximation algorithm, we will ignore the small term f. Our goal in this section is to show how to handle the minimization of h so we will later maximize c in a separate problem. To define *hurdles*, we need some more definitions.

We will use the following geometric representation of G_π: the nodes $\pi_0, \ldots, \pi_n$ are placed counter-clockwise on a circle **C**, each break $\{\pi_i, \pi_{i+1}\}$ is a **C**-arc segment, and a chord $\{i, i+1\}$ is the line segment that joins points i and $i+1$. Note that in this representations numbers are viewed as node names, moreover, if $\pi_{i+1} = \pi_i \pm 1$, then the break $\{\pi_i, \pi_{i+1}\}$ and the short chord $\{\pi_i, \pi_{i+1}\}$ are indeed two different objects. To avoid confusion, we will apply the word *chord* exclusively to the representations of the chords of G_π, while a *chordal segment* is any line segment that connects two (representations of) nodes of G_π.

If two chords e_0 and e_1 intersect in the interior of **C**, they form an *interleaving pair*.

A *chord component* is a connected component of the graph $<$ chords, interleaving pairs $>$. We will assume that there is more than one chord component; otherwise we will have a trivial case.

We define the *area of a chord component* C, denoted by $A(C)$, as follows: for each chord $e \in C$ (viewed as a line segment) we remove the endpoints, then we take the union of these chords, and finally we take the smallest convex set that contains that union.

Observation 1 *$A(C)$ is a convex polygon and its set of nodes is the the set of endpoints of the chords of C. Moreover, the chords of C subdivide $A(C)$ into convex polygons that either have the entire boundary covered by the chords of C, or the entire boundary with the exception of a single segment.*

This observation leads to the next one:

Observation 2 *$A(C)$ cannot intersect a chord e if $e \notin C$.* (Proof omitted.)

A *crescent* $Cr(i, j)$ is an area bounded by counter-clockwise arc from i to j and the chordal segment $\{i, j\}$. If the counterclockwise listing of the nodes of

$A(C)$ is $i_0, i_1, \ldots, i_k = i_0$, then $interior(\mathbf{C}) - A(C)$ is a disjoint union $Cr(i_0, i_1)$, $\ldots, Cr(i_{k-1}, i_k)$; we call them the *neighbor crescents* of C.

The relative positions of different chord components are described in the next observation.

Observation 3 *$A(C')$ is a subset of one of the neighbor-crescents of C for each chord component $C' \neq C$.* (Proof omitted)

Lemma 1 *If $Cr(i, j)$ is a neighbor-crescent of a chord component C, and there exists a chord contained in that crescent, then the chords contained in $Cr(i, j)$, together with the breaks on the $\mathbf{C}$-arc extending from i to j, form an AC.*

Moreover, $\{i, j\}$ is a chord only if it is a short chord, in which case the AC of $Cr(i, j)$ consists of exactly one chord and exactly one break. component of e most be contained in a single neighbor-crescent. (Proof omitted)

The last lemma characterizes neighbor-crescents of C that contain some chords. We say that other neighbor-crescents are *empty*. Obviously, if $Cr(i, j)$ is an empty neighbor crescent, the $\mathbf{C}$-arc from i to j forms a single break, we say that that break is *associated* with C. We define *edge component* of C as the set consisting of the chords of C and the breaks associated with C.

Lemma 2 *Edge components of chord components form a DAC decomposition.* (Proof omitted)

Given a fixed DAC $\mathcal{D}$ of the edge set of G_π into cycles, we define *a cycle component* as a connected component of the graph where the nodes are the edges (chords and breaks) and a pair of edges is connected if either 1) they are in the same edge component, or 2) they belong to the same cycle from $\mathcal{D}$.

Observation 4 *The union of chords of a cycle components, including their endpoints, forms a connected set.* (Proof omitted)

This observation allows us to apply Observations 1, 2, 3, and Lemma 1 to chord set of a cycle component in the same way as to a chord component. In particular, for a chord component C we can define $A(C) = A(C \cap \text{chords})$ and the neighbor-crescents. If C, C_1, C_2 are cycle components and $A(C_1)$ and $A(C_2)$ are contained in two different neighbor-crescents of C, then we say that C *separates* C_1 and C_2.

A cycle is *oriented* if it contains a chord $\{i, i+1\}$ and in each of the crescents $Cr(i, i+1), Cr(i+1, i)$ it contains a break incident to $\{i, i+1\}$. A cycle component is oriented if it contains an oriented cycle, or if it consists of two edges only. Note that a cycle of two edges is always a separate singleton cycle component—this is the case when its chord is short.

A *hurdle* is an unoriented cycle component that does not separate two other unoriented cycle components.

2.2 Breaking Cycle Components into Edge Components

Our cost is the number of breaks minus the number of ACs plus the number of hurdles. To minimize the cost, we need to maximize the number of ACs in our DAC, while simultaneously minimizing the number of hurdles. We would like to separate those two tasks as much as we can.

In our quest for a small number of reversals, we will first maximize the number of cycles found in each edge component. We will show soon that by restricting ourselves to ACs contained in a single edge component, we are not decreasing the number of ACs in a decomposition, and sometimes we can even increase that number. However, this restriction may increase the number of hurdles. The amortized analysis introduced in the next subsection shows how to take care of these extra hurdles.

Let us consider now an optimum DAC $\mathcal{C}$ and a neighbor-crescent $Cr(i,j)$ of a chord component X. By Lemma 1 we can partition all edges into two ACs A and B, where A is the set of edges contained in $Cr(i,j)$. We will modify $\mathcal{C}$ so that every cycle $C \in \mathcal{C}$ will satisfy $C \subset A$ or $C \subset B$.

By definition, i and j are the only nodes that belong simultaneously to edges of A and B. Consider $C \in \mathcal{C}$ such that $C \cap A \neq \emptyset$ and $C \cap B \neq \emptyset$. Because C is connected, without loss of generality we may assume that both $C \cap A$ and $C \cap B$ contain at least one edge that contains i.

We will distinguish now between several cases.

Odd case. Suppose that i belongs to exactly one edge of $C \cap A$. Because every node in $Cr(i,j)$ other than i and j must belong to an even number of edges of C, this means that j also belongs to exactly one edge of $C \cap A$. Note that i belongs to two other edges that in turn belong to another cycle of $D \in \mathcal{C}$, and D has exactly the same properties: i.e. for $\mathbf{C} \in \{C, D\}$, $\mathbf{A} \in \{A, B\}$ and $\mathbf{i} \in \{i, j\}$ there exists exactly one edge of $\mathbf{A} \cap \mathbf{C}$ that is incident to $\mathbf{i}$. We consider two subcases.

Odd Group case. For some $k > 2$ there exists a cycle of nodes $i = i_0, i_1 = j, \ldots, i_k = i_0$ and a sequence of chord components $X_0, \ldots, X_{k_1}$ such that $Cr(i_l, i_{l+1})$ is a neighbor-crescent of X_l. Let Y_l be an AC formed from the edge component of X_l and other edge components that are not contained in $Cr(i_l, i_{l+1})$. By applying above argument inductively, one can see that for $\mathbf{C} \in \{C, D\}$, $\mathbf{A} \in \{Y_0, \ldots, Y_k\}$ and $\mathbf{i} \in \{i_0, \ldots, i_k\}$ there exists exactly one edge of $\mathbf{A} \cap \mathbf{C}$ that is incident to $\mathbf{i}$.

Thus we can replace cycles C and D with k cycles of the form $(C \cup D) \cap Y_l$. We associate $k-2$, the resulting increase in the number of cycles with the convex polygon bounded by the polyline $(i_0, i_1, \ldots i_k = i_0)$.

Odd Pair case. Same as the Odd Group case, but for $k = 2$, so we have no assured increase in the number of cycles.

Even case. Now, both $C \cap A$ and $C \cap B$ is an AC. We can replace C with these intersections and the modified DAC surely has more cycles. We can associate the increase in the number of cycles with node i.

2.3 Amortized Analysis

Throughout the paper, we will use potential analysis to assure that we obtain the promised approximation ratio. For every unit of the cost of the optimum solution we can place $1\tfrac{1}{8}$ of the potential units, and for every unit of the cost of our solution, we place -1 of the potential units. We deliver a desired solution if the sum of the placed potential units is non-negative. At many stages of the analysis, we add and subtract the potential units in various parts of our structure; such a move is valid if we assure that the sum of additions does not exceed the sum of subtractions.

Each break contributes 1 both to the cost of the optimum solution and the cost of the solution obtained by our algorithm. Thus we can place $1\tfrac{1}{8} - 1 = \tfrac{3}{8}$ on each break.

Each AC of the optimum solution contributes -1 to the optimal cost, so we can place $-1\tfrac{1}{8}$ on this cycle. However, when the ACs of the optimum solution span more than one edge component, we break them as described in the previous subsection.

After the break-up, the maximal number of cycles misrepresents the true number in the optimum, to account for that we place the corrective amounts of units. Initially, we place them as follows: in Odd Group case with k chord components we place $1\tfrac{1}{8}(k-2)$ on the polygon that separates these components, in Odd Pair case that increases the number of cycles we place $1\tfrac{1}{8}$ on the chord separating the two components, and in Even Case we place $1\tfrac{1}{8}$ on a separating node. Once the break-up is complete, edge components incident to objects with corrective units share those units evenly. The least possible share occurs for Odd Group case with $k=3$ and it equals to $1\tfrac{1}{24}$.

Because of the break-up, we lost the account of the hurdles in the optimum solution, so we do not take them into account, this can only decrease the total balance of the potential. At each edge component that we estimate to be a hurdle we place -1 unit. The estimation method does have to be correct, the only requirement is that we will have at least as many estimated hurdles as we have the actual ones. In particular, we will estimate every edge component with at least 5 breaks to be a hurdle.

2.3.1 Small Edge Components

We first analyze the case when $\mathcal{C}$ contains only a few edges, in which case our algorithm can find a maximum DAC of this edge component. As a result, on every cycle from the modified optimum solution we may place 1. We will establish the situations when such a maximum DAC of $\mathcal{C}$ is not good enough to assure our approximation ratio, i.e. when the resulting balance of the potential in $\mathcal{C}$ is negative when we estimate that $\mathcal{C}$ is a hurdle.

A case analysis (omitted) shows that this happens in one case only: the cycle component $\mathcal{C}$ consists of exactly one unoriented cycle with 3 breaks. Component $\mathcal{C}$ has three neigbor crescents and at most one neighbor contains an estimated unoriented component.

We total the balance of C and its neigbor crescents that contain oriented components only. If even one of these components, say $\mathcal{D}$, has $w(\mathcal{D}) > 0$, then the sum of balances is at least $3/8 - 1/4 > 0$. IN this case we say that C is *rich*. Thus we can assume that the "oriented" neighbor-crescents contain only 1-components. If an oriented neighbor-crescent contains exactly one such component, we change the solution as shown in [HP96]. If this change turns another component, say $\mathcal{D}$, into a hurdle, than clearly $\mathcal{D}$ is rich.

We are left with the case when each oriented neighbor-crescent contains multiple 1-components. In this case there exists an optimum solution where no cycle overlaps both C and one of this 1-components (this fact follows from the discussion of the Odd Group case and Even Case). If all three neighbor-crescents are oriented, we actually get an optimal solution. Thus were can additionally assume that C has one unoriented neighbor-crescent $Cr(i,j)$ which forms Odd Pair case with C, hence the chordal segment $\{i, j\}$ separates the area of C from the area of another edge component, say $\mathcal{D}$. We say that C is a child of $\mathcal{D}$.

If $w(\mathcal{D}) = 1$, then $\mathcal{D}$ has but one child, and we add the balances of C and $\mathcal{D}$. If $w(\mathcal{D}) = 2$, then $\mathcal{D}$ has at most 4 sides, thus at most 3 children, the balance of $\mathcal{D}$ is $3/4$ and each child has balance $-1/4$, so again, we add the balance of $\mathcal{D}$ to the balances of its children. Thus it remains to consider the case when $w(\mathcal{D}) > 2$ and thus $\mathcal{D}$ is unconditionally estimated to be a hurdle.

Within that parent component $\mathcal{D}$ we must have a plausible single cycle E that was created in $\mathcal{D}$ during the breaking step of $Cr(i, j)$.

We introduce the following terminology for such a situation: cycle E is a *absorber* and the cycle/component C is a *little hurdle*. We say that absorber E absorbs little hurdle C.

2.3.2 Catalogue of Cycles with Negative Potential

We transfer the negative potential of the little hurdles to the respective absorbers. Now the entire potential is contained in the cycles contained in the chord components that are estimated to be hurdles. Our task is to find enough ACs and to absorb enough little hurdles to create the nonnegative balance of the potential. Obviously, we can ignore the cycles of the optimum solution that have non-negative potential. Therefore we need to establish which ACs have negative potential.

Consider an i-cycle from the optimum solution. If it is not an absorber, then its potential is $3/8(i-1)-1$, so for $i = 2, 3, 4, 5$ this potential is $-5/8, -2/8, 1/8$ and $4/8$ respectively. Thus only 2-cycles and 3-cycles have negative potential.

Now consider a (i, j)-*absorber* which we define to be an $(i + j)$-cycle that absorbs j little hurdles. Its balance is $3/8(i + j - 1) - 1/4j - 1 = 3/8(i - 1) + 1/8j - 1$. Note that this covers the case of ordinary i-cycles that will be considered as $(0, i)$-absorbers.

Since (i, j)-absorber of the optimum solution has potential balance $3/8(i-1) + 1/8j - 1$, the absorbers with negative potential are $(1, j)$-absorbers for $j \leq 7$, $(2, j)$-absorbers for $j \leq 4$ and $(3, 1)$-absorbers. As we will show, only $(2, 1)$- and $(3, 1)$-absorbers actually exist.

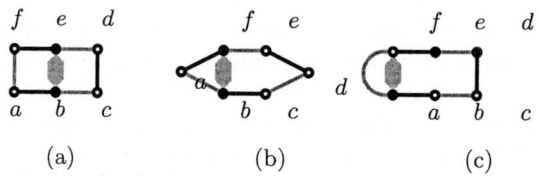

Fig. 1. Conceivable (2,1) absorbers, the hexagons indicate the position of the little hurdle, edge colors indicate chords and breaks. Only (a) can exist, (b) and (c) cannot be decomposed into 2 good cycles

Consider absorber C of j little hurdles $H_1, \ldots, H_j$ there exists a DAC of $C \cup H_1 \cup \ldots H_j$ into $j+1$ ACs, each of them intersecting both C and one or more of the little hurdles. We say that this is a decomposition into *good* ACs.

Consider a good AC D that intersects a little hurdle H. One can see that $D \cap H$ must be a path of three edges, which we will call a *long segment*. The edges of C shall be called *short segments*.

The variety of possible absorbers is restricted by the following two lemmas.

Lemma 3 *A good cycle must contains both kinds of short segments, i.e. at least one break and at least one chord.* (Proof omitted)

Lemma 4 $(1,j)$-*absorbers exist only for $j = 0$ and $(2,j)$-absorbers exist only for $j \leq 1$.* (Proof omitted)

It is worthwhile to note that besides the fact that the variety of possible absorbers is quite restricted, the shapes of (2,1)-absorbers and (3,1)-absorbers are quite restricted as well, as we can see in Fig. 1.

2.3.3 When DAC of a Large Edge Component Is Good Enough

DAC found by our algorithm is good enough if it assures that the balance of the component in question is nonnegative. In a large component, we will require that the sum of balances of ACs of the optimum solutions and ACs found by our algorithm is at least 1, so we can create a hurdle and still have a nonnegative potential.

We define function ϕ such that if C is an AC from the optimum solution, then $-\phi(C)$ is the potential that C contributes to the overall balance of its edge component. We can view a regular i-cycle as a $(i, 0)$-absorber, in the previous subsection we calculated that

$$\phi(C) = 1 - 3/8\,i + 1/8\,j \text{ for an } (i,j)\text{-absorber } C.$$

In turn, when our algorithm finds an (i, j)-absorber C for its solution, this decreases the resulting cost by $1 + j$: 1 for the AC in the decomposition and j for the decrease in the estimate on the number of hurdles in its solution. Thus we can define

$$\gamma(C) = 1 + j.$$

Consequently, if $\mathcal{I}$ is a DAC of the edge component under discussion that is found by our algorithm, and $\mathcal{I}^*$ is an optimal DAC of that component, the condition by sufficiently good $\mathcal{I}$ is

$$\gamma(\mathcal{I}) \geq \phi(\mathcal{I}^*) + 1.$$

Let $n_{i,j}$ be the number of (i,j) absorbers in the optimum DAC. We can rewrite the above condition as follows:

$$\gamma(\mathcal{I}) \geq 1 + {}^5/_8 n_{2,0} + {}^2/_8 n_{3,0} + {}^4/_8 n_{2,1} + {}^1/_8 n_{3,1}.$$

We can show that if an edge component has at least 48 breaks, then it suffices to find a set of non-overlapping ACs that satisfies the above inequality without "1+" term, we use GEDSAC as the acronym for this subproblem. For $2 < m < 48$ breaks it suffices to solve GDoaC problem exactly. We can summarize this section in the form of an algorithm for MIN-SBR problem.

---

1. Input permutation π.
2. Form graph G_π.
3. Decompose G_π into edge components, establish which constitute small hurdles.
4. For each large edge component
   (i) Establish (2,1)- and (3,1)-absorbers.
   (ii) If the component has fewer than 48 breaks, solve exactly GDoaC problem, else
   (iii) Solve GEDSAC problem and add the cycle formed from edges not covered by the solution.
5. Combine DACs of edge component into a single DAC of G_π, re-partition the absorbers and their little hurdles to decrease the number of hurdles.
6. Use this DAC to define a spin π' of π and apply the algorithm of Hannenhalli and Pevzner [HP95] to π'.

---

3 Methods of Solving GEDSAC

Let $\mathcal{V}_{i,j}$ be the set of (i,j)-absorbers, $\mathcal{V}_2 = \mathcal{V}_{2,0} \cup \mathcal{V}_{2,1}$, $\mathcal{V}_3 = \mathcal{V}_{3,0} \cup \mathcal{V}_{3,1}$, $\mathcal{V} = \mathcal{V}_2 \cup \mathcal{V}_3$ and let $\mathcal{E}$ be the set of edge overlapping pairs of cycles from $\mathcal{V}$. Then a solution of GEDSAC is an independent set in $\mathcal{G} = <\mathcal{V}, \mathcal{E}>$. Thus GEDSAC is a kind of maximum independent set problem.

In general, we can easily approximate an maximum independent set in two situations: there is a bound on the number of neighbors that a node may have, or there is a bound on the number of independent neighbors. In the full version we describe an algorithm that simplifies $\mathcal{G}$ so that later independent set techniques will be easier to apply.

This by combining other results of the paper yields our main theorem.

Theorem 2 *There exists a polynomial time approximation algorithms for MIN-SBR problem with approximation ratio 1.375.*

References

[ABIR89] N. Amato, M. Blum, S. Irani and R. Rubinfeld, *Reversing Trains: a turn of the century sorting problem*, J. of Algorithms **10**:413-428, 1989.

[BP93] V. Bafna and P. Pevzner, *Genome rearrangements and sorting by reversals*, Proc. of 34th IEEE FOCS, 148-157, 1993; also in SIAM J. on Computing **25**:272-289, 1996.

[BF94] P. Berman and M. Fürer, *Approximating independent set problem in bounded degree graphs*, Proc. SODA 1994, 365-371.

[BH96] P. Berman and S. Hannenhalli, *Fast Sorting by Reversals*, Proc. of 7th CPM, 168-185, 1996.

[BK99] P. Berman and M. Karpinski, *On some tighter inapproximability results*, Proc. of 26th ICALP, LNCS **1644**:200-209, Springer-Verlag, Berlin, 1999.

[C97] A. Caprara, *Sorting by Reversals is difficult*, Proc. of 1st ACM RECOMB, 75-83, 1997, to appear in SIAM J. of Discr. Math. 2001.

[Ch98] D. A. Christie, *A 3/2 Approximation algorithm for sorting by reversals*, Proc. of 9th ACM-SIAM SODA, 244-252, 1998.

[CB95] D. Cohen and M. Blum, *On the problem of Sorting Burnt Pancakes*, Discrete Appl. Math. **61**:105-125, 1995.

[GP79] W. H. Gates and C. H. Papadimitriou, *Bounds for sorting by prefix reversals*, Discr. Math. **27**:47-57, 1979.

[HY99] M. M. Halldórsson and K. Yoshikara, *Greedy approximation of independent sets in low degree graphs*,

[HP95] S. Hannenhalli and P. Pevzner, *Transforming cabbage into turnip*, Proc. of 27th ACM STOC 1995, 178-189.

[HP96] S. Hannenhalli and P. Pevzner, *To cut... or not to cut*, Proc. of 7th ACM-SIAM SODA 1996, 304-313.

[HS89] C. A. J. Hurkens and A. Schcrijver, *On the size of systems of sets every t of which have an SDR, with an aplication to the worst case ratio of heuristic for Packing Problem*, SIAM J. of Discr. Math. **2**(1):62-72, 1989.

[KST97] H. Kaplan, R. Shamir and R. E. Tarjan, *Faster and simpler algorithm for sorting signed permutations by reversals*, Proc. of 8th ACM-SIAM SODA, 178-187, 1997.

[KS93] J. Kececioglu and D. Sankoff, *Exact and approximation algorithms for the inversion distance between two permutations*, Algorithmica **13**:180-210, 1995.

[P00] P. Pevzner, *Computational Molecular Biology—An Algorithmic Approach*, The MIT Press, Cambridge, 2000.

[SCA90] D. Sankoff, R. Cedergen and Y. Abel, *Genomic divergence through gene rearrangement*, in *Molecular Evolution: Computer Analysis of Protein and Nucleic Acid Sequences*, chapter 26, 428-238, Academic Press, 1990.

[SLA92] D. Sankoff, G. Leduc, N. Antoine, B. Paquin, B. F. Lang and R. Cedergen, *Gene order comparisons for phylogenetic inference: Evolution of the mitochondrial genome*, Proc. Natl. Acad. Sci. USA, **89**:6575-6579, 1992.

Radio Labeling with Pre-assigned Frequencies*

Hans L. Bodlaender[1], Hajo Broersma[2], Fedor V. Fomin[3],
Artem V. Pyatkin[4], and Gerhard J. Woeginger[2]

[1] Institute of Information and Computing Sciences, Utrecht University
P.O. Box 80.089, 3508 TB Utrecht, the Netherlands
hansb@cs.uu.nl
[2] Faculty of Mathematical Sciences, University of Twente
7500 AE Enschede, The Netherlands
{broersma,g.j.woeginger}@math.utwente.nl
[3] Heinz Nixdorf Institute, Paderborn University
Fürstenallee 11, D-33102 Paderborn, Germany
fomin@uni-paderborn.de
[4] Sobolev Institute of Mathematics
pr. Akademika Koptyuga 4, Novosibirsk 630090, Russia
artem@math.nsc.ru

Abstract. A radio labeling of a graph G is an assignment of pairwise distinct, positive integer labels to the vertices of G such that labels of adjacent vertices differ by at least 2. The radio labeling problem (RL) consists in determining a radio labeling that minimizes the maximum label that is used (the so-called span of the labeling). RL is a well-studied problem, mainly motivated by frequency assignment problems in which transmitters are not allowed to operate on the same frequency channel. We consider the special case where some of the transmitters have pre-assigned operating frequency channels. This leads to the natural variants P-RL(l) and P-RL($*$) of RL with l pre-assigned labels and an arbitrary number of pre-assigned labels, respectively.

We establish a number of combinatorial, algorithmical, and complexity-theoretical results for these variants of radio labeling. In particular, we investigate a simple upper bound on the minimum span, yielding a linear time approximation algorithm with a constant additive error bound for P-RL($*$) restricted to graphs with girth ≥ 5. We consider the complexity of P-RL(l) and P-RL($*$) for several cases in which RL is known to be polynomially solvable. On the negative side, we prove that P-RL($*$) is NP-hard for cographs and for k-colorable graphs where a k-coloring is given ($k \geq 3$). On the positive side, we derive polynomial time algorithms solving P-RL($*$) and P-RL(l) for graphs with bounded maximum degree, and for solving P-RL(l) for k-colorable graphs where a k-coloring is given.

---

* The work of HJB, FVF and AVP is partly sponsored by NWO-grant 047.008.006. The work of FVF and HLB is partly sponsored by EC contract IST-1999-14186: Project ALCOM-FT (Algorithms and Complexity - Future Technologies). Part of the work was done while FVF and AVP were visiting the University of Twente. GJW acknowledges support by the START program Y43-MAT of the Austrian Ministry of Science.

Keywords: Radio labeling, frequency assignment, graph algorithms, computational complexity, approximation algorithm, cograph, k-colorling.

1 Introduction

The FREQUENCY ASSIGNMENT problem (FAP) is a general framework focused on point-to-point communication, e.g. in *radio* or *mobile telephone* networks. One of its main threads asks for an assignment of frequencies or frequency channels to transmitters while keeping interference at an acceptable level and making use of the available frequencies in an efficient way. Interference constraints are usually related to the use of the same or similar frequencies at locations within a certain distance (or transmitters within a certain reach) from each other. Due to the scarce resources and the increasing use of frequencies in modern wireless technology, the available frequencies should be used as efficiently as possible. There is usually a trade-off between avoiding interference and the efficient use of frequencies. We will not go deeper into the technical details here.

Graph theoretical issues come into play since possible interference between transmitters is usually modeled by a so-called *interference graph*. Each vertex of the interference graph represents a transmitter. If simultaneous broadcasting of two transmitters may cause an interference, then they are connected by an edge in the interference graph. The frequency channels are usually labeled by positive integers. Frequency channels with 'close' labels are assumed to be 'close' in the spectrum or expected to be 'more likely' to cause interference.

Regarding the assumption that a pair of 'close' transmitters should be assigned different frequencies or frequency channels, the FREQUENCY ASSIGNMENT problem is equivalent to the problem of labeling the interference graph with some constraints on the labeling. In many cases the related labeling problems are variants on what is known as the vertex coloring problem in graph theory. However, Hale [7] observed that the signal propagation may affect the interference even in distant regions (but with decreasing intensity). Hence, not only 'close' transmitters should get different frequencies, but also frequencies used at some distance should be appropriately separated. This leads to a more detailed and complicated modeling of FAP in terms of *distance constrained labeling* of the interference graph. (See for instance the book [8] by Leese.)

In some applications the transmitters are not allowed to operate on the same frequency channel (for example, when every transmitter covers the whole area) while 'close' transmitters should use channels with sufficient separation. In this case non-reusable frequency channels should be assigned to transmitters in a proper way. This leads to the so-called RADIO LABELING problem (RL), i.e. to the problem of assigning distinct labels to the vertices of a graph such that adjacent vertices get labels (positive integers) that differ by at least two. The purpose of RL is to find such a radio labeling with the smallest maximum label.

In this paper we initiate the investigation of two versions of this problem in which some of the transmitters (like military and governmental stations) already have pre-assigned labels corresponding to frequency channels which one is not allowed to change. Then the problem boils down to determining a radio labeling extending a given pre-labeling in a 'best possible' way. In this paper we consider some algorithmical, complexity-theoretical, and combinatorial aspects of these versions of the problem. We do not want to claim that the results in the sequel have immediate practical relevance.

Definitions and preliminary observations. We denote by $G = (V, E)$ a finite undirected and simple graph. The *girth* of G is the length of a shortest cycle in G. For every nonempty $W \subseteq V$, the subgraph of $G = (V, E)$ induced by W is denoted by $G[W]$. A *cograph* is a graph containing no induced path on four vertices. The *(open) neighborhood* of a vertex v in a graph G is $N_G(v) := \{u \in V : \{u, v\} \in E\}$. The degree of a vertex v in G is $d_G(v) := |N_G(v)|$. The maximum degree of G is $\Delta(G) := \max_{v \in V} d_G(v)$. A graph G is *t-degenerate* if each of its subgraphs has a vertex of degree at most t. A *clique* C of a graph $G = (V, E)$ is a nonempty subset of V such that all the vertices of C are pairwise adjacent in G. A nonempty subset of vertices $I \subseteq V$ is *independent* in G if no two of its elements are adjacent in G. The *complement* $\overline{G}$ of $G = (V, E)$ is the graph on V with edge set $\overline{E}$ such that $\{u, v\} \in \overline{E}$ if and only if $\{u, v\} \notin E$.

A *k-coloring* of the vertices of a graph $G = (V, E)$ is a partition $I_1, I_2, \ldots, I_k$ of V into independent sets (in which some of the I_j may be empty); the k sets I_j are called the *color classes* of the k-coloring. The *chromatic number* $\chi(G)$ is the minimum value k for which a k-coloring exists. A *labeling* of the (vertex set of the) graph $G = (V, E)$ is an injective mapping $L: V \to \mathbf{N}^+$ (the set of positive integers). A labeling L of G is called a *radio labeling* of G if for any edge $\{u, v\} \in E$ the inequality $|L(u) - L(v)| \geq 2$ holds; the *span* of such a labeling L is $\max_{v \in V} L(v)$.

The RADIO LABELING problem (RL) is defined as follows: "For a given graph G, find a radio labeling L with the smallest span." The name radio labeling was suggested by Fotakis and Spirakis in [5] but the same notion (under different names) has been introduced independently and earlier by other researchers (see, e.g. Chang & Kuo [1]). Problem RL is equivalent to the special case of the TRAVELING SALESMAN problem TSP(2,1) in which all edge weights (distances) are either one or two. The relation is as follows. For a graph $G = (V, E)$ let K_G be the complete weighted graph on V with edge weights 1 and 2 defined according to E: for every $\{u, v\} \in E$ the weight $w(\{u, v\})$ in K_G is 2 and for $\{u, v\} \notin E$ the weight $w(\{u, v\}) = 1$. The *weight* of a path in K_G is the sum of the weights of its edges. The following proposition can be found in [5].

Proposition 1. *There is a radio labeling of G with span k if and only if there is a hamiltonian path (i.e. a path on $|V|$ vertices) of weight $k - 1$ in K_G.*

Another equivalent formulation of this problem, which was extensively studied in the literature, is the HAMILTONIAN PATH COMPLETION problem (HPC), i.e. the problem of partitioning the vertex set of a graph G into the smallest possible number of sets which are spanned by paths in G.

Now let us turn to the versions of RL with pre-assigned labels. For a graph $G = (V, E)$ a *pre-labeling* L' of a subset $V' \subset V$ is an injective mapping $L' \colon V' \to \mathbf{N}^+$. We say that a labeling L of G *extends* the pre-labeling L' if $L(u) = L'(u)$ for every $u \in V'$. We study the following two problems:

- P-RL(∗): RADIO LABELING WITH AN ARBITRARY NUMBER OF PRE-LABELED VERTICES: For a given graph G and a given pre-labeling L' of G, determine a radio labeling of G extending L' with the smallest span.
- P-RL(l): RADIO LABELING WITH A FIXED NUMBER OF PRE-LABELED VERTICES: For a given graph $G = (V, E)$, a subset $V' \subseteq V$ with $|V'| \leq l$, and a pre-labeling $L' \colon V' \to \mathbf{N}^+$, determine a radio labeling of G extending L' with the smallest span.

Earlier results. As we mentioned above the TRAVELING SALESMAN problem TSP(2,1) (which is equivalent to RL without any pre-labeling) is a well-studied problem. Papadimitriou & Yannakakis [9] proved that this problem is MAX SNP-hard, but gave an approximation algorithm for TSP(2,1) which finds a solution not worse than 7/6 times the optimum solution. Later Engebretsen [3] improved their result by showing that the problem is not approximable within $5381/5380 - \varepsilon$ for any $\varepsilon > 0$.

Damaschke et al. [2] proved that the HAMILTONIAN PATH COMPLETION problem HPC can be solved in polynomial time on cocomparability graphs (complements of comparability graphs). By Proposition 1, the result of Damaschke et al. yields that RL is polynomial time solvable for comparability graphs. Later, this result was rediscovered by Chang & Kuo [1] but under the name of $L'(2,1)$-labeling and only for cographs, a subclass of the class of comparability graphs. Notice that RL is NP-hard for cocomparability graphs because the HAMILTONIAN PATH problem is known to be NP-hard for bipartite graphs which form a subclass of comparability graphs. Recently, Fotakis and Spirakis [5] proved that RL can be solved in polynomial time within the class of graphs for which a k-coloring can be obtained in polynomial time (for some fixed k). Note that, for example, this class of graphs includes the well-studied classes of planar graphs and graphs with bounded treewidth.

We are not aware of any existing results concerning the pre-labeling versions P-RL(∗) and P-RL(l) of RL.

We complete this subsection by mentioning some results concerning the related notion of radio coloring (also known as $L(2,1)$-labeling, $\lambda_{2,1}$-coloring and $\chi_{2,1}$-labeling). A *radio coloring* of a graph $G = (V, E)$ is a function $f \colon V \to \mathbf{N}^+$ such that $|f(u) - f(v)| \geq 2$ if $\{u,v\} \in E$ and $|f(u) - f(v)| \geq 1$ if the distance between u and v in G is 2. So, the difference between radio coloring and radio labeling is that in a radio coloring vertices at distance at least three may have equal labels (or colors). The notion of radio coloring was introduced by Griggs & Yeh [6] under the name $L(2,1)$-labeling. As with radio labeling the span of a radio coloring f of G is $\max_{v \in V} f(v)$. The problem of determining a radio coloring with minimum span has received a lot of attention. However, for only very few graph classes the problem is known to be polynomially solvable. Chang & Kuo [1] obtained a polynomial time algorithm for RC restricted to trees and cographs. The complexity of RC even for graphs of treewidth 2 is a long standing open question. An interesting direction of research was initiated by Fiala et al. [4]. They consider a precolored version of the RADIO COLORING problem, i.e. a version in which some colors are pre-assigned to some vertices. They proved that RC with a given precoloring can be solved in polynomial time for trees.

Our results and organization of the paper. We study algorithmical, complexity-theoretical, and combinatorial aspects of radio labeling with pre-labeled vertices. In Section 2 we give some combinatorial bounds for the minimum span of such labelings. In Section 3, we derive similar results for t-degenerate graphs. In Section 4 we obtain polynomial time algorithms for graphs with bounded degree; these algorithms are based on the results of Section 2.

Section 5 is devoted to the algorithmical study of the radio labeling problem with pre-assigned labels. We study these problems restricted to graphs for which a k-coloring is given, and restricted to cographs, two graph classes for which the radio labeling problem without pre-labeling is known to be solvable in polynomial time. Known and new results on these radio labeling problems are summarized in the following table.

	graphs with a bounded Δ	graphs with a given k-coloring	Cographs
RL	P [5]	P [5]	P [2, 1]
P-RL(l)	P [*]	P [*]	???
P-RL($*$)	P [*]	NPC for $k \geq 3$ [*]	NP [*]

In this table, an entry P denotes solvable in polynomial time, NPC denotes NP-complete, [*] denotes a contribution from this paper, and the sign ??? marks an open problem.

For the results in the middle column, we assume that k is a fixed integer that is not part of the input. Note that the class of graphs with a given k-coloring contains important and well-studied graph classes such as the class of planar graphs and the class of graphs with bounded treewidth.

2 Upper Bounds for the Minimum Span

Let $G = (V, E)$ denote a graph on n vertices, and let $V' \subseteq V$ and $L' : V' \to \mathbf{N}^+$ be a fixed subset of V and a pre-labeling for V', respectively. We define the parameter M which will be very useful in the rest of the paper:

$$M := \max\{n, \max_{v \in V'} L'(v)\}.$$

Clearly, M is straightforward to compute if G and L' are known. And clearly, M is a lower bound on the span of any radio labeling in G extending the pre-labeling L' of G. A natural question is how far M can be away from the minimum span of such a labeling. We will show that the answer to this question heavily relies on the girth of the graph G:

Theorem 1. *Consider a graph G on $n \geq 7$ vertices, and a pre-labeling L' of G. Then there is a radio labeling in G extending L'*

(a) with span $\leq \lfloor (7M-2)/3 \rfloor$
(b) with span $\leq \lfloor (5M+2)/3 \rfloor$ if G has girth at least 4
(c) with span $\leq M+3$ if G has girth at least 5.

All these bounds are best possible. The third bound is even best possible for the class of paths.

Proof. Let us start the proof of Theorem 1 by showing that all the stated bounds indeed are best possible: For (a), let x be a positive integer and y an integer such that $M = 3x+y$ and $-1 \leq y \leq 1$. We consider the complete graph on M vertices in which x vertices are pre-labeled with labels $2, 5, 8, \ldots, 3x-1$, whereas the remaining $2x+y$ vertices are unlabeled. Since we cannot use the labels $1, 2, \ldots, 3x$ at the unlabeled vertices, and the labels at these vertices have to differ by at least two, the span of any radio labeling extending the pre-labeling is at least $3x + 2(2x+y) - 1 = 7x + 2y - 1 = \lfloor (7M-2)/3 \rfloor$.

For (b), let x be a positive integer and y an integer such that $M = 3x+y$ and $0 \leq y \leq 2$. We consider the complete bipartite graph $K_{x+1, 2x+y-1}$ on M vertices. The $x+1$ vertices in the first part of the bipartition are pre-labeled with the labels $2, 5, 8, \ldots, 3x+2$, whereas the $2x+y-1$ vertices in the second part of the bipartition are unlabeled. Note that the pre-labeling forbids the labels $1, \ldots, 3x+y+1$ for the unlabeled vertices. Thus the span of any radio labeling extending the pre-labeling is at least $5x + 2y = \lfloor (5M+2)/3 \rfloor$.

Finally, for (c) we consider the path $v_1 - v_2 - \cdots - v_n$ on $n \geq 7$ vertices. The pre-labeling assigns $L'(v_1) = 1$, $L'(v_3) = 5$, $L'(v_5) = 3$, $L'(v_7) = 7$, and $L'(v_k) = k$ for $k \geq 8$. Then $M = n$, and all but three vertices (v_2, v_4, v_6) are pre-labeled. Since we cannot use the labels $2, 4,$ and 6 at these three vertices, every radio labeling of this path extending the pre-labeling L' has a span of at least $n + 3 = M + 3$.

The greedy preprocessing step. In several of our proofs, we use the same preprocessing step. We denote this step as *greedy preprocessing*.

Greedy preprocessing is done as follows. We extend the pre-labeling by assigning labels from $\{1, \ldots, M\}$ to unlabeled vertices, such that the conditions that all labels are different, and that adjacent vertices have labels that differ by at least two keep being met. This preprocessing step terminates, when we get stuck: then either all vertices have been labeled, or for every unused label $c \in \{1, \ldots, M\}$ and for every unlabeled vertex v, v is adjacent to a vertex labeled with $c-1$ or v is adjacent to a vertex labeled with $c+1$. Note that this greedy preprocessing step does not change the value of M.

Proof of Theorem 1.(a). It is sufficient to consider the case where G is the complete graph on n vertices. Assume that there are $l = |V'|$ pre-labeled vertices and that we start with the available labels $1, 2, \ldots, N = \lfloor (7M-2)/3 \rfloor$. Call a label *blocked*, if it is used in the pre-labeling or if it is adjacent to a (vertex labeled with

a) label in the pre-labeling (and thus cannot be used). Hence there are at most $\min\{3l, M+1\}$ blocked labels. Order the remaining available (nonblocked) labels increasingly, and assign the labels at the odd positions in the ordering to the remaining $n-l$ vertices. Note that we need at most $\min\{3l, M+1\} + 2(n-l) - 1$ labels. If $3l \leq M+1$, then this number is at most

$$3l + 2(n-l) - 1 = 2n + l - 1 \leq 2M + \frac{M+1}{3} - 1 = \frac{7M-2}{3}.$$

If $3l \geq M+1$, then this number is at most

$$M + 1 + 2(n-l) - 1 = M + 2n - 2l \leq M + 2M - 2\frac{M+1}{3} = \frac{7M-2}{3}.$$

So, in both cases we obtain a feasible radio labeling with span at most $\lfloor (7M-2)/3 \rfloor$.

Proof of Theorem 1.(b). We start with $N = \lfloor (5M+2)/3 \rfloor$ available labels $1, 2, \ldots, N$, perform the greedy preprocessing described above, and from now on consider the labeling thus obtained. If all vertices are labeled, then we are done and there is nothing to show. Otherwise, consider some fixed label $c \in \{1, \ldots, M\}$ that is not used in this labeling. Then every unlabeled vertex v must be adjacent to a vertex labeled with $c+1$ or to a vertex labeled with $c-1$. Denote the set of unlabeled vertices adjacent to $c-1$ by A, and denote the set of unlabeled vertices that are adjacent to $c+1$ but not to $c-1$ by B. Then A must be an independent set. (Any edge in A together with the vertex labeled with $c-1$ would induce a triangle, and bring the girth of G down to 3.) And also B must be an independent set.

Assume that l vertices are labeled. Since every label blocks at most two other labels, there are at most $\min\{3l, M+1\}$ blocked labels and we have at least

$$\lfloor (5M+2)/3 \rfloor - \min\{3l, M+1\} \geq n - l$$

available labels for the remaining $n-l$ vertices. The displayed inequality can be seen as follows. If $3l \leq M+1$, then $\lfloor (2M+2)/3 \rfloor \geq 2l$. Adding $M - 3l \geq n - 3l$ yields the desired inequality in this case. If $3l \geq M+2$, then $n - l \leq M - \lceil (M+2)/3 \rceil$. Together with $M - \lceil (M+2)/3 \rceil \leq \lfloor (5M+2)/3 \rfloor - M - 1$ we get the desired inequality also for this second case.

Now we distinguish three subcases. In the first subcase, there are vertices $a \in A$, $b \in B$, such that a is not adjacent to b. We assign the $|A|$ smallest of the $n-l$ available labels to the vertices in A, and the $|B|$ largest of these labels to the vertices in B. This is done in such a way that vertex a receives the largest label in A, and such that vertex b receives the smallest label in B. This gives a radio labeling with span $\leq \lfloor (5M+2)/3 \rfloor$.

In the second subcase, either A or B is empty. In this case, we use the $n-l$ available labels on the $n-l$ vertices in A or B.

In the third subcase, we assume that none of A and B is empty and that they span a complete bipartite graph, i.e. each vertex in A is adjacent to each vertex in B. Consider an arbitrary label $d \in \{1, \ldots, M\}$ that is not used in the pre-labeling. Then every unlabeled vertex $v \in A \cup B$ must be adjacent to the vertex labeled with $d+1$ or to the vertex labeled with $d-1$. There are only two possibilities for this: either all vertices in A are adjacent to the vertex labeled with $d+1$ and all vertices in B are adjacent to the vertex labeled with $d-1$, or all vertices in A are adjacent to the vertex labeled with $d-1$ and all vertices in B are adjacent to the vertex labeled with $d+1$. As a consequence, there are at most $\min\{2l, M+1\}$ blocked labels in this subcase, and at least

$$\lfloor (5M+2)/3 \rfloor - \min\{2l, M+1\} \geq n - l + 1$$

available labels for the remaining $n - l$ vertices. We assign the $|A|$ smallest of these labels arbitrarily to the vertices in A, and the $|B|$ largest of these labels arbitrarily to the vertices in B.

Proof of Theorem 1.(c). We start with the greedy preprocessing described above, and from now on consider the labeling thus obtained. Denote by $C = \{c_1, \ldots, c_k\}$ with $c_1 < c_2 < \cdots < c_k$ the set of all unused labels from $\{1, \ldots, M\}$ in the labeling obtained after the preprocessing. Denote by U the set of unlabeled vertices. It is clear that $|U| \leq |C|$. We prove that the labeling obtained by the greedy preprocessing can always be extended to a radio labeling of G using at most three additional labels $M+1$, $M+2$, and $M+3$.

Without loss of generality, we assume that after the greedy preprocessing there is a vertex of G labeled with M. (If this is not the case, the same proof produces a span $\leq M+2$.) For the sake of convenience we shall identify each labeled vertex with its label. If $|U| = 1$, we just label the only unlabeled vertex with label $M+2$. If $|U| = 2$, then the two unlabeled vertices are either nonadjacent and we can label them with $M+2$ and $M+3$, or at least one of them is not adjacent to M (since G contains no 3-cycles), and we label this vertex with $M+1$ and the other vertex with $M+3$.

From now on we assume $|U| \geq 3$. If C contains two consecutive labels d and $d+1$, then every vertex in U must be adjacent to the vertices labeled with $d-1$ and $d+2$. This yields a cycle of length four in G, and contradicts the assumption on the girth of G. Therefore, there are no consecutive labels in C, and in particular $c_3 - 1 > c_1 + 1$.

Next, we first discuss the case $|U| \geq 4$. Then $|C| \geq 4$ and $c_3 < M$. Suppose for the sake of contradiction that $c_1 = 1$. Then all vertices in U are adjacent to label 2, and at least two vertices of U are adjacent to $c_3 - 1$ or to $c_3 + 1$. This would yield a cycle of length four. This contradiction shows $c_1 \geq 2$. There cannot be more than two vertices of U adjacent to $c_1 - 1$ because otherwise at least two of them would be adjacent to $c_3 + 1$ or to $c_3 - 1$, and we again would obtain a 4-cycle. Similarly, we see that at most two vertices of U are adjacent to $c_1 + 1$. As a consequence, each of the vertices $c_1 - 1$ and $c_1 + 1$ has exactly two neighbors in U. This implies that $|U| = 4$, and that each vertex of U is adjacent to exactly one of $c_1 - 1$ and $c_1 + 1$. Denote the four vertices in U by u_1, u_2, u_3, u_4 such that u_1 and u_2 are the neighbors of $c_1 - 1$, and u_3 and u_4 are the neighbors of $c_1 + 1$. Moreover, we may assume that u_1 and u_3 are adjacent to $c_3 - 1$ and that u_2 and u_4 are adjacent to $c_3 + 1$.

For every label $c \in C \setminus \{c_1, c_3\}$ we have that $|\{c-1, c+1\} \cap \{c_1 - 1, c_1 + 1, c_3 - 1, c_3 + 1\}| = 0$. Otherwise, we either obtain a 4-cycle or can use the label c at one of the vertices in U; as an example, consider, e.g. the case that $c - 1 = c_1 + 1$: then we can use the label c at u_1 or u_2 unless both are adjacent to $c_1 + 3$, yielding a 4-cycle with $c_1 - 1$; the other cases are similar. But then there are only two possibilities for the vertices in U to be the neighbor of $c - 1$ or $c + 1$ without creating a 4-cycle: u_1 and u_4 should be adjacent to one of these labels and u_2 and u_3 to the other. But since $|C \setminus \{c_1, c_3\}| \geq 2$ and C does not contain consecutive integers, we obtain a 4-cycle in G, which is a contradiction. So the case $|U| \geq 4$ can not occur at all.

We are left with the case that $|U| = 3$. By similar arguments as above, we may assume that $c_1 - 1$ is adjacent to u_1, that $c_1 + 1$ is adjacent to u_2 and u_3, that $c_3 + 1$ is adjacent to u_1 and u_2, and that $c_3 - 1$ is adjacent to u_3 (the other cases are analogous). The girth condition implies that the graph induced by the vertices $\{u_1, u_2, u_3\}$ has at most one edge, the edge $\{u_1, u_3\}$. If such an edge exists, then M is nonadjacent to one of these three vertices, and we label this vertex with $M + 1$ and the other one with $M + 3$; in that case the vertex u_2 is labeled with $M + 2$. If there is no such edge, then M is nonadjacent to at least one of the vertices u_i (otherwise we obtain a 4-cycle) and we label this vertex with $M + 1$ and the other two vertices with $M + 2$ and $M + 3$. This completes the proof of Theorem 1.(c). □

Note that the proof of Theorem 1.(c) yields a linear time approximation algorithm determining a radio labeling extending a pre-labeling in graphs with girth ≥ 5 with a span that is at most an additive 3 away from the minimum span.

3 Graphs with Bounded Degeneracy

In this section, we apply the techniques from the proof of Theorem 1 to t-degenerate graphs, i.e. graphs with the property that each of their subgraphs has a vertex of degree at most t. We will also use the easy, well-known fact that a t-degenerate graph on n vertices has at most $t(n-t) + t(t-1)/2$ edges. First we consider the case without any pre-labeling, and obtain the following result which is easy to prove.

Lemma 1. *If G is a t-degenerate graph on n vertices, then it has a radio labeling with span $\leq n + 2t$.*

We now turn to the variant in which a pre-labeling is assumed. We obtain an upper bound for the minimum span of a radio labeling extending the pre-labeling in a t-degenerate graph, depending on M and t only.

Theorem 2. *If G is a t-degenerate graph and L' is a pre-labeling of G, then there exists a radio labeling extending L' with span $\leq M + (4 + \sqrt{3})t + 1$.*

Proof. We again start with the greedy preprocessing step as described in the preceding section, and with available labels $1, 2, \ldots, M + (4 + \sqrt{3})t + 1$. Denote by U the set of unlabeled vertices after this step. We only consider the case where $|U| = 2p$ for some positive integer p; the case with odd $|U|$ is similar and left to the reader. Denote by $C = \{c_1, \ldots, c_{2p}\}$ with $c_1 < \ldots < c_{2p}$ the set of the first $2p$ unused labels. It is clear that C cannot contain three consecutive labels $c - 1, c, c + 1$ (otherwise, we can use the label c at some vertex of U). Thus for every i we have $c_i + 1 < c_{i+2} - 1$. Consider the set C' of vertices $c_i + 1$ and $c_i - 1$ for every odd $i = 1, 3, \ldots, 2p-1$. Then C' contains exactly p pairs of vertices and each unlabeled vertex of U must be adjacent to at least one vertex from every pair. The graph induced by the set $U \cup C'$ has $4p$ vertices. Denote the number of edges of this graph by x. We have $2p^2 \leq x \leq (4p-t)t + (t-1)t/2 = 4pt - t^2/2 - t/2$. Thus $(p-t)^2 \leq 3t^2/4 - t/4 \leq 3t^2/4$, yielding $2p \leq (2 + \sqrt{3})t$. Using Lemma 1 and the fact that $G[U]$ is t-degenerate, we can label the unlabeled vertices of U with the labels $\{M + 2, \ldots, M + (4 + \sqrt{3})t + 1\}$. This completes the proof. □

4 Graphs with a Bounded Maximum Degree

We now turn to graphs with a bounded maximum degree. Using similar proof techniques as in the previous section, we will prove that P-RL(∗) is polynomially solvable within this class of graphs. We first prove the next technical result on t-degenerate graphs.

Theorem 3. *Let $G = (V, E)$ be a t-degenerate graph with maximum degree Δ and let $V' \subseteq V$ be the set of vertices that is pre-labeled by L'. If the number of unlabeled vertices $p = |V \setminus V'| \geq 4\Delta(t + 1)$, then L' can be extended to a radio labeling of G with span M.*

Proof. Let $H = G[V \setminus V']$ be the graph induced by the unlabeled vertices, and start with the available labels $1, 2, \ldots, M$. As long as H has edges we will apply the following labeling procedure. Consider a vertex v in H of minimum positive degree. It has at most t neighbors. We can label them all with unused labels because there are at most $2\Delta - 2$ blocked labels among the unused labels and the number of available labels in every step is at least the number of unlabeled vertices at that moment, which is always at least 4Δ. Adapt H by removing the neighbors of v. In each step we reduce the number of unlabeled vertices by at most t and increase the number of isolated vertices by one. Since $p \geq 4\Delta(t+1)$, we have $q \geq 4\Delta$ isolated vertices in H when H becomes edgeless.

We now show that the set Q of the remaining q isolated vertices of H can be labeled with unused labels from the set $\{1, \ldots, M\}$. Denote the set of such labels by C (it is clear that $|C| \geq q$) and consider the auxiliary bipartite graph G' with vertex partition $Q \cup C$ where an edge $\{v, c\}$ exists if and only if the vertex v of Q can be labeled with the label c. It is sufficient to show that G' has a matching saturating all vertices of Q. Suppose that there is no such matching. Then by standard matching theory there is a set $A \subset Q$ such that $|A| = a$ and $|N(A)| \leq a - 1$ (where $N(A)$ is the neighborhood of A in G'). Let $B = C \setminus N(A)$. We have $|B| \geq q - a + 1$. Note that each label could be forbidden for at most 2Δ vertices and, vice versa, that for every vertex at most 2Δ labels could be forbidden. Therefore, since there are no edges between A and B, we have $a \leq 2\Delta$ and $q - a + 1 \leq |B| \leq 2\Delta$. But this implies $q \leq 4\Delta - 1$, a contradiction. This completes the proof. □

From Theorem 3 we easily obtain the following complexity result for graphs with a bounded maximum degree.

Corollary 1. *Let k be a fixed positive integer. For every graph G with maximum degree $\Delta \leq k$ and pre-labeling L', P-RL(*) can be solved in polynomial time.*

Proof. Each graph with maximum degree Δ is clearly Δ-degenerate. If at most $4\Delta(\Delta + 1)$ vertices of G are not pre-labeled by L', then one can use a brute force algorithm to find a radio labeling extending L' with a minimum span, e.g. by checking all admissible labelings. The time complexity of such a brute force algorithm is $O(n^{4\Delta(\Delta+1)})$. If more than $4\Delta(\Delta+1)$ vertices are unlabeled, then by Theorem 3 there is a radio labeling extending L' with span M (which is clearly the minimum), and from the proof of Theorem 3 it is not difficult to check that such a labeling can be found in polynomial time. □

The above corollary shows that P-RL(l) and P-RL(*) have the same complexity behavior as RL for graphs with a bounded maximum degree, i.e. all three of the problems can be solved in polynomial time. This picture changes if we restrict ourselves to graphs which are k-colorable and for which a k-coloring is given (as part of the input) for some fixed positive integer k. This is the topic of the next section.

5 Graphs with a Bounded Chromatic Number

In this section we concentrate on radio labeling algorithms for graphs with a bounded chromatic number, in particular for the case where a k-coloring of the graph (for a fixed constant k) is provided as part of the input.

Related to Proposition 1 we discussed the useful equivalence between RL and the TRAVELING SALESMAN problem TSP(2,1). The main idea behind the proof of the following theorem is to adapt this equivalence to capture the restrictions of the pre-labeling problem. We omit the proof of the following theorem in this extended abstract.

Theorem 4. *Let $G = (V, E)$ be a graph with a given k-coloring with color classes $I_1, I_2, \ldots, I_k$. Let L' be a pre-labeling of a subset $V' \subseteq V$ with $|V'| = l$. Then a radio labeling L of G extending L' with the smallest possible span can be computed in time $O(n^{4(l+1)k(k-1)})$.*

For each of the graph classes in the following corollary, it is possible to construct a vertex coloring with a constant number of colors in polynomial time. Hence:

Corollary 2. *The radio labeling problem P-RL(l) is polynomially solvable*

- on the class of planar graphs,
- on any class of graphs of bounded treewidth,
- on the class of bipartite graphs.

The above results show that P-RL(l) is solvable in polynomial time for graphs with a bounded chromatic number and a given coloring. This result does not carry over to the more general labeling problem P-RL(*) where the number of pre-labeled vertices is part of the input. The proof of the following theorem is omitted in this abstract.

Theorem 5. *For any fixed $k \geq 3$, problem* P-RL(*) *is NP-hard even when the input is restricted to graphs with a given k-coloring.*

We now turn to the last class of graphs for which RL is known to be polynomially solvable, namely the class of cographs, i.e. graphs without an induced path on four vertices. Using a reduction from 3-PARTITION we show that P-RL(*) is NP-hard for cographs. We omit the proof of the following theorem in this extended abstract.

Theorem 6. *Problem* P-RL(*) *is NP-hard for cographs.*

References

[1] G. J. CHANG AND D. KUO, *The $L(2,1)$-labeling problem on graphs*, SIAM J. Discrete Math., 9 (1996), pp. 309–316.

[2] P. DAMASCHKE, J. S. DEOGUN, D. KRATSCH, AND G. STEINER, *Finding Hamiltonian paths in cocomparability graphs using the bump number algorithm*, Order, 8 (1992), pp. 383–391.

[3] L. ENGEBRETSEN, *An explicit lower bound for TSP with distances one and two*, in Proceedings of the 16th Annual Symposium on Theoretical Aspects of Computer Science (STACS'1999), Springer LNCS 1563, 1999, pp. 373–382.

[4] J. FIALA, J. KRATOCHVÍL, AND A. PROSKUROWSKI, *Distance constrained labelings of precolored trees*, in Proceedings of the 7th Italian Conference on Theoretical Computer Science (ICTCS'2001), Springer LNCS 2202, 2001, pp. 285–292.

[5] D. A. FOTAKIS AND P. G. SPIRAKIS, *A Hamiltonian approach to the assignment of non-reusable frequencies*, in Proceedings of the 18th Conference on Foundations of Software Technology and Theoretical Computer Science (FSTTCS'1998), Springer LNCS 1738, 1998, pp. 18–29.

[6] J. R. GRIGGS AND R. K. YEH, *Labelling graphs with a condition at distance 2*, SIAM J. Discrete Math., 5 (1992), pp. 586–595.

[7] W. K. HALE, *Frequency assignment: Theory and applications*, Proceedings of the IEEE, 68 (1980), pp. 1497–1514.

[8] R. A. LEESE, *Radio spectrum: a raw material for the telecommunications industry*, in Progress in Industrial Mathematics at ECMI 98, Teubner, Stuttgart, 1999, pp. 382–396.

[9] C. H. PAPADIMITRIOU AND M. YANNAKAKIS, *The travelling salesman problem with distances one and two*, Math. Oper. Res., 18 (1993), pp. 1–11.

Branch-and-Bound Algorithms for the Test Cover Problem

Koen M.J. De Bontridder[1], B.J. Lageweg[2], Jan K. Lenstra[3],
James B. Orlin[4], and Leen Stougie[3,5]

[1] Institute of Information and Computing Sciences
Universiteit Utrecht, The Netherlands
koendb@cs.uu.nl,
[2] ORTEC Consultants BV, Gouda, The Netherlands
[3] Department of Mathematics and Computer Science
Technische Universiteit Eindhoven, The Netherlands
{jkl,leen}@win.tue.nl,
[4] Sloan School of Management
Massachussets Institute of Technology, Cambridge, Massachusetts, USA
jorlin@mit.edu
[5] CWI, Amsterdam, The Netherlands
stougie@cwi.nl

Abstract. In the test cover problem a set of items is given together with a collection of subsets of the items, called tests. A smallest subcollection of tests is to be selected such that for every pair of items there is a test in the selection that contains exactly one of the two items. This problem is NP-hard in general. It has important applications in biology, pharmacy, and the medical sciences, as well as in coding theory.
We develop a variety of branch-and-bound algorithms to solve the problem to optimality. The variety is in the definition of the branching rules and the lower bounds to prune the search tree. Our algorithms are compared both theoretically and empirically.

1 Prelude

The input of the *test cover problem* consists of a set of *items*, $\{1, \ldots, m\}$, and a collection $\mathbf{T}$ of *tests*, $T_1, \ldots, T_n \subset \{1, \ldots, m\}$. A test T_l covers or differentiates item pair $\{i, j\}$ if *either* $i \in T_l$ *or* $j \in T_l$, i.e., if $|T_l \cap \{i, j\}| = 1$. A subcollection $\mathcal{T} \subset \mathbf{T}$ of tests is a *test cover* if each of the $m(m-1)/2$ item pairs is covered by at least one test in $\mathcal{T}$. The problem is to find a test cover of minimum cardinality τ^*. The test cover problem is also known as the *minimum test collection* or *minimum test set problem*.

The problem arises naturally in the following general setting of identification problems: Given a set of individuals and a set of binary attributes that may or may not occur in each individual, the goal is to find a minimum subset of attributes (a test cover) such that each individual can be identified uniquely from the information contained in the subset. In this way the incidence vector of any individual with the test cover is a unique binary signature, which distinguishes the individual from any other individual.

The motivation for this research dates back more than 20 years through a request from the Agricultural University in Wageningen, the Netherlands, concerning the identification of potato diseases. Each potato variety is vulnerable to a number of diseases. One wished to have a minimum-size subset of varieties such that each pair of diseases could be distinguished from one another [Lageweg et al. 1980, Kolen & Lenstra 1995]. There are numerous other important applications in biology, pharmacy, and the medical sciences (see e.g., [Moret & Shapiro 1985] and [De Bontridder et al. 2001]), as well as in coding theory (see [De Bontridder 1997]).

The problem is NP-hard [Garey & Johnson 1979]. It remains NP-hard if the number of items in each test is at most 2 [De Bontridder et al. 2002]. Approximation algorithms for the problem are studied by Moret & Shapiro [1985] and De Bontridder et al. [2002].

The main work of Moret & Shapiro [1985] concerns the empirical analysis of branch-and-bound algorithms for the problem. We will use the same basic branch-and-bound scheme: Nodes of the search tree are defined by partial test covers, which are sets of tests that do not cover all item pairs. A descendant node is obtained by adding a test to the partial test cover. There are several criteria to decide on the test to be added. Given a partial test cover, various lower bounds can be defined on the number of tests that need to be added to obtain a test cover. These lower bounds are used in combination with the size of the best feasible solution found to prune the search tree.

The difference with [Moret & Shapiro 1985] is in the different branching rules and lower bounds employed for pruning the search tree. Moreover, we do not only compare the various resulting branch-and-bound algorithms empirically, but we also make a theoretical comparison of the running time required to compute the lower bounds and of their pruning power.

The common ideas behind the algorithms are explained in Section 2. The various branching rules are presented in Section 3. Some of these have appeared in [Moret & Shapiro 1985] and some are new. In Section 4 we present and compare various lower bounds.

From among the variety of branch-and-bound algorithms that can be constructed by combining branching rules and pruning criteria, we make a selection in Section 5. We subject this selection in Section 6 to an empirical performance analysis. As test problems we take a real-life problem instance and a set of random instances. The analysis shows a clear picture in favor of one of the algorithms. We compare our results with the conclusions drawn by Moret & Shapiro [1985]. The results are encouraging for solving larger practical problems.

Throughout the paper log denotes $\log_2$.

2 Partial Test Covers and Equivalence Classes

For solving test cover problems we devise branch-and-bound algorithms, in which the nodes of the search tree correspond to partial test covers and the leaves to test covers, the root being the empty set. The child of a node is defined by adding a test to the partial test cover of that node, and discarding some other tests.

A partial test cover is a collection of tests that does not cover each item pair. Any partial test cover defines an *equivalence relation* on the set of items.

Definition 1. *Given a partial test cover two items are equivalent if there is no test that covers them.*

The equivalence relation corresponding to a partial test cover $\mathcal{T} \subset \mathbf{T}$ induces a partition of the set of all items into subsets $I_1^{\mathcal{T}}, I_2^{\mathcal{T}}, \ldots, I_t^{\mathcal{T}}$, for some t, such that, for each subset, any two items in it are not covered by any test in $\mathcal{T}$. Each item can be in only one set, for if item pairs $\{i_1, i_2\}$ and $\{i_2, i_3\}$ are not covered, then item pair $\{i_1, i_3\}$ is not covered either. Regarded in this way, a test cover is a collection of tests that defines an equivalence relation such that the induced partition consists of exactly m singleton sets.

In the following sections, we use the equivalence relation to define for each node branching rules and criteria to prune the tree at that node. The various branching rules and pruning criteria that we define lead to various algorithms. However, they all start with selecting the so-called *essential* tests.

Definition 2. *A test T is essential if there exists an item pair for which T is the only test in $\mathbf{T}$ that covers it.*

We assume from now on that the root of the search tree includes all essential tests. They can be found in $O(m^2 n)$ time.

3 Quality Criteria and Branching Rules

In this section, we will present criteria to decide on the quality of tests in a node of the search tree, i.e., given that a partial test cover has been selected. We will use these criteria in designing branching schemes and lower bounds for pruning the search tree. As mentioned before, we assume that all essential tests have been included in each node of the search tree.

Let us consider a node represented by the pair $(\mathcal{T}, \mathcal{R})$, with $\mathcal{T}$ the partial test cover and $\mathcal{R}$ the set of discarded tests. Let $I_1^{\mathcal{T}}, I_2^{\mathcal{T}}, \ldots, I_t^{\mathcal{T}}$ be the corresponding partition of the items. $\mathcal{T}$ leaves $\sum_{i=1}^{t} |I_i^{\mathcal{T}}|(|I_i^{\mathcal{T}}| + 1)/2$ pairs of items uncovered. Updating the equivalence classes after adding a test to a partial test cover requires $O(m)$ time, in the most straightforward way.

A test $T \notin \mathcal{T}$ decreases the number of uncovered item pairs instantaneously by

$$D(T|\mathcal{T}) = \sum_{h=1}^{t} |T \cap I_h^{\mathcal{T}}||I_h^{\mathcal{T}} \setminus T|. \tag{1}$$

This leads to the first quality criterion called the *separation criterion* by Moret & Shapiro [1985].

Another criterion called the *information criterion* by Moret & Shapiro [1985], was introduced in a different form in 1961 by Rescigno & Maccacaro [1961], and in the way stated here by Chang et al. [1970]; see also [Moret 1982].

Definition 3. *A partial test cover $\mathcal{T}$ has* information value

$$E(\mathcal{T}) = \log m - \frac{1}{m} \sum_{h=1}^{t} |I_h^{\mathcal{T}}| \log |I_h^{\mathcal{T}}|.$$

Notice that $E(\mathcal{T}) = 0$ if $\mathcal{T} = \emptyset$ and $E(\mathcal{T}) = \log m$ if $\mathcal{T}$ is a test cover. Notice also that E is a monotone function: $E(\mathcal{T}') \leq E(\mathcal{T})$ for any $\mathcal{T}' \subset \mathcal{T}$. The addition of a test T to $\mathcal{T}$ will give an increase in information value of

$$\Delta E(T|\mathcal{T}) = E(\mathcal{T} \cup T) - E(\mathcal{T}), \qquad (2)$$

which defines the quality criterion.

We add here a new quality criterion for a test given a partial test cover, which we call *power* of a test.

Definition 4. *Given a partial test cover $\mathcal{T}$, the* power *of a test $T \notin \mathcal{T}$ is defined by*

$$P(T|\mathcal{T}) = \sum_{h=1}^{t} \min\{|T \cap I_h^{\mathcal{T}}|, |I_h^{\mathcal{T}} \setminus T|\}.$$

A final criterion, also mentioned by Moret & Shapiro [1985], is the so-called *least-separated pair* criterion. It was applied before by microbiologists [Rypka et al. 1967], [Rypka et al. 1978], [Payne 1981].

Definition 5. *Given a partial test cover $\mathcal{T}$, item pair $\{i, j\}$ is* least-separated *if amongst the yet uncovered item pairs it is the one for which the minimum number of covering tests is available. In case there is not a unique least-separated item pair, we choose one of them arbitrarily. Given the least-separated item pair $\{i, j\}$, we define*

$$S(T|\mathcal{T}) = \begin{cases} 1 & \text{if } |T \cap \{i,j\}| = 1; \\ 0 & otherwise. \end{cases}$$

Moret & Shapiro [1985] use the name *nearly essential* tests to clarify the notion of this criterion: at least one of the tests for which $S(T|\mathcal{T}) = 1$ must be contained in a test cover. We follow Moret & Shapiro [1985] by using this criterion only in combination with any of the previous three for breaking the many ties that application of this criterion will meet.

Each of the quality criteria can be incorporated in the following general branching scheme. Let $C \in \{D, \Delta E, P, S\}$. Given a node $(\mathcal{T}, \mathcal{R})$, we order the tests $T \notin \mathcal{T} \cup \mathcal{R}$ according to non-increasing $C(T|\mathcal{T})$ value: $C(T_1|\mathcal{T}) \geq$

$C(T_2|\mathcal{T}) \geq \ldots \geq C(T_l|\mathcal{T})$, with l the number of tests not in $\mathcal{T} \cup \mathcal{R}$. We create the children $(\mathcal{T} \cup T_1, \mathcal{R})$, $(\mathcal{T} \cup T_2, \mathcal{R} \cup T_1)$, $(\mathcal{T} \cup T_3, \mathcal{R} \cup T_1 \cup T_2), \ldots, (\mathcal{T} \cup T_l, \mathcal{R} \cup_{h=1}^{l-1} T_h)$, and explore the children in this order.

Notice that each of the first three branching schemes requires $O(mn \log m)$ in each node of the search tree. The least-separated pair criterion takes less time but will always be combined with one of the other three.

4 Lower Bounds for Pruning

We will use the quality criteria from the previous section here to define pruning criteria for the nodes of the search tree. Again consider a node $(\mathcal{T}, \mathcal{R})$ with a corresponding partition of the items $I_1^\mathcal{T}, I_2^\mathcal{T}, \ldots, I_t^\mathcal{T}$. We define four lower bounds on the number of tests that have to be added to $\mathcal{T}$ to obtain a test cover. For each of the lower bounds we also give the complexity of computing them.

4.1 Lower Bound by Ideal Tests

The first one, which we denote $L_1(\mathcal{T})$, is straightforward and, as the notation suggests, independent of the discarded tests. It assumes the existence of *ideal* or *even splitting* tests, regardless of the availability of such tests in the instance. Given any partition of the items in equivalence classes, such a test contains half of the items in each of the equivalence classes.

Lemma 1. *The number of ideal tests that have to be added to partial test cover $\mathcal{T}$ in order to obtain a test cover is bounded from below by*

$$L_1(\mathcal{T}) = \lceil \log(\max_{h=1,\ldots,t} I_h^\mathcal{T}) \rceil.$$

Proof. Given a set of m items, defining $\frac{1}{2}m(m-1)$ item pairs, any ideal test covers $\lfloor \frac{1}{4}m^2 \rfloor$ of the item pairs. From this observation the lemma follows easily.

Computing this lower bound requires time $O(m)$ in each node of the search tree. This can be improved by using a more efficient data structure for storing the sizes of the equivalence classes.

4.2 Lower Bounds by Power

The lower bound $L_1(\mathcal{T})$ does not use any specific information about the tests available in the problem instance to supplement $\mathcal{T}$ to a test cover. Indeed, ideal tests may not be available at all in the node. The following lower bounds do use specific problem instance and node information. They are based on the power $P(T|\mathcal{T})$ of a test T. Clearly, $1 \leq P(T|\mathcal{T}) \leq m/2$, for any $\mathcal{T}$ and any $T \notin \mathcal{T}$. Moreover, notice that $P(T|\mathcal{T}) \geq P(T|\mathcal{T}')$ for any $\mathcal{T} \subset \mathcal{T}'$ and any $T \notin \mathcal{T}'$. Given node $(\mathcal{T}, \mathcal{R})$, we order the l tests not in $\mathcal{T} \cup \mathcal{R}$ according to non-increasing power, $P(T_1|\mathcal{T}) \geq P(T_2|\mathcal{T}) \geq \ldots \geq P(T_l|\mathcal{T})$.

Let $F(m,n)$ denote the minimum total power that any n tests need to cover m items. Then, given a partial test cover $\mathcal{T}$, with resulting equivalence classes $I_1^\mathcal{T}, I_2^\mathcal{T}, \ldots, I_t^\mathcal{T}$, any k tests should have total power at least $\sum_{h=1}^t F(|I_h^\mathcal{T}|, k)$, in order to supplement $\mathcal{T}$ to a test cover. Using the tests not in $\mathcal{T} \cup \mathcal{R}$, and assuming that their cumulative effect is the sum of their individual effects on covering item pairs, we obtain our second lower bound on the number of tests to be added to $\mathcal{T}$:

$$L_2(\mathcal{T}, \mathcal{R}) = \min\{k | \sum_{j=1}^k P(T_j | \mathcal{T}) \geq \sum_{h=1}^t F(|I_h^\mathcal{T}|, k)\}.$$

Through a smart dynamic programming recursion the list of values for $F(h,k)$, $h = 1, \ldots, m$, $k = 1, \ldots, n$, can be computed in $O(mn)$ time. We omit the description and its correctness proof in this extended abstract. Moreover, notice that the F-values are instance independent. Thus, such a list can be computed once and for all.

Having the list of F-values and the powers of tests in an ordered list, given a partial test cover, computing $L_2(\mathcal{T}, \mathcal{R})$ requires $O(m)$ time.

In fact, we can improve this lower bound by considering subsets of the set of equivalence classes. Let $H \subset \{1, \ldots, t\}$ and $I_H^\mathcal{T} = \cup_{h \in H} I_h^\mathcal{T}$. Define the power of T relative to $I_H^\mathcal{T}$ as $P_H(T|\mathcal{T}) = \sum_{h \in H} \min\{|T \cap I_h^\mathcal{T}|, |I_h^\mathcal{T} \setminus T|\}$. For each subset $I_H^\mathcal{T}$ a lower bound like $L_2(\mathcal{T}, \mathcal{R})$ must hold, i.e., given an ordering of the tests not in $\mathcal{T} \cup \mathcal{R}$ according to non-increasing power relative to $I_H^\mathcal{T}$ ($P_H(T_1|\mathcal{T}) \geq \ldots \geq P_H(T_l|\mathcal{T})$),

$$L_2^H(\mathcal{T}, \mathcal{R}) = \min\{k | \sum_{j=1}^k P_H(T_j|\mathcal{T}) \geq \sum_{h \in H} F(|I_h^\mathcal{T}|, k)\}$$

is also a lower bound on the number of tests to be added to $\mathcal{T}$. Taking the maximum over all possible subsets H yields our third lower bound:

$$L_3(\mathcal{T}, \mathcal{R}) = \max_{H \subset \{1, \ldots, t\}} L_2^H(\mathcal{T}, \mathcal{R}).$$

We do not know if $L_3(\mathcal{T}, \mathcal{R})$ can be computed in polynomial time.

The bounds described so far have been presented in order of increasing quality.

Lemma 2. $L_1(\mathcal{T}) \leq L_2(\mathcal{T}, \mathcal{R}) \leq L_3(\mathcal{T}, \mathcal{R})$.

Proof. The inequalities follow directly from the definitions.

4.3 Lower Bounds by Information Value

Our final lower bound was considered by Moret & Shapiro [1985]) and is based on the information value of a partial test cover. Given a node $(\mathcal{T}, \mathcal{R})$, order the l

tests not in $\mathcal{T} \cup \mathcal{R}$ such that $\Delta E(T_1|\mathcal{T}) \geq \Delta E(T_2|\mathcal{T}) \geq \ldots \geq \Delta E(T_l|\mathcal{T})$. Then,

$$L_4(\mathcal{T},\mathcal{R}) = \min\{k | E(\mathcal{T}) + \sum_{j=1}^{k} \Delta E(T_j|\mathcal{T}) \geq \log m\}$$

is a lower bound on the number of tests needed to raise the information value from $E(\mathcal{T})$ to $\log m$, since $\Delta E(T|\mathcal{T}) \geq \Delta E(T|\mathcal{T}')$ for any $\mathcal{T} \subset \mathcal{T}'$ and any $T \notin \mathcal{T}'$. The latter observation was mentioned by Moret & Shapiro [1985] and proved by De Bontridder [1997]. Given the ordering of the tests on ΔE-value the computation of this lower bound requires $O(m)$ time.

Lemma 3. $L_4(\mathcal{T},\mathcal{R})$ *is not dominated by* $L_3(\mathcal{T},\mathcal{R})$ *and does not dominate* $L_1(\mathcal{T})$.

Proof. The proof is by example. See the full paper for details.

4.4 Pruning Criteria

All lower bounds can be used in the following two criteria to prune nodes. Consider a node $(\mathcal{T},\mathcal{R})$ with lower bound L, and let U be the size of the best solution found so far.

- Prune the node if
  (PC1) $L > |\mathbf{T}| - |\mathcal{T}| - |\mathcal{R}|$, or
  (PC2) $|\mathcal{T}| + L \geq U$.

The second criterion prunes because descendants of the node cannot yield a better solution than the current best one. The first criterion prunes because the node leaves an infeasible instance. Another pruning criterion to cut off infeasible nodes is as follows:

- Prune the node if
  (PC3) a pair of items exists for which no covering test exists in $\mathbf{T}\backslash\mathcal{R}$.

The last criterion, like PC2, prunes a node for being inferior to the best found solution. The advantage of this criterion is that no ordering of tests is required. It relies on the following lemma in which we use the notation $P_I(T|\mathcal{T}) = \min\{|I \cap T|, |I\backslash T|\}$.

Lemma 4. *To cover the item pairs of a set I of $2^n - i$ items, $1 \leq i \leq 2^{n-1} - 1$, with n tests, $T_1, \ldots, T_n$, each of the tests needs to have $P_I(T_j|\emptyset) \geq 2^{n-1} - i$, $j = 1, \ldots, n$.*

Proof. Suppose the theorem is not true, then among the n tests there is at least one, say T, with $P_I(T|\emptyset) < 2^{n-1} - i$. After selection of this test I splits into I_1 and I_2, with $\max\{|I_1|,|I_2|\} > \max\{i, 2^n - i - 2^{n-1} + i\} \geq 2^{n-1}$. Clearly, the remaining $n - 1$ tests are not able to cover the item pairs of a set of more than 2^{n-1} items.

- Prune the node if
  (PC4) given $L = \lceil \log |I_t^\mathcal{T}| \rceil$, $L = U - |\mathcal{T}| - 1$ and there do not exist L tests $T \in \mathbf{T}\backslash(\mathcal{T} \cup \mathcal{R})$ with $P_{I_t^\mathcal{T}}(T|\mathcal{T}) \geq |I_t^\mathcal{T}| - 2^{L-1}$.

5 A Variety of Branch-and-Bound Algorithms

In the previous sections we have presented branching rules and bounding rules as basic ingredients for branch-and-bound algorithms. We have not yet specified a search strategy. We have chosen here for the depth-first search strategy in order to limit the storage of intermediate solutions. Since each node can have quite a lot of children, by the branching rules from Section 3, the number of intermediate solutions in a breadth-first search would become very large, even on shallow trees. Depth-first search has also the well-known advantage of finding feasible solutions rather quickly, which can then be used in defining upper bounds for pruning the search tree. The depth-first search strategy together with the branching and pruning rules give rise to a variety of branch-and-bound algorithms. We made a selection for our empirical tests.

In all of the algorithms we implemented some preprocessing. As mentioned already we add all essential tests directly in the root of the search tree. However, in any node it may happen that only one test remains for covering a yet uncovered item pair. In that case we add this *locally essential* test to the node's partial test cover, before we apply any other branching or bounding to that node. This preprocessing automatically checks pruning criterion PC3.

The selected branch-and-bound algorithms are listed below. They are defined by a triple $\{C, L, x\}$. C defines on which of the quality criteria the branching rule is based: $C \in \{D, P, \Delta E, S(*)\}$ (see Section 3), where $*$ is any of $\{D, P, \Delta E\}$ and $S(*)$ indicates the least-separation criterion combined with $*$ for breaking ties. L defines the type of lower bound used in pruning criteria PC1 and PC2 (see Section 4.4): $L \in \{L_1, L_2, L_3, L_4\}$ (see Sections 4.1, 4.2, and 4.3). Finally, $x \in \{Y, N\}$ indicates whether (Y) or not (N) pruning criterion PC4 is applied (see Section 4.4). In PC4, if applied, the same lower bound is used as the one for L in $\{C, L, Y\}$. We have selected the following algorithms: (P, L_2, N), (D, L_2, N), $(S(P), L_2, N)$, $(S(D), L_2, N)$, (P, L_3, N), (D, L_4, N), $(S(P), L_4, N)$, $(\Delta E, L_4, N)$, $(S(\Delta E), L_4, N)$, (P, L_2, Y), $(S(P), L_2, Y)$, and $(S(D), L_2, Y)$.

6 Empirical Performance Analysis

We have investigated the empirical performance of the algorithms defined above by applying them to one real-world example concerning the identification of potato diseases, and 140 artificial examples. The potato instance has 63 tests and 28 items. The artificial instances are randomly generated and can be characterized by the triple (m,n,p). The parameters m, n, and p denote the number of items, the number of tests, and the probability that an item belongs to a test, respectively. Probability $p \in \{0.1, 0.25, 0.5\}$ is the same for each item and each test. We have chosen $m \in \{25, 50, 100\}$ and $n \in \{25, 50, 100\}$, and generated instances $(50,n,p)$ for all three values of n and p, and $(m,50,p)$ for all three values of m and p. For each triple we generated ten feasible instances. Only for the triple $(50, 25, 0.1)$ we were not able to find such an instance. The experiments were performed on a Sun Ultra 10 Creator3D workstation and all computation times

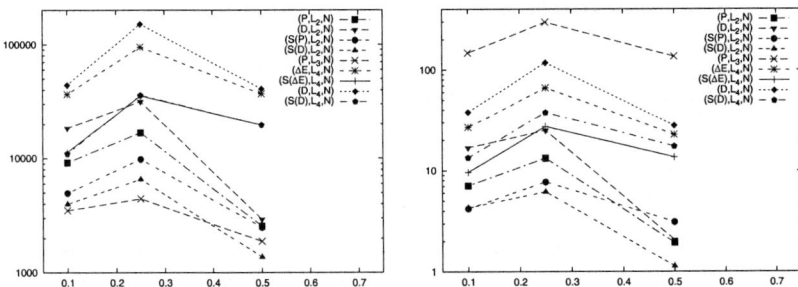

Fig. 1. Average number of branches and average computation time of nine algorithms (logarithmic scale)

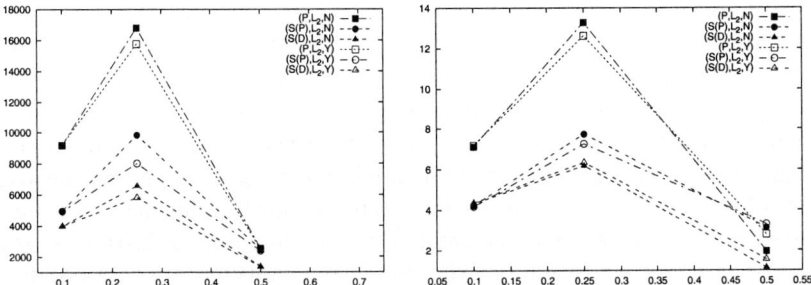

Fig. 2. Average number of branches and average computation time of six algorithms

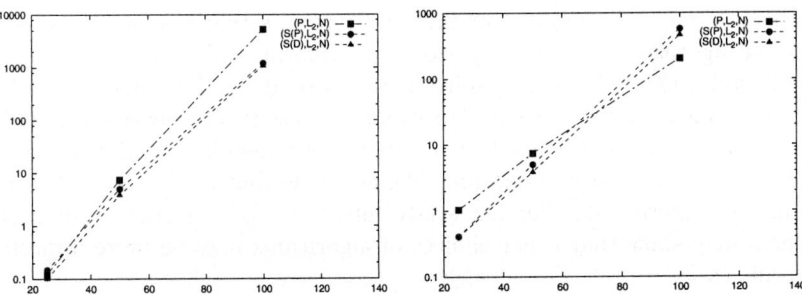

Fig. 3. On the right the influence of the number of items and on the left the influence of the number of tests (logarithmic scale)

are measured in cpu seconds. Below the performance of the various algorithms is compared and conclusions on the relative performance are drawn.

We started with comparing nine algorithms that do not use pruning criterion PC4 on the instances with 50 tests and 50 items. For the three probabilities the number of branches and the average computation time are given in Figure 1. The

Table 1. Results on the potato instance

	(P, L_2, N)	$(S(P), L_2, N)$	$(S(D), L_2, N)$
Number of branches	404	391	334
Computation time	0.2	0.2	0.2

results clearly show that on these instances, using lower bound L_2 is the best choice. The stronger lower bound L_3 is not useful as a consequence of its expensive computation. The algorithms (P, L_2, N), $(S(P), L_2, N)$, and $(S(D), L_2, N)$ appear to be most promising. For these three algorithms we studied the effect of applying pruning criterion PC4 on the same instances. The number of branches and the average computation time are given in Figure 2. As we see, the effect of pruning criterion PC4 is small: the saving on the number of branches is often annihilated by the extra computation time. We did not include PC4 in further experiments.

For (P, L_2, N), $(S(P), L_2, N)$, and $(S(D), L_2, N)$ we also studied the influence of the number of items and the number of tests on the average computation time. In Figure 3 we show the influence of the number of items by depicting the average computation time for the instances with 50 tests and the influence of the number of tests by depicting the average computational time for the instances with 50 items. We see that the results of the power as quality criterion improve when the ratio between the number of tests and the number of items increases.

Finally, we tested the three algorithms on the potato instance that initiated the research over 20 years ago. As we see in Table 6 the potato instance is not a challenge anymore.

Based on our test results, our first conclusion is that our new power-based bounding algorithms clearly outperform the information-based bounding algorithm introduced by Moret & Shapiro [1985]. We endorse the conclusion of Moret & Shapiro that the most efficient algorithms are based on the least-separated pair criterion. We found that the better results are obtained by combining this criterion with the power based bounding. We conclude that $(S(D), L_2, N)$ performs best in our experiments. Further study into the typical structure of problem instances may show that other choices of algorithms may be more appropriate in specific cases.

7 Postlude

We have presented branch-and-bound algorithms for the test cover problem. Some of them are old. The new ones are based on the notion of *power* of tests. Empirical tests indicate that the new lower bound is the most powerful one. The test results show encouraging behavior of the better algorithms on reasonably sized problems. However, they also reveal exponentially increasing running times in the number of tests and the number of items. Especially, the number of items has a quite dramatic effect on the running times. Test results not presented here show that $(50, 100, 0.25)$-instances cannot be solved by algorithms using the information criterion within a few hours. More testing is required to find limits the size of problem instances that are still solvable by the better algorithms.

A practically and theoretically interesting feature we have not considered so far is the use of clever data structures for storing and retrieving information computed. We expect a significant speedup of the algorithms by implementing improved data structures.

References

[1] H. Y. Chang, E. Manning, G. Metze, *Fault diagnosis of digital systems*, Wiley, New York, 1970.
[2] K. M. J. De Bontridder, *Methods for solving the test cover problem*, Master's Thesis, Department of Mathematics and Computer Science, Technische Universiteit Eindhoven, February 1997.
[3] K. M. J. De Bontridder, B. V. Halldórsson, M. M. Halldórsson, C. A. J. Hurkens, J. K. Lenstra, R. Ravi, L. Stougie, Approximation algorithms for the test cover problem, Submitted for publication, 2002.
[4] M. R. Garey, D. S. Johnson, *Computers and Intractability: A Guide to the Theory of NP-completeness*, Freeman, San Francisco, 1979.
[5] A. W. J. Kolen, J. K. Lenstra, Combinatorics in operations research, in R. Graham, M. Grötschel, L. Lovász (eds.), *Handbook of Combinatorics*, Elsevier, Amsterdam, 1995, 1875–1910.
[6] B. J. Lageweg, J. K. Lenstra, A. H. G. Rinnooy Kan, Uit de practijk van de besliskunde, In A. K. Lenstra, H. W. Lenstra, J. K. Lenstra, editors, *Tamelijk briljant; Opstellen aangeboden aan Dr. T. J. Wansbeek.* Amsterdam, 1980.
[7] B. M. E. Moret, Decision trees and diagrams, *Computing Surveys 14*, 1982, 593–623.
[8] B. M. E. Moret, H. D. Shapiro, On minimizing a set of tests. *SIAM Journal on Scientific and Statistical Computing 6*, 1985, 983–1003.
[9] R. W. Payne, Selection criteria for the construction of efficient diagnostic keys. *Journal of Statistical Planning and Information 5*, 1981, 27–36.
[10] A. Rescigno, G. A. Maccacaro, The information content of biological classification, in C. Cherry (ed.), *Information Theory: Fourth London Symposium*, Butterworths, London, 1961, 437–445.
[11] E. W. Rypka, W. E. Clapper, I. G. Brown, R. Babb, A model for the identification of bacteria. *Journal of General Microbiology 46*, 1967, 407–424.
[12] E. W. Rypka, L. Volkman, E. Kinter, Construction and use of an optimized identification scheme. *Laboratory Magazine 9*, 1978, 32–41.

Constructing Plane Spanners of Bounded Degree and Low Weight*

Prosenjit Bose[1], Joachim Gudmundsson[2], and Michiel Smid[1]

[1] School of Computer Science, Carleton University
Ottawa, Canada K1S 5B6
{jit,michiel}@scs.carleton.ca
[2] Institute of Information and Computing Sciences, Utrecht University
P.O.Box 80.089, 3508 TB Utrecht, the Netherlands
joachim@cs.uu.nl

Abstract. Given a set S of n points in the plane, we give an $O(n \log n)$-time algorithm that constructs a plane t-spanner for S, with $t \approx 10.02$, such that the degree of each point of S is bounded from above by 27, and the total edge length is proportional to the weight of a minimum spanning tree of S. These constants are all worst case constants that are artifacts of our proofs. In practice, we believe them to be much smaller. Previously, no algorithms were known for constructing plane t-spanners of bounded degree.

1 Introduction

Given a set S of n points in the plane and a real number $t > 1$, a *t-spanner* for S is a graph G with vertex set S such that any two vertices u and v are connected by a path in G whose length is at most $t \cdot |uv|$, where $|uv|$ is the Euclidean distance between u and v. If this graph has $O(n)$ edges, then it is a sparse approximation of the (dense) complete Euclidean graph on S.

Many algorithms are known that compute t-spanners with $O(n)$ edges that have additional properties such as bounded degree, small spanner diameter (i.e., any two points are connected by a t-spanner path consisting of only a small number of edges), low weight (i.e., the total length of all edges is proportional to the weight of a minimum spanning tree of S), and fault-tolerance; see, e.g., [10, 11, 13], and the surveys [9, 14]. All these algorithms compute t-spanners for any given constant $t > 1$.

In this paper, we consider the construction of *plane* t-spanners. (Obviously, in order for a t-spanner to be plane, t must be at least $\sqrt{2}$.) It is known that the Delaunay triangulation is a t-spanner for $t = 2\pi/(3\cos(\pi/6))$, see [11]. Furthermore, in [4], it is shown that other plane graphs such as the minimum weight triangulation and the greedy triangulation are t-spanners, for some constant t. In [12], it

---

* Research supported in part by the Natural Science and Engineering Research Council of Canada and the Swedish Foundation for International Cooperation in Research and Higher Education.

is shown that, for any real number $r > 0$, a plane graph can be constructed from the Delaunay triangulation that is a t-spanner for $t = (1 + 1/r)2\pi/(3\cos(\pi/6))$ and whose total edge length is at most $2r + 1$ times the weight of a minimum spanning tree of S. Observe that all these spanners may have unbounded degree.

No algorithms are known that compute a plane t-spanner, for some constant t, of bounded degree. In this paper, we show how to compute such a spanner that, additionally, has low weight. That is, the main result of this paper is the following theorem.

Theorem 1. *There is an $O(n \log n)$-time algorithm that computes, when given a set S of n points in the plane and a constant $\varepsilon > 0$, a graph*

1. *that is plane,*
2. *that is a t-spanner for S, for $t = (\pi + 1)2\pi/(3\cos(\pi/6))(1 + \varepsilon)$,*
3. *in which each point of S has degree at most 27,*
4. *and whose weight is bounded from above by a constant times the weight of a minimum spanning tree of S.*

Our algorithm consists of the following steps. First, it computes the Delaunay triangulation of S, $DT(S)$, and a degree-3 spanning subgraph $BDS(S)$ of $DT(S)$ that includes the convex hull $CH(S)$ of S. This graph $BDS(S)$ partitions $CH(S)$ into (possibly degenerate) simple polygons, such that each point of S is on the boundary of at most three polygons. (This part of the algorithm is described in Section 2.)

Then, for each polygon P in the above partition, our algorithm processes the vertices of $DT(S) \cap P$ in breadth-first order, and prunes this part of the Delaunay triangulation such that each vertex of P has low degree. The resulting graph is a plane spanner for the vertices of P in the sense that any two vertices u and v of P are connected by a path whose length is at most a constant times the length of a shortest path between u and v that is completely contained in P. By combining all the spanners for each of the polygons, we get a plane spanner of bounded degree. (This part of the algorithm is described in Section 3.)

In the last step of the algorithm (see Section 4), we run the greedy algorithm (see [10]) on the graph obtained in Section 3. This results in a graph for which the claims in Theorem 1 hold.

2 Computing a Degree-3 Plane Spanning Graph (SPANNINGGRAPH)

The aim of this section is, given a set S of n points in the plane, to construct a low-degree plane spanning graph $BDS(S)$ of S that contains the convex hull of S. The algorithm presented in this section will be used as a pre-processing step in the main algorithm, as described in Section 1. The following lemma explains why we are interested in computing such a spanning graph.

Lemma 1. *Let $BDS(S)$ be a plane spanning graph of the set S that contains the convex hull of S, and let d be the maximum degree of any vertex of $BDS(S)$. The graph $BDS(S)$ partitions the convex hull of S into (possibly degenerate) simple polygons such that*

1. *the union of the vertices of all these polygons is equal to the set S,*
2. *no point of S is contained in the interior of any of these polygons, and*
3. *each point of S is on the boundary of at most d polygons.*

Degree-5 plane spanning subgraph. Let G be the union of the convex hull of S and a minimum spanning tree T of S. Then G is clearly a plane spanning graph. It is well known that each vertex of T has degree at most six, and the angle between any two adjacent edges of T is at least $60°$. Therefore, G has degree less than or equal to six. Since there always exists a minimum spanning tree of degree at most five, the graph G has degree at most five.

Degree-3 plane spanning subgraph. Consider an arbitrary triangulation Δ of S. Barnette [1] has shown that every 3-connected planar graph contains a spanning tree of maximum degree at most three; his proof of this result implies an $O(n \log n)$-time algorithm for computing such a tree. If we apply this to Δ, then we obtain a plane degree-3 spanning tree of S. However, by adding the convex hull of S to this tree, we again obtain a degree-5 plane spanning graph of S that contains the convex hull of S.

We now show how to improve the degree to three. Our construction uses the *canonical ordering* of de Fraysseix et al. [7]. Let Δ be an arbitrary triangulation of S, and let $(v_1, v_2, \ldots, v_n)$ be a numbering of the points of S. For any i, $1 \leq i \leq n$, let Δ_i denote the subgraph of Δ induced by the vertices $v_1, v_2, \ldots, v_i$, and let C_i be the outer face of Δ_i. We say that this numbering is a canonical ordering of Δ, if the following conditions are satisfied:

- v_1 and v_2 are adjacent in Δ and are both on the convex hull of S.
- For all i, $3 \leq i \leq n$, the following holds.
  • the subgraph Δ_i is 2-connected.
  • The point v_i is in the interior of the outer face C_{i-1} of Δ_{i-1}.
  • Let N_i be the set of all vertices of Δ_{i-1} that are connected by an edge, in Δ, to v_i. The set N_i contains at least two elements, and forms a path on the boundary of C_{i-1}.

A canonical ordering always exists. It can be computed, in reverse order, as follows. Start with the complete triangulation Δ, and successively remove a vertex from the boundary of the outer face that is incident to exactly two vertices on the outer face. It is easy to show that such a vertex always exists. The time to compute this canonical ordering is $O(n)$; see [7].

We apply this to an arbitrary triangulation Δ of S. Consider a canonical ordering $(v_1, v_2, \ldots, v_n)$ of Δ. For each i, $3 \leq i \leq n$, we will compute a subgraph G_i of Δ. During this construction, we maintain the following invariant.

Invariant: The graph G_i is a spanning graph for $\{v_1, v_2, \ldots, v_i\}$, its degree is at most three, and it contains all edges of Δ that form the boundary of C_i.

We define the graph G_3 to be the triangle spanned by v_1, v_2, and v_3. Let i be an integer with $4 \leq i \leq n$, and assume we have constructed the graph G_{i-1}. Consider the set N_i of all vertices of Δ_{i-1} that are connected by an edge, in Δ, to v_i. Recall that $|N_i| \geq 2$ and N_i forms a path on the boundary of C_{i-1}. Let $u_1, \ldots, u_k$ be all vertices of N_i, in the order in which they occur on this path. We distinguish two cases.

$k = 2$: We obtain G_i from G_{i-1} by removing the edge (u_1, u_2) and adding the edges (v_i, u_1) and (v_i, u_2). (See the top part of Figure 1.)

$k > 2$: We obtain G_i from G_{i-1} by removing the edges (u_1, u_2) and (u_{k-1}, u_k) and adding the edges (v_i, u_1), (v_i, u_2) and (v_i, u_k). (See the bottom part of Figure 1.)

It is clear that the invariant is maintained during this construction. Hence, the graph $BDS(S) := G_n$ is a plane spanning graph of S of maximum degree at most three. Observe that C_n is equal to the outer face of Δ. Therefore, the boundary of C_n is the convex hull of S. Hence, the invariant implies that $BDS(S)$ contains the convex hull of S.

We denote the entire algorithm for computing $BDS(S)$, when given the triangulation Δ, by SPANNINGGRAPH(Δ). An example is given in Figure 2. The following lemma summarizes the result of this section.

Lemma 2. *Given a triangulation Δ of a set S of n points in the plane, we can compute, in $O(n)$ time, a plane spanning graph $BDS(S)$ that contains the convex hull of S and in which each vertex has degree at most three.*

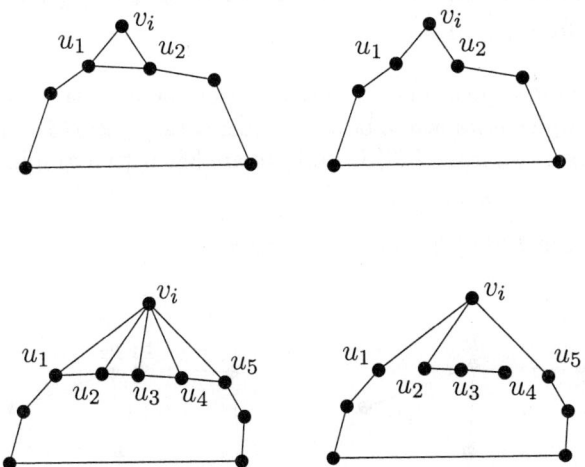

Fig. 1. Illustrating the construction of a degree-3 plane spanning graph

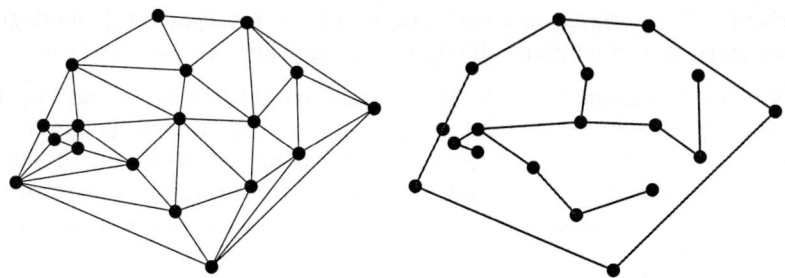

Fig. 2. A triangulation Δ (on the left), and the output of algorithm SPANNINGGRAPH(Δ) (on the right)

2.1 Algorithm TRANSFORMPOLYGON

The polygons obtained from algorithm SPANNINGGRAPH(Δ) may be degenerate, i.e., there may be vertices of degree one or cut vertices in any such polygon. An easy way to handle this problem is to symbolically transform each degenerate polygon P' into a non-degenerate polygon P. This is done by doubling edges adjacent to degenerate vertices of P' and then running an Euler tour. Figure 3 illustrates how this transformation is performed on a degenerate vertex of degree 3. All other degenerate vertices can be handled in a similar fashion.

This transformation does not affect the complexity of the spanner algorithm if the vertices are only split symbolically. We denote the algorithm that transforms a degenerate polygon P' into a non-degenerate polygon P by TRANSFORMPOLYGON(P').

3 Computing a Spanner of a Simple Polygon (POLYGONSPANNER)

The idea behind our algorithm is to start with the Delaunay triangulation $DT(S)$ of a point set S which is known to be a t-spanner with $t \leq 2\pi/(3\cos(\pi/6))$ (shown in [11]). Remove edges from $DT(S)$ until we are left with a graph $PS(S)$ having two properties:

1. every vertex in $PS(S)$ has bounded degree,

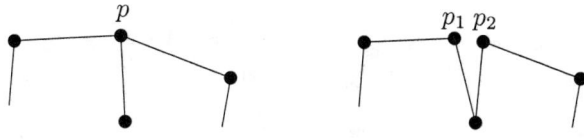

Fig. 3. Illustrating algorithm TRANSFORMPOLYGON

2. for every edge $xy \in DT(S)$, there is a path between x and y in $PS(S)$ whose length is at most $\pi + 1$ times $|xy|$.

These two properties combined imply that $PS(S)$ is a plane graph, every vertex has bounded degree and it is a $((\pi+1)2\pi/(3\cos(\pi/6)))$-spanner. To achieve the first property, we construct the maximum degree-3 spanning subgraph, $BDS(S)$, of the Delaunay triangulation as described in the previous section. Of course, this subgraph is not necessarily a constant spanner, so too many edges have been removed at this point. Therefore, we need to add some carefully chosen edges in order to maintain the first property and achieve the second. Our starting point is $BDS(S)$. As noted previously, $BDS(S)$ decomposes the plane into a set of simple polygons, $\{P_1, \ldots, P_k\}$. Now, each edge $xy \in DT(S)$ that is not in $BDS(S)$ must be a diagonal in some P_i. Thus, it is sufficient to show how to properly approximate all such edges within this set of polygons.

Let P be a simple polygon with vertex set V, and let $t > 1$ be a real number. A graph with vertex set V is called a *t-spanner for V w.r.t. P*, if for every pair of points $u, v \in V$ that form a diagonal of P, there exists a path between u and v in the graph, whose length is at most t times the length of $|uv|$.

In this section, we show how to compute, for some constant $t > 1$, a plane t-spanner for V w.r.t. P, whose degree is bounded by a constant. By constructing such a spanner for each polygon P_i, we get the desired spanner.

Let $G = (V, E)$ be the plane graph, where V is the set of vertices of P and E is the union of the edge set of P and the edges of the Delaunay triangulation of V contained in P. Let v_1 be an arbitrary vertex of V, and let $v_1, v_2, \ldots, v_n$ be the vertices of V according to a clockwise breadth-first ordering (bfs-order) of G with start vertex v_1. A clockwise breadth-first search of a graph is simply a standard breadth-first search with the additional constraint that when the children of a vertex are processed, they are processed in clockwise order. For any two vertices u and v of V, we write $u <_{bf} v$, if u strictly precedes v in the bfs-ordering. We extend this ordering to $\leq_{bf}$ in the obvious way.

We start with some properties of the bfs-numbering.

Let $T(G, v_1)$ be the breadth-first search tree of G rooted at vertex v_1. The level of any vertex v in $T(G, v_1)$ is denoted by $\ell(v)$. (The root is at level zero, i.e., $\ell(v_1) = 0$.)

Lemma 3. *Let v be any vertex in $T(G, v_1)$ such that $v \neq v_1$.*

a) *Any Delaunay edge (v, w) that is not an edge of P partitions P into two simple polygons P_1 and P_2. We may assume w.l.o.g. that v_1 is a vertex of P_1. For every vertex u in P_2 with $u \neq v$ and $u \neq w$, we have $\min(\ell(v), \ell(w)) < \ell(u)$ and $\max(\ell(v), \ell(w)) \leq \ell(u)$.*

b) *The line segment between v and its parent u in $T(G, v_1)$ partitions P into two simple polygons P_1 and P_2 such that all children of u that precede v are in P_1 and all siblings that succede v are in P_2. The vertex v can be adjacent to at most one child of u that precedes v.*

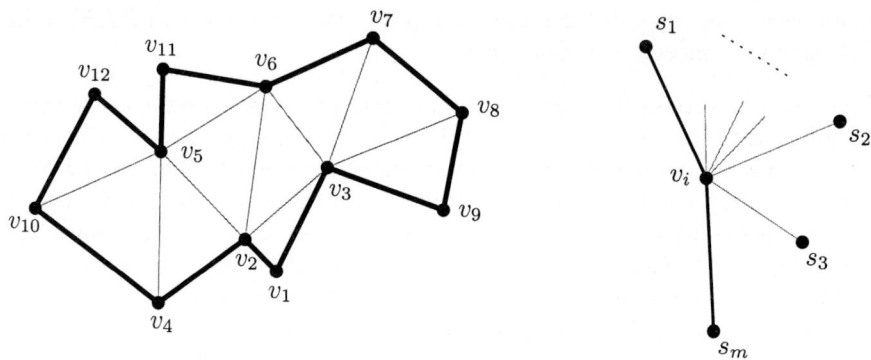

Fig. 4. Illustration of clockwise breadth-first search

Proof. (a) The claim holds since either v or w must be an ancestor to all vertices in P_1. (b) Follows since the siblings of the parent of v are ordered in clockwise order, see Fig. 4. □

The sequence of vertices adjacent to a vertex v, sorted in clockwise order around v, will be denoted by $\mathcal{N}(v)$ and will be called the *ordered neighborhood* of v.

Lemma 4. *The vertices in $\mathcal{N}(v)$ preceding v in the bfs-ordering are the parent of v, denoted $p(v)$, and at most two vertices in $\mathcal{N}(v) \cap \mathcal{N}(p(v))$.*

Proof. Let v_1 and v_2 be the two vertices adjacent to v and $p(v)$. The quadrangle $(p(v), v_1, v, v_2)$ partitions P into five simple polygons, denoted $P_1, \ldots, P_5$ as illustrated in fig. 5. P_2 and P_3 are bounded by $(p(v), v_1)$ and $(p(v), v_2)$ respectively and hence cannot contain any vertices connected to v. Now consider P_4 and P_5 with boundary edge (v_1, v) and (v_2, v) respectively. The proof for both P_4 and P_5 are identical. We consider P_4. There are two cases:

- If $\ell(v) = \ell(v_1)$ then every vertex u in P_4, except v and v_1, must be a successor of v since $\ell(u) > \ell(v)$ according to Lemma 3a.
- If $\ell(v) > \ell(v_1)$ then every point u in P_4 must be a successor of v since $p(v) <_{bf} v_1$, otherwise $p(v)$ could not be the parent of v.

The lemma follows. □

Now we are ready to present the algorithm that computes a bounded-degree plane spanner for V w.r.t. P. Recall that $G = (V, E)$ is the plane graph, where E is the union of the edge set of P and the edge set consisting of all edges of the Delaunay triangulation of V that are contained in P. Our algorithm processes all the vertices of G according to their bfs-ordering starting with v_1.

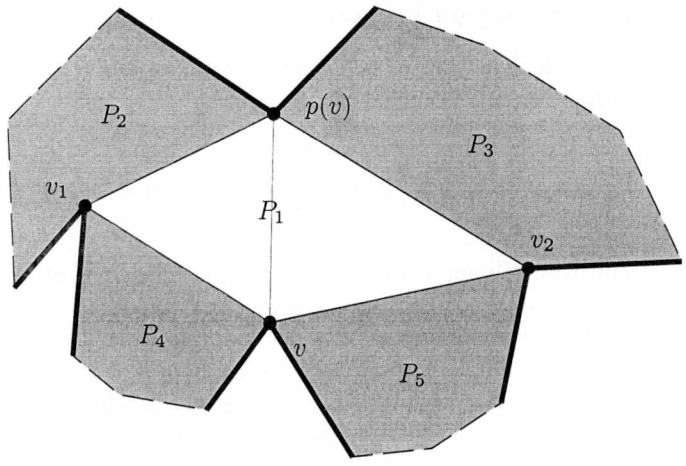

Fig. 5. Illustration for Lemma 4

Initialize E' to be the set of edges of P. At the end of the processing, E' will be the edges of the desired spanner. Let $s_1, s_2, \ldots, s_m$ be the ordered neighborhood $\mathcal{N}(v_1)$ of v. That is, $s_1, s_2, \ldots, s_m$ are all vertices that are connected by an edge of E to v_1. These vertices occur in clockwise order around v_1. We assume that the edges (s_1, v_1) and (s_m, v_1) are edges of the polygon P. The following two steps are performed in the processing of v_1.

First, the interior angle at v_1 is divided into a minimum number of intervals of at most 90°. For each interval, add the shortest edge in E that is connected to v_1 to the edge set E'. Second, add to E' all edges (s_j, s_{j+1}), $1 \leq j < m$. This terminates the processing of vertex v_1.

A generic processing step is similar. Assume that vertex v_i of V is next to be processed. Let $s_1, \ldots, s_m$ be the ordered neighborhood of v_i. We assume that (s_1, v_i) and (s_m, v_i) are edges of P. It follows from Lemma 4 that the parent of v_i has been processed and possibly at most two vertices s_a, s_b in $\mathcal{N}(v_i) \cap \mathcal{N}(p(v_i))$. The remainder of the ordered neighborhood of v_i needs to be processed. Let $\alpha_1(v_i) = \angle s_1 v_i s_a$ and $\alpha_2(v_i) = \angle s_b v_i s_m$. The following two steps are performed.

First, we divide the interior angle $\angle s_1 v_i s_a$ into a minimum number of intervals of at most 90°. The same is done for the angle $\angle s_b v_i s_m$. For each interval, we add to the set E' the shortest edge in E that connects v_i to a vertex that has not been processed. Secondly, we add all edges (s_j, s_{j+1}), $1 \leq j < m$, to E'. Vertex v_i has now been processed.

We continue processing vertices in this way until all vertices have been processed. The output of this algorithm is the graph $G' = (V, E')$. We will refer to this algorithm as POLYGONSPANNER(P).

3.1 Analysis of Algorithm POLYGONSPANNER

In this section, we will show that G' is a spanner with respect to P and that the degree of any vertex of G' is bounded by a constant.

Lemma 5. *Let v be any vertex of P. At the moment when v is being processed by algorithm* POLYGONSPANNER, *there are at most five edges in E' that are incident to v.*

Proof. The proof uses Lemma 4. First observe that the parent $p(v)$ of v is the first vertex incident to v that is being processed. The edges incident to v that the algorithm may add to E' while processing $p(v)$ are the Delaunay edge $(v, p(v))$ and the edges on the ordered neighborhood of $p(v)$; hence a total of three edges. Assume that v is connected to the vertices v_1 and v_2 on the ordered neighborhood of $p(v)$. These are the only two vertices adjacent to v that can be processed before it. However, since $p(v)$ is in $\mathcal{N}(v) \cap \mathcal{N}(v_1)$ and also in $\mathcal{N}(v) \cap \mathcal{N}(v_2)$, at most two edges adjacent to v can be added to E' while processing v_1 and v_2. Hence, when v is being processed, there are at most five edges in E' incident to v. □

Once a vertex has been processed, no additional edges adjacent to it are ever added. Together with the above results, it follows that the degree, in the graph G', of any vertex $v \in V$ is bounded from above by

$$5 + \lceil \alpha_1(v)/90° \rceil + \lceil \alpha_2(v)/90° \rceil + 2 \le 9 + \alpha_1(v)/90° + \alpha_2(v)/90° < 13,$$

since $\alpha_1(v) + \alpha_2 v < 360°$. Hence, the degree is at most 12. This immediately gives a bound of at most 36 if all polygons are processed this way since each vertex can participate in at most 3 polygons. We show how to reduce this upper bound in Section 4.

Lemma 6. *For any edge $(u, v) \in DT(V)$ inside P, there exists a path in G' between u and v whose length is at most $\pi + 1$ times the edge $|uv|$.*

Proof. Consider an arbitrary edge $e = (u, v)$ in the Delaunay triangulation of S that is contained in P. If $e \in G'$, the lemma holds, so assume that $e \notin G'$.

Assume w.l.o.g. that $u <_{bf} v$. It follows from the algorithm that there must exist a vertex v' in the ordered neighborhood of u such that $\angle v'uv < 90°$ and $(v', u) \in E'$. Let $v' = s_1, s_2, \ldots, s_k = v$ be this sequence of vertices in the ordered neighborhood of u from v' to v. Since $(v', u) \in E'$ is shorter than (u, v) by definition, we only need to show that the segment (v, v') is well approximated in G' in terms of (u, v).

Consider the polygon P', which is a sub-polygon of P, consisting of the vertices $u, s_1, \ldots, s_k$. The shortest path in P' from v' to v, denoted $S_{P'}(v', v)$, consists of diagonals of P' and is contained in the triangle u, v, v'. By convexity, the length of $S_{P'}(v', v)$ is at most 2 times $|uv|$.

An edge of $S_{P'}(v', v)$ has as endpoints two vertices in the neighborhood of u. Let (s_i, s_j) be an arbitrary edge in $S_{P'}(v', v)$. Let $D_{P'}(s_i, s_j)$ be the sequence of edges between s_i and s_j in the ordered neighborhood of u. This path is in G'. We bound the length of $D_{P'}(s_i, s_j)$ in terms of $|s_i s_j|$.

In [2], it is shown (by modifying an argument in [8]), that the length of $D_{P'}(s_i, s_j)$ is at most $\pi/2$ times $|s_i s_j|$, provided that (i) the straight-line segment between s_i and s_j lies outside the Voronoi region induced by u, and (ii) that the path lies on one side of the line through s_i and s_j.

Condition (ii) holds trivially. We now show that condition (i) also holds. The vertices u, s_i and s_j form a triangle where $\angle s_i u s_j < 90°$. This implies that for any point x on the segment $[s_i, s_j]$, $\min\{|xs_i|, |xs_j|\} < |xu|$. This means that x cannot be in the Voronoi region induced by u. Therefore, $D_{P'}(s_i, s_j)/|s_i s_j| \leq \pi/2$.

Putting it all together, it follows that the shortest path in G' between v' and v has length at most $\pi|uv|$. This implies that the shortest path in G' between u and v is at most $(\pi+1)|uv|$, hence the lemma follows. □

4 The Complete Algorithm and Its Analysis

In this section, we present the algorithm which will prove Theorem 1. The algorithm is given in Figure 6. Step 7 refers to the greedy algorithm of Gudmundsson et al. [10]. This algorithm takes as input an arbitrary t-spanner $PS(S) = (S, E)$ and a real constant $\varepsilon > 0$, and computes a $(1+\varepsilon)t$-spanner $G = (S, E')$ of $PS(S)$, such that $E' \subseteq E$ and the weight of E' is bounded by a constant times the weight of a minimum spanning tree of S. In the rest of this section, we will prove that for the output of algorithm PLANARSPANNER(S, ε), the claims in Theorem 1 hold.

Let us first consider the running time of algorithm PLANARSPANNER. Step 1 takes $O(n \log n)$ time (see [6]) and, by Lemma 2, Step 2 takes $O(n)$ time. By the results of Section 3, Steps 4 and 5 take $O(|P'|)$ time for each polygon P' in the partition induced by the spanning graph $BDS(S)$. Since each point of S is a vertex of at most three such polygons P' (see Lemma 1), the total time for the for-loop is $O(n)$. Since each graph G_P contains $O(|P|)$ edges, Step 6 takes

---

Algorithm PLANARSPANNER(S, ε)
1. $DT \leftarrow$ Delaunay triangulation of S
2. $BDS(S) \leftarrow$ SPANNINGGRAPH(DT) (see Section 2)
3.   **for each** polygon P' in the partition induced by $BDS(S)$ **do**
4. $P \leftarrow$ TRANSFORMPOLYGON(P') (see Section 2.1)
5. $G_P \leftarrow$ POLYGONSPANNER(P) (see Section 3)
6. $PS(S) \leftarrow$ union of all G_P
7. $G \leftarrow$ GREEDYSPANNER$(PS(S), \varepsilon)$ (see [10])
8.   output G

Fig. 6. The construction of a plane sparse spanner

$O(n)$ time. Finally, by the results in [10], Step 7 takes $O(n \log n)$ time. Hence, the total time of the algorithm is $O(n \log n)$.

Since the output G is a subgraph of $PS(S)$, and $PS(S)$ is a subgraph of the Delaunay triangulation DT of S, it is clear that G is plane, proving the first claim in Theorem 1.

To prove the second claim in Theorem 1, first observe that DT is a $(2\pi/(3\cos(\pi/6)))$-spanner for S, see [11]. Consider any two distinct points u and v of S, and let Q be a path in DT between u and v having length at most $(2\pi/(3\cos(\pi/6))) \cdot |uv|$. Consider an arbitrary edge (x, y) on Q that is not an edge in $PS(S)$. This edge must be a diagonal in some polygon P'. By Lemma 6, there is a path in $PS(S)$ between x and y whose length is at most $(\pi + 1)|xy|$. Therefore, there is a path in $PS(S)$ between u and v of length at most $(\pi + 1)(2\pi/(3\cos(\pi/6)))|uv|$. Hence, $PS(S)$ is a $((\pi + 1)(2\pi/(3\cos(\pi/6))))$-spanner for S. The results in [10] imply that G is a t-spanner for S, for $t = (\pi + 1)(2\pi/(3\cos(\pi/6)))(1 + \varepsilon)$.

Next, we prove the third claim in Theorem 1. As described in Section 2.1, a degenerate polygon P' is transformed in Step 4 to a simple polygon P by doubling some edges. It is easily seen that in this process, any original point u is replaced by at most three new vertices u_1, u_2, and u_3. An analysis similar to the one just above Lemma 6 shows that when these three vertices are merged again, the degree of u in $PS(S)$ is at most $3 \cdot 5 + 3 + \lceil \frac{\alpha_1(u_1)}{90°} \rceil + \lceil \frac{\alpha_2(u_1)}{90°} \rceil + \lceil \frac{\alpha_1(u_2)}{90°} \rceil + \lceil \frac{\alpha_2(u_2)}{90°} \rceil + \lceil \frac{\alpha_1(u_3)}{90°} \rceil + \lceil \frac{\alpha_2(u_3)}{90°} \rceil$. Since all these six interior angles sum up to 360°, the degree of u is at most 27.

The fourth claim in Theorem 1 follows from results in [3, 5]: Since $PS(S)$ is a $((\pi + 1)(2\pi/(3\cos(\pi/6))))$-spanner for S, the graph G produced by the greedy algorithm has a weight that is proportional to the weight of a minimum spanning tree of S.

5 Concluding Remarks

The results in this paper generalize to the *constrained* case, i.e., when the set S is a set of line segments rather than a point set, and edges of the spanner are not allowed to intersect any edge of S. The details will be presented in the full paper.

There are additional properties that would be interesting to consider, for example minimizing the diameter of the spanner. Below we show that this property cannot be obtained for plane spanners.

Lemma 7. *For any real number $t > 1$, there exists a set S of n points and two points $u, v \in S$ such that for any plane t-spanner it holds that any t-spanner path between u and v contains at least $n/2 + 1$ edges.*

Proof. The construction of the point set is illustrated in Figure 7: For simplicity we assume that n is even. Place u and v on a horizontal line such that the horizontal distance between u and v is $t \cdot \frac{n-2}{2}$. □

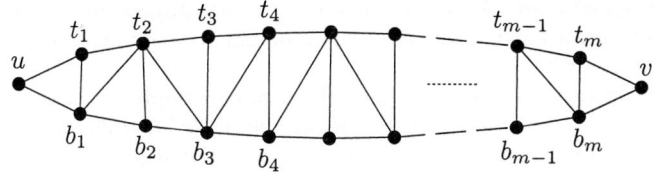

Fig. 7. Illustrating the proof of Lemma 7

Acknowledgements

The authors wish to thank Anil Maheshwari, Pat Morin and David Wood for fruitful discussions on this topic.

References

[1] D. Barnette. Trees in polyhedral graphs. *Canadian Journal of Mathematics*, 18:731–736, 1966.
[2] P. Bose and P. Morin. Online routing in triangulations. In *Proc. 10th Annu. Internat. Sympos. Algorithms Comput.*, volume 1741 of *Lecture Notes Comput. Sci.*, pages 113–122. Springer-Verlag, 1999.
[3] G. Das, P. Heffernan, and G. Narasimhan. Optimally sparse spanners in 3-dimensional Euclidean space. In *Proc. 9th Annu. ACM Sympos. Comput. Geom.*, pages 53–62, 1993.
[4] G. Das and D. Joseph. Which triangulations approximate the complete graph? In *Proc. International Symposium on Optimal Algorithms*, volume 401 of *Lecture Notes Comput. Sci.*, pages 168–192. Springer-Verlag, 1989.
[5] G. Das, G. Narasimhan, and J. Salowe. A new way to weigh malnourished Euclidean graphs. In *Proc. 6th ACM-SIAM Sympos. Discrete Algorithms*, pages 215–222, 1995.
[6] M. de Berg, M. van Kreveld, M. Overmars, and O. Schwarzkopf. *Computational Geometry: Algorithms and Applications*. Springer-Verlag, Berlin, Germany, 2nd edition, 2000.
[7] H. de Fraysseix, J. Pach, and R. Pollack. How to draw a planar graph on a grid. *Combinatorica*, 10:41–51, 1990.
[8] D. P. Dobkin, S. J. Friedman, and K. J. Supowit. Delaunay graphs are almost as good as complete graphs. *Discrete Comput. Geom.*, 5:399–407, 1990.
[9] D. Eppstein. Spanning trees and spanners. In J.-R. Sack and J. Urrutia, editors, *Handbook of Computational Geometry*, pages 425–461. Elsevier Science Publishers, Amsterdam, 2000.
[10] J. Gudmundsson, C. Levcopoulos, and G. Narasimhan. Improved greedy algorithms for constructing sparse geometric sp anners. In *Proc. 7th Scand. Workshop Algorithm Theory*, volume 1851 of *Lecture Notes Comput. Sci.*, pages 314–327, Berlin, 2000. Springer-Verlag.
[11] J. M. Keil and C. A. Gutwin. Classes of graphs which approximate the complete Euclidean gr aph. *Discrete Comput. Geom.*, pages 13–28, 1992.

[12] C. Levcopoulos and A. Lingas. There are planar graphs almost as good as the complete graphs and almost as cheap as minimum spanning trees. *Algorithmica*, 8:251–256, 1992.

[13] C. Levcopoulos, G. Narasimhan, and M. Smid. Improved algorithms for constructing fault-tolerant spanners. *Algorithmica*, 32:144–156, 2002.

[14] M. Smid. Closest point problems in computational geometry. In J.-R. Sack and J. Urrutia, editors, *Handbook of Computational Geometry*, pages 877–935. Elsevier Science Publishers, Amsterdam, 2000.

Eager st-Ordering

Ulrik Brandes

Department of Computer & Information Science, University of Konstanz
ulrik.brandes@uni-konstanz.de

Abstract. Given a biconnected graph $G = (V, E)$ with edge $\{s, t\} \in E$, an st-ordering is an ordering $v_1, \ldots, v_n$ of V such that $s = v_1$, $t = v_n$, and every other vertex has both a higher-numbered and a lower-numbered neighbor. Previous linear-time st-ordering algorithms are based on a preprocessing step in which depth-first search is used to compute lowpoints. The actual ordering is determined only in a second pass over the graph. We present a new, incremental algorithm that does not require lowpoint information and, throughout a single depth-first traversal, maintains an st-ordering of the biconnected component of $\{s, t\}$ in the traversed subgraph.

1 Introduction

The st-ordering of vertices in an undirected graph is a fundamental tool for many graph algorithms, e.g. in planarity testing, graph drawing, or message routing. It is closely related to other important concepts such as biconnectivity, ear decompositions or bipolar orientations.

The first linear-time algorithm for st-ordering the vertices of a biconnected graph is due to Even and Tarjan [2, 3]. Ebert [1] presents a slightly simpler algorithm, which is further simplified by Tarjan [7]. All these algorithms, however, preprocess the graph using depth-first search, essentially to compute lowpoints which in turn determine an (implicit) open ear decomposition. A second traversal is required to compute the actual st-ordering.

We present a new algorithm that avoids the computation of lowpoints and thus requires only a single pass over the graph. It appears to be more intuitive, explicitly computes an open ear decomposition and a bipolar orientation on the fly, and it is robust against application to non-biconnected graphs. Most notably, it can be stopped after any edge traversal and will return an st-ordering of the biconnected component containing $\{s, t\}$ in what has been traversed of the graph until then. The algorithm can thus be utilized in lazy evaluation, for instance when only the ordering of few vertices is required, and on implicitly represented graphs that are costly to traverse more than once.

The paper is organized as follows. In Sect. 2 we recall basic definitions and correspondences between biconnectivity, st-orderings and related concepts. Section 3 is a brief review of depth-first search and lowpoint computation. The new algorithm is developed in Sect. 4 and discussed in Sect. 5.

2 Preliminaries

We consider only undirected and simple graphs $G = (V, E)$. A (simple) *path* $P = (v_0, e_1, v_1, \ldots, e_k, v_k)$ in G is an alternating sequence of vertices $V(P) = \{v_0, \ldots, v_k\} \subseteq V$ and edges $E(P) = \{e_1, \ldots, e_k\} \subseteq E$ such that $\{v_{i-1}, v_i\} = e_i$, $1 \leq i \leq k$, and $v_i = v_j$ implies $i = j$ or $\{i, j\} = \{0, k\}$. The *length* of P is k. A path is called *closed* if $v_0 = v_k$, and *open* otherwise.

A graph G is *connected* if every pair of vertices is linked by a path, and it is *biconnected* if it remains connected after any vertex is removed from G.

We are interested in ordering the vertices of a biconnected graph in a way that guarantees forward and backward connectedness. Determining such an ordering is an essential preprocessing step in many applications including planarity testing, routing, and graph drawing.

Definition 1 (*st-ordering* [5]). *Let $G = (V, E)$ be a biconnected graph and $s \neq t \in V$. An ordering $s = v_1, v_2, \ldots, v_n = t$ of the vertices of G is called an st-ordering, if for all vertices v_j, $1 < j < n$, there exist $1 \leq i < j < k \leq n$ such that $\{v_i, v_j\}, \{v_j, v_k\} \in E$.*

Lemma 1 ([5]). *A graph $G = (V, E)$ is biconnected if and only if, for each edge $\{s, t\} \in E$, it has an st-ordering.*

Several linear-time algorithms for computing st-orderings of biconnected graphs are available [2, 1, 7]. All of these are based on a partition of the graph into oriented paths.

An *orientation* assigns a direction to each edge in a set of edges. An *st-orientation* (also called a *bipolar orientation*) of a graph G is an orientation such that the resulting directed graph is acyclic and s and t are the only source and sink, respectively. The following lemma is folklore.

Lemma 2. *A graph $G = (V, E)$ has an st-orientation if and only if it has an st-ordering. These can be transformed into each other in linear time.*

Proof. An st-ordering is obtained from an st-orientation by topological ordering, and an st-orientation is obtained from an st-ordering by orienting edges from lower-numbered to higher-numbered vertices. □

A sequence $D = (P_0, \ldots, P_r)$ of (open) paths inducing graphs $G_i = (V_i, E_i)$ with $V_i = \bigcup_{j=0}^{i} V(P_j)$ and $E_i = \bigcup_{j=0}^{i} E(P_j)$, $0 \leq i \leq r$, is called an *(open) ear decomposition*, if $E(P_0), \ldots, E(P_r)$ is a partititon of E and for each $P_i = (v_0, e_1, v_1, \ldots, e_k, v_k)$, $1 \leq i \leq r$, we have $\{v_0, v_k\} \subseteq V_{i-1}$ and $\{v_1, \ldots, v_{k-1}\} \cap V_{i-1} = \emptyset$. An ear decomposition *starts* with edge $\{s, t\} \in E$, if $P_0 = (s, \{s, t\}, t)$.

Lemma 3 ([8]). *A graph $G = (V, E)$ is biconnected if and only if, for each edge $\{s, t\} \in E$, it has an open ear decomposition starting with $\{s, t\}$.*

Note that, given an open ear decomposition $P_0, P_1, \ldots, P_r$ starting with edge $\{s,t\}$, it is straightforward to construct an st-orientation. Simply orient P_0 from s to t, and $P_i = (u, \ldots, w)$, $1 \leq i \leq r$, from u to w (from w to u) if u lies before (after) w in the partial ordering induced by $P_0, \ldots, P_{i-1}$. Since the orientation of an ear conforms to the order of its endpoints, no cycles are introduced, and s and t are the only source and sink.

3 Depth-First Search and Biconnectivity

Starting from a *root* vertex s, a *depth-first search* (DFS) of an undirected graph $G = (V, E)$ traverses all edges of the graph, where the next edge is chosen to be incident to the most recently visited vertex that has an untraversed edge. An edge $\{v, w\}$ traversed from v to w is called *tree edge*, denoted by $v \to w$, if w is encountered for the first time when traversing $\{v, w\}$, and it is called *back edge*, denoted by $v \hookrightarrow w$, otherwise. For convenience we use $v \to w$ or $v \hookrightarrow w$ to denote the respective edge as well as the fact that there is such an edge between v and w, or the path $(v, \{v, w\}, w)$. We denote by $v \xrightarrow{*} w$ a (possibly empty) path of tree edges traversed in the corresponding direction. Note that the tree edges form a spanning *DFS tree* $T = T(G)$ rooted at s, i.e. $s \xrightarrow{*} v$ for all $v \in V$ and $v \hookrightarrow w$ implies $w \xrightarrow{*} v$. For $v \in V$ let $T(v)$ be the subtree induced by all $w \in V$ with $v \xrightarrow{*} w$. We will use the graph in Fig. 1 as our running example.

DFS is the basis of many efficient graph algorithms [6], which often make use of the following notion. The *lowpoint* of a vertex $u \in V$ is the vertex w closest to s in $T(G)$ with $w = u$ or $u \xrightarrow{*} v \hookrightarrow w$. If no such path exists, u is its own lowpoint. Lowpoints are important mostly for the following reason.

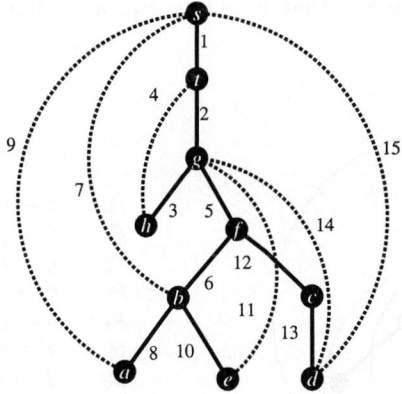

Fig. 1. Running example with edges numbered in DFS traversal order, tree edges depicted solid, and back edges dashed (redrawn from [7])

Lemma 4 ([6]). *A graph $G = (V, E)$ is biconnected if and only if, in a DFS tree $T(G)$, only the root is its own lowpoint and there is at most one tree edge $s \to t$ such that s is the lowpoint of t (in this case, s is the root).*

Previous linear-time st-ordering algorithms first construct a DFS tree and simultaneously compute lowpoints for all vertices. In a second traversal of the graph they use this information to determine an st-ordering. A new biconnectivity algorithm by Gabow [4] that requires only one pass over the graph and in particular does not require lowpoints raised the question whether we can st-order the vertices of a biconnected graph in a similar manner.

4 An Eager Algorithm

We present a new linear-time algorithm for st-ordering the vertices of a biconnected graph. It is eager in the sense that it maintains, during a depth-first search, an ordering of the maximum traversed subgraph for which such ordering is possible without the potential need to modifiy it later on. It is introduced via three preliminary steps that indicate how the algorithm also computes on the fly an open ear decompostion and an st-orientation. Pseudo-code for the complete algorithm is given at the end of this section.

4.1 Open Ear Decomposition

Let $G = (V, E)$ be a biconnected graph with $\{s, t\} \in E$, and let T be a DFS tree of G with root s and $s \to t$. We define an open ear decomposition $D(T) = (P_0, \ldots, P_r)$ using local information only. In particular, we do not make use of lowpoint values.

Let $P_0 = (s, \{s, t\}, t)$ and assume that we have defined $P_0, \ldots, P_i$, $i \geq 0$. If there is a back edge left that is not in E_i, we define P_{i+1} as follows. Let

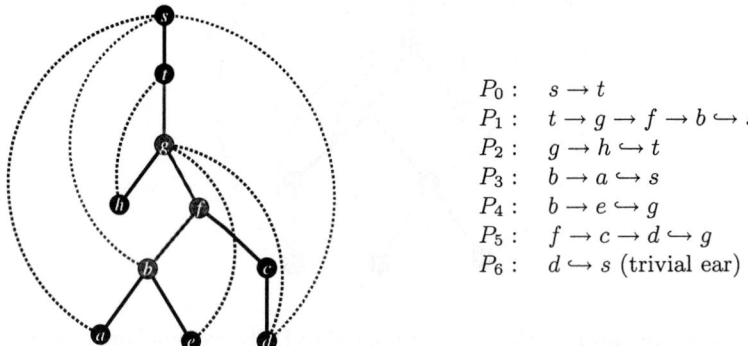

Fig. 2. An ear decomposition $D(T)$ obtained from DFS tree T

$v, w, x \in V$ such that $w, x \in V_i$, $v \hookrightarrow w \notin E_i$, $w \to x$, and $x \xrightarrow{*} v$ (see Fig. 2). Using the last vertex u on the tree path from x to v with $u \in V_i$ (potentially v itself), we set $P_{i+1} = u \xrightarrow{*} v \hookrightarrow w$. Since $w \to x \xrightarrow{*} u$, P_{i+1} is open. It is called *ear of* $v \hookrightarrow w$, and *trivial* if $u = v$.

Theorem 1. $D(T)$ *is an open ear decomposition starting with* $\{s, t\}$.

Proof. Clearly, $D(T) = (P_0, \ldots, P_r)$ is a sequence of edge-disjoint open paths. It remains to show that they cover the entire graph, i.e. $V_r = V$ and $E_r = E$.

First assume there is an uncovered vertex and choose $u \notin V_r$ such that $s \xrightarrow{*} u$ has minimum length. Let w be the lowpoint of u. Since G is biconnected and $u \neq s, t$, Lemma 4 implies that there exist $x, v \in V$ with $w \to x \xrightarrow{*} u \xrightarrow{*} v \hookrightarrow w$ and $x \neq u$. It follows that $w, x \in V_r$ by minimality of u, so that the decomposition is incomplete, since $v \hookrightarrow w$ satisfies all conditions for another ear.

Since all vertices are covered, all tree edges are covered by construction. Finally, an uncovered back edge satisfies all conditions to define another (trivial) ear. □

4.2 *st*-Orientation

We next refine the above definition of an open ear decomposition to obtain an *st*-orientation. We say that a back edge $v \hookrightarrow w$, and likewise its ear, *depends* on the unique tree edge $w \to x$ for which $x \xrightarrow{*} v$. Ears in $D(T)$ are oriented in their sequential order: P_0 is oriented from s to t, whereas P_i, $0 < i \leq r$, is oriented according to the tree edge it depends on. See Fig. 3 for an example.

The following lemma shows that orienting back edges and their ears parallel to the tree edge they depend on nicely propagates into subtrees. If $\{v, w\} \in E_i$, $0 \leq i \leq r$, is oriented from v to w let $v \prec_i w$, $0 \leq i \leq r$.

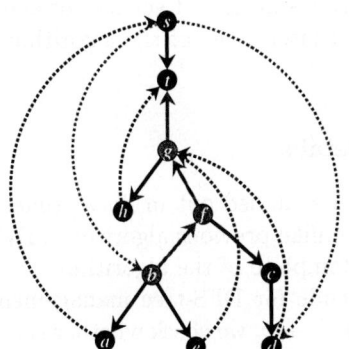

$P_1 = t \xrightarrow{*} b \hookrightarrow s$ depends on $s \to t$
$P_2 = g \xrightarrow{*} h \hookrightarrow t$ depends on $t \to g$
$P_3 = b \xrightarrow{*} a \hookrightarrow s$ depends on $s \to t$
$P_4 = b \xrightarrow{*} e \hookrightarrow g$ depends on $g \to f$
$P_5 = f \xrightarrow{*} d \hookrightarrow g$ depends on $g \to f$
$P_6 = d \xrightarrow{*} d \hookrightarrow s$ depends on $s \to t$

Fig. 3. Orientation of ears in $D(T)$

Lemma 5. *For all $0 \leq i \leq r$, the above orientation of $P_0, \ldots, P_i$ yields an st-orientation of G_i, and $\prec_i$ is a partial order satisfying: If $w \to x \in E_i$ and $w \prec_i x$ ($x \prec_i w$), then $w \prec_i v$ ($v \prec_i w$) for all $v \in T(x) \cap V_i$.*

Proof. The proof is by induction over the sequence $D(T)$. The invariant clearly holds for P_0. Assume it holds for some $i < r$ and let P_{i+1} be the ear of $v \hookrightarrow w$. Let $w \to x \in E_i$ be the tree edge that $v \hookrightarrow w$ depends on and assume it is oriented from w to x (the other case is symmetric). The last vertex $u \in V_i$ on $x \xrightarrow{*} v$ satisfies $w \prec_i u$. All vertices of P_{i+1} except w are in $T(x)$, and since P_{i+1} is oriented like $w \to x$, the invariant is maintained. □

Corollary 1. *The above orientation of $D(T)$ yields an st-orientation of G.*

4.3 st-Ordering

We finally show how to maintain incrementally an ordering of V_i during the construction of $D(T)$. Starting with the trivial st-ordering of P_0, let $P_i = u \xrightarrow{*} v \hookrightarrow w$, $0 < i \leq r$, be the ear of $v \hookrightarrow w$. If P_i is oriented from u to w (w to u), insert the sequence of inner vertices $V(P_i) \setminus \{u, w\}$ of P_i in the order given by the orientation of P_i immediately after (before) u.

Lemma 6. *For all $0 \leq i \leq r$, the ordering of V_i is a linear extension of $\prec_i$.*

Proof. The proof is again by induction over the sequence $D(T)$. The invariant clearly holds for P_0. Assume that it holds for some $i < r$ and that $P_{i+1} = u \xrightarrow{*} v \hookrightarrow w$ depends on $w \to x$. If $w \prec_i x$ (the other case is symmetric), the inner vertices of P_{i+1} are inserted immediately before u. By Lemma 5 and our invariant, u is after w, so that the ordering is also a linear extension of $\prec_{i+1}$. □

Corollary 2. *The above ordering yields an st-ordering of G.*

Note that inserting the inner vertices of an ear $P = u \xrightarrow{*} v \hookrightarrow w$ next to its *destination* w rather than its *origin* u may result in end vertices of another ear being in the wrong relative order during a later stage of the algorithm. An example of this kind is shown in Fig. 4.

4.4 Algorithm and Implementation Details

It remains to show how the above steps can be carried out in linear time. We base our algorithm on depth-first search, but unlike previous algorithms, DFS is not used for preprocessing but rather as the template of the algorithm.

The algorithm is given in Alg. 1, where code for DFS-tree management is implicit. Each time DFS traverses a back edge $v \hookrightarrow w$, we check whether the tree edge $w \to x$ it depends on is already oriented (note that x is the current child of w on the DFS path). If $w \to x$ is oriented, the ear of $v \hookrightarrow w$ is oriented in the same direction and inserted into the ordering. For each tree edge in a newly

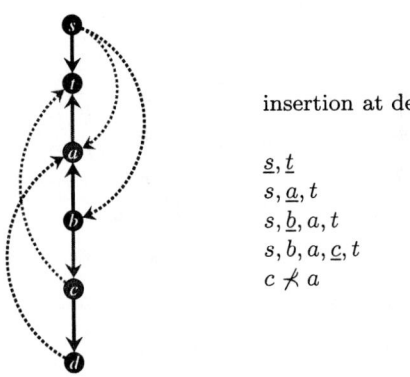

insertion at destination	insertion at origin
$\underline{s}, \underline{t}$	$\underline{s}, \underline{t}$
$s, \underline{a}, t$	$s, \underline{a}, t$
$s, \underline{b}, a, t$	$s, \underline{b}, a, t$
$s, b, a, \underline{c}, t$	$s, b, \underline{c}, a, t$
$c \not< a$	$s, b, c, \underline{d}, a, t$

Fig. 4. Example showing that it is important to insert an ear next to its origin in the tree rather than the destination of its defining back edge

---

Algorithm 1: Eager *st*-ordering

Input: graph $G = (V, E)$, edge $\{s, t\} \in E$
Output: list L of vertices in biconnected component of $\{s, t\}$ (in *st*-order)

process_ears(tree edge $w \to x$) begin
  foreach $v \hookrightarrow w$ *depending on* $w \to x$ **do**
    determine $u \in L$ on $x \xrightarrow{*} v$ closest to v;
    set P to $u \xrightarrow{*} v \hookrightarrow w$;
    if $w \to x$ *oriented from w to x (resp. x to w)* **then**
      orient P from w to u (resp. u to w);
      insert inner vertices of P into L right before (resp. after) u;
    foreach *tree edge* $w' \to x'$ *of* P **do** *process_ears*($w' \to x'$);
  clear dependencies on $w \to x$;
end

dfs(vertex v) begin
  foreach *neighbor w of v* **do**
    if $v \to w$ **then** *dfs(w)*;
    if $v \hookrightarrow w$ **then**
      let x be the current child of w;
      make $v \hookrightarrow w$ depend on $w \to x$;
      if $x \in L$ **then** *process_ears*($w \to x$);
end

begin
  initialize L with $s \to t$;
  dfs(t);
end

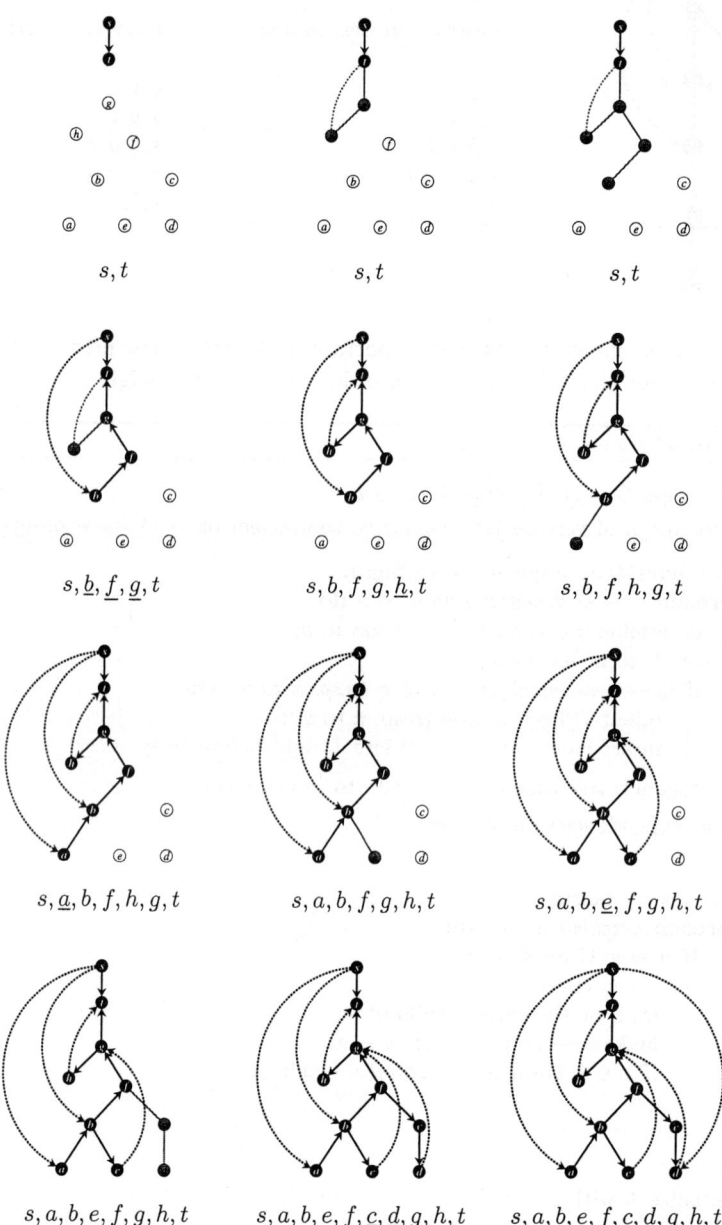

Fig. 5. Intermediate steps of the algorithm

oriented ear, we recursively process the ears of not yet oriented back edges that depend on it.

After each traversal of an edge we are therefore left with an *st*-ordering of the biconnected component of $\{s,t\}$ in the traversed subgraph. Consequently, the algorithm is robust against application to non-biconnected graphs, and the input graph is biconnected if and only if the ordering returned contains all vertices.

Theorem 2. *Algorithm 1 computes an st-ordering of the biconnected component containing $\{s,t\}$ in linear time.*

Proof. Given the discussion above, it is sufficient to show that the algorithm determines the entire open ear decomposition. Each ear is the ear of a back edge; if the back edge depends on an already oriented tree edge, the ear is determined and oriented. If the tree edge is not yet oriented the back edge is added to the set of dependent edges and processed as soon as the tree edge is oriented. It follows from the same argument as in the proof of Theorem 1 that all tree edges in the biconnected component of $\{s,t\}$ will eventually be oriented. □

Figure 5 shows the intermediate steps of the algorithm when applied to the running example. Note that the *st*-ordering is different from that obtained in [7], because, e.g., $d \hookrightarrow g$ is traversed before lowpoint-edge $d \hookrightarrow s$ (otherwise the order of c and d would be reversed).

An efficient implementation of Alg. 1 uses a doubly-linked list L, and stores all edges depending on a tree edge (including the edge itself) in a cyclic list that can be realized with an edge array (since these lists are disjoint). Since only the orientation of tree edges is needed, it is sufficient to store the orientation of the incoming DFS edge at each vertex. In total we need a doubly-linked vertex list, four vertex arrays (incoming edge, orientation of incoming edge, current child vertex, pointer to list position), two vertex stacks (DFS and ear traversal), and one edge array (next dependent edge). The edge array also serves to indicate whether an edge has already been traversed.

5 Discussion

We have presented a simple algorithm to compute an *st*-ordering of the vertices of a biconnected graph. It requires only a single traversal of the graph and maintains, after each step of the DFS, a maximum partial solution for the *st*-ordering, the *st*-orientation, and the open ear decomposition problem on the traversed subgraph.

Without modification, the algorithm can also be used to test for biconnectedness (simply check whether the length of the returned list equals the number of vertices in the graph), and it is readily seen from the inductive proof of Lemma 6 that resulting *st*-orderings have the following interesting property.

Corollary 3. *Algorithm 1 yields st-orderings in which the vertices of every subtree of the DFS tree form an interval.*

It is interesting to note that storing (tree) edge orientations at vertices corresponds to Tarjan's use of +/- labels in [7]. Since they can be interpreted as storing orientations at the parent rather than the child, they need to be updated, though. While the use of lowpoints eliminates the need to keep track of dependencies, lowpoints are known only after traversing the entire graph. Although Tarjan's algorithm [7] remains the most parsimonious, we thus feel that the eager algorithm is more flexible and intuitive.

Finally, the algorithm can be used to determine a generalization of bipolar orientations to non-biconnected graphs, namely acyclic orientations with a number of sources and sinks that is at most one larger than the number of biconnected components, and which is bipolar in each component. Whenever DFS backtracks over a tree edge that has not been orientated, it has completed a biconnected component. We are hence free to orient this tree edge any way we want and thus to determine, by recursively orienting dependent ears, a bipolar orientation of its entire component. The combined orientations are acyclic, and for each additional biconnected component we add at most one source or one sink, depending on how we choose to orient the first edge of that component and whether both its incident vertices are cut vertices.

Acknowledgments

I would like to thank Roberto Tamassia for initiating this research by making me aware of Gabow's work on path-based DFS, and Roberto Tamassia and Luca Vismara for very helpful discussions on the subject. Thanks to three anonymous reviewers for their detailed comments.

References

[1] J. Ebert. *st*-ordering the vertices of biconnected graphs. *Computing*, 30(1):19–33, 1983.

[2] S. Even and R. E. Tarjan. Computing an *st*-numbering. *Theoretical Computer Science*, 2(3):339–344, 1976.

[3] S. Even and R. E. Tarjan. Corrigendum: Computing an *st*-numbering. *Theoretical Computer Science*, 4(1):123, 1977.

[4] H. N. Gabow. Path-based depth-first search for strong and biconnected components. *Information Processing Letters*, 74:107–114, 2000.

[5] A. Lempel, S. Even, and I. Cederbaum. An algorithm for planarity testing of graphs. In P. Rosenstiehl, editor, *Proceedings of the International Symposium on the Theory of Graphs (Rome, July 1966)*, pages 215–232. Gordon and Breach, 1967.

[6] R. E. Tarjan. Depth-first search and linear graph algorithms. *SIAM Journal on Computing*, 1:146–160, 1973.

[7] R. E. Tarjan. Two streamlined depth-first search algorithms. *Fundamenta Informaticae*, 9:85–94, 1986.

[8] H. Whitney. Non-separable and planar graphs. *Transactions of the American Mathematical Society*, 34:339–362, 1932.

Three-Dimensional Layers of Maxima

Adam L. Buchsbaum[1] and Michael T. Goodrich[2]*

[1] AT&T Labs, Shannon Laboratory
180 Park Ave., Florham Park, NJ 07932
alb@research.att.com

[2] Dept. of Information & Computer Science, University of California
Irvine, CA 92697-3425
goodrich@ics.uci.edu

Abstract. We present an $O(n \log n)$-time algorithm to solve the three-dimensional layers-of-maxima problem, an improvement over the prior $O(n \log n \log \log n)$-time solution. A previous claimed $O(n \log n)$-time solution due to Atallah, Goodrich, and Ramaiyer [SCG'94] has technical flaws. Our algorithm is based on a common framework underlying previous work, but to implement it we devise a new data structure to solve a special case of dynamic planar point location in a staircase subdivision. Our data structure itself relies on a new extension to dynamic fractional cascading that allows vertices of high degree in the control graph.

1 Introduction

A point $p \in \Re^d$ *dominates* another point $q \in \Re^d$ if each coordinate of p exceeds that of q. In a set S of n points in $\Re^d$, the *maximum points* are those that are not dominated by any point in S. The *maxima set problem*, to find all the maximum points in S, is a classic problem in computational geometry, dating back to the early days of the discipline [11]. Interestingly, the algorithm presented in the original paper by Kung, Luccio, and Preparata [11] remains the most efficient known solution to this problem; it runs in $O(n \log n)$ time when $d = 2$ or 3 and $O(n \log^{d-2} n)$ time when $d \geq 4$. This problem has subsequently been studied in many other contexts, including solutions for parallel computation models [9], for point sets subject to insertions and deletions [10], and for moving points [7].

The *layers-of-maxima problem* iterates this discovery: after finding the maximum points, remove them from S and find the maximum points in the remaining set, iterating until S becomes empty. The iteration index in which a point is a maximum is defined to be its *layer*. More formally, for $p \in S$, $layer(p) = 1$ if p is a maximum point; otherwise, $layer(p) = 1 + \max\{layer(q) : q \text{ dominates } p\}$. The *layers-of-maxima problem*, which is related to the convex layers problem [4], is to determine $layer(p)$ for each $p \in S$, given S.

With some effort [1], the three-dimensional layers-of-maxima problem can be solved in time $O(n \log n \log \log n)$, using techniques from dynamic fractional

---

* Supported by DARPA Grant F30602-00-2-0509 and NSF Grant CCR-0098068.

cascading [13]. We sketch this result in Section 2. In addition, Atallah, Goodrich, and Ramaiyer [2] claim an $O(n \log n)$-time algorithm, but their presentation appears to have several problems. A simple, linear-time reduction from sorting gives an $\Omega(n \log n)$-time lower bound in the comparison model.

In this paper, we give an $O(n \log n)$-time, $O(n \log n / \log \log n)$-space algorithm for the three-dimensional layers-of-maxima problem. Before we present our algorithm, however, we briefly outline how it relates to the prior algorithm claimed to run in $O(n \log n)$ time [2].

1.1 Relation to the Prior Claim

The previous algorithm [2] was based on the use of two new data structures. The first was a dynamic extension of the point-location structure of Preparata [14] to work in the context of staircase subdivisions, and the second was an extension of the biased search tree data structure of Bent, Sleator, and Tarjan [3] to support finger searches and updates. The second structure was used as an auxiliary structure in the first, and both were analyzed by detailed case analyses. Unfortunately, there are crucial cases omitted in the original analyses of both of these structures, and filling in the details for these omitted cases appears to require increasing the running time so as to negate the claimed time bound of $O(n \log n)$ for the three-dimensional layers-of-maxima problem.

Our solution to the three-dimensional layers-of-maxima problem is also based on a dynamic data structure for staircase subdivisions. But our structure is not an extension of Preparata's point-location approach, nor is it based on any biased data structures. Instead, our data structure exploits a new extension of dynamic fractional cascading, which may be of independent interest.

We sketch the basic approach for our algorithm in Section 2, which follows the space-sweeping framework that provided the basis for prior work [1, 2]. In Section 3, we present our new data structure to solve a special case of dynamic planar point location in a staircase subdivision, which is the key to the basic algorithm. In Section 4, we present an extension of dynamic fractional cascading that allows our data structure to achieve $O(\log n)$ amortized time per point in S, yielding an $O(n \log n)$-time algorithm for three-dimensional layers of maxima.

2 A 3D-Sweep Framework

Assume $S \subset \Re^3$. We use a three-dimensional sweep algorithm to compute the layers-of-maxima on S. Denote by $z(p)$ the z-coordinate of point p. Similarly define $x(p)$ and $y(p)$. Define $S_i(\ell) = \{p \in S : z(p) > i \wedge layer(p) = \ell\}$; $S_{-\infty}(\cdot)$ thus partitions S by layer, and the problem is to compute this partition. We process the points in S in order by decreasing z-coordinate, breaking ties arbitrarily.

Invariant 1 *When processing a point with z-coordinate i, points in $\bigcup_\ell S_i(\ell)$ are correctly labeled with their layers.*

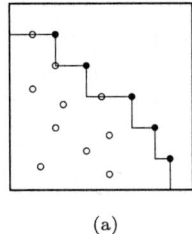 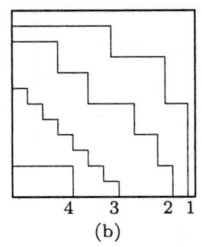

Fig. 1. In this and succeeding pictures, the x-dimension increases left-to-right, and the y-dimension increases bottom-to-top. (a) A set of points in the plane and the staircase bounding their dominance region. Extremal points are shown filled. (b) A staircase subdivision of the plane. Staircases are labeled by their corresponding layers

For each layer, we maintain a subset of the points so far assigned to that layer. Define the *dominance region*, $D(p)$, of a point p to be the set of all points in $\Re^d$ that are dominated by p; for a set X of points, $D(X) = \bigcup_{p \in X} D(p)$. For a point $p \in \Re^3$, let $\pi(p)$ be the projection of p onto the (x, y) plane; for a set X of points, $\pi(X) = \bigcup_{p \in X} \{\pi(p)\}$.

In two dimensions, the dominance region of a set of points is bounded by a *staircase*, which can be identified with its extremal points; see Figure 1(a). Denote by $M_i(\ell)$ the extremal points of $D(\pi(S_i(\ell)))$. That is, of the points so far assigned to each layer ℓ, $M_i(\ell)$ is the set of extremal points of the staircase induced by the dominance region of their two-dimensional projections. We maintain $M_i(\ell)$ for each layer ℓ so far assigned a point, thereby dividing the 2D-plane into staircase regions; see Figure 1(b). We identify layers with their staircases.

Assume Invariant 1 is true before the first point with z-coordinate i is processed. We process the points $P_i = \{p \in S : z(p) = i\}$ as follows. First, we calculate $layer(p)$ for each $p \in P_i$. To do so, we identify where $\pi(p)$ lies in the staircase subdivision defined by $M_i(\cdot)$.

1. If $\pi(p)$ lies on or above staircase 1, then assign $layer(p) = 1$.
2. Otherwise, if $\pi(p)$ lies below the highest-numbered staircase, say s, then p is the first point assigned to $layer(p) = s + 1$.
3. Otherwise, $\pi(p)$ lies between two staircases, ℓ and $\ell + 1$, possibly lying on staircase $\ell + 1$; in this case, assign $layer(p) = \ell + 1$.

(See Figure 2(a).) Assume $M_i(\cdot)$ is correctly maintained. In the first case, no previously processed point dominates p; in the other cases, p is dominated by at least one point in $layer(p) - 1$ but not by any point in layer $layer(p)$ or higher. No point q processed after p dominates p, because for each such q we know that $z(q) \leq z(p)$. Thus, the layer assignment maintains Invariant 1. We need only show how to represent $M_i(\cdot)$ to allow testing of whether a point lies on or above some staircase and how to update $M_i(\cdot)$ to account for the points in P_i.

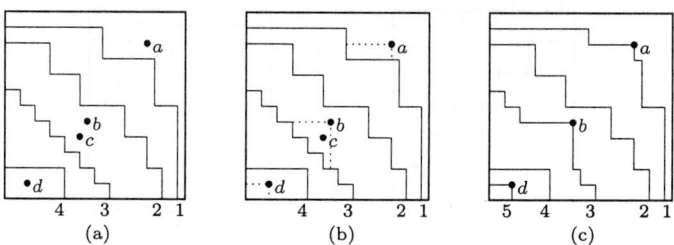

Fig. 2. (a) New points a, b, c, and d are assigned layers: $layer(a) = 1$; $layer(b) = layer(c) = 3$; $layer(d) = 5$. (b) Dotted lines show how a and b affect their staircases and d affects the plane. (c) After updating the staircases; c has no effect

It suffices to maintain each $M_i(\ell)$ as a list of points ordered by decreasing y-coordinate (and hence increasing x-coordinate), including the sentinel points $(-\infty, \infty)$ and $(\infty, -\infty)$, which allows easy determination of whether point $\pi(p)$ lies on or above the staircase ℓ. Let $u, v \in M_i(\ell)$ such that $y(u) > y(p) \geq y(v)$.

1. $\pi(p)$ lies on staircase ℓ if (a) $x(p) = x(u)$ or (b) $x(u) < x(p) \leq x(v)$ and $y(p) = y(v)$;
2. otherwise, $\pi(p)$ lies above staircase ℓ if $x(p) > x(u)$;
3. otherwise, $\pi(p)$ lies below staircase ℓ.

Having determined layers for all points in P_i, we update $M_i(\cdot)$ as follows. In any order, consider each $p \in P_i$ in turn; let $\ell = layer(p)$. If $\pi(p)$ lies on or below staircase ℓ, no further action for p is necessary; $\pi(p)$ can lie below staircase ℓ at this point due to the addition of some other $q \in P_i$ that was previously processed into $M_i(\ell)$. Otherwise, $\pi(p)$ still lies above staircase ℓ and so becomes an extremal point defining the new staircase. In this case, remove all points, if any, from $M_i(\ell)$ that are no longer extremal staircase points. Then, insert $\pi(p)$ into its proper place in $M_i(\ell)$. Thus, $M_i(\cdot)$ becomes $M_j(\cdot)$ for the next z-coordinate j processed by the sweep algorithm.

To find the points in $M_i(\ell)$ that must be removed by the addition of $\pi(p)$, find $u, v \in M_i(\ell)$ such that $y(u) > y(p) \geq y(v)$. If $x(p) \geq x(v)$, then v is no longer extremal, so remove it from $M_i(\ell)$, and iterate at the successor of v in $M_i(\ell)$. If $x(p) < x(v)$, stop: v and its successors are still extremal. At the termination of this iteration, insert $\pi(p)$ as the successor of u in $M_i(\ell)$. See Figure 2(b)–(c).

2.1 Analysis

Let $Q(N)$ be the time to determine the layer of a point. Let $R(N)$ be the time to determine the points u, v in the corresponding staircase that isolate where a new point begins to affect extremal points. Let $I(N)$ and $D(N)$ be the times to insert and delete (rsp.) each extremal point into/from a staircase. Each point p is the subject of one layer query and engenders at most one staircase transformation, to the staircase defining the dominance region of layer(p). While the number of points deleted during this transformation can be large, over the course of the entire procedure each point is inserted and deleted at most once into/from its layer's staircase. The total time is thus $O(n(Q(N) + R(N) + I(N) + D(N))$.

Implementing each $M_i(\ell)$ as a simple ordered list, $Q(n) = O(\log^2 n)$: each query to determine where a point p lies with respect to a staircase takes $O(\log n)$ time, and binary search on the staircases determines layer(p). $R(n) = I(n) = D(n) = O(\log n)$, so the algorithm runs in $O(n \log^2 n)$ total time and $O(n)$ space.

We can improve the running time to $O(n \log n \log \log n)$ by using van Emde Boas trees [16] instead of ordered lists, but the space becomes $O(n^2)$. We can reduce the space back to $O(n)$ with the same $O(n \log n \log \log n)$ time bound using dynamic fractional cascading [13]. We omit the details of this implementation. Instead, we devise a more sophisticated representation for $M_i(\cdot)$, which itself uses an extension of dynamic fractional cascading, and which reduces the running time to $O(n \log n)$ but uses space $O(n \log n / \log \log n)$.

3 Improved Implementation

Since identifying the staircases that determine the layer of a new point is the time consuming part above, we build a data structure to facilitate that query.

Let T be a search tree on the set of y-coordinates in S, i.e., each leaf an element of $\{y(p) : p \in S\}$. T is built a priori and remains unchanged throughout the algorithm. Each internal node u of T spans a *slice* of the y-coordinate space defined by the closed range $[\min(u), \max(u)]$, where $\min(u)$ (rsp., $\max(u)$) is the minimum (rsp., maximum) y-coordinate stored at a leaf descendant of u. Staircases pass through these slices; we say a staircase *enters* a slice at the top (maximum y-coordinate in the slice) and *exits* at the bottom (minimum y-coordinate in the slice). We say a staircase *spans* a slice if it enters and exits at different x-coordinates; staircases that simply pass vertically through a slice do not span that slice. See Figure 3(a). With each internal node u, we store the entry and exit x-coordinates of the staircases that span u's slice, associating with each such record the index of the corresponding staircase.

Say a point p *stabs* a staircase in a slice if an infinite vertical line through p stabs a horizontal section of that staircase inside that slice. See Figure 3(b). By testing a specific set of slices for stabbed staircases, we can identify the layer assigned to a new point. For some slice u and x-coordinate x let $pred_{entry(u)}(x)$ (rsp., $pred_{exit(u)}(x)$) be the predecessor of x among u's entry (rsp., exit) x-coordinates; let $succ_{exit(u)}(x)$ be the successor of x among u's exit x-coordinates; and let $lab(\cdot)$ be the index of the staircase with the corresponding coordinates.

Fact 1 *Point p stabs a staircase in a slice if and only if the entry and exit points of the staircase fall on opposite sides of $x(p)$.*

Lemma 1. *Assume p stabs a staircase in slice u. The highest such staircase, i.e., that with the lowest label, is staircase $lab(pred_{entry(u)}(x(p)))$; the lowest such staircase, i.e., that with the highest label, is staircase $lab(succ_{exit(u)}(x(p)))$.*

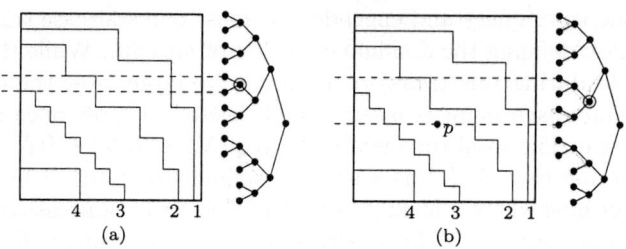

Fig. 3. A staircase subdivision with an associated search tree on the y-coordinates. Each of (a) and (b) depict a slice (between the dashed lines) corresponding to the circled internal node. (a) Staircase 2 spans the slice, but staircase 1 does not. (b) Point p stabs staircase 2 in the slice but stabs none of the other staircases in that slice

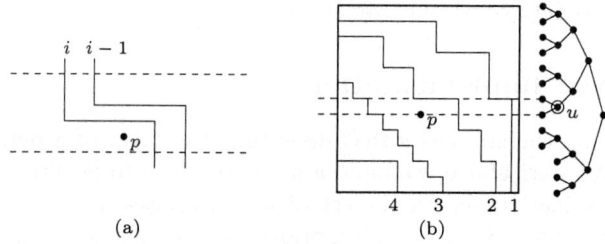

Fig. 4. (a) A slice in which point p stabs staircases i and $i-1$. Staircases do not cross, so $i-1 = lab(pred_{entry(u)}(x(p)))$, or else p stabs a staircase higher than $i-1$ in the slice. Symmetrically, $i = lab(succ_{exit(u)}(x(p)))$, or else p stabs a staircase lower than i in the slice. (b) The nearest ancestor, u, of leaf $y(p)$ whose slice contains a staircase stabbed by p is circled. The slice at the top child of u contains the stabbed staircase

Proof. This follows from Fact 1 and that staircases do not cross. See Figure 4(a).

Lemma 2. *Point p stabs a staircase in some slice u if and only if*

$$lab(pred_{entry(u)}(x(p))) \leq lab(succ_{exit(u)}(x(p))).$$

Proof. Assume $lab(pred_{entry(u)}(x(p))) = i \leq j = lab(succ_{exit(u)}(x(p)))$. Staircases do not cross, so the entry and exit points of both layers i and j must fall on opposite sides of $x(p)$; apply Fact 1. The other direction follows from Lemma 1.

Lemma 2 implies that we can determine if a point stabs a staircase in a given slice in the time it takes to query some dictionary data structure. We use this lemma to determine $layer(p)$ for a new point p as follows. (See Figure 4(b).)

1. Traverse the leaf-to-root path from $y(p)$ in T.

2. Find the nearest ancestor u whose slice contains a staircase stabbed by p. If $u = y(p)$, stop: p lies on a horizontal staircase segment. Else, go to step (3).
3. Determine the nearest child of u whose slice contains a staircase stabbed by p. There must be some child as such, and it is not an ancestor of $y(p)$. Formally, let $c_1, \ldots, c_i$ be the children of u in top-down order ($\min(c_j) > \max(c_{j+1})$). Let c_j be the child of u that is an ancestor of $y(p)$. Find c_a such that p stabs a staircase in slice c_a and either (1) $a < j$ and there exists no b such that $a < b < j$ and p stabs a staircase in slice c_b; or (2) $a > j$ and there exists no b such that $a > b > j$ and p stabs a staircase in slice c_b.

Lemma 3. *If the procedure stops in step (2), $layer(p) = lab(pred_{entry(p)}(x(p)))$. Otherwise, the procedure stops in step (3). In this case, if $a > j$, then $layer(p) = lab(pred_{entry(c_a)}(x(p)))$; if $a < j$, then $layer(p) = lab(succ_{exit(c_a)}(x(p))) + 1$.*

Proof. If (and only if) the procedure stops in step (2), then p lies on a horizontal segment of a staircase, which is identified by $pred_{entry(p)}(x(p))$.

Otherwise, if $a > j$, then c_a is below c_j. By assumption, p stabs a staircase in slice c_a. By Lemma 1, the highest such staircase is given by $pred_{entry(c_a)}(x(p))$. Since p does not stab any staircase before stabbing this one, p lies between it and the next higher one, and so $layer(p) = lab(pred_{entry(c_a)}(x(p)))$.

If $a < j$, then c_a is above c_j, and p must stab the staircase that is identified by $succ_{exit(c_a)}(x(p))$. Since p stabs no staircase before this one, p lies between it and the next lower one, and so $layer(p) = lab(succ_{exit(c_a)}(x(p))) + 1$.

3.1 Updating T

Recall the sweep algorithm framework. Once the points in P_i have been assigned their layers, we must update the staircase extremal points. Let p be a new point assigned to layer ℓ. In addition to the new data structure in T, we still maintain each $M_i(\ell)$ as an ordered list to facilitate finding the points $u, v \in M_i(\ell)$ such that $y(u) > y(p) \geq y(v)$, indicating where $\pi(p)$ begins to affect the staircase. Assume $\pi(p)$ still lies above staircase ℓ, or else p engenders no further update.

While T itself remains static though this process, the sets of staircases spanning various slices change to reflect the addition of $\pi(p)$ to staircase ℓ. As in Section 2, iterate to find each v that is no longer extremal by the addition of $\pi(p)$ to staircase ℓ. Let v' be the first v encountered that remains in $M_i(\ell)$; i.e., after deleting the obviated v's and inserting $\pi(p)$, v' is $\pi(p)$'s successor in $M_i(\ell)$. The semi-closed range $[y(p), y(v'))$ (top-down) defines the extent of p's effect on $M_i(\ell)$, which we call p's *range of influence*. (In Figure 2(b), the dotted lines identify the ranges of influence for points a, b, and d.)

We must now update the slices at ancestors of $y(v)$ in T for each v deleted from $M_i(\ell)$. For each, traverse the $y(v)$-to-root path. For the slice at each node u on the path, there are four cases, depending on the relative positions of $y(p)$ and $\max(u)$ (the top of the slice) and those of $y(v')$ and $\min(u)$ (the bottom of the slice). (See Figure 5.)

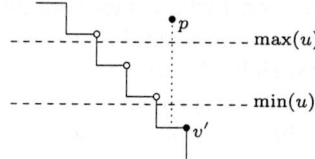

Fig. 5. A staircase, ℓ, being updated by the insertion of p. The dotted line identifies the range of p's influence: $[y(p), y(v'))$. The open circles are extremal points that must be removed from the staircase (and hence from which leaf-to-root traversals are applied). By the assumption that p belongs to layer ℓ, if the top of some slice u (indicated by dashed lines) is inside p's range of influence, then the entry point for staircase ℓ in that slice is given by $pred_{entry(u)}(x(p))$; otherwise, some other staircase would intercede between staircase ℓ and p. Similarly, if the bottom of the slice is inside p's range of influence, then the exit point for staircase ℓ in that slice is given by $pred_{exit(u)}(x(p))$

1. $\max(u) > y(p)$, $\min(u) \leq y(v')$. Both the top and bottom of the slices are outside p's range of influence. This and further ancestor slices are not affected by p, so terminate the traversal.
2. $\max(u) \leq y(p)$, $\min(u) \leq y(v')$. The top of the slice is inside p's range of influence, but the bottom is outside. The entry point of staircase ℓ, which in this case is given by $pred_{entry(u)}(x(p))$, changes to $x(p)$, so delete the old entry point and insert the new one.
3. $\max(u) > y(p)$, $\min(u) > y(v')$. The bottom of the slice is inside p's range of influence, but the top is outside. The exit point of staircase ℓ, which in this case is given by $pred_{exit(u)}(x(p))$, changes to $x(p)$, so delete the old exit point and insert the new one.
4. $\max(u) \leq y(p)$, $\min(u) > y(v')$. Both the top and bottom of the slice are inside p's range of influence. Staircase ℓ no longer spans this slice. If it did before, $pred_{entry(u)}(x(p))$ and $pred_{exit(u)}(x(p))$ are both labeled ℓ and give the old entry and exit points, which are deleted. If $pred_{entry(u)}(x(p))$ and $pred_{exit(u)}(x(p))$ are not labeled ℓ, then staircase ℓ did not span this slice before, and no update is required.

Finally, perform the same operations on the slices encountered during a $y(p)$-to-root traversal of T. Correctness of this procedure follows from the fact that whenever a staircase spans a slice, there exists at least one extremal point of the staircase within the slice. That extremal point earlier caused entry and exit points to be recorded appropriately. Note that any slice whose top or bottom (or both) falls within p's range of influence is an ancestor of $y(p)$ or some $y(v)$ that was deleted above and hence gets updated appropriately.

3.2 Analysis

Straightforward implementations of T do not improve upon the results of Section 2. For example, if we construct T as a balanced, binary search tree and use red-black trees [8] to implement the entry and exit lists, then $Q(N) = O(\log^2 N)$ as before: $O(\log n)$ time at each level of T to perform a stabbing test using Lemma 2. $R(N) = O(\log n)$, and $I(N) = D(N) = O(\log^2 N)$ (argue as with $Q(N)$). Each point in S yields an element in at most one entry and exit list in each level of T. The algorithm thus runs in $O(n \log^2 n)$ time and $O(n \log n)$ space.

Using van Emde Boas trees [16] for the entry and exit lists reduces the running time to $O(n \log n \log \log n)$, but as each slice's entry and exit list requires $O(n)$ space, the total space increases to $O(n^2)$. Using Willard's q-fast tries [17] yields time $O(n \log^{1.5} n)$ and space $O(n \log n)$.

In general, however, if we set the arity of T to some d and use a dictionary that on N items admits predecessor and successor queries, insertions, and deletions in time $\mathcal{D}(N)$, then $Q(N) = O(\mathcal{D}(N)(\log_d N + d))$: $\mathcal{D}(N) \log_d N$ to find the first ancestor slice that contains a stabbed staircase, and $d \cdot \mathcal{D}(N)$ to find the nearest child slice containing the stabbed staircase. $R(N) = O(\log N)$ as before, and $I(N) = D(N) = O(\mathcal{D}(N) \log_d N)$. If $d = O(\sqrt{\log n})$ and $\mathcal{D}(N) = O(\log \log n)$, this would yield an $O(n \log n)$-time algorithm. Again we could use van Emde Boas trees [16], but the space would still be $O(n^2)$. Using Mehlhorn and Näher's [12] modification reduces the space to $O(n)$, but the time becomes expected.

We next show how to apply dynamic fractional cascading with this method to achieve $O(n \log n)$ time (worst case) and $O(n \log n / \log \log n)$ space.

4 Dynamic Fractional Cascading

Let $G = (V, E)$ be a *control graph* with a *catalog* $C(v) \subset \Re^+ = \Re \cup \{-\infty, \infty\}$ associated with each vertex $v \in V$ and a *range* $R(u, v) = R(v, u)$, which is an interval on the open real line, associated with each edge $\{u, v\} \in E$. The *local degree*, $d(v)$, of a vertex v is the maximum number of incident edges whose ranges contain any given element of $\Re^+$. That is, if $N_x(v) = \{u : \{v, u\} \in E \wedge x \in R(v, u)\}$, then $d(v) = \max\{|N_x(v)| : x \in \Re^+\}$. G has *locally bounded degree* d if $d \geq \max\{d(v) : v \in V\}$. Let $N = |V| + |E| + \sum_{v \in V} |C(v)|$.

A *generalized path* in G is a sequence $(v_1, \ldots, v_p)$ of vertices such that for each $1 < j \leq p$, there exists an edge $\{v_i, v_j\} \in E$ for some $1 \leq i < j$. Chazelle and Guibas [5] consider the problem of traversing a generalized path of G, at each vertex v determining the predecessor in $C(v)$ (or at each finding the successor) of a query value $x \in \Re^+$, given that $x \in R(e)$ for each edge e induced by the path. They design the *fractional cascading* data structure, which requires $O(N)$ space and answers each such query in time $O(p \log d + \log N)$ time, where p is the number of vertices in the path and G is of locally bounded degree d. Insertions and deletions from individual catalogs can be performed in $O(\log N)$ amortized time each, but then the query time degrades to $O(p \log d \log \log N + \log N)$.

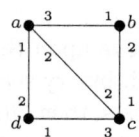 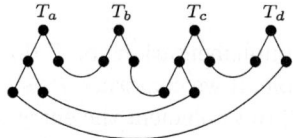

Fig. 6. A control graph G (left) and the associated new graph G' (right). Numbers on edges of G indicate an ordering of the neighbors of each vertex. Leaves of subtrees in G' are implicitly numbered left to right

Mehlhorn and Näher [13] show how to perform each update in $O(\log \log N)$ amortized time if the position of the update is known, i.e., given a pointer to the successor (or predecessor) of the element to be inserted or deleted. The query time becomes $O(p \log \log N + \log N)$. Their results assume that $d = O(1)$. Raman [15] notes that this result is easily extended to yield $O(p(\log d + \log \log N) + \log N)$ query time in the case of arbitrary d and with Dietz goes on [6, 15] to make the update time worst case $O(\log d + \log \log N)$.

The above extension of dynamic fractional cascading to arbitrary d uses an idea of Chazelle and Guibas that replaces each vertex of G by a uniform *star tree*. Using a more local technique, we can show the following slightly more general result. In particular, it allows a (constant) few vertices in the traversal of the path to be of arbitrarily high degree while still maintaining the overall time bound from Mehlhorn and Näher's original result. Let $(v_1, \ldots, v_p)$ denote a generalized path. Let $(\{u_2, v_2\}, \ldots, \{u_p, v_p\})$ be any sequence of edges that can be used to traverse the path; i.e., for each $1 < j \leq p$, $\{u_j, v_j\} \in E$ and there is some $1 \leq i < j$ such that $u_j = v_i$. Redefine $d(v)$ to be the (real) degree of v.

Theorem 1. *An $O(N)$-space dynamic fractional cascading data structure can be built that performs queries in $O(p \log \log N + \log N + \sum_{i=2}^{p}(\log d(u_i) + \log d(v_i)))$ time and insertions and deletions in $O(\log \log N)$ amortized time each, given a pointer to the predecessor or successor of the item to be inserted or deleted.*

Proof. Construct a new graph $G' = (V', E')$ as follows. For each $v \in V$, build a balanced binary tree T_v on $d(v)$ leaves. Let $\eta_v(i)$ be the i'th neighbor of v in G for some arbitrary but fixed ordering of neighbors; and let v_i denote the i'th leaf of T_v in symmetric order. G' is comprised of all the trees T_v plus an edge $\{u_i, v_j\}$, connecting the appropriate leaves of T_u and T_v, for each edge $\{u, v\} \in E$ such that $\eta_u(i) = v$ and $\eta_v(j) = u$. (See Figure 6.) The degree of each vertex in G' is thus two. We maintain a dynamic fractional cascading data structure on G'. Each catalog $C(v)$ is stored at the root of the corresponding T_v. Since $|V'| \leq 4|E|$ and $|E'| \leq 5|E|$, $N' = |V'| + |E'| + \sum_{v \in V} |C(v)| = \Theta(N)$.

Now map paths in G to paths in G'. To start a query at $v \in V$, start at the root of T_v in V'. To traverse edge $\{u, v\}$ in G, traverse in G' the path from the root of T_u to that of T_v, via the appropriate leaves. (Record with each edge $\{u, v\} \in E$ the indices of the appropriate leaves in T_u and T_v.)

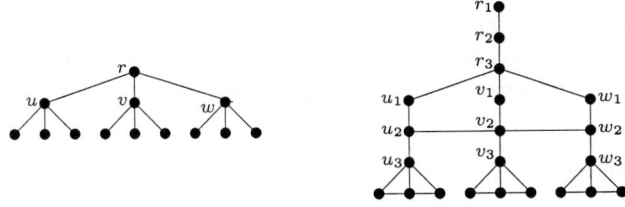

Fig. 7. A 3-ary tree (left) and the derived control graph (right)

Mehlhorn and Näher [13, Lemma 2] show that traversing the new path takes $O(\ell)$ time, where $\ell = \log d(u) + \log d(v)$ is the length of the new path, while querying the catalogs $C(u)$ and $C(v)$ can still be done in $O(\log \log N)$ time each [13, Lemma 3] and insertion and deletion in $O(\log \log N)$ amortized time.

4.1 Application to Layers of Maxima

Theorem 2. *The three-dimensional layers-of-maxima problem can be solved in $O(n \log n)$ time and $O(n \log n / \log \log n)$ space.*

Proof. We use the framework of Section 3. Let T be a static, balanced search tree of arity $\sqrt{\log n}$. (At most one internal child of every node has degree less than $\sqrt{\log n}$; the others have degree $\sqrt{\log n}$.) The depth of T is thus $O(\log n / \log \log n)$. The leaves of T correspond to the y-coordinates of points in S.

Modify T as follows. Transform each internal node u into a length-3 path (u_1, u_2, u_3); let the parent of u_1 correspond to that of u, which will be similarly modified; and let the children of u_3 correspond to those of u, which will be similarly modified unless they were leaves. That is, if v is the parent of u, v_3 becomes the parent of u_1. For each pair (u, v) of consecutive children in the original T, doubly link the corresponding nodes u_2 (u if it is a leaf) and v_2 (v if it is a leaf). See Figure 7. T is no longer a tree, but the notion of levels of T extends easily; in particular, the leaf-to-root paths maintain length $O(\log n / \log \log n)$.

Apply dynamic fractional cascading per Theorem 1 with T as the control graph; for each original node u, store catalogs at the new node u_2 (u if it is a leaf) to record slice u's entry and exit lists. $Q(N) = O(\log n)$: $O(\log \log n)$ time at each level of T to perform a stabbing test using Lemma 2; $O(\sqrt{\log n} \log \log n)$ time to find the appropriate stabbed child of the first stabbed ancestor, using the children links; and $O(\log \log n)$ time per level of T to traverse edges incident to nodes of degree $O(\sqrt{\log n})$. The transformation of the original T allows traversing the children of a node in $O(\sqrt{\log n} \log \log n)$ time, since they are linked at constant-degree nodes. $R(N) = O(\log n)$ using ordered lists of extremal points for each staircase. $I(N) = D(N) = O(\log n)$: each update due to a point p is identified by the predecessor of $x(p)$ in the entry and exit lists of slices encountered in a leaf-to-root traversal of T, which we maintain while traversing the query path in T, so we can update the fractional cascading structure at a cost of $O(\log \log n)$ per level of T. The overall time is thus $O(n \log n)$.

Each point in S engenders an element in at most one entry and exit list in each level of T. The space is thus $O(n \log n / \log \log n)$.

5 Conclusion

We have provided an $O(n \log n)$-time, $O(n \log n / \log \log n)$-space algorithm to solve the three-dimensional layers-of-maxima problem. Our algorithm uses a 3D-sweep on the input point set, maintaining a staircase subdivision of the plane with an improved dynamic fractional cascading data structure that accommodates vertices of high degree in the control graph.

While $\Omega(n \log n)$ is a lower bound on the time to solve this problem in the comparison model, it remains open to achieve that bound using linear or at least $o(n \log n / \log \log n)$ space as well as to provide bounds for higher dimensions.

Our algorithm framework employs a special case of dynamic point location in a staircase subdivision: namely, when all the y-coordinates are known a priori. It remains open to solve this problem in general. It is also open whether there are other applications of the dynamic fractional cascading extension and to combine it with Raman's worst-case update technique.

Acknowledgement

We thank Rajeev Raman for helping set our dynamic fractional cascading extension in context with prior work.

References

[1] P. K. Agarwal. Personal communication, 1992.
[2] M. J. Atallah, M. T. Goodrich, and K. Ramaiyer. Biased finger trees and three-dimensional layers of maxima. In *Proc. 10th ACM SCG*, pages 150–9, 1994.
[3] S. W. Bent, D. D. Sleator, and R. E. Tarjan. Biased search trees. *SIAM J. Comp.*, 14(3):545–68, 1985.
[4] B. Chazelle. On the convex layers of a planar set. *IEEE Trans. Inf. Thy.*, IT-31:509–17, 1985.
[5] B. Chazelle and L. J. Guibas. Fractional cascading: I. A data structure technique. *Algorithmica*, 1(2):133–62, 1986.
[6] P. F. Dietz and R. Raman. Persistence, amortization and randomization. In *Proc. 2nd ACM-SIAM SODA*, pages 78–88, 1991.
[7] P. G. Franciosa, C. Gaibisso, and M. Talamo. An optimal algorithm for the maxima set problem for data in motion. In *Abs. CG'92*, pages 17–21, 1992.
[8] L. J. Guibas and R. Sedgewick. A dichromatic framework for balanced trees. In *Proc. 19th IEEE FOCS*, pages 8–21, 1978.
[9] J.-W. Jang, M. Nigam, V. K. Prasanna, and S. Sahni. Constant time algorithms for computational geometry on the reconfigurable mesh. *IEEE Trans. Par. and Dist. Syst.*, 8(1):1–12, 1997.

[10] S. Kapoor. Dynamic maintenance of maximas of 2-d point sets. In *Proc. 10th ACM SCG*, pages 140–149, 1994.
[11] H. T. Kung, F. Luccio, and P. Preparata. On finding the maxima of a set of vector. *J. ACM*, 22(4):469–76, 1975.
[12] K. Mehlhorn and S. Näher. Bounded ordered dictionaries in $O(\log \log N)$ time and $O(n)$ space. *IPL*, 35(4):183–9, 1990.
[13] K. Mehlhorn and S. Näher. Dynamic fractional cascading. *Algorithmica*, 5(2):215–41, 1990.
[14] F. P. Preparata. A new approach to planar point location. *SIAM J. Comp.*, 10(3):473–482, 1981.
[15] R. Raman. *Eliminating Amortization: On Data Structures with Guaranteed Fast Response Time*. PhD thesis, U. Rochester, 1992.
[16] P. van Emde Boas. Preserving order in a forest in less than logarithmic time and linear space. *IPL*, 6(3):80–2, June 1977.
[17] D. E. Willard. New trie data structures which support very fast search operations. *JCSS*, 28(3):379–94, 1984.

Optimal Terrain Construction Problems and Applications in Intensity-Modulated Radiation Therapy*

Danny Z. Chen, Xiaobo S. Hu**, Shuang Luan,
Xiaodong Wu***, and Cedric X. Yu

Abstract. In this paper, we study several rectilinear terrain construction problems, which model *leaf sequencing problems* in intensity-modulated radiation therapy (IMRT). We present a novel unified approach based on geometric techniques for solving these terrain construction problems. Our approach leads to the first algorithms for several leaf sequencing problems in IMRT that are practically fast and guarantee the optimal quality of the output solutions. Our implementation results show that our terrain construction algorithms run very fast on real medical data.

1 Introduction

In this paper, we study several rectilinear terrain construction problems. A terrain is rectilinear if each of its faces is orthogonal to an axis of the 3-D space. A rectilinear terrain $\mathcal{T}(\cdot)$ is said to be *regular* if it is defined on a regular 2-D grid $\mathcal{RG}$, i.e., each cell c of $\mathcal{T}(\cdot)$ orthogonal to the z-axis is a "lift" of the corresponding cell of $\mathcal{RG}$ by an integer value $\mathcal{T}(c) \geq 0$ in the z direction (e.g., see Figure 1(a)). We assume that $\mathcal{RG}$ consists of n columns of grid cells. An *elementary block* on a column of $\mathcal{RG}$ is a regular terrain of a unit height defined on some consecutive cells of that column. Hence, each elementary block on column i can be represented by a non-empty continuous interval $I_i = [l_i, r_i]$, where l_i and r_i are the starting and ending points of I_i. A *building block* on $\mathcal{RG}$ is a sequence of elementary blocks on a set of contiguous columns of $\mathcal{RG}$, with exactly one elementary block on each of those columns, denoted by $BB = \{I_i, I_{i+1}, \ldots, I_{i+q}\}$ ($1 \leq i \leq n, q \geq 0, i + q \leq n$). In particular, we are interested in two special types of building blocks, called (δ, t)-*building-block* and δ-*building-block*, defined as follows.

---

* This research was supported in part by the NSF under Grant CCR-9988468. The contact information of Chen, Hu, Luan, and Wu is Department of Computer Science and Engineering, University of Notre Dame, Notre Dame, IN 46556, USA. {chen,shu,sluan,xwu}@cse.nd.edu. The contact information of Yu is Department of Radiation Oncology, University of Maryland School of Medicine, Baltimore, MD 21201-1595, USA. cyu002@umaryland.edu.
** The work of this author was supported in part by the NSF under Grant MIP-9701416.
*** Corresponding author. The work of this author was supported in part by a fellowship from the Center for Applied Mathematics, University of Notre Dame.

Definition 1. *Given a separation value $\delta > 0$ and a distance threshold $t > 0$, a building block $BB = \{I_i, I_{i+1}, \ldots, I_{i+q}\}$ is a (δ, t)-building-block (called (δ, t)-block) if the following two conditions hold:*

1. *The length $|I_j| = r_j - l_j$ of each elementary block $I_j \in BB$ satisfies $|I_j| \geq \delta$,*
2. *if $q > 0$, for each $I_j \in BB$ ($i \leq j < i+q$), the distance between I_j and I_{j+1}, defined as $\text{dist}(I_j, I_{j+1}) = \max\{|l_j - l_{j+1}|, |r_j - r_{j+1}|\}$, is no larger than t.*

BB is a δ-building-block (called δ-block) if the following two conditions hold:

1. *For each $I_j \in BB$, $|I_j| \geq \delta$, and*
2. *if $q > 0$, then for each $I_j \in BB$ ($i \leq j < i+q$), the length of the intersection between I_j and I_{j+1} is at least δ (i.e., $|I_j \cap I_{j+1}| \geq \delta$).*

A terrain construction problem seeks a set of building blocks to build a given regular terrain $T(\cdot)$, by stacking up the building blocks. For example, the four δ-blocks in Figures 1(b)–(e) can be used to build the regular terrain in Figure 1(a). Note that when laying a building block on top of a terrain, part of the building block may drop to the lower layers of the terrain.

In this paper, we consider two optimal regular terrain construction problems.

Optimal regular terrain construction with (δ, t)-building-blocks in 3-D (min-# (δ, t)-OTC): Given a regular terrain $T(\cdot)$ on a 2-D regular grid RG with n columns of grid cells, and two integers, a separation value $\delta > 0$ and a distance threshold $t > 0$, build the terrain $T(\cdot)$ with the minimum number of k (δ, t)-blocks. A related problem in 3-D, called **min-d (δ, t)-OTC**, is also useful: Given an integer $\kappa > 0$, find at most κ (δ, t)-blocks to build the terrain $T(\cdot)$, such that the maximum distance d between any two consecutive elementary blocks in the κ (δ, t)-blocks is minimized.

Optimal regular terrain construction with δ-building-blocks in 3-D (δ-OTC): Given a regular terrain $T(\cdot)$ on a 2-D regular grid RG with n columns of grid cells, and an integer separation value $\delta > 0$, find k δ-blocks to build the terrain $T(\cdot)$ such that k is minimized. An interesting special case of the δ-OTC problem is the following 2-D version called **monotone partition using vertical segments** (MPUVS): Given an arbitrary polygon $\mathcal{P}$ (possibly with holes), partition $\mathcal{P}$ into the minimum number of x-monotone subpolygons by adding to $\mathcal{P}$ only vertical segments.

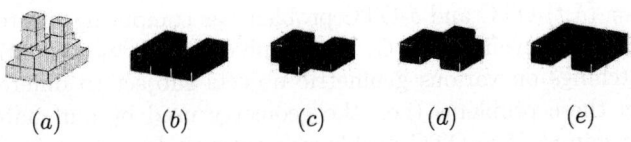

(a)   (b)   (c)   (d)   (e)

Fig. 1. (a) A regular terrain. (b) – (e) Four δ-blocks for building the terrain in (a), with $\delta = 1$

Geometrically, a terrain construction problem can also be viewed as a partition problem: Given a regular terrain $\mathcal{T}(\cdot)$, partition it into (say) the minimum number of specified geometric objects (e.g., the building blocks). For example, each δ-block forms a rectilinear monotone polygon whose "width" is at least δ with respect to a line (e.g., the x-axis). Hence, the 2-D case of the 3-D δ-OTC problem (with the height of each cell of $\mathcal{T}(\cdot)$ being either 1 or 0) is a problem of partitioning a rectilinear polygon (possibly with holes) into the minimum number of this kind of x-monotone polygons.

The above optimal regular terrain construction problems arise in *intensity-modulated radiation therapy* (IMRT). IMRT uses radiation beams to eradicate tumors while sparing surrounding critical structures and healthy tissues. A *multileaf collimator* (MLC) can help deliver intensity-modulated radiation in a dynamic or static fashion, called *dynamic leaf sequencing* or *static leaf sequencing* in medical literature. Actually, the min-# (δ, t)-OTC problem models the dynamic leaf sequencing problem, and the δ-OTC problem models the static leaf sequencing problem; the special structures of the building blocks of each OTC problem reflect the specific medical constraints of the corresponding leaf sequencing problem. Section 6 will discuss in greater detail how these OTC problems arise in medical applications.

Previous work on these terrain construction problems is mainly concerned with leaf sequencing problems. Previous methods for the dynamic leaf sequencing problem (modeled by the min-# (δ, t)-OTC problem) are all heuristics [9, 10, 21, 23, 25, 26]. They attempt to somehow choose a set of elementary blocks for the sub-terrain defined on each column of $\mathcal{RG}$ (see Figure 2(c)) to build that sub-terrain, and then from these elementary blocks form a set of feasible (δ, t)-blocks for building the whole terrain. But, these methods guarantee no good quality of the output (δ, t)-building blocks, and may have a daunting execution time. Likewise, previous methods for the static leaf sequencing problem (modeled by the δ-OTC problem) are also all heuristics [1, 2, 8, 11, 19, 22]. Related to the MPUVS problem, there are geometric results on optimal partition of a polygonal domain into monotone parts [15, 16, 17]. The monotone partitions in [15, 16, 17] are on a simple polygon, and may use non-vertical segments and Steiner points.

Although no proofs have been known, we believe that the general OTC problems in 3-D are computationally intractable. Due to their useful roles in medical practice, it makes sense for us to develop practically efficient OTC algorithms that guarantee a good quality of the output solutions.

We present a novel unified approach based on geometric techniques for solving the (δ, t)-OTC and δ-OTC problems. One of our key ideas is to formulate the min-# (δ, t)-OTC and δ-OTC problems as computing shortest paths in a weighted directed acyclic graph G. The graph G is built by computing optimal bipartite matchings on various geometric objects subject to different medical constraints of these problems (i.e., the geometry used by our matching algorithm for the min-# (δ, t)-OTC problem is different from that for the δ-OTC problem). Further, since we need to compute optimal bipartite matchings on many sets of geometric objects, we use techniques for computing such matchings

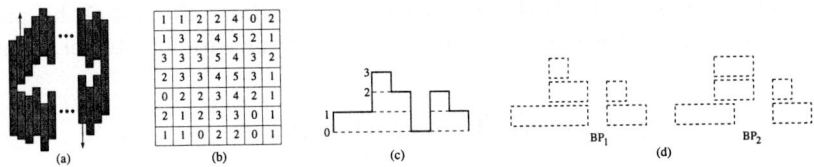

Fig. 2. (a) An MLC. (b) A regular terrain on a 2-D grid RG. (c) A sub-terrain on column one of RG (defined by the solid rectilinear curve). (d) The two block patterns of the sub-terrain in (c)

in a *batch fashion* to speed up these matching computations. In addition, we use some implicit graph and path representation schemes to significantly reduce the space bounds of our algorithms without (greatly) affecting their time bounds.

Consequently, we achieve the first 3-D (δ, t)-OTC and δ-OTC algorithms that guarantee the optimal quality of their output solutions and are practically fast. We implemented our (δ, t)-OTC and δ-OTC algorithms and tested them on real medical data. Our implementation results show that our OTC algorithms run very fast on real medical data (all under one second).

2 Preliminaries

To help present our algorithms, we first discuss some geometric structures. For the given regular terrain $\mathcal{T}(\cdot)$ on a 2-D grid RG, let $\mathcal{T}_i$ denote the sub-terrain of $\mathcal{T}(\cdot)$ defined on the i-th column RG_i of RG. Clearly, we can use a 2-D rectilinear monotone curve to represent $\mathcal{T}_i$, as shown in Figure 2(c).

We need to determine the minimum number of elementary blocks (i.e., intervals), called the *interval number* IN_i, that are used to build each $\mathcal{T}_i$ on RG_i. Suppose we perform a horizontal trapezoidal decomposition of the region under the curve $\mathcal{T}_i$ using all points on $\mathcal{T}_i$ that have an integer y-coordinate (e.g., see Figure 2(c)); then IN_i is the number of rectangles (i.e., blocks) in this trapezoidal decomposition. In Figure 2(c), $IN_i = 5$. For a sub-terrain $\mathcal{T}_i$ with an interval number IN_i, a *block pattern* of $\mathcal{T}_i$ is a group of IN_i intervals (i.e., blocks) on the 1-D RG_i that together build $\mathcal{T}_i$. Observe that a $\mathcal{T}_i$ can have many distinct block patterns (e.g., the sub-terrain in Figure 2(c) has two different block patterns in Figure 2(d)). In the worst case (e.g., a $\mathcal{T}_i$ with the shape of a Tower of Hanoi), $\mathcal{T}_i$ can have $IN_i!$ distinct patterns; but, this is not a common case in medical data. Let Π_i denote the total number of distinct *legal patterns* of $\mathcal{T}_i$ (i.e., those that are correct patterns of $\mathcal{T}_i$ and the length of each interval in such a pattern is $\geq \delta$). We order all distinct legal patterns of each $\mathcal{T}_i$ in an arbitrary way, and denote the k-th pattern in this order by $BP_i(k)$.

Clearly, each legal pattern of $\mathcal{T}_i$ specifies an optimal set of elementary blocks for building the sub-terrain $\mathcal{T}_i$ of $\mathcal{T}(\cdot)$. Of course, one can use more than IN_i elementary blocks (intervals) to build the sub-terrain $\mathcal{T}_i$. Such intervals can be obtained by splitting the IN_i intervals of a legal pattern of $\mathcal{T}_i$ into more intervals.

The splitting points thus used are called *Steiner points*. In the rest of this paper, we focus on solving the OTC problems using patterns without Steiner points.

3 The Min-# (δ, t)-OTC Algorithm

Given a regular terrain $\mathcal{T}(\cdot) = \{\mathcal{T}_1, \ldots, \mathcal{T}_n\}$ defined on a 2-D grid RG, we say that a set of (δ, t)-blocks that build $\mathcal{T}(\cdot)$ forms a (δ, t)-*building-block set* (called (δ, t)-*block set*). We formulate the min-# (δ, t)-OTC problem as finding a shortest path in a weighted directed acyclic graph (DAG) $G = (V, E)$, and build the DAG G by computing maximum cardinality matchings in a set of bipartite graphs. Since we need to compute numerous such maximum bipartite matchings, it is crucial to use an efficient matching algorithm. We exploit the geometric structures of the problem to speed up the matching computations.

Our algorithm has two key steps: (1) build a weighted DAG $G = (V, E)$ to capture the structures of all possible (δ, t)-block sets; (2) find a shortest path in G, which defines an optimal (δ, t)-block set.

3.1 The Graph Construction

To build the DAG G, we need to characterize the structures of the (δ, t)-block sets for $\mathcal{T}(\cdot)$. Let $P = \{R_1, R_2, \ldots, R_q\}$ denote a (δ, t)-block set with q (δ, t)-blocks. Recall that each (δ, t)-block R_i consists of a sequence of non-empty intervals for the $\mathcal{T}_j$'s on some consecutive columns RG_j's of RG, with exactly one interval of R_i per $\mathcal{T}_j$ that is spanned by R_i and the length of each such interval $\geq \delta$. Denote by $I(R_i, \mathcal{T}_j)$ the interval for $\mathcal{T}_j$ in a (δ, t)-block R_i. If R_i does not span $\mathcal{T}_j$, then let $I(R_i, \mathcal{T}_j) = \varnothing$. The interval set $P(\mathcal{T}_j) = \{I(R_1, \mathcal{T}_j), I(R_2, \mathcal{T}_j), \ldots, I(R_q, \mathcal{T}_j)\}$ for a (δ, t)-block set P then forms a legal block pattern of $\mathcal{T}_j$.

Lemma 1. *A (δ, t)-block set P contains exactly one pattern of each sub-terrain $\mathcal{T}_i$ of $\mathcal{T}(\cdot)$.*

Consider two consecutive patterns in P, say $P(\mathcal{T}_j)$ and $P(\mathcal{T}_{j+1})$ for $\mathcal{T}_j$ and $\mathcal{T}_{j+1}$, respectively. Let I_1 (resp., I_2) be an interval $[l_1, r_1]$ (resp., $[l_2, r_2]$). Recall that the distance $dist(I_1, I_2)$ between I_1 and I_2 is defined as $\max\{|l_1 - l_2|, |r_1 - r_2|\}$. Let $P(\mathcal{T}_j)$ (resp., $P(\mathcal{T}_{j+1})$) be the k_j-th (resp., k_{j+1}-th) pattern $BP_j(k_j)$ (resp., $BP_{j+1}(k_{j+1})$) of $\mathcal{T}_j$ (resp., $\mathcal{T}_{j+1}$). For the given threshold t, we define a bipartite graph $G_b^{[t]}(j, k_j, k_{j+1}) = (A, B, E)$ on $P(\mathcal{T}_j)$ and $P(\mathcal{T}_{j+1})$, such that the vertex set A (resp., B) represents the intervals of $P(\mathcal{T}_j)$ (resp., $P(\mathcal{T}_{j+1})$). An edge in $G_b^{[t]}(j, k_j, k_{j+1})$ connects $u \in A$ and $v \in B$ if and only if $dist(I_u, I_v) \leq t$, where I_u (resp., I_v) is the interval for u (resp., v).

A *matching* in a graph H is a subset M of edges in H such that no two distinct edges of M share a common vertex. For convenience, we also view a matching in $G_b^{[t]}(j, k_j, k_{j+1})$ as a matching of $BP_j(k_j)$ and $BP_{j+1}(k_{j+1})$. Let $\mathcal{T}_j \propto_P \mathcal{T}_{j+1}$ denote the pair set $\{(I(R_i, \mathcal{T}_j), I(R_i, \mathcal{T}_{j+1})) \mid I(R_i, \mathcal{T}_j) \neq \varnothing, I(R_i, \mathcal{T}_{j+1}) \neq \varnothing, i = 1, 2, \ldots, q\}$ defined by P. Then each $(I(R_i, \mathcal{T}_j), I(R_i, \mathcal{T}_{j+1})) \in \mathcal{T}_j \propto_P$

$\mathcal{T}_{j+1}$ is an edge in $G_b^{[t]}(j, k_j, k_{j+1})$. Note that for two distinct (δ, t)-blocks R_{i1} and R_{i2} in P, $I(R_{i1}, \mathcal{T}_j)$ and $I(R_{i2}, \mathcal{T}_j)$ are different interval elements in the pattern $P(\mathcal{T}_j)$. Thus, $\mathcal{T}_j \propto_P \mathcal{T}_{j+1}$ forms a matching of $P(\mathcal{T}_j)$ and $P(\mathcal{T}_{j+1})$. Lemma 2 further explores a key property of $\mathcal{T}_j \propto_P \mathcal{T}_{j+1}$.

Lemma 2. *For an optimal (δ, t)-block set P^*, each pair set $\mathcal{T}_j \propto_{P^*} \mathcal{T}_{j+1}$ is a maximum cardinality matching of the corresponding bipartite graph $G_b^{[t]}(j, k_j, k_{j+1})$, and replacing $\mathcal{T}_j \propto_{P^*} \mathcal{T}_{j+1}$ by any other maximum matching of $G_b^{[t]}(j, k_j, k_{j+1})$ does not change the optimality of the resulted (δ, t)-block set.*

Lemma 2 implies that the global optimality of a (δ, t)-block set can be captured locally by a maximum cardinality matching in each such bipartite graph $G_b^{[t]}(j, k_j, k_{j+1})$. Based on Lemmas 1 and 2, if one could somehow find the "right" patterns for an optimal (δ, t)-block set P^*, one pattern per $\mathcal{T}_j$, say $BP_1(k_1), BP_2(k_2), \ldots, BP_n(k_n)$ (recall that $BP_j(k_j)$ denotes the k_j-th block pattern of $\mathcal{T}_j$), then P^* can be computed as follows: For each $j = 1, 2, \ldots, n-1$, build a bipartite graph $G_b^{[t]}(j, k_j, k_{j+1})$; compute a maximum matching M_j in each $G_b^{[t]}(j, k_j, k_{j+1})$. The optimal (δ, t)-block set P^* is then defined by the maximum matchings $M_1, M_2, \ldots, M_{n-1}$. Let $(I', I'') \in M_j$ for intervals $I' \in BP_j(k_j)$ and $I'' \in BP_{j+1}(k_{j+1})$. Then I' and I'' are in the same (δ, t)-block of P^*. Thus, $P^* = \{BP_1(k_1), BP_2(k_2), \ldots, BP_n(k_n)\}$ forms an optimal (δ, t)-block set.

Hence, how to pick the right patterns for an optimal (δ, t)-block set, one pattern per sub-terrain $\mathcal{T}_j$, is a key to our min-# (δ, t)-OTC algorithm. Suppose an optimal (δ, t)-block set P^* contains patterns $BP_j(k_j)$ and $BP_{j+1}(k_{j+1})$ and M_j is a maximum matching of $BP_j(k_j)$ and $BP_{j+1}(k_{j+1})$. We say that a (δ, t)-block R_i of P^* is *introduced* by a pattern $BP_j(k_j)$ in P^* if the first interval of R_i is in $BP_j(k_j)$. Clearly, $BP_1(k_1)$ in P^* introduces IN_1 (δ, t)-blocks of P^*. The following lemma is a cornerstone to construct the desired graph G.

Lemma 3. *For an optimal (δ, t)-block set $P^* = \{BP_1(k_1), BP_2(k_2), \ldots, BP_n(k_n)\}$, $BP_1(k_1)$ introduces IN_1 (δ, t)-blocks of P^*, and $BP_{j+1}(k_{j+1})$ introduces $(IN_{j+1} - |M_j|)$ (δ, t)-blocks of P^* for each $j = 1, 2, \ldots, n-1$.*

Based on Lemmas 1 and 3, we construct the weighted DAG $G = (V, E)$ as follows. Every pattern $BP_j(k)$ of each sub-terrain $\mathcal{T}_j$ on RG_j corresponds to a vertex in V, simply denoted by $BP_j(k)$, for $j = 1, 2, \ldots, n$ and $k = 1, 2, \ldots, \Pi_j$. For any pair of patterns $BP_j(k)$ and $BP_{j+1}(k')$ of two consecutive sub-terrains, we put a directed edge $(BP_j(k), BP_{j+1}(k'))$ in E with a weight $(IN_{j+1} - |M_j(k, k')|)$, where $M_j(k, k')$ is a maximum matching between $BP_j(k)$ and $BP_{j+1}(k')$. In addition, we introduce two special vertices s_0 and s_1 to G. From s_0, a directed edge in E goes to each pattern of $\mathcal{T}_1$ (i.e., $(s_0, BP_1(k)) \in E$ for $k = 1, 2, \ldots, \Pi_1$) with a weight IN_1. Each pattern $BP_n(k)$ of $\mathcal{T}_n$ has a directed edge $(BP_n(k), s_1)$ to s_1 with a weight 0. Then, by computing a shortest path from s_0 to s_1 in G, we obtain an optimal (δ, t)-block set P^* with the minimum number of (δ, t)-blocks.

Lemma 4. *An s_0-to-s_1 shortest path SP in the DAG $G = (V, E)$ defines an optimal (δ, t)-block set P^*.*

To construct G, we need to compute a maximum matching in each bipartite graph $G_b^{[t]}(j, k, k')$, for $j = 1, 2, \ldots, n-1$, $k = 1, 2, \ldots, \Pi_j$, and $k' = 1, 2, \ldots, \Pi_{j+1}$. The best known algorithm for finding a maximum matching in a general N-vertex bipartite graph takes $O(N^{2.5})$ time [14]. For each edge $(BP_j(k), BP_{j+1}(k'))$ in G, we compute a maximum matching in the corresponding bipartite graph $G_b^{[t]}(j, k, k')$. Let m be the maximum interval number among all sub-terrains $\mathcal{T}_j$'s (i.e., $m = \max_{j=1}^n IN_j$). The total time for computing the maximum matchings in all such bipartite graphs is $O(|E| \cdot m^{2.5})$. Finding an s_0-to-s_1 shortest path in G takes $O(|E|)$ time. Since $|E| = O(\sum_{i=1}^{n-1} \Pi_i \Pi_{i+1})$, the time bound of our min-# (δ, t)-OTC algorithm is $O((\sum_{i=1}^{n-1} \Pi_i \Pi_{i+1}) \cdot m^{2.5})$.

3.2 Geometric Matchings

We now exploit the geometry of the min-# (δ, t)-OTC problem for computing maximum matchings. For a value $r > 0$, let $H[r] = (A_h, B_h, E_h)$ denote an N-vertex geometric bipartite graph, where A_h and B_h are each a set of points in the plane, and E_h consists of all edges (a, b) for $a \in A_h$ and $b \in B_h$ such that the distance $dist_h(a, b) \leq r$ between a and b under a given L_p metric. The problem seeks a maximum cardinality matching M_h in $H[r]$.

Lemma 5. *(Efrat, Itai, and Katz [13]) For an N-vertex geometric bipartite graph $H[r]$ and a matching M in $H[r]$, all augmenting paths with respect to M can be found in $O(N \log N)$ time. Further, a maximum cardinality matching M^* in $H[r]$ can be computed in $O(N^{1.5} \log N)$ time.*

Geometry can help the maximum matching computations in our min-# (δ, t)-OTC algorithm. Consider two block patterns $BP_j(k)$ and $BP_{j+1}(k')$ of two consecutive sub-terrains. For each interval $I = [l, r] \in BP_j(k)$ (resp., $I' = [l', r'] \in BP_{j+1}(k')$), we map it to a point $(l, r) \in A_g(k)$ (resp., $(l', r') \in B_g(k')$) in the plane. The L_∞ distance $L_\infty(p_1, p_2)$ between two points $p_1 = (x_1, y_1)$ and $p_2 = (x_2, y_2)$ is $\max\{|x_1 - x_2|, |y_1 - y_2|\}$.

Thus, we map each bipartite graph $G_b^{[t]}(j, k, k')$ to a geometric bipartite graph $G_g^{[t]}(j, k, k') = (A_g(k), B_g(k'), E_g(k, k'))$ in the plane, where $E_g(k, k')$ consists of all pairs (p, p') such that $p \in A_g(k)$, $p' \in B_g(k')$, and $L_\infty(p, p') \leq t$. Based on Lemmas 5, we obtain the following key result.

Lemma 6. *A maximum matching of any two block patterns $BP_j(k)$ and $BP_{j+1}(k')$ can be computed in $O(m^{1.5} \log m)$ time.*

Theorem 1. *The min-# (δ, t)-OTC problem can be solved in $O((\sum_{i=1}^{n-1} \Pi_i \Pi_{i+1}) \cdot m^{1.5} \log m)$ time.*

3.3 Accelerating the Matching Computations Based on Batching

Since we need to compute many maximum geometric matchings, we develop techniques for computing such matchings in a *batch fashion* to accelerate the matching computations. Our batching idea is: For a given pattern $BP_{j+1}(k')$ of $\mathcal{T}_{j+1}$, group together for $\mathcal{T}_j$ the computations of the maximum matchings for all pairs of $BP_j(k)$ and $BP_{j+1}(k')$, $k = 1, 2, \ldots, \Pi_j$.

Let M^* be a maximum matching of $BP_j(k)$ and $BP_{j+1}(k')$, and consider computing a maximum matching for $BP_{j+1}(k')$ and another pattern of $\mathcal{T}_j$, say $BP_j(k'')$. Let $G_g^{[t]}(j, k'', k')$ be the geometric bipartite graph for $BP_j(k'')$ and $BP_{j+1}(k')$, and $A_g(k'')$ be the point set representing $BP_j(k'')$. Define the *difference* $diff(A_g(k), A_g(k''))$ between $A_g(k)$ and $A_g(k'')$ as $(|A_g(k)| - |A_g(k) \cap A_g(k'')|)$ (note that $|A_g(k)| = |A_g(k'')|$). Let M' be the edge set after removing all edges in M^* incident to points in $A_g(k) \setminus A_g(k'')$. Thus, the total number of needed augmenting phases for obtaining a maximum matching of $G_g^{[t]}(j, k'', k')$ from the matching M' is $O(diff(A_g(k), A_g(k'')))$. In each phase, we compute an augmenting path in $G_g^{[t]}(j, k, k')$ for the matching obtained in the previous phase, in $O(m \log m)$ time by Lemma 5.

For each pattern $BP_{j+1}(k')$ of $\mathcal{T}_{j+1}$, we arrange into a sequence the matching computations involving $\mathcal{T}_j$ for all pairs of $BP_j(k)$ and $BP_{j+1}(k')$, $k = 1, 2, \ldots, \Pi_j$, such that the total number of augmenting phases used in that sequence for computing all the Π_j maximum matchings is minimized. We first build a weighted complete graph $G_{mst}(j) = (V_{mst}(j), E_{mst}(j))$, whose vertices represent the Π_j patterns of $\mathcal{T}_j$. Each edge $(u, v) \in E_{mst}(j)$ has a weight $diff(A_g(u), A_g(v))$, where $A_g(w)$ is the point set corresponding to the pattern of $\mathcal{T}_j$ represented by the vertex w of $G_{mst}(j)$, denoted by $BP_j(w)$. Next, by using Chazelle's algorithm [4], we compute a minimum spanning tree $MST(j)$ of $G_{mst}(j)$ in $O(\Pi_j^2 \alpha(\Pi_j^2, \Pi_j))$ time, where $\alpha(\cdot, \cdot)$ is the functional inverse of Ackermann's function. Then, traverse $MST(j)$ in a depth-first search manner, and for each vertex v of $MST(j)$ visited, compute a maximum matching of $BP_j(v)$ and $BP_{j+1}(k')$ using M^* in $O(diff(A_g(v), A_g(u)) \cdot m \log m)$ time, where u is the vertex of $MST(j)$ visited right before v and M^* is a maximum matching of $BP_j(u)$ and $BP_{j+1}(k')$.

Let $W(j)$ denote the total weight of $MST(j)$, called the *MST-difference* of $\mathcal{T}_j$. Note that $W(j) = \Omega(\Pi_j)$. Therefore, the total time to compute all maximum matchings for the whole DAG G is $O(\sum_{i=1}^{n-1} \Pi_{i+1}(W(i) + m^{0.5}) \cdot m \log m)$.

Theorem 2. *The min-# (δ, t)-OTC problem is solvable in $T = \min\{O((\sum_{i=1}^{n-1} \Pi_i \Pi_{i+1}) \cdot m^{1.5} \log m), O(\sum_{i=1}^{n-1} \Pi_{i+1}(W(i) + m^{0.5}) \cdot m \log m)\}$ time.*

3.4 Reducing the Space Bound

The algorithm in Sections 3.1–3.3 separates the construction of the graph G from computing a shortest path in G, thus having to store G explicitly. In extreme cases, this algorithm may suffer severe space problems since explicitly storing G

and all matchings for G uses $O(m \sum_{i=1}^{n-1} \Pi_i \Pi_{i+1})$ space. By exploiting the specific structures of the DAG G and using an implicit graph representation scheme and the actual path reporting technique in [5], an actual path SP in G can be output in a substantially smaller space bound and (almost) the same time bound. The details are left to the full paper.

Theorem 3. *The min-# (δ, t)-OTC problem can be solved in $O(\frac{T \cdot \log n}{\log(h+1)})$ time and $O(\Pi^2 + h\Pi)$ space, where T is defined in Theorem 2 and h is any integer with $1 \leq h \leq n$.*

4 The Min-d (δ, t)-OTC Algorithm

We first consider the set $\mathcal{D}$ of all possible values for t^*. Since a $\mathcal{T}_j$ defines $O(m^2)$ possible intervals, there are seemingly $O(m^4)$ such distances for each pair of $\mathcal{T}_j$ and $\mathcal{T}_{j+1}$. But, a more careful analysis shows that any $\mathcal{T}_j$ and $\mathcal{T}_{j+1}$ actually define only $O(m^2)$ possible distances for t^*. Since each $dist(I, I') = \max\{|l - l'|, |r - r'|\}$ is decided either by the left endpoints of I and I' or by their right endpoints, it is sufficient to consider all possible pairs of left (resp., right) endpoints of the intervals for $\mathcal{T}_j$ and $\mathcal{T}_{j+1}$. The intervals for each $\mathcal{T}_j$ have only $O(m)$ left (resp, right) endpoints. Thus, each pair of $\mathcal{T}_j$ and $\mathcal{T}_{j+1}$ defines $O(m^2)$ possible distances for t^*, and the set $\mathcal{D}$ has $O(nm^2)$ values.

Next, by using an implicit scheme, these $O(nm^2)$ values for t^* can be stored in $O(nm)$ time and space (this part is left to the full paper). Hence, the binary search process of our min-d (δ, t)-OTC algorithm has $O(\log(nm))$ stages, each stage calling our min-# (δ, t)-OTC algorithm once.

Theorem 4. *The min-d (δ, t)-OTC problem can be solved in $O(\frac{T \cdot \log n \cdot \log(nm)}{\log(h+1)})$ time and $O(\Pi^2 + h\Pi + nm)$ space, where T and h are defined as in Theorem 3.*

5 The δ-OTC and MPUVS Algorithms

This section first shows our δ-OTC algorithm. We solve this problem by using the same framework as that for the min-# (δ, t)-OTC problem in Section 3. But, the geometry used by our maximum matching algorithm for the δ-OTC problem is different from that for the min-# (δ, t)-OTC problem.

As in Section 3, we divide the terrain $\mathcal{T}(\cdot)$ into n sub-terrains $\mathcal{T}_1, \mathcal{T}_2, \ldots, \mathcal{T}_n$, with $\mathcal{T}_j$ defined on the j-th column RG_j of RG. Note that for any two consecutive intervals I and I' in the same δ-block, $|I \cap I'| \geq \delta$. For any two patterns $BP_j(k)$ and $BP_{j+1}(k')$ of two consecutive sub-terrains $\mathcal{T}_j$ and $\mathcal{T}_{j+1}$, we construct a bipartite interval graph $G_I(j, k, k') = (A_I(k), B_I(k'), E_I(k, k'))$. The vertices of $A_I(k)$ (resp., $B_I(k')$) represent the intervals in $BP_j(k)$ (resp., $BP_{j+1}(k')$), and an edge $(u, v) \in E_I(k, k')$ connects $u \in A_I(k)$ and $v \in B_I(k')$ iff $|I_u \cap I_v| \geq \delta$, where I_u (resp., I_v) is the interval represented by u (resp., v). Let m be the maximum interval number of all $\mathcal{T}_j$'s. Note that a maximum matching in $G_I(j, k, k')$ can be obtained in $O(m \log \log m)$ time and $O(m)$ space, if the interval endpoints are already given sorted [7]. Note that the endpoints of the intervals in each $BP_j(k)$ are in sorted order since they are from the vertices of the monotone curve $\mathcal{T}_j$.

Theorem 5. *The δ-OTC problem can be solved in $O(\frac{(\sum_{i=1}^{n-1} \Pi_i \Pi_{i+1}) \cdot m \log \log m \cdot \log n}{\log(h+1)})$ time and $O(h\Pi)$ space, where h is any integer with $1 \leq h \leq n$.*

The $O(n \log n)$ time MPUVS algorithm is based on a careful "simulation" of the above 3-D δ-OTC scheme. We leave this algorithm to the full paper.

6 Applications in IMRT

The terrain construction problems that we study arise in radiation treatment planning. Radiation therapy is a major modality for cancer treatments. It uses a set of focused radiation beams to eradicate tumors while sparing surrounding critical structures and healthy tissues [20].

Currently, a device called *multileaf collimator* (MLC) plays an important role in the *intensity-modulated radiation therapy* (IMRT) and can be used to deliver intensity-modulated beams in a dynamic or static fashion [3, 23, 24]. An MLC has multiple pairs of opposite leaves of the same rectangular shape and size (see Figure 2(a)). The leaves can move up and down to form a rectilinear region that is monotone to (say) the x-axis. The cross-section of a cylindrical radiation beam is shaped by this rectilinear monotone region. A common constraint requires such a region to be connected. Another constraint is called the *minimum leaf separation* [8]: If the beam shaping region is not closed at any two opposite MLC leaves of the same pair or two neighboring pairs, then the distance between these two leaves must be $\geq$ a specified separation value δ.

An effective dynamic IMRT method proposed by Yu [23], called *intensity-modulated arc therapy* (IMAT) [9, 10, 21, 25, 26], works as follows. A turned-on beam source is moved along a given *beam path* (typically, a 3-D circular arc around the patient). The beam shape can be changed by an MLC as the beam source is moving. A sequence of the *beam's eye views* is specified on the beam path (usually, every 10 degrees per view). For the i-th view, there is a 2-D tumor region TR_i (as projected on a plane orthogonal to that view) on which a radiation dose distribution $S_i(\cdot)$ is prescribed. Actually, $S_i(\cdot)$ can be viewed as a regular terrain defined on a regular grid on TR_i (e.g., see Figure 2(b)). Let C_i be the center point of TR_i. The beam usually moves in a constant speed. Thus, one may assume each traversal of the beam path by a turned-on beam source delivers a unit dose (called an *intensity level*) to a simple rectilinear x-monotone polygon BR_i on each visited TR_i such that BR_i is the cross-section of the shaped cylindrical beam on TR_i centered at C_i. The beam source may travel the beam path multiple times to deliver the whole distributions. A traversal of a continuous portion of the beam path by a turned-on beam source is called an *arc* [23]. When the beam goes from C_i to C_{i+1}, the beam shape is changed from a BR_i to a BR_{i+1} by moving the MLC leaves up or down. A key constraint is: The maximum moving distance of each leaf for changing any BR_i to BR_{i+1} must be $\leq$ a given "threshold" t. If a leaf must move a distance $> t$ to change a BR_i to BR_{i+1} on an arc, then possibly this leaf movement cannot be finished in the time when the beam source goes from C_i to C_{i+1}. If this occurs, the beam source must be turned off while going from C_i to C_{i+1} (a highly undesirable treatment interruption), and the arc is *infeasible* and must be cut into two arcs between TR_i and TR_{i+1}. A key problem here is to find minimum k feasible arcs (each arc is a sequence of consecutive BR_i's) to deliver the dose distributions. Note that minimizing the number of arcs used means minimizing the treatment interruptions. An important basic case of this problem is called 2-D **dynamic leaf sequencing** (DLS): Only one MLC leaf pair (instead of multiple pairs) is considered and the intensity surface $S_i(\cdot)$ on each (1-D) TR_i is a 2-D rectilinear x-monotone curve called *intensity profile* [23] (e.g., see Figure 2(c)).

It is now easy to see that our min-# (δ, t)-OTC problem is equivalent to the 2-D DLS problem. Each (1-D) TR_i corresponds to a column RG_i of a 2-D regular grid RG. Each sub-terrain T_i of $T(\cdot)$ corresponds to an intensity profile $S_i(\cdot)$ defined on TR_i. Each arc corresponds to a (δ, t)-block. An optimal (δ, t)-block set for building $T(\cdot)$ gives an optimal set of arcs for the 2-D DLS problem.

Static IMRT methods are also used in common practice [8, 11, 22]. In the static setting, only one 2-D tumor region TR is considered and the beam source does not move. A set of beam shaping regions SR_i (called *segments* in medical literature) is used to deliver a prescribed dose distribution (i.e., an intensity surface) $S(\cdot)$ on TR. Each segment SR_i is a simple rectilinear x-monotone polygon on TR that satisfies the minimum leaf separation constraint (specified by δ). When the MLC leaves move to change from one SR_i to another SR_{i+1}, the beam source is turned off. Since the overhead associated with turning the beam source on/off, leaf moving, and verification dominates the total treatment time [11, 22], it must be minimized. This gives rise to the 3-D **static leaf sequencing** (SLS) problem: Given a prescribed dose distribution $S(\cdot)$ on a 2-D tumor region TR, find k segments SR_i on TR for the dose delivery, such that k is minimized. Clearly, our δ-OTC problem is equivalent to the 3-D SLS problem.

7 Implementation and Experiments

We implemented the 3-D min-# (δ, t)-OTC and δ-OTC algorithms using C. The maximum matching subroutine was obtained from DIMACS. Another key component is that of generating all distinct legal block patterns for each sub-terrain T_i. We incorporated our optimal output-sensitive algorithm for the pattern generation into our OTC programs. Our algorithms were tested on a Sun Blade 1000 Workstation using an UltraSparc III Processor with 600MHz and 512 Megabytes of memory. We conducted experiments for both the DLS and SLS algorithms on dose prescriptions for many real medical cases as well as randomly generated dose distributions on tumor regions of various sizes.

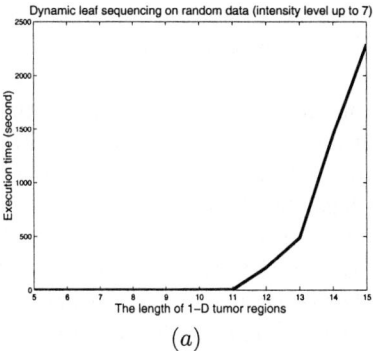
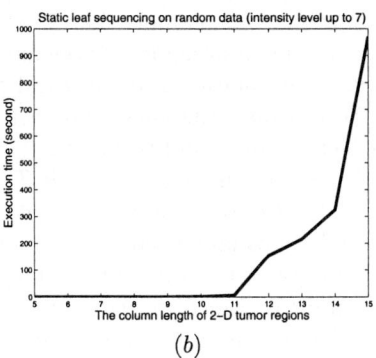

Fig. 3. (a) The execution time versus the length of 1-D tumor region on random dose prescriptions for the DLS problem. (b) The execution time versus the column length of 2-D tumor region on random dose prescriptions for the SLS problem

We tested our min-# (δ, t)-OTC algorithm for the DLS problem on 30 real medical data sets, and our δ-OTC algorithm for the SLS problem on 80 real medical data sets. The optimal output quality of our OTC algorithms was also compared with the previous leaf sequencing heuristics. Our experiments showed that the execution times of our OTC algorithms on real medical data sets are very fast, all under few seconds for a typical prescription with 32 intensity profiles (i.e., sub-terrains $\mathcal{T}_i$). We simulated a sliding-window based DLS approach and a sliding-window based SLS algorithm on real medical data sets. On average, our min-# (δ, t)-OTC algorithm generates 10% less arcs than the DLS approach, and our δ-OTC algorithm produces 15% less segments than the SLS algorithm.

We observe that in common medical data, the maximum intensity level of an intensity surface (i.e., terrain $\mathcal{T}$) for IMAT ranges from 3 to 5, the size of a 2-D tumor region TR ranges from 5 × 5 to 10 × 10, and the number of eye views on a beam path is no more than 32 (for the DLS problem). To further study the feasibility of our OTC algorithms, we tested them on many randomly generated dose prescriptions. Figure 3(a) (resp., Figure 3(b)) gives a plot of the average execution time (in seconds) over 20 runs versus the length of the 1-D TR_i (resp., the column length of the 2-D TR) for the DLS (resp., SLS) problem. The maximum intensity level of each terrain is 7 (which is already higher than the maximum intensity levels seen in common IMAT prescriptions), and the intensity levels of its cells are randomly generated. Each DLS dose prescription uses 32 intensity profiles (one per eye view) on a beam path. For the SLS problem, we considered 2-D tumor regions of a square shape. Figures 3(a) and 3(b) show that the execution times of our OTC algorithms grow rapidly with respect to the length of tumor regions with random data. This is because, as the length of an intensity profile (i.e., sub-terrain) increases, the number of its block patterns increases very quickly. But, the figures also indicate that when the length of a tumor region is no more than 11 (this includes the majority of the real medical prescriptions), our algorithms generally finish in a few seconds. This corresponds to our experiments on real medical data.

We attribute the amazingly fast execution times of our OTC algorithms on real medical data to two factors: (1) In real medical prescriptions, the interval number IN_i of an intensity profile T_i is usually very small, typically ranging from 3 to 5; (2) the total number Π_i of distinct legal block patterns of an intensity profile T_i is usually small too, rarely reaching hundreds. On average, each Π_i is in the order of tens. An interesting observation is that the intervals of a block pattern for real medical data tend to share left or right endpoints considerably. Note that an intensity profile with shared interval endpoints has less *distinct* legal block patterns. Thus, the actual execution times of our OTC algorithms mostly are significantly better than their theoretical worst case bounds.

References

[1] T. R. Bortfeld, J. Stein, K. Preiser, and K. Hartwig, Intensity Modulation for Optimized Conformal Therapy, *Proc. Symp. Principles and Practice of 3-D Radiation Treatment Planning*, Munich, 1996.

[2] A. L. Boyer, Use of MLC for Intensity Modulation, *Med. Phys.*, 21 (1994), p. 1007.

[3] A. L. Boyer and C. X. Yu, Delivery of Intensity-Modulated Radiation Therapy Using Dynamic Multileaf Collimator, *Seminar in Radiat. Oncol.*, 9 (2) (1999).

[4] B. Chazelle, A Minimum Spanning Tree Algorithm with Inverse-Ackermann Type Complexity, *Journal of the ACM*, 47 (2000), pp. 1028–1047.

[5] D. Z. Chen, O. Daescu, X. Hu, and J. Xu, Finding an Optimal Path without Growing the Tree, accepted to the Special Issue of *J. of Algorithms* on Selected Papers from the *6th Annual European Symp. on Algorithms (ESA '98)*.

[6] D. Z. Chen, X. Hu, and X. Wu, Optimal Polygon Cover Problems and Applications, *Proc. 11th Annual International Symp. on Algorithms and Computation*, 2000, *Lecture Notes in Computer Science*, Vol. 1969, Springer Verlag, pp. 564–576.

[7] D. Z. Chen, X. Hu, and X. Wu, Maximum Red/Blue Interval Matching with Applications, *Proc. 7th Annual International Computing and Combinatorics Conf.*, 2001, pp. 150–158.

[8] D. J. Convery and S. Webb, Generation of Discrete Beam-Intensity Modulation by Dynamic Multileaf Collimation under Minimum Leaf Separation Constraints, *Phys. Med. Biol.*, 43 (1998), pp. 2521–2538.

[9] C. Cotrutz, C. Kappas, and S. Webb, Intensity Modulated Arc Therapy (IMAT) with Multi-Isocentric Centrally Blocked Rotational Fields, *Phys. Med.*, 15 (1999), p. 239.

[10] C. Cotrutz, C. Kappas, and S. Webb, Intensity Modulated Arc Therapy (IMAT) with Centrally Blocked Rotational Fields, *Phys. Med. Biol.*, 45 (2000), pp. 2185–2206.

[11] J. Dai and Y. Zhu, Minimizing the Number of Segments in a Delivery Sequence for Intensity-Modulated Radiation Therapy with a Multileaf Collimator, *Phys. Med.*, 28 (2001), pp. 2113–2120.

[12] E. A. Dinitz, Algorithm for Solution of a Problem of Maximum Flow in a Network with Power Estimation, *Soviet Math. Dokl.*, 11 (1970), pp. 248–264.

[13] A. Efrat, A. Itai, and M. J. Katz, Geometry Helps in Bottleneck Matching and Related Problems, *Algorithmica*, 31 (2001), pp 1–28.

[14] J. E. Hopcroft and R. M. Karp, An $n^{5/2}$ Algorithm for Maximum Matching in Bipartite Graphs, *SIAM J. Comput.*, 2 (1973), pp. 225–231.

[15] J. M. Keil, Decomposing a Polygon into Simpler Components, *SIAM J. Comput.*, 14 (4) (1985), pp. 799–817.

[16] J. M. Keil, Polygon Decomposition, *Handbook on Computational Geometry*, J.-R. Sack and J. Urrutia (Eds.), Elsevier Science Publishers, Amsterdam, 1999.

[17] R. Liu and S. Ntafos, On Decomposing Polygons into Uniformly Monotone Parts, *Information Processing Letters*, 27 (1988), pp. 85–89.

[18] A. Schweikard, R. Z. Tombropoulos, and J. R. Adler, Robotic Radiosurgery with Beams of Adaptable Shapes, *Proc. 1st Int. Conf. on Computer Vision, Virtual Reality and Robotics in Medicine, Lecture Notes in Computer Science*, Vol. 905, Springer, 1995, pp. 138–149.

[19] S. Webb, Configuration Options for Intensity-Modulated Radiation Therapy Using Multiple Static Fields Shaped by a Multileaf Collimator, *Phys. Med. Biol.*, 43 (1998), pp. 241–260.

[20] K. R. Winston and W. Lutz, Linear Accelerator as a Neurosurgical Tool for Stereotactic Radiosurgery, *Neurosurgery*, 22 (3) (1988), pp. 454–464.

[21] E. Wong, J. Chen, G. Bauman, E. Yu, A. R. Dar, Plan Evaluations for Clinical Use of Simplified Intensity Modulated Arc Therapy, *Medical Physics*, 28 (6) (2001), p. 1306.

[22] P. Xia and L. J. Verhey, MLC Leaf Sequencing Algorithm for Intensity Modulated Beams with Multiple Static Segments, *Med. Phys.*, 25 (1998), pp. 1424–1434.

[23] C. X. Yu, Intensity-Modulated Arc Therapy with Dynamic Multileaf Collimation: An Alternative to Tomotherapy, *Phys. Med. Biol.*, 40 (1995), pp. 1435–1449.

[24] C. X. Yu, Design Considerations of the Sides of the Multileaf Collimator, *Phys. Med. Biol.*, 43 (5) (1998), pp. 1335–1342.

[25] C. X. Yu, D.-J. Chen, A. Li, L. Ma, D. M. Shepard, and M. Sarfaraz, Intensity-Modulated Arc Therapy: Clinical Implementation and Experience, *Proc. 13th Int. Conf. on the Use of Computers in Radiation Therapy*, W. Schlegel and T. Bortfeld (eds.), Heidelberg, May 2000.

[26] C. X. Yu, A. Li, L. Ma, D. M. Shepard, M. Sarfaraz, T. Holmes, M. Suntharalingham, and C. Mansfield, Clinical Implementation of Intensity-Modulated Arc Therapy, *Elekta Oncology Symposium on IMRT*, Thomas Jefferson University, PA, March 2000.

Geometric Algorithms for Density-Based Data Clustering

Danny Z. Chen[1] *, Michiel Smid[2] **, and Bin Xu[1] ***

[1] Department of Computer Science and Engineering, University of Notre Dame
Notre Dame, IN 46556, USA
{chen,bxu}@cse.nd.edu
[2] School of Computer Science, Carleton University
Ottawa, Ontario, Canada K1S 5B6
michiel@scs.carleton.ca

Abstract. We present new geometric approximation and exact algorithms for the density-based data clustering problem in d-dimensional space $\mathbb{R}^d$ (for any constant integer $d \geq 2$). Previously known algorithms for this problem are efficient only for uniformly-distributed points. However, these algorithms all run in $\Theta(n^2)$ time in the worst case, where n is the number of input points. Our approximation algorithm based on the ϵ-fuzzy distance function takes $O(n \log n)$ time for any given fixed value $\epsilon > 0$, and our exact algorithms take sub-quadratic time. The running times and output quality of our algorithms do not depend on any particular data distribution. We believe that our fast approximation algorithm is of considerable practical importance, while our sub-quadratic exact algorithms are more of theoretical interest. We implemented our approximation algorithm and the experimental results show that our approximation algorithm is efficient on arbitrary input point sets.

1 Introduction

Data clustering is a fundamental problem that arises in many applications (e.g., data mining, information retrieval, pattern recognition, biomedical informatics, and statistics). The main objective of data clustering is to partition a given data set into clusters (i.e., subsets) based on certain criteria. Significant challenges to clustering are that the size of the input data is often very large, and that little *a priori* knowledge about the structure of the data is known. For example, huge sets of spatial data can be generated by satellite images, medical instruments, video cameras, etc.

In this paper, we consider the density-based data clustering problem for spatial data. This problem can be defined as follows [7]. Suppose we are given a set S of n points in the d-D space $\mathbb{R}^d$ (for any constant integer $d \geq 2$) and

---

* The work of these authors was supported in part by Lockheed Martin Corporation and by the National Science Foundation under Grant CCR-9988468.
** The work of this author was supported in part by NSERC.
*** Corresponding author.

two parameters $\delta > 0$ and $\tau > 1$. For any point p in $\mathbb{R}^d$, we denote by $N_\delta(p)$ the sphere centered at p and having radius δ in $\mathbb{R}^d$ (based on some given distance function). This sphere $N_\delta(p)$ is called the δ-*neighborhood* of p.

1. If for a point $p \in S$, there are at least τ points of S (including p) in the sphere $N_\delta(p)$, i.e., $|S \cap N_\delta(p)| \geq \tau$, then all points of $S \cap N_\delta(p)$ belong to the same cluster of S.
2. For two subsets C_1 and C_2 of S, if each of C_1 and C_2 belongs to a cluster and if $C_1 \cap C_2 \neq \emptyset$, then $C_1 \cup C_2$ belongs to the same cluster.
3. A *cluster* of S is a maximal set satisfying the two conditions above.
4. All points of S that do not belong to any cluster are called *noise*.
5. The *density-based clustering (DBC) problem* is to find all clusters of S, and all noise of S.

Note that the shape of a cluster depends on the given distance function.

Intuitively, one can view the DBC problem as finding galaxies and isolated stars in $\mathbb{R}^3$, and cities and rural residences in $\mathbb{R}^2$, based on a certain neighborhood density of the data points.

It is easy to solve the DBC problem in $O(n^2)$ time. Jain and Dubes [8] introduced a density-based method to identify clusters among points in $\mathbb{R}^d$. A heuristic algorithm for determining the parameters δ and τ was given in [7]. Commonly used DBC approaches in the data mining community in general first represent the input point set S by a certain data structure, and then search the δ-neighborhood of each point of S to form clusters. For example, the DBSCAN (Density Based Spatial Clustering of Applications with Noise) algorithms [7, 10] are based on the R*-tree, whereas the FDC (Fast Density-based Clustering) algorithm [11] is based on the k-D tree These DBC algorithms [6, 7, 10, 11] are efficient on uniformly-distributed points when each δ-neighborhood on average contains only a constant number of points of S. However, on non-uniformly distributed points, the running times of such algorithms are all $\Theta(n^2)$ in the worst case. Note that for many database applications, since their data is of enormous size and need not be uniformly distributed, these $\Theta(n^2)$-time algorithms for the DBC problem are highly impractical.

In this paper, we present faster approximation and exact algorithms for the DBC problem in $\mathbb{R}^d$ ($d \geq 2$). The running times and output quality of our algorithms do not depend on any particular data distribution. To achieve nearly linear time DBC algorithms, we consider developing approximate solutions. Our approximation is in the following sense [3]. Let $\epsilon > 0$ be any given constant. For a point $p \in \mathbb{R}^d$, the boundary of our δ-neighborhood $N_\delta(p)$ is *fuzzy* (called ϵ-*fuzzy*), i.e., each point of S whose distance to p is within the range $[(1-\epsilon)\delta, (1+\epsilon)\delta]$ is arbitrarily considered to be in or out $N_\delta(p)$. On the other hand, points whose distances to p are less than $(1-\epsilon)\delta$ (resp., larger than $(1+\epsilon)\delta$) are guaranteed to be in $N_\delta(p)$ (resp., not in $N_\delta(p)$). Studying approximations of this kind is useful because in many practical situations, the input data is imprecise or an approximate solution obtained in a small amount of time is sufficient. Our main results are summarized below.

- An $O(n^{2(1-1/(d+2))}\text{polylog}(n))$ time exact DBC algorithm for the Euclidean distance metric.
- An $O(kn \log n)$ time exact DBC algorithm, if $d = 2$, for the convex distance function based on a regular k-gon.
- An $O(n \log n + n(1/\epsilon)^{d-1})$ time approximate DBC algorithm for the ϵ-fuzzy distance function.
- We have implemented our approximate DBC algorithm and tested it for various clustered point sets. The experimental results show that our approximation algorithm is efficient and does not depend on any particular data distribution (as indicated by our theoretical analysis).

We think our fast approximation algorithm is of considerable practical importance, while our sub-quadratic exact algorithms are more of theoretical interest.

We present an overview of the general algorithmic steps in Section 2, and the efficient exact and approximation theoretical implementations of the general algorithm in Section 3. Some modifications to the approximate DBC approach are given in Section 4 to simplify our approximate DBC algorithm and its implementation. The experimental results of the approximate DBC algorithm based on these modifications are shown in Section 5.

2 Main Ideas and Algorithm Overview

Let $S \subseteq \mathbb{R}^d$ be the input point set, and let $S_\delta(p)$ denote the point set $S \cap N_\delta(p)$ for any point p in $\mathbb{R}^d$. A point $p \in S$ is called a *dense* point if $|S_\delta(p)| \geq \tau$; otherwise, p is called a *sparse point*.

One of our main ideas is to formulate the DBC problem as that of computing special connected components of an undirected graph G, which is defined as follows. Each vertex v_p of G corresponds to exactly one point p of S. An edge connects two distinct vertices v_p and v_q in G if and only if (i) the distance $dist(p, q)$ between the two corresponding points p and q of S (based on a given distance metric $dist$) is less than or equal to δ, and (ii) p or q is a dense point. We say that a connected component of G is *non-trivial* if it contains a vertex v_p such that the corresponding point $p \in S$ is a dense point. In the rest of this paper, we let p denote both a point in S and its corresponding vertex in G. The following lemma is obvious.

Lemma 1. *Each cluster of S corresponds to exactly one non-trivial connected component of G.*

Hence, our goal is to compute all non-trivial connected components of G. But, storing and finding G's connected components in a straightforward manner will take $O(n^2)$ time since G can have $\Theta(n^2)$ edges. Our second idea is to avoid looking at all edges of G when computing its non-trivial connected components. In fact, we seek to examine only $O(n)$ edges of G (i.e., $O(n)$ pairs of points of S) in this computation. To do that, we assume for the rest of this section that we have some data structures that can be used to obtain information on the (dynamically changing) subset of S in the δ-neighborhood $N_\delta(p)$ of any point $p \in S$. In Section 3, we will show how these data structures can actually be theoretically implemented exactly and approximately.

Basically, we find each non-trivial connected component in G by a breadth-first search traversal of G, starting at some dense point. In particular, whenever a point $p \in S$ is included in a cluster, p is removed from S (and hence it is removed from all relevant data structures storing p), so that no subsequent search in G will visit the vertex p again. To achieve this, some care must be taken. Below is an overview of our general algorithm. (The details on how to actually carry out the key steps of this algorithm will be shown in Section 3.)

1. Compute the subsets of all dense and sparse points in S, denoted by D and H, respectively.
2. Build a dynamic data structure T_S (resp., T_D) for the point set S (resp., D). /* Initially, T_S contains all input points, and T_D contains only the dense points. */
3. While $T_D \neq \emptyset$, do the following: /* There is at least one more cluster to be found. */
   (a) Initialize an empty queue DQ, and get a new cluster ID.
   (b) Delete an arbitrary point p from T_D, put p into DQ, and delete p from T_S.
   (c) While $DQ \neq \emptyset$, do the following:
   /* This while loop finds the cluster containing all (dense) points in DQ. */
      i. Remove the first point p from DQ, and assign the cluster ID to p.
      ii. Find the set $CS_\delta(p)$ of all points of $S_\delta(p)$ that are currently in T_S, and delete these points from T_S. /* $CS_\delta(p)$ contains both dense and sparse points. */
      iii. While $CS_\delta(p) \neq \emptyset$, do the following:
         A. Remove a point q from $CS_\delta(p)$, and assign the cluster ID to q.
         /* Each point q in $S_\delta(p)$ belongs to the same cluster as the dense point p. */
         B. If q is a dense point, then put q at the end of DQ, and delete q from T_D. Otherwise (q is a sparse point), find the set $DS_\delta(q)$ of all dense points of $S_\delta(q)$ currently in T_D, delete these points from T_D, and put these points at the end of DQ.
         /* All dense points of $S_\delta(q)$ belong to the same cluster as the sparse point q. */
4. All points of S that are still in T_S are noise.

We now show the correctness of this clustering algorithm. First, we argue that once a point p of S is visited by the breadth-first search procedure, p will not be visited again in the rest of the search. This can be seen easily since whenever a point of S is reported in either T_S or T_D, it is immediately deleted from all the corresponding data structures. This implies that only $O(n)$ edges of G (i.e., $O(n)$ pairs of points of S) are visited by the algorithm.

Next, we argue that each cluster of S (i.e., each non-trivial connected component of G) is obtained correctly by the above breadth-first search procedure (contained in the outermost while loop of Step 3). Note that there are two types of edges in G: (1) those connecting two dense points, and (2) those connecting a dense point and a sparse point. When a dense point p is taken from the queue DQ (called the *dense-point queue*), the edges of both types connecting p and all other unvisited points in $S_\delta(p)$ are visited; this is done inside the innermost while loop of Step 3. After a sparse point q in a cluster is visited, the type (2) edges connecting q and all unvisited dense points in $S_\delta(q)$ are visited, putting these dense points into the queue DQ (also inside the innermost while loop of Step 3). In an inductive fashion, the breadth-first search procedure explores the connectivity of G. Hence, the non-trivial connected component of G being searched is found correctly, as stated by the next lemma.

Lemma 2. *All non-trivial connected components of the graph G are correctly computed by the above clustering algorithm. Furthermore, the algorithm only visits $O(n)$ edges of G.*

Observe that both the data structures T_S and T_D used in our algorithm need not be fully dynamic. What we essentially need are two types of operations on T_S and T_D: (1) searching for the points of S in $N_\delta(p)$ for a point $p \in S$ (i.e., the point set $S_\delta(p)$), and (2) deleting points from the two data structures. Each query point is a point in S, and the insertion operation is not necessary. This fact will be extensively explored by the version of approximation algorithm used in our programming implementation in Sections 4 and 5.

3 Efficient Theoretical Implementations of the Clustering Algorithm

In the previous section, we have given a high-level overview of our clustering algorithm. To obtain fast theoretical implementations for the exact and approximation algorithms, we must carefully choose the data structures that support the search and deletion operations.

3.1 Computing the Dense Points

In Step 1 of the clustering algorithm, the set S is partitioned into D (the dense points) and H (the sparse points). To implement this step, we need to compute, for each point p of S, the number of points in $S \cap N_\delta(p)$. As we will see below, the efficiency of a solution to this problem heavily depends on the metric used and on whether we compute the size of $N_\delta(p)$ exactly or approximately.

We first consider the Euclidean metric. Hence, $N_\delta(p)$ is a ball of radius δ centered at p in $\mathbb{R}^d$. The query can be reduced to a half-space reporting query in $\mathbb{R}^d$ [2]. If we choose the proper parameter, then the running time becomes $O(n^{2(1-1/(d+2))}\text{polylog}(n))$, which is (slightly) sub-quadratic in n.

We now assume that the dimension d is equal to two, and let us use a convex distance function based on a regular k-gon, where $k \geq 3$ is an integer. If k is sufficiently large, then this metric approximates the Euclidean metric arbitrarily closely. For each k-gon K_p centered at point $p \in S$, we partition it into $k-2$ triangles. The number of points contained in K_p can be obtained by efficiently calculating the number of points contained in each of the $k-2$ triangles. In this way, all dense points of S can be computed in $O(kn \log n)$ time.

Observe that in this solution, we approximate the Euclidean circle C of radius δ by a regular k-gon K. Hence, we ignore points of S that are in $C_p \setminus K_p$. In our next solution, we use a more "fuzzy" approach to approximate $|C_p \cap S|$ in $\mathbb{R}^d$ ($d \geq 2$), which is due to Arya and Mount [3]: Any point of S that is within Euclidean distance $(1-\epsilon)\delta$ of p is guaranteed to be counted; any point of S whose Euclidean distance to p is more than $(1+\epsilon)\delta$ is guaranteed not to be counted; any other point of S may or may not be counted. Here, $\epsilon > 0$ is a given fixed real number. Arya and Mount [3] have shown how to preprocess S in $O(n \log n)$ time, such that an approximate range counting query can be answered in $O(\log n + (1/\epsilon)^{d-1})$ time. This leads to an algorithm that computes all points of S that are "approximately" dense, in $O(n \log n + n(1/\epsilon)^{d-1})$ time.

3.2 Steps 2 and 3

To implement Steps 2 and 3 of our clustering algorithm, we need a data structure for the following problem. Let Z be a set of n points in $\mathbb{R}^d$. (In the clustering algorithm, Z is either the entire set S or the dense point set D.) We want to process an on-line operation sequence, in which each operation is

- a query of the form "given a point p of Z, report all points of Z that are contained in the δ-neighborhood $N_\delta(p)$," or
- a deletion of a point from Z.

Let $q_1, q_2, \ldots, q_k$ be all query points, and let A_i be the output for query point q_i, $i \geq 1$. The sequence of operations has the properties that $k \leq n$ and that the sets A_i, $i \geq 1$, are pairwise disjoint. Hence, $\sum_i |A_i| \leq n$.

Let us again consider this problem first for the Euclidean metric. Agarwal, Eppstein, and Matoušek [1] have shown that, for any parameter m, $n \leq m \leq n^{\lfloor (d+1)/2 \rfloor}$, a data structure can be built in $O(m^{1+\nu})$ time such that any query can be answered in $O((n/m^{\lfloor (d+1)/2 \rfloor}) \log n + |A_i|)$ time, and any deletion can be done in $O(m^{1+\nu}/n)$ time. Here, ν is an arbitrarily small positive real constant. Using this result, the entire sequence of operations takes $O(m^{1+\nu} + (n^2/m^{\lfloor (d+1)/2 \rfloor}) \log n)$ time. If we choose $m = n^{2(1-1/\lfloor (d+3)/2 \rfloor)}$, then the entire running time for implementing Steps 2 and 3 is $O(n^{2(1-1/(d+2))} \text{polylog}(n))$, which is the time for computing the dense points (see Section 3.1).

Next we assume that $d = 2$ and use the convex distance function based on the regular k-gon K. Klein et al. [9] have designed a data structure that can be built in $O(kn \log n)$ time, that supports each query in $O(k \log n + |A_i|)$ time and each deletion in $O(k \log n)$ time. (Recall that in a query, we need to report all points of Z that are contained in a translate of K.) Hence, for this metric, Steps 2 and 3 can be implemented in $O(kn \log n)$ time.

Finally, if we use the "fuzzy" notion of neighborhood in $\mathbb{R}^d$, $d \geq 2$ (see Section 3.1), we can use the data structure of Arya and Mount [3]. This data structure answers each query in $O(\log n + (1/\epsilon)^{d-1} + |A_i|)$ time, and supports each deletion in $O(\log n)$ time. (The latter claim can be proved using the same techniques that were used to dynamize the approximate nearest neighbor data structure of Arya et al. [5].) This leads to an $O(n \log n + n(1/\epsilon)^{d-1})$ time implementation of Steps 2 and 3.

3.3 The Final Time Bounds

By combining the results obtained above, we have the following theorem.

Theorem 1. *Let S be a set of n points in $\mathbb{R}^d$, and let $\delta > 0$ and $\tau > 1$ be two parameters. The density-based clustering problem for S (based on δ and τ) can be solved in*

1. *$O(n^{2(1-1/(d+2))} \text{polylog}(n))$ time for the Euclidean metric.*
2. *$O(kn \log n)$ time, if $d = 2$, for the convex distance function based on a regular k-gon.*
3. *$O(n \log n + n(1/\epsilon)^{d-1})$ time for the ϵ-fuzzy distance function, for any real constant $\epsilon > 0$.*

4 Refinements to the Approximate DBC Approach

This section presents some refinements to the approximate DBC approach given in Section 3. There are two reasons for making these refinements: (i) It simplifies the approximate DBC algorithm considerably and makes it easier for our implementation to use existing geometric computing software such as Arya and Mount's ANN library [4] (instead of starting from scratch), hence greatly reducing our programming efforts; (ii) it reduces the execution time and memory usage in the actual implementation (by some constant factors), which is especially meaningful when the data sets are very large. The approximate DBC approach given in this section is based on Arya and Mount's BBD tree data structure for approximate range search in $\mathbb{R}^d$ for $d \geq 2$ [3]. A static BBD tree data structure has been implemented in the ANN library by Arya and Mount [4]. We modified ANN to accommodate deletions.

Hence, we make the following key refinements: 1) Using only one approximate range search data structure, a BBD tree T, for storing both the input point set S and dense point set D, instead of two data structures T_S and T_D as discussed in Section 3 (i.e., T is a combination of T_S and T_D); 2) a fast deletion operation on T. Note that T does not store the dense points twice, but merely marks each point of S as dense or sparse.

4.1 Approximate Range Search Tree T

Our BBD tree T has the same structure as that for approximate range search in [3], with a constant bucket size. In addition, we need to include more information in T for our refinements.

We denote the subtree of T rooted at a node v by $subtree(v)$, the root of a subtree T' of T by $root(T')$, and the height of T' by $height(T')$. Note that each node v in T corresponds to a subregion (denoted by $cell(v)$) in the partition of the space $\mathbb{R}^d$ based on the input point set S, as shown in [3].

T stores all points of S. Each point p of S has a flag indicating that p is either "present" or "deleted", and another flag indicating whether p is a dense or sparse point. (That is, when a deletion is performed on a point p, p is merely marked as "deleted", without being physically removed from T.)

For each node v in T, in addition to the fields that a node has as in [3], v has several other fields: (a) $all\_points\_number(v)$: the total number of (dense and sparse) points of S that are currently stored as "present" in $subtree(v)$; (b) $dense\_points\_number(v)$: the total number of dense points in S that are currently stored as "present" in $subtree(v)$; (c) $parent(v)$: a pointer to the parent of v (if any) in T. These fields are useful for the search and deletion operations on T.

Step 1 of our approximate DBC algorithm (as shown in Section 3.1) simply uses the algorithm for approximate range *counting queries* in [3] to identify approximately dense and sparse points of S. We assume this is already done. Then by using a procedure similar to the one for Lemma 4 of [3], our approximate range search BBD tree T can be constructed in $O(n \log n)$ time and $O(n)$ space. Also, note that by using the field $all\_points\_number(root(T))$ (resp., $dense\_points\_number(root(T))$), it is easy to check in $O(1)$ time whether T contains any point (resp., dense point) of S after many deletions.

4.2 Search and Deletion Operations on T

T needs to support some key operations on the dense point set D: (I) Report all dense points in the δ-neighborhood $N_\delta(q)$ of a point $q \in S$ (q may be a sparse point) that are currently "present" in T; (II) delete a given dense point p from T. We also need from T similar operations on the entire input point set S. Due to their similarity, we only show these operations of T on the dense point set D.

Let $R\_outer(p)$ denote the spherical range of radius $(1+\epsilon)\delta$, and $R\_inner(p)$ denote the spherical range of radius $(1-\epsilon)\delta$, both centered at p. Like the approximate range search algorithm in [3], our approximate range reporting procedure recursively searches the BBD tree T starting from its root, avoiding the nodes whose cells are outside the inner range $R\_inner(p)$ and only considering the nodes whose cells are either inside the outer range $R\_outer(p)$ or whose cells are neither inside $R\_outer(p)$ nor outside $R\_inner(p)$. For a node v whose cell $cell(v)$ intersects $R\_inner(p)$, the procedure searches $subtree(v)$ to find all "present" dense points in $subtree(v)$, by using the fields $dense\_points\_number$ of appropriate nodes in $subtree(v)$. The recursive search at a node v stops if

$dense\_points\_number(v) = 0$. Thus, the search never wastes time on any subtree of T that contains no "present" dense points.

The deletion of a dense point p from T is simple. Suppose p is given (by a reporting procedure) for deletion. Let $leaf(p)$ be the leaf node of T whose cell contains p. We then mark p as "deleted" in T, and trace the path from $leaf(p)$ to $root(T)$ in T, adjusting the values of $all\_points\_number(v)$ and $dense\_points\_number(v)$ for each node v on that path, by following the pointer $parent(v)$.

We have the following lemma about our approximate range reporting and deletion operations on T without proof.

Lemma 3. *On the BBD tree T that is set up to store n input points in $\mathbb{R}^d$, each deletion operation takes $O(\log n)$ time, and each approximate range reporting operation takes $O(m \log n + m(1/\epsilon)^{d-1})$ time for any given fixed value $\epsilon > 0$, where m is the number of points that it reports.*

4.3 The Resulted Approximate DBC Algorithm

Using the approximate range search BBD tree T, we obtain an $O(n \log n + n(1/\epsilon)^{d-1})$ time approximate DBC algorithm in $\mathbb{R}^d$ that is relatively simple to implement. Now we discuss the algorithm based on the steps given in Section 2.

Step 1 is simply carried out by performing n *counting queries* on the approximate range search data structure in [3], one for each point of S. This takes altogether $O(n \log n + n(1/\epsilon)^{d-1})$ time. This step classifies approximately the points of S into dense and sparse points.

Step 2 builds the approximate range search BBD tree T by using an algorithm similar to the data structure construction algorithm in [3], in $O(n \log n)$ time and $O(n)$ space.

Step 3 is carried out by mainly performing approximate range reporting operations and deletion operations on T. Each point of S, once found by a reporting operation, is deleted from T. Since every point of S (dense or sparse) is reported and deleted exactly once in T, at most n approximate range reporting operations and n deletion operations on T are performed. By Lemma 3, the total time of this step is $O(n \log n + n(1/\epsilon)^{d-1})$.

Theorem 2. *The approximate DBC algorithm for n input points in $\mathbb{R}^d$ given in this section takes $O(n \log n + n(1/\epsilon)^{d-1})$ time and $O(n)$ space for any given fixed value $\epsilon > 0$.*

5 Experimental Results for the Approximate DBC Algorithm

To show the good performance of the approximate DBC algorithm presented in Section 4, we implemented it in C++ on a Sun Sparc 20 workstation running Solaris. Our implementation is based on a careful extension of the programming structure of the ANN library created by Arya and Mount [4]. This is because the data structures used in [4] for approximate nearest neighbor queries and our approximate DBC algorithm in Section 4 are both hinged on similar BBD trees. In fact, the availability of the ANN library significantly helped our programming efforts in this study. On the other hand, our implementation work is still quite substantial (with altogether over ten thousand lines of code).

Our experimental study has several goals. The first goal is to find out, in various density-based clustering settings, how the execution time of our approximate DBC algorithm varies as a function of each of several parameters: (1) the input data size $n = |S|$, (2) the accuracy factor ϵ, and (3) the input data dimension d. The effects of the radius δ of the δ-neighborhood $N_\delta(\cdot)$ with respect to these parameters are also examined. Our second goal is to develop software for our approximate DBC algorithm, and make it available to density-based clustering applications. Besides, we know the theoretical upper time bound for our approximate DBC algorithm is $O(2^d n \log n + n(2\sqrt{d}/\epsilon)^{d-1})$, which appears to be quite high for various practical settings. (when d is not a constant). Our third goal thus is to compare the experimental results with this theoretical upper time bound to determine the practical efficiency of our approximate DBC algorithm.

All input data sets used in our experiments were randomly generated. Instead of uniformly distributed input points, we use clustered input data sets generated as follows: 10 "core" points were first chosen from the uniform distribution (in the interval [0, 1]) in the unit hypercube, and then many points based on a Gaussian distribution with a standard deviation of 0.05 centered around each core point were generated in the unit hypercube (see [3, 4] for more information on this method of generating data sets). For this kind of clustered input point sets, the commonly used density-based clustering algorithms [6, 7, 10, 11] will be inefficient because they search the δ-neighborhood of each input point without deleting input points in that neighborhood, thus taking quadratic time.

Each curve in Figures 1, 2, and 3 below represents the average of 20 experiments whose data sets were generated with different seeds for the random number generator that we used.

5.1 Execution Time and Accuracy

We considered the values of ϵ varying from 0 to 0.5, the values of the radius δ of the δ-neighborhood varying from 1/2 to 1/128, and data sets of size 100K. Figure 1 gives the execution time of our approximate DBC algorithm as a function of ϵ with some fixed values of δ, for 2-D data sets.

Figure 1 shows some clear patterns. As ϵ decreases from 0.05 to 0, there are significant increases in execution times for larger ranges (i.e., bigger values of δ), and not so significant increases for smaller ranges (e.g., the execution time increases by 325.3% with a radius of 1/2, and only by 11.9% with a radius of 1/128). The increases of the execution times are rather slow when ϵ varies from 0.5 to 0.05, and the trends tend to be "flat" (e.g., as ϵ decreases from 0.5 to 0.05, which is a "long" interval, the execution time increases by 144.1% with a radius of 1/2, and by 52.1% with a radius of 1/128).

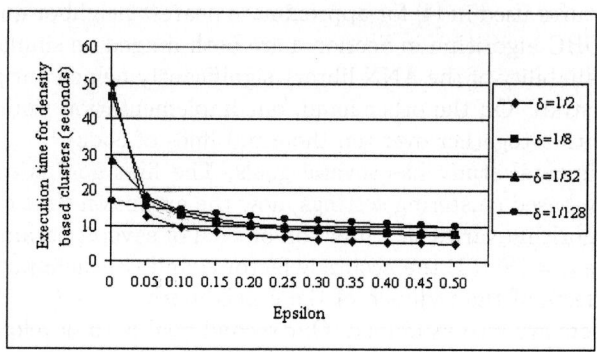

Fig. 1. Relation between the execution time of our approximate DBC algorithm and ϵ, for 2-D data sets of size 100K

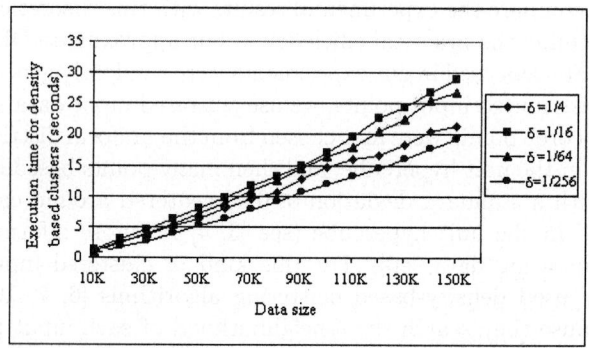

Fig. 2. Relation between the execution time of our approximate DBC algorithm and the data size, for 2-D data sets

These patterns suggest that for a relatively small radius δ, the execution time increases quite slowly with respect to the decrease of the accuracy ϵ, in comparison with the theoretical upper time bound.

5.2 Execution Time and Data Size

We considered input data sets of sizes varying from 50K to 150K, the values of δ varying from 1/4 to 1/256, and ϵ of 0.05. Figure 2 shows some of the results for 2-D data sets.

In Figure 2, the four curves for different values of δ all indicate that the execution times increase almost linearly with respect to the increases of data sizes. Further, the execution times increase very slowly as data sizes increase (with a small positive slope). Comparing to the theoretical upper time bound, the increase of the execution time is much slower as the data sizes increase.

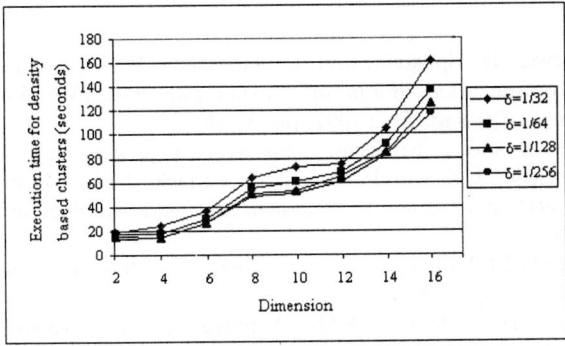

Fig. 3. Relation between the execution time of our approximate DBC algorithm and the dimension, for data sets of size 100K

5.3 Execution Time and Data Dimension

A key issue is how the execution time depends on the dimensionality of the input data set. Since the dimension value d acts as the power parameter in the theoretical upper time bound, d is also a crucial factor to the execution time.

We considered dimensions varying from 2 to 16, the values of δ ranging from $1/32$ to $1/256$, ϵ of 0.05, and data sets of size 100K. As shown in Figure 3, the increases of the execution times are much slower than what the theoretical upper time bound predicts (e.g., for the values of $\delta = 1/32, 1/64, 1/128,$ and $1/256$, when the dimension d changes from 2 to 16, the execution time increases 7.7, 6.7, 7.1, and 7.6 times, respectively, which are much less than the upper bound).

5.4 Discussion of the Experimental Results

In the above experiments, we used clustered input data sets instead of uniformly distributed ones. One can see that comparing to the theoretical upper time bound, the increase of the execution time of our approximate DBC algorithm is quite slow with respect to the increase of the data size and dimension, and with respect to the decrease of the accuracy ϵ.

From the experimental results shown in Section 5.1, we know that by changing the value of ϵ, output solutions with different accuracy can be obtained, and the execution time of our approximate DBC algorithm becomes smaller if we use bigger ϵ. Note that different levels of accuracy may be needed for different real clustering applications, and fast execution time is a key requirement to many such applications. Therefore, we can achieve a good trade-off between the execution time and quality of the output based on the requirements of a specific application, so that we not only can solve a given clustering problem with sufficient accuracy but also complete the computation within a required time period. Hence in real applications, our approximate DBC algorithm based on the approximate range search data structure can be made efficient and practical.

References

[1] P. K. Agarwal, D. Eppstein, and J. Matoušek, Dynamic half-space reporting, geometric optimization, and minimum spanning trees, *Proc. 33rd Annual IEEE Symp. Found. Comput. Sci.*, 1992, pp. 80-89.

[2] P. K. Agarwal and J. Erickson, Geometric range searching and its relatives. In: B. Chazelle, J. E. Goodman and R. Pollack (Eds.), *Advances in Discrete and Computational Geometry*, American Mathematical Society, Providence, RI, 1999, pp. 1–56.

[3] S. Arya and D. M. Mount, Approximate range searching, *Comput. Geom. Theory Appl.*, 17 (2000), pp. 135–152.

[4] S. Arya and D. M. Mount, ANN: A library for approximate nearest neighbor searching, *2nd CGC Workshop on Computational Geometry*, 1997. lso, see http://www.cs.umd.edu/~mount/.

[5] S. Arya, D. M. Mount, N. S. Netanyahu, R. Silverman, and A. Wu, An optimal algorithm for approximate nearest neighbor searching in fixed dimensions, *J. ACM*, 45 (1998), pp. 891–923.

[6] M. Ester, H.-P. Kriegel, J. Sander, M. Wimmer, and X. Xu, Incremental clustering for mining in a data warehousing environment, *Proc. 24th Int. Conf. on Very Large Databases*, 1998, pp. 323-333.

[7] M. Ester, H. P. Kriegel, J. Sander, and X. Xu, A density-based algorithm for discovering clusters in large spatial databases with noise, *Proc. 2nd Int. Conf. on Knowledge Discovery and Data Mining*, 1996, pp. 226-231.

[8] A. K. Jain and R. C. Dubes, *Algorithms for Clustering Data*, Prentice Hall, 1988.

[9] R. Klein, O. Nurmi, T. Ottmann, and D. Wood, A dynamic fixed windowing problem, *Algorithmica*, 4 (1989), pp. 535-550.

[10] J. Sander, M. Ester, H.-P. Kriegel, and X. Xu, Density-based clustering in spatial databases: The algorithm GDBSCAN and its application, *Data Mining and Knowledge Discovery*, 2 (2) (1998), pp. 169-194.

[11] B. Zhou, D. W. Cheung, and B. Kao, A fast algorithm for density-based clustering in large database, *Proc. 3rd Pacific-Asia Conf. on Methodologies for Knowledge Discovery and Data Mining*, 1999, pp. 338-349.

Balanced-Replication Algorithms for Distribution Trees

Edith Cohen[1] and Haim Kaplan[2]

[1] AT&T Labs–Research
Florham Park, NJ 07932 USA
edith@research.att.com

[2] School of Computer Science, Faculty of exact sciences, Tel-Aviv University
Tel Aviv 69978, Israel
haimk@post.tau.ac.il

Abstract. In many Internet applications, requests for a certain object are routed bottom-up over a tree where the root of the tree is the node containing the object. When an object becomes popular, the root node of the tree may become a hotspot. Therefore many applications allow intermediate nodes to acquire the ability to serve the requests, for example by caching the object. We call such distinguished nodes *primed*. We propose and analyse different algorithms where nodes decide when to become primed; these algorithms balance the maximum load on a node and the number of primed nodes.

Many applications require both fully distributed decisions and smooth convergence to a stable set of primed nodes. We first present optimal algorithms which require communication across the tree. We then consider the natural previously proposed THRESHOLD algorithm, where a node becomes primed when the incoming flow of requests exceeds a threshold. We show examples where THRESHOLD exhibits undesirable behavior during convergence. Finally, we propose another fully distributed algorithm, GAP, which converges gracefully.

1 Introduction

The Internet provides a platform for decentralized applications where content or services are requested, stored, and provided by very large number of loosely coupled hosts. In these applications, there is no centralized support. Requests are forwarded from peer to peer, and each peer can become a server for each item.

Many supporting architectures for such applications are such that the search pattern for each request is "routed" that is, every peer has a list of neighbors, and for each request a peer might receive, there is a preferred neighbor that brings the request "closer" to its destination. Therefore, the forwarding pattern of each request forms a tree, with edges directed towards the root. We refer to such a structure as a *distribution tree*.

When a certain request becomes popular, the respective root of its distribution tree can become a hotspot. In this case, hosts further down the tree can "replicate" the service – that is, acquire the ability to satisfy these requests, e.g., by "caching" data or duplicating code. The focus of this paper is proposing and evaluating algorithms for hosts to decide when to become replicas.

Before providing further details on our model, we describe some concrete applications and architectures where this setup arises. A relatively old existing application is the Domain Name Service (DNS). The DNS service already offers caching of records at intermediate nodes; in fact, caching is the default at all "intermediate" servers and DNS records are cached until they expire. A forward looking application is to place HTTP caches (possibly active caches) "near" Internet routers and augment them with the capability to intercept and forward HTTP traffic; such a service can shorten latency and network load. The growing popularity of Peer-to-Peer applications (e.g., [3, 2]) sparked the development of architectures for fully distributed name-lookup, which enable clients to locate objects or services. Two such proposed architectures are Chord [10] (which is based on consistent hashing [4]) and CAN [9] ("Content Addressable Network"). With Chord and CAN, objects have keys which are hashed to points in some metric space S. Each node is responsible for some segment of the space S. Note that the metric of S has no correspondence to the underlying network distances. Each node maintains a small number of pointers $P(v)$ to nodes responsible for other regions of S. When a node receives a request for an object with key k, it forwards it to the node in $P(v)$ responsible for a region containing the point closest to the hash of k (according to a metric on S). Thus, the search paths for each object form a distribution tree, with the root of the tree being the node responsible for the region of the metric space where the hash of the object lies. Note that even though these hashing-based architectures balance the load well across different objects, replication is still necessary for avoiding hotspots caused by highly skewed per-object demands. Thus, to alleviate hotspots, it is proposed in [9, 10] that intermediate nodes cache items (or keys).

In all these applications, replication on intermediate nodes constitutes a tool in avoiding "hotspots" at the roots of distribution trees. Such hotspots can be caused by flash-crowd events or objects with persistent high demand. We would like to iterate that generally, each distribution tree corresponds to a single "item," and thus one might worry about interaction of flows from different trees and that "priming" actions performed at different trees will not balance out across nodes). In many of these applications, however, the balance between trees is taken care of by the process selecting these trees – e.g., in CAN and CHORD, the trees are such that nodes are essentially assigned randomly to their positions in different trees. Thus, it is reasonable to consider the trees independently of each other and focus on a single tree, which is the problem we model and address in this paper.

We say that we *prime* a node when we give it the ability to serve requests. We refer to a set of primed nodes as an *allocation*. We look for algorithms that select an allocation. We model the problem and propose metrics for the quality of a solution. The objective is to balance two parameters: the maximum load of a node, which we would like to be as small as possible, and the total size of the allocation (number of primed nodes), which we also would like to be as small as possible. Other desirable properties pertain to the complexity and communication requirements of the algorithm; and the stability of the allocation while the algorithm converges.

We first relate the worst-case allocation size needed to satisfy a flow of certain size. We provide optimal algorithms which minimize the size of the allocation, given a bound on maximum load a node can process. The algorithms are distributed and requires linear number of messages (two rounds of passing messages up and down the tree).

We then consider a simple natural algorithm proposed with both Chord and CAN: The THRESHOLD algorithm runs locally at each node and primes the node when the incoming flow of requests is above a certain threshold. The THRESHOLD algorithm has obvious advantages as it requires no additional communication as decisions are made only according to the amount of incoming flow. We provide bounds on the size of the allocation obtained and show that it is close to the worst-case upper bound. We show, however, that since priming decisions depend on decisions made by other nodes, THRESHOLD can suffer from very spiky convergence where interim allocation size can be considerably higher than final size.

Finally, we propose a different local algorithm GAP. Like THRESHOLD, GAP also makes its decision based on the amount of incoming flow and also meets the same upper bound. In contrast to THRESHOLD, it considers only the total flow generated below the deciding node and thus priming decisions are independent of decision made at other nodes. It follows that GAP obtains right away a stable allocation.

Related Work

Plaxton, Rajaraman, and Richa proposed a related scheme that also takes into account the underlying topology when overlaying the distribution network [8]. Other related work focused on balancing load through replication, but mainly addressed a different aspect – the layout of the distribution trees (placing caches as "nodes") on top of larger network (e.g. [1, 5]). A conceptually similar problem is that of placing proxy caches in an existing distribution tree as to minimize average number of hops. This problem was explored by Golin et al, which provided a quadratic dynamic programming algorithm [6].

2 Model

A *distribution tree* is a rooted tree with flow generated at the leafs and flowing towards the root. We suggest algorithms that would *prime* some nodes in the tree. A prime node can serve flow coming into it while a nonprime node only forwards flow to its parent. The goal is to prime as few as possible nodes such that all flow is serviced subject to constraints that we define below. As explained in the introduction, flow corresponds to requests, and "priming" corresponds to configuring the node such that it is able to service requests.

Formally, the input to our problem is the tree T and a function $f : v \to R_+$ mapping each leaf v to a non-negative real which describes the flow generated at v. We also assume another input parameter, denoted by C, which is a number that bounds the maximum amount of flow a node can serve. We also assume that each leaf generates at most C units of flow.

An algorithm that solves the problem outputs an *allocation*, which is a subset of the nodes that are primed. The conditions which the allocation has to satisfy vary between the following two service models. These two models differ by whether a primed node is allowed to forward flow or not.

The *Must-Serve* Model: In this model a primed node serves all its incoming flow. An allocation constitutes a solution in this model if when we cut the tree by removing the edge from each primed node to its parent then the flow generated at the leafs of each subtree is at most C (and therefore can be served at the root of this subtree). In other words, an allocation is a good solution if when we process nodes from the leafs up forwarding all the flow into an unprimed node to its parent then each time we reach a primed node it has at most C units of flow coming into it.

The *Serve-or-Forward* Model: In this model a primed node can serve at most C units of the flow coming into it and forwards the rest. Here an allocation constitutes a solution if the following process ends such that the amount of flow reaching the root is at most C if it is primed and 0 if it is not primed. We process the nodes from the leafs up. When an unprimed node is processed we "forward" all its flow to the parent; when a primed node is processed we assign it to serve all the flow it receives, up to C units, and forward the rest to its parent.

In both models an allocation constitutes a solution if there is an assignment of all flow to primed nodes (each unit of flow is assigned to a node on its way to the root) such that the *load* (assigned flow) of each primed node is at most C and the load of other nodes is 0. The *size* of an allocation is the number of nodes it primes. An allocation is *optimal* if it is correct and of minimum size. The objective of minimizing the number of primed nodes models a wide range of applications where being primed consumes resources (e.g., storage space).

The must-serve model is more restrictive than the serve-or-forward model; thus must-serve allocations constitute serve-or-forward allocations. Figure 1 illustrates an example flow serviced in the two service models.

We analyse a static setting where the flow generated by each leaf remains fixed through time.[1] We consider both centralized and distributed algorithms. For distributed algorithms we assume that each node is able to communicate with its parent and children. In the distributed settings, priming decisions of different nodes can depend on each other, and we consider the "convergence"

---

[1] The algorithms, however, are applicable in a dynamic context where flow changes.

process until the nodes decide on a stable allocation. We measure the convergence time in units of the maximum "reaction time" of a node, that is, the maximum elapsed time until it reacts to a new input.

One of our distributed algorithms, GAP, assumes flow reporting, where each child reports to its parent on the flow generated below it (regardless of whether it is served or not). This reporting process is similar to "hit count" reporting in cache hierarchies.

3 What Allocation Size is Necessary?

Let R be the total flow generated in the tree. An obvious lower bound on the required allocation size is R/C. This bound can be further refined if we also consider the maximum number of children Δ. The following lemma show a better bound for the must-serve model. A similar bound also exists for the server-and-forward model.

Lemma 1. *For any C and $\Delta > 4$, there are instances such that $\Delta R/(4C)$ storage is necessary with must-serve.*

Proof. Consider a complete Δ-ary tree of depth d where all leaf nodes generate $2C/\Delta$ flow. Therefore the total flow is $R = 2C\Delta^{d-1}$.

Consider a 2-level subtree of this tree consisting of a node and Δ leaves at the bottom of the tree. At least $\lceil \Delta/2 \rceil$ of its children must be primed, since otherwise the parent receives more than C units of flow. Furthermore, there is an optimal selection that uses exactly $\lceil \Delta/2 \rceil$ children. Otherwise, since the parent node can serve all remaining children, it can replace all "extra" primed children, serving at least the same flow with at most the same coverage.

The number of the remaining unprimed children is $\lfloor \Delta/2 \rfloor$. The combined value of the flow of these unprimed children strictly exceeds $C/2$, therefore at or above the parent level these flows are forwarded or served as complete units,

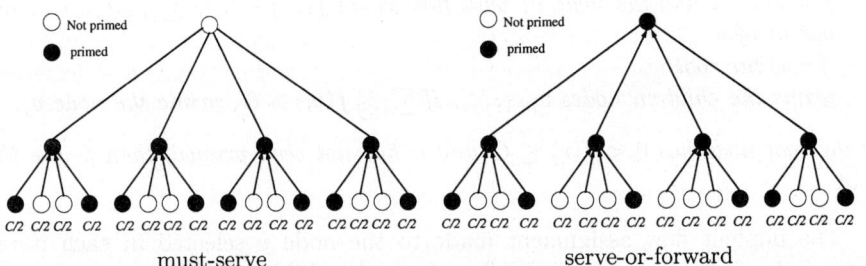

Fig. 1. Example of optimal allocation for the two variants of the problem. The tree has 16 leaves each generating a flow of $C/2$, thus the total flow is 8C. Serve-or-forward primes 11 nodes and must-serve primes 12 nodes

(each primed ancestor node can serve the flow of exactly one of these remaining-children-flows). Thus, there is one primed internal node for each of the Δ^{d-1} sets of remaining children.

We obtain that the size of an optimal solution of this instance is $\Delta^{d-1}(1 + \lceil \Delta/2 \rceil) \geq \frac{R\Delta}{4C}$. □

4 Optimal Algorithms

In this extended abstract we present only the optimal algorithm for the serve-or-forward model (OPT_SoF). The algorithm for the must-serve model is somewhat simpler and will appear in the full version of the paper.

The algorithm OPT_SoF works in two phases. In the first phase it processes the nodes from the leaves upward, and makes some priming decisions. In the second phase additional nodes are primed throughout the tree.

The algorithm makes implicit assignment of flow to primed nodes. Each node primed in the first phase is assigned C units of flow that are generated below it. Nodes primed in the second phase may be assigned less than C units of flow.[2]

Our algorithm utilizes the notion of a *residual problem*. In each iteration of the first phase the algorithm primes a node, and allocates C units of flow for it to serve. Then it generates a residual instance T' by transferring children of the primed node v that have remaining unserved flow to v's parent. In the following we denote by $f(v)$ the flow into node v at the "current" residual problem.

Algorithm 1 (OPT_SoF for serve-or-forward)
Iterate *until all nodes are processed:*
Let v be a lowest node such that $f(v) > C$.
Order the children of v by increasing flow: $v_1, v_2, \ldots$ with $f(v_1) \leq f(v_2) \leq \cdots$.
Let j be the smallest integer such that $\sum_{i \leq j} f(v_i) \geq C$.
prime the node v.

- *if v is not the root (generate the residual problem):*
  transfer the children $v_{j+1}, v_{j+2}, \ldots$ with their respective flow values $f(v_{j+1}), \ldots$ and the node v_j with flow value $f(v_j) - C + \sum_{i<j} f(v_i)$ to the parent of v.
- *if v is the root:*
  prime the children nodes $v_{j+1}, \ldots$. If $\sum_{i \leq j} f(v_i) > C$, prime the node v_j.

If the root node has $0 < f(r) \leq C$ and it has not been primed, then prime the root.

The implicit flow assignment made to the node v selected in each iteration of the first phase is all the flow from the children $v_1, \ldots, v_{j-1}$ and any $C - \sum_{i<j} f(v_i)$ units of flow from v_j. It is easy to see that there is one-to-one

---
[2] Note that there could be several such mappings for each optimal allocation, including mappings where nodes assigned in the first phase obtain less than C units of flow. We consider particular assignments where this holds.

correspondence between solutions of the residual instance T' and solutions of the original instance T that prime v and allocate to it these C units of flow from $v_1, \ldots, v_j$.

We can implement OPT_SOF to run in $O(n \log n)$ time where n is the number of nodes in the tree. To do that we maintain the children of every node in a heap. When processing a node v we do a series of delete-mins on the heap of v until we have accumulated the lightest children whose weight is above C. We may need to reinsert the last deleted child, and finally we meld the heap of v with the heap of the parent of v.

A distributed implementation of OPT_SOF follows the two phases. The first "upward" phase, communication occurs from the leaves to the root with nodes primed along the way. The information of transferred children is transferred up the tree. In the second "downward" phase, the root node "decided" which descendants need to be primed, and this information is passed down the tree to these nodes. The amount of communication passed when children are transferred could be (in the worst case) of the order of the number of descendants of the transferring node. It is not hard to see that the communication can be reduced considerably if an aggregation is performed of all transferred children of similar sizes. In this case, logarithmic size messages yield an approximate solution.

We now establish the correctness and optimality of OPT_SOF.

Theorem 1. OPT_SOF *produce a correct allocation of minimum size.*

We first establish correctness. A residual problem is generated after each iteration of the first phase. Once a node is primed in the first phase it gets exactly C units of flow in the residual problem and therefore no more flow can be assigned to it later on. All unserved flow that was generated below it, is transferred to its parent. The transferring is ok, since any ancestor is able to serve a unit of flow. When the first phase is completed all the children of the root in the residual network generate less than C flow. Therefore having a subset of these children serve the flow they generate, and having the root serve the rest, produces a legal solution.

We now show that the allocation produced is of minimum size. Consider an iteration of the first phase. Let T be the network when the iteration starts, let A be the primed node v along with the flow assigned to it, and let T' be the residual problem of T with respect to A. It is not hard to see that if T has an optimal solution that extends A, then any combination of A with an optimal solution of T' yields an optimal solution of T. Thus by induction on the iterations of the first phase, it suffices to show that the choice of A can be extended to an optimal solution of T. As a base case of the induction we also have to show that the solution of the residual problem obtained by the second phase is optimal, which is straightforward. The following claim shows that a choice A of OPT_SOF can be extended to an optimal solution. We omit the proof from this extended abstract.

Claim. Consider an instance T and let v be a lowest node with flow exceeding C. There exist an optimal solution with v being primed and fully utilized.

It remains to show that the flow assigned to v by OPT_SOF is also consistent with some optimal solution. This is established in the following claim and concludes the optimality proof. We omit the proof from this extended abstract.

Claim. Let v be the first node primed by OPT_SOF. Consider a optimal solution S where v is primed and fully utilized. We can restructure S such that the flow allocated to v is the same as the flow allocated to v by OPT_SOF.

We next consider algorithms where nodes make local decisions on when to become primed, by simply observing incoming flow.

5 The THRESHOLD Algorithm

The THRESHOLD(C, Δ) algorithm runs independently at each node. The only communication it receives is the amount of incoming flow and we assume each node knows Δ, the indegree of its parent node (or some bound on it).

A copy of THRESHOLD running at a node v performs as follows:

Algorithm 2 (THRESHOLD(C, Δ)) *Let $f(v)$ be the flow into the node v. The node v is primed if and only if*

- $f(v) > C/\Delta$, or
- v is the root and $f(v) > 0$.

Initially, the algorithm starts with some configuration of primed nodes (possibly no primed nodes), and each node runs a copy of THRESHOLD. Since each node makes local decisions that depend on decisions made by the nodes below it, it may change its decision multiple times. We argue that eventually, a stable allocation is reached.

In fact, THRESHOLD(C, Δ) converges to the same allocation generated by the following iterative algorithm.

- Prime all nodes such that the flow generated below them exceeds C/Δ, and the flow generated below each of their children does not exceed C/Δ.
- Truncate the tree by removing these newly primed nodes along with all nodes and flow generated below them, and repeat this process.

It is not hard to prove, by induction on the iterations of the procedure above, that THRESHOLD indeed generates the same solution as the algorithm above.

The following lemma follows from this equivalence between these two algorithms.

Lemma 2. THRESHOLD(C, Δ) *eventually reaches a correct stable allocation. The allocation produced is legal in the must-serve model and thus also applies to the serve-or-forward model. The final allocation is of size at most $R\Delta/C$ (since each primed node serves at least C/Δ flow.)*

We define the *convergence time* of THRESHOLD as the time it takes for it to reach the stable state. To bound the convergence time of THRESHOLD we let t be the time it takes for a node to respond to a change in its incoming flow, where by *responding* we mean a change of state from prime to unprime or vice versa and/or updating the outgoing flow. It is easy to prove by induction that if d is the depth of the network T then the convergence time is at most $d * t$.

How far can THRESHOLD be from the optimal solution? We describe an example where the worst-case ratio approaches Δ for the serve-or-forward model. There exists a similar example for the must-serve model that we will describe in the full version of the paper. For an integer i, construct a binary tree of depth i; and then attach Δ children to each of the 2^i leaves of the binary tree. Now suppose that each of the $\Delta 2^i$ leaves of the modified tree generates flow of value $C/\Delta + \epsilon$ ($\epsilon < C/(\Delta 2^i)$). The THRESHOLD algorithm will prime all leaves, and result in an allocation of size $\Delta 2^i$. The algorithm OPT_SoF will prime all depth-i nodes and the root node of the tree, and thus will obtain an allocation of size $2^i + 1$. The ratio of the allocation size obtained by THRESHOLD to the optimal allocation of OPT_SoF approaches Δ as i increases.

Observe that any algorithm that takes independent local decisions at each node would not be able to improve on this approximation ratio. Intuitively, since no information is available to a node on what its siblings send to its parent, the algorithm cannot allow sending more than C/Δ units of flow to the parent and at the same time guarantee that every node receives at most C units of flow.

5.1 Convergence of THRESHOLD

We learned that the worst-case final allocation obtained by THRESHOLD is within Δ of optimal. We now consider the size of "intermediate" allocations obtained during the convergence process.

Consider a tree that consists of a path of n nodes $v_1, \ldots, v_n$, (ordered from the bottom to the root); each path node has $\Delta - 1$ leaves attached to it. The $\Delta - 1$ leaves at each path node generate $\epsilon C/\Delta$ flow in total. (To simplify the presentation we assume that $1/\epsilon$ is integral.) The total flow entering the tree is $R = n\epsilon C/\Delta$. Observe that THRESHOLD will never prime a leaf node, thus only the nodes on the path get primed.

Consider the following possible convergence scenario. (1) Initially, all nodes are not primed; thus the ith path node v_i sees $i\epsilon C/\Delta$ flow. (2) Every v_i for $i \geq 1/\epsilon$ receives more than C/Δ flow and gets primed. (3) Next, all primed nodes above $v_{1+1/\epsilon}$ see only $\epsilon C/\Delta$ flow and unprime themselves. (4) The lowest primed path node $v_{1/\epsilon}$ remains primed from now on, since all nodes below it remain unprimed. (5) The process continues on the new tree obtained after truncating the current tree at $v_{1/\epsilon}$. The new tree has the same structure with $n' = n - 1/\epsilon$ interior path nodes.

It is not hard to see that convergence takes $\epsilon n = R\Delta/C$ such "iterations." The average number of nodes that change state (from primed to unprimed or vice versa) in an iteration is $\Omega(n)$. The largest intermediate allocation occurs at the first iteration and is of size $n - 1/\epsilon = (R\Delta/C - 1)/\epsilon$. That is, there is a factor of $1/\epsilon$ between intermediate and final allocation sizes. [3]

To summarize, we showed that the simple local THRESHOLD algorithm converges to an allocation of size $O(\Delta R/C)$. It can have, however, undesirable convergence patterns. In the next section we propose and analyse a different local algorithm, GAP.

6 The GAP Algorithm

The GAP algorithm runs locally at each node. Similarly to THRESHOLD, decisions are made locally, but unlike THRESHOLD, decisions made at each node are independent of the state of other nodes. The only information GAP relies on at each node is the total amount of flow generated below each of its children (regardless of where it is served). In practice, this can be accomplished by periodic reports from children to parents.

Let $F(v)$ be the total flow generated under a node v and let $F_s(v)$ be the flow generated under its heaviest child, that is $F_s(v) = \max\{F(u)|u \in \text{children}(v)\}$. If v is a leaf then $F_s(v) = 0$.

Algorithm 3 (GAP(C, μ) algorithm) *At each node v, GAP(C, μ) performs the following.*

1. *If $F(v) < \mu C$, then v is not primed.*
2. *If $F(v) \geq \mu C$ then*
   - *if $F_s(v) < \mu C$, v is primed.*
   - *otherwise, v is primed with probability $\min\{1, (F(v) - F_s(v))/(\mu * C)\}$.*
3. *If v is the root, it did not get primed in 2, and it has flow entering it, then prime v.*

In our subsequent analysis, we assume that only leaf nodes have flow below μC. We do that without loss of generality since the operation of GAP on an instance is equivalent to GAP operating on the same instance when maximal subtrees rooted at a node v with $F(v) < \mu C$ are contracted to a single leaf node with flow $F(v)$.

We first bound the expected size of the allocation produced by GAP. (The proof is omitted.)

---

[3] Interestingly, these undesirable properties are not merely an artifact of the "sharp" threshold exhibited by THRESHOLD: a similar convergence pattern with very large size of intermediate allocation and slow convergence time is possible even under a natural modification of THRESHOLD. In this modification we prime a node when the incoming flow exceeds C/Δ and unprime a node when its flow drops to L/Δ or below (for some $\epsilon C < L < C$). This variant of THRESHOLD would result in worse-quality final allocations (which can be C/L times the allocation size obtained by the original version of THRESHOLD.) The worst-case convergence patterns, however, are similar.

Lemma 3. *The expected number of nodes primed by* GAP *at step 2 is at most* $2R/(\mu * C) - 1$, *for* $R \geq \mu C$.

Since GAP is a randomized algorithm, there is positive probability that nodes are assigned flow that exceeds C. The following lemma gives probabilistic bounds on the maximum flow entering a node when we apply GAP. Notice that since each leaf generates at most C units of flow, it is clear that a primed leaf never serves more than C units of flow. So the lemma only considers the maximum flow entering an internal prime node.

Lemma 4. *In the must-serve model, the expected amount of flow serviced by a primed internal node is at most* $\Delta * \mu * C$. *Furthermore, the probability that the serviced flow exceeds* $i * \Delta * \mu * C$ *is at most* $(1 - 1/e)^i$.

Proof. To simplify the calculations, we normalize the flow to be in units of μC. In these units we first have to show that the expected amount of flow serviced by a primed node is at most Δ.

Consider an internal node v_0. Since v_0 is internal $F(v_0) \geq 1$. Consider a child v_1 of v_0. We claim that the expected value of flow that v_0 receives from v_1 is at most 1. Thus the total expected flow passed to v_0 from all children combined is at most Δ.

We now prove the claim. We can assume without loss of generality that $F(v_1) > 1$ (since otherwise the claim trivially follows). Let v_2 the child of v_1 with the largest amount of flow generated below it, and in general let v_{i+1} be the child of v_i with the largest amount of flow generated below it. We call $v_1, v_2, \ldots, v_\ell$ the *heavy path* hanging from v_1. Note that v_ℓ is a leaf. Let $x_i = F(v_i) - F_s(v_i) = F(v_i) - F(v_{i+1})$ ($i \geq 0$) be the difference in flow associated with the edge (v_{i-1}, v_i) of the heavy path. We define a node to be *always primed* if $x_i \geq 1$, or $F(v_i) \geq 1$ and either v_i is the last node on the path or $F(v_{i+1}) < 1$. (Note that the latter can happen only if v_i is the last or next to last node on the path since we assume that $F(v) < 1$ only if v is a leaf.) It is easy to see that v_ℓ or $v_{\ell-1}$ must be always primed. Therefore there is an always primed node on the heavy path. Let k be the maximum such that v_k is not "always primed". Observe that we must have that $x_i < 1$ for $1 \leq i \leq k$. (That is v_{k+1} is the first always primed node on the path.)

Consider a run of dice and let $p \leq k$ be the lowest index such that v_p is primed; define $p = k + 1$ if there is no primed node before v_{k+1} on the path. By definition, the amount of flow v_0 gets from v_1 is at most $\sum_{j \leq p} x_p$. The probability that v_j ($1 \leq j \leq k$) is the lowest-indexed primed node on the path is $x_j \Pi_{b=1}^{j-1}(1 - x_b)$. It follows from the above that the expected amount of flow passed to v_0 from v_1 is at most

$$g_1 = (1-x_1)(x_1 + (1-x_2)(x_2 + (1-x_3)(\cdots + (1-x_{k-1})(x_{k-1} + (1-x_k)x_k)\cdots))).$$

We show that $g_1 \leq 1$. Define $g_i = (1 - x_i)(x_i + (1 - x_{i+1})(x_{i+1} + \ldots))$. ($g_i$ is the expected flow value passed to v_{i-1} from v_i). By definition we have $g_{i-1} = (1 - x_i)(x_i + g_i)$. We show by a backward induction on i that $g_i \leq 1$. It is easy to see that $g_k = (1 - x_k)x_k \le 1$. Assume inductively that $g_i \le 1$, then $g_{i-1} = (1 - x_i)(x_i + g_i) \le (1 - x_i)(x_i + 1) \le 1$.

We now prove the second part of the lemma. We show that the probability, over runs of GAP, that g_i exceeds L is at most $(1 - 1/e)^L$. Let G_i be the random variable that corresponds to the value of g_i in different runs of GAP. We show the claim by downward induction on i.

The base of the induction is the r.v. G_k. The outcome is x_k with probability $1 - x_k$ and 0 otherwise. The likelihood that the outcome exceeds $L > x_k$ is 0. The likelihood that it exceeds $L \le x_k$ is $(1 - x_k) \le (1 - 1/e)^{x_k} \le (1 - 1/e)^L$ (we use the tautology $(1 - a) \le (1 - 1/e)^a$ for $(0 < a < 1)$.)

For the r.v. G_i we have the relation

$$P[G_i > L] \le (1 - x_i)P[G_{i+1} \ge L - x_i] .$$

From the induction hypothesis we have $P[G_{i+1} \ge L - x_i] \le (1 - 1/e)^{L-x_i}$. Substituting the above we obtain that

$$P[G_i > L] \le (1 - x_i)(1 - 1/e)^{L-x_i} \le (1 - 1/e)^L .$$

The last inequality follows again from the tautology $(1 - a) \le (1 - 1/e)^a$ for $(0 < a < 1)$. $\square$

The analysis of GAP implies that if we set $\mu = \Theta(1/(\Delta \log(R\Delta/C)))$, then there is a constant probability that the maximum load served by a primed node is at most C. In the full version of this paper we show how we can use $\mu = \Theta(1/\Delta)$ together with iterative application of GAP in order to obtain better tradeoff.

References

[1] B. Awerbuch, Y. Bartal, and A. Fiat. Distributed paging for general networks. In *Proc. 7th ACM-SIAM Symposium on Discrete Algorithms*, pages 574–583. ACM-SIAM, 1996.
[2] Open Source Community. Gnutella. In *http://gnutella.wego.com/*, 2001.
[3] Napster Inc. The napster homepage. In *http://www.napster.com/*, 2001.
[4] D. Karger, Lehman E., T. Leighton, M. Levine, D. Lewin, and R. Panigraphy. Consistent hashing and random trees: distributed caching protocols for relieving hot spots on the World Wide Web. In *Proc. 29th Annual ACM Symposium on Theory of Computing*, pages 654–663. ACM, 1997.
[5] M. R. Korupolu, C. G. Plaxton, and R. Rajaraman. Placement algorithms for hierarchical cooperative caching. In *Proc. 10th ACM-SIAM Symposium on Discrete Algorithms*. ACM, 1999.
[6] B. Li, M. Golin, G. Italiano, X. Deng, and K. Sohraby. On the optimal placement of Web proxies in the Internet. In *Proceedings of the IEEE Infocom '99 Conference*, 1999.
[7] Q. Lv, P. Cao, E. Cohen, K. Li, and S. Shenker. Search and replication in unstructured peer-to-peer networks. In *Proceedings of the ACM SIGMETRICS'02 Conference*, 2002.

[8] C. G. Plaxton, R. Rajaraman, and A. W. Richa. Accessing nearby copies of replicated objects in a distributed environment. In *Proc. 9th annual ACM Symposium on Parallel Algorithms and Architectures*, pages 311–320. ACM, 1997.
[9] S. Ratnassamy, P. Francis, M. Handley, R. Karp, and S. Shenker. A scalable content-addressable network. In *Proceedings of the ACM SIGCOMM'01 Conference*, 2001.
[10] I. Stoica, R. Morris, D. Karger, and H. Frans Kaashoek, M. ad Balakrishnana. Chord: A scalable peer-to-peer lookup service for internet applications. In *Proceedings of the ACM SIGCOMM'01 Conference*, 2001.

Butterflies and Peer-to-Peer Networks

Mayur Datar

Department of Computer Science, Stanford University
Stanford, CA 94305, USA
datar@cs.stanford.edu

Abstract. Research in Peer-to-peer systems has focussed on building efficient Content Addressable Networks (CANs), which are essentially distributed hash tables (DHT) that support location of resources based on unique keys. While most proposed schemes are robust to a large number of random faults, there are very few schemes that are robust to a large number of adversarial faults. In a recent paper ([2]) Fiat and Saia have proposed such a solution that is robust to adversarial faults.

We propose a new solution based on multi-butterflies that improves upon the previous solution by Fiat and Saia. Our new network, *multi-hypercube*, is a fault tolerant version of the hypercube, and may find applications to other problems as well. We also demonstrate how this network can be maintained dynamically. This addresses the first open problem in the paper ([2]) by Fiat and Saia.

1 Introduction

Peer-to-peer (P2P) systems are distributed systems without (ideally) any centralized control or hierarchical organization, which make it possible to share various resources like music [8, 4], storage [6] etc over the Internet. One approach to building P2P systems is to build a distributed hash table (DHT) that supports location of resources based on their unique key. Such a network is called a content addressable network (CAN) and various solutions ([9, 10, 12]) have been proposed for building efficient CANs.

While most schemes for building CANs are fairly robust against random attacks, a powerful agent like a government or a corporate can attack the system by carefully deleting (making faulty) chosen points or nodes in the system. For instance, the Gnutella [4] file sharing system, while specifically designed to avoid the vulnerability of a central server, has been found (refer [11]) to be highly vulnerable to an attack by removing a very small number of carefully chosen nodes. Thus it is unclear if any of these systems are robust against massive orchestrated attacks.

Recent work by Fiat and Saia [2] presents a CAN with n nodes that is censorship resistant, i.e. fault tolerant to an adversary deleting up to a constant fraction[1] of the nodes. It is clearly desirable for a P2P system to be censorship

---

[1] The paper provides a system that is robust to deletion of up to half the nodes by an adversary. It can be generalized to work for arbitrary fraction.

resistant and to the best of our knowledge this is the first such scheme of its kind. However, a drawback of the solution presented in [2] is that it is designed for a fixed value of n (the number of participating nodes) and does not provide for the system to adapt dynamically as n changes. In fact the first open problem that they mention in their paper (Sect. 6 of [2]) is the following: "*Is there a mechanism for dynamically maintaining our network when large numbers of nodes are deleted or added to the network? ..*"

This paper solves this open problem by proposing a new network that can be maintained dynamically and is censorship resistant. Our new network, *multi-hypercube*, is a fault tolerant version of the hypercube network and may find applications to other problems as well. We first present a static solution that is much simpler than that presented in [2] and improves upon their solution. Next we show how we can dynamically maintain our network as nodes join and leave.

A drawback of our solution is that it requires a reconfiguration after an adversarial attack (details in Sect. 3.1). This does not involve adding new edges or nodes to the network. It only involves sending messages along existing edges to label some nodes as "faulty". This reconfiguration step requires $O(\log n)$ time and $O(n \log n)$ messages are sent, where n is the number of nodes in the network.

We present a table (refer to Table 1) that compares all of these solutions based on some important factors. The parameter n is the number of nodes participating in the network.

Table 1. Comparison of recent solutions

Network	linkage (degree)	query cost (path length)	Messages per query	Fault Tolerance	Dynamic	Data Replication factor
CAN [10]	$O(d)$	$O(n^{1/d})$	$O(n^{1/d})$	?	Yes	$O(1)$
Chord [12]	$O(\log n)$	$O(\log n)$	$O(\log n)$	Random	Yes	$O(1)$
Viceroy [9]	7	$O(\log n)$	$O(\log n)$	?	Yes	$O(1)$
CRN [2]	$O(\log n)$	$O(\log n)$	$O(\log^2 n)$	Adversarial	No	$O(\log n)$
MBN (this paper)	$O(\log n)$	$O(\log n)$	$O(\log n)$	Adversarial	No	$O(1)$
DMBN (this paper)	$O(\log n)$	$O(\log n)$	$O(\log n)$	Adversarial	Yes	$O(1)$

Paper Organization: We begin by briefly reviewing some of the related work in Sect. 2. In Sect. 3 we present a simpler and better censorship resistant network. Section 4 provides a dynamic construction of our network. Finally we conclude with a discussion of open problems in Sect. 5.

2 Related Work

Most CANs are built as an overlay network. The goal is to build a CAN with short query path length since it is directly related to the latency observed by the node that issues the query. Besides a small query path length, other desirable features of a solution include low degree for every node, fewer messages per query, fault tolerance etc.

Recently various solutions – *CAN* ([10]), *Chord* ([12]), *Viceroy* ([9]) etc. – have been proposed to building CANs. Please refer to Table 1 for a comparison of their various performance parameters. A common feature to all of these solutions is the use of an underlying abstract hash space to which nodes and data items are hashed. It is critical to all these schemes that the data items are hashed uniformly and deterministically based on their keys, so that any node can compute the hash value of a data item solely based its unique key.

The *CAN* system designed by Ratnasamy, Francis et al [10] uses a virtual d-dimensional (for a fixed d) Cartesian coordinate space on a d-torus as its hash space. The hash space used by *Chord* [12] and *Viceroy* [9] is identical. It can be viewed as a unit circle $[0, 1)^2$ where numbers are increasing in the clockwise direction. The *Chord* system tries to maintain an approximate hypercube network in a dynamic manner, while the *Viceroy* system tries to maintain an approximate butterfly network in a dynamic and decentralized manner.

The censorship resistant network (henceforth CRN) developed by Fiat and Saia [2] departs from all of the above solutions in trying to provide adversarial fault tolerance. The aim is to build a network such that even after an adversary deletes (makes faulty) up to $n/2$ nodes, $(1 - \epsilon)$ fraction of the remaining nodes should have access to $(1 - \epsilon)$ fraction of the data items, where ϵ is a fixed error parameter. However, their solution assumes that there are n nodes participating in the network, where n is fixed. Their solution can be extended to form a network that is *spam resistant*, i.e. resistant to an adversary that can not only delete a large number of nodes but also make them collude so that they forward arbitrary false data items (or messages) during query routing. In a very recent paper ([3]), the authors have extended their previous work to build a CAN that is *dynamically fault-tolerant*. Their notion of *dynamic* differs from ours and as per their notion of *dynamic fault-tolerance*, the network is built for a certain value n of the number of participating nodes. However, there may be a large turn around of the participating nodes and during any time period for which the adversary deletes γn nodes, δn ($\delta > \gamma$) new nodes are added to the system, always maintaining that there are at least κn live nodes for some fraction κ. At every time instant the network should remain censorship resistant. They build such a system that is an adaptation of CRN and has properties similar to it.

3 Multi-butterfly Network (Multi-hypercube)

In this section we present a censorship resistant network based on multi-butterflies, which we refer to as MBN (Multi-Butterfly Network). Our solution is better than CRN in the following respects:

1. While routing in CRN requires $O(\log^2 n)$ messages, routing in our network requires $O(\log n)$ messages.

---

[2] While the *Chord* paper [12] describes their hash space as "identifier circle modulo 2^m" the two are equivalent

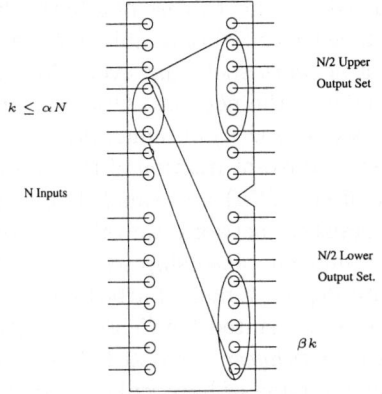 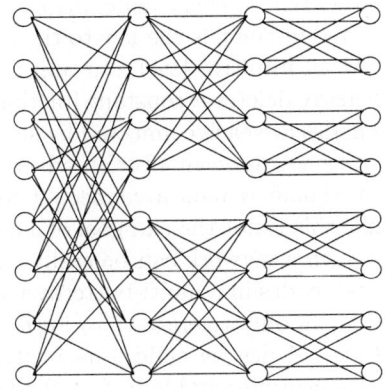

Fig. 1. Splitter with N inputs and N outputs

Fig. 2. Twin Butterfly with 8 inputs

2. The data replication factor in CRN is $O(\log n)$. i.e. every data item is stored at $O(\log n)$ nodes. In our network the data replication factor is $O(1)$.
3. The data availability in our network degrades smoothly with the number of adversarial deletions. No such guarantees are given for CRN.

Similar to [2] we first present a static version, where the number of participating nodes(n) is fixed. Later, we will provide a dynamic construction that maintains an *approximate version* of this network and has similar properties.

Our construction is based on multi-butterflies. Fig. 2 shows a twin-butterfly. Multi-butterfly networks were introduced by Upfal [13] for efficient routing of permutations and were later studied by Leighton and Maggs [7] for their fault tolerance. Please refer to their papers for details. Butterflies and Multi-butterflies belong to the class of splitter networks, whose building block is a splitter (Fig. 1). In an N-input splitter of a multi-butterfly with multiplicity d, every input node is connected to d nodes from the upper and lower output nodes. Similarly every output node has edges from $2d$ input nodes. The splitter is said to have (α, β)-expansion if every set of $k \leq \alpha N$ input nodes is connected to at least βk upper output nodes and βk lower output nodes, where $\alpha > 0$ and $\beta > 1$ are fixed constants (Refer to Fig. 1). Thus, the N input nodes and $N/2$ upper (lower) output nodes form a concentrator with (α, β)-expansion. A multi-butterfly is said to have (α, β)-expansion if all its splitters have (α, β)-expansion. Splitters are known to exist for any $d \geq 3$, and they can be constructed deterministically in polynomial time [13], but randomized wirings will typically provide the best possible expansion. In fact, there exists an explicit construction of a splitter with N inputs and any $d = p+1$, p prime, and $\beta \leq d/(2(d-4)\alpha + 8)$ (Corollary 2.1 in [13]).

In the case of a butterfly network any pair of input and output nodes are connected by a unique (bit correcting) logical path. However, in the case of a multi-butterfly there is a lot of redundancy since there is a choice of d edges to choose from, at every node, instead of a single edge as in the case of a butterfly. This redundancy is the key to the fault tolerance (refer [7]) of a multi-butterfly.

At an intuitive level, our aim is to build a network such that even after an adversary deletes a constant fraction of the nodes in the network, $\Omega(n)$ remaining nodes are each connected by small length paths to $\Omega(n)$ of remaining nodes. In other words, even after an adversary deletes a constant fraction of the nodes, there should remain a connected component of size $\Omega(n)$ and small diameter. While this is not the end goal of a censorship resistant network, we will see later how such a network can be easily enhanced to make it censorship resistant. Although nodes in a multi-butterfly have constant degree, a multi-butterfly with n nodes can only tolerate $O(n/\log n)$ faults and is not suited for Censorship Resistance. Hence we build a new network called *multi-hypercube* that is based on multi-butterflies and is in fact the fault tolerant version of hypercube. In short, a multi-hypercube is to a multi-butterfly as a hypercube is to a butterfly. If the role of all the nodes in a single row of a multi-butterfly is played by a single node then what we get is a multi-hypercube[3]. Consider an N input splitter in a multi-hypercube. In this splitter, upper $N/2$ input nodes are connected to lower $N/2$ output nodes (and vice versa), via an expander of degree d. As a result we get better expansion factor (β) for the same degree as compared to that in a multi-butterfly, where instead of a $(N/2, N/2)$ expander we have a $(N, N/2)$ concentrator. To the best of our knowledge this network has not been studied earlier, neither are we aware of the use of the term multi-hypercube. A formal definition follows:

Multi-hypercube: A multi-hypercube of dimension m and multiplicity d consists of 2^m nodes, where every node has degree $2md$. A node with binary representation $b_1 b_2 \ldots b_m$ is adjacent to $2d$ nodes at each level i ($1 \leq i \leq m$). At level i it has "out-edges" with d nodes whose first i bits are $b_1 b_2 \ldots b_{i-1} \overline{b_i}$. It also has "in-edges" from d nodes belonging to the same set, i.e. nodes with first i bits given by $b_1 b_2 \ldots b_{i-1} \overline{b_i}$. The connections are such that the expansion property holds for every splitter, like in the case of a multi-butterfly.

Thus, a multi-hypercube with n nodes has degree $2d \log n$ for each node. A multi-hypercube is a fault tolerant version of the hypercube network, and turns out to be ideal for censorship resistance. We hope that this network will find other applications as well.

Given n, the network that we build is a multi-hypercube with n nodes and (α, β) expansion. The fault tolerance property that we will prove (Theorem 1) about the multi-hypercube is the exact equivalent of the corresponding property for a multi-butterfly. We refer to this network as the Multi-butterfly network (MBN), since we prefer to visualize it as a multi-butterfly. Data items are deterministically and uniformly hashed onto nodes using their unique keys. Thus the data replication factor is 1 and using consistent hashing, as in [12], we can guarantee that whp the load on any node is at most $O(\log n)$ times the expected

---

[3] The caveat is that in every splitter we only maintain the "cross" edges and not the "straight" edges

average load. It is straightforward to construct our network in a distributed manner. The construction requires 2 broadcasts from every node, with a total of $2n^2$ messages sent and assumes that each node has $O(\log n)$ memory, similar to the creation of CRN. Please refer to the full version of the paper ([1]) for details which are omitted here due to lack of space.

Routing: Every node (source) that wishes to access a data item computes the hash of its key and finds out the index of the node the data item belongs to (destination). Routing between the source and destination is done using the standard, logical *bit-correcting* path as in a hypercube. Due to the redundancy in connections, in a fault free multi-hypercube we will have a choice of d out edges at each level (splitter) of the routing. This choice may be reduced if some of the nodes become faulty, as we shall see later. The number of messages sent for a single query is at most $\log n$ and time taken is also $\log n$.

3.1 Fault Tolerance

In this subsection we prove the fault tolerance for our network. The proof is similar to that presented in [7]. We will view the multi-hypercube as a multi-butterfly where a single node plays the role of all the nodes in a row of the multi-butterfly. In the discussion below we will refer to nodes on level 0 (leftmost level in the figures) as input nodes and nodes on level $\log n$ as output nodes treating them separately. But in reality same node is playing the role of all the nodes in a row. We prove the following theorem:

Theorem 1. *No matter which f nodes are made faulty in the network, there are at least $n - \frac{\beta f}{\beta - 1}$ nodes that still have a $\log n$ length logical path to at least $n - \frac{f}{\alpha(\beta-1)}$ nodes, such that all nodes on the path are not faulty, where (α, β) are the expansion parameters for every splitter.*

We first describe which outputs to remove. Examine each splitter in the multi-butterfly and check if more than $\epsilon_0 = \alpha(\beta - 1)$ fraction of the input nodes are faulty. If so, then "erase" the splitter from the network as well as all descendants nodes, i.e all nodes to the right of the splitter. The erasure of an m-input splitter causes the removal of m multi-butterfly outputs, and accounts for at least $\epsilon_0 m$ faults. Moreover, since a (faulty) node plays the role of all nodes in the same row the output nodes "erased" by it, by virtue of it being in different splitters, are the same. Thus we can attribute the erasure of an output node to a unique largest splitter that "erased" it. Hence, at most $\frac{f}{\epsilon_0} = \frac{f}{\alpha(\beta-1)}$ outputs are removed by this process. We next describe which inputs to remove. Remember, for every splitter each input node is connected to d nodes from either upper or lower outputs depending on its position in the splitter. Working from the $\log n$th level backwards, examine each node to see if all of its outputs lead to faulty nodes that have not been erased. If so, then declare the node as faulty. We prove that at most $f/(\beta-1)$ additional nodes are declared to be faulty at each level of this process. The proofs for the following two lemmas are omitted for lack of space and can be found in the full version of the paper ([1]).

Lemma 1. *In any splitter, at most α fraction of the inputs are declared to be faulty as a consequence of propagating faults backward. Moreover, at most $\alpha/2$ fraction are propagated by faulty upper outputs and at most $\alpha/2$ fraction by faulty lower outputs.*

Lemma 2. *Even if we allow the adversary to make f nodes faulty on every level there will be at most $\frac{f}{\beta-1}$ propagated faults on any level.*

We erase all the remaining faulty nodes. The process of labeling nodes faulty guarantees that an input node that is not faulty has a path to all the output nodes that are not erased. This leaves a network with $n - \frac{\beta f}{\beta-1}$ input nodes and $n - \frac{f}{\alpha(\beta-1)}$ outputs nodes such that every remaining input has a logical path to every remaining output. The process of marking nodes "faulty" is important to the efficient functioning of a CAN after an adversarial attack. It tells every node the set of nodes it should not forward a query, reducing the earlier choice of d nodes it had at each level. While the algorithm above gives an off-line (centralized) algorithm to label nodes faulty what we require is an online algorithm that lets us do this without requiring a central authority. It was shown in [5] that such an algorithm exists. In other words we can reconfigure a faulty network in an online manner with just the live nodes talking to each other. This reconfiguration step requires $O(\log n)$ time and $O(n \log n)$ messages are sent. Please refer to [5] for details. It is required that we do this reconfiguration after an adversarial attack. Since we cannot figure out when the attack has occurred, we suggest that the network does this reconfiguration at regular intervals. During the time interval after an adversary attack and before the reconfiguration is done, it may happen that a node may forward a query to another node that is "faulty", but not yet so marked, and consequently the query may not reach the destination. In such a case the node can retry forwarding the query through another node, after waiting for a "timeout", hoping that it is not faulty. However such retry's can take a lot of time and we may end up sending a lot of messages. This temporary lack in query routing efficiency of the network should be contrasted with the advantage that it has of requiring fewer messages per query.

One possible choice of parameters we may choose for our network are as follows: Choose the multiplicity d such that $\alpha(\beta-1) \geq 2/3$ and $\beta \geq 3$. Substituting these values in the Thm. 1 gives us that no matter which f nodes are made faulty, there are at least $n - \frac{3f}{2}$ nodes that can each reach $n - \frac{3f}{2}$ nodes through a logical path. Note that the guarantee above is deterministic as opposed to whp, as in case of CRN([2]). Moreover we can characterize the "loss" smoothly in our case. Thus if we loose $f = \sqrt{n}$ nodes we know that all but $n - \frac{3\sqrt{n}}{2}$ nodes can reach all but $n - \frac{3\sqrt{n}}{2}$ nodes. Such guarantees are not given in the case of CRN, which does not characterize the behavior when the number of faults is sublinear.

Enhancement: Theorem 1 guarantees that no matter which $n/2$ nodes are made faulty by the adversary, there are $n/4$ remaining nodes (call this set I) that are each connected by $\log n$ length logical paths to $n/4$ nodes (call this set O). However, this by itself is not sufficient to give us a censorship resistant network. We achieve censorship resistance by further enhancing the network as follows: Every node in the network is additionally connected to $k_1(\epsilon)$ (a constant that is function of the error parameter ϵ) random nodes in the network. Moreover, every data item is maintained at $k_2(\epsilon)$ nodes in the network, which are specified by $k_2(\epsilon)$ independent hash functions. First step increases the degree of every node by an additive constant, while the second step makes the data replication factor $k_2(\epsilon)$ (a constant) instead of 1.

Routing in this enhanced network takes place as follows: A node x that is looking for a data item y computes the $k_2(\epsilon)$ nodes (call this set O') that the data item will be maintained at, using to the different hash functions. Let I' denote the set of $k_1(\epsilon)$ random nodes that the node x is connected to as above. For every pair of nodes $(a,b) \in I' \times O'$, from the cross product of I' and O', the node uses a as a proxy to route the query to b. Note that the cross product $I' \times O'$ has a constant size. Using Lemma 4.1 from [2] it is easy to see that for all but ϵ fraction of the nodes the set I' contains a node from I, i.e. $I' \bigcap I \neq \phi$. Similarly, all but ϵ fraction of the data items are maintained at some node in O, i.e $O' \bigcap O \neq \phi$. The above two conditions guarantee a successful query. However, the guarantees in this extension are not deterministic. We can summarize the properties of the network in the following theorem.

Theorem 2. *For a fixed number of participating nodes n, we can build a MBN such that:*

- *Every node has indegree and outdegree equal to $d \log n$.*
- *The data replication factor is $O(1)$.*
- *Query routing requires no more than $\log n$ hops and no more than $\log n$ messages are sent.*
- *Even if f nodes are deleted (made faulty) by any adversary at least $n - \frac{3f}{2}$ nodes can still reach at least $n - \frac{3f}{2}$ nodes using $\log n$ length paths.*
- *This network can be enhanced so that as long as the number of faults is less than $n/2$, whp $(1 - \epsilon)$ fraction of the live nodes can access $(1 - \epsilon)$ fraction of the data items.*

4 Dynamic Multi-butterfly Network

In this section we will describe how to dynamically maintain the MBN described in the earlier section in an "approximate" manner, as n changes over time. The network that we build has the following properties:

- Every node will be connected to $O(\log n)$ other nodes. Query requires $O(\log n)$ time and $O(\log n)$ messages are sent during each query.
- The fault tolerance of the network will be similar to that of MBN. Namely, if at any time there are f adversarial faults, $n - O(f)$ nodes still have $O(\log n)$ length path to $n - O(f)$ of the nodes.

- We assume that there are no adversarial faults while the network builds. While, at every instant the network that is built is fault tolerant to adversarial faults, we cannot add more nodes to the network once adversarial faults happen. In other words our network admits only one round of an adversarial attack. We do however allow random faults as the network builds.

We refer to our dynamic network as DMBN for Dynamic Multi-Butterfly Network. Similar to *Chord* [12] and *Viceroy* [9] we hash the nodes and data items onto a unit circle $[0,1)$ using their ip-address, keys etc. We refer to the hash value as the identifier for the node or data item. We assume that the precision of hashing is large enough to avoid collisions. For $x \in [0,1)$, $Successor(x)$ is defined as the node whose identifier is clockwise closest to x. A data item with identifier y is maintained at the node $Successor(y)$. We also maintain successor and predecessor edges similar to *Chord* and *Viceroy*. In these respects our network is exactly similar to *Chord*. While *Chord* tries to maintain an approximate hypercube we try to maintain an approximate multi-hypercube. In order to do so we need to define an appropriate notion of splitters and levels. Consider a dyadic interval $I = [z, z + 1/2^i)$ ($i \geq 0$). This interval is further broken into two intervals $I_l = [z, z + 1/2^{i+1}), I_u = z + 1/2^{i+1}, z + 1/2^i)$. Let $S = S_l \bigcup S_u$ be the set of nodes whose identifiers belong to the intervals I, I_l, I_u respectively . The sets of nodes S_u, S_l along with the edges between them form a splitter in DMBN. As in MBN, all nodes in S_u, maintain outgoing edges with d random nodes from S_l and vice versa. The index i that determines the width of the dyadic interval defines the level to which this splitter belongs.

4.1 Definitions and Preliminaries

We will refer to a node with identifier x as node x. Let $x = 0.x_1x_2x_3\ldots x_p$ be the binary representation of x, where p is the precision length.

Definition 1. *A dyadic interval pair (DIP) for x ($0 \leq x < 1$) at level i ($i \geq 0$) is defined as $DIP(x,i) = \{[0.x_1x_2\ldots x_i, 0.x_1x_2\ldots x_i 1 = 0.x_1x_2\ldots x_i + 1/2^{i+1}), [0.x_1x_2\ldots x_i 1, 0.x_1x_2\ldots x_i + 1/2^i)\}$.*

Thus $DIP(x,i)$ are two consecutive intervals of length $1/2^{i+1}$ which agree on x on the first i bits. The nodes that belong to $DIP(x,i)$ form a level i splitter of the multi-hypercube that we are trying to maintain in an approximate manner. Every node in this interval pair maintains edges with d random nodes from the other interval in the pair. This other interval is defined below.

Definition 2. *A dyadic interval (DI) for x ($0 \leq x < 1$) at level i ($i \geq 0$) is defined as $DI(x,i) = [0.x_1x_2\ldots x_i\overline{x_{i+1}}, 0.x_1x_2\ldots x_i\overline{x_{i+1}} + 1/2^{i+1})$.*

Lemma 3. *For any x ($0 \leq x < 1$) and $i \geq 0$, let k_u and k_l be the number of nodes whose identifiers belong to the 2 intervals in the dyadic interval pair $DIP(x,i)$. If $k = k_u + k_l \geq c\log n$ for some constant c, then with probability at least $1 - 1/n^2$, $max(\frac{k_u}{k_l}, \frac{k_l}{k_u}) < 2$.*

The Lemma says that if the dyadic interval pair is fairly populated it will be evenly balanced. The proof (refer to the full version [1]) follows from a trivial application of Chernoff bounds along with the observation that any node is equally likely get hashed onto any of the 2 intervals.

If n is the number of participating nodes, whp every node can estimate $\log n$ within a constant factor. Refer to Claim 4.1 from [9].

4.2 Dynamic Construction

Node Join: Similar to *Chord* [12] a typical query in our network is of the form $successor(x)$ where given a value x we return the node $successor(x)$. A node that joins the network has access to some live node in the network, which it will use to issue $successor$ queries as it joins the network. The following steps are executed by every new node with identifier x that joins the network:

- Similar to *Chord* it finds $successor(x)$ and establishes edges to successor and predecessor nodes.
- Similar to *Chord*, all data items that are currently held by $successor(x)$, but have identifiers less than x are transferred to x.
- $i = 0, done = false$
  do
    - Check if the total number of nodes in the interval $DI(x,i)$ exceeds $c \log n$ for some constant c (based on Lemma 3). This can be easily done using a single successor query and then following successor edges till we encounter $c \log n$ nodes or overshoot the interval.
    - If there are less than $c \log n$ nodes in $DI(x,i)$ connect to all of them, set $done = true$.
    - Else choose d random values $(r_1^i, r_2^i, \ldots, r_d^i)$ from the interval $DI(x,i)$. For every value z in this set issue the query $successor(z)$ and maintain an edge with this node.
    - Increment i (Move to the next level splitter)

  while(!$done$)

In short every node establishes edges with d random nodes from the dyadic interval $DI(x,i)$ for $i \geq 0$. It does so till the interval $DI(x,i)$ becomes so small (as i increases)that it contains only $c \log n$ nodes, at that point it maintains a connection to all of them. It is easy to prove that in less than $O(\log n)$ levels the number of nodes that fall into a dyadic interval reduce to $O(1)$ whp. As a result every node will maintain connections to $O(\log n)$ dyadic intervals and have degree $O(\log n)$.

Continuous Update: We will describe in short how these edges are maintained (updated) over time as nodes leave and join. Choosing d random values $(r_1^i, r_2^i, \ldots, r_d^i)$ and maintaining connection with $successor(r_k^i)$ $(1 \leq k \leq d)$ is a mechanism to guarantee that every node x maintains connection with d random nodes from $DI(x,i)$. As nodes join and leave $successor(r_k^i)$ may change. As a result it is necessary for nodes to check if the node that it maintains an edge to is indeed $successor(r_k^i)$. It is sufficient to do this check whenever the number of nodes reduce or grow by a factor 2, i.e. whenever a nodes estimate of $\log n$ changes. A node should also check for other boundary conditions like "does the smallest dyadic interval it maintains edges to have less than $c \log n$ nodes". We could also follow a more pro-active strategy such that whenever a node leaves or joins the network we adjust the connections corresponding to r_k^i's for other nodes. We omit the details for lack of space.

Node Leave: Similar to *Chord* a node that wishes to leave transfers its data items to its successor.

4.3 Routing:

Routing takes place along the usual *bit-correcting* logical path. At every level i the ith bit of the query y (data identifier) is compared with the ith bit of the node x. If the two bits match the query enters the next level. Else it is forwarded to one of the d random nodes that x connects to from $DI(x, i)$, thereby "correcting" the ith bit. Finally, as i increases the interval $DI(x, i)$ becomes so small that it has only $c \log n$ nodes, at which point x connects to all the nodes in that interval and can forward the query to $successor(y)$.

4.4 Fault Tolerance:

Consider a dyadic interval pair A, B that forms a splitter at some level i and has more than $c \log n$ nodes[4]. We know, from Lemma 3, that the interval pair is well balanced (up to a factor 2) whp. In our construction we maintain that every node in A connects to d random edges in B and vice versa. It follows from Lemma 4.1 in [2] that whp (at least $1 - 1/n^2$), the splitter formed by the interval pair A, B will have the crucial expansion property for parameters α, β that satisfy $2\alpha\beta < 1$ [5] The proof for fault tolerance follows exactly as in MBN. We replace splitters with the dyadic intervals and the arguments follow. We have to be slightly careful in our argument due to the slight imbalance in the number of nodes in a dyadic interval pair (Lemma 3). We omit the details due to lack of space. We get the following theorem, which is the equivalent of Theorem 1. Note, the guarantees are no more deterministic but instead probabilistic. The phrase *whp* in the theorem is with respect to the random hashing of nodes and the random connections maintained by nodes in different splitters.

Theorem 3. *No matter which f nodes are made faulty in the network, there are at least $n - \frac{\beta f}{\beta - 1}$ nodes that still have a $O(\log n)$ length path to at least $n - \frac{f}{\alpha(\frac{\beta}{2} - 1)}$ nodes whp.*

---

[4] If the number of nodes is less than $c \log n$ a complete bipartite graph is maintained between A, B.
[5] Factor 2 comes the slight imbalance in the number of nodes in the two intervals.

Similar to the static case(MBN), DMBN must reorganize itself after an adversary attack. In order to make the network censorship resistant we hash the data items multiple times ($k_2(\epsilon)$) and have every node connect to extra $k_1(\epsilon)$ nodes, similar to the enhancements discussed in Sect. 3.1.

It is important to note that the dynamic construction wont work after an adversarial attack. The construction assumes that all $successor(x)$ queries will be answered correctly. This is necessary for new nodes to establish their connections. However once an adversarial attack has taken place such a guarantee cannot be given. After the attack remaining nodes can still query for data and they are guaranteed to have access to most of the data. This follows from the fact that we maintained a fault tolerant network till the time of the attack.

5 Open Problems

Some open problems that remain to be addressed for fault tolerant CANs are:

- Can we build an "efficient" dynamic CAN that is fault tolerant to adversarial faults and allows dynamic maintenance even after an adversary attack, i.e. allows multiple rounds of adversary attack.
- Could multi-butterflies be used in an efficient manner to construct a spam resistant network.
- Are there lower bounds for average degree of nodes, query path length etc. for a network that is fault tolerant to linear number of adversarial faults.

Acknowledgment

The author would like to thank the anonymous referee for various helpful suggestions.

References

[1] M. Datar. Butterflies and Peer-to-Peer Networks. Detailed version available at http://dbpubs.stanford.edu/pub/2002-33.
[2] A. Fiat, and J. Saia. Censorship Resistant Peer-to-Peer Content Addressable Network. In *Proc. Thirteenth Annual ACM-SIAM Symposium on Discrete Algorithms*, Jan. 2002.
[3] J. Saia, A. Fiat, S. Gribble, A. Karlin, and S. Saroiu. Dynamically Fault-Tolerant Content Addressable Networks. In *First International Workshop on Peer-to-Peer Systems*, 2002.
[4] Gnutella website. http://gnutella.wego.com/.
[5] A. Goldberg, B. Maggs, and S. Plotkin. A parallel algorithm for reconfiguring a multi-butterfly network with faulty switches. In *IEEE Transactions on Computers*, 43(3), pp. 321-326, March 1994.

[6] J. Kubiatowicz, D. Bindel, Y. Chen, S. Czerwinski, P. Eaton, D. Geels, R. Gummadi, and S. Rhea. OceanStore: An architecture for global-scale persistent storage. In *Proc. of the Ninth International Conference on Architectural Support for Programming Languages and Operating Systems(ASPLOS 2000), Boston, MA, USA, November 2000.*

[7] T. Leighton, and B. Maggs. Expanders Might be Practical: Fast Algorithms for Routing Around Faults on Multibutterflies. In *Proc. of 30th IEEE Symposium on Foundations of Computer Science, pp. 384-389, Los Alamitos, USA. 1989.*

[8] Napster website. http://www.napster.com/.

[9] D. Malkhi, M. Naor, and D. Ratajczak. Viceroy: A Scalable and Dynamic Lookup Network. In *Proc. of the 21st ACM Symposium on Principles of Distributed Computing (PODC 2002), Monterey, CA, USA, July 2002.*

[10] S. Ratnasamy, P. Francis, M. Handley, R. Karp, and S. Shenker. A Scalable Content-Addressable Network. In *Proc. of ACM SIGCOMM 2001 Technical Conference, San Diego, CA, USA, August 2001.*

[11] S. Saroiu, P. Gummadi, and S. Gribble. A Measurement Study of Peer-to-Peer File Sharing Systems. In *Proc. of Multimedia Computing and Networking, 2002.*

[12] I. Stoica, R. Morris, D. Karger, F. Kaashoek, and H. Balakrishnan. Chord: A Scalable Peer-to-Peer Lookup Service for Internet Applications. In *Proc. of ACM SIGCOMM 2001 Technical Conference* August 2001.

[13] E. Upfal. An $O(\log N)$ deterministic packet-routing scheme. In *Journal of ACM, Vol. 39, No. 1, Jan. 1992, pp 55-70.*

Estimating Rarity and Similarity over Data Stream Windows

Mayur Datar[1,*] and S Muthukrishnan[2]

[1] Department of Computer Science, Stanford University
Stanford, CA 94305
datar@cs.stanford.edu
[2] AT&T Research
Florham Park NJ USA
muthu@research.att.com

Abstract. In the *windowed data stream model*, we observe items coming in over time. At any time t, we consider the *window* of the last N observations $a_{t-(N-1)}, a_{t-(N-2)}, \ldots, a_t$, each $a_i \in \{1, \ldots, u\}$; we are required to support queries about the data in the window. A crucial restriction is that we are only allowed $o(N)$ (often polylogarithmic in N) storage space, so not all items within the window can be archived.

We study two basic problems in the windowed data stream model. The first is the estimation of the rarity of items in the window. Our second problem is one of estimating similarity between two data stream windows using the Jacard's coefficient. The problems of estimating rarity and similarity have many applications in mining massive data sets. We present novel, simple algorithms for estimating rarity and similarity on windowed data streams, accurate up to factor $1 \pm \epsilon$ using space only logarithmic in the window size.

1 Introduction

Consider a *stream* of data elements (possibly infinite) $\ldots, a_{t-2}, a_{t-1}, a_t, \ldots$, where a_t represents the data element observed at time t. In the *windowed data stream model*, at any time t we consider the *window* of the last N observations $a_{t-(N-1)}, a_{t-(N-2)}, \ldots, a_t$, each $a_i \in \{1, \ldots, u\}$. We are allowed to query this data for any function of interest – say, we wish to compute the minimum or the median of these items. Intuitively, in this model, we consider only a recent window within a continuing stream of observations, letting aged data disappear from the computation. The first crucial aspect of this model is that the data under consideration changes over time as new observations get added to the "right" and old observations to the "left" get deleted because the window "moves" past the old items. The second crucial aspect is that in scenarios where this model is applicable, storage is at premium and we can not store the entire window in

---

* This work was done while the author was a DIMACS visitor.

memory. In particular, only $o(N)$ and $o(u)$ memory (often only memory polylogarithmic in N and u) may be used. Clearly, computing most functions *exactly* is impossible in this model[8], and hence, the focus is on approximating functions. This *windowed data stream model* was formalized in a recent paper by Datar et al. [8].

While the above model may, at first glance, appear limiting, it however arises quite naturally in several applications. Data stream of observations are now fundamental to many data processing applications. For example, telecommunication network elements such as switches and Internet routers periodically generate records of their traffic — telephone calls, Internet packet and flow traces — which are data streams [15]. Atmospheric observations — weather measurements, lightning stroke data and satellite imagery — also produce multiple data streams [23]. Emerging sensor networks produce many streams of observations, e.g., highway traffic conditions [20]. Sources of data streams — large scale transactions, web clicks, ticker tape updates of stock quotes, toll booth observations — are ubiquitous in daily life. For more applications and a survey of the recent work in the databases and algorithms community in data streams, refer to Babcock et al. [3]. Two observations hold in practice:

- In many of these applications, despite the exponential growth in the capacity of storage devices, it is not common for such streams to be stored. Instead they must be processed "on the fly" as they are produced, because, the data is produced at a very high rate and it is challenging (if not impossible) to store and manipulate them offline.
- In many of these scenarios, data analysis often involves asking questions about recent time and the goal is to make decisions based on the statistics or models gathered over the "recently observed" data elements. For example: How many distinct customers made a call through a given switch in the past 24 hours? Thus we need to discard "stale data" while gathering statistics and building models.

These two observations naturally lead to the *windowed data stream model* that we focus on in this paper. We study two fundamental problems in this model. The first requires estimating a distributional parameter (rarity) in a windowed stream and the other involves comparing two windowed streams (similarity). Both these problems are novel in this context, and they are well motivated. We present the first known streaming algorithms for both these problems: our algorithms take time and space only logarithmic in N, and provide $1 \pm \epsilon$ approximation of rarity as well as similarity. Our results thus add to the growing body of data stream algorithms. In what follows, we will define the problem and describe our results in more detail.

1.1 Problems

We define the two problems below. Recall that at any time t, we consider the window $a_{t-N-1}, a_{t-N-2}, \ldots, a_t$ of N observations. This is a multiset of items from some Universe set $[u] = \{1, \ldots, u\}$.

1. *(Rarity)* We say an item $i \in a_{t-(N-1)}, a_{t-(N-2)}, \ldots, a_t$ is α-*rare* for integer α if i appears precisely α times, i.e., $a_j = i$ for precisely α indexes j, for $j \in \{t-(N-1), \ldots, t\}$. Let $\#\alpha - rare$ denote the number of such items. Likewise, let $\#distinct$ denote the number of distinct items in $a_{t-(N-1)}, a_{t-(N-2)}, \ldots, a_t$. We define α-*rarity* ρ_α as the ratio: $\rho_\alpha = \frac{\#\alpha - rare}{\#distinct}$.
The problem is to estimate α-rarity at any time t. Thus, 1-rarity is the fraction of items that appear uniquely in the window. More generally, α-rarity is a measure of number of items that repeat exactly α times within the window.

   Previous works have studied various distributional parameters like estimating the number of distinct items or the most frequent item etc. But our measure above is more general, being able to capture the "inverse" (how many items appear x items as opposed to how many times an item i appears) relation completely, for different values of α.

2. *(Similarity)* We study the classical Jacard coefficient of similarity between two streams a and b. Let X_t be the set of distinct items in a windowed data stream $a_{t-(N-1)}, a_{t-(N-2)}, \ldots, a_t$, and let Y_t be that in windowed data stream $b_{t-(N-1)}, b_{t-(N-2)}, \ldots, b_t$. The Jacard similarity between these two windowed data streams at time t is defined as $\sigma_t = \frac{|X_t \cap Y_t|}{|X_t \cup Y_t|}$.
The problem is to estimate the similarity σ_t at any time t.

   This is the classical notion of similarity between two sets. It has been useful in estimating transitive closures [6], web page duplicate detection [2] and data mining [7], among other things.

Rarity ρ finds many data mining applications. For example, consider the data stream of IP-addresses that access any online store like Amazon. The set of rare IP-address (for the appropriate α) in a reasonable time window denotes the set of users who accessed the store website, but did not explore it in depth, for perhaps they were unhappy with their customer experience. A different, somewhat unusual application of rarity is as an indicator of a potential Denial of Service attack. Consider the data stream of IP *flows* (loosely, contiguous packets that comprise a session) that are sent to or received from a particular subnet (range of IP addresses). If the number of flows with a single packet (only the first SYN packet) is large it indicates that there is a Denial of Service attack on one or more of the servers in the subnet.

Similarly, σ too has many data mining applications. For example, let X_t (Y_t) denote the set of distinct IP-addresses that access a particular news website A (B, respectively). Estimating the similarity between these two sets in a short time window indicates if a lot of the readers of the two newspapers look to both for their news coverage. This example can be extended to any other online service.

An important observation about both ρ and σ is that in most applications, we are not interested in calculating them precisely but rather interested in only checking if their values are significantly large. In particular, we are interested in estimating them accurately provided they exceed a certain threshold value.

Thus we will assume that ρ and σ need only be estimated accurately to some additive, prespecified precision p which is a fixed fraction. Another observation is that in applications where ρ_α finds use, α is often small, so we will focus on the case of α being some constant, for most part.

1.2 Our Results

No results are known in windowed stream model for estimating either ρ or σ. We present first known results for these problems. Our results are as follows.

We present an algorithm that uses only $O(\log N + \log u)$ space, and per-item processing time $O(\log \log N)$ and estimates ρ by $\hat{\rho}$ such that $\hat{\rho} \in [1 \pm \epsilon]\rho + \epsilon p$, for any given fraction ϵ, with high probability. For estimating σ we get an identical result.

Notice that if one additionally wished to estimate just $\#\alpha-rare$ or $|X_t \cap Y_t|$ (numerators of ρ and σ respectively), then we can use known results for estimating the number of distinct elements ($\#distinct$) [8, 12] and union size of streams ($|X_t \cup Y_t|$) [13] in addition to our estimates of $\hat{\rho}$, and $\hat{\sigma}$ and get $1 \pm \epsilon$ approximation for them as well, up to relative precision p.

1.3 Related Work

There has been very little work on algorithms for the windowed data stream model. This model was defined and studied in [8] where the authors presented algorithms for keeping various aggregates (the number of 1's, sums etc) as well as the L_p norm of the window. In [5], the authors studied the problem of maintaining a uniform random sample of the items within the current window. These results do not give any bound for estimating rarity or similarity.

A related data stream model is one in which the window size is unbounded. In other words, N is the total number of items we have seen so far, from the beginning of the time. So there is no notion of recency, and the entire past is considered in estimating distributions as well as similarity. We call this the *unbounded data stream model*. Substantial amount of work has been done in this model. This includes computing L_1, L_2 norms [1, 11, 18], computing similarity [7][1], clustering [16], maintaining decision trees [9], maintaining histograms and other synopsis structures [15, 14] etc. For a survey of data stream algorithms, refer to [3]. Our windowed data stream model proves to be more challenging than the unbounded data stream model. In particular, consider the problem of exactly calculating the minimum of the items on the data stream. In the unbounded data stream model, we can maintain the current minimum as new data items arrive, trivially. In contrast, this is provably not possible in the windowed data stream model (see [8]).

There is another data stream model which is based on transactions, wherein items may be inserted or deleted arbitrarily, and we need to estimate functions

---

[1] While the algorithm is not described as a streaming algorithm, it applies to the unbounded streaming model but not for windowed model

in presence of these updates. This is called the *cash register model* in [15]. Here, a delete operation is explicitly specified as an input; in contrast, in the windowed model, the delete corresponding to the element that "falls off" of the window is implicit. Unless we buffer the entire window we have no means to explicitly figure out, at every instant, the element that is deleted in the windowed stream model.

1.4 Overview of Techniques

First consider the problem of estimating ρ. For now focus on the unbounded data stream model. One may be tempted to randomly sample each item, but we need to coordinate the sampling of an item when it applies multiple times, and this calls for sampling from the universe, rather than the individual items in the data stream.[2] We use min-wise hashing, which was introduced in [6, 2], to sample from the universe, and are able to derive an unbiased estimator for rarity. Min-wise hashing, per se, is expensive to compute, but suitable approximations can be generated [19, 21] and we show that these suffice to solve our problem.

Now let us consider the additional complication in the windowed model. Here the crux of the problem can be abstracted as follows. For each item in the data stream, we compute a hash function and we need to maintain the minimum of the hash values. Maintaining the minimum of a data stream, as we noted above, is easy for the unbounded data stream model, but impossible in the windowed model using limited (sublinear) memory. However, we use the observation that the hash values are nearly random and therefore we can keep the entire "chain" of consecutive minimums amongst these hash values in the time increasing order within our space bounds, and obtain our result. The same approach works for σ estimation as well.

We begin with preliminaries in the next section, moving on to prove our main results in the subsequent sections.

2 Preliminaries

A central tool to all our algorithms is the concept of min-wise hashing (also called min-hashing). Let π be a randomly chosen permutation over $[u] = \{1,\ldots,u\}$. For a subset $A \subseteq [u]$ the *min-hash* of A for the given permutation π ($h_\pi(A)$) is defined as : $h_\pi(A) = min_{i \in A} \pi(i)$.

Thus the min-hash of any subset is the minimum element of that subset after we have applied the given permutation to its elements. The crucial (simple to prove) property of min-hash functions is: For any pair of subsets $A, B \subseteq [u]$,

$$Pr[h_\pi(A) = h_\pi(B)] = \frac{|A \cap B|}{|A \cup B|},$$

---

[2] It is quite easy to see using techniques in [22] that random sampling directly will not work.

where the probability above is defined over the random choice of the permutation π. Applications of min-hashing use this property. The following lemma, that is based on this property, is taken from [7]:

Lemma 1. *Let $h_1(A), h_2(A), \ldots, h_k(A)$ ($h_1(B), h_2(B), \ldots, h_k(B)$) be k independent min-hash values for the set A (B, respectively). Let $\hat{S}(A, B)$ be the fraction of the min-hash values that they agree on, i.e.,*

$$\hat{S}(A,B) = |\{l | 1 \leq l \leq k, h_l(A) = h_l(B)\}|/k.$$

For $0 < \epsilon < 1$, $1 > p > 0$, $k \geq 2\epsilon^{-3}p^{-1}\log \tau^{-1}$,

$$\hat{S}(A,B) \in (1 \pm \epsilon)\frac{|A \cap B|}{|A \cup B|} + \epsilon p$$

for a prespecified precision p, with success probability at least $1 - \tau$.

The ideal family of min-hash functions is defined by the set of all permutations over $[u]$. However it requires $O(u \log u)$ bits to represent any random permutation from this family; hence, they are not useful for data stream applications. In [21, 19], a family of *approximate* min-hash functions is presented. A family of hash functions, $\mathcal{H} \subset [n] \to [n]$ (where $[n] = \{1, \ldots, n\}$), is called $\epsilon' - min - wise\ independent$ if for any $X \subset [n]$ and $x \in [n] - X$ we have

$$Pr_{h \in \mathcal{H}}[h(x) < min\ h(X)] = \frac{1}{|X|+1}(1 \pm \epsilon')$$

In particular, [19] presents a family of $\epsilon' - min - wise\ independent$ hash functions such that any function from this family can be represented using $O(\log u \log(1/\epsilon'))$ bits only and each hash function can be computed efficiently in $O(\log(1/\epsilon'))$ time; in addition, any hash function h chosen randomly from this family satisfies

$$Pr[h(A) = h(B)] = \frac{|A \cap B|}{|A \cup B|} \pm \epsilon'$$

In fact, the family proposed in [19] is the set of polynomials over $GF(u)^3$ of degree $O(\log(1/\epsilon'))$. It is now straightforward to observe that Lemma 1 still holds true if we used ϵ'-min-hash functions instead of min-hash functions; only the value of k will be appropriately adjusted in terms of ϵ' and the additive error ϵp changes to $\epsilon p + \epsilon'$. The advantage is that we only use $O(\log u \log(1/\epsilon'))$ bits for storing each hash function and $O(k)$ hashes are needed for the entire estimation which is very efficient in terms of space.

Our results that follow will crucially use min-hashing. We will describe all our results in terms of "perfect" min-hash functions. But it will be implicit that our results can be made to work with ϵ'-min-hash functions with small change in the number of hash functions needed, instead of the perfect min-hash functions; hence, the space bound will be only $O(\log u \log(1/\epsilon'))$ for generating the required hashes [21, 19]. With this implicit understanding, we will work henceforth with perfect min-wise hashing and as a result, the presentation should be simpler.

---
[3] We assume that u is a prime. If not we choose the next biggest prime.

3 Estimating Rarity in Unbounded Data Stream Model

In what follows we will present an algorithm for estimating $\rho_\alpha(t)$ up to precision p, at any time instant t, in the unbounded data stream model.

Consider a multiset S. Let $mult_S(i)$ denote the multiplicity of an element i in the multiset S. Also let D denote the set of distinct elements in the multiset S. For any integer α we can represent D as a union of two disjoint subsets, $D = R_\alpha \cup \overline{R_\alpha}$ where $R_\alpha = \{i | i \in D, mult_S(i) = \alpha\}$ and $\overline{R_\alpha} = D - R_\alpha$. In other words R_α is the set of α-rare elements in S. Our aim is to accurately estimate the ratio $\rho_\alpha = |R_\alpha|/|D|$.

We make the following simple observations: Since $R_\alpha \subseteq D$, $\frac{|R_\alpha \cap D|}{|R_\alpha \cup D|} = \frac{|R_\alpha|}{|D|}$. Using the property of min-hash functions we have $Pr[h(R_\alpha) = h(D)] = \rho_\alpha$. Thus one idea would be to maintain the min-hash values for R_α and D and estimate this ratio. Moreover, $h(R_\alpha) = h(D)$ if and only if the element corresponding to the min-value (after hashing) in D belongs to R_α. This observation precludes the need to maintain a separate min-hash value for both R_α and D. We only need to maintain a min-hash value for D and check if the count of the element corresponding to min-value is no more than α.

Based on these observations we get the following simple algorithm: We choose k min-hash functions $h_1, h_2, \ldots, h_k$ for some parameter k to be determined later. We maintain $h_i^*(t) = min_{a_j, j \leq t} h_i(a_j)$ at each time t, that is, min-hash value of the set $\{\ldots, a_{t-2}, a_{t-1}, a_t\}$. We also maintain k counters $(c_1(t), c_2(t), \ldots, c_k(t))$, one counter corresponding to each min-hash value $h_i^*(t)$. The min-hash values are all initialized to ∞ and counters are all initialized to zero, at $t = 0$. When the new element a_{t+1} is seen, we update these values as follows. For each i, we compute $h_i(a_{t+1})$. If $h_i(a_{t+1})$ is strictly smaller than $h_i^*(t)$, we update the new minimum to be $h_i^*(t+1) = h_i(a_{t+1})$ and set the counter $c_i(t+1)$ to one. If $h_i(a_{t+1}) = h_i^*(t)$, we increment the counter $c_i(t)$ by one to get $c_i(t+1)$. Else we do no further processing for element a_{t+1}, $h_i^*(t+1) = h_i^*$ and $c_i(t+1) = c_i(t)$, and consider the next item in the data stream. We have the following claim whose proof follows trivially from the preceding discussion:

Lemma 2. *Let $\hat{\rho}_\alpha(t)$ be the fraction of the counters $c_i(t)$ that equal α, i.e.,*

$$\hat{\rho}_\alpha(t) = |\{l | 1 \leq l \leq k, c_i(t) = \alpha\}|/k.$$

For $0 < \epsilon < 1$, $1 > p > 0$, $k \geq 2\epsilon^{-3} p^{-1} \log \tau^{-1}$,

$$\hat{\rho}_\alpha(t) \in (1 \pm \epsilon)\rho_\alpha(t) + \epsilon p$$

for a prespecified precision p, with success probability at least $1 - \tau$.

The remainder of the algorithm trivially follows from the claim. We merely return estimate $\hat{\rho}_\alpha(t)$ at time t. The space used is $O(k)$ for the hash values as well as the counters. In addition, we need to store $O(k)$ seeds for the ϵ'-min-hash functions h_i each of which takes $O(\log u \log(1/\epsilon'))$ bits as discussed earlier. Processing each item takes time to update the k hash values and counters which takes time $O(k \log(1/\epsilon'))$ using results in [19]. The appropriate value of k can be determined in terms of ϵ' using the preceding Lemma, but for setting all parameters to be desired constants, $k = O(1)$. That gives the claimed result for estimating $\rho_\alpha(t)$ at any time t, for the unbounded data stream model.

Some additional observations are as follows: Keeping exact counters $c_i(t)$, as described above has the nice property that α need not be fixed apriori. Instead, it can be specified by the user at query time and changed for different queries. In addition, our technique above can work for estimating the size (ratio of sizes) of any "special" subset of the distinct elements provided we can efficiently check if the element corresponding to the min-value belongs to this subset [4]. Moreover, our algorithm is well suited for a distributed scenario. Say we saw a stream A at one site and another stream B at another site and wished to consider the stream $C = A || B$ where $||$ stands for concatenation of the stream. Then, it is easy to see that if we maintain the min-hashes $(h_i^*(A), h_i^*(B))$ along with the counters $(c_i(A), c_i(B))$ separately at the two sites, we can compute the min-hashes and counters for the concatenated (union) stream. The amount of communication required (and the time required for computing) is the size of the min-hashes plus counters, which as we saw is $O(k) = O(1)$.

4 Similarity Estimation

We now consider the similarity estimation problem. Recall from Section 2 that to estimate similarity between two sets X and Y, we need to only compare k min-hash values for each. In the unbounded data stream model, it is trivial to maintain the k min-hash values as new items arrive. However, the challenge for the windowed data stream model, is to maintain $h_i^*(t)$'s for the window $t-(N-1), \ldots, t$ as t progresses. Consider the task of maintaining any single $h_i^*(t)$. At any time t, we would like to maintain the minimum value over the window $t - (N - 1), \ldots, t$ for the "hashed" stream $h_i(a_j)$. An item a_j is *active* at time t iff $j \in \{t - (N - 1), \ldots, t\}$. At time t, consider two active data items d_1 and d_2 with arrival times t_1, t_2 such that $t_1 < t_2$. If it is the case that $h_i(d_1) \geq h_i(d_2)$ then we say that item d_2 *dominates* item d_1. At any time t, we need not store the hash value for d_1 if there exists such a dominating element d_2. This is because the hash value of d_2 is no greater than that of d_1 and for all times in the future whenever d_1 is active, d_2 will also be active since $t_1 < t_2$. Thus $h_i^*(t)$ is not affected by the hash value for d_1 which can be safely discarded.

Based on the observation above, we propose the use of the following linked list $\mathcal{L}_i(t)$. Every element of $\mathcal{L}_i(t)$ consists of the pair $(h_i(a_j), j)$ for some active data item a_j at time t. Thus $\mathcal{L}_i(t)$ looks like

$$\{(h_i(a_{j_1}), j_1), (h_i(a_{j_2}), j_2), \ldots, (h_i(a_{j_l}), j_l)\},$$

---

[4] As a special case, the same algorithm can be used, even if we slightly change the definition of rarity to include all elements with multiplicity less than α (and at least 1).

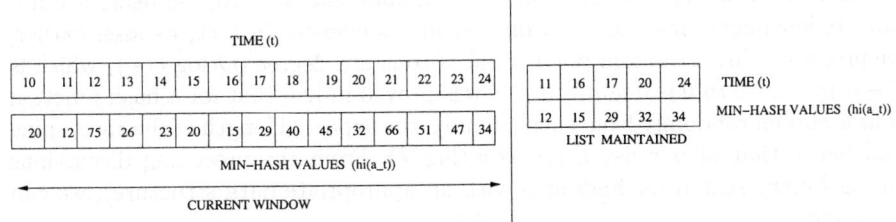

Fig. 1. An example of linked list maintained by the algorithm

where each j_m is the arrival time of active item a_{j_m}. Please refer to Figure 1 for an example. The list $\mathcal{L}_i(t)$ satisfies the property that both the hash values and the arrival times are strictly increasing in the list from left to right. In other words,

$$j_1 < j_2 < \ldots < j_l \ \& \ h_i(a_{j_1}) < h_i(a_{j_2}) < \ldots < h_i(a_{j_l}).$$

Then clearly $h_i^*(t) = h_i(a_{j_1})$.

The algorithm to maintain list $\mathcal{L}_i(t)$ is simple. When we see item a_{t+1}, we compute its hash value $h_i(a_{t+1})$. We find the largest β such that $h_i(a_{j_\beta}) < h_i(a_{t+1})$; here $h_i(a_{j_\beta})$ is from the list $\mathcal{L}_i(t)$. We perform the following steps:

1. We delete all $(h_i(a_{j_k}), j_k)$ from $\mathcal{L}_i(t)$ for which $k > \beta$. Now, $(h_i(a_{j_\beta}), j_\beta)$ is the rightmost item in the list $\mathcal{L}_i(t)$.
2. We insert $(h_i(a_{t+1}), t+1)$ after $(h_i(a_{j_\beta}), j_\beta)$ in list $\mathcal{L}_i(t)$. Hence, $(h_i(a_{t+1}), t+1)$ is now the rightmost element in the list.
3. If the leftmost pair $(h_i(a_{j_1}), j_1)$ is not active at time $t+1$, i.e., $j_1 \notin \{t+1-(N-1), \ldots, t+1\}$, then the leftmost item is deleted, and the second item from left on the list becomes the leftmost.

The list thus obtained is now $\mathcal{L}_i(t+1)$. (We have omitted the description of the boundary cases since they are obvious.)

The difficulty with the algorithm we have described above is that we are explicitly storing $\mathcal{L}_i(t)$ at any time t, and hence, the space used is its length, $|\mathcal{L}_i(t)|$. However,

Lemma 3. *With high probability, over the random choice of min-hash function h_i, we have $|\mathcal{L}_i(t)| = \Theta(H_N)$, where H_N is the Nth Harmonic number, given by $1 + 1/2 + 1/3 + \cdots + 1/N = \Theta(\log N)$.*

Here is a simple, high level argument for proving the lemma above. Consider building a treap on the last N elements where the arrival times are fully ordered (in-order traversal gives increasing arrival times) and hash values ($h_i(a_j)$'s) are heap ordered. Then the list $\mathcal{L}_i(t)$ is the *right spine* of this treap. If the hash values are fully independent, i.e. every hash function is a random permutation, then it is well known (refer [22])that the expected length of the right spine is H_N. In fact, it was proved in [24], the length is $\Theta(\log N)$ with very high probability. However, the hash function families that we propose using are not fully independent, instead have limited independence. In fact, as seen earlier, we propose using a random polynomial of constant degree ($O(\log 1/\epsilon')$) which is $\epsilon' - min - wise\ independent$([19]). It was proved in [24] that for a hash function that is chosen randomly from such a family, the expected length of the right spine and hence that of our list $\mathcal{L}_i(t)$, is $\Theta(\log N)$. From the preceding discussions (using binary search for finding β with an appropriate data structure), we can conclude

Theorem 1. *There is an algorithm for estimating similarity σ between any two windowed data streams, with pre-specified precision p and up to $1 \pm \epsilon$ factor with high probability; the algorithm uses $O(\log N + \log u)$ words of space (or $O((\log N)(\log u))$ bits of space) and takes time $O(\log \log N)$ per data item, with high probability.*

Note that the success probability in the theorem above is predicated on the random choice of the min-hash functions, and *not* over the distribution of the input. In other words, the theorem above holds for an arbitrary (worst case) input.

5 Estimating Rarity over Windowed Data Streams

In this section, we return to the problem of estimating α-rarity, but now we work in the windowed data stream model. Recall that the algorithm for estimating α-rarity in the unbounded data stream model required us to maintain min-hash and counter pairs for k independent min-hash functions. In the windowed data stream model, we follow the same overall algorithm, but modify it as follows.

The main idea here is simple: we maintain for each min-hash value, several (not merely latest) arrival times of items which map to that min-hash value. In particular, we maintain a list $\mathcal{L}_i^t$ of pairs $(h_i(a_{j_k}), list_{i,j_k}^t)$. Here, $list_{i,j_k}^t$ is an ordered list of the latest $\alpha + 1$ time instances ℓ such that $h_i(a_\ell) = h_i(a_{j_k})$; "latest" refers to largest indexes ℓ. (If there are fewer than α such ℓ, we store them all.) Thus $\mathcal{L}_i^t$ now comprises $\{(h_i(a_{j_1}), list_{i,j_1}^t), (h_i(a_{j_2}), list_{i,j_2}^t), \ldots, (h_i(a_{j_l}), list_{i,j_l}^t)\}$. A property of $list$'s worth noting: if we concatenate $list_{i,j_k}^t$'s in the increasing order of j_k (i.e., in the order they appear in $\mathcal{L}_i^t$), then the elements will be strictly increasing. This data structure differs from the one in Section 3 in that here we store $list_{i,j_k}^t$'s, while previously, we only stored a single value for each $h_i(a_{j_1})$, which one can observe will be the largest index in $list_{i,j_k}^t$'s.

When the next item in the data stream, i.e., a_{t+1}, is read, we need to update the lists. This update procedure is similar to the one in Section 4, but there are additional cases to consider. When we see item a_{t+1}, We compute its hash value $h_i(a_{t+1})$. We find the largest β such that $h_i(a_{j_\beta}) \leq h_i(a_{t+1})$, for $h_i(a_{j_\beta})$ in $\mathcal{L}_i(t)$. We perform the following steps:

1. We delete all $(h_i(a_{j_k}), list_{i,j_k}^t)$ pairs from $\mathcal{L}_i(t)$ for which $k > \beta$. Now, $(h_i(a_{j_\beta}), list_{i,j_\beta}^t)$ is the rightmost item in the list $\mathcal{L}_i(t)$.

2. If $h_i(a_{j_\beta}) \neq h_i(a_{t+1})$, we insert $(h_i(a_{t+1}), list_{i,t+1})$ after $(h_i(a_{j_\beta}), list^t_{i,j_\beta})$ in list $\mathcal{L}_i(t)$. Here, $list_{i,t+1}$ consists of the singleton item $t+1$. Hence, $(h_i(a_{t+1}), list_{i,t+1})$ is now the rightmost element in the list $\mathcal{L}_i(t)$.
3. If $h_i(a_{j_\beta}) = h_i(a_{t+1})$, we append $t+1$ to $list^t_{i,j_\beta}$; if the resulting list has more than $\alpha + 1$ items, we delete the smallest index in it.
4. Let x be the smallest index in $list^t_{i,j_1}$; if a_x is not active at time $t+1$, we remove x from $list^t_{i,j_1}$. As a result if $list^t_{i,j_1}$ becomes empty, we remove the pair $(h_i(a_{j_1}), list^t_{i,j_1})$ from $\mathcal{L}_i(t)$.

The list $\mathcal{L}_i(t)$ that remains after the operations above is the list $\mathcal{L}_i(t+1)$. (Again, as in Section 4, the boundary cases have been omitted.) We can argue as in Section 4 that the total size of $\mathcal{L}_i(t)$ is $O(\alpha \log N)$ with high probability, which is $O(\log N)$ for α's of our interest (small constants). Again, as before, we can conclude that this algorithm yields an $1 \pm \epsilon$ approximation, up to precision p, in bounds identical to Theorem 1 for the problem of estimating α-rarity in windowed data streams, giving our result.

Note that in the preceding discussion we assumed that α is a small constant; this is the case that is applicable in the mining scenarios we described in Section 1. However, if one really wishes to solve the problem of estimating ρ_α, for arbitrarily large α, over windowed data streams, we can reduce the memory requirement as follows: Instead of maintaining an ordered list of the latest α time instances we maintain the data structure from [8] for each min-hash value. This data structure can be used to approximately count the number of instances of the hash value in the current window. It uses space $\frac{1}{\tau} \log \alpha$ and maintains a count up to α with accuracy $1 \pm \tau$.

6 Concluding Remarks

Much remains to be done in the windowed data stream model. Our technique of keeping the min-hash value together with a count or list of occurrences has been recently applied to a quite different problem of mining the database structure and building data quality browser [10]. We expect our techniques to find other applications in practice.

References

[1] N. Alon, Y. Matias, M. Szegedy. The space complexity of approximating the frequency moments. In *Proc. Twenty-Eighth Annual ACM Symp. on Theory of Computing*, 1996.
[2] A. Broder. Filtering Near-Duplicate Documents. In *Proc. of FUN*, 1998.
[3] B. Babcock, S. Babu, M. Datar, R. Motwani, and J. Widom. Models and Issues in Data Stream Systems. In *Proc. of Principles of Database Systems*, Madison, Wisconsin, June 3 – 5, 2002.
[4] A. Broder, M. Charikar, A. Frieze, and M. Mitzenmacher. Min-wise Independent Permutations. In *Proc. of STOC*, 1998.

[5] B. Babcock, M. Datar, and R. Motwani. Sampling from a Moving Window over Streaming Data. In *Proc. of Thirteenth Annual ACM-SIAM Symp. on Discrete Algorithms*, 2002.
[6] E. Cohen. Size-Estimation Framework with Applications to Transitive Closure and Reachability. *Journal of Computer and System Sciences* 55 (1997): 441–453.
[7] E. Cohen, M. Datar, S. Fujiwara, A. Gionis, P. Indyk, R. Motwani, J. Ullman, and C. Yang. Finding Interesting Associations without Support Pruning. In *Proc. of the 16th International Conference on Data Engineering, San Diego, USA, 2000*.
[8] M. Datar, A. Gionis, P. Indyk, and R. Motwani. Maintaining Stream Statistics over Sliding Windows. In *Proc. of Thirteenth Annual ACM-SIAM Symp. on Discrete Algorithms*.
[9] P. Domingos, G. Hulten, and L. Spencer. Mining time-changing data streams. In *Proc. of the 7th International Conference on Knowledge Discovery and Data Mining*, 2001.
[10] T. Dasu, T. Johnson, S. Muthukrishnan and V. Shkapenyuk Mining database structure, or to How to build a data quality browser. In *Proc. of the SIGMOD*, 2002.
[11] J. Feigenbaum, S. Kannan, M. Strauss, M. Viswanathan. An Approximate L1-Difference Algorithm for Massive Data Streams. In *Proc. 40th IEEE Symp. on Foundations of Computer Science*, 1999.
[12] P. Flajolet, G. Martin. Probabilistic Counting. In *Proc. 24th IEEE Symp. on Foundations of Computer Science*, 1983.
[13] P. Gibbons and S. Tirthapura. Estimating simple functions on the union of data streams. In *Symp. on Parallel Algorithms and Architectures*, 2001.
[14] A. Gilbert, S. Guha, P. Indyk, Y. Kotidis, S. Muthukrishnan, and M. Strauss. Fast, Small-Space Algorithms for Approximate Histogram Maintenance. In *ACM Symp on Theory of Computing (STOC)*, 2002.
[15] A. Gilbert, Y. Kotidis, S. Muthukrishnan, and M. Strauss. Surfing wavelets on streams: One-pass summaries for approximate aggregate queries. In *Proc. of VLDB*, 2001.
[16] S. Guha, N. Mishra, R. Motwani, L. O'Callaghan. Clustering data streams. In *Proc. 2000 Annual IEEE Symp. on Foundations of Computer Science*, pages 359–366, 2000.
[17] M. R. Henzinger, P. Raghavan, S. Rajagopalan. Computing on data streams. Technical Report TR 1998-011, Compaq Systems Research Center, Palo Alto, California, May 1998.
[18] P. Indyk. Stable Distributions, Pseudorandom Generators, Embeddings and Data Stream Computation. In *Proc. 41st IEEE Symp. on Foundations of Computer Science*, 2000.
[19] P. Indyk A Small Approximately Min-Wise Independent Family of Hash Functions. In *Journal of Algorithms 38(1): 84-90 (2001)*.
[20] S. Madden and M. J. Franklin. Fjording the stream: An architecture for queries over streaming sensor data. In *Proceedings of ICDE*, 2002.
[21] K. Mulmuley An Introduction through Randomized Algorithms. Prentice Hall, 1993.
[22] R. Motwani, P. Raghavan. *Randomized Algorithms*. Cambridge University Press, 1995.
[23] NOAA. U.S. national weather service. http://www.nws.noaa.gov/ .
[24] R. Seidel, and C. Aragon. Randomized Search Trees. In *Algorithmica (1996) 16*, pp 464–497.

Efficient Constructions of Generalized Superimposed Codes with Applications to Group Testing and Conflict Resolution in Multiple Access Channels

Annalisa De Bonis and Ugo Vaccaro

Dipartimento di Informatica ed Applicazioni, Università di Salerno
84081 Baronissi (SA), Italy
{debonis,uv}@dia.unisa.it

Abstract. In this paper we introduce a parameterized generalization of the well known superimposed codes. We give algorithms for their constructions and we provide non-existential results. We apply our new combinatorial structures to the efficient solution of new group testing problems and access coordination issues in multiple access channels.

1 Introduction

Superimposed codes were introduced in the seminal paper by Kautz and Singleton [26]. Since then, they have been extensively studied both in the coding theory community (see [16] and the references quoted therein) and the combinatorics community under the name of *cover free* families [18, 31, 32]. Informally, a collection of subsets of a finite set is r-*cover free* if no subset in the collection is included in the union of r others. One gets the binary vectors of a superimposed code by considering the characteristic vectors of members of an r-cover free family.

Superimposed codes are a very basic combinatorial structure and find application in an amazing variety of situations, ranging from cryptography and data security [10, 17, 33] to computational molecular biology [2, 13, 15], from multiaccess communication [15, 26] to database theory [26], from pattern matching [23] to distributed coloring [28, 34], circuit complexity [7], broadcasting in radio networks [11], and other areas of computer science.

In this paper we introduce a parameterized generalization of superimposed codes. For particular values of the parameters our codes reduce to the classical Kautz and Singleton superimposed codes. Our codes also include a first generalization of superimposed codes introduced in [16], and the combinatorial structures considered in [24]. For our generalized superimposed codes we provide constructions and non existential results, both aspects have relevance to the application areas considered in this paper. Our motivations to introduce the generalization comes from new algorithmic issues in *Combinatorial Group Testing* and *Conflict Resolution in Multiple Access Channels*.

Combinatorial Group Testing. In group testing the task is to determine the *positive* members of a set of objects $\mathcal{O}$ by asking subset queries of the form "does the set $Q \subseteq \mathcal{O}$ contain a positive object?". The very first group testing problem arose almost sixty years ago in the area of chemical analysis [14], where it was employed as a blood test technique to detect the infected members of a population. Since then, combinatorial group testing has exhibited strong relationships with several computer science subjects: algorithms, complexity theory, computational geometry, and computational learning theory, among others. The recent monograph [15] gives an excellent account of these aspects. Combinatorial group testing is also experiencing a renaissance in Computational Molecular Biology where it finds ample applications for screening libraries of clones with hybridization probes [1, 4] and sequencing by hybridization [29, 30]. We refer to [15, 19] for an account of the fervent development of the subject.

More to our points, recent work [19] suggests that classical group testing procedures should take into account also the possibility of the existence of "inhibitory items", that is, items whose presence in the tested set could render the outcome of the test meaningless. In other words, if during the execution of an algorithm we tested a subset $Q \subseteq \mathcal{O}$ containing positive objects *and* inhibitory items, we would get the same answer as Q did not contain any positive object. Similar issues were considered in [12, 21] where additional motivations for the problem were given. In Section 4 we show that our generalized superimposed codes play a crucial role in estimating the computational complexity of this new group testing problem. More precisely, our codes represent both a basic tool in designing efficient algorithmic solutions for the problem at hand, and also in deriving a general lower bound on the number of subset queries to be performed by any algorithm solving it.

Conflict Resolution in Multiple Access Channels. Loosely speaking, with multiple access channels we intend communication media that interconnect a number of users and in which any packet transmission by a single user is broadcasted to all other users connected to the channel. Multiple access channels have been implemented on coaxial cable, fiber optics, packet radio, or satellite transmission media; a well known example is the ETHERNET (see for instance [3] for references in this area).

A model commonly taken as basis for mathematical studies of multiple access channels assumes that the whole system is synchronous, and that at each time instant any number of users can transmit a packet of data over the channel. There is no central control. If just one user transmit its packet in a given time unit, the packet is successfully broadcasted to every other user, if more than one user transmit in a *same* time unit, then all packets are lost because of interference. All users on the channel have the capability of detecting which one of the following events hold: no packet transmission, successful transmission of just one packet, interference due to packet conflict. A key ingredient for an efficient use of multiple access channels is a *conflict resolution algorithm*, that is, a protocol that schedule retransmissions so that each of the conflicting users eventually transmit singly (and therefore successfully) on the channel. A conflict resolution algorithm may be used to coordinate access to the channel in the following way. Access alternates between time instants in which access is unrestricted and time instants in which access is restricted to resolve conflicts. Initially access is unrestricted and all users are allowed to transmit packets at their wish. When a conflict arises, only the involved users execute an algorithm to resolve it and the other users abstain from transmitting. After conflict resolution, access to the channels is again unrestricted (more on this scenario in Section 5).

Conflict resolution in multiple access channels is a source of many challenging algorithmic problems, we refer the reader to the excellent survey paper [9] for a nice account of the vast literature on the topic. The great majority of this body of work assumes the standard hypothesis that conflict arises if more than one user try to transmit at the same time on the channel. However, already in the 80's Tsybakov et al. [35] studied multiple access channels in which simultaneous transmission of up to $c \geq 2$ users is allowed, and conflict arises if strictly more than c users try to transmit at the same time instant. Other examples of such more flexible channels include the adder channel [6] and the Gaussian multiple-access channel [5]. In general, in multiple-frequency wireless channels or in multiple-wavelength optical networks it is indeed possible to carry out simultaneous broadcasting of more than one packet in the same time unit. Also, a somewhat similar scenario has been considered in [20]. It is clear that to fully exploit these new capabilities, new conflict resolutions algorithms must be devised.

The contributions of our paper to this issue are presented in Section 5, where it is essentially shown that our generalized superimposed codes are in a sense equivalent to totally non-adaptive conflict resolution protocols for these more powerful multiple access channels, just as like classical superimposed codes corresponds to totally non adaptive conflict resolution protocols on the standard multiple access channel [26, 27]. Informally, with totally non-adaptive conflict resolution protocols we mean the following: The retransmission schedule of each user is fixed (i.e., does not depend on the time in which collision occurs and on the set of conflicting users), and known beforehand the collision event occurs. Therefore, the behaviour of each user is fixed and does not need to adapt to the behaviour of other users. In contrast, adaptive conflict resolution protocols are more flexible; for instance, they can query other users to find out the identities of the conflicting ones and, on the basis of this acquired knowledge, schedule the retransmissions to solve the conflict. Totally non adaptive conflict resolution protocols have obvious advantages over adaptive ones, of course at the expenses of possibly longer conflict resolution schedules. For completeness, we remark that adaptive conflict resolution protocols in our scenario have been given in [8].

1.1 Structure of the Paper and Summary of Results.

In Section 2 we introduce the basic concepts and we define our generalized superimposed codes; we also point out their relationships with previously known combinatorial structures. In Section 3 we present upper and lower bounds on the length of generalized superimposed codes; this is equivalent to giving algorithms for constructing generalized superimposed codes with "many codewords" and to proving non-existential results.

In Section 4 we present the application of our codes to the design of efficient algorithms for group testing in presence of inhibitors. Moreover, we show that our codes play an important role also in bounding from below the complexity of *any* algorithm for group testing in presence of inhibitors. In Section 5 we formally define the multiple access channel under study, we provide an algorithm for conflict resolution and we estimate its performance in terms of the codeword length of our generalized superimposed codes.

2 Basic Definitions

A set $\mathcal{C} = \{c_1, \ldots, c_n\}$ of n binary vectors of length N is called a *binary code* of size n and length N. Each c_j is called *codeword* and for any i, $1 \leq i \leq N$, $c_j(i)$ denotes the i-th entry of c_j. A binary code $\mathcal{C}$ can be represented by an $N \times n$ binary matrix $C = \|c_j(i)\|$, $i = 1, \ldots, N$ and $j = 1, \ldots, n$, with codewords as columns. A binary code is said k-uniform if all columns have exactly k entries equal to 1.

For each binary vector c_j of length N, let S_{c_j} denote the subset of $\{1, \ldots, N\}$ defined as $S_{c_j} = \{i \in \{1, \ldots, N\} : c_j(i) = 1\}$. Therefore, to any binary code $\mathcal{C} = \{c_1, \ldots, c_n\}$ of length N we can associate a family $\mathcal{F} = \{S_{c_1}, \ldots, S_{c_n}\}$ of subsets of $\{1, \ldots, N\}$. It is clear that this association is invertible, that is, from a family of subsets of $\{1, \ldots, N\}$ one uniquely gets a binary code $\mathcal{C}$ of length N. The set $\{1, \ldots, N\}$ will be called the *ground set* of $\mathcal{F}$.

Given $q > 1$ codewords (columns) $c_{\ell_1}, \ldots, c_{\ell_q}$, we denote with $c_{\ell_1} \vee \ldots \vee c_{\ell_q}$ the boolean sum (OR) of $c_{\ell_1}, \ldots, c_{\ell_q}$. We say that the column c_h is covered by the column c_j if any 1 entry of c_h corresponds to a 1 entry of c_j.

Definition 1. *Let p, r and d be positive integers and let $d \leq r$. We call a binary code $\mathcal{C} = \{c_1, \ldots, c_n\}$, with $n \geq p + r$, (p, r, d)-superimposed if for any distinct $p + r$ codewords $c_{h_1}, \ldots, c_{h_p}, c_{\ell_1}, \ldots, c_{\ell_r}$ there exist $r - d + 1$ distinct indices $j_1, \ldots, j_{r-d+1} \in \{\ell_1, \ldots, \ell_r\}$ such that $c_{h_1} \vee \ldots \vee c_{h_p}$ is not covered by $c_{j_1} \vee \ldots \vee c_{j_{r-d+1}}$. The minimum length of a k-uniform (p, r, d)-superimposed code of size n is denoted by $N(p, r, d, k, n)$, whereas that of an arbitrary (p, r, d)-superimposed code of size n is denoted by $N(p, r, d, n)$.*

Informally, the family of subsets associated to the binary vectors of a (p, r, d)-superimposed code is such that for any p subsets and any r subsets, there exist $r - d + 1$ subsets among the r's such that the union of the p subsets are not included in the union of the $r - d + 1$'s. Notice that (p, r, d)-superimposed codes are a generalization of the superimposed codes introduced by Kautz and Singleton [26] which corresponds to our definition for the case $p = d = 1$. The families of sets associated to such codes are often referred to with the name of r-cover free families and have been extensively studied in the field of Extremal Set

Theory [18, 31]. An r-cover free family is such that no member of the family is contained in the union of any other r members of the family.

Dyachkov and Rykov [16] generalized r-cover families by introducing (p, r)-cover free families. A family is said (p, r)-cover free if the union of any p members of the family is not contained in the union of any other r members of the family. A (p, r)-cover free family corresponds exactly to our codes with parameter $d = 1$. Finally the combinatorial structure considered in [24] coincides with ours for $p = 1$ and $d = r$.

3 Bounds on the Length of (p, r, d)-superimposed Codes

In this section we will present upper and lower bounds on the minimum length $N(p, r, d, n)$ of (p, r, d)-superimposed codes with n codewords.

Theorem 1. *For any $n \geq r \geq d$, it results*

$$N(p, r, d, n) = O\left(\frac{(r+p)\sqrt{(r+1) \cdot (r-d+1)}}{p\sqrt{d}} \log \frac{n}{r+p}\right).$$

Moreover, if $d > \lceil r/2 \rceil$ then

$$N(p, r, d, n) = O\left(\frac{r}{p} \log \frac{n}{r+p}\right).$$

For $p = o(d\sqrt{d})$, the following theorem gives better results than Theorem 1.

Theorem 2. *Let n, p, r and d be positive integers and let $d \leq r$ and $p + r \leq n$. Then, it results $N(p, r, d, n) < 24(\frac{r}{d})^2 (\log_2 n + 2)$.*

It is worth pointing out that by setting $p = d = 1$ in the above two theorems, we recover the best known upper bound $O(r^2 \log n)$ on the length of superimposed codes [18, 22]

We also give a lower bound on the minimum length of (p, r, d)-superimposed codes of size n. Such a lower bound holds for the more general case when a given codeword may occur more than once in the code. Since we will make use of this generalization in the rest of the paper, we will express our lower bound directly in the case of possible multiple occurrences of codewords. We will refer to such codes with the term of multi-codes.

Definition 2. *Let p, r and d be positive integers and let $d \leq r$. A binary multi-code $\tilde{C} = \{c_1, \ldots, c_n\}$, with $n \geq p + r$, is called (p, r, d)-superimposed if for any distinct $p + r$ indices $h_1, \ldots, h_p, \ell_1, \ldots, \ell_r$ there exist $r - d + 1$ distinct indices $j_1, \ldots, j_{r-d+1} \in \{\ell_1, \ldots, \ell_r\}$ such that $c_{h_1} \vee \ldots \vee c_{h_p}$ is not covered by $c_{j_1} \vee \ldots \vee c_{j_{r-d+1}}$. The minimum length of a k-uniform (p, r, d)-superimposed multi-code of size n is denoted by $\tilde{N}(p, r, d, k, n)$ whereas that of an arbitrary (p, r, d)-superimposed multi-code of size n is denoted by $\tilde{N}(p, r, d, n)$.*

Observe that a (p,r,d)-superimposed multi-code may contain at most $d+p-1$ occurrences of a given codeword c. Indeed let $c_{\ell_1},\ldots,c_{\ell_{d+p}}$ be $d+p$ identical codewords, and let $c_{\ell_{d+p+1}},\ldots,c_{\ell_{r+p}}$ be any other codewords. One has that $c_{\ell_1}\cup\ldots\cup c_{\ell_p}$ is covered by the union of any $r-d+1$ codewords belonging to $\{c_{\ell_{p+1}},\ldots,c_{\ell_{r+p}}\}$. Hence, one has that the following theorem holds.

Theorem 3. *Let n, p, r and d be positive integers and let $d \leq r$ and $n \geq p+r$. Then, it results $\tilde{N}(p,r,d,n) \leq N(p,r,d,n) \leq \tilde{N}(p,r,d,(d+p-1)n)$.*

Theorem 4. *Let p, r and d be positive integers and let $d \leq r$ and $n \geq p+r$. Then, it results*

$$\tilde{N}(p,r,d,n) \geq \frac{1}{1+\log_2 e}\left\lfloor\frac{r}{pd}\right\rfloor \log_2 \frac{n-p+1}{rp}.$$

The following corollary follows from Theorems 3 and 4.

Corollary 1. *Let p, r and d be positive integers and let $d \leq r$ and $n \geq p+r$. Then, it results*

$$N(p,r,d,n) \geq \frac{1}{1+\log_2 e}\left\lfloor\frac{r}{pd}\right\rfloor \log_2 \frac{n-p+1}{rp}.$$

4 Efficient Algorithms for Group Testing with Inhibitors

The classical group testing scenario consists of a set $\mathcal{S}=\{s_1,\ldots,s_n\}$ of items p of which are *defective* (positive), while the others are *good* (negative). The goal of a group testing strategy is to identify all defective items. To this aim, items of $\mathcal{S}$ are pooled together for testing. A test yields a positive feedback if the tested pool contains one or more positive members of $\mathcal{S}$ and a negative feedback otherwise. The group testing strategy is said *non-adaptive* if all tests are performed in parallel whereas it is said *adaptive* when the tests are performed sequentially, and which test to perform at a given step may depend on the feedbacks of the previously executed tests. In non-adaptive strategies each test is decided beforehand and does not depend on the feedbacks of previous tests.

We briefly describe how Kautz and Singleton superimposed codes [26] provide a non-adaptive group testing strategy for such a scenario; this will give intuitions for our more complicated case. The correspondence is obtained by associating the columns of the superimposed code with items and the rows with tests. Entry (i,j) of the matrix is 1 if s_j belongs to the pool used for the i-th test, and 0 otherwise. Let y denote the column of length N with $y(i)=1$, $i=1,\ldots,N$, if and only if the response to the i-th test is positive. This column is equal to the boolean sum of the columns associated to the p defective items. The code provides a strategy to uniquely identify the p defectives if the boolean sums of p columns are all distinct. Codes with this property, which is weaker than the cover-free property illustrated in Section 2, have been considered by Kautz and Singleton as well. However, the cover-free property allows a more efficient detection of defective items. Indeed, the feedback column y will cover only the columns associated to the p defective items. For that reason, it will be sufficient to inspect individually the columns associated to the n items instead of inspecting the boolean sums associated to all distinct $\binom{n}{p}$ p-tuples of items.

4.1 Group Testing with Inhibitors

In this section we consider the variation of classical group testing which has been introduced by Farach et al. [19]. A related model was considered in [12]. In this search model, which we call *Group Testing with Inhibitors* (GTI), the input set consists not only of positive items and negative items, but also of a group of r items called *inhibitors*.

A pool tests positive if and only if it contains one or more positive items and no inhibitor. The problem is to identify the set of the positive items. Farach et al. [19] have proved that this problem has the same asymptotical lower bound of the apparently harder problem of identifying both the set of the positives and that of the inhibitors. They have also described a *randomized* algorithm to find the p positives which achieve the information theoretic bound when $p + r \ll n$. In [13] the authors improved on the results given in [19].

4.2 The Threshold Model

We introduce a generalization of the GTI model presented in the previous section. In this new model the presence of positives in a test set can be detected only if the test set contains a number of inhibitors smaller than a fixed threshold d. Our goal is to identify all positive items using as few tests as possible.

Our algorithm consists of four phases.

Phase 1. Find a group of items Q which tests positive.
Phase 2. Find a group of items containing exactly $d-1$ inhibitors and at least one positive item.
Phase 3. Find $r - d + 1$ inhibitors and discard them.
Phase 4. Find all positives.

In the following we will describe how to perform each of the above phases.

Phase 1: The search strategy performed during this phase is provided by a (p, r, d)-superimposed code of size n. We associate the columns of the code with the n items and the rows with the tests. Entry (i, j) of the matrix is 1 if s_j belongs to the pool used for the i-th test, and 0 otherwise. Let y denote the feedback column, i.e., $y(i) = 1$, $i = 1, \ldots, N$, if and only if the response to the i-th test is positive. Hence, one has that $y(i) = 1$, $i = 1, \ldots, N$, if and only if among the items pooled for the i-th test there are at least one positive and no more than $d-1$ inhibitors. Let $c_{h_1}, \ldots, c_{h_p}$ be the p codewords associated to the p defective items and let $c_{\ell_1}, \ldots, c_{\ell_r}$ those associated with the r inhibitors. Then, for any $i = 1, \ldots, N$, one has that $y(i) = 1$ if and only if $(c_{h_1} \vee \ldots \vee c_{h_p})(i) = 1$ and there exist $r-d+1$ indices $j_1, \ldots, j_{r-d+1} \in \{\ell_1, \ldots, \ell_r\}$ such that $(c_{j_1} \vee \ldots \vee c_{j_{r-d+1}})(i) = 0$.

Since the code is (p, r, d)-superimposed then one has that this condition is verified for at least one index $i \in \{1, \ldots, N\}$. Moreover, such an index i exists for any choice of $c_{h_1}, \ldots, c_{h_p}$ and any choice of $c_{\ell_1}, \ldots, c_{\ell_r}$. Consequently, the strategy associated with the (p, r, d)-superimposed code guarantees a positive feedback for any choice of the p defectives and for any choice of the r inhibitors.

Notice that the search strategy performed during this phase is completely non-adaptive, a feature of some interest in practical applications.

Phase 2: In the following we will denote with $Half_L$ ($Half_R$, resp.) a function which takes in input a set $A = \{a_1, \ldots, a_m\}$ and returns the set consisting of the first $\lfloor m/2 \rfloor$ (the last $\lceil m/2 \rceil$, resp.) elements of A. Let Q be the group of items returned by Phase 1. Then, Phase 2 consists of the following procedure.

$\quad A \leftarrow S \setminus Q$
$\quad B \leftarrow Q$
$\quad$ while($|A| > 1$)
$\quad\quad T \leftarrow B \cup Half_L(A)$
$\quad\quad$ if T tests positive then $B \leftarrow T$
$\quad\quad\quad\quad A \leftarrow Half_R(A)$
$\quad\quad\quad$ else $A \leftarrow Half_L(A)$
$\quad$ return(B)

The above procedure preserves the invariant that B contains at most $d - 1$ inhibitors and $B \cup A$ contains at least d inhibitors. Since the algorithm terminates as soon as $|A|$ becomes equal to 1, then it follows that the set B returned by the procedure contains exactly $d - 1$ inhibitors. Moreover, since $Q \subseteq B$, then B contains at least one positive item.

Phase 3: A variant of the classical group testing is used to find the $r - d + 1$ inhibitors $s_{\ell_1}, \ldots, s_{\ell_{r-d+1}}$ contained in $S \setminus B$. The variant consists in adding the items in the set B returned by Phase 2 to each tested group T. This assures a positive feedback if the tested group T contains no inhibitor, and a negative response if the tested group T contains at least 1 inhibitor.

Then, the $r - d + 1$ inhibitors $s_{\ell_1}, \ldots, s_{\ell_{r-d+1}}$ found in $S \setminus B$ are discarded from S.

Phase 4: The set $S \setminus \{s_{\ell_1}, \ldots, s_{\ell_{r-d+1}}\}$ is searched to find the p positives. Since the undiscarded $d - 1$ inhibitors do not interfere with the tests, then a classical group testing algorithm can be applied to find all positives.

Observe that the number of tests executed in Phase 1 is as small as $N(p, r, d, n)$. Phase 2 requires $\lceil \log |S \setminus Q| \rceil \leq \lceil \log(n - 1) \rceil$ tests since the search space reduces by one half at each step. Phase 3, as well as Phase 4, performs a standard group testing strategy. The cost of the optimal group testing strategy (see Chapter 2 of [15]) to find q defectives in a set of size n is $O(q \log(\frac{n}{q}))$. Consequently, Phase 3 requires $O\left((r - d + 1) \log\left(\frac{n}{r-d+1}\right)\right)$ tests, whereas Phase 4 requires $O(p \log(\frac{n}{p}))$ tests. Combining all the above estimates one gets the following theorem.

Theorem 5. *There exists a strategy to find the p positives which uses*

$$N(p,r,d,n) + O\left(\log n + (r-d+1)\log\left(\frac{n}{r-d+1}\right) + p\log\frac{n}{p}\right)$$

tests.

Plugging in the above upper bound either the expression for $N(p,r,d,n)$ given in Theorem 1 or that of Theorem 2 provides an explicit estimate of the cost of our strategy. We remark that for many values of the involved parameters, the leading term in the estimation of the number of tests required by our strategy is just $N(p,r,d,n)$.

The introduced generalization of superimposed codes intervenes in our group testing problem not only in establishing an upper bound on its optimal cost, but also in determining a lower bound on it. Namely, we have the following results.

Theorem 6. *Any strategy to find all positives requires at least $N(p,r,d,n-1)$ tests.*

The following result is an immediate consequence of the information theoretic bound and of Theorems 4 and 6.

Theorem 7. *Any strategy to find all positives requires $\Omega(p\log\frac{n}{p} + \frac{r}{pd}\log\frac{n-p+1}{rp})$ tests.*

5 Conflict Resolution in Multiple Access Channels

In this section we show how our codes can be used for resolving conflicts in multiple access communication when simultaneous transmissions of up to d users on the same channel is allowed. We first define the mathematical model formally.

5.1 The Multiaccess Model

The contemplated scenario consists of a system comprising a set of n users $u_1, \ldots, u_n$ and a single channel which allows up to d users to successfully transmit at the same time. We make the following standard assumptions.

- *Slotted system.* We assume that the time be divided into time slots and that the transmission of a single packet require one time slot. Simultaneous transmissions are those occurring in the same time slot.
- *Threshold conflict.* If no more than d users transmit during the same time slot then their transmissions are *successful*. Collisions arise if more than d users attempt to transmit at the same time.
- *Immediate feedback.* We assume that at the end of each slot, the system provides each user with a feedback which says whether packets have been transmitted during that slot and whether a conflict has occurred.

- *Retransmission of conflicts.* We assume that packets involved in the conflict must be retransmitted until they are successfully received. Users involved in the conflict are said *backlogged*.
- *Bounded number of backlogged users.* We assume that the number of backlogged users does not exceed a given bound q.
- *Blocked access.* We assume that when a conflict occurs only users involved in the conflict are allowed to transmit until the conflict is resolved. Collision resolution algorithms using this assumption are called *blocked access algorithms* and the time employed to resolve the conflict is called *conflict resolution period*. We define the *length* of the conflict resolution period as the number of time slots the conflict resolution period is divided into.

See [9, 3] for an extensive discussion on the implications of the above assumptions.

5.2 Complexity of Non-adaptive Conflict Resolution

A conflict resolution algorithm schedules users' transmissions so that for each user there is a time slot during which her transmission is successful. This property guarantees that a conflict is resolved within the conflict resolution period.

We present a conflict resolution algorithm for the multiaccess model described in the previous section. In our conflict resolution algorithm, each user u_j is permanently associated with a set of time slots. When a new conflict occurs a conflict resolution period starts and the conflict is resolved by having each backlogged user transmit only during the time slots allocated to her. Algorithms like ours are *totally non adaptive*, in contrast to *adaptive* conflict resolution protocols. The latter may query other users to find out the identities of the conflicting ones and, on the basis of this acquired knowledge, schedule the retransmissions to solve the conflict. Totally non adaptive conflict resolution protocols have obvious advantages over adaptive ones.

Our algorithm works as follows. Let $u_{\ell_1}, \ldots, u_{\ell_s}$, $d < s \leq q$, denote the users involved in the conflict. Then, transmissions from users other than $u_{\ell_1}, \ldots, u_{\ell_s}$ are blocked, whereas, for each $h \in \{1, \ldots, s\}$, user u_{ℓ_h} transmits only during the time slots which have been allocated to her. A conflict resolution algorithm should guarantee that for each $h \in \{1, \ldots, s\}$, there is a time slot among those associated to user u_{ℓ_h} during which at most $d-1$ users in $\{u_{\ell_1}, \ldots, u_{\ell_s}\} \setminus \{u_{\ell_h}\}$ are allowed to transmit. We use a $(1, q-1, d)$-superimposed code of size n to construct the time slot subsets to be associated to the n users. To this aim, we associate each user with a distinct codeword of the $(1, q-1, d)$-superimposed code. The time slots assigned to a given user are those corresponding to the 1-entries in the associated codeword. The length N of the $(1, q-1, d)$-superimposed code coincides with the number of slots in the conflict resolution period. It follows that a backlogged user has to wait for no more than N time slots before her packet is successfully transmitted through the channel.

If $s < q$, let $\{u_{\ell_{s+1}}, \ldots, u_{\ell_q}\}$ be any $q-s$ users not involved in the conflict. By definition of $(1, q-1, d)$-superimposed code, one has that for each $h \in \{1, \ldots, s\}$, there exist $q-d$ indices $j_1,\ldots,j_{q-d} \in \{\ell_1,\ldots,\ell_q\} \setminus \{\ell_h\}$ such that $u_{\ell_h}(i) = 1$ and $c_{j_1}(i) = 0, \ldots c_{j_{q-d}}(i) = 0$, for some $i \in \{1,\ldots,N\}$. Consequently, for each $h \in \{1,\ldots,s\}$, there is a time slot $i \in \{1,\ldots,N\}$ among those assigned to user u_{ℓ_h} which has been assigned to at most other $d-1$ users in $\{u_{\ell_1},\ldots,u_{\ell_q}\} \setminus \{u_{\ell_h}\}$. Since $\{u_{\ell_1},\ldots,u_{\ell_s}\} \setminus \{u_{\ell_h}\} \subseteq \{u_{\ell_1},\ldots,u_{\ell_q}\} \setminus \{u_{\ell_h}\}$, then it follows that such time slot i has been assigned to at most other $d-1$ users in $\{u_{\ell_1},\ldots,u_{\ell_s}\} \setminus \{u_{\ell_h}\}$. As a consequence, there is a time slot within the conflict resolution period during which u_{ℓ_h}'s transmission is successful.

We have seen that a generalized superimposed code can be used as a basic tool for a totally non adaptive conflict resolution algorithm in the multiple access channel previously described. Actually we can prove that in our scenario any totally non adaptive conflict resolution algorithm corresponds to a generalized superimposed (multi)code. This will allow us to estimate also from below the complexity of non adaptive conflict resolution algorithms.

To this aim, fix any non-adaptive conflict resolution algorithm and suppose that this algorithm divide the conflict resolution period into t slots. The conflict resolution algorithm allows each user u_j to transmit only at given time slots within the conflict resolution period. For each user u_j, $j = 1,\ldots,n$, let us define a binary column c_{u_j} of length t such that $c_{u_j}(i) = 1$, $i = 1,\ldots,t$, if and only if u_j is allowed to transmit at time slot i. At this point it is rather easy to see that $c_{u_1},\ldots,c_{u_n}$ form a $(1,q-1,d)$ multi-code of size n. Therefore, from the above discussion we get the following result.

Theorem 8. *In any non-adaptive conflict resolution period, the length of the conflict resolution period coincides with the length of a $(1,q-1,d)$ superimposed multi-code of size n.*

From the above theorem one has that the length $\tilde{N}(1,q-1,d,n)$ of the shortest $(1,q-1,d)$-superimposed multi-code provides both upper and lower bound on the number of time slots required to resolve conflict non-adaptively. Therefore, our explicit upper and lower bounds on $\tilde{N}(1,q-1,d,n)$ (recall that $\tilde{N}(\cdot) \leq N(\cdot)$) given in Section 3 yield explicit estimates on the goodness of the conflict resolution protocol presented in this section.

References

[1] E. Barillot, B. Lacroix, and D. Cohen, "Theoretical analysis of library screening using an n-dimensional pooling strategy", *Nucleic Acids Research*, 6241–6247, 1991.

[2] D. J. Balding, W. J. Bruno, E. Knill, and D. C. Torney, "A comparative survey of non-adaptive pooling design" in: Genetic mapping and DNA sequencing, *IMA Volumes in Mathematics and its Applications*, Springer-Verlag, 133–154, 1996.

[3] D. Bertsekas and R. Gallager, *Data Networks*, Prentice Hall, 1992.

[4] W. J. Bruno, D. J. Balding, E. Knill, D. Bruce, C. Whittaker, N. Dogget, R. Stalling, and D. C. Torney, "Design of efficient pooling experiments", *Genomics*, **26**, 21–30, 1995.

[5] G. Caire, E. Leonardi, and E. Viterbo, "Modulation and coding for the Gaussian collision channel", *IEEE Trans on Inform. Theory*, **46**, No. 6, 2007–2026, 2000.

[6] S. C. Chang and E. J. Weldon, "Coding for t-user multiple-access channels", *IEEE Trans. on Inform. Theory*, **25**, 684–691, 1979.

[7] S. Chaudhuri and J. Radhakrishnan, "Deterministic restrictions in circuit complexity", *STOC 96*, 30–36.

[8] R. W. Chen and F. K. Hwang, "K-definite group testing and its application to polling in computer networks", *Congressus Numerantium*, **47**, 145–149, 1985.

[9] B. S. Chlebus, "Randomized communication in radio networks", in: *Handbook of Randomized Computing*, Kluwer Academic Publishers, vol. I, 401-456, 2001.

[10] B. Chor, A. Fiat, and M. Naor, "Tracing traitors", *Crypto 94*, LNCS, **839**, Springer–Verlag, Berlin, 257–270, 1994.

[11] A. E. F. Clementi, A. Monti and R. Silvestri, "Selective families, superimposed codes, and broadcasting on unknown radio networks", *SODA 01*, 709–718.

[12] P. Damaschke, "Randomized group testing for mutually obscuring defectives", *Information Processing Letters*, **67**, 131–5, 1998.

[13] A. De Bonis and U. Vaccaro, "Improved algorithms for group testing with inhibitors", *Information Processing Letters*, **67**, 57–64, 1998.

[14] R. Dorfman, "The detection of defective members of large populations", *Ann. Math. Statist.*, **14**, 436–440, 1943.

[15] D. Z. Du and F. K. Hwang, *Combinatorial Group Testing and its Applications*, World Scientific, 2nd edition, 2000.

[16] A. G. Dyachkov, V. V. Rykov, "A survey of superimposed code theory", *Problems Control & Inform. Theory*, **12**, No. 4, 1–13, 1983.

[17] M. Dyer, T. Fenner, A. Frieze, A. Thomason, "On key storage in secure networks", *J. of Cryptology*, **8**, 189–200, 1995.

[18] P. Erdős, P. Frankl, and Z. Füredi, "Families of finite sets in which no set is covered by the union of r others", *Israel J. of Math.*, **51**, 75–89, 1985.

[19] M. Farach, S. Kannan, E. H. Knill and S. Muthukrishnan, "Group testing with sequences in experimental molecular biology", *Sequences 1997*, IEEE Computer Society, 357–367.

[20] F. Meyer auf der Heide, C. Scheideler, V. Stemann, "Exploiting storage redundancy to speed up randomized shared memory simulations", *STACS 95*, 267–278.

[21] F. K. Hwang, "A Tale of Two Coins", *Amer. Math. Monthly* **94**, 121–129, 1987.

[22] F. K. Hwang and V. T. Sós, "Non adaptive hypergeometric group testing", *Studia Sc. Math. Hungarica*, vol. **22**, 257–263, 1987.

[23] P. Indyk, "Deterministic superimposed coding with application to pattern matching", *FOCS 97*, IEEE Press, 127–136.

[24] G. O. H. Katona and T. G. Tarján, "Extremal problems with excluded subgraphs in the n-cube", *Lecture Notes in Math.*, **1018**, Springer–Verlag, Berlin, 84–93, 1983.

[25] I. Katzela and M. Naghshineh, "Channel assignment schemes for cellular mobile telecommunication systems: a comprehensive survey", *IEEE Personal Communications,* 10–31, June 1996.

[26] W. H. Kautz and R. R. Singleton, "Nonrandom binary superimposed codes", *IEEE Trans. on Inform. Theory*, **10**, 363–377, 1964.

[27] J. Komlós and A. G. Greenberg, "An asymptotically non-adaptive algorithm for conflict resolution in multiple-access channels", *IEEE Trans. on Inform. Theory*, **31**, No. 2, 302–306, 1985.

[28] N. Linial, "Locality in distributed graph algorithms", *SIAM J. on Computing*, **21**, 193–201, 1992.

[29] D. Margaritis and S. Skiena, "Reconstructing strings from substrings in rounds", *FOCS 95*, 613–620.
[30] P. A. Pevzner and R. Lipshutz, "Towards DNA sequencing chips", *MFCS 1994*, Lectures Notes in Computer Science, 143–158, Springer Verlag.
[31] M. Ruszinkó, "On the upper bound of the size of the r-cover-free families", *J. of Combinatorial Theory*, Series A, **66**, 302–310, 1994.
[32] Yu.L. Sagalovich, "Separating systems", *Problems of Information Transmission*, **30**, No. 2, 105–123, 1994.
[33] D. R. Stinson, Tran van Trung and R. Wei, " Secure frameproof codes, key distribution patterns, group testing algorithms and related structures", *Journal of Statistical Planning and Inference*, **86** , 595–617, 2000.
[34] M. Szegedy and S. Vishwanathan, "Locality based graph coloring", *STOC 93*, ACM Press, 201–207.
[35] B. S. Tsybakov, V. A. Mikhailov and N. B. Likhanov, "Bounds for packet transmissions rate in a random-multiple-access system", *Problems of Information Transmission*, **19**, 61–81, 1983.

Frequency Estimation of Internet Packet Streams with Limited Space*

Erik D. Demaine[1], Alejandro López-Ortiz[2], and J. Ian Munro[2]

[1] Laboratory for Computer Science, Massachusetts Institute of Technology
Cambridge, MA 02139, USA
edemaine@mit.edu

[2] Department of Computer Science, University of Waterloo
Waterloo, Ontario N2L 3G1, Canada
{alopez-o,imunro}@uwaterloo.ca

Abstract. We consider a router on the Internet analyzing the statistical properties of a TCP/IP packet stream. A fundamental difficulty with measuring traffic behavior on the Internet is that there is simply too much data to be recorded for later analysis, on the order of gigabytes a second. As a result, network routers can collect only relatively few statistics about the data. The central problem addressed here is to use the limited memory of routers to determine essential features of the network traffic stream. A particularly difficult and representative subproblem is to determine the top k categories to which the most packets belong, for a desired value of k and for a given notion of categorization such as the destination IP address.

We present an algorithm that deterministically finds (in particular) all categories having a frequency above $1/(m+1)$ using m counters, which we prove is best possible in the worst case. We also present a sampling-based algorithm for the case that packet categories follow an arbitrary distribution, but their order over time is permuted uniformly at random. Under this model, our algorithm identifies flows above a frequency threshold of roughly $1/\sqrt{nm}$ with high probability, where m is the number of counters and n is the number of packets observed. This guarantee is not far off from the ideal of identifying all flows (probability $1/n$), and we prove that it is best possible up to a logarithmic factor. We show that the algorithm ranks the identified flows according to frequency within any desired constant factor of accuracy.

1 Introduction

Problem. The goal of this research is to develop algorithms that extract essential characteristics of network traffic streams passing through routers, specifically estimates of the heaviest users and most popular sites, subject to a limited amount

---

* This research is partially supported by the Natural Science and Engineering Research Council of Canada, by the Canada Research Chairs Program, and by the Nippon Telegraph and Telephone Corporation through the NTT-MIT research collaboration.

of memory about previously seen packets. Such characteristics are essential for designing accurate models and developing a general understanding of Internet traffic patterns, which are important for such applications as efficient network routing, caching, prefetching, information delivery, and network upgrades. In addition, information of the load distribution has direct applications to billing users.

As the network stream passes by, we have only a few nanoseconds to react to each packet. This time permits, at best, indexing into one of a small number of registers and storing a new value or incrementing or decrementing a few counters. Memory is limited primarily because it must be on the chip that is handling our processing, in order to keep up.

Ideally, we would like to determine the heaviest k users, for a desired value of k, over some time period. However, because some users may have nearly equal load, answering this question exactly is impossible using little space. Rather, one problem we consider is to determine all users above a given load threshold during some time period. A second case of interest is the weaker requirement of identifying a short list of elements guaranteed to include all of these heavy users. Of course, we would like to be able to solve these problems in the worst case for all possible input sequences, but failing that, we may settle for a probabilistic method provided it is robust (accurate with high probability).

Application. In practice, this frequency estimation information is used both for billing purposes and for traffic engineering decisions. In our particular case, this research is motivated by the need to determine the largest packet flows which most heavily influence the characteristics of a router. The routers in question serve large capacity connections on backbones across the continental United States. In network-administration parlance, we need to determine the flows that "shape" the pipe. The information collected in this scenario is important for short- and long-term traffic engineering and routing decisions on the pipe.

In this application, we augment the router by adding a monitoring system to the router box that collects aggregate statistics on the traffic. This system monitors the packet stream as it passes by, and must collect statistical data in real time. Given the current bandwidth capacities at the network core, the processing time must be on the order of nanoseconds for each packet. This imposes particular restrictions in the nature and amount of operations that can be performed per packet, usually limited to manipulating a small number of registers. Often we can assume the existence of a hardware-based hash-table (associative memory). This table implements a hardware lookup operation using only a few clock cycles. It returns an index associated with the entry if present or an error flag otherwise.

As an example, routers from one of the largest vendors (Cisco) collect perfect statistics on low-bandwidth connections but rely on sampling for higher speeds. The following excerpt from the Cisco *NetFlow* manual [5] illustrates this:

> Forwarding rates on a Gigabit Switch Router... an order of magnitude greater than traditional platforms that support NetFlow. "Touching" every

switched packet for NetFlow accounting becomes a challenge at these high switching rates. However, collecting characteristic statistics on IP traffic being forwarded... is still a necessary tool for managing and planning a network.

In order to scale to higher forwarding rates, NetFlow will now allow the user to sample one out of every "x" IP packets being forwarded... This feature will substantially decrease the CPU utilization needed to account for NetFlow packets.

However, this sampling method is often unsatisfactory given the nature of Internet traffic [9, 23]. Moreover, in many cases, a small percentage of the packet categories account for a large percentage of the traffic. In general, because of the nature and characteristics of Internet traffic and intended routing application, we require counting mechanisms that examine the vast majority of packets using contiguous sampling of packet bursts.

Our results. We consider a general model in which packets have been classified into *categories*. Examples of interesting categorizations include the IP address and/or port of the packet's source and/or destination. We illustrate under a variety of weak models of computation, storage, and network distributions that carefully arranged counting of repetitions of packets' categories can lead to accurate estimates of the most common packet categories above a certain threshold. To give some intuition, a representative example of how counters can be used is the following: when a packet streams by, the process can check whether its category matches any of the currently monitored categories, and if so, increment that counter. The idea is that the category with the highest counter is likely to be the most popular category.

The primary difficulty in counting with very few counters is to know which categories to monitor. If we never reset the counters and start counting newly discovered categories, we may never notice the most popular category, thus never counting them and discovering their popularity. On the other hand, if we reset counters too frequently, we will not gain enough statistics to be sure which counter is significantly higher than the others.

We resolve this trade-off with the following matching upper and lower bounds for monitoring a stream of *unknown* length using m counters:

1. In the worst-case omniscient-adversary model [Section 3]:
   (a) All categories that occur more than $1/(m+1)$ of the time can (in particular) be deterministically reported after a single pass through the stream. However, it is unknown which reported categories have this frequency.
   (b) This result is best possible: if the most common category has frequency of less than $1/(m+1)$, then the algorithm can be forced to report only uniquely occurring elements.
2. In the stochastic model [Section 4]:
   (a) All categories that occur with relative frequency $> (c \ln n)/\sqrt{mn}$ for a constant $c > 0$ can be reported after a single pass through the stream.
   (b) The algorithm estimates the frequencies of the reported categories to within a desired error factor $\varepsilon > 0$ (influencing c).

(c) The results hold *with (polynomially) high probability*, meaning that the probability of failure is at most $1/n^i$ for a desired constant i (also influencing c).

(d) This result is best possible up to constant factors: if the maximum frequency is below $f/\sqrt{nm}$, then the algorithm can be forced to report only uniquely occurring elements with probability at least $(e^{-1+1/e})^f$.

3. Both of these one-pass algorithms can be implemented in a small constant amount of worst-case time per packet.

Related work. Some variants of this problem have been previously considered in the context of one pass analysis of database streams [1, 10, 20], query streams to a search engine [3], and packet data streams [7, 9, 19, 21]. Morris [24] showed that it is possible to approximately count up to n using $\lg\lg n$ bits, and Flajolet [15] gave a detailed analysis of this algorithm. Vitter [26] shows how to sample in a small amount of space and linear time in a single pass. A related problem is computing the spectra (approximate number of distinct values) of a stream which can be achieved in $\lg n$ space [16, 27]. Alon et al. show that the first five moments can be approximated in $\lg n$ space while surprisingly all other (higher) moments require linear space [1].

On the particular issue of estimating frequencies, Fang et al. [10] propose heuristics to compute all values above a threshold. Charikar et al. [3] propose algorithms to compute the top k candidates in a list of length l under a Zipfian distribution. Estan and Varghese [9] identify supersets likely to contain the dominant flows and give a probabilistic estimate of the expect count value in terms of a user selected threshold.

2 Model

This section formalizes the problems and models addressed in this paper, some aspects of which were mentioned in Section 1 in the context of our application. There are three key aspects to the problem and model: what computational power and storage we have to gather statistics about streams, what distributions the streams follow, and what guarantees we make about quality of results. We cover each aspect in the next three subsections.

2.1 Computation and Storage

We use a more restrictive model for the algorithms we develop, and a more powerful model for proving lower bounds, strengthening our results.

2.1.1 Model for Algorithms. Our basic model of computation is that a *statistics-gathering* process watches a stream of n packets passing through an Internet router or similar device. The stream is rapid, so the process can make only one pass through the data, and furthermore can perform little computation per packet. Specifically, we limit the amount of computation to $O(1)$ operations per packet. The storage space available to the process is limited, but a more important limiting factor is that the *working store* of the process is very small: all actively used variables (e.g., counters) must fit in a small cache in order to keep up with the data stream. Thus, in some settings, we may be willing to record a significant amount of data (but still much less than one item per packet) to external storage, and make a final pass through these records at the end of the computation.

A key operation that the statistics gathering process can perform is *counting*. The process is limited to having at most m active counters at any time. Each counter has an associated category that it *monitors*. A counter can be incremented, decremented, or reset to monitor a different category.

Counters can be associatively indexed based on the monitored category. This indexing structure can be implemented in hardware by associative memory, or in software using dynamic perfect hashing [25]. In the latter case, our worst-case running times turn into with-high-probability running times.

We believe that this model of computation captures essentially the entire spectrum of possible algorithms, while capturing all of the important limiting factors in the application. For lower bounds, however, we will consider an even more powerful model, described next.

2.1.2 Model for Lower Bounds. For the purpose of lower bounds, we consider a broad model of computation in which the process can maintain at most m categories in working store at any time, in addition to examining the category of the current packet under consideration. Arbitrary amounts of memory and computation can be used for counters or other structures, but categories must be treated as opaque objects from an arbitrary space with unknown structure, and at most m categories can be stored. The only operation allowed on categories is testing two for equality; in the lower-bound context where we ignore computation time, this operation permits hashing based on categories currently in working store. The process can return candidate most-popular categories only from the m categories that it has in working store.

2.2 Network Traffic Distributions

We propose three broad models of the network traffic distributions that enable us to prove guarantees on quality. All of these models lead to interesting theoretical results which are closely related to the practical problem.

The two most general models are *worst-case distributions*. In this context, the network traffic is essentially arbitrary, and at any moment, an adversary can choose the next packet's category. Algorithms in this model are difficult but surprisingly turn out to be possible. There are two subtly different versions of the model. In the *omniscient adversary* model, the adversary knows everything about the algorithm's execution, and can choose the packet sequence to be the absolute most difficult. In the slightly less powerful but highly natural *oblivious adversary* model, the adversary knows the entire algorithm, but does not know the results of any random coin tosses made by the algorithm. Thus the algorithm can hope to win over the adversary with high probability by using random bits.

Of course, these worst-case models are overly pessimistic, and limit the provable strength of any algorithm. Fortunately, real traffic is not worst-case, but rather follows some sort of distribution. A natural such distribution is the *stochastic* model: an arbitrary probability distribution specifies the relative frequencies of the category, but in what order these categories occur in the packet stream is uniformly random. While this model may not precisely match reality, we feel that it is sufficiently representative to lead to highly practical algorithms. (We plan to evaluate this statement experimentally.)

2.3 Guarantees

It is impossible in general to report the most common category in one pass using less than $\Theta(n)$ storage. For example, such storage is clearly necessary when all categories occur uniquely except for one category that occurs twice. Fortunately, a user of this system is only interested in categories that occur particularly often, i.e., above some frequency threshold.

It turns out that, for each model of network traffic, there is a particular threshold below which it is impossible to accurately detect, but above which it is possible to accurately detect. When we have no extra storage beyond the working store, we can only report m such categories with any confidence. When we have extra storage beyond the working store, we can record more values and make a final pass to choose the largest k frequencies for a desired value of k. In either case we guarantee that, out of the categories whose frequencies are above threshold, the approximately top k are reported. "Approximately" means that the frequency (as opposed to rank) is within a desired constant-factor error.

3   Worst-Case Bounds without Randomization

This section develops an algorithm for the most difficult model, the worst-case omniscient adversary.

3.1   Classic Majority Algorithm

Our starting point is the elegant algorithm [13] for determining whether a value occurs a majority of the time in a stream, i.e., occurs more than $n/2$ times in a stream of length n. The basic model under which this algorithm was developed is that we should make as few passes as possible through the data and as few comparisons as possible, while using the smallest possible amount of space—a single counter.

Algorithm MAJORITY

1. Initialize the counter to zero.
2. For each element in the stream:
   (a) If the counter is zero, define the current element to be the monitored element of the counter.
   (b) If the current element is the monitored element, increment the counter.
   Otherwise, decrement the counter.

If the algorithm terminates with a counter value of zero, then the last monitored element or the last value on the stream could have occurred up to $n/2$ times, though not a majority. On the other hand, if the counter value is positive, the last monitored element is the only value that could have occurred in a majority of the positions. A simple rescan (not permitted in our model) will confirm or deny this hypothesis. With a bit of thought, one sees that the scheme uses at most $\lceil 3n/2 \rceil - 2$ comparisons, which Fischer and Salzberg [13] show to be optimal.

3.2 Generalization

This majority algorithm is a gem, often used in undergraduate lectures and assignments. However, the following generalization does not seem to have appeared. Our initial description ignores issues of data structures required to effectively decrement m counters at once or manage any other aspects of the algorithm; these issues will be addressed later.

Theorem 1. *There is a single-pass algorithm using m counters that determines a set of at most m values including all that occur strictly more than $n/(m+1)$ times in an input stream of length n.*

Proof. The scheme is indeed a generalization of Algorithm MAJORITY:

Algorithm FREQUENT

1. Initialize the counters to zero.
2. For each element in the stream:
   (a) If the current element is not monitored by any counter and some counter is zero, define the current element to be the monitored element of that counter.
   (b) If the current element is the monitored element of a counter, increment the counter. Otherwise, decrement every counter.

The reaction to a value not in a full slate of candidates is admittedly Draconian, but it is effective. To demonstrate this effectiveness, consider any element x that occurs $t > n/(m+1)$ times. Suppose that x is read t_f times when all other candidate locations are full with other values, and t_i times when either it is already present or there is space to add it. Thus, x's counter is incremented t_i times, and $t_f + t_i = t > n/(m+1)$. Furthermore, let t_d denote the number of times that a counter monitoring x is decremented as another value is read. Because a counter never goes negative, $t_i \geq t_d$. If this inequality is strict, then x ends up with a positive count at the end of the algorithm.

With each of the $t_f + t_d$ times decrements occur, we can associate m occurrences of other values along with the occurrence of x, for a total of $m+1$ unique locations in the input steam. Thus, $(m+1)(t_f + t_d) \leq n$. If the final value of x's counter is zero, then $t_d = t_i$, so $t = t_f + t_i = t_f + t_d > n/(m+1)$, i.e., $(m+1)(t_f + t_d) > n$, which is a contradiction. Hence $t_i > t_d$, so x's counter remains positive and x is one of at most m candidates remaining. □

This method thus identifies at most m candidates for having appeared more than $n/(m+1)$ times, and does so with no use of probabilistic methods. Clearly there remains the issue of how to perform the appropriate updates quickly. Most notably, there is the issue of decrementing and releasing several counters simultaneously.

3.3 Data Structures

To support decrementing all counters at once in constant time, we store the counters in sorted order using a differential encoding. That is, each counter actually only stores how much larger it is compared to the next smallest counter. Now incrementing and decrementing counters requires them to move significantly in the total order; to support these operations, we coalesce equal counters (differentials of zero) into common *groups*.

The overall structure is a doubly linked list of groups, ordered by counter value. Each group represents a collection of equal counters, consisting of two parts: (1) a doubly linked list of counters (in no particular order, because they all have the same value), and (2) the difference in value between these counters and the counters in the previous group, or, for the first group, the value itself. Each "counter" no longer needs to store a value, but rather stores its group and its monitored element.

Because of lack of space, we omit the details of Algorithm FREQUENT in combination with these data structures.

Theorem 2. *Algorithm* FREQUENT' *implements Algorithm* FREQUENT *from Theorem 1 in $O(1)$ time per packet.*

3.4 Lower Bound

Algorithm FREQUENT achieves the best possible frequency threshold according to the model presented in Section 2.1.2.

Theorem 3. *For any n and m, and any deterministic one-pass algorithm storing at most m elements at once, there is a sequence of length n, in which one element occurs at least $n/(m+1)-1$ times and the other elements are all unique, and on which the algorithm terminates with only uniquely occurring elements stored.*

Proof. We initially imagine there being n distinct elements, divided by a yet-to-be-determined scheme into $m + 1$ *classes*. We maintain that each element stored by the algorithm is from a different class. At each step, the algorithm examines its at most $m + 1$ elements, discards one, and reads the next element from the stream. The adversary chooses the next element from the same class as the element that was discarded. (At the beginning, the adversary chooses arbitrarily.)

In this way, the algorithm learns only that elements from different classes are different, but does not learn about elements from a common class. Thus, at the end, the adversary is free to choose which elements in a class are equal and which are not. In particular, the adversary can choose the largest class, which must have size at least $n/(m+1)$, to have all its members equal except for possibly one member of the m being returned by the algorithm; and choose all other classes to have all distinct elements. □

4 Probabilistic Frequency Counts

This section develops algorithms for the stochastic model, in which an arbitrary probability distribution specifies the relative frequencies of the categories, but in what order these categories occur in the packet stream is uniformly random. We distinguish two cases according to whether the process is allowed extra storage so long as the working store is small; see Section 2.1.1.

4.1 Overview

The basic algorithm works as follows. We divide the stream into a collection of rounds, carefully sized to balance the counter-reset trade-off described in the first section. At the beginning of each round, the algorithm samples the first m distinct packet categories, which is equivalent to sampling m packets uniformly at random. The algorithm then counts their occurrences for the duration of the round. Applying Chernoff bounds, we prove that the counts obtained during a round are close to the actual frequencies of the categories. The k categories with the maximum counter values at the end of the round are the winners for that round. If extra *nonworking storage* is available to the algorithm, we record these winners and their counts for a final tournament at the end of the algorithm. Otherwise, we reserve a constant fraction of the working storage for the current best winners, and only compare against those. In either case, we prove that with high probability the true frequencies of the final winners are close to the frequencies of the truly most popular categories. The probabilities are slightly higher when extra nonworking space is available.

The ideal choice for the size of a round in this algorithm depends on the length n of the stream and on the probability distribution on categories. Of course, the algorithm does not generally know the probabilities, and may not even know for how long it will be monitoring the stream: imagine a scenario in which the statistics gathering process is running constantly, and at will a networks designer can request the current guess and confidence of the most popular categories; as time passes, the confidence increases. To solve these problems, we harness the algorithm in an adaptive framework that gradually increases the round length until the confidence is determined to suffice. This flexible framework requires monitoring the stream for only slightly longer.

4.2 Algorithm with Extra Nonworking Storage

More precisely, we divide the input stream into rounds of r packets each. The algorithm then works as follows:

Algorithm PROBABILISTIC

1. For each round of r elements:
   (a) Assign the m counters to monitor the first m distinct elements that appear in the round.
   (b) For each element, if the element is being monitored, increment the appropriate counter.
   (c) Store the elements and their counts to the extra nonworking storage.
2. Pass through the elements and counts stored in extra nonworking storage.
3. Return the k distinct elements with the largest counts, for the desired value of k. (If an element appears multiple times in the list, we effectively drop all but its largest count.)

Theorem 4. *Fix any constants $c > 0$ and $\alpha > 1$. Call an element above threshold if it has relative frequency at least $\tau = (c \ln n)/\sqrt{mn}$. Suppose that t elements are above threshold. If c is sufficiently large with respect to α, then with high probability, Algorithm* PROBABILISTIC *($r = \sqrt{mn}$) returns a list of k elements whose first $\min\{k, t\}$ elements are as if we perturbed each element's relative frequency within a factor of α and then took the top $\min\{k, t\}$ elements.*

4.3 Algorithm without Extra Nonworking Storage

A simple modification to Algorithm PROBABILISTIC avoids the use of extra storage by computing the maximum frequencies online at the cost of using some counter space:

> **Algorithm** PROBABILISTIC-INPLACE
> 1. Reserve $m/2$ of the m counters to store the current best candidates.
> 2. For each round of r elements:
>   (a) Assign the $m/2$ unreserved counters to monitor the first $m/2$ distinct elements that appear in the round, and zero these counters.
>   (b) For each element, if the element is being monitored, increment the appropriate counter.
>   (c) Replace the $m/2$ reserved counters with the top out of all m counters.
> 3. Return the $m/2$ reserved counters.

As stated, this algorithm does not run in constant time per packet, incurring a $\Theta(m)$ cost at the end of every round. However, this large cost can be avoided, similar to Algorithm FREQUENT. Again we omit details because of lack of space.

We obtain the same results as in Theorem 4, only with m half as large and k constrained to be at most $m/2$.

Theorem 5. *Suppose that t elements are above threshold, i.e., have relatively frequency at least $(c \ln n)/\sqrt{mn/2}$. If c is sufficiently large with respect to α, then with high probability, Algorithms* PROBABILISTIC-INPLACE *and* PROBABILISTIC-INPLACE' *($r = \sqrt{mn/2}$) return a list of $m/2$ elements whose first $\min\{m/2, t\}$ elements are as if we perturbed each element's relative frequency within a factor of α and then took the top $\min\{m/2, t\}$ elements.*

4.4 Streams of Unknown Length

If the value of n is unknown to the algorithm, we can guess the value of n to be 1 and run the algorithm, then guess consecutively $n = 2, 4, \ldots, 2^j, \ldots$ until the stream is exhausted. At round j, we can find the top elements so long as their probability satisfies $p > j/\sqrt{2^j m}$. This bound is within a factor of roughly $\sqrt{2}$ compared to if we knew n a priori.

4.5 Lower Bound

We prove a matching lower bound for the algorithms above, up to constant factors, in the model of computation presented in Section 2.1.2:

Theorem 6. *Consider the distribution in which one element x has relative frequently (just) below $f/\sqrt{mn}$, and e.g. every other element occurs just once. For any probabilistic one-pass algorithm storing at most m elements at once, the probability of failing to report element x is, asymptotically, at least $(e^{-1+1/e})^f \approx 0.5314636^f$. Consequently, if $f = \Theta(1)$, there is a constant probability of failure, and f must be $\Omega(\lg n)$ to achieve a polynomially small probability of failure.*

5 Conclusion

The main open problem that remains is to consider the more relaxed but highly natural oblivious-adversary worst-case model, which allows randomization internally to the algorithm but assumes nothing about the input stream. We are hopeful that it is possible to achieve results similar to the stochastic model by augmenting our algorithm to randomly perturb the sizes of the rounds. The idea is that such perturbations prevent the adversary from knowing when the actual samples occur.

Acknowledgments

We thank the anonymous referees for their helpful comments and thorough review.

References

[1] N. Alon, Y. Matias and M. Szegedy. "The space complexity of approximating the frequency moments", *STOC*, 1996, pp. 20–29.

[2] B. Bloom. "Space/time trade-offs in hash coding with allowable queries", *Comm. ACM*, 13:7, July 1970, pp. 422–426.

[3] M. Charikar, K. Chen and M. Farach-Colton. "Finding frequent items in data streams", to appear in *ICALP*, 2002.

[4] S. Chaudhuri, R. Motwani and V. Narasayya. "Random sampling for histogram construction: how much is enough", In *SIGMOD*, 1998, pp. 436–447.

[5] Cisco Systems. *Sampled NetFlow*, http://www.cisco.com/univercd/cc/td/doc/product/software/ios120/120newft/120limit/120s/120s11/12s_sanf.htm, April 2002.

[6] K. Claffy, G. Miller, K. Thompson. The nature of the beast: recent traffic measurements from an Internet backbone. In *Proc. 8th Annual Internet Society Conference*, 1998.

[7] M. Datar, A. Gionis, P. Indyk and R. Motwani. "Maintaining stream statistics over sliding windows", In *SODA*, 2002, pp. 635–644.

[8] N. G. Duffield and M. Grossglauser. "Trajectory sampling for direct traffic observation", In *Proc. ACM SIGCOMM*, 2000, pp. 271–282.

[9] C. Estan and G. Varghese. "New directions in traffic measurement and accounting", In *Proc. ACM SIGCOMM Internet Measurement Workshop*, 2001.

[10] M. Fang, N. Shivakumar, H. Garcia-Molina, R. Motwani and J. Ullman. "Computing iceberg queries efficiently", *VLDB*, 1998, pp. 299–310.

[11] J. Feigenbaum, S. Kannan, M. Strauss, and M. Viswanathan. "An approximate L^1-difference algorithm for massive data streams", In *FOCS*, 1999, pp. 501–511.

[12] J. Feigenbaum, S. Kannan, M. Strauss, and M. Viswanathan. "Testing and Spot Checking of Data Streams", In *SODA*, 2000, pp. 165–174.

[13] M. J. Fischer and S. L. Salzberg. "Finding a Majority Among N Votes: Solution to Problem 81-5 (Journal of Algorithms, June 1981)", *J. Algorithms*, 3(4):362–380, 1982.

[14] W. Feller. *An Introduction to Probability Theory and its Applications.* 3rd Edition, John Wiley & Sons, 1968.
[15] P. Flajolet. "Approximate counting: a detailed analysis", *BIT*, 25, 1985, pp. 113–134.
[16] P. Flajolet and G. N. Martin. "Probabilistic counting algorithms", *J. Computer and System Sciences*, 31, 1985, pp. 182–209.
[17] P. B. Gibbons and Y. Matias. "New sampling-based summary statistics for improving approximate query answers", In *Proc. ACM SIGMOD International Conf. on Management of Data*, June 1998, pp. 331–342.
[18] I. D. Graham, S. F. Donelly, S. Martin, J. Martens, and J. G. Cleary. Nonintrusive and accurate measurements of unidirectional delay and delay variation in the Internet. *Proc. 8th Annual Internet Society Conference*, 1998.
[19] P. Gupta and N. Mckeown. "Packet classification on multiple fields", In *Proc. ACM SIGCOMM*, 1999, pp. 147–160.
[20] P. J. Haas, J. F Naughton, S. Seshadri and L. Stokes. "Sampling-Based Estimation of the Number of Distinct Values of an Attribute", In *VLDB*, 1995, pp. 311–322.
[21] P. Indyk. "Stable Distributions, Pseudorandom Generators, Embeddings and Data Stream Computations", In *FOCS*, 2000, pp. 189–197.
[22] J. G. Kalbfleisch, *Probability and Statistical Inference*, Springer-Verlag, 1979.
[23] R. Mahajan and S. Floyd. "Controlling High Bandwith Flows at the Congested Router", In *Proc. 9th International Conference on Network Protocols*, 2001.
[24] R. Morris. "Counting large numbers of events in small registers", *Comm. ACM*, 21, 1978, pp. 840–842.
[25] Rajeev Motwani and Prabhakar Raghavan. *Randomized Algorithms*, Cambridge University Press, 1995.
[26] J. S. Vitter. "Optimum algorithms for two random sampling problems", In *FOCS*, 1983, pp. 65–75.
[27] K.-Y. Whang, B. T. Vander-Zanden, H. M. Taylor. "A Linear-Time Probabilistic Counting Algorithm for Database Applications", *ACM Trans. Database Systems* 15(2):208–229, 1990.

Truthful and Competitive Double Auctions

Kaustubh Deshmukh[1], Andrew V. Goldberg[2],
Jason D. Hartline[1], and Anna R. Karlin[1]

[1] Computer Science Department, University of Washington
{kd,hartline,karlin}@cs.washington.edu
[2] Microsoft Research
1065 La Avenida, SVC 5, Mountain View, CA 94043
goldberg@microsoft.com

Abstract In this paper we consider the problem of designing a mechanism for double auctions where bidders each bid to buy or sell one unit of a single commodity. We assume that each bidder's utility value for the item is private to them and we focus on truthful mechanisms, ones were the bidders' optimal strategy is to bid their true utility. The profit of the auctioneer is the difference between the total payments from buyers and total to the sellers. We aim to maximize this profit. We extend the competitive analysis framework of basic auctions [9] and give an upper bound on the profit of any truthful double auction. We then reduce the competitive double auction problem to basic auctions by showing that any competitive basic auction can be converted into a competitive double auction with a competitive ratio of twice that of the basic auction. In addition, we show that better competitive ratios can be obtained by directly adapting basic auction techniques to the double auction problem. This result provides insight into the design of profit maximizing mechanisms in general.

1 Introduction

Dynamic pricing mechanisms, and specifically auctions with multiple buyers and sellers, are becoming increasing popular in electronic commerce. We consider *double auctions* in which there is one commodity in the market with multiple buyers and sellers each submitting a single bid to either buy or sell one unit of the commodity (E.g. [7]). The numerous applications of double auctions in electronic commerce, including stock exchanges, business-to-business commerce, bandwidth allocation, etc. have led to a great deal of interest in fast and effective algorithms [16, 18].

For double auctions, the auctioneer, acting as a broker, is faced with the task of matching up a subset of the buyers with a subset of the sellers of the same size. The auctioneer decides on a price to be paid to each seller and received from each buyer in exchange for the transfer of one item from each of the selected sellers to each of the selected buyers. The *profit of the auctioneer* is the difference between the prices paid by the buyers and the prices paid to the sellers.

We assume that each bidder has a private utility value for the item. For the buyers this utility value is the most that they are willing to buy the item for, and for the sellers it is the least they are willing to sell for. We focus on double auction mechanisms that are *truthful*: the best strategy of a selfish bidder that is attempting to maximize their own gain is to bid their true utility value.

The traditional economics approach to the study of profit maximizing auctions is to construct the optimal Bayesian auction given the prior distribution from which the bidders' utility values are drawn (e.g., [3, 15]). In contrast, following [10, 6, 9], we attempt to design mechanisms that maximize profit under *any* market conditions. As in competitive analysis of online algorithms, we gauge a truthful double auction mechanism's performance on a particular bid set by comparing it against the profit that would be achieved by an "optimal" auction, $\mathcal{M}_{opt}$, on the same bidders.

If, for every bid set, a particular truthful double auction mechanism $\mathcal{M}$ achieves a profit that is close to that of the optimal $\mathcal{M}_{opt}$, we say that the auction mechanism $\mathcal{M}$ is *competitive* against $\mathcal{M}_{opt}$, or simply competitive. For example, we might be interested in constructing double auctions that are competitive with the optimal single-price omniscient mechanism, $\mathcal{F}$. This is the mechanism which, based on perfect knowledge of buyer and seller utilities, selects a single price b_{opt} for the buyers and a single price s_{opt} for the sellers. It then finds the largest k such that the highest k buyers each bid at least b_{opt} and the lowest k sellers each bid at most s_{opt}. It then matches these buyers and sellers up, paying all the sellers s_{opt} and charging each of the buyers b_{opt}. The profit of the auctioneer is thus $k(b_{opt} - s_{opt})$.

1.1 Results

This paper makes the following contributions:

- We extend the framework for competitive analysis of basic auctions to double auctions. This framework is motivated by a number of results bounding truthful mechanism In particular, we show that no monotone[1] double auction (even a *multi-priced* mechanism) can achieve a higher profit than twice the optimal *single-priced* mechanism $\mathcal{F}$ discussed above.
- We present a reduction from double auctions to basic auctions by showing how to construct a competitive double auction from any competitive basic auction while losing only a factor of two in competitive ratio.
- We show how the basic auction from [8] can be adapted to the double auction problem yielding better competitive ratio than one get by applying the aforementioned reduction. We also discuss the possibility of making similar adaptations for other profit maximizing mechanism design problems.

1.2 Related Work

We study profit maximizing single round double auctions when the utility value of each bidder is private and must be truthfully elicited. When the utilities are

---
[1] See Section 2.2 for the definition of monotonc.

public values this problem becomes trivial. The following variants of the problem have been previously studied.

When the goal is not to maximize profit of the auctioneer, but to find an outcome that is maximizes the *common welfare*, i.e., the sum of the profits of each of the bidders, subject to the constraint that the auctioneer's profit is non-negative, McAffee [13] gives a truthful mechanism that approaches optimal as the number of sold items in the optimal solution grows. Note that the Vickrey-Clarke-Groves [4, 11, 17] mechanism, the only mechanism that always gives the outcome that maximizes the common welfare, always gives a non-positive profit to the auctioneer (assuming *voluntary participation*[2]).

Our results are closely related to the *basic auctions* for a single item available in unlimited supply, e.g., for digital goods [10, 9]. As such, the approach we take in this paper closely parallels that in [9]. Furthermore, as we explain later, the basic auction is a special case of the double auction where all sellers have utility zero.

An "online" version of the double auction, where bids arrive and expire at different times, was considered by Blum, Sandholm, and Zinkevich [2] (also known as a *continuous double auction* [18]). Their mechanism must make decisions without knowing what bids will arrive in the future. They consider the goals of optimizing the profit of the auctioneer and of maximizing the number of items sold. Their solution assumes that bidders are compelled to bid their true utility value despite the fact that the algorithms they develop are not truthful, i.e., the utility values of the bidders are public. An interesting open question left by our work is the problem of a profit maximizing online double auction for the private value model. For private values, an online variant of the basic auction problem was considered by Bar-Yossef et al. [1] in a competitive framework for profit maximization.

Of course, auctions, be they traditional or combinatorial, have received a great deal of attention (see e.g., the surveys [5, 12]).

2 Preliminaries

Throughout the paper we will be using the notation $b_{(i)}$ to represent the ith largest buyer bid and $s_{(i)}$ for the ith smallest seller bid.

Definition 1. *A* single-round sealed-bid double-auction mechanism *is one where:*

- *Given the two bid vectors* $\mathbf{b} = (b_1, \ldots, b_n)$ *and* $\mathbf{s} = (s_1, \ldots, s_n)$, *the mechanism computes a pair of allocation vectors,* $\mathbf{x}$ *and* $\mathbf{y} \in \{0,1\}^n$, *and payment vectors* $\mathbf{p}$ *and* $\mathbf{q} \in \mathbb{R}^n$, *subject to the constraints that:*
  - *The number of winning buyers is equal to the number of winning sellers, i.e.,* $\sum_i x_i = \sum_i y_i$.[3]

---

[2] Defined in Section 2.
[3] We assume that the auctioneer neither has any items for sale nor is willing to purchase any. For this reason, we can also assume that the number of buyer bids equals

- $0 \le p_i \le b_i$ (resp. $s_i \le q_i$) for all winning buyers (resp. sellers) and that $p_i = 0$ (resp. $q_i = 0$) for all losing buyers (resp. sellers). These are the standard assumptions of no positive transfers and voluntary participation. See, e.g., [14].
  - If $x_i = 1$ buyer i wins (i.e., receives the item) and pays price p_i, otherwise we say that buyer i loses. If $y_i = 1$ seller i wins (i.e., sells the item) and gets paid q_i, otherwise we say that seller i loses.
  - The profit of the mechanism is $\mathcal{M}(\mathbf{b}, \mathbf{s}) = \sum_i p_i - \sum_i q_i$.

Note that the *basic auction* problem of [10] can be viewed as a special case of the double auction problem with all sell bids equal to zero.

We say the *mechanism is randomized* if the procedure used to compute the allocations and prices is randomized. Otherwise, the mechanism is *deterministic*. Note that if the mechanism is randomized, the profit of the mechanism, the output prices, and the allocation are random variables.

We use the following private value model for the bidders:
  - Each bidder has a private utility value for the item. We denote the utility value for buyer i by u_i and the utility value for seller i by v_i.
  - Each bidder bids so as to maximize their *profit*: For buyers (resp. sellers) this means they bid b_i (resp. s_i) to maximize profit given by $u_i x_i - p_i$ (resp. $q_i - v_i y_i$).
  - Bidders bid with full knowledge of the auctioneer's strategy. However, the bidding occurs in advance (i.e., without knowledge) of any coin tossing done by a randomized auctions.
  - Bidders do not collude.

Finally, we formally define the notion of truthfulness.

Definition 2. *We say that a deterministic double auction is truthful if, for each bidder i and any choice of bid values for all other bidders, bidder i's profit is maximized by bidding their utility value, i.e., by setting $b_i = u_i$ for buyers and by setting $s_i = v_i$ for sellers.*

Definition 3. *We say that a randomized auction is truthful if it can be described as a probability distribution over deterministic truthful auctions.*

As bidding u_i (resp. v_i) is a dominant strategy for buyer i (resp. seller i) in a truthful auction, in the remainder of this paper, we assume that $b_i = u_i$ and $s_i = v_i$ unless mentioned otherwise.

---

the number of seller bids. If there are any extra buyers or sellers, the auctioneer can earn the same amount of profit by ignoring the extra low bidding buyers or high bidding sellers.

2.1 Bid Independence

We describe a useful characterization of truthful mechanisms using the notion of *bid independence*. Let $\mathbf{b}_{-i}$ denote the vector of bids $\mathbf{b}$ with b_i removed, i.e., $\mathbf{b}_{-i} = (b_1, \ldots, b_{i-1}, ?, b_{i+1}, \ldots, b_n)$. We call such a vector *masked*. Similarly, let $\mathbf{s}_{-i}$ denote the masked vector of bids $\mathbf{s}$ with s_i removed, i.e., $\mathbf{s}_{-i} = (s_1, \ldots, s_{i-1}, ?, s_{i+1}, \ldots, s_n)$. Given bid vectors $\mathbf{b}$ and $\mathbf{s}$, the bid-independent mechanism defined by the randomized functions f and g is defined as follows:

For each buyer i, if $f(\mathbf{b}_{-i}, \mathbf{s}) \leq b_i$, buyer i wins at the price $p_i = f(\mathbf{b}_{-i}, \mathbf{s})$. Otherwise, buyer i loses and makes no payment. Similarly for each seller i, if $s_i \leq g(\mathbf{b}, \mathbf{s}_{-i})$, seller i wins and receives a payment of $q_i = g(\mathbf{b}, \mathbf{s}_{-i})$. Otherwise, seller i loses and receives no compensation.[4]

The following theorem, which is a straightforward generalization of the equivalent result for basic auctions in [9], relates bid independence to truthfulness.

Theorem 1. *A double auction is truthful if and only if it is bid-independent.*

2.2 Monotonicity

We define the notion of monotone double auctions to characterize "reasonable" truthful mechanisms. Using standard terminology, we say that random variable X *dominates* random variable Y if for all x

$$\Pr[X \geq x] \geq \Pr[Y \geq x].$$

Definition 4. *A double auction mechanism is monotone if it is defined by a pair of bid-independent functions f and g (each taking as input a buy vector and a sell vector, where one of the two vectors is masked) such that for any buy and sell vectors $\mathbf{b}$ and $\mathbf{s}$, we have:*

- *For any pair of buyers i and j such that $b_i \leq b_j$ the random variable $f(\mathbf{b}_{-i}, \mathbf{s})$ dominates the random variable $f(\mathbf{b}_{-j}, \mathbf{s})$.*
- *For any pair of sellers i and j such that $s_i \leq s_j$ the random variable $g(\mathbf{b}, \mathbf{s}_{-i})$ dominates the random variable $g(\mathbf{b}, \mathbf{s}_{-j})$.*

To get a feel for this definition, observe that when $b_i \leq b_j$ the bids visible in the masked vector $\mathbf{b}_{-j}$ are the same as those visible in the masked vector $\mathbf{b}_{-i}$ except for the fact that the smaller bid b_i is visible in $\mathbf{b}_{-j}$ whereas the larger bid b_j is visible in $\mathbf{b}_{-i}$. Intuitively, monotonicity means that if buyer bids are increased while keeping the seller bid vector constant, then the threshold prices output by the bid-independent function f increase. Similarly for the sellers.

---

[4] In fact, bid-independence allows the inequalities, $f(\mathbf{b}_{-i}, \mathbf{s}) \leq b_i$ and $s_i \leq g(\mathbf{b}, \mathbf{s}_{-i})$ to be either strict or non-strict at the discretion of the functions f and g.

2.3 Single Price Omniscient Mechanism

A key question is how to evaluate the performance of mechanisms with respect to the goal of profit maximization. Consider the following definitoin:

Definition 5. *The optimal single price omniscient mechanism, $\mathcal{F}$, is the mechanism that uses the optimal single buy price and single sell price. It achieves the optimal single price profit of*

$$\mathcal{F}(\mathbf{b},\mathbf{s}) = \max_{i} i(b_{(i)} - s_{(i)}).$$

A theorem we prove later shows that no reasonable (possibly multi-priced) truthful mechanism can achieve profit above $2\mathcal{F}(\mathbf{b},\mathbf{s})$. This motivates using $\mathcal{F}$ as a performance metric. Unfortunately, it is impossible to be competitive with $\mathcal{F}$. This is shown in [9] for basic auctions, a special case of the double auction, when the $\mathcal{F}$ sells to only the highest bidder. Hence, we compare the performance of truthful mechanisms with the profit of the optimal single price omniscient mechanism that transfers at least two items from sellers to buyers.

Definition 6. *The optimal fixed price mechanism that transfers at least two items, $\mathcal{F}^{(2)}$, has profit*

$$\mathcal{F}^{(2)}(\mathbf{b},\mathbf{s}) = \max_{i \geq 2} i(b_{(i)} - s_{(i)}).$$

2.4 Competitive Mechanisms

We now formalize the notion of a competitive mechanism:

Definition 7. *We say that a truthful double auction $\mathcal{M}$ is β-competitive against $\mathcal{F}^{(2)}$ if, for all bid vectors $\mathbf{b}$ and $\mathbf{s}$ the expected profit of $\mathcal{M}$ satisfies*

$$\mathbf{E}[\mathcal{M}(\mathbf{b},\mathbf{s})] \geq \mathcal{F}^{(2)}(\mathbf{b},\mathbf{s})/\beta.$$

We say that $\mathcal{M}$ is competitive if $\mathcal{M}$ is β-competitive for some constant β.

3 Upper Bound on the Profit of Truthful Mechanisms

In this section, we show that the profit for all monotone double auction mechanisms the is bounded by $2\mathcal{F}(\mathbf{b},\mathbf{s})$. Goldberg et al. showed that for basic auctions this result holds without the factor of two:

Theorem 2. *For input bids $\mathbf{b}$, no truthful monotone basic auction has expected profit more than $\mathcal{F}(\mathbf{b})$ [9].*

We conjecture that this bound holds for double auctions as well, though what we prove below is a factor of two worse.

Lemma 1. *For any value v and buy and sell bids $\mathbf{b}$ and $\mathbf{s}$, define $\mathbf{b}'$ and $\mathbf{s}'$ as $b'_i = b_i - v$ and $s'_i = v - s_i$ for $1 \leq i \leq n$. Then for any monotone double auction, $\mathcal{M}$:*

$$\mathbf{E}[\mathcal{M}(\mathbf{b}, \mathbf{s})] \leq \mathcal{F}(\mathbf{b}') + \mathcal{F}(\mathbf{s}').$$

Proof. Let $\mathbf{x}$, $\mathbf{y}$, $\mathbf{p}$, and $\mathbf{q}$ be the outcome and prices when $\mathcal{M}$ is run on $\mathbf{b}$ and $\mathbf{s}$. Let $X = \{i : x_i = 1\}$ and $Y = \{i : y_i = 1\}$. Note $|X| = |Y|$. Thus,

$$\mathcal{M}(\mathbf{b}, \mathbf{s}) = \sum_i p_i - \sum_i q_i = \sum_{i \in X} p_i - \sum_{i \in Y} q_i$$
$$= \sum_{i \in X} (p_i - v) + \sum_{i \in Y} (v - q_i).$$

Let $\mathcal{A}_{v,\mathbf{s}}$ be the basic auction that on $\mathbf{b}'$ simulates $\mathcal{M}(\mathbf{b}, \mathbf{s})$ to compute prices p_i for each bidder b'_i and then offers them $p_i - v$. It is easy to see that this is truthful, monotone (since $\mathcal{M}$ is), and gives revenue

$$\mathcal{A}_{v,\mathbf{s}}(\mathbf{b}') = \sum_{i \in X} (p_i - v).$$

Using the bound on the revenue of any monotone basic auction (Theorem 2) we get:

$$\mathbf{E}\left[\sum_{i \in X} (p_i - v)\right] = \mathbf{E}[\mathcal{A}_{v,\mathbf{s}}(\mathbf{b}')] \leq \mathcal{F}(\mathbf{b}').$$

Combining this with the analogous argument for $\mathbf{s}'$ we have:

$$\mathbf{E}[\mathcal{M}(\mathbf{b}, \mathbf{s})] = \mathbf{E}\left[\sum_{i \in X} (p_i - v)\right] + \mathbf{E}\left[\sum_{i \in Y} (v - q_i)\right] \leq \mathcal{F}(\mathbf{b}') + \mathcal{F}(\mathbf{s}').$$

□

Theorem 3. *For any bid vectors $\mathbf{b}$ and $\mathbf{s}$, any truthful monotone double auction, $\mathcal{M}$, has expected profit at least $2\mathcal{F}(\mathbf{b}, \mathbf{s})$.*

Proof. Find the largest ℓ such that $b_{(\ell)} \geq s_{(\ell)}$ and choose $v \in [s_{(\ell)}, b_{(\ell)}]$. Now we let $\mathbf{b}'$ and $\mathbf{s}'$ be $b'_i = b_i - v$ and $s'_i = v - s_i$ for $1 \leq i \leq n$ as in Lemma 1 giving $\mathbf{E}[\mathcal{M}(\mathbf{b}, \mathbf{s})] \leq \mathcal{F}(\mathbf{b}') + \mathcal{F}(\mathbf{s}')$ for our choice of v.

Note that $\mathcal{F}(\mathbf{b}, \mathbf{s}) = \max_i i(b_{(i)} - s_{(i)})$. Let k be the number of winners in $\mathcal{F}(\mathbf{b}, \mathbf{s})$. Note that by our choice of v, we have $b_{(k)} \geq v$ and $s_{(k)} \leq v$. This gives:

$$\mathcal{F}(\mathbf{b}') = \max_i i(b_{(i)} - v) \leq \max_i i(b_{(i)} - s_{(i)}) = \mathcal{F}(\mathbf{b}, \mathbf{s}), \text{ and}$$
$$\mathcal{F}(\mathbf{s}') = \max_i i(v - s_{(i)}) \leq \max_i i(b_{(i)} - s_{(i)}) = \mathcal{F}(\mathbf{b}, \mathbf{s}).$$

Thus, $\mathbf{E}[\mathcal{M}(\mathbf{b}, \mathbf{s})] \leq \mathcal{F}(\mathbf{b}') + \mathcal{F}(\mathbf{s}') \leq 2\mathcal{F}(\mathbf{b}, \mathbf{s})$. □

4 Reducing Competitive Double Auctions to Competitive Basic Auctions

In this section we describe a general technique for converting any β-competitive basic auction into a 2β-competitive double auction. Let $\mathcal{A}$ be a basic auction that is β-competitive against $\mathcal{F}^{(2)}(\mathbf{b})$. We assume for the discussion here that $b_{(n)} \geq s_{(n)}$ though it is not difficult remove this assumption. We construct a double auction mechanism $\mathcal{M}_\mathcal{A}$ that is 2β-competitive against $\mathcal{F}^{(2)}(\mathbf{b}, \mathbf{s})$ as follows:

1. If $n = 2$ run the Vickery auction, output its outcome, and halt.
2. Let $\mathbf{b}'$ and $\mathbf{s}'$ be n-dimensional vectors with components by $b'_i = b_i - s_{(n)}$ and $s_i = b_{(n)} - s_i$. Let $\mathbf{b}''$ (resp. $\mathbf{s}''$) be $\mathbf{b}'$ with the smallest (resp. largest) bid deleted.
3. With probability $1/2$ run $\mathcal{A}(\mathbf{b}'')$. If i wins $\mathcal{A}$ at price p''_i, buyer i wins $\mathcal{M}_\mathcal{A}$ at price $p_i = \max(b_{(n)}, p''_i + s_{(n)})$. Let k be the number of winners in $\mathcal{A}(\mathbf{b}'')$. To determine the outcome for the sellers, run the k-Vickery auction on $\mathbf{s}$.
4. Otherwise (with probability $1/2$) run $\mathcal{A}(\mathbf{s}'')$. If i wins $\mathcal{A}$ at price q''_i, seller i wins $\mathcal{M}_\mathcal{A}$ at price $q_i = \min(s_{(n)}, b_{(n)} - q''_i)$. As in Step 2, we run a k-Vickery auction on the buyers to determine the outcome, where k is the number of winners in $\mathcal{A}(\mathbf{s}'')$.

Theorem 4. $\mathcal{M}_\mathcal{A}$ *is truthful.*

We omit the proof.

Theorem 5. *If $\mathcal{A}$ is β-competitive, $\mathcal{M}_\mathcal{A}$ is 2β-competitive against $\mathcal{F}^{(2)}(\mathbf{b}, \mathbf{s})$.*

Proof. If $n = 2$, $\mathcal{M}_\mathcal{A}$ runs Vickrey and is 2-competitive. For the rest of the proof assume $n \geq 3$. Let $k \geq 2$ be the number of items sold by $\mathcal{F}^{(2)}(\mathbf{b}, \mathbf{s})$. Thus,

$$\mathcal{F}^{(2)}(\mathbf{b}, \mathbf{s}) = k(b_{(k)} - s_{(n)}) + k(b_{(n)} - s_{(k)}) - k(b_{(n)} - s_{(n)}).$$

But by definition $\mathcal{F}^{(2)}(\mathbf{b}') \geq k(b_{(k)} - s_{(n)})$ and likewise for $\mathbf{s}'$, therefore

$$\mathcal{F}^{(2)}(\mathbf{b}, \mathbf{s}) \leq \mathcal{F}^{(2)}(\mathbf{b}') + \mathcal{F}^{(2)}(\mathbf{s}') - k(b_{(n)} - s_{(n)}). \tag{1}$$

Note that for the buyers (and similarly for sellers):

$$\mathcal{F}^{(2)}(\mathbf{b}') \leq \mathcal{F}^{(2)}(\mathbf{b}'') + b_{(n)} - s_{(n)}. \tag{2}$$

Because $k \geq 2$, from Equations (1) and (2) we have

$$\mathcal{F}^{(2)}(\mathbf{b}, \mathbf{s}) \leq \mathcal{F}^{(2)}(\mathbf{b}'') + \mathcal{F}^{(2)}(\mathbf{s}'').$$

Note that becaus $\mathcal{A}$ is β-competitive, the expected revenue from Step 3 and Step 4 are $\mathcal{F}^{(2)}(\mathbf{b}'')/2\beta$ and $\mathcal{F}^{(2)}(\mathbf{s}'')/2\beta$ respectively. Thus,

$$\mathbf{E}[\mathcal{M}_\mathcal{A}(\mathbf{b}, \mathbf{s})] \geq \tfrac{1}{2\beta}(\mathcal{F}^{(2)}(\mathbf{b}'') + \mathcal{F}^{(2)}(\mathbf{s}'')) \geq \tfrac{1}{2\beta}\mathcal{F}^{(2)}(\mathbf{b}, \mathbf{s}).$$

□

Plugging in the 4-competitive Sampling Cost Sharing auction [6], we get a double auction with a competitive ratio of 8. Plugging in the 3.39-competitive Consensus Revenue Estimate auction [8] we get a competitive ratio of 6.78. We can do better if we customize mechanisms for the double auction problem.

5 The Revenue Extraction and Estimation Technique

5.1 Revenue Extraction

For the basic auction problem, the cost sharing mechanism of Moulin and Shenker [14] is the basis for auctions with good competitive ratios [6, 8]. The cost sharing mechanism is defined as follows:

CostShare$_C$: Given bids **b**, find the largest k such that the highest k bidders can equally share the cost C. Charge each one of those C/k.

This mechanism is truthful and if $C \leq \mathcal{F}(\mathbf{b})$ then CostShare$_C$ has revenue C, otherwise it has no revenue.

The SCS (Sampling Cost Sharing) auction for the basic problem is defined as follows. First partition the bidders into two sets, $\mathbf{b}'$ and $\mathbf{b}''$, and compute the optimal fixed price revenues from each set, $\mathcal{F}(\mathbf{b}')$ and $\mathcal{F}(\mathbf{b}'')$. Then cost share the optimal revenue for one set on the bids among the other set and vice versa (i.e., run CostShare$_{\mathcal{F}(\mathbf{b}')}(\mathbf{b}'')$ and CostShare$_{\mathcal{F}(\mathbf{b}'')}(\mathbf{b}')$). It is easy to see that profit of the auctioneer is the minimum of the two optimal revenues. The key to the analysis is showing that the expected value of the smaller optimal revenue is at least 1/4 of the optimal revenue for **b**.

We could attempt to follow the same mechanism framework for the double auction problem if we had an equivalent of CostShare$_C$ for double auctions. Unfortunately, there is no exact cost sharing analog.

Lemma 2. *For any value C, there is no truthful mechanism for the double auction problem that always achieves a profit of at least C when C is possible, i.e., $\mathcal{F}(\mathbf{b}, \mathbf{s}) \geq C$.*

Proof. Suppose for a contradiction that such a mechanism $\mathcal{M}_C$ did exist. Consider the single buyer, single seller case with $b_1 = s_1 + C$. Theorem 1 and $\mathcal{M}_C$'s truthfulness implies that the price for b_1 is given by a bid-independent function $f(s_1)$. Since the $C = \mathcal{F}(\mathbf{b}, \mathbf{s})$ is possible and $\mathcal{M}_C$ is assumed to achieve at least C if it is possible, $f(s_1)$ must be $s_1 + C$. Symmetrically, $g(b_1)$ must be $b_1 - C$. Thus, for s_1 and b_1 satisfying $b_1 = s_1 + C$ the profit of $\mathcal{M}_C$ is C as it should be. Now consider inputs $b'_1 = s'_1 + 2C$. Given f and g as above, the buy price is $p'_1 = b'_1 - C$ and the sell price is $q'_1 = s'_1 + C = b'_1 - C$, thus the profit is 0 giving a contradiction. □

Further, the cost sharing problem has requirements that are unnecessary for our application to profit maximizing auctions. We isolate our desired properties in the *revenue extraction problem*.

Definition 8. *Given a target revenue R, we want a truthful mechanism to achieve (or approximate) revenue R if the optimal profit is at least R.*

In the case of double auctions, the optimal profit above is $\mathcal{F}(\mathbf{b},\mathbf{s})$. Note that unlike the cost sharing problem we do not require anything of our mechanism if R is not achievable. Furthermore, we are interested in both exact and approximate solutions.

An exact revenue extractor does not exist for the double auction problem so we define the following approximate revenue extractor that gives revenue $\frac{k-1}{k}R$, where k is the number of winners in $\mathcal{F}^{(2)}(\mathbf{b},\mathbf{s})$. It is a natural hybrid of the k-item Vickrey [17] auction, which sells the k items to the highest k bidders at the $(k+1)$-st highest price, and the Moulin and Shenker cost sharing mechanism above.

> RevenueExtract$_R$: Given bids $\mathbf{b}$ and $\mathbf{s}$, find the largest k such that $k(b_{(k)} - s_{(k)}) \geq R$, i.e., the k extremal buyers and sellers can generate the revenue of R. Sell to the highest $k-1$ buyers at price $b_{(k)}$ and buy from the lowest $k-1$ sellers at price $s_{(k)}$. All other bidders including $b_{(k)}$ and $s_{(k)}$ are rejected.

It is easy to see that RevenueExtract$_R$ has the claimed properties.

One can convert the Sampling Cost Sharing auction for the basic problem into a double auction by using RevenueExtract$_R$ instead of CostShare$_C$. The resulting auction is simple to describe, but its analysis is complicated by the fact that RevenueExtract$_R$ may have no revenue if there is only one item exchanged in the optimal solution. We omit its analysis because the Consensus Revenue Estimate double auction that we present next gives a better competitive ratio. None the less, the sampling cost sharing approach with revenue extraction is interesting because it appears quite general and may work in other contexts.

5.2 Consensus Revenue Estimate

Another application of the RevenueExtract$_R$ leads to the main result of this section: an extension of the Consensus Revenue Estimate (CORE) basic auction to the double auction problem. The resulting CORE double auction is 3.75-competitive against $\mathcal{F}^{(2)}(\mathbf{b},\mathbf{s})$. (The basic CORE auction is 3.39 competitive.)

Next we describe the CORE double auction. The only differences between CORE for basic auctions [8] and CORE for double auctions are the use of RevenueExtract$_R$ instead of CostShare$_C$ and the recomputation of the optimal choice of constants p and c to take into account the fact that RevenueExtract$_R$ is an approximation.

First we describe the consensus estimate problem. For values r and ρ, function g is a ρ-consensus estimate of r if

- g is a *consensus*: for any w such that $r/\rho \leq w \leq r$, we have $g(w) = g(r)$.
- $g(r)$ is a nontrivial lower bound on r, i.e., $0 < g(r) \leq r$.

We define the *payoff*, γ, of a function g on r as

$$\gamma(r) = \begin{cases} g(r) & \text{if } g \text{ is a } \rho\text{-consensus estimate of } r \\ 0 & \text{otherwise.} \end{cases}$$

Definition 9. *The* consensus estimate *problem for ρ is to find a probability distribution, $\mathcal{G}$, over functions such that for all r the expected payoff, $\mathbf{E}[\gamma(r)]$, is big relative to r.*

The solution of [8] chooses $\mathcal{G}$ as the following probability distribution that depends on a parameter $c > \rho$. Let

$$g_u(r) = r \text{ rounded down to nearest } c^{i+u} \text{ for integer } i.$$

and take $\mathcal{G}$ as the the distribution of functions g_U for U uniform $[0, 1]$.

Theorem 6. *[8] For $\mathcal{G}$ as defined above, for all r, $\mathbf{E}[\gamma(r)] = \frac{r}{\ln c}(\frac{1}{\rho} - \frac{1}{c})$.*

It is easy to see that if k, the number of winners in $\mathcal{F}^{(2)}(\mathbf{b}, \mathbf{s})$, is at least three then $\mathcal{F}^{(2)}(\mathbf{b}_{-i}, \mathbf{s})$ and $\mathcal{F}^{(2)}(\mathbf{b}, \mathbf{s}_{-i})$ are in the interval $[\frac{k-1}{k}\mathcal{F}^{(2)}(\mathbf{b}, \mathbf{s}), \mathcal{F}^{(2)}(\mathbf{b}, \mathbf{s})]$.

Our CORE double auction picks g from $\mathcal{G}$ as above and runs the bid-independent auction defined by function f (the bid-independent function for sellers is analogous):

$$f(\mathbf{b}_{-i}, \mathbf{s}) = \text{extract}_{g(\mathcal{F}^{(2)}(\mathbf{b}_{-i}, \mathbf{s}))}(\mathbf{b}_{-i}, \mathbf{s})$$

where extract_R is the bid-independent function defining the RevenueExtract$_R$ mechanism (Theorem 1 implies that extract_R exists).

Combining $\frac{k-1}{k}$-approximate revenue extraction with the expected payoff of consensus estimate for $\rho = \frac{k}{k-1}$ gives the expected revenue of

$$\frac{\mathcal{F}^{(2)}(\mathbf{b}, \mathbf{s})}{\ln c}\left(1 - \frac{k-1}{kc}\right). \quad (3)$$

Thus, we are competitive for $k \geq 3$.

In order to be competitive in general we must also consider the case where the number of winners in $\mathcal{F}^{(2)}$ is $k = 2$ (recall that $k \geq 2$ by definition). In this case the 1-item Vickrey double auction is 2-competitive. To get an auction that is competitive for all $k \geq 2$, we run Vickrey with probability p and the consensus revenue estimate auction otherwise. We optimize the choice of p and c to give the CORE double auction.

Theorem 7. *The CORE double auction is 3.75-competitive against $\mathcal{F}^{(2)}$.*

Proof. We consider the case $k = 2$ and $k \geq 3$ separately.
  $k = 2$: Our expected profit is $p\mathcal{F}^{(2)}/2$.

$k \geq 3$: From Vickrey we get $p\mathcal{F}^{(2)}/k$ and from and the consensus revenue estimate we get $(1-p)$ times the quantity in Equation (3) for a combined expected profit of:

$$\mathcal{F}^{(2)}(\mathbf{b},\mathbf{s})\left(\frac{p}{k} + \frac{1-p}{\ln c}\left(1 - \frac{k-1}{kc}\right)\right).$$

Our choice of p and c optimizes and balances the two cases. Numerical analysis gives $c = 2.62$ and $p = 0.54$ as a near-optimal choice. This choice gives a competitive ratio of 3.75. □

Note that the competitive ratio of the CORE basic auction is better than the competitive ratio of the CORE double auction (3.39 vs. 3.75). This difference is due to the fact that the former uses an exact revenue extractor and the latter uses an approximation.

References

[1] Z. Bar-Yossef, K. Hildrum, and F. Wu. Incentive-compatible online auctions for digital goods. In *Proc. 13th Symp. on Discrete Alg.* ACM/SIAM, 2002.

[2] A. Blum, T. Sandholm, and M. Zinkevich. Online algorithms for market clearing. In *Proc. 13th Symp. on Discrete Alg.*, pages 971–980, 2002.

[3] J. Bulow and J. Roberts. The Simple Economics of Optimal Auctions. *The Journal of Political Economy*, 97:1060–90, 1989.

[4] E. H. Clarke. Multipart Pricing of Public Goods. *Public Choice*, 11:17–33, 1971.

[5] S. DeVries and R. Vohra. Combinatorial Auctions: A survey. Unpublished manuscript, 2000.

[6] A. Fiat, A. V. Goldberg, J. D. Hartline, and A. Karlin. Competitive generalized auctions. In *Proc. 34rd ACM Symposium on the Theory of Computing*. ACM Press, 2002. To appear.

[7] Daniel Friedman and John Rust, editors. *The Double Auction Market: Institutions, Theories, and Evidence*. Addison Wesley, 1993.

[8] A. V. Goldberg and J. D. Hartline. Competitiveness via consensus. Technical Report MSR-TR-2002-73, Microsoft Research, Mountain View, CA., 2002.

[9] A. V. Goldberg, J. D. Hartline, A. Karlin, and A. Wright. Competitive Auctions. Submitted to *Games and Economic Behavior.*, 2001.

[10] A. V. Goldberg, J. D. Hartline, and A. Wright. Competitive Auctions and Digital Goods. In *Proc. 12th Symp. on Discrete Alg.*, pages 735–744. ACM/SIAM, 2001. Also available as InterTrust Technical Report STAR-TR-99.09.01, 1999, http://www.star-lab.com/tr/tr-99-01.html.

[11] T. Groves. Incentives in Teams. *Econometrica*, 41:617–631, 1973.

[12] P. Klemperer. Auction theory: A guide to the literature. In *Journal of Economic Surveys*, pages 227–286. 13(3), 1999.

[13] R. Preston McAfee. A dominant strategy double auction. In *Journal of Economic Theory*, volume 56, pages 434–450, 1992.

[14] H. Moulin and S. Shenker. Strategyproof Sharing of Submodular Costs. to appear in *Economic Theory*.

[15] R. Myerson. Optimal Auction Design. *Mathematics of Operations Research*, 6:58–73, 1981.

[16] T. Sandholm and S. Suri. Market clearability. In *Proc. of the 17th International Joint Conf. on Artificial Intelligence (IJCAI)*, pages 1145–1151, 2001.
[17] W. Vickrey. Counterspeculation, Auctions, and Competitive Sealed Tenders. *J. of Finance*, 16:8–37, 1961.
[18] P. Wurman, W. Walsh, and M. Wellman. Flexible double auctions for electronic commerce: Theory and implementation. In *Decision Support Systems*, volume 24, pages 17–27, 1998.

Optimal Graph Exploration without Good Maps

Anders Dessmark[1,*] and Andrzej Pelc[2,**]

[1] Department of Computer Science, Lund Institute of Technology
Box 118, S-22100 Lund, Sweden
andersd@cs.lth.se

[2] Département d'Informatique, Université du Québec à Hull
Québec J8X 3X7, Canada
pelc@uqah.uquebec.ca

Abstract. A robot has to visit all nodes and traverse all edges of an unknown undirected connected graph, using as few edge traversals as possible. The quality of an exploration algorithm $\mathcal{A}$ is measured by comparing its cost (number of edge traversals) to that of the optimal algorithm having full knowledge of the graph. The ratio between these costs, maximized over all starting nodes in the graph and over all graphs in a given class $\mathcal{U}$, is called the *overhead* of algorithm $\mathcal{A}$ for the class $\mathcal{U}$ of graphs. We construct natural exploration algorithms, for various classes of graphs, that have smallest, or – in one case – close to smallest, overhead. An important contribution of this paper is establishing lower bounds that prove optimality of these exploration algorithms.

1 Introduction

A robot has to visit all nodes and traverse all edges of an unknown undirected connected graph, using as few edge traversals as possible. If the robot has complete knowledge of the explored graph G, i.e., if it has an oriented labeled isomorphic copy of it showing which port at a visited node leads to which neighbor, then exploration with fewest edge traversals starting from node v corresponds to the shortest *covering walk* from v: the shortest, not necessarily simple, path in G starting from v and containing all edges. The length of this shortest covering walk is called the *cost* of G from v, and is denoted $opt(G, v)$. For example, if G is an Eulerian graph then $opt(G, v)$ is the number of edges in G, for any v, and if G is a tree then $opt(G, v) = 2(n-1) - ecc(v)$, where n is the number of nodes in G and $ecc(v)$ is the eccentricity of the starting node v, i.e., the distance from v to the farthest leaf. In this latter case, depth-first search ending in the leaf farthest from the starting node v clearly uses fewest edge traversals.

However, graph exploration is often performed when the explored graph is partially or totally unknown. We consider three scenarios, providing the robot with varying amount of information. Under the first scenario, the robot does not

---

* This research was done during a visit of Anders Dessmark at the Université du Québec à Hull.
** Andrzej Pelc was supported in part by NSERC grant OGP 0008136.

have any *a priori* knowledge of the explored graph. We refer to this scenario as *exploration without a map*. Under the second scenario, the robot has an unlabeled isomorphic copy of the explored graph. We call it an *unanchored map* of the graph. Finally, under the third scenario, the robot has an unlabeled isomorphic copy of the explored graph with a marked starting node. We call it an *anchored map* of the graph. It should be stressed that even the scenario with an anchored map does not give the robot any sense of direction, since the map is unlabeled. For example, when the robot starts the exploration of a line, such a map gives information about the length of the line and distances from the starting node to both ends, but does not tell which way is the closest end. In the case of an $n \times m$ torus, the availability of either type of map is equivalent to the information that the explored graph is an $n \times m$ torus.

In all scenarios we assume that all nodes have distinct labels, and all ports at a node v are numbered $1,..., \deg(v)$ (in the explored graph, not in the map). Hence the robot can recognize already visited nodes and traversed edges. However, it cannot tell the difference between yet unexplored edges incident to its current position. If the robot decides to use such an unexplored edge, the actual choice of the edge belongs to the adversary, as we are interested in worst-case performance. For a given exploration algorithm $\mathcal{A}$, the *cost* $\mathcal{C}(\mathcal{A}, G, v)$ of this algorithm run on a graph G from a starting node v is the worst-case number of edge traversals taken over all of the above choices of the adversary.

For a given graph G and a given starting node v, a natural measure of quality of an exploration algorithm $\mathcal{A}$ is the ratio $\mathcal{C}(\mathcal{A}, G, v)/opt(G, v)$ of its cost to that of the optimal algorithm having complete knowledge of the graph. This ratio represents the relative penalty payed by the algorithm for the lack of knowledge of the environment. For a given class $\mathcal{U}$ of graphs, the number

$$\mathcal{O}_{\mathcal{U}}(\mathcal{A}) = \sup_{G \in \mathcal{U}} \max_{v \in G} \frac{\mathcal{C}(\mathcal{A}, G, v)}{opt(G, v)}$$

is called the *overhead* of algorithm $\mathcal{A}$ for the class $\mathcal{U}$ of graphs. For a fixed scenario, an algorithm is called *optimal* for a given class of graphs, if its overhead for this class is minimal among all exploration algorithms working under this scenario.

Since $\mathcal{C}(DFS, G, v) \leq 2e$, and $opt(G, v) \geq e$, for any graph G with e edges and any starting node v (depth-first search traverses each edge at most twice, and every edge has to be traversed at least once), it follows that the overhead of DFS is at most 2, for any class of graphs. Hence, for any class of graphs, the overhead of an optimal algorithm is between 1 and 2, under every scenario (DFS does not use any information about the explored graph).

The following remark will be useful for proving lower bounds on overhead of exploration algorithms. Suppose that the robot, at some point of the exploration, is at node w, then moves along an already explored edge e incident to w, and immediately returns to w. For any set of decisions of the adversary, an algorithm causing such a pair of moves, when run on a graph G from some starting node v, has cost strictly larger than the algorithm that skips these two moves. Hence, we restrict attention to exploration algorithms that never perform such returns. We call them *regular*.

1.1 Related Work

Exploration and navigation problems for robots in an unknown environment have been extensively studied in the literature (cf. the survey [14]). There are two principal ways of modeling the explored environment. In one of them a geometric setting is assumed, e.g., unknown terrain with convex obstacles [6], or room with polygonal [8] or rectangular [3] obstacles. Another way is to represent the unknown environment as a graph, assuming that the robot may only move along its edges. The graph model can be further specified in two different ways. In [1, 4, 5, 9] the robot explores strongly connected directed graphs and it can move only in the direction from head to tail of an edge, not vice-versa. In [2, 7, 11, 12, 13] the explored graph is undirected and the robot can traverse edges in both directions. The efficiency measure adopted in most papers dealing with exploration of graphs is the cost of completing this task, measured by the number of edge traversals by the robot. In some papers additional restrictions on the moves of the robot are imposed. It is assumed that the robot has either a restricted tank [2, 7], forcing it to periodically return to the base for refueling, or that it is tethered, i.e., attached to the base by a rope or cable of restricted length [11]. It is proved in [11] that exploration can be done in time $O(e)$ under both scenarios. Another direction of research concerns exploration of anonymous graphs [4, 5, 10].

The work most closely related to the present paper is that from [13]. The authors consider exploration of undirected graphs (both arbitrary graphs and trees). The adopted efficiency measure is similar in spirit to our notion of overhead but differs from it in an important way. Similarly as in the present paper, in [13] the authors consider the ratio of the cost of an algorithm lacking some knowledge of the graph to that of the optimal algorithm having this knowledge. (In particular, they study the scenario with an unanchored map.) However, for a given graph, both costs are maximized over all starting nodes and the ratio of these maxima is considered as the performance measure of the algorithm on the graph. (Then the supremum of this ratio is taken over all graphs in the considered class). This approach should be contrasted with our definition of overhead, where the ratio is computed for each starting node individually and then maximized over all possible starting nodes in the graph. In order to see the difference between both approaches, consider the case of the line with availability of an unanchored map (which, for the line, is equivalent to knowing its length). The maximum cost of depth-first search on the line L_n of length n is $2n-1$ (the maximum taken over all starting nodes). On the other hand, $opt(L_n, v) \leq 3n/2$ for *some* starting node v. This gives a ratio close to 4/3 and leads to the conclusion, proved in [13], that DFS is optimal for lines (and in fact for all trees), according to their measure. However, this measure (and hence the obtained result) can be viewed as biased in favor of DFS because for some starting nodes v (close to the endpoints of the line) the ratio of the cost of DFS to $opt(L_n, v)$ is approximately 2. This is captured by our notion of overhead, and in fact leads to the conclusion that, if n is known, there are exploration algorithms of a line better than DFS, according to the overhead measure.

Table 1. Summary of results

	Anchored Map	Unanchored map	No map
Lines	overhead: 7/5 optimal	overhead: $\sqrt{3}$ optimal	Depth-First Search overhead: 2
Trees	overhead: 3/2 optimal	overhead: < 2 lower bound $\sqrt{3}$	
General graphs	Depth-First Search, overhead: 2, optimal		

1.2 Our Results

The aim of this paper is to establish which exploration algorithms have the lowest possible overhead for each of the three scenarios described above: no knowledge of the graph, availability of an unanchored map, and availability of an anchored map. It turns out that some of these algorithms are fairly natural and our main contribution is proving lower bounds that show their optimality.

Depth-first search is among the most natural exploration algorithms in an unknown graph. At every point of the exploration the robot chooses an unexplored edge, if it exists. Otherwise, it backtracks to the most recently visited node with an unexplored incident edge. If no such node exists, exploration is completed. Since DFS traverses every edge at most twice, regardless of adversary's choices, its overhead is at most 2, for all classes of graphs.

While for the class of all (undirected, connected) graphs, depth-first search turns out to be an optimal algorithm for all scenarios, the situation for trees is much different. We show that, under the scenario without any knowledge, DFS is still optimal for trees but this is not the case if a map is available. Under the scenario with an unanchored map, we show that optimal overhead is at least $\sqrt{3}$ but strictly below 2 (and thus DFS, with overhead 2, is not optimal). Under the scenario with an anchored map, we construct an optimal algorithm for trees and show that its overhead is 3/2. We also consider exploration of the class of lines (simple paths). In this case, depth-first search remains optimal for the scenario without any knowledge, with overhead 2. Under the scenario with an unanchored map, we construct an optimal algorithm and show that its overhead is $\sqrt{3}$. Finally, under the scenario with an anchored map, we construct an optimal algorithm and show that its overhead is 7/5.

2 Lines

In this section we construct optimal exploration algorithms for the class $\mathcal{L}$ of lines. A line of length n is a graph $L_n = (V, E)$, where $V = \{v_0, ..., v_n\}$ and

$E = \{[v_i, v_{i+1}] : i = 0, 1, ..., n-1\}$. It turns out that, for all three scenarios, optimal algorithms require at most 2 returns (changes of direction).

2.1 Exploration with an Anchored Map

We consider the scenario in which an anchored map of the line is available to the robot. This is equivalent to knowing the length n of the line and the distances a and b between the starting node and the endpoints. Assume that $a \leq b$.

Algorithm Anchored-Line

- Let $x = 3a + n$ and $y = 2n - a$.
- If $x \leq y$ then
  - go at distance a in one direction, or until an endpoint is reached;
  - if an endpoint is reached then
    >return, go to the other endpoint, and stop

    else
    >return, go to the endpoint, return, go to the other endpoint, and stop

  else
  - go to an endpoint, return, go to the other endpoint, and stop.

Theorem 1. *Algorithm Anchored-Line has overhead 7/5 for the class $\mathcal{L}$ of lines.*

Proof. Denote Algorithm Anchored-Line by $\mathcal{A}$. By definition, $\mathcal{C}(\mathcal{A}, L_n, v) = \min(x, y)$. Since $opt(L_n, v) = a + n$, we have:
if $a \leq \lfloor n/4 \rfloor$ then $\mathcal{C}(\mathcal{A}, L_n, v)/opt(L_n, v) \leq \frac{3\lfloor n/4 \rfloor + n}{\lfloor n/4 \rfloor + n} \leq 7/5$;
if $a \geq \lceil n/4 \rceil$ then $\mathcal{C}(\mathcal{A}, L_n, v)/opt(L_n, v) \leq \frac{2n - \lceil n/4 \rceil}{\lceil n/4 \rceil + n} \leq 7/5$. Hence $\mathcal{O}_\mathcal{L}(\mathcal{A}) \leq 7/5$
Since, for an arbitrary n divisible by 4 and $a = n/4$, we have $\mathcal{C}(\mathcal{A}, L_n, v)/opt(L_n, v) = 7/5$, this proves $\mathcal{O}_\mathcal{L}(\mathcal{A}) = 7/5$.

The next theorem proves that Algorithm Anchored-Line is optimal for the class of lines. The proof will appear in the full version of the paper.

Theorem 2. *Every exploration algorithm with an anchored map has overhead at least 7/5 for the class of lines.*

2.2 Exploration with an Unanchored Map

We now present an algorithm for exploration of lines with an unanchored map (i.e., the length n of the line L_n is known to the robot but the starting node v is unknown) that has overhead $\sqrt{3}$, and show that it is optimal for the class of lines.

Algorithm Unanchored-Line

- Let $a = \lfloor \frac{\sqrt{3}-1}{2} n \rfloor$.
- If the starting node v is an endpoint then
   - go to the other endpoint and stop.
  else
   - go at distance a in one direction or until an endpoint is reached;
   - if an endpoint is reached then
      return, go to the other endpoint, and stop
    else
      return, go to the endpoint, return, go to the other endpoint, and stop.

Theorem 3. *Algorithm Unanchored-Line has overhead not larger than $\sqrt{3}$ for the class $\mathcal{L}$ of lines.*

The proof will appear in the full version of the paper.

Theorem 4. *For all algorithms $\mathcal{A}$ with an unanchored map, $\mathcal{O}_\mathcal{L}(\mathcal{A}) \geq \sqrt{3}$.*

Proof. The proof is divided in two parts. First we show that for every exploration algorithm of the line with an unanchored map, there is an algorithm that does at most two returns and has equal or smaller overhead. In the second part we show that the overhead is at least $\sqrt{3}$ for all algorithms with at most two returns.

For each value of n, all regular algorithms can be classified according to the maximum number of returns performed while exploring the line L_n before reaching an endpoint. Fix $n \geq 4$ and let *type k* be the set of algorithms that always do at most k returns before reaching an endpoint, and that do exactly this many returns for some combination of starting node and (adversary) choice of direction. Notice that one algorithm can be of different types for different values of n, and that algorithm Unanchored-Line is of type 1 for every $n \geq 3$. We now show that for every exploration algorithm $\mathcal{A}$ there exists an algorithm $\mathcal{A}'$ such that $\mathcal{A}'$ is of type 1 and $\max_{v \in L_n}(\frac{\mathcal{C}(\mathcal{A}', L_n, v)}{opt(L_n, v)}) \leq \max_{v \in L_n}(\frac{\mathcal{C}(\mathcal{A}, L_n, v)}{opt(L_n, v)})$.

Depth-first search is the only algorithm of type 0. Since $\mathcal{O}_{\{L_n\}}(\text{DFS}) \geq 7/4 \geq \sqrt{3}$, when $n \geq 4$, it follows that the algorithm Unanchored-Line is an algorithm of type 1 with the required property, for algorithms of type 0.

Consider algorithms of type 2. Every such algorithm $\mathcal{A}$ can be described as follows (assuming that no endpoint is encountered before the first two returns):
- Traverse a edges in one direction.
- Return and traverse $a + b$ edges in the opposite direction.
- Return again and go to the endpoint.
- Return and go to the other endpoint.

We show that $a = 0$ minimizes the overhead for $\mathcal{A}$, which in effect proves that for every algorithm of type 2 there is an algorithm of type 1 with equal or smaller overhead. Let x be the distance from v to the endpoint in the direction of the first traversal. There are three ranges of values of x that are of interest for calculating the overhead of $\mathcal{A}$. When $x \leq a$, then $\mathcal{A}$ makes $x + n$ traversals, we call this case 1. When $a < x < n - b$, then the number of traversals is $2a + 2b + x + n$, this is case 2. Finally, when $x \geq n - b$, then the number of traversals is $2a + 2n - x$, this is case 3. Clearly $opt(L_n, v) = \min(n + x, 2n - x)$. Observe that for $a > \frac{n}{2}$, case 1 is dominated by case 3 (choose $x = n - 1$), and since $\frac{x+n}{\min(n+x, 2n-x)} = 1$ when $a \leq \frac{n}{2}$, it follows that case 1 never dominates and can be discarded from our considerations. In case 3, $opt(L_n, v)$ does not depend on a and $a = 0$ minimizes the ratio in this case. Case 2 is divided into two subcases. When $a < x \leq \frac{n}{2}$, we get $\frac{C(A, L_n, v)}{opt(L_n, v)} = \frac{2a+2b+x+n}{x+n}$ which is maximized for $x = a + 1$, giving $\frac{3a+2b+n+1}{n+a+1}$. When $\frac{n}{2} < x < n - b$, we get $\frac{C(A, L_n, v)}{opt(L_n, v)} = \frac{2a+2b+x+n}{2n-x}$ which is maximized for $x = n - b - 1$, giving $\frac{2a+b+3n-1}{n+b+1}$. In both cases, this maximum value of the ratio $\frac{C(A, L_n, v)}{opt(L_n, v)}$ is minimized for $a = 0$. Hence for every algorithm of type 2 there is an algorithm of type 1 with equal or smaller overhead.

Now we consider algorithms of type $k > 2$. Let $\mathcal{A}$ be such an algorithm. Assuming that no endpoint is encountered before the first 3 returns, the initial behavior of $\mathcal{A}$ can be described as follows:
- Traverse a edges in one direction.
- Return and traverse $a + b$ edges in the opposite direction.
- Return and traverse $b + c$ edges in the first direction.
- Return.

Note that $c > a$ and $b + c \leq n - 2$. Let $\mathcal{A}'$ be the algorithm of type $k - 1$ that behaves exactly like $\mathcal{A}$ but for which $a = 0$. Let x be the distance from v to the closest endpoint. Clearly, $\mathcal{A}'$ performs fewer traversals (in the worst case) than $\mathcal{A}$ when $x > a$ and thus $\frac{C(\mathcal{A}', L_n, v)}{opt(L_n, v)} \leq \frac{C(\mathcal{A}, L_n, v)}{opt(L_n, v)}$ in this case. It remains to show that $\max_{v \in L_n}(\frac{C(\mathcal{A}', L_n, v)}{opt(L_n, v)}) \geq \frac{C(\mathcal{A}', L_n, v')}{opt(L_n, v')}$ for every starting node v' within distance a of an endpoint. Choose a starting node v and an initial direction such that the robot does not encounter an endpoint before returning twice. This gives $\frac{C(\mathcal{A}', L_n, v)}{opt(L_n, v)} \geq \frac{3b+2c+n}{n+\min(b,c)}$. On the other hand $\frac{C(\mathcal{A}', L_n, v')}{opt(L_n, v')} \leq \frac{2b+x+n}{\min(n+x, 2n-x)}$. The inequality $\frac{3b+2c+n}{n+\min(b,c)} \geq \frac{2b+x+n}{\min(n+x, 2n-x)}$ is proved using case-by-case analysis.

Thus, for any algorithm $\mathcal{A}$ of type k, we have shown an algorithm $\mathcal{A}'$ of type $k - 1$ whose overhead for the line L_n is equal to or less than that of $\mathcal{A}$.

It follows by induction that for any regular algorithm $\mathcal{A}$ there exists an algorithm $\mathcal{A}'$ of type 1 such that $\max_{v \in L_n}(\frac{C(\mathcal{A}', L_n, v)}{opt(L_n, v)}) \leq \max_{v \in L_n}(\frac{C(\mathcal{A}, L_n, v)}{opt(L_n, v)})$, which concludes the first part of the proof.

Now it is enough to prove that for any algorithm $\mathcal{A}'$ of type 1 that performs a edge traversals before the first return, there exists a starting node v such that $\frac{C(\mathcal{A}, L_n, v)}{opt(L_n, v)} \geq \sqrt{3} - \frac{c}{n}$ for some positive constant c. First assume that $a \leq \frac{\sqrt{3}-1}{2}n$. Choose the starting node v at distance $a + 1$ from an endpoint p of the line. The adversary chooses the direction of p for the first traversal. The algorithm performs $a+2n-1$ traversals and $opt(L_n, v) = a+n$, giving $\frac{C(\mathcal{A}, L_n, v)}{opt(L_n, v)} = \frac{a+2n-1}{a+n} \geq \frac{a+2n}{a+n} - \frac{c}{n} \geq \sqrt{3} - \frac{c}{n}$, for $a \leq \frac{\sqrt{3}-1}{2}n$ and a sufficiently large n. If, on the other hand, $a > \frac{\sqrt{3}-1}{2}n$, choose v at distance 1 from the closest endpoint p. The adversary chooses the opposite direction of p for the first traversal. The algorithm performs

$2a+1+n$ traversals and $opt(L_n, v) = 1+n$, giving $\frac{C(A,L_n,v)}{opt(L_n,v)} = \frac{2a+1+n}{1+n} \geq \frac{1+\sqrt{3}n}{1+n} \geq \sqrt{3} - \frac{c}{n}$ for $a > \frac{\sqrt{3}-1}{2}n$ and a sufficiently large n.

2.3 Exploration without a Map

We finally consider the scenario when no map is available to the robot. The following theorem shows that depth-first search is optimal in this case. The proof will appear in the full version of the paper.

Theorem 5. *Every exploration algorithm without a map has overhead at least 2 for the class $\mathcal{L}$ of lines.*

3 Arbitrary Trees

In this section we describe an optimal algorithm for tree exploration with an anchored map and prove that its overhead is $3/2$ for the class $\mathcal{T}$ of all trees. As for the scenario with an unanchored map, we show that the optimal algorithm has overhead strictly smaller than 2, and hence it is not DFS. (By Theorem 4, the optimal overhead for the class $\mathcal{T}$ cannot be smaller than $\sqrt{3}$.) Notice that if no map is available then depth-first search is an optimal algorithm for exploring the class of all trees. Its overhead is 2. This follows from Theorem 5.

3.1 Exploration with an Anchored Map

Lemma 1. *For all algorithms $\mathcal{A}$ with an anchored map, $\mathcal{O}_\mathcal{T}(\mathcal{A}) \geq 3/2$.*

Proof. We construct a tree T of arbitrarily large size n, with starting node v_1, for which $\frac{C(A,T,v_1)}{opt(T,v_1)} \geq \frac{3}{2} - \frac{c}{n}$, for some constant c and any algorithm $\mathcal{A}$. The tree T of size $n = 3m+1$ is defined by the set of nodes $V(T) = \{v_1, ..., v_{m+1}, x_1, ..., x_m, y_1, ...y_m\}$ and the set of edges $E(T) = \{[v_i, y_i], [v_i, x_i], [x_i, v_{i+1}] : 1 \leq i \leq m\}$ (see Figure 1).

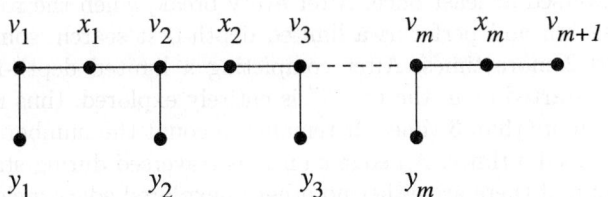

Fig. 1. The tree T

Clearly $opt(T, v_1) = 4m$: every edge $[v_i, y_i]$ is traversed twice and the remaining edges only once. Consider a robot that has an anchored map. When the robot is in v_i and the edges $[v_i, y_i]$ and $[v_i, x_i]$ are both unexplored, the adversary chooses the edge $[v_i, x_i]$ when $\mathcal{A}$ decides to use an unexplored edge. Let $i_0 \leq m$ be the integer such that the robot concludes exploration in y_{i_0} or v_{i_0+1}. For all $i \in \{1, ..., m\} \setminus \{i_0\}$, the robot traverses the edges $[v_i, y_i]$, $[v_i, x_i]$ and $[x_i, v_{i+1}]$ no less than 6 times in total: the edge $[v_i, y_i]$ twice and either $[v_i, x_i]$ 3 times and $[x_i, v_{i+1}]$ once (if it returns immediately), or each of the two edges $[v_i, x_i]$ and $[x_i, v_{i+1}]$ twice (otherwise). For i_0, the number is at least 5, giving the ratio $\frac{C(\mathcal{A}, T, v_1)}{opt(T, v_1)} \geq \frac{6m-1}{4m}$, which proves the lemma.

We now present an optimal algorithm for exploring a tree T with an anchored map. Let D be the eccentricity of the starting node v. Consider all elementary paths of length D starting at v. Two such paths $P_1 = (v_0 = v, v_1, ..., v_D)$ and $P_2 = (v'_0 = v, v'_1, ..., v'_D)$ are called isomorphic if there exists an automorphism of T such that $f(v_i) = v'_i$, for all $i = 0, ..., D$.

Algorithm Anchored-Tree Choose one node on the map of T at distance D from v. Let P be the path on the map from v to this node. Perform a depth-first search with the following adjustments. Suppose that at some point of the exploration the robot, using the map, can determine that its current position corresponds to a node u on a path isomorphic to P, and there is at least one visited node different from u, with unexplored edges. Call this situation a *break*. When a break occurs, continue depth-first search in the subtree T' containing u and resulting from removal of all unexplored edges incident to u. Call this procedure a *limited* depth-first search. When no unexplored edges remain in this subtree, resume "standard" depth-first search, i.e., move to u and continue depth-first search in the rest of the tree, until the next break. (Notice that many breaks can occur during the exploration.) The robot stops when there are no more unexplored edges.

Lemma 2. *Algorithm Anchored-Tree explores any tree T of size n with starting node v, using at most $4(n-1) - 3D$ edge traversals.*

Proof. It is clear that the robot traverses every edge outside of P twice. The edges on P are traversed at least once. After every break, when the robot interrupts depth-first search and performs a limited depth-first search, some edges on P are traversed 2 more times. After completing a limited depth-first search in a subtree T', started at u, the tree T' is entirely explored, thus no edges on P are traversed more than 3 times. It remains to count the number of edges on P that are traversed 3 times. An edge e on P is traversed during standard depth-first search only if there are either no other unexplored edges incident to visited nodes, or if the robot cannot determine that it is on a path isomorphic to P (otherwise a break occurs). In the first case, e will not be traversed again hence the total number of its traversals is 1. For the second case to occur, there must be at least one more edge on the map at the same distance from v, or else the robot could determine that it is on a path isomorphic to P. Only such edges on P can be traversed 3 times. Thus the number of edges on P traversed 3 times is bounded by the number of edges not on P, i.e., by $n-1-D$. Consequently we have three groups of traversals.

1. $n-1-D$ edges outside of P are traversed exactly twice, contributing $2(n-1-D)$ traversals.
2. First traversals of edges on P, a total of D traversals.
3. Two additional traversals of at most $n-1-D$ edges on P, a total of $2(n-1-D)$ traversals.

Thus, the total number of traversals is at most $4(n-1)-3D$.

We use a modified version of Anchored-Tree that runs Anchored-Tree if $D > 2n/3$, and otherwise runs DFS.

Theorem 6. $\mathcal{O}_\mathcal{T}(\texttt{Modified-Anchored-Tree}) = 3/2$.

Proof. If $D \leq 2n/3$, the ratio is $\frac{c(\texttt{DFS},T,v)}{opt(T,v)} \leq \frac{2(n-1)-1}{2(n-1)-2n/3} \leq 3/2$. If $D > 2n/3$, the ratio is $\frac{c(\texttt{Anchored-Tree},T,v)}{opt(T,v)} \leq \frac{4(n-1)-3D}{2(n-1)-D}$ which is maximized for $D = 2n/3$, giving the ratio $\frac{4(n-1)-2n}{2(n-1)-2n/3} < 3/2$.

3.2 Exploration with an Unanchored Map

We now show that an optimal algorithm with an unanchored map has overhead strictly smaller than 2, and thus it is not DFS. We do not make any attempt at optimizing the constant, and show an algorithm with overhead at most 1.99. We first present an algorithm that improves on depth-first search for trees of high diameter and at least 38 nodes.

Algorithm Unanchored-Tree The algorithm explores a tree T of size $n > 37$ and diameter $D \geq 0.99n$. It works in three phases. Let v be the starting node and let $a = \lfloor 0.3n \rfloor$. Let P be a path of length D on the map.

1. Perform depth-first search until a node z is found at distance a from v. Let P' be the path from v to z, and let u be the node on P' at distance $\lceil a - 0.01n \rceil$ from v (see Figure 2). Move to u.
2. Perform depth-first search in the subtree T' of T containing v and resulting from the removal of u. Then return to u.
3. Perform depth-first search in the remaining part of tree T, with the following modification. If the robot, using the map, can identify a nonempty set S of nodes on P such that the current position w corresponds to an element of this set (i.e. the robot "knows" that it is on P), and there is at least one visited node different from w with incident unexplored edges, we say that a *break* occurs. When a break occurs, continue depth-first search in the subtree containing u and resulting from removal of all unexplored edges incident to w. Call this a *limited* depth-first search. When no unexplored edges remain in this subtree, return to w and resume depth-first search in

the remaining part of the tree, until the next break. (Notice that many breaks can occur during the exploration.) Stop when there are no more unexplored edges.

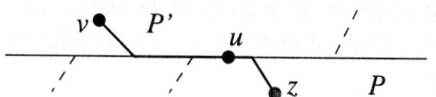

Fig. 2. Nodes visited in phase 1 of Algorithm Unanchored-Tree

Lemma 3. *Let T^* be the class of trees with $n > 37$ and $D \geq 0.99n$. We have $\mathcal{O}_{T^*}(\text{Unanchored-Tree}) \leq 1.99$.*

The proof will appear in the full version of the paper.

It is easy to verify that $\frac{C(\text{DFS},T,v)}{opt(T,v)} \leq 1.99$ for all trees with less than 38 nodes and for all trees of diameter $D < 0.99n$. Together with Lemma 3 this gives the following theorem.

Theorem 7. *There exists an algorithm $\mathcal{A}$ with an unanchored map, for which $\mathcal{O}_T(\mathcal{A}) \leq 1.99$.*

Recall that Theorem 4 implies $\mathcal{O}_T(\mathcal{A}) \geq \sqrt{3}$, for all algorithms with an unanchored map. The construction of an optimal exploration algorithm with an unanchored map, for the class of trees, and establishing the value of the best overhead remains an open problem.

4 Arbitrary Graphs

In this section we consider the class $\mathcal{G}$ of arbitrary undirected connected graphs, and prove that the overhead of any exploration algorithm for this class is at least 2, under all three scenarios. This implies that, for the class $\mathcal{G}$, depth-first search is optimal. Since the scenario with an anchored map provides most information among all three scenarios, it is enough to prove this lower bound under this scenario. To this end, we construct a class of Eulerian graphs S_m of arbitrarily large size, each with a distinguished starting node x, such that for any exploration algorithm $\mathcal{A}$, $C(\mathcal{A}, S_m, x) \geq 2e - o(e)$, where e is the number of edges in S_m. Since $opt(S_m, x) = e$, this will prove our result.

The building blocks of graphs S_m are graphs called *thick lines*, defined in [13]. A thick line L of length n is a graph defined by the set of nodes $V(L) = \{v_0, v_1, \ldots, v_n, x_1, \ldots, x_n, y_1, \ldots, y_n\}$ and the set of edges

$$E(L) = \{[x_i, v_{i-1}], [x_i, v_i], [y_i, v_{i-1}], [y_i, v_i] \ : \ 1 \leq i \leq n\}.$$

The nodes v_0 and v_n are called the *ends* of L. For $i \in \{0, \ldots, n-1\}$, the cycle $[v_i, x_{i+1}, v_{i+1}, y_{i+1}, v_i]$ is called the *cycle connecting v_i and v_{i+1}*. We denote L by $v_0 \diamond v_1 \diamond \ldots \diamond v_n$. Notice that a thick line of length n is an Eulerian graph with $4n$ edges.

Thick lines were used in [13] to prove that DFS is optimal under the scenario with an unanchored map. (Notice that since a thick line L is an Eulerian graph, $opt(L, v)$ does not depend on v, and hence, in this special case, the measure from [13] coincides with our measure of overhead.) In fact, the following lemma is proved in [13].

Lemma 4. *Suppose that the robot starts at node v_0 of a thick line of length n and consider any exploration algorithm. Then there exists an adversary such that, when the robot reaches v_k, $k \in \{2, \ldots, n\}$, for the first time then at least 6 moves have already been performed along the edges of the cycle connecting v_{k-2} and v_{k-1}.*

Lemma 4 was used in [13] to show that, under the scenario with an unanchored map, the cost of every exploration algorithm in a thick line of length n is at least $8n - 12$ which is enough to prove that overhead is at least 2. However, for the scenario with an anchored map, this is not the case. If the distances of the starting node from both ends of the thick line are known, there is a simple exploration algorithm with overhead 15/8. (It is based on the same idea as the algorithm for ordinary lines constructed in Section 2.) Thus, in order to strengthen the result from [13] to the scenario with an anchored map, we need the following, slightly more complicated class of graphs.

A *thick star* of radius m is a graph S_m consisting of m thick lines of length m, which have exactly one common node: one of the ends of each of these lines. Call this node x and consider it to be the starting node of the robot. All thick lines of length m attached to x are called *branches* of S_m.

Lemma 5. *For any exploration algorithm $\mathcal{A}$ with an anchored map, we have $\mathcal{C}(\mathcal{A}, S_m, x) \geq 8m^2 - o(m^2)$.*

Proof. By Lemma 4, at the time when the robot reaches the other end of any branch it must use at least $6(m-1) + 2$ edge traversals in this branch. At least $2m$ additional traversals are needed to return to the starting node x. This must be repeated at least $m - 1$ times (there is no need of returning from the last branch), for a total of $(8m - 4)(m - 1) = 8m^2 - o(m^2)$ edge traversals.

Since every graph S_m is an Eulerian graph with $4m^2$ edges, this proves the following result, implying that DFS is an optimal algorithm for the class $\mathcal{G}$.

Theorem 8. *For any exploration algorithm $\mathcal{A}$, $\mathcal{O}_\mathcal{U}(\mathcal{A}) = 2$, for the class $\mathcal{G}$ of all undirected connected graphs.*

References

[1] S. Albers and M. R. Henzinger, Exploring unknown environments, SIAM Journal on Computing 29 (2000), 1164-1188.

[2] B. Awerbuch, M. Betke, R. Rivest and M. Singh, Piecemeal graph learning by a mobile robot, Proc. 8th Conf. on Comput. Learning Theory (1995), 321-328.

[3] E. Bar-Eli, P. Berman, A. Fiat and R. Yan, On-line navigation in a room, Journal of Algorithms 17 (1994), 319-341.

[4] M. A. Bender, A. Fernandez, D. Ron, A. Sahai and S. Vadhan, The power of a pebble: Exploring and mapping directed graphs, STOC'98, 269-278.

[5] M. A. Bender and D. Slonim, The power of team exploration: Two robots can learn unlabeled directed graphs, FOCS'94, 75-85.

[6] A. Blum, P. Raghavan and B. Schieber, Navigating in unfamiliar geometric terrain, SIAM Journal on Computing 26 (1997), 110-137.

[7] M. Betke, R. Rivest and M. Singh, Piecemeal learning of an unknown environment, Machine Learning 18 (1995), 231-254.

[8] X. Deng, T. Kameda and C. H. Papadimitriou, How to learn an unknown environment I: the rectilinear case, Journal of the ACM 45 (1998), 215-245.

[9] X. Deng and C. H. Papadimitriou, Exploring an unknown graph, Journal of Graph Theory 32 (1999), 265-297.

[10] K. Diks, P. Fraigniaud, E. Kranakis, A. Pelc, Tree exploration with little memory, SODA'2002, San Francisco, U. S. A., January 2002.

[11] C. A. Duncan, S. G. Kobourov and V. S. A. Kumar, Optimal constrained graph exploration, SODA'2001, 807-814.

[12] P. Panaite and A. Pelc, Exploring unknown undirected graphs, Journal of Algorithms 33 (1999), 281-295.

[13] P. Panaite, A. Pelc, Impact of topographic information on graph exploration efficiency, Networks, 36 (2000), 96-103.

[14] N. S. V. Rao, S. Hareti, W. Shi and S. S. Iyengar, Robot navigation in unknown terrains: Introductory survey of non-heuristic algorithms, Tech. Report ORNL/TM-12410, Oak Ridge National Laboratory, July 1993.

Approximating the Medial Axis from the Voronoi Diagram with a Convergence Guarantee

Tamal K. Dey and Wulue Zhao

Ohio State University
Columbus OH 43210, USA
{tamaldey,zhaow}@cis.ohio-state.edu
http://cis.ohio-state.edu/~tamaldey/medialaxis.htm

Abstract. We show that the medial axis of a surface $S \subseteq \mathbb{R}^3$ can be approximated as a subcomplex of the Voronoi diagram of a point sample of S. The subcomplex converges to the true medial axis as the sampling density approaches infinity. Moreover, the subcomplex can be computed in a scale and density independent manner. Experimental results as detailed in a companion paper corroborate our theoretical claims.

1 Introduction

Medial axis provides an alternative compact representation of shapes which have found its use in a number of applications ranging over image processing, solid modeling, mesh generation, motion planning, and many others [12, 13, 14, 15, 16, 17, 18]. Medial axis is defined when the shape is embedded in an Euclidean space and is endowed with a distance function. Informally, it is the set of all points that have more than one closest point on the shape. The shapes in this paper are surfaces embedded in three dimensions.

Application demands have prompted research in the computational as well as the mathematical aspects of the medial axis in recent years. As a mathematical structure they are instable since a small change in a shape can cause a relatively large change in its medial axis, see [11, 19]. They are hard to compute exactly due to numerical instability associated with their computations. Only few algorithms, that too for special class of shapes, have been designed till date to compute the exact medial axis [8, 13]. Consequently, efforts have been made to approximate the medial axis. For polyhedral input Etzion and Rappoport [10] suggest an approximation method based on an octree subdivision of space. Another scheme considered by many uses a set of sample points on the shape and then approximates the medial axis with the Voronoi diagram of these points [3, 4, 7].

We follow the Voronoi diagram approach. It is apt for point clouds that are increasingly being used for geometric modeling over a wide range of applications. It is known that the Voronoi vertices approximate the medial axis of a curve in 2D. In fact, Brandt and Algazi [7] show that if the sample density approaches infinity, the Voronoi vertices in this case converge to the medial axis. Unfortunately, the same is not true in three dimensions. In order to alleviate this problem in the context of surface reconstruction, Amenta and Bern [1] identify some Voronoi vertices called 'poles' that lie close to the medial axis. Boissonnat and Cazals [6] and Amenta, Choi and Kolluri [3] show that the poles converge to the medial axis as the sample density approaches infinity.

The convergence result of poles to the medial axis is a significant progress in the medial axis approximation in 3D. However, many applications require and often prefer a non discrete approximation over a discrete one. In 3D, since poles lie close to the medial axis, Amenta, Choi and Kolluri [3] design an algorithm that connects them with a cell complex from a weighted Delaunay triangulation. This method requires a second Voronoi diagram to compute the medial axis. Although this algorithm maintains the topology of the medial axis, it produces noisy 'spikes' in many cases.

In this paper we propose to approximate the medial axis directly from the Voronoi diagram. Approximating the medial axis straight from the Voronoi diagram in 3D has been proposed in the past, see for example, the work of Attali et al. [4, 5]. They show how to prune the Voronoi diagram with an angle and a length criterion to approximate the medial axis. They do not establish any convergence guarantee and the proposed pruning is scale and density dependent.

Our algorithm also uses two criteria to select a complex from the Voronoi diagram. But, unlike [4, 5], the selection is scale and density independent and has a convergence guarantee. We filter Delaunay edges from the Delaunay triangulation of the sample points and then output their dual Voronoi facets as an approximate medial axis. The approximation distance depends upon a sampling density parameter ε that approaches zero as sampling density on the surface S approaches infinity. Our experiments with different data sets also support our theoretical claims. A detailed description of these experiments appear in a companion paper [9].

2 Preliminaries and Definitions

Let P be a point sample from a smooth compact surface $S \subseteq \mathbb{R}^3$ without boundary. A ball is called *medial* if it meets S only tangentially in at least two points. The medial axis of S is defined as the closure of the set of centers of all medial balls. Each point on S has two medial balls, one touching it from outside and the other touching it from inside.

Obviously, the medial axis of S can be approximated from a sample P only if it is dense enough to carry information about the features of S. Following the recent work [1], we define the local feature size $f()$ as a function $f : S \to \mathbb{R}$ where $f(x)$ is the distance of $x \in S$ to the medial axis. Intuitively, $f()$ measures how complicated S is locally. It is known that the function $f()$ is 1-Lipschitz continuous, i.e., $f(p) \leq f(q) + \|p - q\|$ for any two points p, q in S [1]. A sample is an ε-sample if each point $x \in S$ has a sample point within $\varepsilon f(x)$ distance.

The Voronoi diagram V_P for a point set $P \in \mathbb{R}^3$ is a cell complex consisting of Voronoi cells $\{V_p\}_{p\in P}$ and their facets, edges and vertices, where $V_p = \{x \in \mathbb{R}^3 \mid \|p - x\| \leq \|q - x\|, \forall q \in P\}$. The dual complex, D_P, called the Delaunay triangulation of P, consists of Delaunay tetrahedra and their incident triangles, edges and vertices. A Delaunay tetrahedron is dual to a Voronoi vertex, a Delaunay triangle is dual to a Voronoi edge, a Delaunay edge is dual to a Voronoi facet and a Delaunay vertex is dual to a Voronoi cell. We say $e = \text{Dual}\, g$ if e and g are dual to each other in the two diagrams. It is an important result proved by Amenta and Bern [1] that the Voronoi cells are elongated along the normal direction to the surface at the sample points if the sample is sufficiently dense. They define *poles* to approximate these normals.

Definition 1. *The* pole p^+ *of a sample point p is the farthest Voronoi vertex in the Voronoi cell V_p. If V_p is unbounded, p^+ is taken at infinity. The vector $\mathbf{v}_p = p^+ - p$ is called the* pole vector *for p and its direction is taken as the average of all directions of infinite edges in case V_p is unbounded, see Figure 1.*

Amenta and Bern showed that the pole vector $\mathbf{v}_p$ approximates the normal $\mathbf{n}_p$ to the surface S at p up to orientation [1].

Definition 2. *The* tangent polygon *for a sample point is defined as the polygon in which the plane through p with $\mathbf{v}_p$ as normal intersects V_p. See Figure 1(a) for an illustration.*

Since $\mathbf{v}_p$ approximates $\mathbf{n}_p$, the tangent polygon approximates the tangent plane at p restricted within V_p. We define a dual structure to the tangent polygon from the Delaunay triangulation D_P.

Definition 3. *The* umbrella U_p *for a sample point p is defined as the topological disc made by the Delaunay triangles incident to p that are dual to the Voronoi edges intersected by the tangent polygon. See Figure 1(b).*

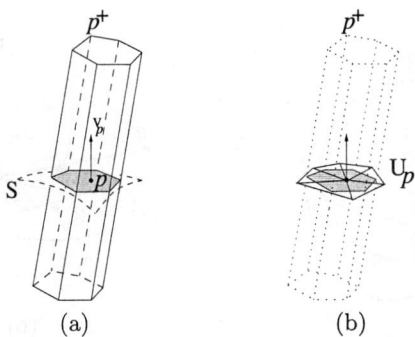

Fig. 1. A Voronoi cell V_p. The corresponding pole, pole vector, tangent polygon (a), and the umbrella (b)

Notations. In what follows we use the following notations. The notation $\angle \mathbf{u}, \mathbf{v}$ denotes the the *acute* angle between the lines supporting two vectors $\mathbf{u}$ and $\mathbf{v}$. The vector going from a point p to q is denoted with $\mathbf{t}_{pq}$. The normal to a triangle σ is $\mathbf{n}_\sigma$ and its circumradius is R_σ.

3 Algorithm

Our aim is to approximate the medial axis M of a smooth surface $S \subseteq \mathbb{R}^3$ from a point sample P. A subset of Voronoi facets chosen as a dual of a set of selected Delaunay edges in D_P constructs this approximation. This means we need some conditions to filter these Delaunay edges from D_P.

3.1 Angle and Ratio Conditions

Let us examine a medial ball B closely to determine which Delaunay edges should we select. Consider Figure 2(a). The segment pq makes an angle θ with the tangent plane at p and q where the medial ball touches the surface S. If B touches S in more than two points, let p and q be such that the angle θ is maximum. We associate each medial axis point m, and also the points where B meets S, with such an angle θ, which we call their *medial angle*. We approximate each medial axis point with the medial angle $\theta > \theta_0$ with an angle condition. For the rest of the medial axis points we apply a ratio condition.

Approximation of the medial angle θ for a medial axis point requires an approximation to the tangent plane at that point. It follows from Lemma 1 that the triangles in the umbrellas necessarily lie flat to S. Therefore, we take umbrella triangles in U_p for approximating the tangent plane at a sample point p and determine all Delaunay edges pq that make relatively large angle with this tangent plane. The angle between an edge pq and a triangle σ is measured by the acute angle $\angle \mathbf{n}_\sigma, \mathbf{t}_{pq}$. We say a Delaunay edge pq satisfies the *angle condition* $[\theta]$ if $\max_{\sigma \in U_p} \angle \mathbf{n}_\sigma, \mathbf{t}_{pq} < \frac{\pi}{2} - \theta$ (see Figure 2(b)).

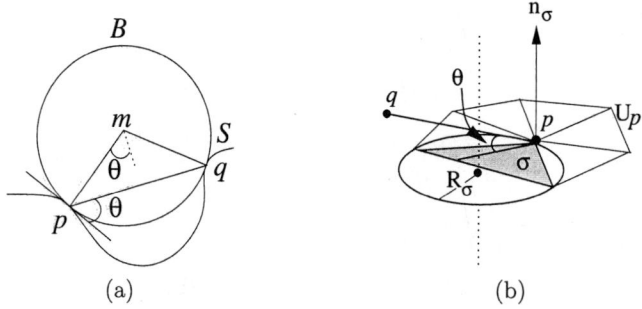

Fig. 2. A medial axis point m, its medial angle θ and the corresponding medial ball (a), illustration for angle and ratio conditions (b)

Only angle condition cannot approximate the medial axis in a density independent manner. If we fix θ for all models, some of the medial axis points with medial angle below θ are not approximated. In that case we cannot hope for the convergence in the limit when density approaches infinity. Also, in practice, it turns out that the value of θ needs to be smaller with increasing sample density [9]. That is why we use the ratio condition to capture those edges whose duals approximate the medial axis points with medial angle below some threshold θ. For this, we compare the length of the Delaunay edges with the circumradii of the umbrella triangles. We say an edge pq satisfies the *ratio condition* $[\rho]$ if $\min_{\sigma \in U_p} \frac{\|p-q\|}{R_\sigma} > \rho$ (see Figure 2(b)).

With the angle and ratio conditions together the thresholds for these conditions can remain fixed. We prove it theoretically. In practice we observe that $\theta = \theta_0 = \frac{\pi}{8}$ and $\rho = \rho_0 = 8$ are appropriate for all data sets that are sufficiently dense. With these two values we enumerate the steps of our algorithm MEDIAL to approximate the medial axis.

MEDIAL(P)
1 Compute V_P and D_P;
2 $F = \emptyset$;
3 for each $p \in P$
4   Compute U_p;
5   for each Delaunay edge $pq \in U_p$
6   if pq satisfies angle Condition $[\theta_0]$
          or ratio condition $[\rho_0]$
7          $F := F \cup \text{Dual } pq$
8   endif
9   endfor
10 endfor
11 output closure(F)

4 Analysis

In this section we establish a convergence result that justifies MEDIAL. For a sample point p we define μ_p, δ_p, η_p and β_p as follows:

$$\mu_p = \text{radius of the larger medial ball at } p, \quad \delta_p = \max_{pqr \in U_p} \frac{R_{pqr}}{\mu_p}, \quad \eta_p = \delta_p \left(\frac{\mu_p}{f(p)}\right),$$

$$\beta_p = \arcsin \frac{\eta_p}{1 - 2\eta_p} + arcsin\left(\frac{2}{\sqrt{3}} \sin(2 \arcsin \frac{\eta_p}{1 - 2\eta_p})\right) + \frac{2\eta_p}{1 - 6\eta_p}.$$

The ratio and angle conditions in MEDIAL are based on the assumption that the umbrella triangles lie flat to the surface and are small compared to the local feature size. These two facts are stated formally in Lemma 1 and 2 which in turn are derived from recent results [1, 2]. We skip the details.

Lemma 1. *Let pqr be any umbrella triangle in* U_p. *We have* $\angle \mathbf{n}_{pqr}, \mathbf{n}_p \leq \beta_p$.

Lemma 2. *Let pqr be any umbrella triangle in* U_p. *The circumradius* $R_{pqr} \leq \left(\frac{\varepsilon}{1-\varepsilon}\right)\left(\frac{1}{\sin(\pi/2 - 3\arcsin \varepsilon/(1-\varepsilon))}\right) f(p)$.

Corollary 1. $\delta_p = O(\varepsilon)\frac{f(p)}{\mu_p}$, $\eta_p = O(\varepsilon)$, *and* $\beta_p \leq 6\eta_p = O(\varepsilon)$ *for small* ε.

4.1 Convergence

The convergence analysis proceeds in part by showing that each point in the output is within $O(\varepsilon^{1/8})\mu_p$ distance from a point in the medial axis for some $p \in P$. In the limit $\varepsilon \to 0$ this distance vanishes. Conversely, we also argue that each point in the medial axis has a nearby point in the output, the distance between which also vanishes as ε reaches zero.

Some of the points on the sampled surface S have infinitely large medial balls. This poses some difficulty in our analysis. To prevent this we enclose S and hence its sample within a sufficiently large bounding sphere. This ensures that μ_p has an upper bound for each point p on S. Of course, the bounding sphere changes the medial axis outside S, but we can keep these changes as far away from S as we wish by choosing a sufficiently large bounding sphere. In particular, the medial axis inside S does not change at all with this modification. In the analysis to follow, we assume that the input point set P samples S as well as the bounding sphere. With the bounding sphere assumption, we have $\Delta \leq \frac{f(p)}{\mu_p} \leq 1$ for any point $p \in S$ where $\Delta > 0$ is a constant dependent on S. It is important that, although Δ depends on S, it remains independent of its sampling.

In the analysis we prove convergence for the subset of the medial axis of S that remain unchanged with the bounding sphere assumption. Let M denote this subset of the medial axis which consists of the centers of the medial balls that touch S but not the bounding sphere. For the ε-sample P, let $\mathcal{L}_\varepsilon$ be the subcomplex of V_P defined by the Voronoi facets and their closures whose dual Delaunay edges satisfy either the ratio condition [$\rho = 8$] or the angle condition [$\theta = \frac{\pi}{8}$], and connect two points in S. We show that the underlying space L_ε of $\mathcal{L}_\varepsilon$ converges to M in the limit $\varepsilon \to 0$. The algorithmic consequence of this result would be to add the sample points of a large bounding sphere to the input sample of S and then filter only from those Delaunay edges that connect sample points of S. However, we observed that the algorithm without this additional sample points work well in practice.

Let m and m' be the centers of the two medial balls at p and μ, μ' their radii respectively. It is a simple observation that m and m' are contained in V_p [1]. Suppose $w \in V_p$ be a point in L_ε so that $\mathbf{t}_{pw} \cdot \mathbf{t}_{pm} > 0$. This means w and m lie on the same side of the tangent plane at p. We will show that, if $\|w - p\| \leq \mu$, the distance between w and its closest point on pm is small. This fact is used to show that, if the Voronoi facet of V_p containing w makes large angle with the line of pm, then w must be near m.

Before we proceed to prove the above fact, we need another geometric property of the umbrella triangles and their circumcircles. For an umbrella triangle

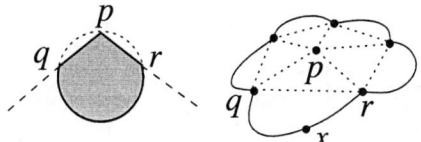

Fig. 3. C_{pqr} of an umbrella triangle pqr (left), and the flower of p

pqr, consider the cone on its plane with p as apex and opening angle $\angle qpr$. Let C_{pqr} denote the intersection of this cone with the circumcircle of pqr, see Figure 3. We define $\mathrm{Fl}_p = \cup_{pqr \in U_p} C_{pqr}$, the *flower* of p.

Lemma 3. V_p *does not contain any point of the boundary of* Fl_p *inside.*

The flower of p lies very flat to the surface and cannot intersect the segment mm' at any point other than p. This means that the above lemma implies that Fl_p intersects V_p completely and separates m and m' on its two sides within V_p. In particular, any segment connecting a point $w \in V_p$ with m, where w and m lies on the opposite sides of Fl_p, must intersect Fl_p.

In the next two lemmas let w° denote the point on the *line* of mm' which is closest to $w \in V_p$.

Lemma 4. *If w° lies in the segment mm', $\|w - w^\circ\| \leq 2\tan(\arcsin 2\eta_p)\mu_p$.*

Proof. Let m' be the center of the medial ball at p so that w and m' lie on opposite sides of the flower at p within V_p. Consider the segment wm'. Let y be the foot of the perpendicular dropped from from p to wm'. Since wm' intersects Fl_p and any point in Fl_p is within $2\delta_p \mu_p$ distance, we must have $\|y-p\| \leq 2\delta_p \mu_p$. Therefore, $\angle pm'y \leq \arcsin \frac{\|y-p\|}{\|m'-p\|} \leq \arcsin \frac{2\delta_p \mu_p}{f(p)} = \arcsin 2\eta_p$. It follows that $\|w - w^\circ\| \leq \|m - m'\|\tan \angle pm'y \leq 2\mu_p \tan(\arcsin 2\eta_p)$.

Lemma 5. *Let $F = \mathrm{Dual}\, pq$ be a Voronoi facet where pq satisfies the angle condition $[\theta]$ with $\theta \geq 2\eta_p + \beta_p$. Any point w in F with $\|w - p\| \leq \mu$ is within $\frac{1}{\sin(\theta - \beta_p)}(2\tan(\arcsin 2\eta_p))\mu_p$ distance from m where m and μ are the center and radius of the medial ball at p with $\mathbf{t}_{pm} \cdot \mathbf{t}_{pw} > 0$.*

Proof. Consider the ball B with radius μ around p. The point w necessarily lies inside B since $\|w - p\| \leq \mu$. Therefore, w° lies in the segment pm. We can apply Lemma 4 to assert $\|w - w^\circ\| \leq 2\tan(\arcsin 2\eta_p)\mu_p$.

Let the plane of F intersect the line of pm at x at an angle α. We have $\alpha > \theta - \beta_p$ since the normal to F makes more than θ angle with the normal of an umbrella triangle (by the angle condition $[\theta]$) which in turn makes an angle less than β_p with the surface normal at p (Lemma 1). With the requirement that $\theta > 2\eta_p + \beta_p$, we have $\alpha > 2\eta_p$.

Also, the plane of F cannot intersect the segment mm'. This is because the medial balls at p are empty of any other sample point and thus both m and m'

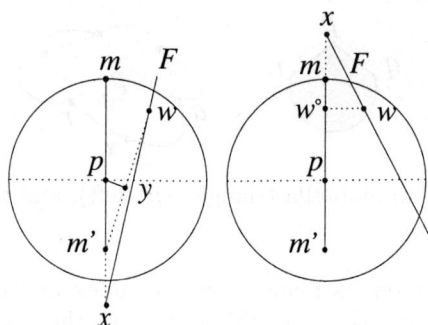

Fig. 4. Illustration for Lemma 5. The picture in the left is not a possible configuration due to the constraint on α

belong to V_p. In particular, the segment mm' must be inside the Voronoi cell V_p. The segment wm' necessarily lies inside V_p and intersects the flower of p, say at y. This means $\|y-p\| \leq 2\delta_p \mu_p$. If x lies below m' as shown in the left picture of Figure 4, we have $\alpha < \angle pxw < \angle pm'y < \tan \frac{\|y-p\|}{\|m'-p\|} < 2\eta_p$. This contradicts the assertion that $\alpha > 2\eta_p$. So, x cannot lie below m'. Instead, it lies above m as shown in the right picture in Figure 4. From the triangle $ww°x$ we have $\|w - m\| \leq \|w - x\| = \frac{\|w-w°\|}{\sin \alpha} \leq \frac{2\tan(\arcsin 2\eta_p)}{\sin(\theta-\beta_p)} \mu_p$.

Now we combine both the angle and ratio conditions to show that the dual Voronoi facets in $\mathcal{L}_\varepsilon$ lie either close to the medial axis or far away from the sample point.

Lemma 6. *Let $pq \in D_P$ with $p \in S$ and $\alpha = \max_{\sigma \in U_p} \left(\frac{\pi}{2} - \angle \mathbf{n}_\sigma, \mathbf{t}_{pq}\right)$. Let w be any point in the Voronoi facet $F = \text{Dual} pq$ and m and μ be the center and radius of the medial ball at p so that $\mathbf{t}_{pm} \cdot \mathbf{t}_{pw} > 0$. If pq satisfies the ratio condition $[\rho = 8]$ then either $\|w - m\| = O(\eta_p^{1/8})\mu_p$ or $\|w - p\| = \Omega(\eta_p^{1/8})\mu$.*

Proof. By definition pq satisfies the angle condition $[\alpha]$. Consider the case when $\alpha > \eta_p^{7/8}$. For sufficiently small ε, we have $\eta_p^{7/8} > 2\eta_p + \beta_p$. If $\|w - p\| \leq \mu$ we can apply Lemma 5 to conclude that

$$\|w - m\| \leq \frac{2\tan(\arcsin 2\eta_p)\mu_p}{\sin 2\eta_p^{7/8}} = O\left(\frac{\eta_p}{\eta_p^{7/8}}\right)\mu_p = O(\eta_p^{1/8})\mu_p.$$

Next consider the case when $\alpha \leq \eta_p^{7/8}$. Let B be the empty ball centered at w and containing p, q on its boundary. Let T denote the tangent plane to B at p. Clearly, $\mathbf{n}_p$ is normal to T. Let ptu be the triangle in U_p to which $\mathbf{t}_{pq}$ project along the direction of $\mathbf{n}_p$ or $-\mathbf{n}_p$. If B does not intersect the circumcircle of ptu at any point other than p, then for sufficiently small ε, $\angle \mathbf{t}_{pq}, \mathbf{n}_p > \angle \mathbf{t}_{pq}, \mathbf{n}_{ptu} - \angle \mathbf{n}_{ptu}, \mathbf{n}_p > \frac{\pi}{2} - \alpha - \beta_p > \frac{\pi}{2} - 2\eta_p^{7/8}$. This means, R, the radius of B satisfies

$$R > \frac{\|p - q\|}{2\sin 2\eta_p^{7/8}} > \frac{\rho_0 \delta_p \mu_p}{4\eta_p^{7/8}} > 2\eta_p^{1/8} f(p).$$

Since $p \in S$, we have $\frac{f(p)}{\mu_p} > \Delta$ for some constant $\Delta > 0$ with the bounding sphere assumption. This gives $R = \Omega(\eta_p^{1/8})\mu_p$.

In the other case when B intersects the circumcircle of ptu at points other than p, let C be the circle of intersection of B with the plane containing pq and $\mathbf{n}_{ptu}$. The segment ps in which C intersects the plane of ptu satisfies two conditions: $\angle \mathbf{t}_{ps}, \mathbf{t}_{pq} \leq \alpha$ and $\|s - p\| \leq 2\delta_p \mu_p$. The first condition follows from the definition α. The second condition follows from the fact that B intersects the plane of ptu in a circle that must be inside its circumcircle since B is an empty ball. The case where $\|p - q\| \leq \|s - p\|$ is not possible since $\|p - q\| > \rho_0 \delta_p \mu_p > 2\delta_p \mu_p \geq \|s - p\|$ for $\rho_0 = 8$. So, consider the case where $\|p - q\| > \|s - p\|$. In this case the radius r of C satisfies:

$$r = \frac{\|q - s\|}{2 \sin \angle qps} > \frac{\|p - q\| - \|s - p\|}{2 \sin \angle qps} > \frac{(\rho_0 - 2)\delta_p \mu_p}{2 \sin \angle qps} = \frac{(\rho_0 - 2)f(p)}{2\alpha}$$
$$> 3\eta_p^{1/8} f(p) = \Omega(\eta_p^{1/8})\mu.$$

The radius of B is at least $r = \Omega(\eta_p^{1/8})\mu$.

To complete the proof of convergence we need the following lemma which says that all points in V_p that are far away from p cannot be too far from a medial axis point. This lemma is extracted from a result of Boissonnat and Cazals ([6], Proposition 18). Although we need slightly different constants, the proof remains same.

Lemma 7. *Let $w \in V_p$ be a point such that $\|w - p\| > \eta\mu$, where μ is the radius of a medial ball at p with the center m and $\mathbf{t}_{pm} \cdot \mathbf{t}_{pw} > 0$, and $\eta \geq \varepsilon^{1/4}$. Then, for sufficiently small $\varepsilon > 0$, $\|w - m\| = O(\varepsilon^{3/4})\mu$ if the medial angle of p is larger than $\varepsilon^{1/3}$.*

Recall that we defined L_ε as the underlying space of a Voronoi subcomplex $\mathcal{L}_\varepsilon \subseteq V_P$ where P is an ε-sample. The Voronoi facets in $\mathcal{L}_\varepsilon$ are dual to Delaunay edges connecting sample points in S and satisfying the angle condition $[\theta = \frac{\pi}{8}]$ or the ratio condition $[\rho = 8]$. We want to show $\lim_{\varepsilon \to 0} L_\varepsilon = M$, where M is the subset of the medial axis defined by the medial balls that touch only S but not the bounding sphere.

Theorem 1. $\lim_{\varepsilon \to 0} L_\varepsilon \subseteq M$.

Proof. Let F be a facet in $\mathcal{L}_\varepsilon$. First, consider the case when $pq = \text{Dual } F$ is selected by the angle condition $[\theta = \frac{\pi}{8}]$. Let $w \in F$ be any point and μ and m are as defined in Lemma 5.

If w is more than μ away from p, we can apply Lemma 7 to conclude $\|w - m\| = O(\varepsilon^{3/4})\mu$ if p has a medial angle $\Omega(\varepsilon^{1/3})$. Otherwise, Lemma 5 applies to assert that $\|w - m\| \leq \frac{2\tan(\arcsin 2\eta_p)\mu_p}{\sin(\pi/8 - \beta_p)} = O(\varepsilon)\mu_p$ (Corollary 1). In both cases w reaches m in the limit $\varepsilon \to 0$.

Next, consider the case when pq is selected by the ratio condition $[\rho_0 = 8]$. Since p is a point in S, we can apply Lemma 6. Then, either $\|w-m\| = O(\eta_p^{1/8})\mu_p$

or $\|w - p\| = \Omega(\eta_p^{1/8})\mu$. In the first case, we get the required convergence as $\lim_{\varepsilon \to 0} \eta_p = 0$ (Corollary 1). In the second case, $\eta_p^{1/8} \gg \varepsilon^{1/4}$ for sufficiently small ε. Thus, the condition to apply Lemma 7 is satisfied if p has a medial angle $\Omega(\varepsilon^{1/3})$. This implies that w reaches m in the limit $\varepsilon \to 0$.

In the next theorem we establish the converse of the previous theorem by showing that a point in M with the medial angle at least $\Omega(\varepsilon)$ has a nearby point in L_ε.

Theorem 2. $M \subseteq \lim_{\varepsilon \to 0} L_\varepsilon$.

Proof. (sketch.) Let $m \in M$ be a medial axis point with the medial angle $\theta > c\frac{\varepsilon}{\Delta}$ for some constant $c > 0$ and μ be the radius of the corresponding medial ball B. Let B meet S tangentially at x and y so that $\angle xmy = \theta$. Grow B centrally till it meets a sample point, say p, for the first time. Without loss of generality assume that $\angle pmx > \angle pmy$ so that $\angle pmx > c\frac{\varepsilon}{2\Delta}$. Now grow B further keeping it always empty and maintaining p on its boundary until its boundary hits a sample point q at least $c_1 \frac{\varepsilon}{\Delta}\mu$ away from p where $0 < c_1 < c$ is another constant chosen appropriately. This growth is always possible. When B has only p on its boundary, grow it by moving its center w along the ray pm. When B has two points p and q on its boundary, grow it by moving w along any ray lying on the bisector plane of p and q. When B has three points on its boundary, move w along the equidistant line whose points are equidistant from all the three points. When B has a set Q of four points on its boundary, move w along the equidistant line of a triangle σ on the convex hull of Q where the plane of σ separates Q and the center of B. Notice that Q cannot enclose the center of B since in that case we have a point q too far from p satisfying the stopping condition. It can be shown that, at any time of this growth the ray along which the center w is moved makes more than $c_2 \frac{\varepsilon}{\Delta}$ angle with the ray xw for an appropriate constant $c_2 > 0$. This in turn implies that the center w has to be moved at most $O(\varepsilon)\mu$ away from m before B encloses a ball of radius $\varepsilon\mu$ around x. By this time B must have hit a sample point more than $c_1 \frac{\varepsilon}{\Delta}\mu$ away from p since x has a sample point within $\varepsilon\mu$ distance and x is sufficiently away from p.

Therefore, when the growth has stopped, the center w of B is only $O(\varepsilon)\mu$ away from m, and we have a Delaunay edge pq whose length is at least $c_1 \frac{\varepsilon}{\Delta}\mu$. By appropriate choice of c and c_1, one can ensure that $c_1 \frac{\varepsilon}{\Delta}\mu > \rho_0 \delta_p \mu_p$. Therefore, pq satisfies the ratio condition $[\rho_0]$ and its dual Voronoi facet is in $\mathcal{L}_\varepsilon$. The center w belongs to this Voronoi facet and hence to L_ε with the condition $\|w - m\| = O(\varepsilon)\mu$. The theorem follows.

Remark: Note that it is essential to assume that no five sample points lie on the boundary of an empty sphere so that the dual facets of all edges with an empty circumscribing sphere exist in the Voronoi diagram. This excludes, for example, sample points on a sphere.

We implemented MEDIAL in C++ using the CGAL library [21] for the Voronoi diagram and the Delaunay triangulation. Details of this experiment and other related ones are reported in [9]. Some sample results are shown in

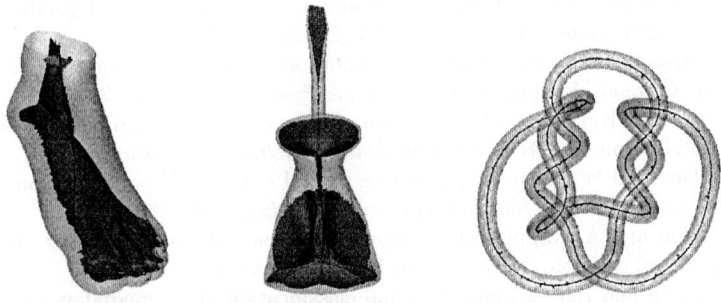

Fig. 5. Medial axis construction with MEDIAL for some data sets

Figure 5. In this picture we show only 'inner' parts of the medial axis which are separated using the reconstruction software COCONE [20].

5 Conclusions

In this paper we present the convergence analysis of an algorithm that approximates the medial axis of a surface from the Voronoi diagram of its sample points. Unlike previous approaches, this algorithm is scale and density independent. Experimental studies suggest that the algorithm computes clean medial axes without any fine tuning of the parameters. Details of this study appear in [9].

Although this is the first result that shows that the medial axis of a surface in 3D can be computed as a Voronoi subcomplex with guaranteed convergence, we could not prove that the approximated medial axis maintains the topology of the original one. We believe that this is true under sufficient density assumption since our experiments with dense data sets do not suggest otherwise. We plan to investigate this aspect in future research.

References

[1] N. Amenta and M. Bern. Surface reconstruction by Voronoi filtering. *Discr. Comput. Geom.* **22** (1999), 481–504.
[2] N. Amenta, S. Choi, T. K. Dey and N. Leekha. A simple algorithm for homeomorphic surface reconstruction. *Internat. J. Comput. Geom. Applications*, **12** (2002), 125–121.
[3] N. Amenta, S. Choi and R. K. Kolluri. The power crust, unions of balls, and the medial axis transform. *Comput. Geom. Theory and Applications* **19** (2001), 127–153.
[4] D. Attali and J.-O. Lachaud. Delaunay conforming iso-surface, skeleton extraction and noise removal. *Comput. Geom. : Theory Appl.*, 2001, to appear.
[5] D. Attali and A. Montanvert. Computing and simplifying 2D and 3D continuous skeletons. *Computer Vision and Image Understanding* **67** (1997), 261–273.

[6] J. D. Boissonnat and F. Cazals. Natural neighbor coordinates of points on a surface. *Comput. Geom. Theory Appl.* **19** (2001), 87–120.

[7] J. W. Brandt and V. R. Algazi. Continuous skeleton computation by Voronoi diagram. *Comput. Vision, Graphics, Image Process.* **55** (1992), 329–338.

[8] T. Culver, J. Keyser and D. Manocha. Accurate computation of the medial axis of a polyhedron. *5th ACM Sympos. Solid Modeling Applications*, (1999), 179–190.

[9] T. K. Dey and W. Zhao. Approximate medial axis as a Voronoi subcomplex. *7th ACM Sympos. Solid Modeling Applications*, (2002), 356–366.

[10] M. Etzion and A. Rappoport. Computing Voronoi skeletons of a 3D polyhedron by space subdivision. *Tech. Report*, Hebrew University, 1999.

[11] P. J. Giblin and B. B. Kimia. A formal classification of 3D medial axis points and their local geometry. *Proc. Computer Vision and Pattern Recognition (CVPR)*, 2000.

[12] L. Guibas, R. Holleman and L. E. Kavraki. A probabilistic roadmap planner for flexible objects with a workspace medial axis based sampling approach. *Proc. IEEE/RSJ Intl. Conf. Intelligent Robots and Systems*, 1999.

[13] C. Hoffman. How to construct the skeleton of CSG objects. *The Mathematics of Surfaces, IVA*, Bowyer and J. Davenport Eds., Oxford Univ. Press, 1990.

[14] R. L. Ogniewicz. Skeleton-space: A multiscale shape description combining region and boundary information. *Proc. Computer Vision and Pattern Recognition*, (1994), 746–751.

[15] D. Sheehy, C. Armstrong and D. Robinson. Shape description by medial axis construction. *IEEE Trans. Visualization and Computer Graphics* **2** (1996), 62–72.

[16] E. C. Sherbrooke, N. M. Patrikalakis and E. Brisson. An algorithm for the medial axis transform of 3D polyhedral solids. *IEEE Trans. Vis. Comput. Graphics* **2** (1996), 44–61.

[17] G. M. Turkiyyah, D. W. Storti, M. Ganter, H. Chen and M. Vimawala. An accelerated triangulation method for computing the skeletons of free-form solid models. *Computer Aided Design* **29** (1997), 5–19.

[18] M. Teichman and S. Teller. Assisted articulation of closed polygonal models. *Proc. 9th Eurographics Workshop on Animation and Simulation*, 1998.

[19] F.-E. Wolter. Cut locus & medial axis in global shape interrogation & representation. *MIT Design Laboratory Memorandum 92-2*, 1992.

[20] http://www.cis.ohio-state.edu/~tamaldey/cocone.html.

[21] http://www.cs.ruu.nl/CGAL.

Non-independent Randomized Rounding and an Application to Digital Halftoning

Benjamin Doerr and Henning Schnieder

Mathematisches Seminar II, Christian–Albrechts–Universität zu Kiel
Ludewig–Meyn–Str. 4, D–24098 Kiel, Germany
{bed,hes}@numerik.uni-kiel.de
http://www.numerik.uni-kiel.de/~{bed,hes}/

Abstract. We investigate the problem to round a given [0, 1]–valued matrix to a 0, 1 matrix such that the rounding error with respect to 2 × 2 boxes is small. Such roundings yield good solutions for the digital halftoning problem as shown by Asano et al. (SODA 2002). We present a randomized algorithm computing roundings with expected error at most 0.6287 per box, improving the 0.75 non-constructive bound of Asano et al. Our algorithm is the first one solving this problem fast enough for practical application, namely in linear time.
Of a broader interest might be our rounding scheme, which is a modification of randomized rounding. Instead of independently rounding the variables (expected error 0.82944 per box in the worst case), we impose a number of suitable dependencies.
Experimental results show that roundings obtained by our approach look much less grainy than by independent randomized rounding, and only slightly more grainy than by error diffusion. On the other hand, the latter algorithm (like all known deterministic algorithms) tends to produce unwanted structures, a problem that randomized algorithms like ours are unlikely to encounter.

Keywords: Randomized rounding, discrepancy, digital halftoning

1 Introduction

In this paper we are concerned with rounding problems and, in particular, the digital halftoning problem. In general form, rounding problems are of the following type: Given some numbers $x_1, \ldots, x_n$, one is looking for roundings $y_1, \ldots, y_n$ such that some given error measures are small. By rounding we always mean that $y_i = \lfloor x_i \rfloor$ or $y_i = \lceil x_i \rceil$. Since there are 2^n possibilities, such rounding problems are good candidates for hard problems. In fact, even several restricted versions like the combinatorial discrepancy problem are known to be NP–hard.

On the other hand, there are cases that can be solved optimally in polynomial time. Knuth [9] for example has shown that there exists a rounding such that all errors $|\sum_{i=1}^{k}(y_i - x_i)|$ and $|\sum_{i=1}^{k}(y_{\pi(i)} - x_{\pi(i)})|$ for a fixed permutation π are at most $\frac{n}{n+1}$. Such roundings can be obtained by computing a maximum flow in a network. A recent generalization to arbitrary totally unimodular rounding problems can be found in [5].

1.1 The Digital Halftoning Problem: A Matrix Rounding Problem

Our study of rounding problems is motivated by an application to digital halftoning. The digital halftoning problem is to convert a continuous-tone intensity image (each pixel may have an arbitrary 'color' on the white-to-black scale) into a binary image (only black and white dots are allowed). An intensity image can be represented by a $[0,1]$-valued $m \times n$ matrix A. Each entry a_{ij} corresponds to the brightness level of the pixel with coordinates (i,j). Since many devices, e.g., laser printers, can only output white and black dots, we have to round A towards a $0,1$ matrix. Naturally, this has to be done in a way that the resulting image looks similar to the original one.

This notion of similarity is a crucial point. From the viewpoint of application, similarity is defined via the human visual system: A rounding is good, if an average human being can retrieve the same information from both the original image and the rounded one. Using this criterion, several algorithms turned out to be useful: Floyd and Steinberg [7] proposed the error diffusion algorithm that rounds the entries one by one and distributes the rounding error over neighboring not yet rounded entries. Lippel and Kurland [10] and Bayer [3] investigated the ordered dither algorithm, which partitions the image into small submatrices and rounds each submatrix by comparing its entries with a threshold matrix of same dimension. One advantage of this approach is that this algorithm can be parallelized easily. Knuth [8] combined ideas of both approaches to get an algorithm called dot diffusion.

What was missing so far from the theoretical point of view is a good mathematical formulation of similarity. Such a similarity measure is desirable for two reasons: Firstly, it allows to compare algorithms without extensive experimental testing. This is particularly interesting, since comparing different halftonings is a delicate issue. For example, it makes a huge difference whether the images are viewed on a computer screen or are printed on a laser printer. Even different printers can give different impressions. Therefore, a more objective criterion would be very helpful. A second reason is that having a good criterion, one would have a clearer indication of how a digital halftoning algorithm should work. Thus developing good algorithms would be easier.

On this year's SODA conference, Asano et al. [2] reported that they made some progress into this direction. Experimental results indicate that good digital halftonings have small error with respect to all 2×2 subregions. More formally, we end up with this problem:

1.2 Problem Statement

Let $A \in [0,1]^{m \times n}$ denote our input matrix. A set $R_{ij} := \{(i,j), (i+1,j), (i,j+1), (i+1,j+1)\}$ for some $i \in [m-1], j \in [n-1]$ is called a 2×2 subregion (or box) in $[m] \times [n]$.[1] Denote by $\mathcal{H}$ the set of all these boxes. We write $A_{R_{ij}}$

---

[1] For a number r we denote by $[r]$ the set of positive integers not exceeding r.

for the 2 × 2 matrix $\begin{pmatrix} a_{i,j} & a_{i,j+1} \\ a_{i+1,j} & a_{i+1,j+1} \end{pmatrix}$ induced by R_{ij}. For any matrix A put $\Sigma A := \sum_{i,j} a_{ij}$.

For a matrix $B \in \{0,1\}^{m \times n}$ — which by definition is a rounding of A — we define the rounding error of A with respect to B by

$$d_{\mathcal{H}}(A,B) := \sum_{R \in \mathcal{H}} |\Sigma A_R - \Sigma B_R|.$$

We usually omit the subscript $\mathcal{H}$ when there is no danger of confusion.

From a broader perspective, this rounding problem has the interesting aspect that the errors regarded depend on few variables (here four) only. Thus the common approach of randomized rounding is not very effective, since it depends on high concentration results for sums of many independent random variables.

1.3 Theoretical Results

Asano et al. [2] exhibited that roundings B such that $d(A, B)$ is small, yield good digital halftonings. They showed that for any A an optimal rounding B^* satisfies $d(A, B^*) \leq \frac{3}{4}|\mathcal{H}|$. They also gave a polynomial time algorithm computing a rounding B such that $d(A, B) - d(A, B^*) \leq \frac{9}{16}|\mathcal{H}| = 0.5625|\mathcal{H}|$. It is easy to see that there are matrices A such that all roundings (and in fact all integral matrices) B have $d(A, B) \geq \frac{1}{2}|\mathcal{H}|$.

A major draw-back of the algorithm given in [2] is that it is not very practical, as it requires the solution of an integer linear program with totally unimodular constraint matrix. As pointed out by the authors, this is too slow for digital halftoning applications.

In this paper we present a randomized algorithm that runs in linear time. It may be used in parallel without problems. The roundings computed by our algorithm have an expected error that exceeds the optimal one by at most $\frac{15}{32}|\mathcal{H}| \leq 0.4688|\mathcal{H}|$ (instead of $0.5625|\mathcal{H}|$). In addition, our roundings satisfy the absolute bound $E(d(A, B)) \leq 0.6287|\mathcal{H}|$ beating the non-algorithmic bound of $\frac{3}{4}|\mathcal{H}|$ of [2].

The distribution of the resulting error is highly concentrated around the expected value. The probability that the error exceeds $(0.6287+\varepsilon)|\mathcal{H}|$ is bounded by $\exp(-\Omega(\varepsilon^2|\mathcal{H}|))$. More concrete, when processing a 4096 × 3072 pixel image, the probability that the error exceeds the expectation by more than $0.01|\mathcal{H}|$ is less than $2.5 \cdot 10^{-7}$. Hence a feasible approach in practice is to compute the random rounding without checking its error.

Another nice feature from the viewpoint of application to digital halftoning is that the roundings computed by our algorithm have small rounding error also with respect to other geometric structures that might indicate good (according to humans' eyes) halftonings. In particular, the expected error with respect to the set $\mathcal{H}_3$ of all 3 × 3 boxes is bounded by $0.82944|\mathcal{H}_3|$, and the error with respect to single entries (1 × 1 boxes) is bounded by $0.5|\mathcal{H}_1|$, which is optimal in the worst-case.

1.4 Non-independent Randomized Rounding

The key idea of our algorithm might also be of a broader interest. We develop a randomized rounding scheme where the individual roundings are not independent. The classical approach of randomized rounding due to Raghavan and Thompson [11, 12] is to round each variable independently with probability depending on the fractional part of its value. This allows to use Chernoff-type large deviation inequalities showing that a sum of independent random variables is highly concentrated around its expectation. Randomized rounding has been applied to numerous combinatorial optimization problems that can be formulated as integer linear programs. Though being very effective in the general case, one difficulty with randomized rounding is to use structural information of the underlying problem. One way to overcome this is to use correlation among the events. This allows to strengthen the classical bounds as shown by Srinivasan [13].

In this paper we try so use the structure in an earlier phase, namely in the design of the random experiment. This leads to randomized roundings where the variables are not rounded independently. There have been a few attempts in this direction (cf. [11, 12, 4]), but they do not go further than translating constraints into dependencies or imposing restrictions on the number of variables rounded up or down. Therefore we feel that the option to design the random experiment in a way that it reflects the structure of the underlying problem has not been exploited sufficiently. In this paper we try to move a step forward into this direction. We impose dependencies that are not necessary in the sense of feasibility, but helpful since they reduce the expected rounding error. For the rounding problem arising from digital halftoning, this improves the bound of $0.82944|\mathcal{H}|$ obtained by independent randomized rounding down to $0.6287|\mathcal{H}|$.

In some sense this result can be seen as an extension of ideas of [1, 6]. These results, however, apply only to a very restricted class of rounding problems: The so-called combinatorial discrepancy problem is to round the vector $x = \frac{1}{2}\mathbf{1}_n$ to a $0,1$ vector y such that $\|A(x - y)\|_\infty$ is small for some $0,1$ matrix A.

Though a general result on non-independent randomized rounding is hard to imagine due to the high dependence on the structure of the target problem, we believe that this work shows the power of non-independent randomized rounding and motivates further research in this area.

1.5 Experimental Results

To estimate the visual quality of our algorithm, we generated roundings of several both real-world and artificial images with existing algorithms and the new one. It turns out that the use of dependencies in the random experiment greatly reduces the graininess of the output images (compared to independent randomized rounding). Still, the error diffusion algorithm remains unbeaten in this category.

On the other hand, since our algorithms is randomized, we have no problems with unwanted structures or textures, a weakness of error diffusion and other deterministic algorithms. Summarizing our experimental results, we feel that our work brings randomized rounding back into a group of feasible approaches for the digital halftoning problem in which neither can claim himself superior to the others.

2 Randomized Rounding

For a number x we write $\lfloor x \rfloor$ for the largest integer not exceeding x, $\lceil x \rceil$ for the smallest being not less that x, and $\{x\} := x - \lfloor x \rfloor$ for the fractional part of x. We say that some random variable X is a *randomized rounding* of x if $P(X = \lfloor x \rfloor + 1) = \{x\}$ and $P(X = \lfloor x \rfloor) = 1 - \{x\}$. In particular, if $x \in [0,1]$, we have $P(X = 1) = x$ and $P(X = 0) = 1 - x$.

We first analyze what can be achieved with independent randomized rounding. The result below is needed not only to estimate the superiority of non-independent randomized rounding, but also in the proofs of two of our results. Note that the proof (omitted for reasons of space) has to be different from typical randomized rounding applications: Since the boxes are small, using a large deviation bound makes no sense and one has to compute the expected error exactly.

We say that B is an *independent randomized rounding* of A if each entry b_{ij} is a randomized rounding of a_{ij} and all these roundings are mutually independent.

Theorem 1. *Let $A \in [0,1]^{m \times n}$, B an independent randomized rounding of A and B^* an optimal rounding of A. Then*

$$E(d(A,B)) \leq 0.82944|\mathcal{H}|,$$
$$E(d(A,B)) \leq 0.75|\mathcal{H}| + d(A,B^*).$$

□

The two bounds of Theorem 1 are tight as shown by matrices with all entries 0.4 and 0.5 respectively.

3 Non-independent Randomized Rounding

In this section we improve the previous bounds by adding some dependencies to the rounding process. We start with an elementary approach called *joint randomized rounding*, which reduces the chance that neighboring matrix entries are both rounded in the wrong way. This leads to a first improvement in Corollary 1. We then add further dependencies leading to what we call *block randomized rounding*.

Definition 1. *Let $a_1, a_2 \in [0,1]$. We say that (b_1, b_2) is a joint randomized rounding of (a_1, a_2) if $P(b_1 = 1 \wedge b_2 = 0) = a_1$, $P(b_1 = 0 \wedge b_2 = 1) = a_2$ and $P(b_1 = 0 \wedge b_2 = 0) = 1 - a_1 - a_2$ holds if $a_1 + a_2 \leq 1$, and if $P(b_1 = 1 \wedge b_2 = 0) = 1 - a_2$, $P(b_1 = 0 \wedge b_2 = 1) = 1 - a_1$ and $P(b_1 = 1 \wedge b_2 = 1) = a_1 + a_2 - 1$ holds in the case $a_1 + a_2 > 1$.*

Immediately, we have

Lemma 1. *Let $a_1, a_2 \in [0,1]$. Then (b_1, b_2) is a joint randomized rounding of (a_1, a_2) if and only if the following two properties are valid:*

(i) For all $i \in [2]$, we have $P(b_i = 1) = a_i$ and $P(b_i = 0) = 1 - a_i$. Hence b_i is a randomized rounding of a_i.

(ii) $b_1 + b_2$ is a randomized rounding of $a_1 + a_2$.

The next lemma shows that two joint randomized roundings are superior to four independent randomized roundings in terms of the rounding error.

Lemma 2. *Let $A = (a_{11}, a_{12}, a_{21}, a_{22})$ be a box. Let (b_{11}, b_{12}) be a joint randomized rounding of (a_{11}, a_{12}), and (b_{21}, b_{22}) one of (a_{21}, a_{22}) independent of the first. Then the expected error of A with respect to $B = (b_{11}, b_{12}, b_{21}, b_{22})$ is at most $E(d(A, B)) = \frac{16}{27}|\mathcal{H}| \leq 0.5926|\mathcal{H}|$. In comparison with an optimal rounding B^* we have $E(d(A, B)) - d(A, B^*) \leq 0.5|\mathcal{H}|$.*

Proof (sketched). $b_{11} + b_{12}$ behaves like a randomized rounding of $a_{11} + a_{12}$, and the same holds for the second row. Hence all we have to do is to bound the expected deviation of a sum of two independent randomized roundings from the sum of the original values. □

The bounds Lemma 2 are sharp as shown by matrices having $a_{11} + a_{12} = a_{21} + a_{22} = \frac{1}{3}$ and $a_{11} + a_{12} = a_{21} + a_{22} = \frac{1}{2}$ respectively. Plastering the grid with joint randomized roundings already yields a first improvement over independent randomized rounding:

Corollary 1. *Let $A \in [0,1]^{m \times n}$. Compute $B \in \{0,1\}^{m \times n}$ by independently obtaining $(b_{i,2j-1}, b_{i,2j})$ from $(a_{i,2j-1}, a_{i,2j})$ by joint randomized rounding for all $i \in [m], j \in [\frac{n}{2}]$, and also independently obtaining b_{in} from $a_{in}, i \in [m]$, by usual randomized rounding, if n is odd. Then*

$$E(d(A, B)) \leq 0.7111|\mathcal{H}|.$$

Proof. At least half of the boxes, namely all R_{ij} such that j is odd, are rounded in the manner of Lemma 2. The remaining ones contain four independent randomized roundings. Thus $E(d(A,B)) \leq \frac{1}{2}0.5926|\mathcal{H}| + \frac{1}{2}0.82944|\mathcal{H}| \leq 0.7111|\mathcal{H}|$ by Theorem 1 and Lemma 2. □

Block Randomized Rounding

Definition 2. *Let $A = (a_{11}, a_{12}, a_{21}, a_{22})$ be a box. We call $B = (b_{11}, b_{12}, b_{21}, b_{22})$ a block randomized rounding of A if*

(i) each single entry of B is a randomized rounding of the corresponding one of A, i.e., $P(b_{ij} = 1) = a_{ij}$ and $P(b_{ij} = 0) = 1 - a_{ij}$ for all $i, j \in [2]$.

(ii) each pair of neighboring entries has the distribution of the corresponding joint randomized rounding, i.e., in addition to (i) we have for all $(i,j), (i',j') \in [2] \times [2]$ such that either $i \neq i'$ or $j \neq j'$,

$$P(b_{ij} + b_{i'j'} = \lfloor a_{ij} + a_{i'j'} \rfloor + 1) = \{a_{ij} + a_{i'j'}\}$$
$$P(b_{ij} + b_{i'j'} = \lfloor a_{ij} + a_{i'j'} \rfloor) = 1 - \{a_{ij} + a_{i'j'}\}.$$

(iii) the box in total behaves like a randomized rounding, i. e., we have
$$P(\Sigma B = \lfloor \Sigma A \rfloor + 1) = \{\Sigma A\}$$
$$P(\Sigma B = \lfloor \Sigma A \rfloor) = 1 - \{\Sigma A\}.$$

By item *(iii)* of this definition, block randomized roundings have low rounding errors. If B is a block randomized rounding of a 2×2 matrix A, then $E(d(A,B)) \leq \frac{1}{2}$ and $E(d(A,B)) - \min_{B^*} d(A,B^*) \leq \frac{1}{8}$. The interesting point is that block randomized roundings always exist:

Lemma 3. *For all boxes $A = (a_{11}, a_{12}, a_{21}, a_{22})$, a block randomized rounding exists and can be generated efficiently.*

Proof. We first show that it is enough to consider boxes A such that $s := \Sigma A = a_{11} + a_{12} + a_{21} + a_{22} \leq 2$. Assume $s > 2$. Define $A' = (a'_{ij})$ by $a'_{ij} := 1 - a_{ij}$. Then $\Sigma A' = 4 - \Sigma A < 2$. Let $B' = (b'_{ij})$ be a block randomized rounding of A'. Define $B = (b_{ij})$ by $b_{ij} := 1 - b'_{ij}$. Now it remains to show that the pair (A, B) fulfills (i) to (iii) of Definition 2, which we leave to the reader.

For the remainder of this proof let us assume that $s \leq 2$. Let $i, i', j, j' \in [2]$. We call $e = \{(i,j), (i',j')\}$ an *edge*, if either $i \neq i'$ or $j \neq j'$, and write $a_e := a_{ij} + a_{i'j'}$. An edge e is *heavy*, if $a_e > 1$. We first note that there cannot be disjoint heavy edges. If e and f were disjoint and heavy, then $s = a_e + a_f > 2$ contradicting our previous assumption. In particular, there can be at most two different heavy edges. We will treat these three cases separately after introducing some more notation. We denote by E_{ij} the 2×2 matrix having all entries zero except a single entry of one on position (i, j). For an edge $e = \{(i,j), (i',j')\}$ let $E_e := E_{ij} + E_{i'j'}$. Finally, put $E_\backslash := E_{11} + E_{22}$ and $E_/ := E_{12} + E_{21}$.

Case 1: No heavy edges. If $s \leq 1$, let B be such that $P(B = E_{ij}) = a_{ij}$ for all $i, j \in [2]$, and $P(B = 0) = 1 - s$. Here as well as in the remaining cases, we leave it to the reader to check that (i) to (iii) of Definition 2 are fulfilled. Assume now that $s > 1$, but there are still no heavy edges. By symmetry, we may assume that a_{11} is a minimal entry of A. Put $p_\backslash = \min\{s - 1, a_{11}\}$ and $p_/ = s - 1 - p_\backslash$. Define B by

$$P(B = E_\backslash) = p_\backslash,$$
$$P(B = E_/) = p_/,$$
$$P(B = E_{ii}) = a_{ii} - p_\backslash, i \in [2],$$
$$P(B = E_{ij}) = a_{ij} - p_/, i, j \in [2], i \neq j.$$

Since a_{11} is minimal and there are no heavy edges, all probabilities defined above are non-negative.

Case 2: One heavy edge. Omitted.
Case 3: Two heavy edges. Omitted. □

Since block randomized roundings have low rounding errors and can be computed efficiently, the following algorithm suggests itself: Partition the input matrix A into 2×2 boxes and round each thereof as in the proof of Lemma 3. To be precise:

Definition 3. Let $A \in [0,1]^{m \times n}$. We say that B is a block randomized rounding of A, if it is computed by the following rounding scheme:

(i) for all $i \in [\frac{m}{2}], j \in [\frac{n}{2}]$, $R := \{2i-1, 2i\} \times \{2j-1, 2j\}$, $B_{|R}$ is a block randomized rounding of $A_{|R}$ as in the proof of Lemma 3,
(ii) if m is odd, then for all $j \in [\frac{n}{2}]$, $(b_{m,2j-1}, b_{m,2j})$ is a joint randomized rounding of $(a_{m,2j-1}, a_{m,2j})$ as in Definition 1,
(iii) if n is odd, then for all $i \in [\frac{m}{2}]$, $(b_{2i-1,n}, b_{2i,n})$ is a joint randomized rounding of $(a_{2i-1,n}, a_{2i,n})$ as in Definition 1,
(iv) if both m and n are odd, then b_{mn} is a randomized rounding of a_{mn}.
(v) All roundings in (i) to (iv) shall be independent.

We shall first analyze the error of block randomized roundings and then turn to the computational aspects of this approach.

Theorem 2. Let B be a block randomized rounding of $A \in [0,1]^{m \times n}$. Then

(i) The expected rounding error is $E(d(A,B)) \leq 0.6287|\mathcal{H}|$.
(ii) If B^* is an optimal rounding, then

$$E(d(A,B)) - d(A,B^*) \leq \tfrac{15}{32}|\mathcal{H}| \leq 0.4688|\mathcal{H}|.$$

(iii) $P(d(A,B) > E(d(A,B)) + \varepsilon|\mathcal{H}|) < 9\exp(-\tfrac{1}{72}\varepsilon^2(m-2)(n-2))$.

Proof (sketched). At least a quarter of all boxes (those R_{ij} where both i and j are odd) are block randomized roundings. Less than half of the boxes contain two independent joint randomized roundings, namely those R_{ij} where either i or j is odd. The remaining boxes, which are a fraction of at most a quarter, have all their entries rounded independently. Thus Theorem 1, Lemma 2 and the remark following Definition 2 yield the bounds (i) and (ii).

The proof of the large deviation bound is an application of the Chernoff inequality. Note that $\mathcal{H}$ can be partitioned into nine sets of boxes such that the errors within the boxes of each partition class are mutually independent random variables. □

We remark that the large deviation bound is not optimal, but by far sufficient for our purposes. The matrix A with all entries 0.3588 has $E(d(A,B)) \geq 0.6173|\mathcal{H}|$ for a block randomized rounding B. Thus our analysis is quite tight.

Algorithmic Properties of Block Randomized Rounding

Computing matrix roundings with the randomized approach above is fast. Each single block randomized rounding of a 2×2 block takes constant time, therefore the whole rounding can be done in linear time. From the view-point of application to digital halftoning, this is a crucial improvement over the algorithm of [2], that roughly has quadratic time complexity (ignoring a polylogarithmic factor). As stated in [2], this is too slow for high-resolution images.

The problem of computing time can be further addressed with parallel computing (and this is actually an issue when discussing digital halftoning algorithms). Since the block randomized roundings of 2×2 blocks are done independently, it is no problem to assign them to different processors.

Further Aspects of Block Randomized Roundings

Let us sketch another advantage of our approach. Suppose that we do not want to find a good rounding with respect to the 2×2 boxes, but with respect to larger structures, say 3×3 boxes. We currently have no hint whether the error with respect to these sets is a better measure for the visual quality of the resulting digital halftoning, but it seems plausible to try this experimentally. Hence we need an algorithm computing such roundings.

Whereas it seem difficult to extend the approach of Asano et al. to larger structures, non-independent rounding does very well: For 3×3 boxes, we may even use the same rounding scheme as before. Doing so, each 3×3 box contains exactly one block randomized rounding, two joint randomized roundings and one single randomized rounding. Since the four values of the block randomized rounding in total behave like a single randomized rounding, and so do the two values of each joint randomized rounding, the expected error of a 3×3 box is just given by Theorem 1. We have

Theorem 3. *With respect to the family $\mathcal{H}_3$ of 3×3 boxes, a block randomized rounding B of A has expected error $d_{\mathcal{H}_3}(A,B) \leq 0.82944|\mathcal{H}_3|$.*

We thus may claim that our approach has a broader range of application than previous results on this problem. This, by the way, works also in the other direction: From the view-point of digital halftoning, it is of course desirable that also the single entries are not rounded to badly, i.e., the error with respect to 1×1 boxes should not be too large. Our algorithm solves this in a very simple manner: If B is a block randomized rounding of A, then each single entry b_{ij} is a randomized rounding of a_{ij}. Hence the expected error of the single entries is at most $\frac{1}{2}mn$, which is, of course, optimal in the worst-case.

4 Experimental Results

We applied the three classical algorithms mentioned in the introduction together with independent randomized rounding and our algorithm to several images. For reasons of space we are not able do give any details concerning the three algorithms except for the remark that all three are deterministic. All image data used 1 byte per pixel resulting in an integer value between 0 and 255. We used two types of input data: Real-world images taken with a digital camera, and artificial images produced with a commercial imaging software. Naturally, the first type is more suitable to estimate how well the algorithm performs in real-world applications, whereas the second is better suited to demonstrate the particular strengths and weaknesses of an algorithm.

For reasons of space the images displayed in this paper are only small parts of the images we processed.[2] These parts have a size of 160×160 pixels, and are displayed in 72 dpi. All printers nowadays can handle higher resolutions, of

---
[2] The full size images can be found at http://www.numerik.uni-kiel.de/~hes/NIRR.htm.

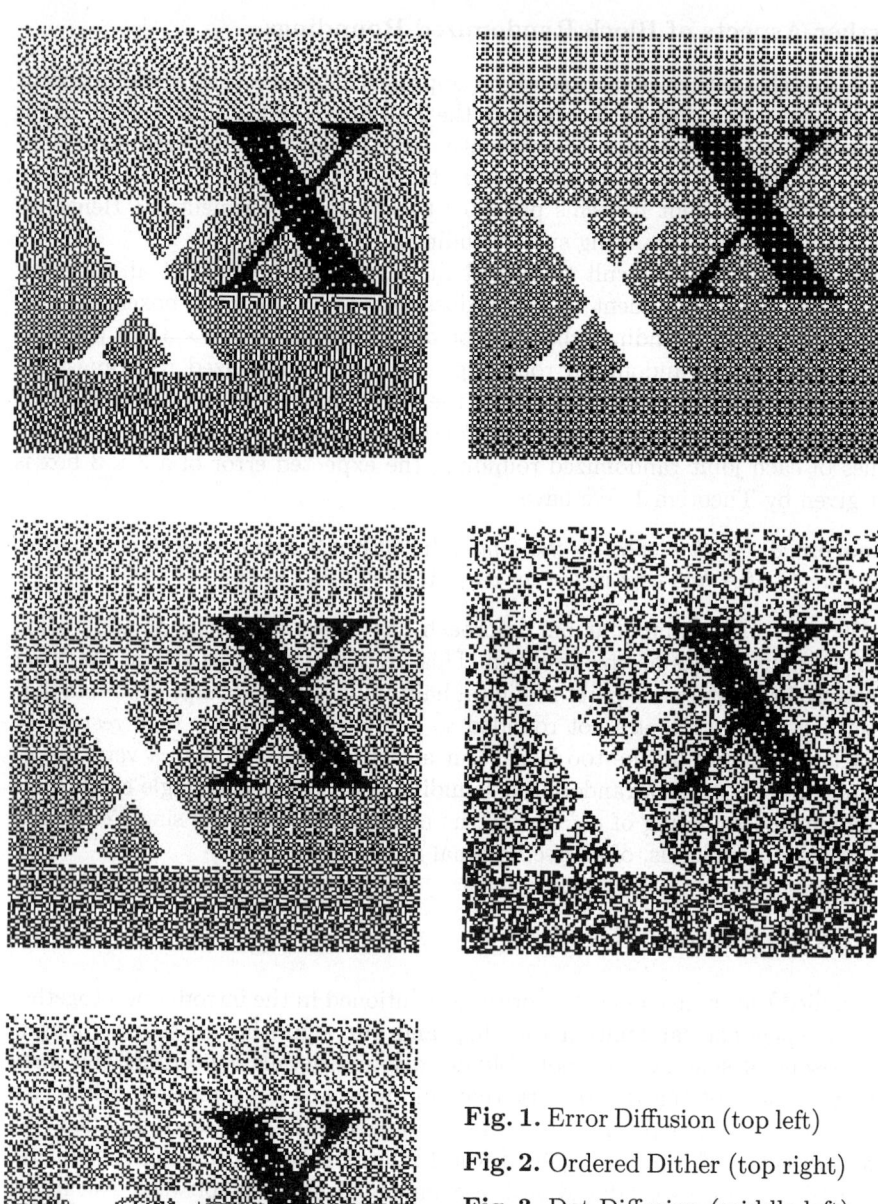

Fig. 1. Error Diffusion (top left)

Fig. 2. Ordered Dither (top right)

Fig. 3. Dot Diffusion (middle left)

Fig. 4. Randomized Rounding (middle right)

Fig. 5. Non-independent Randomized Rounding (bottom)

course, but the single pixels would be harder to recognize, and some unwanted effects like small white dots disappearing in a large black area would spoil the result.

All known algorithms for the digital halftoning problem tend to produce some kind of structures or textures, which draw unwanted attention. Generally two kinds of textures can be observed. First there are regular patterns like snakes, crosses or labyrinths. In particular error diffusion and ordered dither algorithm tend to produce those, as can be seen in Fig. 1 and 2.

The second form of unwanted structures are grains. Grains emerge, if in dark (respectively light) parts of the picture two or more white (respectively black) pixels touch each other and thus build a recognizable block. As observed already in the seventies, randomized rounding is very vulnerable to this problem, which is why it is not used in practice for digital halftoning. On the other end we find error diffusion, which hardly produces any grains. It seems that algorithms that are good concerning graininess tend to produce unwanted structures and vice versa. In this sense, non-independent randomized rounding seems to be a fair compromise: Being by far less grainy than independent randomized rounding on the one hand, it is unlikely to produce unwanted structures on the other.

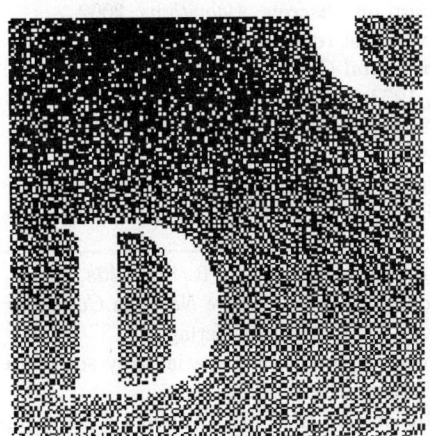 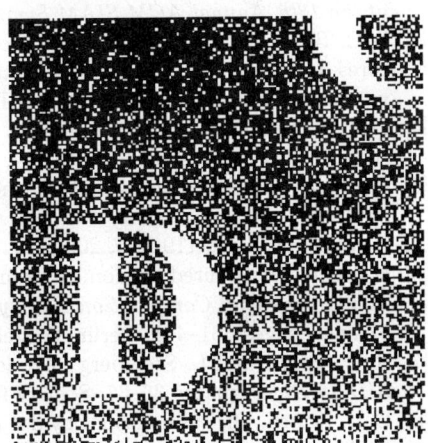

Fig. 6. Non-ind. Rand. Rounding   **Fig. 7.** Randomized Rounding

5 Summary and Outlook

This paper describes a new approach in randomized rounding. By imposing suitable dependencies, we improve the expected rounding error significantly. For a particular problem arising in digital halftoning this improves previous algorithms both according to run-time and rounding error. In particular, we presented the first algorithm solving the rounding problem proposed by Asano et al. fast enough for practical application, namely in linear time.

On the methodological side, this paper shows that non-independent randomized rounding can be very effective if one succeeds in finding the right dependencies. We believe that this is a fruitful approach, in particular for problems where smallish structures do not allow to use large deviation estimates.

Acknowledgments

We would like to thank Tetsuo Asano, Naoki Katoh, Koji Obokata and Takeshi Tokuyama for providing us with a preprint of their SODA 2002 paper and some useful discussion on the topic.

References

[1] N. Alon, B. Doerr, T. Łuczak, and T. Schoen. On the discrepancy of combinatorial rectangles. Tentatively accepted for publication in *Random Structures & Algorithms*, 2001.

[2] T. Asano, N. Katoh, K. Obokata, and T. Tokuyama. Matrix rounding under the L_p-discrepancy measure and its application to digital halftoning. In *Proceedings of the 13th Annual ACM-SIAM Symposium on Discrete Algorithms*, 2002.

[3] B. E. Bayer. An optimum method for two-level rendition of continous-tone pictures. In *Conference Record, IEEE International Conference on Communications*, volume 1, pages (26–11)–(26–15). IEEE, 1973.

[4] D. Bertsimas, C. Teo, and R. Vohra. On dependent randomized rounding algorithms. *Oper. Res. Lett.*, 24:105–114, 1999.

[5] B. Doerr. Lattice approximation and linear discrepancy of totally unimodular matrices. In *Proceedings of the 12th Annual ACM-SIAM Symposium on Discrete Algorithms*, pages 119–125, 2001.

[6] B. Doerr. Structured randomized rounding and coloring. In R. Freivalds, editor, *Fundamentals of Computation Theory*, volume 2138 of *Lecture Notes in Computer Science*, pages 461–471, Berlin–Heidelberg, 2001. Springer Verlag.

[7] R. W. Floyd and L. Steinberg. An adaptive algorithm for spatial grey scale. In *SID 75 Digest*, pages 36–37. Society for Information Display, 1975.

[8] D. E. Knuth. Digital halftones by dot diffusion. *ACM Trans. Graphics*, 6:245–273, 1987.

[9] D. E. Knuth. Two-way rounding. *SIAM J. Discrete Math.*, 8:281–290, 1995.

[10] B. Lippel and M. Kurland. The effect of dither on luminance quantization of pictures. In *IEEE Trans. Commun. Tech. COM-19*, pages 879–888, 1971.

[11] P. Raghavan. Probabilistic construction of deterministic algorithms: Approximating packing integer programs. *J. Comput. Syst. Sci.*, 37:130–143, 1988.

[12] P. Raghavan and C. D. Thompson. Randomized rounding: a technique for provably good algorithms and algorithmic proofs. *Combinatorica*, 7(4):365–374, 1987.

[13] A. Srinivasan. Improved approximation guarantees for packing and covering integer programs. *SIAM J. Comput.*, 29:648–670, 1999.

Computing Homotopic Shortest Paths Efficiently

Alon Efrat[1], Stephen G. Kobourov[1], and Anna Lubiw[2]

[1] Dept. of Computer Science, Univ. of Arizona
{alon,kobourov}@cs.arizona.edu
[2] School of Computer Science, Univ. of Waterloo
alubiw@uwaterloo.ca

Abstract. We give algorithms to find shortest paths homotopic to given disjoint paths that wind amongst n point obstacles in the plane. Our deterministic algorithm runs in time $O(k \log n + n\sqrt{n})$, and the randomized version in time $O(k \log n + n(\log n)^{1+\varepsilon})$, where k is the input plus output sizes of the paths.

1 Introduction

Geometric shortest paths are a major topic in computational geometry; see the survey paper by Mitchell [15]. A shortest path between two points in a simple polygon can be found in linear time using the "funnel" algorithm of Chazelle [3] and Lee and Preparata [13]. A more general problem is to find a shortest path between two points in a polygonal domain. In this case the "rubber band" solution is not unique, or, to put it another way, different paths may have different homotopy types. When the homotopy type of the solution is not specified, there are two main approaches, the visibility graph approach, and the continuous Dijkstra (or shortest path map) approach [15]. In this paper, we address the problem of finding a shortest path when the homotopy type is specified. Colloquially, we have a "sketch" of how the path should wind its way among the obstacles, and we want to pull the path tight to shorten it.

Homotopic shortest paths are used in VLSI routing [5, 11, 14]. A related problem is that of drawing graphs with "fat edges": given a planar weighted graph G, find a planar drawing such that all the edges are drawn as thickly as possible and proportional to the corresponding edge weights. Duncan et al. [7] and Efrat et al. [10] present an $O(kn + n^3)$ algorithm for this problem, where n is the number of edges and k is the maximum of their input and output complexities. Hershberger and Snoeyink [12] give an algorithm for the homotopic shortest path problem. Their algorithm assumes a triangulation of size n of the polygonal domain, and finds a shortest path homotopic to a given path of k edges in time linear in k plus the number of triangles (with repetition) visited by the input path. This can be nk in the worst case, and our aim is to reduce it when output size permits. Cabello et al. [2] consider the related problem of testing if two given paths are homotopically equivalent. Our original work [9] was done independently of theirs and the idea of the first step of the algorithms is the same. The current presentation of our work incorporates their more efficient implementation of this idea.

We now define homotopy, and give a precise description of our problem. Let $\alpha, \beta : [0,1] \longrightarrow \mathbb{R}^2$ be two continuous curves parameterized by arc-length. Then α and β are *homotopic* with respect to a set of obstacles $V \subseteq \mathbb{R}^2$ if α can be continuously deformed into β while avoiding the obstacles; more formally, if there exists a continuous function $h : [0,1] \times [0,1] \to \mathbb{R}^2$ such that:

1. $h(0,t) = \alpha(t)$ and $h(1,t) = \beta(t)$, for $0 \le t \le 1$
2. $h(\lambda, 0) = \alpha(0) = \beta(0)$ and $h(\lambda, 1) = \alpha(1) = \beta(1)$ for $0 \le \lambda \le 1$
3. $h(\lambda, t) \notin V$ for $0 \le \lambda \le 1, 0 < t < 1$

Let $\Pi = \{\pi_1, \pi_2, \ldots, \pi_n\}$ be a set of disjoint, simple polygonal paths and let the endpoints of the paths in Π define the set T of at most $2n$ fixed points in the plane. Note that we allow a path to degenerate to a single point. We call the fixed points of T "terminals," and call the interior vertices of the paths "bends," and use "points" in a more generic sense, e.g. "a point on a path." We assume that no two terminals/bends lie on the same vertical line. Our goal is to replace each path $\pi_i \in \Pi$ by a shortest path σ_i that is homotopic to π_i with respect to the set of obstacles T; see Fig. 1. Note that σ_i is unique. Let $\Sigma = \{\sigma_1, \ldots \sigma_n\}$ be the set of resulting paths. Observe that these output paths may [self] intersect by way of segments lying on top of each other, but will be *non-crossing*.

In order to distinguish how paths go past terminals, we find it convenient to regard each terminal as a small diamond. This actually changes the model of homotopy, because the endpoint of the path is now fixed at one corner of the diamond, and cannot swivel around the terminal. In other words, a path that begins at terminal t and spirals around t before heading off is, with diamond terminals, no longer homotopic to the path that simply heads off. This is the only difference, however, and we can easily remove such spiraling from the final output. Cabello et al. [2] call the above two models the "pin" and "pushpin" models, respectively. They also consider the "tack" model, where the endpoints of the paths are not regarded as obstacles.

Let k_{in} be the number of edges in all the paths of Π. Let k_{out} be the number of edges in all the paths of Σ. Note that k_{in} and k_{out} can be arbitrarily large compared to n, and that $k_{out} \le nk_{in}$. The algorithm of Hershberger and Snoeyink [12] finds homotopic shortest paths in time $O(nk_{in})$. The deterministic algorithm presented in this paper runs in time $O(k_{out} + k_{in} \log n + n\sqrt{n})$, and the randomized algorithm in time $O(k_{out} + k_{in} \log n + n(\log n)^{1+\varepsilon})$. These are improvements except when k_{in} is quite small compared to n.

Our algorithm relies on the algorithm of Bar-Yehuda and Chazelle [1], which uses linear time polygon triangulation and ideas of linear time Jordan sorting, and finds a trapezoidization of n disjoint polygonal chains with a total of k edges in time $O(k + n(\log n)^{1+\varepsilon})$. Replacing this by plane sweep makes our algorithm implementable and yields a running time of $O(k_{out} + (n+k_{in}) \log(n+k_{in}))$ with an extra $O(n\sqrt{n})$ factor for the deterministic version. The algorithm of Cabello et al. [2] tests whether two paths are homotopically equivalent under the pin, pushpin, or tack models in $O((n + k_{in})\log(n + k_{in}))$ time as they opt for the practical plane sweep approach.

In the remainder of this section we give an outline of our algorithm. Although k_{in} can be arbitrarily large compared to n, one easily forms the intuition that, because the paths are simple and disjoint, k_{in} can be large in a non-trivial way only because path sections are repeated. For example, a path may spiral arbitrarily many times around a set of points, but each wrap around the set is the same. Our method makes essential use of this observation. It seems difficult to search explicitly for repeated path sections. Instead we begin in section 2 by applying vertical shortcuts to the paths (homotopically) so that each left and each right local extreme point occurs at a terminal. These terminals must then be part of the final shortest paths, and we have decomposed the paths into "essential" x-monotone pieces with endpoints at terminals. In section 3 we argue that the number of homotopically distinct pieces is at most $O(n)$. We bundle together all the homotopically equivalent pieces. Routing one representative from each such bundle using the straightforward "funnel" technique takes $O(k_{in} + n^2)$ time total. In section 3 we reduce this using a "shielding technique" where we again exploit the fact that the paths are disjoint and use the knowledge gained in routing one shortest path to avoid repeating work when we route subsequent paths. The final step of the algorithm is to unbundle, and recover the final paths by putting together the appropriate pieces. This is straightforward, and we say no more about it. We summarize the main steps of the algorithm below. The remainder of this paper consists of one section for each of steps 1, 2, and 3 (step i in Section $i + 1$).

Summary of the Main Algorithm
1. shortcut paths to divide into essential monotone pieces
2. bundle homotopically equivalent pieces
3. find the shortest path for each bundle
4. unbundle to recover final paths

---

2 Shortcutting to Find Essential Monotone Pieces

We begin by applying vertical shortcuts to reduce each path to a sequence of essential x-monotone path sections; see Fig. 1. A *vertical shortcut* is a vertical line segment ab joining a point a on some path π with a point b also on π, and with the property that the subpath of π joining a and b, π_{ab}, is homotopic to the line segment ab. We only do *elementary vertical shortcuts* where the subpath π_{ab} consists of [portions of] 2 line segments or 3 line segments with the middle one vertical, and the other two non-vertical. We distinguish *left [right] shortcuts* which are elementary vertical shortcuts where π_{ab} contains a point to the left [right] of the line through ab, and *collinear shortcuts* where π_{ab} lies in the line through ab; see Fig. 2. Collinear shortcuts are applied after left/right shortcuts to eliminate consecutive vertical segments. We in fact only apply *maximal* elementary vertical shortcuts, where the subpath π_{ab} cannot be increased.

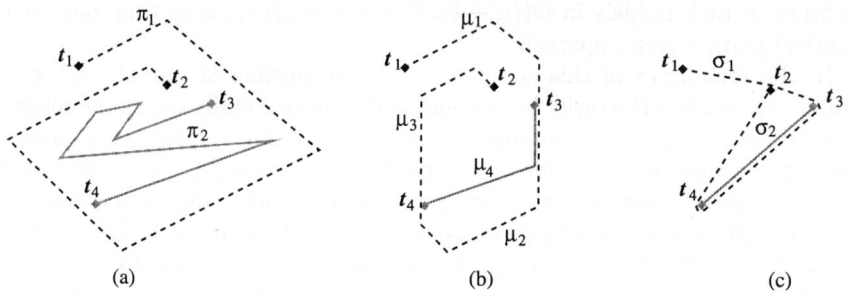

Fig. 1. (a) Paths π_1 and π_2 joining terminals t_1 to t_2 and t_3 to t_4, respectively. (b) After vertical shortcuts, π_1 consists of 3 monotone pieces: μ_1 from t_1 to t_3, μ_2 from t_3 to t_4, and μ_3 from t_4 to t_2; π_2 consists of one x-monotone piece, μ_4, homotopically equivalent to μ_2; (c) Final homotopic shortest paths, σ_1 and σ_2

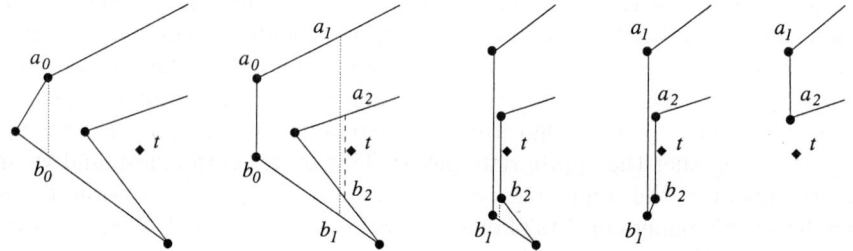

Fig. 2. A sequence of maximal elementary vertical shortcuts (bends are represented by circles and terminals by diamonds). Shortcuts a_0b_0, a_1b_1 and a_2b_2 are left shortcuts; b_2b_1 is a right shortcut; and a_2b_1 and a_1a_2 are collinear shortcuts

This means that for left and right shortcuts, either a or b is a bend, or the line segment ab hits a terminal.

A *local left [right] extreme* of a path is a point or, more generally, a vertical segment, where the x-coordinate of the path reaches a local min [max]. Every left [right] local extreme provides a left [right] shortcut *unless* the local extreme *is locked at* a terminal, meaning that the left [right] extreme contains the left [right] point of the terminal's diamond. The first two pictures in Fig. 2 show left extremes that provide left shortcuts; the middle picture shows left extremes locked at a terminal.

To detect the maximal left shortcut for a local left extreme, l, we use range searching. Let the segments preceding and following l along the path be l_p and l_f. Let their right endpoints be $r(l_p)$ and $r(l_f)$, and let t be the leftmost of these two points. Then t determines the maximal potential shortcut that can be performed at l in the absence of terminals. We must query the triangle/trapezoid bounded by l, l_p, l_f and the vertical line through t to find the leftmost terminal inside it. This determines the actual maximal left shortcut that can be performed at l. As described in section 2.2, we use an idea from Cabello et al. [2] to reduce the range queries to orthogonal range queries, which can be answered more efficiently.

We perform elementary vertical shortcuts until none remain, at which time all local left and right extremes are locked at terminals. We claim that the final shortest paths must be locked in the same way at the same terminals, so we have divided the paths into their *essential monotone pieces*.

Lemma 1 *Let π be a path, and let μ be a result of performing left and right phases of elementary vertical shortcuts on π until no more are possible. Suppose that the local left and right extremes of μ are locked at the terminals $t_{i_1}, \ldots, t_{i_l}$ in that order. Let σ be a shortest path homotopic to π. Then the local left [right] extremes of σ are locked at exactly the same ordered list of terminals, and furthermore, the portion of σ between t_{i_j} and $t_{i_{j+1}}$ is a shortest path homotopic to the portion of μ between those same terminals.*

Proof Sketch. Because π and μ are homotopic, σ is the shortest path homotopic to μ. We can thus go from μ to σ using "rubber band" deformations that only shorten the path, and such deformations cannot loosen a left [right] extreme from the terminal it is locked at. □

Doing elementary vertical shortcuts in an arbitrary order may result in crossing paths; see Fig. 3. To guarantee non-crossing paths we do the elementary vertical shortcuts in alternating *left* and *right phases*. Throughout the algorithm we maintain the sets L and R of local left [right] extremes not locked at terminals. A left phase continues until L is empty. At each step we remove a local left extreme from L, perform the maximal left shortcut there, perform the consequent collinear shortcuts, and update the sets L and R. The new shortcut may destroy one or two elements of R, or create one new element of R or L. These changes occur only at the newly added shortcut and are easy to detect. It is crucial for the correctness of this approach that a left shortcut does not destroy any local left extremes, nor change the shortcuts that will be performed at them. Observe that more than one left and one right phase may be required, since the right phase may add new members to L, for example, in Fig. 2 at least left, right and left phases are needed. Two things remain: to prove that the final paths do not cross, and to analyze the algorithm. We devote a subsection to each.

2.1 Non-crossing Paths

Claim 2 *Each left [right] phase of shortcuts preserves the property that paths are non-crossing.*

It is crucial for steps 2 and 3 of our algorithm that the paths remain non-crossing. By symmetry, we can concentrate on a left phase. Note that a left phase, as described, is non-deterministic. At each stage we choose one member l from the set L of current local left extremes, and use it to perform a left shortcut. To prove the Lemma we first show that for each left phase there is *some* sequence of choices that preserves the property that the paths are non-crossing. We then argue that the end result of a phase does not depend on the choices made.

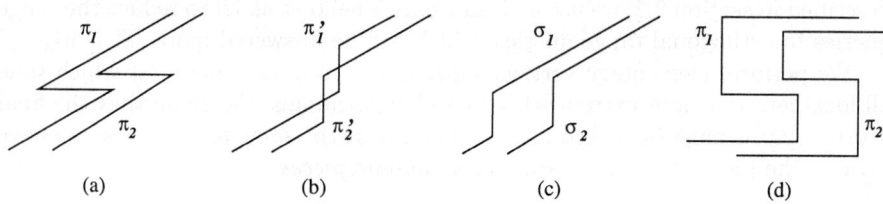

Fig. 3. (a) Disjoint paths π_1 and π_2; (b) Crossing after a left shortcut in π_1 and a right shortcut in π_2. No further shortcuts are possible, hence the crossing cannot be removed; (c) Disjoint paths output by step 1 after two right shortcuts; (d) Rectified version of (a)

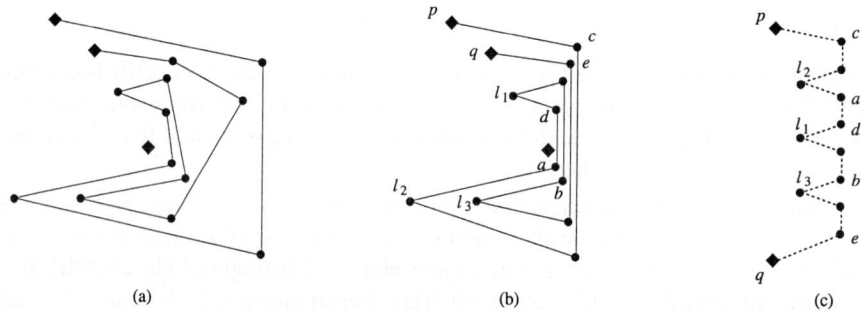

Fig. 4. (a) A path undergoes a right phase; (b) Handling local left extreme l_1 first yields the shortcut ab. Handling the local left extreme l_2 first yields shortcut cd. The final result of handling l_1, l_2 and l_3 in any order is shortcut ce; (c) An "unfolded" version of part (b)

Claim 3 *Performing left shortcuts in rightmost order of the local left extremes leaves the paths non-crossing.*

Observe that we could complete a left phase of shortcuts using this "rightmost" order, at a cost of an extra $\log k_{in}$ factor for maintaining a heap. There is no practical advantage to doing this.

Lemma 4 *The end result of a left phase does not depend on the sequence of choices made during the phase.*

Proof. We begin with the claim that the set $\mathcal{L}$ of all local left extremes that appear in L over the course of the phase is independent of the choices made during the phase. This implies that the set of left shortcuts performed during the course of the phase is also independent of the choices made during the phase. However, the set of collinear shortcuts is *not* independent. In particular, the order in which we perform left shortcuts affects the set of collinear shortcuts; see Fig. 4.

Consider one left phase. Let L_0 be the initial set of local left extremes not locked at terminals. Let $\mathcal{L}$ be the union of L over the course of the phase. Suppose two sequences of choices C_1 and C_2 during a left phase yield sets $\mathcal{L}_1$ and $\mathcal{L}_2$. We would like to show that $\mathcal{L}_1 = \mathcal{L}_2$. Consider $l \in \mathcal{L}_1$. We will prove $l \in \mathcal{L}_2$ by induction on the number of left shortcuts performed in the phase before l enters $\mathcal{L}_1$. If this number is 0 then $l \in L_0$ and we are done. Otherwise, l is added to L as a result of some left shortcut and consequent collinear shortcuts. Any vertical segment used in the collinear shortcuts may, in its turn, have arisen as a result of some left shortcut and consequent collinear shortcuts. Tracing this process, we find that l is formed from a set of left shortcuts linked by vertical segments, all in the same vertical line as l; see Fig. 4. All these left shortcuts arose from local left extremes that entered $\mathcal{L}_1$ before l did, and thus, by induction, are in $\mathcal{L}_2$. They only leave $\mathcal{L}_2$ when we perform their shortcuts, and thus the choice sequence C_2 includes these shortcuts—though possibly in a different order than in C_1. The consequent collinear shortcuts will merge all the verticals forming l, and cannot merge more than that because the segments attached before and after l are not vertical (l is a local left extreme). Thus l is in $\mathcal{L}_2$.

This proves that the set of local left extremes, and thus the set of left shortcuts is independent of choices made during the phase. Any vertical segment that is in the final set of paths output by the phase arises through left shortcuts plus consequent collinear shortcuts. Since any set of choices leads to the same set of left shortcuts, (possibly in different order), the consequent collinear shortcuts will arrive at the same final vertical shortcuts—i.e. the same final paths. □

2.2 Implementation and Run-Time Analysis

We use range queries to identify the maximal shortcut that can be performed at a local extreme of the path. Our original version of this paper [9] used general simplex range searching, which is time-consuming. Cabello et al. [2], in their paper on testing homotopy, independently developed this same idea of shortcutting to divide a path into essential monotone pieces. However, they were able identify shortcuts using *orthogonal* range queries, which can be done much more efficiently. The main insight is that shortcuts depend only on the *aboveness* relation among terminals and x-monotone path sections, and that the aboveness relation forms a partial order. This partial order can be preserved while each monotone path section is modeled as a horizontal line segment, and the whole picture becomes "rectified"; see Fig. 3(d). Shortcuts can now be identified using orthogonal range queries.

We find the aboveness relation from a trapezoidization of the input, which can be computed in time $O(k_{itin} + n(\log n)^{1+\varepsilon})$ using the algorithm of Bar-Yehuda and Chazelle [1]. The partial order can be extended to a total order in linear time. We now assign to each terminal and x-monotone path section, a new y-coordinate equal to its rank in the total order. As justified in [2], shortcuts after this transformation correspond to shortcuts in the original. Each shortcut now requires an orthogonal range query: given an axis-aligned query rectangle, certify that it is empty or else return the rightmost [leftmost] terminal inside it.

Chazelle's data structure for segment dragging [4] solves this with $O(n \log n)$ preprocessing time, $O(n)$ space, and $O(\log n)$ query time. Simpler algorithms require an extra logarithmic factor in space [8, 16]. We need one more ingredient to bound the running time of step 1 of our algorithm by $O(k_{in} \log n + n(\log n)^{1+\varepsilon})$.

Lemma 5 *The number of elementary vertical shortcuts that can be applied to a set of paths with a total of k_{in} segments is at most $2k_{in}$.*

Proof. Assume that no two terminals and/or bends line up vertically. Consider the set of vertical lines through bends and through the left and right sides of each terminal's diamond. An elementary shortcut operates between two of these vertical lines. If a left [right] shortcut has its leftmost [rightmost] vertical at a bend, then after the shortcut, this vertical disappears forever; see Fig. 2. There are thus at most k_{in} such elementary shortcuts. Consider, on the other hand, a left shortcut that has its leftmost vertical at a terminal. This only occurs when two previous left shortcuts are stopped at the terminal, and then combined in a collinear shortcut. See the right hand pictures of Fig. 2. Thus an original edge of the path has disappeared. Note that elementary shortcuts never fragment an edge of the path into two edges, but only shorten it from one end or the other. Thus there are at most k_{in} elementary shortcuts of this type. Altogether, we obtain a bound of $2k_{in}$ elementary shortcuts. □

3 Bundling Homotopically Identical Paths

Let $M = \{\mu_1 \ldots \mu_m\}$ be the set of essential x-monotone paths obtained from step 1 of our algorithm. In the second step of the algorithm we bundle homotopically equivalent paths in M. More precisely, we take one representative path for each equivalence class of homotopically equivalent paths in M. This is justified because the paths in each equivalence class have the same homotopic shortest path. Because the paths in M are non-crossing and x-monotone, it is easier to detect homotopic equivalence: two paths are homotopically equivalent if they have the same endpoints, and, between these endpoints no terminal lies vertically above one path and vertically below the other.

We perform the bundling using a trapezoidization of M; see Fig. 5. We apply the trapezoidization algorithm of Bar-Yehuda and Chazelle [1] to the paths obtained after the shortcuts of step 1, but before these paths are chopped into monotone pieces. We claim that we can perturb the paths output from step 1 so that they become disjoint and the Bar-Yehuda and Chazelle algorithm can be applied. The trapezoidization algorithm takes time $O(k_{in} + n(\log n)^{1+\varepsilon})$. Once we have a trapezoidization of M, we can in linear time bundle homotopically equivalent paths as follows. While scanning each $\mu \in M$, we check if it is homotopically equivalent to the path "below" it, by examining all the edges of the trapezoidization that are incident to μ from below. If all these trapezoidization edges reach the same path μ_j and none pass through a terminal on the way to μ_j, and μ_i and μ_j have the same terminals as endpoints, then we mark μ_i

as a duplicate. Let $R = \{\rho_1 \ldots \rho_r\}$ be the paths of M that are not marked as duplicates.

Lemma 6 *The number of paths in R is bounded by $2n$.*

Proof. Note that every terminal t is either a right endpoint of paths in R or a left endpoint of paths in R, but not both. This is because a terminal cannot have local left extremes and local right extremes locked at it without the paths crossing.

For each terminal $t \in T$, associate its bottommost incident path $\phi(t)$ (its "floor"), and the first path hit by a vertical ray going up from t, $\gamma(t)$ (its "roof"). We claim that every path in R is $\phi(t)$ or $\gamma(t)$ for some $t \in T$. This proves that the number of paths in R is at most $2n$.

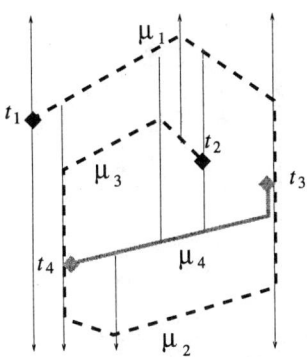

Fig. 5. A trapezoidization helps verify that μ_2 and μ_4 are equivalent.

Consider a path $\rho \in R$ with left and right terminals s and t, respectively. Suppose that ρ is not a roof. Then every vertical ray extending downward from a point of ρ must hit the same path σ. (If two rays hit different paths, then in the middle some ray must hit a terminal.) Furthermore, since ρ is not the bottommost path incident to s, σ must hit s. Similarly σ must hit t. But then σ and ρ are homotopically equivalent. □

4 Shortest Paths

In this section we find shortest paths homotopic to the monotone paths $R = \{\rho_1, \ldots, \rho_r\}$ produced in the previous section. Note that r is $O(n)$. Let ρ'_i denote the shortest path homotopic to ρ_i. We route each path using a funnel technique. The funnel algorithm of [3] and [13] operates on a triangulation of the n points (the terminals in our case), and follows the path through the triangulation maintaining a current "funnel" containing all possible shortest paths thus far. The algorithm takes time proportional to the number of edges in the path plus the number of intersections between the triangulation edges and the path.

Rather than a triangulation, we will use a trapezoidization formed by passing a vertical line through each of the n terminals; see Fig. 6(a). Then, since each path ρ_i is x-monotone, it has $O(n)$ intersections with trapezoid edges, and the funnel algorithm takes time proportional to n plus the number of edges in ρ_i. Since the total number of edges in R is bounded by k_{in}, this gives a total over all paths of $O(n^2 + k_{in})$. (Note that the number of edges in the output is $O(n^2)$, so it doesn't appear explicitly here.)

We improve this by making use of the fact that the paths are disjoint, and re-using information gained when each path is routed. We give a randomized

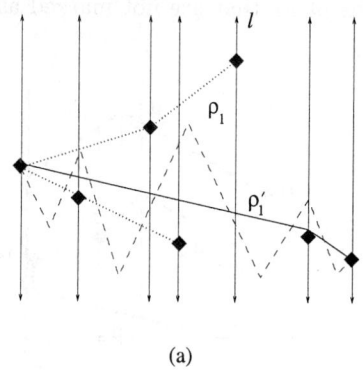 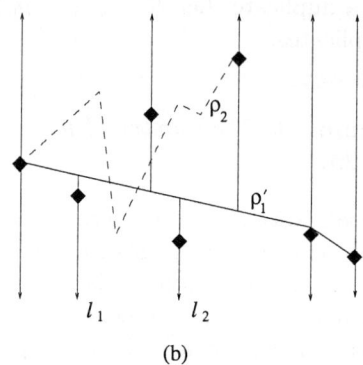

Fig. 6. (a) A path ρ_1 (dashed), its funnel up to line l (dotted) and the final homotopic shortest path ρ'_1; (b) After the shortest path ρ'_1 has been found, shielding allows us to route another path ρ_2 without examining vertical lines l_1 and l_2

algorithm to route the paths of R in time $O(n \log n + k)$, and a deterministic algorithm with $O(n\sqrt{n} + k)$ running time. Recall that $k = k_{in} + k_{out}$.

Both methods use a "shielding" technique. We begin by describing this idea intuitively. First note that the ρ_i's can be routed independently, since none affects the others. The initial paths ρ_i are non-crossing, and so are the final shortest paths, ρ'_i. If ρ_j is below ρ_i, and we have already computed ρ'_j, then ρ'_j behaves as a barrier that "shields" ρ_i from terminals that are vertically below ρ'_j; see Fig. 6(b).

To utilize shielding we will modify the basic trapezoidization described above as we discover shortest paths ρ'_i. In particular, the upward vertical ray through terminal t, $u(t)$, will be truncated at the lowest shortest path ρ'_i that is strictly above t and does not bend at t. The downward vertical ray through terminal t, $d(t)$, will be truncated in an analogous way. To route a new path ρ_i through this modified trapezoidization we use the following algorithm.

Routing ρ_i with Shielding

1. Identify the first trapezoid that ρ'_i will traverse. This can be done in $O(\log n)$ time because the shortest paths observe the same vertical ordering as the original ρ_j's.
2. Traverse from left to right the sequence of trapezoids that ρ'_i will pass through. (Note that ρ_i itself may pass through different trapezoids; see Fig. 6(b).) We construct the funnel for ρ'_i as we do this traversal. Suppose that our path enters trapezoid τ. To leave τ on the right we have two cases. If the right side of τ is a point, then it is a terminal t, and we are locked between two paths that terminate or bend at t. Then the funnel collapses to this point, and we proceed to the next trapezoid if ρ_i continues. Otherwise the right side of τ is a vertical through some terminal t, and (unless ρ_i ends

at t) we have a choice of two trapezoids to enter, the upper one with left side $u(t)$ or the lower one with left side $d(t)$. We follow path ρ_i until it crosses the infinite vertical line through t. If it passes above t then we enter the upper trapezoid, and otherwise we enter the lower trapezoid. We update the funnel to include t.
3. When we reach the right endpoint of ρ_i, the funnel gives us the shortest homotopic path ρ_i'.
4. We update the trapezoidization as follows. For each vertical segment $u(t)$ or $d(t)$ that is intersected by ρ_i' we chop the segment at its intersection point with ρ_i', provided that the intersection point is not t (i.e. that ρ_i' does not bend at or terminate at t).

Without yet discussing the order in which we route the paths, we can say a bit about the timing of this shielding method. As we traverse a path ρ_i we spend time proportional to the size of ρ_i plus the number of trapezoids traversed by ρ_i'. When ρ_i' leaves a trapezoid, say at the line segment $u(t)$ above terminal t, it may happen that ρ_i' will bend at t or terminate at t. In this case t is part of the output path ρ_i', and we can charge the work for this trapezoid to the output size k_{out}. If, on the other hand, ρ_i' does not bend or terminate at t, then it crosses $u(t)$ and we chop $u(t)$ there. In this case we charge the work for this trapezoid to the chop. Thus the total time spent by the algorithm is $O(k+C)$ where C is the total number of chops performed at the n verticals.

For shielding to be effective we need to route paths in an order that makes C grow more slowly than n^2. We first analyze the randomized algorithm where we choose the next path to route at random with uniform probability from the remaining paths.

Lemma 7 *If paths are routed in random order then the expected number of times we chop a segment $u(t)$ or $d(t)$ is $O(\log r)$, and thus C is $O(n \log n)$, and the routing takes time $O(k + n \log n)$.*

Proof. This follows from a standard backward analysis. See for example [6] for other similar proofs. Let l be a vertical line through a point t, and let u denote a ray emerging vertically from t. Assume that m paths intersecting u have been inserted up to now, and we are about to insert a new one. Then u is chopped if and only if the new path creates an intersection point below all existing intersection points on u, but above t itself. Since the order of the insertions is random, the probability of the new intersection point being below all other intersection points is $1/m$. Summing over all paths yields the claimed bound. □

Finally, we describe how to achieve a routing time of $O(k + n\sqrt{n})$ deterministically. Suppose that the paths $\{\rho_1, \ldots, \rho_r\}$ are in order from top to bottom. (We get this for free from step 2 of our algorithm.) Partition the paths into blocks $B_1, \ldots, B_{\sqrt{r}}$ each of size $\sqrt{r}$, and route them one block at a time from $B_{\sqrt{r}}$ to B_1. Within each block we process the paths in order. The largest increasing [decreasing] sequence with this ordering has size $\sqrt{r}$, therefore the number of chops at each vertical is $\sqrt{r}$. Thus C is $O(n\sqrt{n})$, and this routing method takes time $O(k + n\sqrt{n})$.

5 Conclusion and Open Problems

Given n terminal points in the plane, and a set of disjoint paths joining pairs of the terminals, we can find shortest paths homotopic to the originals in time $O(k \log n + n\sqrt{n})$, or with randomization, in time $O(k \log n + n(\log n)^{1+\varepsilon})$, where k is the sum of input and output sizes of the paths. Unless k grows quite slowly compared to n, this is better than the Hershberger-Snoeyink algorithm which runs in time $O(nk)$.

One open problem is to improve the running time of our deterministic algorithm. The bottleneck is the $O(n\sqrt{n})$ arising from the shielding technique in Section 4. There is no reason to think that $O(n\sqrt{n})$ is the "right" answer—it is just the best one can do by ordering the paths while remaining otherwise oblivious to them.

Another open problem is to extend our efficient shortest homotopic path algorithm to the "tack" model, as described in [2]. Under this model the endpoints of paths need not be obstacles; they act as tacks that are pushed all the way into the backing, allowing the rubber bands of the paths to pass freely over them. Cabello et al. [2] successfully use the shortcut technique to find the essential x-monotone pieces so they can test equivalence of two paths under the tack model. However, the problem of finding shortest paths seems more difficult, because the final shortest paths cross in general, and the later steps of our algorithm fail.

References

[1] R. Bar-Yehuda and B. Chazelle. Triangulating disjoint Jordan chains. *International Journal of Computational Geometry & Applications*, 4(4):475–481, 1994.

[2] S. Cabello, Y. Liu, A. Mantler, and J. Snoeyink. Testing homotopy for paths in the plane. In *18th Annual Symposium on Computational Geometry*, pages 160–169, 2002.

[3] B. Chazelle. A theorem on polygon cutting with applications. In *23rd Annual Symposium on Foundations of Computer Science*, pages 339–349, 1982.

[4] B. Chazelle. An algorithm for segment-dragging and its implementation. *Algorithmica*, 3:205–221, 1988.

[5] R. Cole and A. Siegel. River routing every which way, but loose. In *25th Annual Symposium on Foundations of Computer Science*, pages 65–73, 1984.

[6] M. de Berg, M. van Kreveld, M. H. Overmars, and O. Schwarzkopf. *Computational Geometry: Algorithms and Applications*. Springer-Verlag, 2nd edition, 2000.

[7] C. A. Duncan, A. Efrat, S. G. Kobourov, and C. Wenk. Drawing with fat edges. In *9th Symposium on Graph Drawing (GD'01)*, pages 162–177, September 2001.

[8] H. Edelsbrunner. A note on dynamic range searching. *Bulletin of the European Association for Theoretical Computer Science*, 15:34–40, Oct. 1981.

[9] A. Efrat, S. Kobourov, and A. Lubiw. Computing homotopic shortest paths efficiently. Technical report, April 2002. http://www.arXiv.org/abs/cs/0204050.

[10] A. Efrat, S. G. Kobourov, M. Stepp, and C. Wenk. Growing fat graphs. In *18th Annual Symposium on Computational Geometry*, pages 277–278, 2002.

[11] S. Gao, M. Jerrum, M. Kaufmann, K. Mehlhorn, W. Rülling, and C. Storb. On continuous homotopic one layer routing. In *4th Annual Symposium on Computational Geometry*, pages 392–402, 1988.

[12] Hershberger and Snoeyink. Computing minimum length paths of a given homotopy class. *CGTA: Computational Geometry: Theory and Applications*, 4:63–97, 1994.

[13] D. T. Lee and F. P. Preparata. Euclidean shortest paths in the presence of rectilinear barriers. *Networks*, 14(3):393–410, 1984.

[14] C. E. Leiserson and F. M. Maley. Algorithms for routing and testing routability of planar VLSI layouts. In *17th Annual ACM Symposium on Theory of Computing*, pages 69–78, 1985.

[15] J. S. B. Mitchell. Geometric shortest paths and network optimization. In J.-R. Sack and J. Urrutia, editors, *Handbook of Computational Geometry*. 1998.

[16] H. Samet. *The Design and Analysis of Spatial Data Structures*. Addison-Wesley, Reading, MA, 1989.

An Algorithm for Dualization in Products of Lattices and Its Applications*

Khaled M. Elbassioni

Department of Computer Science, Rutgers University
110 Frelinghuysen Road, Piscataway NJ 08854-8003
elbassio@paul.rutgers.edu

Abstract. Let $\mathcal{L} = \mathcal{L}_1 \times \cdots \times \mathcal{L}_n$ be the product of n lattices, each of which has a bounded width. Given a subset $\mathcal{A} \subseteq \mathcal{L}$, we show that the problem of extending a given partial list of maximal independent elements of $\mathcal{A}$ in $\mathcal{L}$ can be solved in quasi-polynomial time. This result implies, in particular, that the problem of generating all minimal infrequent elements for a database with semi-lattice attributes, and the problem of generating all maximal boxes that contain at most a specified number of points from a given n-dimensional point set, can both be solved in incremental quasi-polynomial time.

1 Introduction

Let $\mathcal{L} \stackrel{\text{def}}{=} \mathcal{L}_1 \times \cdots \times \mathcal{L}_n$ be the product of n partially ordered sets (posets). We assume that each poset $\mathcal{L}_i$ is a lattice, i.e., for every pair of elements $x, y \in \mathcal{L}_i$, there exists a unique minimum element called the *meet* and a unique maximum element called the *join*. Throughout we shall denote by $\preceq$ the precedence relation in $\mathcal{L}$, and use, as customary, $\vee$ and $\wedge$ to denote the join and meet operators over $\mathcal{L}$. For $\mathcal{A} \subseteq \mathcal{L}$, denote by $\mathcal{A}^+ = \{x \in \mathcal{L} \mid x \succeq a, \text{ for some } a \in \mathcal{A}\}$ and $\mathcal{A}^- = \{x \in \mathcal{L} \mid x \preceq a, \text{ for some } a \in \mathcal{A}\}$, the ideal and filter generated by $\mathcal{A}$. Any element in $\mathcal{L} \setminus \mathcal{A}^+$ is called *independent of* $\mathcal{A}$. Let $\mathcal{I}(\mathcal{A})$ be the set of all maximal independent elements for $\mathcal{A}$ (also referred to as the *dual* of $\mathcal{A}$):

$$\mathcal{I}(\mathcal{A}) \stackrel{\text{def}}{=} \{p \in \mathcal{L} \mid p \notin \mathcal{A}^+ \text{ and } (q \in \mathcal{L}, q \succeq p, q \neq p \Rightarrow q \in \mathcal{A}^+)\}.$$

Then for any $\mathcal{A} \subseteq \mathcal{L}$, we have the following decomposition of $\mathcal{L}$:

$$\mathcal{A}^+ \cap \mathcal{I}(\mathcal{A})^- = \emptyset, \quad \mathcal{A}^+ \cup \mathcal{I}(\mathcal{A})^- = \mathcal{L}. \tag{1}$$

Call a subset $\mathcal{A} \subseteq \mathcal{L}$ an antichain if no two elements of $\mathcal{A}$ are comparable. Given $\mathcal{A} \subseteq \mathcal{L}$, we consider the problem of incrementally generating all elements of $\mathcal{I}(\mathcal{A})$:

---

* This research was supported in part by the National Science Foundation Grant IIS-0118635, and by DIMACS, the National Science Foundation's Center for Discrete Mathematics and Theoretical Computer Science

$DUAL(\mathcal{L}, \mathcal{A}, \mathcal{B})$: Given an antichain $\mathcal{A} \subseteq \mathcal{L}$ in a lattice $\mathcal{L}$ and a partial list of maximal independent elements $\mathcal{B} \subseteq \mathcal{I}(\mathcal{A})$, either find a new maximal independent element $x \in \mathcal{I}(\mathcal{A}) \setminus \mathcal{B}$, or prove that $\mathcal{A}$ and $\mathcal{B}$ form a dual pair: $\mathcal{B} = \mathcal{I}(\mathcal{A})$.

Clearly, the entire set $\mathcal{I}(\mathcal{A})$ can be generated by initializing $\mathcal{B} = \emptyset$ and iteratively solving the above problem $|\mathcal{I}(\mathcal{A})| + 1$ times. If $\mathcal{L}$ is the Boolean cube, i.e., $\mathcal{L}_i = \{0, 1\}$ for all $i = 1, \ldots, n$, the above dualization problem reduces to the well known *hypergraph transversal* problem, which calls for enumerating all minimal subsets that intersect all edges of a given hypergraph. The complexity of the hypergraph transversal problem is still an important open question. The best known algorithm runs in quasi-polynomial time $poly(n, m) + m^{o(\log m)}$, where $m = |\mathcal{A}| + |\mathcal{B}|$ (see [12]), providing partial evidence that the problem is not NP-hard. In this note, we will extend this result by showing that the problem is still unlikely to be NP-hard when each $\mathcal{L}_i$ is a lattice with bounded width. Specifically, for $x \in \mathcal{L}_i$, denote by $x^\perp$ the set of immediate predecessors of x, i.e., $x^\perp = \{y \in \mathcal{L}_i \mid y \prec x, \nexists z \in \mathcal{L}_i : y \prec z \prec x\}$, and let in-deg$(\mathcal{L}_i) = \max\{|x^\perp| : x \in \mathcal{L}_i\}$. Similarly, denote by $x^\top$ the set of immediate successors of x, and let out-deg$(\mathcal{L}_i) = \max\{|x^\top| : x \in \mathcal{L}_i\}$. Let $d = \max_{i \in [n]} \max\{\text{in-deg}(\mathcal{L}_i), \text{out-deg}(\mathcal{L}_i)\}$, and $\mu = \mu(\mathcal{L}) \stackrel{\text{def}}{=} \max\{|\mathcal{L}_i| : i \in [n]\}$. Finally, denote by $W(\mathcal{L}_i)$ the *width* of $\mathcal{L}_i$, i.e. the maximum size of an antichain in $\mathcal{L}_i$ and let $W \stackrel{\text{def}}{=} \max_{i \in [n]} \{W(\mathcal{L}_i)\}$ be the maximum width of the n lattices. The main result of this paper is the following.

Theorem 1. *Problem $DUAL(\mathcal{L}, \mathcal{A}, \mathcal{B})$ can be solved in $poly(n, \mu) + m^{d \cdot \rho \cdot o(\log m)}$ time, if $\mathcal{L}$ is a product of lattices, where $m = |\mathcal{A}| + |\mathcal{B}|$ and $\rho \stackrel{\text{def}}{=} 2W \ln(W + 1)$.*

Let us remark that the special case of Theorem 1 (with a slightly stronger bound) for integer lattices $\mathcal{L}_i = \{0, 1, 2, \ldots\}$ (where $W = 1$) appears in [3], and implies that the set of minimal integer solutions for a monotone system of linear inequalities can be incrementally generated in quasi-polynomial time.

Let us also note that the result of Theorem 1 can be immediately extended to products of *join semi*-lattices, i.e. under the relaxed requirement that every two elements of $\mathcal{L}_i$ have only a unique maximum element. If, moreover, each semi-lattice $\mathcal{L}_i$ has an acyclic precedence graph (i.e., the underling graph is a rooted tree), then the bound on the running time in Theorem 1 can be further improved to $poly(n, \mu) + m^{o(\log m)}$, see [9] for more details.

In the next section, we motivate Theorem 1 by some applications related to the generation of combinatorial structures described by systems of polymatroid inequalities. In particular, we illustrate that the problem of generating infrequent elements in databases with lattice attributes (data mining) and the generation of maximal boxes that contain at most a certain number of points from a given set of points (geometry), can be reduced in quasi-polynomial time to dualization in products of bounded-width lattices. The proof of Theorem 1 will be given in Section 3.

2 Generating Minimal Feasible Solutions for Systems of Polymatroid Inequalities

Let $\mathcal{L} = \mathcal{L}_1 \times \cdots \times \mathcal{L}_n$ be the product of n lattices, and consider a system of inequalities

$$f_i(x) \geq t_i, \quad i = 1, \ldots, r, \tag{2}$$

over the elements $x \in \mathcal{L}$. We assume that each function f_i, $i = 1, \ldots, r$, is polymatroid, that is, f_i is integer-valued, monotone (i.e., $f(x) \leq f(y)$ whenever $x \preceq y$), submodular:

$$f(x \vee y) + f(x \wedge y) \leq f(x) + f(y) \quad \text{for all } x, y \in \mathcal{L},$$

and $f(l) = 0$, where l is the minimum element of $\mathcal{L}$. Let $\mathcal{F} \subseteq \mathcal{L}$ be the set of minimal elements of $\mathcal{L}$ satisfying (2). Given a set of polynomially computable functions $f_1, \ldots, f_r$, and a set of integer thresholds $t_1, \ldots, t_r$, an interesting problem is to incrementally generate all elements of $\mathcal{F}$:

GEN($\mathcal{L}, \mathcal{F}, \mathcal{X}$): *Given an antichain $\mathcal{X} \subseteq \mathcal{F}$, either find a new element in $\mathcal{F} \setminus \mathcal{X}$, or prove that no such element exists: $\mathcal{X} = \mathcal{F}$.*

In the next two subsections, we consider two examples for such polymatroid systems. An important property of these systems, which enables us to generate their minimal feasible set $\mathcal{F}$ in quasi-polynomial time, is that the size of the dual set $\mathcal{I}(\mathcal{F})$ is upper-bounded by a (quasi-) polynomial $p(\cdot)$ in $|\mathcal{F}|$, and the size of the input description of (2). More precisely, the following result holds (see [4] and [7]).

Theorem 2. *Let $\mathcal{F}$ be the set of all minimal feasible solutions for a system of polymatroid inequalities (2) over a lattice $\mathcal{L}$ and let $\mathcal{X} \subseteq \mathcal{F}$ be an arbitrary subset of $\mathcal{F}$ of size $|\mathcal{X}| \geq 2$. Then*

$$|\mathcal{I}(\mathcal{F}) \cap \mathcal{I}(\mathcal{X})| \leq r|\mathcal{X}|^{(\log t)/c(2\mu(\mathcal{L})n, |\mathcal{X}|)}, \tag{3}$$

where $t = \max\{t_1, \ldots, t_r\}$, and $c = c(\alpha, \beta) \sim \log \log_\alpha \beta$ is the unique positive root of the equation $2^c(\alpha^{c/\log \beta} - 1) = 1$.

Consequently, it is enough for such examples to consider the following problem GEN($\mathcal{L}, \mathcal{F}, \mathcal{I}(\mathcal{F}), \mathcal{A}, \mathcal{B}$) of jointly generating all elements of $\mathcal{F}$ and $\mathcal{I}(\mathcal{F})$:

GEN($\mathcal{L}, \mathcal{F}, \mathcal{I}(\mathcal{F}), \mathcal{A}, \mathcal{B}$): *Given two explicitly listed collections $\mathcal{A} \subseteq \mathcal{F}$ and $\mathcal{B} \subseteq \mathcal{I}(\mathcal{F})$, either find a new element in $(\mathcal{F} \setminus \mathcal{A}) \cup (\mathcal{I}(\mathcal{F}) \setminus \mathcal{B})$, or prove that these collections are complete: $(\mathcal{A}, \mathcal{B}) = (\mathcal{F}, \mathcal{I}(\mathcal{F}))$.*

It was observed in [2, 13] that, if $\mathcal{L} = \{0,1\}^n$ is the Boolean cube, then the above joint generation problem GEN($\mathcal{L}, \mathcal{F}, \mathcal{I}(\mathcal{F}), \mathcal{A}, \mathcal{B}$) can be reduced in polynomial time to problem DUAL($\mathcal{L}, \mathcal{A}, \mathcal{B}$). In fact, it is straightforward to see that the same observation holds for any lattice $\mathcal{L}$. Clearly this observation, together with inequality (3), implies that, for systems of polymatroid inequalities

(2), problem GEN($\mathcal{L}, \mathcal{F}, \mathcal{X}$) is quasi-polynomial time reducible to dualization since, in the process of jointly generating elements of $\mathcal{F}$ and $\mathcal{I}(\mathcal{F})$, the number of intermediate elements generated from $\mathcal{I}(\mathcal{F})$ cannot be very large. On the other hand, it is interesting to note that, for many examples of polymatroid inequalities, the dual problem of incrementally generating $\mathcal{I}(\mathcal{F})$ turns out to be NP-hard [6, 13, 16].

Corollary 1. *All minimal feasible solutions to a monotone system of polymatroid inequalities (2), over products of bounded-width lattices, can be incrementally generated in quasi-polynomial time.*

Examples of combinatorial structures described by polymatroid inequalities can be found in [4, 15, 17]. The next two subsections also describe two more such examples.

2.1 Maximal Frequent and Minimal Infrequent Elements in Databases with Semi-lattice Attributes

The notion of frequent sets in data mining [1] has a natural generalization over products of semi-lattices. Formally, consider a database $\mathcal{D} \subseteq \mathcal{L}$ of transactions, each of which is an n-dimensional vector of attributes over $\mathcal{L}$. For an element $x \in \mathcal{L}$, denote by

$$S(x) = S_\mathcal{D}(x) \stackrel{\text{def}}{=} \{p \in \mathcal{D} \mid p \succeq x\},$$

the set of transactions in $\mathcal{D}$ that *support* x. Note that, by this definition, the function $|S(\cdot)| : \mathcal{L} \mapsto \mathbb{Z}_+$ is an anti-monotone *supermodular* function, and hence, the function $f : \mathcal{L} \mapsto \mathbb{Z}_+$, defined by $f(x) = |\mathcal{D}| - |S(x)|$ is polymatroid.

Given $\mathcal{D} \subseteq \mathcal{L}$ and an integer threshold t, an element $x \in \mathcal{L}$ is called t-frequent if it is supported by at least t transactions in the database, i.e. if $|S_\mathcal{D}(x)| \geq t$. Conversely, $x \in \mathcal{L}$ is said to be t-infrequent if $|S_\mathcal{D}(x)| < t$. Denote by $\mathcal{F}_{\mathcal{D},t}$ the set of all minimal t-infrequent elements of $\mathcal{L}$ with respect to the database $\mathcal{D}$. Then $\mathcal{I}(\mathcal{F}_{\mathcal{D},t})$ is the set of all maximal t-frequent elements. It is clear that $\mathcal{F}_{\mathcal{D},t}$ is the set of minimal feasible solutions of the polymatroid inequality $f(x) \geq |\mathcal{D}| - t + 1$, and therefore can be generated in incremental quasi-polynomial time by Corollary 1.

The separate and joint generation of maximal frequent and minimal infrequent elements of a database are important tasks in knowledge discovery and data mining [1, 14, 18]. In many applications, the data attributes assume values ranging over products of semi-lattices of small width, e.g. quantitative attributes [19], taxonomies [18], and lattices of small sizes in logical analysis of data [8]. The special case of the above result for databases $\mathcal{D}$ of binary attributes can be found in [5, 6].

2.2 Packing Points into Boxes

Given a set of n-dimensional points $\mathcal{S} \subseteq \mathbb{R}^n$, suppose that we want to generate all maximal n-dimensional boxes that contain at most t points of $\mathcal{S}$, where

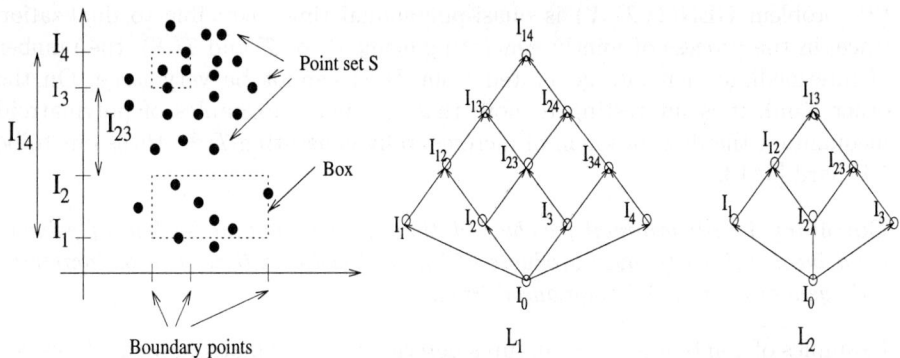

a: A 2-dimensional example. b: The corresponding lattice of intervals $\mathcal{L}_1 \times \mathcal{L}_2$.

Fig. 1. Packing points into boxes: each box has at most $t = 6$ points

$1 \leq t \leq |\mathcal{S}|$ is a given integer threshold. Suppose further that the end points of each candidate box in each dimension, are to be selected from a small set of *boundary* points $\mathcal{P}_i$, for $i = 1, \ldots, n$. This problem has important applications in data mining, see, e.g., [11]. Interestingly, this problem can be cast as of generating minimal feasible solutions of a polymatroid inequality over a product of lattices of bounded width. Indeed, for each set of boundary points $\mathcal{P}_i$, let us construct the corresponding *lattice of intervals* $\mathcal{L}_i$ whose elements are the different intervals defined by these boundary points. The meet of any two intervals is their intersection, and the join is their span, i.e. the minimum interval containing both of them. The minimum element of $\mathcal{L}_i$ is the empty interval I_0. A 2-dimensional example is shown in Figure 1, where we denote the interval between two boundary points $I_j, I_k \in \mathcal{P}_i$ by I_{jk}. Clearly, the product $\mathcal{L} = \mathcal{L}_1 \times \cdots \times \mathcal{L}_n$ is the set of all possible boxes, and the problem is to generate the maximal elements of this product that contain at most t points from the set $\mathcal{S}$. Now consider the function

$$f(x) = |\{q \in \mathcal{S} \mid \text{point } q \text{ is contained inside the box } x\}|,$$

over the elements of $\mathcal{L}$, and observe that this function is supermodular. It follows then that the function $|\mathcal{S}| - f(x)$ is polymatroid over the elements x of the *dual lattice* $\mathcal{L}^*$ (that is, the lattice $\mathcal{L}^*$ with the same set of elements as $\mathcal{L}$, but such that $x \prec y$ in $\mathcal{L}^*$ whenever $x \succ y$ in $\mathcal{L}$). We conclude, therefore, that all maximal boxes that contain at most t points from $\mathcal{S}$, are the minimal solutions $x \in \mathcal{L}^*$ of the polymatroid inequality $|\mathcal{S}| - f(x) \geq |\mathcal{S}| - t$, and hence can be generated in quasi-polynomial time.

3 Dualization in Products of Lattices

In this section we prove Theorem 1. Let $\mathcal{L} = \mathcal{L}_1 \times \cdots \times \mathcal{L}_n$ where each $\mathcal{L}_i$ is a lattice with minimum element l_i and maximum element u_i. Given two antichains $\mathcal{A} \subseteq \mathcal{L}$, and $\mathcal{B} \subseteq \mathcal{I}(\mathcal{A})$, we say that $\mathcal{B}$ is *dual to* $\mathcal{A}$ if $\mathcal{B} = \mathcal{I}(\mathcal{A})$, i.e., if

$\mathcal{B}$ contains all the maximal elements of $\mathcal{L} \setminus \mathcal{A}^+$. Let us remark that, by (1), the latter condition is equivalent to $\mathcal{A}^+ \cup \mathcal{B}^- = \mathcal{L}$.

Given any $\mathcal{Q} \subseteq \mathcal{L}$, let us denote by

$$\mathcal{A}(\mathcal{Q}) = \{a \in \mathcal{A} \mid a^+ \cap \mathcal{Q} \neq \emptyset\}, \qquad \mathcal{B}(\mathcal{Q}) = \{b \in \mathcal{B} \mid b^- \cap \mathcal{Q} \neq \emptyset\},$$

the subsets of $\mathcal{A}, \mathcal{B}$ whose ideal and filter respectively intersect $\mathcal{Q}$. To solve problem DUAL($\mathcal{L}, \mathcal{A}, \mathcal{B}$), we shall use the same general approach used in [12] to solve the hypergraph transversal problem, by decomposing it into a number of smaller subproblems which are solved recursively. In each such subproblem, we start with a sub-lattice $\mathcal{Q} = \mathcal{Q}_1 \times \cdots \times \mathcal{Q}_n \subseteq \mathcal{L}$ (initially $\mathcal{Q} = \mathcal{L}$), and two subsets $\mathcal{A}(\mathcal{Q}) \subseteq \mathcal{A}$ and $\mathcal{B}(\mathcal{Q}) \subseteq \mathcal{B}$, and we want to check whether $\mathcal{A}(\mathcal{Q})$ and $\mathcal{B}(\mathcal{Q})$ are dual in $\mathcal{Q}$. As mentioned before, the latter condition is equivalent to checking whether $\mathcal{Q} \subseteq \mathcal{A}(\mathcal{Q})^+ \cup \mathcal{B}(\mathcal{Q})^-$. To estimate the reduction in problem size from one level of the recursion to the next, we measure the change in the "volume" of the problem defined as $v = v(\mathcal{A}, \mathcal{B}) \stackrel{\text{def}}{=} |\mathcal{A}||\mathcal{B}|$. Since $\mathcal{B} \subseteq \mathcal{I}(\mathcal{A})$ is assumed, the following condition holds, by (1) for the original problem and all subsequent subproblems:

$$a \not\leq b, \quad \text{for all } a \in \mathcal{A}, b \in \mathcal{B}. \tag{4}$$

In the next section we develop several rules for decomposing a given dualization problem into smaller subproblems. The algorithm will select between these rules in such a way that the total volume is reduced from one iteration to the next. The base case for recursion is when one of the sets $\mathcal{A}$ or $\mathcal{B}$ becomes sufficiently small, in which case the problem is easily seen to be polynomially solvable.

Proposition 1. *Suppose* $\min\{|\mathcal{A}|, |\mathcal{B}|\} \leq \text{const}$, *then problem* $DUAL(\mathcal{L}, \mathcal{A}, \mathcal{B})$ *can be solved in* $\text{poly}(n, m, \mu(\mathcal{L}))$ *time for any lattice product* $\mathcal{L}$.

Note that, after performing decomposition, we may end up with a subproblem DUAL($\mathcal{L}', \mathcal{A}', \mathcal{B}'$), where $\mathcal{L}'$ is a sub-poset of $\mathcal{L}$, $\mathcal{A}' \subseteq \mathcal{A}$, $\mathcal{B}' \subseteq \mathcal{B}$, in which some elements of $\mathcal{A}'$, or $\mathcal{B}'$ do not belong to $\mathcal{L}'$. Clearly, if there is an $a \in \mathcal{A}'$ ($b \in \mathcal{B}'$) such that $a^+ \cap \mathcal{L}' = \emptyset$ (respectively, $b^- \cap \mathcal{L}' = \emptyset$) then this element a (respectively, b) can be *eliminated* from the subproblem. On the other hand, if $a \notin \mathcal{L}'$ but $a^+ \cap \mathcal{L}' \neq \emptyset$, then a cannot be eliminated. In such a case, it is necessary to *project* the element a to the poset $\mathcal{L}'$, by replacing it with the set of minimal elements in $a^+ \cap \mathcal{L}'$. In general this minimal set may be large, causing the number of elements of $\mathcal{A}$ or $\mathcal{B}$ to increase exponentially after a succession of decompositions. However, if the sub-poset $\mathcal{L}'$ is a lattice, then each element in $\mathcal{A}'$ or $\mathcal{B}'$ projects only to *a single* element in $\mathcal{L}'$. Thus, we shall always maintain the property that each new subproblem is defined over lattices.

3.1 Decomposition Rules

In general, the algorithm will decompose a given problem by selecting an $i \in [n]$ and decomposing $\mathcal{L}_i$ into a number of sub-lattices, defining accordingly a number

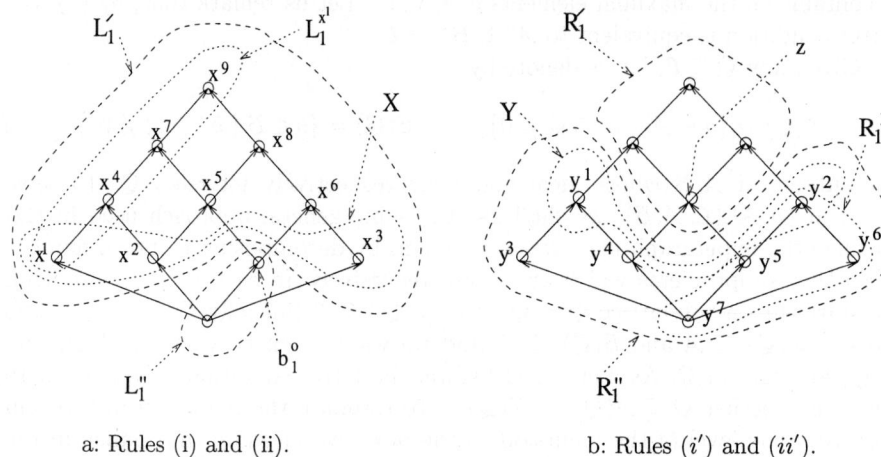

a: Rules (i) and (ii).   b: Rules (i') and (ii').

Fig. 2. Decomposing the lattice $\mathcal{L}_1$

of lattice products. Specifically, let $a^o \in \mathcal{A}$, $b^o \in \mathcal{B}$ be arbitrary. By (4), there exists an $i \in [n]$, such that $a_i^o \not\leq b_i^o$. Let us assume, without loss of generality, that $i = 1$ and set $\mathcal{L}'_1 \leftarrow \mathcal{L}_1 \setminus (b_1^o)^-$, $\mathcal{L}''_1 \leftarrow \mathcal{L}_1 \cap (b_1^o)^-$. (Alternatively, we may set $\mathcal{L}'_1 \leftarrow \mathcal{L}_1 \cap (a_1^o)^+$ and $\mathcal{L}''_1 \leftarrow \mathcal{L}_1 \setminus (a_1^o)^+$.) For brevity, we shall denote by $\overline{\mathcal{L}}$ the product $\mathcal{L}_2 \times \cdots \times \mathcal{L}_n$, and accordingly by $\overline{q}$ the vector $(q_2, \ldots, q_n)$, for an element $q = (q_1, q_2, \ldots, q_n) \in \mathcal{L}$.

Let $\mathcal{X}$ be the set of *minimal* elements in $\mathcal{L}'_1$ and define $\mathcal{L}_1^x = \mathcal{L}'_1 \cap x^+$ for $x \in \mathcal{X}$ (see Figure 2-a). Let further

$$\mathcal{A}'_1 = \{a \in \mathcal{A} \mid a_1 \not\leq b_1^o\}, \qquad \mathcal{A}''_1 = \mathcal{A} \setminus \mathcal{A}'_1,$$
$$\mathcal{B}'_1 = \{b \in \mathcal{B} \mid b_1 \not\leq b_1^o\}, \qquad \mathcal{B}''_1 = \mathcal{B} \setminus \mathcal{B}'_1.$$

Denoting by $\mathcal{L}^x = \mathcal{L}_1^x \times \mathcal{L}_2 \times \cdots \times \mathcal{L}_n$, for $x \in \mathcal{X}$, and by $\mathcal{L}'' = \mathcal{L}''_1 \times \mathcal{L}_2 \times \cdots \times \mathcal{L}_n$ the sub-lattices of $\mathcal{L}$ induced by the above decomposition, we conclude that $\mathcal{A}$ and $\mathcal{B}$ are dual in $\mathcal{L}$ if and only if

$$\mathcal{A}, \mathcal{B}'_1 \text{ are dual in } \mathcal{L}^x, \text{ for all } x \in \mathcal{X}, \tag{5}$$
$$\mathcal{A}''_1, \mathcal{B} \text{ are dual in } \mathcal{L}'', \tag{6}$$

each of which is a dualization problem on lattices. Thus we obtain our first decomposition rule:

Rule (i) Solve $|\mathcal{X}|$ subproblems (5) together with subproblem (6).

Clearly, subproblems (5) and (6) are not independent. Once we know that (6) is satisfied, we gain some information about the solution of (5). The following lemma shows how to utilize such dependence to further decompose (5).

Lemma 1. Given $z \in \mathcal{L}_1$, let $\mathcal{L}'_1 = \mathcal{L}_1 \cap z^+$, $\mathcal{L}''_1 \subseteq \mathcal{L}_1 \cap z^- \setminus \{z\}$ be two disjoint subsets of $\mathcal{L}_1$. Define

$$\mathcal{A}'' = \{a \in \mathcal{A} \mid a_1^+ \cap \mathcal{L}''_1 \neq \emptyset\}, \qquad \mathcal{A}' = \{a \in \mathcal{A} \setminus \mathcal{A}'' \mid a_1^+ \cap \mathcal{L}'_1 \neq \emptyset\},$$
$$\mathcal{B}' = \{b \in \mathcal{B} \mid b_1^- \cap \mathcal{L}'_1 \neq \emptyset\}, \qquad \mathcal{B}'' = \{b \in \mathcal{B} \setminus \mathcal{B}' \mid b_1^- \cap \mathcal{L}''_1 \neq \emptyset\}.$$

Suppose further that we know that $\mathcal{L}'_1 \times \overline{\mathcal{L}} \subseteq (\mathcal{A}' \cup \mathcal{A}'')^+ \cup (\mathcal{B}')^-$, then

$$\mathcal{L}''_1 \times \overline{\mathcal{L}} \subseteq (\mathcal{A}'')^+ \cup (\mathcal{B}' \cup \mathcal{B}'')^- \iff \forall a \in \widetilde{\mathcal{A}} : \mathcal{L}''_1 \times (\overline{\mathcal{L}} \cap \overline{a}^+) \subseteq (\mathcal{A}'')^+ \cup (\mathcal{B}'')^-,$$

where $\widetilde{\mathcal{A}} = \{a \in \mathcal{A}' \cup \mathcal{A}'' \mid a_1 \npreceq z\}$.

By considering the dual lattice of $\mathcal{L}$ (that is, the lattice $\mathcal{L}^*$ with the same set of elements as $\mathcal{L}$, but such that $x \prec y$ in $\mathcal{L}^*$ whenever $x \succ y$ in $\mathcal{L}$), and exchanging the roles of $\mathcal{A}$ and $\mathcal{B}$, we get the following symmetric version of Lemma 1.

Lemma 2. Let $\mathcal{L}''_1 = \mathcal{L}_1 \cap z^-$, $\mathcal{L}'_1 \subseteq \mathcal{L}_1 \cap z^+ \setminus \{z\}$ be two disjoint subsets of $\mathcal{L}_1$ where $z \in \mathcal{L}_1$. Let $\mathcal{A}'', \mathcal{A}', \mathcal{B}'', \mathcal{B}'$ be defined as in Lemma 1, and let $\widetilde{\mathcal{B}} = \{b \in \mathcal{B}' \cup \mathcal{B}'' \mid b_1 \nsucceq z\}$. Suppose we know that $\mathcal{L}''_1 \times \overline{\mathcal{L}} \subseteq (\mathcal{A}'')^+ \cup (\mathcal{B}' \cup \mathcal{B}'')^-$, then

$$\mathcal{L}'_1 \times \overline{\mathcal{L}} \subseteq (\mathcal{A}' \cup \mathcal{A}'')^+ \cup (\mathcal{B}')^- \iff \forall b \in \widetilde{\mathcal{B}} : \mathcal{L}'_1 \times (\overline{\mathcal{L}} \cap \overline{b}^-) \subseteq (\mathcal{A}')^+ \cup (\mathcal{B}')^-.$$

To use Lemma 2, suppose that subproblem (6) has no solution (i.e. there is no $q \in \mathcal{L}'' \setminus [(\mathcal{A}''_1)^+ \cup (\mathcal{B}'_1 \cup \mathcal{B}''_1)^-])$. We proceed in this case as follows. For $x \in \mathcal{L}_1$, let $\widetilde{\mathcal{A}}(x) = \{a \in \mathcal{A} \mid a_1 \npreceq x\}$, $\widetilde{\mathcal{B}}(x) = \{b \in \mathcal{B} \mid b_1 \nsucceq x\}$, and $\mathcal{A}'_1(x) = \{a \in \mathcal{A}'_1 \mid a_1 = x\}$. Let us use $x^1, \ldots, x^k$ to denote the elements of $\mathcal{L}'_1$ and assume, without loss of generality, that they are topologically sorted in this order, that is, $x^j \prec x^h$ implies $j < h$ (see Figure 2-a). Let us decompose (5) (which is equivalent to checking whether $\mathcal{L}'_1 \times \overline{\mathcal{L}} \subseteq \mathcal{A}^+ \cup (\mathcal{B}'_1)^-$) further into the k subproblems

$$\{x^j\} \times \overline{\mathcal{L}} \subseteq \left[\left(\bigcup_{y \in (x^j)^\perp} \widetilde{\mathcal{A}}(y)\right) \cup \mathcal{A}'_1(x^j)\right]^+ \cup (\mathcal{B}'_1)^-, \quad j = 1, \ldots, k. \quad (7)$$

The following lemma, which follows inductively from Lemma 2, will allow us to eliminate the contribution of the set $\mathcal{A}''_1$ in subproblems (7) at the expense of possibly introducing at most $|\mathcal{B}|^d$ additional subproblems.

Lemma 3. Given $x^j \in \mathcal{L}'_i$, suppose we know that $(y^- \cap \mathcal{L}_1) \times \overline{\mathcal{L}} \subseteq \widetilde{\mathcal{A}}(y)^+ \cup \mathcal{B}^-$ for all $y \in (x^j)^\perp$. Then (7) is equivalent to

$$\{x^j\} \times \left[\overline{\mathcal{L}} \cap \left(\bigwedge_{y \in (x^j)^\perp} \overline{b}(y)\right)^-\right] \subseteq \mathcal{A}'_1(x^j)^+ \cup (\mathcal{B}'_1)^-, \quad (8)$$

for all collections $\{b(y) \in \widetilde{\mathcal{B}}(y) \mid y \in (x^j)^\perp\}$.

Informally, Lemma 3 says that, given $x^j \in \mathcal{L}'_1$, if the dualization subproblems for all sub-lattices that lie below x^j have been already verified to have no solution, then we can replace the solution to subproblem (7) by solving at most $\prod_{y \in (x^j)^\perp} |\widetilde{\mathcal{B}}(y)|$ subproblems of the form (8). Observe that it is important to check subproblems (7) in the topological order $j = 1, \ldots, k$ in order to be able to use Lemma 3. Thus we get

Rule (ii) Solve subproblem (6). If it has a solution then we get a point $q \in \mathcal{L} \setminus (\mathcal{A}^+ \cup \mathcal{B}^-)$. Otherwise, we solve subproblems (8), for all collections $\{b(y) \in \widetilde{\mathcal{B}}(y) \mid y \in (x^j)^\perp\}$, for $j = 1, \ldots, k$ (in the topological order).

Suppose finally that we decompose $\mathcal{L}_1$ by selecting an element $z \in \mathcal{L}_1$, letting $\mathcal{R}'_1 \leftarrow \mathcal{R}_1 \cap z^+$, $\mathcal{R}''_1 \leftarrow \mathcal{L}_1 \setminus z^+$, $\mathcal{R}' = \mathcal{R}'_1 \times \overline{\mathcal{L}}$, and $\mathcal{R}'' = \mathcal{R}''_1 \times \overline{\mathcal{L}}$. Let $\mathcal{Y}$ denote the set of *maximal* elements in $\mathcal{R}''_1$ and define $\mathcal{R}^y_1 = \mathcal{R}''_1 \cap y^-$ for $y \in \mathcal{Y}$ (see Figure 2-b). Let further

$$\mathcal{A}'_2 = \{a \in \mathcal{A} \mid a_1 \succeq z\}, \qquad \mathcal{A}''_2 = \mathcal{A} \setminus \mathcal{A}'_2,$$
$$\mathcal{B}'_2 = \{b \in \mathcal{B} \mid b_1 \succeq z\}, \qquad \mathcal{B}''_2 = \mathcal{B} \setminus \mathcal{B}'_2.$$

By exchanging the roles of $\mathcal{A}$ and $\mathcal{B}$ and replacing $\mathcal{L}$ by its dual lattice $\mathcal{L}^*$ in rules (i), (ii) above, we can also derive the following symmetric versions of these rules:

Rule (i') Solve the subproblem

$$\mathcal{A}, \mathcal{B}'_2 \text{ are dual in } \mathcal{R}', \tag{9}$$

and the $|\mathcal{Y}|$ subproblems

$$\mathcal{A}''_2, \mathcal{B} \text{ are dual in } \mathcal{R}^y_1 \times \overline{\mathcal{L}}, \quad \text{for all } y \in \mathcal{Y}. \tag{10}$$

Rule (ii') Solve subproblem (9), and if it does not have a solution, then solve the subproblems

$$\{y^j\} \times \left[\overline{\mathcal{L}} \cap \left(\bigvee_{x \in (y^j)^\top} \overline{a}(x)\right)^+\right] \subseteq (\mathcal{A}''_2)^+ \cup (\mathcal{B}''_2(y^j))^-, \tag{11}$$

for all collections $\{a(x) \in \widetilde{\mathcal{A}}(x) \mid x \in (y^j)^\top\}$, for $j = 1, \ldots, h$, where $y^1, \ldots, y^h$ denote the elements of $\mathcal{R}''_1$ in *reverse* topological order, and $\mathcal{B}''_2(x) = \{a \in \mathcal{B}''_2 \mid b_1 = x\}$.

Finally it remains to remark that all the decomposition rules described above result, indeed, in dualization subproblems over lattices.

3.2 The Algorithm

Given antichains $\mathcal{A}, \mathcal{B} \subseteq \mathcal{L} = \mathcal{L}_1 \times \cdots \times \mathcal{L}_n$ that satisfy the necessary duality condition (4), we proceed as follows:

Step 1. For each $k \in [n]$:
  1. (*eliminate:*) if $a_k^+ \cap \mathcal{L}_k = \emptyset$ for some $a \in \mathcal{A}$ ($b_k^- \cap \mathcal{L}_k = \emptyset$ for some $b \in \mathcal{B}$), then a (respectively, b) can be discarded from further consideration;
  2. (*project:*) if $a_k \notin \mathcal{L}_k$ for some $a \in \mathcal{A}$ ($b_k \notin \mathcal{L}_k$ for some $b \in \mathcal{B}$), set $a_k \leftarrow \bigwedge\{x \mid x \in a_k^+ \cap \mathcal{L}_k\}$ (respectively, set $b_k \leftarrow \bigvee\{x \mid x \in b_k^- \cap \mathcal{L}_k\}$).

Thus we may assume for next steps that $\mathcal{A}, \mathcal{B} \subseteq \mathcal{L}$.

Step 2. If $\min\{|\mathcal{A}|, |\mathcal{B}|\} < \delta = 2$, then dualization can be solved in $poly(n, m, \mu)$ time.

Step 3. Let $a^o \in \mathcal{A}$, $b^o \in \mathcal{B}$. Find an $i \in [n]$ such that $a_i^o \not\leq b_i^o$. Assume, without loss of generality, that $i = 1$ and set $\mathcal{L}_1' \leftarrow \mathcal{L}_1 \setminus (b_1^o)^-$, $\mathcal{L}_1'' \leftarrow \mathcal{L}_1 \cap (b_1^o)^-$. Let $\mathcal{X}, \mathcal{A}_1', \mathcal{A}_1'', \mathcal{B}_1', \mathcal{B}_1''$, and $\widetilde{\mathcal{B}}$ be as defined in the previous subsection, and let

$$\epsilon_1^{\mathcal{A}} = \frac{|\mathcal{A}_1'|}{|\mathcal{A}|}, \qquad \epsilon_1^{\mathcal{B}} = \frac{|\mathcal{B}_1''|}{|\mathcal{B}|}.$$

Observe that $\epsilon_1^{\mathcal{A}} > 0$ and $\epsilon_1^{\mathcal{B}} > 0$ since $a^o \in \mathcal{A}_1'$ and $b^o \in \mathcal{B}_1''$.

Step 4. Define $\epsilon(v) = \rho(W)/\chi(v)$, where $v = v(\mathcal{A}, \mathcal{B})$, $\rho(W) = 2W \ln(W+1)$, and $\chi(v)$ is defined to be the unique positive root of the equation

$$\left(\frac{\chi(v)}{\rho(W)}\right)^{\chi(v)} = \frac{v^d}{(1 - e^{-\rho(W)})(\delta^d - 1)}$$

and observe that $\epsilon(v) < 1$ for $v \geq \delta^2$, $\delta \geq 2$.

If $\min\{\epsilon_1^{\mathcal{A}}, \epsilon_1^{\mathcal{B}}\} > \epsilon(v)$, we use decomposition rule (i), which amounts to solving recursively $|\mathcal{X}|$ subproblems (5), each of which has a volume of at most $v(\mathcal{A}, \mathcal{B}_1') = |\mathcal{A}||\mathcal{B}_1'|$, and subproblem (6) of volume $v(\mathcal{A}_1'', \mathcal{B}) = |\mathcal{A}_1''||\mathcal{B}|$. This gives rise to the recurrence

$$\begin{aligned}
C(v) &\leq 1 + |\mathcal{X}|C(|\mathcal{A}||\mathcal{B}_1'|) + C(|\mathcal{A}_1''||\mathcal{B}|) \\
&\leq 1 + W \cdot C((1 - \epsilon_1^{\mathcal{B}})v) + C((1 - \epsilon_1^{\mathcal{A}})v) \\
&\leq 1 + (W + 1)C((1 - \epsilon(v))v), \quad (12)
\end{aligned}$$

where $C(v)$ denotes the number of recursive calls on a subproblem of volume at most v.

Step 5. If $\epsilon_1^{\mathcal{A}} \leq \epsilon(v)$, we apply rule ($ii$) and get the recurrence

$$\begin{aligned}
C(v) &\leq 1 + C(|\mathcal{A}_1''||\mathcal{B}|) + \sum_{j=1}^{k}\left(\prod_{y \in (x^j)^\perp} |\widetilde{\mathcal{B}}(y)|\right) C(|\mathcal{A}_1'(x^j)||\mathcal{B}_1'|) \\
&\leq 1 + C(|\mathcal{A}_1''||\mathcal{B}|) + |\widetilde{\mathcal{B}}|^d \sum_{j=1}^{k} C(|\mathcal{A}_1'(x^j)||\mathcal{B}_1'|) \\
&\leq 1 + C((1 - \epsilon_1^{\mathcal{A}})v) + |\mathcal{B}|^d C(\epsilon_1^{\mathcal{A}} v) \\
&\leq 1 + C((1 - \epsilon_1^{\mathcal{A}})v) + \frac{v^d}{\delta^d} C(\epsilon_1^{\mathcal{A}} v) \\
&\leq C((1 - \epsilon)v) + \frac{v^d}{\delta^d - 1} C(\epsilon v), \quad \text{for some } \epsilon \in (0, \epsilon(v)], \quad (13)
\end{aligned}$$

where the second inequality follows from the fact that $|(x^j)^\perp| \leq d$, the third inequality follows from $\sum_{j=1}^k C(|\mathcal{A}_1'(x^j)||\mathcal{B}_1'|) \leq C(\sum_{j=1}^k |\mathcal{A}_1'(x^j)||\mathcal{B}_1'|) = C(|\mathcal{A}_1'||\mathcal{B}_1'|)$ since $\{\mathcal{A}_1'(x^j) \mid j = 1,\ldots,k\}$ is a partition of $\mathcal{A}_1'$ and the function $C(\cdot)$ is super-additive, the forth inequality follows from $|\mathcal{B}|^d \leq v(|\mathcal{A}|,|\mathcal{B}|)^d/\delta^d$, and the last inequality follows from the fact that $v \geq \delta^2$ and $\delta \geq 2$.

Step 6. We assume for next steps that $\epsilon_1^A > \epsilon(v)$. Then there exists a point $z \in \mathcal{X}$, such that $|\{a \in \mathcal{A} \mid a_1 \succeq z\}| \geq \epsilon_1^A |\mathcal{A}|/|\mathcal{X}| > \epsilon(v)|\mathcal{A}|/W$. Let $\mathcal{R}_1' \leftarrow \mathcal{R}_1 \cap z^+$, $\mathcal{R}_1'' \leftarrow \mathcal{L}_1 \setminus z^+$, and let $\mathcal{Y}, \mathcal{A}_2', \mathcal{A}_2'', \mathcal{B}_2', \mathcal{B}_2'', \tilde{\mathcal{A}}$ be as defined in the previous subsection. Let also

$$\epsilon_2^A = \frac{|\mathcal{A}_2'|}{|\mathcal{A}|}, \qquad \epsilon_2^B = \frac{|\mathcal{B}_2''|}{|\mathcal{B}|},$$

and observe that $\epsilon_2^A > \frac{\epsilon(v)}{W}$ by our selection of $z \in \mathcal{L}_1$, and that $\epsilon_2^B > 0$ since $b^o \notin \mathcal{B}_2'$.

Step 7. If $\epsilon_2^B > \epsilon(v)$, then we use decomposition rule (i') which gives

$$\begin{aligned}
C(v) &\leq 1 + C(|\mathcal{A}||\mathcal{B}_2'|) + |\mathcal{Y}|C(|\mathcal{A}_2''||\mathcal{B}|) \\
&\leq 1 + C((1-\epsilon_2^B)v) + W \cdot C((1-\epsilon_2^A)v) \\
&\leq 1 + C((1-\epsilon(v))v) + W \cdot C((1-\frac{\epsilon(v)}{W})v),
\end{aligned} \qquad (14)$$

since $\epsilon_2^A > \frac{\epsilon(v)}{W}$ and $\epsilon_2^B > \epsilon(v)$.

Step 8. Finally if $\epsilon_2^B \leq \epsilon(v)$, use rule (ii') and get thus the recurrence

$$C(v) \leq 1 + C(|\mathcal{A}||\mathcal{B}_2'|) + \sum_{j=1}^h \left(\prod_{x \in (y^j)^\top} |\tilde{\mathcal{A}}(x)| \right) C(|\mathcal{A}_2''||\mathcal{B}_2''(y^j)|)$$

$$\leq C((1-\epsilon)v) + \frac{v^d}{\delta^d - 1} C(\epsilon v), \quad \text{for some } \epsilon \in (0, \epsilon(v)]. \qquad (15)$$

One can show by induction on $v = v(\mathcal{A},\mathcal{B}) \geq 4$ that recurrences (12)–(15) imply that

$$C(v) \leq v^{\chi(v)},$$

see [10]. Note that, for $\delta \geq 2$, $d \geq 1$ and $W \geq 1$, we have $(\chi/\rho(W))^\chi < 3(v/\delta)^d$, and thus,

$$\chi(v) < \frac{d\log(v/\delta) + \log 3}{\log(\chi/\rho(W))} \sim \frac{d\rho(W)\log v}{\log d + \log\log v}.$$

As $v(\mathcal{A},\mathcal{B}) < m^2$, we get $\chi(v) = d\rho(W)o(\log m)$. This establishes the bound stated in Theorem 1.

References

[1] R. Agrawal, H. Mannila, R. Srikant, H. Toivonen and A. I. Verkamo, Fast discovery of association rules, in *Advances in Knowledge Discovery and Data Mining*, pp. 307–328, AAAI Press, Menlo Park, California, 1996.

[2] J. C. Bioch and T. Ibaraki, Complexity of identification and dualization of positive Boolean functions, *Information and Computation* 123 (1995) 50–63.
[3] E. Boros, K. Elbassioni, V. Gurvich, L. Khachiyan and K. Makino, On generating all minimal integer solutions to a monotone system of linear inequalities, in *Automata, Languages and Programming, 28-th International Colloquium (ICALP 2001), Lecture Notes in Computer Science* 2076, pp. 92–103, 2001. To appear in *SIAM Journal on Computing*.
[4] E. Boros, K. Elbassioni, V. Gurvich and L. Khachiyan, An inequality for polymatroid functions and its applications, to appear in *Discrete Applied Mathematics*, 2002. DIMACS Technical Report 2001-14, Rutgers University.
[5] E. Boros, V. Gurvich, L. Khachiyan and K. Makino, Dual bounded generation: partial and multiple transversals of a hypergraph, *SIAM Journal on Computing* **30** (6) (2001) 2036–2050.
[6] E. Boros, V. Gurvich, L. Khachiyan and K. Makino, On the complexity of generating maximal frequent and minimal infrequent sets, in *Proc. 19th International Symposium on Theoretical Aspects of Computer Science, (STACS)*, Mrach 2002, *Lecture Notes in Computer Science* 2285, pp. 133–141.
[7] E. Boros, K. Elbassioni, V. Gurvich and L. Khachiyan, Extending the Balas-Yu bounds on the number of maximal independent sets in graphs to hypergraphs and lattices, DIMACS Technical Report 2002-27, Rutgers University.
[8] E. Boros, P. L. Hammer and J. N. Hooker, Predicting cause-effect relationships from incomplete discrete observations, *SIAM J. Discrete Math.* **7** (1994), 481–491.
[9] K. Elbassioni, On dualization in products of forests, in *Proc. 19th International Symposium on Theoretical Aspects of Computer Science (STACS)*, March 2002, *Lecture Notes in Computer Science* 2285, pp. 142–153.
[10] K. Elbassioni, An algorithm for dualization in products of lattices, DIMACS Technical Report 2002-26, Rutgers University.
[11] J. Edmonds, J. Gryz, D. Liang and R. J. Miller, Mining for empty rectangles in large data sets, in *Proc. 8th Int. Conf. on Database Theory (ICDT)*, Jan. 2001, *Lecture Notes in Computer Science* 1973, pp. 174–188.
[12] M. L. Fredman and L. Khachiyan, On the complexity of dualization of monotone disjunctive normal forms, *Journal of Algorithms*, 21 (1996) 618–628.
[13] V. Gurvich and L. Khachiyan, On generating the irredundant conjunctive and disjunctive normal forms of monotone Boolean functions, *Discrete Applied Mathematics*, 96-97 (1999) 363–373.
[14] D. Gunopulos, R. Khardon, H. Mannila, and H. Toivonen, Data mining, hypergraph transversals and machine learning, in *Proc. 16th ACM-SIGACT-SIGMOD-SIGART Symposium on Principles of Database Systems*, (1997) pp. 12–15.
[15] E. Lawler, J. K. Lenstra and A. H. G. Rinnooy Kan, Generating all maximal independent sets: NP-hardness and polynomial-time algorithms, *SIAM Journal on Computing*, 9 (1980) 558–565.
[16] K. Makino and T. Ibaraki, Interior and exterior functions of Boolean functions, *Discrete Applied Mathematics*, 69 (1996) 209–231.
[17] R. C. Read and R. E. Tarjan, Bounds on backtrack algorithms for listing cycles, paths, and spanning trees, *Networks* 5 (1975) 237–252.
[18] R. Srikant and R. Agrawal, Mining generalized association rules, in *Proc. 21st Int. Conf. on Very Large Data Bases*, pp. 407–419, 1995.
[19] R. Srikant and R. Agrawal, Mining quantitative association rules in large relational tables, in *Proc. ACM-SIGMOD 1996 Conf. on Management of Data*, pp. 1–12, 1996.

Determining Similarity of Conformational Polymorphs*

Angela Enosh[1], Klara Kedem[1], and Joel Bernstein[2]

[1] Computer Science Department, Ben-Gurion University of the Negev
84105 Beer Sheva, Israel
{anosh, klara}@cs.bgu.ac.il
[2] Chemistry Department, Ben-Gurion University of the Negev
84105 Beer Sheva, Israel
{yoel}@bgumail.bgu.ac.il

Abstract. Conformational polymorphs are identical molecules that crystallize in different spatial formations. Understanding the amount of difference between the polymorphs might aid drug design as there is a widespread assumption that there exists a direct connection between the conformations in the crystallized form of the molecule and the conformations in the solvent.
We define a measure of similarity between conformational polymorphs and present an algorithm to compute it. For this end we weave together in a novel way our graph isomorphism method and substructure matching. We tested our algorithm on conformational polymorphs from the Cambridge Structural Database. Our experiments show that our method is very efficient in practice and has already yielded an important insight on the polymorphs stored in the data base.

1 Introduction

Conformational polymorphism stands for the existence of more than one crystal structure for a particular molecule [2, 3, 7]. For example, diamond and graphite are conformational polymorphs. The existence of polymorphic crystal structures provides a unique opportunity to study structure-property relationship since the only variable among polymorphic forms is that of structure.

The Cambridge Structural Database [6] (CSD) is a data depository of approximately 250,000 published crystal structures of molecules, and it grows at the rate of about 10% a year. A record of a crystal structure in the CSD contains all the data concerning the molecule, *e.g.*, the atom types, the bonds, the coordinates of the atoms in the crystal structure, and other useful information such as the crystallographic space group. Although there are a few thousand entries in the CSD that are known to be members of polymorphic systems, the number of those which may be described as conformational polymorphs has not

---
* K. Kedem was supported by NSF CISE Research Infrastructure grant #EIA-9972853, by NSF #CCR-9988519, and by a grant from the Israeli Defence Ministry.

been determined, and this in spite of the fact that the phenomenon has been recognized for about 30 years. The reason is both scientific and technological. Scientifically, there has not been any attempt to determine a quantitative measure (or measures) for comparing the conformations and defining when a pair of molecular structures should be considered conformationally different. From the technical aspect, different crystal structures are reported by different authors, with different numbering (i.e. atom labeling) systems. Thus, with the currently available CSD software, a one-to-one comparison of geometric features, in particular the analysis of conformational parameters, is a rather tedious task. This paper describes our successful attempt to overcome these problems and add more insight to the complex process of drug design.

Each entry in the CSD contains one molecule which can be viewed as a graph embedded in $3D$. The vertices of the graph are the atoms and there exists an edge between two vertices if there is a bond between the corresponding atoms. The graphs of two conformational polymorphs are isomorphic but the isomorphism is not known and has to be computed.

In Fig. 1 we present an example of two molecules (conformational polymorphs). Chemically they are the same molecules (C stands for carbon, N for nitrogen, and O for oxygen), but their geometric structure is different and so is the numbering of the atoms. Thus the graphs of two conformational polymorphs are reported in the CSD in random numbering of vertices and ordering of edges. We would like to note here that the molecules in the CSD also have some dangling hydrogen atoms, but since their measurements are very imprecise we ignore them.

How does one define a similarity measure between two geometrically different molecules? The way we suggest here is to find a large connected substructure that is geometrically alike in both molecules (see terms and definitions below), while complying with the graph isomorphism (which is unknown). Next we overlay the molecules so that their common substructures are superimposed on each other and measure the RMS distance between the two molecules. There may be many isomorphic mappings between these two molecules and for each mapping there might be many large connected substructures that are geometrically alike in the molecules. As we describe in detail below, we pick the mapping and the sub-

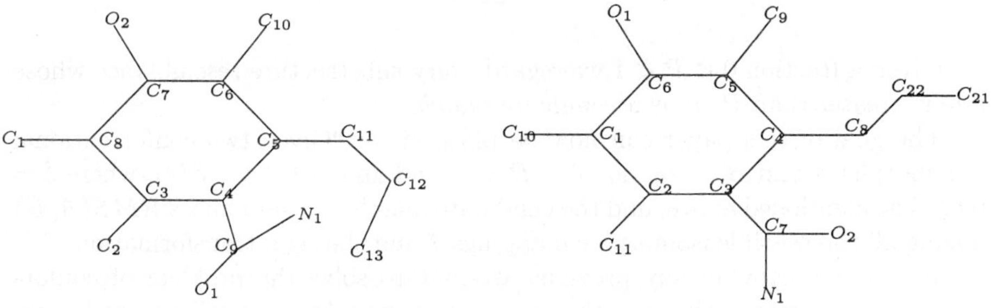

Fig. 1. Example of conformational polymorphs

structure that yield the *smallest RMS* distance between the molecules, and this is what we call *the similarity measure* between the conformational polymorphs.

There are no efficient algorithms known for graph isomorphism in general graphs. What comes to our aid is a combination of observations made on the graph topology of the molecules, such as the small average degree of nodes (see Table 1, Section 4), and the fact that the graphs are mostly planar. For the geometric structure comparison we exploit the fact that atoms of a molecule have to be at least some known distance apart from each other (we denote it by $\delta = 0.68$ Å, see [4]).

Terms and definitions. Conformational polymorphs are *identical* molecules that crystallize in different spatial formations. We represent each molecule as a 3-dimensional labeled undirected graph $G = (V, E)$, where V represents the atoms and E represents the links (bonds) between these atoms. Each vertex (atom) is labeled by the atom coordinates and type (*e.g.*, carbon, nitrogen). Given two conformational polymorphs A and B we construct their underlying graphs $G_A = (V_A, E_A)$ and $G_B = (V_B, E_B)$. For simplicity we slightly abuse notation by referring to G_A and G_B as A and B respectively.

Isomorphic mapping f between graphs A and B is a bijection of V_A onto V_B that preserves atom type and vertex adjacency. Since there is a significant symmetry in molecules, there might be a number of isomorphic mappings f from A to B.

Given a rigid transformation T, an isomorphic mapping f, and an $\varepsilon \geq 0$ we define *substructure resemblance* S_{AB} of A and B as a maximally connected subgraph of A, such that $\rho(T(a), f(a)) \leq \varepsilon$, $a \in A$, where ρ is the Euclidean distance between two points. We call two points, a and b, $\varepsilon-close$ if the distance between them is not greater than ε. We will show later that for each a there exists at most one such point $b = f(a) \in B$. There might be several substructure resemblances between A and B for the same transformation T. The *size* of S_{AB} is defined to be the number of its vertices.

Given f, and assume that A and B have n atoms each, the *RMS distance* (Root Mean Square distance), between the atoms of A and B is:

$$RMS(A, B) = \sqrt{(\sum_1^n ||(a - f(a)||^2/n)}$$

Given a fraction $0 < P < 1$ we regard every substructure resemblance whose size is greater than $P \cdot n$ as a *candidate match*.

The goal of this paper can now be phrased as: "Given two conformational polymorphs A and B, a fraction $0 < P < 1$, and an ε, $(0 < \varepsilon < \delta/2)$, where $\delta = 0.68$ Å as mentioned above, find the candidate match that minimizes $RMS(A, B)$ among all the possible isomorphic mappings f and the rigid transformations T".

We do not know of any previous attempt to solve the problem of conformational polymorphs that we phrase in this paper. In what follows we briefly survey previous work related to the two components of our proposed algorithm, namely, graph isomorphism and molecule substructure resemblance.

Background, graph isomorphism. Graphs $G_1 = (V_1, E_1)$ and $G_2 = (V_2, E_2)$ are *isomorphic* iff there exists a bijection that matches the nodes of V_1 and V_2 in a manner that preserves adjacency [11]. Graph isomorphism is in NP, but its hardness is still an open question.

The problem of graph isomorphism has generated a lot of interest over the years and still draws attention both because of the wide range of its applications and because of the uncertainty of its place in the complexity hierarchy. Though all the attempts to give a polynomial solution for this problem for a general graph have failed, there are classes of graphs for which isomorphism can be solved in polynomial, or even linear time. For instance, it has been shown that the problem can be solved in linear time for trees (see [14]) by using the graph center concept. Graph isomorphism is also solvable in polynomial-time both for graphs with nodes of bounded valence [16] and for graphs with a bounded treewidth [19]. Clearly, the run-time complexity for graphs with bounded properties depends on the value of the bound. See also k-contractible graphs [17]. A polynomial algorithm for planar graph isomorphism can be found in [14]. A good survey appears in [10].

The naive approach, known as "brute force search", solves the graph isomorphism problem by examining all the possible permutations of the nodes in the adjacency matrix of one graph against a single permutation of the other. Each possible permutation defines an ordering of the nodes, and the algorithm simply checks whether one of the possible permutations of the former graph is identical to the chosen permutation of the latter graph. However, the time complexity of generating all the possible permutations of a graph is $O(n!)$, thus using the brute force method without exploiting some other properties of the graph is not doable. *Graph invariants* are structural properties of the graph that are preserved by isomorphism. Strong invariants can reduce the number of mappings that should be tested in order to determine isomorphism. They are used in most of the heuristics to prune the run-time complexity (see, *e.g.*, [10]). Some graph invariants are: the number of vertices of the graph, the number of edges of the graph, the sorted sequence of degrees of the vertices, the determinant of the adjacency matrix. *Vertex invariants* are local properties of vertices that are preserved by isomorphism, *e.g.*, vertex degree, the k-cliques invariant – assign a number to each node v which is the number of different cliques of size k that contain v, and others. We will use invariants in our algorithm.

Background, geometric point matching and molecule resemblance. Identifying common $3D$ substructure in two geometric patterns is an important mission in diverse fields such as computer vision, biology, crystallography, astronomy, and others. Comparing structural features of two molecules is a frequently asked question which comes from the assumption that structure similarity implies functional similarity.

There is a very large bulk of work that deals with comparing ($3D$) pairwise molecule structure and in particular protein structures. The goal is mainly to find a maximal substructure that is common to both molecules. One approach used by Finn *et al* [12] to compare strcutures of molecules is randomized and involves assuming that an α fraction of the atoms of the compared molecules is similar in shape. The mapping between the atoms is known. Thus, by randomly picking three atoms (about $O(1/\alpha^3)$ times) they compute a common substructure very efficiently with high probability. The geometric hashing approach [13, 18] is a very robust method for protein comparison, but since proteins' backbones are just polygonal chains in $3D$ we do not survey works on protein structure resemblance in this paper. A version of geometric hashing has been applied to molecules in [12]. The method described in [1] regards molecules as sets of points in $3D$ and ignore the bonds between the atoms. A *distance graph* C is built based on similar intra-molecule distances: given molecules A and B, each pair (a, b), $a \in A$ and $b \in B$, is a node in C, and there is an edge from (a, b) to (a', b') if the distance between a and a' is very close to the distance between b and b'. The problem of finding a minimal common substructure is reduced to finding a maximal clique in C. The intra-molecule distance approach is also applied in [15] who present an algorithm for detecting occurrence of a $3D$ pattern within a large molecule. A good survey of methods to compute substructure resemblance in molecules can be found in [8].

2 Our Algorithm

Our algorithm consists of two major phases: Phase I finds the set of candidate matches and Phase II finds the isomorphic mapping and the transformation with the lowest RMS distance among the candidates.

2.1 Phase I – Finding Candidate Matches

Let us denote the two molecules by A and B. Their underlying graphs are also denoted by A and B respectively. The trivial approach to find a *superset* of all candidate matches is to go over all possible adjacent (and non-collinear) triplets of atoms in the two molecules. (It is a superset because it does not take into consideration the graph isomorphism.) Each triplet represents a triangle. If the triangles in the two molecules are almost congruent then we compute an appropriate rigid transformation T that superimposes one triangle on the other (using, *e.g.*, the SVD – singular value decomposition). We next find how many points of $T(A)$ are ε-close to B. If ε is less than $\delta/2$, then within a sphere of radius ε about each atom of B there is at most one point of $T(A)$. We count the number of ε-close atoms. If the number of ε-close pairs is larger than $P \cdot n$ then this match is stored in the superset of candidate matches waiting to be verified in Phase II. Since Phase II might be costly in runtime, rejecting candidates at an early stage will speed up the algorithm. Below we incorporate the knowledge of the graph topology to allow fast rejection of false candidates.

We will incorporate vertex invariants in our algorithm to prune the number of computed candidates. The vertex invariants we use are *atom type*, *degree*, *level* and *cycle size invariant*. The names of the first two invariants are self-explanatory. The latter two invariants will be explained after we define the *center of a graph*, and some operations we perform on the graphs of the molecules.

The following definitions and facts can be found in Foulds [11]. Let $G = (V, E)$ be a graph. We define the edge-distance $dist(u, v)$ between two vertices u and v in V to be the number of edges in the shortest path between u and v. The eccentricity, $e(v)$, $v \in V$, is defined as $max_{u \in V} dist(v, u)$. The radius, $r(G)$, of G is equal to the eccentricity of the vertex in V with minimum eccentricity. A vertex $v \in V$ is termed *central* if $e(v) = r(G)$. The *center* of G is the set of all central vertices in G and can be computed by the definitions above.

Assume, for sake of simplicity, that the center of G consists of one vertex $u \in G$, scanning G in a BFS order, starting from u, naturally defines the level invariant of a vertex.

The center of a general graph might consist of an arbitrary number of central vertices, *e.g.*, when all the vertices of the graph compose one simple cycle. However, a tree has a center comprising either of one vertex or of two adjacent vertices [11]. The set of central vertices of a tree can be easily obtained by removing the tree leaves and then recursively removing the leaves from the resulting tree until we are left with one vertex or two adjacent vertices. Thus, if the graph is a tree with two central vertices, we start BFS from both and assign each vertex of the graph the smaller level of the two BFS's.

The graphs of molecules are not trees. In fact, many of them have cycles of 5 – 7 atoms (Benzene, being a very common cycle among the organic molecules, has six carbon atoms in a cycle, but some cycles have a larger number of atoms). In the process described below we show how we convert a molecule's graph into a tree, so as to have 'centers' comprising of one or two central vertices only. Once we get these centers, we can define the level of each vertex in the original graph of the molecule by its BFS level from the center (or two centers).

A *cycle* in G can be an *elementary cycle* which does not contain any other cycle in it, or a *complex cycle* which is composed of more than one elementary cycle (see Fig. 2). We find all the cycles in G by assigning directions to G's edges using any DFS scanning of G, and then applying an algorithm that finds all the strongly connected components in a directed graph [5]. (A strongly connected component C of a directed graph $D = (V, E)$ is a subset of V, such that for every two vertices $u, v \in C$ there is a path from v to u and from u to v in D.) We now define the *cycle size* invariant for a vertex $v \in G$ as the size of the strongly connected component it belongs to. We claim that all the elementary and complex cycles of isomorphic graphs can be found in this way, no matter where we start the DFS scan [5].

We convert the undirected graph G into a tree by collapsing each strongly connected component in the directed graph above into one vertex (see Fig. 4), and compute the center of the tree.

Since one vertex in the tree can be an elementary or a complex cycle in G, the recursive procedure produces a center which may have one of the following sets of vertices of G: (*i*) one vertex, (*ii*) two adjacent vertices, (*iii*) one cycle, (*iv*) a vertex connected to a cycle, (*v*) or two adjacent cycles linked together. By a cycle here we mean elementary or complex cycle. See Fig. 3, where the gray dots represent a vertex of G and an empty circle – a cycle. In cases (*i*)

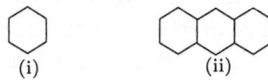

Fig. 2. (i) elementary and (ii) complex cycles

and (ii) the found center vertices are called the center of G. In case (iii), if the center is composed of an elementary cycle then all the vertices of the cycle are regarded as center of G; if it is a complex cycle then, we compute the center of the complex cycle (by the constructive definition of the center above). In cases (iv) and (v) we pick the endpoints of the edge connecting the vertex to the cycle (iv) or connecting the two cycles (v), respectively, as the center of G (see Fig. 3). We remain with a center that is comprised of either one vertex of G, (i), or two, (ii, iv, v), or, if the center was an elementary cycle, then all its vertices remain as a center. We now compute the *level invariant* for the vertices of G. If the center is an elementary cycle we compute the level of each vertex of G as the smallest level among all the BFS levels, when the BFS starts at each vertex of the elementary cycle. In the experiments we have performed on 528 polymorph pairs, the average number of central vertices was 2.

Returning to the graphs of the molecules. We compute the centers of A and B as described above. Since the graphs of the two molecules A and B are isomorphic their centers must be congruent. Each of the vertices of A and B is assigned the four invariants. By $Inv(a) = Inv(b)$ we mean that all the invariants of a are identical to all the corresponding invariants of b. Thus, if the equality does not hold for vertices $a \in A$ and $b \in B$ we deduce that a cannot be mapped to b by any isomorphic mapping.

In the algorithm that finds candidate matches we go over all possible adjacent triplets of vertices (atoms) in A and B respectively. We check for each pair of triplets $(a_1, a_2, a_3) \in A$ and $(b_1, b_2, b_3) \in B$ that all the invariants of a_i are identical to the corresponding invariants of b_i, $i = 1, 2, 3$. If this condition does not hold then this pair is rejected. Otherwise, if these triplets are non-collinear, we observe the triangles Δ_A and Δ_B defined by the atoms, respectively. If Δ_A and Δ_B are almost congruent, in terms of side lengths, then we compute the rigid transformation $T = (\Delta_A, \Delta_B)$ that superimposes Δ_A on Δ_B. Otherwise, the pair is rejected. For each $a \in A$ we compute $T(a)$ and look whether there is

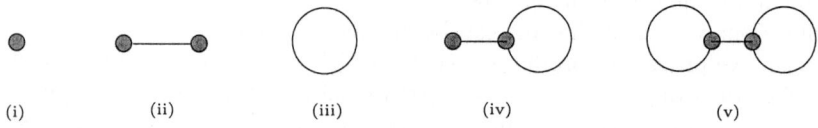

Fig. 3. Examples of different forms of central vertices

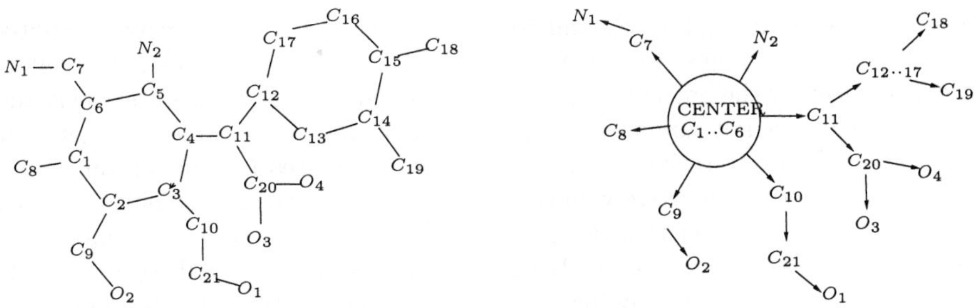

Fig. 4. A molecule and its tree representation; the center, the levels, *e.g.*, the level of atom N_1 is 2 and of atoms C_{17} and C_{13} is 3

an ε-close atom $b \in B$, and if there is one whether $Inv(a) = Inv(b)$. If the answer is 'yes' to both conditions then a and b are ε-close and we add 1 to the number of hits of T, and collect the nodes a and b into sets V'_A and V'_B, respectively.

Let S_A be the graph that has V'_A as vertices and all the edges of A that connect pairs of vertices of V'_A, and let S_B be defined similarly for B. We now compute all the *maximally connected components* in S_A and in S_B. Each maximally connected component whose number of vertices is larger than $P \cdot n$ is stored as a candidate match S_{AB} waiting to be verified in phase II.

The computation of the invariants is $O(n)$ where n is the number of vertices in the graphs. Because of the bounded degree of the vertices the total number of triplet-pairs for which we compute a transformation is $O(n^2)$. Applying the transformation T and finding the ε-close points is done in time $O(n \log n)$. Thus the time complexity of this algorithm is $O(n^3 \log n)$. We note here that by adding geometric hashing in the part that finds the point of B that is ε-close to a point in A the theoretical runtime complexity can be made $O(n^3)$. As we show in Section 3, the fast rejection in our system reduces the number of candidates to $O(n)$ in practice.

2.2 Phase II – Picking the Best Match

The candidate matches computed above are a hybrid between a graph-based match and a pointwise match but do not guarantee that each candidate match complies with the overall isomorphism of the graphs of the molecules. In this phase we verify whether a candidate match is not a false candidate by finding if there exists an isomorphic mapping f from A to B that the candidate complies with. If it does then we compute the RMS distance between A and B, based on f and T. We report the candidate match for which the RMS distance between the molecules is minimal.

Recall that a candidate match is a connected subgraph of A, with a transformation T that transforms each of its vertices to an ε-neighborhood of a vertex of B which has the same vertex invariants. Let us denote the candidate match by $S_A \subset A$ and $S_B \subset B$, the matching subgraphs of A and B, respectively.

We first check whether S_A and S_B are isomorphic. The ε-closness, computed in phase I, determines the correspondence of the graphs' vertices. It is left to confirm whether the links in S_A and S_B between matching vertices are consistent. If S_A and S_B are not isomorphic then we have to check whether a subset of them is isomorphic by iteratively removing extremal vertices (see definition below) from S_A and S_B, respectively, until they comply with the isomorphism or until the size of S_A is smaller than $P \cdot n$. (This procedure is theoretically cumbersome but hardly occurs in practice.) If the candidate is not rejected we proceed to expand the isomorphic mapping to all the vertices in $C_A = A - S_A$ and $C_B = B - S_B$.

We define the *extremal* nodes of S_A (S_B) as all the vertices of S_A (S_B) that have at least one adjacent node in C_A (C_B).

Let $u \in S_A$ be an extremal vertex. We start scanning C_A from u to the leaves of C_A, in a BFS order (here the leaves are the nodes with highest BFS level in the scan). This scan defines many paths from u. Each path is denoted by a string that contains the concatenated list of atom types on the path, *e.g.*, $NCCOC$ stands for nitrogen, carbon, carbon, oxygen, carbon. If, along the BFS scan we reach a vertex twice at the same level (a cycle is detected), then we add a special character (say '#') to the character that represents its atom type, in order to distinguish between paths that pass through cycles and paths that do not. This string is assigned to all the vertices on the path from u to the leaf v of C_A. Clearly, there can be many paths from u to v (if the paths go through cycles). The strings of all the paths that go through a vertex are assigned to the vertex and are sorted lexicographically. The collection of the sorted strings of a vertex is called the *signature* of the vertex.

We create signatures for the vertices of C_A (and C_B). We find the desired match between C_A and C_B by scanning A and B simultaneously in BFS order starting from the extremal vertices of S_A and S_B, towards C_A and C_B respectively. At each BFS level we map vertices of C_A to vertices of C_B that have the same signature. If, at some level, there is more than one vertex of C_A that have the same signature of the vertex v of C_B then we pick the vertex which is closest to v, since at this stage all options preserve isomorphism (by having equal signatures). If, at some level, we do not find a vertex with the same signature we remove this candidate from the list.

Runtime analysis: the number of candidate matches is equivalent to the number of pairs of triplets we have compared in phase I, which is in practice order $O(n)$. Defining a signature for each vertex of C_A (where $|C_A| < (1 - P) \cdot n$) cannot really be bounded, it is the number of paths which in a graph theoretical sense can be exponential, but in our case, as experimentation shows, is close to linear, since the overall runtime is close to quadratic in n.

Table 1. Summarized results: (a) Classifies the molecules by number of atoms (b) Number of molecules in this category (c) Average number of atoms in the center (d) Average node degree (e) Average number of triplets compared (f) Average number of candidates (g) Average runtime in seconds

(a) atoms	(b) mol.	(c) center size	(d) node degree	(e) compared triplets	(f) number of candidates	(g) time
< 10	14	1.36	1.75	16	15	0.06
10 − 50	389	2.02	2.10	64	59	2.00
50 − 100	123	2.18	2.17	183	93	16.00
> 100	2	7.50	2.08	182	68	48.00

3 Experimental Results

We have applied our algorithm on 528 pairs of conformational polymorphs out of about 1500 pairs in the CSD.[1] We ran the program on PC Pentium II processor Intel MMX, 64 MB RAM. The parameter values we used are as follows. $\varepsilon = 0.3$ and $P = 0.2$ (which means that $n \cdot 0.2$ atoms have to be ε-close in order to become a candidate). The performance of the algorithm is summarized in Table 1.

The first column (a) classifies the results by the size of the molecule n (the number of atoms). The second column holds the number of molecules of this size. As shown in the table, the average size of the center (c) is mostly close to 2. The deviation in the fourth row is from an atypically large cycle in one of the samples which contains 13 atoms in an elementary cycle. The average node degree (d) over the whole graph nodes is almost constant. As expected, the number of remaining triplets after applying just graph invariants (e) grows linearly with n, and the number of candidates that comply with the minimum substructure size requirement (f) is naturally smaller than that of the triplets. The runtime presented in (g) is about quadratic in n. It should be noted that when the molecules are assymetric the algorithm runs much faster. We present two examples for conformational polymorphs and the results they yield. Fig. 5 presents two conformational polymorphs of hexanitroazobenzene, a high energy material, and the overlaid of the two molecules. Blue spheres represent the common substructure (ε-close) and the green and red spheres represent the unmatched atoms in A and B respectively. The hexanitroazobenzene molecule has two cycles linked together, from the figure we see that the polymorphs of this compound have one geometric identical cycle and the other cycles are perpendicular to each other. Another example is the virazol (antiviral agent) molecule, an anti-AIDS compound. Fig. 6 describes the two conformations and their overlay. The RMS distance is presented in these figures. In the calculation of the RMS distance we ignored the hydrogen atoms in order to produce more accurate results.

---

[1] There are conformational polymorphs that do not have the same number of atoms. This happens when in the crystallization process some solvent atoms crystallize with the polymorph. Our program is not deal with it but will in the future.

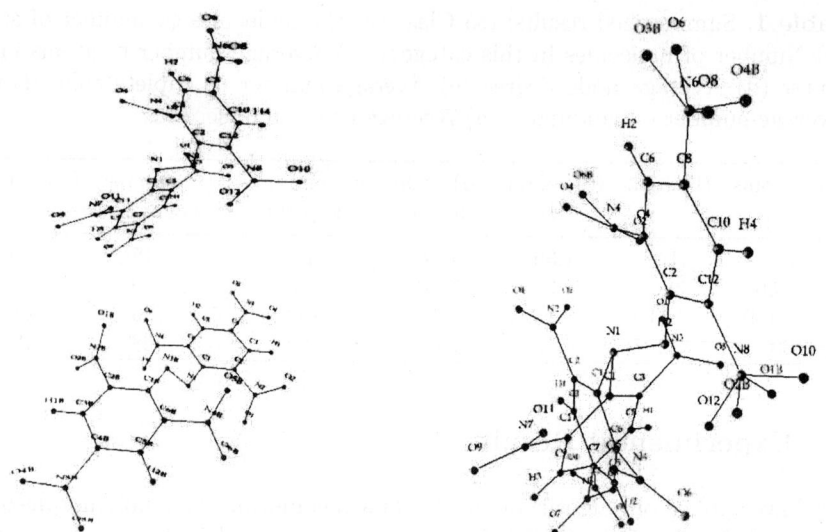

Fig. 5. Two conformational polymorphs of hexanitroazobenzene(left). The overlaid polymorphs of ll hexanitroazobenzene (right), RMS-distance=1.785

4 Conclusion and Suggestions for Further Research:

We proposed, tested and validated an algorithm for polymorph structure alignment. We tested our system on 528 pairs of conformational polymorphs. We have detected a large number (about 10%) of polymorph pairs that are identical in shape. This result in itself is interesting.

Another measure of similarity that we implemented and tested, is reported in [9]. We simply allow in our system to pick the largest correct candidate match as the best match and output the number of hits.

Though the sizes of the molecules are small (up to 200 atoms, mostly below 100 atoms), the experiments show that the runtime does not explode with n and thus this method provides a very useful tool for chemists.

In the Cambridge Structural Database [6] there are over 3000 entries which contain qualifying descriptions as being members of polymorphic systems. Our next goal is to find unique geometric features that are common to various polymorphic forms. This may give us a clue about the existence of polymorphic forms for other compounds.

References

[1] H. G. Barrow and R. M. Burstall, "Subgraph isomorphism, matching relational structures and maximal cliques", *Information Processing Letters*, 4:83-84, 1976.
[2] J. Bernstein, "Conformational polymorphism", Chapter 13 in Organic solid-state Chemistry, G.Desiraju, ed. from Studies in Organic Chemistry, Elsevier, Amsterdam Vol 32, 1987.

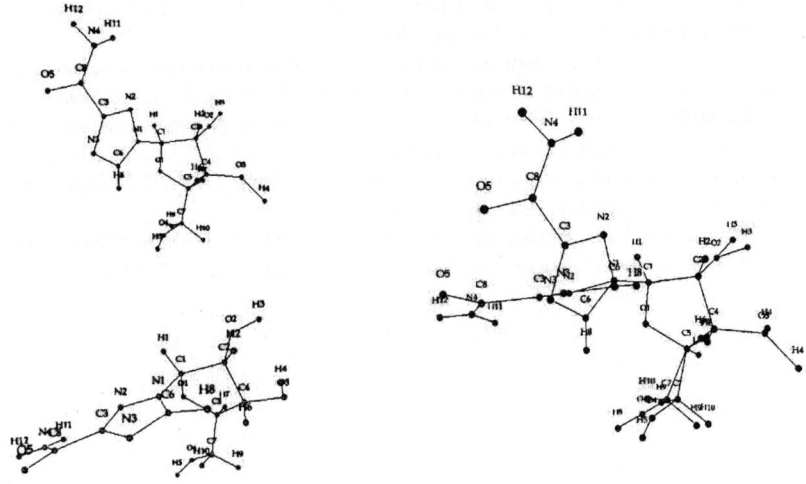

Fig. 6. Two conformational polymorphs of virazole (left). The overlaid polymorphs of virazole (right), RMS-distance=1.892

[3] J. Bernstein, R. E. Davis, L. Shimoni, N. L. Chang, "Patterns in hydrogen bonding: functionality and graph set analysis in crystals", *Angew. Chem. Int. Ed. Engl.* 34:1555-1573, 1995.
[4] R. Blom and A. Haaland, *J. Mol. Struc.* 128:21-27, 1985.
[5] T. H. Cormen, C. E. Leiserson and R. L. Rivest, "Introduction to algorithms", The MIT press, Cambridge, 1990.
[6] CSD: Cambridge Structural Database, http://www.ccdc.cam.ac.uk.
[7] J. D. Dunitz and J. Bernstein, "Disappearing polymorphs", *Accounts of Chemical Research*, 28:193-200 1995.
[8] J. P. Doucet and J. Weber, Computer-aided molecular design: Theory and applications, Academic Press 1996.
[9] A. Enosh, "Determining similarity of conformational polymorphs", M.SC. thesis, Computer Science Department, Ben-Gurion University, Israel.
[10] S. Fortin. "The graph isomorphism problem", Technical Report University of Alberta, 96-20 1996.
[11] L. R. Foulds, Graph theory applications, Springer 1992.
[12] P. Finn, D. Halperin, L. Kavraki, J. Latombe, R. Motwani, C. Shelton, and S. Venkatsubramanian, "Geometric manipulation of flexible ligands", *Applied Computational Geometry: Towards Geometric Engineering*, LNCS 1148:67-78. Springer-Verlag, 1996.
[13] D. Fischer, R. Nussinov and H. J. Wolfson, "3-D substructure matching in protein molecules", In *Proc. 3rd. Annual Symposium on Combinatorial Pattern Matching*, Tucson, Arizona, USA. LNCS 644:136-150. Springer-Verlag, 1992.
[14] J. E. Hopcroft and R. E. Tarjan, "Isomorphism of planar graphs", Complexity of Computer Computation, R. E. Miller and J. W. Thatcher, editors, Plenum, New York, 131-152, 1972.

[15] A. M. Lesk, "Detection of 3-D patterns of atoms in chemical structures", *Communications of ACM*, 22:219-224, 1979.
[16] E. M. Luks, "Isomorphism of graphs of bounded valence can be tested in polynomial time", *Journal of Computer and System Sciences*, 25:42-65, 1982.
[17] G. L. Miller, "Isomorphism of k-contractible graphs. A generalization of bounded valence and bounded genus", *Information and Control*, 56:1-20, 1983.
[18] H. J. Wolfson and I. Rigoutsos, "Geometric hashing", *IEEE Computational Science and Engineering*, 4(4):10-21, 1997.
[19] K. Yamazaki, H. L. Bodlaender, B. Fluiter and D. M. Thilikos, "Isomorphism for graphs of bounded distance width", *Algorithmica*, 24(2):105-127, 1999.

Minimizing the Maximum Starting Time On-line

Leah Epstein[1,*] and Rob van Stee[2,**]

[1] School of Computer Science, The Interdisciplinary Center
Herzliya, Israel
lea@idc.ac.il
[2] Institut für Informatik, Albert-Ludwigs-Universität
Georges-Köhler-Allee, 79110 Freiburg, Germany
vanstee@informatik.uni-freiburg.de

Abstract. We study the scheduling problem of minimizing the maximum starting time on-line. The goal is to minimize the last time that a job starts. We show that while the greedy algorithm has a competitive ratio of $\Theta(\log m)$, we can give a constant competitive algorithm for this problem. We also show that the greedy algorithm is optimal for resource augmentation in the sense that it requires $2m - 1$ machines to have a competitive ratio of 1, whereas no algorithm can achieve this with $2m - 2$ machines.

1 Introduction

In this paper, we study on-line multiprocessor scheduling with a new objective function: the *maximum starting time*. Jobs arrive on-line to be scheduled on m parallel machines. These machines can be either identical or *related*, in which case each machine has a speed that determines how long it takes to run one unit of work.

We study the on-line paradigm where jobs arrive one by one. A job J_j is defined by its size and by its order in the input sequence. Denote the starting time of job J_j by S_j. We denote the cost of an algorithm $\mathcal{A}$ on a job sequence $\sigma = \{J_1, \ldots, J_n\}$ by $\mathcal{A}(\sigma) = \max_j S_j$. An algorithm is required to run the jobs on each machine in the order of arrival.

An example of this situation is the following. There is a loading station where trucks are loaded with goods. These goods need to be delivered to different places, after which the trucks return to the loading station to pick up a new load. At the end of a work day, the station can close as soon as the truck carrying the last load has left, and does not need to wait for the trucks to return. The time it takes to deliver the goods in one truck is the size of the job. (Here we consider a truck load to be "one job", e. g. each truck contains only one item, or items for only one destination (client).

---

* Research supported by Israel Science Foundation (grant no. 250/01)
** This work done while the author was at the CWI, The Netherlands. Research supported by the Netherlands Organization for Scientific Research (NWO), project number SION 612-30-002

We use two measures to study the performance of on-line algorithms. The *competitive ratio* compares an on-line algorithm to an optimal off-line algorithm OPT that knows the job sequence in advance (but can not change the order in which jobs run on a machine, i.e. it also has to run jobs on a machine in the order of their arrival). The competitive ratio $\mathcal{R}(\mathcal{A})$ of an on-line algorithm $\mathcal{A}$ is the infimum value of $\mathcal{R}$ such that for every sequence σ,

$$\mathcal{A}(\sigma) \leq \mathcal{R} \cdot \text{OPT}(\sigma) . \tag{1}$$

The second measure involves *resource augmentation*. Assume the on-line algorithm uses $\tilde{m}$ machines, where $\tilde{m} > m$. What is the minimum value of $\tilde{m}$ such that the cost of the on-line algorithm is bounded by the cost of the optimal off-line algorithm (i.e. the competitive ratio is at most 1)? Resource augmentation was originally introduced by [11], and further widely studied for various scheduling and load balancing problems [5, 7, 11, 12, 14].

Note that if a sequence σ contains at most m jobs, then $\text{OPT}(\sigma) = 0$. By (1), any algorithm with finite competitive ratio needs to have zero cost and run all jobs on different machines in that case.

All previous work assumed that the output needs to be collected by the same system, and hence the last *completion* time was considered in numerous papers [10, 4, 13, 1, 9, 8]. Other papers also considered different functions of the completion times [2] but never the starting times. Resource augmentation for scheduling of jobs one by one was also considered with the maximum completion time goal function [6, 3]. However, to the best of our knowledge, no previous work on the above goal function exists.

We show the following results for the competitive ratio on identical machines:

- The greedy algorithm, which assigns each job to the least loaded machine, has competitive ratio $\Theta(\log m)$.
- The greedy algorithm has optimal competitive ratios for 2 and 3 machines, which are 2 and 5/2 respectively.
- There exists a constant competitive algorithm BALANCE which has competitive ratio 12 for any m (hence the greedy algorithm is far from having optimal competitive ratio for general m).
- No deterministic algorithm for general m has competitive ratio smaller than 4.

For two related machines, we give in the full paper a tight bound of $q+1$ for the competitive ratio, where q is the speed of the fastest machine relative to the slowest. We omit these results from this extended abstract.

We show the following results for resource augmentation on identical machines:

- The greedy algorithm has competitive ratio 1 if it uses $2m-1$ machines (and is compared to an optimal off-line algorithm with m machines).
- Any on-line algorithm which uses $2m-2$ machines has competitive ratio larger than 1, and any on-line algorithm which uses $2m-1$ machines has competitive ratio of at least 1. Hence the greedy algorithm is optimal in this measure.

Note that the off-line version of minimizing the maximum starting time is strongly NP-hard. The off-line problem of minimizing the maximum completion time (minimizing the makespan) is a special case of our problem. A simple reduction from the makespan problem to our problem can be given by adding m very large jobs (larger than the sum of all other jobs) in the end of the sequence. Each machine is forced to have one such job, and the maximum starting time of the large jobs, is the makespan of the original sequence.

We present results on the greedy algorithm in Section 2, the constant competitive algorithm BALANCE in Section 3, lower bounds in Section 4 and results for resource augmentation in Section 5.

2 The Greedy Algorithm

GREEDY always assigns an arriving job on the machine where it can start the earliest (see [10]). In some upper bound proofs we use the following definition: a *final* job is a job that starts as the last job on some machine in OPT's schedule.

Theorem 1. $\mathcal{R}(\text{GREEDY}) = \Theta(\log m)$ *on identical machines.*

Proof. Let $\lambda = \text{OPT}(\sigma)$. Note that all on-line machines are occupied until time $\text{GREEDY}(\sigma)$. We cut the schedule of GREEDY into pieces of time length 2λ starting from the bottom.

If there are less than m final jobs, there are less than m jobs, hence GREEDY is optimal. Suppose there are m final jobs.

Claim: At time $2i\lambda$, at most $m/2^i$ final jobs did not start yet.

Proof: By induction. The claim holds for $i = 0$. Assume it holds for some $i \geq 0$.

A final job is called *missing* if it did not start before time $2\lambda i$. Let k be the number of missing jobs. We have $k \leq m/2^i$ starting at time $2\lambda i$ or later. The total size of non-final jobs running at any time after $2\lambda i$ is at most $k\lambda$. This follows because GREEDY schedules the jobs with monotonically increasing start times, hence if there are k missing final jobs, then all the unstarted jobs must have arrived after the $m-k$-th final job. That job is started before time $2\lambda_i$ and hence the unstarted jobs must be scheduled by OPT on the machines where it runs the last k final jobs. Since OPT completes all these (non-final) jobs no later than at time λ, the total size of these jobs is at most $k\lambda$.

At most $k/2$ machines can be busy with these jobs during the entire time interval $[2\lambda i, 2\lambda(i+1)]$. Hence $k/2$ or more final jobs start in this interval (one for every machine that is not busy with non-final jobs during the entire interval and that was also not running a final job already). At most $k/2$ final jobs will be missing at time $2\lambda(i+1)$, and $k/2 \leq m/2^{i+1}$. □

At time $2\lambda \log_2 m$, only one final job is missing, therefore $\text{GREEDY}(\sigma) \leq 2\lambda \log_2 m + \lambda$, hence $\mathcal{R}(\text{GREEDY}) = O(\log m)$.

To show that $\mathcal{R}(\text{GREEDY}) = \Omega(\log m)$, we use a job sequence that consists of a job of size 1 followed by a job of size M (a large constant, e.g. $M = m$), repeated m times. The optimal algorithm can assign the jobs so that no job starts later than at time 1, whereas GREEDY starts the last job at time $1 + \lfloor \log_2 m \rfloor$. □

We now consider the competitive ratio of GREEDY for $m = 2, 3$. In Section 4, we will show matching lower bounds. Hence, GREEDY is optimal for $m = 2, 3$.

Lemma 1. *On identical machines, $\mathcal{R}(\text{GREEDY}) \le 2$ for $m = 2$, and we have $\mathcal{R}(\text{GREEDY}) \le 5/2$ for $m = 3$.*

Proof. We omit the case $m = 2$ in this extended abstract; it is similar to the case $m = 3$.

For $m = 3$, suppose GREEDY has competitive ratio $\rho > 5/2$ and define σ and J_ℓ as above (taking $\varepsilon = \frac{1}{2}(\rho - 5/2)$). Assume $\text{OPT}(\sigma) = 1$. Denote the total size of all jobs but J_ℓ by V. Note that the size of J_ℓ is irrelevant for the competitive ratio; we may assume it has size 0. Denote the total size of all jobs of size at most 1 by V'. Since $\text{OPT}(\sigma) = 1$, OPT starts all its jobs no later than at time 1; the jobs that it completes before time 1 have total size at most 3.

We have $V \ge 3(\rho - \varepsilon) > 15/2$, since all three of GREEDY's machines are busy until past time $\rho - \varepsilon > 5/2$ when J_ℓ arrives.

- If σ contains no jobs larger than 1, consider the optimal off-line schedule. Two final jobs are of size at most 1, and the third (J_ℓ) is of size 0. The rest of the jobs are completed by time 1, and their total size is at most 3. Hence $V = V' \le 5$, a contradiction.
- If σ contains one job larger than 1, then $V' \le 4$: one final job has size 0, and one must have size at most 1 (since only one can be larger than 1). The rest of the jobs are of size at most 1, and have total size at most 3. Consider the least loaded machine among the two machines that do not run the job larger than 1, at the time J_ℓ arrives. Since $V' \le 4$, it cannot have a load more than 2. But then GREEDY starts J_ℓ no later than at time 2.
- If σ contains two jobs larger than 1, then analogously to the previous cases, $V' \le 3$. Denote the time that GREEDY starts the second large job by t_2. Similarly to in the previous case, we have $t_2 \le 3/2 < 5/2$. At most a volume of 1 of jobs starts after t_2, since OPT has to run all these jobs and J_ℓ on one machine if $\text{OPT}(\sigma) = 1$: two of OPT's machines are already running large jobs and cannot be used anymore.
  - If $t_2 \ge 1/2$, then in the worst case GREEDY assigns all the jobs that arrive after t_2 to one machine and starts J_ℓ no later than at time $5/2$.
  - If $t_2 < 1/2$, then at the time the second large job arrives GREEDY starts no job later than at time $1/2$. Hence the on-line machine that has no large job has load at most $3/2$ at this time, since all jobs on that machine have size at most 1 and GREEDY always uses the least loaded machine. Since after t_2, at most a volume 1 of jobs still arrives, J_ℓ starts no later than at time $5/2$. □

3   Algorithm BALANCE

We give an algorithm for identical machines of competitive ratio 12. This algorithm works in phases and uses an estimate on OPT(σ) which is denoted by λ. A job is called *large* if its size is more than λ, and *small* otherwise; if $\lambda \geq \text{OPT}(\sigma)$, OPT can only run one such job on each machine. Also, once OPT has done this, it cannot use that machine anymore for any job.

A phase of BALANCE ends if it is clear from the small jobs that arrived in the phase, and from the large jobs that exist, that if another job arrives then $\lambda \leq \text{OPT}(\sigma)$. In this case we double λ and start a new phase.

In every phase, BALANCE only uses machines that do not already have large jobs. Each such machine will receive jobs according to one of the two following possibilities.

1. Only small jobs, of total weight in that phase less than 3λ.
2. Small jobs of weight less than 2λ, and one large job on top of them.

A machine that received a large job is called *large-heavy*, a machine that received weight of at least 2λ of small jobs in the current phase is called *small-heavy*. Both small-heavy and large-heavy machines are considered *heavy*. A machine that received more than a weight of λ of small jobs in the current phase but at most 2λ (and no large job) is considered *half-heavy*. Other machines are *non-heavy*. A machine that is not heavy (but possibly half-heavy) is called *active*. The algorithm BALANCE also maintains a set Q that contains the active machines.

Define λ_i as the value of λ in phase i. The algorithm BALANCE starts with phase 0 which is different from the other phases. In phase 0, m jobs arrive that are assigned to different machines. We then set λ_0 equal to the size of the smallest job that has arrived. Then the first of the regular phases starts.

Phases: A new phase starts when $Q = \emptyset$, i. e. there are no active machines anymore. (Phase 1 starts when phase 0 ends.) At the start of phase $i > 0$, we set $\lambda_i = 2\lambda_{i-1}$. Then Q contains all machines that do not have a large job. This holds because no machine has yet received any job in the current phase, so no machine can be small-heavy. Note that such a large job has arrived in some previous phase, but that the definition of large jobs has changed compared to the previous phase. I. e. not all the large jobs from previous phases are still large.

At all times, the algorithm only uses active machines. When the phase starts, all active machines are non-heavy. Each phase consists of two parts. The first part continues as long as there is at least one non-heavy machine among the active machines. As soon as no machine is non-heavy, the second part starts.

Part 1 For small jobs, BALANCE uses the machines in Q in a Next Fit-fashion, moving to the next machine as soon as a machine has received a load of more than λ_i in the current phase. An arriving large job is assigned to a machine that already has weight of more than λ_i. If no such machine exists, it is assigned to the active machine that BALANCE is currently using or going to use for small jobs (there is a unique such machine, and all other non-heavy machines did not receive any jobs in the current phase). A machine that receives a large job becomes large-heavy, and is removed from Q.

Part 2 We again start using the machines in Q in a Next Fit-fashion, moving to the next machine as soon as the machine has received a total load at least $2\lambda_i$ in the current phase. A machine that receives weight of at least $2\lambda_i$ of small jobs in total in this phase becomes small-heavy and hence stops being active (is removed from Q). A machine that receives a large job becomes large-heavy and also stops being active (it is removed from Q).

As long as $|Q| > 0$, there are active machines. When $Q = \emptyset$, a new phase starts. An example of a run of BALANCE can be seen in Figure 1.

We show that as soon as a first job in the new phase arrives, then $\lambda_{i-1} \leq \text{OPT}(\sigma)$. (Note that it is possible that no jobs arrive in a phase; this happens if $Q = \emptyset$ at the beginning of a phase.)

Lemma 2. *In each phase $i > 0$ in which jobs arrive, we have $\text{OPT}(\sigma) \geq \lambda_i/2$, where σ is the sequence of jobs that arrived until phase i, including the first job of phase i.*

Proof. The lemma holds for phase 1, since there is at least one machine of the optimal off-line algorithm that has two scheduled jobs after the first job in phase 1 arrives.

Consider a phase $i > 1$. If phase i starts when phase $i - 1$ is still in its first part, then no machines are small-heavy. Hence in total m jobs have arrived that were considered large in phase $i - 1$ (where some may have arrived before phase $i - 1$). After the first job arrives in phase i, we have $\text{OPT}(\sigma) > \lambda_{i-1} = \lambda_i/2$.

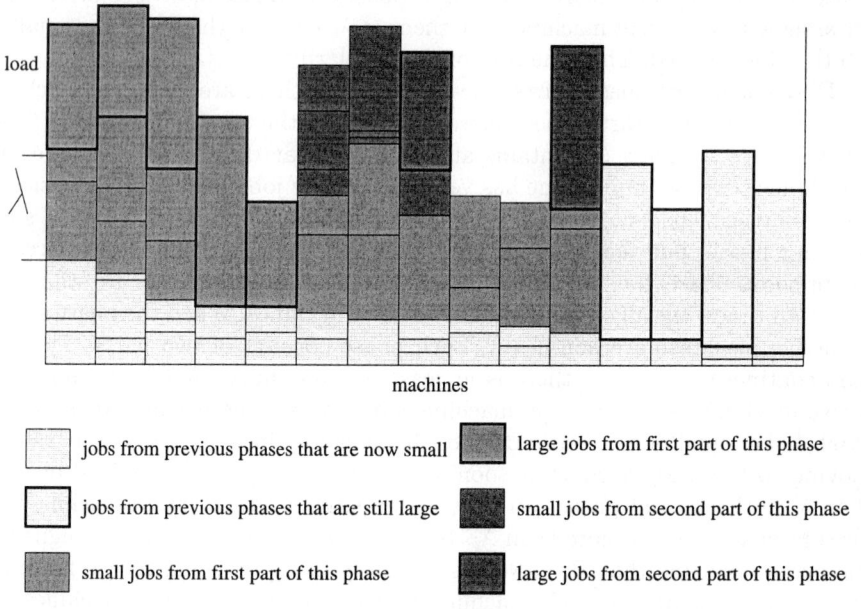

Fig. 1. A run of BALANCE

If phase i starts while phase $i-1$ is in its second part, let K be the set of large jobs that were assigned to non-heavy machines in phase $i-1$. (If no such jobs exist, $K = \emptyset$). The jobs in K arrived in part 1 of phase $i-1$, since in part 2 only half-heavy machines are used. In part 1 of a phase, the active machines that have already been used are half-heavy or large-heavy.

Assume by contradiction that $\text{OPT}(\sigma) < \lambda_{i-1}$. Suppose $K \neq \emptyset$. Denote the last job in K by J_K and denote the set of machines that are still active after J_K has arrived by Q'. Write $q = |Q'|$. There was no half-heavy machines available for J_K, so all the machines that already received jobs in phase $i-1$, including the one that received J_K, are large-heavy at this point (they cannot be small-heavy in part 1). If $K = \emptyset$, define Q' as the set of active machines at the start of phase $i-1$. Clearly, all machines not in Q' are large-heavy at that point.

From this, we have that there exist $m-q$ large jobs after J_K has arrived (or at the start of phase $i-1$): all machines not in Q' either were large-heavy when phase $i-1$ started, or became large-heavy during it. Hence there are $m-q$ machines of OPT with a large job, since OPT cannot put two large jobs on one machine; OPT cannot put any more jobs on those machines if $\text{OPT}(\sigma) < \lambda_{i-1}$. Consider the set Q'_{OPT} of machines of OPT that do not run any of the $m-q$ large jobs that arrived already. We have $|Q'_{\text{OPT}}| = |Q'| = q$.

We calculate how much weight can be assigned by BALANCE to the machines in Q' (or equivalently, by OPT to the machines in Q'_{OPT}) in the remainder of phase $i-1$. In the schedule of OPT, the machines in Q'_{OPT} have some q jobs running last on them. Apart from that they have at most an amount of $\text{OPT}(\sigma) < \lambda_{i-1}$ small jobs.

Let $q_1 \leq q$ be the number of large jobs assigned by BALANCE to machines in Q' in the remainder of phase $i-1$. At the end of phase $i-1$, each machine in Q' is either small-heavy, or has an amount of at least λ_{i-1} small jobs and a large job. The total weight of small jobs assigned in phase $i-1$ to the machines of Q' by BALANCE is at least $(2q - q_1)\lambda_i$.

Suppose we remove the q largest jobs assigned in phase $i-1$ to the machines of the set Q' in the assignment of BALANCE. This means that we remove q_1 large jobs and $q - q_1$ small jobs. By definition, each small removed job has size of at most λ_{i-1}, so we removed at most an amount of $(q-q_1)\lambda_{i-1}$ small jobs. Therefore we are left with total weight of at least $q\lambda_{i-1}$ on the machines in Q', counting only weight from jobs that arrived in this phase.

This implies that even if OPT runs the largest q jobs last on the machines in Q'_{OPT}, it starts at least one of them at time λ_{i-1} or later, by the total weight of the other jobs. This gives a contradiction, already without the first job in phase i. This proves the lemma. □

Theorem 2. *Algorithm* BALANCE *has a competitive ratio of 12.*

Proof. Consider the last phase $\ell > 0$ in which jobs arrived. (If $\ell = 0$, BALANCE is optimal.) Let $\Lambda = \lambda_\ell$. We have $\text{OPT} \geq \Lambda/2$ by Lemma 2. Consider the machines that received jobs in phase ℓ, and for each such machine, consider the total size of jobs below the last job that is run on that machine. (For the machines that did not receive jobs in this phase, we have stronger bounds.) This size consists of three parts:

- The small jobs of phase ℓ
- The small jobs of previous phases
- The large jobs of previous phases

For the computation, for phases $0 < i < \ell$ in which a machine got only small jobs, we replace an amount of $2\lambda_i$ of small jobs from that phase by one (large) job. (Possibly a small job is broken in two parts to get a total of exactly $2\lambda_i$.) Because we only consider machines that received jobs in phase ℓ, the maximum starting time is unaffected by this substitution. As a result, each machine receives at most a weight of $2\lambda_i$ of small jobs in phase i.

In phase ℓ, each machine receives at most 2Λ of small jobs before it receives its last job. The value of λ is doubled between phases, hence the total amount of small jobs from previous phases on a machine is at most $\sum_{i<\ell} 2\lambda_i \leq 2\Lambda$.

We still need to consider the large jobs from previous phases. We count the large jobs not by the phases they arrive; instead, each large job is counted in the first phase where it is not large anymore, and the machine is active again. The large jobs that replace 2Λ worth of small jobs as described above, are always already small in the subsequent phase. For each phase $i \leq \ell$, a machine has at most one job that has just become small. This job is of size at most λ_i. Hence in total the size of all these jobs is at most $\sum_{i \leq \ell} \lambda_i \leq 2\Lambda$. Therefore the total load below the last jobs on any machine is at most $2\Lambda + 2\Lambda + 2\Lambda \leq 6\Lambda$. Since $\Lambda \leq 2\text{OPT}$, we are done. □

4 Lower Bounds

In the following proofs, we take M to be a large constant. If we construct a job sequence σ that contains a job of size M, then we assume that M is larger than $\mathcal{R}$ times the sum of smaller jobs in σ, where $\mathcal{R}$ is the competitive ratio that we want to show. This choice of M ensures that if a machine is assigned a job of size M, it cannot receive any other job after this without violating the competitive ratio. For the proof of the following lemma, we refer to the full paper.

Lemma 3. *Suppose we have a job sequence σ that shows that $\mathcal{R}(\mathcal{A}) \geq \mathcal{R}$ for all on-line algorithms on m_1 machines. Then for any $m > m_1$, $\mathcal{R}(\mathcal{A}) \geq \mathcal{R}$ for all on-line algorithms on m machines, as well.*

Theorem 3. *Take $\alpha = (\sqrt{5}+1)/2 \approx 1.618$ and M a large constant. For all on-line algorithms $\mathcal{A}$, we have the following lower bounds for the competitive ratio.*

Number of machines	Job sequence	$\mathcal{R}$
2	$1, M, 1, M$	2
3	$1/2, 1/2, M, 1, M, 1, M$	$5/2$
4	$\alpha-1, \alpha-1, M, M, 1, M, 1, M$	$\alpha + 1 \approx 2.618$

Moreover, as the number of machines tends to infinity, the competitive ratio tends to at least 4.

Proof. For $m \leq 4$, we use the job sequences described in the table above. For these sequences, any on-line algorithm that has a better competitive ratio than in the last column of the table must assign these jobs in the same way as the greedy algorithm, or violate the competitive ratio. In all cases, after the last job arrives we have $\text{OPT}(\sigma) = 1$ and $\mathcal{A}(\sigma) = \mathcal{R}$.

As an example, for $m = 4$, the first four jobs must be assigned to four different machines, the next two jobs to the machines with the jobs of size $\alpha - 1$, and the last two to the machine that does not have a job of size M yet. The sequence stops as soon as $\mathcal{A}$ assigns a job differently than described here, or after the fourth large job.

For larger m, we use the following job sequence. Assume $m = 2^r$ for some $r \geq 3$, and consider a sequence of real numbers $\{k_i\}_{i=0}^{\infty}$ with properties to be defined later. We will first define the job sequence and then specify for which value of r it works. The job sequence consists of $r + 1$ steps. For $1 \leq i \leq r$, in step i first $m/2^i$ jobs of size k_i arrive, and then $m/2^i$ jobs of size M. In step $r + 1$, one last job of size k_r arrives, followed by a job of size M.

We denote the optimal maximum starting time after step i by OPT_i. If $k_i \leq k_{i+1}$ for all $1 \leq i \leq r-1$, then for $1 \leq i \leq r$, we have $\text{OPT}_i = k_{i-1}$ (we put $k_0 = 0$), which is seen as follows. We describe the optimal schedule after step i. (We note that the optimal schedules after different steps can be very different.) There are $m/2^i$ machines with one job of size k_i, and $m/2^i$ machines with one job of size M. These machines do not have any other jobs. The remaining machines have one job of size k_s for some $s < i$, and after it one job of size M. After the last step we have $\text{OPT}_{r+1} = k_r$. In this case, all machines have one job of size k_s for some $s \leq r$, and after it one job of size M.

We will now define the sequence $\{k_i\}_{i=0}^{\infty}$ in such a way that the on-line algorithm cannot place two jobs on the same machine in one step. By induction we can see that at the start of step i $(1 \leq i \leq r)$, $m(1 - 1/2^{i-1})$ jobs of size M have already arrived. Thus if the on-line algorithm places the $m/2^{i-1}$ jobs from step i on different machines (that moreover do not have a job of size M yet), then also by induction, after every step i $(1 \leq i \leq r)$, every machine of the on-line algorithm either has a job of size M, or it has one job of *each* size k_j, for $1 \leq j \leq i$.

Define $s_i = \sum_{j=1}^{i} k_j$. If the on-line algorithm does put two jobs on the same machine in some step $i \leq r$, then by the above the last job on that machine starts at time $\sum_{j=1}^{i} k_j$ and the implied ratio is

$$\mathcal{R}_i = \frac{\sum_{j=1}^{i} k_j}{k_{i-1}} = \frac{s_i}{k_{i-1}}. \qquad (2)$$

If the on-line algorithm never does this, then in the final step $r + 1$ it has only $m/2^r = 1$ machine left without a job of size M, and this machine has one job of each size k_j for $1 \leq j \leq i$. The on-line algorithm has minimal cost if it places the two jobs from step $i + 1$ on this machine, and the implied competitive ratio is thus $\mathcal{R}_{r+1} = (\sum_{j=1}^{r} k_j + k_r)/k_r = (s_r + k_r)/k_r$. Using (2), we will define the sequence $\{k_i\}_{i=0}^{\infty}$ so that $\mathcal{R}_i = \mathcal{R}$ is a constant for $1 \leq i \leq r + 1$. This implies $k_0 = 0$, $k_1 = 1$, $k_i = \mathcal{R}k_{i-1} - \sum_{j=1}^{i-1} k_j = \mathcal{R}k_{i-1} - s_{i-1}$ for $i > 1$. This proves a competitive ratio of $\mathcal{R}$ if $k_i \le k_{i+1}$ for $1 \le i \le r-1$ and $(s_r + k_r)/k_r \ge \mathcal{R}$ (where this last condition follows from step $r+1$). We have

$$(s_r + k_r)/k_r \ge \mathcal{R} \iff k_r + s_r \ge \mathcal{R} k_r = s_{r+1} \quad \text{using (2)}$$
$$\iff k_r \ge s_{r+1} - s_r = k_{r+1}$$
$$\iff s_{r+1} \ge s_{r+2} \iff k_{r+2} \le 0 \iff s_{r+3} \le 0.$$

Hence it is sufficient to show that the sequence $\{s_i\}_{i=0}^\infty$ has its first nonpositive term s_{r+3} for some $r \ge 1$. This value of r determines for which m this job sequence shows a lower bound of $\mathcal{R}$, since $m = 2^r$. Note that if s_{r+3} is nonpositive, we have to stop the job sequence after step $r+1$ at the latest, because by the above $k_{r+2} \le 0 < k_1 \le k_r$: the sequence is no longer non-decreasing. As stated above, we will in fact give one final job of size k_r in step $r+1$, and a job of size M, and thus not use any value k_i for $i > r$. The sequence $\{s_i\}_{i=0}^\infty$ satisfies the recurrence $s_{i+2} - \mathcal{R}s_{i+1} + \mathcal{R}s_i = 0$. For $\mathcal{R} < 4$, the solution of this recurrence is given by

$$s_i = \frac{2\sin(\theta i)\sqrt{\mathcal{R}}^i}{\sqrt{4\mathcal{R} - \mathcal{R}^2}} \quad \text{where} \quad \cos\theta = -\frac{1}{2}\sqrt{\mathcal{R}} \quad \text{and} \quad \sin\theta = \sqrt{1 - \frac{\mathcal{R}}{4}}.$$

Since $\sin\theta \ne 0$, then $\theta \ne 0$, which implies $s_i < 0$ for some value of i. Furthermore, this value of i tends to ∞ as $\mathcal{R}$ tends to 4 from below. Direct calculations show that for $i = 1, 2, 3, 4$, $s_i > 0$, hence given such minimal integer i, we can define $r = i - 3$. From the calculations it also follows that $\{k_i\}_{i=0}^r$ is non-decreasing.

In conclusion, for any value of $\mathcal{R} < 4$ it is possible to find a value r so that any on-line algorithm has at least a competitive ratio of $\mathcal{R}$ on 2^r machines. By Lemma 3, this implies that for every $\varepsilon > 0$, there exists a value m_1 such that for any on-line algorithm $\mathcal{A}$ on $m > m_1$ machines, $\mathcal{R}(\mathcal{A}) \ge 4 - \varepsilon$. □

Note that this proof does not hold for $\mathcal{R} \ge 4$, because the solution of the recurrence in that case is not guaranteed to be below 0 for any i.

Corollary 1. *On identical machines,* GREEDY *is optimal for* $m = 2, 3$.

Proof. This follows from Lemma 1 and Theorem 3. □

5 Resource Augmentation

We now consider on-line algorithms that have more resources than the off-line algorithm. It turns out that in these changed circumstances, GREEDY is optimal in the sense that it requires the minimum possible number of machines to have a competitive ratio of 1. We only consider identical machines in this section.

Lemma 4. $\mathcal{R}(\text{GREEDY}) = 1$ *if it has at least* $2m - 1$ *machines.*

Proof. Let $h = \text{GREEDY}(\sigma)$ and $h^* = \text{OPT}(\sigma)$. Note that the last job J_ℓ that is assigned at time h by GREEDY is a final job for OPT as well, since this is the very last job in the sequence. Let S be the set of on-line machines of GREEDY that only contain non-final jobs or J_ℓ. Since there are at most m final jobs, $|S| \geq 2m - 1 - (m-1) = m$. All of GREEDY's machines are occupied from 0 to h. The machines in S are occupied during this time by non-final jobs. Let W be the total size of non-final jobs. We have $W \geq mh$. But $W \leq h^*m$. Hence $h \leq h^*$. □

Note that a similar proof shows that the competitive ratio of GREEDY tends to zero as the number of on-line machines tends to ∞.

Lemma 5. *Any algorithm that has at most $2m - 2$ machines has a competitive ratio greater than 1.*

Proof. Suppose $\mathcal{A}$ has a competitive ratio of at most 1. We use a construction in phases, where in each phase the size of the arriving jobs is equal to the total size of all the jobs from the previous phases. Let n_i denote the number of jobs in phase i, and M_i denote the size of the jobs in phase i. We determine the number of phases later. We take $n_0 = m$ and $n_i = 2m - 1$ for $i > 0$. Furthermore, we take $M_0 = 1$, $M_1 = n_0 M_0 = m$ and

$$M_i = \sum_{j=0}^{i-1} n_j M_j = \sum_{j=0}^{i-2} n_j M_j + n_{i-1} M_{i-1} = 2m M_{i-1} \text{ for } i > 1.$$

Claim: After i phases, at least $\min(m + (m-1)(1 - \frac{1}{2^i}), 2m - 2)$ machines are non-empty.

Proof: We use an induction. All jobs from phase 0 have to be assigned to different machines to have a finite competitive ratio, so m machines are non-empty after phase 0.

Consider phase i for $i > 0$. During each phase $i > 0$, the optimal costs are at most M_i: all the jobs from the previous phases go together on one machine, followed by one job of size M_i. All other machines have two jobs of size M_i. In order to have a competitive ratio of 1, $\mathcal{A}$ can assign at most one job of size M_i on each non-empty machine, and at most 2 such jobs on each empty machine. Let x be the number of non-empty machines at the start of phase i. If $x = 2m - 2$ we are done immediately. Else, we have $x \geq m + (m-1)(1 - \frac{1}{2^{i-1}})$ by induction. The number of machines that become non-empty in phase i is at least $(2m - 1 - x)/2$, so after phase i, at least $m - \frac{1}{2} - \frac{1}{2}x + x$ machines are non-empty. By induction, we have $m - \frac{1}{2} + \frac{1}{2}x \geq m - \frac{1}{2} + (m + (m-1)(1 - \frac{1}{2^{i-1}}))/2 = m + (m-1)(1 - \frac{1}{2^i})$. □

Taking $k = \lceil \log_2 m \rceil$, we have that after k phases, $m + (m-1)(1 - \frac{1}{2^k}) \geq m + (m-1)(1 - \frac{1}{m}) > 2m - 2$, hence $\mathcal{A}$ needs more than $2m - 2$ machines to maintain a competitive ratio of 1. □

Note that no algorithm $\mathcal{A}$ which uses $2m - 1$ machines can have competitive ratio less than 1, due to the sequence $1, \ldots, 1$ ($2m$ jobs). At least two jobs run on the same on-line machine, hence $\mathcal{A}(\sigma) = \text{OPT}(\sigma) = 1$.

6 Conclusions

We showed that the greedy algorithm is far from being optimal in one measure (competitive ratio), but optimal in a different measure (amount of resource augmentation). This phenomenon raises many questions. Which of the two measures is more appropriate for this problem? Furthermore, which measure is appropriate for other problems? Is it possible to introduce a different measure that would solve the question: is GREEDY a good algorithm to use?

Acknowledgement

The authors would like to thank Han La Poutré for helpful suggestions.

References

[1] S. Albers. On the influence of lookahead in competitive paging algorithms. *Algorithmica*, 18(3):283–305, Jul 1997.
[2] A. Avidor, Y. Azar, and J. Sgall. Ancient and new algorithms for load balancing in the l_p norm. In *Proc. 9th ACM-SIAM Symp. on Discrete Algorithms*, pages 426–435, 1998.
[3] Y. Azar, L. Epstein, and R. van Stee. Resource augmentation in load balancing. *Journal of Scheduling*, 3(5):249–258, 2000.
[4] Y. Bartal, A. Fiat, H. Karloff, and R. Vohra. New algorithms for an ancient scheduling problem. In *Proc. 24th ACM Symposium on Theory of Algorithms*, pages 51–58, 1992. To appear in *Journal of Computer and System Sciences*.
[5] P. Berman and C. Coulston. Speed is more powerful than clairvoyance. *Nordic Journal of Computing*, 6:181–193, 1999.
[6] M. Brehob, E. Torng, and P. Uthaisombut. Applying extra-resource analysis to load balancing. *Journal of Scheduling*, 3(5):273–288, 2000.
[7] J. Edmonds. Scheduling in the dark. *Theoretical Computer Science*, 235:109–141, 2000.
[8] R. Fleischer and M. Wahl. Online scheduling revisited. *Journal of Scheduling*, 3(5):343–353, 2000.
[9] T. Gormley, N. Reingold, E. Torng, and J. Westbrook. Generating adversaries for request-answer games. In *Proceedings of the Eleventh Annual ACM-SIAM Symposium on Discrete Algorithms*, pages 564–565. ACM-SIAM, 2000.
[10] R. L. Graham. Bounds for certain multiprocessor anomalies. *Bell System Technical Journal*, 45:1563–1581, 1966.
[11] B. Kalyanasundaram and K. Pruhs. Speed is as powerful as clairvoyance. *Journal of the ACM*, 47:617–643, 2000.
[12] Bala Kalyanasundaram and Kirk Pruhs. Maximizing job completions online. In G. Bilardi, G. F. Italiano, A. Pietracaprina, and G. Pucci, editors, *Algorithms - ESA '98, Proceedings Sixth Annual European Symposium*, volume 1461 of *Lecture Notes in Computer Science*, pages 235–246. Springer, 1998. To appear in Journal of Algorithms.
[13] D. R. Karger, S. J. Phillips, and E. Torng. A better algorithm for an ancient scheduling problem. *Journal of Algorithms*, 20:400–430, 1996.
[14] C. A. Phillips, C. Stein, E. Torng, and J. Wein. Optimal time-critical scheduling via resource augmentation. *Algorithmica*, 32(2):163–200, 2002.

Vector Assignment Problems: A General Framework[*]

Leah Epstein[1] and Tamir Tassa[2]

[1] School of Computer Science, The Interdisciplinary Center
P.O.B 167, 46150 Herzliya, Israel
lea@idc.ac.il
[2] Department of Applied Mathematics, Tel-Aviv University
Ramat Aviv, Tel Aviv, Israel
tassa@post.tau.ac.il

Abstract. We present a general framework for vector assignment problems. In such problems one aims at assigning n input vectors to m machines such that the value of a given target function is minimized. While previous approaches concentrated on simple target functions such as max-max, the general approach presented here enables us to design a PTAS for a wide class of target functions. In particular we are able to deal with non-monotone target functions and asymmetric settings where the cost functions per machine may be different for different machines. This is done by combining a graph-based technique and a new technique of preprocessing the input vectors.

1 Introduction

In this paper we present a general framework for dealing with assignment problems in general and vector assignment problems in particular. An assignment problem is composed of the following three ingredients:

- Items: $x^1, \ldots, x^n$;
- Containers: $c^1, \ldots, c^m$;
- A target function: $F : \{1, \ldots, m\}^{\{1, \ldots, n\}} \to \mathcal{R}^+$.

Each item is characterized by a parameter or a set of parameters that reflect the "size" of the item. That size may be a scalar, a vector or whatever the application which gave rise to the problem dictates.
The containers may be characterized by their capacity; that capacity would be a scalar or a vector, in accord with the type of the items to be stored.
The set $\{1, \ldots, m\}^{\{1, \ldots, n\}}$ consists of all possible assignments of items to containers. Each assignment is referred to as *a solution* to the problem. In all assignment problems there is a natural addition operation between items. Hence, given an

---

[*] Research supported in part by the Israel Science Foundation (grant no. 250/01).

assignment (solution) $A \in \{1, \ldots, m\}^{\{1,\ldots,n\}}$, we may compute the *load* in each container as

$$l^k = \sum_{A(i)=k} x^i \, .$$

The target function evaluates for each solution a nonnegative *cost*. That function takes into account the loads l^k and possibly also the container capacities, if such capacities are given.

Such problems are known to be strongly NP-hard. Hence, polynomial time approximation schemes (PTAS) are sought. Such schemes produce, in polynomial time, a solution (i.e., an assignment) whose cost is larger than that of an optimal solution by a factor of no more than $(1 + \mathrm{Const} \cdot \varepsilon)$, where $\varepsilon > 0$ is an arbitrary parameter. Namely, if Φ^o is the optimal cost and $\varepsilon > 0$ is a given parameter, the scheme produces a solution A that satisfies

$$F(A) \leq (1 + \mathrm{Const} \cdot \varepsilon) \cdot \Phi^o \, , \tag{1}$$

where the constant is independent of the input data (n, m and $\mathbf{x}^i$, $1 \leq i \leq n$) and ε, but may depend on the dimension of the vectors.

The above formulation encompasses all problems that were studied in the art so far. However, the chosen target functions in those studies were limited to a narrow class of "natural" functions, as described below. Motivated by an interesting problem that arises in transmitting multiplexed video streams, we suggest here a general framework that includes a much wider class of target functions.

Overview. We focus here on Vector Assignment Problems (VAP), where the items $\mathbf{x}^i$, $1 \leq i \leq n$, and the resulting loads $\mathbf{l}^k$, $1 \leq k \leq m$, are vectors in $(\mathcal{R}^+)^d$. We consider target functions of the form:

$$F(A) = f(g^1(\mathbf{l}^1), \ldots, g^m(\mathbf{l}^m)) \, . \tag{2}$$

Here: corresponding load vectors.
$g^k : (\mathcal{R}^+)^d \to \mathcal{R}^+$ ($1 \leq k \leq m$) are functions that evaluate a cost per container.
$f : (\mathcal{R}^+)^m \to \mathcal{R}^+$ is a function that evaluates the final cost over all containers.

Relation to previously studied problems. This suggested framework includes many problems that are already known in the art. The terminology in those problems may vary. In some scalar problems the containers are referred to as *bins*. In other scalar problems and in most all vector problems the terms *items*, *containers* and *assignment* are replaced with *jobs*, *machines* and *scheduling*, respectively. Since we have in mind applications that do not deal with scheduling, we adopt herein a slightly more general terminology: *vectors*, *machines* and *assignment*.

Herein, we list some of those problems. The first 4 examples are scalar. The last one is a vector problem.

1. The classical problem in this context is the scalar *makespan problem*. In that problem one aims at minimizing the maximal load. It is described by (2) with $d = 1$, $g^k = id$ and $f = \max$. See [11, 12, 13, 15].

2. The ℓ_p minimization problem is given by (2) with $d = 1$, $g^k(x) = x^p$ and $f(y_1, \ldots, y_m) = \sum_{k=1}^{m} y_k^p$. The case $p = 2$ was studied in [2, 4] and was motivated by storage allocation problems. The general case was studied in [1].
3. Problem (2) with $d = 1$, $g^k(x) = h(x)$ for all k, where $h : \mathcal{R}^+ \to \mathcal{R}^+$ is some fixed function, and f is either the maximum or sum of its arguments, was studied in [1]. Other choices for f are the inverse minimum or the inverse sum. By considering those choices, one aims at **maximizing** the minimal or average completion time. See also [6, 16].
4. The Extensible Bin Packing Problem is given by (2) with $d = 1$, $g^k(x) = \max\{x, 1\}$ for all k and $f(y_1, \ldots, y_m) = \sum_{k=1}^{m} y^k$. See [5, 8, 9].
5. The Vector Scheduling Problem, see [3], coincides with (2) with $f = g^k = \max$.

In most of the above examples the target functions were monotone. Namely, when adding an item to a container, the value of the target function increases, or at least does not decrease. Such monotonicity is indeed natural when dealing with bin packing or job scheduling: every item that is stored in a bin decreases the remaining available space in that bin; every job assigned to a machine increases the load on that machine. However, we present in this paper the so called *line-up problem* that arises in video transmission and broadcasting, where the target function has a different nature: it aims at optimizing the quality of the transmitted video. Such functions are not monotone - increasing the size of a vector component may actually decrease the value of the target function.

We note in passing that a related class of problems that we exclude from our discussion is that in which the goal is to minimize the number of containers that are used for packing, subject to some condition (such conditions are usually associated with the capacity of the containers). See [10, 7, 14, 17, 3].

Notation agreements. Throughout this paper we adopt the following conventions:

- Small case letters denote scalars; bold face small case letters denote vectors.
- A superscript of a vector denotes the index of the vector; a subscript of a vector indicates a component in that vector. E.g., 1_j^k denotes the jth component of the vector 1^k.
- If $\gamma(k)$ is any expression that depends on k, then $f(\gamma(k))_{1 \leq k \leq m}$ stands for $f(\gamma(1), \ldots, \gamma(m))$.
- If x is a scalar then $x_+ = \max\{x, 0\}$.
- If $\circ$ is any operation between scalars then $\mathbf{v} \circ c$ is the vector whose jth component (for all values of j) is $\mathbf{v}_j \circ c$. Similarly, if $\propto$ is any relation between scalars, then $\mathbf{v} \propto c$ or $\mathbf{v} \propto \mathbf{w}$ mean that the relation holds component-wise.

2 The Cost Functions

Herein we list the assumptions that we make on the outer cost function $f(\cdot)$ and the inner cost functions $g^k(\cdot)$.

Definition 1.

1. *A function* $h : (\mathcal{R}^+)^n \to \mathcal{R}^+$ *is monotone if*

$$h(\mathbf{x}) \leq h(\mathbf{y}) \quad \forall \mathbf{x}, \mathbf{y} \in (\mathcal{R}^+)^n \text{ such that } \mathbf{x} \leq \mathbf{y} . \tag{3}$$

2. *The function* $h : (\mathcal{R}^+)^n \to \mathcal{R}^+$ *is dominating the function* $\tilde{h} : (\mathcal{R}^+)^n \to \mathcal{R}^+$ *if there exists a constant* η *such that*

$$\tilde{h}(\mathbf{x}) \leq \eta h(\mathbf{x}) \quad \forall \mathbf{x} \in (\mathcal{R}^+)^n . \tag{4}$$

3. *The function* $h : (\mathcal{R}^+)^n \to \mathcal{R}^+$ *is Lipschitz continuous if there exists a constant* M *such that*

$$|h(\mathbf{x}) - h(\mathbf{y})| \leq M \|\mathbf{x} - \mathbf{y}\|_\infty \quad \forall \mathbf{x}, \mathbf{y} \in (\mathcal{R}^+)^n . \tag{5}$$

Assumption 1. *The function* $f : (\mathcal{R}^+)^m \to \mathcal{R}^+$ *is:*

1. *monotone;*
2. *linear with respect to scalar multiplications, i.e.,* $f(c\mathbf{x}) = cf(\mathbf{x})$ *for all* $c \in \mathcal{R}^+$ *and* $\mathbf{x} \in (\mathcal{R}^+)^m$;
3. *dominating the max norm with a domination factor* η_f *that is independent of* m;
4. *Lipschitz continuous with a constant* M_f *that is independent of* m;
5. *recursively computable (explained below).*

Assumption 2. *The functions* $g^k : (\mathcal{R}^+)^d \to \mathcal{R}^+$ *are:*

1. *dominating the* ℓ_∞ *and the* ℓ_1 *norms with a domination factor* η_g *that does not depend on* m;
2. *Lipschitz continuous with a constant* M_g *that does not depend on* m.

By assuming that f is recursively computable we mean that there exists a family of functions $\psi^k(\cdot, \cdot)$, $1 \leq k \leq m$, such that

$$f(g^1, \ldots, g^k, 0, \ldots, 0) = \psi^k \left(f(g^1, \ldots, g^{k-1}, 0, \ldots, 0), g^k \right) \tag{6}$$

(note that $f(0, \ldots, 0) = 0$ in view of Assumption 1-2). For example, if f is a weighted ℓ_p norm on $\mathcal{R}^m$, $1 \leq p \leq \infty$, with weights $(w_1, \ldots, w_m)$, then ψ^k is the ℓ_p norm on $\mathcal{R}^2$ with weights $(1, w_k)$.

Next, we see what functions comply with the above assumptions. Assumption 1 dictates a quite narrow class of outer cost functions. $f = \max$ is the most prominent member of that class (luckily, in many applications this is the only relevant choice of f). Other functions f for which our results apply are the ℓ_p norms taken on the t largest values in the argument vector, where $t = \min(m_0, m)$ for some constant m_0; e.g., the sum of the largest two components. Assumption 1 is not satisfied by any of the usual ℓ_p norms for $p < \infty$ because of the conjunction of conditions 3 and 4: no matter how we rescale an ℓ_p norm, $p < \infty$, one of the parameters η_f (condition 3) or M_f (condition 4) would depend on m.

As for g^k, basically any norm on $\mathcal{R}^d$ is allowed. The most interesting choices are the ℓ_p norms and the Sobolev norms, $\|\mathbf{l}\|_{1,p} := \|\mathbf{l}\|_p + \|\mathbf{\Delta l}\|_p$ where $\mathbf{\Delta l} \in \mathcal{R}^{d-1}$ and $\Delta l_j = l_{j+1} - l_j$, $1 \leq j \leq d-1$. Another natural choice is the "extensible bin" cost function, $g^k(\mathbf{l}^k) = \|\max\{\mathbf{l}^k, \mathbf{c}^k\}\|$; here $\mathbf{c}^k$ is a constant vector reflecting the parameters of the kth machine and the outer norm may be any norm.

It is interesting to note that the set of functions that comply with either Assumption 1 or 2 is closed under positive linear combinations. For example, if f_1 and f_2 satisfy Assumption 1, so would $c_1 f_1 + c_2 f_2$ for all $c_1, c_2 > 0$.

3 A Graph Based Scheme

3.1 Preprocessing the Vectors by Means of Truncation

Let I be the original instance of the VAP. We start by modifying I into another problem instance $\bar{I}$ where the vectors $\bar{\mathbf{x}}^i$ are defined by

$$\bar{\mathbf{x}}^i_j = \begin{cases} \mathbf{x}^i_j & \text{if } \mathbf{x}^i_j \geq \varepsilon \|\mathbf{x}^i\|_\infty \\ 0 & \text{otherwise} \end{cases} \qquad 1 \leq i \leq n,\ 1 \leq j \leq d. \tag{7}$$

Lemma 1. *Let A be a solution to I and let $\bar{A}$ be the corresponding solution to $\bar{I}$. Then*

$$(1 - C_1 \varepsilon) F(\bar{A}) \leq F(A) \leq (1 + C_1 \varepsilon) F(\bar{A}) \qquad \text{where} \quad C_1 = M_g \eta_g. \tag{8}$$

Proof. Let $\mathbf{l}^k$ and $\bar{\mathbf{l}}^k$, $1 \leq k \leq m$, denote the load vectors in A and $\bar{A}$ respectively. In view of (7),

$$\bar{\mathbf{l}}^k \leq \mathbf{l}^k \leq \bar{\mathbf{l}}^k + \varepsilon \sum_{A(i)=k} \|\mathbf{x}^i\|_\infty \tag{9}$$

Since $\|\mathbf{x}^i\|_\infty = \|\bar{\mathbf{x}}^i\|_\infty \leq \|\bar{\mathbf{x}}^i\|_1$ we conclude that $\sum_{A(i)=k} \|\mathbf{x}^i\|_\infty \leq \|\bar{\mathbf{l}}^k\|_1$. Recalling Assumption 2-1 we get that

$$\sum_{A(i)=k} \|\mathbf{x}^i\|_\infty \leq \eta_g g^k(\bar{\mathbf{l}}^k). \tag{10}$$

Therefore, by (9) and (10), $\bar{\mathbf{l}}^k \leq \mathbf{l}^k \leq \bar{\mathbf{l}}^k + \varepsilon \eta_g g^k(\bar{\mathbf{l}}^k)$. Next, by the uniform Lipschitz continuity of g^k we conclude that

$$(1 - C_1 \varepsilon) g^k(\bar{\mathbf{l}}^k) \leq g^k(\mathbf{l}^k) \leq (1 + C_1 \varepsilon) g^k(\bar{\mathbf{l}}^k) \qquad \text{where} \quad C_1 = M_g \eta_g. \tag{11}$$

Finally, we invoke the monotonicity of f and its linear dependence on scalar multiplications to arrive at (8). □

We assume henceforth that the input vectors have been subjected to the truncation procedure (7). To avoid cumbersome notations we shall keep denoting the truncated vectors by $\mathbf{x}^i$ and their collection by I.

3.2 Large and Small Vectors

Let Φ^o denote the optimal cost, let A^o be an optimal solution, $F(A^o) = \Phi^o$, and let $\mathbf{l}^k$, $1 \leq k \leq m$, be the load vectors in that solution. Then, in view of Assumption 1-3 and Assumption 2-1, $\mathbf{l}^k \leq \eta_f \eta_g \Phi^o$, $1 \leq k \leq m$. Consequently, we conclude that all input vectors satisfy the same bound, $\mathbf{x}^i \leq \eta_f \eta_g \Phi^o$, $1 \leq i \leq n$. Hence, we get the following lower bound for the optimal cost:

$$\Phi^o \geq \Phi := \frac{\max_{1 \leq i \leq n} \|\mathbf{x}^i\|_\infty}{\eta_f \eta_g} . \tag{12}$$

This lower bound induces a decomposition of the set of input vectors into two subsets (multi-sets) of large and small vectors as follows:

$$\mathcal{L} = \{\mathbf{x}^i : \|\mathbf{x}^i\|_\infty \geq \Phi \varepsilon^{2d+1}, \ 1 \leq i \leq n\} , \tag{13}$$

$$\mathcal{S} = \{\mathbf{x}^i : \|\mathbf{x}^i\|_\infty < \Phi \varepsilon^{2d+1}, \ 1 \leq i \leq n\} . \tag{14}$$

We present below a technique to replace $\mathcal{S}$ with another set of vectors $\tilde{\mathcal{S}} = \{\mathbf{z}^1, \ldots, \mathbf{z}^{\tilde{\nu}}\}$ where

$$\tilde{\nu} = |\tilde{\mathcal{S}}| \leq \nu = |\mathcal{S}| \quad \text{and} \quad \|\mathbf{z}^i\|_\infty = \Phi \varepsilon^{2d+1} \ \ 1 \leq i \leq \tilde{\nu} . \tag{15}$$

In other words, all vectors in $\tilde{\mathcal{S}}$ are large.

Let $\mathbf{x} \in \mathcal{S}$. Then, in view of the truncation procedure (7),

$$\varepsilon \leq \frac{x_j}{\|\mathbf{x}\|_\infty} \leq 1 \quad \forall x_j > 0, \ 1 \leq j \leq d . \tag{16}$$

Next, we define a geometric mesh on the interval $[\varepsilon, 1]$:

$$\xi_0 = \varepsilon \ ; \quad \xi_i = (1+\varepsilon)\xi_{i-1} \ , \quad 1 \leq i \leq q \ ; \quad q := \left\lfloor \frac{-\lg \varepsilon}{\lg(1+\varepsilon)} \right\rfloor + 1 . \tag{17}$$

In view of the above, every nonzero component of $\mathbf{x}/\|\mathbf{x}\|_\infty$ lies in an interval $[\xi_{i-1}, \xi_i)$ for some $1 \leq i \leq q$. Next, we define

$$\mathbf{x}' = \|\mathbf{x}\|_\infty \mathcal{H}\left(\frac{\mathbf{x}}{\|\mathbf{x}\|_\infty}\right) , \tag{18}$$

where the operator $\mathcal{H}$ retains components that are 0 or 1 and replaces every other component by the left end point of the interval $[\xi_{i-1}, \xi_i)$ where it lies. Hence, the vector $\mathbf{x}'$ may be in one of $s = (q+2)^d - 1$ linear subspaces of dimension 1 in $\mathcal{R}^d$; we denote those subspaces by W^σ, $1 \leq \sigma \leq s$. In view of the above, we define the set

$$\mathcal{S}' = \{\mathbf{x}' : \mathbf{x} \in \mathcal{S}\} . \tag{19}$$

Next, we define for each type $1 \leq \sigma \leq s$

$$\mathbf{w}^\sigma = \sum \{\mathbf{x}' : \mathbf{x}' \in \mathcal{S}' \cap W^\sigma\} ; \tag{20}$$

namely, $\mathbf{w}^\sigma$ aggregates all vectors $\mathbf{x}'$ of type σ. We now slice this vector into large identical "slices", where each of those slices and their number are given by:

$$\tilde{\mathbf{w}}^\sigma = \frac{\mathbf{w}^\sigma}{\|\mathbf{w}^\sigma\|_\infty} \cdot \Phi\varepsilon^{2d+1} \quad \text{and} \quad \kappa_\sigma = \left\lceil \frac{\|\mathbf{w}^\sigma\|_\infty}{\Phi\varepsilon^{2d+1}} \right\rceil . \qquad (21)$$

Finally, we define the set $\tilde{S}$ as follows:

$$\tilde{S} = \cup_{\sigma=1}^{s} \{\mathbf{z}^{\sigma,q} = \tilde{\mathbf{w}}^\sigma \; : \; 1 \le q \le \kappa_\sigma\} . \qquad (22)$$

Namely, the new set $\tilde{S}$ includes for each type σ the "slice"-vector $\tilde{\mathbf{w}}^\sigma$, (21), repeated κ_σ times. As implied by (21), all vectors in $\tilde{S}$ have a max norm of $\Phi\varepsilon^{2d+1}$, in accord with (15). Also, the number of vectors in $\tilde{S}$, $\tilde{\nu} = \sum_{\sigma=1}^{s} \kappa_\sigma$, is obviously no more than ν as the construction of the new vectors implies that $\kappa_\sigma \le |S' \cap W^\sigma|$ (recall that $\|\mathbf{x}'\|_\infty < \Phi\varepsilon^{2d+1}$ for all $\mathbf{x}' \in S'$).

So we have modified the original problem instance I, having n input vectors $\mathcal{L} \cup \mathcal{S}$, into an intermediate problem instance $I' = \mathcal{L} \cup \mathcal{S}'$, see (19), and then to a new problem instance,

$$\tilde{I} = \mathcal{L} \cup \tilde{S} , \qquad (23)$$

see (20)-(22), that has $\tilde{n} = n - \nu + \tilde{\nu}$ input vectors. The following theorem states that those problem instances are close in the sense that interests us.

Theorem 1. *For each solution $A \in \{1, \ldots, m\}^{\{1, \ldots, n\}}$ of I there exists a solution $\tilde{A} \in \{1, \ldots, m\}^{\{1, \ldots, \tilde{n}\}}$ of $\tilde{I}$ such that*

$$(1 - C_1\varepsilon) \cdot \left(F(\tilde{A}) - C_2\Phi\varepsilon\right) \le F(A) \le (1 + C_1\varepsilon) \cdot \left(F(\tilde{A}) + C_2\Phi\varepsilon\right) , \qquad (24)$$

where C_1 is given in (8) and

$$C_2 = M_f M_g . \qquad (25)$$

Conversely, for each solution $\tilde{A} \in \{1, \ldots, m\}^{\{1, \ldots, \tilde{n}\}}$ of $\tilde{I}$ there exists a solution $A \in \{1, \ldots, m\}^{\{1, \ldots, n\}}$ of I that satisfies (24).

Proof. Let A be a solution of I and A' be its counterpart solution of I'. Let $\mathbf{l}^k$ and $\mathbf{l}'^k$, $1 \le k \le m$, denote the load vectors in A and A', respectively. By (18), $1 \le \mathbf{l}^k / \mathbf{l}'^k \le 1 + \varepsilon$. Hence, by Assumption 2-1,

$$\|\mathbf{l}^k - \mathbf{l}'^k\|_\infty \le \varepsilon \eta_g g^k(\mathbf{l}'^k) . \qquad (26)$$

Therefore, by the uniform Lipschitz continuity of g^k,

$$(1 - C_1\varepsilon) g^k(\mathbf{l}'^k) \le g^k(\mathbf{l}^k) \le (1 + C_1\varepsilon) g^k(\mathbf{l}'^k) \quad 1 \le k \le m \qquad (27)$$

where C_1 is as in (8). Applying f on (27) and using Assumptions 1-1 and 1-2, we get that

$$(1 - C_1\varepsilon) F(A') \le F(A) \le (1 + C_1\varepsilon) F(A') . \qquad (28)$$

Next, given a solution A' of I', we construct a solution $\tilde{A}$ of $\tilde{I}$ such that

$$F(\tilde{A}) - C_2 \Phi \varepsilon \leq F(A') \leq F(\tilde{A}) + C_2 \Phi \varepsilon , \qquad (29)$$

with C_2 as in (25). Showing this will enable us to construct for any solution A of I a solution $\tilde{A}$ of $\tilde{I}$ for which, in view of (28) and (29), (24) holds. Then, in order to complete the proof, we shall show how from a given solution $\tilde{A}$ of $\tilde{I}$, we are able to construct a solution A' of I' for which (29) holds.

To this end, we fix $1 \leq \sigma \leq s$ and define for every machine k the following vector:

$$\mathbf{y}^{\sigma,k} = \sum \{\mathbf{x}'^i \ : \ \mathbf{x}'^i \in \mathcal{S}' \cap W^\sigma \ , \ A'(i) = k\} ; \qquad (30)$$

i.e., $\mathbf{y}^{\sigma,k}$ is the sum of small vectors of type σ in I' that are assigned to the kth machine. Recalling (21), $\tilde{\mathcal{S}}$ includes the vector $\tilde{\mathbf{w}}^\sigma$ repeated κ_σ times, where

$$\kappa_\sigma = \left\lceil \sum_{k=1}^m \frac{\|\mathbf{y}^{\sigma,k}\|_\infty}{\Phi \varepsilon^{2d+1}} \right\rceil . \qquad (31)$$

We may now select for each k an integer $t_{\sigma,k}$ such that

$$\left| t_{\sigma,k} - \frac{\|\mathbf{y}^{\sigma,k}\|_\infty}{\Phi \varepsilon^{2d+1}} \right| \leq 1 \qquad (32)$$

and $\sum_{k=1}^m t_{\sigma,k} = \kappa_\sigma$. The integers $t_{\sigma,k}$ can be found in the following manner: We define $t_{\sigma,k}^{low} = \lfloor \|\mathbf{y}^{\sigma,k}\|_\infty / \Phi \varepsilon^{2d+1} \rfloor$ and $t_{\sigma,k}^{high} = \lceil \|\mathbf{y}^{\sigma,k}\|_\infty / \Phi \varepsilon^{2d+1} \rceil$. Clearly, $\sum_{k=1}^m t_{\sigma,k}^{low} \leq \kappa_\sigma$ and $\sum_{k=1}^m t_{\sigma,k}^{high} \geq \kappa_\sigma$. Since $t_{\sigma,k}^{high} - t_{\sigma,k}^{low} \leq 1$ for all $1 \leq k \leq m$, there exists an integer number $0 \leq x \leq m$ such that $\sum_{k=1}^m t_{\sigma,k}^{low} = \kappa_\sigma - x$. Finally, we set

$$t_{\sigma,k} = \begin{cases} t_{\sigma,k}^{high} & 1 \leq k \leq x \\ t_{\sigma,k}^{low} & x < k \leq m \end{cases}$$

With this, the solution $\tilde{A}$ is that which assigns to the kth machine, $1 \leq k \leq m$, $t_{\sigma,k}$ vectors $\tilde{\mathbf{w}}^\sigma$ for all $1 \leq \sigma \leq s$ (and coincides with A' for all large vectors in $\mathcal{L}$). In view of (32) and the definition of $\tilde{\mathbf{w}}^\sigma$, see (21),

$$\|t_{\sigma,k} \cdot \tilde{\mathbf{w}}^\sigma - \mathbf{y}^{\sigma,k}\|_\infty \leq \Phi \varepsilon^{2d+1} . \qquad (33)$$

Therefore, summing (33) over $1 \leq \sigma \leq s$, we conclude that $\tilde{\mathbf{l}}^k$ and $\mathbf{l}'^k$ – the loads on the kth machine in $\tilde{A}$ and A' respectively – are close,

$$\|\tilde{\mathbf{l}}^k - \mathbf{l}'^k\|_\infty \leq s \Phi \varepsilon^{2d+1} . \qquad (34)$$

However, as (17) and the definition of s imply that $s \leq \varepsilon^{-2d}$ for all $0 < \varepsilon \leq 1$. We conclude by that $\|\tilde{\mathbf{l}}^k - \mathbf{l}'^k\|_\infty \leq \Phi \varepsilon$. Finally, the Lipschitz continuity of both g and f imply that (29) holds with C_2 as in (25).

Next, we show how to construct from a solution $\tilde{A}$ of $\tilde{I}$, a solution A' of I' for which (29) holds. The two assignments will coincide for the large vectors $\mathcal{L}$. As for the small vectors, let us fix one vector type $1 \leq \sigma \leq s$, where s is the number of types. $\tilde{\mathcal{S}}$ includes the vector $\tilde{\mathbf{w}}^\sigma$ repeated κ_σ times, (21)-(22). Let $t_{\sigma,k}$ be the number of those vectors that $\tilde{A}$ assigns to the kth machine. The counters $t_{\sigma,k}$ satisfy the bounds on them. We now assign the vectors $\mathbf{x}' \in \mathcal{S}' \cap W^\sigma$, see (20), to the m machines so that the ℓ_∞-norm of their sum in the kth machine is greater than $(t_{\sigma,k} - 1)\Phi\varepsilon^{2d+1}$ but no more than $(t_{\sigma,k} + 1)\Phi\varepsilon^{2d+1}$. In view of (20) and (21), it is easy to see that such an assignment exists: Assign the jobs one by one greedily, in order to obtain in the kth machine, $1 \leq k \leq m$, a load with an ℓ_∞-norm of at least $(t_{\sigma,k} - 1)\Phi\varepsilon^{2d+1}$. Since the ℓ_∞-norm of the sum of all small jobs is at least $(\sum_{k=1}^m t_{\sigma,k} - 1)\Phi\varepsilon^{2d+1}$, this goal can be achieved. Also, as the size of each of those jobs is no more than $\Phi\varepsilon^{2d+1}$, we may perform this assignment in a manner that keeps the load in each machine below $t_{\sigma,k}\Phi\varepsilon^{2d+1}$. After achieving that goal in all machines, we assign the remaining jobs so that the total load in each machine is bounded by $(t_{\sigma,k} + 1)\Phi\varepsilon^{2d+1}$. This is possible given the small size of the jobs and the size of their sum (at most $(\sum_{k=1}^m t_{\sigma,k})\Phi\varepsilon^{2d+1}$). Clearly, if we let $\mathbf{y}^{\sigma,k}$ denote the sum of vectors $\mathbf{x}'$ of type σ thus assigned to the kth machine, then $\mathbf{y}^{\sigma,k}$ satisfies (33). As we saw before, this implies that $\tilde{A}$ and A' satisfy (29). This completes the proof. □

3.3 The Scheme

In view of the previous two subsections, we assume that the original set of input vectors I was subjected to the truncation procedure, along the lines of §3.1, and then modified into a problem instance $\tilde{I}$ where all vectors are large, using the procedure described in §3.2. For convenience, we shall keep denoting the number of input vectors in $\tilde{I}$ by n and the input vectors by $\mathbf{x}^i$, $1 \leq i \leq n$. Hence, all vectors in $\tilde{I}$ satisfy $\|\mathbf{x}^i\|_\infty \geq \Phi\varepsilon^{2d+1}$ for $1 \leq i \leq n$. This, together with (7) on one hand and (12) on the other hand, yield the following lower and upper bounds:

$$\varepsilon^{2d+2} \leq \frac{x_j^i}{\Phi} \leq \eta_f \eta_g \quad \text{for } 1 \leq i \leq n,\ 1 \leq j \leq d \text{ and } x_j^i \neq 0. \quad (35)$$

Next, we define a geometric mesh on the interval given in (35):

$$\xi_0 = \varepsilon^{2d+2}\ ;\ \xi_i = (1+\varepsilon)\xi_{i-1}\ ,\ 1 \leq i \leq q\ ;\ q := \left\lfloor \frac{\lg(\eta_f \eta_g \varepsilon^{-2(d+1)})}{\lg(1+\varepsilon)} \right\rfloor + 1. \quad (36)$$

In view of the above, every nonzero component of $\mathbf{x}^i/\Phi$, $1 \leq i \leq n$, lies in an interval $[\xi_{i-1}, \xi_i)$ for some $1 \leq i \leq q$. We use this in order to define a new set of vectors,

$$\hat{I} = \left\{ \hat{\mathbf{x}}^i = \Phi\mathcal{H}\left(\frac{\mathbf{x}^i}{\Phi}\right) : \mathbf{x}^i \in \tilde{I} \right\}, \quad (37)$$

where the operator $\mathcal{H}$ replaces each nonzero component in the vector on which it operates by the left end point of the interval $[\xi_{i-1}, \xi_i)$ where it lies.

The proof of the following theorem is omitted due to space restrictions.

Theorem 2. *Let $\tilde{A}$ be a solution of $\tilde{I}$ and let $\hat{A}$ be the corresponding solution of $\hat{I}$. Then*

$$(1 - C_1\varepsilon)F(\hat{A}) \leq F(\tilde{A}) \leq (1 + C_1\varepsilon)F(\hat{A}) , \qquad (38)$$

where C_1 is given in (8).

The vectors in $\hat{I}$ belong to the set

$$W = \mathcal{X}^d \quad \text{where} \quad \mathcal{X} = \{0, \xi_0, \ldots, \xi_{q-1}\} . \qquad (39)$$

As the size of W is $s = (q+1)^d$, it may be ordered:

$$W = \{\mathbf{w}^1, \ldots, \mathbf{w}^s\} . \qquad (40)$$

With this, the set of modified vectors $\hat{I}$ may be identified by a configuration vector

$$\mathbf{z} = (\mathbf{z}_1, \ldots, \mathbf{z}_s) \quad \text{where} \quad \mathbf{z}_i = \#\{\hat{\mathbf{x}} \in \hat{I} \ : \ \hat{\mathbf{x}} = \mathbf{w}^i\} , \quad 1 \leq i \leq s . \qquad (41)$$

Next, we may describe all possible assignments of vectors from $\hat{I}$ to the m machines using a layered graph $G = (V, E)$. To that end, assume that $\hat{A} : \hat{I} \to \{1, \ldots, m\}$ is such an assignment. We let $\hat{I}^k$ denote the subset of $\hat{I}$ consisting of those vectors that were assigned to one of the first k machines,

$$\hat{I}^k = \{\hat{\mathbf{x}} \in \hat{I} \ : \ \hat{A}(\hat{\mathbf{x}}) \leq k\} \qquad 1 \leq k \leq m .$$

Furthermore, we define the corresponding *state vector*

$$\mathbf{z}^k = (\mathbf{z}_1^k, \ldots, \mathbf{z}_s^k) \quad 1 \leq k \leq m \quad \text{where} \quad \mathbf{z}_i^k = \#\{\hat{\mathbf{x}} \in \hat{I}^k \ : \ \hat{\mathbf{x}} = \mathbf{w}^i\} , \ 1 \leq i \leq s.$$

We note that

$$\emptyset = \hat{I}^0 \subseteq \hat{I}^1 \subseteq \ldots \subseteq \hat{I}^{m-1} \subseteq \hat{I}^m = \hat{I} \qquad (42)$$

and

$$\mathbf{0} = \mathbf{z}^0 \leq \mathbf{z}^1 \leq \ldots \leq \mathbf{z}^{m-1} \leq \mathbf{z}^m = \mathbf{z} , \qquad (43)$$

where $\mathbf{z}$ is given in (41). In addition, when $0 < k < m$, $\hat{I}^k$ may be any subset of $\hat{I}$ while $\mathbf{z}^k$ may be any vector in $Z = \{\mathbf{y} \ : \ \mathbf{0} \leq \mathbf{y} \leq \mathbf{z}\}$. With this, we define the graph $G = (V, E)$ as follows:

- The set of vertices consists of $m+1$ layers, $V = \cup_{k=0}^m V^k$. If $v \in V$ is a vertex in the kth layer, V^k, then it represents one of the possible state vectors after assigning vectors to the first k machines. Hence $V^0 = \{\mathbf{0}\}$, $V^m = \{\mathbf{z}\}$ and the intermediate layers are $V^k = Z$, $0 < k < m$.
- The set of edges consists of m subsets:

$$E = \cup_{k=1}^m E^k \quad \text{where} \quad E^k = \{(\mathbf{u}, \mathbf{v}) \ : \ \mathbf{u} \in V^{k-1} , \ \mathbf{v} \in V^k , \ \mathbf{u} \leq \mathbf{v}\} . \qquad (44)$$

In other words, there is an edge connecting two vertices in adjacent layers, $\mathbf{u} \in V^{k-1}$ and $\mathbf{v} \in V^k$, if and only if there exists an assignment to the kth machine that would change the state vector from $\mathbf{u}$ to $\mathbf{v}$.

Note that all intermediate layers, V^k, $0 < k < m$, are composed of the same number of vertices, t, given by the number of sub-vectors that $\mathbf{z}$ has:

$$t = |Z| = \prod_{i=1}^{s}(\mathbf{z}_i + 1) \leq (n+1)^s \ . \tag{45}$$

Next, we turn the graph into a weighted graph, using a weight function $w : E \rightarrow \mathcal{R}^+$ that computes the cost that the given edge implies on the corresponding machine: Let $e = (\mathbf{u}, \mathbf{v}) \in E^k$. Then the difference $\mathbf{v} - \mathbf{u}$ tells us how many vectors of each of the s types are assigned by this edge to the kth machine. The weight of this edge is therefore defined as

$$w(e) = g^k(T(\mathbf{v} - \mathbf{u})) \quad \text{where} \quad T(\mathbf{v} - \mathbf{u}) = \sum_{i=1}^{s}(\mathbf{v}_i - \mathbf{u}_i)\mathbf{w}^i \ , \tag{46}$$

$\mathbf{w}^i$ are as in (40). We continue to define a cost function on the vertices, $r : V \rightarrow \mathcal{R}^+$. The cost function is defined recursively according to the layer of the vertex, using Assumption 1-5: $r(v) = 0$, $v \in V^0$;

$$r(v) = \min\left\{\psi^k(r(u), w(e)) \ : \ u \in V^{k-1}, \ e = (u,v) \in E^k\right\} \quad , \quad v \in V^k$$

(the functions ψ^k are as in (6)). This cost function coincides with the cost function of the VAP, (2). More specifically, if $v \in V^k$ and it represents a subset of vectors $\hat{I}^k \subseteq \hat{I}$, then $r(v)$ equals the value of an optimal assignment of the vectors in $\hat{I}^k$ to the first k machines. Hence, the cost of the end vertex, $r(v)$, $v \in V^m$, equals the value of an optimal solution of the VAP for $\hat{I}$.

The goal is to find the shortest path from V^0 to V^m that achieves this minimal cost. Namely, we look for a sequence of vertices $v^k \in V^k$, $0 \leq k \leq m$, such that $e^k := (v^{k-1}, v^k) \in E^k$, $1 \leq k \leq m$ and $f(w(e^1), \ldots, w(e^m)) = r(v^m)$. We may apply a standard algorithm to find this minimal path within $\mathcal{O}(|V| + |E|)$ steps. As $|V| \leq 2 + (m-1) \cdot (n+1)^s$ and $|E| = \sum_{k=1}^{m}|E^k| \leq m \cdot (n+1)^{2s}$ the running time would be polynomial in n and m.

The shortest path thus found represents an assignment of the vectors of the modified set $\hat{I}$, $\hat{A} : \hat{I} = \{\hat{\mathbf{x}}_1, \ldots, \hat{\mathbf{x}}_n\} \rightarrow \{1, \ldots, m\}$. We need to translate this assignment into an assignment of the original vectors, $A : I = \{\mathbf{x}_1, \ldots, \mathbf{x}_n\} \rightarrow \{1, \ldots, m\}$. To that end, let us review all the problem modifications that we performed:

- First modification: I to $\bar{I}$, see (7) and Lemma 1.
- Second modification: $\bar{I}$ to $\tilde{I}$, see (20)-(23) and Theorem 1.
- Third modification: $\tilde{I}$ to $\hat{I}$, see (37) and Theorem 2.

In view of the above, we translate the solution that we found, $\hat{A}$, into a solution $\tilde{A}$ of $\tilde{I}$, then – along the lines of Theorem 1 – we translate it into a solution $\bar{A}$ of $\bar{I}$ and finally we take the corresponding solution A of I.

The proof of the following theorem is omitted due to space restrictions.

Theorem 3. *Let Φ^o be the optimal cost of the original problem instance I. Let A be the solution of I that is obtained using the above scheme. Then A satisfies (1) with a constant that depends only on η_g, M_g and M_f.*

References

[1] N. Alon, Y. Azar, G. Woeginger, and T. Yadid. Approximation schemes for scheduling. In *Proc. 8th ACM-SIAM Symp. on Discrete Algorithms*, pages 493–500, 1997.

[2] A. K. Chandra and C. K. Wong. Worst-case analysis of a placement algorithm related to storage allocation. *SIAM Journal on Computing*, 4(3):249–263, 1975.

[3] C. Chekuri and S. Khanna. On multi-dimensional packing problems. In *Proceedings of the Tenth Annual ACM-SIAM Symposium on Discrete Algorithms*, pages 185–194, 1999.

[4] R. A. Cody and E. G. Coffman, Jr. Record allocation for minimizing expected retrieval costs on drum-like storage devices. *J. Assoc. Comput. Mach.*, 23(1):103–115, 1976.

[5] E. G. Coffman, Jr. and George S. Lueker. Approximation algorithms for extensible bin packing. In *Proceedings of the Twelfth Annual ACM-SIAM Symposium on Discrete Algorithms*, pages 586–588, 2001.

[6] J. Csirik, H. Kellerer, and G. Woeginger. The exact lpt-bound for maximizing the minimum completion time. *Operations Research Letters*, 11:281–287, 1992.

[7] W. F. de la Vega and G. S. Lueker. Bin packing can be solved within $1 + \epsilon$ in linear time. *Combinatorica*, 1(4):349–355, 1981.

[8] P. Dell'Olmo, H. Kellerer, M. G. Speranza, and Zs. Tuza. A 13/12 approximation algorithm for bin packing with extendable bins. *Information Processing Letters*, 65(5):229–233, 1998.

[9] P. Dell'Olmo and M. G. Speranza. Approximation algorithms for partitioning small items in unequal bins to minimize the total size. *Discrete Applied Mathematics*, 94:181–191, 1999.

[10] M. R. Garey, R. L. Graham, D. S. Johnson, and A. C. C. Yao. Resource constrained scheduling as generalized bin packing. *Journal of Combinatorial Theory (Series A)*, 21:257–298, 1976.

[11] R. L. Graham. Bounds for certain multiprocessor anomalies. *Bell System Technical Journal*, 45:1563–1581, 1966.

[12] R. L. Graham. Bounds on multiprocessing timing anomalies. *SIAM J. Appl. Math*, 17:416–429, 1969.

[13] D. S. Hochbaum and D. B. Shmoys. Using dual approximation algorithms for scheduling problems: theoretical and practical results. *Journal of the ACM*, 34(1):144–162, 1987.

[14] N. Karmarkar and R. M. Karp. An efficient approximation scheme for the one-dimensional bin–packing problem. In *Proc. 23rd Ann. Symp. on Foundations of Computer Science*, 1982.

[15] S. Sahni. Algorithms for scheduling independent tasks. *Journal of the Association for Computing Machinery*, 23:116–127, 1976.

[16] G. J. Woeginger. A polynomial time approximation scheme for maximizing the minimum machine completion time. *Operations Research Letters*, 20:149–154, 1997.

[17] G. J. Woeginger. There is no asymptotic PTAS for two-dimensional vector packing. *Information Processing Letters*, 64(6):293–297, 1997.

Speeding Up the Incremental Construction of the Union of Geometric Objects in Practice*

Eti Ezra, Dan Halperin, and Micha Sharir

School of Computer Science, Tel Aviv University
{estere,danha,michas}@post.tau.ac.il

Abstract. We present a new incremental algorithm for constructing the union of n triangles in the plane. In our experiments, the new algorithm, which we call the Disjoint-Cover (DC) algorithm, performs significantly better than the standard randomized incremental construction (RIC) of the union. Our algorithm is rather hard to analyze rigorously, but we provide an initial such analysis, which yields an upper bound on its performance that is expressed in terms of the expected cost of the RIC algorithm. Our approach and analysis generalize verbatim to the construction of the union of other objects in the plane, and, with slight modifications, to three dimensions. We present experiments with a software implementation of our algorithm using the CGAL library of geometric algorithms.

1 Introduction

Computing the union of n triangles in the plane is a fundamental problem in computational geometry with many applications. For example, this problem arises in the construction of the forbidden portions of the configuration space in certain robot motion planning problems, and in hidden surface removal for visibility problems in three dimensions [8, 10].

Computing the union, by constructing the *arrangement* of the triangles (namely, the subdivision of the plane into vertices, edges, and faces induced by the boundary segments of the triangles), may result in a solution which is too slow in practice. This is because it is likely that most vertices of the arrangement lie in the interior of the union, so computing them is wasteful. Naturally, one would like to have an output-sensitive algorithm. However, such an algorithm is unlikely to exist: Even the problem of deciding whether the union of a given

---

* Work reported in this paper has been supported in part by the IST Programme of the EU as a Shared-cost RTD (FET Open) Project under Contract No IST-2000-26473 (ECG - Effective Computational Geometry for Curves and Surfaces), by The Israel Science Foundation founded by the Israel Academy of Sciences and Humanities (Center for Geometric Computing and its Applications), and by the Hermann Minkowski – Minerva Center for Geometry at Tel Aviv University. Micha Sharir has also been supported by NSF Grants CCR-97-32101 and CCR-00-98246, and by a grant from the U.S.-Israeli Binational Science Foundation.

set of triangles in the plane covers another given triangle is a *3SUM-hard* problem [5]. The best known solutions for problems from this family require $\Theta(n^2)$ time in the worst case, even though the size of the output may be only linear or even constant.

1.1 Randomized Incremental Construction

We compare our new algorithm to a randomized incremental algorithm (RIC) for constructing the union, which is *quasi output sensitive*, and which is a variant of a similar algorithm due to Mulmuley [10] (a similar algorithm is also presented by Agarwal and Har-peled [1]).

Given a set T of n triangles in the plane, the RIC algorithm computes their union as follows: We compute a random permutation $D := (\Delta_1, \ldots, \Delta_n)$ of T, and insert the triangles one at a time, in their order in D. In the i'th iteration, we compute the partial union $\bigcup_{j=1}^{i} \Delta_j$. This is accomplished by finding the intersection points of the boundary of the next triangle Δ_i with the boundary of the preceding union $\bigcup_{j<i} \Delta_j$, and by removing all features of the union that lie inside Δ_i. For further details concerning possible implementations of these insertion steps, see [1, 10]. Our DC algorithm also computes the union incrementally, by inserting the triangles one at a time, and it differs from the RIC algorithm in the order in which the triangles are inserted. In our study we use a twofold approach to measuring the cost of the algorithms. Our first cost measure is the number of vertices that the algorithms generate (some of which may not appear on the boundary of the union), and the set of these vertices depends only on the random permutation D. This cost measure allows us to ignore details of the implementation of the algorithms. However, the actual expected cost of the algorithms (in the unit cost model) depends on this number in a more subtle way

Define the *depth* $d(v)$ of a vertex v to be the number of triangles of T that contain v in their interior. Vertices at depth 0 are the vertices of the union, and they have to be constructed by any algorithm that computes the union. We are thus only interested in the *residual cost* of the algorithm, defined as the (expected) number of positive-depth vertices that the algorithm constructs.

Let $\mathcal{A}(T)$ denote the arrangement of T, and let L_i denote the number of vertices of $\mathcal{A}(T)$ that are intersections of triangle boundaries and have depth i, for $1 \leq i \leq n-2$ (the vertices of the triangles themselves are ignored in the analysis). Then the expected number of vertices at positive depth constructed by the RIC algorithm is known to be $\theta(\mathcal{A}(T)) = \sum_{i=1}^{n-2} \frac{2}{(i+1)(i+2)} \cdot L_i$; we refer to this sum as *Mulmuley's theta series* [10].

1.2 Related Work

Agarwal and Har-Peled gave a randomized incremental algorithm for constructing the union of n triangles in the plane, whose analysis is based on Mulmuley's theta series [1]. An earlier variant, due to Mulmuley [10], constructs partial unions for hidden surface removal. If the given triangles are *fat* (every angle of each triangle is at least some constant positive angle), or arise in the union of Minkowski sums of a fixed convex polygon with a set of pairwise disjoint convex polygons (which is the problem one faces in translational motion planning of a convex polygon), then their union has only linear or near-linear complexity [7, 9], and more efficient algorithms, based on either deterministic divide-and-conquer, or on randomized incremental construction, can be devised, and are presented in the above-cited papers.

1.3 Our Results

We present an incremental algorithm for constructing the boundary of the union. The algorithm, which we call the *Disjoint Cover* (DC) algorithm, inserts the triangles one by one in some order. Each insertion is performed exactly as in the RIC algorithm. The difference is in the order in which we process the triangles. The intuition behind our approach is that the random order used in the RIC construction makes sure that deep vertices of the arrangement are very unlikely to be constructed; however, shallow vertices have rather high probability of being created. A typical bad situation is when there exist triangles that cover many shallow vertices. If we could force these triangles to be inserted first, they would have eliminated many vertices that will be constructed under a random insertion order. This is exactly what the new algorithm is trying to achieve.

2 Constructing the Union: Disjoint Cover Algorithm

Define the *weight* $w(v)$ of a vertex v (at positive depth) to be $\frac{1}{d(v)}$. We denote by V^+ the set of vertices of the arrangement $\mathcal{A}(T)$ at positive depth (considering, as above, only intersection points of the triangle boundaries). Suppose that the insertion order of the DC algorithm (to be described shortly) is $(\Delta_1, \ldots, \Delta_n)$. Regard the triangles of T as open triangles. Define $S_{\Delta_j} = V^+ \cap \left(\Delta_j \setminus \bigcup_{i<j} \Delta_i\right)$, namely, the set of vertices in the interior of Δ_j that are not covered by the interior of any previously inserted triangle. The *weight* $W(\Delta_j)$, for $j = 1, \ldots, n$, is then defined to be the sum of the weights of the vertices in S_{Δ_j}. Note that $\{S_\Delta\}_\Delta$ is a partition of V^+ into pairwise disjoint sets.

The DC algorithm chooses an insertion order that aims to maximize the sequence $(W(\Delta_1), \ldots, W(\Delta_n))$ in lexicographical order. In an ideal setting (which is too expensive to implement, and which will therefore be modified shortly), we proceed as follows. Suppose we have already chosen $(\Delta_1, \ldots, \Delta_j)$ to be inserted. For each remaining triangle Δ, we set (temporarily) S_Δ to be the set of all vertices of V^+ in the interior of Δ that are not covered by $\bigcup_{i \leq j} \Delta_i$. We compute the corresponding weights $W(\Delta)$ of all the remaining Δ's, and set Δ_{j+1} to be the triangle with the maximum weight. We proceed in this manner until all triangles have been chosen.

The problem with this approach is that it requires knowledge of all the vertices of $\mathcal{A}(T)$, which in general is too expensive to compute. Instead, we consider a smaller subset R. We fix some parameter r, select r random pairs of triangles from T, construct and collect the intersection points, if any, of the boundaries of each pair. We now estimate each set S_Δ by the corresponding set $S_\Delta \cap R$, which is computed using only the vertices in R, and consequently estimate $W(\Delta)$ by the sum of weights of vertices in $S_\Delta \cap R$. At present, this simplification should be viewed as purely heuristic—the theory of random sampling and ε-approximations (see, e.g., [11]) is not directly applicable to argue that $S_\Delta \cap R$ is a good approximation of S_Δ, because the portion of the plane over which S_Δ is estimated at the $(j+1)$-th step, namely, $\Delta \setminus \bigcup_{i<j} \Delta_i$, may not have constant complexity, which is a (sufficient) condition that is usually needed to be assumed in order to facilitate the application of the random sampling theory. Nevertheless, our experimental results indicate that this heuristic performs very well in practice—see Section 3.

The above discussion is summarized in the following lemma. (Proofs are omitted in this abstract.)

Lemma 1 *Given a set T of n triangles and a vertex set R as above, the construction of the insertion order by the DC algorithm takes $O(n|R| + |R|\log|R|)$ time.*

The following theorem relates the residual cost of the DC algorithm (in its ideal setting) to that of the RIC algorithm:

Theorem 1 *Let T be a collection of n triangles with κ intersection points at positive depth. Then the number of positive-depth vertices generated by the ideal DC algorithm is at most $O(n^{2/3}\kappa^{1/3}M^{1/3})$, where M is the expected number of positive-depth vertices generated by the RIC algorithm.*

Note that when M and κ are $\Theta(n^2)$, both algorithms generate the same quadratic number of vertices asymptotically. If either of these two parameters is strictly subquadratic, then the DC algorithm will produce a strictly subquadratic number of vertices at positive depth. However, the upper bound of Theorem 1 seems rather pessimistic, and we believe that it can be improved (as is strongly suggested by our experimental results).

We observe, though, that there exist (rather pathological) examples in which $\kappa \ll n^2$ and the DC algorithm performs considerably worse than the RIC algorithm. This lower bound is exemplified in Theorem 2. No such examples are known for $\kappa = \Theta(n^2)$, and we conjecture that in this case the residual cost of the DC algorithm is at worst comparable with that of the RIC (and is likely to be small in practice).

Theorem 2 *There exist collections T of n triangles, for arbitrary large n, with $\kappa = O(n^{5/3})$ intersection points at positive depth, in which the number of positive-depth vertices generated by the RIC algorithm and by the DC algorithm is $O(n)$ and $\Omega(n^{4/3})$, respectively.*

3 Experimental Results I: Number of Positive Depth Vertices

In this section we present experimental results comparing the number of positive-depth vertices constructed by the RIC and by the DC algorithms.

The motivation to devise the DC algorithm is practical. We wish to precede the incremental construction of the union with a simple and fast procedure that will speed up the more heavy-duty incremental stage. The incremental stage uses rather involved data structures for representing the topology of the partially constructed union and for searching in it. In comparison, the preprocessing stage of the DC algorithm, where we compute the order of insertion, uses very simple operations (details are omitted in this version). In our experiments the preprocessing time is negligible compared with the time of the incremental construction, even when $|R|$ is linear in the number of input triangles.

3.1 Input Sets

The input data that we used is depicted in Figures 1 and 2. It consists of the following sets:

- **regular** (Figure 1(a)): Arbitrary triangles (each generated from a random triple of vertices) randomly placed inside a square.
- **fat** (Figure 1(b)): Equilateral triangles of a fixed size randomly placed inside a square.
- **fat_with_grid** (Figure 1(c)): A grid-like pattern fully covered by many random fat triangles; half of the triangles form the grid and the other half are the fat triangles.
- **star-shaped_robot_and_obstacles** (Figure 2(a,b)): The triangulated Minkowski sums of a star-shaped robot and triangular obstacles. The resulting arrangement contains star-shaped sets, each such set intersects its adjacent sets in a superlinear number of points.
- **L-shaped_robot_and_obstacles** (Figure 2(c,d)): The triangulated Minkowski sums of an L-shaped robot and rectangular obstacles. A fifth of the rectangular obstacles are long and narrow rectangles, and the rest of them are small squares. The resulting arrangement contains L-shaped sets rotated by 180 degrees, each such set intersects its adjacent sets in a superlinear number of points.

For each type of input, we have experimented with a varying number of triangles, up to 800 per run, except for the two robot data sets, in which we use a single input set of size 2832 and 840, respectively. For each specific input set, we ran each algorithm five times, and the results reported below are the averages over the number of positive-depth vertices created during the construction. The complexity of the union of the *fat* input is almost linear in the number of triangles (see Figure 1), while the union of the *fat_with_grid* input can have a superlinear number of holes. The *star-shaped_robot_and_obstacles* input (resp.

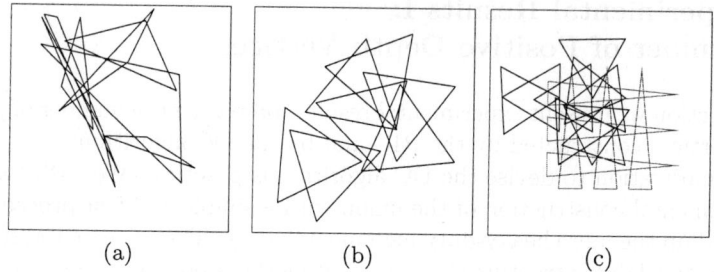

Fig. 1. Regular input (a), fat input (b) and fat_with_grid input (c)

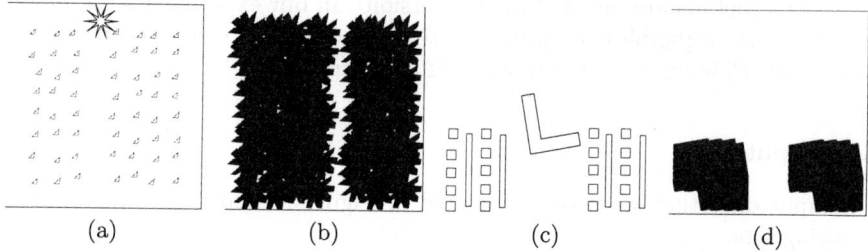

Fig. 2. A star-shape robot and triangular obstacles (a) and the Minkowski sum of the robot (rotated by 180 degrees) with each of the obstacles (b). An L-shape robot and rectangular obstacles (c) and the Minkowski sum of the robot (rotated by 180 degrees) with each of the obstacles (d)

L-shaped_robot_and_obstacles input) arises in translational motion planning of a star-shaped (resp. L-shaped) polygon as robot amid disjoint triangular (resp. rectangular) obstacles which are placed inside a square [8]. See Figure 2. Using our Minkowski sums package [3], we computed the Minkowski sum of the star-shaped (resp. L-shaped) robot with each of the obstacles and triangulated each resulting sum. Then we collected all such triangles to form our data set.

3.2 Results

We present experimental results of applying both algorithms to each of the data sets described in Section 3.1. We present the number of positive-depth vertices created by each of these algorithms, which, as discussed above, is our first yardstick for measuring and comparing the performance of the algorithms.

In all our experiments the DC algorithm performs better, and in several cases significantly better, than the RIC. For each input type we show five graphs. Besides the graph for the RIC algorithm, we present graphs for the DC algorithm where the (maximal) size of R varies between a constant, a logarithmic term in the number of input triangles, a linear term, and R being the full set V, where V is the set of vertices of the underlying arrangement (considering, as above, only intersection points of the triangle boundaries). The results are presented in Fig-

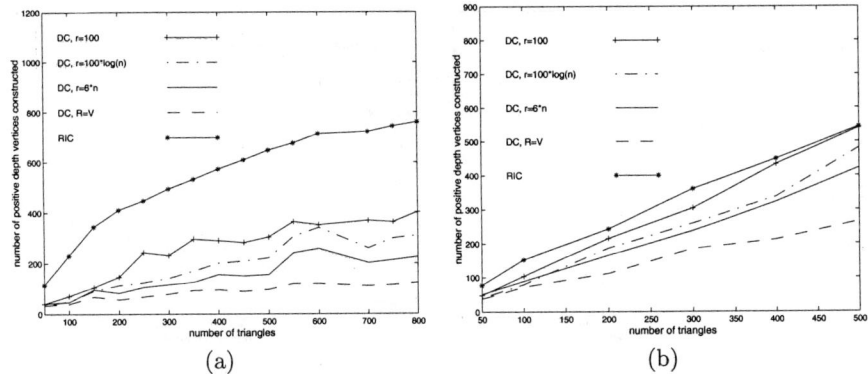

Fig. 3. Number of positive-depth vertices created for the *regular* input (a) and for the *fat* input (b); r denotes the number of pairs of triangles used to construct the sample R

ures 3 and 4 and in Table 1. Note that the results reported in this section are independent of implementation details; they only depend on the insertion order (permutation) determined by the DC and the RIC algorithms.

In all the graphs we see that if we take the whole set V into account in computing the insertion order then the savings in the union construction stage are big. In general, in all our experiments, the DC algorithm performs better than the RIC,[1] and the performance improves as the size of R increases. In some cases, e.g., for the *fat_with_grid* input, even if we use much smaller samples R, e.g., samples of size linear in the input size, then we save the construction of over 9700 vertices (out of about 10040) during the incremental stage when the input consists of 302 triangles (Figure 4). The saving is similar when we use $R = V$, due to the fact that most of the arrangement vertices are contained in the fat triangles covering the grid. Hence, even when choosing a random subset smaller than V, the DC algorithm first inserts most of the fat triangles into the union with high probability.

For the *fat* input (see Figure 3(b)), there is still improvement, but it is less significant than the improvement obtained for the other input sets. It can be shown that the amount of work that the RIC algorithm performs for fat triangles is always close to linear.

For the *fat_with_grid* input (Figure 4) the RIC algorithm performs poorly since the intersection points of the grid tend to be shallow on the average.

For the *star-shaped_robot_and_obstacles* input (Table 1 (left)) and for the *L-shaped_robot_and_obstacles* input, we get an improvement comparable with that obtained for the *regular* input. (The source of this improvement is discussed in the full version).

---

[1] Recall that this may fail to hold in some pathological examples, where $|V| \ll n^2$.

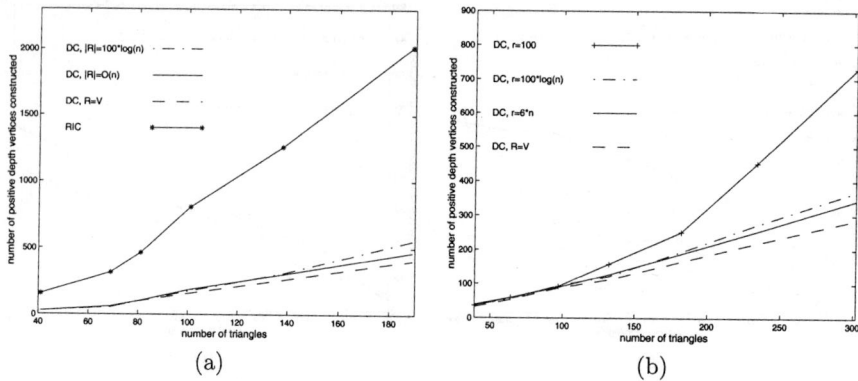

Fig. 4. Number of positive-depth vertices created for the *fat_with_grid* input. Since the differences between the number of positive-depth vertices constructed by the DC algorithm and by the RIC algorithm are highly significant for this input, we compare in (a) the RIC algorithm to the DC algorithm only for $r = 100$ and $r = 6n$, and in (b) we zoom in on the number of positive-depth vertices constructed by the DC algorithm for all different random sample sizes

We remark that the number of vertices $|V|$ in some of our examples is huge (reaching roughly half a million for the *regular* input with 800 triangles), rendering the construction of the union by first computing the entire underlying arrangement unacceptable (when using exact arithmetic). Similarly, our experimenting with $R = V$ is not intended (for such inputs) as a viable implementation, and is used only for measuring performance.

4 Experimental Results II: Running Times

In this section we briefly present our implementation and report some of our experimental results, comparing the running times of the RIC and the DC algorithms. We use the same input sets as in Section 3.1.

In the experiments reported in Section 3 we focused on the number of positive-depth vertices constructed by the RIC and the DC algorithms. This number is a pure estimator in the sense that it is an (expected) lower bound on the time complexity of any implementation of the DC or RIC algorithms, independent of the actual implementation. However, any implementation requires additional operations for constructing the union boundary. Such operations are needed for updating the structure representing the union during the construction. For example, the RIC algorithm can be implemented using trapezoidal decomposition of the complement of the union constructed so far, and corresponding conflict lists between trapezoids and crossing triangles [1, 10]. The expected running time of this implementation is $O(n \log n + \theta_1(\mathcal{A}(T)))$, where $\theta_1(\mathcal{A}(T)) = \sum_{i=0}^{n-2} \frac{1}{(i+1)} \cdot L_i$. The function $\theta_1(\cdot)$ expresses the expected overall

Table 1. The number of positive-depth vertices constructed for the *star-shaped_robot_and_obstacles* input (left) and for the *L-shaped_robot_and_obstacles* input (right)

algorithm	number of positive-depth vertices created, n=2832
DC, $R = V$	328
DC, $r = 6n$	943
DC, $r = 100 \cdot \log n$	1379
DC, $r = 100$	1880
RIC	2037.92

algorithm	number of positive-depth vertices created, n=840
DC, $R = V$	303
DC, $r = 6n$	554.5
DC, $r = 100 \cdot \log n$	654.5
DC, $r = 100$	1070.34
RIC	1423.74

number of conflicts over a random permutation $D := (\Delta_1, \ldots, \Delta_n)$ of the input triangles [1]. Exactly the same implementation can be applied to the DC algorithm.

Our implementation of the union algorithms is based on the CGAL (version 2.3) and LEDA (version 4.3) libraries. The implementation employs the CGAL maps and arrangements packages [4, 6]. In our implementation, which is more practically oriented, we do not maintain trapezoidal decompositions, since, as our experience shows, they incur significant overhead during the construction. Instead, we maintain the union and its complement as a collection of undecomposed faces. In this representation, the additional operations that are needed are point location of vertices of newly-inserted triangles, and traversal of their edges through the current union. In this section, we introduce another estimator for the running time of our algorithms, which is based on the total cost of the edge traversals: Every insertion of a triangle into the partially constructed union is performed by the insertion of the three edges defining the triangle, one at a time. In each such insertion of an edge e, we find the intersection points of e with the boundary of the current union, by traversing its zone in the union, and then insert the portions of e that do not lie inside the present union.

The running time of our implementation is dominated by the total number of edges traversed when inserting the triangles into the union constructed so far. In our experiments, we measure the total number of visited edges by each of the two algorithms. We also compare running times and show that there is a high correlation between the number of traversed edges and the time undertaken for constructing the union by each of the two algorithms.

4.1 Results

In all our experiments, for all five data sets presented in Section 3.1, the DC algorithm performs better than the RIC algorithm. However, (i) the improvement is not as significant as in the number of generated positive-depth vertices, and

(ii) the performance does not significantly improve as the size of the sample R increases. The explanation to these phenomena is: First, both algorithms need to construct the vertices of the union, and this tends to partially hide the savings due to constructing fewer positive-depth vertices. Second, the edge traversal cost may be high when the combinatorial complexity of the visited faces (of the union and its complement) is large. For instance, for the *regular* input the optimal random sample is obtained when $r = 6n$. The ratio between the total number of traversed edges by the DC and by the RIC algorithms in this case, for 800 input triangles, is roughly 10:13. The ratio between the running times, in this case is roughly 10:14. It is conceivable, though, that this cost can be significantly reduced with more careful implementation.

For the *fat* input the DC algorithm only slightly improves the performance. Recall that the RIC algorithm has good performance on the *fat* input, and improving its performance is more difficult (and less feasible) in this case.

For the *fat_with_grid* input the DC algorithm performs significantly better than the RIC algorithm. In this case the RIC algorithm traverses faces with a large number of holes on average. The DC algorithm first inserts the fat triangles, and hence traverses faces with a smaller number of holes. The ratio between the total number of traversed edges by the DC and by the RIC algorithms for 302 input triangles, is roughly 1:50 for all three cases where, $r = 100 \log n$, $r = 6n$, and $R = V$. The explanation to this phenomenon is similar to that given in Section 3.2, where we got a similar number of positive-depth vertices constructed by the DC algorithm, for a random sample of size linear and for $R = V$.

For the *star-shaped_robot_and_obstacles* and the *L-shaped_robot_and_obstacles* inputs, the DC algorithm performs better than the RIC algorithm. The ratio between the running times of the DC algorithm for $r = 6n$ and the RIC algorithm are 10:13 for the *star-shaped_robot_and_obstacles* input, and 10:14 for the *L-shaped_robot_and_obstacles* input.

In most of our experiments, the preprocessing time of the DC algorithm, when taking $R = V$, consumed most of the running time of the union construction. (Recall that we do not propose taking $R = V$ in practice. We carried out these experiments in order to gain a better understanding of what sample size is sufficient to give a good approximation of the ideal DC algorithm). However, already by decreasing R to be linear in the number of triangles (even though the bound of Lemma 1 on the preprocessing cost is $O(n^2 \log n)$) we get a significant improvement in the preprocessing time. For 400 triangles from the *regular* input set, we have that, in all cases, excluding the case $R = V$, the ratio between the preprocessing time and the entire union construction time is less than 1:15. For the *fat, fat_with_grid, star-shaped_robot_and_obstacles* and *L-shaped_robot_and_obstacles* data sets, this ratio is roughly 1:28, 1:70, 1:45 and 1:40, respectively, when taking $r = 6n$.

Note that in all our experiments, when taking $r = 6n$, the total number of edges traversed by the DC algorithm is very close to that number when taking $R = V$.

For lack of space, the full report on the experiments, including running times, is omitted here. These details are provided in the full version of the paper.

5   Conclusions

The experiments reported above demonstrate the practical advantages of the DC algorithm relative to the RIC algorithm. Our results show that the number of positive-depth vertices constructed by the RIC algorithm is larger (and, in many cases, significantly larger) than the number of such vertices constructed by the DC algorithm, for all five kinds of input.

We note that the DC algorithm in its ideal setting (with $R = V$) is *deterministic*. In practice we use randomness, but only to estimate the weights of triangles. After doing so, the insertion order is still computed deterministically (although it is a random variable).

The DC algorithm can be generalized for other geometric objects in the plane, and also can be extended to higher dimensions. Since the calculation of the disjoint cover of each such object deals mostly with counting the number of vertices contained in the interior of that object, the order of insertion can be easily calculated if we are provided with suitable primitives for calculating intersection points and for testing orientations of triples. We are currently in the process applying our algorithm to the union of ellipses and of lenses formed by pairs of intersecting disks.

In another batch of experiments, we experimented with a Divide & Conquer (D&C) algorithm. In the D&C algorithm we first order the triangles of the input in a random permutation, and then recursively construct the union of the triangles in the first half of the permutation, and the union of those in the second half. In the merge step we compute the whole union as the union of the two partial unions constructed so far, using a straightforward sweep-line algorithm [2]. In this construction, one can show that given a collection T of n triangles, the number of positive-depth vertices constructed by the D&C algorithm is $O(\theta(\mathcal{A}(T)) \cdot \log n)$. In other words, the expected cost of the D&C algorithm is worse than that of the RIC algorithm. It is, however, simple to implement.

Nevertheless, through our experiments, the DC algorithm did not decrease performance relative to the RIC algorithm. In some cases, as the case of the *fat_with_grid* input, the improvement in the performance achieved by the DC algorithm was extremely significant. In other cases (the *regular* input and the two *robot_and_obstacles* inputs), the actual improvements were still significant, ranging from 20 to 30 percent. We also note that, in our experiments, the preprocessing time, undertaken for random samples that are logarithmic or even linear in the size of the input, is negligible relative to the entire union construction time, and the improvement in performance obtained in these cases was similar to the improvement we gained when running the DC algorithm in its ideal setting. This implies that, by a small amount of additional work (recall also that the preprocessing stage is very easy to implement), we obtain an algorithm that achieves better performance in practice than the RIC algorithm.

References

[1] P. K. Agarwal and S. Har-Peled. Two randomized incremental algorithms for planar arrangements, with a twist. Manuscript, 2001.

[2] M. de Berg, M. van Kreveld, M. Overmars, and O. Schwarzkopf. *Computational Geometry: Algorithms and Applications*. Springer-Verlag, Berlin, 1997.

[3] E. Flato. Robust and efficient construction of planar Minkowski sums. Master's thesis, Dept. Comput. Sci., Tel-Aviv Univ., 2000. http://www.cs.tau.ac.il/~flato.

[4] E. Flato, D. Halperin, I. Hanniel, O. Nechushtan, and E. Ezra. The design and implementation of planar maps in CGAL. *The ACM Journal of Experimental Algorithmics*, 5, 2000. Also in LNCS Vol. 1668 (WAE '99), Springer, pp. 154–168.

[5] A. Gajentaan and M. H. Overmars. On a class of $O(n^2)$ problems in computational geometry. *Comput. Geom. Theory Appl.*, 5:165–185, 1995.

[6] I. Hanniel and D. Halperin. Two-dimensional arrangements in CGAL and adaptive point location for parametric curves. In *Proc. of the 4th Workshop of Algorithm Engineering*, volume 1982 of *Lecture Notes Comput. Sci.*, pages 171–182. Springer-Verlag, 2000.

[7] K. Kedem, R. Livne, J. Pach, and M. Sharir. On the union of Jordan regions and collision-free translational motion amidst polygonal obstacles. *Discrete Comput. Geom.*, 1:59–71, 1986.

[8] J.-C. Latombe. *Robot Motion Planning*. Kluwer Academic Publishers, Boston, 1991.

[9] J. Matoušek, N. Miller, J. Pach, M. Sharir, S. Sifrony, and E. Welzl. Fat triangles determine linearly many holes. In *Proc. 32nd Annu. IEEE Sympos. Found. Comput. Sci.*, pages 49–58, 1991.

[10] K. Mulmuley. *Computational Geometry: An Introduction through Randomized Algorithms*. Prentice Hall, Englewood Cliffs, New Jersey 07632, U.S.A., 1994.

[11] J. Pach and P. K. Agarwal. *Combinatorial Geometry*. John Wiley & Sons, New York, NY, 1995.

Simple and Fast: Improving a Branch-And-Bound Algorithm for Maximum Clique*

Torsten Fahle

University of Paderborn
Department of Mathematics and Computer Science
Fürstenallee 11, D-33102 Paderborn, Germany
tef@uni-paderborn.de

Abstract. We consider a branch-and-bound algorithm for maximum clique problems. We introduce cost based filtering techniques for the so-called *candidate set* (i.e. a set of nodes that can possibly extend the clique in the current choice point).
Additionally, we present a taxonomy of upper bounds for maximum clique. Analytical results show that our cost based filtering is in a sense as tight as most of these well-known bounds for the maximum clique problem.
Experiments demonstrate that the combination of cost based filtering and vertex coloring bounds outperforms the old approach as well as approaches that only apply either of these techniques. Furthermore, the new algorithm is competitive with other recent algorithms for maximum clique.

Keywords: maximum clique, branch-and-bound, constraint programming, cost based filtering.

1 Introduction

A *Clique* is an undirected graph $C = (V_C, E_C)$ where all nodes in V are pairwise adjacent. The *Maximum Clique Problem (MC)* on a graph $G = (V, E)$ asks for a clique $C = (V_C, E_C)$ in G having a largest node set $|V_C|$. It is known that MC is NP-hard [3].

Despite its complexity, the maximum clique problem is important in many fields. Cliques are used in integer programming for presolving [19] and deriving valid cuts [1]. They appear in coding theory, fault diagnostics, or pattern recognition. For an overview on applications we refer to [7].

In this paper, we introduce some simple cost based domain filtering techniques [13]. We test them on a branch-and-bound algorithm proposed by Carraghan and Pardalos [9]. Their algorithm is still believed to be one of the fastest

---

* Partially supported by the Future and Emerging Technologies programme of the EU under contract number IST-1999-14186 (ALCOM-FT).

clique solver for sparse graphs. It recursively enlarges a clique by adding nodes of a candidate set. Pruning based on simple bounds is used to cut off useless parts of the search tree.

Our cost based filtering tightens the candidate set by removing or fixing certain nodes. We show analytically that this domain filtering in a sense is as tight as typical upper bounds for MC. As we will see later, our filtering dominates 7 out of 8 bounds proposed for MC. The bound that is not dominated (but also does not dominate the filtering) is the vertex coloring bound. We add it to the algorithm and achieve a significant improvement on the DIMACS benchmark set. Within a 6 hours time limit the enhanced algorithm can prove optimal results for 45 out of 66 benchmark instances, whereas the old approach is only able to finalize 33 instances. Furthermore, it turns out that domain filtering improves 56 out of 66 test cases compared to applying coloring bounds alone. (An earlier version of this work was presented at the CPAIOR'02 workshop [10]).

1.1 Literature Review

Cost based filtering [13] aims at fixing variables with respect to the objective function. It typically combines techniques from Optimization with domain filtering stemming from Constraint Programming. It is an important technique when solving complex optimization problems. Optimization constraints for various applications have been addressed e.g. in [13, 14, 16, 11, 12, 18].

The maximum clique problem has attracted researchers in computer science, operations research and other fields for many years. We can therefore only scratch some results and have to refer to some recent overview in [7] for further details.

Balas and Yu [5] presented a new idea for implicitly enumerating cliques. They search for a maximal induced triangulated subgraph T of G in which then a maximum clique C is searched. In a second phase they extend T in a way that does not enlarge the max. clique. Applied in a branch-and-bound scheme their approach was quite successful as the bounds generated turned out to be quite tight. A branch-and-bound algorithm using vertex coloring bounds was presented by Wood [20].

Cuts for an integer programming formulations of the maximum clique problem have been investigated in [4]. The linear independent set formulation, $\max\{\sum_{i=1}^{n} x_i \mid x_i + x_j \leq 1, \forall \{i,j\} \notin E, x \in [0\ldots 1]^n\}$ was used in their approach.

Recently, Caprara, Pisinger and Toth proposed a solver for the quadratic Knapsack problem which was applied also to the maximum clique problem [8].

Enumerating algorithms have been contributed by many researchers. Of certain interests for this work are the methods of Carraghan and Pardalos, and Östergård. We will return to those algorithms in Sec. 2. Before that we will briefly introduce some notation. In Sec. 2.1 we will present the filtering constraints. Some analytical results are given in Sec. 3, empirical studies are presented in Sec. 4. Finally, we conclude.

1.2 Notation

Throughout the paper we denote the neighborhood of a node $v \in V$ as $N(v) = \{u \in V | \{u,v\} \in E\}$ and the degree of a node $v \in V$ as $\delta(v) = |N(v)|$. Given a node set U, the graph $G_U = (U, E \cap (U \times U))$ is the graph *induced* by U. We write $N_U(v) = \{u \in U | \{u,v\} \in (U \times U) \cap E\}$ and the $\delta_U(v) = |N_U(v)|$ if we speak of the neighborhood, or degree, resp. of a node v in the subgraph of G induced by U. For a clique C we call $P = \bigcap_{v \in C} N(v)$ the *candidate set of C*. A clique $\widetilde{C}$ is an *extension* of a clique C if $C \subseteq \widetilde{C}$. We use the notation $\widetilde{C}(C)$ for extensions. To ease the notation, we often identify a clique C with its set of nodes. Finally, $\mathcal{Z}(G) = \{I \subseteq V \mid G_I \text{ is a connected component in } G\}$ contains the *node sets of connected components* of the graph $G = (V, E)$

2 Simple Branch-And-Bound Algorithms for Maximum Clique

Carraghan and Pardalos [9] enumerate the cliques of a graph $G = (V, E)$ according to some lexicographical order. Without pruning, the algorithm finds the largest clique in G_V containing node v_1, then the largest clique in $G_{V \setminus \{v_1\}}$ containing v_2, etc. A candidate set $P = \bigcap_{v \in C} N(v)$ containing all nodes that are adjacent to the clique C currently under construction is used for bounding. A simple depth first search and static variable ordering is used to guide the branch-and-bound (Alg. 1). The well-known dfmax [22] program (D. Applegate and D. Johnson) is an efficient implementation of that idea.

Algorithm 1 A simple Algorithm to find the Maximum Clique C^*

---

function findClique(set C, set P)   **function** main( )

1: **if** ($|C| > |C^*|$) **then**   1: $C^* \leftarrow \emptyset$
2: $\quad C^* \leftarrow C$ 2: findClique($\emptyset$, V)
3: **if** ($|C| + |P| > |C^*|$) **then** 3: **return** C^*
4: $\quad$ **for all** $p \in P$ in predetermined order:
$\quad\quad$ **do**
5: $\quad\quad P \leftarrow P \setminus \{p\}$
6: $\quad\quad C' \leftarrow C \cup \{p\}$
7: $\quad\quad P' \leftarrow P \cap N(p)$
8: $\quad\quad$ findClique(C', P')

Östergård [17] developed a variant of the previous method. Instead of determine the cliques in a decreasing node set, his algorithm starts on $G_{\{v_n\}}$ and determines the largest clique on that graph. Then it considers $G_{\{v_{n-1}, v_n\}}$ and determines the largest clique there. By adding more and more nodes, and determine cliques on each of those sets, it ends up with a largest clique in G. During that incremental process, bounds on the largest cliques containing a certain node

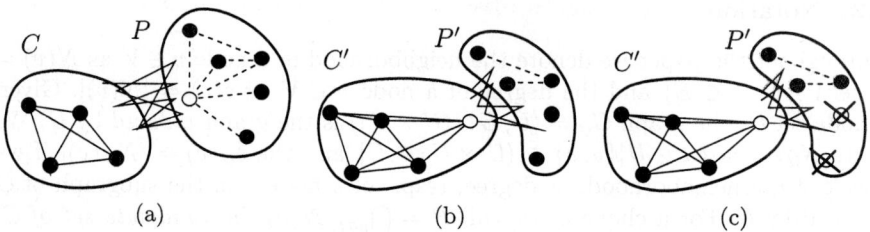

Fig. 1. (a) a clique C and its candidate set P. (b) applying lemma 2 fixes one more node. (c) as a result, two nodes have degree zero now, and can be eliminated according to lemma 1

can be determine as a by-product. With these bounds at hand, Östergård can speed up finding cliques significantly.

2.1 Observations on the Candidate Set

Both algorithms sketched above share the idea of a candidate set $P = \bigcap_{v \in C} N(v)$ containing only those nodes that may extend the clique C currently under construction. P is also used for pruning (line 3), since the largest clique to be discovered in the current part of the search tree can contain $|C| + |P|$ nodes at most. We will show now that two simple observations can help to tighten the candidate set, and thus, potentially reduce the number of choice points explored.

Looking at the nodes in P in more detail, we can characterize those that can never extend a given clique to a maximum clique:

Lemma 1. *Let $G = (V, E)$ be a graph, C be a clique on G, and P be a candidate set, i.e. $P = \bigcap_{v \in C} N(v)$. Furthermore, let $\sigma \in \mathbb{N}$ be a lower bound on the size of a maximum clique in G. Then, for every $v \in P$ such that $|C| + \delta_P(v) < \sigma - 1$, v cannot extend C to a maximum clique of G.*

(proof omitted due to space limitations)

The next lemma identifies nodes that will be members of any extension of the current clique to a maximum one:

Lemma 2. *Let $G = (V, E)$ be a graph, C be a clique on G, and P be a candidate set, i.e. $P = \bigcap_{v \in C} N(v)$. Then, every $v \in P$ such that $\delta_P(v) = |P| - 1$, is contained in any maximum clique of G that also contains C.*

(proof omitted due to space limitations)

Figure 1(a) represents the settings of the algorithms described above, Figure 1(b),(c) sketches the situation described by lemma 1, 2.

2.2 An Improved Branch-And-Bound Algorithm

With these observations at hand we are ready to present an improved enumeration algorithm for MC (Alg. 2, on page 489). Between lines 7 and 8 of the first algorithm we insert some domain filtering technique based on lemma 1 and 2. We know that the next improvement has to be better than the best clique found so far. Hence, we use $(|C^*|+1)$ as a lower bound for the necessary checks.

Domain filtering does not depend on the order in which the two lemmas are applied. The only situation where the lemmas can be in conflict arises when there is $v \in P'$ such that $|P'| - 1 = \delta_{P'}(v)$ and $\delta_{P'}(v) + |C'| < |C^*|$. Then $|P'| - 1 + |C'| < |C^*| \Rightarrow |P'| + |C'| \leq |C^*|$. But in such a case we cannot improve on C^* anymore, and that part of the search tree is pruned in the next recursion level by line 3 of Alg. 2. Furthermore, it is easy to see that the order in which nodes are considered within each lemma does not change the final result. Also, it is clear that when applying lemma 1 first, then lemma 2, no node will fulfill the requirements of lemma 1 again in that branch-and-bound node.

Algorithm 2 An Improved Algorithm for the Maximum Clique Problem

function findClique2(set C, set P)
1:   if $(|C| > |C^*|)$ then
2:     $C^* \leftarrow C$
3:   if $(|C| + |P| > |C^*|)$ then
4:     for all $p \in P$ in predetermined order: do
5:       $P \leftarrow P \setminus \{p\}$
6:       $C' \leftarrow C \cup \{p\}$
7:       $P' \leftarrow P \cap N(p)$
8:       // domain filtering
9:       // reduce possible set
10:      while $(\exists v \in P' : \delta_{P'}(v) + |C'| < |C^*|)$ do
11:         $P' \leftarrow P' \setminus \{v\}$ // lemma 1
12:      // increase required set
13:      while $(\exists v \in P' : \delta_{P'}(v) = |P'| - 1)$ do
14:         $C' \leftarrow C' \cup \{v\}$ // lemma 2
15:         $P' \leftarrow P' \setminus \{v\}$ // lemma 2
16:      // here it holds: $(\forall v \in P' : (|C^*| - |C'|) \leq \delta_{P'}(v) < (|P'| - 1)$
17:      findClique2(C', P')

3 Upper Bounds vs. Domain Filtering: Some Analytical Results

In this section we present some analytical results on the two constraints. Especially, we show that some upper bounds for the MC used in OR methods are already subsumed by the domain filtering.

3.1 Accuracy

We still assume the setting presented in Sec. 2: C is the clique currently under construction, $P = \bigcap_{v \in C} N(v)$ the corresponding candidate set. The following lemma lists some well-known upper bounds $\mathcal{U}_i(C,P)$ for a maximal clique $C^*(C)$ (see also [7]):

Lemma 3 (Upper Bounds for MC). *The following upper bounds hold for MC:*

1. *Only nodes from the candidate set can extend C:* $\mathcal{U}_1(C,P) = |C| + |P|$.
2. *A node with maximum degree in P limits the size of any extension of C to a maximal clique:* $\mathcal{U}_2(C,P) = |C| + \max\{\delta_P(v) + 1 \mid v \in P\}$.
3. *Only one connected component in P can extend C:* $\mathcal{U}_3(C,P) = |C| + |I^{max}|$, where I^{max} denotes the node set of the largest connected component in $\mathcal{Z}(G_P)$
4. *A k-clique requires k nodes with degree at least $k-1$:*
   $\mathcal{U}_4(C,P) = |C| + \max\{k \mid \exists v_1 < \cdots < v_k \in P,\ \delta_P(v_i) \geq k-1\}$
5. *A k-clique has $k(k-1)/2$ edges:* $\mathcal{U}_5(C,P) = |C| + \max\{k \in \mathbb{N} \mid \frac{k(k-1)}{2} \leq |E_P|\}$, where E_P is the edge set of the graph $G_P = (P, E_P)$ induced by P.
6. *Apply $\mathcal{U}_4(C,P)$ on connected components of G_P only:*
   $\mathcal{U}_6(C,P) = |C| + \max_{I \in \mathcal{Z}(G_P)} \max\{k \mid \exists v_1 < \cdots < v_k \in I,\ \delta_P(v_i) \geq k-1\}$
7. *Apply $\mathcal{U}_5(C,P)$ on connected components of G_P only:*
   $\mathcal{U}_7(C,P) = |C| + \max_{I \in \mathcal{Z}(G_P)} \max\{k \in \mathbb{N} \mid \frac{k(k-1)}{2} \leq |E_I|\}$,
   where E_I is the edge set of the graph $G_I = (P, E_I)$ induced by I.
8. *Any k-clique needs k colors in a vertex coloring:* $\mathcal{U}_8(C,P) = |C| + \chi(G_P)$, where $\chi(G_P)$ denotes the (vertex) chromatic number of the graph induced by P.

Proof. For applications of these bounds and further information we refer to [9] for $\mathcal{U}_1$, [5, 20] for $\mathcal{U}_8$, and [7] for some bound similar to $\mathcal{U}_5$. The other bounds are simple extensions of $\mathcal{U}_1$.

Obviously, some of these bounds are contained in others leading to the following relationship (to our knowledge, such a taxonomy has not been presented before, though some relations are obvious and have probably been used earlier):

Lemma 4 (Taxonomy of Bounds). *Let $G = (V,E)$ be a graph, C a (not necessarily maximum) clique in G, and $P = \bigcap_{v \in C} N(v)$ the corresponding candidate set. Then*

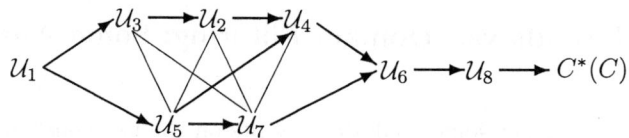

$(\mathcal{U}_i \to \mathcal{U}_j :\Leftrightarrow \mathcal{U}_i \geq \mathcal{U}_j$ and $\mathcal{U}_i - \mathcal{U}_j :\Leftrightarrow$ neither $\mathcal{U}_i \geq \mathcal{U}_j$, nor $\mathcal{U}_i \leq \mathcal{U}_j$).

(proof omitted due to space limitations)

Next, we show that the filters described in Sec. 2.1 already include bounds $\mathcal{U}_1 - \mathcal{U}_7$. I.e. after applying lemma 1, 2 to (C,P) and obtaining a new (C',P'), those bounds cannot prune the current part of the subtree.

Theorem 1. Let $G = (V, E)$ be a graph, C be a clique on G, and P be a candidate set, i.e. $P = \bigcap_{v \in C} N(v)$. Furthermore, let $\sigma \in \mathbb{N}$ be a lower bound on the size of a maximum clique in G. Let $|C| + |P| > \sigma$ and let C' and P' be the node sets after applying the domain filtering of algorithm 2, lines 8 – 16. Then, $\mathcal{U}_i(C', P') > \sigma$ for $i = 1, \ldots, 7$.

Proof. According to lemma 4 it is sufficient to prove $\mathcal{U}_6(C', P') > \sigma$. So let us assume, $\mathcal{U}_6(C', P') \leq \sigma$, i.e. $|C'| + \max_{I \in \mathcal{Z}(G_{P'})} \max\{k \mid \exists v_1 < \cdots < v_k \in I, \delta_{P'}(v_i) \geq k - 1\} \leq \sigma$. After applying lines 8 – 16 of algorithm 2 we have: $\forall v \in P' : (\sigma - |C'|) \leq \delta_{P'}(v)$.
$\Rightarrow \forall v \in P' : |C'| + \delta_{P'}(v) \geq \sigma \geq |C'| + \max_{I \in \mathcal{Z}(G_{P'})} \max\{k \mid \exists v_1 < \cdots < v_k \in I, \delta_{P'}(v_i) \geq k - 1\}$
$\Rightarrow \forall v \in P' : \delta_{P'}(v) \geq \max_{I \in \mathcal{Z}(G_{P'})} \max\{k \mid \exists v_1 < \cdots < v_k \in I, \delta_{P'}(v_i) \geq k - 1\} =: k'$

So the degree of all nodes in P' is at least k', especially, there is a connected component in P' having at least $k' + 1$ nodes with degree larger than k' — being a contradiction to the maximality of $\mathcal{U}_6(C', P')$. □

Bound $\mathcal{U}_8$ is not included in the above theorem. The following two examples show that depending on the situation we may subsume the coloring bound, or it may be more accurate than filtering alone:

Assume, the candidate set after domain filtering is given as sketched in the right graph. Furthermore, we have guessed $\sigma = 4$, and we have already $|C'| = 2$. P' cannot be reduced further by domain filtering, as $\forall v \in P' : \underbrace{\sigma - |C'|}_{=2} \leq \underbrace{\delta_{P'}(v)}_{=2} < \underbrace{|P'| - 1}_{=5}$. Now, as $\chi(G_{P'}) = 2$ we get $\mathcal{U}_8(C', P') = 4 = \sigma$. Thus, applying the coloring bound here can prune parts of the search tree that would still be considered when using filtering only.

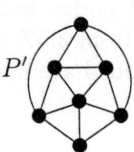

Things change when considering the left graph: Having $\sigma = 3$ and $|C'| = 1$, domain filtering cannot fix any further node in P' as $\forall v \in P' : \underbrace{\sigma - |C'|}_{=2} \leq \underbrace{\delta_{P'}(v)}_{3,4,5} < \underbrace{|P'| - 1}_{=6}$. As $\chi(G_{P'}) = 4$ here, we obtain $\mathcal{U}_8(C', P') = 5 > \sigma$ showing that in this case the coloring bound cannot prune the remaining subtree.

3.2 Computational Complexity

Better accuracy of bounds is usually payed by higher running time when calculating bounds. We briefly compare running time per choice point for calculating bounds or domain filtering, respectively.

As benchmark graphs for cliques are typically rather dense, we assume that the graph G is stored as an adjacency matrix. Also, we assume, that $|C'|$ can be determined in $O(|C'|)$. $\mathcal{U}_1$ requires the size of P, which can be determined in $O(|P|)$ within a reasonable data structure. Also the max. degree, required by $\mathcal{U}_2$, can be determined in that time if degree-information is available (otherwise: $O(|P|^2)$).

A largest connected component in P can be detected by a DFS in a dense graph in time $O(|P|^2)$. However, $\mathcal{U}_4$ being tighter than $\mathcal{U}_3$ can be computed in time $O(|P|\log|P|)$ if degree information is available (or in quadratic time otherwise). Also $\mathcal{U}_6$ and $\mathcal{U}_7$ base their information on connected components, and need therefore time $O(|P|^2)$.

If degree information is at hand, bound $\mathcal{U}_5$ can be determined in $O(|P|)$, otherwise this bound needs to visit G, giving time complexity $O(|P|^2)$.

Vertex coloring, needed for $\mathcal{U}_8$ is NP-hard in general. For bound calculation, however, heuristics are used. They need to visit each node and each edge a constant number of time, giving running time $O(|P|^2)$ in a dense graph (see e.g. [2, 5, 20]).

Our domain filtering, as described in algorithm 2, lines 8 – 16, does perform $|P|$ loops at most, as each loop excludes one node from P. Being dependent on degree information, each exclusion of a node requires an update of the degrees of adjacent nodes. That step requires at most time $O(|P|)$ leading to a total running time of $O(|P|^2)$ for the entire path from the root of the search tree to one of its leaves.

The domain filtering is therefore at least as accurate as bounds $\mathcal{U}_1$–$\mathcal{U}_7$, and requires not more running time asymptotically, than the more effective bounds of these seven candidates.

4 Experiments

Obviously, as the tighter bounds need some more computational effort, there is a trade-off between the effectiveness of pruning techniques and the overall efficiency of the approach. In this section we study the empirical behavior of the new approach. We use the well-established DIMACS benchmark set [15, 21] for the comparison. It consists of 66 instances, ranging from 28 – 3361 nodes, and 420 – 11 million edges. The density of the underlying graphs range from 3.5% – 99.8%. To our knowledge for some of these instances optimal solutions have not been proven so far.

4.1 The Implementation

An efficient implementation has to consider some more "tricks" than those described in the algorithms 1 and 2. When determine $P' = P \cap N(v)$, we use a loop running over all elements $p \in P$. If $\{p,v\} \in E$, p is copied to P'. For non-adjacent elements we decrease a counter m that contains the number of elements still needed in order to obtain $|C'|+|P'| > |C^*|$. Should m ever be decremented below zero we skip that recursion-level immediately. (We refer to [22] for more details).

For the algorithm using domain filtering we also update degree information on the subgraph induced by P. That is, we decrease the degree of a node $p' \in P'$ if an adjacent node $p \in P$ is not copied to P'. By keeping a copy of P in each recursion level we can easily perform incremental updates on the degrees before entering the next recursion level.

In lines 8 – 16 of algorithm 2 we perform domain filtering. We implement the two while loops using a stack to store those nodes, that either have not been copied to P' (lines 5 – 7), or have been removed during domain filtering (lines 8 – 16). In the implementation, we take the next element p from the stack, decrease the degree of all adjacent nodes p', perform the degree tests on those nodes, and in case of removal, push p' onto the stack. We continue until the stack is empty.

For the coloring bounds we use four heuristics: Two of them follow the DSATUR approach of Brelaz [2], e.g. choose the node with a maximum number of colored neighbors and assign the minimal unused color to that node. The other two methods build an independent set for each color class. If the current set cannot be extended anymore, a new set is opened, and the number of sets required is the heuristic chromatic number. As we apply nodes in increasing, and decreasing order of degrees, we obtain four heuristics, and the minimum number calculated is used as the bound.

4.2 Numerical Results

All algorithms were coded in C++ and compiled by the GNU g++ 2.95.3 compiler using full optimization. Our benchmark tests run on a Pentium III-933 PC with 512 MB ram operating Linux 2.4.17. We stopped each run after 6 hours (21 600 sec).

Effects of Domain Filtering In table 2 we compare our implementation[1] of dfmax using only bound $\mathcal{U}_1$ (Alg. 1), an implementation using coloring bounds only (χ), another using domain filtering only (DF), and a forth using both ($\chi + DF$). The latter three implementations correspond to Alg. 2.

As expected, the number of choice points decreased considerably when applying domain filtering. The savings range from 3.6 (mann_a9) to 30.4 (p_hat1500_1) when comparing dfmax with the DF approach. In contrast, running time of the current code is typically a factor of 2 times slower than the dfmax code. Here, domain filtering is computationally too expensive compared to the simple $\mathcal{U}_1$ bound: It is still cheaper to traverse larger useless parts of the search tree than detecting these parts.

The c-fatxxx instances play a special role in this context. They are quite easy to solve using domain filtering. Each instance requires 5 choice points only to prove optimality, with running time being negligible. The dfmax algorithm uses

---

[1] For a fare comparison, we decided to use identical subroutines for all implementations. We adapted dfmax to our settings resulting in a slightly faster algorithm ($< 2\%$) than the original dfmax

dramatically more choice points (a factor of more than one million in the case of c-fat200-5, and more than 10^9 for c-fat500-10). Wood [20] also reports only a few number of choice points (1–27) for his approach on those instances. Hence, we consider this behavior as a drawback of dfmax rather than as an advantage of our filtering.

Using vertex color bounds heavily helps to detect useless parts of the search tree. Out of the 66 benchmark instances, 45 could be solved to optimality when using coloring bounds, whereas only 33 can be finalized by dfmax. Again, it turns out that DF helps to reduce choice points considerable. Additionally, also a positive impact on running time can be seen. Since the coloring bounds a rather expensive to compute, any reduction of the remaining graph helps to decrease overall running time. In 56 out of 66 cases, the overall running time benefits from domain filtering (measured as: (1) better result, (2) (if results equal) being faster, (3) (if results and running time equal) finding best solution earlier).

It should be noted that especially for simple instances running time is negatively affected by more sophisticated approaches (e.g. some of the p_hatxx instances). For those cases some control mechanism that switches off expensive techniques should be used.

Comparison to other Algorithms Though the focus of this paper is not on designing a fastest clique solver, we also compare our results to some recent results presented in the literature. We compare results of Wood [20], Balas and Xue [6], and Östergård [17] against our $\chi + DF$ results. Since the results in column one of table 2 refer to the Carraghan/Pardalos algorithm, we have shown already, that we beat that algorithm in many cases. Unfortunately, for the other approaches different machines where used as well as different time limits. Hence, we can only try to show a general tendency, rather than comparing details. Also, the papers mentioned do not present results on all 66 instances, whereas we have good results on some of the omitted instances. It should be noted here, that we do not use lower bound heuristics, that sometimes give good or even optimal results in the first node of the branch-and-bound tree.[2]

Wood's approach is a branch-and-bound using fractional coloring and lower bound heuristics. It can solve 38 instances to optimality using running times up to 19 hours on a SUN S10. In 10 cases, the heuristics can prove optimality in the root node already (i.e. lower and upper bounds are the same). In the remaining 29 cases our approach always uses less choice points than his fastest method (MC_C), the typical factor being 4–8. When considering MC_D (the approach using the fewest number of choice points), our approach wins in 14 cases, MC_D has less choice points in 15 cases. However, in at least 20 cases the running time is higher than ours.[3].

Balas and Xue developed a heuristics to determine fractional coloring bounds within a branch-and-bound framework. They used a DEC Alpha 300–400 AXP

---

[2] The reason being that we want to show the effects of bounds and filtering clearly
[3] A SUN Sparc 10/51 is about 10 times slower than our computer, and we transformed the running times with this factor

and a 5 hours time limit. It is not noted whether the result given are all proven optimal results, or best results after stopping the algorithm. Again, we omit those cases, where heuristics already prove optimality in the root node. In 19 out of 38 cases we use less choice points than their approach. Assuming that their computer is about 4 times slower than ours, we beat that algorithm in 13 cases.

Östergård's approach was already described in Sec. 2. As no numbers on choice points are given, we can only compare running times. The computer used is a 500 MHz PC, roughly 2 times slower than ours. He considers 38 instances, of which we can solve 8 faster than his approach. Since our domain filtering approach is compatible with the ideas of Östergård, combining both would be an interesting experiment. However, Östergård also mentions that his approach sometimes uses much more time than other solvers, so a detailed experimental evaluation on all DIMACS instances is necessary here.

5 Conclusions and Future Work

We presented some simple constraints that tighten the candidate set used in solvers for maximum clique instances. Using a taxonomy of typical bounds for MC we were able to show that the tightened candidate set is in a sense as least as tight as seven of these bounds.

A brief complexity analysis furthermore showed, that the asymptotic running time is not higher than the time complexity of efficient bounds used. In an experimental study we showed significant reduction of choice points when using the new approach. Introducing some tighter bounds further improves the algorithm, and the combination of tight bounds and domain filtering is the most promising approach tested in this paper. Hence, the combination of a simple branch-and-bound, domain filtering, and coloring bounds defines a simple and fast solver for MC.

Obviously, the bounds' quality is most important for the overall efficiency. They prune the search tree and lead to the promising parts. Domain filtering fixes variables in a choice point. As we thereby avoid branching on those variables, domain filtering helps to stabilize the search within the solution space.

Since the domain filtering approach is rather generic, it can be combined easily with other MC algorithms – like the one of Östergård.

In this paper, we focused on the effects of domain filtering for MC, and the algorithms tested were designed to show these effects rather then to be extremely fast. When considering other techniques as well, running time of the algorithm can be reduced further. As shown by others, lower bound heuristics can prove optimality in the root node for 10 DIMACS instances. Also for the p_hatxx and the mann_axx instances applying lower bound heuristics provides a solution better than the best solution found after 6 hours of pure systematic search. Since our domain filtering depends on a lower bound for the maximum clique, such a heuristic also improves effects of domain filtering.

file	|V|	density	our dfmax |C*|	ChPts	time	|C*|	χ ChPts	time	|C*|	DF ChPts	time	|C*|	χ+DF ChPts	time
brock200_1	200	74.54%	21	38043497	28.82	21	74613	181.82	21	3350353	50.61	21	66042	139.46
brock200_2	200	49.63%	12	54314	0.05	12	530	0.64	12	3539	0.16	12	437	0.46
brock200_3	200	60.54%	15	453265	0.41	15	2487	4.93	15	32170	0.99	15	2332	3.34
brock200_4	200	65.77%	17	2161909	1.73	17	9708	16.86	17	167750	3.64	17	8779	12.27
brock400_1	400	74.84%	≥27	24423311512	≥21600	≥24	6682621	≥21600	≥25	1115779306	≥21600	≥24	7513237	≥21600
brock400_2	400	74.92%	≥29	19952147706	≥21600	≥29	7159353	≥21600	≥29	809348442	≥21600	≥29	6743670	≥21600
brock400_3	400	74.79%	≥31	25092151368	≥21600	≥24	6208742	≥21600	≥24	1142054135	≥21600	≥24	7061823	≥21600
brock400_4	400	74.89%	33	20152485865	20577.7	≥25	5370110	≥21600	≥25	1016877999	≥21600	≥25	5950301	≥21600
brock800_1	800	64.93%	≥21	21193921452	≥21600	≥20	7825966	≥21600	≥21	579955608	≥21600	≥21	9205790	≥21600
brock800_2	800	65.13%	≥21	23396947454	≥21600	≥20	7861777	≥21600	≥20	684885264	≥21600	≥20	9747602	≥21600
brock800_3	800	64.87%	≥21	22077418714	≥21600	≥20	7596314	≥21600	≥20	645967913	≥21600	≥20	9464943	≥21600
brock800_4	800	64.97%	≥26	21514428210	≥21600	≥20	7688179	≥21600	≥21	643687866	≥21600	≥20	9529585	≥21600
c-fat200-1	200	7.71%	12	52	0.01	12	34	0	12	5	0	12	5	0.01
c-fat200-2	200	16.26%	24	444	0	24	70	0.01	24	5	0	24	5	0.01
c-fat200-5	200	42.58%	58	268435599	1307.45	58	172	0.04	58	5	0	58	5	0.01
c-fat500-1	500	3.57%	14	71	0.01	14	40	0.02	14	5	0.01	14	5	0.02
c-fat500-10	500	37.38%	≥124	15086121946	≥21600	126	376	0.3	126	5	0.01	126	5	0.03
c-fat500-2	500	7.33%	26	683	0.01	26	76	0.03	26	5	0	26	5	0.02
c-fat500-5	500	18.59%	64	5703074	5.71	64	190	0.06	64	5	0.01	64	5	0.03
hamming10-2	1024	99.02%	≥512	2963756831	≥21600	512	513	8.92	≥512	902493400	≥21600	512	257	7.74
hamming10-4	1024	82.89%	≥32	35273440082	≥21600	≥32	7782157	≥21600	≥32	2933556637	≥21600	≥32	7677209	≥21600
hamming6-2	64	90.48%	32	42787	0.03	32	33	0	32	1891	0.01	32	17	0.01
hamming6-4	64	34.92%	4	221	0	4	28	0	4	47	0	4	31	0
hamming8-2	256	96.86%	≥128	13603244188	≥21600	128	129	0.14	≥128	1582399395	≥21600	128	65	0.11
hamming8-4	256	63.92%	16	3742143	4.05	16	2063	9.97	16	252418	7.47	16	1950	9.16
johnson16-2-4	120	76.47%	8	4177630	1.51	8	96767	12.48	8	904446	3.57	8	126460	11.87
johnson32-2-4	496	87.88%	≥16	49631406919	≥21600	≥16	133437402	≥21600	≥16	4918222342	≥21600	≥16	139645893	≥21600
johnson8-2-4	28	55.56%	4	90	0	4	13	0	4	21	0	4	15	0
johnson8-4-4	70	76.81%	14	12544	0.01	14	50	0.05	14	1205	0.02	14	39	0.04
keller4	171	64.91%	11	1275236	0.79	11	2664	3.93	11	107086	1.53	11	1771	3.79
keller5	776	75.15%	≥24	23430537443	≥21600	≥24	2694646	≥21600	≥24	737283168	≥21600	≥25	2671681	≥21600
keller6	3361	81.82%	≥43	17888655376	≥21600	≥43	1390946	≥21600	≥43	540201863	≥21600	≥43	1362220	≥21600
MANN_a9	45	92.73%	16	329235	0.1	16	49	0.03	16	92210	0.11	16	31	0.01
MANN_a27	378	99.01%	≥116	15557799773	≥21600	≥126	40001	≥21600	≥116	23120760471	≥21600	126	39351	15223.3
MANN_a45	1035	99.63%	≥220	12119272496	≥21600	≥327	13098	≥21600	≥234	23274505520	≥21600	≥331	10367	≥21600
MANN_a81	3321	99.88%	≥438	5947703016	≥21600	≥994	30709	≥21600	≥467	23035700588	≥21600	≥996	8113	≥21600
p_hat1000-1	1000	24.48%	10	1801019	2.18	10	28375	28.99	10	81573	9.9	10	19430	24.65
p_hat1000-2	1000	49.01%	≥42	19361957667	≥21600	≥44	990436	≥21600	≥41	1149227032	≥21600	≥44	968615	≥21600
p_hat1000-3	1000	74.42%	≥49	21936773057	≥21600	≥50	928896	≥21600	≥47	1603605303	≥21600	≥50	920905	≥21600
p_hat1500-1	1500	25.34%	12	13825580	20.77	12	140943	299.62	12	454914	97.29	12	136620	179.65
p_hat1500-2	1500	50.61%	≥46	24069645860	≥21600	≥52	705644	≥21600	≥46	1893106666	≥21600	≥52	656300	≥21600
p_hat1500-3	1500	75.36%	≥53	26020312163	≥21600	≥55	849096	≥21600	≥53	1920941837	≥21600	≥56	885013	≥21600
p_hat300-1	300	24.38%	8	11634	0.02	8	271	0.17	8	450	0.06	8	254	0.11
p_hat300-2	300	48.89%	25	1553140	1.25	25	1443	5.34	25	182168	2.52	25	1121	4.52
p_hat300-3	300	74.45%	36	1960518988	1614.35	36	210265	1668.4	36	251846120	2561.11	36	171086	1285
p_hat500-1	500	25.31%	9	89908	0.11	9	778	1.28	9	4978	0.43	9	690	0.9
p_hat500-2	500	50.46%	36	292274682	253.76	36	38031	363.22	36	35324032	459.89	36	32413	305.9
p_hat500-3	500	75.19%	≥44	23974852012	≥21600	≥48	1113372	≥21600	≥43	2103038234	≥21600	≥48	993862	≥21600
p_hat700-1	700	24.93%	11	343857	0.44	11	2617	4.7	11	20630	1.94	11	2195	4.01
p_hat700-2	700	49.76%	44	10969888234	10397.8	44	231488	3921.47	44	1394776054	16637.8	44	188823	3129.95
p_hat700-3	700	74.80%	≥49	22686699084	≥21600	≥54	737898	≥21600	≥49	1851669803	≥21600	≥54	653774	≥21600
san1000	1000	50.15%	≥10	11298828624	≥21600	15	29502	4158.17	≥10	207288356	≥21600	15	35189	4566.13
san200_0.7_1	200	70.00%	30	14265131069	5713.23	30	348	2.32	30	3805796652	8848.04	30	301	2.35
san200_0.7_2	200	70.00%	≥18	36793469795	≥21600	18	419	1.07	≥18	5332111203	≥21600	18	394	0.99
san200_0.9_1	200	90.00%	≥48	44733138057	≥21600	70	32148	171.54	≥48	16789724889	≥21600	70	20239	93.92
san200_0.9_2	200	90.00%	≥41	47851554296	≥21600	60	446005	4096.28	≥41	3788397406	≥21600	60	309378	2896.35
san200_0.9_3	200	90.00%	≥44	22287415000	≥21600	44	38240	347.31	≥44	1804345383	≥21600	44	32327	292.44
san400_0.5_1	400	50.00%	13	1414713059	924.5	13	703	8.9	13	155275559	2095.96	13	882	10.11
san400_0.7_1	400	70.00%	≥22	86272548474	≥21600	40	12553	672.43	≥22	17971994383	≥21600	40	11830	638.98
san400_0.7_2	400	70.00%	≥17	60943311431	≥21600	30	21511	218.47	≥17	8165125369	≥21600	30	26818	239.58
san400_0.7_3	400	70.00%	≥22	25369766347	≥21600	22	141744	833.87	≥18	2116314580	≥21600	22	213162	925.61
san400_0.9_1	400	90.00%	≥49	51823450631	≥21600	100	610230	17850.1	≥49	6422756477	≥21600	100	291195	10829.3
sanr200_0.7	200	69.69%	18	7985123	5.96	18	28835	51.19	18	665013	11.55	18	25582	37.68
sanr200_0.9	200	89.76%	≥40	27782611608	≥21600	≥41	2634239	≥21600	≥40	2453566354	≥21600	≥41	2649347	≥21600
sanr400_0.5	400	50.11%	13	4283999	4.07	13	36414	53.04	13	251463	11.74	13	32883	34.64
sanr400_0.7	400	70.01%	21	5624332491	4700.66	21	11010391	31686.3	21	408188818	8765.3	21	9759158	23887.5

Fig. 2. Results on the Dimacs Instances. We compare the number of choice points and running time in seconds of our implementation of dfmax, our algorithm using coloring bounds only (χ), our algorithm using domain filtering only (DF), and a forth using both ($\chi + DF$)

The ordering in which nodes are considered during search can have a severe impact on running time, too. We found in an additional study that some instances benefit from different orderings. For mann_a81 e.g. we find a solution of 1096 within 10 sec. when using a reverse ordering, whereas the best solutions found by the other approaches after 6 hours is still below 1000 (and below 500 when not using coloring bounds). It is not clear though how to detect these effects while searching, and utilize them to improve convergence.

References

[1] A. Atamtürk and G. L. Nemhauser and M. W. P. Savelsberg. Conflict Graphs in Integer Programming. *European Journal of Operations Research*, 121:40–55, 2000.
[2] D. Brelaz. New methods to color the vertices of a graph. *Communcations of the ACM*, 22:251–256, 1979.
[3] M. R. Garey and D. S. Johnson. *Computers and Intractability*. W. H. Freeman & Co., 1979.
[4] E. Balas, S. Ceria, G. Couruéjols and G. Pataki. Polyhedral Methods for the Maximum Clique Problem. in [15, p. 11–28].
[5] E. Balas and C. S. Yu. Finding a Maximum Clique in an Arbitrary Graph. *SIAM Journal Computing*, 14(4):1054–1068, 1986.
[6] E. Balas and Xue. Weighted and Unweighted Maximum Clique Algorithms with Upper Bounds from Fractional Coloring. *Algorithmica*, 15:397–412, 1996.
[7] I. M. Bomze, M. Budinich, P. M. Pardalos, M. Pelillo. *The Maximum Clique Problem*. Handbook of Combinatorial Optimization, volume 4. Kluwer Academic Publishers, 1999.
[8] A. Caprara and D. Pisinger and P. Toth. Exact Solutions on the Quadratic Knapsack Problem. *Informs Journal on Computing*, 11(2):125–137, 1999.
[9] R. Carraghan and P. M. Pardalos. An exact algorithm for the maximum clique problem. *Operations Research Letters* 9:375–382, 1990.
[10] T. Fahle. Cost Based Filtering vs. UpperBounds for Maximum Clique *CP-AI-OR'02 Workshop*, Le Croisic/France, 2002.
[11] T. Fahle, U. Junker, S. E. Karisch, N. Kohl, M. Sellmann, B. Vaaben. Constraint programming based column generation for crew assignment. *Journal of Heuristics* 8(1):59–81, 2002.
[12] T. Fahle and M. Sellmann. Constraint Programming Based Column Generation with Knapsack Subproblems. *Annals of Operations Reserach, Vol 114*, 2003, to appear.
[13] F. Focacci, A. Lodi, M. Milano. Cost-Based Domain Filtering. *Proc. CP'99* LNCS 1713:189–203, 1999.
[14] F. Focacci, A. Lodi, M. Milano. Cutting Planes in Constraint Programming: An Hybrid Approach. *Proceedings of CP'00*, Springer LNCF 1894:187–200, 2000.
[15] D. S. Johnson and M. A. Trick. *Cliques, Colorings and Satisfiability*. 2nd DIMACS Implementation Challenge, 1993. American Mathematical Society, 1996.
[16] U. Junker, S. E. Karisch, N. Kohl, B. Vaaben, T. Fahle, M. Sellmann. A Framework for Constraint programming based column generation. *Proc. CP'99* LNCS 1713:261–274, 1999.
[17] P. R. J. Östergård. A fast algorithm for the maximum clique problem. *Discrete Applied Mathematics*, to appear.

[18] G. Ottosson and E. S. Thorsteinsson. Linear Relaxation and Reduced-Cost Based Propagation of Continuous Variable Subscripts. *CP-AI-OR'00*, Paderborn, 2000, submitted.

[19] M. W. P. Savelsbergh. Preprocessing and probing techniques for mixed integer programming problems. *ORSA Journal on Computing*, 6:445–454, 1994.

[20] D. R. Wood. An algorithm for finding a maximum clique in a graph. *Operations Research Letters*, 21:211–217, 1997.

[21] ftp://dimacs.rutgers.edu/pub/challenge/graph/benchmarks/clique/ Dimacs Clique Benchmark Instances.

[22] dfmax.c. ftp://dimacs.rutgers.edu/pub/challenge/graph/solvers/

Online Companion Caching

Amos Fiat[1], Manor Mendel[2], and Steven S. Seiden[3]*

[1] School of Computer Science, Tel-Aviv University
fiat@tau.ac.il
[2] School of Computer Science, The Hebrew University
mendelma@cs.huji.ac.il
[3] Department of Computer Science, Louisiana State University, Baton Rouge

Steve Seiden died in a tragic accident on June 11, 2002. The other authors would like to dedicate this paper to his memory.

Abstract. This paper is concerned with online caching algorithms for the (n,k)-companion cache, defined by Brehob et. al. [3]. In this model the cache is composed of two components: a k-way set-associative cache and a companion fully-associative cache of size n. We show that the deterministic competitive ratio for this problem is $(n+1)(k+1)-1$, and the randomized competitive ratio is $O(\log n \log k)$ and $\Omega(\log n + \log k)$.

1 Introduction

A popular cache architecture in modern computer systems is the *set-associative cache*. In a k-way set-associative cache, a cache of size s is divided into $m = s/k$ disjoint sets, each of size k. Addresses in main memory are likewise assigned one of m types, and the i'th associative cache can only store memory cells whose address is type i. Special cases includes *direct-mapped caches*, which are 1-way set associative caches, and fully-associative caches, which are s-way set associative caches.

In order to overcome "hot-spots", where the same set associative cache is being constantly accessed, computer architects have designed hybrid cache architectures. Typically such a cache has two or more components. A given item can be placed in any of the components of the cache. Brehob et. al. [3] considered the (n,k) companion cache, which consists of two components: A k-way set associative called the *main cache*, and a fully-associative cache of size n, called the *companion cache* (the names stem from the fact that typically $mk \gg n$). As argued by Brehob et. al. [3], many of the L1-cache designs suggested in recent years use companion caches as the underlying architecture. Several variations on the basic companion cache structure are possible. These include reorganization/no-reorganization and bypassing/no-bypassing. Reorganization is the ability to move an item from one cache component to another, whereas bypassing is the ability to avoid storing an accessed item in the cache. A schematic view of the companion cache is presented in Fig. 1.

---

* This research was partially supported by the Louisiana Board of Regents Research Competitiveness Subprogram and by AFOSR grant No. F49620-01-1-0264.

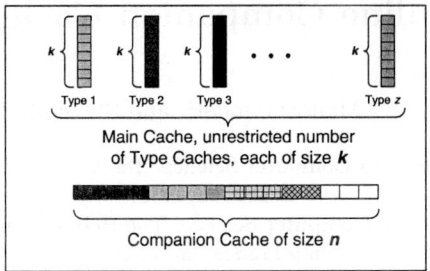

Fig. 1. A schematic description of a companion cache

Since maintenance of the cache must be done online, and this makes it impossible to service requests optimally, we use *competitive analysis*. The usual assumption is that any referenced item is brought into the cache before it is accessed. Since items in the cache are accessed much more quickly than those outside, we associate costs with servicing items as follows: If the referenced item is already in the cache then we say that the reference is a *hit* and the cost is zero. Otherwise, we have a *fault* or *miss* which costs one. Roughly speaking, an online caching algorithm is called r-competitive if for any request sequence the number of faults is at most r times the number of faults of the optimal offline algorithm, allowing a constant additive term.

Previous Results: Maintenance of a fully associative cache of size k is the well known *paging problem* [2]. Sleator and Tarjan [8] proved that natural algorithms such as Least Recently Used are k-competitive, and that this is optimal for deterministic online algorithms. Fiat *et. al.* [4], improved by McGeoch and Sleator [7] and Achlioptas *et. al.* [1], show a tight $\approx \ln k$ competitive randomized algorithm. k-way set associative caches can be viewed as a collection of independent fully associative caches, each of size k, and therefore they are uninteresting algorithmically. Brehob *et. al.* [3] study deterministic online algorithms for $(n, 1)$-companion caches. They investigate the four previously mentioned variants, i.e., bypassing/no-bypassing and reorganization/no-reorganization.

Our Results: This paper studies deterministic and randomized caching algorithms for a (n, k)-companion cache. We consider the version where reorganization is allowed but bypassing is not. We show that the deterministic competitive ratio is exactly $(n + 1)(k + 1) - 1$. For randomized algorithms, we present an upper bound of $O(\log n \log k)$ on the competitive ratio, and a lower bound of $\Omega(\log n + \log k)$. For the special case of $k = 1$ that was studied in [3], our bounds on the randomized competitive ratio are tight up to a constant factor. The results of [3] and those of this paper are summarized and compared in Table 1.

We note that any algorithm for the reorganization model can be implemented (in online fashion) in the no-reorganization model while incurring a cost at most three times larger, and any algorithm for the bypassing model can be implemented (in online fashion) in the no-bypassing model while incurring a cost

Table 1. Summary of the results in [3] and in this paper. The size of the companion cache is n

Previous Results [3] (only for main cache of size $k = 1$):

Bypass	Reorg'	det'/rand'	Upper Bound	Lower Bound
−	−	det	$2n + 2$	$n + 1$
✓	−	det	$2n + 3$	$2n + 2$
✓	✓	det	$2n + 3$	

New Results (main cache of arbitrary size k):

Bypass	Reorg'	det'/rand'	Upper Bound	Lower Bound
−	✓	det	$(n+1)(k+1) - 1$	$(n+1)(k+1) - 1$
−/✓	−/✓	det	$O(nk)$	$\Omega(nk)$
−/✓	−/✓	rand	$O(\log k \log n)$	$\Omega(\log k + \log n)$

at most two times larger. Thus, the competitive ratio (both randomized and deterministic) differs by at most constant factor between the different models.

The techniques we use generalize *phase partitioning* and *marking algorithms* [6, 4].

2 The Problem

In the (n, k)-companion caching problem, there is a slow *main memory* and a fast *cache*. The items in main memory are partitioned into m *types*, the set of types is T ($|T| = m$). The cache consists of a two separate components:

- The Main Cache: Consisting of a cache of size k for each type. I.e., every type t, $1 \leq t \leq m$, has its own cache of size k which can hold only items of type t.
- The Companion Cache: A cache of size n which can hold items of any type.

We refer to these components collectively simply as *the cache*. If an item is stored somewhere in the cache, we say it is *cached*. Our basic assumptions are that there are at least $k+1$ items of every type and that the number of types, m, is greater than the size of the companion cache, n.

A caching algorithm is faced with a sequence of requests for items. When an item is requested it must be cached (*i.e.*, bypassing is not allowed). If the item is not cached, a fault occurs. The goal is to minimize the number of faults. A caching algorithm can swap items of the same type between the main and companion caches without incurring any additional cost (*i.e.*, reorganization is allowed).

We use the competitive ratio to measure the performance of online algorithms. Formally, given an item request sequence σ, the cost of an online algorithm A on σ, denoted by $\text{cost}_A(\sigma)$, is the number of faults incurred by A. An algorithm is called r-competitive if there exists a constant c, such that for any request sequence σ, $E[\text{cost}_A(\sigma)] \leq r \cdot \text{cost}_{\text{Opt}}(\sigma) + c$.

To simplify the analysis later, we mention the following fact (attributed to folklore):

Proposition 1. *We may assume that Opt is lazy, i.e., Opt evicts an item only when a requested item is not cached.*

Straightforward lower bounds follow from the classical paging problem.

Theorem 1. *The deterministic competitive ratio for the (n,k)-companion caching problem is at least $(n+1)(k+1) - 1$. The randomized competitive ratio is at least $H_{(k+1)(n+1)-1} = \Omega(\log n + \log k)$.*

Proof. Consider the situation where there are $(n+1)(k+1)$ items of $n+1$ types, $k+1$ items of each type. In this case, a caching algorithm has $(n+1)k + n = (n+1)(k+1) - 1$ cache slots available. If we compare this situation to the regular paging problem where the virtual memory consists of $(n+1)(k+1)$ pages (items) and the cache is of size $(n+1)(k+1) - 1$, we find the two problems are exactly the same. A companion caching algorithm induces a paging algorithm, and the opposite is also true. Hence a lower bound on the competitive ratio for paging implies the same lower bound for companion caching. We conclude there are lower bounds of $(n+1)(k+1) - 1$ on the deterministic competitive ratio and $H_{(n+1)(k+1)-1} = \Omega(\log n + \log k)$ on the randomized competitive ratio for companion caching. □

3 Phase Partitioning of Request Sequences

In [6, 4] the request sequence for the paging problem is partitioned into *phases* as follows: A phase begins either at the beginning of the sequence or immediately after the end of the previous phase. A phase ends either at the end of the sequence or immediately before the request for the $(k+1)$st distinct page in the phase. Similarly, we partition the request sequence for the companion caching problem into phases. However, the more complex nature of our problem implies more complex partition rules.

Let $\sigma = \sigma_1, \sigma_2, \ldots, \sigma_{|\sigma|}$ denote the request sequence. The indices of the sequence are partitioned into a sequence of disjoint consecutive subsequences $D_1, D_2, \ldots, D_f$, whose concatenation gives $\{1, \ldots, |\sigma|\}$. The indices are also partitioned into a sequence of disjoint (ascending) subsequences $P_1, P_2, \ldots, P_f$.

In Fig. 2 we describe how to generate the sequences D_i and P_i. D_i is a consecutive sequence of indices of requests *issued during* phase i. P_i is a (possibly non-consecutive, ascending) sequence of indices of requests *associated* with phase i.

Given a set of indices A we denote by $\mathbb{I}(A) = \{\sigma_\ell | \ell \in A\}$ the set of items requested in A, and by $\mathbb{T}(A)$ the set of types of items in $\mathbb{I}(A)$. Table 2 shows an example of phase partitioning.

In [6] it is shown that any paging algorithm faults at least once in each complete phase. Here we show a similar claim for companion caching.

```
Pᵢ:        The indices of the requests associated with phase i.
Dᵢ:        The indices of the requests issued during phase i.
N(t):      The indices of requests of type t that have not yet
           been associated with a phase.
M(t)       = {σ_ℓ | ℓ ∈ N(t)}

For every type t ∈ T: M(t) ← ∅, N(t) ← ∅
P₁ ← ∅, D₁ ← ∅
i ← 1
For ℓ ← 1, 2, ...   Loop on the requests
    Let σ_ℓ be the current request and t₀ be its type.
    Let m_t ← { max{0, |M(t)| − k}            t ≠ t₀
              { max{0, |M(t₀) ∪ {σ_ℓ}| − k}   t = t₀
    If Σ_{t∈T} m_t > n then   End of Phase Processing:
        For every type t ∈ T such that m_t > 0 do
            Pᵢ ← Pᵢ ∪ N(t)
            M(t) ← ∅, N(t) ← ∅
        i ← i + 1
        Pᵢ ← ∅, Dᵢ ← ∅
    Dᵢ ← Dᵢ ∪ {ℓ}
    N(t₀) ← N(t₀) ∪ {ℓ}
    M(t₀) ← M(t₀) ∪ {σ_ℓ}
```

Fig. 2. Phase partition rules described as an algorithm

Proposition 2. *For any (online or offline) caching algorithm, it is possible to associate with each phase (except maybe the last one) a distinct fault.*

Proof. Consider the request indices in P_i together with the index j that ends the phase (i.e., $j = \min D_{i+1}$). One of the items in $\mathbb{I}(P_i)$ must be evicted after being requested and before σ_j is served. This is simply because the cache can not hold all these items simultaneously. We associate this eviction with the phase.

We must show that we have not associated the same eviction to two distinct phases. Let i_1 and i_2 be two distinct phases, $i_1 < i_2$. If the evictions associated with i_1 and i_2 are of different items then they are obviously distinct. Otherwise, the evictions associated with i_1 and i_2 are of the same type t, and $t \in \mathbb{T}(P_{i_1}) \cap \mathbb{T}(P_{i_2})$, which means that all indices $\ell \in P_{i_2}$, where σ_ℓ is of type t, must have $\ell > \max D_{i_1}$. Thus, an eviction associated with phase i_2 cannot be associated with phase i_1. □

To help clarify our argument in the proof of Proposition 2, consider the third phase in Table 2. Here $\mathbb{I}(P_3) = \{b_4, b_1, b_2, b_5, d_1, d_2\}$, and the phase ends because of the request to d_3. It is not possible that all these items reside in the cache simultaneously and thus at least one of the items in $\mathbb{I}(P_3)$ must be evicted before or on the request for item d_3. The item evicted can be either some b_i, $i = 1, 2, 4, 5$, or some d_i, $i = 1, 2$. If, for example, the item evicted is some b_i, then this eviction

Table 2. An example for an (n,k)-companion caching problem where $n = 3$ and $k = 2$. The types are denoted by the letters a, b, c, d. The ith item of type $\beta \in \{a, b, c, d\}$ is denoted by β_i

Req. seq.	$a_1b_1d_1c_1a_2a_3b_2a_4b_3c_2$	$b_4a_5c_3d_2b_1c_4a_3a_2$	$a_1a_3b_2b_4b_5$	$d_3\ldots$
Phase	$i = 1$	$i = 2$	$i = 3$	
D_i	$\{1,\ldots,10\}$	$\{11,\ldots,18\}$	$\{19,\ldots,23\}$	
P_i	$\{1,2,5,6,7,8,9\}$	$\{4,10,12,13,\\ 16,17,18\}$	$\{3,11,14,15,\\ 21,22,23\}$	
$\mathbb{T}(P_i)$	$\{a,b\}$	$\{a,c\}$	$\{b,d\}$	

must have occurred after $\max D_1$ — the end of the first phase — and therefore it cannot be an eviction associated with the first phase.

4 Deterministic Marking Algorithms

In a manner similar to [6], based on the phase partitioning of Section 3, we define a class of online algorithms called *marking* algorithms.

Definition 1. *During the request sequence an item $e \in \bigcup_t M(t)$ is called* marked *(see Figure 2 for a definition of $M(t)$). An online caching algorithm that never evicts marked items is called a* marking *algorithm.*

Remarks:

1. The phase partitioning and dynamic update of the set of marked items can be performed in an online fashion (as given in the algorithm of Fig. 2).
2. At any point in time, the cache can accommodate all marked items.
3. Unlike the marking algorithms of [6], it is not true that immediately after $\max D_i$ all marks of the ith phase are erased. Only the marked items of types $t \in \mathbb{T}(P_i)$ will have their markings erased immediately after $\max D_i$.

For a specific algorithm, at any point in time during the request sequence, a type t that has more than k items in the cache is called *represented in the companion cache*. Note that for marking algorithms, a type is in $\mathbb{T}(P_i)$ if and only if it is represented in the companion cache at $\max D_i$ or it is the type of the item that ended phase i.

Proposition 3. *The number of faults of any marking algorithm on requests whose indices are in P_i is at most $n(k+1) + k = (n+1)(k+1) - 1$.*

Proof. Each item e of type t requested in request index $\ell \in P_i$, is marked and is not evicted until after $\max D_i$. We note that $|\mathbb{T}(P_i)| \leq n + 1$ since at most n types are represented in the companion cache, and the type of the item whose request ends the phase may also be in $\mathbb{T}(P_i)$. Thus, $|\mathbb{I}(P_i)| \leq (n+1)k + n$. □

We conclude from Proposition 3 and Proposition 2:

Theorem 2. *Any marking algorithm is $(n+1)(k+1) - 1$ competitive.*

Since the marking property can be realized by deterministic algorithms, we conclude that the deterministic competitive ratio of the (n, k)-companion caching problem is $(n+1)(k+1) - 1$.

5 Randomized Marking Algorithms

In this section we present an $O(\log n \log k)$ competitive randomized marking algorithm. The building blocks of our randomized algorithms are the following three eviction strategies:

On a fault on an item of type t:

Type Eviction. Evict an item chosen uniformly at random among all unmarked items of type t in the cache.

Cache-Wide Eviction. Let T be the set of types represented in the companion cache, let U be the set of all unmarked items in the cache whose type is in $T \cup \{t\}$. Evict an item chosen uniformly at random from U.

Skewed Cache-Wide Eviction. Let T be the set of types represented in the companion cache, let $T' \subset T \cup \{t\}$ be the set of types with at least one unmarked item in the cache. Choose t' uniformly at random from T', let U be the set of all unmarked items of type t', and evict an item chosen uniformly at random from U.

Remarks:

- Type eviction may not be possible as there may be no unmarked items of type t in the cache.
- Cache-wide eviction and skewed cache-wide eviction are always possible, if there are no unmarked pages of types represented in the companion cache and no unmarked pages of type t in the cache then the fault would have ended the phase.

The algorithms we use are:

Algorithm TP_1. Given a request for item e of type t, not in the cache: Update all phase related status variables (as in the algorithm of Figure 2).

- If t is not represented in the companion cache and there are unmarked items of type t, use type-eviction.
- Otherwise — use cache-wide eviction.

Algorithm TP$_2$. Given a request for item e of type t, not in the cache: Update all phase related status variables (as in the algorithm of Figure 2). Let the current request index be $j \in D_i$, $i \geq 1$.

- If t is not represented in the companion cache and there are unmarked items of type t, use type-eviction.
- If t is represented in the companion cache, $e \in \mathbb{I}(P_{i-1})$, and there are unmarked items of type t, use type eviction.
- Otherwise — use skewed cache-wide eviction.

Algorithm TP. If $k < n$ use TP$_1$, otherwise, use TP$_2$.

Theorem 3. *Algorithm TP is $O(\log n \log k)$ competitive.*

Due to lack of space we only give an overview of the proof in the next section.

5.1 Proof Overview

We give an analogue to the definitions of new and stale pages used in the analysis of the randomized marking paging algorithm of [4].

Definition 2. *For phase i and type t, denote by i^{-t} the largest index $j < i$ such that $t \in \mathbb{T}(P_j)$. If no such j exists we denote $i^{-t} = 0$, and use the convention that $P_0 = \emptyset$. Similarly, i^{+t} is the smallest index $j > i$ such that $t \in \mathbb{T}(P_j)$. If no such index exists, we set $i^{+t} = "\infty"$, and use the convention that $P_\infty = \emptyset$.*

Definition 3. *An item e of type t is called new in P_i if $e \in \mathbb{I}(P_i) \setminus \mathbb{I}(P_{i^{-t}})$. We denote by $g_{t,i}$ the number of new items of type t in P_i. Note that if $t \notin \mathbb{T}(P_i)$ then $g_{t,i} = 0$.*

Let i_{end} denote the index of the last *completed* phase.

Definition 4. *For $t \in \mathbb{T}(P_i)$, let $L_{t,i} = \mathbb{I}(P_i) \cap \{$items of type $t\}$. Note that $|L_{t,i}| \geq k$. Define*

$$\ell_{t,i} = \begin{cases} |L_{t,i}| - k & i < i_{\text{end}} \wedge t \in \mathbb{T}(P_i) \setminus \mathbb{T}(P_{i+1}), \\ 0 & \text{otherwise.} \end{cases}$$

We use the above definitions to give an amortized lower bound on the cost to Opt of dealing with the sequence σ:

Lemma 1. *There exist $C > 0$ such that for any request sequence σ,*

$$\text{cost}_{\text{Opt}}(\sigma) \geq C \max\left\{ \sum_{i \leq i_{\text{end}}} \sum_{t \in P_i} g_{t,i}, \sum_{i \leq i_{\text{end}}} \sum_{t \in P_i} \ell_{t,i} \right\}. \tag{1}$$

Proof (sketch). Consider a phase P_j and the set $\mathbb{T}(P_j)$, for every $t \in \mathbb{T}(P_j)$ there exist phases P_{j-t} and P_{j+t}, except for some constant number of phases and types.

The number of new items $g_{j+t,t}$, $t \in \mathbb{T}(P_j)$, can be associated with a cost to opt, as follows: If an item of type t is not present in the adversary cache at the end of phase P_j, then the adversary must have faulted on this item when it was subsequently requested during phase P_{j+t}. If some ℓ of these elements were in the cache at the end of phase P_j (irrespective of their type), then we know that (a) these elements were not requested in P_j — as they are new in P_{j+t} for some t, and (b) under the assumption that opt is lazy — all of these ℓ items have been in opt's cache since (at least) the end of (some) P_{j-t} (where t depends on the specific item in question).

This means that the sequence of requests comprising P_j must have had ℓ items that were in the cache (at least immediately after being requested), yet must have been subsequently evicted before the end of phase P_j.

We also need to avoid overcounting the same evictions multiple times. We argue that we do not do so because these evictions are all accounted for before the end of phase P_j. Now, although in general the requests associated with two phases can be interleaved — requests to items of type $t \in \mathbb{T}(P_j)$ can only occur after the end of phase P_{j-t}.

This (almost) proves the 1st lower bound in (1), with the caveat that the proof above seemingly requires an additive constant. A more refined argument shows that this additive constant is not really required.

The proof of the 2nd lower bound in (1) requires similar arguments and is omitted. □

We will give upper bounds (algorithms) that belong to a restricted family of randomized algorithms, specifically *uniform type preference* algorithms defined below. The main advantage of using such algorithms is that their analysis is simplified as they have the property that while dealing with requests σ_j, $j \in D_i$, the companion cache contains only items of types in $\mathbb{T}(P_i) \cup \mathbb{T}(P_{i-1})$.

Definition 5. *A type preference algorithm is a marking algorithm such that when a fault occurs on an item of a type that is not represented in the companion cache, it evicts an item of the same type, if this is possible.*

Definition 6. *A uniform type preference algorithm is a randomized type preference algorithm maintaining the invariant that at any point in time between request indices $1 + \max D_{i-t}$ and $\max D_i$, inclusive, and any type $t \in \mathbb{T}(P_i)$, all unmarked items of type t in $\mathbb{I}(P_{i-t})$ are equally likely to be in the cache.*

Note that both TP_1 and TP_2 are uniform type preference algorithms.

We use a charge-based amortized analysis to compute the online cost of dealing with a request sequence σ. We charge the expected cost of all but a constant number of requests in σ to at least one of two "charge counts", charge(D_i) and/or charge(P_j) for some $1 \leq i \leq j \leq i_{\text{end}}$. The total cost associated with the online algorithm is bounded above by a constant times

$\sum_{1 \leq i \leq i_{\text{end}}} \text{charge}(D_i) + \sum_{1 \leq i \leq i_{\text{end}}} \text{charge}(P_i)$, excluding a constant number of requests.

Other than a constant number of requests, every request $\sigma_\ell \in \sigma$ has $\ell \in D_{i_1} \cup P_{i_2}$ for some $1 \leq i_1 \leq i_2 \leq i_{\text{end}}$.

We use the following strategy to charge the cost associated with this request to one (or more) of the charge(D_i), charge(P_j):

1. If $\ell \in P_i$ and type(σ_ℓ) $\in \mathbb{T}(P_i) \setminus \mathbb{T}(P_{i-1})$ then we charge the (expected) cost of σ_ℓ to charge(P_i). These charges can be amortized against the cost of **Opt** to deal with σ_ℓ. This amortization is summarized in Proposition 5 (for any uniform type preference algorithm).
2. If $\ell \in D_i$ and type(σ_ℓ) $\in \mathbb{T}(P_{i-1})$ then we charge the (expected) cost of σ_ℓ to charge(D_i). These charges will be amortized against the cost of **Opt** to within a poly-logarithmic factor.

To compute the expected cost of a request σ_ℓ, $\ell \in D_i$, type(σ_ℓ) $\in \mathbb{T}(P_{i-1})$, we introduce an analogue to the concept of "holes" used in [4]. In [4] holes were defined to be stale pages that were evicted from the cache.

Definition 7. *We define the number of holes during D_i, h_i, to be the maximum over the indices $j \in D_i$ of the total number of items of types in $\mathbb{T}(P_{i-1})$ that were requested in $P_{t,i-1}$ but are not cached when the jth request is issued.*

Proposition 4. *Consider a marking algorithm, a phase i, a type $t \in \mathbb{T}(P_i) \setminus \mathbb{T}(P_{i-1})$, and a request index $\max D_{i-t} < j \leq \max D_i$. Let H be the set of items of type t that were requested in P_{i-t} and evicted afterward without being requested again, up to request index j (inclusive). Then $|H| \leq \hat{g}_{t,i} + \ell_{t,i-t}$, where $\hat{g}_{t,i} \leq g_{t,i}$ is the number of new items of type t requested after $\max D_{i-t}$ and up to time j (inclusive).* □

Proposition 5. *For a uniform type preference algorithm, the expected number of faults on request indices in P_i for items of type $t \in \mathbb{T}(P_i) \setminus \mathbb{T}(P_{i-1})$ is at most $(1 + H_{n+k})(g_{t,i} + \ell_{t,i-t})$. I.e., charge($P_i$) $\leq (1 + H_{n+k})(g_{t,i} + \ell_{t,i-t})$.*

Proof. Fix a type $t \in \mathbb{T}(P_i) \setminus \mathbb{T}(P_{i-1})$. There are $g_{t,i}$ faults on new items of type t, the rest of the faults are on items in $L_{t,i-t}$ that were evicted before being requested again. By Proposition 4, the number of items in $L_{t,i-t}$ that are not in the cache at any point of time is at most $\hat{g}_{t,i} + \ell_{t,i-t} \leq g_{t,i} + \ell_{t,i-t}$. For any a, b in $L_{t,i-t}$ that have not been requested after $\max D_{i-t}$, the probability that a has been evicted since $1 + \max D_{i-t}$ is equal to the probability that b has been evicted since $1 + \max D_{i-t}$.

Let r denote the number of items in $L_{t,i-t}$ that have been requested since after $\max D_{i-t}$. There are $|L_{t,i-t}| - r$ unmarked items of $L_{t,i-t}$. The probability that an item of $L_{t,i-t}$ is cached is therefore $(g_{t,i} + \ell_{t,i-t})/(|L_{t,i-t}| - r)$. Thus, the expected number of faults on requests indices in P_i for items in $L_{t,i-t}$ is at most

$$\sum_{r=0}^{|L_{t,i-t}|-1} \frac{g_{t,i} + \ell_{t,i-t}}{|L_{t,i-t}| - r} \leq (g_{t,i} + \ell_{t,i-t}) H_{|L_{t,i-t}|} \leq (g_{t,i} + \ell_{t,i-t}) H_{n+k}. \qquad \Box$$

Proposition 6. *A type preference algorithm has the following properties:*

1. *During D_i, only types in $\mathbb{T}(P_{i-1}) \cup \mathbb{T}(P_i)$ may be represented in the companion cache.*
2. *During D_i, when a type $t \in \mathbb{T}(P_i) \setminus \mathbb{T}(P_{i-1})$ becomes represented in the companion cache, there are no unmarked cached items of type t, and t stays represented in the companion cache until $\max D_i$, inclusive.* □

Proposition 7. *For a type preference algorithm,*

$$h_i \leq \sum_{t \in \mathbb{T}(P_i) \setminus \mathbb{T}(P_{i-1})} (g_{t,i} + \ell_{t,i-t}) + \sum_{t \in \mathbb{T}(P_{i-1})} g_{t,(i-1)+t}. \tag{2}$$

Definition 8. *At any point during D_i, call a type $t \in \mathbb{T}(P_{i-1})$ that has unmarked items in the cache and is represented in the companion cache an* active type. *Call an unmarked item $e \in \mathbb{I}(P_{i-1})$ of an active type an* active item.

Proposition 8. *The following properties hold for type preference algorithms:*

1. *During D_i, the set of active types is monotone decreasing w.r.t. containment.*
2. *During D_i, the set of active items is monotone decreasing w.r.t. containment.*

Proposition 9. *For TP_1, charge(D_i) — The expected number of faults on request indices in D_i to types in $\mathbb{T}(P_{i-1})$ — is at most $h_i(1 + H_{k+1}(1 + H_{(n+1)(k+1)}))$.*

Proof. First, we count the expected number of faults on items in $\cup_{t \in \mathbb{T}(P_{i-1})} L_{t,i-1}$. By Proposition 8, the set of active items is monotone decreasing, where an item becomes inactive either by being marked, or because its type is no longer represented in the companion cache. Let $\langle m_j \rangle_{j=1,\ldots,w}$, be a sequence, indexed by the event index, of the number of active items. An event is either when an active item is requested, or when an active type t becomes inactive by being no longer represented in the companion cache (it is possible that one request generates two events, one from each case).

If the jth event is a request for active item, then $m_{j+1} = m_j - 1$. Otherwise, if the jth event is the event of type t becoming inactive, and before that event there were b active items of type t, then $m_{j+1} = m_j - b$.

In the first case, the expected cost of the request, conditioned on m_j, is at most h_i/m_j. In the second case, there are b items of type t that became inactive, each had probability of $\frac{h_i}{m_j}$ of not being in the cache at that moment. This means that the expected number of items among the up-until now active items of type t, that are not in the cache, at this point in time, is $\frac{h_i b}{m_j}$.

Let g_t denote the number of new items of $P_{(i-1)+t}$ (Def. 3) requested during D_i ($g_t \leq g_{t,(i-1)+t}$). After type t becomes inactive, the number of items among $L_{t,i-1}$ that are not in the cache can increase only when a new item of type t is requested. Therefore the expected number of items among $L_{t,i-1}$ that are not in the cache, after the jth event (the event when t became inactive), is at most $\frac{h_i b}{m_j} + g_t$.

Because of the uniform type eviction property of TP_1, the probability that an item in $L_{t,i-1}$ is not in the cache is the expected number of items among $L_{t,i-1}$, and not in the cache, divided by the number of unmarked items among $L_{t,i-1}$, and therefore the expected number of faults on items of $L_{t,i-1}$ after the jth event is at most

$$\sum_{a=1}^{b}(\frac{h_i b}{m_j}+g_t)\cdot \frac{1}{a} = (\frac{h_i b}{m_j}+g_t)H_b.$$

Note that $b \leq k+1$, and $\sum_{t \in P_{i-1}} g_t \leq h_i$, and so the expected number of faults on items $e \in \cup_{t \in \mathbb{T}(P_{i-1})} L_{t,i-1}$, conditioned on the sequence $\langle m_j \rangle_j$ is at most

$$h_i H_{k+1} + h_i \sum_j \frac{(m_j - m_{j-1})H_{k+1}}{m_j} \qquad (3)$$

The sequence $\langle m_j \rangle_j$ is itself a random variable, but we can give an upper bound on the expected number of faults on items $e \in \cup_{t \in \mathbb{T}(P_{i-1})} L_{t,i-1}$ by bounding the *maximum* of (3) over all feasible sequences $\langle m_j \rangle_j$. The worst case for (3) will be when $\langle m_j \rangle_j = \langle (n+1)(k+1) - j \rangle_{j=1}^{(n+1)(k+1)-1}$. Thus,

$$h_i H_{k+1}(1 + \sum_j \frac{(m_j - m_{j+1})}{m_j}) \leq h_i H_{k+1}(1 + H_{(n+1)(k+1)})$$

We are left to add faults on new items of types in $\mathbb{T}(P_{i-1})$. There are at most $\sum_{t \in \mathbb{T}(P_{i-1})} g_{t,(i-1)+t} \leq h_i$ such faults. □

Lemma 2. TP_1 *is* $O(\log k \max\{\log n, \log k\})$ *competitive.*

Proof. Each fault is counted by either charge(P_i) (Proposition 5) or charge(D_i) (Proposition 9) (faults on request indices in D_i for items of type in $\mathbb{T}(P_{i-1})\setminus \mathbb{T}(P_i)$ are counted twice), and by Lemma 1, we have that the expected number of faults of TP_1 is at most $O(H_{k+1} H_{(n+1)(k+1)})\, \text{cost}_{\text{Opt}}$. □

For algorithm TP_2, we have the following result:

Lemma 3. TP_2 *is* $O(\log n \max\{\log n, \log k\})$ *competitive.* □

The proof, which uses ideas similar to the proof of Lemma 2, has been omitted.

Proof (of Theorem 3). Follows immediately from Lemma 2 and Lemma 3. □

Unfortunately, the competitive ratio of type preference algorithms is always $\Omega(\log n \log k)$. The proof of this claim is omitted in this version for lack of space.

References

[1] D. Achlioptas, M. Chrobak, and J. Noga. Competitive analysis of randomized paging algorithms. *Theoretical Computer Science*, 234:203–218, 2000.

[2] L. A. Belady. A study of replacement algorithms for virtual storage computers. *IBM Systems Journal*, 5:78–101, 1966.

[3] M. Brehob, R. Enbody, E. Torng, and S. Wagner. On-line restricted caching. In *Proceedings of the 12th Symposium on Discrete Algorithms*, pages 374–383, 2001.

[4] A. Fiat, R. Karp, M. Luby, L. A. McGeoch, D. D. Sleator, and N. E. Young. Competitive paging algorithms. *Journal of Algorithms*, 12:685–699, 1991.

[5] N. Jouppi. Improving direct-mapped cache by the addition of small fully-associative cache and prefetch buffer. *Proc. of the 17th International Symposiuom on Computer Architecture*, 18(2):364–373, 1990.

[6] A. Karlin, M. Manasse, L. Rudolph, and D. D. Sleator. Competitive snoopy caching. *Algorithmica*, 3(1):79–119, 1988.

[7] L. McGeoch and D. D. Sleator. A strongly competitive randomized paging algorithm. *J. Algorithms*, 6:816–825, 1991.

[8] D. D. Sleator and R. E. Tarjan. Amortized efficiency of list update and paging rules. *Communication of the ACM*, 28:202–208, 1985.

[9] A. Seznec. A case for two-way skewed-associative caches. In *Proc. of the 20th International Symposuim on Computer Architecture* pages 169–178, 1993.

Deterministic Communication in Radio Networks with Large Labels*

Leszek Gąsieniec[1], Aris Pagourtzis[2,3,**], and Igor Potapov[1]

[1] Department of Computer Science, University of Liverpool
Liverpool L69 7ZF, UK
{leszek,igor}@csc.liv.ac.uk

[2] Institute of Theoretical Computer Science, ETH Zürich
pagour@inf.ethz.ch

[3] Department of Electrical and Computer Engineering
National Technical University of Athens

Abstract. We study **deterministic** gossiping in ad-hoc radio networks, where labels of the nodes are **large**, i.e., they are polynomially large in the size n of the network. A label-free model was introduced in the context of randomized broadcasting in ad-hoc radio networks, see [2]. Most of the work on deterministic communication in ad-hoc networks was done for the model with labels of size $O(n)$, with few exceptions; Peleg [19] raised the problem of deterministic communication in ad-hoc radio networks with large labels and proposed the first deterministic $O(n^2 \log n)$-time broadcasting algorithm. In [11] Chrobak et al. proved that deterministic radio broadcasting can be performed in time $O(n \log^2 n)$; their result holds for large labels.

Here we propose two new deterministic gossiping algorithms for ad-hoc radio networks with large labels. In particular:

– a communication procedure giving an $O(n^{5/3} \log^3 n)$-time deterministic gossiping algorithm for *directed* networks and an $O(n^{4/3} \log^3 n)$-time algorithm for *undirected* networks;
– a gossiping procedure designed particularly for undirected networks resulting in an almost linear $O(n \log^3 n)$-time algorithm.

1 Introduction

Mobile *radio networks* are expected to play an important role in future commercial and military applications [22]. Such networks are particularly suitable for situations where instant infrastructure is needed and no central system administration (such as base stations in a cellular system) is available. Typical applications for this type of peer-to-peer networks include: mobile computing in

---

* Research supported in part by GR/N09855 and GR/R85921 EPSRC grants.
** Part of this work was done while this author was with the Department of Computer Science, University of Liverpool.

remote areas; tactical communications; law enforcement operations; and disaster recovery. Radio networks, as well as any other distributed communication system, demand efficient implementation of communication primitives to carry out more complex communication tasks.

There are two important communication primitives encountered in the process of information dissemination in networks: *broadcasting* and *gossiping*. In the *broadcasting problem*, a message from a distinguished *source* node has to be sent to all other nodes. In the *gossiping problem* each node of the network possesses a unique message that is to be communicated to all other nodes in the network.

The study of communication in *ad-hoc*, i.e. *unknown*, radio networks has been mostly devoted to the broadcasting problem. One natural tool for dealing with uncertainty and conflicts in a distributed setting is randomization and, indeed, most earlier work on broadcasting in radio networks focussed on randomized algorithms. Bar-Yehuda et al. [2] gave a randomized algorithm that achieves broadcast in expected time $O(D \log n + \log^2 n)$, where D denotes the network diameter. This is very close to the lower bound of $\Omega(D \log(n/D))$, by Kushilevitz and Mansour [18], and it matches this lower bound when $D = O(n^{1-\epsilon})$, for any $\epsilon > 0$. Further, if D is constant, it also matches the lower bound of $\Omega(\log^2 n)$ for constant diameter networks, obtained by Alon et al. [1]. In the deterministic case, Bar-Yehuda et al. [2] gave a lower bound $\Omega(n)$ for constant diameter networks. For general networks, the best currently known lower bound of $\Omega(n \log n)$ was obtained by Brusci and Del Pinto in [4]. In [6], Chlebus et al. present a broadcasting algorithm with time complexity $O(n^{11/6})$ – the first sub-quadratic upper bound. This upper bound was later improved to $O(n^{5/3} \log^3 n)$ by De Marco and Pelc [14] and then by Chlebus et al. [8] who used finite geometries to develop an algorithm with time complexity $O(n^{3/2})$. Only recently [11], Chrobak, Gąsieniec and Rytter developed a deterministic algorithm for broadcasting in *ad-hoc* radio networks working in time $O(n \log^2 n)$. However their result is non-constructive. Shortly afterwards, Indyk [17] gave an alternative constructive solution with a similar $O(n \log^{O(1)} n)$ complexity. An alternative approach to radio broadcasting has been followed by Clementi, Monti and Silvestri [12]. They presented deterministic broadcasting algorithm for *ad-hoc* radio networks working in time $O(D\Delta \log^2 n)$, where Δ stands for the maximum in-degree. The same authors have studied deterministic broadcasting in known radio networks in the presence of both static as well as dynamic faults [13]. Another interesting setting, where broadcasting and gossiping coincide, was studied by Chlebus et al. in [7]. They presented several upper and lower bounds on the time complexity of the so called *oblivious* (equivalently, *non-adaptive*) gossiping algorithms in unknown *ad-hoc* radio networks.

Until recently, there has not been much known about gossiping in *ad-hoc* radio networks. A discussion was initiated by Chrobak, Gąsieniec and Rytter in [11], where they proved the existence of a sub-quadratic $O(n^{3/2} \log^2 n)$-time deterministic gossiping algorithm in such networks. Its constructive counterpart was proposed later by Indyk in [17]. In [10], Chrobak, Gąsieniec and Rytter introduced a randomized distributed algorithm that performs radio gossiping in the expected time $O(n \log^4 n)$. Clementi, Monti and Silvestri [12] presented an alternative deterministic gossiping algorithm working in time $O(D\Delta^2 \log^3 n)$. Very recently Gąsieniec and Lingas introduced in [15] a new gossiping paradigm leading to two deterministic algorithms with time complexities $O(n\sqrt{D} \log^2 n)$ and $O(D\Delta^{3/2} \log^3 n)$. Deterministic radio gossiping has been recently studied for a model with messages of bounded size; in [9], Christersson et al. proposed a number of gossiping algorithms for this model. Results on deterministic gossiping in *known* radio networks can be found in [16].

All the above mentioned deterministic algorithms have been introduced for the model with small labels. However, it is possible to adapt the algorithms in [11, 17, 12] so as to work for large labels as well without loss of efficiency. On the other hand, there seem to be no easy way to do so for the gossiping algorithms in [11, 17]; the algorithm in [12] works for large labels but it is efficient only for graphs where the maximum in-degree is small.

An alternative radio model was proposed by Ravishankar and Singh. They presented a distributed gossiping algorithm for networks with nodes placed uniformly randomly on a line [20] and a ring [21]. In [3] they study probabilistic protocols for k-point-to-point communication and k-broadcast in radio networks with known local neighborhoods and bounded-size messages.

Our results. The purpose of this paper is to accelerate a discussion, initiated by Peleg in [19], on communication in *ad-hoc* radio networks with large labels, i.e. labels polynomially bounded in n. It is already known that deterministic radio broadcasting can be performed in time $O(n \log^2 n)$ independently of the size of the labels, see [11]. However, very little is known about efficient (other than naive multiple use of a broadcasting procedure) deterministic gossiping in *ad-hoc* radio networks: the algorithm in [12] works for large labels also; nevertheless it is efficient only for graphs where the maximum in-degree is small. The main problem is that the efficient known deterministic gossiping algorithms whose time complexity depends only on the size of the network rely heavily on the ROUND-ROBIN procedure (all nodes transmit one-by-one) which for large labels is quite costly. Here we propose two new gossiping techniques that avoid the use of ROUND-ROBIN and still result in efficient deterministic algorithms for gossiping in *ad-hoc* radio networks with large labels. In particular:

- a communication procedure which provides an $O(n^{5/3} \log^3 n)$–time deterministic gossiping algorithm in *directed* networks and an $O(n^{4/3} \log^3 n)$–time algorithm in *undirected* networks;
- a gossiping procedure designed particularly for undirected networks which gives an almost linear $O(n \log^3 n)$–time algorithm.

2 Preliminaries

The radio network topology of connections is often modeled as a graph $G = (V, E)$, where V is the set of nodes, $n = |V|$ stands for the size of the network and E is the set of all connections (links) available in G. The graph can be *directed*, i.e. the information can be passed along an edge only in one direction that is consistent with its orientation. In case of an *undirected* (equivalently *symmetric*) graph of connections the information can flow in both directions along every edge, however in one direction at a time. Node w is a *neighbor* of node v if there is a link from w to v. It is assumed (for gossiping to be feasible) that the graph of connections is (strongly) connected. In each step a node may choose to be in one of two modes: either the *receive* mode or the *transmit* mode. A node in the transmit mode sends a message along all its out-going links. A message is delivered to all the recipients during the step in which it is sent. A node in the receive mode attempts to hear the messages delivered along all its in-coming links. A message that is delivered to a node is not necessarily heard by that node. The basic feature of the radio network is that if more than one link brings in messages to a node v during a single step, then v does not receive any of the messages, and all of them are lost as far as v is concerned. Node v can receive a message delivered along a link if this link is the only one bringing in a message. If more than one message arrive at a node at one step then a *collision* at the node is said to occur. We assume that in our model *collision detection* is not available. In particular, nodes cannot distinguish between background noise and noise caused by an occurrence of a collision.

In what follows we assume that labels of all nodes in the network have values not greater than N, where N is polynomially bounded in n, i.e., $N < n^{const}$, where *const* is a known constant. Note that the value of $\log N$ is linear in $\log n$, hence $O(\log N) = O(\log n)$. We will explore this equivalence throughout the rest of the paper. For the purpose of this presentation we assume that the value n is known, however this assumption can be dropped with the help of a standard doubling technique, see [6].

Let us now recall some definitions used in deterministic communication in ad-hoc radio networks. We say that a set S *hits* a set X iff $|S \cap X| = 1$ and that S *avoids* X iff $|S \cap X| = 0$. We say that S hits X on element x if S hits X and $S \cap X = \{x\}$.

Selective family. We say that a family R of subsets of $\{0, 1, \ldots, N\}$ is k-*selective*, for a positive integer k, if for any subset $Z \subseteq \{0, 1, \ldots, N\}$ such that $|Z| \leq k$ there is a set $S \in R$ such that S hits Z. It is known that for each k, N there exists a k-selective family of size $O(k \log N)$, see [12].

Strongly selective family. We say that a family R of subsets of $\{0, 1, \ldots, N\}$ is *strongly k-selective*, for a positive integer k, if for any subset $Z \subseteq \{0, 1, \ldots, N\}$ such that $|Z| \leq k$, and any $z \in Z$, there is a set $S \in R$ such that $S \cap Z = \{z\}$. It is known that for each k, N there exists a strongly k-selective family of size $O(k^2 \log N)$, see [12].

Selector and compound selector. Given a positive integer w, a family R_w of subsets of $\{0, 1, \ldots, N\}$ is called a w-*selector* if it satisfies the following property:

(∗) For any two disjoint sets $X, Y \subseteq \{0, 1, \ldots, N\}$ with $w/2 \leq |X| \leq w$ and $|Y| \leq w$ there exists a set in R_w which hits X and avoids Y.

It was proved in [11] that there exists a w-selector of size $O(w \log N) = O(w \log n)$.

Fact 1 *Let R_w be a w-selector and $Z \subseteq \{0, 1, \ldots, N\}$ such that $|Z| \leq w \leq \frac{3|Z|}{2}$. Let also $Y \subseteq Z$ be such that for all $y \in Y$ there exists a set in R_w which hits Z on y. Then Y contains at least $\frac{1}{4}$ of the elements of Z.*

Proof. We prove this by contradiction. Let $|Y| < \frac{1}{4}|Z| < w$ and $X = Z \setminus Y$ (i.e. no set in R_w hits Z on an element of X); hence, $|X| > \frac{3}{4}|Z| \geq \frac{1}{2}w$. By the definition of w-selector there exists a set S in R_w that hits X, say on element x, and avoids Y. Then Z is also hit by S on x – contradiction.

We now define a *compound w-selector* as follows. For each $j = 0, \ldots, \lceil \log_{3/2} w \rceil$ let $\hat{S}_j = (S_{j,0}, S_{j,1}, \ldots, S_{j,m_j-1})$ be a $\lceil (\frac{3}{2})^j \rceil$-selector with $m_j = O((\frac{3}{2})^j \log n)$ sets. The *compound w-selector* consists of *blocks*, each block having $\lceil \log_{3/2} w \rceil + 1$ *positions*, such that the jth position of block b contains set $S_{j, b \bmod m_j}$. Note that the size of a compound w-selector is comparable with the size of a regular w-selector.

Fact 2 *For any $k = \lceil (\frac{3}{2})^j \rceil \leq w$, a sequence of $O(k \log^2 n)$ consecutive sets in a compound w-selector contains a full occurrence of a compound k-selector.*

The fact follows from the observation that $O(k \log n)$ consecutive blocks contain a full occurrence of an $\lceil (\frac{3}{2})^j \rceil$-selector for all $0 \leq j \leq \log_{3/2} k$; note also that w.l.o.g. we may assume $O(\log w) = O(\log N) = O(\log n)$, since it is useless to have a w-selector over $\{0, \ldots, N\}$ with $w > 2N$.

Throughout the paper, we use selectors and (strongly or not) selective families as deterministic communication algorithms in the following sense: a family $R = \{S_1, \ldots, S_m\}$ of sets of node labels can be seen as such an algorithm consisting of m steps; at each step i, nodes with labels in S_i choose to be in transmit mode while the rest set themselves to receive mode.

2.1 Neighbor Gossip in Stars

One of the most difficult tasks in radio gossiping with large labels is to learn as much as possible about the local neighborhood. The task of learning about the complete neighborhood is known as *neighbor gossip*. Indeed, if all nodes are aware of their neighborhood, gossiping can be performed in almost linear time by selection of a leader in time $O(n \log^3 n)$, see [8], and efficient $O(n)$-time implementation of a DFS traversal presented in [6]. Since familiarizing with the neighborhood appears to be as hard as gossiping itself, the main effort in our algorithms is devoted to obtain a complete or sometimes partial information about the neighbors.

As an example we initially consider the neighbor gossip problem in the case where the graph of connections $G = (V, E)$ is a star, i.e., $V = \{v_0, v_1, \ldots, v_n\}$ and $E = \{(v_0, v_i) : i = 1, \ldots, n\}$ where node v_0 is the center of the star. We assume that the labels of all nodes are large (bounded by N). The main task is to inform the central node v_0 about the labels of all nodes at the ends of the arms of the star, since node v_0 can inform the others in a single step.

By using a w-selector such that $n \leq w \leq \frac{3}{2}n$, according to Fact 1, at least a fraction $\frac{1}{4}$ of labels of arm nodes will be delivered to the central node in time $O(w \log N) = O(n \log n)$.

Note that the process of labels collection in node v_0 can be iterated. I.e., after each delivery node v_0 sends a message asking nodes whose labels have been delivered to remain silent, and during the next round only a fraction of arm nodes is taking part in delivery, etc. The cost of each round is $O(x \log n)$, where x is the current number of arm nodes taking part in the delivery process. Hence the total cost of the whole delivery process is bounded by $O(n \log^2 n)$. Note also that this sequence of decreasing (in size) selectors, can be replaced by a multiple (but a constant number of times) application of a compound n-selector.

2.2 Partial Neighbor Gossip in General Graphs

In general graphs we can use compound selectors to estimate the degree of each node in the network. It follows immediately from Fact 1 and Fact 2 that:

Fact 3 *By applying a compound w-selector (s.t. $w \geq \frac{3}{2}d$) on a graph, all nodes of degree d will learn about at least a fraction $\frac{1}{4}$ of their neighbors in time $O(d \log^2 n)$.*

The above implies that after application of a compound w-selector (s.t. $w \geq \frac{3}{2}n$) on a graph of size n, if a node v has received a number x of labels (of neighbors) then for the degree d of v it holds $x \leq d \leq 4x$.

3 Deterministic Gossiping in Directed Networks

We say that a node v owns message m_v if m_v originates in v. The algorithm consists of 3 stages. During stage 1 we reduce a number of active nodes in the network, s.t., each node has at most k active neighbors. During stage 2 we perform (partial) gossiping on the reduced network. And in stage 3 every node repeats exactly the same pattern of transmissions as used in the stage 1.

Stage 1 The following procedure REDUCE(k) decreases the number of messages that are to be distributed during the gossiping process. The procedure consists of several iterations of a loop. In the beginning of each iteration a number of messages is already distributed to all nodes of the network - these messages are called *secure*, and the remaining messages are called *insecure*. We also assume that in the beginning of each iteration insecure messages are available only at the nodes who own them. We say that a given node v has a k-reach neighborhood if its in-neighbors contain at least k insecure messages. The procedure REDUCE has the following structure.

REDUCE(k);
loop

1. Run compound n-selector, applied only to nodes who own insecure messages (i.e. only owners of insecure messages take part in transmissions), in order to collect at least $\frac{k}{4}$ messages at all nodes with k-reach neighborhood (and possibly at some other nodes with $\frac{k}{4}$-reach neighborhood).
2. Choose a leader λ among nodes who received $\geq \frac{k}{4}$ messages. If such a node does not exist then **exit** the loop.
3. Let $K(\lambda)$ be the set of insecure messages collected by λ in this step. Distribute set $K(\lambda)$ to all nodes of the network securing all messages in $K(\lambda)$.

end loop

Lemma 1. *Procedure* REDUCE*(k) runs in time* $O(\frac{n^2 \log^3 n}{k})$.

Proof. Each step of the loop runs in time $O(n \log^3 n)$. Step 1 corresponds to a single execution of compound n-selector of size $O(n \log^2 n)$. Step 2 is based on a binary search and broadcasting procedure, and it can be implemented in time $O(n \log^3 n)$, for details check [8]. Step 3 corresponds to a single call of a broadcasting procedure running in time $O(n \log^2 n)$. Each chosen leader secures at least $\Omega(k)$ messages thus there are at most $O(\frac{n}{k})$ executions of the body of the loop. Altogether the complexity is bounded by $O(\frac{n^2 \log^3 n}{k})$.

Lemma 2. *After an execution of procedure* REDUCE*(k) each node in the network has in its in-neighborhood at most k insecure messages.*

Proof. Follows from the definition of the procedure. I.e., if there were nodes with k-reach neighborhood, one of them would be chosen as a leader and appropriate messages would become secure.

We say that after the execution of procedure REDUCE each node who owns insecure message remains *active* and all other nodes become *dormant*. For each node v in the network its *active in-neighborhood* $AN(v)$ is defined as the set of active nodes in the in-neighborhood of v. Note that, by Lemma 2, after execution of procedure REDUCE(k) it holds $|AN(v)| < k$, for all nodes v.

Stage 2

Fact 4 *If there are dormant nodes in the graph then there is a path from any active node w to some dormant node that contains only active (internal) nodes and has length at most n.*

Proof. Let v be some dormant node. In the original graph there is a path of length at most n from w to v. If there are dormant nodes in this path, let the first of them be v'. The path from w to v' proves the fact.

If all nodes are active then the graph is of degree $\leq k$; recall also that it is strongly connected.

At this stage we execute a strongly k-selective family n times with only active nodes involved in transmissions, but this time performing regular gossiping, i.e., each node (when it is its time to transmit) transmits its whole current knowledge. Since the distance from any node with insecure message to some dormant node is at most n and the size of k-strongly selective family is bounded by $O(k^2 \log N) = O(k^2 \log n)$ each insecure message will move to some dormant node in time $O(nk^2 \log n)$. In case that all nodes are active (at the beginning of this stage) a complete gossiping is performed.

Stage 3 At this stage every node repeats a pattern of transmissions used in procedure REDUCE, where (eventually) all dormant nodes managed to secure their messages. This is enough to ensure that all messages are eventually distributed to all nodes in the network. The cost of this stage equals to the running cost of procedure REDUCE(k), and it is bounded by $O(\frac{n^2 \log^3 n}{k})$.

Theorem 1. *Deterministic gossiping in directed ad-hoc radio networks with large labels can be completed in time $O(n^{5/3} \log^3 n)$.*

Proof. Take $k = n^{1/3}$. The initial execution of the procedure REDUCE($n^{1/3}$) as well as its final repetition run in time $O(\frac{n^2 \log^3 n}{n^{1/3}}) = O(n^{5/3} \log^3 n)$. The cost of the second stage, when we move remaining insecure messages to dormant nodes works in time $O(nk^2 \log n) = O(n(n^{1/3})^2 \log n) = O(n^{5/3} \log n)$. Thus the total complexity of the gossiping algorithm is $O(n^{5/3} \log^3 n)$.

Let D be the *eccentricity* of graph G. The following holds.

Corollary 1. *Deterministic gossiping in directed ad-hoc radio networks with eccentricity D can be completed in time $O(n^{4/3} D^{1/3} \log^3 n)$.*

Proof. The 1st and the 3rd part of the algorithm take time $O(\frac{n^2 \log^3 n}{k})$, as shown before, while for the second stage $O(Dk^2 \log n)$ time suffices. In order to balance those two complexities we choose $k = \frac{n^{2/3}}{D^{1/3}}$ ending up with total complexity $O(n^{4/3} D^{1/3} \log^3 n)$.

4 Deterministic Gossiping in Undirected Networks

We show here how to implement the algorithm from the last section more efficiently in undirected networks. The following holds.

Theorem 2. *Deterministic gossiping in undirected ad-hoc radio networks with large labels can be completed in time $O(n^{4/3} \log^3 n)$.*

Proof. The difference between the settings in directed and undirected networks is in the implementation of the middle stage, i.e., moving insecure messages to the dormant nodes.

First, note that after completion of the procedure REDUCE(k) the whole graph is split into several connected components containing nodes with insecure messages. Note that each connected component is separated from the other by a boundary consisting of dormant nodes only; thus, computations in each of the connected components can be performed simultaneously and independently (there are no collisions between two different connected components). The second stage is performed as follows.

1. Initially each node is informed about its complete neighborhood. This is done by single application of a k-strongly selective family in time $O(k^2 \log n)$, and then:
2. We choose a leader λ_k in each connected component C_k, with the help of binary search and broadcasting, done in time $O(n \log^3 n)$, for details see [8].
3. We build a DFS spanning tree T_k in C_k rooted in λ_k. While building tree T_k we also collect all insecure messages residing in C_k and we broadcast them later from the root to all other nodes in C_k – thus also to all nodes that are neighbors of dormant nodes in the border. This can be done in time $O(n \log^2 n)$ [6].
4. We run the strongly k-selective family once again in order to transmit all insecure messages to appropriate dormant nodes. A single run of the k-selective family suffices since the active in-neighborhood of dormant nodes is bounded by k. The execution of this step takes $O(k^2 \log n)$ time.

The total cost of this procedure is bounded by $O(n \log^3 n + k^2 \log n)$. And since the cost of the other steps is bounded by $O(\frac{n^2 \log^3 n}{k})$, plugging in $k = n^{2/3}$ we end up with the total complexity $O(n^{4/3} \log^3 n)$.

4.1 Almost Optimal $O(n \log^3 n)$–Time Deterministic Gossiping

In the remaining part of this section we show how to design an almost linear $O(n \log^3 n)$–time deterministic gossiping algorithm for undirected radio networks.

We will use the following definitions. We split the set of nodes into a logarithmic number of groups $V_1, V_2, \ldots, V_k$, where in group V_i there are all nodes with in-degree $> c^i \sqrt{n}$ and perhaps some nodes of degree $> \frac{1}{4} c^i \sqrt{n}$ (due to approximate computation of degrees of nodes) and all their neighbors, where constant $c = \frac{3}{2}$ and $k = O(\log n)$. For any subset $V' \subseteq V$, the border of V' is the subset of V' formed by all nodes that are adjacent to nodes in $V - V'$. Note that $V_k \subseteq V_{k-1} \subseteq \ldots \subseteq V_1$ and that the border of V_i (if non-empty, i.e. if $V_i \neq V$) contains nodes of degree $\leq c^i \sqrt{n}$.

A *map* is a graph with a distinguished node v_c called the center. Nodes in a map are divided into layers, such that nodes in the same layer are at the same distance from the center. It is assumed that all nodes in the same map are aware of the whole topology of the map. Moreover the map is equipped with two operations: COLLATE that allows to collect information available at the border of the map in time $O(r + b)$, where r is the radius of the map and b is the size of the border; and DISTRIBUTE that helps to distribute a message from the source to all nodes in the border of the map.

An implementation of operation COLLATE is done as follows. Information from the border of the map traverses toward the center layer by layer. We use an amortization argument involving two buckets: b, the size of the border, and r, the distance from the border to the center. Assume inductively that border information is available at j nodes $w_1, w_2, \ldots, w_j$ at layer i. If there is any node w_x at layer $i-1$ adjacent to $l > 1$ nodes at layer i, all l nodes but one transmit their messages to w_x sequentially and we amortize this cost using bucket b; we repeat that process until all remaining messages at layer i can be transmitted to layer $i-1$ simultaneously and the cost of this simultaneous transmission is amortized with the help of bucket r.

An implementation of operation DISTRIBUTE is based on reversing all transmissions in operation COLLATE. Hence its complexity is also bounded by $O(r+b)$.

Lemma 3. *Each connected component in set V_i has diameter of size $O(\frac{\sqrt{n}}{c^i})$.*

Proof. Consider nodes belonging to a diameter of some connected component in V_i. It is a known fact that on the shortest path between any two points in an undirected graph of size n there are at most $\frac{4n}{d}$ nodes of degree $\geq d$, otherwise we would have a shortcut. Hence the diameter of the component has at most $\frac{16n}{c^i\sqrt{n}}$ nodes of degree $\geq \frac{1}{4}c^i\sqrt{n}$. Let $X = \{x_1, x_2, \ldots, x_l\}$ be the set of all nodes in the diameter of degree $< \frac{1}{4}c^i\sqrt{n}$. Each node in X must have a neighbor in the component of degree $\geq \frac{1}{4}c^i\sqrt{n}$. If l was greater than $\frac{24\sqrt{n}}{c^i}$ then there would be six nodes in X that would share the same distance-two neighbor, thus forming a shortcut. Hence, the number of nodes on the diameter is bounded by $\frac{16n}{c^i\sqrt{n}} + \frac{24\sqrt{n}}{c^i} = O(\frac{\sqrt{n}}{c^i})$.

The gossiping algorithm works as follows.

1. Initially each node computes its approximate degree and membership in appropriate groups V_i. The approximate degree is computed with the help of a compound n-selector in time $O(n \log^2 n)$, see Fact 3.
2. For $i = k$ downto 1 do
   A Find a leader λ_i^j in each connected component C_i^j of V_i.
   B Build a map of all nodes in C_i^j centered in λ_i^j using $O(\frac{n}{c^i\sqrt{n}})$ steps with the help of maps built in previous stages (of larger index i) as follows. At the beginning of step s we assume that the map $M_i^j(s-1)$ centered in λ_i^j covers all nodes in C_i^j at distance $\leq s-1$ from λ_i^j. We also assume that if a node v belongs to $M_i^j(s-1)$ and v belongs to some $C_{i+1}^{j'}$ (from previous iteration) then all nodes belonging to $C_{i+1}^{j'}$ are also covered by $M_i^j(s-1)$. Hence the border of $M_i^j(s-1)$ is formed by nodes of degree $\leq c^{i+1}\sqrt{n}$.
   In step s we extend the map $M_i^j(s-1)$ by the nodes $v_1, \ldots, v_r$ that belong to V_i and are adjacent to the border of $M_i^j(s-1)$ and by the nodes of all connected components C_{i+1}^* containing nodes $v_1, \ldots, v_r$. This is done as follows.
   (i) Nodes $v_1, \ldots, v_r$ learn about their neighbors in the border of $M_i^j(s-1)$ as well as about the relevant interconnection edges; this is done in

time $O(b_{s-1})$, by letting nodes it the border of $M_i^j(s-1)$ transmit one by one.

(ii) All nodes $v_1, \ldots, v_r$ make an attempt to transmit their whole knowledge (including information about C_{i+1}^*'s containing them) during a single application of a compound $c^{i+1}\sqrt{n}$-selector. Each node at the border of $M_i^j(s-1)$ receives messages from at least a fraction $(\frac{1}{4})$ of its neighbors. This is done in time $O(c^{i+1}\sqrt{n}\log^2 n)$.

(iii) Information gathered at the border is transmitted efficiently to the leader (operation COLLATE), processed and then re-transmitted (operation DISTRIBUTE) to the border. This is done in order to allow each vertex at the border to be aware of whole currently recognized neighborhood. The time complexity of those two operations is bounded by $O(\frac{\sqrt{n}}{c^i} + b_{s-1})$, i.e., the sum of the diameter of V_i and the size of the border b_{s-1} of $M_i^j(s-1)$.

(iv) All nodes on the border of $M_i^j(s-1)$ run $c^{i+1}\sqrt{n}$-selective family to inform their neighbors (in V_i) that all of them, excluding already recognized neighborhood, should make another attempt to transmit their whole knowledge, and when this is finished the algorithm goes to (ii). This is done in time $O(c^{i+1}\sqrt{n}\log n)$.

The whole process is repeated at most logarithmic number of times since the number of unknown edges (connections) between the border of $M_i^j(s-1)$ and its unknown neighborhood drops by a fraction after each iteration (property of a compound selector). The cost of running (i) through (iv) is bounded by $O(c^{i+1}\sqrt{n}\log^2 n + b_{s-1}\log n)$.

(v) When the knowledge of the whole neighborhood of $M_i^j(s-1)$ is finally available at the leader, it generates a new map $M_i^j(s)$ and distributes the map to all nodes in it in time $O(\frac{\sqrt{n}\log^2 n}{c^i})$ by using the algorithm from [5]. Thus the total cost of running (i) through (v) is bounded by $O(c^{i+1}\sqrt{n}\log^2 n + b_{s-1}\log n)$.

Since the diameter of V_j is bounded by $O(\frac{\sqrt{n}}{c^i})$ and we go along the diameter with distance at least one during each phase, (i) through (v), the whole process will be completed in time $O(\frac{\sqrt{n}}{c^i} \times c^{i+1}\sqrt{n}\log^2 n + (b_1 + b_2 + \ldots + b_r)\log n) = O(n\log^2 n + n\log n) = O(n\log^2 n)$.

Since the cost of each iteration (for-loop) is bounded by $O(n\log^2 n)$ the total cost of all iterations is bounded by $O(n\log^3 n)$.

3. At this stage all nodes of degree $\geq \sqrt{n}$ are aware of their whole neighborhood. All other nodes having degree $< \sqrt{n}$ can learn about their neighborhood in time $O(n\log n)$ by application of a strongly $\sqrt{n}$-selective family. When the neighborhoods are known at every node, gossiping is accomplished by a selection of a global leader and application of a DFS method in time $O(n\log^3 n)$ [8, 6]. Thus the total cost of the algorithm is $O(n\log^3 n)$.

Theorem 3. *Deterministic gossiping in undirected radio networks with large labels can be completed in $O(n\log^3 n)$ time.*

References

[1] N. Alon, A. Bar-Noy, N. Linial and D. Peleg, A lower bound for radio broadcast, *Journal of Computer and System Sciences* 43, (1991), pp. 290–298.

[2] R. Bar-Yehuda, O. Goldreich, and A. Itai, On the time complexity of broadcast in multi-hop radio networks: An exponential gap between determinism and randomization, *Journal of Computer and System Sciences*, 45 (1992), pp. 104–126.

[3] R. Bar-Yehuda, A. Israeli, and A. Itai, Multiple communication in multi-hop radio networks, *SIAM Journal on Computing* 22 (1993), pp. 875–887.

[4] D. Brusci, and M. Del Pinto, Lower bounds for the broadcast problem in mobile radio networks, *Distributed Computing*, 10 (1997), pp. 129–135.

[5] I. Chlamtac and O. Weinstein, The Wave Expansion Approach to Broadcasting in Multihop Radio Networks, *IEEE Trans. on Communications*, 39(3), pp. 426–433, 1991.

[6] B. S. Chlebus, L. Gąsieniec, A. M. Gibbons, A. Pelc, and W. Rytter, Deterministic broadcasting in unknown radio networks, In Proc. 11*th ACM-SIAM Symp. on Discrete Algorithms*, (SODA'2000), pp. 861–870.

[7] B. S. Chlebus, L. Gasieniec, A. Lingas, and A. Pagourtzis, Oblivious gossiping in ad-hoc radio networks, In Proc *5th Int. Workshop on Discrete Algorithms and Methods for Mobile Computing and Communications*, (DIALM'2001), pp. 44-51.

[8] B. S. Chlebus, L. Gąsieniec, A. Östlin, and J. M. Robson, Deterministic radio broadcasting, In Proc 27*th Int. Colloq. on Automata, Languages and Programming*, (ICALP'2000), LNCS 1853, pp. 717–728.

[9] M. Christersson, L. Gąsieniec, and A. Lingas, Gossiping with bounded size messages in ad-hoc radio networks, to appear in Proc 29*th Int. Colloq. on Automata, Languages and Programming*, (ICALP'2002), Malaga, July 2002.

[10] M. Chrobak, L. Gąsieniec, and W. Rytter, A randomized algorithm for gossiping in radio networks, In Proc. 7*th Annual Int. Computing and Combinatorics Conference*, (COCOON'2001), pp. 483-492.

[11] M. Chrobak, L. Gąsieniec, and W. Rytter, Fast broadcasting and gossiping in radio networks, In Proc. 41*st IEEE Symp. on Found. of Computer Science*, (FOCS'2000), pp. 575–581. A full version to appear in *Journal of Algorithms*.

[12] A. E. F. Clementi, A. Monti, and R. Silvestri, Selective families, superimposed codes, and broadcasting in unknown radio networks, In Proc. 12*th ACM-SIAM Symp. on Discrete Algorithms*, Washington, DC, 2001, pp. 709–718.

[13] A. E. F. Clementi, A. Monti, and R. Silvestri, Round robin is optimal for fault-tolerant broadcasting on wireless networks, In Proc. 9*th Ann. European Symposium on Algorithms*, (ESA'2001), pp 452–463.

[14] G. De Marco, and A. Pelc, Faster broadcasting in unknown radio networks, *Information Processing Letters*, 79(2), pp 53-56, (2001).

[15] L. Gąsieniec & A. Lingas, On adaptive deterministic gossiping in ad-hoc radio networks, In Proc. 13*th Annual ACM-SIAM Symposium on Discrete Algorithms* (SODA'2002), January 2002, pp 689-690.

[16] L. Gąsieniec and I. Potapov, Gossiping with unit messages in known radio networks, to appear in Proc. *2nd IFIP International Conference on Theoretical Computer Science*, Montreal, August 2002.

[17] P. Indyk, Explicit constructions of selectors and related combinatorial structures with applications, In Proc. 13*th ACM-SIAM Symp. on Discrete Algorithms*, (SODA'2002), January 2002, pp 697–704.

[18] E. Kushilevitz, and Y. Mansour, An $\Omega(D\log(N/D))$ lower bound for broadcast in radio networks, *SIAM J. on Computing*, 27 (1998), pp. 702–712.
[19] D. Peleg, Deterministic radio broadcast with no topological knowledge, 2000, a manuscript.
[20] K. Ravishankar and S. Singh, Asymptotically optimal gossiping in radio networks, *Discrete Applied Mathematics* 61 (1995), pp 61-82.
[21] K. Ravishankar and S. Singh, Gossiping on a ring with radios, *Par. Proc. Let.* 6, (1996), pp 115-126.
[22] S. Tabbane, *Handbook of Mobile Radio Networks*, Artech House Publishers, 2000.

A Primal Approach to the Stable Set Problem*

Claudio Gentile[1], Utz-Uwe Haus[2], Matthias Köppe[2],
Giovanni Rinaldi[1], and Robert Weismantel[2]

[1] Istituto di Analisi dei Sistemi ed Informatica "Antonio Ruberti"– CNR, Roma,
Italy
{gentile,rinaldi}@iasi.rm.cnr.it
[2] Otto-von-Guericke-Universität Magdeburg
Department of Mathematics/IMO, Germany
{haus,mkoeppe,weismant}@imo.math.uni-magdeburg.de

Abstract. We present a new "primal" algorithm for the stable set problem. It is based on a purely combinatorial construction that can transform every graph into a perfect graph by replacing nodes with sets of new nodes. The transformation is done in such a way that every stable set in the perfect graph corresponds to a stable set in the original graph. The algorithm keeps a formulation of the stable set problem in a simplex-type tableau whose associated basic feasible solution is the incidence vector of the best known stable set. The combinatorial graph transformations are performed by substitutions in the generators of the feasible region. Each substitution cuts off a fractional neighbor of the current basic feasible solution. We show that "dual-type" polynomial-time separation algorithms carry over to our "primal" setting. Eventually, either a non-degenerate pivot leading to an integral basic feasible solution is performed, or the optimality of the current solution is proved.

1 Introduction

The *stable set problem* (or *node packing problem*) is one of the most studied problems in combinatorial optimization. It can be defined as follows: Let $(G, \mathbf{c})$ be a weighted graph, where $G = (V, E)$ is a graph with $n = |V|$ nodes and $m = |E|$ edges and $\mathbf{c} \in \mathbf{R}_+^V$ is a node function that assigns a weight to each node of G. A set $S \subseteq V$ is called *stable* if its nodes are pairwise nonadjacent in G. The problem is to find a stable set S^* in G of maximum weight $\mathbf{c}(S^*) = \sum_{v \in S^*} c(v)$. The value $\mathbf{c}(S^*)$ is called the **c**-*weighted stability number* $\alpha_\mathbf{c}(G)$ of the graph G.

This problem is equivalent to maximizing the linear function $\sum_{v \in S} c(v) x_v$ over the *stable set polytope* P_G, the convex hull of the incidence vectors of all the stable sets of G. Thus linear programming techniques can be used to solve the problem, provided that an explicit description of the polytope is given. It

---

* First, second, third, and fifth named authors supported by the European DONET program TMR ERB FMRX-CT98-0202. Second, third, and fifth named authors supported by grants FKZ 0037KD0099 and FKZ 2495A/0028G of the Kultusministerium of Sachsen-Anhalt. Fifth named author supported by a Gerhard-Hess-Preis and grant WE 1462 of the Deutsche Forschungsgemeinschaft.

is nowadays well known that, the stable set problem being NP-hard, it is very unlikely that such a description can be found for instances of arbitrary size. Moreover, even if a partial description is at hand, due to the enormous number of inequalities, it is not obvious how to turn this knowledge into a useful algorithmic tool.

Despite these difficulties, the literature in combinatorial optimization of the last thirty years abounds with successful studies where nontrivial instances of NP-hard problems were solved with a cutting plane procedure based on the generation of strong cuts obtained from inequalities that define facets of certain polytopes.

The idea of using facet defining inequalities in a cutting plane algorithm was first proposed by Padberg in [16]. He proved that the inequality $\sum_{v \in C} x_v \leq \frac{1}{2}(|C| - 1)$ associated to a chordless cycle of odd length (an *odd hole*) is facet-defining for P_G or can be "lifted" to obtain a facet-defining inequality. Later *antihole inequalities* were introduced in [15]. Borndörfer [5] gives an extensive list of references to further facet defining inequalities for which a characterization is known.

It is not a trivial task to exploit this vast amount of knowledge on the stable set polytope to devise an effective cutting plane algorithm that is able to solve non trivial instances of large size. Attempts have been made by Nemhauser and Sigismondi [14] and Balas et al. [1], but, unlike in the case of other NP-hard problems, polyhedrally based cutting plane algorithms for the stable set problem have not yet shown their superiority over alternative methods. On the other hand, several approaches have been tried to solve difficult instances. For a collection of papers on algorithms for the stable set problem and for a recent survey on the subject, see [12] and [4], respectively.

The cutting plane procedure mentioned before has a "dual flavor," in the sense that the current solution is infeasible until the end, when feasibility and hence optimality is reached. For this reason, a "primal" algorithm, i.e., an algorithm that always maintains a feasible solution, is desirable. A primal cutting plane procedure was first proposed by Young [18]: One starts with an integral basic feasible solution, then either pivots leading to integral solutions are performed or cuts are generated that are satisfied by the current solution at equality. Padberg and Hong [17] were the first to propose a similar primal procedure based on strong polyhedral cutting planes. These kinds of algorithms produce a path of adjacent vertices of the polytope associated with the problem.

A profound study of the vertex adjacency for the polytope of the set partitioning problem was produced by Balas and Padberg [2]. They provided the theoretical background for the realization of a primal algorithm that produces a sequence of adjacent vertices of the polytope, ending with the optimal solution. Their basic technique was to replace a column of the current simplex tableau with a set of new columns in order to guarantee the next pivot to lead to an integral basic feasible solution.

These ideas were generalized to the case of general integer programming by Haus, Köppe, and Weismantel [10, 11], who called their method the "Integral Basis Method." This method does neither require cutting planes nor enumeration techniques. In each major step the algorithm either returns an augmenting direction that is applicable at the given feasible point and yields a new feasible point with better objective function value or provides a proof that the point under consideration is optimal. This is achieved by iteratively substituting one column by columns that correspond to irreducible solutions of a system of linear diophantine inequalities. A detailed description of the method is given in the paper [11].

The present paper provides some first graph theoretical tools for a primal algorithm for the stable set problem in the same vein as the work of Balas and Padberg and of Haus, Köppe, and Weismantel.

The cardinality of the largest stable set of a graph G $= (V, E)$ is called the *stability number* of G and denoted by $\alpha(G)$. The minimum number of cliques of G whose union coincides with V is called the *clique covering number* of G and denoted by $\overline{\chi}(G)$. A graph G is *perfect* if and only if $\alpha(G') = \overline{\chi}(G')$ for all subgraphs G' of G induced by subsets of its node set V. For the fundamentals on perfect graphs and balanced matrices and on their connections we refer to, e.g., [7].

A graph is perfect if and only if its clique formulation defines an integral polytope. Moreover, for perfect graphs the stability number can be computed in polynomial time [9]; thus, also the separation problem for P_G is polynomially solvable in this case. Therefore, one can devise a primal cutting plane algorithm for the stable set problem for perfect graphs. We start, for example, with the edge formulation and with a basic feasible solution corresponding to a stable set. Then we perform simplex pivots until either we reach optimality or we produce a fractional solution. In the latter case we add clique inequalities to the formulation that make the fractional solution infeasible, we step back to the previous (integral) basic feasible solution, and we iterate.

Suppose now that the graph is not perfect. We assume that at hand is a graph transformation that in a finite number of steps transforms the original graph into a possibly larger graph that is perfect. Then it may be possible to apply again the previous primal cutting plane procedure as follows: As soon as the fractional solution cannot be cut off by clique inequalities, because other valid inequalities for P_G would be necessary, we make one or more steps until the clique formulation of the current graph makes the fractional solution infeasible. This procedure eventually finds an optimal stable set in the latest generated graph. It can be used for solving the original problem as long as the graph transformation is such that the optimal stable set in this graph can be mapped into an optimal stable set in the original graph.

This procedure provides a motivation for this paper where in Section 2 we define valid transformations that have the desired properties mentioned above; in Section 3 we translate the graph transformations into algebraic operations on the simplex tableaux; in Section 4, we give some algorithmic properties of the proposed transformations.

2 Valid Graph Transformations

Throughout this section, we will denote by $G^0 = (V^0, E^0)$ and $\mathbf{c}^0$ the graph and node-weight function of the original weighted stable set problem, respectively. The purpose of this section is to devise a family of transformations $(G, \mathbf{c}) \mapsto (G', \mathbf{c}')$ with the property $\alpha_{\mathbf{c}}(G) = \alpha_{\mathbf{c}'}(G')$, i.e., transformations maintaining the weighted stability number. After a sequence of those transformations, a perfect graph G^* with a node-weight function $\mathbf{c}^*$ will be produced. In perfect graphs the stability number can be computed in polynomial time [9]. Moreover, the $\mathbf{c}^*$-weighted stable set problem in G^* can be solved with linear programming over the clique formulation of G^*. (Finding the maximal cliques in G^* will be expensive as well. In Section 4, however, we will show that our transformations can only enlarge the size of the cliques and never destroy the property of maximality.)

Typically, one is not only interested in the weighted stability number of a graph but also in a stable set where the maximum is attained. Thus, once the $\mathbf{c}^*$-weighted stable set problem in G^* is solved, one would like to recover a corresponding maximum $\mathbf{c}^0$-weighted stable set in the original graph G^0. For this purpose, we shall attach a *node labeling* $\sigma \colon V \to 2^{V^0}$ to each graph $G = (V, E)$. This labeling assigns a stable set $\sigma(v) \subseteq V^0$ in the original graph to each node $v \in V$. The label of a node also determines its weight by the setting

$$c(v) = \sum_{u \in \sigma(v)} c^0(u) \quad \text{for } v \in V. \tag{1}$$

Now, given a stable set $S \subseteq V$ in G with labeling σ, we intend to reconstruct a stable set $S^0 \subseteq V^0$ in G^0 by $S^0 = \bigcup_{s \in S} \sigma(s)$. For this to work, we need to impose some properties on a labeling.

Definition 1 (valid labeling). *Let $G = (V, E)$ be a graph. A mapping $\sigma \colon V \to 2^{V^0}$ is called a* valid node labeling *of G (with respect to G^0) if the following conditions hold:*

(a) *For $v \in V$, $\sigma(v)$ is a nonempty stable set in G^0.*
(b) *For every two distinct nodes $u, v \in V$ with $\sigma(u) \cap \sigma(v) \neq \emptyset$, the edge (u, v) is in E; i.e., nodes with non-disjoint labels cannot be in the same stable set.*

(c) *Let $u, v \in V$ be distinct nodes. If there exists an edge $(u^0, v^0) \in E^0$ with $u^0 \in \sigma(u)$ and $v^0 \in \sigma(v)$, then the edge (u, v) belongs to E.*

Lemma 1. *Let σ be a valid labeling of a graph $G = (V, E)$ and let $\mathbf{c} \colon V \to \mathbf{R}_+$ be defined by (1). Let S be a stable set in G. Then $S^0 = \bigcup_{s \in S} \sigma(s)$ is a disjoint union, giving a stable set in G^0 with $\mathbf{c}^0(S^0) = \mathbf{c}(S)$.*

Definition 2 (faithful labeling). *Let σ be a valid labeling of a graph $G = (V, E)$ with respect to G^0 and let $\mathbf{c} \colon V \to \mathbf{R}_+$ be defined by (1). We call σ a* faithful labeling *of G if for every stable set S in G that has maximum weight with respect to $\mathbf{c}$, the stable set $S^0 = \bigcup_{s \in S} \sigma(s)$ in G^0 has maximum weight with respect to $\mathbf{c}^0$. A faithfully labeled graph $(G, \mathbf{c}, \sigma)$ is a weighted graph $(G, \mathbf{c})$ with a faithful labeling σ.*

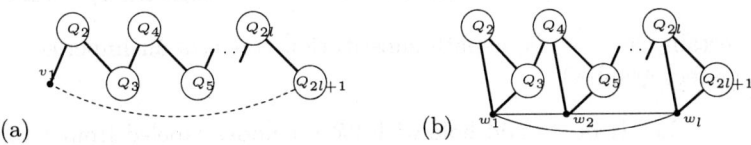

Fig. 1. The clique-path substitution. (a) An odd path of cliques; the dashed line indicates that only some of the edges between v_1 and the nodes of Q_{2l+1} need to present. (b) The result of the substitution in the case of $R = \emptyset$; note that, to unclutter the picture, some edges have been omitted, in fact all nodes w_i are connected with Q_2 and Q_{2l+1}

Definition 3 (valid transformation). *A valid graph transformation is a transformation that turns a faithfully labeled graph* (G, c, σ) *into a faithfully labeled graph* (G', c', σ').

We first consider a very simple transformation. Take a path $P = (v_1, v_2, \ldots, v_{2l}, v_{2l+1})$ of odd length in G that together with the edge (v_{2l+1}, v_1) forms an odd hole. Let S be a stable set in G with $v_1 \in S$. Since there are at most l elements of S in P, there exists an index i such that both v_{2i} and v_{2i+1} are not in S. Therefore, if we replace v_1 by l pairwise adjacent copies $w_1, \ldots, w_l$, where w_i is adjacent to both v_{2i} and v_{2i+1}, for $i = 1, \ldots, l$, it is not difficult to see that any stable set in G corresponds to a stable set in the new graph. The advantage of applying such an operation is, as will be made clear in the following, that in the new graph the odd hole has disappeared. This observation motivates the following definition, where the set of all nodes in G adjacent to a node v is denoted by $N_G(v)$.

Definition 4 (clique-path substitutions). *Let* $G = (V, E)$ *be a graph with a valid node labeling* $\sigma: V \to 2^{V_0}$. *For some* $l > 0$, *let* $P = (Q_1 = \{v_1\}, Q_2, \ldots, Q_{2l+1})$ *be a sequence of cliques of G such that* $Q_{i,i+1} := Q_i \cup Q_{i+1}$ *is a clique in G for all* $i \in \{1, \ldots, 2l\}$. *We call P an odd path of cliques. Now let*

$$R = \{v \in Q_{2l+1} : v \text{ is not adjacent to } v_1 \text{ in } G\}.$$

A clique-path substitution along P that transforms a graph G with a valid labeling into a graph G' with a labeling σ' *is obtained in the following way:*

- *replace v_1 by the clique of new nodes* $W = \{w_1, w_2, \ldots, w_l\} \cup \{t_r : r \in R\}$;
- *for $w \in W$ connect w to all nodes of $N_G(v_1)$;*
- *for $i \in \{1, \ldots, l\}$ connect w_i to all the nodes of $Q_{2i,2i+1}$, then set $\sigma'(w_i) = \sigma(v_1)$;*
- *for $r \in R$ connect t_r to r and all the nodes of $N_G(r)$, then set $\sigma'(t_r) = \sigma(v_1) \cup \sigma(r)$.*

In Figure 1 an odd path of cliques and the result of the substitution are shown. Definition 4 does not require the cliques Q_i to be pairwise disjoint, thus they may share nodes.

Proposition 1. *Clique-path substitutions are valid graph transformations.*

In certain cases, a clique-path substitution converts an imperfect graph G into a perfect graph G'.

Example 1. Let H be an odd hole with $2k+1$ nodes labeled from 1 to $2k+1$. Pick node 1 and consider the path P from 1 to $2k+1$ through all nodes. Apply the clique-path substitution of node 1 along P, and call the resulting graph H'. In Figure 2(a,b) both the original graph H and the transformed graph H' are shown for $k=2$. The constraint matrix of the maximal-clique formulation for H' turns out to be a balanced matrix, and hence H' is perfect; see [7].

Example 2. The same substitution also turns the odd wheel W (Figure 2 c) into a perfect graph W' (Figure 2 d); indeed, the "hub" node h is connected to all other nodes in W', which is otherwise the same as H'. This is in contrast to the cutting-plane approach, where lifting an odd-hole cut is indispensable to make it strong, because odd-hole cuts define facets for the stable set polytope associated with a graph that is the odd hole itself.

Example 3. Now we consider an odd antihole $\overline{H}_{2k+1} = (V, E)$ with $2k+1$ nodes, labeled from 1 to $2k+1$, see Figure 3. The set $V \setminus \{1\}$ can be partitioned into two cliques: Q_{odd} and Q_{even}. The former contains all nodes with an odd label (except node 1), the latter contains all nodes with an even label. So we can consider the odd path of cliques $P = (\{1\}, Q_{even} \setminus \{2\}, \{2\}, Q_{odd} \setminus \{3\}, \{3\})$.

The graph G' resulting from the clique-path substitution of node 1 along P is shown in Figure 3. It has two new nodes 1' and 1'' replacing node 1; node 1' is connected to all nodes but 2, while node 1'' is connected to all nodes but $2k+1$. The complement of G' only contains the simple path $\{(1'', 2), (2, 3), \ldots, (2k, 2k+1), (2k+1, 1')\}$, hence G' is perfect; see [13].

Our examples imply that, since the Strong Perfect Graph Conjecture [3] seems to be proven [6], we can transform every minimally imperfect graph into a perfect graph with a single clique-path substitution.

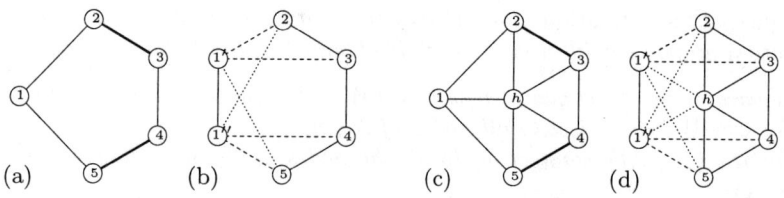

Fig. 2. Applying the clique-path substitution to an odd hole and an odd wheel. (a) The 5-hole H_5. (b) The result of the clique-path substitution of node 1 in H_5 along the path $P = (\{1\}, \{2\}, \{3\}, \{4\}, \{5\})$. (c) The 5-wheel W_5. (d) The result of the clique-path substitution of node 1 of the odd hole in W_5 along the path $P = (\{1\}, \{2\}, \{3\}, \{4\}, \{5\})$

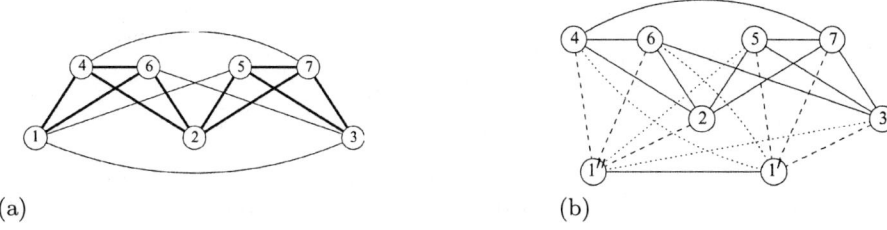

(a) (b)

Fig. 3. Applying the clique-path substitution to an odd antihole. (a) The 7-antihole $\overline{H}_7$; the edges of the complete subgraphs induced by the relevant cliques in the path of cliques $P = (\{1\}, Q_{\text{even}} \setminus \{2\}, \{2\}, Q_{\text{odd}} \setminus \{3\}, \{3\})$ are shown with thick lines. (b) The result of the clique-path substitution of node 1 along P; the edges introduced by the clique-path substitution are drawn with dashed lines, whereas edges merely "inherited" from node 1 are drawn with a dotted line

Moreover, for any connected graph $G = (V, E)$, it is easy to construct a finite sequence of clique-path substitutions leading to a graph that is the disjoint union of complete graphs, hence to a perfect graph: We start with a clique $Q \subseteq V$ that is maximal w.r.t. inclusion. When $Q = V$, we are done. Otherwise, we take a node $v_1 \in V \setminus Q$ such that $Q_2 := Q \cap N_G(v_1) \neq \emptyset$. Let $Q_3 := Q \setminus Q_2$. Now $P = (\{v_1\}, Q_2, Q_3)$ is an odd path of cliques in G. The clique-path substitution along P leads to a graph G', where all the new nodes have been adjoined to the clique Q. We continue with G' and a maximal clique in G' containing the enlarged clique. Since $|V \setminus Q|$ decreases in each step by at least one, the procedure terminates with a complete graph. This construction can be applied independently to all the connected components of an arbitrary graph.

Remark 1 (Comparison to LP-based branch-and-bound procedures). We use a simple example to illustrate the possible advantage of a method based on valid graph transformations, compared to LP-based branch-and-bound procedures. For $k \in \{1, 2, \ldots\}$ and $l \in \{2, 3, \ldots\}$, let the graph C_{2l+1}^k be the disjoint union of k odd holes H_{2l+1}. The maximal clique formulation of the stable set problem in C_{2l+1}^k is

$$\max \sum_{i=1}^{k} \sum_{j=1}^{2l+1} x_{i,j}$$
$$\text{s.t.} \quad x_{i,j} + x_{i,j+1} \leq 1 \quad \text{for } i \in \{1, \ldots, k\} \text{ and } j \in \{1, \ldots, 2l\},$$
$$x_{i,1} + x_{i,2l+1} \leq 1 \quad \text{for } i \in \{1, \ldots, k\},$$
$$x_{i,j} \in \{0, 1\} \quad \text{for } i \in \{1, \ldots, k\} \text{ and } j \in \{1, \ldots, 2l+1\}.$$
(2)

The unique optimal solution to the LP relaxation of (2) is given by $x_{i,j} = \frac{1}{2}$ for $i \in \{1, \ldots, k\}$ and $j \in \{1, \ldots, 2l+1\}$. An LP-based branch-and-bound procedure would now select one node variable, $x_{1,1}$ say, and consider the two subproblems obtained from (2) by fixing $x_{1,1}$ at 0 and 1, respectively. The graph-theoretic interpretation of this variable fixing is that the copy of C_{2l+1} corresponding to the node variables $x_{1,\cdot}$ is turned into a perfect graph in both branches, see Figure 4.

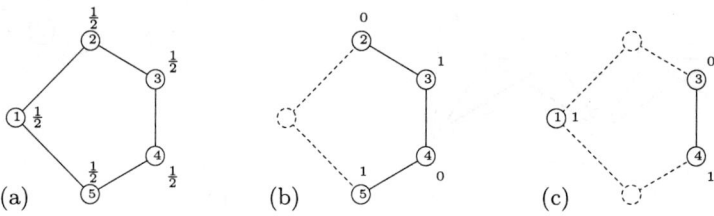

Fig. 4. Graph-theoretic interpretation of the branch operation on a fractional node variable in an LP-based branch-and-bound procedure. The numbers shown aside the nodes are the node variable values in an optimal solution to the LP relaxation. (a) One of the copies of C_5 in the graph C_5^k. (b) Fixing the first node variable at zero. (c) Fixing the first node variable at one

Hence, the optimal basic solutions to the LP relaxations of the subproblems attain integral values there. Since there remain $k-1$ odd holes in both subproblems, the branch-and-bound procedure clearly visits a number of subproblems exponential in k.

On the other hand, a method using clique-path substitutions, which performs the whole enumeration implicitly, can turn the graph C_{2l+1}^k into a perfect, validly labeled graph $(G, \mathbf{c}, \sigma)$ by performing only k substitution steps of the type shown in Figure 2. The substitution steps can be accomplished in a way such that one keeps a maximal-clique formulation of the transformed graph, see Section 4. The optimal solution to the LP relaxation of this formulation is integral, and the corresponding maximum stable set in the original graph C_{2l+1}^k can be computed by means of the node labeling σ.

3 A Primal Integer Programming Approach

Since graph transformations transform weighted stable set problems to weighted stable set problems, they can be used as a tool within any optimization algorithm for the stable set problem. In this section, we will deal with weighted stable set problems in a specific algorithmic framework, namely in a primal integer programming setting in the vein of work of Balas and Padberg [2] and of Haus, Köppe, and Weismantel [10, 11]. It will turn out that the graph transformations discussed in the previous section can be re-interpreted as column operations in an integral simplex tableau.

Again let $(G = (V, E), \mathbf{c})$ be a weighted graph. Let $Q_1, \ldots, Q_k$ be cliques in G that cover all the edges of G. Introducing a slack variable z_{Q_i} for each clique Q_i, the weighted stable set problem is formulated as the integer program

$$\max \mathbf{c}^\mathsf{T} \mathbf{x} : \sum_{v \in Q_i} x_v + z_{Q_i} = 1 \text{ for } i \in \{1, \ldots, k\}, \quad \mathbf{x} \geq 0,\ \mathbf{z} \geq 0, \qquad (3)$$

$$\mathbf{x} \in \mathbf{Z}^V,\ \mathbf{z} \in \mathbf{Z}^k. \qquad (4)$$

We call (3) a *maximal-clique formulation* if all cliques Q_i, $i = \{1, \ldots, k\}$, are maximal. Moreover, if the set $\{Q_i\}_{i=\{1,\ldots,k\}}$ includes all maximal cliques of G, (3) is called the *complete maximal-clique formulation*.

Now let $S \subseteq V$ be a stable set in G. To construct a basic feasible solution associated with S we select for each of the rows in the program (3) a basic variable as follows:

- For every $v \in S$, let i_v be a row index such that $v \in Q_{i_v}$. We select x_v as the basic variable associated with the row i_v of the tableau.
- For each of the remaining clique constraints i we select the slack variable z_{Q_i} as the corresponding basic variable.

Note that the indices i_v are all distinct, and hence the construction yields a basis corresponding to S. As usual, let B and N denote the sets of basic and non-basic variables, respectively. We can now rewrite (3) in tableau form

$$\mathbf{y}_B + \mathbf{A}_B^{-1}\mathbf{A}_N \mathbf{y}_N = \mathbf{b}, \quad (\mathbf{y}_B, \mathbf{y}_N) \in \mathbf{Z}_+^{n+k}.$$

The variables y_j correspond to node variables x_v or slack variables z_{Q_i}. We will henceforth call a tableau obtained by this procedure a *canonical tableau* for S.

Starting from a basic feasible integer solution, the Balas–Padberg procedure [2] and the Integral Basis Method by Haus, Köppe and Weismantel [10, 11] proceed as follows. As long as nondegenerate integral pivots are possible, such steps are performed, improving the current basic feasible integer solution. When a solution is reached that would only permit a degenerate integral pivot, a "column-generation procedure" replaces some nonbasic columns by new "composite" columns, which are nonnegative integral combinations of nonbasic columns in the current tableau. In this way, eventually columns are generated that allow nondegenerate integral pivots, or optimality of the current basic feasible solution is proved.

The Balas–Padberg procedure has not proven to be an efficient algorithm for set partitioning problems. Also the implementation of the Integral Basis Method as described in [11] shows a rather weak computational performance when applied to stable set problems. The reason is that both algorithms generate the composite columns to add to the tableau without making use of the graph-theoretic properties of the problem. Using "strong", "combinatorial" composite columns whenever possible is a key to improved performance.

Definition 5 (odd alternating path of cliques). *Let $(G = (V, E), \mathbf{c}, \sigma)$ be a faithfully labeled weighted graph. Fix an integral simplex tableau for a formulation of the $\mathbf{c}$-weighted stable set problem in G. Let x_{v_1} be a non-basic variable of positive reduced cost, and let $P = (\{v_1\}, Q_2, \ldots, Q_{2l+1})$ be an odd path of cliques in G. Again let $R = \{v \in Q_{2l+1} : v \text{ is not adjacent to } v_1 \text{ in } G\}$. For $i \in \{1, \ldots, l\}$, assume there is a non-basic slack variable z_i for the clique $Q_{2i,2i+1} = Q_{2i} \cup Q_{2i+1}$, and there is $x_{j_i} = 1$ with $j_i \in Q_{2i}$. Moreover, presume that all variables x_v for $v \in R$ are nonbasic. In this setting, we call $(\{v_1\}, Q_{2,3}, Q_{4,5}, \ldots, Q_{2l,2l+1})$ an odd alternating path of cliques.*

In this setting, we are going to remove the column of the non-basic variable x_{v_1} from the tableau and replace it by non-basic integral combinations of other columns of the tableau. We shall use the notation $x_v \wedge x_w$ for a new variable associated with a column that is the sum of the columns for x_v and x_w.

Definition 6. *For an odd alternating path of cliques* $(\{v_1\}, Q_{2,3}, Q_{4,5}, \ldots,$
$Q_{2l,2l+1})$ *we define the corresponding* alternating-path substitution *in the tableau as follows: substitute x_{v_1} by new binary variables according to the following column operations:*

- $u_r = x_{v_1} \wedge x_r$ *for all* $r \in R$,
- $y_i = x_{v_1} \wedge z_i$ *for all* $i \in \{1, \ldots, l\}$,

where all new variables are non-basic and 0/1.

Observation 1. Let $\bar{x}$ and $\bar{z}$ denote the variables **x** and **z**, respectively, in the formulation obtained after the substitution. Then we can map a solution of the new formulation back into a solution of the old formulation via the following relations:

$$x_{v_1} = \sum_{i=1}^{l} y_i + \sum_{r \in R} u_r, \tag{5a}$$
$$x_r = \bar{x}_r + u_r \quad \text{for all } r \in R, \tag{5b}$$
$$z_i = \bar{z}_i + y_i \quad \text{for all } i \in \{1, \ldots, l\}. \tag{5c}$$

For all other variables, we have $x_v = \bar{x}_v$ and $z_i = \bar{z}_i$. In the following we will denote by F a generic formulation of type (3), by $S_P(F)$ the formulation obtained by applying an alternating-path substitution along P to F. Moreover, given a formulation $F' = S_P(F)$, we will denote by $R_P(F')$ the formulation obtained by applying the mapping (5a)–(5c) to F'.

Lemma 2. *The integer program obtained by a sequence of alternating-path substitutions is an integer programming formulation of the stable set problem in G; the optimal solutions translate into the maximum stable sets in G via the iterated mapping* (5).

Proof. Let $(G', \mathbf{c}', \sigma')$ denote the labeled graph obtained from $(G, \mathbf{c}, \sigma)$ by performing the clique-path substitution along $P = (\{v_1\}, Q_2, Q_3, \ldots, Q_{2l}, Q_{2l+1})$. As in Definition 4, let t_r for $r \in R$ and w_i for $i \in \{1, \ldots, l\}$ denote the new nodes arising from the substitution. We show that the problem resulting from the above column operations is a formulation of the $\mathbf{c}'$-weighted stable set problem in G'.

The key is to realize that the new variables correspond to the new nodes in the following way:

(i) For $i \in \{1, \ldots, l\}$, variable $y_i = x_{v_1} \wedge z_i$ corresponds to the new node w_i.
(ii) For $r \in R$, variable $u_r = x_{v_1} \wedge x_r$ corresponds to the new node t_r.

To verify (i), let $i \in \{1, \ldots, l\}$ and note that the original formulation of the **c**-weighted stable set problem in G implies the following inequalities and equations:

$$x_{v_1} + x_v \leq 1 \quad \text{for } v \in N_G(v_1),$$
$$\sum_{v \in Q_{2i,2i+1}} x_v + z_i \leq 1.$$

By (5), we obtain:

$$\sum_{i=1}^{l} y_i + \sum_{r \in R} u_r + \bar{x}_v \leq 1 \quad \text{for } v \in N_G(v_1),$$
$$\sum_{v \in Q_{2i,2i+1}} \bar{x}_v + \sum_{v \in Q_{2i,2i+1} \cap R} u_v + \bar{z}_i + y_i = 1.$$

Hence, for all variables x_t corresponding to the neighbors $t \in N_{G'}(w_i)$, as given by Definition 4, the new formulation implies an inequality $x_t + y_i \leq 1$. The correspondence (ii) can be verified analogously.

4 Algorithmic Properties

It has already been pointed out that, when all the maximal cliques in G are employed, the integrality constraints in the formulation (3) can be dropped if the underlying graph G is perfect.

Proposition 2. *Let G be a graph and consider a maximal-clique formulation F for the associated stable set problem, then the formulation $S_P(F)$, obtained after an alternating-path substitution for P, is a maximal-clique formulation for the modified graph G'.*

Proof. All new nodes arising from the substitution belong to each (maximal) clique Q such that $v_1 \in Q$. These cliques are extended to maximal cliques as the variables associated with all the new nodes are inserted in the corresponding inequalities. Moreover, all the nodes w_i coming from the sum of the columns of x_{v_1} and of a maximal clique $Q_{i,i+1}$ belong also to the clique $Q_{i,i+1} \cup \{t_i\}$, that now replaces the clique $Q_{i,i+1}$ in the formulation. Finally, all the nodes t_r coming from the sum of the columns of x_{v_1} and of x_r, for $r \in R$, belong also to the maximal cliques containing either r or v_1, that are extended accordingly. All the other cliques are not extended by this operation, so the corresponding inequalities are still facet-defining.

Now we consider the *identification problem* for odd alternating paths of cliques. First we show that an alternating-path substitution "cuts off" the fractional point $(\mathbf{x}^F, \mathbf{z}^F)$ obtained by a single pivoting step applied to a basic integer solution $(\mathbf{x}^I, \mathbf{z}^I)$. Moreover, such an alternating path with $R = \emptyset$ can be found in polynomial time if it exists.

Definition 7. *Let $(\mathbf{x}^I, \mathbf{z}^I)$ be a basic integer solution and let v_i be a nonbasic variable. If the basic solution obtained by pivoting in x_{v_1} is fractional, we call it a fractional neighbor of $(\mathbf{x}^I, \mathbf{z}^I)$ and denote it by $(\mathbf{x}^F, \mathbf{z}^F)$.*

Lemma 3. *Let F be a formulation for the graph G, $(\mathbf{x}^I, \mathbf{z}^I)$ be a basic integer solution, and x_{v_1} a non-basic variable. Assume that pivoting x_{v_1} into the basis would produce a fractional neighbor $(\mathbf{x}^F, \mathbf{z}^F)$. Consider an alternating-path substitution along $P = (v_1, Q_{2,3}, \ldots, Q_{2l,2l+1})$ and the corresponding formulation $F' = S_P(F)$. Then the solution $(\mathbf{x}^F, \mathbf{z}^F)$ is not feasible for the mapping $R_P(F')$ on the space of the initial variables (where R is the mapping defined in Observation 1).*

Observation 2. *Lemma 3 is valid as long as the variables z_i for $i \in \{1, \ldots, l\}$ and x_r for $r \in R$ are equal to zero for both the solutions $(\mathbf{x}^I, \mathbf{z}^I)$ and $(\mathbf{x}^F, \mathbf{z}^F)$. Therefore, these variables do not have to be necessarily nonbasic.*

In the following, we shall consider the alternating-path substitution in the case of $R = \emptyset$. In this case the path of cliques is a cycle. In fact, an easy computation shows that the alternating-path substitution along $P = (\{v_1\}, Q_{2,3}, \ldots, Q_{2l,2l+1})$ is at least as strong as adding the inequality

$$\sum_{i=1}^{l} \left(\sum_{j \in Q_{2i,2i+1}} x_j \right) + x_{v_1} + \sum_{i=1}^{l} \bar{z}_i = l, \tag{6}$$

to the problem, which is a lifted odd-hole inequality resulting from the odd hole of length $2l+1$ of nodes $v_1, v_2, v_3, v_4, \ldots, v_{2l}, v_{2l+1}$, where v_i can be chosen as any node in Q_i for $i \in \{2, \ldots, 2l+1\}$. This enables us to translate polynomial-time "dual separation" results into "primal separation" results.

Theorem 1. *Let F be a formulation and let $(\mathbf{x}^I, \mathbf{z}^I)$ denote a basic integer solution. Suppose that there exists an alternating-path substitution along $P = (\{v_1\}, Q_{2,3}, \ldots, Q_{2l,2l+1})$ such that the inequality corresponding to each of the cliques $Q_{2,3}, \ldots, Q_{2l,2l+1}$ coincides with or is dominated by one of the clique inequalities in F and $R = \emptyset$. Then one can find such a substitution in polynomial time in the size of F. Moreover, the fractional neighbor $(\mathbf{x}^F, \mathbf{z}^F)$ that would have been obtained by pivoting x_{v_1} into the basis is infeasible for the formulation $R_P(S_P(F))$.*

In a dual-type method, one is interested in finding the cut from a given class that is *most violated* by the current solution. Modifying the standard algorithm for separating odd-hole inequalities [8], we can solve the analogous problem in our primal setting:

Proposition 3. *The problem of finding the alternating-path substitution of the type as in Theorem 1, whose corresponding inequality (6) is the most violated by a fractional neighbor $(\mathbf{x}^F, \mathbf{z}^F)$ of a basic integral solution $(\mathbf{x}^I, \mathbf{z}^I)$, can be solved in polynomial time in the size of F.*

We conclude this section with the following easy observation:

Observation 3. *If F contains an inequality for each clique of G of size at most h, then Proposition 3 gives an exact polynomial time primal separation procedure for all the inequalities of type (6). Note that for $h = 3$, inequalities of type (6) include all the odd-hole and the odd-wheel inequalities.*

References

[1] Egon Balas, Sebastián Ceria, Gérard Cornuéjols, and Gabor Pataki, *Polyhedral methods for the maximum clique problem*, In Johnson and Trick [12], pp. 11–28.

[2] Egon Balas and Manfred W. Padberg, *On the set-covering problem: II. An algorithm for set partitioning*, Operations Research **23** (1975), 74–90.

[3] Claude Berge, *Färbung von Graphen, deren sämtliche bzw. deren ungerade Kreise starr sind*, Wiss. Z. Martin-Luther-Univ. Halle-Wittenberg, Math.-Natur. Reihe (1961), 114–115.

[4] Immanuel M. Bomze, Marco Budinich, Panos M. Pardalos, and Marcello Pelillo, *The maximum clique problem*, Handbook of Combinatorial Optimization (Supplement Volume A) (D.-Z. Du and P. M. Pardalos, eds.), vol. 4, Kluwer Academic Publishers, Boston, MA, 1999.

[5] Ralf Borndörfer, *Aspects of set packing, partitioning, and covering*, Dissertation, Technische Universität Berlin, 1998, published by Shaker-Verlag, Aachen.

[6] Maria Chudnovski, Neil Robertson, Paul Seymour, and Robin Thomas, Talk given at the Oberwolfach meeting on Geometric Convex Combinatorics, June 2002.

[7] Gérard Cornuéjols, *Combinatorial optimization: Packing and covering*, CBMS-NSF Regional Conference Series in Applied Mathematics, no. 74, SIAM, Philadelphia, 2001.

[8] A. M. H. Gerards and A. Schrijver, *Matrices with the Edmonds-Johnson Property*, Combinatorica **6** (1986), no. 4, 365–379.

[9] Martin Grötschel, László Lovász, and Alexander Schrijver, *Geometric algorithms and combinatorial optimization*, Algorithms and Combinatorics, vol. 2, Springer, Berlin, 1988.

[10] Utz-Uwe Haus, Matthias Köppe, and Robert Weismantel, *The Integral Basis Method for integer programming*, Mathematical Methods of Operations Research **53** (2001), no. 3, 353–361.

[11] _____, *A primal all-integer algorithm based on irreducible solutions*, To appear in *Mathematical Programming Series B*, preprint available from URL http://www.math.uni-magdeburg.de/~mkoeppe/art/haus-koeppe-weismantel-ibm-theory-rr.ps, 2001.

[12] D. S. Johnson and M. A. Trick (eds.), *Clique, coloring, and satisfiability: Second DIMACS implementation challenge, DIMACS*, vol. 26, American Mathematical Society, 1996.

[13] L. Lovász, *Normal hypergraphs and the weak perfect graph conjecture*, Topics on perfect graphs, North-Holland, Amsterdam, 1984, pp. 29–42.

[14] G. L. Nemhauser and G. L. Sigismondi, *A strong cutting plane / branch and bound algorithm for node packing*, Journal of the Operational Research Society (1992), 443–457.

[15] G. L. Nemhauser and L. E. Trotter, *Properties of vertex packing and independent system polyhedra*, Mathematical Programming **6** (1973), 48–61.

[16] M. W. Padberg, *On the facial structure of set packing polyhedra*, Mathematical Programming **5** (1973), 199–215.

[17] Manfred W. Padberg and Saman Hong, *On the symmetric travelling salesman problem: a computational study*, Math. Programming Stud. (1980), no. 12, 78–107.

[18] Richard D. Young, *A simplified primal (all-integer) integer programming algorithm*, Operations Research **16** (1968), no. 4, 750–782.

Wide-Sense Nonblocking WDM Cross-Connects

Penny Haxell[1]*, April Rasala[2]**, Gordon Wilfong[3], and Peter Winkler[3]

[1] Department of Combinatorics and Optimization, University of Waterloo
Waterloo, Ont. Canada, N2L 3G1
pehaxell@math.uwaterloo.ca
[2] MIT Laboratory for Computer Science
Cambridge, MA, USA 02139
arasala@theory.lcs.mit.edu
[3] Bell Labs, Lucent Technologies
Murray Hill NJ, USA 07974
{gtw,pw}@research.bell-labs.com

Abstract. We consider the problem of minimizing the number of wavelength interchangers in the design of wide-sense nonblocking cross-connects for wavelength division multiplexed (WDM) optical networks. The problem is modeled as a graph theoretic problem that we call *dynamic edge coloring*. In dynamic edge coloring the nodes of a graph are fixed but edges appear and disappear, and must be colored at the time of appearance without assigning the same color to adjacent edges.
For wide-sense nonblocking WDM cross-connects with k input and k output fibers, it is straightforward to show that $2k-1$ wavelength interchangers are always sufficient. We show that there is a constant $c > 0$ such that if there are at least ck^2 wavelengths then $2k-1$ wavelength interchangers are also necessary. This improves previous exponential bounds. When there are only 2 or 3 wavelengths available, we show that far fewer than $2k-1$ wavelength interchangers are needed. However we also prove that for any $\varepsilon > 0$ and $k > 1/2\varepsilon$, if the number of wavelengths is at least $1/\varepsilon^2$ then $2(1-\varepsilon)k$ wavelength interchangers are needed.

1 Introduction

A wavelength division multiplexed (WDM) network employs multiple wavelengths in order to carry many channels in an optical fiber. A WDM network contains places at which multiple fibers come together. At these places, channels that have previously been routed along the same fiber may each need to be moved to different fibers and possibly also change wavelengths. Switching is, ideally, done by a WDM cross-connect that allows each incoming input channel to be routed to any (unused) output channel. To do this the cross-connect requires, among other things, expensive components called *wavelength interchangers* that permute the wavelengths on a fiber in any desired manner.

---
* Partially supported by NSERC.
** Supported by a Lucent GRPW Fellowship.

Over time, the demands on the network change. Some connections are no longer needed and requests are made to add new connections. Any fiber in the network with an available wavelength can handle the addition of a new request by routing it on the unused wavelength. However, at cross-connects the interactions between requests can be more complicated. A cross-connect is said to be *wide-sense nonblocking* if there is an on-line algorithm that assures that it can always meet demands. (This is weaker than *strictly nonblocking* where the demands are never blocked even when previous demands have been routed arbitrarily).

Our goal here is to minimize the number of wavelength interchangers in the design of a wide-sense nonblocking cross-connect with k input fibers and k output fibers. It is easily seen that $2k-1$ wavelength interchangers suffice, even with greedy routing; we show that there is a constant $c > 0$ so that with ck^2 wavelengths, $2k-1$ wavelength interchangers are necessary as well, *regardless* of the routing algorithm. This improves previous exponential bounds.

On the positive side, in the case where there are only 2 or 3 wavelengths there is a significant reduction in the number of wavelength interchangers required. However, we also show that for any $\varepsilon > 0$ and $k > 1/2\varepsilon$, if there are at least $1/\varepsilon^2$ wavelengths then $2(1-\varepsilon)k$ wavelength interchangers are necessary.

This WDM cross-connect problem is shown to be equivalent to a dynamic edge coloring problem for bipartite multigraphs and the results are stated and derived in terms of this edge coloring problem.

2 Wavelength Division Multiplexing

In wavelength division multiplexing (WDM) an optical fiber or other medium carries many channels at once, subject to the constraint that each employs a different wavelength from some fixed set of Λ wavelengths. WDM systems greatly increase the available bandwidth of existing facilities, and are rapidly proliferating; systems with 80 wavelengths are becoming commonplace and systems with thousands are being contemplated.

Optimal use of bandwidth in a WDM network requires switches that can change the wavelength, as well as the fiber, on which a channel is carried [13, 14, 7, 16, 17]. A $k \times k$ *WDM cross-connect* should in theory be able to dynamically route up to Λk incoming channels on k fibers in any specified way onto k outgoing fibers, subject to the constraint that no two channels of the same wavelength are output on the same fiber. When the cross-connect is in operation, "demands" arrive and depart, and must be handled without knowledge of the future; each demand consists of an input channel (that is, an input fiber and a wavelength) and an output channel to which it must be linked.

The cost of a WDM cross-connect is dominated by the cost of the components, called "wavelength interchangers", that permute wavelengths on a fiber [23]. Thus our goal is to study WDM cross-connect designs that minimize the number of wavelength interchangers required to achieve certain nonblocking properties.

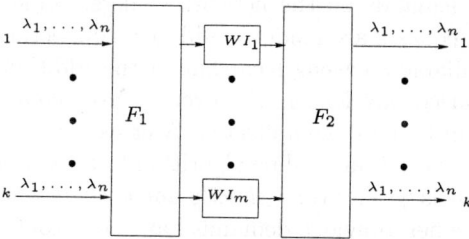

Fig. 1. A WDM split cross-connect with k input fibers and k output fibers

We consider an important class of WDM cross-connects known as *split cross-connects*, illustrated in Figure 1. In such cross-connects, any input channel can be routed to any wavelength interchanger not currently servicing a demand with the same input wavelength; and similarly any output channel can be routed from any wavelength interchanger not servicing a demand of the same output wavelength, regardless of how any previous demands have been routed [19, 20]. Thus in order to satisfy a demand the only decision necessary is to choose which available wavelength interchanger to use.

A demand for a connection from input channel I to output channel O is said to be *valid* if neither I nor O is part of an already routed demand. Demands for connections are requested and withdrawn over time. The nonblocking properties of a split WDM cross-connect are said to be *rearrangeably*, *wide-sense* or *strictly* nonblocking where

(i) "rearrangeably nonblocking" means that there exists an available wavelength interchanger to service any valid demand although the wavelength interchangers assigned to currently routed demands might have to be changed;
(ii) "wide-sense nonblocking" means that there exists an algorithm that always finds an available wavelength interchanger to service a valid demand assuming that all current assignments of wavelength interchangers to demands have been done using the same algorithm; and
(iii) "strictly nonblocking" means that there always exists an available wavelength interchanger to service any valid demand irrespective of how the previous assignments of wavelength interchangers to demands was performed.

The question of whether weakening the nonblocking constraint on traditional (i.e. non-WDM) cross-connects allows for less complex designs has been well studied. In traditional cross-connect design, the goal is to minimize the size of the cross-connect (i.e. the number of edges in the directed graph representing the connectivity in the cross-connect), and nonblocking properties are concerned with being able to route a valid demand avoiding edges used by previous demands. In this case, it has been shown that $\Omega(k \log k)$ is a lower bound on the size of a wide-sense nonblocking cross-connect [21] (in fact, it is actually shown to be a lower bound for the weaker rearrangeably nonblocking constraint). Also, it is known that $O(k \log k)$ is an upper bound for strictly nonblocking cross-connects [2] (and hence also an upper bound for wide-sense nonblocking cross-connects). That is, these bounds are tight (up to a constant factor) for both wide-sense and strictly nonblocking cross-connects and so there is no reduction in the size of a cross-connect to be gained by relaxing the nonblocking constraint to wide-sense. However, for more general kinds of demands (e.g. multicast demands, as in [3, 9]), there is a reduction in the required size of a wide-sense nonblocking cross-connect compared to a strictly nonblocking cross-connect.

Thus in the case of WDM cross-connects, it is again natural to study whether the weaker nonblocking properties allow for more efficient designs. It has been shown that for rearrangeably nonblocking split WDM cross-connects, k wavelength interchangers are necessary and sufficient [23]. The stronger property of strictly nonblocking was shown to have upper and lower bounds of $2k-1$ wavelength interchangers [19]. We consider the question of where the bounds for the intermediate case of wide-sense nonblocking lie. We begin by defining an equivalent graph edge coloring problem and then present our technical results in terms of the edge coloring problem.

3 The Graph-Theoretic Model

A *graph* here consists of a set of nodes and a multiset of edges; if there is at most one edge for each pair of nodes, we say that the graph is *simple*. In a *dynamic graph* edges appear and disappear over time (see Section 5 for the precise definition). An algorithm for edge coloring a dynamic graph must assign a color to each new edge presented without any knowledge of future additions or deletions. At all times adjacent edges must have different colors. The goal of the algorithm is to minimize the total number of colors ever assigned to edges of the graph.

Determining the number of necessary wavelength interchangers in a wide-sense nonblocking cross-connect can be cast as an edge coloring problem for dynamic bipartite graphs of fixed maximum degree. The set A of nodes on the left side of the graph represents the set of wavelengths available on each of the input fibers; the set B of nodes on the right side represents the set of wavelengths available on the output fibers. We define $n = \max(|A|, |B|)$ to be the *size* of the graph. (Normally the set of input wavelengths and the set of output wavelengths are the same, thus the bipartition is balanced.) An edge $\{u, v\}$ is present if there is currently a request on the cross-connect from some input fiber to some output fiber such that the request starts on wavelength u and ends on wavelength v. The color assigned to the edge represents the wavelength interchanger that the demand is routed through.

Since the inexpensive part of the cross-connect can route channels to and from the wavelength interchangers arbitrarily, the identities of the input and output fibers for a particular demand are not needed for the graph model. However, the *number* of input (or output) fibers is critical because it bounds the number of channels of a given wavelength, thus the degree of the graph.

Figure 2 shows a 2×2 cross-connect with 3 wavelength interchangers handling 4 wavelengths. Figure 3 shows the corresponding graph, with $n = 4$ nodes

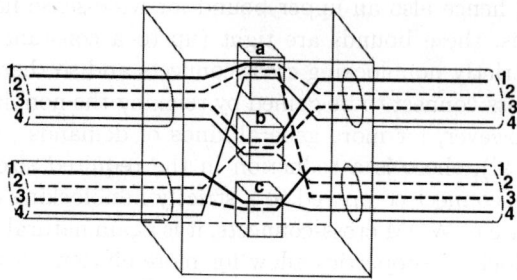

Fig. 2. A 2 × 2 cross-connect with 4 wavelengths and 3 wavelength interchangers

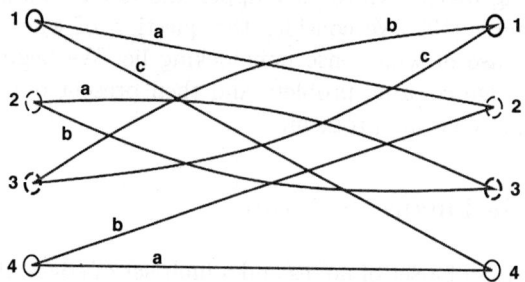

Fig. 3. The graph corresponding to the cross-connect in Figure 2

on a side and maximum degree $\Delta = 2$. We have labeled the wavelengths (and nodes) numerically, and the wavelength interchangers (and colors) by letters a, b and c. Table 1 shows a mapping of terminology and notation between WDM cross-connects and the graph model.

Obviously at least Δ colors will be necessary to edge color a dynamic graph if a node can have as many as Δ incident edges; on the other hand with $2\Delta - 1$ colors the colorer can employ any strategy and will never be stymied. We show in Section 6.1 that for every edge coloring algorithm there exists a dynamic graph with only $O(\Delta^2)$ nodes that requires $2\Delta - 1$ colors. This result holds also for simple graphs.

This raises the question of whether smaller dynamic graphs require fewer than $2\Delta - 1$ colors. In Section 6.2 we consider bipartite dynamic graphs with very few nodes. These graphs can be edge colored with substantially fewer than $2\Delta - 1$ colors. In particular, dynamic graphs with 2 nodes on each side of the bipartition can be edge colored with $3\Delta/2$ colors and this bound is tight. For bipartite dynamic graphs with 3 nodes on each side, $15\Delta/8$ colors suffice and there is a lower bound of $7\Delta/4$. Due to space constraints, we omit all proofs in Section 6.2.

This leaves open the question of how many colors are necessary when the number of nodes in the graph is between a small constant and $O(\Delta^2)$. In Section 6.3 we provide a lower bound of $2(1-\varepsilon)\Delta$ colors for graphs with at least $1/\varepsilon^2$ nodes.

Table 1. Mapping of terminology and notation of WDM cross-connects and graph model

WDM cross-connects	graph model
$\Lambda = \#$ of wavelengths	$n = \max(|A|, |B|)$
demand	edge
$k = \#$ input/output fibers	$\Delta =$ maximum node degree
$\#$ wavelength interchangers	$\#$ edge colors

4 Edge Coloring of Graphs

A proper edge coloring of a graph G requires that adjacent edges be assigned distinct colors. The minimum number of colors needed to color the edges of a graph, usually called the chromatic index, is a classical graph parameter that has been studied extensively; see e.g. [4] or [12]. König [15] showed that every bipartite graph with maximum degree Δ has chromatic index at most Δ; Vizing [22] proved that every *simple* graph with maximum degree Δ has chromatic index either Δ or $\Delta+1$. Even so, for many classes of graphs, determining the exact chromatic index is NP-complete [11, 5, 10].

More recently this problem has been considered in settings in which the entire graph is not known in advance. One such body of work considers *constrained* edge colorings in which a partially colored graph is given as input. The remaining edges must be legally colored without ever re-coloring any edges [13, 14, 7, 16, 17, 6]. In the more standard version of the on-line edge coloring problem, the graph is presented one edge (or node) at a time and each edge must be colored by the algorithm as it is presented. Favrholdt and Nielsen [8] consider on-line edge coloring with a fixed number of colors, the goal being to color as many edges as possible. Bar-Noy et al. [1] provide a graph of maximum degree Δ and $O(\binom{2\Delta-1}{\Delta})$ nodes for which any on-line edge coloring algorithm requires $2\Delta-1$ colors; this implies that for graphs with n nodes and $O(\log n)$ maximum degree one can do no better than the greedy on-line coloring algorithm in the worst-case.

5 Problem Definitions

Let $\delta_E(v)$ be the degree of node v given a set E of edges. Given a set N of nodes and a maximum degree Δ define a set of edges E to be *valid* if and only if

1. for any $\{u, v\} \in E$, $u, v \in N$;
2. for any $v \in N$, $\delta_E(v) \leq \Delta$.

Let the *edge sequence* $\mathcal{E} = (E_1, E_2, \ldots)$ be a sequence of valid edge sets E_i. We define a dynamic graph $\mathcal{G}(N, \Delta, \mathcal{E})$ to be the sequence of graphs $(G_1, G_2 \ldots)$ such that $G_t = (N, E_t)$. We assume that $\mathcal{G}$ starts as just the set N of nodes and define $E_0 = \emptyset$ to be the initial set of edges.

A coloring C assigns a color $C(e)$ to each edge e. We define $\mathcal{C} = (C_1, C_2 \ldots)$ to be a *proper coloring* of $\mathcal{G}$ if for any i

1. C_i is a proper coloring of G_i;
2. for any e such that $e \in E_{i-1}$ and $e \in E_i$, $C_{i-1}(e) = C_i(e)$.

If the node set N consists of "left nodes" from a set A and "right nodes" from a set B, and all edges connect a node from A with a node from B, then the dynamic graph is bipartite and we write $\mathcal{G}(A, B, \Delta, \mathcal{E})$ instead of $\mathcal{G}(N, \Delta, \mathcal{E})$. All of the dynamic graphs we construct here are bipartite, on account of our intended application.

Our lower bounds apply to any algorithm, deterministic or randomized. However, the dynamic graph that witnesses the lower bound may depend on past choices made by the algorithm. Thus it is possible that there exists a randomized algorithm whose *expected* performance is not governed by our lower bounds.

6 New Results

We begin by showing that for every edge coloring algorithm there exists a dynamic graph with $O(\Delta^2)$ nodes that requires $2\Delta - 1$ colors.

6.1 A Lower Bound for Polynomial Size Graphs

We show that for any edge coloring algorithm there is a dynamic bipartite graph $\mathcal{G}(A, B, \Delta, \mathcal{E})$ with $\max(|A|, |B|) = n = (\frac{1}{4} + o(1))\Delta^2$ such that the algorithm must use $2\Delta - 1$ colors to edge color $\mathcal{G}$.

Define the *spectrum* $S(v)$ of a node v to be the set of colors of its incident edges. We say that a node is *full* if its degree is Δ, i.e. $|S(v)| = \Delta$.

We begin by showing that if there is some edge sequence that at some time results in a graph with a particular property then for any edge coloring algorithm we can construct an edge sequence that will require $2\Delta - 1$ colors.

Lemma 1. *Let $\mathcal{G} = \mathcal{G}(A, B, \Delta, \mathcal{E})$ be a dynamic bipartite graph that has been colored so that at a certain stage j, $G_j(A, B, E_j)$ has $m = 1 + \lceil \log_2 \Delta \rceil$ full nodes with the same spectrum S where*

1. *one of the m points (say, x) is on one side of the bipartition, the rest $(y_1, \ldots, y_{m-1})$ are all on the other side;*
2. *there are currently no edges between x and any of the y_i.*

Then for any edge coloring algorithm there is an edge sequence $\mathcal{E}' = (E'_1, E'_2, \ldots)$ where $E'_r = E_r$ for $r = 1, 2, \ldots, j$ such that the algorithm will require $2\Delta - 1$ colors to edge color $\mathcal{G}(A, B, \Delta, \mathcal{E}')$.

Proof. The edge sequence E' is defined for E_r, $1 \le r \le j$ and now we describe which edges are to be deleted and which edges are to be added to progress from E'_i to E'_{i+1} for $i \ge j$. Consider any edge coloring algorithm and let C'_i be the coloring of the edges of E'_i by that algorithm.

Define $S_i(v)$ to be the spectrum of node v according to C'_i. Let $T_i := S_i(x) \cap S$ and $t_i := \lfloor |T_i|/2 \rfloor$. Let X_i and Y_i be disjoint subsets of T_i of size t_i. To define

E'_{i+1} from E'_i, delete the edges of E'_i incident to x whose colors lie in X_i, and the edges of E'_i incident to y_{i-j} whose colors lie in Y_i. Notice that even after removing these edges, the union of the current spectrums of x and y_{i-j} still contains S. Replace the removed edges by t_i new edges from x to y_{i-j}. Notice that all of their colors must lie outside the set S, thus $S_{i+1}(x)$ now has t_i new colors.

Thus C'_{i+1} is such that $|T_{i+1}| = \lceil |T_i|/2 \rceil$ so that after precisely $m-1 = \lceil \log_2 \Delta \rceil$ stages we have $|T_{j+m-1}| = 1$, thus $|S_{j+m-1}(x) \cup S| = 2\Delta - 1$. The total number of nodes originally required is thus $m = 1 + \lceil \log_2 \Delta \rceil$. □

Now what remains for us to show is that for any edge coloring algorithm we can define $E_1, E_2, \ldots, E_j$ so that the edge coloring algorithm will either use all $2\Delta - 1$ colors or force the conditions stated in Lemma 1. In what follows, when we speak of edges being colored we mean by some arbitrary edge coloring algorithm.

We choose opposing nodes x and y, connect by Δ edges, and name the resulting colors 1 through Δ. These will be called the *light* colors. Colors Δ through $2\Delta - 1$ are said to be *dark*. Note that color Δ is both light and dark and thus there are at most $\Delta - 1$ colors that are only light and at most $\Delta - 1$ colors that are only dark.

Next we choose Δ new points on each side, say u_1 through u_Δ on the left and v_1 through v_Δ on the right, and connect completely to form a copy of the complete $\Delta \times \Delta$ bipartite graph, $K_{\Delta,\Delta}$. Suppose first that at least 3/4 of the colors of these Δ^2 edges are dark.

Then we proceed to construct a point with completely dark spectrum, which, in combination with the $\{x, y\}$ edges, uses all colors. To do this, we direct our attention to new nodes $w_1, w_2, \ldots, w_k$ on the right, and successively connect all the u_i's to w_1, then to w_2 etc., each time discarding the light edges and keeping the dark ones. Since there are only $\Delta - 1$ colors that are not dark and we add Δ edges to each w_i, at least one new dark edge must appear each time. Hence, after $k = \Delta^2/4$ iterations some u_i must be full and dark, and so all $2\Delta - 1$ colors are being used.

If, on the other hand, more than 1/4 of the $K_{\Delta,\Delta}$ edge colors are light, we construct instead a light point. Again we successively connect all the u_i's to $w_1, w_2, \ldots, w_k$, this time discarding the dark edges and keeping the light ones. Here, however, we gain at least 3 light edges per w_i since if w_i has only one light incident edge we already have all the colors, and if it has just two light colors then we do the following: we remove one of the light edges $\{x, y\}$ that has the same color as one of the edges incident with w_i. We delete the other edge incident with w_i having a light color. Note that the spectra of x and w_i are disjoint and together consist of $2\Delta - 2$ colors. We then add a new edge $\{x, w_i\}$ and this edge must be given a color different from the $2\Delta - 2$ colors in the spectra of x and w_i and so all $2\Delta - 1$ colors will be used.

Hence after only $k = (\frac{1}{3})(3\Delta^2/4)$ iterations we have a light point, say u_1, among the u_i's; replace u_1 by u'_1 and continue until there are $1 + \lceil \log_2 \Delta \rceil$ light points. Each time a light point is replaced Δ light edges are lost from the current batch of u_i's and u'_i's, so altogether $\Delta^2/4 + \Delta \lceil \log_2 \Delta \rceil$ w_i's may be needed.

Either way we have $n = (\frac{1}{4} + o(1))\Delta^2$.

Combining this construction with Lemma 1 (using x or y as the "lonely" point, as necessary) allows us to conclude the following.

Theorem 1. *For any edge coloring algorithm, there is a dynamic bipartite graph $\mathcal{G}(A, B, \Delta, \mathcal{E})$ with $\max(|A|, |B|) = n$ where $n = (\frac{1}{4} + o(1))\Delta^2$ such that the edge coloring algorithm must use at least $2\Delta - 1$ colors to color $\mathcal{G}$. (Thus, in particular, a $k \times k$ wide-sense nonblocking cross-connect in an n-wavelength system requires a full complement of $2k-1$ wavelength interchangers.)*

A similar argument gives the same result for simple graphs, however the constant factor in $n = O(\Delta^2)$ is larger than $1/4 + o(1)$.

Given the result of Theorem 1 it is natural to ask whether for all edge coloring algorithms and all n, there exists a dynamic graph of size n with maximum degree Δ that requires $2\Delta - 1$ colors to be edge colored. The following results show that for small graphs strictly fewer than $2\Delta - 1$ colors are sufficient.

6.2 Small Bipartite Graphs Need Fewer Colors

We address the case of dynamic bipartite graphs $\mathcal{G}(A, B, \Delta, \mathcal{E})$ with $n = |A| = |B|$ nodes where n is small, i.e. $n = 2$ or $n = 3$. For $n = 2$ there is an algorithm that uses at most $3\Delta/2$ colors and for $n = 3$ there is an algorithm that uses at most $15\Delta/8$ colors. We then consider lower bounds for these two cases. We assume throughout this section that Δ is divisible by 8. We omit all proofs in this section due to space constraints.

Theorem 2. *If $\mathcal{G}(A, B, \Delta, \mathcal{E})$ has $|A| = |B| = 2$ then it can be edge colored with $3\Delta/2$ colors.*

Theorem 3. *If $\mathcal{G}(A, B, \Delta, \mathcal{E})$ has $|A| = |B| = 3$, then it can be edge colored with $15\Delta/8$ colors.*

The next results show that our upper bound on the number of colors sufficient to edge color 2×2 dynamic bipartite graphs is tight whereas there remains a gap for 3×3 dynamic bipartite graphs.

Theorem 4. *For any edge coloring algorithm, there exists a dynamic bipartite graph $\mathcal{G}(A, B, \Delta, \mathcal{E})$ where $|A| = |B| = 2$ for which the edge coloring requires at least $3\Delta/2$ colors.*

Theorem 5. *For any edge coloring algorithm, there exists a dynamic bipartite graph $\mathcal{G}(A, B, \Delta, \mathcal{E})$ where $|A| = |B| = 3$ such that the edge coloring algorithm must use at least $7\Delta/4$ colors.*

The results of this section lead to the following question: How many colors are necessary and sufficient to edge color all dynamic graphs with maximum degree Δ and $o(\Delta^2)$ nodes?

6.3 A Lower Bound

Theorem 6. *For any edge coloring algorithm, any $\varepsilon > 0$ and $\Delta > 1/2\varepsilon$, there exists a dynamic bipartite graph with fewer than $1/\varepsilon^2$ nodes on each side that requires the algorithm to use more than $2(1-\varepsilon)\Delta$ colors.*

Note that the lower bound on Δ is necessary, since if $\Delta \leq 1/2\varepsilon$ then $2(1-\varepsilon)\Delta \geq 2\Delta - 1$, and we can never force more than $2\Delta - 1$ colors.

Proof. In view of Theorems 4 and 5 we may assume $\varepsilon < \frac{1}{8}$, thus $\Delta > 4$ and $2(1-\varepsilon)\Delta > 2(1 - \frac{1}{8})4 > 7$. Let us fix ε and Δ accordingly, put $q = \lceil \varepsilon \Delta \rceil$ and let C be a set of $\lfloor 2(1-\varepsilon)\Delta \rfloor$ colors. We will construct a dynamic graph whose edge coloring from C leads to a contradiction. The method employs a variation of part of the proof of Theorem 1, and indeed by letting Δ be a function of ε one could deduce a weaker form of that theorem.

We consider first the case in which $\lceil 2\varepsilon\Delta \rceil$ is even, thus equal to $2q$; then $|C| = 2\Delta - 2q$. Let $m = \lceil 1/\varepsilon \rceil < 8\varepsilon/7$. Let $X := \{x_1, \ldots, x_m\}$ be a set of m left nodes and $Y := \{y_1, \ldots, y_m\}$ a set of right nodes, each x_i matched by Δ parallel edges to y_i. When these edges have been colored we choose $A \subset C$ with $|A| = |C|/2 = \Delta - q$ so as to maximize the number of edges colored by A. All edges *not* colored from A are now removed.

Let s be the sum of the degrees of the nodes in X, so that $s \geq m\Delta|A|/|C| \geq m\Delta/2$ by choice of A. Since no node can have more than $\Delta - q$ incident edges colored by A, we also have $s \leq m(\Delta - q) \leq m\Delta - \Delta$; thus there are at least Δ places for new edges to be introduced, incident to nodes in X.

Next we consider a new node z_1 on the right, adding a full complement of Δ edges between z_1 and X and then deleting all edges that did not get colored by colors in A. Since at least $\Delta - (|C| - |A|) = q$ of the edges incident to z_1 must have been A-colored, the degree sum s will increase by at least q.

We now repeat the operation with more right-hand nodes $z_2, z_3, \ldots, z_t$ where $t = \lceil 4/7\varepsilon^2 \rceil \geq m/2\varepsilon$. Then

$$s \geq m\Delta/2 + tq$$
$$\geq m\Delta/2 + \frac{m}{2\varepsilon}\varepsilon\Delta$$
$$\geq m\Delta > m\Delta - \Delta ,$$

an impossibility.

When $\lceil 2\varepsilon\Delta \rceil$ is odd, thus equal to $2q-1$, we have to give away a bit more. Then $|C| = 2\Delta - 2q + 1$; and $q \geq 2$ since by assumption $2\varepsilon\Delta > 1$, and it follows that $q - 1 > \frac{2}{3}\varepsilon\Delta$.

Let $m = \lceil 3/2\varepsilon \rceil \leq \frac{13}{8}\varepsilon$, and select A as above but with $|A| = \Delta - q + 1 > |C|/2$. We now have $s > m\Delta/2$ and $s \leq m|A| = m\Delta - m(q-1) \leq m\Delta - (3/2\varepsilon)(\frac{2}{3}\varepsilon\Delta) \leq m\Delta - \Delta$, so again there are at least Δ places for new edges to be introduced.

As before $\Delta - (|C| - |A|) = q$ of the edges incident to each successive z_i must be colored by A, so the degree sum increases by at least q each time. Taking $t = 13/16\varepsilon^2 > 2m/\varepsilon$ now causes the contradiction in the same manner as above. □

A simpler version of the proof of Theorem 6 above shows that when Δ is large relative to $1/\varepsilon$, we get a stronger but asymptotic result, namely that $(1 + o(1))/2\varepsilon^2$ nodes on a side suffice to force more than $2(1-\varepsilon)\Delta$ colors.

A similar construction shows that for any edge coloring algorithm, there is a simple dynamic bipartite graph with $|A| = |B| \geq \max\left[2(1-\varepsilon)\Delta, (1 + \frac{(1-\varepsilon)}{\varepsilon}\Delta)\right]$ that requires the algorithm to use $2(1-\varepsilon)\Delta$ colors.

7 Conclusions and Future Work

We have presented a variety of results concerning the number of wavelength interchangers needed in a wide-sense nonblocking $k \times k$ WDM cross-connect by considering the problem of edge coloring dynamic graphs with maximum degree $\Delta = k$. In particular, for the case of 2 wavelengths, the necessary and sufficient number of wavelength interchangers is $3k/2$. When there are 3 wavelengths, a lower bound of $7k/4$ and an upper bound of $15k/8$ was given. However we showed that if there are about $k^2/4$ or more wavelengths then $2k-1$ wavelength interchangers are necessary. Thus in this case, the greedy algorithm is optimal. Furthermore, this implies that weakening the nonblocking capability from strictly nonblocking to wide-sense nonblocking does not reduce the number of wavelength interchangers needed and hence does not reduce the cost of the cross-connect. We have also shown that for any $\varepsilon > 0$ and $k > 1/2\varepsilon$, if there are at least $1/\varepsilon^2$ wavelengths then $2(1-\varepsilon)k$ wavelength interchangers are necessary.

The major remaining question would be to determine the number of wavelength interchangers necessary and sufficient for such cross-connects supporting $o(k^2)$ wavelengths. In particular, it would be interesting to know the smallest number of wavelengths so that any wide-sense nonblocking $k \times k$ WDM cross-connect supporting these wavelengths would require $2k-1$ wavelength interchangers. We would also like to know whether it is true that for every fixed number of wavelengths, say Λ, there is some $c < 2$ such that there is a wide-sense nonblocking $k \times k$ WDM cross-connect supporting Λ wavelengths with only ck wavelength interchangers.

Along the lines of [18] one could consider using less powerful wavelength interchangers (e.g. those with the ability to swap only two wavelength channels leaving the others fixed) and ask how using such weaker wavelength interchangers affects the number of wavelength interchangers needed in a wide-sense nonblocking cross-connect.

References

[1] A. Bar-Noy, R. Motwani, and J. Naor. The greedy algorithm is optimal for on-line edge coloring. *Information Processing Letters*, 44(5):251–253, 1992.

[2] L. A. Bassalygo and M. S. Pinsker. Complexity of an optimal nonblocking switching network without reconnections. *Problems Inform. Transmission*, 9:64–66, 1974.

[3] L. A. Bassalygo and M. S. Pinsker. Asymptotically optimal networks for generalized rearrangeable switching and generalized switching without rearrangement. *Problemy Peredachi Informatsii*, 16:94–98, 1980.
[4] C. Berge. *Graphs and Hypergraphs*. North Holland, Amsterdam, 1973.
[5] L. Cai and J. A. Ellis. NP-completeness of edge-coloring some restricted graphs. *Discrete Appl. Math.*, 30:15–27, 1991.
[6] I. Caragiannis, C. Kaklamanis, and P. Persiano. Edge coloring of bipartite graphs with constraints. *Theoretical Computer Science*, 270(1-2):361–399, 2002.
[7] T. Erlebach, K. Jansen, C. Kaklamanis, M. Mihail, and P. Persiano. Optimal wavelength routing on directed fiber trees. *Theoretical Computer Science*, 221(1-2):119–137, 1999.
[8] L. Favrholdt and M. Nielsen. On-line edge coloring with a fixed number of colors. In *Foundations of Software Technology and Theoretical Computer Science*, pages 106–116, Dec. 2000.
[9] P. Feldman, J. Friedman, and N. Pippenger. Wide-sense nonblocking networks. *SIAM J. Disc. Math.*, 1(2):158–173, 1988.
[10] M. R. Garey and D. S. Johnson. *Computers and Intractability: A Guide to the Theory of NP-Completeness*. W. H. Freeman and Co., San Francisco, CA, 1979.
[11] I. Holyer. The NP-completeness of edge-coloring. *SIAM Journal of Computing*, 10:718–720, 1981.
[12] T. R. Jensen and B. Toft. *Graph Colouring Problems*. Wiley, New York, 1995.
[13] C. Kaklamanis and P. Persiano. Efficient wavelength routing on directed fiber trees. In *Proceedings of the 4th European Symposium on Algorithms (ESA '96)*, pages 460–470, 1996.
[14] C. Kaklamanis, P. Persiano, T. Erlebach, and K. Jansen. Constrained bipartite edge coloring with applications to wavelength routing. In *Proceedings of the 24th International Colloquium on Automata, Languages, and Programming (ICALP '97)*, pages 493–504, 1997.
[15] D. König. Graphok és alkalmazásuk a determinánsok és a halmazok elméletére (in Hungarian). *Mathematikai és Természettudományi Értesítő*, 34:104–119, 1916.
[16] V. Kumar and E. J. Schwabe. Improved access to optical bandwidth. In *Proceedings of 8th ACM-SIAM Symposium on Discrete Algorithms(SODA '97)*, pages 437–444, 1997.
[17] M. Mihail, C. Kaklamanis, and S. Rao. Efficient access to optical bandwidth. In *Proceedings of the 36th Annual IEEE Symposium on the Foundations of Computer Science (FOCS '95)*, pages 548–557, 1995.
[18] R. Ramaswami and G. H. Sasaki. Multiwavelength optical networks with limited wavelength conversion. In *Proceedings of IEEE INFOCOM*, volume 2, pages 489–498, 1997.
[19] A. Rasala and G. Wilfong. Strictly non-blocking WDM cross-connects. In *Proceedings of Symposium on Discrete Algorithms (SODA '00)*, pages 606–615, 2000.
[20] A. Rasala and G. Wilfong. Strictly non-blocking WDM cross-connects for heterogreous networks. In *Proceedings of Symoposium on Theory of Computation (STOC '00)*, 2000.
[21] C. E. Shannon. Memory requirements in a telephone exchange. *Bell System Tech. J.*, 29:343–349, 1950.
[22] V. G. Vizing. On an estimate of the chromatic class of a p-graph (in Russian). *Diskret. Analiz*, 3:23–30, 1964.
[23] G. Wilfong, B. Mikkelsen, C. Doerr, and M. Zirngibl. WDM cross-connect architectures with reduced complexity. *Journal of Lightwave Technology*, pages 1732–1741, October 1999.

Efficient Implementation of a Minimal Triangulation Algorithm

Pinar Heggernes and Yngve Villanger

Department of Informatics, University of Bergen
N-5020 Bergen, Norway
{pinar,yngvev}@ii.uib.no

Abstract. LB-triang, an algorithm for computing minimal triangulations of graphs, was presented by Berry in 1999 [1], and it gave a new characterization of minimal triangulations. The time complexity was conjectured to be $O(nm)$, but this has remained unproven until our result. In this paper we present and prove an $O(nm)$ time implementation of LB-triang, and we call the resulting algorithm LB-treedec. The data structure used to achieve this time bound is tree decomposition. We also report from practical runtime tests on randomly generated graphs which indicate that the expected behavior is even better than the proven bound.

1 Introduction

Many important graph problems are concerned with adding edges to a given arbitrary graph to obtain a chordal supergraph, and the resulting chordal graph is called a *triangulation* of the given graph. A triangulation H of G is *minimal* if no subgraph of H is a triangulation of G. The first algorithms for computing minimal triangulations appeared more than 25 years ago [12], [15] with $O(nm)$ time complexity, where n is the number of vertices and m is the number of edges of the input graph. After a time gap of 20 years, the problem began to be restudied [3], [4], [7], [14] in the following sandwich version: given an arbitrary triangulation H of G, compute a minimal triangulation M of G with $G \subseteq M \subseteq H$. The best known theoretical time complexity of computing minimal triangulations, with or without the sandwich property, has remained as $O(nm)$.

Algorithm LB-triang [1] presented a new characterization of minimal triangulations, and also solved the sandwich problem, allowing the order in which the vertices are processed as input. The time complexity of LB-triang was conjectured to be $O(nm)$, however this remained unproven since its presentation. (The straight forward implementation suggested in [1] requires $O(nm')$ time, where m' is the number of edges in the resulting triangulation).

In this paper, we prove that Algorithm LB-triang can be implemented in $O(nm)$ time using a a tree decomposition of the input graph. In the start the tree decomposition consists of only one tree node containing all the vertices of G. At each step, the tree decomposition is refined by encountering and inserting minimal separators of G into the data structure. At the end, the data structure contains a clique tree of the computed minimal triangulation of G. We call this new $O(nm)$ time and $O(m')$ space algorithm LB-treedec. In addition to a theoretical time complexity analysis of our algorithm, we also present runtime results from a practical implementation.

This extended abstract is organized as follows. In the next section we give the necessary background. Section 3 presents Algorithm LB-treedec and contains the main results of this paper. Section 4 concludes the paper and mentions open questions and future research directions.

2 Background

A graph is denoted $G = (V, E)$, with $n = |V|$, and $m = |E|$. All graphs that we work on are connected and simple. $G(A)$ is the subgraph induced by a vertex set $A \subset V$, but we often denote it simply by A when there is no ambiguity. A *clique* is a set of vertices that are all pairwise adjacent. For all the following definitions, we will omit subscript G when it is clear from the context which graph we work on. The *neighborhood* of a vertex x in G is $N_G(x) = \{y \neq x \mid xy \in E\}$. The neighborhood of a set of vertices A is $N(A) = \cup_{x \in A} N(x) - A$, and we define $N[A] = N(A) \cup A$.

For a connected graph $G = (V, E)$ with $X \subseteq V$, $\mathcal{C}_G(X)$ denotes the set of connected components of $G(V - X)$. $S \subset V$ is called a *separator* if $|\mathcal{C}(S)| \geq 2$, an xu-*separator* if vertices x and u are in different connected components of $\mathcal{C}(S)$, a *minimal xu-separator* if S is an xu-separator and no proper subset of S is an xu-separator, and a *minimal separator* if there is some pair $\{x, u\}$ such that S is a minimal xu-separator. Equivalently, S is a minimal separator if there exist C_1 and C_2 in $\mathcal{C}(S)$ such that $N(C_1) = N(C_2) = S$. Two separators S and S' are *crossing* if there exist two components $C_1, C_2 \in \mathcal{C}(S)$, $C_1 \neq C_2$, such that $S' \cap C_1 \neq \emptyset$ and $S' \cap C_2 \neq \emptyset$.

A *chord* of a cycle is an edge connecting two non-consecutive vertices of the cycle. A graph is *chordal*, or *triangulated*, if it contains no chordless cycle of length ≥ 4. Edges that are added to an arbitrary graph G to obtain a triangulation H of G are called *fill edges*, with $m' = |E(H)|$.

Theorem 1. (Lekkerkerker and Boland [11]) *A graph G is chordal iff for every vertex x in G, every minimal separator contained in $N_G(x)$ is a clique.*

Algorithm LB-triang (Berry [1])
Input: A graph $G = (V, E)$, and an order $v_1, v_2, ..., v_n$ on the vertices of G.
Output: A minimal triangulation $H = (V, E + F)$ of G.
begin
    $H = G$; $F = \emptyset$;
    for $i = 1$ **to** n **do**
        for each connected component C in $\mathcal{C}_G(N_H[v_i])$ **do**
            Make $N_G(C)$ into a clique by adding edge set F';
            $F = F + F'$; $H = (V, E + F)$;
end

It is shown in [2] that the set of minimal separators included in the neighborhood of a vertex x is exactly $\{N(C) \mid C \in \mathcal{C}(N[x])\}$. Thus with help of Theorem 1 it can easily be shown that LB-triang produces a triangulation. The proof that the computed triangulation is minimal relies on results from [10] and [13], and involves showing that the computed minimal separators are pairwise non-crossing. It should be noted that the set of minimal separators of G processed in this way by LB-triang is exactly the set of all minimal separators of the resulting minimal triangulation H.

The interesting part for the purpose of this paper is the time complexity of LB-triang. It is mentioned in [1] that $\mathcal{C}_H(S) = \mathcal{C}_G(S)$ and $N_H(C) = N_G(C)$ for each $C \in \mathcal{C}_G(S)$; thus the component search can be done in G rather than in H, as described in Algorithm LB-triang. Despite this, proving an $O(nm)$ time complexity for LB-triang turned out to be a more difficult task than first expected. The most time consuming part is making the encountered minimal separators into cliques by adding fill edges. These fill edges are not needed in the component search, but they are needed in finding $N_H[v_j]$ at later steps j. The problem is that each minimal separator, and thus each fill edge, can be encountered and inserted several times, which gives the need to keep a sorted list of the minimal separators or the fill edges so that redundant copies can be removed. Such an approach requires $O(nm')$ time.

Our algorithm does not store the edges of each computed minimal separator, but rather stores the minimal separators as vertex sets, since these are all cliques. Computing $N_H[v_i]$ is not straight forward in this setting, and requires scanning of some of the computed minimal separators. Our practical implementation of Algorithm LB-triang relies on two important structures, called tree decompositions and clique trees.

Definition 1. *A* tree-decomposition *of a graph* $G = (V, E)$ *is a pair*
$$(\{X_i \mid i \in I\}, \ T = (I, M))$$
where $\{X_i \mid i \in I\}$ *is a collection of subsets of* V, *and* T *is a tree, such that:*
- $\bigcup_{i \in I} X_i = V$,
- $(u, v) \in E \Rightarrow \exists i \in I$ *with* $u, v \in X_i$, *and*
- *for all vertices* $v \in V$, $\{i \in I \mid v \in X_i\}$ *induces a connected subtree of* T.

Thus each tree node corresponds to a vertex subset X_i, also called a *bag* (in which the graph vertices are placed). We will not distinguish between vertex subsets and their corresponding tree nodes. Consequently, we will refer to the tree T when we mention the corresponding tree decomposition. More about tree decompositions and their importance can be found in [6]. In our implementation, we will let each edge (X, Y) of T contain the set of vertices in $X \cap Y$. Thus we will often refer to edges of T also as vertex subsets.

For chordal graphs, tree decompositions exist where the bags are exactly the maximal cliques of the graph [8]. Such tree decompositions are called *clique trees* [5]. One important property of clique trees which is related to our implementation is the following.

Lemma 1. (Ho and Lee [9]) *Let T be a clique tree of a chordal graph G. Every edge of T is a minimal separator of G, and for every minimal separator S in G, there is an edge $(K_i, K_j) = K_i \cap K_j = S$ in T.*

A chordal graph has at most n maximal cliques and $n-1$ minimal separators, and hence the number of nodes and edges in a clique tree is $O(n)$ [9].

3 LB-Treedec: An $O(nm)$ Time Implementation of LB-Triang

At each step i, Algorithm LB-triang identifies and makes into cliques the minimal separators of G (and H) included in $N_H(v_i)$, where H is the partially filled graph so far. In our implementation these computed minimal separators are stored as vertex lists. Thus computing $N_H(v_i)$ is not straight forward as the fill edges are not actually added. However, at the beginning of step i, it is sufficient to consider $N_G(v_i)$ and the minimal separators computed so far, in order to compute $N_H(v_i)$, since an edge is a fill edge of the transitory H if and only if its endpoints belong to a previously computed separator. Our general approach will be as follows. We start with a tree decomposition T of G consisting of only one bag containing all the vertices of G. At each step, whenever we encounter a new minimal separator S, we check whether S can be inserted into T as an edge to refine the tree decomposition. We will show that if S can be inserted then there is only one bag containing the vertices belonging to S and the vertices that S minimally separates. We split this bag into two bags, insert the separator as an edge between the two new bags, correctly couple the two new bags to the rest of the tree through the neighbors of the old bag, and maintain in this way a tree decomposition of G where the bags get smaller and smaller. This is the intuition behind the implementation, and we will now give the formal details. For the following discussions, let T_x denote the subtree of T induced by all tree nodes containing graph vertex x.

3.1 Data Structures and Implementation Details

We will first rewrite Algorithm LB-triang to operate on the tree decomposition data structure. The new algorithm that thus results is called LB-treedec, where the neighbors of a vertex are found through original edges and the already computed separators, using the tree structure to extract this information. At step i of the algorithm, let $U(i)$ be the union of all minimal separators computed at earlier steps containing vertex v_i, and let U_A be the union of all minimal separators computed so far. Every vertex in $U(i)$ is a neighbor of v_i in the transitory graph H since these minimal separators are cliques in H. Fill edges of H appear only within minimal separators, thus $U(i) \cup N_G(v_i)$ is the set of all neighbors of v_i at step i, including v_i itself if $U(i) \neq \emptyset$. In fact, it can be shown that no fill edge of H incident to v_i is created after step i, thus $N_H[v_i] = U(i) \cup N_G(v_i) \cup \{v_i\}$. Consequently, at every step i, the final adjacency set of v_i in H is computed, and can be inserted into H directly.

Algorithm LB-treedec
Input: A graph $G = (V, E)$, and an order $v_1, v_2, ..., v_n$ on the vertices of G.
Output: A minimal triangulation $H=(V, E+F)$ of G, and a clique tree T of H;
begin
    $H = G$; $T = (\{V\}, \emptyset)$; $U_A = \emptyset$;
    for $i = 1$ **to** n **do**
(1)   Compute $U(i)$ using the separator information stored in the edges of T;
            $N_H[i] = N_G(v_i) \cup U(i) \cup \{v_i\}$;
        for each connected component C in $\mathcal{C}_G(N_H[i])$ **do**
            $S = N_G(C)$; $U_A = U_A \cup S$;
(2)   **if** there is a bag X in T containing v_i and S and a subset of C **then**
                $X_1 = S \cup (X \cap C)$; $X_2 = X \cap (G - C)$;
(3)   Replace X with X_1, X_2, and edge $(X_1, X_2) = S$ in T;
end

Steps of this algorithm marked as (1) - (3) need to be explained further. Step (1) is more involved, and we will describe this step last. For the time being, assume that the neighborhood of v_i in H at step i of the algorithm is correctly found.

Let $S = N_G(C)$ for a component $C \in \mathcal{C}_G(N_H[i])$ found at step i. S is a minimal separator of G and of H, separating v_i and a subset of C. If there is a tree node X containing v_i and S and any vertex of C, then this tree node can be split into two tree nodes and S can be inserted as an edge between the new tree nodes.

Lemma 2. *Let $S = N_G(C)$ for a component $C \in \mathcal{C}_G(N_H[i])$ found at step i. Then there is at most one bag X in the current tree T containing both $(\{v_i\} + S)$ and a subset of C.*

Proof. Remember first that S is a minimal separator separating v_i from C. For every vertex $u \in C$, we know that no previously computed minimal separator S' contains both u and v_i. Otherwise S and S' would be crossing separators, and this would contradict the correctness of Algorithm LB-triang.

Assume now on the contrary that there are two bags X_1 and X_2 that both contain v_i and S and at least one vertex belonging to C. Let u be a vertex of C that belongs to X_1 and x a vertex of C that belongs to X_2. If $u = x$, then we have a contradiction since every tree edge between X_1 and X_2 is a minimal separator containing u and v_i. Thus $u \in X_1 - X_2$, and $x \in X_2 - X_1$. No tree edge on the path between X_1 and X_2 in T contains both u and x, and every tree edge on this path contains S and v_i. Since u and x do not appear together in any bag on this path, there must exist at least one edge S' that does not contain any of u and x. This means that there is a minimal ux-separator S' such that $(\{v_i\} + S) \subseteq S'$. Since S' separates u and x in G, S' must contain some vertex of C. Since S' contains both v_i and a vertex of C, S and S' are crossing, leading to the desired contradiction.

Invariant 1. *At each step of the algorithm, the tree T is a tree decomposition of G and of the resulting minimal triangulation H.*

Proof. We will prove this invariant by induction. The base case is true since a tree with only one tree node containing all vertices of G is a tree decomposition of both G and H.

Let $S = N_G(C)$ for a component $C \in \mathcal{C}_G(N_H[i])$ found at step i. If there is no bag containing v_i and S and a subset of C, no changes need to be done in the tree, and we have still a tree decomposition. Otherwise, let X be the bag of T containing v_i and S and a subset of C. Assume that the invariant is true for T. We will show, by explaining how X is replaced by X_1, X_2 and the edge $(X_1, X_2) = S$, that the tree T' resulting from this operation is a tree decomposition of G and H.

After the tree node X is removed and the tree nodes X_1 and X_2 are inserted, the tree nodes that were previously incident to X must be connected to X_1 or X_2 instead. It can be easily shown that for any edge $(X, Y) = X \cap Y = S'$ previously incident to X, S' is a subset of X_1 or X_2 or both. Thus we simply connect each such edge to that of the new tree nodes that it is a subset of (arbitrary one can be chosen if it is a subset of both; a good idea is to choose the one that results in a smaller diameter of the tree). We leave it to the reader to verify that the new tree T' fulfills the requirements of a tree decomposition, both for G and H.

We will now describe how to search for the tree node X efficiently. We start from a tree node U containing a vertex u of C and do a depth first search until we find a tree node X that contains v_i. If X also contains a vertex of C, then by Lemma 2 we have found the unique tree node that we want to split. If X does not contain a vertex of C, then by the connected subtree property of a tree decomposition, we know that no tree nodes of T further away from U can contain a vertex of C, and we can conclude that the tree does not need to be updated. Every graph vertex u has a pointer to a tree node that contains u. Thus the starting point of the search is decided in constant time. We will in the next section prove that searching for v_i in a tree node at step i is an amortized constant time operation. In addition, the starting points of the searches at step i belong to disjoint components of G, and thus the searches can be done in disjoint subtrees of T by marking the already traversed paths and adding pointers as shortcuts. What now remains to be explained is how to decide whether or not X contains a vertex of C.

Let $S = N_G(C)$ for a component $C \in \mathcal{C}_G(N_H[i])$ found at step i, and let U be a tree node of T containing a vertex of C. Let $(Y, X) = S'$ be the last edge in a depth first search that starts from U in T and reaches a tree node X containing v_i. Then certainly, X contains a vertex of C iff S' contains a vertex of C. Thus, in addition to the above searches, we also need to check the last edges leading to the tree nodes containing v_i when the searches stop. These edges are edges of T with exactly one endpoint in T_{v_i}. Let us denote this set of tree edges by $Border(i)$. The sum of the sizes of these edges is $O(m)$ as will be shown during the time complexity analysis. Therefore, at each step i we can create characteristic vectors for all these edges in $O(m)$ time and space, and delete these at the end of each step. Then checking each vertex of each component C for membership in the desired $Border(i)$ is a constant time operation per check.

We will now describe how to compute the union $U(i)$. The main reason for using the tree data structure is to be able to compute $N_H(v_i)$ efficiently. Note that T_x is a connected subgraph of T for every x at each step of the algorithm, due to Invariant 1. From the construction of T, it is clear that $U(i)$ can be found by computing the union of all the edges of T_{v_i}. However, this may require $O(m')$ time at each step, which we cannot afford. We have to compute this union in a different way. The main idea is that the sum of the sizes of the edges of T_{v_i} is $O(m')$, whereas the sum of the sizes of the edges belonging to $Border(i)$ is $O(m)$. Now we show how these border edges can be used to compute $U(i)$.

Observation 1. *Let u and v be two vertices of G. A separator S containing both u and v is present as an edge in T iff T_u and T_v share an edge $\equiv S$ in T.*

Observation 1 gives an alternative definition of the desired union: $U(i) = \{u \mid T_u \text{ shares an edge with } T_{v_i}\}$. We define the following disjoint vertex sets for step i of the algorithm:

- $Inner(i) = \{u \neq v_i \mid \text{every edge of } T_u \text{ is an edge of } T_{v_i}\}$
- $InnerOuter(i) = \{u \mid T_u \text{ has at least one edge that is an edge of } T_{v_i} \text{ and at least one edge that is not an edge of } T_{v_i}\}$
- $BorderOuter(i) = \{u \mid T_u \text{ has no edges in common with } T_{v_i} \text{ and } T_u \text{ has a node containing } v_i\}$
- $Outer(i) = \{u \mid T_u \text{ has no edges in common with } T_{v_i} \text{ and no node of } T_u \text{ contains } v_i\}$

We can see that $U(i) = Inner(i) \cup InnerOuter(i)$. The reason why we have partitioned the vertices into these subsets is that some of these subsets are less time consuming to compute than others, and we will therefore not compute $Inner(i) \cup InnerOuter(i)$ directly, but through set operations on the listed subsets. The following connection should be clear: $U_A = Inner(i) \cup InnerOuter(i) \cup BorderOuter(i) \cup Outer(i)$.

$Border(i)$ is the set of edges (X, Y) in T such that X contains v_i, and Y does not contain v_i. $Border(i)$ can be computed readily during the depth first searches of step i. The union of all vertices belonging to the tree edges (minimal separators) in $Border(i)$ gives us exactly $(InnerOuter(i) \cup BorderOuter(i))$. However, we need to separate $InnerOuter(i)$ from $BorderOuter(i)$, and also compute $Inner(i)$, since our goal is to compute $U(i) = Inner(i) \cup InnerOuter(i)$.

Let (X, Y) be an edge in $Border(i)$ where X contains v_i, and let u be a graph vertex belonging to $X \cap Y$. For each such vertex u, we need to decide whether $u \in InnerOuter(i)$ or $u \in BorderOuter(i)$. A naive and straight forward approach would be to scan every edge in T_{v_i} incident to X to decide what kind of subtree T_u is. If u does not appear on any such edge of T_{v_i} then $u \in BorderOuter(i)$. Otherwise, $u \in InnerOuter(i)$. Since we cannot afford to scan all the edges of T_{v_i}, we will avoid this by including enough information in each tree node so that scanning edges that belong to $Border(i)$ is enough. Each tree node X containing vertex u has a variable N_u which is the number of the neighboring tree nodes in T that also contain u. (See Figure 1.) Another variable S_u is initialized to -1 when X is created. When the edge (X,Y) is being examined, for each vertex u belonging to this edge, if $S_u < i$ then S_u is updated to i, and C_u is updated to equal $N_u - 1$. If $S_u = i$ then C_u is decremented. Clearly, if C_u reaches 0 during the scanning of the edges in $Border(i)$, u belongs to $BorderOuter(i)$. Otherwise u belongs to $InnerOuter(i)$. After all vertices u belonging to the edges of $Border(i)$ are processed, we will have identified the sets $InnerOuter(i)$ and $BorderOuter(i)$.

It remains to compute $Inner(i)$. Observe that $(Inner(i) \cup Outer(i))$ is readily computed since $Inner(i) \cup Outer(i) = U_A - (InnerOuter(i) \cup BorderOuter(i))$. Since every tree node that an $Inner(i)$ vertex belongs to contains v_i, and no tree node that an $Outer(i)$ vertex belongs to contains v_i, $Inner(i)$ is easily separated from $Outer(i)$. The computation of $U(i)$ at step i of LB-treedec is summarized in the following algorithm:

$U(i) = U_A$; $Inner(i) = U(i)$;
Compute T_{v_i} and $Border(i)$ by depth first search in T;
for each edge $S = (X,Y)$ in $Border(i)$ with $v_i \in X$ **do**
    for each u in S **do**
        if $X.S_u < i$ **then**
            $X.S_u = i$; $X.C_u = X.N_u - 1$;
        else
            $X.C_u = X.C_u - 1$;
            if $X.C_u = 0$ **then** Remove u from $U(i)$;
            Remove u from $Inner(i)$;
for each u in $Inner(i)$ **do**
    Let U be any tree node containing u;
    if $v_i \notin U$ **then** Remove u from $Inner(i)$ and $U(i)$;

We use characteristic vectors to implement $U_A, U(i)$, and $Inner(i)$, so that membership can be tested and changed in constant time. Each of these vectors require $O(n)$ space. Observe that $U(i)$ and $Inner(i)$ are cleared and reused at each step of Algorithm LB-treedec.

We have thus explained the details of Algorithm LB-treedec. Note that all the minimal separators of H are inserted into T during the algorithm by the correctness of LB-triang [1]. From this and from the proof of Invariant 1 it follows that T at the end of the algorithm is actually a clique tree of H.

The data structure details of Algorithm LB-treedec are illustrated in Figure 1. Given the discussion of the implementation and of the complexity analysis, it should be clear how the shown structures work. We would like to stress that tree nodes and tree edges are *not* implemented as characteristic vectors in general. Thus each tree node and each tree edge has simply a sorted list of the vertices it contains. To start with the single tree node requires $\Theta(n)$ space. Every time a tree node is split into two tree nodes because of the insertion of a new minimal separator S, the space requirement increases by $O(|S|)$. Since every inserted

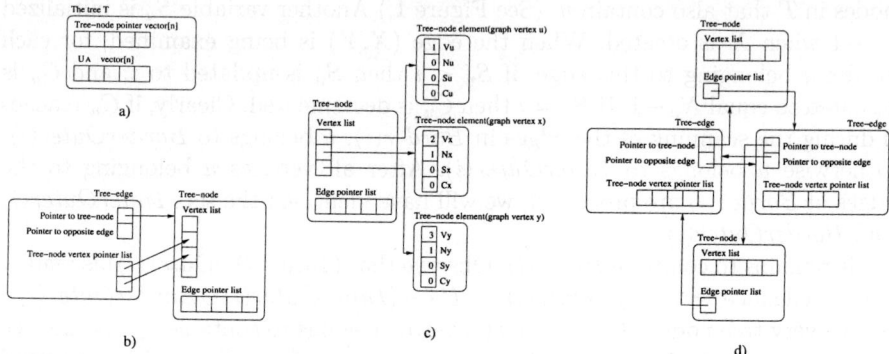

Fig. 1. a) The tree data object where each graph vertex has a pointer to a tree node containing it. b) The interaction between a tree edge and a tree node. c) The relationship between a tree node and the graph vertices contained in it. d) An edge between two tree nodes. Only vectors that end with [n] have lenght exactly n. The length of each of the other lists is the number of vertices it actually contains

minimal separator of H has to be the neighborhood of a distinct component, the sum of the sizes of all the separators of H is $O(m')$. Thus the total space required is $O(m')$ including the mentioned characteristic vectors, assuming that $n \leq m'$.

3.2 Time Complexity Analysis

Observation 2. *The total number of tree nodes created during Algorithm LB-treedec is less than $2n$.*

Lemma 3. *Let k be the total number of all requests asking whether or not a given vertex belongs to a given tree node in T during Algorithm LB-treedec. Then the total time for all requests is $O(n^2 + k)$.*

Proof. In our data structures, each tree node X contains a sorted list of the vertices that belong to X. In addition, there is a pointer p_X in X that points to the most recently requested element in X. Each time we receive a request to check whether or not v_i belongs to X, we scan the vertices in the locations between the previous position of this pointer until we reach v_i or a vertex v_j with $j > i$. The pointer is moved during this scan. Since element v_i is never requested at other steps than i, each tree node X is scanned exactly once. Following requests involving vertex v_i at step i are handled in constant time by simply checking whether p_X is pointing to v_i or to a larger vertex. When a tree node is split into two new nodes, the pointers of the new tree nodes are moved to the beginning of the lists again. Thus in the worst case, each tree node ever created during the algorithm is scanned once. Since at most $2n$ tree nodes are created by Observation 2, the total time is $O(n^2 + k)$.

The number of such requests during the whole algorithm in $O(n^2)$ since the depth first tree searches at each step i are done in disjoint subtrees and require $O(n)$ requests asking whether or not v_i belongs to the tree nodes traversed by the searches. $Border(i)$ is also computed during the same search. However, we need to show that the sum of the sizes of the minimal separators (or tree edges) contained in $Border(i) = O(m)$. Recall that at step i, both deciding which tree nodes that contain v_i should be split, and the computation of $U(i)$ require the scanning of $Border(i)$ tree edges.

Let $I = \{x \mid x$ belongs to a tree node of $T_{v_i}\}$. Consider the graph $G(V - I)$. The number of minimal separators belonging to $Border(i)$ is at most the number of connected components of this graph. Remember that each minimal separator S in $Border(i)$ is defined as $S = N_G(C)$ for distinct connected components C of $G(V - I)$. Thus the sum of the sizes of all involved minimal separators is at most twice the number of edges in G, $O(m)$. Consequently, at step i, the scanning of $Border(i)$ edges and the creation of the mentioned characteristic vectors can be done in $O(m)$ time. In addition, at each step i, we might have to check every vertex of every computed component C for membership in the appropriate $Border(i)$ edge. Since these components are disjoint subgraphs of G, and membership test can be done in constant time for $Border(i)$ edges, this is an $O(n)$ operation for each step i.

We have thus proven the following.

Theorem 2. *Algorithm LB-treedec requires $O(nm)$ time.*

3.3 Practical Runtime Tests

The runtime tests of Algorithm LB-treedec exhibit interesting properties. We have run the code for the implementation on many random graphs and have observed that LB-treedec behaves in an $O(n^2)$ fashion rather than $O(nm)$ in practice. The implementation is coded using C++ and STL, and the practical tests are run on a machine with Intel Pentium III 1GHz processor and 512 MB RAM. Our code can be obtained via anonymous ftp from ftp.ii.uib.no at directory pub/pinar/LB-treedec/.

A classical $O(nm)$ time algorithm is Lex-M [15]. We have also coded this algorithm in C++, and tested its runtime against our LB-treedec. The first chart in Figure 2 shows the results of these tests. We have tested random graphs, all with the same number of vertices, and increasing the number of edges. The x-axis shows the number of edges of each graph as a percentage of all possible edges. As we can see from the first chart in Figure 2 the runtime of LB-treedec is superior to Lex-M except for very sparse graphs. Our runtime is highest when there are few edges in the input graph. As the input graph gets denser, the runtime decreases.

We have then examined when the maximum runtime occurs, and our tests indicate that the graphs that require the maximum runtime have $O(n)$ edges. This is illustrated by the second chart of Figure 2. To do this, we generated, for each number n of vertices shown on the x-axis, connected random graphs

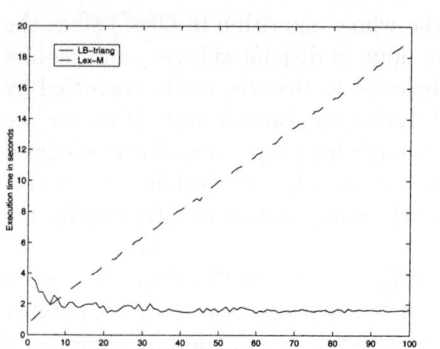
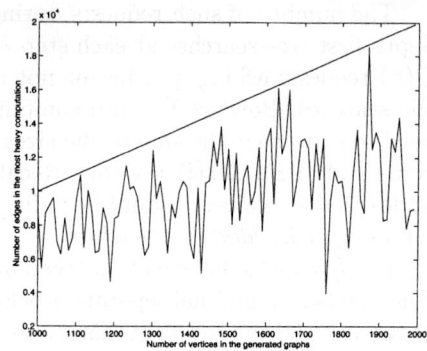

Fig. 2. The chart on the left shows the runtimes of Lex-M and LB-treedec compared on random graphs all with 1000 vertices and increasing number of edges. The chart on the right shows, for numbers $n \in [1000, 2000]$ of vertices, the numbers of edges in the graph on n vertices that requires maximum LB-treedec runtime

with all possible numbers of edges, and examined for each n the number of edges in the graph that required the highest runtime. As can be seen from the chart, the worst runtime occurs when the input graph is sparse. The straight line corresponds to $10n$.

4 Conclusion and Further Work

We have presented Algorithm LB-treedec which is an $O(nm)$ time new version of Algorithm LB-triang. LB-treedec achieves this time bound by utilizing a tree decomposition data structure and carefully placing links between elements that are related. The space complexity of LB-treedec is $O(m')$ where m' is the number of edges of the output graph H. In practical tests, Algorithm LB-treedec exhibits an interesting $O(n^2)$ runtime pattern on randomly generated graphs. In particular, it is faster than the classical Lex-M algorithm of the same theoretical time complexity. However, special graphs on which LB-Treedec requires $\Theta(nm)$ time can be generated, thus an important open problem is whether or not Algorithm LB-triang can be implemented within $O(n^2)$ theoretical time bound. It is still an open question whether or not minimal triangulations can be computed in $O(n^2)$ time in general.

The original Algorithm LB-triang is on-line, which means that the order in which the vertices are processed can be chosen during the course of the algorithm. Algorithm LB-treedec can be made on-line at the cost of more space. If the tree nodes are implemented as characteristic vectors of length n, then membership of each graph vertex in a tree node can be tested in $O(1)$ time. This on-line version requires $O(nm)$ time and $O(n^2)$ space. An interesting question is whether an on-line version can be implemented in $O(nm)$ time and $O(m')$ space.

Every tree decomposition of a graph G corresponds to a triangulation of G. Many important graph theoretical problems search for triangulations with various properties. One of the important NP-hard problems is deciding treewidth, which is equivalent to finding a triangulation where the largest clique is as small as possible. Algorithm LB-treedec gives interesting tools that might be useful for new heuristics for this kind of problems.

Acknowledgment

The authors thank Genevieve Simonet for her useful comments.

References

[1] A. Berry. A wide-range efficient algorithm for minimal triangulation. In *Proceedings of the 10th Annual ACM-SIAM Symposium on Discrete Algorithms*, 1999.
[2] A. Berry, J-P. Bordat, and P. Heggernes. Recognizing weakly triangulated graphs by edge separability. *Nordic Journal of Computing*, 7:164–177, 2000.
[3] J. R. S. Blair, P. Heggernes, and J. A. Telle. Making an arbitrary filled graph minimal by removing fill edges. In R. Karlsson and A. Lingas, editors, *Algorithm Theory - SWAT '96*, pages 173–184. Springer Verlag, 1996. Lecture Notes in Computer Science 1097.
[4] J. R. S. Blair, P. Heggernes, and J. A. Telle. A practical algorithm for making filled graphs minimal. *Theoretical Computer Science*, 250:125–141, 2001.
[5] J. R. S. Blair and B. W. Peyton. An introduction to chordal graphs and clique trees. In J. A. George, J. R. Gilbert, and J. W. H. Liu, editors, *Graph Theory and Sparse Matrix Computations*, pages 1–30. Springer Verlag, 1993. IMA Volumes in Mathematics and its Applications, Vol. 56.
[6] H. L. Bodlaender. A tourist guide through treewidth. *Acta Cybernetica*, 11:1–21, 1993.
[7] E. Dahlhaus. Minimal elimination ordering inside a given chordal graph. In R. H. Möhring, editor, *Graph Theoretical Concepts in Computer Science - WG '97*, pages 132–143. Springer Verlag, 1997. Lecture Notes in Computer Science 1335.
[8] F. Gavril. The intersection graphs of subtrees in trees are exactly the chordal graphs. *J. Combin. Theory Ser. B*, 16:47–56, 1974.
[9] C-W. Ho and R. C. T. Lee. Counting clique trees and computing perfect elimination schemes in parallel. *Inform. Process. Lett.*, 31:61–68, 1989.
[10] T. Kloks, D. Kratsch, and J. Spinrad. On treewidth and minimum fill-in of asteroidal triple-free graphs. *Theoretical Computer Science*, 175:309–335, 1997.
[11] C. G. Lekkerkerker and J. C. Boland. Representation of a finite graph by a set of intervals on the real line. *Fundamenta Mathematicae*, 51:45–64, 1962.
[12] T. Ohtsuki. A fast algorithm for finding an optimal ordering in the vertex elimination on a graph. *SIAM J. Comput.*, 5:133–145, 1976.
[13] A. Parra. *Structural and algorithmic aspects of chordal graph embeddings*. PhD thesis, Technische Universität Berlin, 1996.
[14] B. W. Peyton. Minimal orderings revisited. *SIAM J. Matrix Anal. Appl.*, 23(1):271–294, 2001.
[15] D. J. Rose, R. E. Tarjan, and G. S. Lueker. Algorithmic aspects of vertex elimination on graphs. *SIAM J. Comput.*, 5:266–283, 1976.

Scheduling Malleable Parallel Tasks: An Asymptotic Fully Polynomial-Time Approximation Scheme*

Klaus Jansen

Institut für Informatik und praktische Mathematik
Christian-Albrechts-Universität zu Kiel
24 098 Kiel, Germany
kj@informatik.uni-kiel.de

Abstract. A malleable parallel task is one whose execution time is a function of the number of (identical) processors allotted to it. We study the problem of scheduling a set of n independent malleable tasks on an arbitrary number m of parallel processors and propose an asymptotic fully polynomial time approximation scheme. For any fixed $\epsilon > 0$, the algorithm computes a non-preemptive schedule of length at most $(1 + \epsilon)$ times the optimum (plus an additive term) and has running time polynomial in n, m and $1/\epsilon$.

1 Introduction

In this paper, we study the following scheduling problem. Suppose there is given a set of n tasks $\mathcal{T} = \{T_1, \ldots, T_n\}$ and a set of m identical processors $\{1, \ldots, m\}$. Let $D \subset \{1, \ldots, m\}$ be the set of possible numbers of processors for any task. For example, for a linear array of processors or a PRAM model, $D = \{1, 2, \ldots, m\}$ and for a hypercube with $m = 2^r$ processors, $D = \{2^0, 2^1, \ldots, 2^r\}$ where each number 2^i is the size of a subcube. Each task T_j has an associated function $p_j : D \to \mathbb{R}^+$ that gives the execution time $p_j(\ell) \leq 1$ of that task in terms of the number $\ell \in D$ of processors that are assigned to T_j. Given β_j processors allotted to task T_j, these β_j processors are required to execute task T_j in union and without preemption, i.e. they all have to start processing task T_j at some starting time τ_j, and complete it at $\tau_j + p_j(\beta_j)$. A feasible non-preemptive schedule for the PRAM model consists of a processor allotment $\beta_j \in D$ and a starting time $\tau_j \geq 0$ for each task T_j such that for each time step τ, the number of active processors does not exceed the total number of processors, i.e.

$$\sum_{j : \tau \in [\tau_j, \tau_j + p_j(\beta_j))} \beta_j \leq m.$$

---

* Supported by EU Project APPOL I + II, Approximation and Online Algorithms, IST-1999-14084 and IST-2001-32012, by EU Research Training Network ARACNE, Approximation and Randomized Algorithms in Communication Networks, HPRN-CT-1999-00112 and by DAAD French-German Exchange Program Procope, Scheduling Malleable Tasks.

For other models like linear arrays or hypercubes, the processor assignment for the tasks has to be specified in addition. The objective is to find a feasible non-preemptive schedule that minimizes the overall makespan

$$\max\{\tau_j + p_j(\beta_j) : j = 1, \ldots, n\}.$$

This problem is called malleable parallel task scheduling (MPTS) and denoted by $P|fctn|C_{max}$, and has recently been studied in several papers, see e.g. [2, 8, 14, 22, 27]. The problem of non-malleable parallel task scheduling (NPTS) denoted by $P|size_j|C_{max}$ is a restriction of MPTS in which the processor allotments are known a priori, i.e. for each task both the number of assigned processors and its execution time are given as part of the input. Closely related problems to NPTS are rectangle packing, strip packing (see e.g. [1, 5, 18, 25, 26]) and resource constrained scheduling (see e.g. [3, 11, 10]). Since both NPTS and MPTS are strongly NP-hard even for a fixed number ($m \geq 5$) of processors [8], it is natural to ask how well the optimum for restricted variants can be approximated. Jansen and Porkolab [14] studied the case when there is only a constant number of processors and presented polynomial-time approximation schemes for both MPTS and NPTS which compute for any fixed $\epsilon > 0$ schedules in $O(n)$ time with absolute performance ratio $(1 + \epsilon)$.

Regarding the complexity, the preemptive variant $P|size_j, pmtn|C_{max}$ of NPTS is also NP-hard [6, 8], however the question whether it is strongly NP-hard was open [7, 8]. The problem $Pm|size_j, pmtn|C_{max}$ (with constant number of processors) can be solved in polynomial time [4] by formulating it as a linear program with n constraints and $O(n^m)$ nonnegative variables and computing an optimal solution by any polynomial-time linear programming algorithm. Jansen and Porkolab [15] focused on computing (exact) optimal solutions and presented a linear time algorithm for $Pm|size_j, pmtn|C_{max}$. Furthermore, they proposed an algorithm for solving the general problem $P|size_j, pmtn|C_{max}$ and showed that this algorithm runs in $O(n) + poly(m)$ time, where $poly(.)$ is a univariate polynomial. This gives a strongly polynomial complexity for any m that is polynomially bounded by n, and hence also implies that $P|size_j, pmtn|C_{max}$ cannot be strongly NP-hard, unless P=NP. Furthermore, they focused on preemptive schedules for $P|fctn_j, pmtn|C_{max}$ where migration is allowed, that is task T_j may be assigned to different processor sets of $size_j$ during different execution phases [4, 7, 8]. They extended the algorithm above also to malleable tasks with running time polynomial in m and n. These results are based on methods by Grötschel et al. [13] and use the ellipsoid method. Recently, Jansen and Porkolab [16] found a more practical fully polynomial time approximation scheme for $P|fctn_j, pmtn|C_{max}$ that runs in $O(n^2 m(\epsilon^{-2} + \ln n) \ln(n\epsilon^{-1})(\ln m + n\epsilon^{-1}))$ time.

For a particular input I, let $OPT(I)$ be the minimum makespan, and let $A(I)$ denote the makespan obtained by algorithm A. The absolute performance ratio of A is $sup_I A(I)/OPT(I)$. The asymptotic performance ratio of A is $limsup_{Opt(I) \to \infty} A(I)/OPT(I)$. A fully polynomial approximation algorithm is an approximation algorithm A that runs in time polynomial not only in the length of I but also in $\frac{1}{\epsilon}$. The best currently known absolute approximation ratios for NPTS and MPTS for the PRAM model, linear arrays and hypercubes are 2 [9, 22, 25, 26]. On the other hand 3/2 is the best known absolute approximation ratio for the monotonic variant MMPTS [23] (where $p_j(\ell)\ell \le p_j(\ell')\ell'$ and $p_j(\ell) \ge p_j(\ell')$ holds for each task T_j and pair $\ell \le \ell'$).

Kenyon and Remila [18] proposed an asymptotic polynomial time approximation scheme for strip packing (i.e. packing of rectangles in a strip of width 1 while minimizing the total height) where each rectangle has height at most 1. Since strip packing is equivalent to NPTS with consecutive processor addresses, this implies also an AFPTAS for NPTS where each task has processing time at most 1. In this paper, we focus on asymptotic polynomial time approximation schemes for the non-preemptive malleable parallel task scheduling problem $P|fctn_j|C_{max}$. We show how to use preemptive schedules (with migration) for $P|fctn_j, pmtn|C_{max}$ and a new rounding technique in order to generalize the result by Kenyon and Remila to malleable tasks for the PRAM, linear array and hypercube model. Our main result is the following:

Theorem 1. *There is an algorithm A which, given an instance I with n tasks, m processors and execution times $p_j(\ell) \le 1$ depending on the numbers $\ell \in D \subset \{1,\ldots,m\}$ of processors allotted to tasks T_j, and a positive number $\epsilon > 0$, produces a non-preemptive schedule of the tasks with length*

$$A(I) \le (1+\epsilon)OPT(I) + O(1/\epsilon^2).$$

The running time of A is polynomial in n and $1/\epsilon$ with a preprocessing time of $O(n|D|)$.

The algorithm for MPTS in the PRAM model works as follows. In the first phase we solve approximately the preemptive scheduling problem (where we also allow migration). This can be done by solving a linear program approximately in $O(n(\epsilon^{-2}+\ln n)\ln(n\epsilon^{-1})(n|D|\ln|D|+n^2|D|\epsilon^{-1}))$ time similar to [16]. To improve this running time we use a new preprocessing step where we generate a table of size $O(\min(n|D|, n^2/\epsilon))$ for a given accuracy $\epsilon > 0$. Using the preprocessing step we obtain an approximation scheme to compute a preemptive schedule with running time $O(n(\ln n + \epsilon^{-2})\min(n|D|, n^2\epsilon^{-1})n\epsilon^{-1}\ln(n\epsilon^{-1}) + n|D|)$.

The main new idea is an algorithm to convert the approximate preemptive schedule into an approximate non-preemptive schedule (via computing an unique allotment for almost all tasks). To do this we construct a sequence of wide and narrow rectangles (or fractions of task pieces). The wide rectangles are packed on a stack in non-increasing order of their widths. For the narrow rectangles we store the total area. Then we split the stack into a constant number of groups and use a rounding technique to determine the unique allotment (i.e. unique processor numbers) for almost all tasks. A constant number of tasks may have a fractional assignment; and these tasks are moved to the end of the schedule. In this way we obtain an instance of a fractional strip packing problem for the wide tasks. We show that the fractional strip packing problem can be solved approximately in $O(M(\epsilon^{-2} + \ln M)(M + \epsilon^{-3})\ln(\epsilon^{-1}))$ time where $M = O(\epsilon^{-2})$ is number of different widths of rectangles. Finally we insert the narrow tasks using the next fit decreasing height (NFDH) heuristic. The second phase can be solved in time $O(n^2(\ln n + \epsilon^{-2})\ln(n\epsilon^{-1}) + \epsilon^{-4}\max(\mathcal{M}(\epsilon^{-2}), \epsilon^{-3}\ln(\epsilon^{-1})))$ where $\mathcal{M}(N)$ is the time to invert an $(N \times N)$ matrix. Interestingly, the final non-preemptive schedule computed via preemptive scheduling with migration, rounding and strip packing is approximately close to the length of an optimum non-preemptive schedule.

2 Preemptive Schedules

In this section we study the preemptive multiprocessor task scheduling problem (for the PRAM model), where a set $\mathcal{T} = \{T_1, \ldots, T_n\}$ of n tasks has to be executed by m processors such that each processor can execute at most one task at a time and each task must be processed simultaneously by several processors. In the preemptive model, each task can be interrupted any time at no cost and restarted later possibly on a different set of processors. We will focus on those preemptive schedules where migration is allowed, that is where each task may be assigned to different processor sets during different execution phases [4, 7, 8].

2.1 Linear Programming Formulation

The preemptive version of the multiprocessor task scheduling problem denoted by $P|fctn_j, pmtn|C_{max}$ can be formulated as the following linear program [15], where D denotes the set of different cardinalities that processor sets executing task T_j can have and each configuration f is a mapping $f : \{1, \ldots, m\} \to \{1, \ldots, n\} \cup \{0\}$ between processors and tasks (where 0 denotes a dummy task to model idle times of processors).

$$\text{Min} \sum_{f \in F} x_f$$
$$\text{s.t.} \sum_{\ell \in D} \frac{1}{p_j(\ell)} \sum_{f \in F : |f^{-1}(j)| = \ell} x_f \geq 1, \quad j = 1, \ldots, n, \quad (1)$$
$$x_f \geq 0, \quad \forall f \in F.$$

The set F denotes the set of all configurations. Furthermore, the variable x_f in the linear program indicates the length of configuration f in the schedule and $|f^{-1}(j)|$ is the number of processors allotted to task T_j in configuration f. Let $Lin_M(I)$ be the optimum value of the linear program (1). Clearly $Lin_M(I)$ is the length of an optimal preemptive schedule (with migration). Furthermore, $Lin_M(I) \leq OPT_{non-pre}(I)$ where $OPT_{non-pre}(I)$ is the length of an optimum non-preemptive schedule.

2.2 Solving the Linear Program

In this subsection we briefly describe how to compute a preemptive schedule of length at most $(1 + \epsilon)Lin_M(I)$ for any $\epsilon > 0$ similar to [16]. The underlying technique is used also for the fractional strip packing problem in Section 3 and for the modified linear programs in the full paper.

We can solve (1) by using binary search on the optimum value and testing in each step the feasibility of a system of (in-) equalities for a given $r \in$

$[d_{max}, nd_{max}]$ where $d_{max} = \max_{1 \le j \le n} d_j$ and $d_j = \min_{\ell \in D} p_j(\ell)$. Notice that the length of an optimal preemptive schedule $Lin_M(I)$ satisfies $d_{max} \le Lin_M(I) \le OPT_{non-pre} \le nd_{max}$. The system of inequalities is given as follows:

$$\sum_{\ell \in D} \frac{1}{p_j(\ell)} \sum_{f \in F: |f^{-1}(j)| = \ell} x_f \ge 1, \quad j = 1, \ldots, n, \quad (x_f)_{f \in F} \in B$$

where

$$B = \{ (x_f)_{f \in F} : \sum_{f \in F} x_f = r, \; x_f \ge 0, \; f \in F \}.$$

This can be done approximately by computing an approximate solution for the following problem:

$$\lambda^* = \text{Max}\{ \lambda : \sum_{\ell \in D} \frac{1}{p_j(\ell)} \sum_{f \in F: |f^{-1}(j)| = \ell} x_f \ge \lambda, \; j = 1, \ldots, n, \quad (x_f)_{f \in F} \in B \}. \tag{2}$$

The latter problem can also be viewed as a fractional covering problem with one block B, and n coupling constraints. Let the n coupling (covering) constraints be represented by $Ax \ge \lambda$. We can compute an $(1-\rho)$ - approximate solution for (2) by $O(n(\rho^{-2} + \ln n))$ iterations (coordination steps) [12], each requiring for a given price vector $y = (y_1, \ldots, y_n)$ an $(1-\rho')$ approximate solution (with $\rho' = \rho/6$) of the problem,

$$\text{Max} \{ y^T Ax : x \in B \}. \tag{3}$$

Since B is just a simplex, the optimum of this linear program is attained at a vertex $\tilde{x}$ of B that corresponds to a (single) configuration $\tilde{f}$ with $\tilde{x}_{\tilde{f}} = r$ and $\tilde{x}_f = 0$ for $f \ne \tilde{f}$. A similar idea was used for the bin packing problem by Plotkin at al. [24]. Thus, it suffices to find a configuration $\tilde{f}$ with tasks that can be executed in parallel and has the largest associated profit value $c_{\tilde{f}}$ in the profit vector $c^T = y^T A$.

For each task T_j we have now different values in D and hence in the corresponding Knapsack Problem the profit $y_j/p_j(\ell)$ depends on the cardinality $\ell \in D$, while the capacity of the knapsack is m. The subroutine corresponds to a generalized Knapsack Problem with different choices for tasks (items). The problem we have to solve (approximately) for a given n-vector $(y_1, \ldots, y_n)$ can be formulated as follows:

$$\begin{array}{ll} \text{Max} & \sum_{j=1}^n \sum_{\ell \in D} \frac{y_j}{p_j(\ell)} \cdot x_{j\ell} \\ \text{s.t.} & \sum_{j=1}^n \sum_{\ell \in D} \ell \cdot x_{j\ell} \le m, \\ & \sum_{\ell \in D} x_{j\ell} \le 1, \quad j = 1, \ldots, n, \\ & x_{j\ell} \in \{0, 1\}, \quad \ell \in D, \; j = 1, \ldots, n. \end{array} \tag{4}$$

This is the Multiple-choice Knapsack Problem. Lawler [20] showed that an $(1-\rho')$-approximate solution of this Knapsack Problem can be computed in $K(n, D, \rho') = O(n|D| \ln |D| + n^2|D|/\rho')$ time. The overhead (i.e. the numerical calculations) per iteration in [12] is bounded by $O(n \ln \ln(n/\rho)) = O(n^2 + n\rho^{-1})$

that is less than the running time required by the knapsack subroutine. By binary search on r we can obtain a solution $(x_f)_{f \in F}$ with $\sum_{f \in F} x_f \leq (1+\epsilon/4) Lin_M(I)$ and $\sum_{\ell \in D} \frac{1}{p_j(\ell)} \sum_{f \in F : |f^{-1}(j)| = \ell} x_f \geq (1-\rho)$, where $Lin_M(I)$ is the length of an optimal preemptive schedule. Now one can define $\tilde{x}_f = x_f(1+4\rho)$ and obtain $\sum_{\ell \in D} \frac{1}{p_j(\ell)} \sum_{f \in F : |f^{-1}(j)| = \ell} \tilde{x}_f \geq (1-\rho)(1+4\rho) = (1+3\rho-4\rho^2) \geq 1$ for $\rho \leq 3/4$. In this case the length of the generated schedule is $\sum_{f \in F} \tilde{x}_f \leq Lin_M(I)(1+4\rho)(1+\epsilon/4) = Lin_M(I)(1+4\rho+\epsilon/4+\rho\epsilon) \leq Lin_M(I)(1+\epsilon) \leq OPT_{non-pre}(1+\epsilon)$ by choosing $\epsilon \leq 1$ and $\rho = 3\epsilon/20$. Since the optimum of (1) lies within interval $[d_{max}, nd_{max}]$, the overall running time of the first phase to compute a $(1+\epsilon)$ approximate solution is bounded by $O(n^2|D|(\epsilon^{-2}+\ln n)\ln(n\epsilon^{-1})(\ln|D|+n\epsilon^{-1}))$ time. In the full paper we show how to improve the running time of the approximation scheme above to

$$O(n^2 \epsilon^{-1}(\epsilon^{-2} + \ln n) \min(n|D|, n^2 \epsilon^{-1}) \ln(n \epsilon^{-1}) + n|D|).$$

This gives an improvement in the running time for $|D| > n\epsilon^{-1}$. We notice that the number of configurations in the solution can be reduced to n by solving several systems of equalities. Furthermore the number of processor numbers per task T_j can be also reduced to $O(\min(n, |D|))$. The running time of such a procedure is at most $O(n(\epsilon^{-2}+\ln n)\mathcal{M}(n))$ where $\mathcal{M}(N)$ is the time to invert a $(N \times N)$ matrix. This step is helpful to reduce the number of preemptions, but in the next rounding step we may use also the larger number of configurations.

3 Converting the Preemptive Schedule

In this section we show how to convert a preemptive schedule into a non-preemptive schedule. First we compute an unique processor allotment for almost all tasks. The key idea is to construct an instance of a strip packing problem with several rectangles for any task (based on the solution (x_f) of the linear program) and to apply a new rounding technique. After the rounding we remove a constant number of tasks with non-unique processor numbers and obtain a instance of strip packing with one rectangle for any remaining task. We show that an approximate solution of this strip packing problem gives us an approximate solution for our scheduling problem $P|fctn_j|C_{max}$ (without preemption).

3.1 Computing an Allotment

Let $p_{max} = \max_j \max_{\ell \in D} p_j(\ell)$. For each task $T_j \in \mathcal{T}$ and each processor number $\ell \in D$ we define the fraction

$$x_{j,\ell} = \frac{1}{p_j(\ell)} \sum_{f \in F : |f^{-1}(j)| = \ell} x_f \in [0,1].$$

Then we construct an instance of a strip packing problem with rectangles $(x_{j,\ell} \cdot p_j(\ell), \ell)$ of height $x_{j,\ell} p_j(\ell)$ and width ℓ for $x_{j,\ell} > 0$. Let L_{wide} be the set of rectangles with width $\ell > \epsilon' m$ and let L_{narrow} be the set of rectangles with width $\ell \leq \epsilon' m$. We place the wide rectangles of L_{wide} on a left-justified stack by order of non-increasing widths. Such a stack packing was introduced in [18] for the strip packing problem. Let $H := H(L_{wide})$ be the height of the stack. Now consider the horizontal lines $y = iH/M$ where M is a later specified constant and $i \in \{1, \ldots, M-1\}$. Each rectangle that intersects with a horizontal line is divided into two rectangles. After that we have M groups of rectangles where each group has total height H/M. Let a_i, b_i be the smallest and largest width of group i. Let $W_{j,i}$ be the set of widths in group i corresponding to task T_j. For the set L_{narrow} we compute the total area $Area_{narrow} = \sum_{j=1}^{n} \sum_{\ell \in D, \ell \leq \epsilon' m} x_{j,\ell} p_j(\ell) \ell$ and place the corresponding rectangles in group 0. For each group $i \in \{0, \ldots, M\}$ and task T_j let $z_{j,i} = \sum_{w: w \in W_{j,i}} y_{j,i}(w)$ be the fraction of task T_j executed in group i (where $y_{j,i}(w)$ is the fraction of task T_j with width w executed in group i). Notice that tasks with the same width could have pieces in different groups.

The goal is now to get unique processor numbers for almost all tasks. To obtain such an assignment we do the following steps:

(1) for each group $i \in \{1, \ldots, M\}$ and task T_j with at least two widths in group i: compute the smallest processing time $p_j(\ell)$ among all processor numbers $\ell \in D \cap [a_i, b_i]$. Let $\ell_{j,i}$ be a corresponding processor number. Now replace the rectangles corresponding to task T_j in group i by $(z_{j,i} p_j(\ell_{j,i}), \ell_{j,i})$.
(2) for each task T_j with at least two widths in group 0: compute the smallest area $p_j(\ell) \ell$ among all processor numbers $\ell \in D \cap [1, \epsilon' m]$. Let $\ell_{j,0}$ be a corresponding processor number. Then replace all rectangles corresponding to task T_j in group 0 by $(z_{j,0} p_j(\ell_{j,0}), \ell_{j,0})$.
(3) round all tasks over the groups using a general assignment problem:

$$\begin{array}{ll} \sum_{j=1}^{n} z_{j,0} p_j(\ell_{j,0}) \ell_{j,0} \leq Area_{narrow} & \\ \sum_{j=1}^{h} z_{j,i} p_j(\ell_{j,i}) \leq H/M & i = 1, \ldots, M \\ \sum_{i=0}^{M} z_{j,i} = 1 & j = 1, \ldots, n \\ z_{j,i} \geq 0 & j = 1, \ldots, n, \ i = 1, \ldots, M \end{array} \quad (5)$$

This problem formulation is closely related to scheduling tasks on unrelated machines; i.e. with $M+1$ machines and makespan as objective function [21]. One can round now the variables $z_{j,i}$ such that there are at most M fractional variables. Let $\mathcal{F}$ be the set of tasks T_j with fractional variables $z'_{j,i} \in (0,1)$ after the rounding. After removing these tasks we have an unique processor number for each remaining task. The tasks in $\mathcal{F}$ will be executed later at the end of the schedule with execution time bounded by $M p_{max} \leq M$ (using $p_{max} \leq 1$). We have obtained now a rectangle packing instance with a set $L'_{wide} = \{(p_j(\ell_{j,i}), \ell_{j,i}) | z'_{j,i} = 1, i > 0\}$ of wide rectangles and a set $L'_{narrow} = \{(p_j(\ell_{j,0}), \ell_{j,0}) | z'_{j,0} = 1\}$ of narrow rectangles.

The rounding step (3) can be done by constructing a bipartite graph $G = (V_1, V_2, E)$ where V_1 is the task set, $V_2 = \{0, \ldots, M\}$ is the machine or group set and $E = \{e = (T_j, i) | z_{j,i} > 0, j = 1, \ldots, n, i = 0, \ldots, M\}$. Each assignment given by (5) can be transformed into another assignment that satisfies the constraints above and that corresponds to a forest [24]. The transformation can be done in $O(|E|M) = O(\min(nM, n|D|)M)$ time. In our case $\epsilon' = \delta/(\delta + 2)$, $\delta = \epsilon/8$ and $M = O(1/(\epsilon')^2) = O(1/\epsilon^2)$ and therefore the running time of the rounding step (3) is bounded by $O(\min(n\epsilon^{-2}, n|D|)\epsilon^{-2})$.

The stack packing can be computed in $O(n^2(\epsilon^{-2} + \ln n)\ln(n\epsilon^{-1}))$ time by sorting the rectangles (an instance with $O(n^2(\epsilon^{-2} + \ln n))$ rectangles). Furthermore, we can compute the values a_i, b_i and $z_{j,i}$ in $O(n^2(\epsilon^{-2} + \ln n))$ time. This implies a running time of $O(n^2(\epsilon^{-2} + \ln n)\ln(n\epsilon^{-1}))$ for the steps (1, 2). We notice that the running time for the rounding steps (1 − 3) is less than the running time for the computation of the approximate preemptive schedule. In general, the heights and widths of rectangles in a group are modified. For example the height of a group could be extremely smaller than H/M.

For the next step we use two sets of instances $inf(L)$ and $sup(L)$ for a given set L of rectangles [18]. First construct a stack packing for L with M groups. Then rounding each rectangle in group i up to b_i (up to m for the first group) generates $sup(L)$, and rounding each rectangle in group i down to b_{i+1} (down to 0 for the last group) generates $inf(L)$. We say that $L \leq L'$ if the stack associated to L viewed as a region in the plane is contained in the stack associated to L'.

Our algorithm computes $sup(L'_{wide})$ (by producing a new stack packing). In this way an instance with at most M distinct widths is generated. We note that the height H' of the new stack packing is bounded by H.

3.2 Fractional Strip Packing

Suppose now that we have a list L of n rectangles that have M distinct widths (number of processors) $w'_1 > w'_2 > \ldots > w'_M > \epsilon'm$. A configuration C_j is defined as a multi-set of widths $(\alpha_{1,j}, \ldots, \alpha_{M,j})$ where $\alpha_{i,j}$ is the number of rectangles of width w'_i in C_j. Let q be the number of configurations, and let β_i denote the total height of all rectangles of width w'_i. The fractional strip packing problem is defined as follows:

$$\begin{aligned}
\text{Min} \quad & \sum_{j=1}^{q} x_j \\
\text{s.t.} \quad & \sum_{j=1}^{q} \alpha_{i,j} x_j \geq \beta_i, \quad i = 1, \ldots, M, \\
& x_j \geq 0, \quad j = 1, \ldots, q.
\end{aligned} \quad (6)$$

Let $Lin(L)$ be the optimum value of this fractional strip packing problem, and $Area(L)$ the total area of all rectangles in L. Fractional strip packing is closely related to fractional bin packing where the right hand side is a vector $(n_1, \ldots, n_M)$ with integer coordinates. The fractional bin packing problem can be solved approximately with additive tolerance t in

$$O(M^6 \ln^2 \frac{Mn}{a\,t} + \frac{M^5 n}{t} \ln \frac{Mn}{a\,t})$$

time using the algorithm by Karmarkar and Karp [17, 19] (where n is the number of items, M is the number of distinct sizes, and a is the size of the smallest item).

We have to solve now the fractional strip packing problem for $L = sup(L'_{wide})$ approximately. This can be done in time polynomial in M, $\sum_i \beta_i \leq n$ and $\frac{1}{t}$ using the method by Karp and Karmarkar (since it does not use the fact that the vector $(\beta_1, \ldots, \beta_M)$ is integer). But for our purpose it is sufficient to get a solution with value $\leq (1+\delta)Lin(sup(L'_{wide}))$. Using the method described in the previous section, we can compute a $(1+\delta)$ approximate solution of the fractional strip packing problem. The subproblem here to solve is a Knapsack problem with unbounded variables. The number of iterations (calls to the block solver) is $O(M(\delta^{-2} + \ln M))$ and the number of rounds in the binary search is at most $O(\ln(\delta^{-1}))$. The latter bound follows from a computation of an integral strip packing (by using longest task first (LTF)) [5, 27] with length

$$LTF(L) \leq \frac{2}{m}\sum_{i=1}^{M}\beta_i w'_i + \max_{1 \leq i \leq M}\beta_i.$$

Notice that the total area divided by the width of the stripe $\frac{1}{m}\sum_{i=1}^{M}\beta_i w'_i$ and the maximum value $\epsilon' \max_{1 \leq i \leq M}\beta_i$ are both lower bounds on the minimum height of a fractional strip packing. Therefore the length $LTF(L)$ of this integral strip packing can be bounded by $(2 + 1/\epsilon')Lin(L)$. In other words, $Lin(L) \in [LTF(L)/(2+1/\epsilon'), LTF(L)]$ and the number of binary search steps is bounded by $O(\ln(\delta^{-1}))$. The knapsack problem with M different types can be solved in $O(M + (\rho')^{-3})$ time [20]. Using $\rho' = \theta(\delta)$ and counting the overhead of $O(M \ln \ln(M\delta^{-1}))$ per iteration we obtain:

Lemma 1. *The fractional strip packing problem with M different widths $w'_1 > w'_2 > \ldots > w'_M > \epsilon'm$ and $\epsilon' = \theta(\delta)$ can be solved approximately with ratio $(1+\delta)$ in time $O(M(\delta^{-2} + \ln M)\ln(\delta^{-1})\max(M + \delta^{-3}, M \ln \ln(M\delta^{-1})))$.*

Using $M = O(1/\epsilon^2)$ and $\rho' = \theta(\epsilon)$, we obtain $O(\epsilon^{-3})$ as running time for the knapsack problem (including the overhead). The number of non-zero basic variables is bounded by $O(M(\epsilon^{-2} + \ln M))$ (in each iteration we include one configuration). This implies that the number of iterations and the number of non-zero variables is $O(\epsilon^{-4})$. Therefore, the running time of this step is $O(\epsilon^{-7}\ln(\epsilon^{-1}))$.

We can reduce the number $O(M(\epsilon^{-2} + \ln M))$ of non-zero basic variables to M by solving again a series of systems of linear equalities. We use the fact that we have only M equalities. By solving $Bz = 0$ for a submatrix B with $M+1$ columns we can choose z to eliminate one coordinate in each iteration. We need $O(\epsilon^{-4})$ rounds with $O(\epsilon^{-2})$ variables and equalities. Each round can be done in $\mathcal{M}(\epsilon^{-2})$ time. Therefore, this rounding step can be done in $O(\epsilon^{-4}\mathcal{M}(\epsilon^{-2}))$ time.

In the next step we place the wide rectangles into the space generated by the non-zero basic variables. Finally, we use a modified version of next fit decreasing height ($MNFDH$) to place the narrow rectangles into the remaining gaps (see also Kenyon and Remila [18]) and schedule the tasks in $\mathcal{F}$ at the end of the generated schedule (using an arbitrary number $\ell \in D$ of processors).

3.3 Description of the Algorithm

The algorithm to convert a preemptive schedule corresponding to an $(1+\delta)$ approximate solution (x_f) of the linear program (1) into a non-preemptive schedule works as follows:

(0) set $M = (1/\epsilon')^2$, $\epsilon' = \delta/(\delta+2)$, $\delta \leq \min(1, \epsilon/8)$,
(1) compute the values $x_{j,\ell} = 1/p_j(\ell) \sum_{f \in F: |f^{-1}(j)|=\ell} x_f \in [0,1]$ for each task $T_j \in \mathcal{T}$ and $\ell \in D$,
(2) construct an instance of strip packing with rectangles $(x_{j,\ell} p_j(\ell), \ell)$ for $x_{j,\ell} > 0$, and let $L_{wide} = \{(x_{j,\ell} p_j(\ell), \ell) | \ell > \epsilon' m, \ell \in D\}$ and $L_{narrow} = \{(x_{j,\ell} p_j(\ell), \ell) | \ell \leq \epsilon' m, \ell \in D\}$,
(3) use the rounding technique to obtain a strip packing instance $L'_{wide} = \{(p_j(\ell_{j,i}), \ell_{j,i}) | z_{j,i} = 1, i > 0\}$, $L'_{narrow} = \{(p_j(\ell_{j,0}), \ell_{j,0}) | z_{j,0} = 1$ and set $\mathcal{F} = \{T_j | z_{j,i} \in (0,1)$ for at least one $i \in \{0, 1, \ldots, M\}\}$,
(4) construct instance $sup(L'_{wide})$ (via stack packing) with a constant number M of distinct widths,
(5) solve the fractional strip packing problem for $sup(L'_{wide})$ approximately with ratio $(1+\delta)$ and round the solution to obtain only M non-zero variables x_j,
(6) place the wide rectangles of L'_{wide} into the space generated by the non-zero variables x_j,
(7) insert the narrow rectangles of L'_{narrow} using modified next fit decreasing height,
(8) schedule the tasks in $\mathcal{F}$ at the end of the schedule.

We obtain now the following result for our scheduling problem.

Theorem 2. *For $\delta \leq \epsilon/8$ and $\epsilon \in (0,1)$, any preemptive schedule of length $(1+\delta) Lin_M(I)$ can be converted in polynomial time into a non-preemptive schedule of length $(1+\epsilon) OPT_{non-pre}(I) + O(\epsilon^{-2})$. The total running time of this algorithm is $O(n^2(\ln n + \epsilon^{-2}) \ln(n\epsilon^{-1}) + \epsilon^{-4} \max(\mathcal{M}(\epsilon^{-2}), \epsilon^{-3} \ln(\epsilon^{-1})))$.*

The proof of this Theorem and the complete analysis of our algorithm can be found in the full version of our paper. For any $\epsilon > 0$ the **main algorithm** works as follows:

(1) set $\delta = \min(1, \epsilon/8)$,
(2) compute a $(1+\delta)$ approximate preemptive schedule (see Section 2),
(3) convert the preemptive schedule into a non-preemptive schedule (see Section 3).

Theorem 2 shows us that the main algorithm is in fact a AFPTAS. This proves our Theorem 1.

4 Further Results

The running time of the first phase can be improved by further ideas. First we can reduce the number of binary search steps from $O(\ln(n\epsilon^{-1}))$ to $O(\ln(\epsilon^{-1}))$. Then we can construct a new linear program for the first phase with a smaller number of variables. In fact we show that an approximate schedule for the wide tasks and a good bound for the total area of all tasks is sufficient as a starting point for the second phase. In this way the computation of the preemptive schedule can be avoided. As result we obtain an improved running time for the first phase of $O(n(\epsilon^{-2} + \ln n) \max[\min(n|D|, n\epsilon^{-2})\epsilon^{-2}, n \ln\ln(n\epsilon^{-1})] \ln(\epsilon^{-1}) + n|D|)$.

Furthermore, we show how the approximation scheme can be generalized to the case where the processors are arranged in a line with consecutive processors for tasks or where the processors are interconnected by a hypercube and each task requires a subcube. Finally, we simplify our method for monotonic malleable tasks while improving the running time of the approximation scheme to $O(n^2(\epsilon^{-2} + \ln n) \max(\ln\ln(n\epsilon^{-1}), s\epsilon^{-2}) \ln(\epsilon^{-1}) + \epsilon^{-2} s \max(\mathcal{M}(s), \epsilon^{-3} \ln(\epsilon^{-1})))$ where $s = O(\log_{1+\epsilon}(\epsilon^{-1}))$ and reducing the additive constant from $O(\epsilon^{-2})$ to $O(\log_{1+\epsilon}(\epsilon^{-1})) = O(\epsilon^{-1} \log(\epsilon^{-1}))$.

The details of these improvements and the algorithms are given in the full version of the paper.

Acknowledgement

The author thanks Evripidis Bampis for many helpful discussions during a research visit at the University of Evry.

References

[1] B. Baker, E. Coffman and R. Rivest, Orthogonal packings in two dimensions, *SIAM Journal on Computing*, 9 (1980), 846-855.

[2] K. Belkhale and P. Banerjee, Approximate algorithms for the partitionable independent task scheduling problem, *International Conference on Parallel Processing* (1990), Vol. 1, 72-75.

[3] J. Blazewicz, W. Cellary, R. Slowinski, and J. Weglarz, Scheduling under resource constraints - deterministic models, *Annals of Operations Research* 7 (1986).

[4] J. Blazewicz, M. Drabowski, and J. Weglarz, Scheduling multiprocessor tasks to minimize schedule length, *IEEE Transactions on Computers*, C-35-5 (1986), 389-393.

[5] E. Coffman, M. Garey, D. Johnson and R. Tarjan, Performance bounds for level-oriented two dimensional packing algorithms, *SIAM Journal on Computing*, 9 (1980), 808-826.

[6] M. Drozdowski, On the complexity of multiprocessor task scheduling, *Bulletin of the Polish Academy of Sciences*, 43 (1995), 381-392.

[7] M. Drozdowski, Scheduling multiprocessor tasks - an overview, *European Journal on Operations Research*, 94 (1996), 215-230.

[8] J. Du and J. Leung, Complexity of scheduling parallel task systems, *SIAM Journal on Discrete Mathematics*, 2 (1989), 473-487.

[9] M. R. Garey and R. L. Graham, Bounds for multiprocessor scheduling with resource constraints, *SIAM Journal on Computing* 4 (1975), 187-200.
[10] M. R. Garey, R. L. Graham, D. S. Johnson and A. C.-C. Yao, Resource constrained scheduling as generalized bin packing, *Journal Combinatorial Theory A*, 21 (1976), 251-298.
[11] M. Garey and D. S. Johnson, Complexity results for multiprocessor scheduling under resource constraints, *SIAM Journal on Computing* 4 (1975), 397-411.
[12] M. D. Grigoriadis, L. G. Khachiyan, L. Porkolab and J. Villavicencio, Approximate max-min resource sharing for structured concave optimization, *SIAM Journal on Optimization*, 11 (2001), 1081-1091.
[13] M. Grötschel, L. Lovász and A. Schrijver, Geometric Algorithms and Combinatorial Optimization, Springer Verlag, Berlin, 1988.
[14] K. Jansen and L. Porkolab, Linear-time approximation schemes for scheduling malleable parallel tasks, *Algorithmica* 32 (2002), 507-520.
[15] K. Jansen and L. Porkolab, Computing optimal preemptive schedules for parallel tasks: Linear Programming Approaches, *Proceedings 11th Annual International Symposium on Algorithms and Computation* (2000), LNCS 1969, Springer Verlag, 398-409, and to appear in: *Mathematical Programming*.
[16] K. Jansen and L. Porkolab, On preemptive resource constrained scheduling: polynomial-time approximation schemes, *Proceedings 9th International Conference on Integer Programming and Combinatorial Optimization* (2002), LNCS 2337, Springer Verlag, 329-349.
[17] N. Karmarkar and R. M. Karp, An efficient approximation scheme for the one-dimensional bin packing problem, *Proceedings 23rd IEEE Symposium on Foundations of Computer Science* (1982), 312-320.
[18] C. Kenyon and E. Remila, A near-optimal solution to a two-dimensional cutting stock problem, *Mathematics of Operations Research* 25 (2000), 645-656.
[19] B. Korte and J. Vygen, Combinatorial optimization: Theory and Algorithms, Springer Verlag, Berlin, 2000.
[20] E. Lawler, Fast approximation algorithms for knapsack problems, *Mathematics of Operations Research*, 4 (1979), 339-356.
[21] J. K. Lenstra, D. B. Shmoys and E. Tardos, Approximation algorithms for scheduling unrelated parallel machines, *Mathematical Programming*, 24 (1990), 259-272.
[22] W. Ludwig and P. Tiwari, Scheduling malleable and nonmalleable parallel tasks, *Proceedings 5th ACM-SIAM Symposium on Discrete Algorithms* (1994), 167-176.
[23] G. Mounie, C. Rapine and D. Trystram, Efficient approximation algorithms for scheduling malleable tasks, *Proceedings 11th ACM Symposium on Parallel Algorithms and Architectures* (1999), 23-32.
[24] S. A. Plotkin, D. B. Shmoys, and E. Tardos, Fast approximation algorithms for fractional packing and covering problems, *Mathematics of Operations Research*, 20 (1995), 257-301.
[25] I. Schiermeyer, Reverse-Fit: A 2-optimal algorithm for packing rectangles, *Proceedings 2nd European Symposium of Algorithms* (1994), LNCS 855, 290-299.
[26] A. Steinberg, A strip-packing algorithm with absolute performance bound two, *SIAM Journal on Computing*, 26 (1997), 401-409.
[27] J. Turek, J. Wolf and P. Yu, Approximate algorithms for scheduling parallelizable tasks, *Proceedings 4th ACM Symposium on Parallel Algorithms and Architectures* (1992), 323-332.

The Probabilistic Analysis of a Greedy Satisfiability Algorithm*

Alexis C. Kaporis, Lefteris M. Kirousis, and Efthimios G. Lalas

University of Patras
Department of Computer Engineering and Informatics
University Campus, 265 04, Patras, Greece
{kaporis,kirousis,lalas}@ceid.upatras.gr

Abstract. Consider the following simple, greedy Davis-Putnam algorithm applied to a random 3-CNF formula of constant density c: Arbitrarily set to TRUE a literal that appears in as many clauses as possible, irrespective of their size (and irrespective of the number of occurrences of the negation of the literal). Reduce the formula. If any unit clauses appear, then satisfy their literals arbitrarily, reducing the formula accordingly, until no unit clause remains. Repeat. We prove that for $c < 3.42$ a slight modification of this algorithm computes a satisfying truth assignment with probability asymptotically bounded away from zero. Previously, algorithms of increasing sophistication were shown to succeed for $c < 3.26$. Preliminary experiments we performed suggest that $c \simeq 3.6$ is feasible running algorithms like the above, which take into account not only the number of occurrences of a literal but also the number of occurrences of its negation, irrespectively of clause-size information.

1 Introduction

We consider random 3-CNF formulas in the standard $G_{n,m}$ model. This means that a random formula is constructed by choosing m clauses uniformly, independently and with replacement from the set of clauses on n variables that contain exactly three literals. The *density* c of a formula is the ratio m/n, i.e. c is the scaled, with respect to n, number of clauses. It is conjectured that there exists a critical threshold density $r_3 \simeq 4.2$ such that almost all formulas of densities (greater) less than r_3 are (un)satisfiable. Friedgut [14], proved the existence of a sequence of threshold values $r_3(n)$ for random 3-CNF formulas, but the convergence of this sequence still remains open. Upper bounds to r_3 have been proven using probabilistic counting arguments, see [12, 13, 17, 18, 20, 21, 22]. The current best upper bound is $r_3 \leq 4.506$ by Dubois, Boufkhad and Mandler in [13].

From [14], to prove that $r_3 > c$ it suffices to show with bounded away from zero probability that a satisfying truth assignment exists for random formulas

---

* Research supported by the University of Patras Research Committee Project Carathéodory under contract no. 2445 and by the Computer Technology Institute

of initial density c. In this vain, Davis-Putnam algorithms of increasing sophistication were rigorously analyzed see [1, 5, 7, 8, 9, 15]. The best lower bound for the satisfiability threshold thus obtained is $r_3 > 3.26$ by Achlioptas and Sorkin in [5]. These algorithms (with the exception of the "plain" pure literal algorithm, which succeeds for formulas of density up to 1.63 [7, 23]) take into account the size of the clauses were the literal to be set appears.

We describe here a greedy Davis-Putnam algorithm that in selecting the literal to be set takes into account only how many clauses, irrespectively of their size, will be satisfied. Its probabilistic analysis is based on the method of differential equations studied by Wormald in [27]. As the algorithm is rather simple, its analysis is also relatively easy. This simplicity, contrasted with the improvement over the lower bounds of previous analyses, is an interesting, we think, aspect of this work.

The algorithm is applied to a 3-CNF formula with n variables and density c. The *degree* of a literal is the number of its occurrences in the formula. If the formula contains clauses of varying size, then any occurrence of a literal, irrespectively of the size of the clause where it appears, counts towards the degree. Let h be an a priori decided integer parameter. At a first phase, the algorithm arbitrarily selects and sets to TRUE literals of degree *at least* h, until they are exhausted. At subsequent phases, it continues with literals of degree *exactly* $h-1$ etc. in decreasing order of the degree. Unit clauses, whenever they appear, are given priority. In the numerical computations, we take $h = 10$ (a larger h gives a larger lower bound, but only with respect to its second decimal digit). Let $\mathcal{X}_j, j = 0, \ldots, h-1$, be the current collection of literals of degree j, and let $\mathcal{X}_h$ be the current collection of literals of degree at least h (although this notation is not uniform for $j < h$ and for h, it is convenient). Literals in $\mathcal{X}_h$ are called h*eavy*. Del&Shrink is the usual procedure of any Davis-Putnam algorithm where clauses that contain the literal that is set to TRUE are deleted, whereas clauses that contain its negation are shrunk by deleting it from them. Formally, the algorithm, is the following:

Greedy Algorithm
begin:
$j \leftarrow h$;
while unset literals exist do:
   while $\mathcal{X}_j \neq \emptyset$ do:
      set an arbitrary literal in $\mathcal{X}_j$ to TRUE, its negation to FALSE;
      Del&Shrink;
      while unit clauses exist do:
         set an arbitrary literal of a unit clause to TRUE, its negation to FALSE;
         Del&Shrink;
      end do;
   end do;
   $j \leftarrow j - 1$;
end do;
if an empty clause is generated then report failure;
else report success;
end;

We were motivated to work with literals rather than variables from the approach in [23] (about the decoupling of a literal from its negation see also Section 3). We were motivated to give priority to large degrees from [2, 5], where the need to capitalize on variable-degree information was pointed out and from [4], where, in the context of the 3-coloring problem, the Brélaz heuristic [6] was analyzed. According to this heuristic, vertices of maximum degree are given priority, but only in case they can be legally colored by 2 out of the 3 colors. The maximum degree of such vertices is not necessarily decreasing, whereas in our case the degree of any literal is always decreasing, a fact that simplifies the analysis. We were motivated to put together all heavy literals from [25], where light vertices are deleted in order to find the k-core of a random graph.

To put in use the above ideas in the probabilistic analysis, we introduce in Section 3 a model for random formulas that we call the *truncated configuration model*. It is a variant of the configuration model for random graphs with a given degree sequence (see e.g. [16]). The name comes from a variant of the Poisson distribution we use, which is called the truncated Poisson distribution. A special case (for $h = 1$) of this distribution was used in [7], in the analysis of the pure literal heuristic. Section 4 outlines our basic tool: Wormald's method of analyzing random processes with differential equations. In Sections 2, 5 and 6 we elaborate on certain aspects particular to analyses of the type we make in this paper. The same or similar techniques as those in the above three sections have been used before (see, in this respect, [1, 2, 5, 7, 8, 23] and also, with respect to algorithms for the coloring problem, see [3, 4]). Although the basic ideas involved are essential to understand the method, the reader may gloss over the technicalities of these three sections. Section 7 contains the main result and Section 8 concern numerical computations and experiments.

2 Rounds

The algorithm proceeds in *rounds*. A round consists of one *free step*, i.e. a step where a literal in X_j is set to TRUE ($j = h, \ldots, 0$), followed by a number of *forced steps* i.e. steps where unit clauses are satisfied (the steps of the inner loop in the pseudo-code above).

As in [5], in the analysis of the evolution of the algorithm we consider as discrete time the number of rounds rather than the number of individual steps, which, for distinction, are to be called *atomic steps*. To explain why this choice of time is made, take into account that as the solution to the differential equations will show (for $c = 3.42$ and $h = 10$), the expected number of unit clauses generated at any atomic step is bounded below 1 (this, as we will show in Section 5, guarantees that the probability of success of the algorithm is bounded above 0). But then, during the course of algorithm the number of unit clauses is equal to 0 unboundedly many times (this happens at the end of each round; all rounds have $O(1)$ atomic steps, so there are $O(n)$ of them, see Section 5). As a consequence, if time corresponds to atomic steps, the evolution of the number of unit clauses cannot be analyzed by the method of differential equations. This is so because to apply this method, the rate of change, from a current step to the next, of the parameter under examination should be given by a smooth function (Lipschitz continuous function, see [27]) of the current scaled value of the parameter. This is not possible for the number of unit clauses, as its rate of change when there is at least one unit clause is discontinuously different from its rate of change when there is none (in the former case we deterministically delete one unit clause). See, for more details, Sections 4 and 6.

3 The Randomness Model

For a 3-CNF formula with n variables, let its *truncated*, scaled (with respect to n) degree sequence be the sequence $x_j, j = 0, \ldots, h$, where for $0 \leq j < h$, x_j is the scaled number of literals of degree exactly j, whereas x_h is the scaled number of heavy literals, i.e., literals of degree at least h. Notice that the degree sequence refers not to variables but to literals. Also, let ℓ denote the scaled total number of literals ($\ell = 2$ in the beginning of the algorithm).

It is immediate that for a uniformly random formula with n variables and density c, each of the $3cn$ literal positions has probability to be occupied by any specified literal equal to $1/(2n)$. But then, for large n, the probability of a literal having degree j follows a Poisson distribution with parameter $3c/2$. So the expected value of the scaled number of literals of degree j is $\ell e^{-(3/2)c}((3/2)c)^j/j!$. From this and from a simplified version of Azuma's inequality (see for example the "Simple Concentration Bound" in [24]) it follows that almost all uniformly random formulas with n variables and density c have truncated scaled degree sequence given by $x_j = \ell e^{-(3/2)c}((3/2)c)^j/j! + o(1)$, for $j = 0, \ldots, h-1$ (then necessarily $x_h = \ell - \sum_{j=0}^{h-1} x_j + o(1)$). So to prove that a result holds a.s., in the large n limit, for uniformly random formulas with n variables and scaled number of clauses c, we only have to prove it for formulas that are additionally conditional on that their truncated scaled degree sequence is the one given above. In the sequel, we assume that the Davis-Putnam algorithm is initialized with such a formula.

Notice that although we start with a 3-CNF formula, at the end of each round we get a $\{3,2\}$-CNF formula, i.e. a formula that may contain 3-clauses and 2-clauses (unit clauses appear only during the course of each round; empty clauses are the clauses we aim at showing that do not appear). We study the evolution of the following $h+3$ scaled stochastic parameters during the course of the algorithm: ℓ, the scaled number of unset literals; $c_i, i = 3, 2$, the scaled number of i-clauses; and x_j, $0 \leq j \leq h-1$. The scaling at all steps of the algorithm is with respect to n, the initial number of variables. However, we will also consider the current densities $\rho_i = 2c_i/\ell, i = 2,3$, of i-clauses, i.e., the current number of i-clauses divided by the current number of unset variables.

We denote by $\mathcal{S}$ the vector whose coordinates are the current unscaled (i.e. multiplied with n) values of the $h+3$ parameters above, i.e.

$$\mathcal{S} = n(\ell, c_3, c_2, x_0, \ldots, x_{h-1}). \tag{1}$$

It can be shown, by the technique in [19], that during the evolution of the algorithm, the formula remains random conditional on the current number of these parameters.

The model we use for formulas that are random conditional on the values of the parameters in S, which we call the *truncated configuration model*, is described as follows: think of the ℓn literals as bins and of the $(3c_3+2c_2)n$ clause positions as balls. Randomly select h subsets of the balls of cardinalities $jx_j n$, $j = 0, \ldots, h-1$, respectively. Then for each $j = 0, \ldots, h-1$ randomly throw the balls of the corresponding subset into the $x_j n$ bins of literals of degree j, in a away that every bin gets exactly j balls. Then randomly throw the remaining $(\ell - \sum_{j=0}^{h-1} j x_j)n$ balls into the bins of heavy literals, but so that each heavy bin gets at least h balls. Although this model allows the same literal, or literals of the same variable, to occur in a clause, such phenomena violating the simplicity of the formula are, for the large n limit, of bounded below 1 probability (the term "simplicity" comes from simple graphs, i.e. graphs without loops or multiple edges). Therefore, if a result, like our main theorem in Section 7, is shown to hold a.s. for the possibly non-simple formulas in this model, it also holds a.s. for random simple formulas.

An important aspect of the algorithm is that at any atomic step, the literal to be set to TRUE is selected on the basis of information about itself and irrespective of properties of its negation. To describe this situation, we say that literals are decoupled from their negation. As a consequence, the literal set to FALSE at any atomic step is always uniformly random over all literals (the restriction that it has to be different from the literal set to TRUE introduces an $o(1)$ discrepancy which is neglected). It is because of this that we can work with a degree sequence based on literals and not, as it is usually the case, with a 2 dimensional degree sequence that at (i,j) gives the number of variables that have i positive and j negative occurrences. From the fact that the literals that are set to FALSE are uniformly random literals, it immediately follows that the expected number of unit clauses generated at any atomic step is the expected number of occurrences in 2-clauses of a random literal. This number is trivially the current density ρ_2 of 2-clauses at that atomic step.

Let now p_h denote the scaled number of literal positions that are occupied with heavy literals. Then $p_h = 3c_3 + 2c_2 - \sum_{j=0}^{h-1} j x_j$. Then the probability mass of the number of literals of degree i, for any fixed $i \geq h$, follows a truncated Poisson distribution. This means that for any $i \geq h$, the probability that a heavy literal has degree i is:

$$\frac{\mu^i}{\left(e^\mu - \sum_{j=0}^{h-1} \frac{\mu^j}{j!}\right) i!}, \tag{2}$$

where μ is the solution, with respect to x, of the following equation:

$$\frac{p_h}{x_h} = \frac{x\left(e^x - \sum_{j=0}^{h-2} \frac{x^j}{j!}\right)}{e^x - \sum_{j=0}^{h-1} \frac{x^j}{j!}}. \tag{3}$$

This result about the truncated Poisson distribution is proved for $h = 1$ and for 3-CNF formulas in [7]. Its generalization to arbitrary h and for $\{3, 2\}$-CNF formulas in the truncated configuration model is not hard. We omit the proof.

4 Wormald's Theorem

As we already pointed out, our analysis is based on the method of differential equations. For an exposition of how the relevant Wormald's theorem is applied to the satisfiability problem see [2]. Roughly, the situation is as follows: suppose that $Y_j, j = 1, \ldots, a$ are stochastic parameters related to a formula, like e.g. the number of clauses with a specified size, or the number of literals with a specified degree. In our case, the Y_j's are the $h + 3$ parameters in $\mathcal{S}$. We want to estimate the evolution of the parameters Y_j during the course of a Davis-Putnam algorithm. The formula initially is a 3-CNF formula with n variables and uniformly random conditional on given initial values $Y_j(0), j = 1, \ldots, a$. These initial values in our case are constant multiples of n (in general, they may be random numbers). As the formula is random with respect to the names (labels) of the literals, we assume that the Davis-Putnam algorithm selects at any atomic step the first literal (in some arbitrary ordering of the labels) that is subject to the restrictions of the algorithm. In other words, the algorithm is assumed to be deterministic and the sample space is determined by the initial formula, only. Notice that, conveniently, a Davis-Putnam algorithm is a Markov process, although Wormald's theory is applicable to more general processes.

Suppose that the expected change of each Y_j from an instance t to the next, conditional on the values $Y_j(t), j = 1, \ldots, a$ of the parameters at t, is, for all possible values of the random parameters $Y_j(t)$, given by

$$\mathbf{E}[Y_j(t+1) - Y_j(t) \mid Y_1(t), \ldots, Y_a(t)] = f_j(t/n, Y_1(t)/n, \ldots, Y_a(t)/n) + o(1), \quad (4)$$

where the f_js are Lipschitz continuous functions. Suppose also that the change of each of the parameters Y_j from t to $t + 1$ is well concentrated on its expected value (see [27] for the exact statement). Then the solution of the system of differential equations:

$$dy_j(x)/dx = f_j(x, y_1(x), \ldots, y_a(x)), j = 1, \ldots, a, \quad (5)$$

with the initial conditions $y_j(0) = Y_j(0)/n$, gives, at any time t and a.s. (in the large n limit), the scaled value of Y_j, i.e.

$$y_j(t/n) = (1/n)Y_j(t) + o(1), j = 1, \ldots, a. \quad (6)$$

In applications, an open, connected and bounded domain D that contains

$$(Y_1(0)/n, \ldots, Y_a(0)/n)$$

and a time interval $[0, t_f)$ are considered, and it is assumed that the hypotheses of the theorem hold up to the last time instant $T < t_f$ such that for all $t \in [0, T]$,

$(Y_1(t)/n, \ldots, Y_a(t)/n) \in D$ (T is a random variable). Then the conclusion of the theorem holds, for large enough n, up to any $t < t_f$ such that for all $x \in [0, t/n]$, $(y_1(x), \ldots, y_a(x)) \in D$. In this context, it is sufficient that the Lipschitz continuity of the f_js holds over $[0, t_f/n] \times D$. The above is only a rough outline of Wormald's theorem, not in its full generality, but restricted to the purposes of our particular problem.

5 Inside a Round and Probability of Success

Assume, for the moment, that the density of 2-clauses remains constant during a round (we will elaborate on this point below). Then the generation of the unit clauses in the forced steps of the round follows the pattern of a Galton-Watson branching process (see [11]).

The number of offspring in a Galton-Watson tree may follow an arbitrary distribution whose expected value is known as the Malthus parameter μ. It is known that if the Malthus parameter is < 1 then, irrespective of other characteristics of the distribution of the offspring, the population certainly becomes extinct, and its expected size equals: $1 + \mu + \mu^2 + \ldots = 1/(1 - \mu)$.

In our case now, as we have shown in Section 3 when referring to the decoupling of a literal from its negation, the Malthus parameter is equal to the density ρ_2 of 2-clauses. Suppose that for a given initial scaled number of clauses c and predetermined h, ρ_2 remains bounded below 1 a.s. during the course of the algorithm. Then, the total number of unit clauses generated during the round is, with high probability, finite. Moreover, their expected number is $m_1 = \rho_2/(1 - \rho_2)$ (recall that we start not with one, but with expectedly ρ_2 unit clauses generated at the free step). In [11, Theorem 5.8], it is proved that $y(s) = sf(y(s))$, where $y(s)$ is the probability generating function of the size of the tree and $f(s)$ is the probability generating function of the number of offspring. From this, by differentiation, we can easily compute the second moment m_2 of the size of a subcritical Galton-Watson tree in terms of the Malthus parameter and the second moment μ_2 of the number of offspring. In our case, μ_2 is equal to the second moment of the number of occurrences in 2-clauses of a random literal occurrence (see Section 7). All we need here is that μ_2 is $O(1)$. Therefore, m_2 as well is $O(1)$, a fact we use below.

Condition now on the actual size $|T|$ of the Galton-Watson tree at a round. In order to derive the differential equations of Wormald's theorem, we need to compute the total expected change during all steps of a round of each parameter in $\mathcal{S}$. First notice for any parameter Y_j as in Wormald's theorem, if its expected change from one atomic step to the next obeys a change law like in Relation (4), then its expected change in $|T|$ consecutive steps, by the smoothness of f_j, is given by

$$\mathbf{E}[Y_j(t + |T|) - Y_j(t) \mid Y_1(t), \ldots, Y_a(t)] = f_j(t/n, Y_1(t)/n, \ldots, Y_a(t)/n)|T| \\ + |T|^2 o(1). \qquad (7)$$

Taking now the expectation of the above relation with respect to $|T|$ we get:

Claim. The expected change of a parameter in $\mathcal{S}$ during a round, unconditionally on the number of unit clauses generated at the round, is equal to its expected change at the free step plus its expected change at the first forced step multiplied with $\rho_2/(1-\rho_2)$. Moreover, to compute the expected change at the first forced step, we may assume, by continuity, that the value of the scaled parameters in $\mathcal{S}$ are the same at the beginning of the free step and at the beginning of the forced steps. The total error term is $m_2 o(1)$, i.e. $o(1)$, as required by Wormald's theorem. We strongly use this claim to derive the differential equations.

By the same technique of conditioning on the size of the tree and then unconditioning with passing to the moments, we can show that the change of a parameter from round to round is concentrated around its mean, as sharply as required by Wormald's theorem.

At a fixed round again, conditional on the size $|T|$ of the Galton-Watson tree, it is easy to see that the probability that both a literal and its negation appear in the unit clauses is $O(|T|^2/n)$ (the decoupling of a literal from its negation is needed here). Therefore we immediately conclude that the unconditional probability of contradiction during the round is $O(m_2/n)$ (when the tree is subcritical). Therefore, for all rounds, the probability that no contradiction occurs is $(1-O(m_2/n))^n = e^{-O(m_2)}$, so the probability of success of the algorithm, as long as the generation of unit clauses is subcritical, is bounded above zero.

It may seem that our arguments above are circular, as we use the subcriticality of the unit-clauses to derive the differential equations, and then from them to prove this sub-criticality. The way out is to apply Wormald's theorem for a domain D which includes only values for which ρ_2 is bounded below 1. Then the conclusion of the theorem remains true as long as the solutions of the differential equations stay in D. Notice that the rôle of D is to force the parameters in $\mathcal{S}$ to a.s. remain inside a desirable path.

6 Transition between Phases and Termination of the Algorithm

The jth $(j = h, \ldots, 0)$ *phase* of the algorithm is comprised by all rounds whose free steps set to TRUE a literal of degree j. In other words, a phase corresponds to the middle loop of the pseudo-code of the algorithm. Fix now a phase j. Can the domain of application of Wormald's theorem $[0, t_f) \times D$ be extended to include the vector $(T_0/n, \mathcal{S}(T_0)/n)$, where T_0 is the instant where the phase j terminates and $\mathcal{S}(T_0)/n$ is the vector of the scaled values of the parameters in $\mathcal{S}$ at T_0? The answer is no, because the expected change of $|\mathcal{X}_j|$ before T_0 (i.e. when $\mathcal{X}_j \neq \emptyset$) and the expected change of $|\mathcal{X}_j|$ at T_0 (i.e. when $\mathcal{X}_j = \emptyset$) cannot both be expressed by the same Lipschitz continuous function, as this expected change changes discontinuously when we pass from values before T_0 to T_0 (before T_0 we always set to TRUE a literal from $\mathcal{X}_j$, whereas we cease doing so at T_0).

This creates a problem with respect to the computation of the scaled value of parameters in $\mathcal{S}$ at the very end of the phase j. We can overcome this difficulty by continuity arguments, i.e. by finding the limiting values of these parameters as we approach the end of the phase (see in this respect [27, Note 5, p. 1220]). An alternative way out is to assume that the transition from phase j to $j-1$ occurs when the scaled parameter x_j becomes not greater than a predetermined $\epsilon = (1/2)10^{-5}$, so the leftover literals remain $\Omega(n)$ and the domain restrictions of Wormald's theorem are satisfied. Since the degree of any literal is monotonically decreasing the total number of leftover literals is decreasing and do not substantially affect the results of the numerical calculations.

A similar as above problem arises with the termination of the algorithm: we carry on the algorithm until the size of the formula is small enough, but remains $\Omega(n)$, and then we apply a criterion for satisfiability of formulas with a given degree sequence given by Cooper et al. [10, Theorem 1(a)]. By simple calculations, it follows from this result in [10] that a random $\{3,2\}$-CNF formula with a given degree sequence is a.s. satisfiable if the sum of the densities of the 3-clauses and 2-clauses $\rho_3 + \rho_2$ is bounded below 1. Notice that the result in [10] refers to 2-CNF formulas; to apply it to $\{3,2\}$-CNF formulas, we randomly delete a literal from each 3-clause.

7 Putting Everything Together

Working with the randomness model described earlier, and using the results about the truncated Poisson distribution and the Claim in Section 5, the expected changes, from a round to the next, of the $h+3$ parameters in S can be computed by elementary probabilistic calculations. Thus, we derive the system of differential equations:

(a) $\dfrac{d\ell}{dt} = -2 - 2\dfrac{\rho_2}{1-\rho_2},$

(b) $\dfrac{dc_3}{dt} = -3\left(\dfrac{d^j_{fr}c_3}{p} + \dfrac{\rho_3}{2}\right) - 3\left(\dfrac{d_{fo}c_3}{p} + \dfrac{\rho_3}{2}\right)\dfrac{\rho_2}{1-\rho_2},$

(c) $\dfrac{dc_2}{dt} = \dfrac{3\rho_3}{2} - \dfrac{2d^j_{fr}c_2}{p} - \rho_2 + \left(\dfrac{3\rho_3}{2} - \dfrac{2d_{fo}c_2}{p} - \rho_2\right)\dfrac{\rho_2}{1-\rho_2},$

(d) $\dfrac{dx_s}{dt} = (6c_3 + 2c_2)\dfrac{(s+1)x_{s+1} - sx_s}{p^2} d^j_{fr} - \dfrac{x_s}{\ell} - \delta_{s,j}$

$+ \left((6c_3 + 2c_2)\dfrac{(s+1)x_{s+1} - sx_s}{p^2} d_{fo} - \dfrac{x_s}{\ell} - \dfrac{sx_s}{p}\right)\dfrac{\rho_2}{1-\rho_2},$

for $s = 0, \ldots, h-2,$

(e) $\dfrac{dx_{h-1}}{dt} = (6c_3 + 2c_2)\dfrac{hy_h - (h-1)x_{h-1}}{p^2} d^j_{fr} - \dfrac{x_{h-1}}{\ell} - \delta_{h-1,j}$

$+ \left((6c_3 + 2c_2)\dfrac{(hy_h - (h-1)x_{h-1})}{p^2} d_{fo} - \dfrac{x_{h-1}}{\ell} - \dfrac{(h-1)x_{h-1}}{p}\right)\dfrac{\rho_2}{1-\rho_2},$

where:

$$\delta_{s,j} = \begin{cases} 1 \text{ if } j = s, \\ 0 \text{ otherwise,} \end{cases} \quad s = 0, \ldots, h-1,$$

$$d^j_{fr} = \begin{cases} p_h/x_h \text{ if } j = h, \\ j \quad\quad \text{ if } 0 \le j \le h-1, \end{cases} \text{ and: } \frac{p_h}{x_h} = \frac{\mu\left(e^\mu - \sum_{s=0}^{h-2} \frac{\mu^s}{s!}\right)}{e^\mu - \sum_{s=0}^{h-1} \frac{\mu^s}{s!}},$$

$$d_{fo} = \frac{\sum_{s=0}^{h-1} s^2 x_s}{p} + \frac{x_h \left(\mu^2 e^\mu + \mu e^\mu - \sum_{s=0}^{h-1} \frac{s^2 \mu^s}{s!}\right)}{\lambda p} - 1.$$

$$y_h = \frac{x_h \mu^h}{\lambda h!}, \text{ and: } \lambda = e^\mu - \sum_{s=0}^{h-1} \frac{\mu^s}{s!}$$

Initial conditions:

$$\ell = 2, \ c_3 = c, \ c_2 = 0, \ x_s = \frac{2e^{-3c/2}(3c/2)^s}{s!}, \text{ for } s = 0, \ldots, h-1.$$

It turns out that for initial density $c = 3.42$, the density $\rho_2 < 1$ all the way until $\rho_3 + \rho_2 < 1$. At this stage we apply the criterion in [10] to conclude that the remaining formula is satisfiable (see Section 5).

Theorem 1. *Random 3-CNF formulas with scaled number of clauses less than 3.42 (the scaling is with respect to the number n of variables) are a.s., in the large n limit, satisfiable.*

8 Experiments

The simulation of the algorithm was implemented on C. For the generation of random 3-CNF formulas, we made use of the code freely distributed at SAT–The Satisfiability Library [26]. Our implementation was influenced and makes use of the code for the implementation of GSAT, also available in the above site. The simulation was implemented for 5×10^5 variables.

The simulation results for the parameters in $\mathcal{S}$ are very close to the corresponding values obtained from the numerical solution of the differential equations as can be seen from Table 1. In this table, line initiated with "d.e." contains the vector solution of the system of differential equations while each following "sim." line contains the corresponding experimental values. Finally, preliminary experiments we conducted suggest that by greedy algorithms that take into account not only the number of clauses satisfied by the selection of a literal but also the number of clauses that are shrunk (but ignore the size of the clause where this literal appears), the value 3.42 can be raised significantly, up to 3.6 and perhaps more.

Acknowledgment

Discussions of the second author with D. Achlioptas were crucial in developing the ideas in this work.

Table 1. Comparison of values obtained from the numerical solution of the differential equations with the values given by the simulation

	t	ℓ	ρ_2	ρ_3	x_0	x_1	x_2	...	x_8	x_9	x_{10}
d.e.	.000000	2.000000	.000000	3.420000	.011833	.060703	.155705	...	.140772	.080240	.073582
sim.	.000000	2.000000	.000000	3.420000	.012078	.060154	.155992	...	.141078	.079986	.073526
d.e.	.010000	1.979478	.051372	3.294281	.013127	.065913	.165471	...	.131459	.072808	.055457
sim.	.010000	1.979436	.051455	3.294243	.013392	.065444	.165650	...	.131904	.072540	.055550
d.e.	.020000	1.957801	.102948	3.165518	.014615	.071725	.176000	...	.121703	.064449	.039038
sim.	.020000	1.957622	.102767	3.165561	.014962	.071308	.175780	...	.122222	.064282	.039098
d.e.	.030000	1.934832	.154631	3.033612	.016330	.078224	.187345	...	.111393	.055419	.024384
sim.	.030000	1.934700	.154814	3.033580	.016772	.077558	.187116	...	.111658	.055340	.024436
d.e.	.040000	1.910425	.206383	2.898391	.018320	.085510	.199558	...	.100470	.045974	.011536
sim.	.040000	1.910484	.206123	2.898846	.018770	.084898	.199070	...	.100874	.045974	.011542
d.e.	.050000	1.884366	.258116	2.759672	.020639	.093699	.212686	...	.088941	.036379	.000520
sim.	.050000	1.884208	.258151	2.759493	.021228	.092942	.212402	...	.089040	.036536	.000508
d.e.	.050510	1.882988	.260752	2.752500	.020767	.094143	.213381	...	.088338	.035890	.000017
sim.	.050510	1.882800	.260922	2.752160	.021382	.093422	.213088	...	.088480	.036042	.000000
d.e.	.060000	1.856401	.310999	2.632164	.022984	.101647	.224767	...	.077376	.020061	.000015
sim.	.060000	1.856256	.311209	2.631981	.023624	.100960	.224676	...	.077812	.020238	.000000
d.e.	.070000	1.826188	.364454	2.500700	.025703	.110515	.237584	...	.064064	.006658	.000012
sim.	.070000	1.826040	.364570	2.500637	.026374	.109922	.237944	...	.064476	.006812	.000000
d.e.	.076105	1.806463	.397338	2.417882	.027596	.116482	.245831	...	.055433	.000008	.000011
sim.	.076105	1.806744	.396826	2.418985	.028210	.115870	.246294	...	.055834	.000120	.000000
d.e.	.080000	1.793301	.418885	2.367798	.028785	.120138	.250699	...	.046740	.000008	.000010
sim.	.080000	1.793068	.419080	2.367642	.029412	.119574	.251230	...	.047058	.000000	.000000
d.e.	.090000	1.757114	.474996	2.234567	.032230	.130388	.263740	...	.027271	.000007	.000008
sim.	.090000	1.756488	.475355	2.233755	.032886	.130000	.263852	...	.027436	.000000	.000000
d.e.	.104000	1.669236	.555691	2.034841	.038258	.147221	.283259	...	.006595	.000005	.000005
sim.	.104000	1.699116	.556006	2.035145	.038954	.146628	.283420	...	.006798	.000000	.000000
d.e.	.109955	1.671313	.590879	1.944219	.041385	.155450	.291943	...	.000005	.000005	.000004
sim.	.109955	1.671960	.590224	1.946347	.042202	.154384	.292072	...	.000180	.000000	.000000
d.e.	.120000	1.617999	.654258	1.790959	.047159	.169702	.305322	...	.000005	.000004	.000003
sim.	.120000	1.617592	.654565	1.790711	.047866	.169326	.305842	...	.000000	.000000	.000000
d.e.	.130000	1.553796	.721099	1.620394	.054604	.186559	.318659	...	.000004	.000003	.000002
sim.	.130000	1.554480	.720315	1.622145	.055456	.186224	.318190	...	.000000	.000000	.000000
d.e.	.140000	1.470975	.793518	1.421630	.064857	.207177	.330796	...	.000003	.000002	.000001
sim.	.140000	1.475280	.789439	1.430790	.065046	.206406	.329776	...	.000000	.000000	.000000
d.e.	.150000	1.346948	.880189	1.170968	.080346	.232620	.336419	...	.000002	.000001	.000000
sim.	.150000	1.346224	.880747	1.169965	.080994	.232604	.335538	...	.000000	.000000	.000000
d.e.	.155000	1.235824	.936604	.979773	.094423	.249925	.330070	...	.000001	.000000	.000000
sim.	.155000	1.234012	.937041	.978266	.095532	.249156	.329176	...	.000000	.000000	.000000
d.e.	.170000	.600820	.803173	.222643	.179484	.225146	.137641	...	.000000	.000000	.000000
sim.	.170000	.595608	.797578	.219674	.179668	.223954	.135280	...	.000000	.000000	.000000
d.e.	.171605	.585324	.782149	.207836	.182467	.221233	.130492	...	.000000	.000000	.000000
sim.	.171605	.579056	.774584	.203503	.183088	.219382	.127930	...	.000000	.000000	.000000

References

[1] Achlioptas, D.: Setting two variables at a time yields a new lower bound for random 3-SAT. Proc. 32nd Annual ACM Symposium on Theory of Computing (STOC '00) 28–37

[2] Achlioptas, D.: Lower bounds for random 3-SAT via differential equations. Theoretical Computer Science **265** (1-2) (2001) 159–185

[3] Achlioptas, D., Molloy, M.: The analysis of a list-coloring algorithm on a random graph. Proc. 38th Annual Symposium on Foundations of Computer Science (FOCS '97) 204–212

[4] Achlioptas, D., Moore, C.: Almost all graphs with average degree 4 are 3-colorable. Proc. 34th Annual ACM Symposium on Theory of Computing (STOC '02) 199–208

[5] Achlioptas D., Sorkin G. B.: Optimal myopic algorithms for random 3-SAT. Proc. 41st Annual Symposium on Foundations of Computer Science (FOCS '00) 590–600

- [6] Brélaz, D.: New methods to color the vertices of a graph. Communications of the ACM **22** (1979) 251–256
- [7] Broder, A., Frieze, A., Upfal, E.: On the satisfiability and maximum satisfiability of random 3-CNF formulas. Proc. 4th ACM-SIAM Symposium on Discrete Algorithms (SODA '93) 322–330
- [8] Chao, M. T., Franco, J.: Probabilistic analysis of two heuristics for the 3-satisfiability problem. SIAM J. of Comp. **15** (4) (1986) 1106–1118
- [9] Chvátal, V., Reed, B.: Mick gets some (the odds are on his side). Proc. 33rd Annual Symposium on the Foundation of Computer Science (FOCS '92) 620–627
- [10] Cooper, C., Frieze, A., Sorkin, G. B.: A note on random 2-SAT with prescribed literal degrees. Proc. 13th ACM-SIAM Symposium on Discrete Algorithms (SODA '02) 316–320
- [11] Devroye, L.: Branching processes and their applications in the analysis of tree structures and tree algorithms. In: Habib, M., McDiarmid, C., Alfonsin, R., Reed, B. (eds.): Probabilistic Methods for Algorithmic Discrete Mathematics. Lecture Notes in Computer Science, Vol. . Springer-Verlag, Berlin (1998) 249–314
- [12] Dubois, O., Boufkhad, Y.: A general upper bound for the satisfiability threshold of random r-SAT. J. of Algorithms **24** (1997) 395–420
- [13] Dubois, O., Boufkhad, Y., Mandler, J.: Typical random 3-SAT formulae and the satisfiability threshold. In: Proc. 11th Symposium on Discrete Algorithms (SODA '00) 126–127
- [14] Friedgut, E., (Appendix by Bourgain, J.): Sharp thresholds of graph properties, and the k-SAT problem. J. AMS **12** (1997) 1017–1054
- [15] Frieze A., Suen S.: Analysis of two simple heuristics for random instances of k-SAT. J. Algorithms. **20** (1996) 312–355
- [16] Janson, S., Łuczak, T., Ruciński, A.: Random Graphs. Wiley (2000)
- [17] Janson, S., Stamatiou, Y. C., Vamvakari, M.: Bounding the Unsatisfiability Threshold of Random 3-SAT. Random Structures and Algorithms **17** (2000) 103–116
- [18] Kamath, A., Motwani, R., Palem, K., Spirakis, P.: Tail bounds for occupancy and the satisfiability threshold conjecture. Random Structures and Algorithms **7** 1995 59–80
- [19] Kaporis, A. C., Kirousis, L. M., Stamatiou, Y. C.: How to prove conditional randomness using the Principle of Deferred Decisions. TR, CTI, Greece, (2002) Available at: www.ceid.upatras.gr/faculty/kirousis/kks-pdd02.ps
- [20] Kaporis, A. C., Kirousis, L. M., Stamatiou, Y. C., Vamvakari, M., Zito, M.: The Unsatisfiability Threshold Revisited. To appear in Discrete Mathematics
- [21] Kirousis, L. M., Kranakis, E., Krizanc, D., Stamatiou, Y. C.: Approximating the Unsatisfiability Threshold of Random Formulas. Random Structures and Algorithms **12** (1998) 253–269
- [22] Maftouhi, M., Fernandez, V.: On random 3-SAT. Combinatorics, Probability and Computing **4** (1995) 190–195
- [23] Mitzenmacher, M.: Tight thresholds for the pure literal rule. TR, Digital Equipment Corporation, (1997) Available at: www.research.compaq.com/SRC/
- [24] Molloy, M.: The probabilistic method. In: Habib, M., McDiarmid, C., Alfonsin, R., Reed, B. (eds.): Probabilistic Methods for Algorithmic Discrete Mathematics. Lecture Notes in Computer Science, Vol. . Springer-Verlag, Berlin (1998) 1–35
- [25] Pittel, B., Spencer, J., Wormald, N. C.: Sudden emergence of a giant k-core in a random graph. J. Combinatorial Theory, Series B. **67** (1996) 111–151
- [26] SAT–The Satisfiability Library, www.intellektik.informatik.tu-darmstadt.de/SATLIB/
- [27] Wormald, N. C. Differential equations for random processes and random graphs. The Annals of Applied Probability. **5** (4) (1995) 1217–1235

Dynamic Additively Weighted
Voronoi Diagrams in 2D[*]

Menelaos I. Karavelas and Mariette Yvinec

INRIA Sophia-Antipolis, Project PRISME
2004 Route des Lucioles, BP 93
06902 Sophia-Antipolis Cedex, France
{Menelaos.Karavelas,Mariette.Yvinec}@sophia.inria.fr

Abstract. In this paper we present a dynamic algorithm for the construction of the additively weighted Voronoi diagram of a set of weighted points in the plane. The novelty in our approach is that we use the dual of the additively weighted Voronoi diagram to represent it. This permits us to perform both insertions and deletions of sites easily. Given a set $\mathcal{B}$ of n sites, among which h sites have a non-empty cell, our algorithm constructs the additively weighted Voronoi diagram of $\mathcal{B}$ in $O(nT(h) + h \log h)$ expected time, where $T(k)$ is the time to locate the nearest neighbor of a query site within a set of k sites. Deletions can be performed for all sites whether or not their cell is empty. The space requirements for the presented algorithm is $O(n)$. Our algorithm is simple to implement and experimental results suggest an $O(n \log h)$ behavior.

1 Introduction

One of the most well studied structures in computational geometry is the Voronoi diagram for a set of sites. Applications include retraction motion planning, collision detection, computer graphics or even networking and communication networks. There have been various generalizations of the standard Euclidean Voronoi diagram, including generalizations to L_p metrics, convex distance functions, the power distance, which yields the power diagram, and others. The sites considered include points, convex polygons, line segments, circles and more general smooth convex objects.

In this paper we are interested in the *Additively Weighted Voronoi diagram* or, in short, AW-Voronoi diagram. We are given a set of points and a set of weights associated with them. Let $d(\cdot, \cdot)$ denote the Euclidean distance. We define the distance $\delta(p, B)$ between a point p on the Euclidean plane $\mathbb{E}^2$ and a weighted point $B = \{b, r\}$ as $\delta(p, B) = d(p, b) - r$. If the weights are positive, the additively weighted Voronoi diagram can be viewed geometrically as the Voronoi diagram for a set of circles, the centers of which are the points and the radii of which are

---

[*] Work partially supported by the IST Programme of the EU as a Shared-cost RTD (FET Open) Project under Contract No IST-2000-26473 (ECG - Effective Computational Geometry for Curves and Surfaces).

the corresponding weights. Points outside a circle have positive distance with respect to the circle, whereas points inside a circle have negative distance. The Voronoi diagram does not change if all the weights are translated by the same quantity. Hence, in the sequel we assume that all the weights are positive. In the same context we use the term *site* to denote interchangeably a weighted point or the corresponding circle. We also define the distance between two sites $B_1 = \{b_1, r_1\}$, $B_2 = \{b_2, r_2\}$ on the plane to be $\delta(B_1, B_2) = d(b_1, b_2) - r_1 - r_2$. Note that $\delta(B_1, B_2)$ is negative if the two sites (circles) intersect at two points or if one is inside the other. If we assign every point on the plane to its closest site, we get a subdivision of the plane into regions. The closures of these regions are called *Voronoi cells*. The one-dimensional connected sets of points that belong to exactly two Voronoi cells are called *Voronoi edges*, whereas points that belong to at least three Voronoi cells are called *Voronoi vertices*. The collection of cells, edges and vertices is called the *Voronoi diagram*. A *bisector* between two sites is the locus of points that are equidistant from both sites. Unlike the case of the Euclidean Voronoi diagram of points in an AW-Voronoi diagram the cell of a given site may be empty. Such a site is called *trivial*. We can actually fully characterize trivial sites: a site is trivial if it is fully contained inside another site.

The first algorithm for computing the AW-Voronoi diagram appeared in [1]. The running time of the algorithm is $O(nc^{\sqrt{\log n}})$, where c is a constant, and it works only in the case of disjoint sites. The same authors presented in [2] another algorithm for constructing the AW-Voronoi diagram, which runs in $O(n \log^2 n)$ time. This algorithm uses the divide-and-conquer paradigm and works again only for disjoint sites. A detailed description of the geometric properties of the AW-Voronoi diagram, as well as an algorithm that treats intersecting sites can be found in [3]. The algorithm runs in $O(n \log^2 n)$ time, and also uses the divide-and-conquer paradigm. A sweep-line algorithm is described in [4] for solving the same problem. The set of sites is first transformed to a set of points by means of a special transformation, and then a sweep-line method is applied to the point set. The sweep-line algorithm runs in time $O(n \log n)$. Aurenhammer [5] suggests a lifting map of the two-dimensional problem to three dimensions, and reduces the problem of computing the AW-Voronoi Voronoi diagram in 2D to computing the power diagram of a set of spheres in 3D. The algorithm runs in $O(n^2)$ time, but it is the first algorithm for constructing the AW-Voronoi diagram that generalizes to dimension $d \geq 3$. If we do not have trivial sites, every pair of sites has a bisector. In this case, the AW-Voronoi diagram is a concrete type of an *Abstract Voronoi diagram* [6], for which optimal divide-and-conquer $O(n \log n)$ algorithms exist. Incremental algorithms that run in $O(n \log n)$ expected time also exist for abstract Voronoi diagrams (see [7, 8]). The algorithm in [8] allows the insertion of sites with empty Voronoi cell. However, it does not allow for deletions and the data structures used are a bit involved. More specifically, the Voronoi diagram itself is represented as a planar map and a history graph is used to find the conflicts of the new site with the existing Voronoi diagram. Finally, an off-line algorithm that constructs the Delaunay triangulation of the centers of the sites and then performs edge-flips in order to restore the AW-Delaunay graph is presented in [9]. Again this algorithm does not allow the deletion of sites, and moreover it does not handle the case of trivial sites.

In this paper we present a fully dynamic algorithm for the construction of the AW-Voronoi diagram. Our algorithm resembles the algorithm in [8], but, it also has several differences. Firstly, we do not represent the AW-Voronoi diagram, but rather its dual. The dual of the AW-Voronoi diagram is a planar graph of linear size [3], which we call the *Additively Weighted Delaunay graph* or AW-Delaunay graph, for short. Moreover, under the non-degeneracy assumption that there are no points in the plane equidistant to more than 3 sites, the AW-Delaunay graph has triangular faces. Our algorithm requires no assumption about degeneracies. It implicitly uses a perturbation scheme which simulates non-degeneracies and yields an AW-Delaunay graph with triangular faces. Hence, representing the AW-Voronoi diagram can be done in a much simpler way compared to [8]. In [8] the insertion is done in two stages. First the history graph is used to find the conflicts of the new site with the existing AW-Voronoi diagram. Then both the planar map representation of the AW-Voronoi diagram and the history graph are updated, in order to incorporate the Voronoi cell of the new site. In our algorithm the insertion of the new site is done in three stages. The first stage finds the nearest neighbor of the new site in the existing AW-Voronoi diagram. Using the nearest neighbor, the second stage determines if the new site is trivial. Starting from the nearest neighbor of the new site, the third stage finds all the Voronoi edges in conflict with the new site and then reconstructs the AW-Voronoi diagram. This is done using the dual graph. Another novelty of our algorithm is that it permits the deletion of sites, which is not the case in [8].

The remainder of the paper is structured as follows. In Section 2 we define the AW-Voronoi and its dual. We review some of the known properties of the AW-Voronoi diagram and provide definitions used in the remainder of the paper. In Section 3 we show how to insert a new site, and discuss how we deal with trivial sites. In Section 4 we describe how to delete sites. In Section 5 we briefly discuss the predicates involved in our algorithm and present experimental results. A more detailed analysis of the predicates and how to compute them can be found in [10]. Section 6 is devoted to conclusions and directions for further research. Due to space limitations the proofs of the lemmas and theorems are omitted. this version of the paper. The interested reader can find these proofs in [11].

2 Preliminaries

Let $\mathcal{B}$ be a set of sites B_j, with centers b_j and radii r_j. For each $j \neq i$, let $H_{ij} = \{y \in \mathbb{E}^2 : \delta(y, B_i) \leq \delta(y, B_j)\}$. Then the (closed) Voronoi cell V_i of B_i is defined to be $V_i = \cap_{i \neq j} H_{ij}$. The connected set of points that belong to exactly two Voronoi cells are called *Voronoi edges*, whereas points that belong to at least three Voronoi cells are called *Voronoi vertices*. The *AW-Voronoi diagram* $\mathcal{V}(\mathcal{B})$ of $\mathcal{B}$ is defined as the collection of the Voronoi cells, edges and vertices. The *Voronoi skeleton* $\mathcal{V}_1(\mathcal{B})$ of $\mathcal{B}$ is defined as the union of the Voronoi edges and

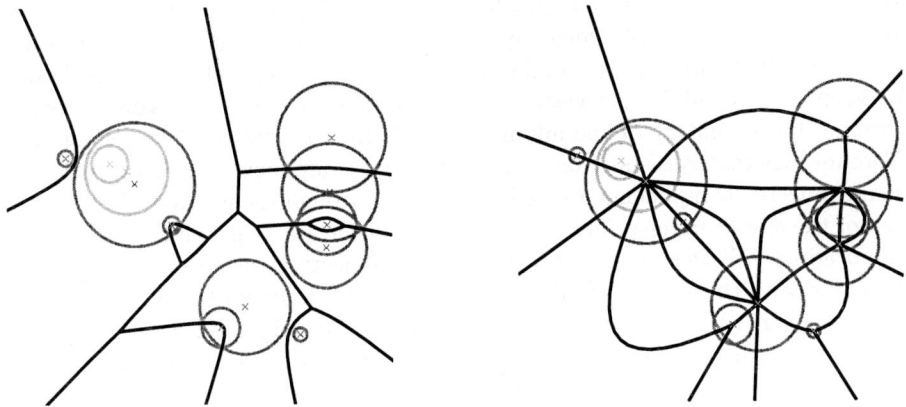

Fig. 1. Left: the AW-Voronoi diagram for a set of 12 sites, among which 2 sites are trivial. Non-trivial sites are shown in gray. Trivial sites are shown in light gray. The Voronoi skeleton is shown in black. Right: a planar embedding of the AW-Delaunay graph of the same set of sites. The edges of the AW-Delaunay graph are shown in black

Voronoi vertices of $\mathcal{V}(\mathcal{B})$. The AW-Voronoi diagram is a subdivision of the plane [3, Property 1]. The Voronoi edges are straight or hyperbolic arcs and each cell is star-shaped with respect to the center of the corresponding site [3, Properties 3 and 4]). In the case of the usual Euclidean Voronoi diagram for a set of points, every point has a non-empty Voronoi cell. In AW-Voronoi diagrams there may exist sites with empty Voronoi cells. In particular, the Voronoi cell V_i of a site B_i is empty if and only if B_i is contained in another site B_j [3, Property 2]. A site whose Voronoi cell has empty interior is called *trivial*, whereas a site whose Voronoi cell has non-empty interior is called *non-trivial* (see Fig. 1(left)).

We call AW-Delaunay graph, and denote by $\mathcal{D}(\mathcal{B})$, the dual graph of the AW-Voronoi diagram $\mathcal{V}(\mathcal{B})$. There is a vertex in $\mathcal{D}(\mathcal{B})$ for each non-trivial site B_i in $\mathcal{B}$. Let B_i and B_j be two sites whose Voronoi cells V_i and V_j are adjacent. We denote by α_{ij}^{kl} the Voronoi edge in $V_i \cap V_j$ whose endpoints are the Voronoi vertices equidistant to B_i, B_j, B_k and B_i, B_j, B_l, respectively. There exists an edge e_{ij}^{kl} in $\mathcal{D}(\mathcal{B})$ connecting B_i and B_j for each edge α_{ij}^{kl} of $\mathcal{V}(\mathcal{B})$ in $V_i \cap V_j$. The fact that we have a planar embedding of linear size for the AW-Delaunay graph [3, Property 7] immediately implies that the size of the AW-Voronoi diagram is $O(n)$. The Voronoi skeleton may consist of more than one connected component [3, Property 9], whereas the dual graph is always connected. If we do not have any degeneracies, the AW-Delaunay graph has the property that all but its outer face have exactly three edges. However, it may contain vertices of degree 2, i.e., we have triangular faces with two edges in common. If the Voronoi skeleton consists of more than one connected component the AW-Delaunay graph may also have vertices of degree 1, which are the dual of Voronoi cells with no vertices (e.g., the Voronoi cell at the top left corner of Fig. 1(left)). To simplify the representation of the AW-Delaunay graph we add a fictitious site called the site at infinity. This amounts to adding a Voronoi vertex on each unbounded edge of $\mathcal{V}_1(\mathcal{B})$. These additional vertices are then connected through Voronoi edges forming the boundary of the infinite site cell. In this compactified version, the Voronoi skeleton consists of only one connected component, and the previously non-connected components are now connected through the edges of the Voronoi cell of the site at infinity. The compactified AW-Delaunay graph corresponds to the original AW-Delaunay graph plus edges connecting the sites on the convex hull of $\mathcal{B}$ with the site at infinity. In the absence of degeneracies, all faces of the compactified AW-Delaunay graph have exactly three edges, but this graph may still have vertices of degree 2. From now on when we refer to the AW-Voronoi diagram or the AW-Delaunay graph, we refer to their compactified versions (see Fig. 1(right)). Degenerate cases arise when there are points equidistant to more than three sites. Then, the AW-Delaunay graph has faces with more than three edges. This is entirely analogous to the situation for the usual Delaunay diagram for a set of points with subsets of more than three cocircular points. In such a case, a graph with triangular faces can be obtained from the AW-Delaunay graph through an arbitrary triangulation of the faces with more than three edges. Our algorithm uses an implicit perturbation scheme and produces in fact such a triangulated AW-Delaunay graph.

Let B_i and B_j be two sites such that no one is contained inside the other. A circle tangent to B_i and B_j that neither contains any of them nor is contained in any of them is called an *exterior bitangent Voronoi circle*. A circle tangent to B_i and B_j that lies in $B_i \cap B_j$ is an *interior bitangent Voronoi circle*. Similarly, a circle tangent to three sites B_i, B_j and B_k is an *exterior tritangent Voronoi circle* if it neither contains any of B_i, B_j and B_k nor is contained in any of them. A circle tangent to B_i, B_j and B_k is called an *interior tritangent Voronoi circle* if it is included in $B_i \cap B_j \cap B_k$. A triple of sites B_i, B_j and B_k can have up to two tritangent Voronoi circles, either exterior or interior. This is equivalent to stating that the AW-Voronoi diagram of three sites can have up to two Voronoi vertices [3, Property 5]. Let π_{ij} denote the bisector of the sites B_i and B_j. As we already mentioned π_{ij} can be a line or a hyperbola. We define the orientation of π_{ij} to be such that b_i is always to the left of π_{ij}. The orientation of π_{ij} defines an ordering on the points of π_{ij}, which we denote by $\prec_{ij}$. Let o_{ij} be the intersection of π_{ij} with the segment $b_i b_j$. We can parameterize π_{ij} as follows: if $o_{ij} \prec_{ij} p$ then $\zeta_{ij}(p) = \delta(p, B_i) - \delta(o_{ij}, B_i)$; otherwise $\zeta_{ij}(p) = -(\delta(p, B_i) - \delta(o_{ij}, B_i))$. The function $\zeta_{ij}(\cdot)$ is a 1–1 and onto mapping from π_{ij} to $\mathbb{R}$. The *shadow region* $S_{ij}(B)$ of a site B with respect to the bisector π_{ij} of B_i and B_j is the locus of points c on π_{ij} such that $\delta(B, C_{ij}(c)) < 0$, where $C_{ij}(c)$ is the bitangent Voronoi circle of B_i and B_j centered at c. Let $\tilde{S}_{ij}(B)$ denote the set of parameter values $\zeta_{ij}(c)$, where $c \in S_{ij}(B)$. It is easy to verify that $\tilde{S}_{ij}(B)$ can be of the form $\emptyset$, $(-\infty, \infty)$, $(-\infty, a)$, (b, ∞), (a, b) and $(-\infty, a) \cup (b, \infty)$, where $a, b \in \mathbb{R}$.

Let α_{ij}^{kl} be an edge of $\mathcal{V}(\mathcal{B})$ on the bisector π_{ij}. Let C_{ijk} and C_{ijl} be the tritangent Voronoi circles associated with the endpoints of α_{ij}^{kl}. We denote by c_{ijk}

(resp. c_{ijl}) the center of C_{ijk} (resp. C_{ijl}). Under the mapping $\zeta_{ij}(\cdot)$, α_{ij}^{kl} maps to the interval $\tilde{\alpha}_{ij}^{kl} = [\xi_{ijl}, \xi_{ijk}] \subset \mathbb{R}$. We define the *conflict region* $R_{ij}^{kl}(B)$ of B with respect to the edge α_{ij}^{kl} to be the intersection $R_{ij}^{kl}(B) = \alpha_{ij}^{kl} \cap S_{ij}(B)$. We say that B *is in conflict* with α_{ij}^{kl} if $R_{ij}^{kl}(B) \neq \emptyset$. Under the mapping by $\zeta_{ij}(\cdot)$, the conflict region $R_{ij}^{kl}(B)$ maps to the intersection $\tilde{R}_{ij}^{kl}(B) = \tilde{\alpha}_{ij}^{kl} \cap \tilde{S}_{ij}(B)$. $\tilde{R}_{ij}^{kl}(B)$ can be one of the following types : (1) $\tilde{R}_{ij}^{kl}(B) = \emptyset$. We say that B is *not in conflict* with α_{ij}^{kl}. (2) $\tilde{R}_{ij}^{kl}(B)$ consists of two disjoint intervals, including respectively ξ_{ijk} and ξ_{ijl}. We say that B *is in conflict with both vertices of* α_{ij}^{kl}. (3) $\tilde{R}_{ij}^{kl}(B)$ consists of a single connected interval. We further distinguish between the following cases: (a) $\tilde{\alpha}_{ij}^{kl} = \tilde{R}_{ij}^{kl}(B)$. We say that B *is in conflict with the entire edge* α_{ij}^{kl}. (b) $\tilde{R}_{ij}^{kl}(B)$ contains either ξ_{ijk} or ξ_{ijl}, but not both. We say that B *is in conflict with one vertex of* α_{ij}^{kl}. (c) $\tilde{R}_{ij}^{kl}(B) \neq \emptyset$ but contains neither ξ_{ijk} nor ξ_{ijl}. We say that B *is in conflict with the interior of* α_{ij}^{kl}. Finally we define the *conflict region* $R_{\mathcal{B}}(B)$ of B with respect to $\mathcal{B}$ as the union $R_{\mathcal{B}}(B) = \bigcup_{\alpha_{ij}^{kl} \in \mathcal{V}(\mathcal{B})} R_{ij}^{kl}(B)$. It is easy to verify that $R_{\mathcal{B}}(B) = V_{\mathcal{B} \cup \{B\}}(B) \cap \mathcal{V}_1(\mathcal{B})$, where $V_{\mathcal{B} \cup \{B\}}(B)$ denotes the Voronoi cell of B in $\mathcal{V}(\mathcal{B} \cup \{B\})$.

3 Inserting a Site Incrementally

In this section we present the incremental algorithm. Let again $\mathcal{B}$ be our set of n sites and let us assume that we have already constructed the AW-Voronoi diagram for a subset $\mathcal{B}_m$ of $\mathcal{B}$. Here m denotes the number of sites in $\mathcal{B}_m$. We now want to insert a site $B \notin \mathcal{B}_m$. The insertion is done in the following steps : (i) locate the nearest neighbor $NN(B)$ of B in $\mathcal{B}_m$; (ii) test if B is trivial; (iii) find the conflict region of B and repair the AW-Delaunay graph. We postpone the discussion on the location of the nearest neighbor until the end of this section.

The first test we have to do is to determine whether B is trivial or not. This can easily be done once we know the nearest neighbor $NN(B)$ of B, since B is trivial if and only if $B \subset NN(B)$ [11, Lemma 1]. Let $R_m(B)$ be the conflict region of B with respect to $\mathcal{B}_m$. Let $\partial R_m(B)$ denote the boundary of $R_m(B)$. $R_m(B)$ is a subset of $\mathcal{V}_1(\mathcal{B}_m)$ and $\partial R_m(B)$ is a set of points on edges of $\mathcal{V}_1(\mathcal{B}_m)$. Points in $\partial R_m(B)$ are the vertices of the Voronoi cell V_B of B in $\mathcal{V}(\mathcal{B}_{m+1})$, where $\mathcal{B}_{m+1} = \mathcal{B}_m \cup \{B\}$. It has been shown in [8, Lemma 1] that $R_m(B)$ is connected. Thus, the aim is to discover the boundary $\partial R_m(B)$ of $R_m(B)$, since then we can repair the AW-Voronoi diagram in exactly the same way as in [8]. The idea is to perform a *depth first search* (DFS) on $\mathcal{V}_1(\mathcal{B}_m)$ to discover $R_m(B)$ and $\partial R_m(B)$, starting from a point on the skeleton that is known to be in conflict with B. Let L denote the boundary of the currently discovered portion of $R_m(B)$. Initially $L = \emptyset$. We are going to represent points in L by the Voronoi edges that contain them. We want the points of $\partial R_m(B)$ to appear in L in the order that they appear on the boundary of the Voronoi region V_B of B in $\mathcal{V}(\mathcal{B}_{m+1})$. Without loss of generality we can choose this order to be the counter-clockwise ordering of the vertices on the boundary of V_B.

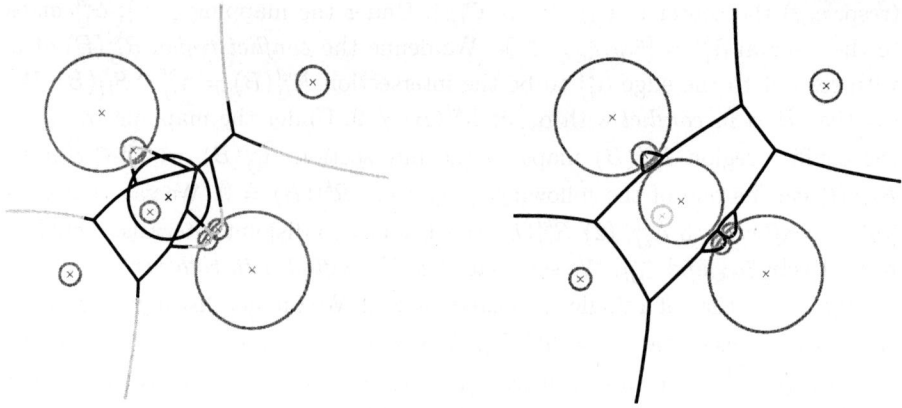

Fig. 2. Left: The AW-Voronoi diagram for a set of sites (gray) and the conflict region (black) of a new site (also black). The portion of the Voronoi skeleton that does not belong to the conflict region of the new site is shown in light gray. Right: The AW-Voronoi diagram after the insertion of the new site. Non-trivial sites, including the new site, are shown in gray. The site in light gray is inside the new site and has become trivial. The Voronoi skeleton is shown in black

As we mentioned in the previous paragraph, we need to find a first point on the Voronoi skeleton $\mathcal{V}_1(\mathcal{B}_m)$, that is in conflict with B. This point is going to serve as the starting point for the DFS. It can be shown that if B is a non-trivial site, then B has to be in conflict with at least one of the edges of the Voronoi cell $V_{NN(B)}$ of $NN(B)$ in $\mathcal{V}(\mathcal{B}_m)$ [11, Lemma 2]. Hence, we can simply walk on the boundary of $V_{NN(B)}$, until we find a Voronoi edge in conflict with B. Let α be the first edge, of the boundary of $V_{NN(B)}$ that we found to be in conflict with B. If B is in conflict with the interior of α, we have discovered the entire conflict region $R_m(B)$. In this case L consists of two copies of α with different orientations. Otherwise, B has to be in conflict with at least one of the two Voronoi vertices of α. In this case we set L to be the edges adjacent to that Voronoi vertex in counter-clockwise order. The DFS will then recursively visit all vertices in conflict with B. Suppose that we have arrived at a Voronoi vertex v (which is a node on the Voronoi skeleton). Firstly, we mark it. Then we look at all the Voronoi edges α adjacent to it. Let v' be the Voronoi vertex of α that is different from v. We consider the following two cases : (1) $\underline{v'\text{ has not been marked}}$. If B is in conflict with the entire edge α, then we replace α in L by the remaining Voronoi edges adjacent to v', in counter-clockwise order. We then continue recursively on v'. If B is not in conflict with the entire edge α, we have reached an edge α containing a point of $\partial R_m(B)$. The list L remains unchanged and the DFS backtracks. (2) $\underline{v'\text{ has already been marked}}$. If B is in conflict with the entire edge α, have found a cycle in $R_m(B)$, or equivalently, B contains a site in $\mathcal{B}_m$, which will become trivial. Since α currently belongs to L, but does not contain any points of $\partial R_m(B)$, we remove it from L. The DFS then backtracks. If B is not in conflict with the entire edge α, then B is in conflict with both vertices of α. Hence α contains two points of $\partial R_m(B)$ in its interior. The list L remains unchanged and the DFS backtracks. Note that in this case α appears twice in L, once per point in $\partial R_m(B)$ that it contains. Fig. 2(top left) shows an example of a conflict region which triggers all the possible cases of the above search algorithm. In our case, the AW-Voronoi diagram is represented through its dual AW-Delaunay graph. It is thus convenient to restate the algorithm for finding the boundary $\partial R_m(B)$ of the conflict region $R_m(B)$ in terms of the AW-Delaunay graph. This is done in the long version of this paper (cf. [11]).

In case of degeneracies, the algorithm uses a perturbation scheme described by the following lazy strategy. Any new site which is found tangent to a tritangent Voronoi circle is considered as not being in conflict with the corresponding Voronoi vertex. Then any Voronoi vertex remains a degree 3 vertex and the dual AW-Delaunay graph is always triangular. This graph, however, is not canonical, but depends on the insertion order of the sites.

During the insertion procedure trivial sites can appear in two possible ways. Either the new site B to be inserted is trivial, or B contains existing sites, which after the insertion of B will become trivial. When deletion of sites is allowed, B may contain other sites which will become non-trivial if B is deleted. For this reason we need to keep track of trivial sites. Since a site is trivial if and only if it is contained inside some other site, there exists a natural parent-child relationship between trivial and non-trivial sites. In particular, we can associate every trivial site to a non-trivial site that contains it. If a trivial site is contained in more than one non-trivial sites, we can choose the parent of the trivial site arbitrarily. A natural choice for storing trivial sites is to maintain, for every non-trivial site, a list containing all trivial sites that have the non-trivial site as their parent. Let $\mathcal{B}_m^+$ be the subset of non-trivial sites of $\mathcal{B}_m$, and let $\mathcal{B}_m^- = \mathcal{B}_m \setminus \mathcal{B}_m^+$. For some $B' \in \mathcal{B}_m^+$, we define $L_{tr}(B')$ to be the list of trivial sites in $\mathcal{B}_m^-$ that have B' as their parent. We note by $\mathcal{L}_m$ the set of all lists $L_{tr}(B')$ for $B' \in \mathcal{B}_m^+$, and correspondingly $\mathcal{L}_{m+1}$ the set of all $L_{tr}(B')$ for $B' \in \mathcal{B}_{m+1}^+$. When a new site B is inserted and B is found to be trivial, we simply add B to $L_{tr}(NN(B))$. If B is non-trivial, let $\mathcal{B}_m^-(B)$ be the set of sites in $\mathcal{B}_m^+$ that are contained in B. Since after the insertion of B all sites in $\mathcal{B}_m^-(B)$ become trivial, we add every $B'' \in \mathcal{B}_m^-(B)$ to $L_{tr}(B)$. Moreover, for every $B'' \in \mathcal{B}_m^-(B)$ we move all sites in $L_{tr}(B'')$ to $L_{tr}(B)$. The following theorem subsumes the run time analysis of our algorithm. A detailed proof can be found in [11].

Theorem 1. *Let $\mathcal{B}$ be a set of n sites among which h are non-trivial. We can construct the AW-Voronoi diagram incrementally in $O(nT(h) + h \log h)$ expected time, where $T(k)$ is the time to locate the nearest neighbor of a query site within a set of k sites.*

We now turn our discussion on the location of the nearest neighbor. The nearest neighbor location of B in fact reduces to the location of the center b of B in $\mathcal{V}(\mathcal{B}_m)$. We can do that as follows. Select a site $B' \in \mathcal{B}_m$ at random.

Look at all the neighbors of B' in the AW-Delaunay graph. If there exists a B'' such that $\delta(B, B'') < \delta(B, B')$, then B' cannot be the nearest neighbor of B. In this case we replace B' by B'' and restart our procedure. If none of the neighbors of B' is closer to B than B', then $NN(B) = B'$. The time to find the nearest neighbor using the above procedure is trivially $O(h)$, where h is the number of non-trivial sites in $\mathcal{B}$. However, we can speed-up the nearest-neighbor location by maintaining a hierarchy of AW-Delaunay graphs as is done in [12] for the Delaunay triangulation for points. The details can be found in [11]. The randomized time analysis for the location and insertion of a point in the Delaunay hierarchy has been given in [12]. Unfortunately, this analysis does not generalize to the AW-Delaunay hierarchy. Our experimental results, however, show that we do get a speed-up and that in practice the nearest-neighbor location is done in time $O(\log h)$, which gives a total running time of $O(n \log h)$ (see Section 5).

4 Deleting a Site

Suppose that we have been given a set $\mathcal{B}$ of sites for which we have already constructed the AW-Voronoi diagram $\mathcal{V}(\mathcal{B})$. Let also $B \in \mathcal{B}$ be a site that we want to delete from $\mathcal{V}(\mathcal{B})$. In this section we describe how to perform the deletion.

Suppose that B is non-trivial. Let $\mathcal{B}_\gamma$ be the set of neighbors of V_B in $\mathcal{D}(\mathcal{B})$. Let also $L_{tr}^+(B)$ be the set of sites in $L_{tr}(B)$ that become non-trivial after the deletion of B. Finally, let $L_{tr}^-(B) = L_{tr}(B) \setminus L_{tr}^+(B)$, $\mathcal{B}_s = \mathcal{B}_\gamma \cup L_{tr}(B)$ and $\mathcal{B}_s^+ = \mathcal{B}_\gamma \cup L_{tr}^+(B)$. The main observation is that the second nearest neighbor of each point in V_B is one of the sites in $\mathcal{B}_s^+$. Moreover, every site in $L_{tr}^-(B)$ is inside one of the sites in $\mathcal{B}_s^+$ [11, Lemma 6]. Consequently, the AW-Voronoi diagram after the deletion of $\mathcal{B}$ can be found by constructing the AW-Voronoi diagram of $\mathcal{B}_\gamma \cup L_{tr}(B)$. More precisely, if b is a degree 3 vertex in $\mathcal{D}(\mathcal{B})$ and $|L_{tr}(B)| = 0$, we simply remove from $\mathcal{D}(\mathcal{B})$ the vertex corresponding to B as well as all its incident edges. If b is a degree 2 vertex in $\mathcal{D}(\mathcal{B})$ and $|L_{tr}(B)| = 0$, we again remove from $\mathcal{D}(\mathcal{B})$ the vertex v_B corresponding to B as well as all its incident edges. In addition, we collapse the edges e and e', where e and e' are the two edges of the star of v_B that are not incident to v_B. If the degree of b in $\mathcal{D}(\mathcal{B})$ is at least 4 or if $|L_{tr}(B)| > 0$, we construct $\mathcal{V}(\mathcal{B}_s)$ and then we find the nearest neighbor of B in $\mathcal{B}_s$. Once the nearest neighbor has been found we compute the conflict region of B in $\mathcal{B}_s$ by means of the procedure described in Section 3. Let $\partial^* R_s(B)$ be the representation, by means of the dual edges, of the conflict region of B in $\mathcal{B}_s$. The edges in $\partial^* R_s(B)$ are the dual of the AW-Voronoi edges in $\partial R_s(B)$. The triangles inside $\partial^* R_s(B)$ are the triangles that must appear in the interior of the boundary of the star of B when B is deleted from $\mathcal{D}(\mathcal{B})$. Therefore we can use these triangles to construct $\mathcal{D}(\mathcal{B} \setminus \{B\})$, or equivalently $\mathcal{V}(\mathcal{B} \setminus \{B\})$. Finally, all lists in $\mathcal{L}(\mathcal{B}_s^+)$ must be merged with their corresponding lists in $\mathcal{L}(\mathcal{B} \setminus \{B\})$.

Suppose that B is trivial. In this case we have to find the non-trivial site B' such that $B \in L_{tr}(B')$ and then delete B from $L_{tr}(B')$. Since $B \subset NN(B)$, B must be in the list of some B', which is in the same connected component of the union of sites as $NN(B)$. It has been shown that the subgraph $\mathcal{K}(\mathcal{B})$ of $\mathcal{D}(\mathcal{B})$ that consists of all edges of $\mathcal{D}(\mathcal{B})$ connecting intersecting sites, is a spanning subgraph of the connectivity graph of the set of sites [13, Chapter 5]. The deletion of a trivial site can, thus, be done as follows : (i) find the nearest neighbor $NN(B)$ of B; (ii) walk on the connected component $\mathcal{C}$ of $NN(B)$ in the graph $\mathcal{K}(\mathcal{B})$ and for every site $B' \in \mathcal{C}$ that contains B, test if $B \in L_{tr}(B')$; (iii) once the site B', such that $B \in L_{tr}(B')$, is found, delete B from $L_{tr}(B')$. The following theorem discusses the cost of deleting a site. A detailed analysis can be found in [11].

Theorem 2. *Let $\mathcal{B}$ be a set of n sites, among which h are non-trivial. Let $B \in \mathcal{B}$, and let $L_{tr}(B)$ be the list of trivial sites whose parent is B. If B is non-trivial, it can be deleted from $\mathcal{V}(\mathcal{B})$ in expected time $O((d+t)T(d+t') + (d+t')\log(d+t'))$, where d is the degree of B in $\mathcal{D}(\mathcal{B})$, t is the cardinality of $L_{tr}(B)$ and t' is the number of sites in $L_{tr}(B)$ that become non-trivial after the deletion of B. If B is trivial it can be deleted from $\mathcal{L}(\mathcal{B})$ in worst case time $O(n)$.*

5 Predicates and Implementation

For the purposes of computing the algebraic degree of the predicates used in our algorithm, we assume that each site is given by its center and its radius. The predicates that we use are the following :

1. Given two sites B_1 and B_2, and a query site B, determine if B is closer to B_1 or B_2. This is equivalent to comparing the distances $\delta(b, B_1)$ and $\delta(b, B_2)$. This predicate is used during the nearest neighbor location phase and it is of algebraic degree 4 in the input quantities.
2. Given a site B_1 and a query site B, determine if $B \subset B_1$. This is equivalent to the expression $\delta(B, B_1) < -2r$, where r is the radius of B. This predicate is used during the insertion procedure in order to determine whether the query site is trivial. The algebraic degree of the predicate is 2.
3. Given two sites B_1 and B_2 determine if they intersect. This predicate is used during the deletion of a trivial site, and its algebraic degree is 2.
4. Given two sites B_1 and B_2 and a tritangent Voronoi circle C_{345} determine the result of the orientation test $\text{CCW}(b_1, b_2, c_{345})$, where b_1, b_2 and c_{345} are the centers of B_1, B_2 and C_{345}, respectively. This predicate is used in order to find the first conflict of a new site B given its nearest neighbor $NN(B)$. The evaluation of this predicate is discussed in [10], where is it also shown that its algebraic degree is 14.
5. Given a Voronoi edge α and a query site B, determine the type of the conflict region of B with α. This predicate is used in order to discover the conflict region of B with respect to the existing AW-Voronoi diagram. A method for evaluating this predicate is presented in [10]. The corresponding algebraic degree is shown to be 16 in the input quantities, using techniques from Sturm sequences theory.

We have implemented two versions of our algorithm, which differ only on how the nearest neighbor location is done. The first one does the nearest neighbor

Table 1. The running times of the two algorithms as a function of the size n of the input set and the number of non-trivial sites h. T_1 indicates the time for the algorithm with one level of the AW-Voronoi diagram and T_2 indicates the running time for an hierarchy of AW-Voronoi diagrams. Unless otherwise indicated, both T_1 and T_2 are given in seconds. The experiments were performed on a Pentium-III 1GHz running Linux

n	h	h/n	T_1	T_2	$T_1/(n\log h)$	$T_2/(n\log h)$
10 000	10 000	1.00	4.75	3.59	1.18×10^{-4}	0.90×10^{-4}
10 000	7 973	0.80	4.46	3.65	1.14×10^{-4}	0.94×10^{-4}
10 000	5 017	0.50	3.64	3.02	0.98×10^{-4}	0.82×10^{-4}
100 000	99 995	1.00	85.17	38.42	1.70×10^{-4}	0.77×10^{-4}
100 000	79 861	0.80	83.37	38.52	1.70×10^{-4}	0.79×10^{-4}
100 000	49 614	0.50	67.15	32.19	1.43×10^{-4}	0.68×10^{-4}
1 000 000	999 351	1.00	> 36 min	425.38	–	0.71×10^{-4}
1 000 000	800 290	0.80	2 130.49	445.58	3.61×10^{-4}	0.75×10^{-4}
1 000 000	497 866	0.50	1 715.94	386.47	3.01×10^{-4}	0.68×10^{-4}

location using the simple procedure described in Section 3. The second implementation maintains a hierarchy of AW-Delaunay graphs. The predicates are evaluated exactly and have been implemented using two scenarios. The two scenarios are adapted, respectively, to number types that support the operations $\{+, -, \times, /, \sqrt{\ }\}$ and $\{+, -, \times\}$ exactly. Both algorithms were implemented in C++, following the design of the library CGAL [14]. Our C++ code also supports CGAL's dynamic filtering [15], which is also used in our experiments. Finally, our experimental results were produced using the implementations of the predicates that do not use square roots for their evaluation. The two algorithms were tested on random circle sets of size $n \in \{10^4, 10^5, 10^6\}$ (see Table 1). The centers were uniformly distributed in the square $[-M, M] \times [-M, M]$, where $M = 10^6$. The radii of the circles were uniformly distributed in the interval $[0, R]$, were R was chosen appropriately so as to achieve different ratios h/n. In particular, we chose R so that the ratio h/n is approximately equal to the values in the set $\{1.00, 0.80, 0.50\}$. The last column of Table 1 suggests that our algorithm runs in time $O(n \log h)$ if we use the AW-Delaunay hierarchy. The algorithm with one level of the AW-Delaunay graph performs well for small inputs, but it is not a good choice for data sets where n is large and $h = \Theta(n)$.

6 Conclusion

This paper proposes a dynamic algorithm to compute the additively weighted Voronoi diagram for a set of weighted points in the plane. The algorithm represents the AW-Voronoi diagram through its dual graph, the AW-Delaunay graph and allows the user to perform dynamically insertions and deletions of sites. Given a set of n sites, among which h have non-empty cell, our algorithm constructs the AW-Voronoi diagram in expected time $O(nT(h) + h \log h)$, where $T(k)$ is the time to locate the nearest neighbor of a site within a set of k sites with non-empty Voronoi cell. Two methods are proposed to locate the nearest neighbor of a given site. The first one uses no additional data structure, performs a simple walk in the AW-Delaunay graph and locates the nearest neighbor in $O(h)$ worst case time. The second method maintains a hierarchy of AW-Delaunay graphs, analog to the Delaunay hierarchy, and uses this hierarchy to perform the nearest neighbor location. Although the analysis of the Delaunay hierarchy does not extend to the case of the AW-Delaunay hierarchy, experimental results suggest that such a hierarchy allows to answer a nearest neighbor query in $O(\log h)$ time. Our algorithm performs deletions of non-trivial sites in almost optimal time. However, deletions of trivial sites are not done very efficiently and this point should be improved in further studies.

Further works also include generalization of our method to more general classes of objects, such as convex objects. More generally, one can think of characterizing classes of abstract Voronoi diagrams that can be computed using the method proposed here, i.e., without using a history or conflict graph. Another natural direction of future research is the generalization of the presented algorithm for the construction of AW-Voronoi diagrams in higher dimensions.

References

[1] Drysdale, III, R. L., Lee, D. T.: Generalized Voronoi diagrams in the plane. In: Proc. 16th Allerton Conf. Commun. Control Comput. (1978) 833–842
[2] Lee, D. T., Drysdale, III, R. L.: Generalization of Voronoi diagrams in the plane. SIAM J. Comput. **10** (1981) 73–87
[3] Sharir, M.: Intersection and closest-pair problems for a set of planar discs. SIAM J. Comput. **14** (1985) 448–468
[4] Fortune, S.: A sweepline algorithm for Voronoi diagrams. In: Proc. 2nd Annu. ACM Sympos. Comput. Geom. (1986) 313–322
[5] Aurenhammer, F.: Power diagrams: properties, algorithms and applications. SIAM J. Comput. **16** (1987) 78–96
[6] Klein, R.: Concrete and Abstract Voronoi Diagrams. Volume 400 of Lecture Notes Comput. Sci. Springer-Verlag (1989)
[7] Mehlhorn, K., Meiser, S., Ó'Dúnlaing, C.: On the construction of abstract Voronoi diagrams. Discrete Comput. Geom. **6** (1991) 211–224
[8] Klein, R., Mehlhorn, K., Meiser, S.: Randomized incremental construction of abstract Voronoi diagrams. Comput. Geom. Theory Appl. **3** (1993) 157–184
[9] Kim, D. S., Kim, D., Sugihara, K.: Voronoi diagram of a circle set constructed from Voronoi diagram of a point set. In Lee, D. T., Teng, S. H., eds.: Proc. 11th Inter. Conf. ISAAC 2000. Volume 1969 of LNCS., Springer-Verlag (2000) 432–443
[10] Karavelas, M. I., Emiris, I. Z.: Predicates for the planar additively weighted Voronoi diagram. Technical Report ECG-TR-122201-01, INRIA Sophia-Antipolis (2002)
[11] Karavelas, M. I., Yvinec, M.: Dynamic additively weighted Voronoi diagrams in 2D. Technical Report No. 4466, INRIA Sophia-Antipolis (2002)
[12] Devillers, O.: Improved incremental randomized Delaunay triangulation. In: Proc. 14th Annu. ACM Sympos. Comput. Geom. (1998) 106–115

[13] Karavelas, M.: Proximity Structures for Moving Objects in Constrained and Unconstrained Environments. PhD thesis, Stanford University (2001)
[14] The CGAL Reference Manual. (2001) Release 2.3, http://www.cgal.org.
[15] Pion, S.: Interval arithmetic: An efficient implementation and an application to computational geometry. In: Workshop on Applications of Interval Analysis to systems and Control. (1999) 99–110

Time-Expanded Graphs for Flow-Dependent Transit Times*

Ekkehard Köhler, Katharina Langkau, and Martin Skutella

Technische Universität Berlin
Fakultät II — Mathematik und Naturwissenschaften
Institut für Mathematik, Sekr. MA 6–1
Straße des 17. Juni 136,
D–10623 Berlin, Germany
{ekoehler,langkau,skutella}@math.tu-berlin.de
http://www.math.tu-berlin.de/~{ekoehler,langkau,skutella}

Abstract. Motivated by applications in road traffic control, we study flows in networks featuring special characteristics. In contrast to classical static flow problems, time plays a decisive role. Firstly, there are transit times on the arcs of the network which specify the amount of time it takes for flow to travel through a particular arc; more precisely, flow values on arcs may change over time. Secondly, the transit time of an arc varies with the current amount of flow using this arc. Especially the latter feature is crucial for various real-life applications of flows over time; yet, it dramatically increases the degree of difficulty of the resulting optimization problems.

Most problems dealing with flows over time and constant transit times can be translated to static flow problems in time-expanded networks. We develop an alternative time-expanded network with flow-dependent transit times to which the whole algorithmic toolbox developed for static flows can be applied. Although this approach does not entirely capture the behavior of flows over time with flow-dependent transit times, we present approximation results which provide evidence of its surprising quality.

1 Introduction

Flows over Time with Fixed Transit Times. Flows over time were introduced more than forty years ago by Ford and Fulkerson [8, 9]. Given a network with capacities and transit times on the arcs, they study the problem of sending a maximal amount of flow from a source node s to a sink node t within a prespecified time horizon T. Ford and Fulkerson show that this problem can be

---

* Extended abstract; information on the full version of the paper can be obtained via the authors' WWW-pages. This work was supported in part by the EU Thematics Network APPOL I+II, Approximation and Online Algorithms, IST-1999-14084 and IST-2001-30012, by the European graduate program 'Combinatorics, Geometry, and Computation', Deutsche Forschungsgemeinschaft, grant GRK 588/1, and by the Bundesministerium für Bildung und Forschung (bmb+f), grant no. 03-MOM4B1.

solved by one minimum-cost flow computation, where transit times of arcs are interpreted as cost coefficients. A problem closely related to the one considered by Ford and Fulkerson is the *quickest s-t-flow problem*. Here, instead of fixing the time horizon T and asking for a flow over time of maximal value, the value of the flow (demand) is fixed and T is to be minimized. This problem can be solved in polynomial time by incorporating the algorithm of Ford and Fulkerson in a binary search framework. Burkard, Dlaska, and Klinz [2] present an algorithm which solves the quickest flow problem in strongly polynomial time.

Time-Expanded Networks. Another important contribution of Ford and Fulkerson's work [8, 9] are *time-expanded networks*. A time-expanded network contains one copy of the node set of the underlying 'static' network for each discrete time step (building a *time layer*). Moreover, for each arc with transit time τ in the static network, there is a copy between each pair of time layers of distance τ in the time-expanded network. Thus, a flow over time in the static network can be interpreted as a static flow in the corresponding time-expanded network. Since this interrelation works bilateral, the concept of time-expanded networks allows to solve a variety of time-dependent flow problems by applying algorithmic techniques developed for static network flows; see, e.g., Fleischer and Tardos [7]. Notice, however, that one has to pay for this simplification of the considered flow problem in terms of an enormous increase in the size of the underlying network. In particular, the size of the time-expanded network is only pseudo-polynomial in the input size. Nevertheless, Fleischer and Skutella [6] show that the time-expanded network can be reduced to polynomial size at the cost of a slightly degraded solution space.

Flows over Time with Flow-Dependent Transit Times. So far we have considered flows over time with fixed transit times on the arcs. In this setting, the time it takes to traverse an arc does not depend on the current flow situation on the arc. Everybody who was ever caught up in a traffic jam knows that the latter assumption often fails to capture essential characteristics of real-life situations. In many applications, such as road traffic control, production systems, and communication networks (e.g., the Internet), the amount of time needed to traverse an arc of the underlying network increases as the arc becomes more congested. Examples and further applications can be found in the survey articles of Aronson [1] and Powell, Jaillet, and Odoni [15].

A fully realistic model of flow-dependent transit times on arcs must take density, speed, and flow rate evolving along an arc into consideration; see, e.g., the book of Sheffi [17] and the report by Gartner, Messer, and Rathi [10] for details in the case of traffic flows. It is a highly nontrivial and open problem to find an appropriate and, above all, tractable mathematical network flow model. There are hardly any algorithmic techniques known which are capable of providing reasonable solutions even for networks of rather modest size. For problem instances of realistic size (as those occurring in real-life applications), already the solution of mathematical programs relying on simplifying assumptions is in general still beyond the means of state-of-the-art computers. We refer to [1, 12, 15, 16] for more details and examples.

In practical applications such as traffic flows, precise information on the behavior of the transit time τ_e of an arc e can usually only be found for the case of static flows that do not vary over time. In this case, τ_e is given as a function of the constant flow rate y_e on arc e. Examples are 'Davidson's function' and a function developed by the U.S. Bureau of Public Roads for traffic flow applications; for details we refer to [17].

Models Known from the Literature. A non-linear and non-convex program with discretized time steps is proposed by Merchant and Nemhauser [13]. In their model, the outflow out of an arc in each time period solely depends on the amount of flow on that arc at the beginning of the time period. Unfortunately, the non-convexity of their model causes analytical and computational problems. Merchant and Nemhauser [14] and Carey [3] describe special constraint qualifications which are necessary to ensure optimality of a solution in this model. Moreover, Carey [4] introduces a slight revision of the model of Merchant and Nemhauser yielding a convex problem instead of a non-convex one.

Köhler and Skutella [11] consider a different model of flow-dependent transit times. There, the pace on an arc depends on its current load, i.e., the entire amount of flow which is currently traveling along that arc. For static flows, the transit time τ_e can be interpreted as a function of the load. This observation is carried over to the non-static case, where flow is allowed to vary over time. In this model, approximate solutions to the quickest s-t-flow problem can be computed very efficiently by one static convex-cost flow computation, similar to the ingenious algorithm of Ford and Fulkerson mentioned above.

Carey and Subrahmanian [5] introduce a generalized time-expanded network for flow-dependent transit times. For each time period, there are several copies of an arc of the underlying 'static' network corresponding to different transit times. However, in order to enforce flow-dependent transit times, special capacity constraints are introduced which give rise to a dependency between the flow on all copies of an arc corresponding to one time step. As a consequence, the resulting static problem on the generalized time-expanded graph can no longer be solved by standard network flow techniques but requires a general linear programming solver. This constitutes a serious drawback with regard to the practical efficiency and applicability of this model.

A New Model. In this paper we propose a closely related approach which overcomes the latter drawback. As in the model of Carey and Subrahmanian, we consider a generalized time-expanded network with multiple copies of each arc for each time step. However, in contrast to their model, we introduce additional 'regulating' arcs which enable us to enforce flow-dependent transit times without using generalized capacity constraints. As a result, we can apply the whole algorithmic toolbox developed for static network flows to the generalized time-expanded network. In the following, we refer to this time-expanded network as *fan graph*.

The underlying assumption for our approach is that at any moment of time the transit time on an arc solely depends on the current rate of inflow into that arc. In the following, we will therefore speak of *flows over time with inflow-*

dependent transit times, emphasizing that transit times are considered as functions of the rate of inflow. Thus, in contrast to the load-dependent model developed in [11], the flow units traveling on the same arc at the same time do not necessarily experience the same pace, as the transit time and thus the pace of every unit of flow is determined when entering the arc and remains fixed throughout. Of course, this model is only a rough approximation of the actual behavior of flow in real-life applications. For example, flows over time with inflow-dependent transit times do in general not obey the *first-in-first-out* (*FIFO*) property[1] on an arc. However, this phenomenon does not occur in the solutions which are generated for the quickest s-t-flow problem below.

The generalized time-expanded network presented in this paper does not entirely capture the behavior of flows over time with inflow-dependent transit times. In fact, the solution space defined by static flows on the fan graph can be seen as a relaxation of the space of flows over time with inflow-dependent transit times. Based on this observation, we present approximation results providing strong evidence of the quality of this relaxation and thus justifying its consideration for practical purposes. We show that an optimal solution to the quickest s-t-flow problem in the fan graph can be turned into a feasible flow over time with inflow-dependent transit times while only losing a factor of $2 + \epsilon$ in the value of the objective function. This result holds for arbitrary non-decreasing transit time functions τ_e. Furthermore, if the transit time functions τ_e are concave, the performance ratio can be improved to $3/2 + \epsilon$.

Comparison to Earlier Work. The approximation results presented in this paper are inspired by the work of Fleischer and Skutella [6] and the subsequent results of Köhler and Skutella [11]. The latter paper presents a $(2 + \epsilon)$-approximation algorithm for the quickest s-t-flow problem in the setting of load-dependent transit times described above. Interestingly, the analysis in [11] can easily be carried over to the model of flows over time with inflow-dependent transit times. In fact, one can check that the generated solutions are $(2 + \epsilon)$-approximations for both models, the load-dependent model and the inflow-dependent model. This implies a surprisingly close relationship between these rather different flow models for the case of quickest s-t-flows. If all transit time functions τ_e are convex and non-decreasing, then the time horizon of a quickest s-t-flow in the load-dependent model and the time horizon of a quickest s-t-flow in the inflow-dependent model cannot differ by more than a factor of 2.

Notice, however, that the algorithm in [11] can only be applied if all transit time functions are convex, while we allow a considerably richer family of transit time functions in this paper. Also, the stronger performance ratio for the case of concave transit time functions does not seem to translate to the model of flows over time with load-dependent transit times.

Basic Definitions and Notation. Let $G = (V, E)$ be a directed graph with node set V and arc set E. Each arc $e \in E$ has associated with it a positive capacity u_e, which is interpreted as an upper bound on the rate of flow enter-

---
[1] The FIFO property requires that flow units are entering and leaving an arc in the same order.

ing e at any moment of time, and a non-negative transit time function τ_e, which determines the time it takes for flow to traverse arc e. Finally, there is a source node $s \in V$ and a sink node $t \in V$. An *s-t-flow over time* (also known as *dynamic s-t-flow* or *time-dependent s-t-flow* in the literature) with time horizon T is given by functions $f_e : [0, T) \to \mathbb{R}^+$, for $e \in E$, where $f_e(\theta)$ defines the rate of flow (per unit time) entering arc e at time θ. This flow arrives at the head node of e at time $\theta + \tau_e(f_e(\theta))$.

In the setting of flows over time, flow conservation in a node $v \in V \setminus \{s\}$ at time $\theta \in [0, T)$ requires that the total inflow into node v until time θ is an upper bound on the total outflow out of node v until time θ. If storage of flow at intermediate nodes is not allowed, equality must hold for all $v \in V \setminus \{s, t\}$ and all $\theta \in [0, T)$. In the following, we mostly assume that storage of flow at nodes is allowed. Finally, all arcs must be empty at time T, i.e., $\tau_e(f_e(\theta)) < T - \theta$ for all $e \in E$ and $\theta \in [0, T)$, and flow must not remain in any node other than the sink at time T.

The value of a flow over time f is the total outflow out of node s until time T minus the total inflow into s until time T. Notice that, due to flow conservation, this amount, which is denoted by $|f|$, is equal to the total inflow into node t until time T minus the total outflow out of t until time T.

In this extended abstract, we omit all details in the last three sections due to space restrictions.

2   A Model for Flows over Time with Inflow-Dependent Transit Times

In this section we define a generalized time-expanded graph—the *fan graph* G^F— in which transit times depend indirectly on the flow rate. As a result, classical network algorithms for static flow problems can be applied to G^F. In the following sections, we demonstrate the strength of this approach by computing approximate quickest s-t-flows with inflow-dependent transit times in G.

Let us, for the moment, turn back to the case of fixed transit times. Assume that we are given a positive, integral transit time τ_e for every arc $e \in E$. Given an integral time horizon T, the time-expanded graph of G introduced by Ford and Fulkerson [8, 9] contains one copy of every node in V for every integral time step, i.e., its node set is $\{v_\theta : v \in V, \theta \in \{0, 1, \ldots, T-1\}\}$. For every arc $e = (v, w) \in E$, and for every $0 \leq \theta < T - \tau_e$, it contains an arc e_θ from v_θ to $w_{\theta+\tau_e}$ of the same capacity as e. Additionally, there are arcs with infinite capacity from v_θ to $v_{\theta+1}$, for all $v \in V$ and $\theta = 0, \ldots, T-2$, which model the possibility to hold flow at a node.

A static flow in the time-expanded graph can be interpreted as a flow over time in G. Take the flow on arc e_θ as the flow rate into arc e during the time interval $[\theta, \theta+1)$. Similarly, any flow over time f with time horizon T corresponds to a flow in the time-expanded graph; for every arc e_θ, take the average flow rate of f on e over the time interval $[\theta, \theta+1)$ as the flow value on e_θ. This relationship allows to solve many time-dependent flow problems in G by transforming them to static flow problems in the time-expanded graph; see, e.g., [7].

2.1 The Fan Graph

In order to model flow-dependent transit times, we take a similar approach, i.e., we expand the graph according to the transit time functions in such a way that transit times indirectly depend on the current flow rate. For the moment, we will assume that the transit time function $\tau_e(x_e)$ of an arc e is given as a piecewise constant, non-decreasing, and left-continuous function with only integral values[2]. To stress the step function character of this transit time function, we will denote it by $\tau_e^s(x_e)$. Later we will use the fact that general transit time functions can be approximated by step functions within arbitrary precision.

The *fan graph* G^F is defined on the same set of nodes $\{v_\theta : v \in V, \theta \in \{0, 1, \ldots, T-1\}\}$ as the time-expanded graph. An illustration of the definition is given in Fig. 1, where a single arc $e = (v, w)$ is expanded according to its transit time function. For every $\theta \in \{0, 1, \ldots, T-1\}$, we define a 'fan' of arcs leaving v_θ, which represent all possible transit times of arc e. In the example, there are three possible transit times on arc e, namely transit times 1, 3, and 6; see Fig. 1. The fan leaving $v(0)$ is shown in Fig. 1 (middle). The fan consists of capacitated horizontal arcs and uncapacitated arcs pointing upwards. While the latter arcs model the different possible transit times, the capacities of the horizontal arcs try to control the distribution of flow according to the transit time function. More precisely, if flow is being sent into the fan at rate x_e, the transit time assigned to the slowest portion of this flow is forced to be at least $\tau_e^s(x_e)$. In the example, the capacities are chosen in such a way that at most x_1 units of flow can travel with transit time $\tau_1 = 1$, at most x_2 units of flow can travel with transit time $\tau_2 = 3$ or faster, and so on.

Figure 1 (right) shows the fan graph of a single arc e. As for the time-expanded graph, there is an arc with infinite capacity from v_θ to $v_{\theta+1}$, for all $v \in V$ and $\theta = 0, \ldots, T-2$, which models the possibility to hold flow at node v.

2.2 The Bow Graph

A major drawback of the fan graph is that it may become very large, and is thus not practical to work with for many real-world applications. Even if all transit times are constant, implying that the fan graph and the time-expanded graph essentially coincide, this graph is only of pseudo-polynomial size. Ford and Fulkerson [8, 9] prove that, in order to compute a maximal flow over time with constant transit times, it is not necessary to explicitly work with the time-expanded graph. Instead, they present a polynomial time algorithm which solely works on the original graph. Motivated by this result, we give a slightly different view of the fan graph, as the time-expansion of a smaller graph, the so-called

---

[2] Of course, the latter assumption can easily be relaxed to allow arbitrary rational values if time is scaled by an appropriate factor.

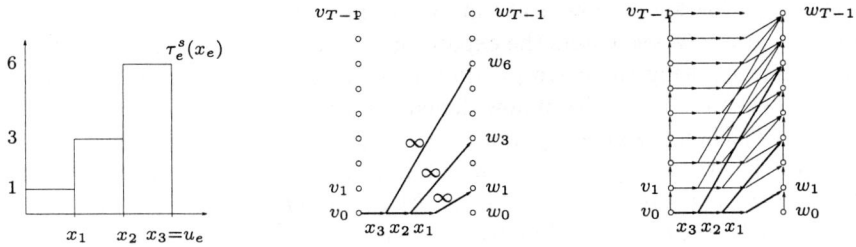

Fig. 1. Definition of the fan graph

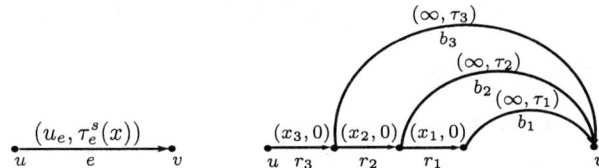

Fig. 2. Definition of the bow graph

bow graph. We will explain how flows over time with inflow-dependent transit times in G relate to a certain class of flows over time in the bow graph.

Again, let us assume that all transit time functions are given as piecewise constant, non-decreasing, and left-continuous functions τ_e^s, $e \in E$. The bow graph, denoted $G^B = (V^B, E^B)$, arises from the original graph by expanding each arc $e \in E$ according to its transit time function. In G^B, every arc $a \in E^B$ has a capacity u_a and a constant transit time $\tau_a \in \mathbb{R}^+$.

For the definition, let us consider a particular arc $e \in E$. For arc e, we are given breakpoints $0 = x_0 < x_1 < \cdots < x_k = u_e$ and corresponding transit times $\tau_1 < \cdots < \tau_k$. Flow entering at rate $x \in (x_{i-1}, x_i]$ needs τ_i time units to traverse arc e. Consider the example in Fig. 2, where an arc is expanded according to the step function in Fig. 1. Arc e is replaced by arcs of two types; *bow arcs* denoted $b_1, \ldots, b_k$ and *regulating arcs* denoted $r_1, \ldots, r_k$. The bow arcs are uncapacitated, they represent all possible transit times of arc e. The transit time of arc b_i is given by τ_i, $i = 1, \ldots, k$; in the example we have $\tau_1 = 1, \tau_2 = 3$, and $\tau_3 = 6$. The regulating arcs have zero transit time, they limit the amount of flow entering the bow arcs. Their capacities are chosen according to the breakpoints of transit time function $\tau_e^s(x_e)$; more precisely, the capacity of arc r_i is set to x_i, $i = 1, \ldots, k$. We will denote the set of bow arcs and regulating arcs of an arc e by E_e^B. Note that the size of this *expansion* E_e^B of arc e is linear in the number of breakpoints of $\tau_e^s(x)$. Also note that the fan graph (see Fig. 1 (right)) is the time-expanded graph of the bow graph (see Fig. 2).

An arbitrary flow over time with inflow-dependent transit time functions τ_e^s in G can be interpreted as a flow over time with constant transit times in the bow graph G^B (and thus as a static flow in the fan graph G^F). The idea is the following. Flow entering arc e at time θ and traversing e with transit time τ_i in the original graph G also enters the expansion of arc e in G^B at time θ and traverses it in time τ_i, using the corresponding bow arc b_i. To be more precise, let f be a flow over time in G with inflow-dependent transit time functions τ_e^s and time horizon T. Then f is given by flow rates $f_e : [0, T) \to \mathbb{R}^+$ for all $e \in E$. According to the transit time functions, flow entering arc e at time θ will need $\tau_e^s(f_e(\theta))$ time to traverse arc e. Let $i \in \{1, \ldots, k\}$ be chosen such that $f_e(\theta) \in (x_{i-1}, x_i]$ and consider the corresponding transit time $\tau_i = \tau_e^s(f_e(\theta))$. We define a flow over time f^B on the expansion of arc e by setting $f_a^B(\theta) := f_e(\theta)$ if a is either the bow arc b_i or a regulating arc r_j with $j \geq i$. For all other arcs we set $f_a^B(\theta) = 0$. Notice that f^B obeys capacity constraints and flow conservation constraints at all intermediate nodes. Such flows over time in G^B which simulate flows over time with inflow-dependent transit times in G are called *feasible in G*. We can draw the following conclusion.

Observation 1. *Let f be a flow over time with inflow-dependent transit time functions τ_e^s in G, which sends D units of flow from s to t within time T. Then, there is a flow over time with constant transit times in G^B which sends D units of flow from s to t within time T.*

Hence, every flow over time in G can be regarded as a flow over time in G^B; however, the converse is not true. By our definition of flow-dependent transit times, flow units entering arc $e \in E$ at the same time, simultaneously arrive at the head node of e. In the bow graph, however, flow units entering the expansion of an arc simultaneously, do not necessarily travel through the expansion at the same pace. The flow is allowed to split up and use bow arcs representing different transit times. Only a portion of the flow is traversing the expansion of arc e at the speed prescribed by the transit time function or possibly slower. The rest might travel at a faster speed. Therefore, the bow graph is a relaxation of our original model of inflow-dependent transit times. The advantage of this relaxation is that it relies on a graph having constant transit times on the arcs. This implies that one can apply algorithms that are known for this much simpler model of flows over time. In particular, this provides the opportunity to use standard network flow algorithms on the time-expanded fan graph G^F of G^B.

3 Approximate Solutions to the Quickest Flow Problem

The purpose of this section is to demonstrate the strength and the usefulness of the bow and the fan graph presented in the last section for solving time-dependent flow problems with inflow-dependent transit times. This is done by presenting approximation algorithms for the quickest flow problem with inflow-dependent transit times that are based on optimal solutions to the relaxation given by G^B and G^F.

3.1 Ford and Fulkerson's Algorithm

A problem closely related to the quickest flow problem is the problem of computing a maximal s-t-flow over time with fixed time horizon T. For the case of fixed transit times τ_e, $e \in E$, Ford and Fulkerson [8] show how this problem can be solved by only one (static) minimum cost flow computation in G. Since we will employ some of the underlying basic insights of this algorithm, we give a brief description of it here.

The algorithm of Ford and Fulkerson is based on the concept of *temporally repeated flows*. Let $(y_e)_{e \in E}$ be a feasible static s-t-flow in G with path decomposition $(y_P)_{P \in \mathcal{P}}$ where $\mathcal{P}$ is a set of s-t-paths. If the transit time $\tau_P := \sum_{e \in P} \tau_e$ of every path $P \in \mathcal{P}$ is bounded from above by T, the static s-t-flow y can be turned into a *temporally repeated s-t-flow* f as follows. Starting at time zero, f sends flow at constant rate y_P into path $P \in \mathcal{P}$ until time $T - \tau_P$, thus ensuring that the last unit of flow arrives at the sink before time T. Feasibility of f with respect to capacity constraints immediately follows from the feasibility of the underlying static flow y. Notice that the value of f is given by

$$|f| = \sum_{P \in \mathcal{P}} (T - \tau_P) y_P = T|y| - \sum_{e \in E} \tau_e y_e . \qquad (1)$$

Here, $|y|$ denotes the value of the static s-t-flow y.

The algorithm of Ford and Fulkerson computes a static s-t-flow y maximizing the right hand side of (1), then determines a path decomposition of y, and finally returns the corresponding temporally repeated flow f. Ford and Fulkerson show that this temporally repeated flow is in fact a maximal s-t-flow over time.

3.2 An Approximation Algorithm for the Quickest Flow Problem

In this section an approximation algorithm is presented for the case of piecewise constant, non-decreasing, and left-continuous transit time functions. Using the algorithm of Burkard et al. [2], one can determine a quickest flow over time in the bow graph G^B in strongly polynomial time. The computation yields the minimal time horizon $\overline{T}^B$ and a static flow y^B in G^B such that the value of the resulting temporally repeated flow f^B in G^B is

$$|f^B| = \overline{T}^B |y^B| - \sum_{a \in E^B} \tau_a y_a^B = D . \qquad (2)$$

Since the quickest flow problem on G^B can be seen as a relaxation of the quickest flow problem with inflow-dependent transit times on G, the value $\overline{T}^B$ is a lower bound on the optimal time horizon $\overline{T}$ in G.

For the following lemma, we consider the bow graph expansion of a particular arc $e \in E$ to bow arcs $b_1, \ldots, b_k$ and regulating arcs $r_1, \ldots, r_k$, where the capacity of r_j is given by x_j.

Lemma 1. *If y^B sends flow along the ith bow arc b_i, then for all bow arcs b_j with index j smaller than i we have $y^B_{b_j} = x_j - x_{j-1}$. Less formally, y^B fills the bows from bottom to top.*

Eventually, we are interested in a flow over time in G^B which yields a feasible flow over time in G. Unfortunately, as discussed above, f^B does in general not satisfy this requirement. One reason is that two flow units entering the expansion of an arc $e \in E$ simultaneously might in fact experience different transit times in f^B due to different transit times on bow arcs in E^B_e.

Therefore, the static flow y^B is rerouted to make sure that it does not split among bows representing different transit times of the same arc $e \in E$. This is achieved by pushing flow from "fast" bow arcs up to the "slowest" flow-carrying bow arc in y^B. More formally, for the expansion of every arc $e \in E$, the modified static flow $\tilde{y}^B$ is defined by setting $\tilde{y}^B_{b_i} := y^B_{r_i}$, if $y^B_{b_i} > 0$ and $y^B_{b_j} = 0$ for all $j > i$, and $\tilde{y}^B_{b_i} := 0$ otherwise; furthermore, $\tilde{y}^B_{r_i} := y^B_{r_i}$, if $y^B_{b_j} = 0$ for all $j > i$, and $\tilde{y}^B_{r_i} := 0$ otherwise. Notice that the value of the flow remains unchanged, i.e., $|\tilde{y}^B| = |y^B|$.

We show that the modified flow $\tilde{y}^B$ yields a flow over time in G with value D and time horizon at most $2\overline{T}^B$. Decompose $\tilde{y}^B$ into flows on s-t-paths $P \in \tilde{\mathcal{P}}$ in G^B with flow values $\tilde{y}^B_P$. This path-decomposition induces a temporally repeated flow $\tilde{f}^B$ in G^B for any time horizon $T \geq \overline{T}^B$. It is easy to determine T such that $|\tilde{f}^B| = T |\tilde{y}^B| - \sum_{a \in E^B} \tau_a \tilde{y}^B_a = D$.

Lemma 2. *The value of T is bounded from above by $2\overline{T}^B$.*

Finally, we show that $\tilde{f}^B$ can be interpreted as a flow over time with inflow-dependent transit times in G. Here, essentially the same argument is used that was already applied in [11] in a similar context. For any arc $e \in E$, the static flow $\tilde{y}^B$ uses at most one bow arc $b^e \in E^B_e$. Thus, $\tilde{f}^B$ naturally induces a flow over time f in G by setting $f_e(\theta) := \tilde{f}^B_{b^e}(\theta)$ for any time θ.

Lemma 3. *The flow over time f in G is feasible, that is, it fulfills the flow conservation constraints.*

Notice that, in contrast to the temporally repeated flow $\tilde{f}^B$ in G^B, the flow over time f in G uses storage of flow at intermediate nodes. More precisely, the flow arriving via arc e at node $v \in V$ at time $\theta + \tau^s_e(f_e(\theta))$ in f waits there for $\tau_{b^e} - \tau^s_e(f_e(\theta))$ time units until the corresponding flow in $\tilde{f}^B$ has also arrived. We call f a *temporally repeated flow (with inflow-dependent transit times)*. The underlying static flow y in G is given by $y_e := \tilde{y}^B_{b^e}$. Putting everything together, we can state the following result.

Theorem 1. *Consider an instance of the quickest flow problem with inflow-dependent transit times where all transit time functions are non-decreasing step functions. If there is a flow over time with inflow-dependent transit times sending D units of flow from s to t within time $\overline{T}$, then there exists a temporally* repeated flow with inflow-dependent transit times satisfying demand D within time horizon at most $2\overline{T}$. Moreover, such a flow can be computed in strongly polynomial time.

Proof. It follows from the description above that the temporally repeated flow f can be computed in strongly polynomial time. Moreover, $|f| = |\tilde{f}^B| = D$ and the time horizon of f is equal to T, the time horizon of $\tilde{f}^B$. Finally, by Lemma 2, $T \leq 2\overline{T}^B \leq 2\overline{T}$. □

Notice that in our analysis we have compared the value T of the computed solution f to the lower bound $\overline{T}^B$ given by an optimal solution f^B to the relaxation of the problem defined by the bow graph G^B. This yields the following result on the quality of this relaxation and thus on the quality of the time-expanded fan graph G^F.

Corollary 1. *The relaxation of the quickest flow problem on the bow graph G^B is a 2-relaxation, that is, the value of an optimal solution to the quickest flow problem with inflow-dependent transit times is within a factor of 2 of the value of an optimal solution to this relaxation.*

4 An Approximation for General Transit Time Functions

So far we have derived a 2-approximation for the case that all transit time functions are non-decreasing step functions. In this section, the approach is generalized to arbitrary non-decreasing transit time functions. The idea is to approximate them by step functions and then to proceed with the algorithm described in the last section. In order to do this, we need the technical requirement that the transit time functions are left-continuous.

Theorem 2. *There exists a $(2+\epsilon)$-approximation algorithm with strongly polynomial running time for the quickest flow problem with inflow-dependent transit times.*

The proof of the theorem is contained in the full version of the paper. Here, we only summarize the following main steps of the algorithm.

1) Construct a relaxed instance by approximating and replacing all transit time functions τ_e by lower step functions τ_e^s.
2) Further relax the problem by considering the corresponding bow graph G^B with constant transit times on the arcs and compute a quickest flow there. Let y^B denote the underlying optimal static flow in G^B.
3) Turn y^B into a static flow y in the original graph G by setting y_e to the amount of flow sent through the expansion of arc e in y^B, for all $e \in E$.
4) Determine the time horizon T such that $T|y| - \sum_{e \in E} \tau_e(y_e) y_e = D$ and output the corresponding temporally repeated flow with inflow-dependent transit times in G.

5 Improved Result for Concave Transit Time Functions

For the special case of non-negative, non-decreasing, and concave transit time functions, the algorithm discussed at the end of the last section achieves a better performance ratio.

Theorem 3. *Consider an instance of the quickest flow problem with inflow-dependent transit times where the transit time function of every arc is non-negative, non-decreasing, and concave. Then, the above algorithm achieves performance ratio $3/2 + \epsilon$.*

There exist bad instances showing that the algorithm has a performance ratio not better than $\sqrt{2}$. For such an example and for the proof of Theorem 3, we refer to the full version of the paper.

Acknowledgments

The authors are much indebted to Lisa Fleischer, Rolf Möhring, and Günter Rote for helpful comments and interesting discussions on the topic of this paper.

References

[1] J. E. Aronson. A survey of dynamic network flows. *Annals of OR*, 20:1–66, 1989.
[2] R. E. Burkard, K. Dlaska, and B. Klinz. The quickest flow problem. *ZOR — Methods and Models of Operations Research*, 37:31–58, 1993.
[3] M. Carey. A constraint qualification for a dynamic traffic assignment model. *Transp. Science*, 20:55–58, 1986.
[4] M. Carey. Optimal time-varying flows on congested networks. *OR*, 35:58–69, 1987.
[5] M. Carey and E. Subrahmanian. An approach for modelling time-varying flows on congested networks. *Transportation Research B*, 34:157–183, 2000.
[6] L. Fleischer and M. Skutella. The quickest multicommodity flow problem. In *Proc. of IPCO'02*, 2002.
[7] L. Fleischer and É. Tardos. Efficient continuous-time dynamic network flow algorithms. *Operations Research Letters*, 23:71–80, 1998.
[8] L. R. Ford and D. R. Fulkerson. Constructing maximal dynamic flows from static flows. *Operations Research*, 6:419–433, 1958.
[9] L. R. Ford and D. R. Fulkerson. *Flows in Networks*. Princeton University Press, Princeton, NJ, 1962.
[10] N. Gartner, C. J. Messer, and A. K. Rathi. Traffic flow theory: A state of the art report. http://www-cta.ornl.gov/cta/research/trb/tft.html, 1997.
[11] E. Köhler and M. Skutella. Flows over time with load-dependent transit times. In *Proc. of ACM-SIAM Symposium on Discrete Algorithms (SODA)*, 2002.
[12] H. S. Mahmassani and S. Peeta. System optimal dynamic assignment for electronic route guidance in a congested traffic network. In *Urban Traffic Networks. Dynamic Flow Modelling and Control*, pages 3–37. Springer, Berlin, 1995.

[13] D. K. Merchant and G. L. Nemhauser. A model and an algorithm for the dynamic traffic assignment problems. *Transp. Science*, 12:183–199, 1978.
[14] D. K. Merchant and G. L. Nemhauser. Optimality conditions for a dynamic traffic assignment model. *Transp. Science*, 12:200–207, 1978.
[15] W. B. Powell, P. Jaillet, and A. Odoni. Stochastic and dynamic networks and routing. Handb. in OR and Man.Sc. vol. 8, pages 141–295. North–Holland, 1995.
[16] B. Ran and D. E. Boyce. *Modelling Dynamic Transportation Networks*. Springer, Berlin, 1996.
[17] Y. Sheffi. *Urban Transportation Networks*. Prentice-Hall, New Jersey, 1985.

Partially-Ordered Knapsack and Applications to Scheduling

Stavros G. Kolliopoulos[1]* and George Steiner[2]**

[1] Department of Computing and Software, McMaster University
stavros@mcmaster.ca
[2] Management Science and Information Systems, McMaster University
steiner@mcmaster.ca

Abstract. In the *partially-ordered knapsack* problem (*POK*) we are given a set N of items and a partial order $\prec_P$ on N. Each item has a size and an associated weight. The objective is to pack a set $N' \subseteq N$ of maximum weight in a knapsack of bounded size. N' should be precedence-closed, i.e., be a valid prefix of $\prec_P$. *POK* is a natural generalization, for which very little is known, of the classical Knapsack problem. In this paper we advance the state-of-the-art for the problem through both positive and negative results. We give an FPTAS for the important case of a *2-dimensional* partial order, a class of partial orders which is a substantial generalization of the series-parallel class, and we identify the first non-trivial special case for which a polynomial-time algorithm exists. We also characterize cases where the natural linear relaxation for *POK* is useful for approximation and we demonstrate its limitations. Our results have implications for approximation algorithms for scheduling precedence-constrained jobs on a single machine to minimize the sum of weighted completion times, a problem closely related to *POK*.

1 Introduction

Let a partially-ordered set (poset) be denoted as $P = (N, \prec_P)$, where $N = \{1, 2, ..., n\}$. A subset $I \subseteq N$ is an *order ideal* (or *ideal* or *prefix*) of P if $b \in I$ and $a \prec_P b$ imply $a \in I$. In the *partially-ordered knapsack* problem (denoted *POK*), the input is a tuple $(P = (N, \prec_P), w, p, h, l)$ where P is a poset, $w : N \to R^+$, $p : N \to R^+$, and h, l are scalars in $(1, +\infty)$. For a set $S \subseteq N$, $p(S)$ ($w(S)$) denotes $\sum_{i \in S} p_i$ ($\sum_{i \in S} w_i$). We can think of being given a knapsack of capacity $p(N)/l$. The sought output is an ideal N' that fits in the knapsack, i.e., $p(N') \leq p(N)/l$, and for which $w(N') \geq w(N)/h$. In the optimization version of *POK* only the knapsack capacity $p(N)/l$ is given and one seeks to maximize $w(N')$. The two formulations are polynomial-time equivalent from the algorithmic perspective. In our exposition we adopt each time the one which is more convenient. A ρ-approximation algorithm, $\rho < 1$, finds an ideal N' such

---
* Research partially supported by NSERC Grant 227809-00
** Research partially supported by NSERC Grant OG0001798

that $p(N') \leq p(N)/l$ and $w(N')$ is at least ρ times the optimum. For simplicity we shall sometimes denote a POK instance as a triple (N, h, l) with $\prec_P, w, p$ implied from the context. Similarly, we occasionally abuse notation and denote a poset by N, or omit P from $\prec_P$ when this leads to no ambiguity.

POK is a natural generalization of the Knapsack problem. An instance of the latter is a POK instance with an empty partial order. Johnson and Niemi [13] view POK as modeling, e.g., an investment situation where every investment has a cost and a potential profit and in which certain investments can be made only if others have been made previously. POK is strongly NP-complete, even when $p_i = w_i$, $\forall i \in N$, and the partial order is bipartite [13] and hence does not have an FPTAS, unless $\mathcal{P} = \mathcal{NP}$. To our knowledge, this is the only hardness of approximation result known. However, the problem is believed to have much more intricate structure. There is an approximation-preserving reduction from the densest k-subgraph problem (DkS) to POK. No NP-hardness of approximation result exists for DkS but the best approximation ratio currently known is $O(n^\delta)$, $\delta < 1/3$ [8]. Very recently, Feige showed that DkS is hard to approximate within some constant factor under an assumption about the average-case complexity of 3SAT [7]. In terms of positive results for POK, there is very little known. In 1983 Johnson and Niemi gave an FPTAS for the case when the precedence graph is a directed out-tree [13]. Recently there has been revived interest due to the relevance of POK for scheduling. It is known that an $O(1)$-factor approximation for a special type of POK instances leads to a $(2-\delta)$-approximation for minimizing average completion time of precedence-constrained jobs on a single machine, a problem denoted as $1|prec|\sum w_j C_j$. Improving on the known factor of 2 for the latter problem (see, e.g., [3], [4], [10], [15]) is one of the major open questions in scheduling theory [18]. The relationship between POK and $1|prec|\sum w_j C_j$ was explored in a recent paper by Woeginger [21].

Due to the scheduling connection, we adopt scheduling terminology for POK instances: items in N are *jobs*, function w assigns *weights* and function p *processing time*. An instance (N, h, l) is a *weight-majority* instance if $h < l$, and *time-majority* if $h > l$. To our knowledge, Woeginger gave after many years the first new results for POK by showing pseudopolynomial algorithms for the cases where the underlying partial order is an interval order or a bipartite convex order [21]. We use the notion of convex orders ourselves so we proceed to define it. The comparability graph of a *bipartite poset* $(X, Y; \prec)$ is a bipartite graph $G = (X, Y; E)$ and by convention, the set of maximal elements of the partial order is Y. A bipartite poset is *bipartite convex* if its comparability graph is *convex*: the vertices can be ordered so that the set of predecessors of any job from Y forms an interval within the jobs in X.

Our contribution. In this paper we advance the state-of-the-art for POK through both positive and negative results. Our positive results are based on structural information of posets which has not been exploited before in approximation algorithms. One of the main applications of posets is in scheduling problems but there are only a few relevant results (e.g., [5], [10],). Moreover, these results are usually derived either by simple greedy scheduling or by relying on an LP solution to resolve the ordering. However, a large amount of combinatorial theory exists for posets. Tapping this source can only help in designing approximation algorithms. Following this approach, we obtain combinatorial algorithms for comprehensive classes of POK instances. These lead to improved approximation algorithms for the corresponding cases of $1|prec|\sum w_j C_j$. Perhaps ironically, we then show that the natural LP relaxation for POK provides only limited information.

The first part of our paper deals with the complexity of POK on two comprehensive classes of partial orders. First, we give an FPTAS for POK when the underlying order is 2-dimensional. Second, we give a polynomial-time algorithm for a special class of bipartite orders.

2-dimensional orders. In Section 2 we provide a bicriteria FPTAS for POK when the underlying order is *2-dimensional*. It achieves a simultaneous $(1 \mp \varepsilon)$-approximation for weight and processing time. We proceed to give background on 2-dimensional orders. A *linear extension* of a poset $P = (N, \prec_P)$ is a linear (total) order L with $a \prec_P b$ implying $a \prec_L b$ for $a, b \in N$. Every poset P can be defined as the intersection of its linear extensions (as binary relations) [20]. The minimum number of linear extensions defining P in this way is the *dimension of P*, denoted by $dim P$. It is well known that $dim P = 2$ exactly when P can be embedded into the Euclidean plane so that $a \prec_P b$ for $a, b \in N$ if and only if the point corresponding to a is not to the right and not above the point corresponding to b. 2-dimensional posets were first characterized by Dushnik and Miller [6] and they can be recognized and their two defining linear extensions can be found in polynomial time. However, recognizing whether $dim P = k$ for any $k \geq 3$ is NP-complete [22]. POK is NP-complete on 2-dimensional partial orders, since the empty partial order is also of dimension 2. It is well known that every directed out-tree poset is series-parallel and that every series-parallel poset is also of dimension 2, but the class of 2-dimensional posets is substantially larger. For example, while the class of series-parallel posets can be characterized by a single forbidden subposet, posets of dimension 2 cannot be defined by a finite list of forbidden substructures. Thus our FPTAS for 2-dimensional POK represents a substantial addition to previously known positive results on directed out-trees [13] and other classes [21]. For a review of the extensive literature on 2-dimensional posets, we refer the reader to [16].

Complement of chordal bipartite orders. A POK instance is called *Red-Blue* if $\forall a \in N$, either $w_a = 0$ (a is red) or $p_a = 0$ (a is blue). Red-Blue bipartite instances of POK are of particular interest, because any $O(1)$-approximation to these would yield improved approximation results for $1|prec|\sum w_j C_j$ [3, 21]. Note also that solving any POK instance can be reduced in an approximation-preserving manner to solving a Red-Blue bipartite instance (cf. Sec. 6). In Section 3 we give a polynomial-time algorithm for POK on Red-Blue bipartite instances where the comparability graph has the following property: its bipartite complement is chordal bipartite. *Chordal bipartite graphs* are bipartite graphs in which every cycle of length 6 or more has a chord. They form a large class of perfect graphs, containing, for example, convex and biconvex bipartite graphs, bipartite permutation graphs (the comparability graphs of bipartite posets of dimension 2), bipartite distance hereditary graphs and interval bigraphs and they can be recognized in polynomial time [17]. For an excellent overview of these graph classes, the reader is referred to [1]. To the best of our knowledge, our result identifies the *first nontrivial* special case of POK which is solvable in polynomial time. All other solvable cases, e.g., when the poset is a rooted tree [13], an interval order [21], a convex bipartite poset [21] or a series-parallel poset, include the case when the partial order is empty, i.e., the classical Knapsack problem. Hence their best algorithm can only be pseudopolynomial, unless $\mathcal{P} = \mathcal{NP}$. In contrast, the class we define does *not* include Knapsack; moreover it is the "maximal" possible in $\mathcal{P}$ since without the Red-Blue constraint the problem becomes NP-hard even on these restricted posets. We also give an FPTAS for POK with general w and p functions and comparability graph whose bipartite complement is chordal bipartite.

As a corollary to our POK results, we obtain in Section 4 an 1.61803-approximation for $1|prec|\sum w_j C_j$ when the partial order of the jobs falls in one of the two classes we described above. Our derivation uses machinery developed by Woeginger in [21].

In the second part of our paper we turn our attention to the problem with a general partial order. We study the natural linear relaxation and provide insights on the structure of optimal solutions. On the positive side we provide in Section 5 a bicriteria-type approximation for weight and processing time on weight-majority instances. On the negative side, we show that the LP solution is inadequate in the general case. Let the *rank* of an ideal I be defined as $w(I)/p(I)$. Poset N is *indecomposable* if its maximum rank ideal is the entire set. It is well-known that in order to improve on the 2-approximation for $1|prec|\sum w_j C_j$ one needs only to consider indecomposable instances [3, 19]. As part of our contribution, we show that indecomposability affects POK, although in a different manner. In Section 6 we show that if the input is indecomposable the LP-relaxation provides essentially no information since the optimum solution assigns the same value to all the variables (cf. Theorem 8). We base our analysis on the fact that the dual LP is related to a transportation problem. We also exploit the flow connection to show non-constructively an upper bound on the integrality gap of the LP-relaxation that holds regardless of indecomposability.

Our guarantee for the weight in our bicriteria-type result is not an approximation ratio in the classical sense (cf. Theorem 7) and as said applies only for the weight-majority case. We give evidence in Section 7 that both these limitations of our algorithm reflect rather the difficulty of the problem itself. We show that an $O(1)$-approximation algorithm for the POK problem restricted to weight-majority instances, would imply a breakthrough $(2-\delta)$-approximation for $1|prec|\sum w_j C_j$. This is a rather surprising fact given that any weight-majority instance defined on an indecomposable poset is by definition infeasible. As mentioned, the hard case for $1|prec|\sum w_j C_j$ is precisely the one where the underlying poset is indecomposable.

2   POK on 2-Dimensional Orders

Consider a POK instance $(P = (N, \prec_P), h, l)$ where $\prec_P$ is 2-dimensional. For simplicity we denote the knapsack capacity $p(N)/l$ by b and $w(N)$ by W. Without loss of generality we assume that processing times, weights and the knapsack capacity are all integers. Define the $b \times W$ *feasibility matrix* $M(N, b, W)$ as follows: For each integer $p \in [1, b]$ and $w \in [1, W]$, the corresponding element $m_{pw} = 1$ if there is an ideal I such that $p(I) = p \leq b$ and $w(I) = w$; otherwise $m_{pw} = 0$. It is clear that any POK problem can be solved by computing its matrix $M(N, b, W)$ and then finding $\max\{w \mid m_{pw} = 1\}$. For any $S \subseteq N$, define $M(S, b, w(S))$ as the feasibility matrix for the induced POK problem on S. For $k \in S$ define further $M'(S, k, b, w(S))$ as the feasibility matrix for the induced POK problem on S which is restricted to ideals containing k as their highest numbered element. When no ambiguity arises, we omit from the matrix notation b and $w(S)$. The (p, w)-th element of matrix $M(S)$ $(M'(S, k))$ is then abbreviated as $m_{pw}(S)$ $(m'_{pw}(S, k))$.

We assume, without loss of generality, that the elements of N have been numbered so that $L_1 = 1, 2, ..., n$ is one of two defining linear extensions of P, while the other one is denoted by L_2. Accordingly $i \prec_P k$ iff $i \prec_{L_1} k$ and $i \prec_{L_2} k$. We will use the notation $i || k$ if $i < k$ as numbers and i is not comparable to k in P. The *principal ideals* are defined by $B_k = \{i \mid i \preceq_P k\}$ and their 'complements' by $\overline{B}_k = \{i \mid i || k\}$ for $k = 1, 2, ..., n$. Partition the ideals of P by their highest numbered element, and let $\mathcal{I}_k$ be the set of ideals with highest numbered element k. If $I \in \mathcal{I}_k$, then we clearly must have $B_k \subseteq I$ and $I \setminus B_k$ must be an ideal in the induced subposet $\overline{B}_k$. Therefore, each ideal $I \in \mathcal{I}_k$ is in a one-to-one correspondence with the ideal $I \setminus B_k$ of the subposet $\overline{B}_k$ and $p(I) = p(I \setminus B_k) + p(B_k)$ and $w(I) = w(I \setminus B_k) + w(B_k)$. Thus, we have proved the following lemma.

Lemma 1. *Consider a POK problem on the poset P with knapsack capacity b and total item value W. If $N_k = \{1, 2, ..., k\}$, and $M'(N_k, k, b, w(N_k))$ $M(\overline{B}_k, b, w(N_k) - w(B_k))$ are the two feasibility matrices of interest, then for their corresponding elements we have $m'_{pw}(N_k, k) = m_{p-p(B_k), w-w(B_k)}(\overline{B}_k)$ for $1 \leq p \leq b$ and $1 \leq w \leq w(N_k)$.*

Since $M(N, b, W)$ contains a 1 in position (p, w) if and only if there is a matrix $M'(N_k, k, b, w(N_k))$ containing a 1 in the same position, $M(N, b, W)$ can be obtained in $O(bWn)$ time if we know the $M'(N_k, k, b, w(N_k))$ matrices for $k = 1, 2, ..., n$. If the computation in Lemma 1 was recursively applied to the subposets $\overline{B}_k$, this would yield an exponential in n computation of $M(N, b, W)$ for general posets P. As the following theorem shows, however, the recursion does not need to go beyond the second level if $dim P = 2$, i.e., it results in a polynomial-time computation.

Theorem 1. *If $dim P = 2$ for a POK problem with knapsack capacity b and total item value W, then $M(N, b, W)$ can be computed by a pseudopolynomial algorithm in $O(n^2 bW)$ time.*

Proof. Partition the set of ideals in P by their highest numbered element and apply Lemma 1. Accordingly we obtain a general element of $M(N, b, W)$ by

$$m_{p,w}(N) = \max_{k=1,2,...,n} m'_{pw}(N_k, k) = \max_{k=1,2,...,n} m_{p-p(B_k),w-w(B_k)}(\overline{B}_k)$$

for $1 \leq p \leq b$ and $1 \leq w \leq W$. We will show that a single computation of $M(\overline{B}_k, b, w(N_k) - w(B_k))$ for a fixed k can be carried out in $O(nbW)$ time. Then the entire computation for $M(N, b, W)$ needs no more than $O(n^2 bW)$ time. Let $C_{kj} \doteq \overline{B}_k \cap B_j = \{i \mid i \in \overline{B}_k, i \preceq_P j\}$ and $\overline{C}_{kj} \doteq \overline{B}_k \cap \overline{B}_j = \{i \mid i \in \overline{B}_k, i < j, i \not\prec_P j\}$ for $k = 1, 2, ..., n$ and $j || k$.

We claim that $dim P = 2$ and $j || k$ imply $\overline{C}_{kj} = \overline{B}_j$: If $i \in \overline{C}_{kj}$, then it can easily be seen that $i \in \overline{B}_j$. For the other direction, if $i \in \overline{B}_j$, then $i < j$ and $i \not\prec_P j$, i.e., $j \prec_{L_2} i$. Furthermore, $j || k$ implies $j < k$ and $j \not\prec_P k$, i.e., $k \prec_{L_2} j$ too, so that by transitivity of L_2 $k \prec_{L_2} i$ also holds, which implies $i \in \overline{B}_k$, and thus $i \in \overline{C}_{kj}$.

Let us compute $M(\overline{B}_k, b, w(N_k) - w(B_k))$, i.e., the feasibility matrix for the ideals of $\overline{B}_k$, in ascending order of $k = 1, 2, ..., n$, For this it suffices to calculate the matrices $M'(\overline{B}_k, j)$ for all $j : j || k$, since $m_{pw}(\overline{B}_k) = \max_{j:j||k}\{m'_{pw}(\overline{B}_k, j)\}$. Applying Lemma 1 to the poset $\overline{B}_k$ shows that $M'(\overline{B}_k, j)$ can be derived from $M(\overline{C}_{kj})$ by $m'_{pw}(\overline{B}_k, j) = m_{p-p(C_{kj}),w-w(C_{kj})}(\overline{C}_{kj})$. But $\overline{C}_{kj} = \overline{B}_j$ therefore we need only the previously computed matrices $M(\overline{B}_j)$. Combining these observations, we obtain

$$m_{pw}(\overline{B}_k) = \max_{j:j||k}\{m'_{pw}(\overline{B}_k, j)\} = \max_{j:j||k}\{m_{p-p(C_{kj}),w-w(C_{kj})}(\overline{C}_{kj})\}$$

which equals $\max_{j:j||k}\{m_{p-p(C_{kj}),w-w(C_{kj})}(\overline{B}_j)\}$. Since the quantities $p(B_k)$, $w(B_k)$, $p(C_{kj})$ and $w(C_{kj})$ can all be calculated in $O(n^3)$ time in a preprocessing step, the claimed complexity of the algorithm follows. □

The algorithm given in Theorem 1 is polynomial in W and b. We show how to compress the table size and find a near-optimal solution instead of the exact optimum. We will obtain a bicriteria FPTAS, i.e., an algorithm that yields an ideal of weight and processing time within $(1 \mp \varepsilon)$ of the desired targets, in time polynomial in $1/\varepsilon$.

We will scale the coefficients in a manner similar to the one used for standard Knapsack (cf. [11]). The difference is that in our case both b and W appear in the running time so we have to scale both weights and processing times. Let $w_{\max} = \max_j w_j$. Let $k_w = \varepsilon w_{\max}/n$, $k_p = \frac{\varepsilon p(N)}{(n+1)l}$. Set $w'_j = \lfloor w_j/k_w \rfloor$, $p'_j = \lfloor p_j/k_p \rfloor$ $\forall j \in N$. In the scaled instance set the knapsack capacity to be $b' = \lceil p(N)/(k_p l) \rceil$. It can be shown that solving the scaled instance yields the following theorem.

Theorem 2. *For a POK problem with knapsack capacity b on poset $P = (N, \prec_P)$ where $dim P = 2$ there is an FPTAS with the following properties. For any $\varepsilon > 0$ it produces, in time $O(n^5/\varepsilon)$, an ideal of processing time at most $(1 + \varepsilon)b$ and weight at least $(1 - \varepsilon)$ times the optimum.*

3 POK on Bipartite Partial Orders

Let $(X, Y; \prec)$ be a bipartite poset with comparability graph $G = (X, Y; E)$. Its bipartite complement $\overline{G} = (X, Y; \overline{E})$ is defined by $\overline{E} = X \times Y \setminus E$. An ideal $X' \cup Y'$, where $X' \subseteq X$ and $Y' \subseteq Y$, is Y-*maximal* if there is no $y \in Y$ such that $X' \cup Y' \cup \{y\}$ is also an ideal. The ideal (X', Y') is X-*minimal* if there is no $x \in X'$ such that $(X' \setminus x, Y')$ is also an ideal.

Each ideal I of a poset is uniquely defined by its *maximal elements* $\max I = \{a \in I \mid \not\exists b \in I$ such that $a \prec b\}$: If we know $\max I$, then $I = \{a \mid \exists b \in \max I$ such that $a \preceq b\}$. The elements of $\max I$ always form an antichain in the poset, and an antichain generates an ideal in this sense. Let $\mathcal{M} = \{I \mid I$ is an ideal such that $\max I$ is a maximal antichain$\}$. It is well known that $(\mathcal{M}, \subseteq)$ is a lattice which is isomorphic to the lattice of maximal antichains.

Lemma 2. *The following statements are equivalent in a bipartite poset $(X, Y; \prec)$ with comparability graph $G = (X, Y; E)$:*
1. *$X' \cup Y'$ is a Y-maximal and X-minimal ideal*
2. *$Y' \cup (X \setminus X')$ is a maximal antichain*
3. *The induced subgraph $\overline{G}[X \setminus X', Y']$ is a maximal complete bipartite subgraph of the bipartite complement $\overline{G}$.*

Proof. 1.$\implies$ 2. Suppose $X' \cup Y'$ is a Y-maximal and X-minimal ideal. We cannot have any edge between $X \setminus X'$ and Y' in G, since (X', Y') is an ideal. Thus $Y' \cup (X \setminus X')$ is an antichain. Since Y' is Y-maximal, every $y \in Y \setminus Y'$ must have a predecessor in $X \setminus X'$, so there is no $y \in Y \setminus Y'$ such that $Y' \cup (X \setminus X') \cup y$ would also be an antichain. Similarly, every $x \in X'$ must have a successor $y \in Y'$, since (X', Y') is X-minimal, so there is no $x \in X'$ for which $Y' \cup (X \setminus X') \cup x$ would also be an antichain. This proves the maximality of the antichain $Y' \cup (X \setminus X')$.

2.$\implies$ 3. obvious.

3.$\implies$ 1. Let $\overline{G}[X \setminus X', Y']$ be a maximal complete bipartite subgraph of $\overline{G}$. This implies that no $x \in X \setminus X'$ can be a predecessor for any $y \in Y'$, thus (X', Y') is an ideal in $(X, Y; \prec)$. Since $\overline{G}[X \setminus X', Y']$ is a maximal complete bipartite subgraph of $\overline{G}$, no $y \in Y \setminus Y'$ is connected to every element of $X \setminus X'$ in $\overline{G}$, so there is no $y \in Y \setminus Y'$ extending the ideal $X' \cup Y'$, i.e., it is Y-maximal. Similarly, no $x \in X'$ is connected to every element of Y' in $\overline{G}$, so $X' \cup Y'$ is also X-minimal. □

It is clear that in order to solve a Red-Blue bipartite instance of POK, we need to search through only the ideals which are Y-maximal and X-minimal, i.e., the ideals in $\mathcal{M}$. By the above lemma, these ideals of a bipartite poset $(X, Y; \prec)$ are in a one-to-one correspondence with the maximal complete bipartite subgraphs of the bipartite complement of its comparability graph. If we have such a subgraph $\overline{G}[U, Z]$, then the corresponding ideal $X \setminus U \cup Z$, its weight $w(X \setminus U \cup Z)$ and processing time $p(X \setminus U \cup Z)$ can all be computed in $O(n)$ time. Furthermore, if $\overline{G}$ is chordal bipartite, then it has only at most $|\overline{E}|$ maximal complete bipartite subgraphs and Kloks and Kratsch [14] found an algorithm which lists these in $O(|X \cup Y| + |\overline{E}|)$ time if $\overline{G}$ is given by an appropriately ordered version of its bipartite adjacency matrix. This proves the following theorem.

Theorem 3. *Consider a* Red-Blue *instance of POK on a bipartite poset* $(X, Y; \prec)$ *with comparability graph* G. *If* $\overline{G}$ *is chordal bipartite then there is an algorithm which solves this POK problem in* $O(n^3)$ *time.*

We proceed now to lift the restriction that the bipartite instance is Red-Blue. The resulting problem is NP-hard since it contains classical Knapsack. Let $I = B \cup C$ be an arbitrary ideal in the bipartite poset $(X, Y; \prec)$, where $B \subseteq X$ and $C \subseteq Y$. It is clear that $\max I$ partitions into $C \cup (\max I \cap B)$. Let $X' = B \setminus \max I$ and $Y' = \{y \in Y \mid \nexists x \in X \setminus X' \text{ such that } x \prec y\}$. It is clear that $C \subseteq Y'$ and that $X' \cup Y'$ is Y-maximal and X-minimal ideal. Then by Lemma 2, $Y' \cup (X \setminus X')$ is a maximal antichain containing $\max I$. The ideal generated by this maximal antichain is $Y' \cup X \in \mathcal{M}$. Furthermore, $I \subseteq (Y' \cup X)$ and $(Y' \cup X) \setminus I$ is an antichain contained in $Y' \cup (X \setminus X')$. Thus I can be *derived* from $(X \cup Y')$ by the deletion of an appropriate unordered subset of $Y' \cup (X \setminus X')$. If we considered for deletion all such subsets of $Y' \cup (X \setminus X')$, and repeated this for all $(X \cup Y') \in \mathcal{M}$, then we would derive *every* ideal of the poset $(X, Y; \prec)$ (some of them possibly several times.)

Consider now an arbitrary instance of POK on a bipartite poset $(X, Y; \prec)$ with knapsack size b and optimal solution weight W^*. If there exists an $(X \cup Y') \in \mathcal{M}$ such that $p(X \cup Y') \leq b$, then clearly $W^* = \max\{w(X \cup Y') \mid p(X \cup Y') \leq b, (X \cup Y') \in \mathcal{M}\}$. Otherwise, consider an infeasible $(X \cup Y') \in \mathcal{M}$, i.e., $p(X \cup Y') > b$. Define $X' = \{x \in X \mid \exists y \in Y' \text{ such that } x \prec y\}$. We can find the largest-weight feasible ideal derivable from $(X \cup Y')$ by the above process by solving the auxiliary Knapsack problem $\{\max w(J) \mid J \subseteq ((X \setminus X') \cup Y')), p(J) \leq b - p(X')\}$. The optimal solution J^* of this can be found by a pseudopolynomial algorithm in $O(nW)$ time, and $X' \cup J^*$ is the largest-weight feasible ideal that can be derived from this $(X \cup Y')$ for the original POK instance. As we have discussed earlier, if $(X, Y; \prec)$ is a bipartite poset whose comparability graph has a bipartite complement $\overline{G} = (X, Y; \overline{E})$ which is chordal bipartite, then $|\mathcal{M}| \leq |\overline{E}|$, so that we need to call the pseudopolynomial algorithm for the solution of at most $|\overline{E}|$ auxiliary problems. This proves the following.

Theorem 4. *Consider an instance of POK on a bipartite poset* $(X, Y; \prec)$ *with comparability graph* G. *If* $\overline{G}$ *is chordal bipartite then there is a pseudopolynomial algorithm which solves this POK problem in* $O(n^3 W)$ *time.*

It is easy to see that invoking as a Knapsack oracle the FPTAS in [12], instead of a pseudopolynomial algorithm, yields a $(1 - \varepsilon)$ weight-approximation to the original problem.

Theorem 5. *Consider an instance of POK on a bipartite poset* $(X, Y; \prec)$ *with comparability graph* G. *If* $\overline{G}$ *is chordal bipartite then there is an FPTAS which for any* $\varepsilon > 0$, *solves this POK problem in* $O(n^3 \log(1/\varepsilon) + n^2(1/\varepsilon^4))$ *time.*

4 Applications to Scheduling

In this section we show how the above pseudopolynomial algorithms for POK given in Theorems 1 and 4 lead to improved polynomial-time approximation algorithms for special cases of the scheduling problem $1|prec|\sum w_j C_j$. In a recent paper, Woeginger [21] defined the following auxiliary problem, which is a special case of *POK*.

PROBLEM: GOOD INITIAL SET (IDEAL)
INSTANCE: An instance of $1|prec|\sum w_j C_j$ with nonnegative integer processing times p_j and weights w_j, a real number γ with $0 < \gamma \leq 1/2$.
QUESTION: Is there an ideal I in the poset P representing the precedence constraints for which $p(I) \leq (1/2 + \gamma)p(N)$ and $w(I) \geq (1/2 - \gamma)w(N)$?

The following theorem shows the strong connection between the solvability of *POK* and the approximability of $1|prec|\sum w_j C_j$. Its derivation uses 2-dimensional Gantt charts as introduced in [9].

Theorem 6. *[21] If C is a class of partial orders on which GOOD INITIAL SET is solvable in pseudopolynomial time, then for any $\varepsilon > 0$, the restriction of the scheduling problem to C, i.e., $1|prec, C|\sum w_j C_j$ has a polynomial-time $(\Phi + \varepsilon)$-approximation algorithm, where $\Phi = 1/2(\sqrt{5}+1) \approx 1.61803$.*

Theorems 1, 4 and 6 yield the following corollaries (proofs omitted).

Corollary 1. *For any $\varepsilon > 0$, $1|prec, dimP = 2|\sum w_j C_j$ has a polynomial-time $(\Phi + \varepsilon)$-approximation algorithm.*

Corollary 2. *For any $\varepsilon > 0$, the problem $1|prec, \prec_P | \sum w_j C_j$ where $P = (X, Y; \prec_P)$ is such that the bipartite complement of its comparability graph is a chordal bipartite graph, has a polynomial-time $(\Phi + \varepsilon)$-approximation algorithm.*

5 The LP-Relaxation for General *POK*

In this section we examine the natural integer program for *POK* and the associated linear relaxation. We give an algorithm for rounding a fractional solution which satisfies weight-majority. Consider the following integer program with parameter $l > 1$:

$$\max \{\sum_{j \in N} w_j x_j \mid \sum_{i \in N} p_i x_i \leq p(N)/l, \ x_j \leq x_i \ i \prec j, \ x_i \in \{0,1\} \ i \in N\}.$$

Relaxing the integrality constraint gives a linear relaxation which we denote by $LP(l)$. Let $\bar{x}$ be a solution to $LP(l)$ with objective value $\sum_{j \in N} w_j \bar{x}_j - w(N)/h$, where $h < l$. We show how to round it to an integral solution, while relaxing by a small factor the right hand side of the packing constraint. The guaranteed value of the objective will also be less than $w(N)/h$, therefore producing a bicriteria approximation. The algorithm itself is simple. Let $\alpha \in (0, 1)$ be a parameter of our choice. We apply filtering based on α. We omit the analysis in this version.

Theorem 7. *Let a POK instance be such that the linear relaxation $LP(l)$ has a solution $\bar{x}$ of value $w(N)/h$ with $1/l < 1/h$. Define $N' \doteq \{j \in N \mid \bar{x}_j \geq \alpha\}$. N' is an ideal and for any $\alpha \in (1/l, 1/h)$ $w(N') \geq \frac{1-h\alpha}{h(1-\alpha)} w(N)$ and $p(N') \leq \frac{p(N)}{\alpha l}$.*

Theorem 7 raises two questions. First, when can we expect the optimal fractional solution to meet the weight-majority condition $1/l < 1/h$? Second, the guarantee on the weight does not conform to the standard definition of a multiplicative ρ-approximation. Can we obtain a proper ρ-approximation on the weight with or without relaxing the upper bound on the processing time? We discuss these two questions in the next two sections.

6 The Effect of Decomposability

In this section we address the question: when can one hope to obtain a weight majority solution to $LP(l)$ and hence apply Theorem 7? We show that decomposability of the input is a necessary condition by gaining insight into the structure of LP solutions. Finally we show, non-constructively, an upper bound on the integrality gap of $LP(l)$.

To simplify our derivation we assume the input instance is bipartite Red-Blue. This is without loss of generality: given a general input N_0 one can transform it to a bipartite input by having for each job $i_0 \in N_0$ a vertex i, i' on each side X, Y of the partition. The vertex i on the X-side assumes the processing time and i' the weight of i_0. Precedence constraint $i_0 \prec j_0$ for $i_0, j_0 \in N_0$ translates to $i \prec j'$. Moreover $i \prec i'$ for all i. Without loss of generality we can assume that in any ideal of N, inclusion of i implies inclusion of i'. Thus there is a one-to-one correspondence between ideals of N and N_0, with the total processing time and weight being the same.

For any $l > 1$ formulation $IPOK$, defined below, is the integer linear programming formulation of the associated partially-ordered knapsack instance.

$$\max \{ \sum_{j \in Y} w_j x_j \mid \sum_{i \in X} p_i x_i \leq p(X)/l, \ x_j - x_i \leq 0 \ i \prec j, \ x_i \in \{0,1\} \ i \in X \cup Y \}.$$

In the linear programming relaxation, $LPOK$, the integrality constraints are relaxed. The proofs of the following two theorems are omitted.

Theorem 8. *The solution $\bar{x}_i = 1/l$ for $i \in X \cup Y$ is optimal for $LPOK$ with knapsack capacity $p(X)/l$ if and only if the underlying poset is indecomposable.*

Theorem 8 shows that when the input is indecomposable, the same fraction of weight and processing time will be assigned to the ideal by the optimal solution to $LPOK$, hence the algorithm from Theorem 7 cannot apply. However, Theorem 8 does not express only the limitations of our algorithm but the limitations of the relaxation as well. The linear program $LPOK$ can be seen as maximizing the rank of the ideal that meets the processing-time packing constraint. When the balanced solution in which all variables are equal is optimal, in the case of indecomposability, the rank computed is equal to $w(Y)/p(X)$, which is equivalent to selecting the entire set $X \cup Y$ as the ideal. Therefore no information is really obtained from $LPOK$ in this case.

Theorem 9. *The fractional optimum of $LPOK$ is at most $(\lambda/l)w(Y)$ on an input with maximum-rank ideal of value $\lambda w(Y)/p(X)$, $\lambda \geq 1$.*

Finally we observe that the valid inequalities for Knapsack (without precedence constraints) recently proposed by Carr et al. [2] do not seem to help in the difficult case, namely indecomposability. If we augment $LPOK$ with these inequalities, it is easy to see that the balanced solution is feasible for the new formulation as well.

7 Evidence of Hardness

The positive result of Theorem 7 has two limitations: first it applies only to fractional solutions satisfying weight-majority and second it does not give a ρ-approximation for the weight in the classical multiplicative sense. In this section we show that overcoming these limitations would lead to improved approximations for both the time-majority case and $1|prec|\sum w_j C_j$. This would make progress on a longstanding open problem. For the purposes of this section a ρ-approximation algorithm, $\rho < 1$, for a POK instance (N, h, l) yields, if the instance is feasible, an ideal of weight at least $\rho w(N)/h$ and processing time at most $p(N)/l$. The idea behind the upcoming theorem is that even if a given set N has no ideal N' such that $w(N')/w(N) > p(N')/p(N)$, we can add auxiliary weight so that a feasible weight-majority instance I' can be defined. If in turn we have an algorithm to extract from I' an ideal with large enough weight, this answer will contain a good fraction of the original weight from N. The proof of the theorem is omitted.

Theorem 10. *Let δ be any constant in $(0, 1)$. If there is a δ-approximation algorithm $\mathcal{A}$ for weight-majority knapsack instances, then there exists an $O(1)$-approximation algorithm for a time-majority instance (N, h, l) where h and l are constants depending on δ.*

Theorem 10 shows that finding an $O(1)$-approximation for the weight-majority case is as hard as finding an $O(1)$-approximation for a class of time-majority instances with small but constant $1/h$ and $1/l$ parameters. This in turn is a highly non-trivial problem. Its solution would lead to a $2 - \beta$ approximation for some fixed β for $1|prec|\sum w_j C_j$, however small $1/h$ and $1/l$. The latter statement can easily be shown using 2-D Gantt charts [9] and methods similar to the ones used in [21]. We omit the details.

Corollary 3. *Let δ be any constant in $(0, 1)$. If there is a δ-approximation algorithm for weight-majority knapsack instances, there is a $2 - \beta$ approximation for some fixed β for $1|prec|\sum w_j C_j$.*

Acknowledgment

Stavros Kolliopoulos thanks Gerhard Woeginger for useful discussions.

References

[1] A. Brandstädt, V. B. Le, and J. P. Spinrad. *Graph Classes: A Survey*. SIAM, Philadelphia, PA, 1999.

[2] R. D. Carr, L. K. Fleischer, V. J. Leung, and C. A. Phillips. Strengthening integrality gaps for capacitated network design and covering problems. In *Proc. 11th ACM-SIAM Symposium on Discrete Algorithms*, 2000.

[3] C. Chekuri and R. Motwani. Precedence constrained scheduling to minimize sum of weighted completion times on a single machine. *Discrete Applied Mathematics*, 98:29–38, 1999.

[4] F. A. Chudak and D. S. Hochbaum. A half-integral linear programming relaxation for scheduling precedence-constrained jobs on a single machine. *Operations Research Letters*, 25:199–204, 1999.

[5] F. A. Chudak and D. B. Shmoys. Approximation algorithms for precedence-constrained scheduling problems on parallel machines that run at different speeds. In *Proc. 8th ACM-SIAM Symposium on Discrete Algorithms*, 581–590, 1997.

[6] B. Dushnik and E. W. Miller. Partially ordered sets. *Amer. J. Math.*, 63:600–610, 1941.

[7] U. Feige. Relations between average case complexity and approximation complexity. In *Proc. 34th ACM Symposium on Theory of Computing*, 534–543, 2002.

[8] U. Feige, G. Kortsarz, and D. Peleg. The dense k-subgraph problem. *Algorithmica*, 29:410–421, 2001.

[9] M. X. Goemans and D. P. Williamson. Two-dimensional Gantt charts and a scheduling algorithm of Lawler. *SIAM J. on Discrete Mathematics*, 13:281–294, 2000. Preliminary version in Proc. SODA 1999.

[10] L. A. Hall, A. S. Schulz, D. B. Shmoys, and J. Wein. Scheduling to minimize average completion time: Off-line and on-line approximation algorithms. *Math. Oper. Res.*, 22:513–544, 1997.

[11] D. S. Hochbaum. Approximating covering and packing problems: set cover, vertex cover, independent set, and related problems. In D. S. Hochbaum, editor, *Approximation Algorithms for NP-hard problems*, 94–143. PWS, 1997.

[12] O. H. Ibarra and C. E. Kim. Fast approximation algorithms for the Knapsack and sum of subset problems. *J. ACM*, 22:463–468, 1975.

[13] D. S. Johnson and K. A. Niemi. On knapsacks, partitions, and a new dynamic programming technique for trees. *Math. Oper. Res.*, 8(1):1–14, 1983.

[14] T. Kloks and D. Kratsch. Computing a perfect edge without vertex elimination ordering of a chordal bipartite graph. *IPL*, 55:11–16, 1995.

[15] F. Margot, M. Queyranne, and Y. Wang. Decompositions, network flows and a precedence-constrained single machine scheduling problem. TR 2000-29, Dept. of Mathematics, University of Kentucky, Lexington. Communicated in 1997.

[16] R. H. Möhring. Computationally tractable classes of ordered sets. In I. Rival, editor, *Algorithms and Order*, 105–194. Kluwer, Boston, MA, 1989.

[17] R. Paige and R. E. Tarjan. Three partition refinement algorithms. *SIAM J. Computing*, 6:973–989, 1987.

[18] P. Schuurman and G. J. Woeginger. Polynomial time approximation algorithms for machine scheduling: ten open problems. *J. of Scheduling*, 2:203–213, 1999.

[19] J. B. Sidney. Decomposition algorithms for single machine sequencing with precedence relations and deferral costs. *Operations Research*, 23:283–298, 1975.

[20] W. T. Trotter. *Combinatorics of Partially Ordered Sets*. Johns Hopkins University Press, Baltimore, Md., 1992.

[21] G. J. Woeginger. On the approximability of average completion time scheduling under precedence constraints. In *Proc. 28th ICALP*, 887–897, 2001.
[22] M. Yannakakis. On the complexity of partial order dimension problem. *SIAM J. Alg. Discr. Methods*, 3:351–358, 1982.

A Software Library for Elliptic Curve Cryptography[*]

Elisavet Konstantinou[1,2], Yiannis Stamatiou[1,2], and Christos Zaroliagis[1,2]

[1] Computer Technology Institute
P.O. Box 1122, 26110 Patras, Greece
[2] Department of Computer Engineering and Informatics
University of Patras, 26500 Patras, Greece
{konstane,stamatiu,zaro}@ceid.upatras.gr

Abstract. We present an implementation of an EC cryptographic library targeting three main objectives: portability, modularity, and ease of use. Our goal is to provide a fully-equipped library of portable source code with clearly separated modules that allows for easy development of EC cryptographic protocols, and which can be readily tailored to suit different requirements and user needs. We discuss several implementation issues regarding the development of the library and report on some preliminary experiments.

1 Introduction

Cryptographic systems based on elliptic curves were introduced independently by Koblitz [15] and Miller [20] in 1985 as an alternative to conventional public key cryptosystems such as RSA [23] and DSA [14]. The main advantage of elliptic curve (EC) cryptosystems is that they use smaller parameters (e.g., encryption key) than the conventional cryptosystems. The reason is that the underlying mathematical problem on which their security is based, the EC discrete logarithm problem (ECDLP), appears to require more time to solve than the analogous problem in groups generated by prime numbers on which the conventional cryptosystems are based. For groups defined on ECs (where the group elements are the points of the EC), the best algorithm for attacking this problem takes time exponential in the size of the group, while for groups generated by prime numbers there are algorithms that take subexponential time. This implies that one may use smaller parameters for the EC cryptosystems than the parameters used in RSA or DSA, obtaining the same level of security. A typical example is that a 160-bit key of an EC cryptosystem is equivalent to RSA and DSA with a 1024-bit modulus. As a consequence, smaller keys result in faster implementations, less storage space, as well as reduced processing and bandwidth requirements.

---

[*] This work was partially supported by the IST Programme of EU under contracts no. IST-1999-14186 (ALCOM-FT) and no. IST-1999-12554 (ASPIS), and by the Human Potential Programme of EU under contract no. HPRN-CT-1999-00104 (AMORE).

There are many important decisions that one should make before starting implementing an elliptic curve cryptosystem (ECC). These include the type of the underlying finite field, the algorithms for implementing the basic algebraic operations, the type of the elliptic curve to be used as well as its generation, and finally the elliptic curve protocols. The fields usually used are either prime fields (denoted by F_p, where p is a prime) or binary fields. Selecting a prime field (which will be our concern here) implies proper choice of p, since all basic operations will be modulo that prime and the security of the system will depend on its size. The larger the prime, the more secure but slower the cryptosystem. The basic algebraic operations of the EC group that must be implemented are addition of points on an EC, and scalar multiplication (multiplication of a point by an integer). The latter is the operation upon which the ECDLP is based. There should also be a way for generating secure elliptic curves and create random points in them. The generation of such curves can be accomplished with three methods, namely the point counting method (see [3]), the method based on the constructive Weil descent [8], and the Complex Multiplication method (see [3]). The latter two methods build ECs of a *suitable order* (where order refers to the number of elements in the group defined by the EC), i.e., the order satisfies certain conditions, necessary for the cryptographic strength of the EC. The former method does not necessarily produce ECs of a suitable order. Once all these basic operations have been implemented, one can start developing cryptographic protocols.

Several studies have been conducted in the past and many papers have been written on software implementations of ECCs. The majority of them focuses either on the efficient implementation of the basic algebraic operations (see e.g., the excellent surveys of [18] and [19]), or on the efficient implementation of a single protocol for one particular finite field. For example, in [4] a nice study of the performance of the Elliptic Curve Discrete Signature Algorithm (ECDSA) over prime fields is presented using the NIST recommended ECs, while in [10] a similar study is presented regarding binary fields.

In this paper, we present an implementation of an EC cryptographic library targeting three main objectives: portability, modularity, and ease of use. Our goal is to provide a fully-equipped library of portable source code with clearly separated modules which allow for the easy development of EC cryptographic protocols, and which can be readily tailored to suit different requirements and user needs. The library has been implemented in ANSI C using the GNUMP [9] library. We have chosen prime fields as the underlying field of our implementations, because of their simplicity in representation and in performing algebraic operations. Our library includes all the basic operations of an EC group, several cryptographic protocols based on ECs (including the EC Discrete Signature Algorithm – ECDSA), and the Complex Multiplication (CM) method for generating secure elliptic curves. The implementation of the latter method included several engineering challenges due to the fact that it heavily relies on the ability to perform unlimited precision computations on complex numbers, and in addition requires the use of special trigonometric and exponentiation functions.

The reason for high precision stems from a crucial step of the method, namely the construction of the so-called Hilbert or Weber polynomials. The GNUMP library for high-precision floating point arithmetic lacks both an implementation of the required functions as well as of high-precision complex number arithmetic, and hence we had to implement them from scratch. For the implementation of the functions, we had to resort to their Taylor series expansion and study the interrelationship between the precision of basic complex number (floating point) arithmetic and the number of terms in the Taylor series necessary to produce correct results. Clearly, the number of terms depended on the required precision. Another problem was that the roots of the Hilbert polynomials are absolutely necessary in order to find the EC, but their construction is a considerably high burden, because of the very high precision they require. On the other hand, the construction of Weber polynomials does not require such a high precision and consequently turns out to be incredibly faster (cf. Section 5), but their roots are not appropriate for the CM method. Hence, we had to find a practical way to transform Weber roots to the roots of the corresponding Hilbert polynomial (something that was not adequately addressed in the bibliography).

At the current stage of the library, we are making no claims that the implemented algorithms are the best or fastest possible (they certainly are not, although preliminary experiments are rather encouraging). We hope, however, that our work will be valuable to those interested in developing easily protocols based on elliptic curve cryptography by providing, in a nutshell, basic and advanced cryptographic primitives and by uncovering many of the pitfalls and their treatment, inherent in their implementation. Our library is publicly available from http://www.ceid.upatras.gr/faculty/zaro/software/ecc-lib/.

To the best of our knowledge, there are two other libraries which offer primitives for EC cryptography: LiDIA [16] and MIRACL [21]. Both are efficient, general purpose, number-theoretic libraries implemented in C++ that offer a vast number of cryptographic primitives. Although there seem to exist implementations of EC cryptographic protocols (e.g., ECDSA) and EC generation methods (CM method) based on these libraries, these implementations are either not publicly offered (LiDIA), or are partly provided (MIRACL).

2 Basic Concepts of Elliptic Curve Algebra

In this section we review some basic concepts regarding elliptic curves and their definition over finite fields. The interested reader may find additional informations, for example, in [3, 25]. We also assume some familiarity with elementary number theory (see e.g., [5]).

The elliptic curves are usually defined over *binary fields* F_{2^m} ($m \geq 1$), or over *prime fields* F_p, $p > 3$. The case $p = 3$ for prime fields requires separate attention but it is, generally, not difficult to transfer results obtained for F_p, $p > 3$, to the case $p = 3$. In our library we use elliptic curves defined over prime fields.

An *elliptic curve* (EC) over the prime field F_p, denoted by $E(F_p)$, is the set of points $(x, y) \in F_p$ (represented by affine coordinates) which satisfy the equation

$$y^2 = x^3 + ax + b \tag{1}$$

where $4a^3 + 27b^2 \neq 0$ (this condition guarantees that Eq. (1) does not have multiple roots in F_p), along with a special point denoted by $\mathcal{O}$, called the *point at infinity*. An addition operation $+$ is defined over $E(F_p)$ such that $(E(F_p), +)$ defines an Abelian group, called the *EC group*, with $\mathcal{O}$ acting as its identity. The addition operation on $E(F_p)$ is defined as: $P+Q = \mathcal{O}$, if $P = -Q$; $P+Q = Q$, if $P = \mathcal{O}$; and $P+Q = R$, otherwise, where, if $P = (x_1, y_1)$ and $Q = (x_2, y_2)$, then the coordinates of the point $R = (x_3, y_3)$ are given by $x_3 = z^2 - x_1 - x_2$, $y_3 = z(x_1 - x_3) - x_3 - y_1$ with $z = \frac{y_1 - y_2}{x_1 - x_2}$, if $P \neq Q$, and $z = \frac{3x_1^2 + a}{2y_1}$, if $P = Q$ (a is the coefficient from Eq. (1)). The negative $-Q$ of a point $Q = (x, y)$ is the point $(x, p-y)$.

A fundamental operation of cryptographic protocols based on EC is the *scalar (or point) multiplication*, i.e., the multiplication of a point P by an integer k (an operation analogous to the exponentiation in multiplicative groups), that produces another point $Q = kP$ (the point resulting by adding P to itself for k times). Several algorithms exist for the fast and efficient implementation of the scalar multiplication operation. Most of them are based on the binary representation of the integer k.

The *order m of an elliptic curve* is the number of points in $E(F_p)$. Hasse's theorem (see e.g., [3, 25]) gives upper and lower bounds for m that are based on the order p of F_p:

$$p + 1 - 2\sqrt{p} \leq m \leq p + 1 + 2\sqrt{p}. \tag{2}$$

The *order of a point P* is the smallest positive integer n for which $nP = \mathcal{O}$. Application of Langrange's theorem (stating that the exponentiation of any group element to the power of the group's order gives the identity element) on $E(F_p)$, gives that $mP = \mathcal{O}$ for any point $P \in E(F_p)$, which in turn implies that the order of a point cannot exceed the order of the elliptic curve.

3 Generation of Elliptic Curves and the Complex Multiplication Method

Many of the security properties of elliptic curve cryptosystems depend on the order of the EC group and this is determined by the generated EC. If this order is *suitable*, i.e., it obeys some specific good properties, then there is a guarantee for a high level of security. The order m of an EC is called *suitable*, if the following conditions are satisfied:

1. m must have a sufficiently large prime factor (greater than 2^{160}).
2. $m \neq p$.
3. For all $1 \leq k \leq 20$, it should hold that $p^k \neq 1 \bmod m$.

The above conditions ensure the robustness of cryptosystems based on the *discrete logarithm problem for EC groups* (ECDLP), since it is very difficult for all known attacks to solve this problem efficiently, if m obeys the above properties. ECDLP asks for determining the value of t when two points P, Q in $E(F_p)$ are given such that P is of order n and $Q = tP$, where $0 \leq t \leq n - 1$.

As it was mentioned in the introduction, there are three methods for generating ECs: the point counting method (see [3]), the method based on the constructive Weil descent [8], and the Complex Multiplication (CM) method (see [3]). The point counting method does not necessarily construct an EC of suitable order, but it may achieve this by repeated applications of the method. The other two methods construct ECs of a suitable order. In [8] it was shown that the method based on the constructive Weil descent suffers from a major drawback that is not easy to handle: it samples from a very small subset of the set of possible elliptic curves. For this reason, we didn't implement this particular method in our library. We only implemented the CM method and compared it to an implementation of the point counting method given in [12]. The point counting method is based on Schoof's algorithm for the exact counting of rational points on an elliptic curve (see [3] for details as well as some improvements on Schoof's algorithm). In the rest of this section we shall briefly review the CM method.

3.1 A High-Level Description of the CM Method

The Complex Multiplication method generates an EC of a suitable order and also computes the coefficients a, b which determine the EC. The method starts by creating an EC of a suitable order and then proceeds to determine the coefficients a, b. To accomplish the latter, the roots of the so-called Hilbert or Weber polynomials have to be computed. Each root of such a polynomial determines two possible elliptic curves. However, only one of them has the desired suitable order.

Let $E(F_p)$ be an elliptic curve of order m defined over the prime field F_p. Hasse's theorem (Eq. (2)) implies that the quantity $Z = 4p - (p + 1 - m)^2$ is positive, and that it exists a unique factorization $Z = Dv^2$, where D is a square-free positive integer and v is some integer. In addition, there exists an integer u such that:

$$4p = u^2 + Dv^2 \qquad (3)$$

where u satisfies $m = p + 1 \pm u$. We say that D is a *CM discriminant* for the prime p and that $E(F_p)$ has a *CM by D*. The Complex Multiplication method takes as input some discriminant D. The basic steps of the method are as follows:

1. Pick a random prime p and check whether Eq. (3) has a solution (u, v), with u, v integers. One algorithm that can be used to solve this equation in u, v is Cornacchia's algorithm [6]. If there is no solution, then another prime p is chosen and the step is repeated. The prime number p is going to be the order of the underlying finite field F_p.
2. There are only two possible orders for the group $E(F_p)$, namely $m = p+1-u$ and $m = p + 1 + u$. Check whether (at least) one of them is suitable. If this

is the case, then proceed to Step 3 (m is the order of the elliptic curve that we will generate). Otherwise, return to Step 1.
3. Construct either the Hilbert or the Weber polynomial, using the discriminant D. It should be noted that both types of polynomials lead to the construction of the same elliptic curve.
4. Compute the roots (modulo p) of either polynomial (this is accomplished by using a slight modification of Berlekamp's algorithm [2]). To further proceed, however, the roots of the Hilbert polynomial are required. If in Step 3 the construction of the Weber polynomial was chosen, then transform their roots (cf. Subsection 3.2) to the roots of the corresponding Hilbert polynomial (constructed using the same discriminant D). From every (Hilbert) root, two elliptic curves will be generated, but only one has the desired order m. Given a root j, the first curve is given by the equation $y^2 = x^3 + ax + b$, where $a = 3k$, $b = 2k$ and $k = \frac{j}{1728-j}$ (all operations are modulo p). The second curve, called the *twist* of the first, is given by $y^2 = x^3 + ac^2x + bc^3$, where c is any quadratic non-residue in F_p.
5. Since only one of the curves has the required suitable order m, we can find the particular one using a simple procedure that is based on Langrange's theorem (for any group point P, it should hold that $mP = \mathcal{O}$): repeatedly pick random points P on each elliptic curve, until a point is found for which $mP \neq \mathcal{O}$. Then, we are certain that the other curve is the one we seek.

3.2 The Hilbert and Weber Polynomials

One major difficulty associated with the construction of Hilbert and Weber polynomials for a given discriminant D, denoted by $H_D(x)$ and $W_D(x)$ respectively, is the need for implementing high precision, complex arithmetic. Especially for the Hilbert polynomials, whose coefficients can become huge for large values of D, the need for high precision is more apparent. On the other hand, the Weber polynomials have smaller coefficients than the equivalent Hilbert polynomials, and their construction requires less time. Due to space limitations, we cannot provide the (rather lengthy) definitions of these polynomials, but will give an example of the two polynomials for $D = 292$ in order to get an idea on the size of the coefficients and hence on the required high precision.

$$W_{292}(x) = x^4 - 5x^3 - 10x^2 - 5x + 1$$
$$H_{292}(x) = x^4$$
$$-206287709860428304608 00x^3$$
$$-9369362251192903875949706611 2000000x^2$$
$$+4552155138637938536962996838 4000000000x$$
$$-3802594610425124047799906426 88000000000000$$

In our library, we have implemented both Hilbert and Weber polynomials, mainly in order to perform a comparative study of how their requirements for time and precision scale in connection with certain parameters such as the discriminant D.

If in Step 3 of the CM method we choose to implement the Weber polynomials, we must transform their roots into the roots of the corresponding Hilbert polynomials (generated from the same discriminant D). A detailed analysis of how this transformation can be carried out is partly presented in [26], where a different representation of the polynomials is used. By working out the details and taking into account the representation of polynomials that we use, we came up with an easy to implement transformation.

4 Implementation Considerations

In this section, we will discuss several implementation issues regarding the development of our library. Our major design goals were to develop a portable, modular, and easy-to-use library. Before designing the library components, several decisions had to be made regarding the choice of the field on which to base the library, its size, as well as the libraries for generating and performing arithmetic with large numbers (e.g., primes).

To address portability, we have chosen to implement our library in ANSI C, using the (ANSI C) GNU Multiple Precision arithmetic library [9] for integer and floating point arithmetic with infinite precision.

We have chosen prime fields F_p mainly due to their simplicity in representation and in performing the basic algebraic operations. For the representation of numbers in F_p, we use the representation of large numbers provided by GNUMP. GNUMP represents integers and floating point numbers using a number of units called *limbs*, each consisting of 32 bits (with 2 limbs being the minimum precision required for any computation). Although GNUMP supports infinite precision computations with integers, we had to enhance it with some useful integer functions such as factorization, primitive root location, etc., which were not provided[1]. We also had to augment GNUMP's floating point capabilities with implementations of the basic complex number algebraic operations (addition, multiplication, exponentiation, and squaring) as well as with a high precision floating point implementation of functions such as $\cos(x)$, $\sin(x)$, $\exp(x)$, $\ln(x)$, $\arctan(x)$ and $\sqrt{x}$ required by other modules of the library (e.g., by the one which implements the Complex Multiplication method) and which are not offered by GNUMP. We implemented these functions using their Taylor series expansion, suitably truncated for our computation needs.

To address modularity, we have built the library around a set of modules that are organized in a bottom-up fashion. There are four major modules: the *Kernel*, the *EC operations* module, the *EC generation* module, and the *Applications* module. The software modules of our final design, along with the most important components that they include and their relationships, are shown in Figure 1.

---
[1] We had initially used the LIP library [17] for the generation and processing of large integers mainly because it provided all these functions. However, we discovered that the LIP implementations were a source of considerable time consumption. Our implementation of these functions turned out to be much more efficient.

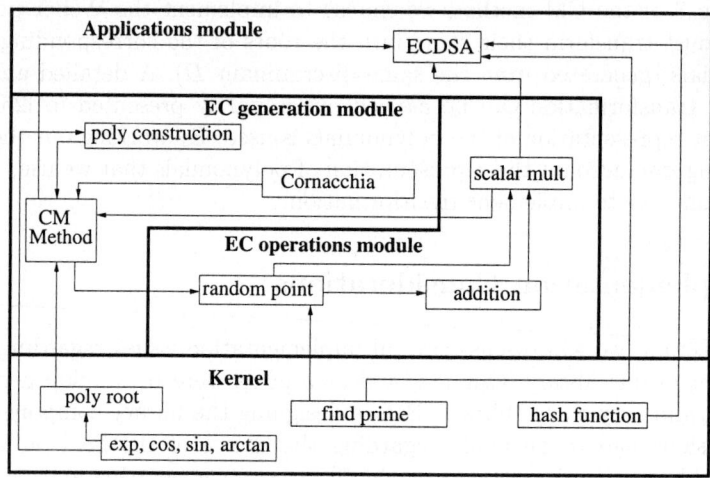

Fig. 1. The architecture of the library

The *Kernel* includes several components that perform the basic algebraic operations as well as trigonometric and exponentiation function computations using infinite precision integer and floating point arithmetic. In addition, it includes components that implement a number of more advanced (and complicated) operations, that we had to build from scratch using the primitives of the basic operations. These advanced operations include components for the manipulation of integer coefficients in polynomials (algebraic operations), and finding roots of polynomials modulo a prime number. All the components in the kernel form the core of our library and they are independent of components in other modules, at a higher level, which use them. This means that they can be optimized independently in order to tune the performance of the basic and advanced arithmetic operations.

The *EC operations* module includes several components that implement the elliptic curve related algebraic operations. We have a component in which the elliptic curve data type is defined (which is an array composed of two infinite precision numbers representing the coefficients a and b in Eq. (1)) along with the structure of a curve point (represented as a pair of two infinite precision integers). There are also components that generate random points on elliptic curves, perform addition of two points of the EC, perform scalar multiplication, and create a base point in the EC.

The *EC generation* module is one of the most important and algorithmically challenging of our library. It is composed of a number of components, including Cornacchia's algorithm [6] for solving a special class of Diophantine equations, that implement the Complex Multiplication (CM) method using both Hilbert and Weber polynomials. Two main issues had to be addressed in the implementation of the CM method: (i) the implementation of complex, high precision, floating point arithmetic as well as the use of special mathematical functions (the handling of this issue was carried out by resorting to the Taylor series expansion of these functions as was discussed in the Kernel module); (ii) the demand for increased floating point arithmetic precision as the value of the discriminant D increases. For example, the construction of the Hilbert polynomial $H_{40}(x)$ required a precision of 2 limbs (and 19 terms in the Taylor series expansion), while the construction of the polynomial $H_{292}(x)$ required a precision of 5 limbs (and 90 terms in the Taylor series expansion). To determine the required precision, we start with some initial value and then check whether the resulting polynomial is computed correctly. If not, the precision is increased and the process is repeated.

The *Applications* module contains a number of cryptographic primitives and high level protocols based on elliptic curves including the Diffie-Hellman key exchange protocol, public/private key generation, data encryption, the Elliptic Curve Digital Signature Algorithm (ECDSA) as well as a simple one-pad encryption scheme. One may easily build a richer set of cryptographic protocols using the operations offered by the other modules of the library with only knowledge the interface with these functions (calling convention and required parameters).

The modular structure reflects the distinct nature and physical separation of the software modules of our library. It is easy to replace one method by another (perhaps more efficient), by simply implementing the new method and plugging it into our library; or one may discard, for example, the complex multiplication code and produce ECs with another method, or use a concrete, precomputed set of elliptic curves such as the ones proposed by NIST (see [7]).

5 Experiments

In this section we report on the performance of the basic components of our library based on some preliminary experiments we conducted. The experiments were carried out on a Pentium III (933 MHz) with 256 MB of main memory, using the GNU multiple precision library, and the ANSI C gcc-2.95.2 compiler. In the following, let p denote a prime, $|p|$ its size, and let $p_{|p|}$ denote a prime p of size $|p|$. We conducted experiments on three different fields with representative prime sizes, namely $F_{p_{175}}$, $F_{p_{192}}$, and $F_{p_{224}}$. Note that Hasse's theorem (Eq. (2)) as well as the first suitability condition (cf. Section 3), imply that $|p|$ must be at least 160 bits long.

We first considered the times required by the scalar multiplication operation and by the EC cryptographic protocols included in the library. We note that no attempt has been made for code optimization, or hard-coding (e.g., writing parts of the library in assembly, etc) as is customary in implementations of EC cryptographic protocols (see [18, 19]). The CPU times are shown in Table 1. As it was expected the timings increase as the size of the field increases.

We next turn to the evaluation of the CM method. To evaluate its efficiency we made an experimental comparison with the point counting method (cf. Section 3). The goal was to investigate the relationship between the time required by repeated applications of the point counting method in order to achieve an EC of suitable order, and the burden of constructing the polynomials in the CM

Table 1. Timing estimates in msecs for various modules of the library

$|p|$	175 bits	192 bits	224 bits
Scalar multiplication	13.6	15.7	19.5
Key generation	19.6	23.9	30.8
ECDH protocol	27.2	31.4	39
ECES encryption	28.8	36.5	46
ECES decryption	13.5	16.3	19.1
Signature	19.1	22.7	30.6
Verification	24.5	28.3	36.8

method. In all experiments, we have chosen two representative values for $|p|$, namely 175 and 192 bits.

We started with the point counting method for which we conducted several experiments. For each experiment we first generated a random prime number p, and subsequently we generated uniformly at random two integers a and b such that $1 \leq a, b \leq p$. These two numbers represent the coefficients in the equation of the elliptic curve (Eq. (1)). We then used an implementation of Schoof's algorithm from [12] to find the order of the curve. In all experiments we conducted, we observed that this is a rather time consuming method having a big variance both w.r.t. the number of repetitions to construct a suitable curve and w.r.t. to the time required. The fastest experiment took 1 repetition and 25 minutes to construct a suitable EC, while the slowest took 21 repetitions and more than 3 hours.

We next experimented with the CM method. Recall that the only input to the CM method is the discriminant D. For our experiments, we have chosen three representative values for D, namely 40, 292, and 472, that produce Hilbert and Weber polynomials of degree 2, 4, and 6, respectively. In Table 2, we report on the outcomes of our experiments for $|p| = 192$ (similar timings were reported for $|p| = 175$). It turns out that the most time consuming step concerns the construction of Hilbert or Weber polynomials (Step 3 of the CM method). In the tables, we report on the number of repetitions to find a suitable order (i.e., number of repetitions required by Steps 1 and 2), on the time required to construct the Hilbert ($T[H]$), or the Weber ($T[W]$) polynomial, and on the time required by all steps excluding $T[H]$ or $T[W]$, denoted by $T_{tot} - T[H]$ and $T_{tot} - T[W]$, respectively. All reported times are averages over 50 experiments, while the number of repetitions are maximum over all experiments.

We would like to make two comments: (1) The construction of the polynomials may seem as a drawback in a first place. However, notice that both Hilbert and Weber polynomials depend only on D (and not on p). This implies that one can generate *off-line* the polynomials for the various values of D considered, and have them handy for the generation of an EC using the CM method. This is a major advantage of this method, and hence the important times are the one appearing in the last two columns of each table. (2) Constructing Weber poly-

Table 2. The CM method for $|p| = 192$ bits. All times are in sec, unless stated otherwise

		$|p| = 192$ bits			
	#repetitions	$T[H]$	$T[W]$	$T_{tot} - T[H]$	$T_{tot} - T[W]$
$D = 40$	7	0.75	0.09	1.66	1.57
$D = 292$	7	37.97	1.00	2.04	1.98
$D = 472$	7	1 h 23 min 1 sec	2.20	3.19	2.56

nomials and then transform their roots to those of the corresponding Hilbert polynomial is incredibly faster than constructing directly Hilbert polynomials.

We conclude by commenting on the efficiency of our implementation of the CM method. We are aware of three other implementations. In [12], a (publicly available) implementation in C++ is given that uses the MIRACL library [21]. The implementation follows the IEEE standard defined in [11]. It is different from ours as it takes as input a prime p and then decides D. Moreover, this implementation uses only Weber polynomials and proceeds with a different method to construct the EC, by avoiding the conversion of Weber roots to Hilbert roots. In [1] and [24] two other (non-publicly available) implementations are given which follow a method similar to ours. The former is implemented in C++ using the LiDIA 2.0 library [16]; the latter uses the NTL library [22]. Preliminary experiments we performed with the implementation in [12] showed that our implementation was by a factor of 1.5 slower, while in some cases (e.g., $D = 40$) it was slightly faster. We believe that this is acceptable, given the fact that no effort regarding any kind of optimization has been made.

6 Concluding Remarks

We have presented the implementation of a library that supports the construction of robust elliptic curve cryptosystems based on F_p, with p a prime larger than 3. The library includes implementations of the basic algebraic operations as well as a variety of cryptographic protocols. One of the strengths of our library is that it includes an implementation of the complex multiplication method for producing elliptic curves of suitable order that ensures robustness of cryptosystems based on ECDLP. We believe that the portability and modularity of the library can be exploited in order to develop more complex protocols as well as to address efficiency issues. The later is a target which we next plan to attack.

References

[1] H. Baier, and J. Buchmann, *Efficient construction of cryptographically strong elliptic curves*, in *Progress in Cryptology – INDOCRYPT 2000*, LNCS 1977 (Springer, 2000), pp. 191-202.

[2] E. R. Berlekamp, *Factoring polynomials over large finite fields*, Math. Comp. 24, 111 (1970), pp. 713-735.

[3] Ian Blake, Gadiel Seroussi, and Nigel Smart, *Elliptic curves in cryptography*, London Math. Society Lecture Note Series 265, Cambridge University Press, 1999.
[4] M. Brown, D. Hankerson, J. Lopez, and A. Menezes, Software Implementation of the NIST Elliptic Curves over Prime Fields, in *Topics in Cryptology* – CT-RSA 2001, LNCS 2020 (Springer, 2001), pp. 250-265.
[5] D. Burton, *Elementary Number Theory*, McGraw Hill, 4th edition, 1998.
[6] G. Cornacchia, *Su di un metodo per la risoluzione in numeri interi dell' equazione* $\sum_{h=0}^{n} C_h x^{n-h} y^h = P$, Giornale di Matematiche di Battaglini 46 (1908), 33–90.
[7] CSRC, Recommended elliptic curves for federal government use, July 1999. Available at: http://csrc.nist.gov/csrc/fedstandards.html.
[8] S. Galbraith, Limitations of constructive Weil descent, in *Public-Key Cryptography and Computational Number Theory*, (Alster and Kazimierz eds.) 2001, pp. 59-70.
[9] GNU Multiple Precision library, ed. 3.1.1, Sept 2000, http://www.swox.com/gmp.
[10] D. Hankerson, J. Lopez, and A. Menezes, Software Implementation of Elliptic Curve Cryptography over Binary Fields, in *Cryptographic Hardware and Embedded Systems* – CHES 2000, LNCS 1965 (Springer, 2000), pp. 1-24.
[11] IEEE P1363/D13, *Standard Specifications for Public-Key Cryptography*, ballot draft, 1999. http://grouper.ieee.org/groups/1363/tradPK/draft.html.
[12] Implementations of Portions of the P1363 Draft. http://grouper.ieee.org/groups/1363/P1363/implementations.html.
[13] Don Johnson, and Alfred Menezes, *The Elliptic Curve Digital Signature Algorithm (ECDSA)*, Tech. Report CORR 99-06, Dept of Combinatorics and Optimization, Univ. of Waterloo, 1999. Available at: http://www.cacr.math.uwaterloo.ca/.
[14] D.W. Kravitz, Digital Signature Algorithm, U.S. Patent #5,231,668, 27 July 1993.
[15] N. Koblitz, *Elliptic curve cryptosystems*, Math. of Comp., 48 (1987), pp.203-209.
[16] LiDIA. *A library for computational number theory*, Technical University of Darmstadt, Germany. Available from http://www.informatik.tudarmstadt.de/TI/LiDIA/Welcome.html.
[17] LIP (Large Integer Package). Available through ftp at the directory texttt/usr/spool/ftp/pub/lenstra at server flash.bellcore.com.
[18] J. López and R. Dahab, *Performance of Elliptic Curve Cryptosystems*, Tech. Report, IC-00-08, May 2000. Available at: http://www.dcc.unicamp.br/ic-main/publications-e.html.
[19] J. López and R. Dahab, *An Overview of Elliptic Curve Cryptography*, Tech. Report, IC-00-10, May 2000. Available at: http://www.dcc.unicamp.br/ic-main/publications-e.html.
[20] V. Miller, Uses of elliptic curves in cryptography, in *Advances in Cryptology* – Crypto '85, LNCS 218 (Springer, 1986), pp. 417-426
[21] Multiprecision Integer and Rational Arithmetic C/C++ Library, http://indigo.ie/ mscott/.
[22] NTL: A Library for doing Number Theory, http://shoup.net/ntl/.
[23] R. Rivest, A. Shamir, and L. Adleman, A Method for Obtaining Digital Signatures and Public-Key Cryptosystems, *Comm. of the ACM*, 21(1978), pp.120-126.
[24] Erkay Savas, Thomas A. Schmidt, and Cetin K. Koc, Generating Elliptic Curves of Prime Order, *Cryptographic Hardware and Embedded Systems* – CHES 2001, LNCS 2162 (Springer, 2001), pp. 145-161.
[25] J.H. Silverman, *The Arithmetic of Elliptic Curves*, Springer, GTM 106, 1986.
[26] Thomas Valente, *A distributed approach to proving large numbers prime*, Rensselaer Polytechnic Institute Troy, New York, Thesis, August 1992.

Real-Time Dispatching of Guided and Unguided Automobile Service Units with Soft Time Windows

Sven O. Krumke*, Jörg Rambau, and Luis M. Torres

Konrad-Zuse-Zentrum für Informationstechnik Berlin
Department Optimization, Takustr. 7, 14195 Berlin-Dahlem, Germany
{krumke,rambau,torres}@zib.de

Abstract. We investigate a real-world large scale vehicle dispatching problem with strict real-time requirements, posed by our cooperation partner, the German Automobile Association. We present computational experience on real-world data with a dynamic column generation method employing a portfolio of acceleration techniques. Our computer program ZIBDIP yields solutions on heavy-load real-world instances (215 service requests, 95 service units) in less than a minute that are no worse than 1% from optimum on state-of-the-art personal computers.

1 Introduction

The German Automobile Association *ADAC* (*Allgemeiner Deutscher Automobil-Club*), second in size only to AAA, the largest such organization in the world, maintains a heterogeneous fleet of over 1600 service vehicles in order to help people whose cars break down on their way. All service vehicles (*units*, for short) are equipped with GPS, which helps to exactly locate each unit in the fleet. In five ADAC help centers (*Pannenhilfezentralen*) spread over Germany, human operators (*dispatcher*) constantly assign units to incoming help requests (*events*, for short) so as to provide for a good quality of service (i.e., waiting times of less than 20–60 minutes depending on the system load) and low operational costs (i.e., short total tour length and little overtime costs). Moreover, about 5000 units of service contractors (*conts*, for short)—not guided by ADAC—can be employed to cover events that otherwise could not be served in time. This manual dispatching system is now subject to automation.

1.1 Informal Problem Description

Given a snapshot in the continuously running planning process, the task of the dispatcher in one of the help centers is to assign a unit or a contractor to each event and a tour to each unit such that every event is served by a unit or contractor that is capable of this service and such that a certain cost function

---
* Research supported by the German Science Foundation (DFG, grant GR 883/10)

is minimized. The result of this planning process is a (tentative) dispatch. The overall goal is to design an automatic online dispatching system that guarantees small waiting times for events and low operational costs when regarded over a larger period of time. Among the many management decisions that impose constraints on the possible dispatches, we mention ADAC's decision to impose a soft deadline on the service of an event that may be missed at the cost of a linearly increasing lateness penalty (*soft time windows*).

A basic building block for such a system is a fast *offline optimization module* that is able to produce an optimal or near-optimal dispatch under strict real-time requirements. More specifically, the optimization module must provide a reasonable answer in less than a second and must have the ability to improve on that solution whenever by some circumstances more time is granted to the optimization process. Given the fact that in an average snapshot 200 yet unserved events have to be assigned to tours for about 100 vehicles at distinct positions, the real-time aspect requires special attention.

1.2 Related Work

See [1] for evidence that the ability of fast reoptimization can help to improve the performance of dynamic dispatching systems. A possible method to organize the dynamics of a dispatching system in the language of *agents* can be found in [10]. Whether or not precomputed routes should be subject to sudden change are discussed in [4], where the optimization part is done by tabu search.

The basic problem can be modeled as a *multi depot vehicle routing problem with soft time windows* MVRPSTW, where each guided vehicle is a depot of its own and the contractors maintain a depot of a certain capacity. Many algorithms, heuristic and exact (i.e., where a performance guarantee can be given a-posteriori), have been proposed for the related *vehicle routing problem with time windows*, where the time windows have to be respected in any feasible dispatch (see [3] and references therein for a survey of various problem types and exact algorithms; see [5] for recent progress in efficiency of exact algorithms; see [9] for a tabu search approach dealing with soft time windows; see [7] for one approaches based on genetic algorithms).

To the best of our knowledge none of the exact algorithms was ever reported to *predictably* meet strict real time requirements in a large scale real-world application, i.e., produce reasonable answers very early (after five seconds) in the course of the optimization process. On the other hand, the heuristic methods cannot guarantee a certain quality of the delivered solution.

1.3 Our Contribution

We will show in this paper that we can employ a custom-made dynamic column generation method (we call it *Dynamic Pricing Control*) to this problem that delivers good solutions after a fraction of a second and yields a provably optimal or near-optimal solution in less than five seconds in all real-world data sets with about 200 events and about 100 units provided by ADAC. The behaviour remains stable even for extremal load problems (artificially augmented real-world problems) with up to 770 events and 200 units: the solution quality was already within 12% from optimum after 5 seconds, within 5% from optimum after 15 seconds, and within 2% after one minute.

We had the chance to compare an implementation of our algorithm with an experimental prototype using meta-heuristics based on genetic algorithms and hill-climbing, which was produced by our industrial partner with serious effort; it turned out that four variants of the code based on meta-heuristics are clearly outperformed by our exact method on real-world data.

1.4 Outline

The rest of the paper is organized as follows: In the next section we introduce the exact setting of the VDP. Section 3 is devoted to the mathematical model that is the basis for the column generation approach. In Section 4 we describe our real-time compliant algorithm. Computational results on real-world data in Section 5 evaluate the effectiveness of various algorithmic tuning concepts. Section 6 summarizes the key points of this paper.

As a remark, we mention that presenting the complete model for ADAC's particular vehicle dispatching problem would not allow any additional insight. We have therefore used a simplified core model for our computational experiments that is sufficient to reveal the problem structure. All missing constraints can be incorporated in the tour generation since they are independent for each vehicle.

2 Problem Specification

In the following, we specify the exact form of VDP that is tackled by our algorithm. An instance of the VDP consists of a set of units, a set of contractors, and a set of events.

Each unit u has a current position o_u, a home position d_u, a logon time t_u^{start}, a shift end time t_u^{end}, and a set of capabilities F_u. Moreover, the costs related to using this unit are specified by values for costs per time unit for each of the following actions: driving c_u^{drv}, serving c_u^{svc}, and overtime c_u^{ot}.

Each contractor v has a home position d_v and a set of capabilities F_v. Moreover, the costs for booking the contractor are specified by a value for costs per service c_v^{svc}.[1]

---

[1] For example, a cab partner (*Taxi*) charges a fixed price for every jump start. For some contractors, in reality, the service costs depend on the contractor-event pairing, but in a tame way: for example, some partners charge three different prices depending on the distance to the request. Incorporating this into the computation of the cost coefficient for a contractor tour does not affect the performance of the algorithm. Our model can even handle cost structures with rebates, because we use a tour-based formulation. We here discuss the simplified model since these features were missing in the competing code so that the benchmarks could only be run with constant constractor costs.

Each event e has a position x_e, a release time θ_e^r, a deadline θ_e^d, a service time δ_e, and a set of required capabilities F_e. Moreover, extra costs related to serving this event are specified by the value of a lateness coefficient c_e^{late} meaning that a cost of c_e^{late} times the delay w.r.t. the deadline of the event is incurred.

A feasible solution of the VDP (a *dispatch*) is an assignment of events to units and contractors capable of serving them, as well as a tour for each unit such that all events are assigned, the service of events does not start before their release times (waiting of units is allowed at no extra-costs), and all tours for all units start at their current positions not before their logon times and end at their home positions. The costs of a dispatch are the sum of all unit costs, contractor costs, and event costs.

3 Modeling

We will use a model based on tour variables. Models of this type are by now well-established in the vehicle routing literature (see, e.g., [3]).

Let $\mathcal{R}$ be the set of all feasible *tours*. This set splits into the sets $\mathcal{R}_u$ of feasible tours for each unit u. A tour in $\mathcal{R}_u$ can be described by an ordered sequence $(u, e_1, e_2, \ldots, e_k)$ of k distinct events visited by u in that order. We will use the sequence (u) to denote the *go-home* tour, i.e., the tour in which u travels from its current position directly to its home position. Feasibility means that the capabilities of the unit are sufficient for e_i, i.e., $F_{e_i} \subseteq F_u$, for all $i = 1, \ldots, k$. Notice that this sequence also fixes the arrival times of u at each event, since we may assume that starting services as early as possible does not increase the cost of a tour (recall that for units there are no waiting costs involved).

For all $R \in \mathcal{R}_u$ we introduce binary variables x_R with the following meaning: $x_R = 1$ if and only if the route R is chosen to be in the dispatch.

The cost of the route R is denoted by c_R and computed as follows. Let δ_u^{ef} be the driving time of unit u from event e to event f. Moreover, let $\delta_u^{o_u e}$ resp. $\delta_u^{ed_u}$ be the driving times of unit u from its current position to event e resp. from event e to its home position d_u. By t_R^e we denote the arrival time at event e in route R. The arrival time of u at its home position be $t_R^{d_u}$. Then the cost c_R of route $R = (u, e_1, e_2, \ldots, e_k)$ can be computed as

$$c_R = c_u^{\text{drv}} \delta_u^{o_u e_1} + \sum_{i=2}^{k} c_u^{\text{drv}} \delta_u^{e_{i-1} e_i} + c_u^{\text{drv}} \delta_u^{e_k d_u} \quad \text{(driving)}$$

$$+ \sum_{i=1}^{k} c_u^{\text{svc}} \delta_{e_i} \quad \text{(service)}$$

$$+ c_u^{\text{ot}} \max\{(t_R^{d_u} - t_u^{\text{end}}), 0\} \quad \text{(overtime)}$$

$$+ \sum_{i=1}^{k} c_{e_i}^{\text{late}} \max\{(t_R^{e_i} - \theta_{e_i}^d), 0\} \quad \text{(lateness)}$$

A feasible "route" S for a contractor v can be written as a set $\{e_1, e_2, \ldots, e_k\}$ of events that this contractor may be assigned to serve, i.e., $F_{e_i} \subseteq F_v$ for all $i = 1, \ldots, k$.

Let t_v^e be the time by which contractor v can have reached event e with one of his vehicles. The cost c_S of such a "tour" S can be computed as follows:

$$c_S = c_v^{\text{svc}} |S| \qquad \text{(service)}$$

$$+ \sum_{i=1}^{k} c_{e_i}^{\text{late}} \max\{(t_v^{e_i} - \theta_{e_i}^{\text{d}}), 0\} \qquad \text{(lateness)}$$

Since this cost is linear in the events served by this contractor, every "tour" S of a contractor can be combined from elementary contractor tours containing each only a single event.[2] Let $\mathcal{S}^v$ be the set of elementary feasible "tours" for v, and let $\mathcal{S}$ be their union over all contractors $v \in V$.

For all $S \in \mathcal{S}^v$ we introduce binary variables x_S with the following meaning: $x_S = 1$ if and only if the elementary "tour" S is chosen to be in the dispatch.

The VDP can now be formulated as a set partitioning problem as follows. Let a_{Re}, b_{Se} be binary coefficients with $a_{Re} = 1$ (resp. $b_{Se} = 1$) if and only if event e is served in tour R (resp. in elementary contractor "tour" S).

$$\min \sum_{R \in \mathcal{R}} c_R x_R + \sum_{S \in \mathcal{S}} c_S x_S \qquad \text{(IP)}$$

subject to

$$\sum_{S \in \mathcal{S}} b_{Se} x_S + \sum_{R \in \mathcal{R}} a_{Re} x_R = 1 \qquad \forall e \in E; \quad (1)$$

$$\sum_{R \in \mathcal{R}_u} x_R = 1 \qquad \forall u \in U; \quad (2)$$

$$x_R \in \{0, 1\} \qquad \forall R \in \mathcal{R}; \quad (3)$$

$$x_S \in \{0, 1\} \qquad \forall S \in \mathcal{S}. \quad (4)$$

Our method is based on solving the linear programming relaxation (LP) of (IP) by dynamic column generation.

We would like to estimate during the column generation process how far we are still away from the optimal solution of (LP). To this end, we use a bound [6] coming from the Langrangean relaxation of (LP) w.r.t. the constraints (1).

Lemma 3.1. *Let (π_e^*, π_u^*) be an optimal solution to the dual of the reduced LP, and let $(x_R^*, x_S^*)^T$ be the corresponding primal solution. Then the cost c_{LP}^{opt} of an*

---

[2] So far, there are no data about the capacities of contractors available to the ADAC. Thus, an infinite capacity is assumed. Later, in the dynamic optimization process, a contractor that declines a request will be removed form the dispatching system for some time.

optimal solution of the LP satisfies

$$c_{LP}^{opt} \geq \sum_{R \in \widetilde{\mathcal{R}}} c_R x_R^* + \sum_{S \in \mathcal{S}} c_S x_S^* + \sum_{u \in U} \min_{R \in \mathcal{R}_u} \left(c_R - \sum_{e \in E} a_{Re} \pi_e^* - \pi_u^* \right) \quad (5)$$

□

This lower bound is useful in the course of a column generation algorithm since its main terms have to be computed during the column generation process anyway.

4 The Algorithm

The input of the top level algorithm in ZIBDIP is an instance of the VDP. The initial LP consists of all elementary tours for all contractors plus a tour for each unit from its current position to its home position (*go-home-tour*). This way, both the initial LP and the initial IP are feasible.

The search for additional columns is done in a *depth-first-search branch&bound tree* (*search tree*, for short) for each unit. Each node in the search tree corresponds to a tour starting at the current position of a unit and ending at the position of the last event served by the unit. A node can be completed to a feasible tour by appending the tour from the position of the last event in the node to the unit's home position. The *pre-cost* of a node in the search tree is the *reduced cost* [2, 8] of the corresponding tour (without returning to the unit's home position and overtime). The *cost* of a node in the search tree is defined as the reduced cost of the corresponding feasible tour (including the costs for returning to the home position and overtime). The *dual prices* for the events and units are taken from the previous run of the LP solver.

The root node r of the search tree corresponds to the empty tour. Given a node v in the tree, the children of v are obtained by appending one event to v that is not yet in v.

We generate columns for each unit in loops with increasing values for the maximal search depth (inner loop) and the maximal search degree (outer loop). The values for the maximal search depth are increased until no progress has been made in the previous step provided the search depth was sufficiently large, at latest when the depth equals the number of events. The search degree is increased until an optimality criterion is met or the search degree has reached the number of events.

While we are adding columns to the LP we fix the upper bound to a negative acceptance threshold: all columns that have reduced costs smaller than the acceptance threshold are added to the LP. This acceptance threshold is updated after each iteration depending on the number of columns produced.

This search on a dynamically growing space ensures that

- the effort of finding new columns is small in the beginning, when the dual variables are not yet in good shape
- the dual variables are updated often in the beginning
- this update is fast since the number of columns in the LP is still small

- we can enforce the output of a feasible integer solution early
- the search is exact later in the run when the dual information is reliable

Whenever a new integral solution is found we output the corresponding dispatch.

5 Computational Results on Real-World Data: Effect of Dynamic Pricing Control

In order to show the effectiveness of the ZIBDIP algorithm, we compared the following setups on the real-world data sets of Table 1:

- an *unmodified* column generation procedure, where the pricing step is done on the whole search space and all columns with negative reduced costs are included in the new reduced LP (all-off)
- *forced early stop* in the pricing step: whenever one column with negative reduced costs is found then the pricing step is interrupted, and the new column is added to the reduced LP (this method was successfully applied by [5] on modified instances of some Solomon problems (earlystop),
- the default settings of ZIBDIP (default).

Table 1. Overview over the data sets used in the evaluation

Data File	System Load	# events	# units
2702_high	high	215	98
Xtreme	simulated extreme load	775	211
prob700	randomized extreme load	700	100

We have plotted over time the development of the optimal solutions of the reduced LPs (RLP) and the corresponding IPs (RIP) as well as the accumulated number of columns.

The results in Figure 1 show that a vanilla column generation algorithm is not capable to meet the real time requirements: far too many generated columns lead to an unacceptable solution time both in pricing and in LP solving. This, by the way, does not change if the column generation step is interrupted after the optimal column was found. It can be seen that *forced early stop* is for our special kind of problem not the proper approach to deal with this diffculty: too many iterations are wasted by adding columns with inferior quality, and so *forced early stop* is clearly outperformed by ZIBDIP's default settings.

Observe in Figure 1 that a successful optimality check for the default setting of ZIBDIP made the program terminate already within 15 seconds whereas the all-off and the earlystop settings need a great deal longer to nail down the optimal LP solution.

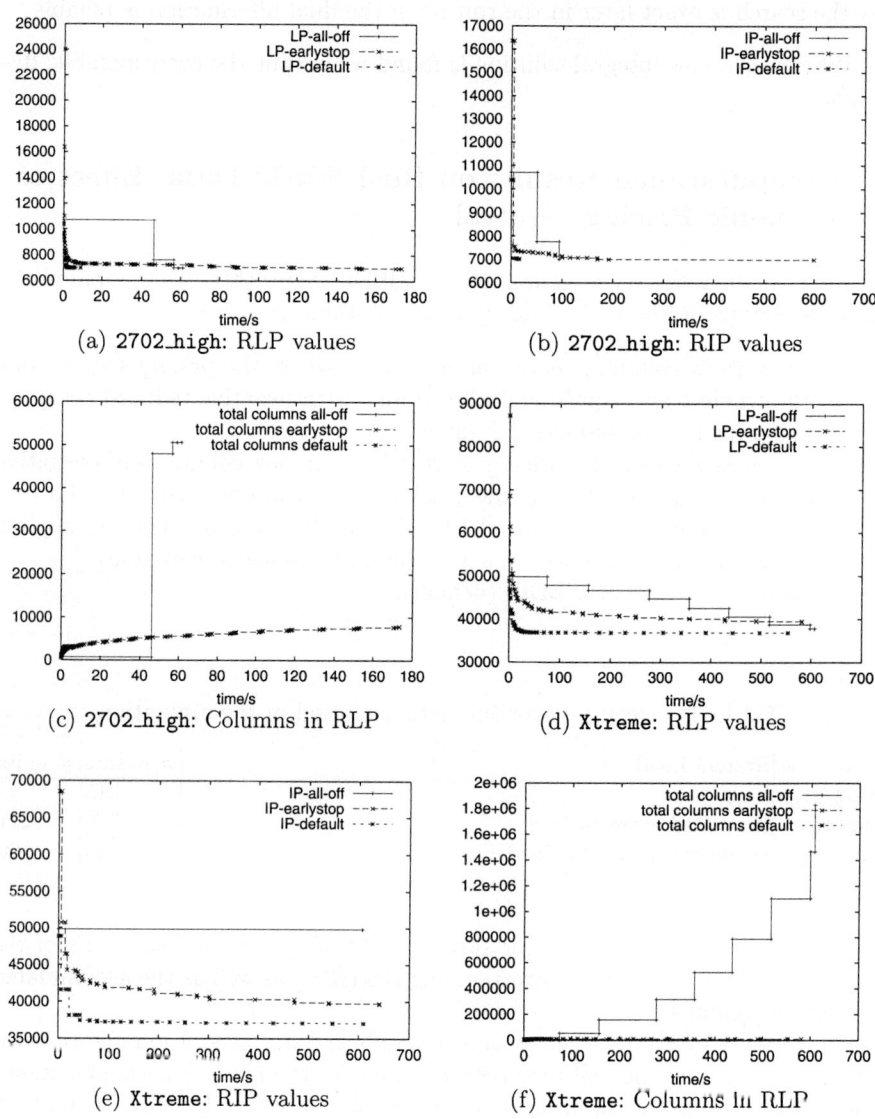

Fig. 1. Results for 2702_high and Xtreme

We have run this test on all of our data with the same results, except that for the low load instances the differences were not equally substancial.

For a more detailed look at the effects of specific algorithmic features of ZIBDIP, we compared the performances of

- *forced early stop* (earlystop),

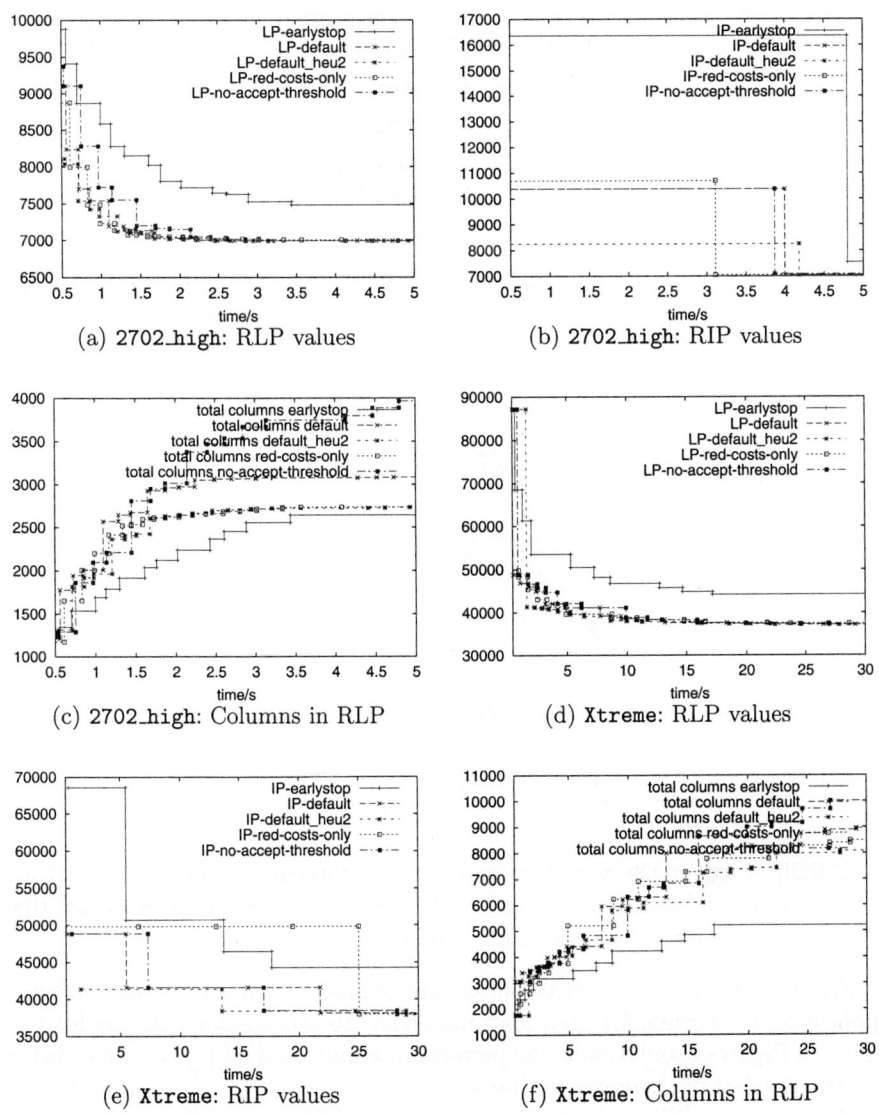

Fig. 2. Results for 2702_high and Xtreme

- ZIBDIP in default settings (default),
- ZIBDIP with a node-2-OPT start heuristics (default_heu2),
- ZIBDIP the acceptance threshold is set to zero throughout, thus all tours with negative reduced costs found in the restricted searchspace are accepted (no-accept-threshold),

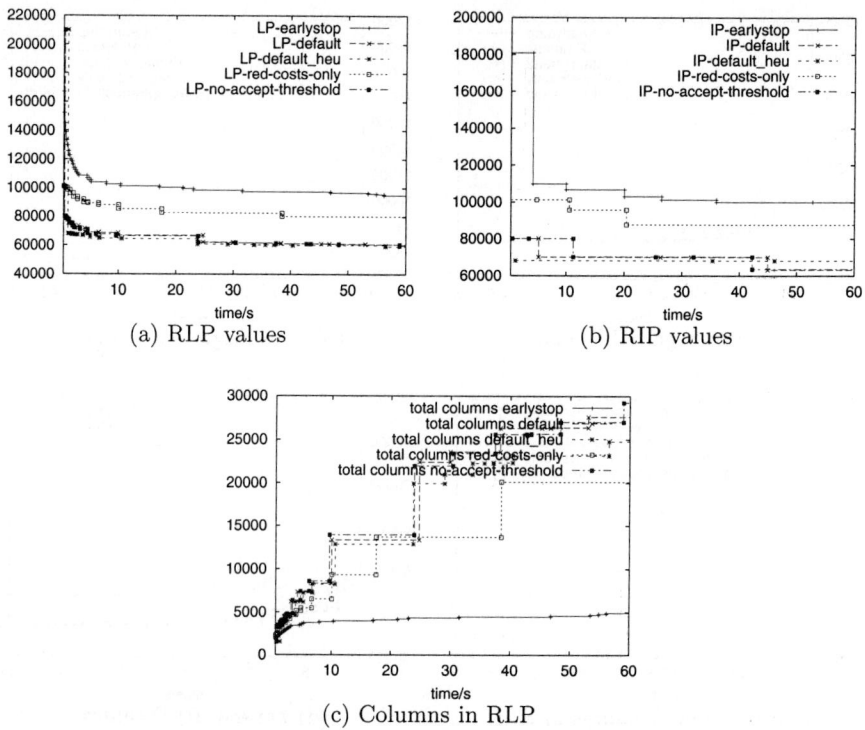

Fig. 3. Results for prob700

- ZIBDIP without the variation of sorting criteria in the pricing step: tours are always extended in the order of increasing reduced costs rather than according to alternating sorting criteria (red-costs-only).

We concentrate on the early phase that is most important for the realtime-application (5 seconds for high load instances, 15 seconds for extreme load instances). Figures 2 and 3 show the performance indicated by the objective values and the number of generated columns.

We add one additional set of input data prob700 with 700 events and 100 units. It was created randomly and has release times in the future. It serves as a stability checker for our methods since we expect that its uniform structure and relatively small late penalties allow for longer tours and make it harder for ZIBDIP to find the optimal tours.

The results in Figures 2 and 3 show that using the starting heuristics ususally improves the IP solutions of ZIBDIP during the first five seconds. The convergence of the further column generation procedure is, however, not affected by the start heuristics.

Dropping the dynamic acceptance threshold leads to more columns resulting in a slightly worse performance in the speed of LP convergence. This behaviour is, however, by far not as significant as the restriction of the search space: once the search space is restricted acceptance control of new columns is a fine tuning issue.

The influence of the variation of sorting criteria, while neglectable on the real-world data, is strong in the random data example. This is plausible because a greedy search by reduced costs is more promising in the absence of long tours. Since prob700 has tours of length 9 in its near-optimal solution while for Xtreme there are only tours of length 4, a pure greedy search is to "narrow-minded" to find good tours early.

Anyway: since varying the sorting criterion does not harm in the other test cases it seems advisable to use it in order to be weaponed against pathologic input data.

6 Conclusion

We have presented the specialized column generation algorithm ZIBDIP that solves a real-world large scale vehicle dispatching problem with soft time windows under realtime requirements. The problem arises as a subproblem in an online-dispatching task that was proposed to us by the German Automobile Association (ADAC). The algorithm clearly outperforms an experimental prototype code based on primal heuristics and genetic algorithms provided by our industrial partner. Moreover, ZIBDIP is able to provide a lower bound based on an optimal LP solution in seconds for all real-world instances provided by ADAC. It was shown that the concept of Dynamic Pricing Control can significantly speed up convergence of the column generation process, thereby making a method that has proven to be effective for large scale offline problems ready for the use in online-algorithms under realtime requirements. The practical impact of this work is that ZIBDIP is being reimplemented into the new commercial standard automatic dispatch system distributed by IPS, one of the main providers of dispatching software, superseding the former code based on primal meta-heuristics. Moreover: this product will finally be used in ADAC's help centers.

References

[1] Julien Bramel and David Simchi-Levi, *On the effectiveness of set covering formulations for the vehicle routing problem with time windows*, Operations Research **45** (1997), no. 2, 295–301.
[2] Vasek Chvatal, *Linear programming*, Freeman, New York, 1983.
[3] Jaques Desrosiers, Yvan Dumas, Marius M. Solomon, and François Soumis, *Time constraint routing and scheduling*, Network Routing (Michael Ball, Tom Magnanti, Clyde Monma, and George Newhauser, eds.), Handbooks in Operations Research and Management Science, vol. 8, Elsevier, Amsterdam, 1995, pp. 35–140.
[4] Soumia Ichoua, Michel Gendreau, and Jean-Yves Potvin, *Diversion issues in real-time vehicle dispatching*, Transportation Science **34** (2000), no. 4, 426–438.

[5] Jesper Larsen, *Vehicle routing with time windows—finding optimal solutions efficiently*, DORSnyt (engl.) (1999), no. 116.

[6] Leon S. Lasdon, *Optimization theory for large systems*, Macmillan, New York, 1970.

[7] Sushil J. Louis, Xiangying Yin, and Zhen Ya Yuan, *Multiple vehicle routing with time windows using genetic algorithms*, 1999 Congress on Evolutionary Computation (Piscataway, NJ), IEEE Service Center, 1999, pp. 1804–1808.

[8] D. G. Luenberger, *Linear and nonlinear programming*, 2 ed., Addison-Wesley, 1984.

[9] Éric Taillard, P. Badeau, Michel Gendreau, F. Guertin, and Jean-Yves Potvin, *A tabu search heuristic for the vehicle routing problem with soft time windows*, Transportation Science **31** (1997), 170–186.

[10] Kenny Qili Zhu and Kar-Loon Ong, *A reactive method for real time dynamic vehicle routing problem*, Proceedings of the 12th ICTAI, 2000.

Randomized Approximation Algorithms for Query Optimization Problems on Two Processors

Eduardo Laber[1]*, Ojas Parekh[2], and R. Ravi[3]**

[1] Puc-Rio, Rio de Janeiro, Brasil
laber@inf.puc-rio.edu
[2] Carnegie Mellon University
odp@andrew.cmu.edu
[3] Carnegie Mellon University
ravi@andrew.cmu.edu

Abstract. Query optimization problems for expensive predicates have received much attention in the database community. In these situations, the output to the database query is a set of tuples that obey certain conditions, where the conditions may be expensive to evaluate computationally. In the simplest case when the query looks for the set of tuples that simultaneously satisfy two expensive conditions on the tuples and these can be checked in two different distributed processors, the problem reduces to one of ordering the condition evaluations at each processor to minimize the time to output all the tuples that are answers to the query. We improve upon a previously known deterministic 3-approximation for this problem: In the case when the times to evaluate all conditions at both processors are identical, we give a 2-approximation; In the case of non-uniform evaluation times, we present a $\frac{8}{3}$-approximation that uses randomization. While it was known earlier that no deterministic algorithm (even with exponential running time) can achieve a performance ratio better than 2, we show a corresponding lower bound of $\frac{3}{2}$ for any randomized algorithm.

1 Introduction

The main goal of query optimization in databases is to determine how a query over a database must be processed in order to minimize the user response time. A typical query extracts the tuples in a relational database that satisfy a set of conditions (predicates in database terminology). For example, consider the set of tuples $\{(a_1, b_1), (a_1, b_2), (a_1, b_3), (a_2, b_1)\}$ and a query which seeks to extract the subset of tuples (a_i, b_j) for which a_i has property $p_1(a)$ and b_j has property $p_2(b)$.

---

* This works was supported by FAPERJ through Jovem Cientista Program (E-26/150.362/2002) and by CNPQ though Kit Enxoval 68.0116/01-0
** This work was supported by an NSF-CNPQ collaborative research grant CCR-9900304

In many scenarios, the time to evaluate the property $p(x)$ for element x in the tuple can be assumed to be constant (i.e., $O(1)$) and hence the solution involves scanning through all the tuples in turn and checking for the properties. However, in the case of evaluating expensive predicates, when a relation contains complex data as images and tables, this assumption is not necessarily true. In fact, image processing and table manipulation may be very time consuming. In this case, the time spent by those operations must be taken into account when designing a query optimization algorithm. There is some work in this direction [1, 2, 3, 5, 7, 8]. In [7], it is proposed the Cherry Picking (CP) approach which reduces the dynamic query evaluation problem to DBOP (Dynamic Bipartite Ordering Problem), a graph optimization problem, which is the focus of our improvement in this paper.

1.1 Problem Statement

An instance of DBOP consists of a bipartite graph $G = (V, E)$ with bipartition (A, B), a function $\delta : V \to \{0, 1\}$ and a function $w : V \to R^+$. Define the function $\gamma : E \to \{0, 1\}$ by $\gamma(uv) = \delta(u) \times \delta(v)$.

The idea is that $w(v)$ is the estimated time required to evaluate the function δ over v and the the goal is to compute the value of $\gamma(e)$ for every $e \in E$ spending as little as possible with evaluations of δ. The only allowed operation is to evaluate the function δ over a node.

In this formulation, note that sets A and B correspond to different attributes of the relation that is queried – the nodes correspond to distinct attribute values and the edges to tuples in the relation. Figure 1(a) shows a graph corresponding to the set of tuples $\{(a_1, b_1), (a_1, b_2), (a_1, b_3), (a_2, b_1)\}$.

As in [7], we assume a distributed scenario with two available processors: D_1 and D_2 that are used to evaluate the function δ over the nodes of A and B respectively. The goal is to minimize some sort of "makespan," that is, the maximum time by which both D_1 and D_2 finish all their evaluations of δ so as to determine all the answers to the given query. Figure 1(b) shows an instance for DBOP. The value inside each node indicates the value of the hidden function δ. Assume that $w(a) = 1$ for every $a \in A$ and $w(b) = 3$ for every $b \in B$. In this case, a greedy algorithm that evaluates the nodes with largest degree first has makespan 6 since D_1 evaluates a_1 followed by a_2 and a_4, while D_2 evaluates b_1 followed by b_4. Note that after these evaluations, given the shown δ values, we can conclude that we have determined the value $\gamma(e)$ for every edge e even without having examined many nodes (such as a_3 or b_2). Thus the crux of the problem is to choose dynamically (based on the given w's and the revealed δ values) an order of evaluation of the nodes for the two processors to minimize makespan. We will call an algorithm which chooses such an order as an *ordering algorithm*.

For an ordering algorithm Alg on a DBOP instance I, let $c(Alg, I)$ denote the time for Alg to evaluate γ on I, that is, $c(Alg, I) = \max\{c(Alg, D_1, I), c(Alg, D_2, I)\}$, where $c(Alg, D_1, I)$ and $c(Alg, D_2, I)$ indicate, respectively, the time in which D_1 and D_2 finish their evaluations. In the complexity measure c,

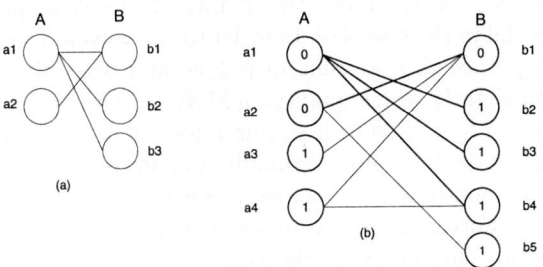

Fig. 1. An instance of DBOP

only the time incurred by the evaluation of δ over the nodes of the graph is charged. When the context is clear we may drop some of the c arguments. Solving DBOP may require some graph manipulation such as as computing the degree of some vertices, determining a vertex cover etc, so we denote by $t(Alg)$ the total running time of Alg in a worst case instance I of size n. We would like t to be a low-order polynomial on the size of the bipartite graph G.

Establishing a measure for the performance of a DBOP algorithm is a subtle issue. For example, a worst case analysis for c is not adequate since any DBOP algorithm should evaluate all nodes when $\delta(v) = 1$ for every $v \in V$. Motivated by this fact, we define the verification problem associated to an instance $I = (G, \delta)$ by a triplet (G, δ, β), where $\beta : E \to \{0, 1\}$ is a boolean function. The goal is to determine, spending as little as possible with δ's evaluation, if for every edge $e \in E$, $\beta(ab) = \delta(a)\delta(b)$, where a and b are the endpoints of e. Note that the verifier knows β and G and it must pay $w(a)$ to evaluate $\delta(a)$. Given an instance $I = (G, \delta)$, we use Opt to denote the fastest verifier for the triplet (G, δ, γ). Furthermore, we use $c(Opt, I)$ to denote the time spent by Opt to verify (G, δ, γ). We observe that $c(Opt, I) \leq c(Alg, I)$ since every algorithm that solves I can be used to verify the corresponding triplet.

Define the quality (performance ratio) of an Algorithm Alg for instance I as

$$q(Alg, I) = \frac{c(Alg, I)}{c(Opt, I)},$$

Furthermore, define the *absolute quality* (quality for short) of an Algorithm Alg as

$$q(Alg) = \max_{I \in Inst} \left\{ \frac{c(Alg, I)}{c(Opt, I)} \right\},$$

where $Inst$ is the set of all possible instances for DBOP .

1.2 Our Results

In [7], it was shown that no deterministic algorithm has quality smaller than 2. A simple algorithm called MBC (Minimum Balanced Cover) was proposed in that paper, which is optimal under the quality metric (has quality 2) but has $t(MBC)$ exponential in the size of G. In order to circumvent this problem, a variant, MBC^*, was proposed, that runs in polynomial time, that is $t(MBC^*) = O(|G|^3)$, but with slightly worse quality: $q(MBC^*) = 3$.

In this paper, we first present a simple linear time deterministic algorithm for DBOP which has the best-possible quality of 2 in the case when all weights w are uniform. In the non-uniform case, we present a randomized algorithm that achieves quality $\frac{8}{3}$. Furthermore, we prove that $\frac{3}{2}$ is a lower bound on the expected quality of any (even exponential-time) randomized strategy.

Our paper is organized as follows. In Section 2, we present lower bounds on $c(Opt, I)$. In section 3, we present a deterministic algorithm for uniform weights. In section 4, we present the rMBC algorithm, a randomized version of MBC^*. This algorithm is used as a subroutine for 2rMBC , the $\frac{8}{3}$-expected quality algorithm proposed in Section 5. In Section 6, we present a lower bound on the quality of any randomized algorithm for DBOP . Finally, in Section 7 we present our conclusions and indicate some open problems.

2   Lower Bounds

Let $G = (A \cup B, E)$ be a simple bipartite graph with node weights $w(\cdot)$. A (vertex) cover for G is the union of two subsets X and Y, $X \subset A$ and $Y \subset B$, such as that for every edge $e \in E$, at least one of e's endpoints is in $X \cup Y$. Given $V' \subset V$, define $w(V')$ to be $\sum_{v \in V'} w(v)$.

Given a graph G and non-negative numbers g_1 and g_2, the MinMax Cover $MMC(g_1, g_2)$ problem is to find a cover $X \cup Y$ for G that minimizes $\max\{g_1 + w(X), g_2 + w(Y)\}$, among all possible covers for G. This problem is NP-complete as shown in [7]. Furthermore, it admits the following natural integer programming formulation.

$$\begin{aligned} &\text{Minimize } t \\ &\text{subject to} \\ &g_1 + \sum_{i \in A} w(a_i) a_i \leq t \\ &g_2 + \sum_{j \in B} w(b_j) b_j \leq t \\ &a_i + b_j \geq 1 \quad \forall (i,j) \in E \\ &a_i \in \{0,1\} \quad \forall i \in A \\ &b_j \in \{0,1\} \quad \forall j \in B \end{aligned}$$

In the above program, a variable a_i (b_j) is assigned to 1 if the node a_i belongs to the cover and it is set to 0, otherwise.

We also consider the well known Minimum Bipartite Vertex Cover (MBVC) problem, which consists of finding a cover $X \cup Y$ for G, with $X \subset A$ and $Y \subset$

B, that minimizes $w(X) + w(Y)$. This problem can be solved in $O(|E|(|A| + |B|) \log(|E|))$ time through a max st flow algorithm [4, 6]. We have the following lemmas.

Lemma 1. *Let $I = (G, \delta)$ be an instance of DBOP and let $X^* \cup Y^*$ be a solution of MMC(0,0) for G. Then $c(Opt, I) \geq \max\{w(X^*), w(Y^*)\}$.*

Proof: For every edge $e \in E$, at least one of its endpoints must be evaluated, otherwise it is not possible to verify whether $\gamma(e) = \delta(a)\delta(b)$, being $e = (e, b)$. Let $X_{opt} \cup Y_{opt}$ be the set of nodes evaluated by Opt when it solves I. Clearly, $X_{opt} \cup Y_{opt}$ is a cover for G, otherwise there is an edge $e \in E$ such that none of its endpoints is evaluated by Opt. Since $X^* \cup Y^*$ is an optimal solution of MMC(0,0) for G, it follows that
$c(opt, I) \geq \max\{w(X_{opt}), w(Y_{opt})\} \geq \max\{w(X^*), w(Y^*)\}$. □

Lemma 2. *Let $I = (G, \delta)$ be an instance of DBOP and let $\overline{X} \cup \overline{Y}$ be a solution of MBVC for G. Then, $c(Opt, I) \geq (w(\overline{X}) + w(\overline{Y}))/2$ and $c(Opt, I) \geq w(z)$, for every $z \in \overline{X} \cup \overline{Y}$*

We defer the proof of this lemma for the extended version of this paper *Proof:* □

For the next lower bound, we need additional notation. Let V' be a subset of V. We define $T(V')$ (denoting the "True" nodes of V') by $T(V') = \{v \in V' | \delta(v) = 1\}$. Moreover, we use $N(V')$ to indicate the neighborhood of V'. Finally, we define $NT(V')$ by $N(T(V'))$.

Lemma 3. *Let $I = (G, \delta)$ an instance of DBOP. Then,*

$$c(Opt, I) \geq \max\{w(NT(B)), w(NT(A))\}$$

Proof: Let $a \in NT(B)$. It implies that there is $b \in B$, with $\delta(b) = 1$ such that $(a, b) \in E$. Hence, in order to verify whether $\gamma(a, b) = \delta(a)\delta(b)$ or not, the node a must be evaluated. Hence, every node in $NT(B)$ must be evaluated by any algorithm for I. A symmetric argument shows that every node in $NT(A)$ must be evaluated. □

We now present our last lower bound that can be viewed as a generalization of the lower bounds presented in Lemmas 1 and 3. We need to introduce the concept of a essential set.

Definition 1. *A set $V' \subset V$ is a essential set if and only if $v \in NT(V)$ for all $v \in V'$.*

We observe that if V' is a a essential set, then all of its nodes must be evaluated by any algorithm, and in particular, by Opt.

Lemma 4. *Let $I = (G, \delta)$ an instance of DBOP. Furthermore, given two essential sets A' and B', with $A' \subset A$ and $B' \subset B$, let G' be the graph induced by $V \setminus (A' \cup B')$. If $\overline{A} \cup \overline{B}$, with $\overline{A} \subset A$ and $\overline{B} \subset B$, solves MMC($w(A')$, $w(B')$) for the graph G', then $c(Opt, I) \geq \max\{w(A') + w(\overline{A}), w(B') + w(\overline{B})\}$*

Proof: Let $X_{opt} \cup Y_{opt}$, with $X_{opt} \subset A$ and $Y_{opt} \subset B$ be the set of nodes evaluated by Opt when it verifies the triplet (G, δ, γ). Since A' and B' are essential sets, it follows that $A' \subset X_{opt}$ and $B' \subset Y_{opt}$. Furthermore, $(X_{opt} \setminus A') \cup (Y_{opt} \setminus B')$ is a cover for G', otherwise one of the edges in G' cannot be verified. Since $\overline{A} \cup \overline{B}$ solves MMC(w(A'),w(B')) for the graph G',

$$\max\{w(A') + w(\overline{A}), w(B') + w(\overline{B})\} \le$$
$$\max\{w(A') + w(X_{opt} \setminus A'), w(B') + w(Y_{opt} \setminus B')\} =$$
$$\max\{w(X_{opt}), w(Y_{opt})\} \le c(Opt, I) . \quad \square$$

3 An Algorithm for Uniform Weights

In this section we present a simple deterministic algorithm which runs in linear time and delivers a quality of 2 in the case when the values of the query weights are identical for all nodes in V.

We assume w.l.o.g that $w(v) = 1$ for all $v \in V$. Let M be some maximal matching of G, and let $M_A = M \cap A$ and $M_B = M \cap B$. Finally we set $K_A = NT(M_B) \setminus M_A$ and $K_B = NT(M_A) \setminus M_B$.

In the first phase of the algorithm we evaluate the function δ in parallel for the nodes in M_A and M_B. This takes time $|M| = |M_A| = |M_B|$. In the second phase we evaluate δ for K_A and K_B in parallel, requiring time $\max\{|K_A|, |K_B|\}$. Since $M_A \cup M_B$ is a cover of G, δ is evaluated for at least one endpoint of each edge in the first phase. After the second phase, we have either evaluated both endpoints of an edge, or we have that δ is 0 on one of its endpoints, hence the algorithm is correct.

For the analysis of the quality guarantee we appeal to Lemma 4. Let G' be the graph induced by $V \setminus (K_A \cup K_B)$, and let $\overline{A} \cup \overline{B}$ be an optimizer for $MMC(|K_A|, |K_B|)$ in the graph G' induced by the bipartition $(A \setminus K_A, B \setminus K_B)$. By Lemma 4, we have that $c(Opt, I) \ge \max\{|K_A| + |\overline{A}|, |K_B| + |\overline{B}|\}$. Thus

$$q \le \frac{|M| + \max\{|K_A|, |K_B|\}}{\max\{|K_A| + |\overline{A}|, |K_B| + |\overline{B}|\}} \le 2 \frac{|M| + \max\{|K_A|, |K_B|\}}{|K_A| + |\overline{A}| + |K_B| + |\overline{B}|}$$

$$\le 2 \frac{|M| + \max\{|K_A|, |K_B|\}}{|M| + |K_A| + |K_B|} \le 2,$$

where the penultimate inequality follows from the fact that since G' contains M, and $\overline{A} \cup \overline{B}$ is a cover of G', $|\overline{A}| + |\overline{B}| \ge |M|$.

4 A Randomized Algorithm

In this section we present rMBC, a randomized algorithm for DBOP. This algorithm is used as a subroutine by the 2rMBC algorithm presented in the next section. rMBC can be viewed as a randomized version of the MBC^* algorithm proposed in [7].

Step 1 : Solve MBVC for the graph G obtaining the sets A' and B'.

Do in Parallel
   Processor D_1:
      $Ne(A') \leftarrow \emptyset$
      For $i := 1$ to $|A'|$
         Select randomly a node a from A' that has not been evaluated yet
         Evaluate $\delta(a)$
         $Ne(A') \leftarrow Ne(A') \cup NT(\{a\})$
      While there exists a node in $Ne(B')$ that has not been evaluated
      or D_2 has not processed all the nodes from B'
         Let a be a node from $Ne(B')$ that has not been evaluated yet
         Evaluate $\delta(a)$

Fig. 2. rMBC Algorithm

Before presenting rMBC , we briefly outline the MBC^* algorithm. MBC^* is divided into three steps. In the first step, the MBVC problem is solved for the input graph G. Let $A^* \cup B^*$, with $A^* \subset A$ and $B^* \subset B$, be the solution for MBVC obtained in the first step. In the second step, the function δ is evaluated for the vertices in $A^* \cup B^*$. At the end of this step, at least one endpoint of every edge has already been computed. Let $e = (a, b)$. If one of the e endpoints, say a, evaluated at the second phase is such that $\delta(a) = 0$, then the evaluation of the other endpoint b is not necessary since $\gamma(e) = 0$. Therefore, only those vertices adjacent to at least one vertex with δ value equal to 1 are evaluated in the third step. The rMBC algorithm is similar to MBC^*, except for the fact that it evaluates the cover nodes following a randomly selected order, while MBC^* follows a fixed order. Figure 2 shows the pseudo-code for rMBC . We only present the code executed in processor D_1. The code executed in D_2 is obtained by replacing every occurrence of A', a and B' by B', b and A', respectively.

From now, we assume w.l.o.g. that $w(A') \geq w(B')$. In this case, the processor D_1 does not become idle during the while loop, and as a consequence,

$$c(D_1, I) = w(A') + w(NT(B') \setminus A') \tag{1}$$

We start the analyzis of rMBC presenting a technical lemma.

Lemma 5. *Let $\pi(A')$ be a random permutation of the nodes in A' and let $\pi(i)$ be the ith node in this permutation. Moreover, let $\overline{\pi}$ be the reverse permutation of π, that is, $\overline{\pi}(i) = \pi(n - i + 1)$, for $i = 1, \ldots, n$. Being $c(D_2, \pi)$ the time in which D_2 ends its task, assuming an execution of rMBC over a permutation π, then*

$$c(D_2, \pi)/c(Opt, I) \leq 3 \tag{2}$$

and

$$c(D_2, \pi) + c(D_2, \overline{\pi})/c(Opt, I) \leq 5 \tag{3}$$

Proof: We defer the proof for an extended version of this paper. □

Theorem 1. *Let $I = (G, \delta)$ be an instance of DBOP. If $c(D_1, I)/c(Opt, I) \leq 7/3$, then $E[q(rMBC, I)] \leq 8/3$*

Proof: Let Π be the set of all possible permutations for the nodes in A'. By definition,

$$E[q(rMBC,I)] = \frac{1}{|A'|!} \sum_{\pi \in \Pi} \max\left\{\frac{c(D_1,I)}{c(Opt,I)}, \frac{c(D_2,\pi)}{c(Opt,I)}\right\} \leq$$

$$\frac{1}{|A'|!} \sum_{\pi \in \Pi} \max\left\{\frac{7}{3}, \frac{c(D_2,\pi)}{c(Opt,I)}\right\} =$$

$$\frac{1}{2|A'|!} \sum_{\pi \in \Pi} \left(\max\left\{\frac{7}{3}, \frac{c(D_2,\pi)}{c(Opt,I)}\right\} + \max\left\{\frac{7}{3}, \frac{c(D_2,\overline{\pi})}{c(Opt,I)}\right\}\right)$$

Hence, in order to establish the result, it suffices to prove that

$$\max\left\{\frac{7}{3}, \frac{c(D_2,\pi)}{c(Opt,I)}\right\} + \max\left\{\frac{7}{3}, \frac{c(D_2,\overline{\pi})}{c(Opt,I)}\right\} \leq 16/3, \quad (4)$$

for every $\pi \in \Pi$. We assume w.l.o.g. that $c(D_2, \pi) \geq c(D_2, \overline{\pi})$. If $c(D_2, \pi) \leq 8/3$, we are done. Hence, we assume that $c(D_2, \pi) > 8/3$. The correctness of (4) can be established by applying the inequality (2) in the cases where $c(D_2, \overline{\pi})/c(Opt, I) \leq 7/3$ and the inequality (3) in the case where $c(D_2, \overline{\pi})/c(Opt, I) > 7/3$. □

5 2rMBC Algorithm

At this section, we present the 2rMBC Algorithm, a polynomial time algorithm with expected quality 8/3.

First, the rMBC algorithm is executed until D_2 finishes evaluating the nodes in B' (we are assuming w.l.o.g. that $w(B') \leq w(A')$). At this point, the execution is interrupted and 2rMBC analyzes the information achieved so far. Based on this analysis, it takes one of the two following decisions: Either to resume the execution of rMBC or to execute a second, new algorithm called 2ndCover. Thus, the algorithm has two phases. The phase 1 is divided into two steps. Let $K = NT(B') \setminus A'$. In step 1, 2rMBC checks if

$$\frac{c(D_1, rMBC, I)}{c(Opt, I)} \leq \frac{w(A') + w(K)}{\max\{w(NT(B')), (w(A') + w(B'))/2\}} \leq \frac{7}{3} \quad (5)$$

If the ratio is not larger than 7/3, then it follows from Theorem 1 that the expected quality of 2rMBC (rMBC) for the current instance is not larger than 8/3. In this case, the execution of rMBC is resumed.

If the test in step 1 fails, then step 2 is executed. In this step, 2rMBC computes a new lower bound on $c(Opt, I)$. Let G' the bipartite graph induced by

$V \setminus K$. It is easy to check that K is a essential set. It follows from Lemma 4 that if $\overline{A} \cup \overline{B}$ solves $MMC(w(K), 0)$ for the graph G', then $max\{(w(K) + w(\overline{A}), w(\overline{B})\}$ is a lower bound on $c(Opt, I)$. The main problem is that solving MMC may require exponential time. Hence, instead of solving $MMC(w(K), 0)$ on G', 2rMBC solves a linear programming (LP) relaxation of the integer programming formulation for $MMC(w(K), 0)$ presented in Section 2. The LP relaxation is obtained replacing the constraints $a_i \in \{0, 1\}$ and $b_j \in \{0, 1\}$ by $0 \leq a_i \leq 1$ and $0 \leq b_j \leq 1$ respectively. Let (a^*, b^*, t^*) the optimum solution for the LP relaxation. The end of step 2 consists in checking whether

$$\frac{c(D_1, rMBC, I)}{c(Opt, I)} \leq \frac{w(A') + w(K)}{t^*} \leq \frac{7}{3} \qquad (6)$$

In the positive case, the execution of rMBC is resumed. Otherwise, the following 2ndCover algorithm is executed.

5.1 The 2ndCover Algorithm

First, the solution of the LP relaxation solved in Step 2 is rounded to obtain a feasible cover $A2 \cup B2$ for G', where $A2 \subset A \setminus K$ and $B2 \subset B$. Let $x = \sum_{i \in A \setminus K} a_i^* w(a_i)$ and $y = \sum_{j \in B} b_j^* w(b_j)$.

The cover is obtained through the following procedure:

0) $A2 \leftarrow \emptyset$; $B2 \leftarrow \emptyset$
1) **For** every $i \in A \setminus K$ such that $a_i^* \geq x/(y+x)$ do $A2 \leftarrow A2 \cup \{a_i\}$
2) **For** every $j \in B$ such that $b_j^* \geq y/(y+x)$ do $B2 \leftarrow B2 \cup \{b_j\}$

Now the processors D_1 and D_2 are used to evaluate the nodes in $A2$ and $B2$ respectively. Finally D_1 and D_2 evaluate the nodes from $NT(B' \cup B2)$ and $NT(A2)$, respectively, that have not yet been processed.

5.2 Algorithm Analysis

In this section, we analyze the performance of 2rMBC.

Theorem 2. *2rMBC is correct.*

Proof: If either inequality (5) or inequality (6) holds, then 2rMBC 's correctness follows from that of rMBC, which in turn follows from the correctness of MBC^* outlined in the introduction of Section 4. On the other hand, if neither inequality (5) nor inequality (6) hold then 2ndCover is executed. Since $A' \cup B'$ is a cover of G, B' must contain the neighbors of K, hence the algorithm processes both endpoints of any edge with an endpoint in K. If $A2 \cup B2$ is a cover of $G' = G[V \setminus K]$ then again an argument analogous to the one used to establish the correctness of MBC^* shows that upon termination of the algorithm, γ may be evaluated for the edges of G'. Thus the only thing which remains to be shown is that $A2 \cup B2$ indeed covers G'. For an edge ij of G', we have that $a_i^* + b_j^* \geq 1 = x/(y+x) + y/(y+x)$, hence at least one of i and j must belong to $A2 \cup B2$. □

Lemma 6. Let $A2 \cup B2$ be the cover for G' obtained by 2ndCover algorithm. Then, $\max\{w(A2), w(B2)\} \leq x + y$.

Proof: It follows from the rounding scheme that

$$w(A2) = \sum_{i \in A2} w_i \leq \sum_{i \in A \setminus K} a_i^* w_i (y+x)/x \leq \frac{y+x}{x} \sum_{i \in A \setminus K} a_i^* w_i = y + x.$$

The analysis for $B2$ is symmetric. □

Theorem 3.

$$E[q(2rMBC))] \leq 8/3$$

Proof: First, we consider the case when the execution of rMBC is resumed after the interruption. In this case, either the test at Step 1 or the test at Step 2 is positive, which implies that

$$\frac{w(A') + w(K)}{c(Opt, I)} \leq \frac{7}{3}.$$

Therefore, it follows from theorem 1 that $E[q(rMBC, I)] \leq 8/3$.

Now, we assume that 2ndCover is executed. In this case, the tests performed at Step 1 and 2 fail, that is, inequalities (5) and (6) do not hold. First, using the observation that inequality (5) fails, we have that

$$w(K) \geq \frac{w(A')}{6} + \frac{7w(B')}{6}. \tag{7}$$

Using the fact that $w(K) + x \leq t^*$, and the fact that (6) does not hold, we have that

$$7x \leq 3w(A') - 4w(K) \tag{8}$$

Replacing the upper bound on $w(K)$ given by inequality (7) in the inequality above, we conclude that

$$x \leq w(A')/3 - 2w(B')/3 \tag{9}$$

We have the following upper bound on $c(2rMBC, I)$.

$$c(2rMBC, I) \leq w(B') + \max\{w(A2), w(B2)\} + \max\{w(NT(A)), w(NT(B))\}.$$

Since $w(A2) \leq x + y$, $w(B2) \leq x + y$ and $y \leq t^*$, we have that

$$c(2rMBC, I) \leq w(B') + x + t^* + \max\{w(NT(A)), w(NT(B))\}, \tag{10}$$

The result follows from inequality (9) and from the fact that $w(NT(A))$, $w(NT(B))$, $(w(A) + w(B))/2$ and t^* are lower bounds on $c(Opt, I)$. □

6   A 3/2 Lower Bound for Randomized Algorithms

In order to give a lower on the expected quality of any randomized algorithm for DBOP , we apply Yao's minmax principle [9].

We consider a distribution D for the instances of DBOP , where the only instances with nonzero probability are all equally likely and have the following properties.

(i) The input graph $G = (V, E)$ has bipartition (A, B), where $A = \{a_1, \ldots, a_{2n}\}$ and $B = \{b_1, \ldots, b_{2n}\}$. Moreover, $E = \{(a_i, b_i) | 1 \leq i \leq 2n\}$ and $w(v) = 1$ for every $v \in V$.

(ii) $|T(A)| = |T(B)| = n$ and $\gamma(e) = 0$ for every $e \in E$.

Observe that there are exactly $\binom{2n}{n}$ instances with these properties. In distribution D, all of them have probability $1/\binom{2n}{n}$. By Yao's principle, the average quality of an optimal deterministic algorithm for D is a lower bound on the expected quality of any randomized algorithm for DBOP .

Our first observation is that $c(Opt, I) = n$ for every instance I with nonzero probability, once Opt evaluates in parallel the nodes in $A \backslash T(A)$ and $B \backslash T(B)$. Let Opt_D be the optimal deterministic algorithm for distribution D. The expected quality of Opt_D is given by the expected cost of Opt_D divided by n.

In order to evaluate the expected cost of Opt_D, we first calculate the expected cost of the optimal sequential algorithm for distribution D. Next, we argue that the optimal parallel algorithm for D cannot overcome the sequential one by a factor larger than 2. The sequential algorithm uses only one processor to evaluate δ over the nodes of the graph. Hence, its cost is given by the number of evaluated nodes. Let $k_1 \geq k_2$. We use $P(k_1, k_2)$ to denote the expected cost of an optimal sequential algorithm for a distribution D_{k_1, k_2}, where the only instances with nonzero probability are all equally likely and have the following properties:

(i) The input graph $G = (V, E)$ has bipartition (A, B), where $A = \{a_1, \ldots, a_{k_1+k_2}\}$ and $B = \{b_1, \ldots, b_{k_1+k_2}\}$. Moreover, $E = \{(a_i, b_i) | 1 \leq i \leq k_1 + k_2\}$ and $w(v) = 1$, for every $v \in V$.

(ii) $\max\{A \setminus |T(A)|, B \setminus |T(B)|\} = k_1$, $\min\{A \setminus |T(A)|, B \setminus |T(B)|\} = k_2$ and $\gamma(e) = 0$ for every $e \in E$.

Clearly, the optimal algorithm for distribution D_{k_1,k_2} firstly evaluates a node from the side with k_1 false nodes. If it succeeds to find a false node, say v, then it does not need to evaluate the node adjacent to v; otherwise, it has. We can formulate the following recursive equation when $k_1 \geq k_2 > 0$.

$$P(k_1, k_2) = \frac{k_1}{k_1 + k_2}[1 + P(\max\{k_1 - 1, k_2\},$$

$$\min\{k_1 - 1, k_2\})] + \frac{k_2}{k_1 + k_2}[2 + P(k_1, k_2 - 1)],$$

In addition, $P(k_1, 0) = k_1$, since the optimal algorithm just need to evaluate the nodes from one side of the graph. It is possible to verify that

$\lim_{n\to\infty} P(n,n)/n = 3$. Since, the optimal parallel algorithm Opt_D cannot overcome the sequential one by a factor larger than 2, we have that $\lim_{n\to\infty} E[c(Opt_D)]/n \geq 3/2$. Therefore, we can state the following theorem.

Theorem 4. *Let Alg be a randomized algorithm for DBOP. Then, $E[q(Alg)] \geq 1.5$.*

7 Conclusion

In this paper we have addressed the problem of evaluating queries in the presence of expensive predicates. We presented a randomized polynomial time algorithm with expected quality 8/3 for the case of general weights. For the case of uniform weights, we show that a very simple deterministic linear time algorithm has quality 2, which is the best possible quality achievable by a deterministic algorithm [7]. On the other hand, we gave a lower bound of 1.5 on the expected quality of any randomized algorithm. Two major questions remain open:

1. Is there a polynomial time deterministic algorithm with quality 2 for non uniform weights ?

2. Where exactly in between 1.5 and 2.667 is the exact value of the quality of the best possible polynomial randomized algorithm for DBOP ?

References

[1] L. BOUGANIM, F. FABRET, F. PORTO, AND P. VALDURIEZ, *Processing queries with expensive functions and large objects in distributed mediator systems*, in Proc. 17th Intl. Conf. on Data Engineering, April 2-6, 2001, Heidelberg, Germany, 2001, pp. 91–98.

[2] S. CHAUDHURI AND K. SHIM, *Query optimization in the presence of foreign functions*, in Proc. 19th Intl. Conf. on Very Large Data Bases, August 24-27, 1993, Dublin, Ireland, 1993, pp. 529–542.

[3] D. CHIMENTI, R. GAMBOA, AND R. KRISHNAMURTHY, *Towards on open architecture for LDL*, in Proc. 15th Intl. Conf. on Very Large Data Bases, August 22-25, 1989, Amsterdam, The Netherlands, 1989, pp. 195–203.

[4] A. V. GOLDBERG AND R. E. TARJAN, *A new approach to the maximum flow problem*, in Proceedings of the Eighteenth Annual ACM Symposium on Theory of Computing, Berkeley, California, 28–30 May 1986, pp. 136–146.

[5] J. M. HELLERSTEIN AND M. STONEBRAKER, *Predicate migration: Optimizing queries with expensive predicates*, in Proc. 1993 ACM SIGMOD Intl. Conf. on Management of Data, May 26-28, 1993, Washington, D. C., USA, 1993, pp. 267–276.

[6] D. S. HOCHBAUM, *Approximating clique and biclique problems*, Journal of Algorithms, 29 (1998), pp. 174–200.

[7] E. S. LABER, F. PORTO, P. VALDURIEZ, AND R. GUARINO, *Cherry picking: A data-dependency driven strategy for the distributed evaluation of expensive predicates*, Tech. Rep. 01, Departamento de Informática, PUC-RJ, Rio de Janeiro, Brasil, April 2002.

[8] T. MAYR AND P. SESHADRI, *Client-site query extensions*, in Proc. 1999 ACM SIGMOD Intl. Conf. on Management of Data, June 1-3,1999, Philadelphia, Pennsylvania, USA, 1999, pp. 347–358.

[9] A. C. YAO, *Probabilistic computations : Toward a unified measure of complexity*, in 18th Annual Symposium on Foundations of Computer Science, Long Beach, Ca., USA, Oct. 1977, IEEE Computer Society Press, pp. 222–227.

Covering Things with Things*

Stefan Langerman[1] and Pat Morin[2]

[1] School of Computer Science, McGill University
3480 University Street, Montreal, H3A 2A7, Canada
sl@cgm.cs.mcgill.ca
[2] School of Computer Science, Carleton University
1125 Colonel By Drive, Ottawa, K1S 5B6, Canada
morin@cs.carleton.ca

Abstract. An abstract NP-hard covering problem is presented and fixed-parameter tractable algorithms for this problem are described. The running times of the algorithms are expressed in terms of three parameters: n, the number of elements to be covered, k, the number of sets allowed in the covering, and d, the combinatorial dimension of the problem. The first algorithm is deterministic and has running time $O'(k^{dk}n)$. The second algorithm is also deterministic and has running time $O'(k^{d(k+1)} + n^{d+1})$. The third is a Monte-Carlo algorithm that runs in time $O'(k^{d(k+1)} + c2^d k^{\lceil (d+1)/2 \rceil \lfloor (d+1)/2 \rfloor} n \log n)$ time and is correct with probability $1 - n^{-c}$. Here, the O' notation hides factors that are polynomial in d. These algorithms lead to fixed-parameter tractable algorithms for many geometric and non-geometric covering problems.

1 Introduction

This paper pertains to the parametrized complexity of some geometric covering problems, many of which are NP-hard. A prototype for these problems is the following HYPERPLANE-COVER problem: Given a set S of n points in $\mathbb{R}^d$, does there exist a set of k hyperplanes such that each point of S is contained in at least one of the hyperplanes? It it known that HYPERPLANE-COVER is NP-hard, even in $\mathbb{R}^2$ (i.e., covering points with lines) [12].

Parametrized complexity [5] is one of the recent approaches for finding exact solutions to some restricted versions of NP-hard problems. With this approach, one tries to identify one or more natural numerical parameters of the problem which, if small, make the problem tractable.

With many problems, such as deciding if a graph with n vertices contains a clique of size k, it is easy to devise algorithms with running time of the form $O(n^k)$, so the problem is polynomial for any constant k. While this is somewhat helpful, it is not completely satisfying since the degree of the polynomial grows quickly as k grows. Fixed-parameter tractable algorithms have more strict requirements. Namely, the running time must be of the form $O(f(k)n^c)$, where f

---
* This research was partly funded by NSERC, MITACS, FCAR and CRM.

is an arbitrarily fast growing function of k, and c is a constant not depending on k.

A typical example is the VERTEX-COVER problem: Given a graph $G = (V, E)$ with n vertices, does G have a subset $V' \subseteq V$ of k vertices such that every edge is incident to at least one vertex of V'? After much work on this problem, its running time has been reduced to $O(1.27^k + kn)$ which, with current computing technology allows for the efficient solution of problem with values of k as large as 200 and arbitrarily large n [4].

This article is an attempt to apply the general techniques of parametrized complexity to geometric NP-hard problems. Indeed many geometric problems have natural parameters that can be used to limit their complexity. The most common of such a parameter is the geometric dimension of the problem.

This has been observed, for example, in the study of linear programming in R^d for low dimension d. For this problem, there exists an algorithm with running time (e.g.) $O(2^{2^d} n)$, so the problem is fixed-parameter tractable. Another remarkable example of this phenomenon is k-*piercing of rectangles*: Given a set of n rectangles in the plane, is there a set of k points such that each rectangle contains at least one of the points? This problem has $O(n \log n)$ algorithms for $k = 1, \ldots, 5$ [13, 15], which makes it believable that it admit an algorithm with running time of the form $O(f(k) n \log n)$ for arbitrary values of k.

In this paper we define an abstract set covering problem called DIM-SET-COVER that includes HYPERPLANE-COVER and many similar problems. As part of this definition, we introduce a new notion of geometric set system hierarchy, in many points similar to the range spaces of Vapnik and Chervonenkis [16], along with a concept of dimension for each set system, which is always at least as large as its VC-dimension. We then give *fixed-parameter tractable* algorithms for DIM-SET-COVER. These are algorithms that have running times of the form $O(f(k, d) n^c)$ where c is a constant, d is the combinatorial dimension of the problem and $f(k, d)$ is an arbitrarily fast growing function of k and d.

We give three algorithms for DIM-SET-COVER. The first is deterministic and runs in $O'(k^{dk} n)$ time. The second is deterministic and runs in $O'(k^{d(k+1)} + n^{d+1})$ time. The third is a Monte-Carlo algorithm that runs in $O'(k^{d(k+1)} + c2^d k^{\lceil (d+1)/2 \rceil \lfloor (d+1)/2 \rfloor} n \log n)$ time and returns a correct answer with probability at least $1 - n^{-c}$. Here and throughout, the O' notation is identical to the standard O notation except that it hides factors that are polynomial in d.

The remainder of the paper is organized as follows. In Section 2 we introduce the DIM-SET-COVER problem. In Section 3 we give a deterministic algorithm for DIM-SET-COVER. In Section 4 we present two more algorithms for DIM-SET-COVER. In Section 5 we describe several covering problems that can be expressed in terms of DIM-SET-COVER. Finally, in

2 Abstract Covering Problems

In this section we present an abstract covering problem that models many geometric and non-geometric covering problems. Throughout this section we illustrate all concepts using the HYPERPLANE-COVER example.

Let U be a (possiblity infinite) universe and let $S \subseteq U$ be a *ground set* of size $n < \infty$. The *covering set* R contains subsets of U and can be covered with d sets $R_0, \ldots, R_{d-1}$ with the properties defined below. For convenience, we also define $R_{-1} = \emptyset$ and $R_d = \{U\}$. Given a set $A \subseteq U$ we define the *dimension* of A as

$$\dim(A) = \min\{i : \exists r \in R_i \text{ such that } A \subseteq r\}$$

and we define the *cover* of A as

$$\text{cover}(A) = \{r \in R_{\dim(A)} \text{ such that } A \subseteq r\} \ .$$

Since $R_d = \{U\}$, it is clear that $\dim(A)$ and $\text{cover}(A)$ are well defined and $\text{cover}(A)$ is non-empty for any $A \subseteq U$. We require that the sets $R_0, \ldots, R_{d-1}$ also satisfy the following property.

Property 1. For any $r_1 \in R_i$ and $r_2 \in R_j$ such that $r_1 \not\subseteq r_2$ and $r_2 \not\subseteq r_1$, $\dim(r_1 \cap r_2) < \min\{i,j\}$.

> In our HYPERPLANE-COVERexample, U is $\mathbb{R}^d$ and S is a set of n points in $\mathbb{R}^d$. We say that an *i-flat* is the affine hull of $i+1$ affinely independent points [14]. Then, the set R_i is the set of all i-flats in $\mathbb{R}^d$. The dimension $\dim(A)$ of a set A is the dimension of the affine hull of A. The set $\text{cover}(A)$ is a singleton containing the unique $\dim(A)$-flat that contains A. Property 1 says that the intersection of an i-flat and a j-flat, neither of which contains the other, is an l-flat for some $l < \min\{i,j\}$. (We consider the empty set to be a (-1)-flat.)

Property 1 has some important implications. The following lemma says that $\text{cover}(A)$ really only contains one interesting set.

Lemma 1 *For any $A \subset U$ and any finite $S \subseteq U$ there is a $\text{set}(A) \in \text{cover}(A)$ such that $r \cap S \subseteq \text{set}(A) \cap S$ for all $r \in \text{cover}(A)$.*

Proof. Suppose, by way of contradiction, that there is no such set in $\text{cover}(A)$. Then, there must exist two sets $r_1, r_2 \in \text{cover}(A)$ such that $r_1 \cap S \not\subseteq r_2 \cap S$ and $r_2 \cap S \not\subseteq r_1 \cap S$. Then, it must be that $r_1 \not\subseteq r_2$ and $r_2 \not\subseteq r_1$ so, by Property 1, $\dim(r_1 \cap r_2) < \dim(A)$. But this is a contradiction since $A \subseteq r_1 \cap r_2$.

> In our HYPERPLANE-COVER example, we could also have defined R_i as the set of all point sets in $\mathbb{R}^d$ whose affine hull has dimension i. Then Lemma 1 says that for any $A \subseteq \mathbb{R}^d$ whose affine hull has dimension i, there exists one i-flat F containing every point set whose affine hull has dimension i and that contains A.

The following lemma says that we can increase the dimension of A by adding a single element to A.

Lemma 2 *For any $A \subseteq U$ and $p \in U \setminus \mathrm{set}(A)$, $\dim(A \cup \{p\}) > \dim(A)$.*

Proof. By definition, $\dim(A \cup \{p\}) \geq \dim(A)$. Suppose therefore, by way of contradiction, that $\dim(A \cup \{p\}) = \dim(A)$. Then there is a set $r' \in R_{\dim(A)}$ that contains $A \cup \{p\}$. But then $r' \in \mathrm{cover}(A)$ and by Lemma 1, $r' \subseteq \mathrm{set}(A)$, contradicting $p \in U \setminus \mathrm{set}(A)$.

> In our HYPERPLANE-COVER example, Lemma 2 implies that if we have a set A whose affine hull is an i-flat and take a point p not contained in the affine hull of A then the $A \cup \{p\}$ is not contained in any i-flat.

The next lemma says that for any $A \subseteq S$, $\mathrm{set}(A)$ has a *basis* consisting of at most $\dim(A) + 1$ elements.

Lemma 3 *For any $r \in R$ such that $r = \mathrm{set}(A)$ for some $A \subseteq U$, there exists a basis $A' \subseteq A$ such that $r = \mathrm{set}(A')$ and $|A'| \leq \dim(A) + 1$.*

Proof. Any such set r can be generated as follows: Initially, set $A' \leftarrow \emptyset$, so that $\dim(A') = -1$. While, $\mathrm{set}(A') \neq \mathrm{set}(A)$, repeatedly add an element of $A \setminus \mathrm{set}(A')$ to A'. By Lemma 2, each such addition increases the dimension of A' by at least 1, up to a maximum of $\dim(A)$. Therefore, A' contains at most $\dim(A) + 1$ elements and $\mathrm{set}(A') = r$, as required.

> In our HYPERPLANE-COVER example, Lemma 3 is equivalent to saying that any i-flat is the affine hull of some set of $i+1$ points.

A k-cover of S is a set $r_1, \ldots, r_k \in R$ such that $S \subseteq \bigcup_{i=1}^{k} r_i$. The DIM-SET-COVER problem is that of determining whether S has a k-cover. Throughout this paper we assume, usually implicitly, that $k \geq 2$, otherwise the problem is not very interesting.

The sets U, R, and $R_0, \ldots, R_{d-1}$ may be infinite, so they are usually represented implicitly. We assume that any algorithm for DIM-SET-COVER accesses these sets using two operations. (1) Given a set $r \in R_i$ and an element $p \in S$, the algorithm can query if p is contained in r. Such a query takes $O'(1)$ time. (2) Given a set $A \subseteq U$, the algorithm can determine $\mathrm{set}(A)$ and $\dim(A)$ in $O'(|A|)$ time. It follows that the only sets accessible to an algorithm are those sets $r \in R$ such that $r = \mathrm{set}(S')$ for some $S' \subseteq S$. We call such sets *accessible sets*. Throughout the remainder of this paper we consider only k-covers that consist of accessible sets.

> In our HYPERPLANE-COVER example, we can determine in $O'(1)$ time if a point is contained in an i-flat. Given $A \subseteq \mathbb{R}^d$, we can compute the affine hull of A in $O'(|A|)$ time.

3 A Deterministic Algorithm

Next we give a deterministic fixed-parameter tractable algorithm for DIM-SET-COVER. The algorithm is based on the *bounded search tree* method [5]. The algorithm works by trying to partition S into sets $S_1, \ldots, S_k$ such that $\dim(S_i) < d$ for all $1 \leq i \leq k$.

Initially, the algorithm sets $S' \leftarrow S$ and $S_1 \leftarrow S_2 \leftarrow \cdots \leftarrow S_k \leftarrow \emptyset$. The algorithm always maintains the invariants that (1) $S', S_1, \ldots, S_k$ form a partition of S and (2) $\dim(S_i) < d$ for all $1 \leq i \leq k$. At the beginning of every recursive invocation the algorithm checks if S' is empty and, if so, outputs *yes*, since $\text{set}(S_1), \ldots, \text{set}(S_k)$ is a k-cover of S. Otherwise, the algorithm chooses an element $p \in S'$. For each $1 \leq i \leq k$, if $\dim(S_i \cup \{p\}) < d$ the algorithm calls itself recursively on $S' \setminus \text{set}(S_i \cup \{p\})$, and $S_1, \ldots, S \cap \text{set}(S_i \cup \{p\}), \ldots, S_k$. If none of the recursive calls gives a positive result the algorithm returns *no*. This is described by the following pseudocode.

BST-DIM-SET-COVER$(S', S_1, \ldots, S_k)$

1: **if** $S' = \emptyset$ **then**
2:    **output** *yes* and quit
3: **else**
4:    choose $p \in S$
5:    **for** $i = 1, \ldots k$ **do**
6:      **if** $\dim(S_i \cup \{p\}) < d$ **then**
7:         BST-DIM-SET-COVER$(S' \setminus \text{set}(S_i \cup \{p\}), S_1, \ldots, S \cap \text{set}(S_i \cup \{p\}), \ldots, S_k)$
8:      **end if**
9:    **end for**
10: **end if**
11: **output** *no*

Theorem 1 *Algorithm* BST-DIM-SET-COVER *correctly solves the* DIM-SET-COVER *problem in* $O'(k^{dk}n)$ *time.*

Proof. We begin by proving the correctness of the algorithm. To do this we consider a *restriction* of the problem. A RESTRICTED-SET-COVER instance is a $(k+1)$-tuple $(S, S_1, \ldots, S_k)$ where $S_i \subset S$ and $\dim(S_i) < d$ for each $1 < i < k$. A solution to the instance consists of a k-cover $r_1, \ldots, r_k$ of S such that $S_i \subseteq r_i$ for all $1 \leq i \leq k$. The *degree of freedom*, of a RESTRICTED-SET-COVER instance is defined as $DF(S_1, \ldots, S_k) = (d-1)k - \sum_{i=1}^{k} \dim(S_i)$. Note that RESTRICTED-SET-COVER is equivalent to DIM-SET-COVER when $S_1 = \cdots = S_k = \emptyset$, in which case $DF(S_1, \ldots, S_k) = dk$.

We claim that a call to BST-DIM-SET-COVER$(S, S_1, \ldots, S_k)$ correctly solves the RESTRICTED-SET-COVER instance $(S, S_1, \ldots, S_k)$. To prove this, we use induction on $DF(S_1, \ldots, S_k)$. If $DF(S_1, \ldots, S_k) = 0$ then $\dim(S_i) = d - 1$ for all $1 \leq i \leq n$, and, by Lemma 2 the RESTRICTED-SET-COVER instance has a solution if and only if S' is empty. Since line 1 of the algorithm checks this condition, the algorithm is correct in this case.

Next suppose that $DF(S_1,\ldots,S_k) = m > 0$. If S' is empty then $S_1,\ldots,S_k$ form a partition of S and line 1 ensures that the algorithm correctly answers *yes*. Otherwise, by construction, any $p \in S'$ is not contained in any set(S_i). If the answer to the RESTRICTED-SET-COVER instance is yes then, in a restricted k-cover $r_1,\ldots,r_k$, p is contained in some set r_i. Therefore, the RESTRICTED-SET-COVER instance $(S, S_1,\ldots, S_i \cup \{p\},\ldots, S_k)$ also has a solution. On the other hand, if the correct answer is no, then none of the restrictions $(S, S_1,\ldots, S_i \cup \{p\},\ldots, S_k)$ for any $1 \leq i \leq k$ have a solution. By Lemma 2, $DF(S_1,\ldots, S_i \cup \{p\},\ldots, S_k) < m$ for all $1 \leq i \leq k$ so, by induction, a call to BST-DIM-SET-COVER$(S_1,\ldots, S_i \cup \{p\},\ldots, S_k)$ correctly solves the RESTRICTED-SET-COVER instance $(S_1,\ldots, S_i \cup \{p\},\ldots, S_k)$. Since the algorithm checks all these k RESTRICTED-SET-COVER instances, it must answer correctly.

Finally, we prove the running time of the algorithm. Each invocation of the procedure results in at most k recursive invocations and, by Lemma 2, each recursive call decreases the degree of freedom by at least one. Therefore, the recursion tree has at most k^{dk} leaves and $\sum_{i=0}^{dk-1} k^i = O(k^{dk-1})$ internal nodes. The work done at each leaf is $O'(k)$ and the work done at each internal node is $O'(kn)$. Therefore the overall running time is $O'(k^{dk}n)$, as required.

4 Kernelization

In this section we give two more algorithms for DIM-SET-COVER that have a reduced dependence on k. The first algorithm is deterministic and runs in $O'(k^{d(k+1)} + n^{d+1})$ time. The second is a Monte-Carlo algorithm that runs in $O'(k^{d(k+1)} + c2^d k^{\lceil (d+1)/2 \rceil \lfloor (d+1)/2 \rfloor} n \log n)$ time and outputs a correct answer with high probability. Both algorithms work by reducing the given DIM-SET-COVER instance to an equivalent *kernel* instance that has size $O(k^d)$ and then solving the new instance using the BST-DIM-SET-COVER algorithm. We begin with a structural lemma.

Lemma 4 *Suppose $|r \cap S| \leq m$ for all $r \in \bigcup_{j=0}^{i-1} R_j$ and there exists an accessible set $r' \in R_i$ such that $|r' \cap S| > km$. Then any k-cover of S contains a set r'' such that $r' \subseteq r''$.*[1]

Proof. By Property 1, any accessible set $r \in R$ that does not contain r' has $\dim(r \cap r') < i$ and therefore $|(r \cap r') \cap S| \leq m$, i.e., r contains at most m elements of $r' \cap S$. Therefore, k such sets contain at most $km < |r' \cap S|$ elements of $r' \cap S$. However, in a k-cover, all elements of $r' \cap S$ must be covered. We conclude that any k-cover must contain a set r'' such that $r' \subseteq r''$.

Lemma 4 is the basis of our *kernelization* procedure. The procedure works by finding sets in R that contain many elements of S and grouping those elements

---
[1] The reader is reminded that we are only considering k-covers consisting of accessible sets.

of S together. Given a subset $S' \subseteq S$, a grouping of the DIM-SET-COVER instance S by S' creates a new DIM-SET-COVER instance as follows: (1) The elements of S' are removed from S and a new element s' is added to S. (2) For every set $r \in R$ such that $S' \subseteq r$, the element s' is added to r.

It is clear that Property 1 is preserved under the grouping operation, so that Property 1 holds for the new DIM-SET-COVER instance. Furthermore, Lemma 3 implies that the new element s' can be represented as a list of at most d elements of S. Thus, operations on grouped instances of DIM-SET-COVER can be done with a runtime that is within a factor of d of non-grouped instances, i.e., the overhead of working with grouped instances is $O'(1)$.

The following lemma shows that, if the group is chosen carefully, the new instance is equivalent to the old one.

Lemma 5 *Suppose $|r \cap S| \leq m$ for all $r \in \bigcup_{j=0}^{i-1} R_j$ and there exists an accessible set $r' \in R_i$ with size $|r' \cap S| > km$. Then the grouping of S by $S' = r' \cap S$ has a k-cover if and only if S has a k-cover.*

Proof. If S has a k-cover $r_1, \ldots, r_k$ then, by Lemma 4, there exists an r_j such that $r' \subseteq r_j$. Consider the sets $r'_1, \ldots, r'_k$ in the grouped instance that correspond to the sets $r_1, \ldots, r_k$. For each $1 \leq i \leq k$, $r_i \subseteq r'_i$. Therefore, $S \setminus S' \subseteq S \subseteq \bigcup_{i=1}^{k} r'_i$. Furthermore, r'_j contains s'. Therefore $r'_1, \ldots, r'_k$ is a k-cover for the grouped instance.

On the other hand, if the grouping of S by S' has a k-cover $r'_1, \ldots, r'_k$, then the corresponding sets $r_1, \ldots, r_k$ in the original instance are a k-cover for S. This is due to the facts that, for each $1 \leq i \leq k$, $r_i = r'_i \setminus \{s'\}$, $s' \notin S$ and $s' \in r'_i$ implies $S' \subseteq r_i$.

The following procedure is used to reduce an instance of DIM-SET-COVER involving a set S of size n into a new instance with size at most k^d.
KERNELIZE(S)
1: **for** $i = 0, \ldots, d-1$ **do**
2:     **while** R_i contains an accessible set r such that $|r \cap S| > k^i$ **do**
3:        group S by $r \cap S$
4:     **end while**
5: **end for**

Lemma 6 *A call to KERNELIZE(S) produces a new instance of DIM-SET-COVER that has a k-cover if and only if the original instance has a k-cover.*

Proof. By the time the procedure considers set R_i in lines 2–4, every r in $\bigcup_{j=0}^{i-1} R_j$ has size $|r \cap S| \leq k^{i-1}$. Therefore, by Lemma 5, any grouping operation performed in line 3 results in a DIM-SET-COVER instance that has a k-cover if and only if the original DIM-SET-COVER instance has a k-cover.

After a call to KERNELIZE(S) we get a new instance of DIM-SET-COVER for which $|r \cap S| \leq k^{d-1}$ for all $r \in R$. One consequence of this is that, if the new instance has more than k^d elements that need to be covered, then we can be immediately certain that it does not have a k-cover. Therefore, the instance that we have left to solve has size $n' \leq k^d$ and can be solved in time $O'(k^{d(k+1)})$ using the BST-DIM-SET-COVER algorithm.

Lemma 7 *The* DIM-SET-COVER *problem can solved in* $O'(k^{d(k+1)} + K(n,k,d))$ *time where* $K(n,k,d)$ *is the running time of the* KERNELIZE *procedure.*

All that remains is to find an efficient implementation of the KERNELIZE procedure. Lemma 3 implies that any accessible set in R_i can be generated as $\text{set}(S')$ where $S' \subseteq S$ has size at most $i+1$. It follows that all the accessible sets in R_i can be generated in $O(n^{i+1})$ time, and each one can be checked in $O(n)$ time. This gives us the following brute force result.

Theorem 2 *The* DIM-SET-COVER *problem can be solved in* $O'(k^{d(k+1)} + n^{d+1})$ *time by a deterministic algorithm.*

Although Theorem 2 implies a faster algorithm for some values of k, d and n, it is not entirely satisfactory. In fact, it does not even satisfy the definition of fixed-parameter tractability since d appears in the exponent of n. To obtain a faster algorithm we make use of randomization.

Define a *heavy covering set* as an accessible set $r \in R_i$ such that $|r \cap S| > n/2k^{d-i}$. Consider the following alternative implementation of KERNELIZE.

HALVING-KERNELIZE(S)
1: **while** $|S| > 2k^d$ **do**
2: $\quad n \leftarrow |S|$
3: $\quad$ **for** $i = 0, \ldots, d-1$ **do**
4: $\quad\quad$ **while** R_i contains a heavy covering set r **do**
5: $\quad\quad\quad$ group S by $r \cap S$
6: $\quad\quad$ **end while**
7: $\quad$ **end for**
8: $\quad$ **if** $|S| > n/2$ **then**
9: $\quad\quad$ output no
10: $\quad$ **end if**
11: **end while**

It is easy to verify, using Lemma 5, that each grouping operation results in an instance of DIM-SET-COVER that has a k-cover if and only if the original instance has a k-cover. Now, after each iteration of the outer while loop, $|r \cap S| \leq n/2k$ for all $r \in S$. Therefore, if $|S|$ is greater than $n/2$ we can be sure that S has no k-cover, so the algorithm only outputs no in line 9 when S has no k-cover. If this is not the case, then $|S|$ decreases by a factor of at least 2 during each iteration.

It seems that HALVING-KERNELIZE is no easier to implement than the original KERNELIZE procedure. However, the difference between the two is that HALVING-KERNELIZE attempts to find heavy covering sets, which contain relatively large fractions of S. This helps, because if we choose $i+1$ elements of S at random, there is a good chance that they will all belong a heavy covering set in R_i, if a heavy covering set exists.

Suppose that $|r \cap S| \le n/(2k^{d-j})$ for all $r \in R_j$, $j < i$. Let $r \in R_i$ be a heavy covering set and consider the following experiment. Initially we set our *sample* S' equal to the empty set. We then repeatedly choose an element p from S at random and add it to our sample S'. If p is contained in $(r \cap S) \setminus \text{set}(S')$ we are successful and we continue, otherwise we are unsuccessful and we stop. The experiment ends when we are either unsuccessful or $\text{set}(S') = r$. In the former case, we call the experiment a *failure*. In the latter case we call it a *success*.

If $\dim(S') = -1$, i.e., $S' = \emptyset$, then the probability that we are successful is at least

$$\frac{|r \cap S|}{n} \ge \frac{1}{2k^{d-i}}$$

Otherwise, if $0 \le \dim(S') = j < i$, then the probability that we are successful is at least

$$\frac{|r \cap S| - |\text{set}(S') \cap S|}{n} \ge \frac{1}{2k^{d-i}} - \frac{1}{2k^{d-j}} = \frac{k^i - k^j}{2k^d},$$

and each successful step increases $\dim(S')$ by at least 1. Therefore, the probability that the entire experiment is a success is at least

$$p_i \ge \frac{1}{2k^{d-i}} \times \prod_{j=0}^{i-1} \left(\frac{k^i - k^j}{2k^d} \right) \qquad (1)$$

$$\ge \frac{1}{2^{i+1} k^{d+di-i-i^2}} \times \prod_{j=1}^{\infty} \left(1 - k^{-j} \right) \qquad (2)$$

$$\ge \frac{1}{2^{i+3} k^{d+di-i-i^2}} \qquad (3)$$

$$\ge \frac{1}{2^{d+2} k^{\lceil (d+1)/2 \rceil \lfloor (d+1)/2 \rfloor}} . \qquad (4)$$

Inequality (3) follows from Euler's pentagonal number theorem [6, Chapter 16] (c.f., Andrews [2]). If we repeat the above experiment x times, then the probability that all of the experiments are failures is at most

$$(1 - p_i)^x$$

Setting $x = 2^{d+2} k^{\lceil (d+1)/2 \rceil \lfloor (d+1)/2 \rfloor}$, this probability is less than $1/2$. If we repeat this procedure $cx \log n$ times, the probability that all experiments are failures is no more than n^{-c}. Thus, we have an algorithm that runs in $O'(cxn \log n)$ time and finds r with probability at least $1 - n^{-c}$.

Lemma 3 implies that the total number of accessible sets in R_i, and hence the total number of heavy covering sets in R_i is at most $\binom{n}{i+1} \le n^{i+1}$. Thus, if we choose $c = i + 1 + c'$ then the probability that there exists a heavy covering set in R_i that is not found by repeating the above sampling experiment $cx \log n$ times is at most $n^{-c'}$. Therefore, one execution of line 4-6 of the HALVING-KERNELIZE algorithm can be implemented to run in $O'(c'x|S| \log n)$ time and fails with probability at most $n^{-c'}$. Since $|S|$ is halved during each iteration of the outer loop, the entire algorithm runs in time $O'(c'xn \log n)$ and the algorithm is correct with probability at least $1 - n^{-c'} d \log n$. Choosing c' large enough yields our main result on randomized algorithms.

Theorem 3 *There exists a Monte-Carlo algorithm for the* DIM-SET-COVER *problem that runs in* $O'(k^{d(k+1)} + c2^d k^{\lceil (d+1)/2 \rceil \lfloor (d+1)/2 \rfloor} n \log n)$ *time and answers correctly with probability at least* $1 - n^{-c}$.

5 Applications

In this section we present a number of covering problems, both geometric and non-geometric, that can be modeled as instances of DIM-SET-COVER and hence solved by the algorithms of the previous two sections. For the geometric applications, the value of d is closely related to the dimension of the geometric space.

5.1 Covering Points with Hyperplanes and Vice-Versa

Given a set S of n points in $\mathbb{R}^d$, does there exist a set of k hyperplanes such that each point of S is contained in at least one hyperplane?

This problem has received considerable attention in the literature and appears to be quite difficult. It is known to be NP-hard even when $d = 2$ [12]. The optimization problem of finding the smallest value of k for which the answer is yes has recently been shown to be APX-hard [3, 11], so unless P = NP there does not exist a $(1 + \epsilon)$-approximation algorithm. Algorithms for restricted versions and variants of this problem have been considered by Agarwal and Procopiuc [1] and Hassin and Megiddo [8].

We have seen, in our running example, that this problem can be modelled as follows: $U = \mathbb{R}^d$ and, for each $0 \leq i \leq d-1$, R_i is the set of i-flats embedded in $\mathbb{R}^d$. To see that this fits into our model, observe that the intersection of an i-flat and a j-flat, neither of which contains the other is an l-flat, for some $l < \min\{i, j\}$ so this system satisfies Property 1. Thus, this is a DIM-SET-COVER instance in which the parameter d is equal to the dimension of the underlying Euclidean space.

In a dual setting, we are given a set S of n hyperplanes in $\mathbb{R}^d$ and asked if there exists a set of k points such that each hyperplane in S contains at least one of the points. By a standard point-hyperplane duality, this problem is equivalent to the previous problem and can be solved as efficiently.

At this point we remark that in the plane, i.e., covering points with lines, it is possible to achieve a deterministic algorithm that performs slightly better than our randomized algorithm.[2] This is achieved by replacing our randomized procedure for finding heavy covering sets with an algorithm of Guibas et al [7] that

---
[2] This result can be generalized to $\mathbb{R}^d$ when the point set S is *restricted*, i.e., no $i+1$ elements of S lie on a common i-flat, for any $0 \leq i < d-1$. However, since any input that has a k-cover necessarily has many points on a common $(d-1)$-flat, this does not seem like a reasonable assumption.

can find all lines containing at least m points of S in time $O((n^2/m)\log(n/m))$. Using this yields a deterministic algorithm with running time $O(k^{2k+2} + kn)$.

5.2 Covering Points with Spheres

Given a set S of n points in $\mathbb{R}^d$, does there exist a set of k hyperspheres such that each point in S is on the surface of at least one of the hyperspheres?

For this problem we take $R_{d+1} = U = \mathbb{R}^d$. The set R_i, $0 \le i \le d$ consists of all i-spheres, so that R_0 consists of points, R_1 consists of pairs of points, R_2 consists of circles, and so on. Again, the intersection of an i-sphere and a j-sphere, neither of which contains the other, is an l sphere for some $l < \min\{i,j\}$, so this system satisfies Property 1. This yields an instance of DIM-SET-COVER whose dimension is 1 greater than that of the underlying Euclidean space.

5.3 Covering Points with Polynomials

Given a set $S = \{(x_1, y_1), \ldots, (x_n, y_n)\}$ of n points in $\mathbb{R}^2$ does there exist a set of k polynomial functions $f_1, \ldots, f_k$, each with maximum degree d, such that for each $1 \le i \le n$, $y_i = f_j(x_i)$ for some $1 \le j \le k$? In other words, is every point in S contained in the graph of at least one of the functions?

This problem is a generalization of the problem of covering points with lines. As such, it is NP-hard and APX-hard, even when $d = 1$.

We say that two points in $\mathbb{R}^2$ are x-distinct if they have distinct x-coordinates. For this problem, R_0 is the set of all points, R_1 is the set of all x-distinct pairs of points, R_2 is the set of all x-distinct triples of points, and so on, until R_{d+1} which is the set of all degree d polynomials. Since there exists a degree d polynomial that contains any set of $d+1$ or fewer x-distinct points, each set in R corresponds to points that can be covered by a degree d polynomial (possibly with some coefficients set to 0).

The intersection of a finite set of i points and a (possibly infinite) set of $j \ge i$ points, neither of which contains the other results in a set of $l < \min\{i, j\}$ points. The intersection of two non-identical degree d polynomials is a set of at most d points. Therefore, the above system of sets satsifies Property 1 and can be solved with our algorithms.

5.4 Covering by Sets with Intersection at Most d

Given a set S of size n and a set C containing subsets of S such that no two elements of C have more than d elements in common, is there a set of k elements in C whose union contains all the elements of S?

This problem is a special case of the classic SET-COVER problem, which was one of the first problems shown to be NP-hard [10]. The version when $d = 1$ is a generalization of covering points by lines and is therefore APX-hard. In fact, finding an $o(\log n)$ approximation for the corresponding optimization problem is not possible unless NP $\subseteq$ ZTIME($n^{O(\log \log n)}$) [11]. On the other hand, Johnson [9] shows that, even with no restrictions on d, an $O(\log n)$-approximation can be achieved with a simple greedy algorithm.

To model this problem as a DIM-SET-COVER instance, we let R_i, $0 \leq i < d$ be all subsets of S of size $i + 1$ that are contained in 1 element of C and we let $R_d = C$. It is easy to verify that these sets satisfy Property 1 and hence we have an instance of DIM-SET-COVER that can be solved with our algorithms. If the sets are given explicitly, the Kernelize procedure runs in $O(n')$ time where $n' = \sum_{r \in C} |r|$ is the input size. Thus, for this case we obtain an algorithm that runs in time $O(n' + k^{d(k+1)})$.

References

[1] P. K. Agarwal and C. M. Procopiuc. Exact and approximation algorithms for clustering. In *Proceedings of the 9th ACM-SIAM Symposium on Discrete Algorithms (SODA 1998)*, pages 658–667, 1998.

[2] George E. Andrews. *The Theory of Partitions*. Addison-Wesley, 1976.

[3] B. Brodén, M. Hammar, and B. J. Nilsson. Guarding lines and 2-link polygons is APX-hard. In *Proceedings of the 13th Canadian Conference on Computational Geometry*, pages 45–48, 2001.

[4] J. Chen, I. Kanj, and W. Jia. Vertex cover: further observations and further improvements. *Journal of Algorithms*, 41:280–301, 2001.

[5] R. G. Downey and M. R. Fellows. *Parameterized Complexity*. Springer, 1999.

[6] Leonhardo Eulero. *Introductio in Analysin Infinitorum*. Tomus Primus, Lausanne, 1748.

[7] L. J. Guibas, M. H. Overmars, and J.-M. Robert. The exact fitting problem for points. *Computational Geometry: Theory and Applications*, 6:215–230, 1996.

[8] R. Hassin and N. Megiddo. Approximation algorithms for hitting objects by straight lines. *Discrete Applied Mathematics*, 30:29–42, 1991.

[9] D. S. Johnson. Approximation algorithms for combinatorial problems. *Journal of Computer Systems Sciences*, 9:256–278, 1974.

[10] R. M. Karp. *Reducibility Among Combinatorial Problems*, pages 85–103. Plenum Press, 1972.

[11] V. S. A. Kumar, S. Arya, and H. Ramesh. Hardness of set cover with intersection 1. In *Proceedings of the 27th International Colloquium on Automata, Languages and Programming*, pages 624–635, 2000.

[12] N. Megiddo and A. Tamir. On the complexity of locating linear facilities in the plane. *Operations Research Letters*, 1:194–197, 1982.

[13] D. Nussbaum. Rectilinear p-piercing problems. In *ISSAC '97. Proceedings of the 1997 International Symposium on Symbolic and Algebraic Computation, July 21-23, 1997, Maui, Hawaii*, pages 316–323, 1997.

[14] Franco P. Preparata and Michael Ian Shamos. *Computational Geometry: An Introduction*. Springer-Verlag, 1985.

[15] M. Sharir and E. Welzl. Rectilinear and polygonal p-piercing and p-center problems. In *Proceedings of the Twelfth Annual Symposium On Computational Geometry (ISG '96)*, pages 122–132, 1996.

[16] V. N. Vapnik and A. Y. A. Chervonenkis. On the uniform convergence of relative frequencies of events to their probabilities. *Theory of Probability and its Applications*, 16:264–280, 1971.

On-Line Dial-a-Ride Problems under a Restricted Information Model

Maarten Lipmann[1], X. Lu[1,2], Willem E. de Paepe[3],
Rene A. Sitters[1], and Leen Stougie[1,4]

[1] Department of Mathematics and Computer Science
Technische Universiteit Eindhoven
P.O.Box 513, 5600 MB Eindhoven, The Netherlands
{m.lipmann,x.lu,r.a.sitters,l.stougie}@tue.nl
[2] East China University of Science and Technology
Shanghai 200237, China
xwlu@ecust.edu.cn
[3] Department of Technology Management
Technische Universiteit Eindhoven
P.O.Box 513, 5600 MB Eindhoven, The Netherlands
w.e.d.paepe@tm.tue.nl
[4] CWI, P.O. Box 94079, 1090GB Amsterdam, The Netherlands
stougie@cwi.nl

Abstract. In on-line dial-a-ride problems, servers are traveling in some metric space to serve requests for rides which are presented over time. Each ride is characterized by two points in the metric space, a *source*, the starting point of the ride, and a *destination*, the end point of the ride. Usually it is assumed that at the release of such a request complete information about the ride is known. We diverge from this by assuming that at the release of such a ride only information about the source is given. At visiting the source, the information about the destination will be made available to the servers. For many practical problems, our model is closer to reality. However, we feel that the lack of information is often a *choice*, rather than inherent to the problem: additional information *can* be obtained, but this requires investments in information systems. In this paper we give mathematical evidence that for the problem under study it pays to invest.

1 Introduction

In dial-a-ride problems servers are traveling in some metric space to serve requests for rides. Each ride is characterized by two points in the metric space, a *source*, the starting point of the ride, and a *destination*, the end point of the ride. The problem is to design routes for the servers through the metric space, such that all requested rides are made and some optimality criterion is met.

Dial-a-ride problems have been studied extensively in the literature of operations research, management science, and combinatorial optimization. Traditionally, such combinatorial optimization problems are studied under the assumption that the input of the problem is known completely to the optimizer.

In a natural setting of dial-a-ride problems requests for rides are presented over time while the servers are enroute serving other rides, making the problem an on-line optimization problem. Examples of such problems in practice are taxi and minibus services, courier services, and elevators. In their on-line setting dial-a-ride problems have been studied in [1] and [4]. In these papers single server versions of the problem are studied, as we will do here. The two papers study the problem in which the rides are specified completely upon presentation, i.e., both the source and the destination of the ride become known at the same time. We diverge from this setting here.

In many practical situations complete specification of the rides is not realistic. Think for example of the problem to schedule an elevator. Here, a ride is the transportation of a person from one floor (the source) to another (the destination), and the release time of the ride is the moment the button on the wall outside the elevator is pressed. The destination of such a ride is revealed only at the moment the person enters the elevator and presses the button inside the elevator.

In this paper we study the on-line single server dial-a-ride problem in which only the source of a ride is presented at the release time of the ride. The destination of a ride is revealed at visiting its source. We call this model the *incomplete ride information model* and refer to the model used in [1] and [4] as the *complete ride information model*. As objective we minimize the time by which the server has executed all the rides and is back in the origin.

We distinguish two versions of the on-line dial-a-ride problem under the incomplete ride information model. In the first version the server is allowed to preempt any ride at any point, and proceed the ride later. In particular the server is allowed to visit the source of a ride and learn its destination without executing the ride immediately. This version we call the *preemptive version*. In the second version, the *non-preemptive* version, a ride has to be executed as soon as the ride has been picked up in the source. In this version we do allow the server to pass a source without starting the ride, in which case he does not learn the destination of the ride at passing the source. We study each version of the problem under various *capacities* of the server. The capacity of a server is the number of rides the server can execute simultaneously. A formal problem definition is given in Section 2.

We perform competitive analysis of deterministic algorithms for the on-line dial-a-ride problems described above. Competitive analysis measures the performance quality of an algorithm for an on-line problem by the worst-case ratio over all possible input sequences of the objective value it produces and the optimal off-line solution value, the so-called *competitive ratio* of the algorithm. For a detailed explanation of competitive analysis and many examples we refer to [3]. For an overview of results on on-line optimization problems we refer to [5]. Typically there are lower bounds on the competitive ratio achievable by any algorithm (even allowing exponential computing time). We derive such lower bounds for deterministic algorithms for the various versions of the on-line dial-

Table 1. Overview of lower bounds (LB) and upper bounds (UB) on the competitive ratio of deterministic algorithms for on-line dial-a-ride problems

	capacity	LB	UB
complete ride information			
preemption	$1, c, \infty$	2 [2]	2 [1]
no preemption	$1, c, \infty$	2 [2]	2 [1]
incomplete ride information			
preemption	$1, c, \infty$	3	3
no preemption	1	$1 + \frac{3}{2}\sqrt{2}$	4
	c	$\max\{1 + \frac{3}{2}\sqrt{2}, c\}$	$2c + 2$
	∞	3	3

a-ride problem under the incomplete ride information model and design and analyze algorithms for their solution.

In [1] a 2-competitive deterministic algorithm is given for the on-line dial-a-ride problem under the complete ride information model, independent of the capacity of the server. In this paper preemption of rides is not allowed. However, the lower bound of 2 comes from a sequence of rides with zero length, an instance of the on-line traveling salesman problem [2], whence the bound also holds for the problem with preemption. We show that under the incomplete ride information model no deterministic algorithm can have a competitive ratio smaller than 3, even if preemption is allowed, and independent of the capacity of the server. For the preemptive version, we design an algorithm with competitive ratio matching the lower bound of 3, independent of the capacity of the server. These results are presented in Section 3.

If preemption is not allowed, we derive a lower bound of $\max\{c, 1 + \frac{3}{2}\sqrt{2}\}$ on the competitive ratio of any deterministic algorithm, where c is a given fixed capacity of the server. We present a $2c + 2$-competitive algorithm for the non-preemptive version. These results are presented in Section 4.

We notice that there is no difference between the preemptive version and the non-preemptive version of the problem if the server has infinite capacity, whence we inherit the matching lower and upper bound of 3 of the preemptive version for this case. An overview of the results obtained in this paper is given in Table 1.

The results in our paper (combined with those from [1]) show the effect of having complete knowledge about rides on worst-case performance for on-line dial-a-ride problems. This is an important issue, since in practice complete information is often lacking. Investments in information systems can help to obtain more complete information, and mathematical support is essential in justifying such investments. Our results concern the objective of minimizing the time by which the server has done all rides and is back in the origin. An interesting question is if similar results can be obtained for other objectives.

We conclude this introduction by referring back to the elevator scheduling problem. We have seen that the typical elevator with only a request button at the wall outside the elevator fits our incomplete ride information model. In an alternative construction of an elevator, the destination buttons could be build outside the elevator, fitting the complete ride information model. Notice that minimizing the latest completion time is not the most natural objective for an elevator.

2 Problem Definition

An instance of the on-line single server dial-a-ride problem (OLDARP) is specified by a metric space $M = (X, d)$ with a distinguished origin $O \in X$, a sequence $\sigma = \sigma_1, \ldots, \sigma_m$ of requests for rides, and a capacity for the server. A server is located at the origin O at time 0 and can move at most at unit speed. We assume that M has the property that for any pair of points $\{x, y\} \in X$ there is a continuous path $p \colon [0, 1] \to X$ in X with $p(0) = x$ and $p(1) = y$ of length $d(x, y)$ (see [2] for a thorough discussion of this model). We add explicitly the assumption that d is symmetric and satisfies the triangle inequality for those readers who do not see this as being implicit in the definition of metric space.

Each *ride* is a triple $\sigma_i = (t_i, s_i, d_i)$, where $t_i \in \mathbb{R}_0^+$ is the time at which ride σ_i is released, $s_i \in X$ is the source of the ride, and $d_i \in X$ is the destination of the ride. Every ride $\sigma_i \in \sigma$ has to be executed (served) by the server; that is, the server has to visit the source, pick up the ride, and end it at the destination. The *capacity* of the server is an upper bound on the number of rides the server can execute simultaneously. We consider unit capacity, constant capacity $c \geq 2$, and infinite capacity for the server. The objective in the OLDARP is to minimize the completion time of the server, that is, the time when the server has served all rides and has returned to the origin.

In this paper we consider the preemptive and non-preemptive versions of the OLDARP, under the *incomplete ride information model* for different capacities of the server.

Definition 1. *Under the* incomplete ride information model *only the source s_i of ride σ_i is revealed at time t_i. The destination of the ride becomes known only at picking up the ride in the source.*

We assume that the sequence $\sigma = \sigma_1, \ldots, \sigma_m$ of rides is given in order of non-decreasing release times, and that the on-line server has neither information about the time when the last ride is released, nor about the total number of rides.

For $t \geq 0$ we denote by $\sigma_{\leq t}$ the set of rides in σ released no later than time t. An on-line algorithm for the OLDARP must determine the behavior of the server at any moment t as a function of t and $\sigma_{\leq t}$, whereas the off-line algorithm knows σ at time 0. A feasible on-line/off-line solution is a route for the server that starts and ends in the origin O and serves all requested rides regarding that each ride is picked up at the source not earlier than the time it is released.

Let $\mathrm{ALG}(\sigma)$ denote the completion time of the server moved by algorithm ALG on the sequence σ of rides and $\mathrm{OPT}(\sigma)$ denote the optimal off-line algorithm's completion time. The competitive ratio of algorithm ALG is

$$\max_{\sigma \in \Sigma} \frac{\mathrm{ALG}(\sigma)}{\mathrm{OPT}(\sigma)},$$

with Σ the class of all possible request sequences.

3 The Preemptive Version

We describe our algorithm SNIFFER, which preempts rides only immediately at the source, just to learn the destinations of the rides: it "sniffs" the rides. Upon visiting the source of a ride for the second time, the ride is completed right away. The algorithm is an adaption of the 2-competitive algorithm for the on-line traveling salesman problem (OLTSP), described in [2]. The proof of 3-competitiveness of SNIFFER borrows parts of the proof in the latter paper. The algorithm is described completely by the actions it takes at any moment t at which the server either arrives in the origin or receives a new request. We use $|T|$ to denote the length of a tour T.

Algorithm SNIFFER

(1) **The server is in the origin at t.**
If the set S of yet unvisited sources is non-empty, compute the optimal traveling salesman tour $T_{\mathrm{TSP}}(S)$ on the points in S, and start following $T_{\mathrm{TSP}}(S)$. Just learn the destinations of the rides with sources in S, without starting to execute any of these rides.
If $S = \emptyset$ and the set R of rides yet to be executed is non-empty, compute the optimal dial-a-ride tour $T_{\mathrm{DAR}}(R)$ on the rides in R. Also compute the optimal dial-a-ride tour $T_{\mathrm{DAR}}(\sigma_{\leq t})$ on all rides requested in $\sigma_{\leq t}$. If $t = 2|T_{\mathrm{DAR}}(\sigma_{\leq t})|$, start following $T_{\mathrm{DAR}}(R)$. If $t < 2|T_{\mathrm{DAR}}(\sigma_{\leq t})|$, remain idle. If no new requests arrive before time $2|T_{\mathrm{DAR}}(\sigma_{\leq t})|$ start following $T_{\mathrm{DAR}}(R)$ at time $2|T_{\mathrm{DAR}}(\sigma_{\leq t})|$.

(2) **The server is on a tour $T_{\mathrm{TSP}}(S)$ at t when a new ride is released.**
Let p_t denote the location of the server at time t. If the new ride, say $\sigma_k = (t, s_k, ?)$ (the question mark indicating the unknown destination), is such that $d(s_k, O) > d(p_t, O)$, then return to the origin via the shortest path, ignoring all rides released while traveling to the origin.
If $d(s_k, O) \leq d(p_t, O)$, ignore the new ride until the origin is reached again and proceed on $T_{\mathrm{TSP}}(S)$.

(3) **The server is on a tour $T_{\mathrm{DAR}}(R)$ at t when a new ride is released.**
Return to the origin as soon as possible via the shortest path, and ignore rides released in the mean time. If the server is executing a ride, the ride is finished before returning to the origin.

We emphasize that the optimal tours that SNIFFER computes are always starting and finishing in the origin.

Theorem 1. *Algorithm* SNIFFER *is 3-competitive for the preemptive* OLDARP *under the incomplete ride information model, independent of the capacity of the server.*

Proof. Let $T_{\text{DAR}}(\sigma)$ be the optimal tour over all rides of σ. It is sufficient to prove that for any sequence σ the server can always be in the origin at time $2|T_{\text{DAR}}(\sigma)|$ to start the final tour on the yet unserved rides. He will then always finish this tour before time $3|T_{\text{DAR}}(\sigma)| \leq 3\text{OPT}(\sigma)$. This is obviously true for any sequence σ consisting of only one ride. We assume it holds for any sequence of $m-1$ rides, and prove that then it also holds for any sequence σ of m rides. Let $\sigma_m = (t_m, s_m, d_m)$ be the last ride in σ (notice that the destination d_m is not given to the on-line algorithm until the moment the source s_m is visited).

(1) Suppose the server is in O at t_m, and $S \neq \emptyset$. He starts $T_{\text{TSP}}(S)$ and returns to O at time $t_m + |T_{\text{TSP}}(S)| \leq 2|T_{\text{DAR}}(\sigma)|$.

(2) Suppose the server is in p_{t_m} following $T_{\text{TSP}}(S)$. If $d(O, s_m) \leq d(O, p_{t_m})$, σ_m is added to a set Q of rides ignored since the last time the server was in O. Let $s_q \in Q$ be the source of a ride visited first in an optimal solution. Since this ride was ignored, the server was at least $d(O, s_q)$ away from the origin at time t_q, and hence had moved at least this distance on $T_{\text{TSP}}(S)$. Thus, the server returns in O before $t_q + |T_{\text{TSP}}(S)| - d(O, s_q)$. Back in O the server commences on $T_{\text{TSP}}(Q)$. Let $P_q(Q)$ be the path of minimum length that starts in s_q, ends in O, and visits all sources in Q. Obviously, $|T_{\text{TSP}}(Q)| \leq d(O, s_q) + |P_q(Q)|$. Hence the server is back in the origin after visiting all sources no later than $t_q + |T_{\text{TSP}}(S)| - d(O, s_q) + d(O, s_q) + |P_q(Q)| = t_q + |T_{\text{TSP}}(S)| + |P_q(Q)| \leq 2|T_{\text{DAR}}(\sigma)|$, since, clearly, $|T_{\text{DAR}}(\sigma)| \geq t_q + |P_q(Q)|$, and $|T_{\text{DAR}}(\sigma)| \geq |T_{\text{TSP}}(S)|$.
If $d(O, s_m) > d(O, p_{t_m})$, the server returns to O immediately, arriving there at $t_m + d(O, p_{t_m}) < t_m + d(O, s_m) \leq |T_{\text{DAR}}(\sigma)|$. Back in O the server computes and starts following an optimal TSP tour over the yet unvisited sources, which has a length of at most $|T_{\text{DAR}}(\sigma)|$. Hence the server is back in O again before time $2|T_{\text{DAR}}(\sigma)|$.
If the server was already moving towards the origin because a ride was released before σ_m that was further away from the origin than the on-line server, then the arguments above remain valid.

(3) Suppose the server is on $T_{\text{DAR}}(R)$ at time t_m, or moving towards O because of another ride released before t_m. Let $t(R)$ be the time at which the server started $T_{\text{DAR}}(R)$. Then $R \subset \sigma_{\leq t(R)}$ and $t(R) = 2|T_{\text{DAR}}(\sigma_{t(R)})|$, by induction. Thus, the server is back in O before time $3|T_{\text{DAR}}(\sigma_{t(R)})|$. There, it starts $T_{\text{TSP}}(S)$ over a set S of unvisited sources, being again back in O before time $3|T_{\text{DAR}}(\sigma_{t(R)})| + |T_{\text{TSP}}(S)| = \frac{3}{2}t(R) + |T_{\text{TSP}}(S)|$. We need to show that this is not greater than $2|T_{\text{DAR}}(\sigma)|$.
Let s_q be the first ride from S served in an optimal solution and $P_q(S)$ the shortest path starting in $s_q \in S$, ending in O, and visiting all sources in S. Clearly, $|T_{\text{DAR}}(\sigma)| \geq t_q + |P_q(S)|$ and $|T_{\text{TSP}}(S)| \leq 2|P_q(S)|$. Since all rides in S are released after $t(R)$, $t_q \geq t(R)$. Therefore, $|T_{\text{DAR}}(\sigma)| \geq t(R) + |P_q(S)|$

and
$$\frac{3}{2}t(R) + |T_{\text{TSP}}(S)| \leq 2t(R) + 2|P_q(S)| \leq 2|T_{\text{DAR}}(\sigma)|.$$

We show that SNIFFER is a best possible deterministic algorithm for the preemptive version of the OLDARP, even though SNIFFER uses preemption only at the source of rides.

Theorem 2. *No deterministic algorithm can have a competitive ratio smaller than $3 - \epsilon$ for the OLDARP under the incomplete ride information model, independent of the capacity of the server, where ϵ is arbitrarily small.*

Proof. For the proof of this theorem we use a commonly applied setting of a two-person game, with an adversary providing a sequence of rides, and an on-line algorithm serving the rides (see [3]). Typically, the outcome of the algorithm is compared by the solution value the adversary achieves himself on the sequence, which usually is the optimal off-line solution value. We consider the OLDARP under the incomplete ride information model where the on-line server has infinite capacity. Let ALG be a deterministic on-line algorithm for this problem. We will construct an adversarial sequence σ of requests for rides. We restrict the adversary by giving his server capacity 1. We will prove that ALG can not be better than $3 - \epsilon$-competitive for this restricted adversary model, where ϵ is arbitrary small.

The metric space $M = (X, d)$ contains the set of points $\{x_1, x_2, \ldots, x_{n^2}\} \cup O$ and the distance function d, where $d(O, x_i) = 1$ and $d(x_i, x_j) = 2$ for all x_i, x_j. To facilitate the exposition we denote point x_i by i.

At time 0 there is one ride in each of the n^2 points in $x_1, x_2, \ldots, x_{n^2}$. If the on-line server visits the source i of a ride at time t with $t \leq 2n^2 - 1$, then the destination turns out to be i as well, and at time $t + 1$ a new ride with source i is released.

In this way, the situation remains basically the same for the on-line server until time $2n^2$. We may assume that at some moment t^*, with $2n^2 - 1 < t^* \leq 2n^2$, there is exactly one ride $\sigma_i = (t_i, i, d_i)$ in each of the points i. Without loss of generality we assume that the vertices i are labelled in such a way that $t_1 \leq \cdots \leq t_{n^2}$.

Thus, at time t^* the on-line server still has to complete exactly n^2 rides. We partition the set of n^2 vertices into n sets: $I_k = \{(k-1)n + 1, \ldots, kn\}$, $k = 1, \ldots, n$. Within each of these sets we order the vertices by the on-line server's first visit to them after time t^*. Let b_{kj}, $j \in \{1, \ldots, n\}$ be the jth vertex in this order in I_k. For all $k \in \{1, \ldots, n\}$ we define $d_{b_{k1}} = b_{k1}$ and $d_{b_{kj}} = b_{k,j-1}$ for all $j \in \{2, \ldots, n\}$. Notice that the destination of ride σ_i only depends on the tour followed by the on-line server until he picks up the ride to look at its destination. For the on-line server this means that n of the n^2 rides can be served immediately since the source equals the destination. For the other $n^2 - n$ rides the server finds out that the destination of the rides he just picked up is another point that he already visited after time t^*. Therefore, $n^2 - n$ points will have to be visited by the on-line server at least twice after time t^*. Hence, the completion time for the on-line server is at least $t^* + 4(n^2 - n) - 1 + 2n > 6n^2 - 2n - 2$.

We will now describe the tour made by the adversary. Given our definition of t^* we have that $t_{n^2} \leq t^* \leq 2n^2$. Since the on-line server needs at least 2 time units to move from a point i to another point i', it follows that $t_i \leq 2i$, for all $i \in \{1, \ldots, n^2\}$. The adversary waits until time $2n$ and then starts to serve the rides $\sigma_1, \ldots, \sigma_n$, by visiting the sources in reversed order of $b_{11}, \ldots, b_{1n}$. The rides with equal source and destination are served immediately at arrival in the point. This takes the adversary $2n$ time units. At time $4n$ the adversary starts serving the rides $\sigma_{n+1}, \ldots, \sigma_{2n}$, and then at time $6n$ the rides $\sigma_{2n+1}, \ldots, \sigma_{3n}$, etc. Continuing like this the server has completed all the rides and is back in the origin at time $2n^2 + 2n$.

Hence, the competitive ratio is bounded from below by $(6n^2 - 2n - 2)/(2n^2 + 2n)$, which can be made arbitrarily close to 3 by choosing n large enough.

4 The Non-preemptive Version

For the non-preemptive version we design an algorithm, called BOUNCER, because the server always "bounces" back to the source, once a ride is completed. This algorithm uses as a subroutine the 2-competitive algorithm for the OLTSP problem from [2].

Algorithm BOUNCER

Perform the OLTSP algorithm on the sources of the rides. This algorithm outputs a tour T. The BOUNCER server follows T, until a source is visited. There he executes the ride, and returns to the source via the shortest path. As soon as the server arrives in the source again, he continues to follow T.

Theorem 3. *Algorithm BOUNCER is $(2c + 2)$-competitive for the OLDARP under the incomplete ride information model, where c is the capacity of the server.*

Proof. Consider any request sequence σ. Since the OLTSP algorithm is 2-competitive, and in any solution for the OLDARP all sources have to be visited, $\text{OPT}(\sigma) \geq |T|/2$. Let $D = \sum_{i:\sigma_i \in \sigma} d(s_i, d_i)$, then also $\text{OPT}(\sigma) \geq D/c$. The completion time of the BOUNCER server is at most $T + 2D \leq 2\text{OPT}(\sigma) + 2c\text{OPT}(\sigma)$.

Corollary 1. *Algorithm BOUNCER is 4-competitive for the OLDARP under the incomplete ride information model, if the capacity of the server is 1.*

Theorem 4. *No non-preemptive deterministic on-line algorithm can have a competitive ratio smaller than $c - \epsilon$ for the OLDARP under the incomplete ride information model, where the capacity of the server c is a constant, and where $\epsilon > 0$ is arbitrarily small.*

Proof. Consider an instance of the OLDARP on the edges of a star graph with $K \gg c$ leaves, where the origin is located in the center of the star, and each leaf has distance 1 to the origin. At time 0, cK rides are released, all with their source in the origin, and each of the leaves being destination of c rides. Thus, there are K sets of c identical rides each, hence the instance has an optimal solution value of $2K$.

Any on-line server can carry at most c rides at a time. The instance is constructed in such a way that, until time $2(cK - c^2)$, any time the on-line server is in the origin he has all different rides (rides with different destinations). It is clear that this can indeed be arranged, given that the on-line server can not distinguish between the rides until he picks them up at the source and he has no possibility to preempt, not even in the source. At time $2(cK - c^2)$ the on-line server can have served at most $cK - c^2$ rides, and hence at least c^2 rides remain yet to be served, requiring an extra of at least $2c$ time units. Hence the completion time of any on-line server is at least $2(cK - c^2) + 2c$.

Therefore, the competitive ratio is bounded from below by

$$\frac{2(cK - c^2) + 2c}{2K} = \frac{cK - c^2 + c}{K}.$$

For any $\epsilon > 0$ we can choose K large enough for the theorem to hold.

Together with Theorem 1 this theorem shows that for servers with capacity greater than 3, the best possible deterministic on-line algorithm for the non-preemptive version of the problem has a strictly higher competitive ratio than SNIFFER for the preemptive problem, in which the server can have any capacity. The following theorem shows that this phenomenon also occurs for lower capacities of the server.

Theorem 5. *No non-preemptive deterministic algorithm can have a competitive ratio smaller than $1 + \frac{3}{2}\sqrt{2} - \epsilon \approx 3.12$ for the OLDARP under the incomplete ride information model, where the capacity of the server c is a constant and ϵ is arbitrarily small.*

Proof. First we consider the OLDARP under the incomplete ride information model when the on-line server has capacity 1. Then we will sketch how to extend the proof for any capacity c.

Let ALG be a non-preemptive deterministic on-line algorithm for this problem. The metric space is defined by the edges of a star graph on $2n+1$ vertices. All edges have length 1. The center of the star is the origin O and the leaves are denoted by a_i ($i = 1 \ldots n$), and b_i ($i = 1 \ldots n$). The point at distance α ($0 < \alpha < 1$) from point a_i, is denoted by a'_i, $i = 1, \ldots, n$, where α is a fixed number that we choose appropriately later.

We give the following sequence σ of rides. At time zero there are three rides in each point a_i and b_i, $i = 1 \ldots n$. If the on-line server visits a source, then the destination turns out to be the same as the source. This kind of rides are called empty rides. One time unit later the ride is replaced by a new ride with the same source. Every source that is visited by the on-line server before time $4n$ sees the same fate. Sources visited after this time are not replaced.

Let t_i^a (resp. t_i^b) be the last moment before time $4n - 4$ that the on-line server is in point a_i (resp. b_i). We set $t_i^a = 0$ (resp. $t_i^b = 0$) if a_i (resp. b_i) is not visited before $4n-4$. Without loss of generality we assume that $t_i^a \leq t_j^a$, $t_i^b \leq t_j^b$, $t_i^a \leq t_j^b$, and $t_i^b \leq t_j^a$, for all $1 \leq i < j \leq n$. Any server needs at least 2 time units to travel from one leaf to another, implying that $t_i^a \leq 4(i-1)$ and $t_i^b \leq 4(i-1)$, for all $i \in \{1, \ldots, n\}$.

We refer to the three rides that were released latest in a leaf as the *decisive rides* and define them as follows. In each point b_i two of the decisive rides are empty and one has destination a_i. In point a_i one of the decisive rides is empty, one has destination a_i', and one is either empty or has destination O. With these rides, the on-line server is unable to distinguish between the points a_i and b_i, and since we did not distinguish between these points before, we may assume that, after time $4n-4$, the on-line server visits point a_i before point b_i. The first ride that the server picks up in point a_i is the ride to a_i'. The first ride the server picks up in point b_i is the ride to a_i. We distinguish between two cases. In the first case the on-line server executes the ride from b_i to a_i before it picks up the second ride in a_i. In this case the second ride is a ride to the origin. Otherwise, the second ride being picked up in a_i is empty.

In the first case the on-line server needs at least 10 time units (from origin to origin) to serve all the rides connected to each pair a_i, b_i, whereas in the optimal off-line solution only $4 + 2\alpha$ time units are required. In the second case the on-line server needs at least $8 + 2\alpha$ time units, whereas in the optimal off-line solution 4 time units are required. The optimal off-line strategy starts a tour at time 0, first serving a_1 and b_1, then serving a_2 and b_2, etc. Empty rides are served without taking any extra time.

The on-line server cannot start with the decisive rides until time $4n - 4$. Assume that he is in the origin at time $4n - 5$ and then moves to point a_1. Now we consider the contribution of a pair a_i, b_i in the total time needed for the on-line and the off-line server and we take the ratio of the two. For fixed α ($0 < \alpha < 1$) this ratio becomes at least

$$\min\left\{\frac{10+(4n-5)/n}{4+2\alpha}, \frac{8+2\alpha+(4n-5)/n}{4}\right\} > \min\left\{\frac{14}{4+2\alpha}, \frac{12+2\alpha}{4}\right\} - \frac{2}{n}.$$

Optimizing over α yields $\alpha = -4 + 3\sqrt{2}$. Hence, for the competitive ratio we find

$$\frac{\text{ALG}(\sigma)}{\text{OPT}(\sigma)} > 1 + \frac{3}{2}\sqrt{2} - \frac{2}{n}.$$

For any ϵ we can choose n large enough for the theorem to hold in the case of a unit capacity server. If the capacity of the server is c, with $c \geq 1$, we just give c copies of the same sequence σ simultaneously. An on-line server cannot benefit from this extra capacity in combining rides from different pairs a_i, b_i. The on-line server will have to do the rides in a specific point in the same order as before. For example the first c rides that the on-line server picks up in a_i are rides to a_i'. Hence, the completion time for the on-line server cannot be smaller than in the capacity 1 case, and the off-line server can complete in exactly the same time.

Corollary 2. *No non-preemptive deterministic on-line algorithm can have a competitive ratio smaller than* $\max\{1 + \frac{3}{2}\sqrt{2}, c\} - \epsilon$ *for the* OLDARP *under the incomplete ride information model, where the capacity of the server c is a constant and* $\epsilon > 0$ *is arbitrarily small.*

5 Discussion

In [1] and [4] the competitive ratio measures the cost of having no information about the release times of future rides. In this discussion we show how we can measure the cost of having no information about the destinations of the rides through the competitive ratio.

Suppose that at time 0 the release times and the location of the sources of the rides are given, but the information about the destinations is again revealed only at visiting the sources.

Both SNIFFER and BOUNCER use the on-line algorithm of Aussiello et al. [2] for a TSP tour along the sources. In case all sources of the rides and the release times are known, an optimal TSP tour over the sources, that satisfies the release time constraints, can be computed (disregarding complexity issues). In this way SNIFFER and BOUNCER gain an additive factor of 1 on their competitive ratio, making SNIFFER 2-competitive and BOUNCER $2c + 1$-competitive.

Notice that the lower bound of $c - \epsilon$ on the competitive ratio for the non-preemptive problem in Theorem 4 is obtained through a sequence of rides all with release time 0. Thus, this lower bound is completely due to the lack of information about the destinations of the rides.

The rides in the sequence giving the lower bound of $1 + \frac{3}{2}\sqrt{2}$ for the non-preemptive problem in Theorem 5 have release times no larger than $4n - 5$. Taking the unserved rides at time $4n - 5$ as an instance given at time 0, shows that the competitive ratio is at least $\min\{\frac{10}{4+2\alpha}, \frac{8+2\alpha}{4}\}$. Optimizing over α yields a lower bound of $\frac{1}{2} + \frac{1}{2}\sqrt{11} \approx 2,15$. Thus, due to the lack of information about destinations only, any algorithm will not be able to attain a ratio of less than $\max\{\frac{1}{2} + \frac{1}{2}\sqrt{11}, c - \epsilon\}$.

In the lower bound construction for the preemptive problem in Theorem 2 the adversary stops giving requests at time $2n^2$. Take the set of rides unserved by any on-line algorithm at that time as an instance with release time 0. Following the proof of Theorem 2 any on-line algorithm will need $4n^2 - 2n$, whereas an optimal tour takes $2n^2$, yielding a lower bound of 2.

Notice that the above lower bounds are established on sequences where all rides have release time 0. For the preemptive version of the problem this is clearly sufficient, since the performance of SNIFFER matches the lower bound. However, for the non-preemptive version higher lower bounds might be obtained using diverse release times of rides.

References

[1] N. Ascheuer, S. O. Krumke, and J. Rambau, *Online dial-a-ride problems: Minimizing the completion time*, Proceedings of the 17th International Symposium on Theoretical Aspects of Computer Science, Lecture Notes in Computer Science, vol. 1770, Springer, Berlin, 2000, pp. 639–650.

[2] G. Ausiello, E. Feuerstein, S. Leonardi, L. Stougie, and M. Talamo, *Algorithms for the on-line traveling salesman*, Algorithmica **29** (2001), 560–581.

[3] A. Borodin and R. El-Yaniv, *Online computation and competitive analysis*, Cambridge University Press, Cambridge, 1998.

[4] E. Feuerstein and L. Stougie, *On-line single server dial-a-ride problems*, Theoretical Computer Science **268(1)** (2001), 91–105.

[5] A. Fiat and G. J. Woeginger (eds.), *Online algorithms: The state of the art*, Lecture Notes in Computer Science, vol. 1442, Springer, Berlin, 1998.

Approximation Algorithm for the Maximum Leaf Spanning Tree Problem for Cubic Graphs[*]

Krzysztof Loryś and Grażyna Zwoźniak

Institute of Computer Science, Wrocław University

Abstract. The problem of finding spanning trees with maximal number of leaves is considered. We present a linear-time algorithm for cubic graphs that achieves approximation ratio 7/4. The analysis of the algorithm uses a kind of the accounting method, that is of independent interest.

1 Introduction

Given a connected undirected graph G, the maximum leaf spanning tree problem is to find in G a spanning tree with the maximum number of leaves. Using reduction from the dominating set problem Garey and Johnson have shown [7] that this problem is NP-complete and then Galbiati, Maffioli and Morzenti established its MAX-SNP completeness [6]. Therefore there exists some constant $\epsilon > 0$ such that there is no $(1 + \epsilon)$-approximation for maximum leaf spanning tree unless $P = NP$ [2], [3].

First approximation solutions to the problem were obtained by Lu and Ravi. In [13] they presented two algorithms based on a local optimization strategy, that achieve performance ratios 3 and 5 and run in time $O(n^7)$ and $O(n^4)$, respectively. This stands in contrast with the dominating set problem, which cannot be approximated within any constant ratio. Later the same authors, using a novel approach based on a notion of leafy trees, obtained another 3-approximation algorithm for this problem that works in near linear time [15]. Recently, their results have been improved with respect to both approximation ratio and running time by Solis-Oba [16]. His algorithm is 2-approximation and works in linear time.

There are several papers that focus on finding leafy spanning trees in restricted graphs. It is known [9], [12], [17] that every connected graph with n vertices and minimum degree $k = 3$ has a spanning tree with at least $n/4 + 2$ leaves. In [12] authors gave the bounds for the number of leaves: $(2n + 8)/5$ for $k = 4$, and $(1 - \Omega(\ln k/k))n$ for arbitrary k. The last bound was improved in [5] to $(k - 5)/(k + 1)2^k + 2$ for the special case of a hypercube of dimension k. These lower bounds are typically proved by constructing spanning trees with desired number of leaves, so presented algorithms can be used to approximate the optimal solution to the maximum leaf spanning tree problem for graphs with minimum degree k.

---

[*] partially supported by Komitet Badań Naukowych, grant 8 T11C 044 19

In this paper we concentrate on cubic graphs, i.e. regular graphs whose all nodes have degree equal 3. This kind of graphs plays a special role of a "boundary" class of graphs. Very often it is the simplest class of graphs for which a problem is as difficult as in the general case (see [8] and [1]). The maximum leaf spanning tree problem is not an exception here. It was shown that it remains NP-hard even if the input is restricted to d-regular graphs for any fixed $d \geq 3$ (see [14] and [7]), so in particular to cubic graphs. There are other reasons for studying the case of cubic graphs. Despite of its restricted nature they appear in a great number of applications. Moreover, studying the problem restricted to cubic graphs helps to understand deeper the unrestricted case.

The main contribution of this paper is a 7/4-approximation algorithm working in linear time. Note that till now the ratios obtained in [15] and [16] for arbitrary graphs were also the best known for 3-regular graphs. Our algorithm looks very simple. Firstly it constructs a forest F of disjoint trees using some rules for adding new vertices, similar to those in [16]. In the second step the algorithm links the trees of F and vertices that do not belong to the forest to form a spanning tree for G. The crucial for the algorithm performance is the way in which it chooses roots of the trees in F and order in which it tests the vertices while extending the trees.

A naive analysis of the algorithm could compare the lower bound on the number of leaves of the forest F created by the algorithm with the upper bound on the number of leaves in the full binary trees (i.e. binary trees with no vertices of degree 2). We note that it would be sufficient only in very specific situation, for example when the forest F would consists of the only one tree. However, when it contains more trees we have to connect them finally and each such operation kills two leaves. Therefore we have to take into account some other factors that either increase the number of leaves in F or cause that the optimal tree cannot have as many leaves as the full binary tree. We relate these factors and reduce the analysis to proving that the loss caused by each operation of connecting trees from F can be covered by savings we have due to some of the factors.

The rest of the paper is organized as follows. In Section 3 we present our approximation algorithm. In Section 4 we show that the performance ratio of the algorithm is 7/4, and in Section 5 we give the proof of the lemma which is essential in our analysis of the algorithm. Due to lack of space, technical proofs of some facts and lemmata used in our analysis are presented in full version of this paper, available from the authors.

2 Preliminaries

Let G be a connected undirected graph. We use $V(G)$ to denote the set of vertices in G and $E(G)$ to denote the set of edges in G. For a vertex $v \in V(G)$ let $\Gamma_G(v)$ denote the set of vertices $\{w : (v, w) \in E(G)\}$. The degree of v in G, $deg_G(v)$, is the number of edges incident to v in G. We denote by $L(G)$ the set of vertices that have degree one in G. If T is a rooted tree, then we use $\text{LCA}_T(u, w)$ to denote the lowest common ancestor of vertices u, w in T.

3 Algorithm

Let G be a connected undirected cubic graph. In the first step our algorithm builds a forest F for G by using three rules (see Fig. 1). Two of them are applied to the leaves of the current tree T_i; they add vertices and edges to F in such a way that as large as possible fraction of vertices in F are leaves. Rule 1 puts to the tree T_i two vertices $u, w \notin V(F)$ adjacent to a leaf $v \in V(T_i)$ with edges (v, u) and (v, w). Rule 2 puts to the tree T_i a vertex $u \notin V(F)$ adjacent to a leaf $v \in V(T_i)$ together with two vertices $w_1, w_2 \notin V(F)$ adjacent to u; this rule adds $(v, u), (u, w_1), (u, w_2)$ to the set of edges $E(T_i)$. Both rules add two leaves to the tree. We name these leaves left and right son of their father. This allows us to impose the depth-first order in which the algorithm visits vertices trying to extend the tree (by these two rules) or to look for a root of a new tree (using rule 3 described below).

Rule 3 initiates a new tree $T_j, j > i$; if $v \in V(T_i)$, $w \in \Gamma_G(v)$, $\Gamma_G(u) = \{w, z_1, z_2\}$ and $w, u, z_1, z_2 \notin V(F)$ yet, then Rule 3 starts to build T_j rooted in w and adds u, z_1, z_2 to the set of vertices $V(T_j)$ and $(w, u), (u, z_1), (u, z_2)$ to the set of edges $E(T_j)$. We refer to T_i as the father of T_j. This relation determines a partial order in F, so we use some other related terms as e.g. ancestor of the tree.

In the second step the algorithm adds edges to connect each vertex which does not belong to F with one of its neighbours in $L(F)$ and each root of the tree in F – except the root of the first tree – with a leaf of another tree.

THE ALGORITHM

1. Construct the forest F
    (a) $F \leftarrow \emptyset$
    (b) $V(T_0) \leftarrow \{r_0, v_1, v_2, v_3\}$, where r_0 is any vertex of G and $v_1, v_2, v_3 \in \Gamma_G(r_0)$;
        $E(T_0) \leftarrow \{(r_0, v_1), (r_0, v_2), (r_0, v_3)\}$; let r_0 be a root of T_0;
        $i \leftarrow 0$
    (c) **if it is possible**
            find the leftmost leaf in T_i that can be expanded by the Rule 1 and expand it; go to the step 1c;
        else
            go to the step 1d;
    (d) **if it is possible**
            find the leftmost leaf in T_i that can be expanded by the Rule 2 and expand it; go to the step 1c
        else
            $F \leftarrow F \cup T_i$ and go to the step 1e
    (e) **if it is possible**
            find the leftmost leaf v in T_i such that Rule 3 can be applied to v and apply this rule to v; $i \leftarrow i + 1$;
            let T_i be a new tree created in this step; go to the step 1c with T_i
        else
            go to the step 1e with the father of T_i
2. Add edges to F to make it a spanning tree for G. □

Note that roots of the trees constructed by our algorithm – except the root of T_0 – have degree one in F. Hence, $L(F) = R \cup \bar{L}(F)$, where R is the set of roots of the trees $T_1, \ldots, T_k$ and $\bar{L}(F)$ is the set of leaves of $T_0, \ldots, T_k$.

Definition 1. *A vertex $v \in V(G)$ is an* exterior vertex *if $v \notin V(F)$. Let EX denote the set of all exterior vertices in G.*

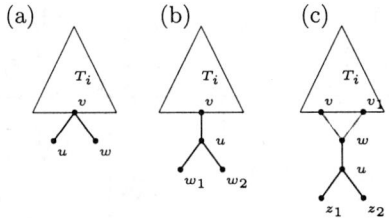

Fig. 1. (a) Rule 1. (b) Rule 2. (c) Rule 3

4 Analysis of the Algorithm

In this section we show that the performance ratio of the algorithm is at most 7/4. Let us fix a terminology used in our analysis. $F = \{T_0, \ldots, T_k\}$ is the forest built by the algorithm for a cubic graph G. In Section 4.1 we give some properties of F. In Section 4.2 we show that the number of leaves in a maximum leaf spanning tree for G (denoted by T_{opt}) is at most one-half of the number of all vertices in G. Finally, in Section 4.3 we estimate the approximation ratio.

4.1 Properties of the Forest Constructed by the Algorithm

Fact 1 *For each $T_i \in F$ there is at least one vertex $v \in V(T_i)$ such that $deg_{T_i}(v) = 3$.*

Fact 2 *Let $v \in V(T_i), T_i \in F$ and $deg_{T_i}(v) = 2$. If $x \in \Gamma_{T_i}(v)$ then $deg_{T_i}(x) = 3$.*

Fact 3 *Let $v \in V(T_i), T_i \in F$ and $deg_{T_i}(v) = 2$. If $w \in \Gamma_G(v)$ then $w \in V(T_i)$.*

Proof. Let us assume that $w \notin V(T_i)$. If $w \in EX$ or $w \in V(T_j), j > i$, then Rule 1 would have been applied to the vertex v. If $w \in V(T_j), j < i$ then Rule 2 or Rule 1 would have been applied to the vertex w. Hence, $w \in V(T_i)$. □

Fact 4 *Let $v \in V(T_i), T_i \in F$ be adjacent to the node $w \in EX$. If $u \neq w$ and $u \in \Gamma_G(v)$ then $u \in V(T_i)$.*

Fact 5 *Let $v \in V(T_i), T_i \in F$ and $u, w \in \Gamma_G(v)$. If $u, w \notin V(T_i)$ then $u, w \in V(T_j), j < i, T_j \in F$, and T_j is an ancestor of T_i, see Fig. 2(d).*

Note that if v is a root of T_i, $i > 0$, $T_i \in F$, then $deg_{T_i}(v) = 1$, so v has two neighbours in $T_j \in F$, where T_j is a father of T_i.

Fact 6 *Let $v \in V(G)$ and $u, w, z \in \Gamma_G(v)$. If $v \in EX$ then two its neighbours, let us say u, w, belong to $\bar{L}(T_i)$ for some $T_i \in F$, and the third one satisfies one of the following conditions:*

- *$z \in \bar{L}(T_i)$, see Fig. 2(a)*
- *$z \in \bar{L}(T_j), j > i, T_j \in F, T_i$ is the ancestor of T_j, see Fig. 2(b)*

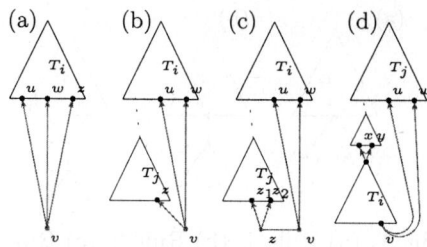

Fig. 2. (a),(b),(c) The neighbours of the exterior vertex v. (d) Two neighbours $\notin T_i$ of a vertex $\in V(T_i)$

- $z \in EX$ and its neighbours $z_1, z_2 \in \bar{L}(T_j)$, $T_j \in F$, where $i = j$, or T_i is an ancestor of T_j, or T_j is an ancestor of T_i, see Fig. 2(c).

Note that from Facts 5 and 6 we have that there are no edges or paths going through exterior vertices between vertices of trees T_i, T_j, $i \neq j$ such that neither T_j is an ancestor of T_i nor T_i is an ancestor of T_j.

From Fact 6 we have that every exterior vertex v has at least two neighbours in one tree $T_i \in F$. We call any two of them *ex-leaves*, and denote by $\bar{L}_{ex}(T_i)$ the set of all ex-leaves in T_i.

Definition 2. *A leaf $v \in \bar{L}(T_i)$, $T_i \in F$ is a* free leaf *if it is not an ex-leaf and for any $w \in \Gamma_G(v)$ we have that $w \notin R$. $\bar{L}_f(T_i)$ denotes the set of all free leaves in T_i.*

Definition 3. *An edge $e = (u, v)$ in a tree $T_i \in F$ is a* free edge, *if $deg_{T_i}(u) = deg_{T_i}(v) = 3$. Let $E_f(T_i)$ denote the set of all free edges in the tree T_i.*

4.2 Number of Leaves in the Maximum Leaf Spanning Tree

We can easily derive the upper bound on the number of leaves in the maximum leaf spanning trees for cubic graphs; it is the number of leaves of a tree that has no internal vertices of degree two.

Fact 7 *Let T be such a tree that $|V(T)| \geq 4$ and for each vertex $v \in V(T)$ $deg_T(v) = 3$ or $deg_T(v) = 1$. The number of leaves $|L(T)|$ equals $\frac{1}{2}(V(T) + 2)$.*

The number of vertices of degree two in the tree affects the number of leaves.

Lemma 1. *Let G be a cubic graph, $|V(G)| \geq 4$, and let T_{opt} be a maximum leaf spanning tree for G. If d vertices have degree 2 in T_{opt}, then $|L(T_{opt})| = \frac{1}{2}(|V(G)| - d + 2)$.*

4.3 Performance Ratio of the Algorithm

The following lemma and corollary ensures that at least one-third of the vertices in the forest F are leaves.

Lemma 2. *Let $T \in F$, where F is the forest built by the algorithm for a cubic graph G, $|V(G)| \geq 4$. Then $|V(T)| = 3|L(T)| - |E_f(T)| - 5$.*

Proof. Note that each node v such that $deg_T(v) = 2$ is adjacent to the vertices u, w such that $deg_T(u) = deg_T(w) = 3$ (see Fact 2). Let us replace each path u, v, w by a new edge (u, w) and remove v with edges adjacent to it to form the new tree T'. T' has $|V(T)| - d$ vertices and $|V(T)| - d - 1$ edges, where d is the number of vertices of degree 2 in T. Note that d equals the number of new edges in T', and there are $|L(T)|$ old edges adjacent to the leaves and $|E_f(T)|$ old edges between vertices of degree 3. Hence, $|V(T)| - d - 1 = d + |L(T)| + |E_f(T)|$, so $d = \frac{|V(T)|-1-|L(T)|-|E_f(T)|}{2}$. The number of leaves $|L(T')| = |L(T)| = \frac{1}{2}(|V(T)|-d+2)$ (see Lemma 1) so

$$2|L(T)| = |V(T)| - d + 2 = |V(T)| - \frac{|V(T)|-1-|L(T)|-|E_f(T)|}{2} + 2$$

$$3|L(T)| = |V(T)| + 5 + |E_f(T)| \quad \text{and} \quad |V(T)| = 3|L(T)| - |E_f(T)| - 5.$$

□

Corollary 1. *Let F be the forest built by the algorithm for a cubic graph G, $|V(G)| \geq 4$. If F is composed of $k+1$ disjoint subtrees $T_0, \ldots, T_k$, then $|V(F)| = 3|L(F)| - 5k - 5 - |E_f(F)|$.*

The following lemma is essential to proving Theorem 1. It says that the number of the trees in F can be bounded by a combination of free edges, free leaves and vertices of degree two in the optimal tree.

Lemma 3. *Let T_{opt} be a maximum leaf spanning tree and let F be a forest built by the algorithm for a cubic graph G, $|V(G)| \geq 4$. Then $2|E_f(F)| + |\bar{L}_f(F)| + 2d + 6 > k$, where d is the number of vertices of degree 2 in T_{opt} and $k+1$ is the number of trees in F.*

The proof of this lemma is very technical and we postpone it to the next section.

Theorem 1. *Let T_{opt} be a maximum leaf spanning tree for a cubic graph G, $|V(G)| \geq 4$, and let T be a spanning tree for G constructed by the algorithm. Then*

$$\frac{|L(T_{opt})|}{|L(T)|} < \frac{7}{4}.$$

Proof. The algorithm constructs the tree T with $|L(F)| - 2k$ leaves, where $k+1$ is the number of trees in the forest F. If d vertices have degree 2 in T_{opt}, then the number of leaves $|L(T_{opt})| = \frac{1}{2}(V(G) - d + 2)$ (see Lemma 1). Moreover,

$V(G) = |V(F)| + |EX|$. We know that $|V(F)| = 3|L(F)| - 5k - 5 - |E_f(F)|$ (see Corollary 1). Let l_r be the number of leaves in F adjacent in G to the roots in R. $|R| = k$, so $l_r = 2k$ (see Fact 5).

$$|L(F)| = |\bar{L}_{ex}(F)| + |\bar{L}_f(F)| + |R| + l_r = |\bar{L}_{ex}(F)| + |\bar{L}_f(F)| + 3k.$$

It follows from definition of $\bar{L}_{ex}(F)$ that $|\bar{L}_{ex}(F)| = 2|EX|$. Hence,

$$|EX| = \frac{1}{2}|\bar{L}_{ex}(F)| = \frac{L(F)}{2} - \frac{|\bar{L}_f(F)|}{2} - \frac{3}{2}k$$

$$\frac{|L(T_{opt})|}{|L(T)|} = \frac{1}{2} \frac{3|L(F)| - 5k - 5 - |E_f(F)| + \frac{L(F)}{2} - \frac{|\bar{L}_f(F)|}{2} - \frac{3}{2}k - d + 2}{|L(F)| - 2k}$$

$$= \frac{1}{4} \frac{7|L(F)| - 13k - 2|E_f(F)| - |\bar{L}_f(F)| - 2d - 6}{|L(F)| - 2k}.$$

From Lemma 3 we have that $2|E_f(F)| + |\bar{L}_f(F)| + 2d + 6 > k$, so $\frac{|L(T_{opt})|}{|L(T)|} < \frac{7}{4}$.
□

5 The Proof of Lemma 3

We have to show that $2|E_f(F)| + |\bar{L}_f(F)| + 2d + 6 > k$ where d is the number of vertices that have degree two in T_{opt}.

Note that k is the number of edges added to F in the second step of the algorithm to connect the trees of F. It is also the number of such pairs of leaves in F which are connected by edges from outside F with roots from the set R. We will assign weights to the free edges, free leaves and vertices which have degree two in T_{opt}. Then we will partition these weights and move parts of them to some leaves of F. Finally, we will show that each pair of leaves connected with root from R received weight at least 1, so the sum of weights moved to these leaves is greater than k.

Let us assign weights to the free edges and free leaves of F, and some additional weights to the root and one edge of T_0, in the following way:

1. $weight(r_0) = 2$
2. $weight(e_0) = 4$ where e_0 - the leftmost edge incident to r_0
3. $weight(e) = 2$ if $e \neq e_0$ is a free edge
4. $weight(v) = 1$ if v is a free leaf
5. $weight(w) = 2$ if $deg_{T_{opt}}(w) = 2$, $w \neq r_0$

Note that

$$2|E_f(F)| + |\bar{L}_f(F)| + 2d + 6 \geq \sum_{e \in E_f(F), e \neq e_0} weight(e) +$$

$$+ \sum_{v \in \bar{L}_f(F)} weight(v) + \sum_{w: deg_{T_{opt}}(w) = 2} weight(w) + weight(r_0) + weight(e_0).$$

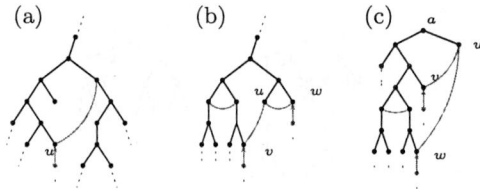

Fig. 3. (a) u and w are best friends. (b) v is the only friend of u. (c) v and w are friends of u

5.1 Some Definitions

Recall that every tree in the forest F is a rooted tree. Our proof takes advantage of the F's structure. We introduce some definitions which characterize how some vertices of F are connected in G. Let T be a tree in F.

Definition 4. *Two vertices $u, w \in V(T)$ are* twins *if they have the same father in T.*

Definition 5. *Vertices $f, u, w \in V(T)$ form a* triangle $\Delta(f, u, w)$ *if u, w are twins, f is their father in T and $(u, w) \in E(G)$.*

Definition 6. *An edge $(u, w) \in E(G)$ is an* arrow *if $u \in R \cup EX$ and $w \in \bar{L}(F)$. A pair of arrows is a pair of edges $(u, w), (u, v) \in E(G)$ where $u \in R \cup EX$ and $v, w \in \bar{L}(T)$.*

Note that if (u, w) is an arrow, then $(u, w) \notin E(F)$.

Definition 7. *Two vertices $u, w \in V(T)$ are* colleagues *if u, w are not twins and $(u, w) \in E(G) \setminus E(T)$.*

Definition 8. *Two vertices $u, w \in V(T)$ are* best friends *if u, w are colleagues and for $x \in \{u, w\}$ we have that $\deg_T(x) = 2$ or $x \in \bar{L}(T)$ is the endpoint of an arrow, see Fig. 3(a).*

Two following definitions concern connections between vertices that belong to the set of leaves $\bar{L}(T)$.

Definition 9. *A vertex $v \in \bar{L}(T)$ is the* only friend *of a free leaf $u \in \bar{L}(T)$ if*

1. *v is the endpoint of an arrow and u, v are colleagues*
2. *there is no vertex $w \in \bar{L}(T), w \neq v$, which satisfies the first condition.*

Definition 10. *Vertices $v, w \in \bar{L}(T)$ are* friends *of a free leaf $u \in \bar{L}(T)$ if v, w are endpoints of arrows, u, v are colleagues and u, w are colleagues.*

The following lemmata show that there is a free edge for each pair of best friends and for each pair of friends of a free leaf.

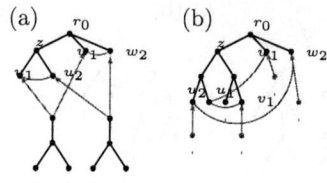

Fig. 4. r_0 as the endpoint of a free edge (r_0, z)

Lemma 4. *Let $u, w \in V(T)$ be best friends, $a = LCA_T(u, w)$ and let u lies on the left from v in T. Then w is the right son of a in T and the edge (w', a) is a free edge, where w' is the left son of a.*

Lemma 5. *Let $v, w \in V(T)$ be friends of a free leaf $u \in \bar{L}(T)$. Let $x \in \{u, w, v\}$ be the vertex which was added to T as a first one. Then $(x', a) \in E(T)$ is a free edge, where a is a father of x in T and x' is a twin of x in T.*

5.2 Distribution of Weights of the Free Edges

Let us consider a vertex $v \in V(T)$ which is the endpoint of an arrow. Recall that the leftmost edge incident to the root of r_0 has $weight(e_0) = 4$ and every free edge $e \neq e_0$ has $weight(e) = 2$.

1. If there is a vertex $w \in V(T)$, such that v, w are best friends, z is the left son of a, where $a = LCA_T(v, w)$, then $e_f = (z, a)$ is a free edge (see Lemma 4); we assign weight to v: $weight(v) := \frac{1}{2} weight(e_f) = 1$, if $e_f \neq e_0$, and $weight(v) := \frac{1}{4} weight(e_f) = 1$, if $e_f = e_0$;
2. If there are vertices $w, u \in \bar{L}(T)$ such that v, w are friends of a free leaf u, $x \in \{u, v, w\}$ is the vertex which was added to T as a first one, x' is a twin of x in T and a is their father in T, then $e_f = (x', a)$ is a free edge (see Lemma 5); we assign weight to v: $weight(v) := \frac{1}{2} weight(e_f) = 1$, if $e_f \neq e_0$, and $weight(v) := \frac{1}{4} weight(e_f) = 1$, if $e_f = e_0$.

Note that e_0, the leftmost edge incident to r_0, could give weights to at most four vertices, because the leftmost son of r_0 has two twins, see Fig. 4. Each of the remaining free edges could give weights to at most two vertices, because each vertex $a \neq r_0$ has at most two children in T, so the leftmost son of $a \neq r_0$ has no more than one twin.

5.3 Breaking T_{opt}

Now we would like to define some particular areas of leaves in T_{opt}.

Definition 11. *A triangle $\Delta(f, u, w)$ breaks T_{opt} if for $x \in \{f, u, w\}$ we have $deg_{T_{opt}}(x) \in \{1, 3\}$ and $f \neq r_0$.*

Definition 12. *A pair of arrows* (u, w), (u, v) *breaks* T_{opt} *if both w and v are in triangles that break* T_{opt} *and* $deg_{T_{opt}}(u) \in \{1, 3\}$.

The condition about u's degree in T_{opt} has nothing to do with intuitive definition of breaking the tree, but it simplifies the analysis of the algorithm.

5.4 Distribution of Weights of Vertices Which Have Degree Two

Recall that every vertex w of degree two in T_{opt} has $weight(w) = 2$, and the root of T_0 has $weight(r_0) = 2$.

Let us consider a vertex $v \in V(T)$ which is the endpoint of an arrow and a twin in a triangle $\Delta(f, v, w)$. $\Delta(f, v, w)$ does not break T_{opt} iff $f = r_0$ or $deg_{T_{opt}}(f) = 2$ or $deg_{T_{opt}}(v) = 2$ or $deg_{T_{opt}}(w) = 2$. In this case we assign weight to v in the following way:

1. if $deg_{T_{opt}}(v) = 2$ then $weight(v) := \frac{1}{2}weight(v) = 1$
2. else if $deg_{T_{opt}}(w) = 2$ then $weight(v) := \frac{1}{2}weight(w) = 1$
3. else if $deg_{T_{opt}}(f) = 2$ or $f = r_0$ then $weight(v) := \frac{1}{2}weight(f) = 1$

It is easy to see that in the special case when we have two triangles with a father r_0, only two vertices may be the endpoints of the arrows. In the other cases a vertex $v \in V(T)$ may be a twin in exactly one triangle, and f may be a father in exactly one triangle. Hence, a weight of $x \in \{v, w, f\}$ may be distributed to at most two vertices.

In the case when $\Delta(f, v, w)$ breaks T_{opt} and $deg_{T_{opt}}(r) = 2$, where $r \in R$ is the endpoint of an arrow (r, v), then $weight(v) := \frac{1}{2}weight(r) = 1$. Note that weight of r may be distributed between two endpoints of the same pair of arrows.

5.5 Pairs of Arrows Which Break T_{opt}

In the case when a pair of arrows breaks T_{opt} we will find weights for its endpoints in some free leaf or in the endpoints of another pair of arrows.

Lemma 6. *One can uniquely assign to each pair of arrows* (r, w), (r, v), $r \in R$ *that breaks* T_{opt}

- *a free leaf u' which has two neighbours in another tree or*
- *a pair of arrows (u', w'), (u', v'), $u' \in EX$ that does not break* T_{opt}.

5.6 Distribution of Weights of the Free Leaves

Now we can move weights of the free leaves. Recall that every free leaf u has $weight(u) = 1$. Let us consider a vertex $v \in V(T)$ which is the endpoint of an arrow.

1. If there is a free leaf $u \in \bar{L}(T)$ such that $v \in V(T)$ is the only friend of $u \in \bar{L}(T)$, then we assign to v the weight of u; $weight(v) := weight(u) = 1$.

2. If v is the endpoint of the pair of arrows that breaks T_{opt}, then we can give v a half of the weight of u, where u is a free leaf assigned to this pair of arrows (see Lemma 6); $weight(v) := \frac{1}{2}weight(u) = \frac{1}{2}$.

Note that if v is the only friend of u, then by definition there is no vertex $w \in V(T)$, $w \neq v$ which is the only friend of u. Note also that if $u \in V(T_i)$, $i \in \{1, \ldots, k\}$ is a free leaf assigned to the pair of arrows that breaks T_{opt}, then u has two neighbours in T_j, $j < i$, $j \in \{0, \ldots, k-1\}$, so there is no vertex $w \in V(T_i)$ which is the only friend of u. In this case weight of u is distributed between two endpoints of the same pair of arrows.

5.7 Counting Weights

The following lemma shows how an endpoint of an arrow may be connected with other vertices in the same tree.

Lemma 7. *Let (u, w) be an arrow, $u \in R \cup EX$, $w \in \bar{L}(T)$. Then one of the following conditions is true:*

1. *w is a twin in a triangle*
2. *there is a vertex $v \in V(T)$, such that w and v are best friends*
3. *there are vertices $v, z \in \bar{L}(T)$, where v is a free leaf, such that w and z are friends of v*
4. *there is a free leaf $v \in \bar{L}(T)$, such that w is the only friend of v*

Now we are ready to prove Lemma 3 as follows.

Proof. (of Lemma 3) Let us consider a pairs of arrows (u_i, w_i), (u_i, v_i), $u_i \in R \cap T_i$, $i = 1, \ldots, k$. If w_i is a twin in a triangle and v_i is a twin in a triangle and the pair of arrows (u_i, w_i), (u_i, v_i) breaks T_{opt}, then there is a free leaf assigned to this pair, or a pair of arrows (u'_i, w'_i), (u'_i, v'_i), $u'_i \in EX$ that does not break the tree T_{opt}. If x_i is a free leaf assigned to (u_i, w_i), (u_i, v_i), then $weight(v_i) = weight(w_i) = \frac{1}{2}$ (see Section 5.6). Otherwise we consider vertices u'_i, w'_i, v'_i and move weights assigned to w'_i, v'_i to the vertices w_i, v_i; to simplify the analysis let us rename (in this proof only) vertices u'_i, w'_i, v'_i to u_i, w_i, v_i.

If (u_i, w_i), (u_i, v_i) does not break T_{opt}, then (1) at least one of $\{u_i, v_i\}$ is not in the triangle which breaks T_{opt}, so its weight is at least 1 (see Lemma 7 and Sections 5.2, 5.6, 5.4) or (2) $deg_{T_{opt}}(u_i) = 2$; in this case $weight(w_i) = weight(v_i) = \frac{1}{2}weight(u_i) = 1$ (see Section 5.4). So in this case at least one endpoint from each pair of arrows has weight at least 1. Let us assign $weight(v_i) := 0$ and $weight(w_i) := 0$ to these endpoints of pairs of arrows (u_i, w_i), (u_i, v_i), $u_i \in R \cap T_i$, $i = 1, \ldots, k$ that have not got weights yet.

$$\sum_{e \in E_f(F), e \neq e_0} weight(e) + \sum_{v \in \bar{L}_f(F)} weight(v) + \sum_{w: deg_{T_{opt}}(w) = 2} weight(w) + weight(r_0) +$$

$$+ weight(e_0) > \sum_{v_i} weight(v_i) + \sum_{w_i} weight(w_i) \geq k. \qquad \square$$

6 Conclusions

Our algorithm can be easily extended to d-regular graphs with $d \geq 4$. However, it seems that simple modifications of its analysis would be insufficient as they do not prevent from considering a huge number of specific cases. Therefore finding approximation algorithms with a ratio less than 2 remains a problem for $d \geq 4$.

References

[1] P. Alimonti, V. Kann, *Some APX-completeness results for cubic graphs*, Theoretical Computer Science, 237, pages 123–134, 2000.

[2] S. Arora, C. Lund, R. Motwani, M. Sudan, M. Szegedy *Proof verification and the hardness of approximation problems*, Proceedings of the Thirty-third Annual IEEE Symposium on Foundations of Computer Science, pages 14–23, 1992.

[3] S. Arora, S. Safra *Probabilistic checking of proofs: A new characterization of NP*, Proceedings of the Thirty-third Annual IEEE Symposium on Foundations of Computer Science, pages 2–13, 1992.

[4] E. W. Dijkstra *Self-stabilizing systems in spite of distributed control*, Communications of ACM 17, pages 643–644, 1974.

[5] W. Duckworth, P. E. Dunne, A. M. Gibbons, M. Zito *Leafy spanning trees in hypercubes*, Technical Report CTAG-97008, University of Liverpool, 1997.

[6] G. Galbiati, F. Maffioli, A. Morzenti *A short note on the approximability of the maximum leaves spanning tree problem*, Information Processing Letters 52, pages 45–49, 1994.

[7] M. R. Garey, D. S. Johnson *Computers and Intractability: A guide to the theory of NP-completeness*, W. H. Freeman, San Francisco 1979.

[8] R. Greenlaw, R. Petreschi *Cubic graphs*, ACM Computing Surveys, 27, pages 471-495, 1995.

[9] J. R. Griggs, D. J. Kleitman, A. Shastri *Spanning trees with many leaves in cubic graphs*, Journal of Graph Theory 13, pages 669–695, 1989.

[10] S. Guha, S. Khuller *Approximation algorithms for connected dominating sets*, Proceedings of Fourth Annual European Symposium on Algorithms, pages 179–193, 1996.

[11] S. Guha, S. Khuller *Improved methods for approximating node weight Steiner trees and connected dominating sets*, Technical Report CS-TR-3849, University of Maryland, 1997.

[12] D. J. Kleitman, D. B. West *Spanning trees with many leaves*, SIAM Journal of Discrete Mathematics 4, pages 99–106, 1991.

[13] H. Lu, R. Ravi *The power of local optimization: approximation algorithms for maximum-leaf spanning tree*, Proceedings of the Thirtieth Annual Allerton Conference on Communication, Control and Computing, pages 533–542, 1992.

[14] P. Lemke *The maximum leaf spanning tree problem for cubic graphs is NP-complete*, IMA Preprint Series nr. 428, Minneapolis, 1988.

[15] H. Lu, R. Ravi *A near-linnear time approximation algorithm for maximum leaf spanning tree*, Journal of Algorithms, Vol. 29, No. 1, pages 132–141, 1998.

[16] R. Solis-Oba *2-approximation algorithm for finding a spanning tree with maximum number of leaves*, Proceedings on the 6th Annual European Symposium on Algorithms, LNCS 1461, pages 441–452, 1998.

[17] J. A. Storer *Constructing full spanning trees for cubic graphs*, Information Processing Letters 13, pages 8–11, 1981.

Engineering a Lightweight Suffix Array Construction Algorithm
(Extended Abstract)*

Giovanni Manzini[1,2] and Paolo Ferragina[3]

[1] Dipartimento di Informatica, Università del Piemonte Orientale
I-15100 Alessandria, Italy
manzini@mfn.unipmn.it
[2] Istituto di Informatica e Telematica, CNR
I-56100 Pisa, Italy
[3] Dipartimento di Informatica, Università di Pisa
I-56100 Pisa, Italy
ferragin@di.unipi.it

1 Introduction

We consider the problem of computing the *suffix array* of a text $T[1,n]$. This problem consists in sorting the suffixes of T in lexicographic order. The suffix array [16] (or PAT array [9]) is a simple, easy to code, and elegant data structure used for several fundamental string matching problems involving both linguistic texts and biological data [4, 11]. Recently, the interest in this data structure has been revitalized by its use as a building block for three novel applications: **(1)** the Burrows-Wheeler compression algorithm [3], which is a provably [17] and practically [20] effective compression tool; **(2)** the construction of succinct [10, 19] and compressed [7, 8] indexes; the latter can store both the input text and its full-text index using roughly the same space used by traditional compressors for the text alone; and **(3)** algorithms for clustering and ranking the answers to user queries in web-search engines [22]. In all these applications the construction of the suffix array is the computational bottleneck both in time and space. This motivated our interest in designing *yet another* suffix array construction algorithm which is fast and "lightweight" in the sense that it uses small space.

The suffix array consists of n integers in the range $[1,n]$. This means that in theory it uses $\Theta(n \log n)$ bits of storage. However, in most applications the size of the text is smaller than 2^{32} and it is customary to store each integer in a four byte word; this yields a total space occupancy of $4n$ bytes. For what concerns the cost of constructing the suffix array, the theoretically best algorithms run in $\Theta(n)$ time [5]. These algorithms work by first building the suffix tree and then obtaining the sorted suffixes via an in-order traversal of the tree. However, suffix tree construction algorithms are both complex and space consuming since they occupy at least $15n$ bytes of working space (or even more, depending on the

---

* Partially supported by Italian MIUR project on "Technologies and services for enhanced content delivery".

text structure [14]). This makes their use impractical even for moderately large texts. For this reason, suffix arrays are usually built using algorithms which run in $O(n \log n)$ time but have a smaller space occupancy. Among these algorithms the current "leader" is the qsufsort algorithm by Larsson and Sadakane [15]. qsufsort uses $8n$ bytes[1] and despite the $O(n \log n)$ worst case bound it is faster than the algorithms based on suffix tree construction.

Unfortunately, the size of our documents has grown much more quickly than the main memory of our computers. Thus, it is desirable to build a suffix array using as small space as possible. Recently, Itoh and Tanaka [12] and Seward [21] have proposed two new algorithms which only use $5n$ bytes. From the theoretical point of view these algorithms have a $\Theta(n^2 \log n)$ worst case complexity. In practice they are faster than qsufsort when the average LCP is small (the LCP is the length of the longest common prefix between two consecutive suffixes in the suffix array). However, for texts with a large average LCP these algorithms can be slower than qsufsort by a factor 100 or more.[2]

In this paper we describe and extensively test a new lightweight suffix sorting algorithm. Our main idea is to use a very small amount of extra memory, in addition to $5n$ bytes, to avoid the degradation in performance when the average LCP is large. To achieve this goal we make use of engineered algorithms and *ad hoc* data structures. Our algorithm uses $5n + cn$ bytes, where c is a user tunable parameter (in our tests c was at most 0.03). For files with average LCP smaller than 100 our algorithm is faster than Seward's algorithm and roughly two times faster than qsufsort. The best algorithm in our tests uses $5.03n$ bytes and is faster than qsufsort for all files except for the one with the largest average LCP.

2 Definitions and Previous Results

Let $T[1, n]$ denote a text over the alphabet Σ. The suffix array [16] (or PAT array [9]) for T is an array $SA[1, n]$ such that $T[SA[1], n], T[SA[2], n]$, etc. is the list of suffixes of T sorted in lexicographic order. For example, for $T = $ babcc then $SA = [2, 1, 3, 5, 4]$ since $T[2, 5] = $ abcc is the suffix with lower lexicographic rank, followed by $T[1, 5] = $ babcc, followed by $T[3, 5] = $ bcc and so on.[3]

Given two strings v, w we write LCP(v, w) to denote the length of their longest common prefix. The average LCP of a text T is defined as the average length of the longest common prefix between two consecutive suffixes. The average LCP is a rough measure of the difficulty of sorting the suffixes: if the average LCP is large we need in principle to examine "many" characters in order to establish the relative order of two suffixes.

---

[1] Here and in the following the space occupancy figures include the space for the input text, for the suffix array, and for any auxiliary data structure used by the algorithm.
[2] This figure refers to Seward algorithm [21]. We are in the process of acquiring the code of the Itoh-Tanaka algorithm and we hope we will be able to test it in the final version of the paper.
[3] Note that to define the lexicographic order of the suffixes it is customary to append at the end of T a special end-of-text symbol which is smaller than any symbol in Σ.

Table 1. Files used in our experiments sorted in order of increasing average LCP

Name	Ave. LCP	Max. LCP	File size	Description
bible	13.97	551	4,047,392	The file bible of the Canterbury corpus
e.coli	17.38	2,815	4,638,690	The file E.coli of the Canterbury corpus
wrld	23.01	559	2,473,400	The file world192 of the Canterbury corpus
sprot	89.08	7,373	109,617,186	Swiss prot database (sprot34.dat)
rfc	93.02	3,445	116,421,901	Concatenation of RFC text files
howto	267.56	70,720	39,422,105	Concatenation of Linux Howto text files
reuters	282.07	26,597	114,711,151	Reuters news in XML format
linux	479.00	136,035	116,254,720	Linux kernel 2.4.5 source files (tar archive)
jdk13	678.94	37,334	69,728,899	html and java files from the JDK 1.3 doc.
chr22	1,979.25	199,999	34,553,758	Assembly of human chromosome 22
gcc	8,603.21	856,970	86,630,400	gcc 3.0 source files (tar archive)

Since this is an "algorithmic engineering" paper we make the following assumptions which correspond to the situation most often faced in practice. We assume $|\Sigma| \leq 256$ and that each alphabet symbol is stored in one byte. Hence, the text $T[1, n]$ takes precisely n bytes. Furthermore, we assume that $n \leq 2^{32}$ and that the starting position of each suffix is stored in a four byte word. Hence, the suffix array $SA[1, n]$ takes precisely $4n$ bytes. In the following we use the term "lightweight" to denote a suffix sorting algorithm which use $5n$ bytes plus some small amount of extra memory (we are intentionally giving an informal definition). Note that $5n$ bytes are just enough to store the input text T and the suffix array SA. Although we do not claim that $5n$ bytes are indeed required, we do not know of any algorithm using less space.

For testing the suffix array construction algorithms we use the collection of files shown in Table 1. These files contain different kind of data in different formats; they also display a wide range of sizes and of average LCP's.

2.1 The Larsson-Sadakane qsufsort Algorithm

The qsufsort algorithm [15] is based on the doubling technique introduced in [13] and first used for the construction of the suffix array in [16]. Given two strings v, w and $t > 0$ we write $v <_t w$ if the length-t prefix of v is lexicographically smaller than the length-t prefix of w. Similarly we define the symbols $\leq_t, =_t$ and so on. Let s_1, s_2 denote two suffixes and assume $s_1 =_t s_2$ (that is, $T[s_1, n]$ and $T[s_2, n]$ have a length-t common prefix). Let $\hat{s}_1 = s_1 + t$ denote the suffix $T[s_1 + t, n]$ and similarly let $\hat{s}_2 = s_2 + t$. The fundamental observation of the doubling technique is that

$$s_1 \leq_{2t} s_2 \iff \hat{s}_1 \leq_t \hat{s}_2. \tag{1}$$

In other words, we can derive the $\leq_{2t}$ order between s_1 and s_2 by looking at the rank of $\hat{s}_1$ and $\hat{s}_2$ in the $<_t$ order.

The algorithm qsufsort works in rounds. At the beginning of the ith round the suffixes are already sorted according to the $\leq_{2^i}$ ordering. In the ith round the algorithm looks for groups of suffixes sharing the first 2^i characters and sorts them according to the $\leq_{2^i}$ ordering using Bentley-McIlroy ternary quicksort [1]. Because of (1) each comparison in the quicksort algorithm takes $O(1)$ time. After at most $\log n$ rounds all the suffixes are sorted. Thanks to a very clever data organization qsufsort only uses $8n$ bytes. Even more surprisingly, the whole algorithm fits in two pages of clean and elegant C code.

The experiments reported in [15] show that qsufsort outperforms other suffix sorting algorithm based on either the doubling technique or the suffix tree construction. The only algorithm which runs faster than qsufsort, but only for files with average LCP less than 20, is the Bentley-Sedgewick multikey quicksort [2]. Multikey quicksort is a *direct comparison* algorithm since it considers the suffixes as ordinary strings and sorts them via a character-by-character comparison without taking advantage of their special structure.

2.2 The Itoh-Tanaka two-stage Algorithm

In [12] Itoh and Tanaka describe a suffix sorting algorithm called two-stage suffix sort (two-stage from now on). two-stage only uses the text T and the suffix array SA for a total space occupancy of $5n$ bytes. To describe how it works, let us assume $\Sigma = \{a, b, \ldots, z\}$. Using counting sort, two-stage initially partitions the suffixes into $|\Sigma|$ *buckets* $B_a, \ldots, B_z$ according to their first character. Note that a bucket is nothing more than a set of consecutive entries in the array SA which now is sorted according to the $\leq_1$ ordering. Within each bucket two-stage distinguishes between two types of suffixes: Type A suffixes in which the second character of the suffix is smaller than the first, and Type B suffixes in which the second character is larger than or equal to the first suffix character. The crucial observation of algorithm two-stage is that when all Type B suffixes are sorted we can derive the ordering of Type A suffixes. This is done with a single pass over the array SA; when we meet suffix $T[i, n]$ we look at suffix $T[i-1, n]$: if it is a Type A suffix we move it to the first empty position of bucket $B_{T[i-1]}$.

Type B suffixes are sorted using textbook string sorting algorithms: in their implementation the authors use MSD radix sort [18] for sorting large groups of suffixes, Bentley-Sedgewick multikey quicksort for medium size groups, and insertion sort for small groups. Summing up, two-stage can be considered an "advanced" direct comparison algorithm since Type B suffixes are sorted by direct comparison whereas Type A suffixes are sorted by a much faster procedure which takes advantage of the special structure of the suffixes.

In [12] the authors compare two-stage with three direct comparison algorithms (quicksort, multikey quicksort, and MSD radix sort) and with an earlier version of qsufsort. two-stage turns out to be roughly 4 times faster than quicksort and MSD radix sort, and 2 to 3 times faster than multikey quicksort and qsufsort. However, the files used for the experiments have an average LCP of at most 31, and we know that the advantage of doubling algorithms (like qsufsort) become apparent for much larger average LCP's.

2.3 Seward copy Algorithm

Independently of Itoh and Tanaka, in [21] Seward describes a lightweight algorithm, called copy, which is based on a concept similar to the Type A/Type B suffixes used by algorithm two-stage.

Using counting sort, copy initially sorts the array SA according to the $\leq_2$ ordering. As before we use the term *bucket* to denote the contiguous portion of SA containing a set of suffixes sharing the same first character. Similarly, we use the term *small bucket* to denote the contiguous portion of SA containing suffixes sharing the first two characters. Hence, there are $|\Sigma|$ buckets each one consisting of $|\Sigma|$ small buckets. Note that one or more (small) buckets can be empty.

copy sorts the buckets one at a time starting with the one containing the fewest suffixes, and proceeding up to the largest one. Assume for simplicity that $\Sigma = \{a, b, \ldots, z\}$. To sort a bucket, let us say bucket B_p, copy sorts the small buckets $b_{pa}, b_{pb}, \ldots, b_{pz}$. The crucial point of algorithm copy is that when bucket B_p is completely sorted, with a simple pass over it copy sorts all the small buckets $b_{ap}, b_{bp}, \ldots, b_{zp}$. These small buckets are marked as sorted and therefore copy will skip them when their "parent" bucket is sorted. As a further improvement, Seward shows that even the sorting of the small bucket b_{pp} can be avoided since its ordering can be derived from the ordering of the small buckets $b_{pa}, \ldots, b_{po}$ and $b_{pq}, \ldots, b_{pz}$. This trick is extremely effective when working on files containing long runs of identical characters.

Algorithm copy sorts the small buckets using Bentley-McIlroy ternary quicksort. During this sorting the suffixes are considered atomic, that is, each comparison consists of the complete comparison of two suffixes. The standard trick of sorting the larger side of the partition last and eliminating tail recursion ensures that the amount of space required by the recursion stack grows, in the worst case, logarithmically with the size of the input text. In [21] Seward compares a tuned implementation of copy with the qsufsort algorithm on a set of files with average LCP up to 400. In these tests copy outperforms qsufsort for all files but one. However, Seward reports that copy is much slower than qsufsort when the average LCP exceeds a thousand, and for this reason he suggests the use of qsufsort as a fallback when the average LCP is large.[4]

Since the source code of both qsufsort and copy is available[5], we have tested both algorithms on our suite of test files which have an average LCP ranging from 13.97 to 8603.21 (see Table 1). The results of our experiments are reported in the top two rows of Table 2 (for a AMD Athlon processor) and Table 3 (for a Pentium III processor). In accordance with Seward's results, copy is faster than

---

[4] In [21] Seward describes another algorithm, called cache, which is faster than copy for files with larger average LCP. However, algorithm cache uses $6n$ bytes.

[5] Algorithm copy was originally conceived to split the input file into 1MB blocks. We modified it to allow the computation of the suffix array for the whole file.

qsufsort when the average LCP is small, and it is slower when the average LCP is large. The turning point appears to be when the average LCP is in the range 100-250. However, this is not the complete story. For example for all files the running time of qsufsort on the Pentium is smaller than the running time for the Athlon; this is not true for copy (see for example files *jdk13* and *gcc*). We conjecture that a difference in the cache architecture and behavior could explain this difference, and we plan to investigate it in the full paper (see also Section 4). We can also see that the difference in performance between the two algorithms does not depend on the average LCP alone. The DNA file *chr22* has a very large average LCP, nevertheless the two algorithms have similar running times. The file *linux* has a much greater average LCP than *reuters* and roughly the same size. Nevertheless, the difference in the running times between qsufsort and copy is smaller for *linux* than for *reuters*.

The most striking data in Tables 2 and 3 are the running times for *gcc*: for this file algorithm copy is 150-200 times slower than qsufsort. This is not acceptable since *gcc* is not a pathological file built to show the weakness of copy, on the contrary it is a file downloaded from a very busy site and we can expect that there are other files like it on our computers.[6] In the next section we describe a new algorithm which uses several techniques for avoiding such catastrophic behavior and at the same time retaining the nice features of copy: the $5n$ bytes space occupancy and the good performance for files with moderate average LCP.

3 Our Contribution: Deep-Shallow Suffix Sorting

Our starting point for the design of an efficient suffix array construction algorithm is Seward copy algorithm. Within this algorithm we replace the procedure used for sorting the small buckets (i.e. the groups of suffixes having the first two characters in common). Instead of using Bentley-McIlroy ternary quicksort we use a more sophisticated technique. More precisely, we sort the small buckets using Bentley-Sedgewick multikey quicksort and we stop the recursion when we reach a predefined depth L (that is, when we have to sort a group of suffixes with a length-L common prefix). At this point we switch to a different string sorting algorithm. This approach has several advantages: **(1)** it provides a simple and efficient mean to detect the groups of suffixes with a long common prefix; **(2)** because of the limit L, the size of the recursion stack is bounded by a predefined constant which is independent of the size of the input text and can be tuned by the user; **(3)** if the suffixes in the small bucket have common prefixes which never exceed L, all the sorting is done by multikey quicksort which is an extremely efficient string sorting algorithm.

We call this approach to suffix sorting *deep-shallow* sorting since we mix an algorithm for sorting suffixes with small LCP (*shallow sorter*) with an algorithm (actually more than one, as we shall see) for sorting suffixes with large LCP (*deep*

---

[6] As we have already pointed out, algorithm copy was conceived to work on blocks of data of size at most 1MB. The reader should be aware that we are using an algorithm outside its intended domain!

Table 2. Running times (in seconds) for a 1400 MHz AMD Athlon processor, with 1GB main memory and 256Kb L2 cache. The operating system was Debian GNU/Linux Debian 2.2. The compiler was gcc ver. 2.95.2 with options -O3 -fomit-frame-pointer. The table reports (user + system) time averaged over five runs. The running times do not include the time spent for reading the input files

	bible	e.coli	wrld	sprot	rfc	howto	reuters	linux	jdk13	chr22	gcc
qsufsort	5.44	5.21	3.27	313.41	321.15	80.23	391.08	262.26	218.14	55.08	199.63
copy	3.36	4.55	1.69	228.47	201.06	68.25	489.75	297.26	450.93	48.69	28916.45
ds0 $L=500$	2.64	3.21	1.29	157.54	139.71	50.36	294.57	185.79	227.94	33.38	2504.98
ds0 $L=1000$	2.57	3.22	1.29	157.04	140.11	50.26	292.25	185.00	235.74	33.27	2507.15
ds0 $L=2000$	2.66	3.23	1.29	157.00	139.93	50.30	292.35	185.46	237.75	33.29	2511.50
ds0 $L=5000$	2.66	3.23	1.31	156.90	139.87	50.36	291.47	185.53	239.48	33.23	2538.78
ds1 $L=200$	2.51	3.21	1.29	169.68	149.12	41.76	301.10	150.35	148.14	33.44	343.02
ds1 $L=500$	2.51	3.22	1.28	161.94	147.35	40.62	309.97	140.85	177.28	33.32	295.70
ds1 $L=1000$	2.51	3.22	1.29	157.60	145.12	40.52	298.23	138.11	202.28	33.27	289.40
ds1 $L=2000$	2.50	3.19	1.27	157.19	140.93	41.10	291.18	139.06	202.30	33.18	308.41
ds1 $L=5000$	2.51	3.18	1.28	157.09	139.73	42.76	289.95	145.74	212.77	33.21	372.35
ds2 $d=500$	2.64	3.19	1.35	157.09	139.34	37.48	292.22	121.15	164.97	33.33	230.64
ds2 $d=1000$	2.55	3.19	1.28	157.33	139.14	38.50	284.50	124.07	184.86	33.30	242.99
ds2 $d=2000$	2.50	3.18	1.27	156.93	139.81	39.67	286.56	128.26	191.71	33.25	266.27
ds2 $d=5000$	2.51	3.19	1.28	157.05	139.65	41.94	289.78	137.08	210.01	33.31	332.55

sorter). In the next sections we describe several deep sorting strategies, i.e. algorithms for sorting suffixes which have a length-L common prefix.

3.1 Blind Sorting

Let $s_1, s_2, \ldots, s_m$ denote a group of m suffixes with a length-L common prefix that we need to deep-sort. If m is small (we will discuss later what this means) we sort them using an algorithm, called *blind sort*, which is based on the *blind trie* data structure introduced in [6, Sect. 2.1] (see Fig. 1). Blind sorting simply consists in inserting the strings $s_1, \ldots, s_m$ one at a time in an initially empty blind trie; then we traverse the trie from left to right thus obtaining the strings sorted in lexicographic order.

The insertion of string s_i in the trie requires a first phase in which we scan s_i and simultaneously traverse the trie until we reach a leaf ℓ. Then we compare s_i with the string associated to leaf ℓ and we determine the length of their common prefix. Finally, we update the trie adding the leaf corresponding to s_i (see [6] for details and for the merits of blind tries with respect to standard compacted tries). Obviously in the construction of the trie we ignore the first L characters of each suffix because they are identical.

Our implementation of the blind sort algorithm uses at most $36m$ bytes of memory. Therefore, we use it when the number of suffixes to be sorted is less

Table 3. Running times (in seconds) for a 1000MHz Pentium III processor, with 1GB main memory and 256Kb L2 cache. The operating system was GNU/Linux Red Hat 7.1. The compiler was gcc ver. 2.96 with options -O3 -fomit-frame-pointer. The table reports (user + system) time averaged over five runs. The running times do not include the time spent for reading the input files

	bible	e.coli	wrld	sprot	rfc	howto	reuters	linux	jdk13	chr22	gcc
qsufsort	4.96	4.63	3.16	230.52	245.02	64.71	290.20	213.81	168.59	42.69	162.68
copy	3.34	4.63	1.76	230.10	197.73	70.82	532.02	324.29	519.10	47.33	35258.04
ds0 $L=500$	2.28	2.96	1.24	122.84	114.50	47.60	246.87	192.02	218.85	26.02	3022.47
ds0 $L=1000$	2.21	2.80	1.20	122.29	114.10	47.62	243.98	191.32	221.65	26.07	3035.91
ds0 $L=2000$	2.21	2.80	1.19	121.80	113.80	48.51	242.71	192.33	222.68	26.04	3026.48
ds0 $L=5000$	2.28	2.94	1.23	121.55	113.72	47.80	242.17	186.40	225.59	25.99	3071.17
ds1 $L=200$	2.29	2.99	1.26	137.27	124.75	36.07	253.79	140.10	127.52	26.28	383.03
ds1 $L=500$	2.18	2.85	1.20	127.29	122.13	34.49	262.66	126.61	150.68	26.10	331.50
ds1 $L=1000$	2.20	2.85	1.20	122.76	119.27	34.60	248.58	123.38	174.54	26.06	325.91
ds1 $L=2000$	2.18	2.79	1.19	121.50	114.85	35.23	240.59	124.24	175.30	25.93	344.11
ds1 $L=5000$	2.19	2.80	1.20	121.80	113.53	37.39	240.42	132.50	190.77	26.05	410.72
ds2 $d=500$	2.18	2.79	1.20	121.66	112.74	30.95	239.44	102.51	133.37	25.95	249.23
ds2 $d=1000$	2.18	2.79	1.19	121.44	112.57	32.16	232.45	105.79	152.49	25.92	262.48
ds2 $d=2000$	2.22	2.81	1.20	121.65	113.35	33.99	235.94	111.35	162.75	26.05	287.02
ds2 $d=5000$	2.23	2.82	1.20	121.54	113.25	36.86	239.99	121.88	186.84	25.99	353.70

than $B = \frac{n}{2000}$. Thus, the space overhead of using blind sort is at most $\frac{9n}{500}$ bytes. If the text is 100MB long, this overhead is 1.8MB which should be compared with the 500MB required by the text and the suffix array.[7]

If the number of suffixes to be sorted is larger than $B = \frac{n}{2000}$ we sort them using Bentley-McIlroy ternary quicksort. However, with respect to algorithm copy, we introduce the following two improvements:

1. As soon as we are working with a group of suffixes smaller than B we stop the recursion and we sort them using blind sort;
2. during each partitioning phase we compute L_S (resp. L_L) which is the longest common prefix between the pivot and the strings which are lexicographically smaller (resp. larger) than the pivot. When we sort the strings which are smaller (resp. larger) than the pivot we can skip the first L_S (resp. L_L) characters since we know they constitute a common prefix.

We have called the above algorithm ds0 and its performances are reported in Tables 2 and 3 for several values of the parameter L (the depth at which we

---

[7] Although we believe this is a small overhead, we point out that the limit $B = \frac{n}{2000}$ was chosen somewhat arbitrarily. Preliminary experimental results show that there is only a marginal degradation in performance when we take $B = \frac{n}{3000}$, or $B = \frac{n}{4000}$. In the future we plan to better investigate the space/time tradeoff introduced by this parameter and its impact on the cache performance.

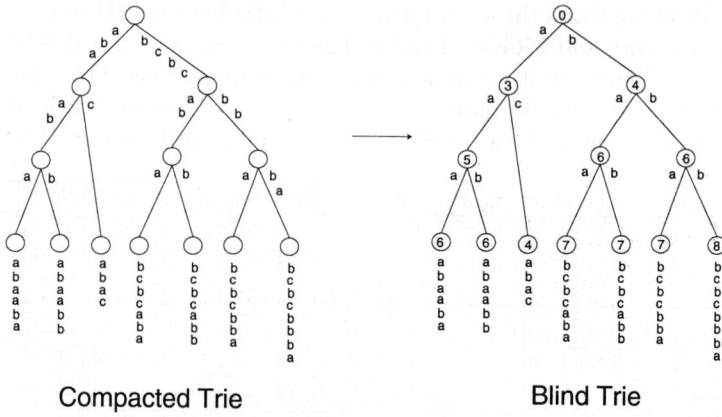

Fig. 1. A standard compacted trie (left) and the corresponding blind trie (right) for the strings: abaaba, abaabb, abac, bcbcaba, bcbcabb, bcbcbba, bcbcbbba. Each internal node of the blind trie contains an integer and a set of outgoing labelled arcs. A node containing the integer k represent a set of strings which have a length-k common prefix and differ in the $(k+1)$st character. The outgoing arcs are labelled with the different characters that we find in position $k+1$. Note that since the outgoing arcs are ordered alphabetically, by visiting the trie leaves from left to right we get the strings in lexicographic order

stop multikey quicksort and we switch to the blind sort/quicksort algorithms). We can see that algorithm ds0 is slower than qsufsort only for the files *jdk13* and *gcc*. If we compare copy and ds0 we notice that our deep-shallow approach has reduced the running time for *gcc* by a factor 10. This is certainly a good start. As we shall see, we will be able to reduce it again by the same factor taking advantage of the fact that the strings we are sorting are all suffixes of the same text.

3.2 Induced Sorting

One of the nice features of two-stage and copy algorithms is that some of the suffixes are not sorted by direct comparison: instead their relative order is derived in constant time from the ordering of other suffixes which have been already sorted. We use a generalization of this technique in the deep-sorting phase of our algorithm. Assume we need to sort the suffixes $s_1, \ldots, s_m$ which have a length-L common prefix. We scan the first L characters of s_1 looking at each pair of consecutive characters (e.g. $T[s_1]T[s_1+1]$, $T[s_1+1]T[s_1+2]$, up to $T[s_1+L-2]T[s_1+L-1]$). As soon as we find a pair of characters, say $\alpha\beta$, belonging to an already sorted small bucket $b_{\alpha\beta}$ the ordering of $s_1, \ldots, s_m$ can be derived from the ordering of $b_{\alpha\beta}$ as follows.

Assume $\alpha = T[s_1+t]$ and $\beta = T[s_1+t+1]$ for some $t < L-1$. Since $s_1, \ldots, s_m$ have a length-L common prefix, every s_i contains the pair $\alpha\beta$ starting from position t. Hence $\mathsf{b}_{\alpha\beta}$ contains m suffixes corresponding to $s_1, \ldots, s_m$ (that is, $\mathsf{b}_{\alpha\beta}$ contains the suffixes starting at $s_1 + t, s_2 + t, \ldots, s_m + t$). Note that these suffixes are not necessarily consecutive in $\mathsf{b}_{\alpha\beta}$. Since the first $t - 1$ characters of $s_1, \ldots, s_m$ are identical, the ordering of $s_1, \ldots, s_m$ can be derived from the ordering of the corresponding suffixes in $\mathsf{b}_{\alpha\beta}$. Summing up, the ordering is done as follows:

1. We sort the suffixes $s_1, \ldots, s_m$ according to their starting position in the input text $T[1, n]$. This is done so that in Step 3 we can use binary search to answer membership queries in the set $s_1, \ldots, s_m$.
2. Let $\hat{s}$ denote the suffix starting at the text position $T[s_1 + t]$. We scan the small bucket $\mathsf{b}_{\alpha\beta}$ in order to find the position of $\hat{s}$ within $\mathsf{b}_{\alpha\beta}$.
3. We scan the suffixes preceding and following $\hat{s}$ in the small bucket $\mathsf{b}_{\alpha\beta}$. For each suffix s we check whether the suffix starting at the position $T[s - t]$ is in the set $s_1, \ldots, s_m$; if so we mark the suffix s.[8]
4. When m suffixes in $\mathsf{b}_{\alpha\beta}$ have been marked, we scan them from left to right. Since $\mathsf{b}_{\alpha\beta}$ is sorted this gives us the correct ordering of $s_1, \ldots, s_m$.

Obviously there is no guarantee that in the length-L common prefix of $s_1, \ldots, s_m$ there is a pair of characters belonging to an already sorted small bucket. In this case we simply resort to the quicksort/blind sort combination. We call this algorithm ds1 and its performances are reported in Tables 2 and 3 for several values of L. We can see that ds1 with $L = 500$ runs faster than qsufsort for all files except *gcc*. In general, ds1 appears to be slightly slower than ds0 for files with small average LCP but it is clearly faster for the files with large average LCP: for *gcc* it is 8-9 times faster.

3.3 Anchor Sorting

Profiling shows that the most costly operation of induced sorting is the scanning of the small bucket $\mathsf{b}_{\alpha\beta}$ in search of the position of suffix $\hat{s}$ (Step 2 above). We now show that we can avoid this operation if we are willing to use a small amount of extra memory. For a fixed $d > 0$ we partition the text $T[1, n]$ into n/d segments of length d: $T[1, d], T[d+1, 2d]$ and so on up to $T[n-d+1, n]$ (for simplicity let us assume that d divides n). We define two arrays Anchor[·] and Offset[·] of size n/d such that, for $i = 1, \ldots, n/d$:

- Offset[i] contains the position of leftmost suffix which starts in the ith segment and belongs to an already sorted small bucket. If in the ith segment does not start any suffix belonging to an already sorted small bucket then Offset[i] = 0.
- Let $\hat{s}_i$ denote the suffix whose starting position is stored Offset[i]. Anchor[i] contains the position of $\hat{s}_i$ within its small bucket.

---
[8] The marking is done setting the most significant bit of s. This means that we can work with texts of size at most 2^{31}. The same restriction holds for qsufsort as well.

The use of the arrays Anchor[·] and Offset[·] is fairly simple. Assume that we need to sort the suffixes $s_1, \ldots, s_m$ which have a length-L common prefix. For $j = 1, \ldots, m$, let t_j denote the segment containing the starting position of s_j. If $\hat{s}_{t_j}$ (that is, the leftmost already sorted suffix in segment t_j) starts within the first L characters of s_j (that is, $s_j < \hat{s}_{t_j} < s_j + L$) then we can sort $s_1, \ldots, s_m$ using the induced sorting algorithm described in the previous section. However, we can skip Step 2 since the position of $\hat{s}_{t_j}$ within its small bucket is stored in Anchor[t_j]. Obviously, it is possible that for some j $\hat{s}_{t_j}$ does not exist or cannot be used. However, since the suffixes $s_1, \ldots, s_m$ usually belong to different segments, we have m possible candidates. In our implementation among the available sorted suffixes $\hat{s}_{t_j}$'s we use the one whose starting position is closest to the corresponding s_j (that is, we choose j which minimizes $\hat{s}_{t_j} - s_j$; this helps Step 3 of induced sorting). If there is no available sorted suffix, then we resort to the blind sort/quicksort combination.

For what concerns the space occupancy of anchor sorting, we note that in Offset[i] we can store the distance between the beginning of the ith segment and the leftmost sorted suffix. Hence Offset[i] $< d$. If we take $d < 2^{16}$ the array Offset requires $2n/d$ bytes of storage. Since each entry of Anchor requires four bytes, the overall space occupancy is $6n/d$ bytes. In our tests we used at least $d = 500$ which yields an overhead of $\frac{6n}{500}$ bytes. If we add the $\frac{9n}{500}$ bytes required by blind sorting with $B = \frac{n}{2000}$, we get a maximum overhead of at most $\frac{3n}{100}$ bytes. Hence, for a 100MB text the overhead is at most 3MB, which we consider a "small" amount compared with the 500MB used by the text and the suffix array.

In Tables 2 and 3 we report the running times of anchor sorting (under the name ds2) for d ranging from 500 to 5000 and $L = d + 50$. We see that for the files with moderate average LCP ds2 with $d = 500$ is significantly faster than copy and roughly two times faster than qsufsort. For the files with large average LCP ds2 is faster than qsufsort for all files except gcc. For gcc ds2 is 15% slower than qsufsort on the Athlon and 50% slower on the Pentium. In our opinion this slowdown on a single file is an acceptable price to pay in exchange for the reduction in space occupancy achieved over qsufsort (5.03n bytes vs. 8n bytes). We believe that the possibility of building suffix arrays for larger files has more value than a greater efficiency in handling files with a very large average LCP.

4 Conclusions and Further Work

In this paper we have presented a novel algorithm for building the suffix array of a text $T[1, n]$. Our algorithm uses 5.03n bytes and is faster than any other tested algorithm. Only on a single file our algorithm is outperformed by qsufsort which however uses 8n bytes.

For pathological inputs, i.e. texts with an average LCP of $\Theta(n)$, all lightweight algorithms take $\Theta(n^2 \log n)$ time. Although this worst case behavior does not occur in practice, it is an interesting theoretical open question whether we can achieve $O(n \log n)$ time using $o(n)$ space in addition to the space required by the input text and the suffix array.

References

[1] J. L. Bentley and M. D. McIlroy. Engineering a sort function. *Software – Practice and Experience*, 23(11):1249–1265, 1993.

[2] J. L. Bentley and R. Sedgewick. Fast algorithms for sorting and searching strings. In *Proceedings of the 8th ACM-SIAM Symposium on Discrete Algorithms*, pages 360–369, 1997.

[3] M. Burrows and D. Wheeler. A block sorting lossless data compression algorithm. Technical Report 124, Digital Equipment Corporation, 1994.

[4] M. Crochemore and W. Rytter. *Text Algorithms*. Oxford University Press, 1994.

[5] M. Farach-Colton, P. Ferragina, and S. Muthukrishnan. On the sorting-complexity of suffix tree construction. *Journal of the ACM*, 47(6):987–1011, 2000.

[6] P. Ferragina and R. Grossi. The string B-tree: A new data structure for string search in external memory and its applications. *Journal of the ACM*, 46(2):236–280, 1999.

[7] P. Ferragina and G. Manzini. Opportunistic data structures with applications. In *Proc. of the 41st IEEE Symposium on Foundations of Computer Science*, pages 390–398, 2000.

[8] P. Ferragina and G. Manzini. An experimental study of an opportunistic index. In *Proc. 12th ACM-SIAM Symposium on Discrete Algorithms*, pages 269–278, 2001.

[9] G. H. Gonnet, R. A. Baeza-Yates, and T. Snider. New indices for text: PAT trees and PAT arrays. In B. Frakes and R. A. Baeza-Yates and, editors, *Information Retrieval: Data Structures and Algorithms*, chapter 5, pages 66–82. Prentice-Hall, 1992.

[10] R. Grossi and J. Vitter. Compressed suffix arrays and suffix trees with applications to text indexing and string matching. In *Proc. of the 32nd ACM Symposium on Theory of Computing*, pages 397–406, 2000.

[11] D. Gusfield. *Algorithms on Strings, Trees, and Sequences: Computer Science and Computational Biology*. Cambridge University Press, 1997.

[12] H. Itoh and H. Tanaka. An efficient method for in memory construction of suffix arrays. In *Proceedings of the sixth Symposium on String Processing and Information Retrieval, SPIRE '99*, pages 81–88. IEEE Computer Society Press, 1999.

[13] R. Karp, R. Miller, and A. Rosenberg. Rapid Identification of Repeated Patterns in Strings, Arrays and Trees. In *Proceedings of the ACM Symposium on Theory of Computation*, pages 125–136, 1972.

[14] S. Kurtz. Reducing the space requirement of suffix trees. *Software—Practice and Experience*, 29(13):1149–1171, 1999.

[15] N. J. Larsson and K. Sadakane. Faster suffix sorting. Technical Report LU-CS-TR:99-214, LUNDFD6/(NFCS-3140)/1-43/(1999), Department of Computer Science, Lund University, Sweden, 1999.

[16] U. Manber and G. Myers. Suffix arrays: a new method for on-line string searches. *SIAM Journal on Computing*, 22(5):935–948, 1993.

[17] G. Manzini. An analysis of the Burrows-Wheeler transform. *Journal of the ACM*, 48(3):407–430, 2001.

[18] P. M. McIlroy and K. Bostic. Engineering radix sort. *Computing Systems*, 6(1):5–27, 1993.

[19] K. Sadakane. Compressed text databases with efficient query algorithms based on the compressed suffix array. In *Proceeding of the 11th International Symposium on Algorithms and Computation*, pages 410–421. Springer-Verlag, LNCS n. 1969, 2000.

[20] J. Seward. The BZIP2 home page, 1997. http://sourceware.cygnus.com/bzip2/index.html.
[21] J. Seward. On the performance of BWT sorting algorithms. In *DCC: Data Compression Conference*, pages 173–182. IEEE Computer Society TCC, 2000.
[22] O. Zamir and O. Etzioni. Grouper: A dynamic clustering interface to web search results. *Computer Networks*, 31(11-16):1361–1374, 1999.

Complexity of Compatible Decompositions of Eulerian Graphs and Their Transformations*

Jana Maxová and Jaroslav Nešetřil

Department of Applied Mathematics
and Institute for Theoretical Computer Science (ITI)
Charles University Malostranské nám. 25, 11 800 Praha 1, Czech Republic
{jana,nesetril}@kam.ms.mff.cuni.cz

Abstract. In this paper some earlier defined local transformations between eulerian trails are generalized to transformations between decompositions of graphs into (possibly more) closed subtrails. For any graph G with a forbidden partition system F, we give an efficient algorithm which transforms any F-compatible decomposition of G into closed subtrails to another one, and at the same time it preserves F-compatibility and does not increase the number of subtrails by more than one. From this, several earlier results for eulerian trails easily follow. These results are embedded into the rich spectrum of results of theory of eulerian graphs and their applications. We further apply this statement to digraphs and discuss the time complexity of enumeration of all F-compatible decompositions (resp. of all F-compatible eulerian trails) in both graphs and digraphs.

1 Introduction

A closed trail in a graph G is a sequence $v_0, e_1, v_1, \ldots, v_{t-1}, e_t, v_t$ of its vertices and edges such that $v_t = v_0$, for every i, e_i is incident with v_{i-1}, v_i, and all edges $e_1, \ldots, e_t$ are distinct. A closed trail which contains all edges of G is called an *eulerian trail*. Eulerian trails belong to the very classical combinatorial structures, yet they play a prominent role in many modern branches of graph theory, such as network flows, graph polynomials, coloring problems, nowhere-zero flows, cycle double cover conjecture, to name just a few (see [6] and [9] for many more examples). Eulerian trails provide also some background structure for problems in areas as diverse computational biology [17, 16] and knot theory [2, 22]. Although the basic questions related to eulerian trails (such as existence, number and various optimization problems) are well-known and were clarified long time ago, the new applications produce specific problems which are interesting (and sometimes hard). One such area of problems is the study of eulerian trails which satisfy special local conditions such as forbidding special consecutive pairs of edges. This study was initiated by A. Kotzig [11, 1] and continued by G. Sabidussi and H. Fleischner who related it to the Cycle double cover conjecture and other well-known problems, see [6, 18].

---
* The research was supported by project LN00A056 of the Ministry of Education of the Czech Republic

In this paper we address this problem in full generality and consider the set $\mathcal{S}(G,F)$ of all decompositions of G into closed trails with respect to a given set F of forbidden transitions at every vertex (see Section 2 for definitions). In section 4 we give an efficient algorithm for transforming any two F-compatible decompositions of $\mathcal{S}(G,F)$ (F a partition system in G) to one another with preserving the F-compatibility and not increasing the number of trails in the decomposition by more than one (see Theorem 1). This result also implies and gives short (and algorithmic) proofs of several results obtained earlier: Kotzig's theorem on transformations between eulerian trails [1, 11] (Corollary 2) and theorem of Fleischner et al. concerning transformations between F-compatible eulerian trails [7] (Corollary 1).

Our approach has been partly inspired by chord-diagrams which play an important role in knot theory (see for example [12, 15] for papers related to Vassiliev invariants). Chord diagrams are defined as a circle with additional set of disjoint chords (i.e. a matching). For a given knot, the chords correspond to the crossing points (in the plane projection) and thus to the double occurrence of a vertex in the eulerian trail. The graphs in knot theory are 4-regular planar graphs (or 2-in 2-out planar digraphs).

This can be naturally generalized to *diagrams* of arbitrary eulerian trail $v_0, e_1, \ldots, e_m, v_m = v_0$ in graph $G = (V, E)$ as follows: We consider a cycle of length m with vertices $v_1, \ldots v_m$ and with extra added disjoint complete graphs $\{K_v\}_{v \in V}$. These complete graphs correspond to vertices of G; $V(K_v) = \{v_i \mid v_i = v\}$. The complete graph K_v is specified by the polygon inscribed in the cycle (these polygons may be degenerated). For distinct vertices $u \neq v$ in G, the polygons K_u, K_v have disjoint vertices. The diagrams of eulerian trails could be also called *polygonal diagrams*.

Some questions related to diagrams of eulerian trails are much harder than for chord diagrams (i.e. diagrams of 4-regular planar graphs). One such problem is e.g. the Gauss problem which is already open for graphs with maximum degree at most six (see [8, 5] for the solution for 4-regular graphs). Still the diagram is a useful tool which motivated some of our constructions below. For example, the three main transformations are schematicly depicted in Fig. 1 (we skip the detailed description).

2 Terminology

We use a standard graph theory notation, see e.g. [13]. $G = (V, E)$ will always denote an (undirected) graph with the vertex-set V and the edge-set E.

$E(v)$ denotes the set of edges incident with v. Any two-element subset of $E(v)$ is called a *transition* through v. $\mathcal{T}(v)$ denotes the set of all transitions through v and $\mathcal{T}(G) = \bigcup_{v \in V} \mathcal{T}(v)$ the set of all transitions of G.

If $X(v)$ is a partition of $E(v)$ into two-element classes (i.e. each class of $X(v)$ is a transition through v), then $X(v)$ is called a *transition system through* v and $X = \bigcup_{v \in V} X(v)$ is called a *transition system of* G.

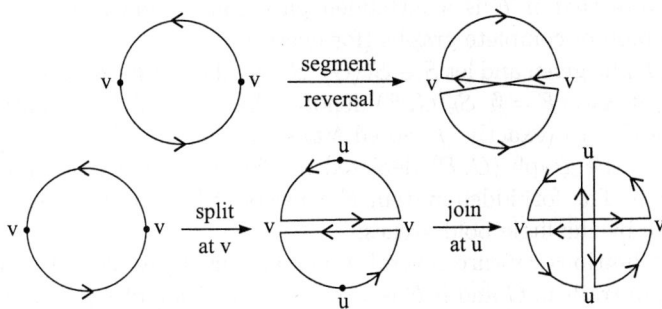

Fig. 1. Diagrams of transformations

$\mathcal{S}_k(G)$ denotes the set of all decompositions of G into exactly k (edge-disjoint) closed trails. In particular, $\mathcal{S}_1(G)$ denotes the set of all (closed) eulerian trails of G. We set $\mathcal{S}(G) = \bigcup_{k=1}^{\infty} \mathcal{S}_k(G)$. Let us remark that two closed trails in an (undirected) graph are considered as equal if one can be obtained from another by a cyclic permutation and possibly by a complete reversal.

Let $S = \{S_1, \ldots, S_k\} \in \mathcal{S}_k(G)$, X_{S_i}, is the set of transitions which correspond to a subsequence of S_i. Then $X_S = \bigcup_{i=1}^{k} X_{S_i}$ is a transition system of G. And conversely, every transition system X of G defines a (unique) $S \in \mathcal{S}(G)$, i.e. there is a unique S such that $X = X_S$. $n(X)$ denotes the number of closed trails in the decomposition of G defined by X (i.e. if $X = X_S$ for $S \in \mathcal{S}_k(G)$, then $n(X) = k$). If there is no danger of confusion, we will not distinguish between the transition system and the decomposition into closed trails defined by it. Let us stress that this correspondence is not between eulerian trails and transitions systems. Of course, any decomposition of the edge set of a connected graph into closed trails may be combined to a single eulerian trail, however at the expense that the transition system is changed. We are interested here in those operations which do not change transition system too much. This is captured by the notion of compatibility. Its origin lies in questions first raised by A. Kotzig for eulerian trails and by G. Sabidussi and H. Fleischner for cycle decompositions.

Let G be a graph and $F \subseteq \mathcal{T}(G)$ any subset of its transitions. $F = \bigcup_{v \in V} F(v)$, where $F(v) \subseteq \mathcal{T}(v)$. Any transition t of G such that $t \in F$ is said to be *forbidden*, any transition t of G such that $t \notin F$ is said to be *admissible* (or also *compatible*). (G, F) will denote a graph with a given forbidden system F. If for every v, $F(v)$ is a transitive relation on $E(v)$ we say that $F = \bigcup_{v \in V} F(v)$ is a *forbidden partition system of G*. This is consistent with definitions given in [6, 7]. This special type of forbidden systems is also used in our algorithm below. It is not hard to see that Theorem 1 does not hold for arbitrary forbidden systems.

Another possibility how to view a forbidden system of transition is the following: To every $F(v)$, there is a corresponding graph $G_{F(v)}$ on the set of neighbors of v with edges corresponding to forbidden transitions through v. More formally, $V(G_{F(v)}) = \{u \in V(G) \mid \{u, v\} \in E(G)\}$, $E(G_{F(v)}) = \{\{u, u'\} \mid \{\{u, v\}, \{u', v\}\}$

$\in F(v)$}. Note that if F is a forbidden partition system in G, then $G_{F(v)}$ is a disjoint union of complete graphs (for every v).

Let (G, F) be given and let $S \in \mathcal{S}_k(G)$, S is said to be F-*compatible* (or simply *compatible*) if $X_S \cap F = \emptyset$. $\mathcal{S}_k(G, F)$ denotes the set of all F-compatible decompositions of G into (exactly) k closed trails. Again, $\mathcal{S}(G, F) = \bigcup_{k=1}^{\infty} \mathcal{S}_k(G, F)$. For exmaple, the graph (G, F) depicted in Fig. 2 satisfies $\mathcal{S}_1(G, F) = \emptyset$ while $\mathcal{S}_2(G, F) \neq \emptyset$. The forbidden system F consists of four transistions represented in the figure by the little bold arches.

These definitions capture several important instances. $\mathcal{S}_1(G)$ is just the set of all eulerian trails in G and if F is a red-blue partition of $E(G)$, then $\mathcal{S}_1(G, F)$ denotes the set of all eulerian trails in G which are alternating in colors [17, 7].

All the above defined notions can be easily modified for directed graphs. In this paper, $D = (V, A)$ always denotes a directed graph, $A^+(v)$ (resp. $A^-(v)$) denotes the set of all arcs outgoing from (resp. incoming to) v. A *transition* through v in D is any two-element subset of $A(v)$ with one element in $A^+(v)$ and the other one in $A^-(v)$. We skip precise definitions of other notions, they are similar to the undirected case.

Our aim now is to define some local transformations between elements of $\mathcal{S}(G, F)$ which preserve F-compatibility. As we will see, our generalization to decompositions into more closed trails is very natural. For simplicity we define the transformation between transition systems, but of course, they uniquely determine also transformations between elements of $\mathcal{S}(G)$.

Definition 1. *Let X, X' be any two transition systems of a graph G such that $|X \setminus X'| = 2$, that is X and X' differ in only two transitions.*

1. *If $n(X') = n(X)$, then we say that X' arises from X be a* **segment reversal***.*
2. *If $n(X') = n(X) + 1$, then we say that X' arises from X be a* **split***.*
3. *If $n(X') = n(X) - 1$, then we say that X' arises from X be a* **join***.*

Note that whenever X and X' differ in only two transitions, then these are transitions through the same vertex (i.e. there is a vertex $v \in V(G)$ such that $X(u) = X'(u)$ for every $u \neq v$) and furthermore $n(X') \in \{n(X), n(X) \pm 1\}$. Hence, $|X \setminus X'| = 2$, if and only if X' arises from X by a segment reversal, by a split or by a join.

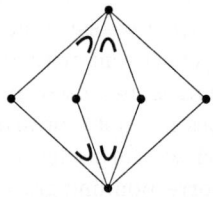

Fig. 2. (G, F)

The above defined transformations have their natural meanings, *segment reversal* corresponds to reversing a segment (i.e. a part of one trail between two occurrences of the same vertex), *split* corresponds to dividing one trail of S into two closed trails and *join* corresponds to joining two elements of S into one closed trail.

Later we will need the following technical definition. Let X be a transition system in a graph G such that $\{g, g'\}, \{h, h'\} \in X(v)$ and $\{g, h\} \notin F(v)$. We define a new transition system $Y = join(X, g, h)$ by:

$$Y = \begin{cases} X \setminus \{\{g,g'\}, \{h,h'\}\} \cup \{\{g,h\}, \{g',h'\}\} & \text{if } \{g',h'\} \notin F(v) \\ X \setminus \{\{g,g'\}, \{h,h'\}\} \cup \{\{g,h'\}, \{g',h\}\} & \text{otherwise} \end{cases}$$

Note that if F is a partition system and $X \cap F = \emptyset$ then also $Y \cap F = \emptyset$.

3 Formulation of Main Problems and Results

We are interested in the possibility of transforming two F-compatible transition systems into one another by means of the transformations defined above such that the F-compatibility is preserved throughout this process and the number of closed trails of any decomposition is not increased by more than one. More formally, we define:

REACHABILITY PROBLEM

INPUT: (G, F) a graph with a forbidden system
X, X' two F-compatible transition systems of G.
QUESTION: Does there exist a sequence $X = X_0, X_1, \ldots, X_n = X'$ such that for all $0 = 1, \ldots, n$:
1. X_i is F-compatible transition system of G,
2. $n(X_i) \leq \max\{n(X), n(X')\} + 1$,
3. $|X_i \setminus X_{i+1}| = 2$,
4. if $n(X_i) = \max\{n(X), n(X')\} + 1$, then $n(X_{i+1}) = n(X_i) - 1$.

Theorem 1. *Let F be a forbidden partition system in G. Then for any transition system X, X' compatible with F, there is a polynomial algorithm which finds the transforming sequence $X = X_0, X_1, \ldots, X_n = X'$ satisfying conditions 1.-4. above.*

In particular, Theorem 1 implies that when F is a partition system, the answer to REACHABILITY PROBLEM is always "yes". A polynomial algorithm for the REACHABILITY PROBLEM ifor partition systems is presented below.

In Section 6 we discuss algorithmic complexity of the following decision and evaluation problems:

GRAPH COMPATIBLE DECOMPOSITION PROBLEM (GCD)

INPUT: (G, F) a graph with a forbidden system
QUESTION: Is $\mathcal{S}(G, F) \neq \emptyset$?

♯GRAPH COMPATIBLE DECOMPOSITION PROBLEM (♯GCD)

INPUT: (G, F) a graph with a forbidden system
QUESTION: Determine $|\mathcal{S}(G, F)|$.

Similarly, we can define GRAPH COMPATIBLE TRAIL PROBLEM (GCT) and ♯GRAPH COMPATIBLE TRAIL PROBLEM (♯GCT) concerning the existence and number of F-compatible eulerian trails. For the lack of space we skip definitions of analogous problems for digraphs: DIGRAPH COMPATIBLE DECOMPOSITION PROBLEM (DCD), ♯DIGRAPH COMPATIBLE DECOMPOSITION PROBLEM (♯DCD), DIGRAPH COMPATIBLE TRAIL PROBLEM (DCT), and ♯DIGRAPH COMPATIBLE TRAIL PROBLEM (♯DCT).

4 Proof of Theorem 1

We prove that for a graph G with a forbidden partition system F and any two F-compatible transition systems X, X', the transforming sequence constructed by the algorithm given below satisfies the required properties 1. – 4. Because of the lack of space, let us only stress some most important parts of the proof: Note that at least one of Y_1 and Y_2 (and similarly also at least one of Y_3 and Y_4) is F-compatible. All X_i's and X_i''s are F-compatible, but it might happen that $n(X_{i+1}) = \ell + 1$ or $n(X'_{i+1}) = \ell + 1$. Still, after Step 3, the current pair X_{i+1} and X'_{i+1} satisfies $\min\{n(X_{i+1}), n(X'_{i+1})\} \leq \ell$ and $|X_{i+1} \cap X'_{i+1}|$ is increased. After Step 4, the current pair X_{i+1} and X'_{i+1} satisfies $n(X_{i+1}), n(X'_{i+1}) \leq \ell$ and $|X_{i+1} \cap X'_{i+1}|$ is again increased or remains the same. Thus, $X = X_0, X_1, \ldots X_k = X'_k \ldots, X'_0 = X'$ (after possibly deleting repetitive consecutive members of the sequence) satisfies conditions 1. – 4. required by Theorem 1.

As to the time complexity of the algorithm, the number of repetitions of Steps 2 to 4 is at most $|X \setminus X'| \leq m$, where m is the number of edges of G. Furthermore (by brute force counting) it is quite clear that each Step 1 to 5 can be done in time $O(m)$. Thus, the time complexity of our algorithm is $O(m^2)$ in the worst case.

Algorithm for REACHABILITY PROBLEM for partition systems

INPUT: (G, F) a graph with a forbidden partition system
$X \neq X'$ transition systems in G such that

$$X \cap F = \emptyset, \quad X' \cap F = \emptyset \text{ and } n(X), n(X') \leq \ell$$

OUTPUT: A sequence $X = X_0, X_1, \ldots, X_k = X'_k, \ldots, X'_1, X'_0 = X'$.

Step 1:
Set $i := 0$, $X_i := X$ and $X'_i := X'$.

Step 2:
Choose a vertex v such that $X_i(v) \neq X'_i(v)$ and edges $a,b,c,d,e \in E(v)$ (possibly $d = e$) such that $\{a,b\},\{c,d\} \in X_i \setminus X'_i$ and $\{a,c\},\{b,e\} \in X'_i \setminus X_i$. Set
$Y_1 := X_i \setminus \{\{a,b\},\{c,d\}\} \cup \{\{a,c\},\{b,d\}\},$
$Y_2 := X_i \setminus \{\{a,b\},\{c,d\}\} \cup \{\{a,d\},\{b,c\}\},$
$Y_3 := X'_i \setminus \{\{a,c\},\{b,e\}\} \cup \{\{a,b\},\{c,e\}\},$
$Y_4 := X'_i \setminus \{\{a,c\},\{b,e\}\} \cup \{\{a,e\},\{b,c\}\}.$

Step 3:
3.1. If $Y_1 \cap F = \emptyset$, then $X_{i+1} := Y_1$, $X'_{i+1} := X'_i$ and go to Step 4.
3.2. If $Y_3 \cap F = \emptyset$, then $X_{i+1} := X_i$, $X'_{i+1} := Y_3$ and go to Step 4.
3.3. If $\min\{n(Y_2), n(Y_4)\} \leq \ell$, then $X_{i+1} := Y_2$, $X'_{i+1} := Y_4$ and go to Step 4.
3.4. If $n(Y_2) = n(Y_4) = \ell + 1$, then
   if a and e appear in different trails of X, then $X'_{i+2} := X'_{i+1} := Y_4$, $X_{i+1} := Y_2$, $X_{i+2} := join(Y_2, a, e)$ $i := i+1$ and go to Step 4,
   otherwise $X_{i+1} := Y_2$, $X_{i+2} := join(Y_2, b, e)$, $X'_{i+2} := X'_{i+1} := X'_i$, $i := i+1$ and go to Step 4.

Step 4:
If $n(X_{i+1}) = \ell+1$, then choose edges g, h such that $\{g,h\} \in X'_{i+1}$ and g, h appear in different trails of X_{i+1}. Set $X_{i+2} := join(X_{i+1}, g, h)$, $X'_{i+2} := X'_{i+1}$, $i := i+1$.

If $n(X'_{i+1}) = \ell+1$, then choose edges g, h such that $\{g,h\} \in X_{i+1}$ and g, h appear in different trails of X'_{i+1}. Set $X'_{i+2} := join(X'_{i+1}, g, h)$, $X_{i+2} := X_{i+1}$, $i := i+1$.

Step 5:
If $X_{i+1} = X'_{i+1}$ then STOP, otherwise $i := i+1$ and go to Step 2.

5 Applications to Eulerian Trails

As F-compatible eulerian trails are special cases of F-compatible decompositions, Theorem 1 can be applied also to eulerian trails. In particular, as its consequence for $k = \ell = 1$, we obtain results of Fleischner et al. [7] and Kotzig [11]. Our corollaries are even stronger than the original results which proved only existence of the transforming sequence, we give a polynomial algorithm which constructs it.

Corollary 1. *Let F be a forbidden partition system in a graph G and let $T, T' \in S_1(G, F)$. Then there is a polynomial algorithm which constructs a sequence $X_0 = X_T, X_1, \ldots, X_n = X_{T'}$ such that for every $i = 0, \ldots, n$:*

1. *X_i defines F-compatible decomposition of G into at most 2 trails*
2. *$|X_i \setminus X_{i+1}| = 2$,*
3. *if $n(X_i) = 2$ then $n(X_{i+1}) = 1$.*

Corollary 2. *Let $T, T' \in \mathcal{S}_1(G)$. Then there is a polynomial algorithm which constructs a sequence $T_0 = T, T_1, \ldots, T_n = T'$ such that for every $i = 1, \ldots, n$, T_i arises from T_{i-1} by a segment reversal.*

Corollary 1 is an immediate consequence of Theorem 1 for $n(X) = n(X') = 1$. For the proof of Corollary 2 one has to prove a little more, namely that a split and a join (performed consecutively) can be replaced by a finite sequence of segment reversals (which is easy to see).

Theorem 1 can be also applied to digraphs (note that we cannot use segment reversals):

Definition 2. *Let X, X' be two transition systems of a digraph D such that $|X \setminus X'| = 2$.*

1. *If $n(X') = n(X) + 1$, then we say that X' arises from X be a* **split**.
2. *If $n(X') = n(X) - 1$, then we say that X' arises from X be a* **join**.

Again, whenever X and X' differ in only two transitions, then these are transitions through the same vertex and furthermore $n(X') = n(X) \pm 1$. Hence, for digraphs, $|X \setminus X'| = 2$, if and only if X' arises from X by a split or by a join.

Let D be a digraph and G its underlying graph. The orientation of D defines a natural forbidden transition system F in G ($F(v)$ consists of two equivalence classes corresponding to $A^+(v)$ and $A^-(v)$). Clearly there is a 1-1 correspondence between elements of $\mathcal{S}_k(G, F)$ and $\mathcal{S}_k(D)$. Thus, if we apply Theorem 1 and observe that segment reversals can be never used we obtain the following corollary. The statement for $n(X) = n(X') = 1$ was also observed in [17, 7].

Corollary 3. *For any transition system X, X' in a digraph D there is a polynomial algorithm which constructs a sequence $X = X_0, X_1, \ldots, X_n = X'$ such that for every i:*

1. *X_i transition system of D,*
2. *$n(X_i) \leq \max\{n(X), n(X')\} + 1$,*
3. *$|X_i \setminus X_{i+1}| = 2$,*
4. *if $n(X_i) = \max\{n(X), n(X')\} + 1$, then $n(X_{i+1}) = n(X_i) - 1$.*

6 Complexity Remarks

In this section we consider the complexity of enumeration problems ♯GCD (resp. ♯DCD, ♯GCT, ♯DCT) for eulerian graphs and digraphs. First let us consider the seemingly simplest case when there are no forbidden transitions.

As there is a 1-1 correspondence between elements of $\mathcal{S}(G)$ and transition systems in G (and the same holds also for digraphs) and $|X(v)|$ equals to the number of all perfect matching of a complete graph on $\deg_G(v)$ vertices (which is $(\deg_G(v) - 1)!!$ if $\deg_G(v)$ is even) we have

$$|\mathcal{S}(G)| = \prod_{v \in V(G)} (\deg_G(v) - 1)!!.$$

Similarly, by counting the number of perfect matchings in a complete bipartite graph, one gets
$$|\mathcal{S}(D)| = \prod_{v \in V(D)} (\deg_D^+(v))!.$$

This should be compared with the difficulty of determination of $|\mathcal{S}_1(G)|$. It is an open (and well known) problem whether this problem is a $\sharp P$-complete or not (see [14]). It is known that for graphs with maximum degree at most six, there is a fully polynomial randomized approximation scheme (FPRAS) and a fully polynomial almost uniform sampling (FPAUS) [20]. Perhaps some of our results will lead to a better understanding of this. The number $|\mathcal{S}_1(D)|$ (of eulerian trails in a digraph D) is of course counted by the classical BEST-formula [4, 19].

For general forbidden systems of transitions we have:

Theorem 2. *Problem $\sharp$GCD (for a general forbidden system) is $\sharp P$-complete.*

Proof: In fact we prove even more. We prove that $\sharp$GCD is equivalent to the problem of counting of all perfect matchings in a graph (which is known to be $\sharp P$-complete [21]).

Given a graph $H = (V, E)$, we define G by
$$V(G) = V \times \{+1, -1\} \cup \{0\},$$
$$E(G) = \{\{0, (v, i)\} \mid i = \pm 1, v \in V\} \cup \{\{(v, +1), (v, -1)\} \mid v \in V\}.$$

Since all vertices of G except 0 are of degree 2, the only forbidden transitions could be through vertex 0. We define $F = F(0)$ as follows:

$$\{\{(v, +1), 0\}, \{(v', -1), 0\}\} \in F(0) \; \forall v, v' \in V(H)$$

$$\{\{(v, +1), 0\}, \{(v', +1), 0\}\} \in F(0) \Leftrightarrow \{v, v'\} \notin E(H)$$

$$\{\{(v, -1), 0\}, \{(v', -1), 0\}\} \in F(0) \Leftrightarrow \{v, v'\} \notin E(H)$$

It is easy to check that $|\{M \mid M \text{ is a perfect matching in } H\}|^2 = |\mathcal{S}(G, F)|$ (transition system of each element of $|\mathcal{S}(G, F)|$ corresponds to an ordered pair of two perfect matchings in H, one on $(v, +1)$- and one on $(v, -1)$-vertices).

Conversely, for a given (G, F), we define a graph H such that the number of perfect matchings in H equals to the number of F-compatible decompositions of (G, F): First, for every vertex $v \in V(G)$ we define a graph H_v on the neighbors of v (in G) with edges corresponding to admissible transitions through v in G. More formally,
$$V(H_v) = \{u \in V(G) \mid \{u, v\} \in E(G)\}$$
$$E(H_v) = \{\{u, u'\} \mid \{\{u, v\}, \{u', v\}\} \notin F(v)\}$$

We put H to be the disjoint union of H_v, $v \in V(G)$. It is clear that $|\{M \mid M \text{ is a perfect matching in } H\}| = |\mathcal{S}(G, F)|$. □

Similarly, one can also prove that ♯DCD is a ♯P-complete problem. The question whether problems ♯GCD and ♯DCD have FPRAS is clearly related to the FPRAS for counting perfect matchings and thus presently open. We have seen that the exact evaluation problems as ♯GCD and ♯DCD are hard while the decision problems GCD and DCD are easy (since the question whether a graph has a perfect matching can be answered in polynomial time). For the decision problems GCT and DCT, the situation is harder. As we will see below, to decide whether a graph G has an F-compatible eulerian trail is an NP-complete problem (and the same holds also for digraphs). Moreover, from the reduction will follow that the corresponding evaluation problems ♯GCT and ♯DCT are ♯P-complete. These seem to be the first NP-completeness and ♯P-completeness results related to the compatible eulerian trails.

However, if we restrict the input to (G, F), F is a forbidden partition system, then testing whether $S_1(G, F) \neq \emptyset$ can be done in polynomial time (for undirected graphs). This follows immediately from the fact that $S_1(G, F) \neq \emptyset \Leftrightarrow S(G, F) \neq \emptyset$, if F is a partition system [11]. Moreover, assuming $S(G, F) \neq \emptyset$, one can find an F-compatible eulerian trail by a polynomial algorithm (see [3]).

Theorem 3. *Problem* DCT *(i.e. testing $S_1(D, F) \neq \emptyset$ for general forbidden system F in a digraph D) is NP-complete.*

Proof: Let D be a given digraph, we construct a digraph D'' and a forbidden system F in D'' such that D has a hamiltonian cycle if an only if (D'', F) has an F-compatible eulerian trail. This finishes the proof as testing hamiltonicity is an NP-complete problem, see [10].

First of all we define a digraph D' by $V(D') = \{v^1, v^2 \mid v \in V(D)\}$ and $A(D') = A^1(D') \cup A^2(D')$, where $A^1(D') = \{(v^1, v^2) \mid v \in V(D)\}$ and $A^2(D') = \{(u^2, v^1) \mid (u, v) \in A(D)\}$. Further, let D'' arises from D' by contracting all arcs in $A^2(D')$.

Note that $|A(D'')| = |V(D)|$ (arcs in $A^1(D')$ correspond to vertices of D) and if $(u, v) \in A(D)$ then $u^2 = v^1$ in D''. Hence if D is hamiltonian, then D'' is eulerian. But not conversely. For the other implication we need to define a forbidden system F in D''. We set $F = \{\{(u^1, u^2), (v^1, v^2)\} \mid u^2 = v^1$ in D'' and $(u, v) \notin A(D)\}$. It follows from the choice of F that (D'', F) has an F-compatible eulerian trail if and only if D has a hamiltonian cycle. It may happen (in fact it mostly happens) that D'' is a directed multigraph. If we insist on D'' being a simple digraph, we subdivide each arc of D'' with a vertex of degree two. □

Corollary 4. *Problem* GCT *(for general forbidden system F) is NP-complete.*

Proof: We will show that DCT can be reduced to our problem. This together with the previous Theorem 3 will finish the proof.

Suppose that (D, F) (a digraph with an arbitrary forbidden system F) is given. We will construct an undirected graph G and a forbidden system F' in G such that $S_1(D, F) \neq \emptyset \Leftrightarrow S_1(G, F') \neq \emptyset$.

We set G to be the corresponding underlying graph of D and let $E^+(v)$ (resp. $E^-(v)$) denote the edges of G corresponding to $A^+(v)$ (resp. $A^-(v)$). If $e \in E(G)$, then the corresponding arc in D is denoted by $\overrightarrow{e}$. We set:

$$F_1(v) = \{\{e,f\} \mid \{e,f\} \subseteq E^-(v) \text{ or } \{e,f\} \subseteq E^+(v)\},$$

$$F_2(v) = \{\{e,f\} \mid \{\overrightarrow{e}, \overrightarrow{f}\} \in F(v)\},$$

$$F' = F_1 \cup F_2.$$

It is easy to see that there is a 1-1 correspondence between F-compatible eulerian trails in D and F'-compatible eulerian trails in G. Indeed, if T is an F-compatible eulerian trail in D, then it is also an F'-compatible eulerian trail in G (since $F_2 \subseteq F'$). Conversely, if T is an F'-compatible eulerian trail in G, then the choice of F_1 guarantees that it is also an eulerian trail in D and the choice of F_2 that it is F-compatible. Hence, $\mathcal{S}_1(D,F) \neq \emptyset \Leftrightarrow \mathcal{S}_1(G,F') \neq \emptyset$. □

It follows immediately from the proof of Theorem 3, that there is a 1-1 correspondence between hamiltonian circuits in the digraph D and F-compatible eulerian trails in D''. Thus the evaluation problem $\sharp$DCT (and by the same argumentation also $\sharp$GCT) can be reduced to counting the number of hamiltonian circuits in a digraph (which is obviously equivalent to counting the number of hamiltonian circuits in an undirected graph, and thus a $\sharp P$-complete problem).

Corollary 5. $\sharp$DCT *and* $\sharp$GCT *(for general forbidden systems F) are $\sharp P$-complete problems.*

References

[1] J. Abrham, A. Kotzig: Transformations of Euler tours, Annals of Discrete Mathematics, 8 (1980), 65-69
[2] P. N. Balister, B. Bollobás, O. M. Riordan, A. D. Scott: Alternating knot Diagrams, euler circuits and the interlace polynomial, Europ. J. Combinatorics (2001) 22, 1-4
[3] A. Benkouar, Y. G. Manoussakis, V. T. Paschos, R. Saad: On the complexity of some hamiltonian and eulerian problems in edge-coloured completegraphs, ISA '91, LNCS Vol 557, Springer-Verlag, 1991, 190-198
[4] N. G. de Bruijn, T. van Aerdenne-Ehrenfest: Circuits and trees in oriented graphs, Simon Stevin 28 (1951), 203-217
[5] H. Crapo, P. Rosenstiehl: On lacets and their manifolds, Discrete Mathematics 233 (2001), 299-320
[6] H. Fleischner: Eulerian graphs and related topics, Annals of Discrete Mathematics 45, North-Holland, 1990
[7] H. Fleischner, G. Sabidussi, E. Wengner: Transforming eulerian trails, Discrete Mathematics 109 (1992), 103-116
[8] H. de Fraysseix, P. Ossona de Mendez: On a characterization of Gauss codes, Discrete and Computational Geometry, 2(2), 1999
[9] Handbook of Combinatorics, edited by R. Graham, M. Grötschel and L. Lovász, North Holland, 1995

[10] R. M. Karp: Reducibility among combinatorial problems, Complexity of Computer Computations, Plenum Press, 1972
[11] A. Kotzig: Eulerian lines in finite 4-valent graphs and their transformations, in P. Erdős and G. Katona (eds.), Theory of Graphs, Akademiai Kiado, Budapest, 1968, 219-230
[12] S. K. Lando: On Hopf algebra in graph theory, J. Combin. Theory, Ser. B 80, 2000, 104-121
[13] J. Matoušek, J. Nešetřil: Invitation to discrete mathematics, Oxford University Press, New York, 1998
[14] M. Mihail, P. Winkler: On the number of eulerian orientations of a graph, Algorithmica (1996) 16, 402-414
[15] S. D. Noble, D. J. A. Welsh: A weighted graph polynomial from chromatic invariants of knots, Ann. Inst. Fourier, Grenoble, (1999) 49, 1057-1087
[16] P. Pančoška, V. Janota, J. Nešetřil: Novel matrix descriptor for determination of the connectivity of secondary structure segments in proteins, Discrete Mathematics 235 (2001), 399-423
[17] P. A. Pevzner: DNA physical mappings and alternating eulerian cycles in colored graphs, Algorithmica (1995) 13, 77-105
[18] G. Sabidussi: Eulerian walks and local complementation, D. M. S. no. 84-21, Univ. de Montréal, 1984
[19] C. A. B. Smith, W. T. Tutte: On unicursal paths in a network of degree 4, American Mathematical Monthly 48 (1941), 233-237
[20] P. Tetali, S. Vempala: Random sampling of euler tours, Algorithmica (2001) 30, 376-385
[21] L. G. Valiant: The complexity of computing permanent, Theoretical Computer Science, 1979, 189-201
[22] D. J. A. Welsh: Complexity: Knots, colourings and counting, London Mathematical Society, Lecture Note Series 186, Cambridge University Press, 1993

External-Memory Breadth-First Search with Sublinear I/O*

Kurt Mehlhorn and Ulrich Meyer

Max-Planck-Institut für Informatik
Stuhlsatzenhausweg 85, 66123 Saarbrücken, Germany

Abstract. Breadth-first search (BFS) is a basic graph exploration technique. We give the first external memory algorithm for sparse undirected graphs with sublinear I/O. The best previous algorithm requires $\Theta(n + \frac{n+m}{D \cdot B} \cdot \log_{M/B} \frac{n+m}{B})$ I/Os on a graph with n nodes and m edges and a machine with main-memory of size M, D parallel disks, and block size B. We present a new approach which requires only $\mathcal{O}(\sqrt{\frac{n \cdot (n+m)}{D \cdot B}} + \frac{n+m}{D \cdot B} \cdot \log_{M/B} \frac{n+m}{B})$I/Os. Hence, for $m = \mathcal{O}(n)$ and all realistic values of $\log_{M/B} \frac{n+m}{B}$, it improves upon the I/O-performance of the best previous algorithm by a factor $\Omega(\sqrt{D \cdot B})$. Our approach is fairly simple and we conjecture it to be practical. We also give improved algorithms for undirected single-source shortest-paths with small integer edge weights and for semi-external BFS on directed Eulerian graphs.

1 Introduction

Breadth-First Search (**BFS**) is a basic graph-traversal method. It decomposes the input graph G of n nodes and m edges into at most n levels where level i comprises all nodes that can be reached from a designated source s via a path of i edges. BFS is used as a subroutine in many graph algorithms; the paradigm of breadth-first search also underlies shortest-paths computations. In this paper we focus on BFS for general undirected graphs and sparse directed Eulerian graphs (i.e., graphs with a cycle that traverses every edge of the graph precisely once).

External-memory (**EM**) computation is concerned with problems that are too large to fit into main memory (internal memory, **IM**). The central issue of **EM** computation is that accessing the secondary memory takes several orders of magnitude longer than performing an internal operation. We use the standard model of **EM** computation [16]. There is a main memory of size M and an external memory consisting of D disks. Data is moved in blocks of size B consecutive words. An I/O-operation can move up to D blocks, one from each disk. For graphs with n nodes and m edges the semi-external memory (**SEM**) setting assumes $c \cdot n \leq M < m$ for some appropriate constant $c \geq 1$.

---

* Partially supported by the Future and Emerging Technologies programme of the EU under contract number IST-1999-14186 (ALCOM-FT) and the DFG grant SA 933/1-1.

A number of basic computational problems can be solved I/O-efficiently. The most prominent example is **EM** sorting [2, 15]: sorting x items of constant size takes $\text{sort}(x) = \Theta(\frac{x}{D \cdot B} \cdot \log_{M/B} \frac{x}{B})$ I/Os. BFS, however, seems to be hard for external-memory computation (and also parallel computation). Even the best **SEM** BFS algorithms known require $\Omega(n)$ I/Os.

Recall the standard $\mathcal{O}(n+m)$-time internal-memory BFS algorithm. It visits the vertices of the input graph G in a one-by-one fashion; appropriate candidate nodes for the next vertex to be visited are kept in a FIFO queue Q. After a vertex v is extracted from Q, the *adjacency list* of v, i.e., the set of neighbors of v in G, is examined in order to update Q: unvisited neighboring nodes are inserted into Q. Running this algorithm in external memory will result in $\Theta(n + m)$ I/Os. In this bound the $\Theta(n)$-term results from the unstructured accesses to the adjacency lists; the $\Theta(m)$-term is caused by m unstructured queries to find out whether neighboring nodes have already been unvisited.

The best **EM** BFS algorithm known (Munagala and Ranade [14]) overcomes the latter problem; it requires $\Theta(n + \text{sort}(n + m))$ I/Os on general undirected graphs. Still, the Munagala/Ranade algorithm pays one I/O for each node.

In this paper we show how to overcome the first problem as well: the new algorithm for undirected graphs needs just $\mathcal{O}(\sqrt{\frac{n \cdot (n+m)}{D \cdot B}} + \text{sort}(n + m))$ I/Os. Our approach is simple and has a chance to be practical. We also discuss extensions to undirected single-source shortest-paths (SSSP) with small integer edge weights and semi-external BFS on directed Eulerian graphs.

This paper is organized as follows. In Section 2 we review previous work and put our work into context. In Section 3 we outline a randomized version of our new approach. The details are the subject of Section 5. We start with a review of the algorithm of Munagala and Ranade (Section 4) and then discuss our improvement (Sections 5.1 and 5.2). Section 6 presents a deterministic version of our new approach. In Section 7 we sketch an extension to some single-source shortest-paths problem. Another modification yields an improved semi-external BFS algorithm for sparse directed Eulerian graphs (Section 8). Finally, Section 9 provides some concluding remarks and open problems.

2 Previous Work and New Results

Previous Work. I/O-efficient algorithms for graph-traversal have been considered in, e.g., [1, 3, 4, 7, 8, 9, 10, 11, 12, 13, 14]. In the following we will only discuss results related to BFS. The currently fastest BFS algorithm for general undirected graphs [14] requires $\Theta(n + \text{sort}(m))$ I/Os. The best bound known for directed **EM** BFS is $\mathcal{O}(\min\{n + \frac{n}{M} \cdot \frac{m}{D \cdot B}, (n + \frac{m}{D \cdot B}) \cdot \log_2 \frac{n}{D \cdot B}\})$ I/Os [7, 8, 9]. This also yields an $\mathcal{O}(n + \frac{m}{D \cdot B})$-I/O algorithm for **SEM** BFS.

Faster algorithms are only known for special types of graphs: $\mathcal{O}(\text{sort}(n))$ I/Os are sufficient to solve **EM** BFS on trees [7], grid graphs [5], outer-planar graphs [10], and graphs of bounded tree width [11]. Slightly sublinear I/O was known for undirected graphs with bounded maximum node degree d: the algorithm [13] needs $\mathcal{O}(\frac{n}{\gamma \cdot \log_d(D \cdot B)} + \text{sort}(n \cdot (D \cdot B)^\gamma))$ I/Os and $\mathcal{O}(n \cdot (D \cdot B)^\gamma)$

external space for an arbitrary parameter $0 < \gamma \leq 1/2$. Maheshwari and Zeh [12] proposed I/O-optimal algorithms for a number of problems on planar graphs; in particular, they show how to compute BFS on planar graphs using $\mathcal{O}(\text{sort}(n))$ I/Os.

SSSP can be seen as the weighted version of BFS. Consequently, all known **EM** SSSP algorithms do not perform better than the respective **EM** BFS algorithms. The best known lower bound for BFS is $\Omega(\min\{n, \text{sort}(n)\} + \frac{n+m}{D \cdot B})$ I/Os. It follows from the respective lower bound for the list-ranking problem [8].

New Results. We present a new **EM** BFS algorithm for undirected graphs. It comes in two versions (randomized and deterministic) and requires $\mathcal{O}(\sqrt{\frac{n \cdot (n+m)}{D \cdot B}} + \text{sort}(n+m))$ I/Os (expected or worst-case, respectively). For sparse graphs with $m = \mathcal{O}(n)$ and realistic machine parameters, the $\mathcal{O}(\sqrt{\frac{n \cdot (n+m)}{D \cdot B}})$-term in the I/O-bound will be dominant. In that case our approach improves upon the I/O-performance of the best previous algorithm [14] by a factor of $\Omega(\sqrt{D \cdot B})$. More generally, the new algorithm is asymptotically superior to the old algorithm for $m = o(\frac{D \cdot B \cdot n}{\log_{M/B} n/B})$; on denser graphs both approaches need $\mathcal{O}(\text{sort}(n+m))$ I/Os.

A simple extension of our new BFS algorithm solves the SSSP problem on undirected graphs with integer edge-weights in $\{1, \ldots, W\}$ for small W: it requires $\mathcal{O}(\sqrt{\frac{W \cdot n \cdot (n+m)}{D \cdot B}} + W \cdot \text{sort}(n+m))$ I/Os. After another modification we obtain an improved algorithm for **SEM** BFS on sparse directed Eulerian graphs: it achieves $\mathcal{O}(\frac{n+m}{(D \cdot B)^{1/3}} + \text{sort}(n+m) \cdot \log n)$ I/Os. A performance comparison for our BFS algorithms is depicted in Figure 1.

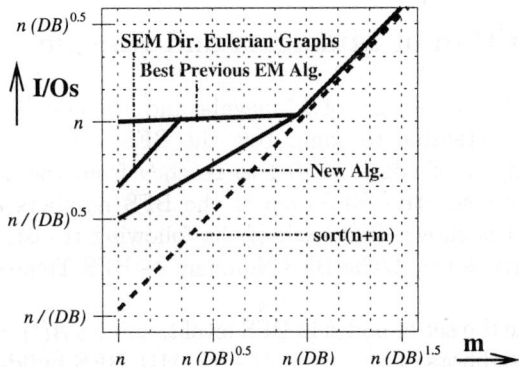

Fig. 1. Comparison: I/O-performance of the new BFS algorithms

3 High-Level Description of the New BFS Algorithm

Our algorithm refines the algorithm of Munagala and Ranade [14] which constructs the BFS tree level-by-level. It operates in two phases. In a first phase it preprocesses the graph and in the second phase it performs BFS using the information gathered in the first phase.

The preprocessing partitions the graph into disjoint subgraphs $\mathcal{S}_i$, $0 \leq i \leq K$ with small average internal shortest-path distances. It also partitions the adjacency lists accordingly, i.e., it constructs an external file $\mathcal{F} = \mathcal{F}_0 \mathcal{F}_1 \ldots \mathcal{F}_i \ldots \mathcal{F}_{K-1}$ where $\mathcal{F}_i$ contains the adjacency lists of all nodes in $\mathcal{S}_i$. The randomized partition is created by choosing seed nodes independently and uniformly at random with probability μ and running a BFS starting from all seed nodes. Then the expected number of seed nodes is $K = \mathcal{O}(\mu \cdot n)$ and the expected shortest-path distance between any two nodes of a subgraph is at most $\mathcal{O}(1/\mu)$. The expected I/O-bound for constructing the partition is $\mathcal{O}(\frac{n+m}{\mu \cdot D \cdot B} + \text{sort}(n+m))$.

In the second phase we perform BFS as described by Munagala and Ranade with one crucial difference. We maintain an external file $\mathcal{H}$ (= hot adjacency lists) which is essentially the union of all $\mathcal{F}_i$ such that the current level of the BFS-tree contains a node in $\mathcal{S}_i$. Thus it suffices to scan $\mathcal{H}$ (i.e., to access the disk blocks of $\mathcal{H}$ in consecutive fashion) in order to construct the next level of the tree. Each subfile $\mathcal{F}_i$ is added to $\mathcal{H}$ at most once; this involves at most $\mathcal{O}(K + \text{sort}(n+m))$ I/Os in total. We prove that after an adjacency list was copied to $\mathcal{H}$, it will be used only for $\mathcal{O}(1/\mu)$ steps on the average; afterwards the respective list can be discarded from $\mathcal{H}$. We obtain a bound of $\mathcal{O}(\mu \cdot n + \frac{n+m}{\mu \cdot D \cdot B} + \text{sort}(n+m))$ on the expected number of I/Os for the second phase. Choosing $\mu = \min\{1, \sqrt{\frac{n+m}{n \cdot D \cdot B}}\}$ gives our bound.

4 The Algorithm of Munagala and Ranade

We review the BFS algorithm of Munagala and Ranade [14], MR_BFS for short. We restrict attention to computing the BFS *level* of each node v, i.e., the minimum number of edges needed to reach v from the source. For undirected graphs, the respective BFS tree or the BFS numbers can be obtained efficiently: in [7] it is shown that each of the following transformations can be done using $\mathcal{O}(\text{sort}(n+m))$ I/Os: BFS Numbers → BFS Tree → BFS Levels → BFS Numbers.

Let $L(t)$ denote the set of nodes in BFS level t, and let $A(t) := N(L(t-1))$ be the multi-set of neighbors of nodes in $L(t-1)$. MR_BFS builds $L(t)$ as follows: $A(t)$ is created by $|L(t-1)|$ accesses to the adjacency lists of all nodes in $L(t-1)$. This causes $\mathcal{O}(|L(t-1)| + |A(t)|/(D \cdot B))$ I/Os. Observe that $\mathcal{O}(1 + x/(DB))$ I/Os are needed to read a list of length x. Then the algorithm removes duplicates from $A(t)$. This can be done by sorting $A(t)$ according to the node indices, followed by a scan and compaction phase; hence, the duplicate elimination takes $\mathcal{O}(\text{sort}(|A(t)|))$ I/Os. The resulting set $A'(t)$ is still sorted.

Now MR_BFS computes $L(t) := A'(t) \setminus \{L(t-1) \cup L(t-2)\}$. Filtering out the nodes already contained in the sorted lists $L(t-1)$ or $L(t-2)$ is possible by parallel scanning. Therefore, this step can be done using $\mathcal{O}((|A(t)| + |L(t-1)| + |L(t-2)|)/(D \cdot B))$ I/Os. Since $\sum_t |A(t)| = \mathcal{O}(m)$ and $\sum_t |L(t)| = \mathcal{O}(n)$, MR_BFS requires $\mathcal{O}(n + \text{sort}(n+m))$ I/Os. The $\Theta(n)$ I/Os result from the unstructured accesses to the n adjacency lists.

The correctness of this BFS algorithm crucially depends on the input graph being undirected: assume inductively that levels $L(0), \ldots, L(t-1)$ have already been computed correctly and consider a neighbor v of a node $u \in L(t-1)$. Then the distance from the source node s to v is at least $t-2$ because otherwise the distance of u would be less than $t-1$. Thus $v \in L(t-2) \cup L(t-1) \cup L(t)$ and hence it is correct to assign precisely the nodes in $A'(t) \setminus \{L(t-1) \cup L(t-2)\}$ to $L(t)$.

Theorem 1 ([14]). *Undirected BFS requires $\mathcal{O}(n + \text{sort}(n+m))$ I/Os.*

5 The New Approach

We show how to speed-up the Munagala/Ranade approach (MR_BFS) of the previous section. We refer to the resulting algorithm as FAST_BFS. We may assume w.l.o.g. that the input graph is connected (otherwise we may run the randomized $\mathcal{O}(\text{sort}(n+m))$-I/O connected-components algorithm of [1] and only keep the nodes and edges of the component C_s that contains the source node; all nodes outside of C_s will be assigned BFS-level infinity, and the BFS computation continues with C_s). We begin with the randomized preprocessing part of FAST_BFS:

5.1 Partitioning a Graph into Small Distance Subgraphs

As a first step, FAST_BFS restructures the adjacency lists of the graph representation: it grows disjoint connected subgraphs $\mathcal{S}_i$ from randomly selected nodes s_i and stores the adjacency lists of the nodes in $\mathcal{S}_i$ in an external file $\mathcal{F}$. The node s_i is called the *master node* of subgraph S_i. A node is selected to be a master with probability $\mu = \min\{1, \sqrt{\frac{n+m}{n \cdot D \cdot B}}\}$. This choice of μ minimizes the total cost of the algorithm as we will see later. Additionally, we make sure that the source node s of the graph will be the master of partition $\mathcal{S}_0$. Let K be the number of master nodes. Then $E[K] = 1 + \mu n$.

The partitioning is generated "in parallel": in each round, each master node s_i tries to capture all unvisited neighbors of its current sub-graph $\mathcal{S}_i$. If several master nodes want to include a certain node v into their partitions then an arbitrary master node among them succeeds.

At the beginning of a phase, the adjacency lists of the nodes lying on the boundaries of the current partitions are active; they carry the label of their master node. A scan through the set of adjacency lists removes these labeled lists, appends them in no particular order to the file $\mathcal{F}$, and forms a set of requests for the involved target nodes. Subsequently, the requests are sorted. A parallel scan of the sorted requests and the *shrunken* representation for the unvisited parts of the graph allows us to identify and label the new boundary nodes (and their adjacency lists). Each adjacency list is active during at most one phase. The partitioning procedure stops once there are no active adjacency lists left. The expected I/O-performance of the preprocessing step depends on the speed with which the graph representation shrinks during the partitioning process.

Lemma 1. *Let $v \in G$ be an arbitrary node; then v is assigned to some subgraph (and hence is removed from the graph representation) after an expected number of at most $1/\mu$ rounds.*

Proof: Consider a shortest path $\mathcal{P} = \langle s, u_j, \ldots, u_2, u_1, v \rangle$ from the source node s to v in G. Let k, $1 \leq k \leq j$, be the smallest index such that u_k is a master node. Then v is assigned to a subgraph in or before the k-th round. Due to the random selection of master nodes, we have $\mathbf{E}[k] \leq 1/\mu$. □

Corollary 1. *Consider an arbitrary node $v \in \mathcal{S}_i$ and let s_i be the master node of the subgraph $\mathcal{S}_i$. The expected shortest-path distance between v and s_i in $\mathcal{S}_i$ is at most $1/\mu$.*

By Lemma 1, the expected total amount of data being processed during the partitioning is bounded by $X := \mathcal{O}(\sum_{v \in V} 1/\mu \cdot (1 + \text{degree}(v))) = \mathcal{O}((n+m)/\mu)$. However, sorting only occurs for active adjacency lists. Thus the preprocessing requires $\mathcal{O}((n+m)/(\mu \cdot D \cdot B) + \text{sort}(n+m))$ expected I/Os.

After the partitioning phase each node knows the (index of the) subgraph to which it belongs. With a constant number of sort and scan operations we can partition the adjacency lists into the format $\mathcal{F}_0 \mathcal{F}_1 \ldots \mathcal{F}_i \ldots \mathcal{F}_{|S|-1}$, where $\mathcal{F}_i$ contains the adjacency lists of the nodes in partition $\mathcal{S}_i$; an entry $(v, w, \mathcal{S}(w), f_{\mathcal{S}(w)})$ from the list of $v \in \mathcal{F}_i$ stands for the edge (v,w) and provides the additional information that w belongs to subgraph $\mathcal{S}(w)$ whose subfile $\mathcal{F}_{\mathcal{S}(w)}$ starts at position $f_{\mathcal{S}(w)}$ within $\mathcal{F}$. The edge entries of each $\mathcal{F}_i$ are lexicographically sorted. In total, $\mathcal{F}$ occupies $\mathcal{O}((n+m)/B)$ blocks of external storage (spread over the D disks in round-robin fashion). $\mathcal{F}$ consists of K subfiles with $\mathbf{E}[K] = 1 + \mu \cdot n$. The size of the subfiles may vary widely. Some spread out over several disk blocks and some may share the same disk block. The following lemma summarizes the discussion.

Lemma 2. *The randomized preprocessing of* FAST_BFS *requires* $\mathcal{O}(\frac{n+m}{\mu \cdot D \cdot B} + \text{sort}(n+m))$ *expected I/Os.*

5.2 The BFS Phase

We construct the BFS levels one by one as in the algorithm of Munagala and Ranade (MR_BFS). The novel feature of our algorithm is the use of a sorted external file $\mathcal{H}$. We initialize $\mathcal{H}$ with $\mathcal{F}_0$. Thus, in particular, $\mathcal{H}$ contains the adjacency list of the source node s of level $L(0)$. The nodes of each created BFS level will also carry identifiers for the subfiles $\mathcal{F}_i$ of their respective subgraphs $\mathcal{S}_i$.

When creating level $L(t)$ based on $L(t-1)$ and $L(t-2)$, FAST_BFS does not access single adjacency lists like MR_BFS does. Instead, it performs a parallel scan of the sorted lists $L(t-1)$ and $\mathcal{H}$. While doing so, it extracts the adjacency lists of all nodes $v_j \in L(t-1)$ that can be found in $\mathcal{H}$. Let $V_1 \subseteq L(t-1)$ be the set of nodes whose adjacency lists could be obtained in that way. In a second step, FAST_BFS extracts from $L(t-1)$ the partition identifiers of those nodes in $V_2 := L(t-1) \setminus V_1$. After sorting these identifiers and eliminating duplicates, FAST_BFS knows which subfiles $\mathcal{F}_i$ of $\mathcal{F}$ contain the missing adjacency lists. The respective subfiles are concatenated into a temporary file $\mathcal{F}'$ and then sorted. Afterwards the missing adjacency lists for the nodes in V_2 can be extracted with a simple scan-step from the sorted $\mathcal{F}'$ and the remaining adjacency lists can be merged with the sorted set $\mathcal{H}$ in one pass.

After the adjacency lists of the nodes in $L(t-1)$ have been obtained, the set $N(L(t-1))$ of neighbor nodes can be generated with a simple scan. At this point the augmented format of the adjacency lists is used in order to attach the partition information to each node in $N(L(t-1))$. Subsequently, FAST_BFS proceeds like MR_BFS: it removes duplicates from $N(L(t-1))$ and also discards those nodes that have already been assigned to BFS levels $L(t-1)$ and $L(t-2)$. The remaining nodes constitute $L(t)$. The constructed levels are written to external memory as a consecutive stream of data, thus occupying $\mathcal{O}(n/(D \cdot B))$ blocks striped over the D disks.

Since FAST_BFS is simply a refined implementation of MR_BFS, correctness is preserved. We only have to reconsider the I/O-bounds:

Lemma 3. *The BFS-phase of* FAST_BFS *requires* $\mathcal{O}(\mu \cdot n + \frac{n+m}{\mu \cdot D \cdot B} + \text{sort}(n+m))$ *expected I/Os.*

Proof: Apart from the preprocessing of FAST_BFS (Lemma 2) we mainly have to deal with the amount of I/Os needed to maintain the data structure $\mathcal{H}$. For the construction of BFS level $L(t)$, the contents of the sorted sets $\mathcal{H}$, $L(t-2)$, and $L(t-1)$ will be scanned a constant number of times. The first $D \cdot B$ blocks of $\mathcal{H}$, $L(t-2)$, and $L(t-1)$ are always kept in main memory. Hence, scanning these data items does not necessarily cause I/O for each level. External memory access is only needed if the data volume is $\Omega(D \cdot B)$. In that case, however, the number of I/Os needed to scan x data items over the whole execution of FAST_BFS is bounded by $\mathcal{O}(x/(D \cdot B))$.

Unstructured I/O happens when $\mathcal{H}$ is filled by merging subfiles $\mathcal{F}_i$ with the current contents of $\mathcal{H}$. For a certain BFS level, data from several subfiles $\mathcal{F}_i$ may be added to $\mathcal{H}$. However, the data of each single $\mathcal{F}_i$ will be merged with $\mathcal{H}$ at most once. Hence, the number of I/Os needed to perform the mergings can be split between (a) the adjacency lists being loaded from $\mathcal{F}$ and (b) those already being in $\mathcal{H}$. The I/O bound for part (a) is $\mathcal{O}(\sum_i (1 + \frac{|\mathcal{F}_i|}{D \cdot B} \cdot \log_{M/B} \frac{n+m}{B})) = \mathcal{O}(K + \text{sort}(n+m))$ I/Os, and $\mathbf{E}[K] = 1 + \mu \cdot n$.

With respect to (b) we observe first that the adjacency list A_v of an arbitrary node $v \in S_i$ stays in $\mathcal{H}$ for an expected number at most of $2/\mu$ rounds. This follows from the fact that the expected shortest-path distance between any two nodes of a subgraph is at most $2/\mu$: let $L(t')$ be the BFS level for which $\mathcal{F}_i$ (and hence A_v) was merged with $\mathcal{H}$. Consequently, there must be a node $v' \in S_i$ that belongs to BFS level $L(t')$. Let s_i be the master node of subgraph S_i and let $d(x, y)$ denote the number of edges on the shortest path between the nodes x and y in S_i. Since the graph is undirected, the BFS level of v will lie between $L(t')$ and $L(t' + d(v', s_i) + d(s_i, v))$. As soon as v becomes assigned to a BFS level, A_v is discarded from $\mathcal{H}$. By Corollary 1, $\mathbf{E}[d(v', s_i) + d(s_i, v)] \leq 2/\mu$. In other words, each adjacency list is part of $\mathcal{H}$ for expected $\mathcal{O}(2/\mu)$ BFS-levels. Thus, the expected total data volume for (b) is bounded by $\mathcal{O}((n+m)/\mu)$. This results in $\mathcal{O}((n+m)/(\mu \cdot D \cdot B))$ expected I/Os for scanning $\mathcal{H}$ during merge operations. By the same argumentation, each adjacency list in $\mathcal{H}$ takes part in at most $\mathcal{O}(1/\mu)$ scan-steps for the generation of $N(L(\cdot))$ and $L(\cdot)$. Similar to MR_BFS, scanning and sorting all BFS levels and sets $N(L(\cdot))$ takes $\mathcal{O}(\text{sort}(n+m))$ I/Os. □

Combining Lemmas 2 and 3 and making the right choice of μ yields:

Theorem 2. *External memory BFS on arbitrary undirected graphs can be solved using* $\mathcal{O}(\sqrt{\frac{n \cdot (n+m)}{D \cdot B}} + \text{sort}(n+m))$ *expected I/Os.*

Proof: By our lemmas the expected number of I/Os is bounded by $\mathcal{O}(\mu \cdot n + \frac{n+m}{\mu \cdot D \cdot B} + \text{sort}(n+m))$. The expression is minimized for $\mu^2 \cdot n \cdot D \cdot B = n + m$. Choosing $\mu = \min\{1, \sqrt{n \cdot D \cdot B/(n+m)}\}$ the stated bound follows. □

6 The Deterministic Variant

In order to obtain the result of Theorem 2 in the worst case, too, it is sufficient to modify the preprocessing phase of Section 5.1 as follows: instead of growing subgraphs around randomly selected master nodes, the deterministic variant extracts the subfiles $\mathcal{F}_i$ from an Euler Tour around the spanning tree for the connected component C_s that contains the source node s. Observe that C_s can be obtained with the deterministic connected-components algorithm of [14] using $\mathcal{O}((1 + \log \log(D \cdot B \cdot n/m)) \cdot \text{sort}(n+m))$ I/Os. The same amount of I/O suffices to compute a (minimum) spanning tree T_s for C_s [3].

After T_s has been built, the preprocessing constructs an Euler Tour around T_s using a constant number of sort- and scan-steps [8]. Then the tour is broken at the source node s; the elements of the resulting list can be stored in consecutive order using the deterministic list-ranking algorithm of [8]. This causes $\mathcal{O}(\text{sort}(n))$ I/Os. Subsequently, the Euler Tour can be chopped into pieces of size $1/\mu$ with a simple scan step. These Euler Tour pieces account for subgraphs S_i with the property that the distance between any two nodes of S_i in G is at most $1/\mu - 1$. Observe that a node v of degree d may be part of $\Theta(d)$ different subgraphs $\mathcal{S}_i$. However, with a constant number of sorting steps it is possible to remove duplicates and hence make sure that each node of C_s is part of exactly one subgraph $\mathcal{S}_i$, for example of the one with the smallest index; in particular, $s \in \mathcal{S}_0$. Eventually, the reduced subgraphs $\mathcal{S}_i$ are used to create the reordered adjacency-list files $\mathcal{F}_i$; this is done as in the old preprocessing and takes another $\mathcal{O}(\text{sort}(n+m))$ I/Os.

The BFS-phase of the algorithm remains unchanged; the modified preprocessing, however, guarantees that each adjacency list will be part of the external set $\mathcal{H}$ for at most $1/\mu$ BFS levels: if a subfile $\mathcal{F}_i$ is merged with $\mathcal{H}$ for BFS level $L(t)$, then the BFS level of any node v in $\mathcal{S}_i$ is at most $L(t) + 1/\mu - 1$. The bound on the total number of I/Os follows from the fact that $\mathcal{O}((1 + \log\log(D \cdot B \cdot n/m)) \cdot \text{sort}(n+m)) = \mathcal{O}(\sqrt{\frac{n \cdot (n+m)}{D \cdot B}} + \text{sort}(n+m))$.

Theorem 3. *There is a deterministic algorithm that solves external memory BFS on undirected graphs using $\mathcal{O}(\sqrt{\frac{n \cdot (n+m)}{D \cdot B}} + \text{sort}(n+m))$ I/Os.*

7 Extension to Some SSSP Problem

We sketch how to modify FAST_BFS in order to solve the Single-Source Shortest-Paths (SSSP) problem on undirected graphs with integer edge-weights in $\{1, \ldots, W\}$ for small W. Due to the "BFS-bottleneck" all previous algorithms for SSSP required $\Omega(n)$ I/Os. Our extension of FAST_BFS needs $\mathcal{O}(\sqrt{\frac{W \cdot n \cdot (n+m)}{D \cdot B}} + W \cdot \text{sort}(n+m))$ I/Os. Thus, for sparse graphs and constant W the resulting algorithm FAST_SSSP requires $\mathcal{O}(\frac{n}{\sqrt{D \cdot B}} + \text{sort}(n))$ I/Os.

For integer weights in $\{1, \ldots, W\}$, the maximum shortest-path distance of an arbitrary reachable node from the source node s is bounded by $W \cdot (n-1)$. FAST_SSSP subsequently identifies the set of nodes with shortest-path distances $1, 2, \ldots$, denoted by levels $L(1), L(2), \ldots$; for $W > 1$ some levels will be empty. During the construction of level $L(t)$, FAST_SSSP keeps the first $D \cdot B$ blocks of each level $L(t - W - 1), \ldots, L(t + W - 1)$ in main memory. The neighbor nodes $N(L(t-1))$ of $L(t-1)$ are put to $L(t), \ldots, L(t + W - 1)$ according to the edge weights. After discarding duplicates from the tentative set $L(t)$, it is checked against $L(t - W - 1), \ldots, L(t-1)$ in order to remove previously labeled nodes from $L(t)$. For the whole algorithm this causes at most $\mathcal{O}(W \cdot \text{sort}(n+m))$ I/Os.

Using the randomized preprocessing of FAST_BFS, the expected length of stay for an arbitrary adjacency list in $\mathcal{H}$ is multiplied by a factor of at most W (as compared to FAST_BFS): the expected shortest-path distance between any two nodes $u, v \in \mathcal{S}_i$ is at most $W \cdot \mathbf{E}[d(u, s_i) + d(s_i, v)] \leq 2 \cdot W/\mu$. Hence, the expected number of I/Os to handle $\mathcal{H}$ is at most $\mathcal{O}(W/\mu \cdot (n+m)/(D \cdot B) + W \cdot \text{sort}(n+m))$. The choice $\mu = \min\{1, \sqrt{\frac{W \cdot (n+m)}{n \cdot D \cdot B}}\}$ balances the costs of the various phases and the stated bound results.

8 Semi-External BFS on Directed Eulerian Graphs

The analysis for FAST_BFS as presented in the previous sections does not transfer to *directed* graphs. There are two main reasons: (i) In order to detect previously labeled nodes during the construction of BFS level $L(t)$ it is usually not sufficient to check just $L(t-1)$ and $L(t-2)$; a node in $A'(t)$ may already have appeared before in any level $L(0),\ldots,L(t-1)$. (ii) Unless a subgraph $\mathcal{S}_i$ is strongly connected it is not guaranteed that once a node $v \in \mathcal{S}_i$ is found to be part of BFS level $L(t)$, all other nodes $v' \in \mathcal{S}_i$ will belong to BFS some levels $L(t')$ having $t' < t + |\mathcal{S}_i|$; in other words, adjacency lists for nodes in $\mathcal{S}_i$ may stay (too) long in the data structure $\mathcal{H}$.

Problem (i) can be circumvented in the **SEM** setting by keeping a lookup table in internal memory. Unfortunately, we do not have a general solution for (ii). Still we obtain an improved I/O-bound for **SEM** BFS on sparse directed *Eulerian graphs*[1]. The preprocessing is quite similar to the deterministic undirected variant of Section 6. However, instead of grouping the adjacency lists based on an Euler Tour around a *spanning tree*, we partition them concerning an Euler Circuit for the *whole graph*:

The PRAM algorithm of [6] yields an Euler Circuit in $\mathcal{O}(\log n)$ time; it applies $\mathcal{O}(n+m)$ processors and uses $\mathcal{O}(n+m)$ space. Hence, this parallel algorithm can be converted into an **EM** algorithm which requires $\mathcal{O}(\log n \cdot \text{sort}(n+m))$ I/Os [8]. Let $\langle v_0, v_1, \ldots, v_{m-1}, v_m \rangle$ denote the order of the nodes on an Euler Circuit for G, starting from one occurrence of the source node, i.e., $v_0 = s$. Let the subgraph $\mathcal{S}_i$ contain the nodes of the multi-set $\{v_{i \cdot (D \cdot B)^{1/3}}, \ldots, v_{(i+1) \cdot (D \cdot B)^{1/3} - 1}\}$. As in the deterministic preprocessing for the undirected case (Section 6), a node v may be part of several subgraphs $\mathcal{S}_i$; therefore, v's adjacency list will only be kept in exactly one subfile $\mathcal{F}_i$. We impose another additional restriction: the subfiles $\mathcal{F}_i$ only store adjacency lists of nodes having outdegree at most $(D \cdot B)^{1/3}$; these nodes will be called *light*, nodes with outdegree larger than $(D \cdot B)^{1/3}$ are called *heavy*. The adjacency lists for heavy nodes are kept in a standard representation for adjacency lists.

The BFS-phase of the directed **SEM** version differs from the fully-external undirected approach in two aspects: (i) The BFS level $L(t)$ is constructed as $A'(t) \setminus \{L(0) \cup L(1) \cup \ldots \cup L(t-1)\}$, where $L(0), L(1), \ldots, L(t-1)$ are kept in internal memory. (ii) The adjacency list of each *heavy* node v is accessed separately using $\mathcal{O}(1 + outdegree(v)/(D \cdot B))$ I/Os at the time the lists needs to be read. Adjacency lists of heavy nodes are *not* inserted into the data structure $\mathcal{H}$. Each such adjacency list will be accessed at most once. As there are at most $m/(D \cdot B)^{1/3}$ heavy nodes this accounts for $\mathcal{O}(m/(D \cdot B)^{1/3})$ extra I/Os.

Theorem 4. *Semi-external memory BFS on directed Eulerian graphs requires* $\mathcal{O}(\frac{n+m}{(D \cdot B)^{1/3}} + \text{sort}(n + m) \cdot \log n)$ *I/Os in the worst case.*

---

[1] An Euler Circuit of a graph is a cycle that traverses every edge of the graph precisely once. A graph containing an Euler Circuit is called Eulerian. If a directed graph is connected then it is Eulerian provided that, for every vertex v, $indegree(v) = outdegree(v)$.

Proof: As already discussed before the modified preprocessing can be done using $\mathcal{O}(\text{sort}(n + m) \cdot \log n)$ I/Os. The amount of data kept in each $\mathcal{F}_i$ is bounded by $\mathcal{O}((D \cdot B)^{2/3})$. Hence, accessing and merging all the $m/(D \cdot B)^{1/3}$ subfiles $\mathcal{F}_i$ into $\mathcal{H}$ during the BFS-phase takes $\mathcal{O}(m/(D \cdot B)^{1/3} + \text{sort}(m))$ I/Os (excluding I/Os to scan data already stored in $\mathcal{H}$).

A subfile $\mathcal{F}_i$ is called *regular* if none of its adjacency lists stays in $\mathcal{H}$ for more than $2 \cdot (D \cdot B)^{1/3}$ successive BFS levels; otherwise, $\mathcal{F}_i$ is called *delayed*. The total amount of data kept and scanned in $\mathcal{H}$ from regular subfiles is at most $\mathcal{O}(m/(D \cdot B)^{1/3} \cdot (D \cdot B)^{2/3} \cdot (D \cdot B)^{1/3}) = \mathcal{O}(m \cdot (D \cdot B)^{2/3})$. This causes $\mathcal{O}(m/(D \cdot B)^{1/3})$ I/Os.

Now we turn to the delayed subfiles: let $\mathcal{D} := \{\mathcal{F}_{i_0}, \mathcal{F}_{i_1}, \ldots, \mathcal{F}_{i_k}\}$, $k \leq m/(D \cdot B)^{1/3}$, be the set of all delayed subfiles, where $i_j < i_{j+1}$. Furthermore, let t_{i_j} be the time (BFS level) when $\mathcal{F}_{i_j}$ is loaded into $\mathcal{H}$; similarly let t'_{i_j} be the time (BFS level) after which all data from $\mathcal{F}_{i_j}$ has been removed from $\mathcal{H}$ again.

Recall that the source node s is the first node on the Euler Circuit $\langle v_0, v_1, \ldots, v_{m-1}, v_m, v_0 \rangle$. Hence, node v_i has BFS level at most i. Furthermore, if v_i belongs to BFS level $x \leq i$ then the successive node v_{i+1} on the Euler Circuit has BFS level at most $x + 1$. As $\mathcal{F}_{i_0}$ contains (a subset of) the adjacency lists of the light nodes in the multi-set $\{v_{i_0 \cdot (D \cdot B)^{1/3}}, \ldots, v_{(i_0+1) \cdot (D \cdot B)^{1/3} - 1}\}$, we find $t'_{i_0} \leq (i_0 + 1) \cdot (D \cdot B)^{1/3}$. More generally, $t'_{i_j} \leq t_{i_{j-1}} + (i_j - i_{j-1} + 1) \cdot (D \cdot B)^{1/3}$.

The formula captures the following observation: once all data of $\mathcal{F}_i$ has been loaded into $\mathcal{H}$, the data of $\mathcal{F}_{i+l}$ will have been completely processed after the next $(l + 1) \cdot (D \cdot B)^{1/3}$ BFS levels the latest. As each $\mathcal{F}_i$ contains at most $\mathcal{O}((D \cdot B)^{2/3})$ data, the total amount of data scanned in $\mathcal{H}$ from delayed $\mathcal{F}_i$ is at most $Z = \sum_{j=0}^{k}(t'_{i_j} - t_{i_j}) \cdot (D \cdot B)^{2/3} \leq (i_0 + 1) \cdot (D \cdot B) + \sum_{j=1}^{k}(t_{i_{j-1}} - t_{i_j} + (i_j - i_{j-1} + 1) \cdot (D \cdot B)^{1/3}) \cdot (D \cdot B)^{2/3}$. The latter sum telescopes, and Z is easily seen to be bounded by $t_k \cdot (D \cdot B)^{2/3} + (i_k + k + 1) \cdot (D \cdot B)$. Using $k, i_k \leq m/(D \cdot B)^{1/3}$ and $t_k \leq n$ this implies another $\mathcal{O}((n + m)/(D \cdot B)^{1/3})$ I/Os. □

Theorem 4 still holds under the weaker memory condition $M = \Omega(n/(D \cdot B)^{2/3})$: instead of maintaining an **IM** boolean array for *all* n nodes it is sufficient to remember subsets of size $\Theta(M)$ and adapt the adjacency lists in **EM** whenever the **IM** data structure is full [8]. This can happen at most $\mathcal{O}((D \cdot B)^{2/3})$ times; each **EM** adaption of the lists can be done using $\mathcal{O}((n + m)/(D \cdot B))$ I/Os.

Theorem 4 also holds for graphs that are *nearly Eulerian*, i.e., $\sum_v |\text{indegree}(v) - \text{outdegree}(v)| = \mathcal{O}(m/n)$: a simple preprocessing can connect nodes with unbalanced degrees via paths of n dummy nodes. The resulting graph G' is Eulerian, has size $\mathcal{O}(n + m)$, and the BFS levels of reachable nodes from the original graph will remain unchanged.

9 Conclusions

We have provided a new BFS algorithm for external memory. For general undirected sparse graphs it is much better than any previous approach. It may facilitate I/O-efficient solutions for other graph problems like demonstrated for some SSSP problem. However, it is unclear whether similar I/O-performance can be achieved on arbitrary directed graphs. Furthermore, it is an interesting open question whether there is a stronger lower-bound for external-memory BFS. Finally, finding an algorithm for depth-first search with comparable I/O-performance would be important.

Acknowledgements

We would like to thank the participants of the GI-Dagstuhl Forschungsseminar "Algorithms for Memory Hierarchies" for a number of fruitful discussions on the "BFS bottleneck".

References

[1] J. Abello, A. Buchsbaum, and J. Westbrook. A functional approach to external graph algorithms. *Algorithmica*, 32(3):437–458, 2002.

[2] A. Aggarwal and J. S. Vitter. The input/output complexity of sorting and related problems. *Communications of the ACM*, 31(9):1116–1127, 1988.

[3] L. Arge, G. Brodal, and L. Toma. On external-memory MST, SSSP and multi-way planar graph separation. In *Proc. 8th Scand. Workshop on Algorithmic Theory*, volume 1851 of *LNCS*, pages 433–447. Springer, 2000.

[4] L. Arge, U. Meyer, L. Toma, and N. Zeh. On external-memory planar depth first search. In *Proc. 7th Intern. Workshop on Algorithms and Data Structures (WADS 2001)*, volume 2125 of *LNCS*, pages 471–482. Springer, 2001.

[5] L. Arge, L. Toma, and J. S. Vitter. I/O-efficient algorithms for problems on grid-based terrains. In *Proc. Workshop on Algorithm Engineering and Experiments (ALENEX)*, 2000.

[6] M. Atallah and U. Vishkin. Finding Euler tours in parallel. *Journal of Computer and System Sciences*, 29(30):330–337, 1984.

[7] A. Buchsbaum, M. Goldwasser, S. Venkatasubramanian, and J. Westbrook. On external memory graph traversal. In *Proc. 11th Symp. on Discrete Algorithms*, pages 859–860. ACM-SIAM, 2000.

[8] Y.-J. Chiang, M. T. Goodrich, E. F. Grove, R. Tamasia, D. E. Vengroff, and J. S. Vitter. External memory graph algorithms. In *6th Annual ACM-SIAM Symposium on Discrete Algorithms*, pages 139–149, 1995.

[9] V. Kumar and E. J. Schwabe. Improved algorithms and data structures for solving graph problems in external memory. In *Proc. 8th Symp. on Parallel and Distrib. Processing*, pages 169–177. IEEE, 1996.

[10] A. Maheshwari and N. Zeh. External memory algorithms for outerplanar graphs. In *Proc. 10th Intern. Symp. on Algorithms and Computations*, volume 1741 of *LNCS*, pages 307–316. Springer, 1999.

[11] A. Maheshwari and N. Zeh. I/O-efficient algorithms for graphs of bounded treewidth. In *Proc. 12th Ann. Symp. on Discrete Algorithms*, pages 89–90. ACM-SIAM, 2001.

[12] A. Maheshwari and N. Zeh. I/O-optimal algorithms for planar graphs using separators. In *Proc. 13th Ann. Symp. on Discrete Algorithms*, pages 372–381. ACM-SIAM, 2002.

[13] U. Meyer. External memory BFS on undirected graphs with bounded degree. In *Proc. 12th Ann. Symp. on Discrete Algorithms*, pages 87–88. ACM–SIAM, 2001.
[14] K. Munagala and A. Ranade. I/O-complexity of graph algorithms. In *Proc. 10th Symp. on Discrete Algorithms*, pages 687–694. ACM-SIAM, 1999.
[15] M. H. Nodine and J. S. Vitter. Greed sort: An optimal sorting algorithm for multiple disks. *Journal of the ACM*, 42(4):919–933, 1995.
[16] J. S. Vitter and E. A. M. Shriver. Algorithms for parallel memory I: Two level memories. *Algorithmica*, 12(2–3):110–147, 1994.

Frequency Channel Assignment on Planar Networks*

Michael Molloy** and Mohammad R. Salavatipour***

Department of Computer Science, University of Toronto
10 King's College Rd., Toronto M5S 3G4, Canada
{molloy,mreza}@cs.toronto.edu

Abstract. For integers $p \geq q$, a $L(p,q)$-labeling of a network G is an integer labeling of the nodes of G such that adjacent nodes receive integers which differ by at least p, and nodes at distance two receive labels which differ by at least q. The minimum number of labels required in such labeling is $\lambda_q^p(G)$. This arises in the context of frequency channel assignment in mobile and wireless networks and often G is planar. We show that if G is planar then $\lambda_q^p(G) \leq \frac{5}{3}(2q-1)\Delta + 12p + 144q - 78$. We also provide an $O(n^2)$ time algorithm to find such a labeling. This provides a $(\frac{5}{3} + o(1))$-approximation algorithm for the interesting case of $q = 1$, improving the best previous approximation ratio of 2.

1 Introduction

The problem of *frequency assignment* arises when different radio transmitters which operate in the same geographical area interfere with each other when assigned the same or closely related frequency channels. This situation is common in a wide variety of real world applications related to mobile or general wireless networks and is best modeled using graph coloring where the vertices of a graph represent the transmitters and adjacencies indicate possible interferences.

There has been recently much interest in the $L(2,1)$-labeling problem, which is the problem of assigning radio frequencies (integers) to transmitters such that transmitters which are close (at distance 2 apart in the graph) to each other receive different frequencies and transmitters which are very close (adjacent in the graph) receive frequencies that are at least two apart. To keep the frequency bandwidth small, we are interested in minimizing the difference of the smallest and largest integers assigned as labels to the vertices of the graph. The minimum range of frequencies is called λ_1^2. In many applications, the differences between

---

* A full version of this paper is available with title "A Bound on the Chromatic Number of the Square of a Planar Graph" at http://www.cs.toronto.edu/~mreza/research.html
** Supported by NSERC, a Sloan Research Fellowship, and a Premier's Research Excellence Award.
*** Supported by Research Assistantship, Department of computer science, and University open fellowship, University of Toronto.

the frequency channels must be at least some specific given numbers. So we study $L(p,q)$-labelings, which are frequency assignments such that the transmitters that are adjacent in the graph get labels that are at least p apart and those at distance two get labels that are at least q apart; λ_q^p is defined similarly.

Several papers have studied this problem and different bounds and approximation algorithms for λ_1^2 have been obtained for various classes of graphs [2, 4, 5, 6, 7, 8, 10, 11, 14, 16], most of them based on Δ, the maximum degree of the graph. Much of this study has been focussed on planar graphs. For example, this problem is proved to be NP-complete for planar graphs in [2, 10]. Jonas [9] showed that for planar graphs $\lambda_1^2 \leq 8\Delta - 13$. This was the best known bound until recently, when Van den Heuvel and McGuinness [13] showed that $\lambda_q^p \leq (4q-2)\Delta + 10p + 38q - 24$. In this paper we improve this bound asymptotically by showing that:

Theorem 1. *For any planar graph G and positive integers $p \geq q$: $\lambda_q^p(G) \leq \frac{5}{3}(2q-1)\Delta + 12p + 144q - 78$.*

Although the proof of this result is lengthy and non-trivial, it yields an easy to implement $O(n^2)$ algorithm to find such a labeling.

For simplicity of exposition, we present the case $p = q = 1$. The proof of the general case is nearly identical. For this special case, we have the following longstanding conjecture:

Conjecture 1 (Wegner [15]). For a planar graph G:

$$\lambda_1^1(G) \leq \begin{cases} \Delta + 5 & \text{if } 4 \leq \Delta \leq 7, \\ \lceil \frac{3}{2}\Delta \rceil + 1 & \text{if } \Delta \geq 8. \end{cases}$$

This conjecture, if true, would be the best possible bound in terms of Δ, as shown by Wegner [15]. S.A. Wong [17] showed that $\lambda_1^1(G) \leq 3\Delta+5$. Recently, Van den Heuvel and McGuinness [13] proved that $\lambda_1^1(G) \leq 2\Delta + 25$. For large values of Δ, Agnarsson and Halldórsson [1] have a better asymptotic bound for $\lambda_1^1(G)$. They prove that if G is a planar graph with $\Delta \geq 749$, then $\lambda_1^1(G) \leq \lceil \frac{9}{5}\Delta(G) \rceil + 1$, but as they noted, this is the best asymptotic bound that they can get via their approach. Recently, Borodin et al. [3] have been able to extend these results to $\lambda_1^1(G) \leq \lceil \frac{9}{5}\Delta(G) \rceil + 1$ for planar graphs with $\Delta(G) \geq 47$. We improve all these results asymptotically by showing that:

Theorem 2. *For a planar graph G, $\lambda_1^1(G) \leq \frac{5}{3}\Delta(G) + 78$.*

Remark 1. For planar graphs G with $\Delta \geq 241$ we can actually obtain $\lambda_1^1(G) \leq \frac{5}{3}\Delta(G) + 24$, but we do not present the proof for this result here, because of space limits.

In section 5 we explain how to modify the proof of Theorem 2 to prove Theorem 1. The technique we use is inspired by that used by Sanders and Zhao [12] to obtain a similar bound on the cyclic chromatic number of planar graphs.

In [2] Bodlaender et al. have given approximation algorithms to compute λ_1^2 for some classes of graphs and noted that the result of Jonas [9] yields an

8-approximation algorithm for planar graphs. In [5] Fotakis et al. use the result of [13] to obtain a $(2 + o(1))$-approximation algorithm for λ_1^1 on planar graphs. They state that a major open problem is to get a polynomial time approximation algorithm of approximation ratio < 2. Agnarsson and Halldórsson [1] also give a 2-approximation. The results of this paper yield a $(\frac{5}{3} + o(1))$-approximation algorithm for λ_1^p and in general a $(\frac{5}{3q}(2q-1) + o(1))$-approximation algorithm for λ_q^p, for planar graphs. The reason for this is that for a planar graph G with maximum degree Δ: $\lambda_q^p(G) \geq q\Delta + p$. So, the algorithm we obtain for Theorem 1 will have approximation ratio of $\frac{5}{3q}(2q-1) + o(1)$ for λ_q^p.

The organization of the paper is as follows: In the next section we give an overview of the algorithms we obtain. In Sections 3 and 4 we give some preliminary definitions and (the sketch of) the proof of Theorem 2. Section 5 contains (the sketch of) the proof of Theorem 1. Finally, in Section 6 we explain the algorithm and talk about the asymptotic tightness of the results.

2 Overview of the Algorithms

We use G^2 to denote the *square* of G, i.e. the graph formed by joining all pairs of vertices which are at distance at most 2 in G. It is convenient to note that $\lambda_1^1(G) = \chi(G^2)$. Thus, our proof of Theorem 2 is simply a proof that the square of any planar graph can be colored with at most $\frac{5}{3}\Delta + 78$ colors.

An edge (u, v) of G is *reducible* if it has the following properties:

(i) H, the graph obtained from G by contracting (u, v) has maximum degree $\Delta(G)$.
(ii) Any $(\frac{5}{3}\Delta(G) + 78)$-coloring of H^2 can "easily" be extended to a coloring of G^2.

The exact meaning of "easily" is made clear in the proofs of Lemmas 8 and 9. For now, it suffices to say that this extension can be done in $O(\Delta(G))$ time. We prove that every planar graph has a reducible edge. Furthermore, that edge can be found in $O(n)$ time. This yields an $O(n^2)$ recursive algorithm for finding the coloring. We elaborate more on this in section 6.

We find a reducible edge using the Discharging Method, which was first used to prove the Four Color Theorem. We start by assigning an initial charge (that will be defined in the proof) to each vertex such that the sum of the charges is negative. Then we move the charges among the vertices based on the 12 discharging rules given in Section 4. This process preserves the sum of the charges, and so at least one vertex will have negative charge. We will show that any vertex with negative charge will have a reducible edge in its neighborhood. Applying the charges and discharging rules and then searching for a negative charge vertex and then finding the associated reducible edge can be done in $O(n)$, as required. We can use exactly the same procedure to develop an algorithm for Theorem 1.

3 Preliminaries

We assume that the given graph is a simple connected planar graph with at least 8 vertices. The vertex set and edge set of a graph G are denoted by $V(G)$ and $E(G)$, respectively. The length of a path between two vertices is the number of edges on that path. We define the distance between two vertices to be the length of the shortest path between them. The square of a graph G, denoted by G^2, is a graph on the same vertex set such that two vertices are adjacent in G^2 iff their distance in G is at most 2. The degree of a vertex v is the number of edges incident with v and is denoted by $d_G(v)$ or simply $d(v)$ if it is not confusing. We denote the maximum degree of a graph G by $\Delta(G)$ or simply Δ. If the degree of v is i, at least i, or at most i we call it an i-vertex, a $\geq i$-vertex, or a $\leq i$-vertex, respectively. By $N_G(v)$, we mean the open neighborhood of v in G, which contains all those vertices that are adjacent to v in G. The closed neighborhood of v, which is denoted by $N_G[v]$, is $N_G(v) \cup \{v\}$. We usually use $N(v)$ and $N[v]$ instead of $N_G(v)$ and $N_G[v]$, respectively.

A vertex k-coloring of a graph G is a mapping $C : V \longrightarrow \{1, \ldots, k\}$ such that any two adjacent vertices u and v are mapped to different integers. The minimum k for which a coloring exists is called the chromatic number of G and is denoted by $\chi(G)$. A vertex v is called *big* if $d_G(v) \geq 47$, otherwise we call it a *small* vertex.

From now on assume that G is a minimum counter-example to Theorem 2.

Lemma 1. *For every vertex v of G, if there exists a vertex $u \in N(v)$, such that $d_G(v) + d_G(u) \leq \Delta(G) + 2$ then $d_{G^2}(v) \geq \frac{5}{3}\Delta(G) + 78$.*

Proof. Assume that v is such a vertex. Contract v on edge (v, u). The resulting graph has maximum degree at most $\Delta(G)$ and because G was a minimum counter-example, the new graph can be colored with $\frac{5}{3}\Delta(G) + 78$ colors. Now consider this coloring induced on G, in which every vertex other than v is colored. If $d_{G^2}(v) < \frac{5}{3}\Delta(G) + 78$ then we can assign a color to v to extend the coloring to v, which contradicts the definition of G. □

If we define H to be the graph obtained from G by contracting (u, v), the above proof actually shows how to extend a coloring of H^2 to a coloring of G^2. Therefore, we have the following:

Type 1 reducible edge: *An edge (u,v) where $d_G(u) + d_G(v) \leq \Delta(G) + 2$ and $d_{G^2}(v) < \frac{5}{3}\Delta(G) + 78$.*

As we mentioned before, Van den Heuvel and McGuinness [13] showed that $\chi(G^2) \leq 2\Delta + 25$. So:

Observation 1. *We can assume that $\Delta(G) \geq 160$, otherwise $2\Delta(G) + 25 \leq \frac{5}{3}\Delta(G) + 78$.*

Lemma 2. *Every ≤ 5-vertex in G must be adjacent to at least two big vertices.*

Corollary 1. *Every vertex of G is a ≥ 2-vertex.*

The proof of Theorem 2 becomes significantly simpler if we can assume that the underlying graph is triangulated, i.e. all faces are triangles, and has minimum degree at least 4. To be able to make this assumption, we modify graph G in two phases and get a contradiction from the assumption on G. In the first phase we make a (simple) triangulated graph G', by adding edges to every non-triangle face of G.

Observation 2. *For every vertex v, $N_G(v) \subseteq N_{G'}(v)$.*

It is also easy to verify that:

Lemma 3. *All vertices of G' are $\geq$3-vertices.*

Lemma 4. *Each $\geq$4-vertex v in G' can have at most $\frac{1}{2}d(v)$ neighbors which are 3-vertices.*

In the second phase we transform graph G' into another triangulated graph G'', whose minimum degree is at least 4. Initially G'' is equal to G'. As long as there is any 3-vertex v we do the following *switching* operation: let x, y, z be the three neighbors of v. At least two of them, say x and y, are big in G' by Lemma 2 and Observation 2. Remove edge (x, y). Since G' (and also G'') is triangulated this leaves a face of size 4, say $\{x, v, y, t\}$. Add edge (v, t) to G''. This way, the graph is still triangulated.

Observation 3. *If v is a small vertex in G then $N_G(v) \subseteq N_{G''}(v)$.*

Lemma 5. *If v is a big vertex in G then $d_{G''}(v) \geq 24$.*

So a big vertex v in G will not be a $\leq$23-vertex in G''. Let v be a big vertex in G and $x_0, x_2, \ldots, x_{d_{G''}(v)-1}$ be the neighbors of v in G'' in clockwise order. We call $x_a, \ldots, x_{a+b}$ (where addition is in mod $d_{G''}(v)$) a *sparse segment* in G'' iff:

- $b \geq 2$,
- Each x_i is a 4-vertex.

In the next two lemmas let's assume that $x_a, \ldots, x_{a+b}$ is a maximal sparse segment of v in G'', which is not equal to all the neighborhood of v. Also assume that x_{a-1} and x_{a+b+1} are the neighbors of v right before x_a and right after x_{a+b}, respectively.

Lemma 6. *There is a big vertex other than v, that is connected to all the vertices of $x_a, \ldots, x_{a+b}$.*

We use u to denote the big vertex, other than v, that is connected to all $x_a, \ldots, x_{a+b}$.

Lemma 7. *All the vertices $x_{a+1}, \ldots, x_{a+b-1}$ are connected to both u and v in G. If x_{a-1} is not big in G then x_a is connected to both u and v in G. Otherwise it is connected to at least one of them. Similarly, if x_{a+b+1} is not big in G then x_b is connected to both u and v in G, and otherwise it is connected to at least one of them.*

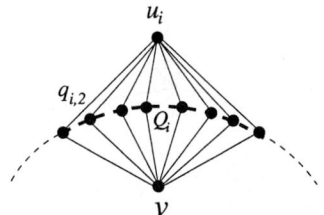

Fig. 1. The configuration of lemma 8

We call $x_{a+1}, \ldots, x_{a+b-1}$ the *inner* vertices of the sparse segment, and x_a and x_{a+b} the *end* vertices of the sparse segment. Consider vertex v and let's call the maximal sparse segments of it $Q_1, Q_2, \ldots, Q_m$ in clockwise order, where $Q_i = q_{i,1}, q_{i,2}, q_{i,3}, \ldots$. The next two lemmas are the key lemmas in the proof of Theorem 1.

Lemma 8. $|Q_i| \leq d_G(v) - \frac{2}{3}\Delta - 73$, for $1 \leq i \leq m$.

Proof. We prove this by contradiction. Assume that for some i, $|Q_i| \geq d_G(v) - \frac{2}{3}\Delta - 72$. Let u_i be the big vertex that is adjacent to all the inner vertices of Q_i (in both G and G'''). (See Figure 1).

For an inner vertex of Q_i, say $q_{i,2}$, we have:

$$d_{G^2}(q_{i,2}) \leq d_G(u_i) + d_G(v) + 2 - (|Q_i| - 3)$$
$$\leq \Delta(G) + d_G(v) - |Q_i| + 5$$
$$\leq \tfrac{5}{3}\Delta(G) + 77.$$

If $q_{i,2}$ is adjacent to $q_{i,1}$ or $q_{i,3}$ in G then it is contradicting Lemma 1. Otherwise it is only adjacent to v and u_i in G, therefore has degree 2, and so along with v or u_i contradicts Lemma 1. □

Therefore, if lemma 8 fails for some Q_i, then there must be a type 1 reducible edge with an end point in Q_i.

Lemma 9. *Consider G and suppose that u_i and u_{i+1} are the big vertices adjacent to all the inner vertices of Q_i and Q_{i+1}, respectively. Furthermore assume that t is a vertex adjacent to both u_i and u_{i+1} but not adjacent to v (see Figure 2) and there is a vertex $w \in N_G(t)$ such that $d_G(t) + d_G(w) \leq \Delta(G) + 2$. Let $X(t)$ be the set of vertices at distance at most 2 of t that are not in $N_G[u_i] \cup N_G[u_{i+1}]$. If $|X(t)| \leq 4$ then:*

$$|Q_i| + |Q_{i+1}| \leq \frac{1}{3}\Delta(G) - 69.$$

Proof. Again we use contradiction. Assume that $|Q_i| + |Q_{i+1}| \geq \frac{1}{3}\Delta(G) - 68$. Using the argument of the proof of Lemma 1, by contracting (t, w), we can color every vertex of G other than t. Note that $d_{G^2}(t) \leq d_G(u_i) + d_G(u_{i+1}) + |X(t)| \leq 2\Delta(G) + 4$. If all the colors of the inner vertices of Q_i have appeared on the

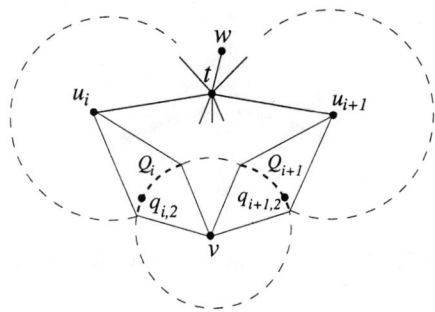

Fig. 2. The configuration of lemma 9

vertices of $N_G[u_{i+1}] \cup X(t) - Q_{i+1}$ and all the colors of inner vertices of Q_{i+1} have appeared on the vertices of $N_G[u_i] \cup X(t) - Q_i$ then there are at least $|Q_i|-2+|Q_{i+1}|-2$ repeated colors at $N_{G^2}(t)$. So the number of colors at $N_{G^2}(t)$ is at most $2\Delta(G) + 4 - |Q_i| - |Q_{i+1}| + 4$, which is at most $\frac{5}{3}\Delta(G) + 76$ and so there is still one color available for t, which is a contradiction.

Therefore, without loss of generality, there exists an inner vertex of Q_{i+1}, say $q_{i+1,2}$, whose color is not in $N_G[u_i] \cup X(t) - Q_i$. If there are less than $\frac{5}{3}\Delta(G) + 77$ colors at $N_{G^2}(q_{i+1,2})$ then we could assign a new color to $q_{i+1,2}$ and assign the old color of it to t and get a coloring for G. So there must be $\frac{5}{3}\Delta(G) + 77$ different colors at $N_{G^2}(q_{i+1,2})$. ¿From the definition of a sparse segment, we have: $N_G(q_{i+1,2}) \subseteq \{v, u_{i+1}, q_{i+1,1}, q_{i+1,3}\}$. There are at most $d_G(u_{i+1})+5$ colors, called the *smaller* colors, at $N_G[u_{i+1}] \cup X(t) \cup N_G[q_{i+1,1}] \cup N_G[q_{i+1,3}] - \{v\} - \{q_{i+1,2}\}$ (note that t is not colored). So there must be at least $\frac{2}{3}\Delta(G)+72$ different colors, called the *larger* colors, at $N_G[v] - Q_{i+1}$. Since $|N_G[v]| - |Q_i| - |Q_{i+1}| \leq \Delta(G) + 1 - \frac{1}{3}\Delta(G) + 68 \leq \frac{2}{3}\Delta(G) + 69$, one of these *larger* colors must be on an inner vertex of Q_i, which without loss of generality, we can assume is $q_{i,2}$. Because t is not colored, we must have all the $\frac{5}{3}\Delta(G) + 78$ colors at $N_{G^2}(t)$. Otherwise we could assign a color to t. As there are at most $\Delta(G) + 4$ colors, all from the *smaller* colors, at $N_G[u_{i+1}] \cup X(t)$, all the *larger* colors must be in $N_G[u_i]$ too. Therefore the larger colors are in both $N_G[v]$ and $N_G[u_i]$. Note that $|N_{G^2}(q_{i,2})| \leq |N_G[v]| + |N_G[u_i]| \leq 2\Delta(G)$. So there are at most $2\Delta(G) - \frac{2}{3}\Delta(G) - 72 = \frac{4}{3}\Delta(G) - 72$ different colors at $N_{G^2}(q_{i,2})$ and so we can assign a new color to $q_{i,2}$ and assign the old color of $q_{i,2}$, which is one of the *larger* colors and is not in $N_{G^2}(t) - \{q_{i,2}\}$, to t and extend the coloring to G, a contradiction. □

Another way of looking at the proof of Lemma 9 is that we showed the following is a reducible edge:

Type 2 reducible edge: (t, w), under the assumptions of Lemma 9.

4 Discharging Rules

We give an initial charge of $d_{G''}(v) - 6$ units to each vertex v. Using the Euler formula, $|V| - |E| + |F| = 2$, and noting that $3|F(G'')| = 2|E(G'')|$:

$$\sum_{v \in V}(d_{G''}(v) - 6) = 2|E(G'')| - 6|V| + 4|E(G'')| - 6|F(G'')| = -12 \quad (1)$$

By these initial charges, the only vertices that have negative charges are 4- and 5-vertices, which have charges -2 and -1, respectively. The goal is to show that, based on the assumption that G is a minimum counter-example, we can send charges from other vertices to $\leq$5-vertices such that all the vertices have non-negative charge, which is of course a contradiction since the total charge must be negative by equation (1).

We call a vertex v *pseudo-big* (in G'') if v is big (in G) and $d_{G''}(v) \geq d_G(v) - 11$. Note that a pseudo-big vertex is also a big vertex, but a big vertex might or might not be a pseudo-big vertex. Before explaining the discharging rules, we need a few more notations.

Suppose that $v, x_1, x_2, \ldots, x_k, u$ is a sequence of vertices such that v is adjacent to x_1, x_i is adjacent to x_{i+1}, $1 \leq i < k$, and x_k is adjacent to u.

Definition: By "*v sends c units of charge through $x_1, \ldots, x_k$ to u*" we mean v sends c units of charge to x_1, it passes the charge to x_2... etc, and finally x_k passes the charge to u. In this case, we also say "*v sends c units of charge through x_1*" and "*u gets c units of charge through x_k*".

Note that in order to simplify the calculations of the total charges on vertex x_i, $1 \leq i \leq k$, we do not take into account the charges that only pass through x_i. We say v *saves* k units of charge on a set of size l of its neighbors, if the total charge sent from v to or through them minus the total charge sent from or through them to v is at most $l - k$ units. So, for example, if v is sending nothing to u and is getting $\frac{1}{2}$ through u then v saves $\frac{3}{2}$ on u.

In discharging phase, a big vertex v of G:

1) Sends 1 unit of charge to each 4-vertex u in $N_{G''}(v)$.
2) Sends $\frac{1}{2}$ unit of charge to each 5-vertex u in $N_{G''}(v)$.

In addition, if v is a big vertex and u_0, u_1, u_2, u_3, u_4 are consecutive neighbors of v in clockwise or counter-clockwise order, where $d_{G''}(u_0) = 4$, then:

3) If $d_{G''}(u_1) = 5$, u_2 is big, $d_{G''}(u_3) = 4$, $d_{G''}(u_4) \geq 5$, and the neighbors of u_1 in clockwise or counter-clockwise order are v, u_0, x_1, x_2, u_2 then v sends $\frac{1}{2}$ to x_1 through u_2, u_1.
4) If $d_{G''}(u_1) = 5$, $5 \leq d_{G''}(u_2) \leq 6$, $d_{G''}(u_3) \geq 7$, and the neighbors of u_1 in clockwise or counter-clockwise order are v, u_0, x_1, x_2, u_2 then v sends $\frac{1}{2}$ to x_1 through u_3, u_2, u_1.
5) If $d_{G''}(u_1) = 5$, u_2 is big, $d_{G''}(u_3) \geq 5$, and the neighbors of u_1 in clockwise or counter-clockwise order are v, u_0, x_1, x_2, u_2 then v sends $\frac{1}{4}$ to x_1 through u_2, u_1.

6) If $d_{G''}(u_1) = 6$, $d_{G''}(u_2) \leq 5$, $d_{G''}(u_3) \geq 7$, and the neighbors of u_1 in clockwise or counter-clockwise order are $v, u_0, x_1, x_2, x_3, u_2$ then v sends $\frac{1}{2}$ to x_1 through u_1.
7) if $d_{G''}(u_1) = 6$, $d_{G''}(u_2) \geq 6$, and the neighbors of u_1 in clockwise or counter-clockwise order are $v, u_0, x_1, x_2, x_3, u_2$ then v sends $\frac{1}{4}$ to x_1 through u_1.

if $7 \leq d_{G''}(v) < 12$ then:

8) If u is a big vertex and $u_0, u_1, u_2, v, u_3, u_4, u_5$ are consecutive neighbors of u where all $u_0, u_1, u_2, u_3, u_4, u_5$ are 4-vertices then v sends $\frac{1}{2}$ to u.
9) if u_0, u_1, u_2, u_3 are consecutive neighbors of v, such that $d_{G''}(u_1) = d_{G''}(u_2) = 5$, u_0 and u_3 are big, and t is the other common neighbor of u_1 and u_2 (other than v), then v sends $\frac{1}{2}$ to t.

Every ≥ 12-vertex v of G'' that was not big in G:

10) Sends $\frac{1}{2}$ to each of its neighbors.

A ≤ 5-vertex v sends charges as follows:

11) if $d_{G''}(v) = 4$ and its neighbors in clockwise order are u_0, u_1, u_2, u_3, such that u_0, u_1, u_2 are big in G and u_3 is small, then v sends $\frac{1}{2}$ to each of u_0 and u_2 through u_1.
12) If $d_{G''}(v) = 5$ and its neighbors in clockwise order are u_0, u_1, u_2, u_3, u_4, such that $d_{G''}(u_0) \leq 11$, $d_{G''}(u_1) \geq 12$, $d_{G''}(u_2) \geq 12$, $d_{G''}(u_3) \leq 11$, and u_4 is big, then v sends $\frac{1}{2}$ to u_4.

It can be proved that after applying the discharging rules:

Lemma 10. *Every vertex v that is not big in G will have non-negative charge.*

Lemma 11. *Every big vertex v that is not pseudo-big will have non-negative charge.*

Lemma 12. *Every pseudo-big vertex v has non-negative charge.*

We omit the proofs of Lemmas 10, 11, and 12 because of a lack of space.

Proof of Theorem 2: By Lemmas 10, 11, and 12 every vertex of G'' will have non-negative charge, after applying the discharging rules. Therefore the total charge over all the vertices of G'' will be non-negative, but this is contradicting equation (1). This disproves the existence of G, a minimum counter-example to the theorem.

5 Generalization to Bound λ_q^p

The general steps of the proof of Theorem 1 are very similar to those of Theorem 2. The case $q = 0$ reduces to the Four Color Theorem. So let's assume that $q \geq 1$. Let G be a planar graph which is a minimum counter-example to Theorem 1. Recall that a vertex is big, if $d_G(v) \geq 47$. The proof of the following lemmas and observations are very similar to the corresponding ones for Theorem 2.

Lemma 13. *Suppose that v is a ≤ 5-vertex in G. If there exists a vertex $u \in N(v)$, such that $d_G(v) + d_G(u) \leq \Delta + 2$ then $d_{G^2}(v) \geq d_G(v) + \frac{5}{3}\Delta + 73$.*

Since the bound $2(2q-1)\Delta + 10p + 38q - 24$ is already proved in [13]:

Observation 4. *We can assume that $\Delta \geq 160$.*

Lemma 14. *Every ≤ 5-vertex must be adjacent to at least 2 big vertices.*

Now construct graph G' from G and then G'' from G' in the same way we did in the proof of Theorem 2. Also, we define the sparse segments in the same way. Consider vertex v and let's call the maximal sparse segments of it $Q_1, Q_2, \ldots, Q_m$ in clockwise order, where $Q_i = q_{i,1}, q_{i,2}, q_{i,3}, \ldots$.

Lemma 15. $|Q_i| \leq d_G(v) - \frac{2}{3}\Delta - 70.$

Lemma 16. *Suppose that u_i and u_{i+1} are the big vertices adjacent to all the vertices of Q_i and Q_{i+1}, respectively. Furthermore assume that t is a ≤ 6-vertex adjacent to both u_i and u_{i+1} but not adjacent to v (see Figure 2) and there is a vertex $w \in N(t)$ such that $d_G(t) + d_G(w) \leq \Delta(G) + 2$. Let $X(t)$ be the set of vertices at distance at most 2 of t that are not in $N[u_i] \cup N[u_{i+1}]$. If $|X(t)| \leq 4$ then:*

$$|Q_i| + |Q_{i+1}| \leq \frac{1}{3}\Delta(G) - 69. \qquad (2)$$

The rest of the proof is almost identical to that of Theorem 2. We apply the same initial charges and discharging rules, and use Lemmas 14, 15, and 16, instead of Lemmas 2, 8, and 9, respectively.

6 The Algorithm and the Asymptotic Tightness of the Result

Now we describe an algorithm that can be used for each of Theorems 1 or 2 to find such colorings. Consider a planar graph G. One iteration of the algorithm is to reduce the size of the problem by finding a reducible edge in G, contracting it, coloring the new smaller graph recursively, and then extending the coloring to G. At each iteration, first we check to see if every ≤ 5-vertex is adjacent to at least 2 big vertices or not. If not, then that vertex along with one of its small neighbors will be a type 1 reducible edge by Lemma 2. Otherwise, we construct the triangulated graph G'' and apply the initial charges and the discharging rules. As the total charge is negative, we can find a big vertex with negative charge. This vertex must have at least one of the configurations of Lemmas 8 and 9 or 15 and 16 for the cases of Theorem 2 or 1, respectively. If it has the configuration of Lemma 8 then one of the inner vertices of the sparse segment along with one of its two big neighbors will be a type 1 reducible edge. Otherwise, if it has the configuration of Lemma 9 then (t, w) will be a type 2 reducible edge (recall t and w from Lemma 9). In either of these cases we reduce the size of the

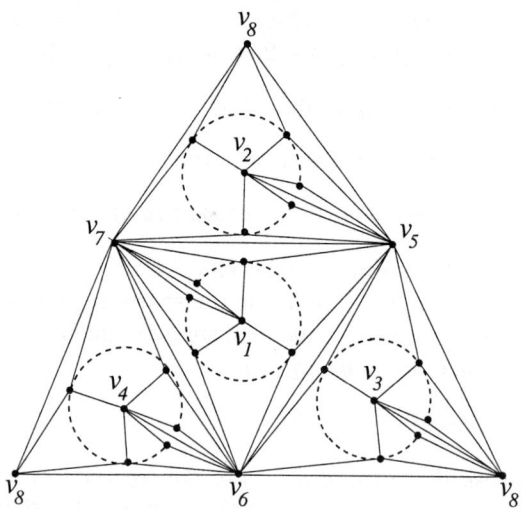

Fig. 3. The extremal graph for Theorem 2

graph by one. Thus, we can iterate until the size of the graph is small enough, i.e constant. In that case we can find the required coloring in constant time.

To see if there is a $\leq$5-vertex with less than 2 big neighbors we spend at most $O(n)$ time, where n is the number of vertices in G. Also, applying the initial charges and discharging rules takes $O(n)$ time. After finding a vertex with negative charge, finding the suitable edge and then contracting it can be done in $O(n)$. Since there are $O(n)$ iterations of the main procedure, the total running time of the algorithm would be $O(n^2)$.

Now we show that these theorems are asymptotically tight, if we use this proof technique. The results of [1] and [3] are essentially based on showing that in a planar graph G, there exists a vertex v such that $d_{G^2}(v) \leq \lfloor \frac{9}{5}\Delta(G) \rfloor + 1$. This also leads to a greedy algorithm for coloring G^2. However, as pointed out in [1], this is the best possible bound. That is, there are planar graphs in which every vertex v satisfies $d_{G^2}(v) \geq \lceil \frac{9}{5}\Delta(G) \rceil$. See [1] for an example. For the moment, let's just focus on the asymptotic order of the bounds and denote the additive constants by C. The reducible configuration in Lemma 8, after modifying the coefficient from $\frac{5}{3}$ to $\frac{9}{5}$, is the only configuration needed in obtaining the bound $\chi(G^2) \leq \frac{9}{5}\Delta(G) + C$. The extremal graph of [1] is actually an extremal graph for this lemma, and this is the reason that we need another reducible structure, like the one in Lemma 9, to improve previously known results asymptotically. But there are graphs that are extremal for both of these lemmas. For an odd value of k, one of these graphs is shown in Figure 3. In this graph G, which is obtained based on a tetrahedron, we have to join the three copies of v_8 and remove the multiple edges (we draw the graph in this way for clarity). The dashed lines represent sequences of consecutive 4-vertices. Around each of $v_1, \ldots, v_4$ there are $3k-5$ of such vertices. It is easy to see that G does not have the configuration of

Lemma 9. So the minimum degree of the vertices in G^2 is of order of $\frac{5}{3}\Delta(G)$ and G does not have the configuration of Lemma 9. One can easily check that if we are going to use only $\frac{3}{2}\Delta(G)+C$ colors, we can get a variation of Lemma 9 in which the coefficient of Δ is $\frac{3}{2}$, instead of $\frac{5}{3}$. But even that configuration does not appear in the graph of Figure 3. Therefore, using these two reducible configurations the best asymptotic bound that we can achieve is $\frac{5}{3}\Delta(G)$, and we probably need another reducible configuration to improve this result asymptotically.

References

[1] G. AGNARSSON AND M. M. HALLDÓRSSON, Coloring powers of planar graphs, *To appear in SIAM J. Disc. Math.*, Earlier version appeared in *Proc. of the 11th annual ACM-SIAM Symp. on Disc. Alg.*, pages 654-662, 2000.

[2] H. L. BODLAENDER, T. KLOKS, R. B. TAN, AND J. VAN LEEUWEN, Approximations for λ-Coloring of Graphs, In *Proc. of 17th Annual Symp. on Theo. Aspc. of Comp. Sci.* pages 395-406, Springer 2000.

[3] O. BORODIN, H. J. BROERSMA, A. GLEBOV, AND J. VAN DEN HEUVEL, Colouring at distance two in planar graphs, *In preparation* 2001.

[4] J. CHANG AND KUO, The $L(2,1)$-labeling problem on graphs, *SIAM J. Disc. Math.* 9:309-316, 1996.

[5] D. A. FOTAKIS, S. E. NIKOLETSEAS, V. G. PAPADOPOULOU, AND P. G. SPIRAKIS, Hardness results and efficient approximations for frequency assignment problems: radio labeling and radio coloring, *J. of Computers and Artificial intelligence*, 20(2):121-180, 2001.

[6] J. P. GEORGES AND D. W. MAURO, On the size of graphs labeled with a condition at distance two, *J. Graph Theory*, 22:47-57, 1996.

[7] J. P. GEORGES AND D. W. MAURO, Some results on λ_j^i-numbers of the products of complete graphs, *Congr. Numer.*, 140:141-160, 1999.

[8] J. R. GRIGGS AND R. K. YEH, Labeling graphs with a condition at distance 2, *SIAM J. Disc. Math.*, 5:586-595, 1992.

[9] T. K. JONAS, Graph coloring analogues with a condition at distance two: $L(2,1)$-labelings and list λ-labelings, Ph.D. Thesis, University of South Carolina, 1993.

[10] S. RAMANATHAN AND E. L. LLOYD, On the complexity of distance-2 coloring, In *Proc. 4th Int. Conf. Comput. and Inform.* pages 71-74, 1992.

[11] S. RAMANATHAN AND E. L. LLOYD, Scheduling algorithms for multi-hop radio networks, *IEEE/ACM Trans. on Networking*, 1(2):166-172, 1993.

[12] D. P. SANDERS AND Y. ZHAO, A new bound on the cyclic chromatic number, *J. of Comb. Theory Series B* 83:102-111, 2001.

[13] J. VAN DEN HEUVEL AND S MCGUINNESS, Colouring the Square of a Planar Graph, *Preprint*.

[14] J. VAN DEN HEUVEL, R. A. LEESE, AND M. A. SHEPHERD, Graph labeling and radio channel assignment, *J. Graph Theory*, 29:263-283, 1998.

[15] G. WEGNER, Graphs with given diameter and a coloring problem, Technical report, University of Dortmund, 1977.

[16] A. WHITTLESEY, J. P. GEORGES, AND D. W. MAURO, On the λ-number of Q_n and related graphs, *SIAM J. Disc. Math.*, 8:499-506, 1995.

[17] S. A. WONG, Colouring Graphs with Respect to Distance, M.Sc. Thesis, Department of Combinatorics and Optimization, University of Waterloo, 1996.

Design and Implementation of Efficient Data Types for Static Graphs*

Stefan Näher and Oliver Zlotowski

Fachbereich IV – Informatik
Universität Trier, 54296 Trier, Germany
{naeher,zlotowski}@informatik.uni-trier.de

Abstract. The LEDA library contains a very general and powerful graph type. This type provides the operations for all graphs of the library in one single interface. In many situations, however, this general interface is not needed and more specialized implementations of graphs could be used to improve the efficiency of graph algorithms. In this paper we present a design and an implementation of a family of special and efficient static graph types which fit into the LEDA environment of graphs and algorithms. They allow to speed up existing programs written with LEDA's standard graph type without changing the existing code.

1 Introduction

The LEDA platform for combinatorial and geometric computing [8, 7, 6] contains a very powerful and flexible graph data type **graph**. This type is used to represent *all* kinds of graphs used in the library, such as directed, undirected, and bidirected graphs, networks, planar maps, and geometric graphs. It provides the operations for all these graphs in one single (sometimes called *fat*) interface. The data type **graph** allows to write programs for graph problems in a form which is very close to the typical text book presentation of algorithms. This is achieved by providing macros for the typical iteration loops of the algorithms and by offering array types for associating data with the nodes and edges of a graph. For a complete description and examples of algorithms see [8] and [6].

The elegant way of turning algorithms into executable programs is one of the main advantages of LEDA and has made it very popular as a platform for the implementation of graph algorithms. However, it is not hard to see, that in many situations the general interface is not necessary and that smaller and more efficient data structures could be used. This is especially true if only one particular graph problem has to be solved, e.g., a maxflow problem.

The main result of this paper is the design and the implementation of a family of more special graph data types which fit into the LEDA environment of graphs and algorithms by providing a subset of the general interface and still supporting the convenient way of implementing graph algorithms. These graphs can speed

---

* This work was supported in part by DFG-Grant Na 303/1-2, Forschungsschwerpunkt "Effiziente Algorithmen für diskrete Probleme und ihre Anwendungen".

up existing graph algorithms written for the LEDA standard graph considerably without changing the existing code. So, all users of LEDA can benefit from the improvements presented in this paper.

The design of the graph types is based on the following two important observations. Most graph algorithms do not change the underlying graph, i.e., they work on a constant or static graph, and secondly, different algorithms are based on different models or categories of graphs. A graph traversal algorithm, for example, needs only to iterate over all nodes adjacent to given node, whereas network algorithms are often based on bidirected or bidirectional graphs for representing the residual network.

The structure of the remainder of this paper is as follows. In Section 2 we introduce the concepts of graph categories and structures. Section 3 gives the specification and implementation of our new static graph data structures. Section 4 describes different methods for associating data with the nodes and edges of a graph. Section 5 proposes a design for writing graph algorithms and Section 6 reports on the experiments we made with different graph types and data access methods. Finally, we give some conclusions and indicate future work.

2 Graph Categories and Structures

The interface of a graph data type consists of two parts. The first part is the representation of the sets V and E of nodes and edges and we call it the *structure* of the graph. The second part is the representation of the actual node and edge objects and we call it the *category* of the graph. The category of a graph defines the mathematical model of the graph. In particular, it describes the relations between the nodes and edges, for instance, that an edge has a target node or that every node has a first incident edge.

2.1 Graph Categories

In the current implementation we support five graph categories which have been identified by inspecting the requirements of various graph and network algorithms and graphs used in computational geometry.

1. Directed Graphs (`directed_graph`)
   This is the most fundamental graph category. It represents the mathematical concept of a directed graph by providing the ability to iterate over all edges incident to a given node v and to ask for the target node of a given edge e.

2. Bidirectional Graphs (`bidirectional_graph`)
   This category extends the basic directed graph category by supporting in addition iterations over all edges ending in a given node v and to ask for the source node of a given edge e. This allows algorithms to distinguish between out-going and in-coming edges of a node which is used in flow algorithms working on a bidirectional representation of the residual network (see [8] Section 7.10 for details).

3. Opposite Graphs (`opposite_graph`)
   This is a variant of the bidirectional graph category that does not support explicit computation of the source or target node of a given edge but supports to walk from one terminal v of an edge e to the other *opposite* one. This allows to implement the edges in a more compact way by storing the bitwise exclusive or of target and source representation in one single data field.

4. Bidirected Graphs (`bidirected_graph`)
   Bidirected graphs are directed graphs containing for each edge (v, w) also the *reversal* edge (w, v). In addition to the operations of directed graphs this category provides an operation for retrieving the reversal edge of a given edge e. This category of graphs is often used in network algorithms for an explicit representation of the residual network as a bidirected graph.

5. Undirected Graphs (`undirected_graph`)
   This category represents undirected graphs where every edge is an unordered pair $\{v, w\}$ incident to both v and w. It supports iteration over all edges incident to a given node v and the opposite operation, i.e., traveling along an edge e from a given endpoint v.

Remark. If only the adjacency relation between nodes is important and edges are not considered individual objects undirected graphs can be represented as bidirected graphs. Indeed, this definition of undirected graphs is often used in theory. However, as soon as data is associated with the edges of a graph (e.g. in a shortest path algorithm) there is a big difference between bidirected and undirected graphs.

2.2 Graph Data Types

Graph structures and categories can be combined arbitrarily to define a *graph data type*. So, for example, we might want to use a fixed-degree bidirected graph, or a static undirected graph, or a dynamic bidirectional graph, etc. In our design these combinations are supported by realizing each graph data type as a parameterized type (C++ class template) that implements the structure of the graph and takes a graph category `category` as its main type argument.

In this paper we concentrate on static graph structures and the corresponding parameterized data type `st_graph<category,node_data,edge_data>`. We have also implemented a prototype of a fixed degree graph, called `fd_graph`, and used it for computing delaunay triangulations and voronoi diagrams in computational geometry (see [10] for a first implementation and experimental evaluation). Different variants of dynamic graph structures (singly-linked and doubly-linked sequences of objects) are currently under development.

3 Static Graphs

A *static* graph consists of a fixed sequence of nodes and edges. The parameterized data type `st_graph<category,node_data,edge_data>` is used to represent static graphs. The first template parameter `category` defines the graph category. The last two parameters are optional and can be used to define user-defined data structures to be included into the node and edge objects (see section 4). For readability we will omit these two parameters in the remainder of this section.

An instance G of the parameterized data type `st_graph<category>` contains a sequence V of nodes and a sequence E of edges. New objects (nodes or edges) can be appended to both sequences only in a construction phase which has to be started by calling `G.start_construction()` and terminated by `G.finish_construction()`. For every node or edge x we define $index(x)$ to be equal to the rank of x in its sequence. During the construction phase, the sequence of the source node index of all inserted edges must be non-decreasing. After the construction phase both sequences V and E are fixed.

The interface of `st_graph<category>` consists of two parts. The first part is independent of the actual graph category and specifies operations which are common to all static graphs. These operations include methods for constructing a static graph and macros to iterate over all nodes and edges of the graph. The second part of the interface is different for every category and typically contains macros to iterate over incident edges or adjacent nodes and methods for traversing a given edge, e.g., by retrieving its target node.

The default constructor creates an empty graph. The construction phase of G starts by calling the operation `G.start_construction(n,m)`. The parameter n defines an upper bound for the number of nodes and m specifies an upper bound for the number edges of G. During the construction phase up to n nodes can be added to G by calling `G.new_node()`.

Some other operations of all static graphs

`node G.new_node()`	appends a new node to V and returns it.
`edge G.new_edge(node v,node w)`	appends a new edge (v,w) to E and returns it. *Precondition:* all edges (u,x) of G with $index(u) < index(v)$ have been created before.
`void G.finish_construction()`	terminates the construction phase.
`int G.index(node/edge x)`	returns the index of x.
`forall_nodes(v,G)`	v iterates over the node sequence.
`forall_edges(e,G)`	e iterates over the edge sequence.

Static graphs are implemented by two arrays V and E storing the node and edge structures of the actual graph category. The local types `node` and `edge` are defined as pointers into array V and array E, respectively.

The following example program constructs a static graph representation of the graph shown in Figure 1. Note, that the interface for the construction of a static graph does not depend upon the actual graph category.

```
typedef st_graph<category> static_graph;
typedef static_graph::node node;
typedef static_graph::edge edge;

static_graph G;
array<node> v(4);
array<edge> e(4);

G.start_construction(4,4);
for (int i = 0; i < 4; i++) v[i] = G.new_node();
e[0] = G.new_edge(v[0],v[1]);
e[1] = G.new_edge(v[0],v[2]);
e[2] = G.new_edge(v[1],v[2]);
e[3] = G.new_edge(v[3],v[1]);
G.finish_construction();
```

3.1 Static Directed Graphs (st_graph<directed_graph>)

For category directed_graph the interface of st_graph is extended by the operations

```
node G.target(edge e)       returns the target node of e.
int  G.outdeg(node v)       returns the out-degree of v.
forall_out_edges(e,v)       e iterates over all edges with source(e) = v.
```

The implementation of static directed graphs is simple. Every node v stores a pointer to its first incident edge and every edge e stores a pointer to its target node. Then G.target(e) just returns the value of the target data field of e. The operation G.outdeg(v) returns the difference between (v+1)->first_out and v->first_out, and the forall_out_edges(e,v) macro can be implemented by a simple array traversal from v->first_out to (v+1)->first_out. Note that we use a sentinel node v_stop and an sentinel edge e_stop at the end of both arrays to keep the iteration as simple as possible. Figure 1 shows an example graph and its internal representation.

3.2 Static Bidirectional Graphs (st_graph<bidirectional_graph>)

For category bidirectional_graph the interface of st_graph is extended by the operations

```
node G.target(edge e)       returns the target node of e.
node G.source(edge e)       returns the source node of e.
int  G.outdeg(node v)       returns the out-degree of v.
int  G.indeg(node v)        returns the in-degree of v.
forall_out_edges(e,v)       e iterates over all edges with source(e) = v.
forall_in_edges(e,v)        e iterates over all edges with target(e) = v.
```

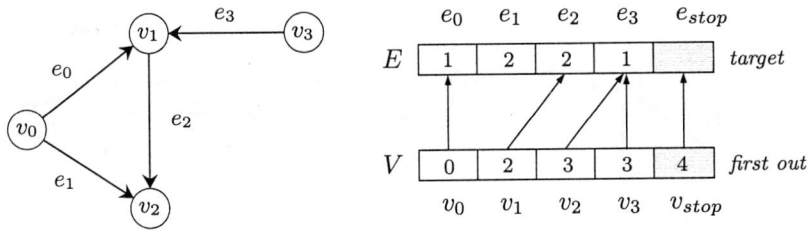

Fig. 1. A sample graph and the corresponding static directed graph representation

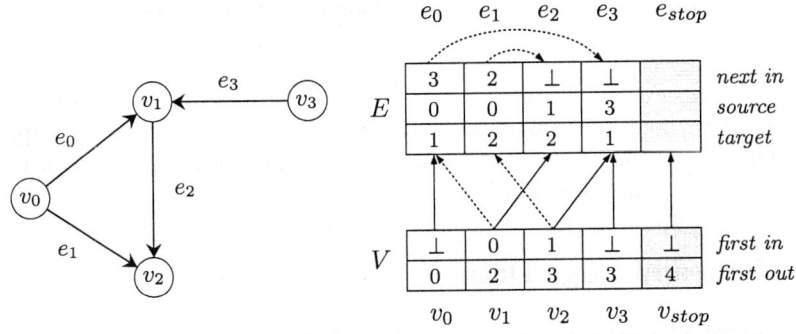

Fig. 2. The internal representation of a static bidirectional graph

In the implementation of static bidirectional graphs every node v stores an additional pointer to the first edge ending in v (NULL if no such edge exists) and every edge e stores a pointer to its source node u and a pointer to its successor in the list of edges ending in u (NULL for the last edge). Then the forall_in_edges(e,v) macro can be implemented by a simple singly-linked list traversal. Figure 2 shows the internal data structure.

3.3 Static Opposite Graphs (st_graph<opposite_graph>)

For category opposite_graph the interface of st_graph is extended by the operations

```
node G.opposite(edge e,node v)    returns the node opposite to v along e.
int  G.outdeg(node v)              returns the out-degree of v.
int  G.indeg(node v)               returns the in-degree of v.
forall_out_edges(e,v)              e iterates over edges with source(e) = v.
forall_in_edges(e,v)               e iterates over edges with target(e) = v.
```

The implementation of opposite graphs is very similar to the implementation of bidirectional graphs with one exception instead of storing the source and target

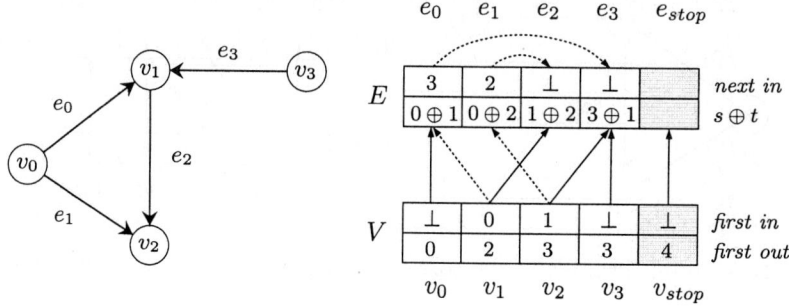

Fig. 3. The internal representation of a static opposite graph

nodes for every edge separately we store the bitwise *xor* of source and target in a single data field. Then G.opposite(e,v) can be computed by e->opposite $\oplus$ v. Of course, v must be either equal to the source or the target of e. Figure 3 shows the internal representation.

4 Node and Edge Data

Many graph algorithms need to associate information with the nodes and edges of a graph. This is part of the interface of most algorithms, e.g., when passing *cost* values for edges and returning *dist* values for nodes in a shortest path algorithm. Furthermore, it is often necessary to associate temporary data, such as labels, potentials, etc. items with the nodes and edges of a graph.

4.1 Node and Edge Arrays

The LEDA standard method for associating information with graph objects are node arrays and edge arrays (see [6] and [8] for a detailed description). They can be viewed as arrays indexed by the nodes and edges of a graph. In their general form they are in fact implemented as arrays and data access is realized by a normal array access with the *index* of the corresponding object. Since data is stored outside of the graph we call this method of assigning information to nodes and edges the *external array* implementation. It is the most flexible way of associating information with graph objects since an arbitrary number of arrays can be used with any graph defining an index operation for its nodes and edges. However, this flexibility has to be paid with a large run-time overhead due to the fact that information is not stored locally (as shown in the experiments in Section 6).

Node and edge arrays support a second, more efficient way of associating data with nodes and edges. In LEDA, nodes and edges of a graph can have additional *slots* for storing arbitrary data. The number of these slots is specified for the standard graph type at run-time in a special constructor and for our new static graph types at compile time by two optional template parameters `data_slots<int>`: e.g., `st_graph<directed_graph,data_slots<3>,data_slots<1>>` defines a static directed graph with three additional node slots and one additional edge slot (see also Section 3). Node and edge arrays can use these data slots instead of allocating an external array. This is discussed in detail in chapter 6.13 of the LEDA book [8]. Since the assignment of the slots to the arrays is done at run-time, we call this method of data assignment run-time or *dynamic slot assignment*. Sometimes the external array method is also called *horizontal* data assignment and the data slot method is called *vertical* data assignment [5].

4.2 Node and Edge Slots

Dynamic slot assignment is very flexible. However there is a run time penalty for storing the slot index in a data member. Although this slot index is constant during the life-time of the array compilers cannot treat it as a constant. This is due to the fact that C++ allows to "cast away constness" of any variable or data member. So even if we declare the slot index data member to be a constant the compiler cannot take any advantage from this information. An inspection of the assembly code shows that the slot index will be loaded from memory for every array access instead of keeping it in a machine register. This slows down data access considerably as shown in the experiments in Section 6.

This observation motivated us to implement a variant of node and edge arrays with compile-time or *static slot assignment*. The resulting data types are called `node_slot` and `edge_slot`. They always use slots and the mapping to the actual slot index is part of the type, i.e., done at compile time by using a slot template parameter. So, node and edge slots take three template parameters: the element type of the array, an integer slot number, and the type of the graph.

As an example for using static slot assignment in a graph algorithm, we present one of the opposite graph types used in the maxflow experiments of Section 6. It uses three node slots for storing the distance, excess, and a successor node, and two edge slots for storing the flow and capacity.

```
typedef st_graph<opposite_graph,data_slots<3>,data_slots<2>> graph;
node_slot<node,0,graph> succ;
node_slot<int, 1,graph> dist;
node_slot<edge,2,graph> excess;
edge_slot<int, 0,graph> flow;
edge_slot<int, 1,graph> cap;
```

4.3 Customizable Node and Edge Types

In our new design we offer a third method for data assignment. Instead of using the pre-defined `data_slot<int>` type for allocating data slots in the nodes and edges of a graph a user can pass any structure derived from this type as second or third template argument. This implements a way to extend nodes and edges by *named* data members. Note that these members are added in addition to data slots specified in the base type. In the example given below we define customized node end edge types for a preflow-push algorithm. Note that the node structure contains one (unnamed) data slot in addition to its (named) excess and level data members.

```
struct flow_node : data_slots<1>      struct flow_edge : data_slots<0>
{ int excess;                         { int flow;
  int level;                            int cap;
};                                    };

typedef st_graph<bidirectional_graph,flow_node,flow_edge> graph;
graph::node v;
forall_nodes(v,G) v->excess = 0;
```

5 Graph Algorithm Templates

To make graph algorithms work with all types of graphs and any data assignment they have to be written as C++ templates leaving the actual graph type and data access method open.

5.1 A Maxflow Algorithm Template

As an example we give the global structure of the maxflow algorithm template used in the experiments of Section 6. Note that we do not give any details of the algorithms here. We adapted the maxflow code described in Chapter 7.10 of the LEDA book [8].

The formal type parameters for the class template are the number type NT, the graph type graph_t, and optional array types for the locally used node data assignment method. If omitted the LEDA standard node array type is used. The run member function template is parameterized by the data assignment methods used for the capacity and flow arrays in the interface of the algorithm.

```
template<class NT, class graph_t,
         class succ_array   = node_array<graph_t::node,graph_t>,
         class excess_array = node_array<NT, graph_t>,
         class dist_array   = node_array<int,graph_t> >
class maxflow {
public:
  typedef graph_t::node node;
  typedef graph_t::edge edge;

  template<class cap_array, class flow_array>
  NT run(const graph_t& G, node s, node t,
         const cap_array& cap, flow_array& flow) { ... }

  // maxflow checker etc.
};
```

5.2 Maxflow Test Programs

The following program shows the use of the maxflow template with a static opposite graph (see Section 3) and static slot assignment for both the temporary and the interface data. The network is read in the DIMACS format [2].

```
typedef st_graph<opposite_graph,data_slots<3>,data_slots<2>> graph;

graph G;
graph::node s,t;
egge_slot<int,0,graph> cap(G);
edge_slot<int,1,graph> flow(G);
read_dimacs_mf(cin,G,s,t,cap);

typedef node_slot<node,0,graph> succ_array;
typedef node_slot<int, 1,graph> excess_array;
typedef node_slot<int, 2,graph> dist_array;

max_flow<int,graph,succ_array,excess_array,dist_array> MF;
MF.run(G,s,t,cap,flow);
```

The next program shows the use of the maxflow template with LEDA's standard graph type and standard node and edge arrays. Note that we do not have to provide template arguments for the local node data since node arrays are the default method.

```
graph G(3,2);
node s,t;
edge_array<int>   cap(G);
edge_array<int>   flow(G);
read_dimacs_mf(cin,G,s,t,cap);
max_flow<int,graph>  MF;
MF.run(G,s,t,cap,flow);
```

6 Experimental Results

The basis of our experiments is the preflow-push code for solving the maxflow problem as presented in the LEDA book in Chapter 7.10. All experiments were executed on a Sun UltraSPARC IIi 440 Mhz with 256 Mbytes of main memory running Solaris 2.7. The test programs have been compiled with gcc-3.0.3. To be able to test the effect of the different static graph data structures and data assignment methods we had to introduce additional template arguments and we changed the global structure of the code as indicated in the previous section. The underlying algorithms and heuristics have not been changed at all.

To measure the influence of the different graph types (LEDA standard graph, static bidirectional graph, static opposite graph) and the different data assignment methods, we performed a total number of nine experiments. In addition, we put our results into relation to the running times of two public available maxflow

Table 1. Running times for maximum flow algorithms on LEDA graph, static bidirectional graph and opposite graph in combination with different methods of data assignment. All problem instances were generated by the ak-generator

Generator: ak(n)	5.000	10.000	15.000	20.000	25.000	30.000	35.000	40.000
External Arrays								
LEDA Graph	7.6	31.6	72.5	130.4	236.4	304.9	464.1	591.2
Static Bidirectional Graph	7.5	30.2	67.0	122.7	191.8	281.9	378.4	525.7
Static Opposite Graph	6.8	27.1	60.7	111.0	174.6	257.0	350.4	454.6
Dynamic Slot Assignment								
LEDA Graph	6.4	27.1	70.9	134.7	184.7	291.4	475.3	580.0
Static Bidirectional Graph	5.7	22.6	51.3	107.1	149.3	205.0	284.4	427.8
Static Opposite Graph	5.4	21.2	50.0	90.0	143.9	204.2	275.4	364.5
Static Slot Assignment								
LEDA Graph	3.7	15.4	48.2	94.8	117.4	205.7	382.4	435.2
Static Bidirectional Graph	2.3	9.4	21.3	48.9	58.4	85.4	118.0	196.8
Static Opposite Graph	2.3	9.3	22.1	38.5	67.0	85.4	113.1	150.6
Other Packages								
Cherkassky/Goldberg	4.8	19.0	43.5	82.9	119.3	173.7	301.1	334.6
BGL push-relabel maxflow	34.8	143.3	323.9	580.4	921.5	–	–	–
Space Requirements [Mbyte]								
LEDA Graph	2.1	4.4	6.5	8.7	10.1	13.0	15.2	17.4
Static Bidirectional Graph	0.9	1.9	2.9	3.8	4.8	5.7	6.7	7.6
Static Opposite Graph	0.8	1.7	2.5	3.4	4.2	5.0	5.9	6.7

implementations: the *hi_pr* code of Cherkassky and Goldberg [1] and the preflow-push implementation of the BGL library [9]. Note, that is is not meant to be a systematic evaluation of different maxflow implementations. Since the *hi_pr* and BGL implementations are based on different preflow push heuristics they might perform better on different problem instances.

All networks were generated with the *ak*-generator of Cherkassky and Goldberg [3, 1, 8]. This generator produces maxflow instances which are hard to solve for preflow-push algorithms. The size of the generated network depends on a parameter i. More precisely, an ak(i)-network consists of $4i+6$ nodes and $6i+7$ edges. For a network of n nodes and m edges the space requirement is $12n+11m$ words for the LEDA graph $5n+5m$ words for the static bidirectional graph, and $5n+4m$ words for the static opposite graph.

Table 1 shows all results of our experiments. The combination of a static opposite graph and static slot assignment is, as expected, faster than any other combination. One can also see, that the effect of choosing different data assignment methods, is larger for the small graphs (bidirectional and opposite) than for the large LEDA graph.

7 Conclusions and Current Work

In this paper we presented the design, the implementation, and an experimental evaluation of a collection of static graph types for the LEDA library. The design is based on the observation that the interface of the general graph data type can be separated into two parts: The global representation of the sets V and E of the nodes and edges, which we call the *structure* of the graph, and the implementation of the actual node and edge objects and their relations to each other which we call the *category* of the graph. All graph types presented in this paper fit into the LEDA environment of graphs and graph algorithms by providing a subset of the interface of the general graph and they can be used together with the various methods for assigning information to the nodes and edges of a graph.

Furthermore, we presented a new graph data structure, the *opposite graph*, for a very compact representation of bidirectional graphs and showed its practical efficiency by using it in LEDA's preflow-push code. Currently, we are working on the implementation of special variants of the static graph structure, such as fixed-degree graphs for applications in computational geometry (see [10] for a prototype implementation), and on dynamic graph types using singly- or doubly-linked lists for the representation of the adjacency lists.

References

[1] Andrew V. Goldberg's Home Page. http://www.avglab.com/andrew/soft.html.
[2] Dimacs implementation file format. http://dimacs.rutgers.edu/Challenges/.
[3] B. V. Cherkassky and A. V. Goldberg. On Implementing Push-Relabel Method for the Maximum Flow Problem. *Algorithmica*, (19):390–410, 1997.
[4] A. V. Goldberg and R. E. Tarjan. A New Approach to the Maximum Flow Problem. *ACM Symposium on Theory of Computing*, (18):136–146, 1986.
[5] Dietmar Kühl. Design patterns for the implementation of graph algorithms, July 1996.
[6] K. Mehlhorn, S. Näher, M. Seel, and C. Uhrig. *The LEDA User Manual Version 4.3*. Algorithmic Solutions GmbH, Saarbrücken, 2001.
[7] Kurt Mehlhorn and Stefan Näher. Leda: a library of efficient data structures and algorithms. *Communications of the ACM*, (38):96–102, 1995.
[8] Kurt Mehlhorn and Stefan Näher. *LEDA: A Platform for Combinatorial and Geometric Computing*. Cambridge University Press, 1999.
[9] Jeremy G. Siek, Lie-Quan Lee, and Andrew Lumsdaine. *The Boost Graph Library*. Addison Wesley, 2001.
[10] O. Zlotowski and M. Bäsken. An experimental comparision of two-dimensional triangulation packages. In *CGAL User Workshop*, 2002. to appear.

An Exact Algorithm for the Uniformly-Oriented Steiner Tree Problem

Benny K. Nielsen, Pawel Winter, and Martin Zachariasen

Department of Computer Science, University of Copenhagen
DK-2100 Copenhagen Ø, Denmark
{benny,pawel,martinz}@diku.dk

Abstract. An exact algorithm to solve the Steiner tree problem for uniform orientation metrics in the plane is presented. The algorithm is based on the two-phase model, consisting of full Steiner tree (FST) generation and concatenation, which has proven to be very successful for the rectilinear and Euclidean Steiner tree problems. By applying a powerful canonical form for the FSTs, the set of optimal solutions is reduced considerably. Computational results both for randomly generated problem instances and VLSI design instances are provided. The new algorithm solves most problem instances with 100 terminals in seconds, and problem instances with up to 10000 terminals have been solved to optimality.

1 Introduction

Suppose that we are given a set P of n points in the plane. We are interested in finding a Steiner minimum tree (SMT) for P under the assumption that the edges are permitted to have only a limited number $\lambda \geq 2$ of equally-spaced orientations. We can assume that the permissible orientations are defined by straight lines making angles $i\omega$, $i = 0, 1, \ldots, \lambda - 1$, with the positive x-axis, where $\omega = \pi/\lambda$. Finding such a tree, called a λ-SMT (Fig. 1), is in general NP-hard[1] since the problem is NP-hard for $\lambda = 2$ [6].

Our interest in λ-SMTs is motivated by recent developments in VLSI technology. It will soon be possible to manufacture chips with wires having more than two orientations. In particular, the $\lambda = 4$ case which corresponds to adding diagonal routing has recently received significant attention [16]; note that a λ-SMT for $\lambda = 4$ can be almost 30% shorter than one with $\lambda = 2$.

Previous research on the Steiner tree problem with uniform orientations includes the early work by Sarrafzadeh and Wong [12] on heuristics for the problem. More recently, Li, Leung and Wong [9] presented comprehensive experimental results for an improved heuristic. Also, Li et al. [8] gave some important structural results for the $\lambda = 3$ case. Brazil, Thomas and Weng [3, 4] presented several fundamental results for the general case, including a Melzak-like algorithm for constructing a λ-SMT for a given topology; a linear-time algorithm for

---

[1] It has not been proved that the problem is NP-hard for *every* value of λ, but we strongly conjecture this to be the case.

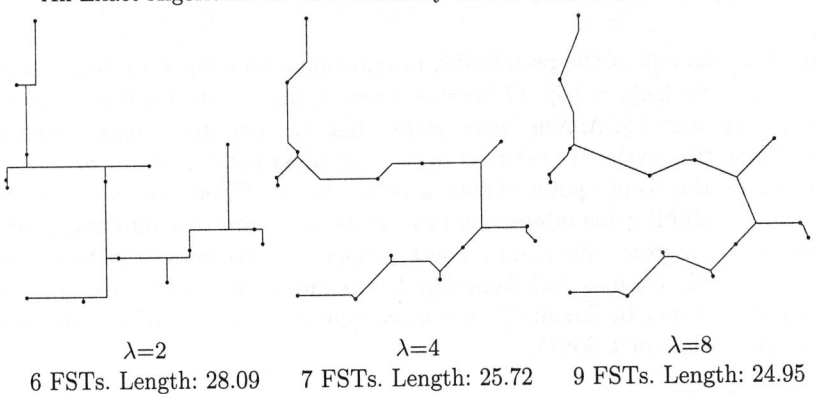

$\lambda=2$ $\lambda=4$ $\lambda=8$
6 FSTs. Length: 28.09   7 FSTs. Length: 25.72   9 FSTs. Length: 24.95

Fig. 1. λ-SMTs with varying λ for the same set of 15 points. Note that the topologies are *not* identical

this problem was given by Nielsen, Winter and Zachariasen [11] for the $\lambda = 3m$ case (where $m \geq 1$ is an integer). Recently, Brazil et al. [5] gave a linear-time algorithm for constructing a λ-SMT for a given topology for all values of λ.

In this paper we describe an exact algorithm for the Steiner tree problem with λ uniform orientations, where $2 < \lambda < \infty$. The exact algorithm is based on the two-phase model which has proven to be very efficient in connection with the rectilinear ($\lambda = 2$) and the Euclidean ($\lambda = \infty$) Steiner minimum tree problems [13, 14, 15, 17]. In the *generation phase* we compute a small set $\mathcal{F}$ of so-called full Steiner trees (FSTs) while ensuring that a union of a subset of $\mathcal{F}$ gives a λ-SMT for the entire set of terminals. Finding this subset of FSTs is called the *concatenation phase*. While the concatenation phase is essentially the same as the concatenation phase in the Euclidean and the rectilinear Steiner tree problems, the generation phase is quite different, and it is the main subject of the first part of this paper. In the second part of the paper we will provide some computational results. They will in particular emphasize how the lengths of λ-SMTs decrease as λ grows.

2 Basic Properties of λ-SMTs

Some of the basic properties for Euclidean SMTs ($\lambda = \infty$) are also true for all λ-SMTs. Most importantly, every λ-SMT is a union of *full Steiner trees* where each terminal has degree 1. The number of Steiner points that can appear in any λ-SMT is still at most $n - 2$. Each Steiner point has degree 3 — apart from a few special cases for $\lambda = 2, 3, 4$ and 6. In the following we will therefore assume that Steiner points have degree 3 (the special handling of degree 4 Steiner points is discussed in the full paper).

Other properties are not shared with the Euclidean SMTs. For instance, edges connecting terminals and Steiner points are in general not straight line segments. Consider a pair of points p and q in the plane. If the Euclidean line segment pq has one of the permissible orientations, then pq is a *straight edge* in λ-geometry and $|pq|_\lambda = |pq|$. Otherwise, there is an infinite number of shortest paths joining p and q. Among these paths, there are two involving precisely one bend. These two paths are referred to as *bent edges* bending at a *corner point*. Let r denote the corner point of such a bent edge pq. Then $|pq|_\lambda = |pr| + |rq|$.

λ-SMTs exhibit some interesting properties which are very important for the efficiency of our exact algorithm. These properties were proved in the seminal paper by Brazil, Thomas and Weng [3]. The reader is referred to this paper and to the survey paper by Brazil [2] for a more systematic and in-depth coverage of various properties of λ-SMTs.

Theorem 1. *The minimum angle at a Steiner point is $\lceil 2\lambda/3 - 1 \rceil \omega$ while the maximum angle is $\lfloor 2\lambda/3 + 1 \rfloor \omega$.*

Theorem 2. *For any set of points there exists a λ-SMT such that each of its FSTs contains at most one bent edge.*

3 Generation of FSTs

The goal of the FST generation phase is to (implicitly) enumerate all FSTs spanning from 2 to n terminals. Most FSTs are discarded since it can be shown that they cannot appear in any λ-SMT. The generation algorithm is based on the fact that there exists an optimal solution where each FST contains at most one bent edge (Theorem 2). In the following a *half FST* is a maximal subtree of an FST with straight edges and one "dangling" *extension ray*. The extension ray extends from the *root* of the half FST.

Every FST is a combination of two half FSTs. When the FST has a bent edge, the extension rays of the two half FSTs intersect at the corner point of the bent edge — otherwise the two extension rays overlap. Furthermore, every half FST with two or more terminals is also a combination of two half FSTs. These straightforward observations are essential since they permit us to focus on generating half FSTs with straight edges only. A typical (half) FST is depicted in Fig. 2. Notice the corner point r of the bent edge. If we split the FST at r, we obtain two half FSTs.

The *size* of an (half) FST is the number of terminals spanned by it. The smallest half FST is a terminal with an extension ray (note that for a given terminal we therefore associate 2λ different half FSTs). Half FSTs (and FSTs) of size 2 are obtained by combining two half FSTs of size 1. A lot of combinations are infeasible as the extension rays do not intersect. Those that do intersect can be tested with a range of different methods to avoid non-optimal/canonical FSTs and half FSTs (these methods will be described in Section 4). Some combinations will survive and they will be added to a list of surviving FSTs and/or a list of half FSTs of size 2.

When all half FSTs of size 2 have been generated, FSTs and half FSTs of size 3 are generated by combining half FSTs of size 1 with half FSTs of size 2. More generally, FSTs and half FSTs of size k are generated by combining a half

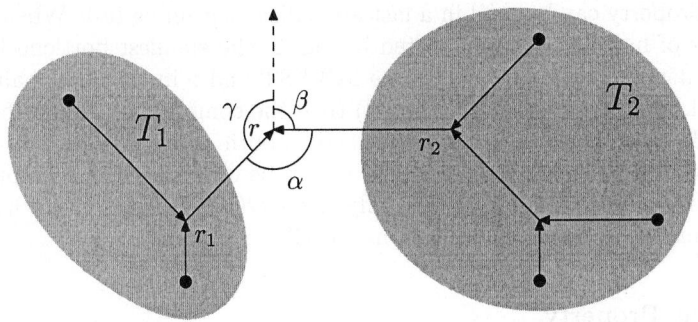

Fig. 2. Two half FSTs T_1 and T_2 which are combined into a new tree T. The dashed arrow is a possible extension of T

FST of size j to one of size $k - j$, for $j = 1, \ldots, k - 1$, such that a given pair of half FSTs is only combined once. The generation of half FSTs of size n is not required, but FSTs of size n must be generated.

4 Pruning Techniques

To avoid combinatorial explosion when generating FSTs and half FSTs, it is necessary to apply some effective pruning techniques. Some of the methods used in the Euclidean and the rectilinear cases can also be applied in other λ-metrics although most of them need some changes. Furthermore, some new techniques, applicable in λ-metrics only, prove to be very efficient.

4.1 Meeting Angles

Let α denote the smaller of two angles which are formed at the intersection point r of the extension rays when a pair of half FSTs is combined (Fig. 2). Assume that the extension ray of the resulting half FST makes angles β and γ with the other extension rays. By Theorem 1, all three angles α, β and γ must be within the interval $\lceil 2\lambda/3 - 1 \rceil \omega$ to $\lfloor 2\lambda/3 + 1 \rfloor \omega$. As a consequence, at most three extension ray candidates exist for the combined half FST.

When the extension rays for a pair of half FSTs meet at an angle of $\pi - \omega$ or the two rays overlap, the combined tree is an FST candidate.

4.2 Bottleneck Steiner Distances

The bottleneck Steiner distance $b(u, v)$ for two terminals u and v is the longest edge on the path between u and v in a minimum spanning tree (MST) for the terminals. No edge on the path between two terminals u and v in an SMT can be longer than $b(u, v)$ [7].

This property can be used in a fast and efficient pruning test. When combining a pair of half FSTs as illustrated in Fig. 2, the smallest bottleneck Steiner distance, denoted by $b^*(u,v)$, u in one half FST and v in the other half FST, is computed. If $b^*(u,v) < \max(|r_1 r|, |r_2 r|)$ then the combination cannot be part of any λ-SMT and therefore it can be pruned as both a half FST and an FST.

Note that bottleneck Steiner distances can be determined in a preprocessing phase. This means that each check only takes $O(n_1 n_2)$ time, where n_1 and n_2 are the number of terminals in each half FST.

4.3 Lune Property

Consider an edge uv in a λ-SMT. Then there cannot exist a terminal t so that $|ut|_\lambda < |uv|_\lambda$ and $|vt|_\lambda < |uv|_\lambda$ — otherwise a shorter tree could be constructed by deleting the edge uv and adding an edge from t to either u or v (whichever reconnects the subtrees obtained by deleting uv). This can be used for another efficient pruning test when combining half FSTs. As with the bottleneck Steiner distance test, we examine the two new edges $r_1 r$ and $r_2 r$ (see Fig. 2). If there exists a terminal whose distances to r_1 and r (resp. r_2 and r) are both shorter than $|r_1 r|_\lambda$ (resp. $|r_2 r|_\lambda$) then the combined half FSTs cannot be in any λ-SMT.

4.4 Upper Bounds

Whenever an (half) FST T is generated by combining two half FSTs T_1 and T_2 at a root r, it is important to compute a tight upper bound on the length of the λ-SMT for the root and terminals in T. If a shorter (heuristic) tree interconnecting the root and the terminals exists, the (half) FST cannot be part of any λ-SMT. The length of some heuristic tree is denoted an *upper bound*.

Probably the simplest upper bound is to determine a terminal t_1 in T_1 closest to the root r, and a terminal t_2 in T_2 closest to the root r. Then $|T_1| + |T_2| + |t_1 r|_\lambda + |t_2 r|_\lambda$ is a valid upper bound.

Another upper bound is to replace the greater of the terms $|t_1 r|_\lambda$ and $|t_2 r|_\lambda$ by the smallest bottleneck Steiner distance between a terminal in T_1 and a terminal in T_2. A third upper bound can be obtained by determining an MST for r and terminals in T.

4.5 Forbidden Subpaths and Canonical Forms

An interesting corrolary of the so-called *forbidden subpath theorem* [3] was recently given in [5]. Assume that we label the nodes of an FST T using labels "odd" and "even" such that the endpoints of any edge in T have distinct labels — thus orienting each edge from the odd to the even endpoint vertex. The *direction* of a straight (directed) edge uv of T is the angle of the vector uv with the positive x-axis. The directions of the two half edges that form a bent edge are defined similarly. Now we have the following theorem:

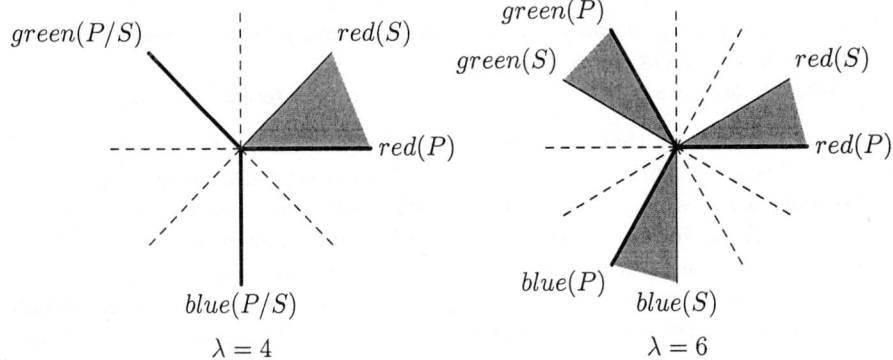

Fig. 3. Coloring of edges in an FST for $\lambda = 4$ and $\lambda = 6$. The primary (P) and secondary (S) directions are also given for each color

Theorem 3. *Let T be an FST in a λ-SMT. Up to rotation, if $\lambda \neq 3m$ then the edges in T have at most 4 different directions: $0, \omega, \lceil 2\lambda/3 \rceil \omega, \lceil 4\lambda/3 \rceil \omega$. Similarly, if $\lambda = 3m$ then the edges in T have at most 6 different directions: $0, \omega, (2\lambda/3)\omega, (2\lambda/3 + 1)\omega, (4\lambda/3)\omega, (4\lambda/3 + 1)\omega$.*

Another way to characterize the result of Theorem 3 is to say that FST edges can be divided into three classes of edges, denoted by *red*, *green* and *blue* edges (Fig. 3). These three classes correspond to edges having directions being approximately $2\pi/3 = (2\lambda/3)\omega$ apart. For $\lambda \neq 3m$, the red edges can have two different directions separated by the angle ω; all blue edges have the same direction and so do all the green edges. For $\lambda = 3m$, each class has edges having two directions separated by the angle ω.

If an FST contains at least 4 directions, then we may denote each straight edge as being either *primary* or *secondary*, depending on which of the two directions of its color class it uses. We assume w.l.o.g. that the primary direction is followed counter-clockwise by the secondary direction as illustrated in Fig. 3. A bent edge uses both the primary and secondary direction of its class. Note that an edge is either a primary edge, a secondary edge or a bent edge.

Let T be any fulsome FST in a λ-SMT, that is, T cannot be split into two or more FSTs having the same total length. (Clearly there exists a λ-SMT for which all FSTs are fulsome.) It turns out that we can freely move primary or secondary edge "material" around in T without changing its length. More precisely, we say that an edge uv in T has a primary (resp. secondary) *component*, if uv is either a bent edge or a primary (resp. secondary) edge. The following powerful theorem was proved in [5]:

Theorem 4. *Let T be a fulsome FST in a λ-SMT. Let e_1 be an edge in T having a secondary component and let e_2 be another edge in T having a primary component. Then there exists a perturbation of the Steiner points in T such that*

- the resulting tree T' has the same length as T,
- all primary (resp. secondary) edges in T remain primary (resp. secondary) edges in T', except for e_1 and e_2,
- either e_1 becomes a primary edge or e_2 becomes a secondary edge.

Thus we may say that we move primary edge material from e_2 to e_1 until either e_1 becomes a primary edge or e_2 becomes a secondary edge (Fig. 4). By performing a series of such so-called zero-shifts we can therefore transform T into a tree T_c having some canonical form: For any numbering $1, \ldots, m$ of the edges in T, there exist a number l, $1 \leq l \leq m$, such that all edges numbered 1 to $l-1$ are primary edges and all edges numbered $l+1$ to m are secondary edges. The l'th edge may be a bent edge. The numbering scheme — and thus the canonical form — that we will choose is the following:

- Every FST is rooted at its lowest index terminal (as given by the input to the algorithm) and the edges are numbered according to a *depth-first* traversal from this root.
- The children of a node u are visited in the order given by their geometric location in T such that the leftmost child is visited before the rightmost child (see Fig. 4). Note that the perturbation used in Theorem 4 preserves this ordering, that is, the chosen edge ordering would be exactly the same when based on T_c (see [5] for more details).

In order to characterize the resulting canonical form using the notion of half FSTs, we need some definitions. A half FST is *clean* if it contains at most three directions (including the extension ray) that form a valid set of purely primary or secondary edges. (For $\lambda = 3m$ a clean half FST has all edges meeting at an angle of $2\pi/3$ at Steiner points.) Otherwise, the half FST is *mixed*.

Using this notation, we may characterize the chosen canonical form as follows (see also Fig. 4). As usual, an FST T is a combination of two half FSTs T_1 and T_2. Let us assume that the terminal t having minimum index is in T_1. Then half FST T_2 is a clean half FST containing only secondary edges. The path $t, s_1, \ldots, s_k$ in T_1 from t to the bent edge (or extension ray of T_1) consists of primary edges, and the subtrees connected to the Steiner points $s_1, \ldots, s_k$ correspond to clean half FSTs. Furthermore, the half FSTs connected to the Steiner points on the left (when traversing from t to the bent edge) contain only primary edges while the half FSTs on the right contain only secondary edges.

This canonical form has a major impact on the combination of half FSTs, since the minimum index terminal t in a mixed half FST must necessarily also be the minimum index terminal in any FST constructed from it. In fact, the path from t to the extension ray will be a prefix of the same path from t to the bent edge, since the bent edge will appear "outside" the half FST. By storing the clean/mixed status and the set of directions used by a half FST, several powerful constant time pruning tests can be applied:

1. A mixed half FST can only be combined with a clean half FST, and the minimum index terminal must reside in the mixed half FST. This is denoted the mixed/clean test.

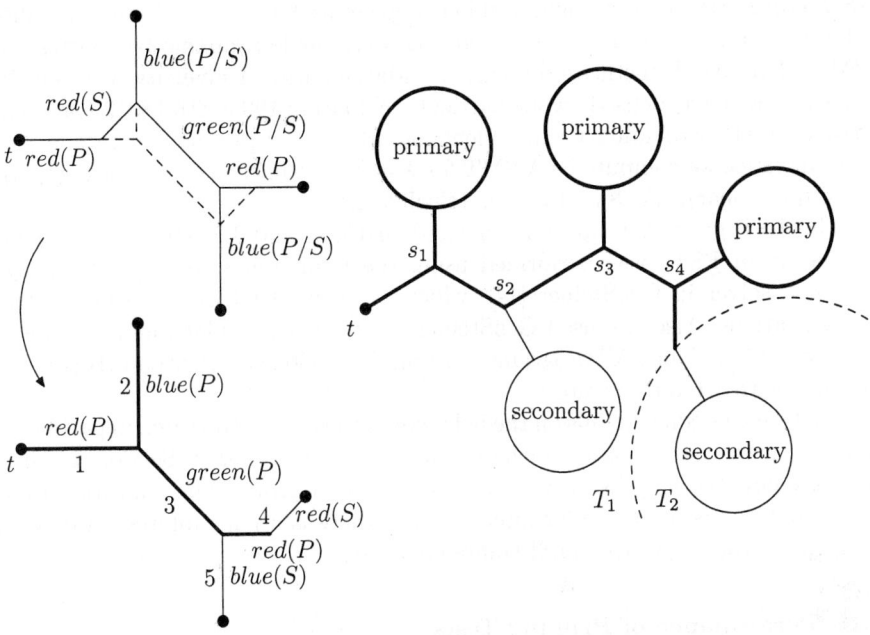

Fig. 4. Left: A zero-shift for an FST. Right: Canonical form for FSTs, where terminal t has minimum index

2. When combining a mixed half FST with a clean half FST, the directions used by the clean half FST must correspond to primary (resp. secondary) edges in the mixed half FST when the clean half FST is on the left (resp. right) as seen from the mixed half FST; in addition, only one extension ray is possible, corresponding to a primary edge.
3. When combining two clean half FSTs into a mixed half FST, the combined half FST must have the required canonical form; e.g., there can be no primary edge in a right subtree on the path from the minimum index terminal to the new root.

For $\lambda = 3m$ it is easy to identify primary and secondary edges in a half FST, since a pair of edges meeting at an angle distinct from $2\pi/3$ uniquely identifies the primary/secondary edge. For $\lambda \neq 3m$ more bookkeeping is necessary, e.g., it is necessary to know all the directions contained in a mixed half FST in order to decide which edges are primary and which are secondary.

5 Experiments

The new exact algorithm for computing uniformly-oriented Steiner trees in the plane was evaluated on three sets of problem instances: VLSI design instances,

OR-Library instances [1], and randomly generated instances (uniformly distributed in a square). The VLSI instances were made available by courtesy of IBM and Research Institute for Discrete Mathematics, University of Bonn. In this study we have focused on one particular chip from 1996 with 180129 problem instances/nets. Preliminary experiments on other chips gave similar results. For every instance we computed a λ-SMT for $\lambda = 2$ (rectilinear), $\lambda = 3$ (hexagonal), $\lambda = 4$ (octilinear), $\lambda = 8$ and $\lambda = \infty$ (Euclidean).

The FST generator was implemented in C++ and LEDA [10]. The concatenation of FSTs was performed using the minimum spanning tree in hypergraph solver in GeoSteiner [14], which again used CPLEX 7.0 for solving LP-relaxations. Also, we used GeoSteiner for solving Euclidean and rectilinear Steiner tree problems. All programs were run on a 930 MHz Pentium III (Linux) machine with 1 Gb of memory.

First we give some details on the behavior of the exact algorithm, in particular concerning the performance of the pruning tests presented in Section 4. Then we show how the lengths of λ-SMTs decrease as λ grows. The transition from $\lambda = 2$ to $\lambda = 4$ is of particular interest, since there is a huge interest in moving to diagonal routing in the VLSI-community [16].

5.1 Performance of Pruning Tests

In Table 1 we give some details on the performance of the pruning tests presented in Section 4. Statistics for three randomly generated instances with 10, 20 and 50 terminals are presented ($\lambda = 4$). The first line gives the number of attempts to combine two half FSTs, and the following lines give the number of remaining combinations after each of the pruning tests. Clearly, the order of execution of the tests is essential, and the given order was found to be superior to other alternatives.

For $n = 10$ (resp. $n = 20$ and $n = 50$) only 0.43% (resp. 0.18% and 0.06%) of the proposed combinations survive all the basic pruning tests. Note the efficiency of the BSD test for the larger instances. The surviving combinations are then extended in all possible directions and they are added to the pool of half FSTs. Finally, the combinations go through various FST tests such as checking for the existence of shorter FSTs for the same sets of terminals.

5.2 Length Reduction When Compared to Rectilinear SMTs

The reductions in length when compared to the rectilinear SMT ($\lambda = 2$) were computed for both VLSI and random instances. The 180129 nets in the chip range from 2 to 86 terminals; however, 99.5% of the nets have 20 or fewer terminals. We solved all the nets in the chip in about an hour ($\lambda = 4$).

The total wire-length reduction for the complete chip is 5.9% for $\lambda = 3$, 10.6% for $\lambda = 4$, 13.4% for $\lambda = 8$ and 14.3% for $\lambda = \infty$. The reduction in percent decreases for larger instances. Chip instances show a smaller reduction when compared to randomly generated instances; this is due to the fact that many of the nets have several axis-aligned terminals. This improves the relative

Table 1. Pruning test statistics for three particular instances for size 10, 20 and 50 ($\lambda = 4$). The left sub-column in each column is the number of combinations surviving each test and the right sub-column is the corresponding percentage. The bottom lines give the final number of FSTs generated and the number of FSTs used in an SMT

Pruning tests	$n = 10$ #	%	$n = 20$ #	%	$n = 50$ #	%
Number of combinations	58759	100.00	349866	100.00	2968266	100.00
Mixed/clean test	45549	77.52	247977	70.88	2140068	72.10
Disjoint combinations	22705	38.64	173996	49.73	1873572	63.12
Simple angle test	11416	19.43	87125	24.90	937147	31.57
Ray intersection	2626	4.47	16318	4.66	172694	5.82
BSD test	731	1.24	2681	0.77	6505	0.22
Lune test	559	0.95	1629	0.47	4477	0.15
Upper bounds	252	0.43	622	0.18	1872	0.06
Number of surviving FSTs	22	0.04	47	0.01	120	0.00
Number of FSTs used in SMT	5	0.01	10	0.00	32	0.00

Table 2. Running times in seconds for the exact algorithm including FST concatenation. For each λ, the column VLSI gives the average running time (in seconds) for the chip instances while the Rand column gives the average running time for the randomly generated instances. GeoSteiner was used for $\lambda = 2$ and $\lambda = \infty$

Size	$\lambda = 2$ VLSI	Rand	$\lambda = 3$ VLSI	Rand	$\lambda = 4$ VLSI	Rand	$\lambda = 8$ VLSI	Rand	$\lambda = \infty$ VLSI	Rand
5	0.002	0.001	0.017	0.018	0.018	0.019	0.018	0.018	0.003	0.004
10	0.004	0.005	0.021	0.024	0.027	0.032	0.035	0.052	0.024	0.041
20	0.011	0.015	0.061	0.069	0.121	0.153	0.262	0.263	0.320	0.232
50	-	0.073	-	0.435	-	1.171	-	2.262	-	1.419
100	-	0.273	-	2.147	-	5.981	-	17.601	-	4.345

quality of the rectilinear SMT. The average reduction for large random problem instances is approximately 4.5% for $\lambda = 3$, 8.6% for $\lambda = 4$, 11.3% for $\lambda = 8$ and 12.1% for $\lambda = \infty$.

5.3 Running Times

The running times of the exact algorithm are presented in Table 2 and Table 3. These running times include FST concatenation, but this is only a small fraction of the total running time for this problem size range. The number of generated FSTs is highest for the rectilinear problem and decreases with increasing λ. Approximately $4n$ FSTs are generated for $\lambda = 2$ and approximately $2n$ for $\lambda = \infty$, where n is the number of terminals.

Table 3. Running times in seconds and average improvements compared to MST in percent (OR-library instances). Each entry is an average of 15 instances. GeoSteiner was used for $\lambda = 2$ and $\lambda = \infty$

Size	$\lambda = 2$		$\lambda = 3$		$\lambda = 4$		$\lambda = 8$		$\lambda = \infty$	
	Sec	Impr	Sec	Impr	Sec	Impr	Sec	Impr	Sec	Impr
10	0.005	10.656	0.028	4.680	0.042	4.523	0.055	3.646	0.034	3.251
20	0.021	11.798	0.061	3.744	0.156	4.482	0.269	3.566	0.208	3.156
30	0.044	11.552	0.154	4.296	0.364	4.202	0.724	3.443	0.564	3.067
40	0.069	10.913	0.343	4.599	0.887	4.393	1.633	3.542	1.056	3.139
50	0.157	10.867	0.355	4.053	1.147	4.220	2.246	3.380	1.335	3.033
60	0.147	11.862	0.634	4.221	2.016	4.656	3.883	3.669	1.729	3.275
70	0.154	11.387	0.891	4.267	2.529	4.406	5.407	3.471	2.447	3.110
80	0.237	11.301	1.491	4.092	3.474	4.404	7.941	3.419	3.071	3.039
90	0.293	11.457	1.180	4.383	4.319	4.460	10.879	3.522	3.323	3.120
100	0.366	11.720	2.195	4.685	6.895	4.656	19.515	3.668	4.785	3.269
250	2.841	11.646	18.087	4.290	80.063	4.500	197.227	3.601	15.339	3.207
500	13.159	11.631	102.193	4.514	386.071	4.652	905.087	3.712	40.764	3.326
1000	202.483	11.655	2386.574	4.502	1618.026	4.649	3784.987	3.712	108.399	3.312
Avg.	-	11.419	-	4.333	-	4.477	-	3.565	-	3.177

The running times for the new general exact algorithm are comparable to the specialized (and highly optimized) algorithms for the rectilinear and Euclidean problems. The average running time growth is reasonable, in particular for small values of λ, and we are (on average) able to solve 100 terminal problem instances in less than 10 seconds. This is even more impressive when compared to the results reported by Li et al. [9] on their *heuristic* method. On average they spend more than 20 minutes on finding a heuristic solution for $n = 100$ and $\lambda = 4$ (SunUltra machine). Finally, we have also tried to solve the largest instance from the OR-library for $\lambda = 4$. This instance has 10000 terminals and it was solved in less than two days.

6 Conclusions

We presented the first exact algorithm for the uniformly-oriented Steiner tree problem in the plane. Our results indicate that the two-phase approach which has been used successfully for the rectilinear and Euclidean Steiner tree problems also can be used to efficiently solve this generalized problem. We plan to incorporate the algorithm into GeoSteiner [14] and to make further investigations into the applicability of octilinear Steiner trees in VLSI design.

Acknowledgments

The third author would like to thank Marcus Brazil, Doreen Thomas and Jia Weng for many fruitful and inspiring discussions during his visit at the University of Melbourne. The authors would like to thank Bernhard Korte and Sven Peyer, University of Bonn, for providing VLSI instances. The research was partially supported by the Danish Natural Science Research Council (contract number 56648).

References

[1] J. E. Beasley. OR-Library: Distributing Test Problems by Electronic Mail. *Journal of the Operational Research Society*, 41:1069–1072, 1990.
[2] M. Brazil. Steiner Minimum Trees in Uniform Orientation Metrics. In D.-Z. Du and X. Cheng, editor, *Steiner Trees in Industries*, pages 1–27. Kluwer Academic Publishers, 2001.
[3] M. Brazil, D. A. Thomas, and J. F. Weng. Minimum Networks in Uniform Orientation Metrics. *SIAM Journal on Computing*, 30:1579–1593, 2000.
[4] M. Brazil, D. A. Thomas, and J. F. Weng. Forbidden Subpaths for Steiner Minimum Networks in Uniform Orientation Metrics. *Networks*, 39:186–202, 2002.
[5] M. Brazil, D. A. Thomas, J. F. Weng, and M. Zachariasen. Canonical Forms and Algorithms for Steiner Trees in Uniform Orientation Metrics. *In preparation*.
[6] M. R. Garey and D. S. Johnson. The Rectilinear Steiner Tree Problem is NP-Complete. *SIAM Journal on Applied Mathematics*, 32(4):826–834, 1977.
[7] F. K. Hwang, D. S. Richards, and P. Winter. *The Steiner Tree Problem*. Annals of Discrete Mathematics 53. Elsevier Science Publishers, Netherlands, 1992.
[8] Y. Y. Li, S. K. Cheung, K. S. Leung, and C. K. Wong. Steiner Tree Constructions in λ_3-Metric. *IEEE Transactions on Circuits and Systems II: Analog and Digital Signal Processing*, 45(5):563–574, 1998.
[9] Y. Y. Li, K. S. Leung, and C. K. Wong. Efficient Heuristics for Orientation Metric and Euclidean Steiner Tree Problems. *Journal of Combinatorial Optimization*, 4:79–98, 2000.
[10] K. Mehlhorn and S. Näher. LEDA - A Platform for Combinatorial and Geometric Computing. Max-Planck-Institut für Informatik, Saarbrücken, Germany, http://www.mpi-sb.mpg.de/LEDA/leda.html, 1996.
[11] B. K. Nielsen, P. Winter, and M. Zachariasen. On the Location of Steiner Points in Uniformly-Oriented Steiner Trees. *Information Processing Letters*, 83:237–241, 2002.
[12] M. Sarrafzadeh and C. K. Wong. Hierarchical Steiner Tree Construction in Uniform Orientations. *IEEE Transactions on Computer-Aided Design*, 11:1095–1103, 1992.
[13] D. M. Warme, P. Winter, and M. Zachariasen. Exact Algorithms for Plane Steiner Tree Problems: A Computational Study. In D.-Z. Du, J. M. Smith, and J. H. Rubinstein, editors, *Advances in Steiner Trees*, pages 81–116. Kluwer Academic Publishers, Boston, 2000.
[14] D. M. Warme, P. Winter, and M. Zachariasen. GeoSteiner 3.1. Department of Computer Science, University of Copenhagen (DIKU), http://www.diku.dk/geosteiner/, 2001.
[15] P. Winter and M. Zachariasen. Euclidean Steiner Minimum Trees: An Improved Exact Algorithm. *Networks*, 30:149–166, 1997.
[16] X Initiative Home Page. http://www.xinitiative.com, 2001.
[17] M. Zachariasen. Rectilinear Full Steiner Tree Generation. *Networks*, 33:125–143, 1999.

A Fast, Accurate and Simple Method for Pricing European-Asian and Saving-Asian Options

Kenichiro Ohta[1,2], Kunihiko Sadakane[1], Akiyoshi Shioura[1], and Takeshi Tokuyama[1]

[1] GSIS Tohoku University
{ken,sada,shioura,tokuyama}@dais.is.tohoku.ac.jp
[2] DC Card Co., Ltd., Tokyo, Japan

Abstract. We present efficient and accurate approximation algorithms for computing the premium price of Asian options. First, we modify an algorithm developed by Aingworth et al. in SODA 2000 for pricing the Europian-Asian option and improve its accuracy (both theoretically and practically) by transforming it into a randomized algorithm. Then, we present a new option named Saving-Asian option, whose merit is in the middle of European-Asian and American-Asian options, and show that our method works for its pricing.

1 Introduction

Background

Options are popular financial derivatives. Options give the right, but not the obligation, to buy or sell something (we consider a stock in this paper) at some point in the future for a specified price (called *strike price*).

A simple option permits buying a stock at the end of the year for a predetermined price. If the stock is worth more than that price, then you can use the option to buy the stock for less than you otherwise could. The price of the option (called *premium* of the option) is usually much less than the underlying price of the stock. Use of options hedges risk more cheaply than using only stocks, and cheaply provides a chance to get large profit if one's speculation is good.

For example, if you are interested in a stock of a current price $200, and forecast that it will possibly go up beyond $300 in the year-end. You may buy 1000 units of the stock (probably falling in debt), and if your forecast comes true, you will gain $100,000; however, if the stock price goes down to $100, you will unfortunately lose $100,000, which you will not be able to afford. Instead, suppose that you can buy at the premium $8 an option that gives you the right to buy the stock at the strike price $220. If the stock price goes up to $300, you will obtain $80 extra (called *payoff*) for each unit by exercising the option and selling the stock at the market price. Thus, if you buy 1250 units of this option, you have a chance to gain total payoff of $100,000 (without considering the debt for the premium) reducing the maximum loss to be $10,000 (just the total premium). You may buy 2000 units of another option that has the strike price $250 and the premium $2, and dream to gain $100,000 with the maximum loss $4,000. But you will see that the latter option is not always better than the former one. Here, you must ask the question whether the option premiums $8 and $2 are fair or not. Therefore, pricing the options is a central topic in financial engineering.

A standard method (Black-Scholes model) is to model the movement of the underlying financial asset as Brownian motion with drift and then to construct an arbitrage portfolio. This yields a stochastic differential equation, and its solution gives a premium of the option. However, it is often difficult to solve such a differential equation, and indeed no closed-form solution is known for the Asian option discussed in this paper.

Therefore, it is widely practiced to simulate the Brownian motion by using a combinatorial model, and to obtain a solution on the model, which we call the *combinatorial exact premium price* or the *exact premium price*, as an approximation of the premium price obtained from the differential equation. The binomial (or trinomial) model is a combinatorial model, in which the time period is decomposed into n time steps, and the Brownian motion is modeled by using a biased random walk on a directed acyclic graph named *recombinant binary (or trinary) tree* of depth n with $n(n+1)/2$ (or n^2) nodes. Although our algorithms and analysis can be easily adjusted to work on the trinomial model, we focus on the binomial model for simplicity. The binomial model is a very popular model since the combinatorial exact premium price converges to the premium price given by the differential equation if we enlarge the size of the model.

In the binomial model, the process of price-movement of a stock (or any financial asset on which the option is based) is represented by a path in the binary recombinant tree. An option is called *path-dependent* if its value at the time of *exercise* depends not only the current price but also the path representing the process. An option is called American type if it permits early exercise. Path-dependency is necessary for designing an option that is secure against the risk caused by sudden change of the market, and also right of early exercise is convenient for users. However, an option with both functions is often difficult to analyze.

Our Problems and Results

The Asian option is a kind of path-dependent options. If we simulate the Black-Scholes model accurately by using a binomial model, the size often becomes large. Unfortunately, it is known to be #P-hard to compute the exact premium price on the binomial model for a path-dependent option in general [5]. Therefore, we need to design an efficient approximation algorithm with a provable high accuracy.

The European Asian option is the simplest Asian option. A naive method (*full-path* method) for computing the exact premium price of an European Asian option is to enumerate all the paths in the model; unfortunately, there are exponential number of paths. Thus, a random sampling method is a popular way to obtain an approximate solution; however, taking a polynomial number of samples naively is not enough to assure a theoretically probable accuracy. There are several polynomial-time approximation algorithms for pricing European-Asian options [5, 6], based on sampling method. For path-dependent call options, howver, the approximation error of a sampling method with a polynomial number of samples has a lower bound that depends on the volatility of the random process represented by the binomial model; moreover, $O(n^4)$ time is necessary to attain the accuracy matching the lower bound.

Recently, Aingworth-Motwani-Oldham [1] gave a breakthrough idea which avoids the influence of volatility to the theoretical error bound. The idea is to aggregate (exponential number of) high-payoff paths by using mathematical formulae during running an approximate aggregation algorithm based on dynamic programming. They proposed an $O(n^2 k)$ time algorithm (referred to AMO algorithm), and proved that its error is bounded by nX/k, where X is the strike price of the option, and k is a parameter giving the time-accuracy tradeoff. Akcoglu et al.[2] presented efficient methods for the pricing of European Asian option, and by using a recursive version of AMO algorithm they reduce the error bound to $n^{\frac{1+\epsilon}{2}} X/k$ spending the same time complexity under the condition that the volatility of the stock is small.

In this paper, we first give a randomized algorithm with an $O(n^2 k)$ time complexity and an $O(\sqrt{n} X/k)$ error bound for which we do not need a volatility condition. The algorithm is indeed a variation of AMO algorithm. The modification itself is quite small, and looks almost trivial at a glance. However, by this modification, the algorithm can be regarded as a variant of the sampling method (without limit of the above mentioned lower bound), as well as that of AMO algorithm. Thus, the algorithm can enjoy advantages of both methods simultaneously. Although algorithms on a uniform model has been mainly considered in the literature [1, 2] in algorithm theory, our algorithm and analysis work on a non-uniform model where the transition probabilities of the stock price may depend on the state of the graph modeling the process, and also work on a trinomial model. Moreover, the error bound can be improved to $O(n^{1/4} X/k)$ for the uniform case unless the transition probability p is extremely close to 1 or 0.

Our idea is the following: By considering a novel random variable, the aggregation process of the algorithm can be regarded as a Martingale process with n random steps. The expected value of its output equals the combinatorial exact price, and the error of its single step is bounded by X/k. Thus, we can apply Azuma's inequality [4] on the Martingale process to obtain the error bound. We show practical quality of our algorithm by an experiment: Indeed, its accuracy (for $n = 30$) is better by a factor nearly 100 than that of Aingworth-Motwani-Oldham's original algorithm.

Inspired from the analysis, we propose an intermediate option between American-Asian and European-Asian options, and show that our method also works for this option. Our option, which we name *Saving-Asian option*, permits early exercise, but the payoff system is different from the American option, so that we can anticipate the action of users and compute the expected payoff accurately. The payoff depends on the average stock price, and hence secures against sudden change of the market. Moreover, compared to the American-Asian option, the Saving-Asian option reduces the risk for the seller; thus the premium is cheaper. Therefore, we believe that our new option and its analysis will be useful in both theory and practice.

2 Preliminaries

We divide the period from the purchase date to the expiration date of an option into n time periods, and the t-th time step is the end of the t-th time period. Let S_t ($t = 0, 1, 2, \ldots, n$) be a random variable representing a stock price at the t-th time step, where S_0 is a constant known as the initial price.

Let X be the strike price of the option. *Payoff* is the value of the option, which is a random variable. In Black-Scholes' theory, the option premium is computed from the expected value of the payoff by subtracting the interest on the premium during the period, and hence it suffices to compute (or approximate) the expected value of the payoff.

2.1 Options

We only consider *call options* in this paper, although pricing of corresponding *put options* can be similarly (and more easily) done. We adopt a convention to write F^+ for $\max\{F, 0\}$.

The **European Call Option** is the most basic option, and its payoff $(S_n - X)^+ = \max\{S_n - X, 0\}$ is determined by the stock price of the expiration date (i.e., at the n-th time step). Note that S_n above is the real stock value that is revealed on the expiration date.

It is quite easy to compute the expected value of the payoff of the European call option if we use the binomial model. A drawback of the European option is that the payoff may be changed drastically by the movement of the stock price just before the expiration date; thus, even if the stock price goes very high during most of the period, it may happen that the option does not make money at the end.

The **European-Asian Option** is an option that can resolve the above-mentioned drawback of European option. The payoff of the European-Asian option is $(A_n - X)^+$, where $A_n = (\sum_{i=1}^{n} S_i)/n$ is the average of the stock prices during the period. Let $T_j = \sum_{i=1}^{j} S_i$ be the running total of the stock price up to the j-th time step. If $T_j > nX$, we will surely exercise the option at the expiration date, and the payoff is at least $T_j/n - X$. We call that the option is *in-the-money* if this happens. Thus, the European-Asian option is more reliable than the European option for buyers.

The **American-Asian Option** permits a buyer to exercise the option in any time period. A buyer receives $A_i - X$ if the option is exercised at the i-th time period, where $A_i = T_i/i$. Apparently, the option is much advantageous for buyers, and hence its premium should be more expensive. One difficulty of this option is that the action of a buyer is highly path-dependent. Even after the status of the option becomes in-the-money, a buyer must decide whether he/she exercises the option immediately; it should depend on both T_i and the current stock price. Thus, its pricing with provable accuracy seems quite difficult (see Section 4).

We propose a new option, named **Saving-Asian Option**[1]. In the Saving-Asian option, a buyer can exercise the option at any time period, and receive $e^{-(n-i)r_0/n}(T_i - iX)/n$ if the option is exercised at the i-th time period, where e^{r_0} is the risk-free interest rate for the whole period. Thus, it is an American type option, but different from a standard American-Asian option since it restricts the payoff for early exercise.

For a buyer, this option is clearly advantageous to the European Asian option, since he/she has a choice to keep the option until the expiration date in which case the payoff is $(A_n - X)^+$ that is exactly same as that of European Asian option. On the other hand, if a buyer exercises at the i-th period and re-invest the money, he/she will have $(T_i - iX)/n$ at the n-th step, which might be larger than $A_n - X = (T_n - nX)/n$. Therefore, if the stock price will drastically go down after enjoying some high-price period, a buyer can exercise early to avoid reduction of his/her profit. Moreover, early exercise has an advantage that a buyer can get money for urgent need.

Intuitively, this option simulates accumulative investment permitting discontinuation, in which a buyer has right to buy $1/n$ unit of the stock by X/n dollars for selling it by the market price every time period, and can stop at the i-th step after investing iX/n dollars to receive the profit obtained so far. Apparently the payoff is path-dependent, and thus the option is not in the category of Markovian-American option given in [5].

Similarly to the American-Asian option, the action of a buyer seems to be path-dependent. However, it is easier to analyze the best action assuming that a buyer has the same model of the stock price movement as the seller. In particular, in the uniform model (defined in the next subsection), once the status of the option becomes in-the-money, a buyer should sell the option in the i-th step if the expectation of the running total after the $(i+1)$-st step is less than $(n-i)X$; thus, the decision depends on the current stock price and the model, but is independent of the history of the movement of the stock. We remark that in our convention in this paper, in-the-money always means $T_i > nX$, although it is common that in-the-money means a buyer can get profit if he/she exercises immediately. We note that a buyer may exercise the option before it becomes in-the-money, and for that case the decision also depends on the current running total; however, this kind of path-dependency can be treated efficiently. We remark that the payoff function can be customized and our analysis still applies (Section 5).

---

[1] This option might be proposed before, although the authors do not know.

2.2 Binomial Model

Let us consider a discrete random process simulating the movement of the price of a stock. The fundamental assumption in the binomial model (and the Black-Scholes model) is that in each time step the stock price S either rises to uS or falls to dS, where $u > d$ are predetermined constants.

Thus, we can model stock price movement by a recombinant binary tree. A *recombinant binary tree*[2] G is a leveled directed acyclic graph whose vertices have at most two parents and two sons. We label the nodes (i,j) where i denotes the level and j denotes the numbering of the nodes in the i-th level ($0 \leq j \leq i$). The node (i,j) has two sons $(i+1,j)$ and $(i+1,j+1)$ if $i \leq n-1$. Therefore, the node (i,j) has parents $(i-1,j)$ and $(i-1,j-1)$ if $i \neq 0$ and $1 \leq j \leq i-1$. Each of the nodes (i,i) and $(i,0)$ has one parent. Intuitively, the graph looks like the structure of Pascal's triangle.

In the model, if we are at a node $v = (i,j)$ and the current stock price is S, we move to $(i+1,j)$ with probability p_v and the stock price rises to uS. With probability $1-p_v$, we move to $(i+1, j+1)$ and the stock price falls to dS. Thus, if we are at the node (i,j), the stock price must be $u^{i-j}d^j S_0$.

The model is called *uniform* if $p_v = p$ for every node v; otherwise it is non-uniform. The uniform model is widely considered [1, 2, 5, 6] since p is uniquely determined under the non-arbitrage condition of the underlying financial object; however, non-uniform model is often useful to customize an option. We consider the uniform model first, and will show later how to deal with the non-uniform model. Our method also works for the trinomial model where each node (except those in the n-th level) has three sons, and stock price moves to one of uS, S, and $u^{-1}S$, although we omit details in this paper. In the uniform model, the probability that the random walk reaches to (i,j) is $\binom{i}{j} p^{i-j}(1-p)^j$. We define $r = up + d(1-p) - 1$, which corresponds to the risk-neutral interest rate for one time period in the risk-neutral model.

Our task is to compute the expected value $E((A_n - X)^+)$ of the payoff. A simple method is to compute the running total $T_n(\mathbf{p})$ of the stock value for each path $\mathbf{p}$ in the graph G together with the probability $prob(\mathbf{p})$ that the path occurs, and exactly compute $E((A_n - X)^+) = \sum_{\mathbf{p}}(prob(\mathbf{p})(T_n(\mathbf{p})/n - X)^+)$. The expected value U of the payoff computed as above is called the *exact value* of the expected pay-off.

However, this needs exponential time complexity with respect to n, since there are 2^n different paths. Random sampling of paths is a popular method to reduce the computation time, although we need to have huge number of paths in order to have a small provable error bound if we naively sample paths.

---

[2] Often called binomial lattice

3 Our Algorithm for Pricing European Asian Option

3.1 AMO Algorithm

We give a brief overview of AMO algorithm (see [1] for details). AMO algorithm is based on dynamic programming and has an $O(n^2 k)$ time complexity with a provable error bound of nX/k, where k is a parameter to give the time-error tradeoff.

For a path **p** from the root to a node of level t, its *stamp* is the pair of its current stock price and the running total. Note that the current stock price corresponds to the node. The basic idea of AMO algorithm is to approximate the running totals appropriately so that the number of different stamps is at most $(t+1)k$, and to store the approximate stamps at the t-th time step into a table with $(t+1)$ rows and k columns. Moreover, if the running total T_t exceeds nX for a path **p** (i.e., the option is in-the-money), the expectation of the payoff of paths containing **p** as a prefix is analytically computed (see Section 3.3), and the stamp corresponding to the path **p** is pruned away from the table.

The row index corresponds to the stock prices. The stock price takes one of $(t+1)$ values $u^i d^{t-i} S_0$ for $i = 0, 1, .., t$ in the binomial model, and hence naturally we assign paths with the stock price $S_t(i) = u^i d^{t-i} S_0$ to the $(i+1)$-st row. The column index corresponds to the running total T_t of paths. Running totals of unpruned paths are assorted into k buckets $B(s)$ for $s = 1, 2, \ldots, k$, where $B(s)$ represents the interval $[b_{s-1}, b_s) = [(s-1)nX/k, snX/k)$.

A cell in the table is indicated by a pair of a stock value and a bucket. Suppose that many stamps are assorted into the cell $C(S_t(i), B(s))$. Then, the algorithm approximates them as a stamp with the current stock price $S_t(i)$ and running total b_{s-1}. The stamp has a weight $w_t(s, i)$, where $w_t(s, i)$ is the summation of the probability that each of the paths occurs.

Since the error caused in one step of the process is bounded by nX/k for $(T_n - nX)^+$, the error contribution to $(A_n - X)^+$ is at most X/k for one step. Thus, the accumulated errors in the final running total will be $n^2 X/k$, and the error in the estimation of the average stock value is bounded by nX/k. More precisely, if $U = E((A_n - X)^+)$ is the exact value of the expected payoff in the binomial model and Φ is the payoff computed by the algorithm, we have $U \geq \Phi \geq U - nX/k$.

3.2 Modified Algorithm

Our modification of AMO algorithm is quite simple. In order to represent the stamps in a cell $C(S_t(i), B(s))$, we apply random sampling so that a stamp with weight w is selected with a probability $w/w_t(s, i)$, and give a weight $w_t(s, i)$ to it. Let Ψ be the payoff value computed by the algorithm. At a glance, it looks merely like a heuristic, and does not improve the theoretical bound.

Indeed, in the worst case, the error caused in one step is only bounded by X/k, and hence we can only prove that the worst case error bound $|U - \Psi|$ is nX/k. However, the algorithm can be also viewed as a sampling method of paths, since stamps stored in the table are real stamps of some suitable paths. We can observe that the selection of paths is smartly done during the runtime of the algorithm: In each step the path-prefixes are clustered by using the table, and a path-prefix is selected from each cluster to continue.

Since the algorithm is randomized, Ψ is a random variable depending on the coin-flips to choose representatives of stamps in the table. Let Y_i be a random variable giving the "exact" payoff value after running the algorithm up to the i-th time step; in other words, after the choice of representatives in all the cells of the table has been determined up to the i-th time step, we consider all the possible suffixes of the representing paths to compute Y_i exactly in the binomial model. Of course, the computation time is exponential, and thus Y_i is only used in the analysis of the performance of our modified AMO algorithm. By definition, $Y_0 = U$ and $Y_n = \Psi$.

Lemma 1. $E(Y_i|Y_0, Y_1, Y_2, .., Y_{i-1}) = Y_{i-1}$ for $i = 1, 2, \ldots, n$. In particular, $E(Y_n) = U$.

Proof. Consider the set $\{a_1, a_2, \ldots, a_q\}$ of stamps in a cell at the i-th time step before selecting its representative. Suppose $Y(a_j)$ is the estimated payoff (exactly computed from the model) for a path with the stamp a_j, and $w(a_j)$ is the weight of a_j. Let $W = \sum_{j=1}^{q} w(a_j)$. If a stamp a_j is selected, it contributes $Y(a_j)W$ to the payoff. Thus, the expectation of the contribution of the stamps after the selection is $\sum_{j=1}^{q}(w(a_j)/W)Y(a_j)W = \sum_{j=1}^{q} w(a_j)Y(a_j)$, which equals the expected contribution before the selection. □

Lemma 1 shows that the sequence $Y_0, Y_1, \ldots, Y_n$ is a *Martingale sequence*. From the argument in the previous section, $|Y_i - Y_{i-1}| < X/k$. Thus, we apply famous Azuma's inequality [4] (also see [3, 7].)

Theorem 1 (Azuma's inequality). *Let $Z_0, Z_1, \ldots$ be a Martingale sequence such that for each i, $|Z_i - Z_{i-1}| < c_i$, then for all $t \geq 0$ and any $C \geq 0$,*

$$Pr[|Z_t - Z_0| \geq C(\sum_{i=1}^{t} c_i^2)^{1/2}] \leq 2exp(-C^2/2).$$

In our case, $c_i = X/k$, and hence, we have the following:

Theorem 2. $|\Psi - U| < C\sqrt{n}X/k$ *with probability* $1 - 2e^{-C^2/2}$.

3.3 Aggregation When the Status Is In-The-Money

It is shown in [1] that in the uniform model, if the status is in-the-money at the t-th time step and the current stock price and running total are S and T, the expectation of payoff is $\{T + \sum_{j=1}^{n-t-1}(1+r)^j S\} - X$. Recall that $r = pu + (1-p)d - 1$. Indeed, $\sum_{j=1}^{n-t-1}(1+r)^j S$ is the expectation of extra running total after the $(t+1)$-st step.

In the non-uniform model, we can compute the value of extra running total for an in-the-money case by using a bottom-up computation as follows: We compute a real value $h(v)$ representing the extra running total for each node of the recombinant binary tree defined as follows: $h(v) = 0$ if v is a leaf (i.e., a node in level n), and $h(v) = p_v(h(w)+S(w))+(1-p_v)(h(w')+S(w'))$ if w and w' are sons of v and $S(w)$ and $S(w')$ are the stock values associated with the nodes w and w', respectively. The expectation of payoff is $T + h(v) - X$ if we are at v, the current running total is T, and the option is in-the-money. The computation time for $h(v)$ for all v is $O(n^2)$, and hence the total asymptotic time complexity does not increase. Therefore, we have the following:

Theorem 3. *Our algorithm approximates the expectation of pay-off in $O(n^2 k)$ time, and its error from the exact expectation is at most $c\sqrt{n}X/k$ with probability $1 - 2e^{-c^2/2}$ for any positive value c.*

3.4 Experimental Performance

Figure 1 gives the comparison of three methods: 1. random sampling, 2. original AMO algorithm, and 3. our algorithm. Here, we consider a uniform model where $S_0 = X = 100$, $u = 1.1$, $d = 1/u$, $pu + (1-p)d = (1.06)^{1/n}$, and we set $k = 1000$. The premium is computed by multiplying $(1.06)^{-1}$ to the expected payoff value. In the random sampling method, we take $20n$ sample paths for one trial and take average over 1000 trials[3]. For our randomized algorithm, we only run single trial.

The graphs show the error from the exact premium computed by using the full-path method. The running time is approximately the same for the three methods in the range $10 \le n \le 30$, and about 0.08 second for $n = 30$, whereas the full-path method takes 1092 seconds. The error of our algorithm is always less than $0.03 = 0.3X/1000$, and smaller than the other methods with factors up to about 100. Also the error tends to decrease if n is increased, and its average is about 0.005 when $25 \le n \le 30$; therefore, it is much better than the theoretical bound $c\sqrt{n}X/1000$. At $n = 30$, the exact premium value is 11.5474 and the error ratio to the premium value is less than 0.0005.

We also run the random sampling algorithm spending 100 times of running time, but the accuracy is still not competitive to our algorithm. Note that we do not implement AMO algorithm with flexible bucket size (a heuristic that is reported to be better than the original AMO algorithm [1]), since its performance depends on tuning of parameters and also the heuristic can be combined with our algorithm, too.

Figure 2 gives the premium prices computed by the algorithms for the case $50 \le n \le 80$, where we use a version of AMO algorithm in which we take the upper value in each bucket in order to give an upper bound of the premium price. The full-path method is not feasible for such a large n. It can be observed that the premium price computed by our algorithm is quite stable.

---
[3] Taking $20000n$ samples at once is oversampling for a small n.

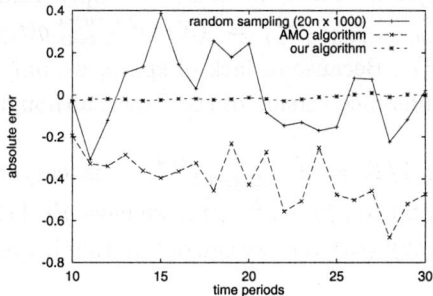

Fig. 1. Errors of the computed premium by three algorithms from the exact value

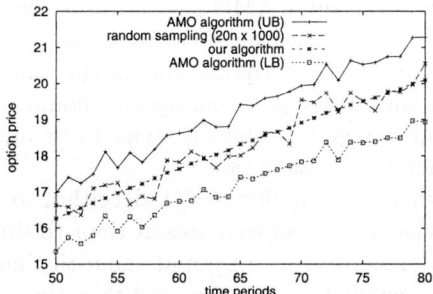

Fig. 2. Premiums computed by the algorithms

3.5 More Precise Analysis for the Uniform Case

The experimental result shows that the analysis in the previous sections overestimates the error. In this subsection, we prove the following theorem:

Theorem 4. *For a uniform model satisfying that $1 - \alpha < p < \alpha$ for a constant $\alpha < 1$, our algorithm approximates the expected pay-off in $O(n^2 k)$ time, and its error from the exact expectation is $O(n^{1/4} X/k)$ with probability $1 - 2e^{-c^2}$, where c is any given positive constant.*

We refine the analysis of the random process in the following way: nodes of the t-th level are processed one-by-one, and coin flips for the cells associated with a node is grouped into one random process; in other words, processing of each row of the table corresponds to one step of the random process.

The total weight $w(t, j)$ of the paths corresponding to the node (t, j) is $\binom{t}{j} p^j (1-p)^{t-j}$. Let $Y_{t,j}$ be the random variable giving the "exact" payoff value just after the algorithm processes the j-th node in the t-th level. Thus, we have a random process with $\sum_{t=0}^{n-1}(t+1) = n(n+1)/2$ total steps. We can easily see that this gives a Martingale process, and $|Y_{t,j} - Y_{t,j+1}| < w(t,j) X/k$ for $j \neq t$, and $|Y_{t,t} - Y_{t+1,0}| < w(t,t) X/k$.

Let $c_{t,j} = w(t,j)X/k$. Then, in order to apply Azuma's inequality, we estimate $\Gamma(p) = \sum_{0 \le j \le t \le n-1} c_{t,j}^2 = (X/k)^2 \sum_{t=0}^{n-1} g(t,p)$, where $g(t,p) = \sum_{j=1}^{t} (\binom{t}{j} p^j (1-p)^{t-j})^2$. Because of lack of space, we only give the estimation for $p = 1/2$ in this conference version of the paper, although it is not difficult to generalize it.

By definition, $g(t, 1/2) = 2^{-t} \sum_{j=1}^{t} \binom{t}{j}^2$. It is easy to see $\sum_{j=1}^{t} \binom{t}{j}^2 = \sum_{j=1}^{t} \binom{t}{j}\binom{t}{t-j} = \binom{2t}{t}$. Since $\binom{2t}{t} \sim 2^{2t}/\sqrt{\pi t}$, we have $g(t, 1/2) \sim 1/sqrt\pi t$. Thus, we have $\Gamma(p) = O((X/k)^2 \sqrt{n})$, and we can obtain the theorem easily by applying Azuma's inequality.

4 Computing Payoff of Saving Asian Option

In [1], it is claimed that a variant of AMO algorithm also works for the American-Asian option. However, it is based on a claim (or assumption) that the early exercise always occurs before the status becomes in-the-money. This is not always true, since a buyer of the American-Asian option should hold the option while the stock price continues to go up. Thus, it seems to be difficult to apply AMO algorithm for the American Asian option.

On the other hand, we can modify AMO algorithm to work on the Saving-Asian option. Recall that if a buyer exercises at the i-th time step and re-invest the money, he/she will receive $(T_i - iX)/n$ at the n-th time step. Suppose that the status is in-the-money at the i-th step, and thus the advantage that a user need not exercise the option is no more valid. In the uniform model, the decision merely depends on whether $T_i - iX$ is larger than the expected value of $T_n - nX$ (knowing the current stock price S) or not; In other words, we should exercise early if and only if $E(T_n - T_i | S_i = S) = S \sum_{j=1}^{n-i} (1+r)^j < (n-i)X$, which means that the expectation of the average stock price after the i-th step is less than X. This condition is path-independent except the path-dependent assumption that the status is in-the-money.

In the non-uniform model, the situation is a little more complicated, since even if $T_i - iX$ is larger than the conditional expectation of $T_n - nX$, it may happen that we should wait for a while. For example, we may postpone to exercise during the X'mas week if a particular stock (e.g. a stock of a department store) is expected to go up during the week in the model provided that the current stock price S is in a certain range.

For each node v in the recombinant binary tree, we define real values $f(v)$ and $g(v)$ as follows: If v is a leaf, $f(v) = 0$ and $g(v) = S(v) - X$, where $S(v)$ is the stock price associated with v. If v has sons w and w' such that w is selected with probability p_v, $f(v) = \max\{p_v g(w) + (1-p_v)g(w'), 0\}$ and $g(v) = S(v) - X + f(v)$.

The value $p_v g(w) + (1-p_v)g(w')$ is the expectation of extra (possibly negative) payoff obtained by postponing the exercise of the option at v, and thus $f(v)$ is the value of the right of postponing the exercise. Indeed, if we postpone and the process goes to w, $S(w) - X$ is added to the current payoff (including the interest) and $f(w)$ gives the value of the right of postponing at w.

We call a node v in the i-th level of the recombinant binary tree a *pseudo-exercise node* if $f(v) = 0$. The values $f(v)$ and $g(v)$ can be computed in a bottom-up fashion in $O(n^2)$ time for all nodes v. The following is the key lemma:

Lemma 2. *If the status is in-the-money, one should exercise the option at the first pseudo-exercise node that is encountered.*

Now, we can run a dynamic programming algorithm that is basically the same as the algorithm in the previous section for the European Asian option (we omit details because of space limitation).

Theorem 5. *Our algorithm approximates the expected pay-off of the Saving-Asian option in $O(n^2 k)$ time, and its error from the exact expectation is at most $c\sqrt{n}X/k$ with probability $1 - 2e^{-c^2/2}$. Moreover, in the uniform model with $1 - \alpha \leq p \leq \alpha$ for a constant $\alpha < 1$, the error bound becomes $O(n^{1/4}X/k)$.*

5 Concluding Remarks

Our algorithm works for a path-dependent option with the following three conditions: 1). The payoff of early exercise at a node v in the t-th level after the path $\mathbf{p}$ of movement on the binomial model is written as $(\gamma(t)F(\mathbf{p}) - G(v))^+$, where $0 < \gamma(t) \leq 1$ is a functions on t, $G(v)$ is a nonnegative function, and $F(\mathbf{p})$ is a summation $\sum_{i=0}^{t} f_i(S_i)$ of the values of nondecreasing functions f_i on the path $\mathbf{p} = S_0, S_1, \ldots, S_t$ of stock movement. 2). There is a threshold value L such that once $F(\mathbf{p}) > L$, the payoff will be destined to be positive (in-the-money). 3). Once $F(\mathbf{p}) > L$, the difference between the payoff (including interest) of immediate exercise and the expected payoff obtained by delaying the exercise is path-independent.

For our Saving-Asian option, $\gamma(t) = e^{-(n-t)r_0/n}$, $f_i(S_i) = S_i/n$, $L = X$, and $G(v) = t(v)X/n$, where $t(v)$ is the level of v. Thus, we may add a path-independent function to the payoff of our Saving-Asian option without losing the accuracy. For example, we may design a variation in which we pay-back a portion of premium for an early exercise satisfying some path-independent conditions.

The experimental performance implies that our theoretical analysis is not tight: Indeed, we may further refine the random process so that each coin flip at a cell of the DP table is one step of a Martingale process. Intuitively, if n is large enough, this will further improve the error analysis by a factor up to $\sqrt{k}$, which explains why the experimental performance is so good; however, the theoretical analysis seems to be complicated and difficult.

It will have industrial value to investigate the applicability of AMO method (or its extension) to several existing options, and design new useful options based on the insight.

Acknowledgement

The authors thank Prof. Syoichi Ninomiya of Center of Res. in Adv. Financial Tech., Tokyo Institute of Technology for his counsel on financial engineering.

References

[1] D. Aingworth, R. Motwani, and J. D. Oldham, Accurate Approximations for Asian Options, *Proc. 11th ACM-SIAM SODA* (2000), pp. 891-901.
[2] K. Akcoglu, M.-Y. Kao, and S. V. Raghavan, Fast Pricing of European Asian Options with Provable Accuracy: Single-Stock and Basket Option, *Proc. ESA 2001, Springer LNCS 2162* (2001), pp. 404-415.
[3] N. Alon, J. Spencer, P. Erdős, *The Probabilistic Method*, Jhon Wiley & Sons, 1992.
[4] K. Azuma, Weighted Sum of Certain Dependent Random Variables, *Tohoku Mathematical Journal* **19** (1967) pp.357-367.
[5] P. Chalasani, S. Jha, and I. Saias, Approximate Option Pricing, *Algoirthmica* **25** (1999), pp.2-21.
[6] P. Chalasani, S. Jha, and A. Varikooty, Accurate Approximation for Europian Asian Options, *Journal of Computational Finance* **1** (1999), pp.11-29.
[7] R. Motwani and P. Raghavan, *Randomized Algorithms*, Cambridge University Press, 1995.

Sorting 13 Elements Requires 34 Comparisons

Marcin Peczarski

Institute of Informatics, Warsaw University
ul. Banacha 2, 02-097 Warszawa, Poland
M.Peczarski@students.mimuw.edu.pl

Abstract. We prove that sorting 13 elements requires 34 comparisons. This solves a long-standing problem posed by D. E. Knuth in his famous book *The Art of Computer Programming, Volume 3, Sorting and Searching*. The result is due to an efficient implementation of an algorithm for counting linear extensions of a given partial order. We present also some useful heuristics which allowed us to decrease the running time of the implementation.

1 Introduction

The problem of finding optimal sorting algorithms is one of fundamental and fascinating problems in the theory of sorting. The problem of finding minimum-comparison sorting algorithms is especially interesting. D. E. Knuth devoted to this problem a part of his famous book *The Art of Computer Programming, Volume 3, Sorting and Searching*.

Let $S(N)$ be the minimum number of comparisons that will suffice to sort N elements. The *information-theoretic lower bound* tells us that

$$S(N) \geq \lceil \lg N! \rceil = c_N \ .$$

Lester Ford, Jr. and Selmer Johnson [2] discovered an algorithm, called *merge insertion*, which nearly and sometimes even exactly matches the theoretical lower bound. Let F_N be the number of comparisons required to sort N elements by merge insertion. We have

N = 1 2 3 4 5 6 7 8 9 10 11 12 13 14 15 16 17 18 19 20 21 22
c_N = 0 1 3 5 7 10 13 16 19 22 26 29 33 37 41 45 49 53 57 62 66 70
F_N = 0 1 3 5 7 10 13 16 19 22 26 30 34 38 42 46 50 54 58 62 66 71

One can see that $S(N) = c_N = F_N$ for $N \leq 11$ and for $N = 20, 21$. After the discovery of merge insertion the first unknown value was $S(12)$. Mark Wells discovered, carrying an exhaustive computer search, that S(12)=30 [7]. Donald Knuth posed the problem of finding the next value $S(13)$ in his classic book [3, Chap. 5.3.1, Exercise 35]. He conjectured that $S(13) = 33$ [4]. In this paper we show that $S(13) = 34$. Our result is based on a smart implementation of the algorithm of Mark Wells.

2 Preliminaries

Let $U = \{u_0, u_1, \ldots, u_{N-1}\}$ be an N-element set to be sorted. Sorting U can be viewed as a sequence of partially ordered sets (U, r), where r is a partial order over U. If U is fixed we will identify the poset (U, r) with r. The sequence starts from the total disorder $r_0 = \{(u, u) : u \in U\}$ and ends with a linearly ordered set $\{u_{i_0} < \ldots < u_{i_{N-1}}\}$. Every subsequent poset is obtained from the previous one as the result of a comparison between two elements of U. Let r be a poset after performing $c - 1$ comparisons and suppose that elements u_j and u_k are being compared in the next step. W.l.o.g. we can assume that $(u_j, u_k) \notin r$ and $(u_k, u_j) \notin r$. If the answer to the comparison is *element u_j is less than element u_k* then the transitive closure of the set $r \cup \{(u_j, u_k)\}$ is the new poset. We will denote it by $r + u_j u_k$.

Definition 1. *If r and l are posets such that $r \subseteq l$ and l is a linear order then l is called a* linear extension *of r.*

Every poset has at least one linear extension. Let $e(r)$ denote the number of linear extensions of a poset r. In particular every totally unordered n-element set has $n!$ linear extensions and every linearly ordered set has exactly one linear extension. Every poset r can be sorted using c comparisons only if c comparisons are sufficient to obtain a linear order from r. The following lemma generalizes the information-theoretic lower bound to the posets.

Lemma 1. *If a poset r can be sorted using c comparisons then $e(r) \leq 2^c$.*

For a poset r the poset $r^* = \{(u, v) : (v, u) \in r\}$ is called the *dual poset* to r.

Definition 2. *Two posets r_1 and r_2 are* congruent *if r_1 and r_2 or r_1 and r_2^* are isomorphic.*

It is easy to see that:

Lemma 2. *Congruent posets need the same number of comparisons to be sorted.*

Let (U, r) be a poset and let A be a subset of U. The poset $(A, r \cap A \times A)$ will be denoted by $r|A$. If $v \in U$ then $r[v]$ will denote the set of elements which are larger than v, precisely $r[v] = \{u : (v, u) \in r \wedge u \neq v\}$. Then we have $r^*[v] = \{u : (u, v) \in r \wedge u \neq v\}$.

3 The Method of Wells

In this section we describe briefly the method which was used to discover that there is no 29-step sorting procedure for 12 elements [7].

The set S_c contains posets which are obtained after c comparisons. The function $search(S_c, r)$ returns TRUE iff the set S_c contains a poset congruent to the poset r.

$S_0 := \{r_0\}$, where $r_0 = \{(u_0, u_0), (u_1, u_1), \ldots, (u_{N-1}, u_{N-1})\}$
for $c := 1$ to c_N do
  $S_c := \emptyset$
  for each $r \in S_{c-1}$ do
    for $j := 0$ to $N - 2$ do
      for $k := j + 1$ to $N - 1$ do
        if $(u_j, u_k) \notin r$ and $(u_k, u_j) \notin r$ then
          $r_1 := r + u_j u_k$
          $r_2 := r + u_k u_j$
          if $search(S_c, r_1) =$ FALSE and $search(S_c, r_2) =$ FALSE then
            $t_1 := e(r_1)$
            $t_2 := e(r_2)$
            if $t_1 \leq 2^{c_N - c}$ and $t_2 \leq 2^{c_N - c}$ then
              if $t_1 \geq t_2$ then
                $S_c := S_c \cup \{r_1\}$
              else
                $S_c := S_c \cup \{r_2\}$

We begin from the set S_0 containing only one, totally unordered poset r_0. In step c every poset $r \in S_{c-1}$ is examined for every unrelated pair u_j and u_k in r in order to verify whether it can be sorted in the remaining $c_N - c + 1$ comparisons. As the result of the comparison of u_j and u_k one can get one of two posets $r_1 = r + u_j u_k$ and $r_2 = r + u_k u_j$. If the number of linear extensions of r_1 (r_2) exceeds $2^{c_N - c}$ then r_1 (r_2) can't be sorted in the remaining $c_N - c$ comparisons (by Lemma 1). It follows that in this case, in order to finish sorting in $c_N - c + 1$ comparisons, elements u_j and u_k should not be compared in step c. If the numbers of linear extensions of both r_1 and r_2 don't exceed $2^{c_N - c}$ then we store one of them for a further analysis. In principle we can choose any of them. It doesn't influence the correctness of the method. Since the poset with the larger number of linear extensions seems to be harder to sort we keep precisely this one. Tights can be broken arbitrarily. Observe that if S_c contains a poset congruent to one of the posets r_1 or r_2 then we don't need to keep any of them (by Lemma 2). This reduces substantially the number of posets to be considered.

If the final set S_{c_N} doesn't contain a linearly ordered set then we conclude that sorting N elements requires more than c_N comparisons. On the other hand if a linearly ordered set belongs to the set S_{c_N} it doesn't mean that it is always possible to sort N elements using c_N comparisons. This happens in the case $N = 13$ where S_{33} contains a linear order and the method fails.

4 The Function Sortable

In this section we propose an improvement of the method described in the previous section. The presented modification enables to overcome with the case where the set S_{c_N} contains a linear order. To begin with we prove a simple lemma.

Lemma 3. *If N elements can be sorted using c_N comparisons then for every c, $0 \leq c \leq c_N$, the set S_c contains at least one poset which can be sorted using $c_N - c$ comparisons.*

Proof. Induction on c. The lemma is obviously true for the initial set S_0. Let r be a poset in S_{c-1} which can be sorted in $c_N - c + 1$ comparisons. It follows that there exists a sorting procedure which sorts r in $c_N - c + 1$ steps. Suppose that the procedure applying to r compares elements u_j and u_k in step c. Hence both $r_1 = r + u_j u_k$ and $r_2 = r + u_k u_j$ can be sorted using $c_N - c$ comparisons. But S_c contains a poset congruent to r_1 or r_2. Hence S_c contains a poset sortable in $c_N - c$ comparisons by Lemma 2.

In the sequel we will extensively use the following corollary.

Corollary 1. *If for some c the set S_c doesn't contain a poset sortable in $c_N - c$ comparisons then there is no sorting procedure for N elements using c_N comparisons.*

Let us consider the recursive function *sortable* described below. The function verifies whether a given poset r can be sorted using c comparisons.

boolean function *sortable*(r, c)
   if $e(r) \leq 2$ **then**
      return TRUE
   for $j := 0$ **to** $N - 2$ **do**
      for $k := j + 1$ **to** $N - 1$ **do**
         if $(u_j, u_k) \notin r$ **and** $(u_k, u_j) \notin r$ **then**
            $r_1 := r + u_j u_k$
            $r_2 := r + u_k u_j$
            if $e(r_1) \leq 2^{c-1}$ **and** $e(r_2) \leq 2^{c-1}$ **then**
               if *sortable*$(r_1, c - 1)$ **and** *sortable*$(r_2, c - 1)$ **then**
                  return TRUE
   return FALSE

We keep the following invariant: *The parameters r and c satisfy $e(r) \leq 2^c$.* The invariant ensures that the recursion always stops because c decreases in each subsequent call and the number of linear extensions is a positive integer.

If $e(r) \leq 2$ then either $e(r) = 1$ or $e(r) = 2$. In the first case r is a linear order. In the latter case $c \geq 1$ and there are two linear extensions of the poset r. Denote them r' and r'' respectively. Let us consider the minimal position where r' and r'' differ and let u' and u'' be the elements on this position in r' and r'' respectively. W.l.o.g. we can assume that $u' < u''$ in r' and $u' > u''$ in r''. Making a single comparison one can determine the proper linear extension and sort r. Therefore the function *sortable* returns TRUE for $e(r) \leq 2$.

If the number of linear extensions of r is larger than 2, one considers posets $r_1 = r + u_j u_k$ and $r_2 = r + u_k u_j$ for each pair of unrelated elements u_j and u_k in r. If the number of linear extensions of r_1 or r_2 exceeds 2^{c-1} then comparing u_j and u_k we cannot sort r in c comparisons by Lemma 1. In the latter case we verify recursively whether r_1 and r_2 are sortable in $c - 1$ comparisons. Observe that the invariant is maintained. If both r_1 and r_2 are sortable then r is also sortable and the function returns TRUE. If there is no pair of sortable posets r_1 and r_2 then the function returns FALSE.

An essential property of this method is that the function *sortable* verifies always correctly whether it is possible to sort r in c comparisons. Unfortunately we can't use the function *sortable* directly to verify if sorting is possible in case $N = 13$ and $c = 33$ because of its exponential complexity in c. But using Corollary 1 we can apply the function *sortable* to improve the method of Wells.

5 Useful Heuristics

Counting the linear extensions is the most time-consuming operation in the method described above. In order to reduce the number of counting operations we apply a few heuristics.

Recursive calls in the function *sortable* should be performed in the lazy manner. We can suppose that among the posets r_1 and r_2 the poset with the larger number of linear extensions is harder to sort. Therefore a possibility that only one call will be sufficient is larger when the function sortable is called first for the poset with the larger number of linear extensions. But if the first call returns TRUE then the second call is still necessary. One can show a poset r and elements u_j and u_k such that $e(r_1) > e(r_2)$, r_1 is sortable but r_2 is not sortable (using the same number of comparisons).

For a poset (U, r) let $V = \{v \in U : \text{there exists } u \in U, u \neq v, (u, v) \in r \text{ or } (v, u) \in r\}$. We have:

- for $u, v \in U \setminus V$ and $w \in V$ the posets $r + uw$, $r + vw$ are congruent and the posets $r + wu$, $r + wv$ are congruent as well;
- for $u, v, x, y \in U \setminus V$ and $u \neq v$, $x \neq y$ the posets $r + uv$, $r + xy$ are congruent.

Therefore if $V = \{u_0, u_1, \ldots, u_{n-1}\}$ it is sufficient to compare only the following pairs of elements:

- u_j and u_k, for $0 \leq j \leq n-2$ and $j+1 \leq k \leq n-1$;
- u_j and u_n, for $0 \leq j < n < N$;
- u_n and u_{n+1}, for $n < N - 1$.

Furthermore we can start the method of Wells from the second step. All posets obtained after the first comparison are congruent and set S_1 always contains only one poset. We can assume that this is the poset $r_0 + u_1 u_0$.

It is known that $e(r_1) + e(r_2) = e(r)$ [3, 7]. The number $e(r)$ is computed in step before. Hence one can store $e(r)$ and it is sufficient to compute only one value: $e(r_1)$ or $e(r_2)$. In practice one of them can be calculated faster. Because counting of linear extensions can take time proportional to their number (see Sect. 6) then intuitively the smaller value should be faster to compute. We can accept that the poset with 'better ordering' has the smaller number of the linear extensions. If $|r^*[j]| + |r[k]| \geq |r[j]| + |r^*[k]|$ then r_1 has 'not worse ordering' than r_2 and in this case we compute

$$t_1 = e(r_1), \qquad t_2 = e(r) - t_1 .$$

We compute

$$t_2 = e(r_2), \qquad t_1 = e(r) - t_2$$

in the other case. Furthermore we know already that if $V = \{u_0, u_1, \ldots, u_{n-1}\}$ and $n < N - 1$ then the posets $r + u_{n+1}u_n$ and $r + u_n u_{n+1}$ are congruent and hence $e(r + u_{n+1}u_n) = e(r + u_n u_{n+1}) = e(r)/2$.

The next improvement follows the lemma:

Lemma 4. *If $(x, u) \in r$, $(v, y) \in r$ and $(v, u) \notin r$ then $e(r + xy) \geq e(r + uv)$.*

Proof. Observe that $r + uv$ and $(r + xy) + uv$ are identical. Since a comparison can't increase the number of linear extensions we have

$$e(r + uv) = e((r + xy) + uv) \leq e(r + xy) .$$

It follows that if $e(r + u_j u_k) > 2^c$ for some u_j and u_k then $e(r + xy) > 2^c$ for every x and y such that $(x, u_j) \in r$ and $(u_k, y) \in r$.

6 Counting Linear Extensions

The problem of counting linear extensions of a given poset is #P-complete [1]. This indicates that the counting linear extensions is probably not easier than generating them. Therefore we can't expect a polynomial-time algorithm for counting, but nevertheless we will show in the next section that the method described here is the fastest existing.

The method is based on the theorem given in [7]. In order to state the theorem precisely we need some new notion.

Definition 3. *Let (U, r) be a poset, $D \subseteq U$ and $d \in D$. The pair (A, B) is called an* admissible partition *of D with respect to element d when the following conditions are satisfied:*

- $A \cup B = D \setminus \{d\}$ and $A \cap B = \emptyset$;
- *if $(a, d) \in r$ and $a \neq d$ then $a \in A$;*
- *if $(d, b) \in r$ and $b \neq d$ then $b \in B$;*
- $(b, a) \notin r$ *for all $a \in A$ and $b \in B$.*

Theorem 1. *Let (U, r) be a poset.*

1. *If $A, B \subseteq U$, $A \cap B = \emptyset$ and $(a, b) \in r$ for $a \in A$ and $b \in B$ then*

$$e(r|A \cup B) = e(r|A) \cdot e(r|B) . \tag{1}$$

2. If $A, B \subseteq U$, $A \cap B = \emptyset$ and $(a,b) \notin r$, $(b,a) \notin r$ for $a \in A$ and $b \in B$ then

$$e(r|A \cup B) = e(r|A) \cdot e(r|B) \cdot \binom{|A|+|B|}{|A|} . \tag{2}$$

3. If $D \subseteq U$ and $d \in D$ then

$$e(r|D) = \sum_{A,B} e(r|A) \cdot e(r|B) , \tag{3}$$

where the sum is taken over all admissible partitions of D with respect to d.

The main idea of the algorithm is to divide the problem into 'simpler' subproblems and to solve them recursively. First, a given poset is partitioned into connected components (in the graph sense). Next, the number of all linear extensions is calculated independently for each component and the results are combined using (2). In order to compute the number of linear extensions of a connected component D we apply point (3) of Theorem 1. We take d such that the number of admissible partitions of D is the smallest possible. It turns out that for such d the number of related elements in $r|D$ is the largest possible. We get a set of pairs of simpler subproblems in this way. Each such subproblem is solved in the recursive manner and the results are combined using (3).

In order to improve the performance of the above algorithm we stop the recursion as early as possible. To this aim we don't partition small components D such that $|D| \leq 5$. If $|D| \leq 2$ then the poset $r|D$ is a linear order (because it is connected) and $e(r|D) = 1$. If D consists of 3, 4, or 5 elements we store the number of linear extensions in 3 auxiliary tables. In order to get the number of linear extension of $r|D$ we compute the index $\sigma(r|D)$ in the appropriate table. If $|D| = 3$ then

$$\sigma(r|D) = \sum_{v \in D} |r[v] \cap D| .$$

If $|D| = 4$ then

$$\sigma(r|D) = \sum_{v \in D} \theta_4(|r[v] \cap D|) ,$$

where the function θ_4 is defined as follows:

$$i = 0 \ 1 \ 2 \ 3$$
$$\theta_4(i) = 0 \ 1 \ 4 \ 10$$

For $|D| = 5$ we have

$$\sigma(r|D) = \sum_{v \in D} \theta_5(|r[v] \cap D|, |r^*[v] \cap D|) ,$$

where the function θ_5 is defined as follows:

$$\begin{array}{rcccccc}
i = & 0 & 1 & 2 & 3 & 4 \\
\theta_5(0,i) = & & 0 & 1 & 3 & 10 \\
\theta_5(1,i) = & 0 & 6 & 8 & 5 \\
\theta_5(2,i) = & 1 & 8 & 1 \\
\theta_5(3,i) = & 3 & 5 \\
\theta_5(4,i) = & 10
\end{array}$$

We have that if $|D_1| = |D_2|$ and $\sigma(r|D_1) = \sigma(r|D_2)$ then $e(r|D_1) = e(r|D_2)$.

7   Results of Experiments

As it was already mentioned the method of Wells produces for $N = 13$ and $c_N = 33$ the sequence of sets $S_1, S_2, \ldots, S_{33}$. The set S_{33} contains a linear order. Therefore we can't say anything about the existence of a sorting procedure for 13 elements using 33 comparisons. Applying the function *sortable* we get that there is no poset in S_{15} which can be sorted using 18 comparisons. By Corollary 1 we have that there is no procedure sorting 13 elements and using 33 comparisons. Since the Ford's and Johnson'n algorithm sorts 13 elements using 34 comparisons we get finally that $S(13) = 34$. Generating the sets $S_1, S_2, \ldots, S_{33}$ took about 2 hours and the analysis of S_{15} took about 8.5 hours.

To ensure correctness of our implementation we checked whether the program can determine properly the known values of $S(N)$ for $N \leq 12$. Supposing we use at most $S(N) - 1$ comparisons we obtained that there is always $c \leq S(N) - 1$ for which the set S_c is empty. This confirm that $S(N) - 1$ comparisons isn't enough to sort N elements. For $N \leq 11$ we checked whether $S(N)$ comparisons is sufficient. We obtained that each set in the sequence $S_1, \ldots, S_{S(N)}$ contains at least one sortable poset, which is consistent with Lemma 3.

There are other algorithms which can be used for computing the exact number of linear extensions of a given poset. The algorithm of Varol and Rotem [8] runs in time $O(n \cdot e(r))$ and the algorithm of Pruesse and Ruskey [5] has the time complexity $O(e(r))$. Both algorithms are designed to generate all linear extensions of a given poset. Pruesse and Ruskey present also a modification that spares the computation time if one wants to compute only the number of linear extensions. The asymptotic time complexity of the algorithm presented in the previous section isn't still known. The following table compares implementations of all three algorithms.

n	$e(r)$	Sect. 6	Varol, Rotem	Puesse, Ruskey
12	2 702 765	0.0006	0.5	0.07
13	22 368 256	0.0014	3.9	0.46
14	199 360 981	0.0033	37	3.4
15	1 903 757 312	0.0073	320	28
16	19 391 512 145	0.018		
21	4 951 498 053 124 096	1.1		
22	69 348 874 393 137 901	2.8		

The poset under consideration was an n-element fence. The computation time is given in seconds. All algorithms are implemented in the C language and compiled using the same compiler with the same optimisation options. Varol's and Rotem's algorithm is very simple. Its source code has about 20 lines. The source code of Pruesse's and Ruskey's algorithm (written by Kenny Wong and Frank Ruskey) is accessible via the Internet [8]. Tests were run on PC with 233MHz processor and 64MB memory.

8 Some Aspects of Implementation

Our algorithms perform a lot of set operations on subsets of the index set $I_N = \{0, 1, 2, \ldots, N-1\}$. Our intention was to perform the following operations as fast as possible: $A \cup B$, $A \cap B$, $A \setminus B$, $|A|$, creating one-element set $\{i\}$ and choosing an arbitrary element from a set A for $A, B \subseteq I_N$, $i \in I_N$. There is also a lot of operations concerning set scanning. Our goal was to implement those operations to run in time proportional to the size of a given set.

A subset $A \subseteq I_N$ is represented as the unsigned integer with bit i set iff $i \in A$. The union, intersection and subtraction of sets are implemented as logical operations: A|B, A&B, A&~B. Creating a single element set, choosing an arbitrary element from a set (elements with the minimal indices are derived) and counting the cardinality of a set are implemented as macros: pof2(a), min_element(A), cardinality(A). Operation min_element(A) is undefined when A represents the empty set. It turns out that the fastest implementation of the macros can be done by storing the appropriate values in arrays.

```
#define pof2(x)         pof2_tbl[x]
#define min_element(x)  min_element_tbl[x]
#define cardinality(x)  cardinality_tbl[x]
typedef char            MIN_EL_TBL_EL;
typedef char            CARD_TBL_EL;
SET                     pof2_tbl[N + 1];
MIN_EL_TBL_EL           min_element_tbl[1 << N];
CARD_TBL_EL             cardinality_tbl[1 << N];
```

The tables min_element_tbl and cardinality_tbl occupy 2^N memory cells each. If it is unacceptable we can apply the following implementation.

```
#define HALF_N          ((N + 1) >> 1)
#define HALF_SET        ((1 << HALF_N) - 1)
#define min_element(x)  ((x) & HALF_SET ?                              \
                         min_element_tbl[(x) & HALF_SET] :             \
                         min_element_tbl[(x) >> HALF_N] + HALF_N)
#define cardinality(x)  (cardinality_tbl[(x) & HALF_SET] +             \
                         cardinality_tbl[(x) >> HALF_N])
MIN_EL_TBL_EL           min_element_tbl[1 << HALF_N];
CARD_TBL_EL             cardinality_tbl[1 << HALF_N];
```

This reduces memory requirements to only $2^{\lceil N/2 \rceil}$ cells and slow down the set operations only by a constant. The implementation of loop 'for each $a \in A$' use the macros defined above.

```
while (A) {
  a~= min_element(A);
  A~~= pof2(a);
  ...
}
```

A poset is represented in the tables $r[0..N-1]$ and $r^*[0..N-1]$ where $r[j]$ and $r^*[j]$ are the sets of indices of all elements larger and smaller than u_j respectively. The congruence of posets is checked using hash functions and Ullman's graph isomorphism algorithm [6]. The complete source code can be downloaded from [9].

Acknowledgement

I would like to thank Krzysztof Diks for his comments, suggestions and encouragement.

References

[1] Brightwell, G., Winkler, D.: Counting Linear Extensions. Order **8** (1991) 225–242
[2] Ford, L., Johnson, S.: A Tournament Problem. American Mathematical Monthly **66** (1959) 387–389
[3] Knuth, D. E.: The Art of Computer Programming, Vol. 3. Sorting and Searching, 2nd ed. Addison-Wesley (1998)
[4] Knuth, D. E., Kaehler E. B.: An Experiment in Optimal Sorting. Information Processing Letters **1** (1972) 173–176
[5] Pruesse, G., Ruskey F.: Generating Linear Extensions Fast. SIAM Journal on Computing **23** (1994) 373–386
[6] Ullman, J. R.: An Algorithm for Subgraph Isomorphism. Journal of the ACM **23** (1976) 31–42
[7] Wells, M.: Elements of Combinatorial Computing. Pergamon Press (1971)
[8] The Combinatorial Object Server: http://www.theory.csc.uvic.ca/~cos
[9] Author's Home Page: http://rainbow.mimuw.edu.pl/~mp2020

Extending Reduction Techniques for the Steiner Tree Problem

Tobias Polzin[1] and Siavash Vahdati Daneshmand[2]

[1] Max-Planck-Institut für Informatik
Stuhlsatzenhausweg 85, 66123 Saarbrücken, Germany
polzin@mpi-sb.mpg.de

[2] Theoretische Informatik, Universität Mannheim
68131 Mannheim, Germany
vahdati@informatik.uni-mannheim.de

Abstract. Reduction methods are a key ingredient of the most successful algorithms for the Steiner problem. Whereas classical reduction tests just considered single vertices or edges, recent and more sophisticated tests extend the scope of inspection to more general patterns. In this paper, we present such an extended reduction test, which generalizes different tests in the literature. We use the new approach of combining alternative- and bound-based methods, which substantially improves the impact of the tests. We also present several algorithmic contributions. The experimental results show a large improvement over previous methods using the idea of extension, leading to a drastic speed-up in the optimal solution process and the solution of several previously unsolved benchmark instances.

Keywords: Steiner Problem, Reductions

1 Introduction

The Steiner problem in networks is the problem of connecting a subset of the vertices of a weighted graph at minimum cost. This is a classical $\mathcal{NP}$-hard problem [5] with many important applications in network design in general and VLSI design in particular. For background information on this problem see [4].

A key ingredient of the most successful algorithms [1, 7, 10] for the Steiner problem are reduction techniques, i.e., methods to reduce the size of a given instance while preserving at least one optimal solution (or the ability to efficiently reconstruct one). While classical reduction tests just inspected simple patterns (vertices or edges), recent and more sophisticated tests extend the scope of inspection to more general patterns (like trees).

In this paper, we present such an extended reduction test, which generalizes different tests in the literature. We use the new approach of combining alternative- and bound-based methods, which substantially improves the impact of the tests. We also present several algorithmic improvements, especially for the computation of the necessary information.

Since reduction techniques cannot be properly evaluated using worst-case analysis, we study the behaviour of the used techniques using the well-established benchmark library SteinLib [6]. The experimental results show a substantial improvement over all previous reduction methods (including our previous results in [10]). In particular, the effect on VLSI-instances is impressive. We achieve far more reductions than in any previous work in relatively short running time. As a consequence, we can now solve all VLSI-instances in SteinLib (including several previously unsolved instances) in tolerable time (few seconds in most cases).

The outline of this paper is as follows. After some preliminaries in the next two subsections, in Section 2 we first describe the test in a generic form. The generic algorithm is substantiated by presenting the applied test conditions (Section 2.1) and criteria for guiding and truncation of expansion (Section 2.2). A very critical issue for the success of such a test is the computation of the necessary information; this and other implementation issues are discussed in Section 2.3. Different design decisions lead to different variants of the test, as described in Section 2.4. Computational experiments on the impact of the tests are reported in Section 3. Finally, Section 4 contains some concluding remarks.

1.1 Definitions

The Steiner problem in networks can be stated as follows (see [4] for details): Given an (undirected, connected) network $G = (V, E, c)$ with vertices $V = \{v_1, \ldots, v_n\}$, edges E and edge weights $c_{ij} > 0$ for all $(v_i, v_j) \in E$, and a set R, $\emptyset \neq R \subseteq V$, of **required vertices** (or **terminals**), find a minimum weight tree in G that spans R (called a **Steiner minimal tree**). If we want to stress that v_i is a terminal, we will write z_i instead of v_i.

We also look at a reformulation of this problem using the (bi-)directed version of the graph: Given $G = (V, E, c)$ and R, find a minimum weight arborescence in $G = (V, A, c)$ ($A := \{[v_i, v_j], [v_j, v_i] \mid (v_i, v_j) \in E\}$, c defined accordingly) with a terminal (say z_1) as the root that spans $R_1 := R \setminus \{z_1\}$.

For any two vertices v_i and v_j, the **distance** $d_{ij} = d(v_i, v_j)$ between v_i and v_j is the length of a shortest path between v_i and v_j. A **Steiner bottleneck** of a path P_{ij} between v_i and v_j is a longest subpath with (only) endpoints in $R \cup \{v_i, v_j\}$; and the **Steiner bottleneck distance** s_{ij} between v_i and v_j is the minimum Steiner bottleneck length over all P_{ij}.

For every tree T in G, we denote by $L(T)$ the leaves of T, by $V(T)$ the vertices of T, and by $c(T)$ the sum of the costs of edges in T. Let T' be a subtree of T. The **linking set** between T and T' is the set of vertices $v_i \in V(T')$ with at least one fundamental path from v_i to a leaf of T not containing any edge of T'. If the linking set between T and T' is equal to $L(T')$, T' is said to be **peripherally contained** in T (Figure 1). A set $L' \subseteq V(T), |L'| > 1$, induces a subtree $T_{L'}$ of T containing for every two vertices $v_i, v_j \in L'$ the fundamental path between v_i and v_j in T. We define L' to be a **pruning set** if L' contains the linking set between T and $T_{L'}$.

A **key node** in a tree T is a node that is either a a nonterminal of degree at least 3 or a terminal. A **tree bottleneck** between $v_i \in T$ and $v_j \in T$ is a longest

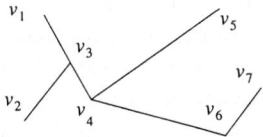

For the depicted tree T, let T' be the subtree of T after removing the edges (v_4, v_6) and (v_6, v_7). The linking set between T and T' is $\{v_1, v_2, v_4, v_5\}$, and therefore T' is not peripherally contained in T. But if we add (v_4, v_6) to T', it is.

Fig. 1. Depiction of some central notions

subpath on the path between v_i and v_j in T in which only the endpoints may be key nodes; and t_{ij} denotes the length of such a tree bottleneck.

1.2 Preliminaries for Reduction Tests

We distinguish between two major classes of reduction tests, namely the alternative-based tests and the bound-based tests [10].

The **alternative-based** tests use the existence of alternative solutions. In case of exclusion tests, it is shown that for any solution containing a certain part of the graph (e.g., a vertex or an edge) there is an alternative solution of no greater cost without this part; the inclusion tests use the converse argument.

A basic alternative-based test is the so-called SD-test [3] (here called s-test): Any edge (v_i, v_j) with $s_{ij} < c_{ij}$ can be excluded. (This test can be extended to the case of equality, if a path corresponding to s_{ij} does not contain (v_i, v_j).)

The **bound-based** tests use a lower bound for the value of an optimal solution under the assumption that a certain part of the graph is contained (in case of exclusion tests) or is not contained (in case of inclusion tests) in the solution. These tests are successful if such a constrained lower bound exceeds a known upper bound, typically the value of a (not necessarily optimal) Steiner tree. But it is usually too costly to recompute a strong lower bound from scratch for each constraint. Here one can use the following simple, but very useful observation:[1]

Lemma 1. *Let $G = (V, A, c)$ be a (directed) network (with a given set of terminals) and $c'' \leq c$. Let $lower'$ be a lower bound for the value of any (directed) Steiner tree in $G' = (V, A, c')$ with $c' := c - c''$. For each incidence vector x representing a feasible Steiner tree for G, it holds: $lower' + c'' \cdot x \leq c \cdot x$.*

A typical application of this lemma is an approach based on Linear Programming: Any linear relaxation can provide a dual feasible solution of value $lower'$ and reduced costs c''. We can use a fast method to compute a constrained lower bound $lower''_{con}$ with respect to c''. Using Lemma 1, it easily follows that $lower_{con} := lower' + lower''_{con}$ is a lower bound for the value of any solution satisfying the constraint.

As an example for such a relaxation, consider the (directed) cut formulation P_C [16]: The Steiner problem is formulated as an integer program by introducing a binary x-variable for each arc (in case of undirected graphs, the

---
[1] Proofs for the lemmas can be found in [9].

bidirected counterpart is used, fixing a $z_1 \in R$ as the root). For each cut separating the root from another terminal, there is a constraint requiring that the sum of x-values of the cut arcs is at least 1. In this way, every feasible binary solution represents a feasible solution for the Steiner problem and each minimum solution (with value $v(P_C)$) a Steiner minimal tree. Now we can use a dual feasible solution of value $lower'$ for LP_C, the linear relaxation of P_C, to apply the method described above. For example, a lower bound for any Steiner tree with the additional constraint that it must contain a certain nonterminal v can be computed by adding $lower'$ to the length of a shortest path with respect to the reduced costs c'' from z_1 via v to another terminal, because any optimal Steiner tree including v must contain such a path.

If a test condition is satisfied, a test action is performed that modifies the current graph. For the tests in this paper, two such modifications are relevant: the elimination of an edge, which is straightforward, and replacing a nonterminal. The latter action is performed once it is established that this vertex has degree at most two in an optimal Steiner tree. The graph is modified by deleting the vertex and the incident edges and introducing a clique over the adjacent vertices in which the cost of each edge is the sum of the costs of the two edges it replaces. Note that in case of parallel edges, only one of minimum cost is kept.

2 Extending Reduction Tests

The classical reduction tests for the Steiner problem inspected just simple patterns (a single vertex or a single edge). There have been some approaches in the literature for extending the scope of inspection [2, 14, 15]. The following function *EXTENDED-TEST* describes in pseudocode a general framework for many of these approaches. The argument of *EXTENDED-TEST* is a tree T that is expanded recursively. For example, to eliminate an edge e, T is initialized with e. The function returns 1 if the test is successful, i.e., it is established that there is an optimal Steiner tree that does not peripherally contain T.

In the pseudocode, the function *RULE-OUT(T, L)* contains the specific test conditions (see Section 2.1): *RULE-OUT(T, L)* returns 1 if it is established that T is not contained with linking set L in at least one optimal Steiner tree.

The correctness can be proven easily by induction using the fact that if T' is a subtree of an optimal Steiner tree T^* and contains no inner terminals, all leaves of T' are connected to some terminal by paths in $T^* \setminus T'$.

Clearly the decisive factor in this algorithm is the realization of the functions *RULE-OUT, TRUNCATE* and *PROMISING*.

Using this framework, previous extension approaches can be outlined easily:

- In [15] the idea of expansion was introduced for the rectilinear Steiner problem.
- In [14] this idea was adopted to the Steiner problem in networks. This variant of the test tries to replace vertices with degree three; if this is successful, the newly introduced edges are tested again with an expansion test. The

expansion is performed only if there is a single possible extension at a vertex, thus eliminating the need for backtracking.
- In [2] backtracking was explicitly introduced, together with a number of new test conditions to rule out subnetworks, dominating those mentioned in [14].
- In [10], we introduced a different test that tries to eliminate edges. Expansion is performed only if there is at most one possible extension (thus inspecting a path) and only if the elimination of one edge implies the elimination of all edges of the path.

$EXTENDED\text{-}TEST(T)$:
    (returns 1 only if T is not contained peripherally in an optimal Steiner tree)
1   **if** $RULE\text{-}OUT(T, L(T))$:
2   **return** 1   (test successful)
3   **if** $TRUNCATE(T)$:
4   **return** 0   (test truncated)
5   **forall** *leaves* v_i *of* T :
6   **if** $v_i \notin R$ **and** $PROMISING(v_i)$:
7            $success := 1$
8   **forall** *nonempty extension* $\subseteq \{(v_i, v_j)$:
                not $RULE\text{-}OUT(T \cup \{(v_i, v_j)\}, L(T) \cup \{v_j\})\}$:
9   **if not** $EXTENDED\text{-}TEST(T \cup extension)$:
10                  $success := 0$
11   **if** *success* :
12   **return** 1   (no acceptable extension at v_i)
13   **return** 0   (in all inspected cases, there was an acceptable extension)

All previous approaches use only alternative-based methods. We present an expansion test that explicitly combines the alternative-based and bound-based methods. This combination is far more effective than previous tests, because the two approaches have complementary strengths. Intuitively speaking, the alternative-based method is especially effective if there are terminals in the vicinity, because it uses the Steiner bottleneck distances. On the other hand, the bound-based method is especially effective if there are no close terminals, because it uses the distances (with respect to reduced costs) to terminals. Furthermore, for the expansion test to be successful, usually many possible extensions must be considered and it is often the case that not all of them can be ruled out using exclusively the alternative- or the bound-based methods, whereas an explicit combination of both methods can do the job.

Although the pseudocode of $EXTENDED\text{-}TEST$ is simple, designing an efficient and effective implementation requires many algorithmic ideas and has to be done carefully, taking the interaction between different actions into account, which is highly nontrivial. Since writing down many pages of pseudocode would be less instructive, we prefer to explain the main building blocks. In the following, we first describe the test conditions for ruling out trees (the function $RULE\text{-}OUT$), using the results of [2] and introducing new ideas. Then we explain the criteria used for truncation and choice of the leaves for expansion (the functions *TRUNCATE* and *PROMISING*). Finally, we will address some implementation issues, particularly data structures for querying different types of distances.

2.1 Test Conditions

For the following test conditions we always consider a tree T with $V(T) \cap R \subseteq L(T)$, i.e., terminals may appear only as leaves of the tree. A very general formulation of the alternative-based test condition is the following:

Lemma 2. *Consider a pruning set L' for T. If $c(T_{L'})$ is larger than the cost of a Steiner tree T' in $G' = (V, \binom{V}{2}, s)$ with L' as terminals, then there is an optimal Steiner tree that does not peripherally contain T. This test can be strengthened to the case of equality if there is a vertex v in $T_{L'}$ that is not in any of the paths used for defining the s-values of the edges of T'.*

A typical choice for L' is $L(T)$, often with some leaves replaced by vertices added to T in the first steps of the expansion. If computing an optimal Steiner tree T' is considered too expensive, the cost of a minimum spanning tree for L' with respect to s can be used as a valid upper bound.

A relaxed test condition compares Steiner bottlenecks with tree bottlenecks:

Lemma 3. *If $s_{ij} < t_{ij}$ for any $v_i, v_j \in T$, there is an optimal Steiner tree that does not peripherally contain T. Again, the test can be strengthened to the case of equality if a path corresponding to s_{ij} does not contain a tree bottleneck of T between v_i and v_j.*

The bound-based test condition uses a dual feasible solution for LP_C of value $lower'$ and corresponding reduced costs c'' (with resulting distances d''):

Lemma 4. *Let $\{l_1, l_2, \ldots, l_k\} = L(T)$ be the leaves of T. Then $lower_{con} := lower' + min_i \{d''(z_1, l_i) + c''(\boldsymbol{T}_i) + \sum_{j \neq i} min_{z_p \in R_1} d''(l_j, z_p)\}$ defines a lower bound for the cost of any Steiner tree under the assumption that it peripherally contains T, where $\boldsymbol{T}_i$ denotes the directed version of T when rooted at l_i.*

In the context of replacement of nonterminals, one can use the following lemma.

Lemma 5. *Let e_1 and e_2 be two edges of T in a reduced network. If both edges originate from a common edge e_3 by a series of replacements, then no optimal Steiner tree for the reduced network that corresponds to an optimal Steiner tree in the unreduced network contains T.*

The conditions above cover the calls $RULE\text{-}OUT(T, L)$ with $L = L(T)$. In case other vertices than the leaves need to be considered in the linking set (as in line 8 of the pseudocode), one can easily establish that all lemmas above remain valid if we treat all vertices of L as leaves.

2.2 Criteria for Expansion and Truncation

The basic truncation criterion is the number of backtracking steps, where there is an obvious tradeoff between the running time and the effectiveness of the test. A typical number of backtracking steps in our implementations is five.

Additionally, there are other criteria that guide and limit the expansion:

1. If a leaf is a terminal, we cannot easily expand over this leaf, because we cannot assume anymore that an optimal Steiner tree must connect this leaf to a terminal by edges not in the current tree. However, if all leaves are terminals (a situation in which no expansion is possible for the original test), we know that at least one leaf is connected by an edge-disjoint path to another terminal (as long as not all terminals are spanned by the current tree). This can be built into the test by another level of backtracking and some modifications of the test conditions. But we do not describe the modifications in detail, because the additional cost did not pay off in terms of significantly more reductions.
2. If the degree deg of a leaf is large, considering all $2^{(deg-1)} - 1$ possible extensions would be too costly and the desired outcome, namely that we can rule out all of these extended subtrees, is less likely. Therefore, we limit the degree of possible candidates for expansion by a small constant, e.g. 8.
3. It has turned out that a depth-first realization of backtracking is quite successful. In each step, we consider only those leaves for expansion that have maximum depth in T when rooted at the starting point. In this way, the bookkeeping of the inspected subtrees becomes much easier and the whole procedure can be implemented without recursion. A similar idea was already mentioned (but not explicitly used) in [2].
4. In case we do not choose the depth-first strategy, a tree T could be inspected more than once. As an example, consider a tree T resulting from an expansion of T' at leaf v_i and then at v_j. If T cannot be ruled out, it is possible that we return to T', expand it at v_j and then at v_i, arriving at T again. This problem can be avoided by using a (hashing-based) dictionary.

2.3 Implementation Issues

Precomputing (Steiner) Distances: A crucial issue for the implementation of the test is the calculation of Steiner bottleneck distances. An exact calculation of all s_{ij} needs time $O(|V|(|E| + |V|\log|V|))$ [2] and space $\Theta(|V|^2)$, which would make the test impractical even for mid-size instances. So we need a good approximation of these distances and some appropriate data structures for retrieving them. Building upon a result of Mehlhorn [8], Duin [1] gave a nice suggestion for the approximation of the Steiner bottleneck distances, which needs preprocessing time $O(|E| + |V|\log|V|)$ (mainly three shortest paths calculations to determine the three nearest terminals for each nonterminal and a minimum spanning tree calculation) and a small running time for each query (the query time can even be constant, if all necessary queries are known in advance [10]). Although the resulting approximate values $\hat{s}_{ij}$ produce quite satisfactory results for the original s-test of Section 1.2, for the extended test the results are much worse than with the exact values. But we observed that $\tilde{s} = \min\{\hat{s}, d\}$ is almost always equal to the exact s-values, and therefore can be used in the extended test as well. Still there remains the problem of computing the d-values: For each vertex v_i we compute and store in a neighbor list the distances to a constant number (e.g. 50) of nearest vertices. But we consider only vertices v_j with $d_{ij} < \hat{s}_{ij}$. This is justified by the observation that in case $d_{ij} \geq \hat{s}_{ij}$, for all descendants v_k of v_j in the shortest paths tree with the root v_i there is a path P_{ik} of Steiner bottleneck length s_{ik} containing at least one terminal, and in such cases s_{ik} is usually quite well approximated by $\hat{s}_{ik}$.

Now, we use different methods for different variants of the test:

1. If we only replace vertices in an expansion test, the d- and s-values do not increase and we can use the precomputed neighbor lists during the whole test using binary search.
2. If we limit the number of (backtracking) steps, then we can confine the set of all possible queries in advance. When a vertex is considered for replacement by the expansion test, we first compute the set of possibly visited neighbors (adjacent vertices, and vertices adjacent to them, and so on, up to the limited depth). Then we compute a distance matrix for this set according to the $\tilde{s}$-values. Using this matrix, each query can be answered in constant time.
3. If we also want to delete edges, we have to store for each vertex in which neighbor list computations it was used. When an edge is deleted, we can redo the $\hat{s}$ computation (or at least those parts that may have changed due to the edge deletion) and restore the affected neighbor lists. This can even be improved by a lazy calculation of the neighbor lists.

Tree Bottlenecks: The tree bottleneck test of Lemma 3 can be very helpful, because every distance between tree vertices calculated for a minimum spanning tree or a Steiner minimal tree computation can be tested against the tree bottleneck; and in many of the cases where a tree can be ruled out, already an intermediate bottleneck test can rule out this tree, leading to a shortcut in the computation. This is especially the case if there are long chains of nodes with degree two in the tree. We promote the building of such chains while choosing a leaf for extension: We first check whether there is a leaf at which the tree can be expanded by only one edge. In this case we immediately perform this expansion, without creating a new key node and without the need of backtracking through all possible combinations of expansion edges.

The tree bottleneck test can be sped up by storing for each node of the tree the length of a tree bottleneck on the path to the starting vertex. For each two nodes v_i and v_j in the tree, the maximum of these values gives an upper bound for the actual tree bottleneck length t_{ij}. Only if this upper bound is greater than the (approximated) Steiner bottleneck distance, an exact tree bottleneck computation is performed.

Computations for the Bound-Based Tests: An efficient method for generating the dual feasible solution needed for the bound-based test of Lemma 4 is the DUAL-ASCENT algorithm described in [10]. We improve the test by calculating a lower bound and reduced costs for different roots. Although the optimal value of the dicut relaxation does not change with the choice of the root, this is not true concerning the value of the dual feasible solution generated by DUAL-ASCENT and, more importantly, the resulting reduced costs can have significantly different patterns, leading to a greater potential for reductions. Even more reductions can be achieved by using stronger lower bounds, as computed with a row generating algorithm [10]. Concerning the tests for the replacement of vertices, we use only the result of the final iteration, which provides an optimal dual solution of the underlying linear relaxation. The dual feasible solutions of the intermediate iterations are used only for the tests dealing with the deletion of edges, because the positive effect of the replacement of a vertex (Section 1.2) cannot be translated easily into linear programs.

Replacement History: Our program package can transform a tree in a reduced network back into a tree in the original instance. For this purpose, we assign a unique ID number to each edge. When a vertex is replaced, we store for each newly inserted edge a triple with the new ID and the two old IDs of the replaced edges. We use this information to implement the test described in Lemma 5. First we do some preprocessing, determining for each ID the edges it possibly originate from (here called ancestors); this can be done in time and space linear in the number of IDs. Later, a test for a conflict between two edges (i.e., they originate from the same edge) can be performed by marking the IDs of the ancestors of one edge and then checking the IDs of ancestors of the other edge; so each such test can be done in time linear in the number of ancestors. We perform this test each time the current tree T is to be extended over a leaf v_i (with (v_k, v_i) in T) by an edge (v_i, v_j). Then we check for a conflict between (v_i, v_j) and (v_k, v_i). This procedure implements an idea briefly mentioned in [2], where a coloring scheme was suggested for a similar purpose. Our scheme has the advantage that it may even discover conflicts in situations where an edge is the result of a series of replacements.

2.4 Variants of the Test

A general principle for the application of reduction tests is to perform the faster tests first so that the stronger (and more expensive) tests are applied to (hopefully) sufficiently reduced graphs. In the present context, different design decisions (e.g., trying to delete edges or replace vertices) lead to different consequences for an appropriate implementation and quite different versions of the test, some faster and some stronger.

We have implemented four versions of expansion tests and integrated them into the reduction process described in [10]. Some details of the corresponding implementations were already given in Section 2.3.

1. For a fast preprocessing we use the linear time expansion test that eliminates paths, as described in [10].
2. A stronger variant tries to replace vertices, but only expands at leaves that are most edges away from the starting vertex.
3. Even stronger but more time-consuming is a version that performs full backtracking.
4. The most time-consuming variant tries to eliminate edges.

3 Computational Results

In this paper, we concentrate on the VLSI-Instances of the benchmark library SteinLib [6]. In a different context, we report in [11] results on the so-called geometric instances of SteinLib using (among other techniques) the methods described in this paper. Other groups of instances could be either solved fairly fast already without these new techniques (as we reported in [10]), or (as for some groups of instances added meanwhile to SteinLib) are deliberately constructed to be hard for the known methods, with the consequence that the impact of the described methods on them is not decisive.

We report results both for the extended reduction techniques described in this paper and for the optimal solution using these techniques. All these results are, to our knowledge, by far the best results achieved for the corresponding instances. In particular, we could solve all VLSI-instances of SteinLib, including several previously unsolved instances, thereby substantially improving the solution times for the previously solved instances (up to some orders of magnitude for the larger instances).

Each of the test series was performed with the same parameters for all instances of all groups. Although in some cases individual parameter tuning could lead to some improvements, the used methods turn out to be fairly robust, so not much is lost by using the same parameters for all instances. All tests were performed on a PC with an AMD Athlon 1.1 GHz processor and 768 MB of main memory, using the operating system Linux 2.2.13. We used the GNU egcs 1.1.2 compiler and CPLEX 7.0 as LP-solver.

In Table 1, we report reduction results for different groups of VLSI-instances of SteinLib. The last group (LIN) was not present in the older versions of SteinLib and included several unsolved instances that we could solve with the help of the techniques described in this paper. For each group, three kinds of results are given:

1. using classical (not extended) reduction tests, as already described in [10];
2. using fast variants of the extended tests (variants 1 and 2 in Section 2.4 in addition to 1 above);
3. using strong variants of the extended tests (variants 3 and 4 in Section 2.4 in addition to 1 and 2 above).

For each group of instances, we give the average values for running time (in seconds) and the fraction of edges remaining after the corresponding reductions.

Detailed results for single instances can be found in [9]. For the sake of comparison, we also provide the best other results [13, 14] we are aware of. Those other results were produced on a Sun ULTRA 167 MHz, but even if one divides the corresponding running times by 10, our strong variants are still much faster, while already our fast variants achieve better reduction results in all cases.

Table 1. Comparison of Reduction Results (Averages)

instance group	classical reductions		extended red., fast variants		extended red., strong variants		Uchoa et. al. [14]	
	time	% of $\|E\|$ remaining	time	% of $\|E\|$ remaining	time	% of $\|E\|$ remaining	time	% of $\|E\|$ remaining
ALUE	2.89	34.50	10.06	0.63	10.36	0.62	1310.93	14.25
ALUT	4.02	38.27	32.62	0.75	33.81	0.49	1806.44	11.76
DIW	0.35	11.57	0.88	0.00	0.88	0.00	214.67	4.10
DMXA	0.06	8.59	0.07	0.00	0.07	0.00	4.14	2.55
GAP	0.13	3.21	0.31	0.00	0.31	0.00	60.54	2.62
MSM	0.13	7.76	0.24	0.05	0.24	0.05	12.90	2.66
TAQ	0.39	23.83	0.98	0.18	0.99	0.16	69.29	11.33
LIN	3.52	27.04	106.17	5.16	1329.21	1.58	—	—

In Table 2, we report the average times (in seconds) for the optimal solution of different groups of instances. The solution method is the same as in [10]; here we have additionally used the extended reduction methods described in this paper (all variants) together with some new methods we described in [12]. Again, for comparison we have included the best other results we are aware of [13], where a branch-and-cut algorithm was used after the reduction phase.

Table 2. Comparison of Times for Exact Solution (Averages)

instance group	our results	Uchoa [13]
ALUE	13	12560
ALUT	35	275817
DIW	0.89	243
DMXA	0.07	4.2
GAP	0.31	61
MSM	0.24	15
TAQ	0.99	163
LIN	20032	—

4 Concluding Remarks

In this paper, we described a generic algorithm and some concrete variants of it for extending the scope of reduction tests. The new approach of explicitly combining alternative- and bound-based methods, together with many algorithmic ideas, lead to substantial improvement over previous tests that used the idea of expansion.

A very important issue in the context of reduction tests is the interaction between different methods. This is especially important for the tests described in this paper, because they tend to transform instances not only in their size, but also in their type. In particular, the success of the described tests for replacing nonterminals, in cooperation with the edge-elimination tests, tend to transform graphs of high connectivity to graphs with many small vertex separators, often consisting of terminals alone. This prepares the ground for another group of reduction tests, which we describe in [12].

References

[1] C. W. Duin. *Steiner's Problem in Graphs*. PhD thesis, Amsterdam University, 1993.
[2] C. W. Duin. Preprocessing the Steiner problem in graphs. In D-Z. Du, J. M. Smith, and J. H. Rubinstein, editors, *Advances in Steiner Trees*, pages 173–233. Kluwer Academic Publishers, 2000.
[3] C. W. Duin and T. Volgenant. Reduction tests for the Steiner problem in graphs. *Networks*, 19:549–567, 1989.
[4] F. K. Hwang, D. S. Richards, and P. Winter. *The Steiner Tree Problem*, volume 53 of *Annals of Discrete Mathematics*. North-Holland, Amsterdam, 1992.
[5] R. M. Karp. Reducibility among combinatorial problems. In R. E. Miller and J. W. Thatcher, editors, *Complexity of Computer Computations*, pages 85–103. Plenum Press, New York, 1972.
[6] T. Koch and A. Martin. SteinLib. http://elib.zib.de/steinlib, 1997.
[7] T. Koch and A. Martin. Solving Steiner tree problems in graphs to optimality. *Networks*, 32:207–232, 1998.
[8] K. Mehlhorn. A faster approximation algorithm for the Steiner problem in graphs. *Information Processing Letters*, 27:125–128, 1988.
[9] T. Polzin and S. Vahdati Daneshmand. Extending reduction techniques for the Steiner tree problem: A combination of alternative- and bound-based approaches. Research Report MPI-I-2001-1-007, Max-Planck-Institut für Informatik, Stuhlsatzenhausweg 85, 66123 Saarbrücken, Germany, 2001.
[10] T. Polzin and S. Vahdati Daneshmand. Improved algorithms for the Steiner problem in networks. *Discrete Applied Mathematics*, 112:263–300, 2001.
[11] T. Polzin and S. Vahdati Daneshmand. On Steiner trees and minimum spanning trees in hypergraphs. Research Report MPI-I-2001-1-005, Max-Planck-Institut für Informatik, Stuhlsatzenhausweg 85, 66123 Saarbrücken, Germany, 2001.
[12] T. Polzin and S. Vahdati Daneshmand. Partitioning techniques for the Steiner problem. Research Report MPI-I-2001-1-006, Max-Planck-Institut für Informatik, Stuhlsatzenhausweg 85, 66123 Saarbrücken, Germany, 2001.

[13] E. Uchoa. *Algoritmos Para Problemas de Steiner com Aplicações em Projeto de Circuitos VLSI (in Portuguese)*. PhD thesis, Departamento De Informática, PUC-Rio, Rio de Janeiro, April 2001.

[14] E. Uchoa, M. Poggi de Aragò, and C. C. Ribeiro. Preprocessing Steiner problems from VLSI layout,. Technical Report MCC. 32/99, Departamento de Informática, PUC-Rio, Rio de Janeiro, Brasil, 1999.

[15] P. Winter. Reductions for the rectilinear Steiner tree problem. *Networks*, 26:187–198, 1995.

[16] R. T. Wong. A dual ascent approach for Steiner tree problems on a directed graph. *Mathematical Programming*, 28:271–287, 1984.

A Comparison of Multicast Pull Models

Kirk Pruhs* and Patchrawat Uthaisombut

Computer Science Department, University of Pittsburgh
kirk,utp@cs.pitt.edu

Abstract. We consider the setting of a web server that receives requests for documents from clients, and returns the requested documents over a multicast/broadcast channel. We compare the quality of service (QoS) obtainable by optimal schedules under various models of the capabilities of the server and the clients to send and receive segments of a document out of order. We show that allowing the server to send segments out of order does not improve any reasonable QoS measure. However, the ability of the clients to receive data out of order can drastically improve the achievable QoS under some, but not all, reasonable/common QoS measures.

1 Introduction

We consider the setting of a web server that receives requests for documents from clients, and returns the requested documents over a multicast/broadcast channel. Such a setting might arise for several reasons. The server and the client may be on the same local area network, or wireless network, where broadcasting is the basic form of communication. Or the server may be using multicast communication to provide better scalability. For example, to provide scalable data service the Hughes DirecPC system[5] uses a broadcast satellite system and Digital Fountain [4] uses IP-multicast. If multiple clients request the same document at approximately the same time, the server may aggregate these requests, and multicast the document only once. One would expect that the ability to aggregate requests would improve Quality of Service (QoS) for the same reason that proxy caches improve QoS, that is, it is common for different users to make requests to the same document.

We formalize this problem in the following manner. Request R_i, $1 \le i \le n$, is for document $D_{\sigma(i)} = D_j$, $1 \le j \le m$. We assume that at some level, say at the transport layer, the document D_j is partitioned into p_j equal sized segments. We assume that the transport layer can transmit one segment from the server to all the clients in one unit of time. We assume that request R_i is made by a client, and immediately communicated to the server, at some time r_i. The completion time $C_i(S)$ for this request in a schedule S is the earliest time that all of the segments of D_j have arrived at the requesting client after time r_i.

---

* Supported in part by NSF grant CCR-0098752, NSF grant ANIR-0123705, and by a grant from the US Air Force.

There are two standard QoS measures for request R_i in a schedule S. The *flow time*, or equivalently user perceived latency, for this request is $F_i(S) = C_i(S) - r_i$. The *stretch* for this request is $ST_i(S) = F_i(S)/p_j$. The motivation for considering stretch is that a human user may have some feeling for the size of the requested document (for example the user may know that video documents are generally larger than text documents) and may be willing to tolerate a larger response time for larger documents. Usually the QoS measures for the requests are combined using an L_p norm to get a QoS measure for the collection of requests. The L_p norm of the flow times for a particular schedule S is $F^{(p)}(S) = \left(\sum_{i=1}^n F_i^p\right)^{1/p}$, where $1 \leq p \leq \infty$. The most commonly used measure of system performance is $F^{(1)}(S)/n$, the average flow time. Also of particular interest is $F^{(\infty)}(S)$, the maximum flow time. The L_p norms for stretch, the average/total stretch, and maximum stretch are similarly defined, and are denoted by $ST^{(p)}(S)$, $ST^{(1)}(S)$, $ST^{(\infty)}(S)$, respectively. When the schedule S is understood we will drop it from our notation.

In the multicast pull setting, a new client might request a document in the middle of the server's transmission of that document to some other clients. Thus potentially this new client could buffer the end of the document at the transport layer. For example, TCP buffers out of order segments. In this case, after finishing the initial transmission of the document, the server need only retransmit the start of the document to satisfy the request of the new client. It is obvious that such client-side buffering can improve the QoS provided by the system.

However, this client buffering of the document may be difficult or impossible for the client to accomplish. One example is the Go-Back-N recovery method that is implemented in the HDLC protocol [12]; In Go-Back-N the receiver discards out of order packets in order to simplify its responsibilities. In the HTTP protocol the client could not generally buffer the end of a broadcast because the identification of the content of the broadcast is in the header. We assume that the client-side transport layer guarantees in order delivery of segments to the application layer. When the client-side transport layer receives a segment, it may either discard the segment, or pass up to the application layer this segment along with any buffered segments that do not violate the transport layer's in order guarantee. There seem to be three natural models of client-side buffering, which we list from least powerful to most powerful:

no buffering: When the client transport layer receives a segment it must either immediately pass the segment up to the application layer, or discard the segment.

cyclic buffering: When the client transport layer receives a segment while its buffer is empty, it may buffer the segment. When the client transport layer receives segment k of a document while its buffer is not empty, it may buffer the segment if segment $k - 1 \bmod p_i$ is also buffered.

arbitrary buffering: When the client transport layer receives a segment, it may buffer the segment.

For the no buffering and arbitrary buffering models, the best algorithm on a client is fixed. If a client cannot buffer, the best algorithm on the client is to discard all segments until it receives segment 1, which it can then pass up to the application layer. After segment i has been passed up to the application layer, the client must discard all segments until it receives segment $i+1$, and so on. If a client can buffer arbitrarily, the best algorithm on the client is to buffer all segments of the requested document that it receives. If a client can buffer cyclically, the best algorithm on the client is unclear; it depends on the server model and/or server algorithm. After a client buffers a segment of a document, the sequence of segments that the client can buffer becomes fixed. Thus, a client may choose not to buffer a segment when its buffer is empty in the hope that a later sequence of segments may satisfy the request faster.

In principle a server could transmit any segment of a document at any point of time. In practice, a server that transmits documents sequentially would be easier to implement, have less overhead, and thus have higher throughput. In the traditional unicast setting, where the server can not aggregate responses, it is easy to see that the ability to transmit a document in an arbitrary order would not improve any reasonable QoS measure. In the multicast setting, it is not at all clear whether the ability of the server to transmit segments out of order can improve some reasonable QoS measure. There seem to be three natural models of server-side transmission ordering, which we list from least powerful to most powerful:

cyclic server ordering: The first segment of a document D_j that the server sends must be the first segment of that document. If the last segment of a document that the server sent was segment i, then the next segment of this document that the server sends must be $(i+1) \bmod p_j$.

cyclic ordering with restart: The first segment of a document D_j that the server sends must be the first segment of D_j. If the last segment sent was segment i, then the next segment must be either segment $(i+1) \bmod p_j$ or the first segment. That is the server may restart sending the document again, say if many new requests arrived for the document.

arbitrary server ordering: The server may send any segment at any time.

There is no obvious choice of the "right" model of server-side transmission ordering and client-side buffering. Different models are appropriate under different settings, and researchers have analyzed algorithms for multicast pull scheduling in various models. For example, in [3] it is claimed that the server scheduling algorithm First-Come-First-Served (FCFS) is $O(1)$-competitive for the objective function of minimizing the maximum flow time, under the assumption of cyclic server ordering and cyclic client buffering. As another example, in [6] it was shown that the server scheduling algorithm BEQUI, which broadcasts each document at a rate proportional to the number of requests for that document, is an $O(1)$-speed $O(1)$-approximation algorithm for average flow time assuming no client-side buffering and cyclic server-side order. In this context, an *s-speed c-approximation* algorithm A has the property that for all inputs, the value of the objective function of the schedule that A produces with speed s server is at most c times the optimal value of the objective function for a speed 1 server [9, 13]. Intuitively an $O(1)$-speed $O(1)$-competitive algorithm is guaranteed to perform well if the system is not too heavily loaded.

1.1 Our Results

Informally, the question that we address in this paper is, "How do the various models of server-side and client-side capabilities affect the QoS provided by the system?" We primarily focus on comparing the optimal schedules in the various models, which in some sense measures the inherent limit of the QoS obtainable from the system. For example, one concrete question we address is "How much better can the average flow time be in an optimal schedule when you have arbitrary server-side ordering and arbitrary client-side buffering as compared to the average flow time for the optimal schedule with cyclic server-side ordering and no client side buffering?" We give a reasonably complete comparison of the QoS achievable in the various models of server/client capabilities for QoS measures that are L_p norms of flow time and stretch.

One natural question that arises when one reads the results in [3, 6] is whether these results carry over to other models, e.g "Is FCFS still $O(1)$ competitive if you have arbitrary server-side ordering and arbitrary client-side buffering?" The results of our comparisons of optimal schedules can answer these questions for the results in [3, 6], and for some reasonable collection of future results that researchers might obtain. For example, if the QoS of the optimal schedule in a more powerful model (that is, where the server/client have additional capabilities) is not much better then the QoS of the optimal schedule in a weaker model, then one can conclude that a competitive online algorithm in the weaker model is competitive in the stronger model. Obviously a competitive online algorithm A in a stronger model is competitive in a weaker model, provided that A is implementable in this weaker model. Where it is possible to do so, we give a general method for converting a general online algorithm A in a stronger model, to a online algorithm B in a weaker model, in such a way that there is little or no deterioration in the QoS for all inputs. We also note when such a general transformation is not possible.

In section 2 we show that in all client models, it is the case that the optimal schedule with cyclic server ordering dominates the optimal schedule with arbitrary server ordering; A schedule S *dominates* a schedule T if the completion time of every request in S is no later than the completion time of that request in T. Additionally, for cyclic client-side buffering and arbitrary client-side buffering, we can convert any online server algorithm using arbitrary ordering to one that is just as good but uses only cyclic server ordering. Hence, arbitrary server ordering is of no more benefit to an optimal server than cyclic server ordering under all reasonable QoS measures (those that are monotone in the completion times of the requests). We thus assume cyclic server ordering for the rest of the paper. Given cyclic server ordering, it is easy to see that the optimal schedule with cyclic client-side buffering dominates the optimal schedule with arbitrary client-side buffering. Thus arbitrary buffering offers no advantages over

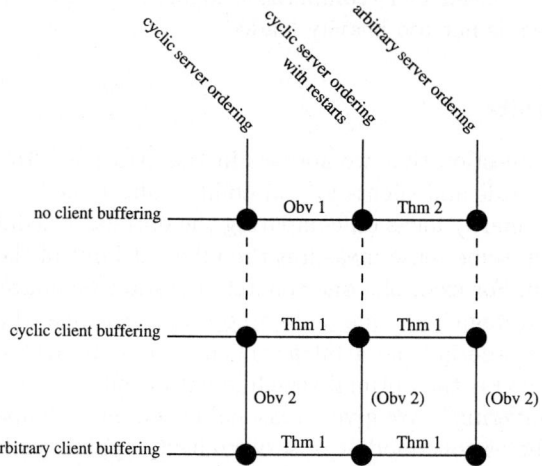

Fig. 1. Equivalence of the optimal schedule among models

cyclic buffering. The rest of paper is then devoted to comparing cyclic client-side buffering with no client-side buffering.

We summarize these results in figure 1. In figure 1, each small circle represents a model with characteristics described by the corresponding vertical and horizontal descriptions. A solid edge between two models signifies that the optimal schedule in the two models are equivalent. The label of each edge tells where the result is stated and proved.

In section 3 we show that the optimal maximum flow time obtainable with client-side buffering is no less than half of the optimal maximum flow time that is obtainable without client-side buffering. We also show that the same result holds when the QoS measure is maximum stretch. Thus client-side buffering doesn't drastically improve the optimal maximum flow time or maximum stretch. And as a consequence of this result, we can conclude that the result from [3], that FCFS is O(1)-competitive with respect to maximum flow time, extends to all models considered here. We also show that this bound of half is tight, that is, there are schedules where buffering reduces the value of the maximum flow time and the value of the maximum stretch by a factor of two.

In contrast we show, in section 4, that for finite p, the L_p norm of the flow times and stretches in the optimal schedule with client-side buffering can be a factor of $n^{\Theta(1/p)}$ less than those in the optimal schedule with no client-side buffering. Thus for these QoS measures, client-side buffering can vastly improve the QoS provided by the system. In particular we show that the value of $F^{(1)}$, $ST^{(1)}$ and $ST^{(2)}$ with client-side buffering can be at least a factor of $n^{1/3}$ smaller than those with no client-side buffering. Further, we show that $F^{(p)}$, $p > 1$, can be at least a factor of $n^{1/2p}$ better, and that $ST^{(p)}$, $p > 2$ can be at least a factor of $n^{1/p}$ better, with client-side buffering than without client-side buffering. We show that for these QoS measures, client-side buffering can improve the achievable QoS by at most a factor of $n^{1/p}$. It is at least of academic interest to close the gap between these upper and lower bounds. While we are not able to fully close the gap here, we make some progress for average flow time. That is for the case that $p = 1$, where the obvious upper bound is n, we show that client side buffering can improve the average flow time by at most a factor of $n^{1/2}$. We summarize these results in table 1.1. Due to space limitations, some proofs have been omitted.

	$p = 1$ (total flow time or total stretch)	$p \geq 2$	$p = \infty$ (max flow time or max stretch)
flow time	$\Omega(n^{1/3})$ (Thm 9) $\leq \sqrt{n} + 1$ (Thm 13)	$\Omega(n^{1/2p})$ (Thm 9) $O(n^{1/p})$ (Thm 12)	$\geq 2 - \frac{2}{n-2}$ (Thm 6) ≤ 2 (Thm 5)
stretch	$\Omega(n^{1/3})$, $p = 1, 2$ (Thm 11) $O(n)$ (Thm 12)	$\Omega(n^{1/p})$, $p \geq 3$ (Thm 11) $O(n^{1/p})$ (Thm 12)	$\geq 2 - \frac{2}{n-2}$ (Thm 6) ≤ 2 (Thm 5)

In section 5 we show that the optimal schedule with a 2-speed server (that is, it can transmit two segments in unit time) assuming no client-side buffering dominates the optimal schedule for a 1-speed server assuming client-side cyclic buffering. Thus there is at most an 2-speed difference between any of the models. As a consequence of this result, we can conclude that the result from [6], that BEQUI is an $O(1)$-speed $O(1)$-approximation algorithm, extends to all models considered here.

1.2 Additional Related Results

A nice introduction to multicast pull scheduling, and some experimental results, can be found in [1]. In [3] it is shown that minimizing the maximum flow time can be approximated well offline, under the assumption of cyclic server order and cyclic client buffering. There are also results in the literature concerning the problem of minimizing the average flow time of non-preemptable unit jobs, in part because unit jobs are easier to deal with combinatorially. This problem was shown to be NP-hard in [7]. An offline $O(1)$-speed $O(1)$-approximation algorithm can be found in [10]. The constants are improved in [7, 8]. A online variation of BEQUI is shown to be an $O(1)$-speed $O(1)$-approximation algorithm in [6]. There has been a fair amount of research into push-based broadcast systems, sometimes called broadcast disks, where the server pushes information to the clients without any concept of a request, ala TV broadcast [2, 11]. It seems that good strategies for push-based systems have little to do with good strategies for pull-based systems.

2 Server-Side Ordering

We show that for all client models, the optimal schedule in the weakest server model dominates the optimal schedule in the strongest server model. For client-side cyclic buffering and client-side arbitrary buffering this is relatively straightforward to see.

Theorem 1. *If the clients use either cyclic buffering, or arbitrary buffering, then the optimal schedule with cyclic server ordering dominates the optimal schedule with arbitrary server ordering.*

Proof. It is sufficient to explain how to turn a schedule S, with arbitrary server ordering, into a schedule T, with cyclic server ordering, in such a way that none of the completion times increase. For all times t, the document that T is broadcasting at the time t is identical to the document that S is broadcasting at the time t, although the segment within the document that S and T broadcast at time t may be different. The first time that S broadcasts any segment of a document D_j, T broadcasts the first segment of D_j. This fixes T's schedule. To see that T doesn't increase any completion time note that a request R_i for document D_j can not complete in S until p_j segments of D_j have been broadcast, which will be when R_i completes in T. □

Note that the above construction can be carried out online. That is, if clients have some form of buffering, then an online server algorithm A that uses arbitrary ordering can be converted to a online server algorithm B that uses only cyclic ordering, with no loss of QoS on all inputs. Assuming that the server uses cyclic ordering, the best client-side algorithm with cyclic buffering is now apparent; A client should start buffering when it receives any segment of the requested document.

We now turn our attention to the case when there is no client-side buffering. It is easy to see that allowing restarts will not improve the optimal offline schedule with cyclic ordering (any partial attempts to broadcast a document before a restart may be omitted without increasing any completion time).

Observation 1. *If the clients cannot buffer, then the optimal schedule, with cyclic server ordering, dominates the optimal schedule, with cyclic server ordering with restarts.*

We are thus left to prove that arbitrary server ordering is no better than cyclic ordering with restarts.

Theorem 2. *If the clients cannot buffer, then the optimal schedule, with cyclic server ordering with restarts, dominates the optimal schedule with arbitrary server ordering.*

Proof. Consider an arbitrary schedule S with arbitrary server ordering. Let t be the first time that S violates the cyclic server ordering with restart property. We show how to iteratively transform S into a new schedule T such that T

doesn't violate the cyclic server ordering property up through time t (including time t) and the completion time of every request in T is no later than the completion time of that request in S. The proof follows by iteratively repeating this transformation.

Assume that at time t the schedule S is multicasting the kth segment of a document D_j. Let the requests for D_j that have been released but not completed by time t be $R_1, \ldots, R_a$, ordered by increasing release time. Let the number of segments that the clients issuing these requests have received just before time t be $N_1, \ldots N_a$, respectively. Note that since the requests are ordered by release times it must be the case that $N_i \geq N_{i+1}$, $1 \leq i \leq a - 1$. For the same reason, no request for D_j was completed between time that the client requesting R_1 received its first segment and time t. If $k-1$ does not equal N_i, for some $i \in [1, a]$, then all clients discard this segment since they do not have buffers; We then construct T by deleting the broadcast of the segment k of D_j at time t. So now assume that $k = N_\ell + 1$ for some ℓ where $1 \leq \ell \leq a$. Let s be the latest time, strictly before time t, that segment $k-1$ of D_j was broadcast. Let $u_1, \ldots, u_b$ be the times during $[s+1, t]$ when S was broadcasting a segment from D_j. We now construct the schedule T from S as follows. In T segment k of D_j is broadcast at time u_1. Note that this preserves cyclic ordering. At time u_{i+1}, $1 \leq i \leq b-1$, T broadcast the segment that S broadcast at time u_i. Note that at time u_1 there was a restart for document D_j in S. Thus the restart at time u_2 in T is legal under the cyclic ordering with restart rules. The subsequent broadcasts in T at the u_i times are legal under the cyclic ordering with restart rules because they were legal at time u_{i-1} in S.

We now claim that the state of all clients is the same just after time t in both S and T. Those clients with requests R_i, $i \in [1, a]$, with $N_i > k-1$ received segments at the same time in S and T. Those clients with requests R_i, $i \in [1, a]$, with $N_i = k-1$ received segments at the same time in S and T except that in T they received segment k at time s instead of at time t, as in S. Those clients with requests R_i, $i \in [1, a]$, with $N_i < k-1$, merely had their segments delayed slightly in T. □

We can now conclude that in every client model, the QoS of the optimal schedule in the strongest server model is no better than the QoS of the optimal schedule in the weakest server model. Thus for the rest of the paper we assume cyclic server ordering. It is then easy to see that the optimal schedule with cyclic client buffering dominates the optimal schedule for arbitrary client buffering. To see this note that since the documents are broadcast cyclically, the client doesn't even have the opportunity to buffer an out of order segment.

Observation 2. *If the server uses cyclic ordering, then the optimal schedule, with cyclic client-side buffering dominates the optimal schedule, with arbitrary client-side buffering.*

Thus the rest of the paper is devoted to comparing cyclic client-side buffering (which we will simply refer to as buffering) and no client-side buffering.

Before proceeding on to the next section we need to introduce some terms and notation. We define the *lifespan* of a request R_i in schedule S to be the time interval $L_i(S) = [r_i, C_i(S)]$. Note that if there are two requests R_i and R_j for the same document with $R_i < R_j$, then $C_i(S) \leq C_j(S)$ since the segments that the client that issued R_i has seen is a strict superset of the segments that the client that issued R_j has seen. Hence no lifespan of a request for a document can properly contain the lifespan of another request for that document.

In a schedule with no client buffering, the transmissions of segments of a document D_j in a schedule S can be partitioned into *broadcasts of D_j*, where a broadcast begins with the transmission of the first segment of D_j and ends at the next time that the last segment of D_j is transmitted. Thus the flow time for a request for D_j in an unbuffered schedule is the time until the start of the next broadcast of D_j plus the duration of that broadcast.

3 Cyclic Buffering vs. No Buffering in L_∞ Norms

In this section, we show that lack of client buffering does not greatly affect the QoS if the QoS measure is maximum flow time or maximum stretch. More precisely, it at most doubles the maximum flow time and maximum stretch. Further we show that this bound is tight.

Theorem 3. *For any schedule S with cyclic client buffering, there exists an unbuffered schedule T such that $F^{(\infty)}(T) \leq 2F^{(\infty)}(S)$, and such that $ST^{(\infty)}(T) \leq 2ST^{(\infty)}(S)$.*

Proof. Consider some arbitrary document D_j. We determine when T broadcasts D_j by iteratively repeating the following process. Let R_l be the latest request to D_j that we have not previously considered. There must be exactly p_j times in $L_l(S)$ when S is broadcasting segments of D_j; Let k_i, $1 \leq i \leq p_j$, be the k_ith such time. Then at time k_i, T is broadcasting segment i of D_j. Let $\mathcal{R}$ be the set of requests to document D_j, such that the life span of these requests overlap with $L_l(S)$. Request R_l and the requests in $\mathcal{R}$ are removed from further consideration.

To see that the schedule T satisfies the desired conditions, consider a request R_l for D_j that was selected at some iteration. Notice that $C_l(T) = C_l(S)$ by construction, and hence $F_l(T) = F_l(S)$, and $ST_l(T) \leq ST_l(S)$. Further for each request $R_a \in \mathcal{R}$ it is the case that $C_a(S) \leq C_l(S)$ since $r_a \leq r_l$ by the definition of R_l. Since $L_a(S)$ and $L_l(S)$ overlap, it must then be the case that $F_a(T) \leq F_a(S) + F_l(S)$. Since requests R_a and R_l are for the same document $ST_a(T) \leq ST_a(S) + ST_l(S)$. Hence, the flow time (stretch) of every request in T is at most the sum of the flow times (stretches) of two requests in S. The claimed results then immediately follow. □

Next we show a matching lower bound.

Theorem 4. *For all even n, there exists an instance I_n such that the maximum flow time of the optimal unbuffered schedule is at least $2 - \frac{2}{n+2}$ times the maximum flow time of the optimal buffered schedule. Similarly, the maximum stretch of the optimal unbuffered schedule is at least $2 - \frac{2}{n+2}$ times the maximum stretch of the optimal buffered schedule. Here n is the number of requests in I.*

It should be noted that one can improve the maximum stretch lower bound to $2 - 1/2^{n-2}$ at the cost of complicating the construction slightly.

4 Cyclic Buffering vs. No Buffering in L_p Norms

In this section we first show that if p if finite, then the L_p norms of the flow times and stretches of buffered schedules can be a factor of $n^{\Theta(1/p)}$ better than those of unbuffered schedules.

4.1 Lower Bounds

All lower bound instances to be presented have the same structure, which we now define formally.

Definition 1. *Let $(k_1, 1, p_1 ; k_2, 1, p_1 ; k_3, m_3, p_3 ; k_4, m_4, p_4)$ denote an input instance for the broadcast scheduling problem where the requests are partitioned into four groups Q_1, Q_2, Q_3 and Q_4 as follows:*

- *There are k_1 requests in Q_1, all of which arrive at time 0, and are to a document D_1 which has length p_1.*
- *There are k_2 requests in Q_2, all of which arrive at time $p_1/2$, and are to document D_1.*
- *There are $k_3 m_3$ requests in group Q_3. From time p_1 to $p_1 + p_3(m_3 - 1)$, there are k_3 requests to a new document of length p_3 every p_3 time units.*
- *There are $k_4 m_4$ requests in group Q_4. From $p_1 + p_3 m_3 + p_1/2$ to time $p_1 + p_3 m_3 + p_1/2 + p_4(m_4 - 1)$ there are k_4 requests to a new document of length p_4 every p_4 time units.*

Note that all requests that arrive at different times are to distinct documents with the exception of requests in Q_1 and Q_2 which are to the same document D_1. Consider the optimal schedule S. Except for the requests for document D_1 at time $p_1/2$, S starts broadcasting a document as soon as the requests for that document arrive, and continues broadcasting the document without interruption until the requests are satisfied. The requests for document D_1 at time $p_1/2$ arrive while document D_1 is being broadcast. The clients that initiate these requests begin receiving and buffering the second half of the document immediately. The first half of the document is broadcast again in S between time $p_1 + m_3 p_3$ and $p_1 + m_3 p_3 + p_1/2$.

Lemma 1. *If $I = (k_1, 1, p_1 ; k_2, 1, p_1 ; k_3, m_3, p_3 ; k_4, m_4, p_4)$, then for any unbuffered schedule S' for I and for any $p \geq 1$, then $F^{(p)}(S')$ is Ω of the minimum of $k_1^{1/p} m_3 p_3$, $(k_3 m_3)^{1/p} p_1$, $k_2^{1/p} m_4 p_4$, $(k_3 p_1/p_3)^{1/p} m_4 p_4$, and $(k_4 m_4)^{1/p} p_1$.*

We now apply this lemma to get our lower bounds for flow time.

Theorem 5. *There exists an input instance I_1 such that $\frac{F(S')}{F(S)} = \Omega(m^{1/3})$. For any $p > 1$, there exists an input instance I_p such that $\frac{F^{(p)}(S')}{F^{(p)}(S)} = \Omega(n^{1/2p})$. Here S' is the optimal unbuffered broadcast schedule for I_p and S is the optimal buffered broadcast schedule for I_p.*

We now turn our attention to the stretch QoS measures.

Lemma 2. *If $I = (k_1, 1, p_1 \,;\, k_2, 1, p_1 \,;\, k_3, m_3, p_3 \,;\, k_4, m_4, p_4)$, then for any unbuffered schedule S' for I and for any $p \geq 1$, $ST^{(p)}(S')$ is Ω of the minimum of $\frac{k_1^{1/p} m_3 p_3}{p_1}$, $\frac{(k_3 m_3)^{1/p} p_1}{p_3}$, $\frac{k_2^{1/p} m_4 p_4}{p_1}$, $\frac{(k_3 p_1/p_3)^{1/p} m_4 p_4}{p_3}$, and $\frac{(k_4 m_4)^{1/p} p_1}{p_4}$.*

We now apply lemma 2 to obtain a lower bound for our stretch QoS measures.

Theorem 6. *For $p = 1, 2$, there exists a input instance I_p such that $\frac{ST^{(p)}(S')}{ST^{(p)}(S)} = \Omega(m^{1/3})$. For any $p \geq 3$, there exists a input instance I_p such that $\frac{ST^{(p)}(S')}{ST^{(p)}(S)} = \Omega(n^{1/p})$. Here, S' is the optimal unbuffered broadcast schedule for I_p and S is the optimal buffered broadcast schedule for I_p.*

4.2 Upper Bounds

Theorem 7. *For any $p \geq 1$ and for any buffered broadcast schedule S for an input instance I with n requests, there exists an unbuffered broadcast schedule T for I such that $F^{(p)}(T) = O(n^{1/p})\, F^{(p)}(S)$ and $ST^{(p)}(T) = O(n^{1/p})\, ST^{(p)}(S)$.*

Proof. The construction of T from S is the same as in Theorem 3. □

We now show that client-side buffering can improve the average flow time by at most a factor of $n^{1/2}$.

Theorem 8. *For any buffered broadcast schedule S for an input instance I with n requests, there exists an unbuffered broadcast schedule T for I such that $F(T) \leq (\sqrt{n} + 1) F(S)$.*

5 Resource Augmentation

In this section, we show that there is at most a 2-speed difference between any of the models.

Theorem 9. *The optimal unbuffered schedule T with a 2-speed server dominates the optimal buffered schedule S for a 1-speed server.*

Proof. We will assume that S broadcasts one segment every two time units, and T broadcasts one segment every time unit. We construct T from S in the following manner. At every time, T broadcasts the document that S is broadcasting. Since the server is broadcasting cyclically this fixes T. Note that even though S and T broadcast the same document at any time, they may be broadcasting different segments of the document since T has a faster server. □

Acknowledgments

The origination of the questions considered in this paper arose from discussions with Jeff Edmonds. The observation that arbitrary server ordering is of no benefit when there is client-side buffering is due to Jiri Sgall.

References

[1] S. Aacharya, and S. Muthukrishnan, "Scheduling on-demand broadcasts: new metrics and algorithms", ACM/IEEE International Conference on Mobile Computing and Networking, 1998.
[2] A. Bar-Noy, R. Bhatia, J. Naor, and B. Schieber, "Minimizing service and operation costs of periodic scheduling", 11-20, SODA 1998.
[3] Y. Bartal, and S. Muthukrishnan, "Minimizing maximum response time in scheduling broadcasts", 558-559, SODA, 2000.
[4] Digital Fountain website, http://www.digitalfountain.com.
[5] DirecPC website, http://www.direpc.com.
[6] J. Edmonds and K. Pruhs, "Broadcast scheduling: when fairness is fine", SODA, 2002.
[7] T. Erlebach and A. Hall, "Hardness of broadcast scheduling and inapproximability of single-source unsplittable min-cost flow", SODA, 2002.
[8] R. Gandhi, S. Khuller, Y.A. Kim, and Y.C. Wan, "Algorithms for Minimimzing Response Time in Broadcast Scheduling", To appear in IPCO, 2002.
[9] B. Kalyanasundaram, and K. Pruhs, "Speed is as powerful as clairvoyance", *JACM*, 2000.
[10] B. Kalyanasundaram, K. Pruhs, and M. Velauthapillai, "Scheduling broadcasts in wireless networks", *European Symposium on Algorithms* (ESA), 2000.
[11] C. Kenyon, N. Schabanel and N. Young, "Polynomial-time approximation schemes for data broadcast", 659-666, STOC 2000.
[12] J. Kurose and K. Ross, *Computer Networking: A Top Down Approach Featuring the Internet*, Addison Wesley Longman, 2001.
[13] C. Phillips, C. Stein, E. Torng, and J. Wein "Optimal time-critical scheduling via resource augmentation", *ACM Symposium on Theory of Computing*, 140 – 149, 1997.

Online Scheduling for Sorting Buffers*

Harald Räcke[1], Christian Sohler[1], and Matthias Westermann[2]

[1] Heinz Nixdorf Institute and Department of Mathematics and Computer Science
Paderborn University, D-33102 Paderborn, Germany
{harry,csohler}@uni-paderborn.de
[2] International Computer Science Institute
Berkeley, CA 94704, USA
marsu@icsi.berkeley.edu

Abstract. We introduce the online scheduling problem for sorting buffers. A service station and a sorting buffer are given. An input sequence of items which are only characterized by a specific attribute has to be processed by the service station which benefits from consecutive items with the same attribute value. The sorting buffer which is a random access buffer with storage capacity for k items can be used to rearrange the input sequence. The goal is to minimize the cost of the service station, i.e., the number of maximal subsequences in its sequence of items containing only items with the same attribute value. This problem is motivated by many applications in computer science and economics.

The strategies are evaluated in a competitive analysis in which the cost of the online strategy is compared with the cost of an optimal offline strategy. Our main result is a deterministic strategy that achieves a competitive ratio of $O(\log^2 k)$. In addition, we show that several standard strategies are unsuitable for this problem, i.e., we prove a lower bound of $\Omega(\sqrt{k})$ on the competitive ratio of the First In First Out (FIFO) and Least Recently Used (LRU) strategy and of $\Omega(k)$ on the competitive ratio of the Largest Color First (LCF) strategy.

1 Introduction

In the online scheduling problem for sorting buffers, we are given a service station and a sorting buffer. An input sequence of items which are only characterized by a specific attribute has to be processed by the service station which benefits from consecutive items with the same attribute value. The sorting buffer which is a random access buffer with limited storage capacity can be used to rearrange the input sequence. Whenever a new item arrives it has to be stored in the sorting buffer. Accordingly, items can also be removed from the sorting buffer and then assigned to the service station. Hence, the service station has

---

* The first two authors are partially supported by the DFG-Sonderforschungsbereich 376, DFG grant 872/8-1, and the Future and Emerging Technologies program of the EU under contract number IST-1999-14186 (ALCOM-FT). The third author is supported by a fellowship within the ICSI-Postdoc-Program of the German Academic Exchange Service (DAAD).

to process a sequence of items that is a partial rearrangement of the input sequence. The goal is to minimize the cost of the service station, i.e., the number of maximal subsequences of items with the same attribute value in the rearranged sequence. This problem is motivated by many applications in computer science and economics. In the following we give some examples.

In computer graphics, the process of displaying a representation of 3D objects, i.e., polygons, is denoted as *rendering*. Each polygon is characterized by several attributes, e.g., color and texture. In the rendering process, a sequence of polygons is transmitted to the graphic hardware. A change of attributes for consecutive polygons in this sequence is denoted as *state change*. One of the determining factors for the performance of a rendering system is the number of state changes in the sequence of polygons. Of course, unless you want to model a black ninja during night, there have to be some state changes. However, the number of state changes can be reduced by preceding a rendering system with a sorting buffer.

Communication in computer systems connected by a network is usually realized by the transfer of data streams. The sending of a data stream by a computer is denoted as *startup*. The overhead induced by the startup process (inclusive the overhead of the receiving computer) is denoted as *startup cost*. Sending many small data streams has a negative impact on the performance of the network, since in this case the startup cost dominates the total cost. The startup cost can be reduced by combining several data streams that are directed to the same destination. Hence, with the assistance of sorting buffers at the computers on the network the performance of the network can be improved.

File server are computer and high-capacity storage devices which each computer on a network can access to retrieve files. Hence, a file server receives a sequence of read and write accesses to files from the computers on the network. A file is denoted as *open*, if it is ready to be accessed. Otherwise, it is denoted as *closed*. By technical reasons, the number of open files on a file server is limited. Since the overhead induced by the opening and closing process takes a lot of time for a file server, it is preferable if as many as possible accesses to an open file can be processed before closing it. The number of opening and closing procedures can be minimized by preceding a file server with a multi service sorting buffer which is a generalization of our sorting buffer. A *multi service sorting buffer* has m identical service stations and each item can be processed by any of these stations. In this example, m is the maximum number of open files on a file server. Note that the multi service sorting buffer problem in which the sorting buffer has storage capacity for only one item is equivalent to the classical paging problem (see, e.g., [2, 10]).

In the painting center of a car plant, a sequence of cars traverses the final layer painting where each car is painted with its own top coat. If two consecutive cars have to be painted in different colors then a *color change* is required. Such a color change causes cost due to the wastage of paint and the use of cleaning chemicals. These costs can be minimized by rearranging the sequence of cars in such a way that cars of the same color preferably appear in consecutive positions.

For this purpose, the final layer painting is preceded by a queue sorting buffer which is a generalization of our sorting buffer. A *queue sorting buffer* consists of several queues each with limited storage capacity. Whenever a new car arrives, it has to be transferred to the tail of any of the queues. Accordingly, cars at the head of any queue can also be removed and then assigned to the final layer painting.

1.1 The Model

We are given a *service station* and a *sorting buffer*. An *input sequence* $\sigma = \sigma_1 \sigma_2 \cdots \sigma_n$ of items which are only characterized by a specific attribute has to be processed by the service station which benefits from consecutive items with the same attribute value. To simplify matters, we suppose that the items are characterized by their color. The sorting buffer which is a random access buffer with storage capacity for k items can be used to rearrange the input sequence in the following way.

The *current input item* σ_i, i.e., the first item of σ that is not handled yet, can be stored in the sorting buffer, or items currently in the sorting buffer can be removed and then assigned to the service station. These removed items result in an *output sequence* $\rho = \rho_1 \rho_2 \cdots \rho_n$ which is a partial rearrangement of σ. We suppose that the sorting buffer is initially empty and, after processing the whole input sequence, the buffer has to be empty again. In addition, we use the following notations. Let the *current input color* denote the color of the current input item. Further, let the *current output item* denote the item that was last assigned to the service station and let the *current output color* denote the color of this item.

The goal is to rearrange the input sequence in such a way that items with the same color preferably appear in consecutive positions in the output sequence. Let each maximal subsequence of the output sequence containing only items with the same color denote as *color block*. Between two different color blocks there is a *color change* at the service station. Let the *cost $C(\sigma)$ of the scheduling strategy* denote the number of color blocks in the output sequence. Then, the goal is to minimize the cost $C(\sigma)$.

The notion of an online strategy is intended to formalize the realistic scenario, where the scheduling strategy does not have knowledge about the whole input sequence in advance. Instead, it learns the input piece by piece, and has to react with only partial knowledge of the input. Online strategies are typically evaluated in a competitive analysis. In this kind of analysis which was introduced by Sleator and Tarjan [10] the cost of the online strategy is compared with the cost of an optimal offline strategy.

In order to obtain a simple, unambiguous model, we assume that an adversary initiates the input sequence $\sigma = \sigma_1 \sigma_2 \cdots \sigma_n$ of items. The online strategy has to serve these items one after the other, that is, it is assumed that σ_{i+1} is not issued before σ_i is not stored in the sorting buffer. For a given sequence σ, let $C_{\mathrm{op}}(\sigma)$ denote the minimum cost produced by an optimal offline strategy. An online strategy is said to be *c-competitive* if it produces cost at most $c \cdot C_{\mathrm{op}}(\sigma) + a$, for each sequence σ, where a is a term that does not depend on σ. The value c is also called the *competitive ratio* of the online strategy.

W.l.o.g., we only consider *lazy* scheduling strategies, i.e., strategies that fulfill the following two properties:

- If an item whose color is equal to the current output color is stored in the sorting buffer, a lazy strategy does not make a color change.
- If there are items in the input sequence that can be stored in the sorting buffer, a lazy strategy does not remove an item from the sorting buffer.

Note that an optimal offline strategy can be transformed into a lazy strategy without increasing the cost.

1.2 Our Contribution

We introduce the online scheduling problem for sorting buffers and present deterministic scheduling strategies for this problem. The strategies aim to rearrange an input sequence in such a way that items of the same color preferably appear in consecutive positions in the output sequence. They are evaluated in a competitive analysis.

At first, we show in Section 2 that several standard strategies are unsuitable for this problem, i.e., we prove a lower bound of $\Omega(\sqrt{k})$ on the competitive ratio of the First In First Out (FIFO) and Least Recently Used (LRU) strategy and of $\Omega(k)$ on the competitive ratio of the Largest Color First (LCF) strategy. Our main result which we present in Section 3 is the deterministic Bounded Waste strategy which achieves a competitive ratio of $O(\log^2 k)$. We believe that the Bounded Waste strategy is well suited for the application in practice because the strategy is simple and has a provably good performance. The main part of the proof concerning the upper bound of the Bounded Waste strategy is given in Section 4.

1.3 Previous Work

Scheduling is continuously an active research area, reflecting the changes in theoretical computer science. When the theory of NP-completeness was developed, many scheduling problems have been shown to be NP-complete (see, e.g., [4]). After the NP-completeness results, the focus shifted to the design of approximation algorithms (see, e.g., [6, 8]). Many natural heuristics for scheduling are in fact online strategies. Hence, when the study of online strategies using competitive analysis became usual, this approach was naturally and quite successfully applied to scheduling (see, e.g., [9]).

The online bin-coloring problem is related to our scheduling problem. The goal in the bin-coloring problem is to pack unit size colored items into bins, such that the maximum number of different colors per bin is minimized. The packing process is subject to the constraint that at any moment at most k bins are partially filled. Moreover, bins may only be closed if they are filled completely.

Krumke et al. [7] present a deterministic strategy that achieves a competitive ratio of $(2k+1)$. In addition, they give an $\Omega(k)$ lower bound on the competitive ratio of any deterministic strategy.

Feder et al. [1] study an online caching problem that is similar to our scheduling problem. In their r-reordering problem, a caching strategy can reorder the sequence of requests $\sigma = \sigma_1 \sigma_2 \cdots$ as long as no request is delayed inordinately, i.e., in the new ordering a request σ_i has to be served before a request σ_j, if $i + r \leq j$. For cache size one, they present a deterministic greedy strategy that achieves competitive ratio two and a lower bound of 1.5 on the competitive ratio of any deterministic strategy. For the case that lookahead l is in addition possible, they give a deterministic strategy that achieves competitive ratio $1 + O(r/l)$ and a lower bound of $1 + \Omega(r/l)$ on the competitive ratio of any strategy.

The following variant of an offline problem for queue sorting buffers is well studied. Suppose we are given a sorting buffer with k queues of infinite storage capacity and all items of the input sequence are already stored in the queues. Now, it remains only to determine a sequence of remove operations to empty the queues. This problem is equivalent to the shortest common super-sequence problem. Fraser and Irving [3] present a polynomial time algorithm for this problem that calculates a $(4k + 12)$-approximation. Further, Jiang and Li [5] show that this problem is not in APX, i.e., for this problem exists no polynomial time algorithm with constant approximation ratio, unless P = NP.

In several practical work, heuristic scheduling strategies are used for the sorting buffer problem (see, e.g., [11]). This work is almost always motivated by the demand for efficient strategies for sorting buffers in manufacturing control. Spieckermann and Voss [11] evaluate some simple heuristic strategies by simulation and come to the conclusion that there is a lack of efficient strategies.

2 Lower Bounds

In this section, we give lower bounds on the competitive ratio of several strategies that have previously been applied to other scheduling problems.

First In First Out (FIFO). Each time an item σ_i is stored in the sorting buffer the FIFO strategy checks whether there is another item of the same color stored in the sorting buffer. If there is no such item, the color of σ_i gets a time stamp. If there is no item of the current output color in the sorting buffer, FIFO selects the color with the "oldest" time stamp.

Theorem 1. *The competitive ratio of the FIFO strategy is at least* $\Omega(\sqrt{k})$.

Proof. W.l.o.g., we assume that $\ell = \sqrt{k-1}$ is integral and even. We consider the colors $c_1, \ldots c_\ell, x, y$ and the sequence $\sigma = (c_1 \cdots c_\ell x^k c_1 \cdots c_\ell y^k)^{\ell/2}$.

At first, we show that σ can be processed with 2ℓ color changes. Consider the sequence $(xy)^{\ell/2} c_1 \cdots c_\ell$ of 2ℓ color changes. During the first ℓ color changes of this sequence the items of colors $c_1, \ldots c_\ell$ are accumulated in the sorting buffer. Note that the total number of these items is $2\ell \cdot \ell/2 = k - 1$.

Now it remains to count the color changes if σ is processed by the FIFO strategy. Whenever a block of x^k or y^k appears FIFO empties the whole sorting buffer, since all other colors stored in it are "older". For each block, this produces ℓ color changes. Hence, if σ is processed by FIFO, at least $2\ell \cdot \ell/2 = k - 1$ color changes occur. □

Least Recently Used (LRU). Similar to FIFO the LRU strategy assigns to each color a time stamp. LRU updates the time stamp of a color each time a new item of that color is stored in the sorting buffer. Thus the time stamp of a color is the time when the most recent item of that color was stored. If there is no item of the current output color in the sorting buffer, LRU selects the color with the "oldest" time stamp.

Theorem 2. *The competitive ratio of the LRU strategy is at least $\Omega(\sqrt{k})$.*

Proof. It is easy to see that for the sequence defined in the proof of Theorem 1 the LRU strategy produces the same output sequence as FIFO. □

Largest Color First (LCF). Another fairly natural strategy for the sorting buffer problem is to free as many locations in the sorting buffer as possible, if the strategy has to make a color change. If there is no item of the current output color in the sorting buffer, the LCF strategy selects a color that has the most items in the sorting buffer.

Theorem 3. *The competitive ratio of the LCF strategy is at least $\Omega(k)$.*

Proof. W.l.o.g., we assume that $k \geq 4$ is even. We consider the colors $c_1, \ldots c_{k-2}, x, y$ and the sequence $\sigma = c_1 \cdots c_{k-2}(xxyy)^{n \cdot k}$, for some integer n.

At first, we show that σ can be processed with $k - 2 + 8n$ color changes. Consider the sequence $c_1 \cdots c_{k-2}(xy)^{4n}$ of $k-2+8n$ color changes. After the items of colors $c_1, \ldots c_{k-2}$ have been removed from the sorting buffer every further color change removes at least $k/2$ items.

Now it remains to count the color changes if σ is processed by the LCF strategy. LCF does not remove the items of colors $c_1, \ldots c_{k-2}$ from the sorting buffer before the items of colors x and y are processed. Hence, if σ is processed by LCF, $2n \cdot k + k - 2$ color changes occur. □

3   Bounded Waste Strategy

How should a good scheduling strategy for sorting buffers look like? On the one hand, no item should be kept in the sorting buffer for a too long, possibly infinite, period of time. This would waste valuable storage capacity that could be used for efficient scheduling otherwise. For example the LCF strategy from the previous section fails to achieve a good competitive ratio because some items are kept in the sorting buffer for nearly the whole sequence.

On the other hand, there is a benefit from keeping an item in the sorting buffer if items of the same color arrive in the near future. Thus, a strategy may fail as well if it removes items too early. Good examples for this phenomenon are the LRU and FIFO strategy from the previous section. These strategies tend to remove items too early from the sorting buffer and, hence, cannot build large color blocks if additional items of the same color arrive.

We need a trade-off between the space wasted by items of a color and the chance to benefit from future items of the same color. Such a trade-off is provided by the *Bounded Waste strategy*. This strategy continues to remove items of the current output color from the sorting buffer as long as this is possible, i.e., until all items in the sorting buffer have a color different from the current output color. Then the strategy has to decide which color is removed next from the sorting buffer. For this purpose we introduce, for each color c, the *penalty* P_c. The penalty P_c for color c is a measure of the space that has been wasted by all items of color c that are currently in the sorting buffer. Initially, the penalty for each color is zero. At each color change, the penalty for each color c is increased by the number of items of color c currently stored in the sorting buffer. Then a color c' with maximal penalty $P_{c'}$ is chosen, an item of color c' is removed from the sorting buffer, and $P_{c'}$ is reset to zero.

For the competitive analysis, we have to compare the Bounded Waste strategy with an optimal offline strategy, for each input sequence σ. In the following we assume that an optimal offline strategy and an arbitrary input sequence σ is fixed. Then the sequence of color changes of the Bounded Waste and optimal offline strategy are fixed as well.

In order to compare both strategies, we introduce the notation of waste. At first, a color change of the Bounded Waste or optimal offline strategy is called *online* or *offline color change*, respectively. We say that an (online or optimal offline) strategy *produces waste* w for color c at an online color change if it has w item of color c in its sorting buffer at this color change. Note that the waste of the optimal offline strategy is also produced at online color changes. Further, we define the (total) *online* and *optimal offline waste* for color c as the total waste produced for color c at all online color changes by the online and optimal offline strategy, respectively. The waste for color c is strongly related to the penalty for c: The penalty for c is equivalent to the waste produced for c by the online strategy since the most recent online color change to c.

In the following we describe how the notion of waste can be used to derive an upper bound on the competitive ratio of the Bounded Waste strategy. Let W_{on}^c and W_{op}^c denote the online and optimal offline waste for color c, respectively, and let C_{on} and C_{op} denote the number of color blocks in the output sequence of the Bounded Waste strategy and the optimal offline strategy, respectively. Then

$$\sum_{\text{color } c} W_{\text{op}}^c \leq k \cdot C_{\text{on}} \quad \text{and} \quad \sum_{\text{color } c} W_{\text{on}}^c \geq k \cdot C_{\text{on}} - k^2 \;,$$

since at all but the last k color changes the online as well as the optimal offline strategy have k items in the sorting buffer, because both strategies are lazy, i.e., they do not make a color change if they have free space in their sorting buffer and remaining items in the input sequence. Let $\Delta^c = W_{op}^c - W_{on}^c$ denote the difference between online and optimal offline waste for color c. Then

$$\sum_{\text{color } c} \Delta^c \leq k^2 \; .$$

The main technical contribution of this paper is to show the following lemma. Its proof is postponed to the next section.

Lemma 1 (Main Lemma). *For each color c,*

$$\Delta^c \geq W_{on}^c - C_{op}^c \cdot O(k \cdot \log^2 k) \; ,$$

with C_{op}^c denoting the number of blocks with color c in the output sequence of an optimal offline strategy.

The following theorem gives an upper bound of the competitive ratio of the Bounded Waste strategy.

Theorem 4. *The Bounded Waste strategy achieves a competitive ratio of $O(\log^2 k)$.*

Proof. With the Main Lemma, we can conclude

$$k^2 \geq \sum_{\text{color } c} \Delta^c$$
$$\geq \sum_{\text{color } c} \left(W_{on}^c - C_{op}^c \cdot O(k \cdot \log^2 k) \right) \geq k \cdot C_{on} - k^2 - C_{op} \cdot O(k \cdot \log^2 k) \; .$$

Hence, $C_{op} \cdot O(\log^2 k) + 2k \geq C_{on}$, which implies the theorem. □

4 Proof of the Main Lemma

In this section, we present the postponed proof of Lemma 1 (Main Lemma).

Proof (of Lemma 1 (Main Lemma)). At first, we show that, for each color c, the penalty P_c is at most $O(k \cdot \log k)$. By the relationship between waste and penalty, this means that the waste produced by the Bounded Waste strategy for color c between two consecutive online color changes to c is at most $O(k \cdot \log k)$. Let $P_c(i)$ denote the penalty for color c directly after the i-th online color change and define $P_c(0) = 0$.

Lemma 2 (Bounded Waste). *For each color c and each online color change i, $P_c(i) = O(k \cdot \log k)$.*

Proof. For the analysis, we use a potential function $\Phi(i)$ that depends on the penalties currently assigned to all colors. The intuition behind this potential function is as follows. We assume that we have $P_c(i)$ units of waste for color c. These units are put on a stack (one stack for all units of the same color). If a unit is at position j on the stack then it contributes with the value $\phi(j)$ to the potential function. The function ϕ is monotonously increasing. Hence, the higher the stack the more expensive become the units of waste.

We define the potential function as follows:

$$\Phi(i) = \sum_{\text{color } c} \sum_{j=1}^{P_c(i)} \phi(j) , \quad \text{with} \quad \phi(j) = 2^{\lfloor \frac{j}{k} \rfloor} + \frac{j}{k} .$$

Our goal is to show that $\Phi(i) = O(k^2)$. This immediately implies that, for each color c, $P_c(i) = O(k \cdot \log k)$. For this purpose, we need the following two propositions.

Proposition 1. *If, for each colors c, $P_c(i) \leq 5k$, then $\Phi(i) = O(k^2)$.*

Proof. It is easy to verify that $\phi(5k)$ is a constant. Hence, it remains to prove that $\sum_{\text{color } c} P_c(i) = O(k^2)$. This follows from the fact that there are at most k colors c with $P_c(i) > 0$. □

Proposition 2. *If there exists a color c, with $P_c(i) \geq 4k$, then $\Phi(i+1) < \Phi(i)$.*

Proof. Let $\Delta\Phi(i+1) = \Phi(i+1) - \Phi(i)$, i.e., the change in the potential at the $(i+1)$-th online color change. Recall that the Bounded Waste strategy first increases the penalty for each color. Then it chooses a color c' with maximal penalty $P_{c'}$, removes an item of color c' from the sorting buffer, and resets $P_{c'}$ to zero.

Let m be the maximal penalty $P_{c'}$ after it has been increased and before it is reset to zero. Then

$$\Delta\Phi(i+1) \leq k \cdot \phi(m) - \sum_{j=1}^{m} \phi(j)$$

$$= k \cdot 2^{\lfloor \frac{m}{k} \rfloor} + m - \sum_{j=1}^{m} 2^{\lfloor \frac{j}{k} \rfloor} - \sum_{j=1}^{m} \frac{j}{k}$$

$$\leq k \cdot 2^{\lfloor \frac{m}{k} \rfloor} - k \cdot \sum_{j=1}^{\lfloor \frac{m}{k} \rfloor - 1} 2^j + m - \frac{m \cdot (m+1)}{2k}$$

$$= k - \frac{m \cdot (m - 2k + 1)}{2k} .$$

Hence, $\Delta\Phi(i+1) < 0$ for $m \geq 4k$ which implies the proposition. □

We prove by induction on i that $\Phi(i) = O(k^2)$. Obviously $\Phi(0) = 0$. For the induction step suppose $\Phi(i) = O(k^2)$. We distinguish between two cases according to the maximal penalty. First, suppose that there exists a color c with $P_c(i) \geq 4k$. Then, we can conclude with Proposition 2 that $\Phi(i+1) < \Phi(i) = O(k^2)$. Now, suppose that, for each colors c, $P_c(i) < 4k$. Then, after increasing the penalties, $P_c(i+1) < 5k$, for each color c. Hence, Proposition 1 yields that $\Phi(i+1) = O(k^2)$. This completes the proof of Lemma 2 (Bounded Waste). □

Now, we introduce the notion of online and offline intervals. Consider the sequence of all online color changes. The offline color changes to color c induce a partition of this sequence into *offline c-intervals*. In addition, the online color changes to color c induce a partition of this sequence into *online c-intervals*. Obviously, there are $C_{op}^c + 1$ offline c-intervals and $C_{on}^c + 1$ online c-intervals, with C_{op}^c and C_{on}^c denoting the total number of color changes to color c made by the optimal offline and Bounded Waste strategy, respectively.

Fix a color c. We show for each offline c-interval I that

$$\Delta^c(I) = W_{op}^c(I) - W_{on}^c(I) \geq W_{on}^c(I) - O(k \cdot \log^2 k) , \qquad (1)$$

with $W_{op}^c(I)$ and $W_{on}^c(I)$ denoting the optimal offline and online waste for color c, respectively, produced in the offline c-interval I. Then the Main Lemma follows immediately, since $\sum_{\text{off. interval } I} W_{op}^c(I) = W_{op}^c$, $\sum_{\text{off. interval } I} W_{on}^c(I) = W_{on}^c$, and there are $C_{op}^c + 1$ offline intervals.

In the following, we fix an offline c-interval I. We partition this interval into two *phases* as follows: The first phase lasts from the beginning of I until the first offline color change to a color different from c. The second phase contains the remaining part of the interval I.

In the remaining part of the proof, we show that an inequality being analogous to Inequality 1 holds for each phase. We will denote the waste produced for color c in a phase of the interval with W_{on}^c and W_{op}^c, if there is no ambiguity.

Lemma 3 (Phase 1). *Let W_{on}^c and W_{op}^c denote the waste produced for color c in the first phase of an offline c-interval. Then*

$$\Delta^c = W_{op}^c - W_{on}^c \geq W_{on}^c - O(k \cdot \log^2 k) .$$

Proof. In the following, the numeration of online intervals and online color changes is with respect to the first phase of the offline c-interval, e.g., if we mention the i-th online color change to color c, we refer to the i-th online color change to color c within this phase. Then let $W_{on}^c(i)$ and $W_{op}^c(i)$ denote the waste for color c produced by the Bounded Waste and optimal offline strategy, respectively, in the i-th online interval. Further, let $n_{on}^c(i)$ and $n_{op}^c(i)$ the number of items of color c in the sorting buffer of the Bounded Waste and optimal offline strategy, respectively, at the i-th online color change to color c. Note that $n_{op}^c(i)$ is monotonously decreasing, since the optimal offline strategy always removes items of color c in the first phase.

In the following proposition, we prove that either an online c-interval satisfies the inequality in Lemma 3 or $n_{op}^c(i)$ decreases by at least a factor of $1/2$.

Proposition 3. *Let i denote the number of an online c-interval that is completely contained in the first phase of an offline c-interval. Then*

$$\Delta^c(i) = W_{\text{op}}^c(i) - W_{\text{on}}^c(i) \geq W_{\text{on}}^c(i) \quad \text{or} \quad n_{\text{op}}^c(i+1) \geq 2 \cdot n_{\text{op}}^c(i+2) \ .$$

Proof. We distinguish between the two cases $n_{\text{op}}^c(i+1) \geq 2 \cdot n_{\text{on}}^c(i+1)$ and $n_{\text{op}}^c(i+1) < 2 \cdot n_{\text{on}}^c(i+1)$. In the first case, we immediately get $W_{\text{op}}^c(i) \geq 2 \cdot W_{\text{on}}^c(i)$. In the second case, we get $n_{\text{op}}(i+1) \geq 2 \cdot n_{\text{op}}^c(i+2)$, since in the $(i+1)$-th online c-interval at least $n_{\text{on}}^c(i+1)$ items with a color different from c have to be stored in the sorting buffer. □

Since the sorting buffer has capacity k, there exists at most $O(\log k)$ online c-intervals that do not satisfy the first inequality in Proposition 3 (including the at most two online c-intervals that are not completely contained within the phase). By Lemma 2 (Bounded Waste) we know that in each online c-interval at most $O(k \cdot \log k)$ waste is produced. Hence, Lemma 3 (Phase 1) follows. □

Lemma 4 (Phase 2). *Let W_{on}^c and W_{op}^c denote the waste produced for color c in the second phase of an offline c-interval. Then*

$$\Delta^c = W_{\text{op}}^c - W_{\text{on}}^c \geq W_{\text{on}}^c - O(k \cdot \log^2 k) \ .$$

Proof. Let C_{on}^c denote the number of online color changes to color c within the second phase of the offline c-interval. We consider the online c-intervals that are completely contained within the phase. There are at most $C_{\text{on}}^c - 1$ such intervals. We define the length ℓ_i of the i-th online c-interval in the phase as the total number of online color changes within this interval. Furthermore, we define the distance d_i of the i-th online c-interval to the end of the phase as the number of online color changes in the phase that take place after the end of the i-th online c-interval.

In the following, we compare the waste that is produced by items appearing during a certain interval. Let N_i^c denote the set of items of color c that appear in the input sequence during the i-th online c-interval of the phase. Note, that these items are all removed from the sorting buffer of the Bounded Waste strategy at the end of the online c-interval while they remain in the sorting buffer of the optimal offline strategy until the end of the phase. This holds, because the phase starts with an offline color change to a color different from c and the next offline color change to c is at the end of the phase.

Now, we study the additional waste that is produced by the optimal offline strategy for items in N_i^c during the phase. Let $W_{\text{on}}^c(N_i^c)$ and $W_{\text{op}}^c(N_i^c)$ denote the online and optimal offline waste for color c, respectively, produced for items in N_i^c. Then

$$W_{\text{op}}^c(N_i^c) - W_{\text{on}}^c(N_i^c) = |N_i^c| \cdot d_i \ ,$$

since after the i-th online c-interval is finished, there follow d_i further color changes in the phase, and at each of these color changes the optimal offline strategy produces waste $|N_i^c|$ for items in N_i^c while the online strategy produces waste zero for items in N_i^c.

Obviously, $W_{\text{on}}^c(N_i^c) = W_{\text{on}}^c(i)$, i.e., the online waste for color c produced for items in N_i^c is the same as the online waste for color c produced during the i-th online c-interval. Unfortunately, this equality does not hold for the optimal offline waste $W_{\text{op}}^c(N_i^c)$ and $W_{\text{op}}^c(i)$. However, the total optimal offline waste for color c produced during the phase is $\sum_i W_{\text{op}}^c(N_i^c) = \sum_i W_{\text{op}}^c(i)$.

We call an online c-interval *problematic* if

$$\Delta^c(N_i^c) = W_{\text{op}}^c(N_i^c) - W_{\text{on}}^c(N_i^c) < W_{\text{on}}^c(N_i^c) ,$$

i.e., the additional waste produced by the optimal offline strategy is less than the online waste for the respective items. The following proposition gives an upper bound on the total number of problematic intervals in the phase.

Proposition 4. *There are at most $O(\log k)$ problematic online c-intervals in the second phase of an offline c-interval.*

Proof. In order to proof the proposition we make the following observation. Suppose all problematic online c-intervals are numbered according to their appearance. Let s and t, with $s < t$, denote the numbers of two problematic online c-intervals. Then $d_s > 2 \cdot d_t$.

Assume for contradiction that $d_t \geq d_s/2$. Then the length ℓ_t of the t-th online c-interval is at most $d_s/2$. Hence, for the the online waste for color c in this interval holds $W_{\text{on}}^c(N_t) \leq \ell_t \cdot |N_t| \leq d_s/2 \cdot |N_t|$. But, since the t-th online c-interval is problematic, it follows that $W_{\text{on}}^c(N_t) > |N_t| \cdot d_t$. This is a contradiction.

It remains to give an upper bound on the distance of an problematic online c-interval. Let i denote the number of a problematic online c-interval. According to Lemma 2 (Bounded Waste), $W_{\text{on}}^c(N_i^c) = O(k \cdot \log k)$. In addition, $W_{\text{on}}^c(N_i) > d_i \cdot |N_i|$, since the i-th online c-interval is problematic. Then $d_i = O(k \cdot \log k)$.

Because the distance of two consecutive problematic online c-intervals increase by a factor of two and the distance is bounded by $O(k \cdot \log k)$, there can only be $O(\log(k \cdot \log k)) = O(\log k)$ such intervals. This yields the proposition. □

Now, we can show the lemma as follows. We first sum up the waste produced by items of problematic online c-intervals

$$\sum_{\text{probl. interval } i} (W_{\text{op}}^c(N_i^c) - W_{\text{on}}^c(N_i^c)) \geq \sum_{\text{probl. interval } i} W_{\text{on}}^c(N_i^c) - O(k \cdot \log^2 k) .$$

The above inequality holds, since there are at most $O(\log k)$ problematic intervals and for each problematic interval i holds $W_{\text{on}}^c(N_i^c) = O(k \cdot \log k)$, according to Lemma 2 (Bounded Waste). For each non-problematic online c-interval i that is completely contained in the phase, $W_{\text{op}}^c(N_i^c) - W_{\text{on}}^c(N_i^c) \geq W_{\text{on}}^c(N_i^c)$, by definition of problematic interval.

It only remains to consider the waste produced by items of the at most two online c-intervals that are not completely contained within the phase. According to Lemma 2 (Bounded Waste), the online waste for color c produced in these intervals is at most $O(k \cdot \log k)$. Hence, $W_{\text{op}}^c(N^c) - W_{\text{on}}^c(N^c) \geq W_{\text{on}}^c(N^c) - O(k \cdot \log k)$, with N^c denoting the set of items of color c that occur in these intervals.

Finally, we can sum up the waste produced by items of all online c-intervals, which yields Lemma 4 (Phase 2). □

This completes the proof of Lemma 1 (Main Lemma). □

References

[1] T. Feder, R. Motwani, R. Panigrahy, and A. Zhu. Web caching with request reordering. In *Proceedings of the 13th ACM-SIAM Symposium on Discrete Algorithms (SODA)*, pages 104–105, 2002.

[2] A. Fiat, R. M. Karp, M. Luby, L. A. McGeoch, D. D. Sleator, and N. E. Young. Competitive paging algorithms. *Journal of Algorithms*, 12(2):685–699, 1991.

[3] C. B. Fraser and R. W. Irving. Approximation algorithms for the shortest common supersequence. *Nordic Journal of Computing*, 2:303–325, 1995.

[4] M. J. Garey and D. S. Johnson. *Computers and intractability: A guide to the theory of NP-completeness*. Freeman, 1979.

[5] T. Jiang and M. Li. On the approximation of shortest common supersequences and longest common subsequences. In *Proceedings of the 21st International Colloquium on Automata, Languages and Programming (ICALP)*, pages 191–202, 1994.

[6] D. Karger, C. Stein, and J. Wein. Scheduling algorithms. In M. J. Atallah, editor, *Handbook of Algorithms and Theory of Computation*. CRC Press, 1997.

[7] S. O. Krumke, W. E. De Paepe, J. Rambau, and L. Stougie. Online bin-coloring. In *Proceedings of the 9th European Symposium on Algorithms (ESA)*, pages 74–85, 2001.

[8] E. L. Lawler, J. K. Lenstra, A. H. G. Rinnooy Kan, and D. B. Shmoys. Sequencing and scheduling: Algorithms and comlexity. In S. C. Graves, A. H. G. Rinnooy Kan, and P. Zipkin, editors, *Handbooks in Operations Research and Management Science, Vol. 4: Logistics of Production and Inventory*, pages 445–552. North-Holland, 1993.

[9] J. Sgall. On-line scheduling. In A. Fiat and G. Woeginger, editors, *Online Algorithms: The State of the Art*, volume 1442, pages 196–231. Springer LNCS, 1998.

[10] D. D. Sleator and R. E. Tarjan. Amortized efficiency of list update and paging rules. *Communications of the ACM*, 28(2):202–208, 1985.

[11] S. Spieckermann and S. Voss. Paint shop simulation in the automotive industry. In *Proceedings of the Workshop for Simulation and Animation in Planning, Education, and Presentation*, pages 367–380, 1996.

Finding the Sink Takes Some Time
An Almost Quadratic Lower Bound for Finding the Sink of Unique Sink Oriented Cubes

Ingo Schurr[*] and Tibor Szabó [**]

Theoretical Computer Science, ETH Zürich
CH-8092 Zürich, Switzerland
{schurr,szabo}@inf.ethz.ch

Abstract. We give a worst-case $\Omega(\frac{n^2}{\log n})$ lower bound on the number of vertex evaluations a deterministic algorithm needs to perform in order to find the (unique) sink of a unique sink oriented n-dimensional cube. We consider the problem in the vertex-oracle model, introduced in [17]. In this model one can access the orientation implicitly, in each vertex evaluation an oracle discloses the orientation of the edges incident to the queried vertex. An important feature of the model is that the access is indeed arbitrary, the algorithm does *not* have to proceed on a directed path in a simplex-like fashion, but could "jump around". Our result is the first super-linear lower bound on the problem. The strategy we describe works even for acyclic orientations. We also give improved lower bounds for small values of n and fast algorithms in a couple of important special classes of orientations to demonstrate the difficulty of the lower bound problem.

1 Introduction

Notation, Definitions. For our purposes a cube is the power set of a set. More precisely, for $A \subseteq B$ finite sets the cube $\mathfrak{C} = \mathfrak{C}^{[A,B]}$ is the edge labeled graph with vertex set $V(\mathfrak{C}) := [A, B] := \{X \mid A \subseteq X \subseteq B\}$, edge set

$$E(\mathfrak{C}) := \{\{v, v \oplus \{l\}\} \mid v \in V(\mathfrak{C}), l \in B \setminus A\}$$

and edge labeling

$$\lambda(\{v, v \oplus \{l\}\}) := l \; ,$$

where the symbol $\oplus$ denotes the symmetric difference of two sets. We will say that an edge e is *l-labeled* if $\lambda(e) = l$, and for a subset L of the labels we say *L-labeled edges* for the set of all edges which are l-labeled with some $l \in L$.

---

[*] Supported by the joint Berlin/Zürich graduate program Combinatorics, Geometry, and Computation (CGC), financed by German Science Foundation (DFG) and ETH Zurich.
[**] Supported by the joint Berlin/Zürich graduate program Combinatorics, Geometry, and Computation (CGC), financed by German Science Foundation (DFG) and ETH Zurich, and by NSF grant DMS 99-70270.

From a purely combinatorial point of view, the cube $\mathfrak{C}^{[A,B]}$ is sufficiently described by its *label set* carr $\mathfrak{C}^{[A,B]} := B \setminus A$. Up to graph-isomorphism even $\dim \mathfrak{C}^{[A,B]} := |B \setminus A|$, the *dimension*, determines the cube. For most of the time we shall work with cubes $\mathfrak{C}^B := \mathfrak{C}^{[\emptyset,B]}$, i.e. power sets. In case carr $\mathfrak{C}$ does not play any role we even abandon the superscript and write $\mathfrak{C}$.

The additional parameter A is helpful in naming the subcubes of a cube. A *subcube* of a cube $\mathfrak{C}^{[A,B]}$ is a cube $\mathfrak{C}^{[X,Y]}$ with $A \subseteq X \subseteq Y \subseteq B$. These subcubes correspond to the *faces* of a geometric realization of $\mathfrak{C}^{[A,B]}$.

Instead of writing down X and Y it is often more convenient to describe a subcube by a vertex and a set of labels generating it: we write $\mathfrak{C}(v, L)$ for the smallest subcube of $\mathfrak{C}$ containing v and $v \oplus L$. In fact for $X = v \cap \bar{L}$ and $Y = v \cup L$ we have $\mathfrak{C}(v, L) = \mathfrak{C}^{[X,Y]}$, where $\bar{L}$ denotes the complement of L with respect to carr $\mathfrak{C}$.

Given an orientation of a cube (i.e. an orientation of its edges), a vertex is called a *sink* if all its incident edges are incoming. An orientation ψ of $\mathfrak{C}$ is called a *unique sink orientation* or USO, if ψ restricted to any subcube has a unique sink. We usually don't distinguish between a USO and the cube equipped with that USO, and refer to the cube as USO as well. An orientation is called *acyclic* if it contains no directed cycle. Acyclic unique sink orientations are abbreviated by AUSO.

The Problem. Easily stated: find the sink of a USO. Following [17] we assume that the orientation is given implicitly, i.e. we can access an *arbitrary* vertex of the USO through an oracle, which then reveals the orientation of the edges incident to the requested vertex. This basic operation is called *vertex evaluation* (sometimes we refer to it as *step* or *query*), and we are interested in evaluating the (unique) sink of a cube by as few vertex evaluations as possible. Formally, let eval($\mathbf{A}, \psi$) be the number of vertex evaluation it takes for a deterministic algorithm $\mathbf{A}$ to evaluate the sink of a USO (or AUSO) ψ, and define $t(n)$ (or $t_{\text{acyc}}(n)$) to be $\min_{\mathbf{A}} \max_{\psi}$ eval($\mathbf{A}, \psi$). Obviously, $t(n) \geq t_{\text{acyc}}(n)$. Let $\tilde{t}(n)$ and $\tilde{t}_{\text{acyc}}$ be the corresponding functions for randomized algorithms; the expected number of evaluations a fastest randomized algorithm takes until it finds the sink of any USO (or AUSO).

Unique sink orientations provide a common framework for several seemingly different problems. Related abstractions were considered earlier, mainly to deal with geometric problems. LP-type problems, Abstract Objective Functions and Abstract Optimization Problems [3, 8, 9, 10, 12, 13] have a rich literature and provide the fastest known algorithms for several geometric problems in the *unit cost model*. As it is always the case with abstractions, it is very well possible that the model of USO is in fact too general for being of any use, but there are a number of results suggesting otherwise.

At first it is not even clear how to find the sink of a USO in $o(2^n)$ evaluations; whether unique sink orientations have enough structure, which distinguishes them from just any orientation of the cube. Indeed, in order to find the sink in an *arbitrary* orientation (with a sink), one needs at least $2^{n-1} + 1$ vertex evaluations [18].

In [17] a deterministic algorithm running in 1.61^n evaluations and randomized algorithm running in 1.44^n evaluations (using that $\tilde{t}(3) = \frac{4074633}{1369468}$ [15]) were given for evaluating the sink of a USO. These algorithms indicate that USOs have some, actually quite rich, structure. Another result, quantitatively pointing to this direction, is due to Matoušek [11]; he showed that the number of USOs is $2^{\Theta(\log n 2^n)}$, which is *significantly less* than $2^{n2^{n-1}}$, the number of all orientations. A bonus for optimists, that within the class of orientations providing the matching lower bound for Matoušek's upper bound the sink can be found in *five*(!) steps.

Motivation. The most obvious and actually quite widely investigated appearance of USOs is the special case of linear programming; i.e. when the polyhedron in question is a slanted geometric cube. Then a linear objective function canonically defines an AUSO on the cube and finding the sink of this orientation obviously corresponds to finding the minimum of the objective function on the polyhedron. The importance of this question is well-demonstrated by the number of papers investigating the running time of specific randomized simplex-like algorithms, like RandomEdge or RandomFacet, on (sometimes even specific) acyclic orientations. The RandomEdge algorithm proceeds on a directed path by choosing uniformly at random among the outgoing edges at each vertex. The RandomFacet algorithm chooses a random facet of the cube where the smaller dimensional algorithm is run and after finding the sink of the facet (and in case that sink is not the global sink) it proceeds to the antipodal facet to find its sink by another smaller dimensional RandomFacet algorithm. Gärtner [4, 5] showed that the expected number of evaluations it takes for the RandomFacet algorithm to evaluate the sink of an AUSO is at most $e^{2\sqrt{n}}$. Gärtner, Henk, and Ziegler [6] analyzed the behavior of RandomEdge and RandomFacet on a specific, particularly interesting orientation, the so-called Klee-Minty cubes.

USOs also appear in less obvious ways, where the correspondence between the problem and the orientation is more abstract. Stickney and Watson [16] defined a combinatorial correspondence between the solution of certain linear complementarity problems [2] (corresponding to so-called P-matrices) and finding the sink in an appropriate USO. In this correspondence the appearance of cycles in the USO is possible.

Similarly, certain quadratic optimization problems, like finding the smallest enclosing ball of $d+1$ affinely independent points in the Euclidean d-space can be reduced to finding the sink in an appropriate USO. For each such point-set, there is a corresponding USO, where the sink corresponds to the smallest enclosing ball of the point set [7]. Here again cycles can arise. The general problem of n points in d-space is known to be reducible to the case when n is small (i.e. around d^2) compared to d. Then one can place the ambient d-space into $\mathbb{R}^{n-1}$ and perturb the points for affine independence.

Note that every linear programm can be translated into a smallest enclosing ball problem. Via this detour unique sink orientations can be seen as an abstraction of linear programming in general (and not only for cube-like polyhedron).

Results. The main finding of our paper is a lower bound of order $\frac{n^2}{\log n}$ for the number of evaluations a deterministic algorithm needs in order to find the sink of an n-dimensional AUSO. Our result is the first non-linear lower bound for the function $t(n)$. On the way we organize our current knowledge about producing USOs. We also prove better lower bounds for small values of n. To motivate our results we look at a couple of simpler classes of orientations.

Lower bounds were found earlier for several related problems or special cases. Aldous [1] considered orientations given canonically by an ordering of the vertices of the cube, which have a unique sink (but not necessarily every subcube has a unique sink). These orientations are acyclic by definition. For them he proves a highly nontrivial exponential lower bound of $\sqrt{2}^{n(1+o(1))}$, using the hitting times of a random walk on the cube. He also provides a simple randomized algorithm which is essentially best possible. Since his model is in some sense weaker and in some sense stronger than our USOs, his results un(?)fortunately do not imply anything for our problem.

Other known lower bounds are related to specific randomized simplex-like algorithms, for example RandomEdge or RandomFacet. On AUSOs Matoušek [12] proved an $e^{\Omega(\sqrt{n})}$ lower bound for the expected running time of RandomFacet, which nicely complements the $e^{O(\sqrt{n})}$ upper bound of Gärtner [4, 5]. Gärtner, Henk and Ziegler [6] showed that the expected running time of RandomEdge on the Klee-Minty cubes is $\Omega(\frac{n^2}{\log n})$. A negative result of Morris [14] could also be interpreted as a lower bound for general USOs. He constructs USOs on which the expected running time of RandomEdge is $((n-1)/2)!$, more than the number of vertices.

Our paper is organized as follows. In Sect. 2 we collect notations, definitions and several known facts about USOs we will make use of. In Sect. 3 we establish our basic building blocks: two different ways of obtaining new USOs from old ones. In Lemma 3 we define a certain product construction of USOs, while in Lemma 4 we describe circumstances under which a local change in an existing USO produces another USO. In order to motivate our lower bound on the general problem, in Sect. 4 we discuss two important smaller classes of USOs for which very fast algorithms exist. Decomposable orientations are built recursively by taking two decomposable orientations of dimension $n-1$ on two disjoint facets of an n-cube and orient all edges between them the same direction. Wiliamson Hoke [19] gave a simplex-like algorithm on them in $O(n^2)$ time. In Prop. 1 we observe that our extra power of being able to "jump around" lets us find the sink in only $n+1$ evaluations. We also observe that this is best possible; i.e. on the class of decomposable orientations we know *a* fastest algorithm. In Prop. 2 we consider the class of the so-called *matching-flip* orientations. This class was introduced by Matoušek and Wagner to give a lower bound for the number of all USOs. Matoušek [11] proved that the number of all USOs is $2^{O(\log n 2^n)}$. The class of matching-flip orientations is in fact so rich that it provides a matching lower bound for the order of magnitude of the logarithm of the number of all USOs. Thus it is somewhat surprising that for this class there is an algorithm finding the sink in five steps.

In Sect. 5 we provide a strategy, which forces every deterministic algorithm to at least $\Omega(\frac{n^2}{\log n})$ evaluations while finding the unique sink of an n-dimensional cube with an acyclic unique sink orientation. We also give a matching lower bound for the 4-dimensional algorithm of [17].

In Sect. 6 we list several open problems.

2 Preliminaries

For an orientation ψ of a cube $\mathfrak{C}^{[A,B]}$ the *outmap* s_ψ *of* ψ is the map assigning to every vertex v the set of labels of outgoing edges, i.e. $s_\psi : \mathfrak{C} \to 2^{\operatorname{carr} \mathfrak{C}}$ with

$$s_\psi(v) = \{l \mid \{v, v \oplus \{l\}\} \text{ is outgoing from } v\} \ .$$

Obviously, a vertex v_0 is a sink iff $s_\psi(v_0) = \emptyset$.

An important property of outmaps of USOs was proved in [17].

Lemma 1. *[17, Lemma 2.2] Let ψ be a USO. Then s_ψ is a bijection.*

Its proof relies on a simple observation, which is also a tool to produce new USOs from old ones. If ψ is an orientation of $\mathfrak{C}$ and $L \subseteq \operatorname{carr} \mathfrak{C}$, then let $\psi^{(L)}$ be the orientation of $\mathfrak{C}$ which agrees with ψ on $\bar{L}$-labeled edges and differ on L-labeled edges.

Lemma 2. *[17, Lemma 2.1] For any $L \subseteq \operatorname{carr} \mathfrak{C}$, if ψ is a USO, then $\psi^{(L)}$ is a USO as well.*

Let us remark here that acyclicity does not necessarily survives this "label-flip"; there are examples of AUSOs ψ and labels $l \in \operatorname{carr} \mathfrak{C}$ (already for $\dim \mathfrak{C} = 3$), such that $\psi^{(\{l\})}$ is *not* an AUSO.

In [17, Lemma 2.3] outmaps of USOs were characterized; $s : \operatorname{V}(\mathfrak{C}) \to 2^{\operatorname{carr} \mathfrak{C}}$ is the outmap of a unique sink orientation, iff

$$(s(v) \oplus s(w)) \cap (v \oplus w) \neq \emptyset \text{ for all } v \neq w \in \operatorname{V}(\mathfrak{C}).$$

We call such outmaps *unique sink outmaps* and – as the acronyms conveniently coincide – we abbreviate them by USO as well. This usually does not cause any confusion since orientations and outmaps determine each other uniquely.

Note that for cubes $\mathfrak{C}^B$ every outmap is a permutation of 2^B. An important example of a unique sink outmap on every $\mathfrak{C}^B$ is the identity. More generally, for every $w \in \operatorname{V}(\mathfrak{C})$ the map ι_w defined by $\iota_w(v) := v \oplus w$ (for any $v \in \operatorname{V}(\mathfrak{C})$) is a USO (note, that $\iota_\emptyset$ is the identity). The corresponding orientation directs every edge towards w. We refer to such orientations as *uniform orientations*. Note that the sink of ι_w is w and the source is $\bar{w} = B \setminus w$.

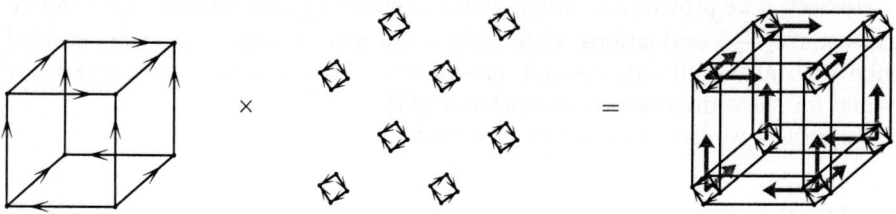

Fig. 1. A product of a three-dimensional USO with two-dimensional ones

3   Basic Constructions

Take a unique sink orientation and replace every vertex by a unique sink oriented cube of a fixed dimension. This construction defines a *product structure* for USOs. Figure 1 tries to illustrate that: In a three-dimensional unique sink orientation the vertices are replaced by two-dimensional unique sink orientations.

The statement of the next lemma provides the formal definition of this product construction, and shows that indeed it produces a unique sink orientation.

Lemma 3. *Let A be a set of labels, $B \subseteq A$ and $\bar{B} = A \setminus B$. For a USO $\tilde{s}$ on $\mathfrak{C}^B$ and for USOs s_u on $\mathfrak{C}^{\bar{B}}$, $u \in V(\mathfrak{C}^B)$, the map s on $\mathfrak{C}^A$ defined by*

$$s(v) = \tilde{s}(v \cap B) \cup s_{v \cap B}(v \cap \bar{B})$$

is a USO.

Furthermore, if $\tilde{s}$ and all s_u are acyclic, then so is s.

Proof. See journal version.

For $|B| = 1$, Lemma 3 says that two $(n-1)$-dimensional unique sink orientations can be combined to an n-dimensional one by placing them in two disjoint facets and directing all edges in between in the same direction.

The other extreme case $|\bar{B}| = 1$ shows that if a cube contains two opposite facets with the same $(n-1)$-dimensional unique sink orientation, the edges between these facets can be directed arbitrarily.

It is easy to see that we can direct all edges of one label arbitrarily only if the unique sink orientations in the two facets not containing edges with this label are the same. On the other hand, by placing a unique sink orientation in one facet and the orientation which has all edges flipped in the other facet, we are forced to direct all edges in between in the same direction. Therefore Lemma 3 is best possible in some sense, one cannot expect to get a more general product construction without further exploring $\tilde{s}$ and/or some s_u.

A simple consequence of Lemma 3 is that in an AUSO sink and source could be placed anywhere, independently of each other. This fact is stated in the next corollary and will be utilized in the proof of our main result.

Corollary 1. *For any two distinct vertices $u, v \in \mathrm{V}(\mathfrak{C}^A)$, there is an acyclic unique sink outmap with sink u and source v.*

Proof. See journal version.

Another way to construct new unique sink orientations is by local modification. Given a n-dimensional unique sink orientation one can replace the orientation of a subcube under certain conditions. It is clear that the replacing orientation has to be a USO, but that does not suffice.

Lemma 4. *Let s be a unique sink outmap on a cube $\mathfrak{C}^A$ and $\mathfrak{C}_0$ be a subcube with label set $B \subseteq A$. If $s(v) \cap \bar{B} = \emptyset$ for all $v \in \mathrm{V}(\mathfrak{C}_0)$ and s_0 is a unique sink outmap on $\mathfrak{C}^B$, then the map $s' : 2^A \to 2^A$ defined by $s'(v) = s_0(v \cap B)$ for $v \in \mathrm{V}(\mathfrak{C}_0)$ and $s'(v) = s(v)$ otherwise is a unique sink outmap on $\mathfrak{C}^A$.*

If s and s_0 are acyclic, then s' is acyclic as well.

For example in Fig. 2 the lower three-dimensional subcube can be directed arbitrarily, since all edges incident to it are incoming.

Proof. See journal version.

Corollary 2. *Let s be a unique sink outmap on a cube $\mathfrak{C}^A$ and $\mathfrak{C}_0$ be a subcube with label set $B \subseteq A$. Suppose that $s(v) \cap \bar{B} = s(v') \cap \bar{B}$ for all $v, v' \in \mathrm{V}(\mathfrak{C}_0)$. If s_0 is a unique sink outmap on $\mathfrak{C}^B$, then the map $s' : 2^A \to 2^A$ defined by $s'(v) = s_0(v \cap B)$ for $v \in \mathrm{V}(\mathfrak{C}_0)$ and $s'(v) = s(v)$ otherwise is a unique sink outmap on $\mathfrak{C}^A$.*

Proof. See journal version.

Let us remark that unlike in Lemma 4, here acyclicity does not necessarily carry over to s'.

Again, the special case $|B| = 1$ is of some interest. In this scenario $\mathfrak{C}_0$ is a single edge and by the preceding corollary, we see that an edge can be flipped, if the outmaps of the two adjacent vertices only differ in the label of the edge.

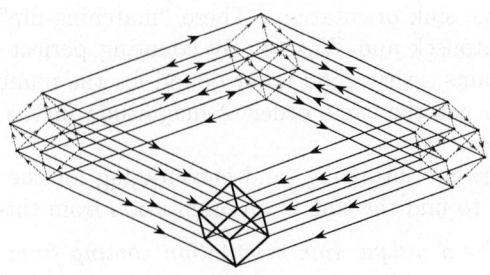

Fig. 2. Three-dimensional flippable subcube

Since flipping an edge solely affects the outmap of the adjacent vertices the converse also holds. Therefore an edge $\{v,w\}$ is flippable iff $s(v) \oplus s(w) = v \oplus w$.

In particular, if ι_w is the outmap of a uniform orientation, then one can flip the edges of an arbitrary matching of the cube and the result is still a USO. As we shall see in the next section this construction is particularly interesting.

4 Interlude: Two Simple Classes

Let us recall that in the special case $|B| = 1$ of Lemma 3 we obtained an n-dimensional USO out of two (possibly distinct) $(n-1)$-dimensional USOs. The new orientation is special in the sense, that all edges of one label point in the same direction. This leads us to the notion of combed orientations; an orientation is called *combed*, if there is a label l for which all l-labeled edges are oriented towards the same facet.

Iterating the construction we get the class of *decomposable* unique sink orientations. Let $\mathcal{D}_1$ be the class of all 1-dimensional unique sink orientations and denote by $\mathcal{D}_{n+1}$ the class of all orientations constructed from two orientations from $\mathcal{D}_n$ using Lemma 3. We call $\mathcal{D}_n$ the class of decomposable unique sink orientations in dimension n. Equivalently, an orientation is decomposable iff every nonzero-dimensional subcube is combed.

In general it is not easy to check, whether a unique sink orientation is decomposable or even combed (one has to to evaluate half of the vertices), but in the class of decomposable unique sink orientations it is easy to find the sink.

Proposition 1. *For a decomposable USO of dimension n one needs at most $n + 1$ vertex evaluations to evaluate the sink. Moreover, $n + 1$ is best possible; i.e. for every deterministic algorithm there is a decomposable USO on which the algorithm needs at least $n + 1$ evaluations.*

Proof. See journal version.

Our second example in this section arises from Lemma 2. Consider a uniform orientation and a matching of the cube. According to Lemma 2 every edge in our matching is flippable. Since flipping one of them does not affect the flippability of other edges of the matching, we can flip any number of them simultaneously.

In consequence, for every matching and every uniform orientation we obtain different unique sink orientations. These "matching-flip" USOs were first constructed by Matoušek and Wagner. By counting perfect matchings of the hypercube one obtains a very good lower bound for the number of unique sink orientations, which has the same order of magnitude in the logarithm as the number of all USOs.

Thus it is somewhat surprising (and encouraging for the general problem) that it is very easy to find the sink of an orientation from this very large class.

Proposition 2. *For a unique sink orientation coming from a uniform orientation by flipping a matching one needs at most 5 steps to find the sink. The value 5 here is best possible.*

Proof. See journal version.

5 The Strategy

In this section we prove the main result of the paper. In order to show lower bounds on $t(n)$ we consider ourselves the oracle playing against a deterministic algorithm, we call Al. We try to make sure that our answers to Al's evaluation requests (i) force Al to evaluate many vertices before the sink, and (ii) could be extended to an AUSO of the whole cube.

Given a sink-finding algorithm Al, we construct an acyclic unique sink orientation for which Al needs an almost-quadratic number of queries. For the first $n - \lceil \log_2 n \rceil$ inquiries we maintain a partial outmap $s : W \to \operatorname{carr} \mathfrak{C}^{[n]}$ on the set $W \subseteq \mathrm{V}(\mathfrak{C}^{[n]})$ of queried vertices, containing the answers we gave Al so far. We also maintain a set L of labels and an acyclic unique sink outmap $\tilde{s}$ on $\mathfrak{C}^L$. This smaller dimensional outmap $\tilde{s}$ is our "building block" which enables us to extend our answers to a global USO at any time.

Before the first inquiry we set $L = W = \emptyset$. After each inquiry we answer to Al by revealing the value of the outmap s at the requested vertex. Then we update L and $\tilde{s}$ such, that the following conditions hold.

(a) $|L| \leq |W|$,
(b) $w' \cap L \neq w'' \cap L$ for every $w' \neq w'' \in W$ and
(c) $s(w) = \tilde{s}(w \cap L) \cup \bar{L}$ for every $w \in W$.

Informally, condition (b) means that the projections of the queried vertices to $\mathfrak{C}^L$ are all distinct. This we shall achieve by occasionally adding a label to L, if the condition would be violated. Condition (c) is hiding two properties of our answers to the inquiries of Al. First that our answers are consistent with $\tilde{s}$ on L-labeled edges, and then that all $\bar{L}$-labeled edges, i.e. the edges leaving $\mathfrak{C}(w, L)$, are outgoing.

Suppose now that Al requests the evaluation of the next vertex u. We can assume that $u \notin W$. Depending on whether there is a $w \in W$ with $u \cap L = w \cap L$ or not we have to distinguish two cases.

Case 1. For every $w \in W$ we have $w \cap L \neq u \cap L$.

Then we answer $s(u) = \tilde{s}(u \cap L) \cup \bar{L}$ and leave L and $\tilde{s}$ unchanged. $(a) - (c)$ all hold trivially by the definition of the updates and the assumption of Case 1.

Case 2. There is a $v \in W$, such that $u \cap L = v \cap L$.

Let us immediately note that by condition (b), there is exactly one such $v \in W$. The assumption of Case 2 and $u \neq v$ implies that we can fix a label $l \in \bar{L}$ such that $l \in u \oplus v$.

We answer $s(u) = s(v) \setminus \{l\}$. Note that as $l \notin L$, $l \in s(v)$.

Now we have to update L and $\tilde{s}$. Our new L we get by adding l. To define the new orientation $\tilde{s}$ we take two copies of the old $\tilde{s}$ on the two facets determined by l. Then by Lemma 3, we can define the orientation of the edges going across arbitrarily, so we make them such that the condition (c) is satisfied. More formally, let

$$\tilde{s}(z) = \begin{cases} \tilde{s}(v \cap L) & \text{if } z = u \cap (L \cup \{l\}) \\ \tilde{s}(w \cap L) \cup \{l\} & \text{if } z = w \cap (L \cup \{l\}) \\ & \text{for some } w \in W, w \neq u \\ \tilde{s}(z) \cup (z \cap \{l\}) & \text{otherwise.} \end{cases}$$

Condition (a) still holds, because we added one element to each of W and L. Since L just got larger, we only need to check condition (b) for the pairs of vertices containing u. By the uniqueness of v, it is actually enough to check (b) for the pair u, v. Since l was chosen from $u \oplus v$ and now is included in L, $u \cap L \neq v \cap L$. Condition (c) is straightforward from the definitions.

We proceed until $|W| = n - \lceil \log_2 n \rceil$, and then change the strategy. By condition (a), $|L| \leq n - \lceil \log_2 n \rceil$. We choose an arbitrary superset $L' \supseteq L$ of L such that $|L'| = n - \lceil \log_2 n \rceil$. The set of labels $\bar{L}'$ determine at least $2^{|\bar{L}'|} \geq n$ disjoint subcubes generated by L'. As $|W| < n$, we can select one of them, say $\mathfrak{C}_0$, which does not contain any point evaluated so far.

Our plan is to apply Lemma 4 with $\mathfrak{C}_0$ and an orientation s, which is consistent with the outmaps of the vertices evaluated so far. The lemma then will enable us to reveal s on $V(\mathfrak{C}^A) \setminus V(\mathfrak{C}_0)$ to Al and still be able to start a completely "new game" on a cube of relatively large dimension. To construct s satisfying the conditions of Lemma 4 we use Lemma 3 twice.

First we define an orientation $\bar{s}$ of $\mathfrak{C}^{L'}$ using Lemma 3 with $A = L', B = L, \tilde{s}$ as defined above, and outmaps s_u on $\mathfrak{C}^{L' \setminus L}$ with the property that for every $w \in W$ the map $s_{w \cap L}$ has its source at $w \cap (L' \setminus L)$. This last requirement can be fulfilled because of condition (b).

Thus the resulting outmap $\bar{s}$ is consistent with the evaluated vertices in the sense that $s(w) \cap L' = \bar{s}(w \cap L')$ for each $w \in W$.

Next we apply Lemma 3 again with L', so we have to construct USOs s_u of $\mathfrak{C}^{\bar{L}'}$ for every $u \in V(\mathfrak{C}^{L'})$. In doing so, we only take care that the sinks of all these orientations are in $\mathfrak{C}_0$, and if $u = w \cap L'$ for some $w \in W$ then $w \cap \bar{L}'$ is the source of s_u. (By condition (b) there can be at most one such vertex w for each u.) The appropriate s_u is constructed in Corollary 1. Now Lemma 3, applied with L', $\bar{s}$ and these s_u's, provides us with an orientation s which agrees with our answers given for the evaluated vertices, and all $\bar{L}'$-labeled edges are incoming into $\mathfrak{C}_0$.

By Lemma 4 we can reveal s on $V(\mathfrak{C}) \setminus V(\mathfrak{C}_0)$ to Al, and still be able to place any orientation on $\mathfrak{C}_0$, a subcube of dimension $n - \lceil \log_2 n \rceil$. Therefore we just proved

$$t_{acyc}(n) \geq n - \lceil \log_2 n \rceil + t_{acyc}(n - \lceil \log_2 n \rceil) .$$

Theorem 1. *A deterministic algorithm needs $\Omega(\frac{n^2}{\log n})$ many steps to find the sink of an acyclic unique sink orientation on a n-dimensional cube.*

Proof. See journal version.

For small dimensions $t(n)$ is easy to check; $t(0) = 1$, $t(1) = 2$, $t(2) = 3$ and $t(3) = 5$. In the following proposition we give a matching lower bound for the SevenStepsToHeaven algorithm of [17] in dimension 4. We note that with similar methods a lower bound of $2n-1$ could be given in any dimension, which is better for small values than our asymptotic lower bound from Theorem 1.

Proposition 3. $t(4) = 7$.

Proof. See journal version.

6 Comments and Open Problems

Of course in an ideal world one could hope to determine the functions $t(n)$, $\tilde{t}(n)$, $t_{acyc}(n)$ and $\tilde{t}_{acyc}(n)$ exactly. There are more realistic goals, though, for the nearer future. A natural question concerning the strategy in our paper is whether the AUSOs we use (i.e. the orientations arising from Lemma 3 and Lemma 4) can be realized geometrically, i.e. as a linear program. A positive answer would be of great interest.

The lone lower bound we are able to give for $\tilde{t}(n)$ is a mere $n/2$, which works even on the set of decomposable orientations. Any nonlinear lower bound would be interesting. An extension of our method for the deterministic lower bound seems plausible.

There is ample space for improvement on the upper bounds as well. The fastest known deterministic algorithm does work for general USOs; a first goal would be to exploit acyclicity in order to separate $t(n)$ and $t_{acyc}(n)$ from each other.

The *only* known nontrivial randomized algorithms for the general USO problem work in a product-like fashion. An improvement of $\tilde{t}(n)$ for a small value of n implies better algorithms for large n. Unfortunately the determination of $\tilde{t}(n)$ becomes unmanageable relatively soon; already the determination of $\tilde{t}(3)$ by Rote [15] presented a significant computational difficulty, and $\tilde{t}(4)$ is still not known. It would be really desirable to create nontrivial randomized algorithms working by an idea *different* from the Product Algorithm of [17]. A possible approach would be to find a way to randomize the deterministic FibonacciSeesaw algorithm [17].

RandomEdge and RandomFacet are very natural randomized algorithms, their analysis is extremely inviting but might be hard. The behavior of RandomFacet is well-understood for AUSO, for general USOs nothing is known. By the construction of Morris [14] RandomEdge is not a good choice to find the sink of a general USO, but for AUSOs not much is known about its behavior.

Acknowledgment

We are grateful to Alan Frieze, Jirka Matoušek, Uli Wagner and Emo Welzl for useful comments and conversations. We also would like to thank Emo Welzl for suggesting the algorithm for decomposable orientations in Prop. 1.

References

[1] David Aldous. Minimization algorithms and random walk on the d-cube. *The Annals of Probability*, 11:403–413, 1983.
[2] Richard W. Cottle, Jong-Shi Pang, and Richrad E. Stone. *The Linear Complementary Problem*. Academic Press, 1992.
[3] Bernd Gärtner. A subexponential algorithm for abstract optimization problems. *SIAM J. Comput.*, 24:1018–1035, 1995.
[4] Bernd Gärtner. Combinatorial linear programming: Geometry can help. In *Proc. 2nd Int. Workshop on Randomization and Approximation Techniques in Computer Science*, volume 1518, pages 82–96, 1998. Lecture Notes in Computer Science.
[5] Bernd Gärtner. The random-facet simplex algorithm on combinatorial cubes. submitted, 2001.
[6] Bernd Gärtner, Martin Henk, and Günter M. Ziegler. Randomized simplex algorithms on Klee-Minty cubes. *Combinatorica*, 18(3):349–372, 1998.
[7] Bernd Gärtner, Hiroyuki Miyazawa, Lutz Kettner, and Emo Welzl. From geometric optimization problems to unique sink orientations of cubes. manuscript in preparation, 2001.
[8] Bernd Gärtner and Emo Welzl. Linear programming – randomization and abstract frameworks. In *Proc. 13th Ann. ACM Symp. Theoretical Aspects of Computer Science*, volume 1046 of *Lecture Notes Comput. Sci.*, pages 669–687. Springer-Verlag, 1996.
[9] Bernd Gärtner and Emo Welzl. Explicit and implicit enforcing – randomized optimization. In *Lectures of the Graduate Program Computational Discrete Mathematics*, volume 2122 of *Lecture Notes in Computer Science*, pages 26–49. Springer-Verlag, 2001.
[10] Gil Kalai. Linear programming, the simplex algorithm and simple polytopes. *Math. Programming*, 79:217–233, 1997.
[11] Jiří Matoušek. manuscript.
[12] Jiří Matoušek. Lower bounds for a subexponential optimization algorithm. *RSA: Random Structures & Algorithms*, 5(4):591–607, 1994.
[13] Jiří Matoušek, Micha Sharir, and Emo Welzl. A subexponential bound for linear programming. *Algorithmica*, 16:498–516, 1996.
[14] Walter D. Morris. Randomized principal pivot algorithms for P-matrix linear complementarity problems. to appear, 2001.
[15] Günter Rote. personal communication.
[16] Alan Stickney and Layne Watson. Digraph models of bard-type algorithms for the linear complementary problem. *Mathematics of Operations Research*, 3:322–333, 1978.
[17] Tibor Szabó and Emo Welzl. Unique sink orientations of cubes. In *Proc. 42nd IEEE Symp. on Foundations of Comput. Sci.*, pages 547–555, 2000.
[18] Emo Welzl. personal communication.
[19] Kathy Williamson Hoke. Completely unimodal numberings of a simple polytope. *Discrete Appl. Math.*, 20:69–81, 1988.

Lagrangian Cardinality Cuts and Variable Fixing for Capacitated Network Design*

Meinolf Sellmann, Georg Kliewer, and Achim Koberstein

Department of Mathematics and Computer Science, University of Paderborn
Fürstenallee 11, D-33102 Paderborn
{sello,geokl,akober}@upb.de

Abstract. We present a branch-and-bound approach for the Capacitated Network Design Problem. We focus on tightening strategies such as variable fixing and local cuts that can be applied in every search node. Different variable fixing algorithms based on Lagrangian relaxations are evaluated solitarily and in combined versions. Moreover, we develop cardinality cuts for the problem and evaluate their usefulness empirically by numerous tests.

1 Introduction

When solving discrete optimization problems to optimality, really two tasks have to be considered. First, an optimal solution must be constructed, and second, the algorithm must prove its optimality. Optimal or at least near optimal solutions can often be found quickly by heuristics or by approximation algorithms, both specially tailored for the given problem. In contrast to the construction of a high quality solution, the algorithmic optimality proof requires the investigation of the entire search space, which in general is much harder than to partly explore the most promising regions only. By eliminating parts of the search space that do not contain improving solutions, tightening strategies can help with respect to both aspects of discrete optimization.

In this paper, we focus on local tightening strategies that can be applied in every search node of a branch-and-bound tree and that may only be valid in the current subtree. We review bound computation algorithms based on Lagrangian relaxation that have been proposed for the CNDP and evaluate their performance in practice.

It is important to note that the algorithms used for bound computations within a branch-and-bound algorithm should not only be measured in terms of quality and computation time. In many successful approaches, they are also used for the selection of the branching cut that should favorably be introduced

---

* This work was partly supported by the German Science Foundation (DFG) project SFB-376, the project "Optimierung in Netzwerken" under grant MO 285/15-1, and by the UP-TV project, partially funded by the IST program of the Commission of the European Union as project number 1999-20 751, and by the IST Programme of the EU under contract number IST-1999-14186 (ALCOM-FT).

in the next branching step, and sometimes they can also be used to tighten the problem formulation within a search node by variable fixing. Or, more generally, by generating local cuts that may only be valid for the subtree rooted by the current node. We embed our algorithms for the computation of the linear continuous relaxation bound of the CNDP in a branch-and-bound framework. We investigate independent variable fixing algorithms and a coupling technique for variable fixing algorithms based on Lagrangian relaxations to tighten the problem formulation within a search node. Additionally, we derive local Lagrangian Cardinality Cuts and evaluate their usefulness in practice.

The paper is structured as follows: In Section 2, we introduce the Capacitated Network Design Problem (CNDP). To solve the problem, we use bounds, variable fixing algorithms and local cardinality cuts based on Lagrangian relaxation as described in Section 3. The entire branch-and-bound-approach is described in Section 4. Finally, in Section 5, we give numerical results. Generally, because of space restrictions we omit all proofs. A full version of the paper can be found in [15].

2 The Capacitated Network Design Problem

The Capacitated Network Design Problem consists of finding an optimal subset of edges in a network $G = (V, E)$ such that we can transport a given demand of goods (so called *commodities*) at optimal total cost. The latter consists of two components: the flow costs and the design costs. The flow cost is the sum of costs for the routing of each commodity, whereby for each arc (i, j) and commodity k a scalar c_{ij}^k determines the cost of routing one unit of commodity k via (i, j). The design costs are determined by the costs of installing the chosen arcs, whereby for each arc (i, j) we are given a fixed edge installation cost f_{ij}. Additionally, there is a capacity u_{ij} on each arc that limits the total amount of flow that can be routed via (i, j).

For all edges $(i,j) \in E$ and commodities $1 \leq l \leq K$, let $b_{ij}^l = \min\{|d^l|, u_{ij}\}$. Using variables $x \in \mathcal{R}_+^{|E|}$ for the flows and $y \in \{0,1\}^{|E|}$ for the design decisions, the mixed-integer linear optimization problem for the capacitated network design is defined as follows:

$$\text{Minimize} \quad L_{CNDP} = \sum_l (c^l)^T x^l + f^T y$$
$$\begin{align}
\text{subject to} \quad & Nx^l = d^l & & (1) \\
& \sum_l x_{ij}^l \leq u_{ij} y_{ij} & \forall\, (i,j) \in E & (2) \\
& x_{ij}^l \leq b_{ij}^l y_{ij} & \forall\, (i,j) \in E,\ 1 \leq l \leq K & (3) \\
& x \geq 0 & & (4) \\
& y \in \{0,1\}^{|E|} & & (5)
\end{align}$$

For ease of notation, we refer to the above LP with L_{CNDP}, which is also used to denote the optimal objective value. The network flow constraints (also called *mass balance constraints*) (1) are defined by the node-arc-incidence matrix $N = (n_{ia})_{i \in V, a \in E}$ and a demand vector $d^k \in \mathcal{R}^{|V|}$ for all commodities k, whereby $n_{ia} = 1$ iff $a = (h,i)$, $n_{ia} = -1$ iff $a = (i,h)$, and $n_{ia} = 0$ otherwise, and $d_i^k > 0$ iff node $i \in V$ is a demand node and $d_i^k < 0$ iff node i is a supply node for commodity k. Without loss of generality, we may assume that there is exactly one demand node and one supply node for each commodity [11].

The total flow on an arc (i,j) is constrained by the capacity u_{ij} (so called *capacity* or *bundle constraints* (2)). The set of *upper bound constraints* (3) is redundant to the problem formulation and provides a tighter LP relaxation of the MIP.

2.1 State of the Art

In several research papers, Crainic, Frangioni, and Gendron develop lower bounding procedures for the CNDP [4]. The main insights are the following: Tight approximations of the so called *strong* LP-relaxation (see L_{CNDP} including the redundant constraints (3)) can be found much faster by Lagrangian relaxation than by optimizing the LP using standard LP-solvers. The authors investigate so called *shortest path* and *knapsack relaxations* (see Section 3). When solving the Lagrangian dual, bundle methods converge faster than ordinary subgradient methods and are more robust. Motivated by this successful work, we evaluate several Lagrangian relaxations in the context of branch-and-bound.

In [11], Holmberg and Yuan present a method to compute exact or heuristic solutions for the CNDP. They use the Lagrangian *knapsack relaxation* in each node of the branch-and-bound tree to efficiently compute lower bounds. Special penalty tests were developed which correspond to variable fixing strategies presented in the paper at hand. An evaluation of the following components is given: subgradient search procedure for solving the Lagrangian dual, primal heuristic for finding feasible solutions, interplay between branch-and-bound and the subgradient search. On top of that work, a heuristic is developed that is embedded in the tree search procedure. That heuristic is able to provide near-optimal solutions on CNDP instances which are far beyond the range of exact methods like Lagrangian relaxation based branch-and-bound or branch-and-cut (represented e.g. by the Cplex implementation).

In [2], Bienstock et al. describe two cutting-plane algorithms for a variant of the CNDP with multi-edges (i.e., an edge can be inserted multiple times). One of them is based on the multicommodity formulation of CNDP and uses cutset and three-partition inequalities. The other one adds the following cutting planes: total capacity, partition and rounded metric inequalities. In a branch-and-cut framework, both variants provide sound results on a benchmark of realistic data. A substantial improvement to this procedure is proposed by Bienstock in [3]. The branch-and-cut algorithm based on ϵ-approximations of linear programs performs better on the same benchmark data.

3 Lagrangian Relaxation Bounds

The CNDP can be viewed as a mixture of a continuous and a discrete optimization problem. The latter is obviously constituted by the design variables, whereas the first is a min cost multi-commodity flow problem (MMCF) that evolves when the design variables are fixed. For the MMCF, besides linear programming solvers, especially cost decomposition approaches based on Lagrangian relaxation have been applied successfully [6]. The bounds we will use for the CNDP will be based on those cost decomposition approaches for the MMCF.

Used for more than 30 years now, Lagrangian relaxation can well be referred to as a standard technique for the bound computation of combinatorial optimization problems. The pioneering work was done by Held and Karp [9, 10] who introduced the new idea when tackling the traveling salesman problem. By omitting some hard constraints and incorporating them in the objective function via a penalty term, upper bounds on the performance (that is, for the CNDP, lower bounds on the costs) can be computed.

Regarding the MMCF and also for the CNDP, we are left with two promising choices of which hard constraints should be softened:

- the bundle constraints ("shortest path relaxation"), or
- the mass balance constraints ("knapsack relaxation").

In the following, we discuss the knapsack relaxation in more detail. For an in depth presentation of the shortest path relaxation, we refer to [15].

3.1 Knapsack Relaxation

For the mass balance constraints to be relaxed, we introduce Lagrangian multipliers μ_i^l for all $1 \leq l \leq K$ and $i \in V$. We get the following linear program:

$$\begin{aligned}
\text{Minimize} \quad & L_{KP}(\mu) = \sum_l \sum_{ij} (c_{ij}^l + \mu_i^l - \mu_j^l)^T x_{ij}^l + f^T y + \mu^T d \\
\text{subject to} \quad & \sum_l x_{ij}^l \leq u_{ij} y_{ij} && \forall\, (i,j) \in E \\
& x_{ij}^l \leq b_{ij}^l y_{ij} && \forall\, (i,j) \in E \\
& x \geq 0 \\
& y \in \{0,1\}^{|E|}
\end{aligned}$$

Whereas the shortest path relaxation decomposes the Lagrangian subproblem by the different commodities, here we achieve an edge-wise decomposition. To solve the above LP, for each $(i,j) \in E$ we consider the following linear program, that is similar to the linear continuous relaxation of a knapsack problem:

$$\begin{aligned}
\text{Minimize} \quad & L_{KP}^{(i,j)}(\mu) = \sum_l \bar{c}_{ij}^l \bar{x}_{ij}^l \\
\text{subject to} \quad & \sum_l \bar{x}_{ij}^l \leq u_{ij} \\
& \bar{x}_{ij}^l \leq b_{ij}^l && \forall\, 1 \leq l \leq K \\
& \bar{x} \geq 0
\end{aligned}$$

where $\bar{c}_{ij}^l = c_{ij}^l + \mu_i^l - \mu_j^l$. For each $(i,j) \in E$, we set $x_{ij}^l = \bar{x}_{ij}^l$ for all $1 \leq l \leq K$, and $y_{ij} = 1$, iff $f_{ij} + L_{KP}^{(i,j)}(\mu) < 0$. Otherwise, we set $x_{ij}^l = 0$ for all $1 \leq l \leq$

K, and $y_{ij} = 0$. Obviously, this setting provides us with an optimal solution for $L_{KP}^{(i,j)}(\mu)$. Thus, the main effort is to solve the problems $L_{KP}^{(i,j)}(\mu)$. But this is an easy task (compare with [12, 13]): first, we can eliminate all variables with positive cost coefficients, i.e., we set $\overline{x}_{ij}^l = 0$ for all $1 \leq l \leq K$ with $\overline{c}_{ij}^l \geq 0$. Next, we sort the $\overline{x}_{ij}^l$ according to increasing cost coefficients $\overline{c}_{ij}^l$, that is, from now on we may assume that $\overline{c}_{ij}^l < \overline{c}_{ij}^{l+1} < 0$ for all $1 \leq l < s \leq K$, where s is the number of negative objective coefficients. Let $k \in \mathbb{N}$ denote the *critical item* with $k = \min\{l \leq s \mid \sum_{h \leq l} b_{ij}^h > u_{ij}\} \cup \{s+1\}$. We obtain $L_{KP}^{(i,j)}(\mu)$ by setting $\overline{x}_{ij}^h = b_{ij}^h$ for all $h < k$, $\overline{x}_{ij}^h = 0$ for all $h > \min\{k, s\}$, and, in case of $k < s+1$, $\overline{x}_{ij}^k = u_{ij} - \sum_{h<k} b_{ij}^h$. Thus, the knapsack subproblem can be solved in time $O(|E|(K \log K))$.

Note, that both relaxations have the integrality property. Thus, the bound we achieve in both settings equals the linear continuous relaxation bound of the CNDP [4].

3.2 Lagrangian Cardinality Cuts

In the presence of a near optimal solution to the CNDP with associated objective value B, in each Lagrangian subproblem we can infer restrictions on the number of edges that need to be installed in any improving solution.

Before we state the idea idea more formally, to ease the notation we introduce some identifiers. For the knapsack relaxation we set $\hat{f}_{ij} = f_{ij} + L_{KP}^{(i,j)}(\mu)$ for all $(i,j) \in E$ and $\hat{B} = B + \mu^T d$ for Lagrangian multipliers μ. Further, denote with L_R the current Lagrangian subproblem

Theorem 1. *Denote with $e_1, \ldots, e_{|E|}$ an ordering of the edges in E such that $i < j$ implies $\hat{f}_{e_i} \leq \hat{f}_{e_j}$. It holds, if $L_R < B$, then*

a) there exist values

$$F = \operatorname{argmin}_{0 \leq u \leq |E|}\{\sum_{h \leq u} \hat{f}_{e_h} < \hat{B}\} \text{ and} \qquad (1)$$

$$U = \operatorname{argmax}_{0 \leq u \leq |E|}\{\sum_{h \leq u} \hat{f}_{e_h} < \hat{B}\}. \qquad (2)$$

b) And, in any improving solution (x, y), it holds that $F \leq \sum_{(i,j) \in E} y_{ij} \leq U$.

Theorem 1 allows us to add cardinality cuts on the number of edges to be used without loosing improving solutions. We will evaluate the effect of local cardinality cuts on the solution process in Section 5.

3.3 Variable Fixing

A big advantage of Lagrangian relaxation based bound computations is that they can be used for variable fixing in a very efficient way. In the presence of an optimal or at least high quality upper bound $B \in \mathbb{R}$ for the CNDP, it is an easy task to check whether a variable y_{ij} can still be set to either of its bounds without worsening the lower bound too much. More formally, given the Lagrangian multipliers μ in the current knapsack subproblem, a value $l \in \{0, 1\}$ and any edge $(i, j) \in E$, we can set

$$y_{ij} = l \quad \text{if } (2l - 1)(f_{ij} + L_{KP}^{(i,j)}(\mu)) > B - L_{KP}(\mu) \tag{3}$$

A similar statement holds for the shortest path relaxation. Using these implications, we can derive two variable fixing algorithms for the two different Lagrangian subproblems. Of course, we could just choose one of the two alternatives (for example the one for which the Lagrangian dual can be solved more quickly) and apply the corresponding variable fixing algorithm. But when using a coupling method for variable fixing algorithms that was published in [14], we can do even more: with the help of dual values gained in the solution process of the Lagrangian subproblem, in every Lagrange iteration we can apply both variable fixing algorithms.

When using the knapsack relaxation, the idea of the coupling method consists in using dual values as Lagrangian multipliers for the shortest path subproblem next. Thus, we need to provide dual values for the bundle as well as the upper bound constraints. Given the current Lagrangian multipliers μ, we solve $|E|$ knapsack subproblems as described in Section 3.1. Again, when given any edge $(i, j) \in E$, we assume that $s \in \mathbb{N}$ denotes the number of negative cost coefficients in $L_{KP}^{(i,j)}(\mu)$, that the remaining variables $\overline{x}_{ij}^l$ are ordered with respect to increasing cost coefficients, and that $k \leq s + 1$ is the critical item.

In case of $k < s+1$, we set $\overline{\lambda}_{ij} = \overline{c}_{ij}^k$, $\overline{\nu}_{ij}^l = \overline{c}_{ij}^l - \overline{c}_{ij}^k$ for all $l < k$ and $\overline{\nu}_{ij}^l = 0$ for all $l \geq k$. And for $k = s+1$, we set $\overline{\lambda}_{ij} = 0$, $\overline{\nu}_{ij}^l = \overline{c}_{ij}^l$ for all $l < k$ and $\overline{\nu}_{ij}^l = 0$ for all $l \geq k$.

Theorem 2. *The vectors $\overline{\lambda}$ and $\overline{\nu}$ define optimal dual values for $L_{KP}(\mu)$.*

Now, if we choose to use the knapsack relaxation, in every Lagrangian subproblem we solve $|E|$ linear continuous knapsack problems and achieve a lower bound for the CNDP. If that bound is worse than B, we can prune the current choice point. Otherwise, we fix variables according to Equation 3. Next, we set up the Lagrangian shortest path subproblems that evolves when using the optimal dual values of the knapsack subproblem in Theorem 2 as Lagrangian multipliers. Then, we fix variables with respect to this substructure.

As our experiments show, it is favorable to use the knapsack relaxation to solve the Lagrangian dual quickly. As one would expect, solving K shortest path problems in every Lagrangian subproblem in addition to the $|E|$ knapsack subproblems is rather costly and slows down the solution process considerably. The following Theorem helps to cope with this situation more efficiently.

Theorem 3. *Given Lagrangian multipliers μ in the knapsack relaxation, denote with $\overline{\lambda} \leq 0$ and $\overline{\nu} \leq 0$ optimal dual values for $L_{KP}(\mu)$. Then,*

$$L_{SP}(\overline{\lambda}, \overline{\nu}) \geq L_{KP}(\mu), \qquad (4)$$

where $L_{SP}(\overline{\lambda}, \overline{\nu})$ denotes the value of the next shortest path subproblem.

This Theorem allows to fix variables with respect to the shortest path relaxation without having to solve the corresponding shortest path subproblems, which improves on the running time of our variable fixing algorithm, but makes it also less effective. Unfortunately, as we shall see in Section 5, even in its strong version the shortest path variable fixing algorithm is already almost ineffective, and therefore this idea cannot be used to improve on the running time of our CNDP solver.

4 A Branch-and-Bound Algorithm

After having described the bound computation and possible tightening strategies based on Lagrangian relaxation, now we sketch the decisions taken in the tree search.

Dominance Cut-Off Rule Apart from the lower bound exceeding the upper bound, the search in the current node can be pruned if the min-cost routing of all commodities only uses edges that have already been decided to be installed. Thus, in every choice point we use a column generation approach to solve the min-cost multicommodity flow problem on the subset of edges with associated y_{ij} that have a current upper bound of 1. And if that routing only uses edges (i, j) with y_{ij} with lower bound 1, we can prune the search and backtrack.

Branching Variable Selection The previous discussion also induces a rule for the selection of the branching variable: it is clearly favorable to choose a variable for branching that is being used by the current optimal min-cost multicommodity flow. Of course, there may be more than just one such variable. Then, we can choose the one with minimal or maximal reduced costs $|\hat{f}_{ij}|$ in the Lagrangian subproblem with the best associated multipliers. The different choices will be evaluated in Section 5.

Tree Traversal A simple depth first search procedure is used to choose the next search node. This allows to find feasible solutions quickly and eases the reuse of Lagrangian multipliers.

Primal Heuristic To find reasonably good and near optimal solutions quickly, in every search node we apply a Lagrangian heuristic that was suggested by Holmberg and Yuan. It works by computing multicommodity flow solutions on a subset of the edges and de-assigning all arcs that carry no flow. For further details, we refer to [11].

Variable Fixing Heuristic Because the Network Design Problem is very hard to be solved exactly, we may decide to search for relatively good solutions quickly. The exact approach can be transformed in a heuristic for the problem by fixing variables more optimistically. Holmberg and Yuan [11] developed the so called α-heuristic for this purpose: While solving the Lagrangian dual, we protocol how often a variable is set to one or to zero. And if one of the values is dominant with respect to a given parameter, the variable is simply set to this value.

5 Numerical Results

We report on our computational experience with the algorithms developed in this paper. The section is structured as follows: first, we introduce the benchmark data used in the experiments. Then, we define the possible parameter settings that activate and deactivate different algorithmic components. And finally, we compare the variants when solving the CNDP from scratch, in the optimality proof, and when using the approach as a heuristic.

All tests were carried out on systems with AMD Athlon, 600MHz processors, and 256 MByte main memory. The code was compiled with the GNU g++ 2.95 compiler using optimization level O3.

5.1 Benchmark Data

Surprisingly, in spite of the theoretical interest the CNDP has drawn and the large number of research groups that have dealt with the problem, apparently there has been no benchmark set established on which researchers can compare algorithms that solve the CNDP exactly. Much work has been done with respect to the computation of lower bounds and the heuristic solution of the problem. Benchmarks used for this purpose (to be found in [4, 11], for example) are still too large to allow the computation of optimal solutions. For variations of the problem (such as the multi-edge CNDP, Network Loading, etc.) benchmark data exists, but it is not straight forward to see how it could be converted into meaningful instances for the pure CNDP as we consider it here.

Thus, we decided to base our comparison on a benchmark of 48 instances generated by a CNDP generator developed by Crainic et al. and described in [4]. It appears as a generator that is used by different research groups, and it was enhanced with a stable random number generator by A. Frangioni. We generated graphs with 12, 18, and 24 nodes with 50 to 440 arcs and 50 to 160 commodities. For the heuristic comparison we use the benchmark set from [4, 5]. The exact details about the benchmark sets we use is given in [15]. There, the exact data regarding our experiments can be found as well, that is left out in the paper because of space restrictions.

5.2 Algorithm Variants Considered in the Experiments

The optimization system developed consists of several parts. The ones compared and evaluated in the experiments are: different Lagrangian relaxation algorithms based on the shortest path (SP) or the knapsack relaxation (KP), respectively; a branch-and-bound algorithm using bounds based on those relaxations, where the branching variable is chosen according to minimal (BR0) or maximal (BR1) absolute reduced cost values $\hat{f}_{ij}$; two different variable fixing algorithms based on the shortest path relaxation (SF) and the knapsack relaxation (KF); and finally, the cardinality interval tightening algorithm that adds Lagrangian cardinality cuts to the problem (CIT).

5.3 Evaluation

With the first experiments we performed we wanted to find out which type of Lagrangian relaxation was preferable. In accordance to the results reported in [11], we found that the knapsack relaxation is clearly superior both with respect to the number of subgradient iterations needed to solve the Lagrangian dual as well as the time needed to solve the Lagrangian subproblems. Because of the space restrictions, we omit a detailed comparison here, and start right away with an evaluation of the impact of Lagrangian cardinality cuts when solving the CNDP using the knapsack relaxation. Table 1 shows a comparison of lower bound routines using the Lagrangian knapsack relaxation with and without cardinality cuts. And Table 2 shows a comparison of two different strategies for the selection of the branching variable. Comparing two variants, the tables give the average percentage of the second variant when compared to the first (that is always set to 100%) with respect to running times and the number of search nodes visited in the branch-and-bound trees. Moreover, we specify minima, maxima, and the variance of those percentages.

Table 1. Impact of cardinality interval tightening using knapsack relaxation with fixation based on knapsack relaxation for solving CNDP. Mean, minimum, maximum values and variance of running time and nodes in the branch-and-bound tree are given

	BR0 → BR0-CIT	BR1 → BR1-CIT
time	**93.7%**	**25.3%**
min	4.72%	0.28%
max	353%	131.5%
variance	62.1%	11.5%
nodes	**38.6%**	**10.1%**
min	0.73%	0.02%
max	120.7%	78.4%
variance	14.5%	2.9%

Clearly, choosing a branching variable with minimal reduced costs is favorable, no matter if cardinality cuts are introduced or not. This result contradicts the recommendation given in [11]. Actually, this result is not very surprising. Intuitively, the variable with the minimal absolute reduced costs is least likely to be set to either of its bounds by variable fixing. It is the variable we have the least knowledge about, and therefore it is a good choice to base a case distinction on it. In contrast, the variable with the largest absolute reduced costs is most likely to be set by variable fixing, and therefore it is no good idea to double the effort by using this variable for branching.

Regarding the introduction of Lagrangian cardinality cuts, Table 1 shows that they have a great impact on the number of search nodes that have to be investigated. Cardinality cuts are also favorable with respect to the total running time, but the gains are not as large as with respect to the size of the search tree. The trade off is caused by the additional effort that is necessary to sort the edges with respect to the current reduced costs $\hat{f}_{ij}$.

When looking at the data more precisely, we find that the primal heuristic works much better in the presence of cardinality cuts. The result of this positive effect is clear: high quality upper bounds are found much earlier in the search, pruning and variable fixing work much better, and the number of search nodes is greatly reduced, which explains the numbers in Table 1.

Table 2. Impact of branching variable selection using knapsack relaxation with fixation based on knapsack relaxation for solving CNDP

	BR0 → BR1	BR0-CIT → BR1-CIT
time	**1817.7%**	**235.1%**
min	68.03%	30.91%
max	7445.5%	1221.7%
variance	37944.6%	604.7%
nodes	**2750.4%**	**311.1%**
min	89.832%	15.636%
max	19415.3%	1427%
variance	172454%	1163.3%

We conjecture that the primal heuristic works so well in the presence of cardinality cuts because they provide a good estimate on the number of edges that need to be installed in order to improve the current solution. Thus, the right amount of edges is opened for the heuristic, and it is able to compute near optimal solutions at a higher rate.

Table 3. Impact of additional shortest path fixing using knapsack relaxation with fixation based on knapsack relaxation. Branching strategy BR0 and cardinality interval tightening are used

	SOLVE: KF → KF-SF	OPT: KF → KF-SF
time	**148.6%**	**144.1%**
min	96.59%	51.87%
max	466%	271.3%
variance	46.1%	13.5%
nodes	**133.8%**	**94.9%**
min	71.42%	20%
max	677.1%	180.3%
variance	166.3%	7.5%

Next, we evaluate the use of the coupling method for variable fixing algorithms for the CNDP. Table 3 shows a comparison of runs when using shortest path variable fixing in addition to the knapsack variable fixing algorithm. The results are very disappointing: not only is the coupled approach inferior with respect to the total running time. On top of that, the reduction of choice points is negligible, and therefore the additional effort taken is almost worthless.

Note, that the number of search nodes when using the coupling method sometimes even exceeds the value when using knapsack variable fixing only. This is caused by differences when building up the search tree: the Lagrangian dual usually stops with different Lagrangian multipliers that have a severe impact on the variable selection. Moreover, the generation of primal solutions differs, which makes the comparison particularly difficult, because variable fixing is highly sensitive to the quality of upper bounds. Thus, to eliminate the last perturbation, we repeated the experiment on the algorithmic optimality proof. That is, in the experiments we present in the following, we provide the algorithm with an optimal solution and let it prove its optimality only.

Table 3 shows the results, that still reveal the poor performance of the additional application of shortest path variable fixing. The reason for this is, that the shortest path variable fixing algorithm is much less effective than the one based on the knapsack relaxation. We tried to improve on the effectiveness of the algorithm by adding node-capacity constraints. If a node is a source for some commodities, its out-capacity must be large enough to push the corresponding supply into the network. Similarly, if a node is a sink node for some commodities, its in-capacity must be large enough to let the required demand in. In contrast to the knapsack relaxation, where the x- and y-variables are not independent, the shortest path relaxation allows to incorporate those constraints very easily. However, even this strengthening did not result in a filtering algorithm that was effective enough to be worth applying.

Next, in Table 4 we compare the performance of the algorithm we developed and the standard MIP-solver Cplex 7.5 when solving the CNDP and when proving the optimality of a given solution. Clearly, using LP-bounds improved by several kinds of cuts that Cplex adds to the problem results in a huge reduction of search nodes. However, Lagrangian relaxation allows to compute lower bounds much faster, so that the approach presented here is still competitive when solving the CNDP. And it achieves an improvement on the running time in the optimality proof.

Regarding the fact that we set up our system for the evaluation of variable fixing algorithms and local Lagrangian cardinality cuts, and taking into account that no sophisticated methods (like, e.g., Bundle methods) for the optimization of the Lagrangian dual are used, and, most importantly, that no global cuts are introduced yet to strengthen the lower bounds computed, we consider these results as very encouraging.

Finally, we compare the non-exact version of our approach (using the α-fixing heuristic) with other heuristic approaches that have been developed for the CNDP (see [5, 7, 8]). In Figure 1, we give the percentage of instances in a benchmark set (set C in [4, 5], containing 31 instances) that have been solved within a given solution quality (in percent, compared with the best known solution). Not only are the α-fixing with and without cardinality cuts clearly superior with respect to the achieved solution quality. On top of that, the heuristic vari-

Table 4. Comparison of CPLEX branch-and-cut algorithm against knapsack relaxation with fixation based on knapsack relaxation and cardinality interval tightening

	OPT: CPLEX → KP-BR0-KF-CIT	SOLVE: CPLEX → KP-BR0-KF-CIT
time	**73.5%**	**229.2%**
min	9.63%	22.48%
max	259%	753.5%
variance	36.5%	356.5%
nodes	**1148.1%**	**3014.6%**
min	196.666%	100%
max	5250%	10279.5%
variance	10762.4%	73762.5%

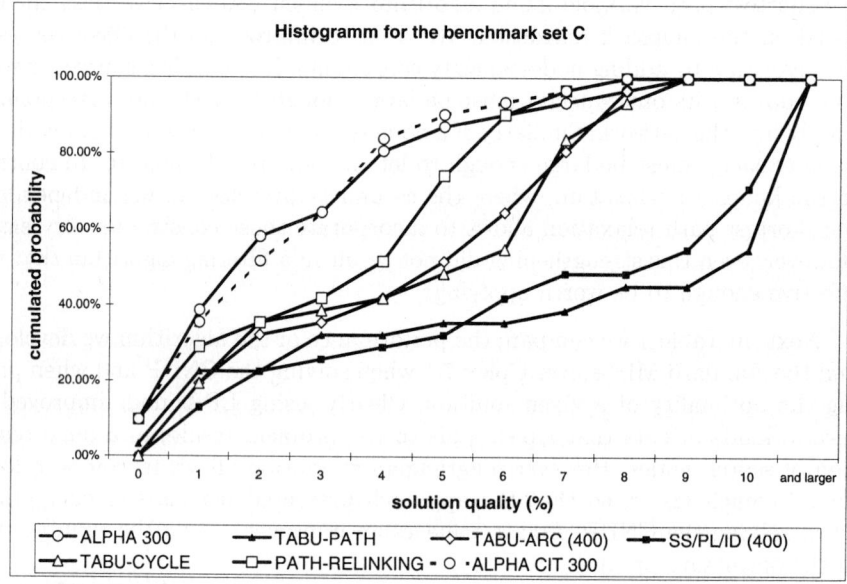

Fig. 1. Comparison of different heuristic solvers for the CNDP

able fixing approach was stopped after at most 300 seconds cpu time. On this benchmark set, heuristic variable fixing is on average about 6 times faster than TABU-PATH and 23 times faster than PATH-RELINKING (using SPECint values to make different architectures comparable).

6 Conclusions and Future Work

We have presented an approach for the solution of the Capacitated Network Design Problem. It is based on a tree search where lower bounds based on Lagrangian relaxation are used for pruning. Two kinds of relaxation are considered, the shortest path and the knapsack relaxation. The latter is clearly favorable with respect to the convergence of the subgradient algorithm that optimizes the Lagrangian dual.

Two different variable fixing algorithms have been proposed in the literature that belong to the kind of relaxation that is chosen. When using the knapsack relaxation, we have shown how variables can also be fixed with respect to shortest path considerations by using dual values in the Lagrangian knapsack subproblem. However, even in combination with node-capacity constraints the shortest path variable fixing algorithm is too ineffective to justify the additional effort that is necessary for its application.

To tighten the problem formulation in a search node, we introduced the idea of local Lagrangian Cardinality Cuts. Experimental results show that their application improves on the overall running time, even though the time per search node increases considerably when they are applied.

Finally, we compared the heuristic variable fixing approach with other heuristic approaches developed for the CNDP. The results show, that the tree search approach we implemented clearly outperforms other heuristics both with respect to the cpu time needed and the quality achieved.

As a subject of further research, methods for the strengthening of the shortest path variable fixing that incorporate the routing costs may improve on the effectivity. Moreover, other tightening strategies such as global cuts can be incorporated to improve on the relaxation gap, which may improve on the overall performance of a Lagrangian relaxation based CNDP solver.

References

[1] R. K. Ahuja, T. L. Magnati, J. B. Orlin. *Network Flows*. Prentice Hall, 1993.
[2] D. Bienstock, O. Günlük, S. Chopra, C. Y. Tsai. Mininum cost capacity installation for multicommodity flows. *Mathematical Programming*, 81:177-199, 1998.
[3] D. Bienstock. Experiments with a network design algorithm using epsilon-approximate linear programs. CORC Report 1999-4.
[4] T. G. Crainic, A. Frangioni, B. Gendron. Bundle-based relaxation methods for multicommodity capacitated fixed charge network design. *Discrete Applied Mathematics*, 112: 73-99, 2001.
[5] T. G. Crainic, M. Gendreau, and J. M. Farvolden. A simplex-based tabu search method for capacitated network design. *INFORMS Journal on Computing*, 12(3):223-236, 2000.
[6] A. Frangioni. Dual Ascent Methods and Multicommodity Flow Problems. Ph.D. Dissertation TD 97-5, Dip. di Informatica, Univ. di Pisa, 1997.
[7] I. Ghamlouche, T. G. Crainic, M. Gendreau. Cycle-based neighbourhoods for fixed-charge capacitated multicommodity network design. Technical report CRT-2001-01. Centre de recherche sur les transports, Université de Montréal.
[8] I. Ghamlouche, T. G. Crainic, M. Gendreau. Path relinking, cycle-based neighbourhoods and capacitated multicommodity network design Technical report CRT-2002-01. Centre de recherche sur les transports, Université de Montréal.
[9] M. Held and R. M. Karp. The travelling-salesman problem and minimum spanning trees. *Operations Research*, 18:1138-1162, 1970.

[10] M. Held and R. M. Karp. The travelling-salesman problem and minimum spanning trees: Part II. *Mathematical Programming*, 1:6–25, 1971.
[11] K. Holmberg and D. Yuan. A Lagrangean Heuristic Based Branch-and-Bound Approach for the Capacitated Network Design Problem. *Operations Research*, 48: 461-481, 2000.
[12] S. Martello and P. Toth. An upper bound for the zero-one knapsack problem and a branch and bound algorithm. *European Journal of Operational Research*, 1:169–175, 1977.
[13] S. Martello and P. Toth. Knapsack Problems – Algorithms and Computer Implementations. *Wiley Interscience*, 1990.
[14] M. Sellmann and T. Fahle. Coupling Variable Fixing Algorithms for the Automatic Recording Problem. *9th Annual European Symposium on Algorithms (ESA 2001)*, Springer LNCS 2161: 134–145, 2001.
[15] M. Sellmann, G. Kliewer, A. Koberstein. Lagrangian Cardinality Cuts and Variable Fixing for Capacitated Network Design. Technical report tr-ri-02-234. University of Paderborn.

Minimizing Makespan and Preemption Costs on a System of Uniform Machines

Hadas Shachnai[1]*, Tami Tamir[2], and Gerhard J. Woeginger[3]

[1] Bell Laboratories, Lucent Technologies
600 Mountain Ave. Murray Hill, NJ 07974
hadas@research.bell-labs.com
[2] Department of Computer Science, Technion
Haifa 32000, Israel
tami@cs.technion.ac.il
[3] Faculty of Mathematical Sciences, Univ. of Twente
P.O. Box 217, 7500 AE Enschede, The Netherlands
woeginge@math.utwente.nl

Abstract. It is well known that for preemptive scheduling on uniform machines there exist polynomial time exact algorithms, whereas for non-preemptive scheduling there are probably no such algorithms. However, it is not clear how many preemptions (in total, or per job) suffice in order to guarantee an optimal polynomial time algorithm. In this paper we investigate exactly this hardness gap, formalized as two variants of the classic preemptive scheduling problem. In *generalized multiprocessor scheduling (GMS)*, we have *job-wise* or *total* bound on the number of preemptions throughout a feasible schedule. We need to find a schedule that satisfies the preemption constraints, such that the maximum job completion time is minimized. In *minimum preemptions scheduling (MPS)*, the only feasible schedules are preemptive schedules with smallest possible makespan. The goal is to find a feasible schedule that minimizes the overall number of preemptions. Both problems are NP-hard, even for two machines and zero preemptions.
For GMS, we develop *polynomial time approximation schemes*, distinguishing between the cases where the number of machines is *fixed*, or given as part of the input. For MPS, we derive matching lower and upper bounds on the number of preemptions required by *any* optimal schedule.

1 Introduction

The problem of preemptive scheduling on uniform machines so as to minimize the overall completion time (or *makespan*), is well-known to be solvable in polynomial time. However, for some instances, any optimal schedule requires $\Omega(m)$ preemptions, where m is the number of machines [7]. While in traditional multiprocessor scheduling the cost of preemptions is relatively small, in the modern distributed computing environment, preemptions typically involve communication, and sometimes require job migration over a network. This can significantly

---

* On leave from the Department of Computer Science, Technion, Haifa 32000, Israel.

increase the cost of the schedule. Therefore it is natural to seek schedules that minimize the overall completion time, while incurring a small number of preemptions. We consider the resulting variants of the classic preemptive scheduling problem. Formally, suppose we are given a set of jobs, $J_1, \ldots, J_n$, with processing times $t_1, \ldots, t_n$, and m uniform machines, $M_1, \ldots, M_m$; the machine M_i has the rate $u_i \geq 1$.

In the *generalized multiprocessor scheduling (GMS)* problem, each job J_j has an associated parameter, a_j, which bounds the number of times J_j can be preempted throughout a feasible schedule. We need to find a schedule that satisfies the preemption constraints, such that the maximum job completion time is minimized.[1]

In the *minimum preemptions scheduling (MPS)* problem, the only feasible schedules are preemptive schedule with smallest possible makespan. The goal is to find a feasible schedule that minimizes the overall number of preemptions.

A straightforward reduction from the *Partition* problem [5] shows that all these problems are NP-hard, even for two machines and zero preemptions. The two classical problems of preemptive and non-preemptive scheduling on uniform machines, which are special cases of GMS, were extensively studied. It is well known that for preemptive scheduling there exist polynomial time exact algorithms, whereas for non-preemptive scheduling there are probably no such algorithms. However, it is not clear how many preemptions (in total, or per job) suffice in order to guarantee an optimal polynomial time algorithm. In this paper we investigate exactly this hardness gap. We give proofs of hardness for some special cases of GMS, and we develop *polynomial time approximation schemes (PTAS)* for this problem.

Our results for generalized multiprocessor scheduling yield important distinction between the identical and uniform machine environments. For the two fundamental scheduling problems that we generalize here, similar solvability/approximability results were obtained in these two environments. In particular, the preemptive scheduling problem is optimally solvable on both identical and uniform machines, and the non-preemptive scheduling problem is strongly NP-hard on both. Yet, the two environments differ already when we allow each of the jobs to be preempted *at most once*. While on identical machines we can use McNaughton's rule [12] to obtain the minimum makespan, on uniform machines the problem is strongly NP-hard (see in Section 2.2).

1.1 Our Results

We now describe our main results for GMS and MPS. We give (in Section 2) proofs of hardness for GMS in the following special cases: (i) each of the jobs can be preempted at most *once* throughout the schedule, and (ii) the overall number of preemptions is bounded by k, for some $1 \leq k < 2(m-1)$.[2] We then

---

[1] Alternatively, we are given a bound Tot on the total number of preemptions.
[2] As shown in [7], for $k \geq 2(m-1)$ the problem is polynomially solvable.

develop (in Section 3.1) a PTAS for any instance of GMS in which the number of machines is *fixed*. Our scheme has *linear* running time, and can be applied also for instances where jobs have release dates, for GMS on *unrelated* machines, and for instances with *arbitrary* preemption costs. In Section 3.2, we give PTASs for instances of GMS with arbitrary number of machines and a bound on the total number of preemptions.

For MPS, we derive (in Section 4) matching lower and upper bounds on the number of preemptions required by *any* optimal schedule. Our results hold for any instance in which a job J_j, $1 \leq j \leq n$, can be processed simultaneously by ρ_j machines, for some $1 \leq \rho_j \leq m$. In particular, we show that the lower bound for the overall number of preemptions, in a schedule that yields the minimum makespan, is $m + \lfloor m/b \rfloor - 2$, where $b = \min_{1 \leq j \leq n} \rho_j$. We give a polynomial time algorithm that achieves this bound. For the special case where $\rho_j = 1 \ \forall j$, our algorithm uses $2(m-1)$ preemptions, as the algorithm presented in [7]; however, our algorithm and its analysis are simpler.

Our main technical contribution is the extension of the approximation technique introduced in [16] for open shop scheduling (and later extended in [11] to other *shop* scheduling problems) to obtain approximation schemes for generalized multiprocessor scheduling, where the number of machine is fixed. We exemplify this in our scheme for GMS on uniform machines. However, the technique can be applied for other variants of the problem, as discussed in Section 3.

1.2 Related Work

The GMS problem generalizes the two classical problems of scheduling on uniform machines to minimize the makespan. When $\forall j, a_j$ is unbounded, we get the preemptive scheduling problem, denoted in standard scheduling notation $Q \mid pmtn \mid C_{max}$. Horvath et al. [10] gave the first optimal algorithm for this problem. When $a_j = 0 \ \forall j$, we get the non-preemptive scheduling problem, $Q \mid\mid C_{max}$, which is strongly NP-hard. For this problem, it was shown in [6] that algorithm *longest processing time (LPT)* yields a ratio of 2 to the optimal makespan. In Section 2.3 we extend this result for the GMS and MPS problems. Hochbaum and Shmoys [8, 9], and later Epstein and Sgall [4], developed PTASs for $Q \mid\mid C_{max}$. However, these schemes cannot be adapted for the GMS problem, since they assume no dependencies among the scheduled jobs. When we allow preemptions in the schedule, each job becomes a set of segments, only one of which can be processed at any point of time.

The MPS problem was studied in [7], in the case where each job can be processed at any time by a *single* machine. The paper shows that there are instances for which any optimal algorithm uses at least $2(m-1)$ preemptions. The paper presents an algorithm that achieves this bound. Other previous works on *parallel* jobs (see, e.g., [1, 3]) assume an *even* partition of the processing of a job J_j among machines that run J_j in parallel, while we do not use this assumption (see in Section 2.1).

Due to space constraints we state some of the results without proofs.[3]

2 Preliminaries

2.1 Minimum Makespan on Uniform Machines

Let I be an instance of MPS, where job J_j has processing time t_j for $1 \leq j \leq n$. J_j can be processed simultaneously on ρ_j machines; the processing of J_j can be shared arbitrarily among the machines. We partition J_j to ρ_j jobs, $J_{j,1}, \ldots, J_{j,\rho_j}$; the processing time of $J_{j,l}$ is t_j/ρ_j. The resulting instance is I'. In the following, we use the instance I' to define the makespan of any optimal schedule of I.

Assume that the jobs in I' are sorted such that $\frac{t_1}{\rho_1} \geq \frac{t_2}{\rho_2} \geq \ldots \geq \frac{t_n}{\rho_n}$, and the machines are sorted such that $u_1 \geq u_2 \geq \ldots \geq u_m$. Let

$$w = \max\{\frac{T_n}{U_m}, \max_{1 \leq j \leq m} \frac{T'_j}{U_j}\}, \qquad (1)$$

where $U_j = \sum_{i=1}^{j} u_i$, T_n is the total length of the jobs in I, and T'_j is the total length of the first j jobs in I'. We omit the proofs of the next results.

Lemma 1. *The makespan of any optimal schedule of I is $w_{OPT}(I) = w$.*

Lemma 2. *Let $\mathcal{A}_{opt}$ be an algorithm that solves the preemptive scheduling problem for I', in $f(n)$ steps. Then $\mathcal{A}_{opt}$ can be adapted to yield the optimal makespan for I, in $f(\sum_{j=1}^{n} \min(\rho_j, m))$ steps.*

2.2 Hardness Results

Consider the *single preemption* scheduling problem, where each of the jobs can be preempted at most *once* throughout the schedule.

Theorem 1. *The single preemption scheduling problem is strongly NP-hard.*

When the overall number of preemptions is bounded by k, for some $1 \leq k \leq 2(m-1)$, we draw the line between hard and easy instances. Specifically, an algorithm that achieves the optimal makespan and uses at most $2(m-1)$ preemptions was given in [7].

Theorem 2. *For any $k < 2(m-1)$, the problem of finding a minimum makespan schedule with at most k preemptions is strongly NP-hard.*

The proofs of Theorems 1 and 2 are given in [14].

---
[3] The detailed proofs are given in [14].

2.3 LPT and the Power of Unlimited Preemptions

Consider the LPT algorithm, that assigns jobs to machines in order of non-increasing processing times. Each job is assigned to the machine that will complete its execution first. When $\rho_j > 1$ for some $1 \leq j \leq n$, we apply this rule to the instance I' (introduced in Section 2.1). Clearly, the resulting schedule is non-preemptive and the parallelism constraints are preserved. In [6], it was shown that LPT yields a ratio of 2 to the optimal *non-preemptive* makespan. A closer analysis of LPT implies the following. Let w denote the optimal preemptive schedule (as defined in Equation (1)).

Theorem 3. *Any LPT schedule yields a 2-approximation to w.*

Thus, LPT is a non-preemptive schedule that achieves 2-approximation for GMS. It can be shown that this bound is tight (see in [14]).

3 Approximation Schemes for GMS

3.1 Fixed Number of Machines

In this section we describe a polynomial time approximation scheme for GMS, where the number of machines, m, is a fixed constant. Note that in this case we may assume that a_j, the maximal number of preemptions of J_j, is fixed and bounded by $2(m-1)$, for all $1 \leq j \leq n$.

We combine in our scheme the technique of [16] for open shop scheduling, with the *configuration graph* approach used in [9, 4], for scheduling on uniform machines. As in [16], we partition the jobs to subsets, by their processing times. The *big* jobs are handled first. The scheme finds a feasible *preemptive* schedule of these jobs, which is within factor $(1+\varepsilon)$ from the optimal. The *small* jobs have non-negligible processing times, yet, the number of possible schedules for these jobs may be exponentially large. Thus, we select this set so that the *overall* processing time of the jobs is small relative to an optimal solution for the problem. This allows us to schedule these jobs *non-preemptively* on the machines (although some or all of these jobs must be preempted, in any optimal solution). Finally, we use a simple algorithm for adding to the schedule the *tiny* jobs, which are processed *non-preemptively*. Since their running times are negligible, so is their contribution to the overall completion time of the schedule. Our scheme proceeds in the following steps. For a given $\varepsilon > 0$, let $\delta = \alpha \varepsilon^3 / (m(a_{max}+1)^2)$.

1. Guess the minimum makespan, $T_\mathcal{O}$, to within factor $1+\varepsilon$.
2. Given $T_\mathcal{O}$, guess P, the maximal load (in processing units) on any machine.
3. Partition the jobs by their processing times to *big*, *small* and *tiny*, given by the sets B, S and T, respectively. Let $\alpha \in (0,1]$ be a constant (to be determined), and $a_{max} = \max_{1 \leq j \leq n} a_j$. Then,

$$B = \{J_j \mid t_j > \alpha \varepsilon P\}, \quad S = \left\{ J_j \mid \frac{\delta \varepsilon}{m} P < t_j \leq \alpha \varepsilon P \right\}, \quad T = \left\{ J_j \mid t_j \leq \frac{\delta \varepsilon}{m} P \right\}$$

4. For any $J_j \in B$, guess a partition of J_j to at most $a_j + 1$ segments, such that each segment is of length at least $\alpha \varepsilon^2 P/(a_{max}+1)^2$, rounded up to the closest integral multiple of $\delta \varepsilon P/m$.
5. Find an optimal schedule of all the segments of the jobs in B, such that the starting time of each segment is an integral multiple of $\delta T_\mathcal{O}$.
6. Schedule the jobs in $S \cup T$ non-preemptively on the machines. First, assign greedily all the jobs in S to *holes* on the machines: for any job $J_j \in S$, we look for the minimal hole in which J_j can be scheduled non-preemptively. The jobs that are scheduled in each hole are processed with non-idle times. If no hole fits for scheduling J_j, then we add J_j at the end of the schedule, on the machine that will complete its execution first. We then proceed to schedule the jobs in T. Starting with the first hole on M_1 (the time interval between the end of the schedule on any machine and $T_\mathcal{O}$ is also considered as a 'hole'), we fill each of the holes, by scheduling sequentially jobs in T. If the hole is too small for the next job, we move this job to the end of the schedule on the same machine. We mark that hole as 'full' and move to the next hole on the same machine. Once all the holes on M_i are marked as full, we move to the first hole on M_{i+1}.

In Step 5. we construct a *configuration graph* to find a partial schedule of the segments of big jobs on the first i machines, $1 \leq i \leq m$.[4]

Analysis We use the next lemmas to show that the schedule output by the scheme is of length at most $T_\mathcal{O}(1+\varepsilon)$.

Lemma 3. *Setting the starting times of the segments of the jobs in B to be integral multiples of $\delta \varepsilon P/m$, can increase the overall length of the schedule at most by factor $1 + \varepsilon$.*

Lemma 4. *Any schedule of length $T_\mathcal{O}$ of the segments of jobs in B, can be transformed into one of length at most $(1+\varepsilon)T_\mathcal{O}$, in which (i) the minimal length of any segment of J_j is at least $\alpha \varepsilon^2 P/(a_{max}+1)^2$, and (ii) the length of each segment is an integral multiple of $\delta \varepsilon P/m$.*

Lemma 5. *There exists $\alpha \in [(\varepsilon^3/(m^2 \cdot (a_{max}+1)^2))^{m-1}, 1]$, such that the set S of small jobs, selected with α, satisfies $\sum_{J_j \in S} t_j \leq \varepsilon P$.*

Theorem 4. *The above scheme yields a $(1+\varepsilon)$-approximation to the minimum makespan, in $O(n + (\frac{1}{\alpha \varepsilon^2})^{1/\alpha \varepsilon^3})$ steps.*

Extensions Our scheme can be applied without change (except for the initial guess for $T_\mathcal{O}$) to instances where jobs have release dates, and to GMS on unrelated machines.

When we are given a bound, Tot, on the total number of preemptions in the schedule, we can apply the above scheme, except that in Step 4. we guess

---
[4] We give the details in [14].

a partition of the big jobs to at most $|B|+Tot$ segments, such that each segment is of length at least $\alpha\varepsilon^2 P/(a_{max}+1)^2$, rounded up to the closest integral multiple of $\delta\varepsilon P/m$.

Arbitrary Preemption Costs Consider now the case where the cost of preemption for J_j on the machine M_i is $c_{ij} \geq 1$, for $1 \leq i \leq m$, $1 \leq j \leq n$. Given $C > 1$, we need to find a preemptive schedule in which the makespan is minimized, and the overall preemption cost is at most C. Our scheme can be applied to this case as follows. For any $1 \leq j \leq n$, let $a_j = 2(m-1)$. (i) Guess $T_\mathcal{O}$ and P. (ii) Partition the jobs (as before) to the sets B, S and T. (iii) Guess the maximal total cost of preemptions on any machine, $\hat{C}$. (iv) Partition the big jobs to segments, as before; round the value c_{ij} up to the next integral multiple of $\varepsilon\hat{C}/(m \cdot (a_{max}+1)^2/(\alpha\varepsilon^2))$; partition the jobs to classes, such that all jobs in a class have the same segment configuration, and the same cost of preemption on each machine. (v) Find a schedule that minimizes preemption cost for the segments of all jobs in B, such that the overall completion time is at most $(1+\varepsilon)T_\mathcal{O}$. (vi) The jobs in $S \cup T$ are scheduled as before.

In Step (v) we find the *minimal cost* of any final assignment of the job segments to the machines. Note that the last segment of any job does not incur preemption cost; thus, given a schedule of the segments, we need to find an assignment of the last segments of the jobs, that minimizes the total preemption cost. This can be done in polynomial time, since the number of machines is fixed. We summarize in the next result.

Theorem 5. *For any fixed $m > 1$, there is a polynomial time approximation scheme for GMS, on unrelated machines, with release dates and arbitrary preemption costs.*

3.2 Arbitrary Number of Machines

Suppose that the number of machines is given as part of the input, and we are given a bound, Tot, on the overall number of preemptions.

Identical Machines In any preemptive schedule, there are *preempted* and *non-preempted* jobs. The processing of a preempted job is split into two or more segments. Note that for identical machines McNaughton schedule incurs at most $m-1$ preemptions (of the last job on each of the first $m-1$ machines).

Definition 1. *A preemptive schedule for a GMS instance is called primitive, if it satisfies the following two conditions. (i) Every preempted job is preempted exactly once. These two segments are processed on two different machines, and (ii) Every machine processes at most two segments of preempted jobs. These segments are either processed first or last on the machine.*

Lemma 6. *For every instance of GMS with an arbitrary number of identical machines and any value of $Tot \geq 1$, there exists an optimal schedule that is primitive.*

Next, we sketch a PTAS for approximating the optimal primitive schedule. Let T denote the makespan of the LPT-schedule; hence, T is a 2-approximation of the optimal makespan. Let $\varepsilon > 0$ be the desired precision of approximation. A job is *big*, if $t_j \geq \varepsilon T$; otherwise, the job is *small*. We introduce a *rounded instance* that corresponds to instance I. For every big job J_j, we define a corresponding rounded job whose processing time equals to t_j rounded down to the next integer multiple of $\varepsilon^2 T$. Rounded big jobs can only be preempted once to form two segments, whose lengths are integer multiples of $\varepsilon^2 T$ and $\geq \varepsilon T$. Let P_s denote the total processing time of all small jobs; then, the rounded instance contains a number of small jobs of length εT, such that their total size equals P_s rounded down to the next integer multiple of $\varepsilon^2 T$. These small jobs must not be preempted at all.

An optimal primitive schedule for the rounded instance can be computed in polynomial time by dynamic programming or via integer programming in fixed dimension. Moreover, an optimal primitive schedule for the rounded instance can be translated into a primitive schedule for the original instance that is a $(1+10\varepsilon)$-approximation of the optimal primitive schedule for the original instance.

Uniform Machines Assume now that the machines are uniform, and $1 \leq Tot < 2(m-1)$ is some constant; then, initially, we guess the optimal completion time, $T_\mathcal{O}$. Next, we guess the set of machines, $\mathcal{M}_p$, on which some jobs will be preempted. Let $m' = |\mathcal{M}_p|$ denote the size of this set; then, $1 \leq m' \leq Tot$. The jobs scheduled on the remaining machines, the set $\mathcal{M}_{np}$, will run non-preemptively. These jobs can be identified efficiently and scheduled to yield a makespan of at most $T_\mathcal{O}(1+\varepsilon)$, by using a transformation to the *multiple knapsack* problem (see in [14]). The makespan of the resulting schedule is bounded by $T_\mathcal{O}(1+\varepsilon)$. Finally, we use the scheme in Section 3.1 to obtain a feasible schedule of the jobs on the machines in $\mathcal{M}_p$, whose overall completion time is at most $T_\mathcal{O}(1+\varepsilon)$.

4 Minimizing the Number of Preemptions

In this section we consider the MPS problem. For convenience, we refer below to the number of *segments*, $N(I)$, generated for the instance I.[5] Lemma 2 implies that we can find efficiently a schedule that minimizes the makespan for a given instance of MPS. However, the schedules generated by adopting an optimal algorithm for $Q|pmtn|C_{max}$, may have an excessive number of job segments. In fact, we may get $\sum_j (\rho_j - 1)$ extra job-segments in the schedule of I.

We derive below a lower bound for $N_s(I)$, for any algorithm that achieves the minimal makespan; then, we present an algorithm, $\mathcal{A}_p$, that achieves this bound.

Theorem 6. *For any m, b, $1 \leq b \leq m$, there exists an instance, I, in which $\rho_j \geq b, \forall j$, and in any optimal schedule of I, $N_s(I) \geq m + n + \lfloor \frac{m}{b} \rfloor - 2$.*

---
[5] The number of segments of J_j is the number of preemptions incurred plus one.

4.1 An Upper Bound on $N_s(I)$

Let w be the length of an optimal schedule for I (By Lemma 1, w can be calculated efficiently using equation (1)). We now describe the algorithm $\mathcal{A}_p$, which achieves the makespan w and also minimizes $N_s(I)$. Assume that the jobs are sorted such that $\frac{t_1}{\rho_1} \geq \ldots \geq \frac{t_n}{\rho_n}$, and the machines are sorted such that $u_1 \leq \ldots \leq u_m$.

As in [7], we define a *disjoint processor system* (DPS) to be a union of disjoint *idle-segments* of r machines with non-decreasing rates, such that the union of the idle-segments is the interval $[0, w]$. Formally, let $M_k^1, \ldots, M_k^r$ be a set of r machines with non-decreasing rates. Then, these r machines form a DPS if M_k^h is idle exactly from τ_k^{h-1} to τ_k^h, where $\tau_k^0 = 0, \tau_k^r = w$ and $\tau_k^{h-1} < \tau_k^h$, for all $1 \leq h \leq r$ (see Figure 1). Note that a DPS is defined relative to a given (partial) schedule.

For each DPS, D_k, let Q_{D_k} denote the *potential* of D_k, that is, the number of processing units that D_k can allocate. Initially, each of the m machines forms a DPS with $r = 1$ and $Q_{D_k} = w u_k$. In general, Q_{D_k} is the weighted-average of the processing potential of the machines that form D_k. Formally, let r_k denote the number of machines that compose D_k, and let u_k^h denote the rate of the machine M_k^h, then $Q_{D_k} = \sum_{h=1}^{r_k} u_k^h(\tau_k^h - \tau_k^{h-1})$. The algorithm $\mathcal{A}_p$ maintains a list, L, of the DPSs, sorted by their potential in non-decreasing order. That is, $Q_{L[1]} \leq Q_{L[2]} \leq \ldots$, where $L[i]$ is the DPS at position i in L. The list L is updated during the execution of the algorithm, according to the current available DPSs. Given a pair of DPSs D_a, D_b, we say that D_a is weaker (stronger) than D_b, if $Q_{D_a} \leq Q_{D_b}$ ($Q_{D_a} \geq Q_{D_b}$).

The jobs, sorted by their processing ratios, are scheduled one after the other. $\mathcal{A}_p$ uses two scheduling procedures:

Greedy-Schedule: This procedure schedules a job, J_j, on the machines that form the DPSs, one after the other, starting from the first machine in the weakest DPS, until J_j is allocated exactly t_j processing units.

Moving-Window: This procedure schedules a job, J_j, on ρ_j DPSs whose total potential is *exactly* t_j. This set of ρ_j DPSs consists of $\rho_j - 1$ DPSs from L and one DPS that is formed from two DPSs in L. To find this set, we scan the list using a window of size $\rho_j + 1$. Initially, the window covers the set of the weakest $\rho_j + 1$ DPSs (given by $L[1], \ldots, L[\rho_j + 1]$). In each iteration we replace the weakest DPS in the window by the next DPS in L, until the total potential of the ρ_j strongest DPSs in the window is large enough to complete J_j. Let $D_{k_1}, \ldots, D_{k_{\rho_j+1}}$ be the set of $\rho_j + 1$ DPSs in the window. We first allocate to J_j all the potential of

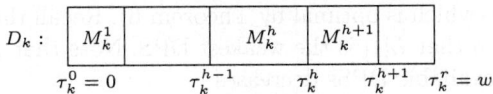

Fig. 1. A DPS D_k

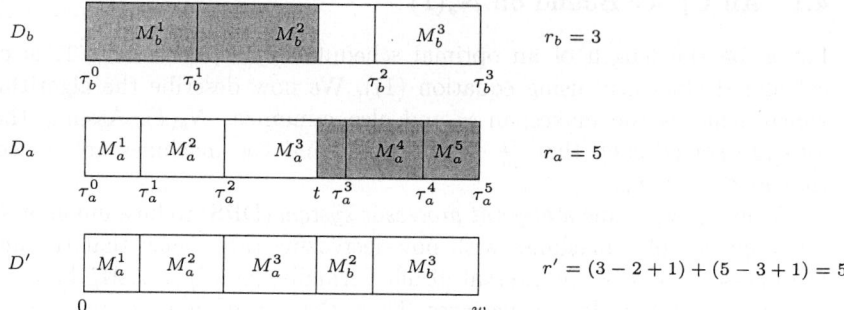

Fig. 2. The schedule of J_j on consecutive DPSs

the DPSs $D_{k_2}, \ldots, D_{k_{\rho_j}}$ (if $\rho_j = 1$ this is an empty set), and complete J_j by allocating to it also some of the potential of length of the schedule of J_j on these two DPSs is exactly w, and the unused intervals of $D_{k_1}, D_{k_{\rho_j+1}}$ form a new DPS.

Figure 2 gives a schedule on two consecutive DPSs: the shaded intervals are allocated to J_j. The non-used intervals of D_a, D_b form a new DPS, D', with potential Q' which is the weighted-average of the non-used intervals. The DPSs D_a, D_b are removed from L and D' is added to L. Since $Q_{D_b} \leq Q' \leq Q_{D_a}$, D' is positioned in L in place of D_a and D_b. We now give a formal description of the algorithm.

Algorithm $\mathcal{A}_p$: The jobs, sorted by the ratios t_j/ρ_j, are scheduled sequentially. The following rules are used when $\mathcal{A}_p$ schedules a job J_j.

- If there are at most ρ_j DPSs in L, or if the total potential of the ρ_j weakest DPSs in L is at least t_j, schedule sequentially J_j and all the remaining jobs, using the greedy-schedule procedure.
- Otherwise, schedule J_j using the moving-window procedure.

We first show that $\mathcal{A}_p$ generates a legal schedule of length w for I; then, we compute the resulting number of job-segments. We distinguish between two phases of $\mathcal{A}_p$: (i) Jobs are scheduled using the moving-window procedure. (ii) Jobs are scheduled using the greedy-schedule procedure.

For simplicity, we analyze $\mathcal{A}_p$ assuming that $\forall j, \rho_j = 1$. In the sequel, we explain how the analysis can be applied for arbitrary ρ_j's. We first show that during the first phase we can always schedule J_j on two consecutive DPSs, such that the idle-segments of these DPSs that remained unused, form a DPS. Next, we show that when we proceed to the greedy-schedule stage, we can schedule all the remaining jobs. Finally, we show that the total number of segments is at most $n + 2(m-1)$ (which is optimal by Theorem 6). Recall that the list L of the DPSs is sorted such that $L[1]$ is the weakest DPS. Note that L becomes shorter as the number of available DPSs decreases.

Claim 1. If $Q_{L[1]} < t_j$ then there exist two consecutive DPSs $D_{L[i]}, D_{L[i+1]}$ such that $Q_{L[i]} \leq t_j$ and $Q_{L[i+1]} \geq t_j$.

Claim 2. Given two DPSs, D_a, D_b, from L, such that $Q_{D_a} \leq t_j$ and $Q_{D_b} \geq t_j$, we can always schedule J_j on D_a, D_b such that the length of the schedule of J_j is exactly w.

The next Claim implies that greedy is suitable for the second phase.

Claim 3. If $Q_{L[1]} \geq t_j$, then we can complete the schedule of the remaining jobs greedily.

We now consider the number of job segments generated by $\mathcal{A}_p$.

Lemma 7. *The total number of segments is at most $n + 2(m-1)$.*

Proof: For a given schedule, define a *busy-segment* as a maximal time interval in which a machine is processing some job without preemptions. Throughout the algorithm, for each machine, M_i, the w-interval of M_i may consist of busy-segments, and one *idle-segment*, which belongs to some DPS (Note that M_i belongs to at most one DPS). Each idle-segment may be partitioned to many busy-segments by the end of the schedule.

Denote by X_0 the initial number of idle or busy segments. Clearly, initially we have one idle-segment of length w on each machine, and no busy-segments. Thus, $X_0 = m$. Denote by X_j the total number of segments (of both types) after the job J_j was scheduled, and by X_n^B the total number of busy-segments after the last job, J_n, was scheduled. That is, $X_n^B = N_s(I)$.

We show that in any schedule that uses the moving-window procedure, the total number of segments may increase by at most two, and in any greedy-schedule, the total number of segments may increase by at most one. Then, we bound the number of moving-window schedules, in order to bound the maximal possible number of segments by the end of the schedule.

Claim 4. If J_j is scheduled in the first phase then $X_j \leq X_{j-1} + 2$.

Claim 5. If $J_j, j < n$, is scheduled in the second phase then $X_j \leq X_{j-1} + 1$.

Claim 6. $X_n^B \leq X_{n-1}$.

Denote by n_w, n_g the number of jobs scheduled in the first and the second phase respectively. Clearly, $n_g = n - n_w$. Combining Claims 4-6, we get:

$$X_n^B \leq X_{n-1} \leq X_0 + 2n_w + n_g - 1 = m + 2n_w + n_g - 1 = m + n + n_w - 1. \quad (2)$$

To bound X_n^B, we bound n_w. Note that each job scheduled during the first phase, unites two DPSs. Thus, after we schedule at most $m - 1$ jobs, we have a single DPS on which we can schedule greedily the remaining jobs. (In some cases, the greedy phase starts when some job J_j with $j < m$ is scheduled.) Thus, $n_w \leq m - 1$. From Equation (2), $X_n^B \leq m + n + n_w - 1 \leq n + 2(m-1)$. This completes the proof of the lemma. □

The analysis of $\mathcal{A}_p$ for instances with arbitrary ρ_j's is similar. Note that when we schedule a job J_j with $\rho_j > 1$ using the moving-window procedure, J_j runs on $\rho_j + 1$ DPSs $D_{k_1}, \ldots, D_{k_{\rho_j+1}}$. All the potential of the DPSs $D_{k_2}, \ldots, D_{k_{\rho_j}}$ is allocated to J_j, and Claim 2 holds for D_{k_1} and $D_{k_{\rho_j+1}}$. The extension of Claims 1 and 3 is straightforward: we consider the set of ρ_j weakest DPSs instead of the weakest one. Finally, Claims 4 - 6 are valid also for the general case: in any moving-window schedule, independent of ρ_j, only the first and last DPSs in the window are united, and in any greedy schedule only the last busy-segment may split (thus adding a single idle-segment). Hence, as in the case where $\forall j, \rho_j = 1$, the resulting number of segments satisfies $X_n^B \leq m + n + n_w - 1$. However, n_w can be bounded by the minimal k for which $\sum_{j=1}^{k} \rho_j \geq m - \min_j \rho_j$ (Note that now, in any moving-window schedule we remove ρ_j DPSs from the DPSs list, and in the worst case, the greedy phase starts with $\min_j \rho_j$ DPSs).

Indeed, sorting the jobs by the ratios t_j/ρ_j, may not provide the minimal k; however, it guarantees that once $\mathcal{A}_p$ reaches the greedy phase, all the remaining jobs can be completed. Thus, for some instances, a different order of the jobs may result with fewer preemptions. Given that $\exists\, b\, \forall\, j$, $\rho_j \geq b$, we get that $n_w \leq \lfloor \frac{m-b}{b} \rfloor$, thus, $N_s(I) = X_n^B \leq m + n + \lfloor \frac{m-b}{b} \rfloor - 1 = m + n + \lfloor \frac{m}{b} \rfloor - 2$. This fits the lower bound given in Theorem 6.

Algorithm $\mathcal{A}_p$ can be implemented in $O(\max(m \log m, n \log n))$ steps. Specifically, $O(m \log m) + O(n \log n)$ steps are required for sorting the lists and calculating w. Given that the lists are sorted, the total time for scheduling the jobs is $O(m + n) + O(n \log m)$. The first phase of the algorithm, in which we schedule the jobs by scanning the list L with the moving-window, can be implemented in $O(m + n) + O(n \log m)$ steps (using skip-lists). The second (greedy) phase requires $O(m + n)$ steps. We summarize our discussion of algorithm $\mathcal{A}_p$ in the next theorem.

Theorem 7. *For any m, b, and any instance I in which $\rho_j \geq b$ for all j, $\mathcal{A}_p$ finds in $O(\max(m \log m, n \log n))$ steps an optimal schedule of I with $N_s(I) \leq m + n + \lfloor \frac{m}{b} \rfloor - 2$.*

References

[1] J. Blazewick, M. Drabowski, and J. Weglarz. Scheduling multiprocessor tasks to minimize schedule length. *IEEE Trans. Comput*, 35(C):389–393, 1986.
[2] C. Chekuri and S. Khanna. A PTAS for the multiple knapsack problem. In *Proc. of SODA*, 213–222, 2000.
[3] X. Deng, N. Gu, T. Brecht and K. Lu. Preemptive scheduling of parallel jobs on multiprocessors. In SODA'96, pp. 159–167.
[4] L. Epstein and J. Sgall. Approximation schemes for scheduling on uniformly related and identical parallel machines. In *Proc. of ESA* LNCS 1643, 151–162, 1999.
[5] M. R. Garey and D. S. Johnson. *Computers and Intractability: A Guide to the Theory of NP-Completeness*. W. H. Freeman, 1979.
[6] T. Gonzalez, O. H. Ibarra and S. Sahni. Bounds for LPT schedules on uniform processors. *SIAM J. on Comp.*, 6, 155–166, 1977.

[7] T. Gonzalez and S. Sahni. Preemptive scheduling of uniform processor systems. *Journal of the ACM*, 25:92–101, 1978.
[8] D.S. Hochbaum and D.B. Shmoys. Using dual approximation algorithms for scheduling problems: Practical and theoretical results. *J. of the ACM*, 34(1):144–162, 1987.
[9] D.S. Hochbaum and D.B. Shmoys. A polynomial approximation scheme for scheduling on uniform processors: Using the dual approximation approach. *SICOMP*, 17(3):539–551, 1988.
[10] E.G. Horvath, S. Lam, and R.Sethi. A level algorithm for preemptive scheduling. *Journal of the ACM*, 24:32–43, 1977.
[11] K. Jansen, M. Sviridenko and R. Solis-Oba. Makespan minimization in job shops: a polynomial time approximation scheme. In *Proc.of STOC*, 394-399. 1999
[12] R. McNaughton. Scheduling with deadlines and loss functions. *Manage. Sci.*, 6:1–12, 1959.
[13] H. Shachnai and T. Tamir. Multiprocessor scheduling with machine allotment and parallelism constraints. 2001. *Algorithmica*, Vol. 32:4, 651–678, 2002.
[14] H. Shachnai, T. Tamir and G.J. Woeginger. Minimizing Makespan and Preemption Costs on a System of Uniform Machines, http://www.cs.technion.ac.il/~hadas/PUB/ucosts.ps.
[15] J.I. Munro, T. Papadakis and R. Sedgewick. Deterministic skip lists. SODA'92.
[16] S.V. Sevastianov and G.J. Woeginger. Makespan minimization in open shops: A polynomial time approximation scheme. *Math. Programming* 82, 191–198, 1998.
[17] D.P. Williamson, L.A. Hall, J.A. Hoogeveen, C.A.J. Hurkens, J.K. Lenstra and D.B. Shmoys. Short shop schedules. Unpublished manuscript, 1994.

Minimizing the Total Completion Time On-line on a Single Machine, Using Restarts

Rob van Stee[1],* and Han La Poutré[2,3]

[1] Institut für Informatik, Albert-Ludwigs-Universität
Georges-Köhler-Allee, 79110 Freiburg, Germany
vanstee@informatik.uni-freiburg.de
[2] Centre for Mathematics and Computer Science (CWI)
Kruislaan 413, NL-1098 SJ Amsterdam, The Netherlands
[3] Eindhoven University of Technology, School of Technology Management
Den Dolech 2, 5600 MB Eindhoven, The Netherlands
Han.La.Poutre@cwi.nl

Abstract. We give an algorithm to minimize the total completion time on-line on a single machine, using restarts, with a competitive ratio of 3/2. The optimal competitive ratio without using restarts is 2 for deterministic algorithms and $e/(e-1) \approx 1.582$ for randomized algorithms. This is the first restarting algorithm to minimize the total completion time that is proved to be better than an algorithm that does not restart.

1 Introduction

We examine the scheduling problem of minimizing the total completion time (the sum of completion times) on-line on a single machine, using restarts. Allowing restarts means that the processing of a job may be interrupted, losing all the work done on it. In this case, the job must be started again later (restarted), until it is completed without interruptions. We study the on-line problem, where algorithms must decide how to schedule the existing jobs without any knowledge about the future arrivals of jobs.

We compare the performance of an on-line algorithm $\mathcal{A}$ to that of an optimal off-line algorithm OPT that knows the entire job sequence σ in advance. The total completion time of an input σ given to an algorithm ALG is denoted by $\text{ALG}(\sigma)$. The competitive ratio $\mathcal{R}(\mathcal{A})$ of an on-line algorithm $\mathcal{A}$ is defined as $\mathcal{R}(\mathcal{A}) = \sup_\sigma \mathcal{A}(\sigma)/\text{OPT}(\sigma)$. The goal is to minimize the competitive ratio.

Known Results For the case where all jobs are available at time 0, the shortest processing time algorithm SPT [7] has an optimal total completion time. This algorithm runs the jobs in order of increasing size. Hoogeveen and Vestjens [4] showed that if jobs arrive over time and restarts are not allowed, the optimal competitive ratio is 2, and they gave an algorithm DSPT ('delayed SPT') which

---

* Work supported by the Deutsche Forschungsgemeinschaft, Project AL 464/3-1, and by the European Community, Projects APPOL and APPOL II.

maintained that competitive ratio. Two other optimal algorithms were given by Phillips, Stein and Wein [5] and Stougie [8].

Using randomization, it is possible to give an algorithm of competitive ratio $e/(e-1) \approx 1.582$ [1] which is optimal [9]. Vestjens showed a lower bound of 1.112 for deterministic algorithms that can restart jobs [12]. This was recently improved to 1.211 by Epstein and Van Stee [2].

We are aware of three previous instances where restarts were proven to help. Firstly, in [6] it was shown that restarts help to minimize the makespan (the maximum completion time) of jobs with unknown sizes on m related machines. Here each machine has its own speed, which does not depend on the job it is running. The algorithm in [6] obtains a competitive ratio of $O(\log m)$. Without restarts, the lower bound is $\Omega(\sqrt{m})$.

Secondly, [10] shows that restarts help to minimize the maximum delivery time on a single machine, obtaining an (optimal) competitive ratio of $3/2$ while without restarts, $(\sqrt{5}+1)/2$ is the best possible. In this problem, each job needs to be delivered after completing, which takes a certain given extra time.

Thirdly, in [3] it is shown that restarts help to minimize the number of *early* jobs (jobs that complete on or before their due date) on a single machine, obtaining an (optimal) competitive ratio of 2 while without restarts, it is not possible to be competitive at all (not even with preemptions).

Our results Until now, it was not known how to use restarts in a deterministic algorithm for minimizing the total completion time on a single machine to get a competitive ratio below 2, whereas a ratio of 2 can be achieved by an algorithm that does not restart. We give an algorithm RSPT ('restarting SPT') of competitive ratio $3/2$. This ratio cannot be obtained without restarts, even with the use of randomization.

Our algorithm is arguably the simplest possible algorithm for this problem that uses restarts: it bases the decision about whether or not it will interrupt a running job J for an arriving job J' solely on J and J'. It ignores, for example, all other jobs that are waiting to be run. We show in section 3 that the analysis of our algorithm is tight and that all "RSPT-like" algorithms, that use a parameter α in stead of the value $2/3$, have a competitive ratio of at least 1.467845. This suggests that a more complicated algorithm would be required to get a substantially better ratio, if possible.

2 Algorithm RSPT

We present our on-line algorithm RSPT for the problem of minimizing the total completion time on a single machine, using restarts. See Figure 1. This algorithm has the following properties (where J, x, s, r and w are defined as in Figure 1). OPT is any optimal off-line algorithm (there can be more than one).

R1 RSPT only interrupts a job J for jobs that are smaller and that can finish earlier than J (i.e. $r + w < s + x$).

> RSPT maintains a queue Q of unfinished jobs. A job is put into Q when it arrives. A job is removed from Q when it is completed. For any time t, RSPT deals first with all arrivals of jobs at time t before starting or interrupting any job.
>
> At any time t where either RSPT completes a job, or one or more jobs arrive while RSPT is idle, RSPT starts to run the smallest (remaining) job in Q. If $Q = \emptyset$, RSPT is idle (until the next job arrives).
>
> Furthermore, if at time r a job J is running that started (most recently) at time s and has size x, and if at time r a new job J' arrives with size w, then RSPT interrupts J and starts to run J' if and only if
>
> $$r + w \leq \frac{2}{3}(s+x). \tag{1}$$
>
> Otherwise, RSPT continues to run J (and J' is put into Q).

Fig. 1. The algorithm RSPT

R2 If RSPT does not interrupt J for a job of size w that arrives at time r, then $r + w > \frac{2}{3}(s+x)$. In this case, if RSPT is still running J at time $r+w$, it runs J until completion.

R3 Suppose that $s < t \leq \frac{2}{3}(s+x)$, and RSPT has been running J continuously from time s until time t. Then at time t, OPT has completed at most one job that RSPT has not completed, and such a job has size at least x.

R4 At any time t, RSPT only interrupts jobs that it cannot finish before time $\frac{3}{2}t$. Hence, RSPT does not interrupt any job with a size of at most half its starting time.

3 RSPT-like Algorithms

It can be seen that the competitive ratio of RSPT is not better than 3/2: consider a job of size 1 that arrives at time 0, and N jobs of size 0 that arrive at time $2/3 + \varepsilon$. RSPT will run these jobs in order of arrival time and have a total completion time of $N + 1$. However, it is possible to obtain a total completion time of $(2/3 + \varepsilon)(N+1) + 1$ by running the jobs of size 0 first. By letting N grow without bound, the competitive ratio tends to 3/2 for $\varepsilon \to 0$.

We define an algorithm RSPT(α) as follows: RSPT(α) behaves exactly like RSPT, but (1) is replaced by the condition $r + w \leq \alpha(s+x)$. It may be possible that RSPT(α) outperforms RSPT for some value of α. However, we show that the improvement could only be very small, if any. To keep the analysis manageable, we analyze only RSPT.

Lemma 1. *For all $0 < \alpha < 1$, $\mathcal{R}(\text{RSPT}(\alpha)) \geq 1.467845$.*

Proof. Similarly to above, we have $\mathcal{R}(\text{RSPT}(\alpha)) \geq 1/\alpha$.

Consider the following job sequence. A job J_1 of size 1 arrives at time 0, a job J_2 of size α at time ε, a job J_3 of size 0 at time α (causing an interruption). For $\varepsilon \to 0$, the optimal cost for this sequence tends to $3\alpha + 1$ (using the order J_2, J_3, J_1). However, RSPT(α) pays $5\alpha + 1$.

Now consider the same sequence where after job J_3, at time $\alpha + \varepsilon$ one final job J_4 of size $\alpha(2\alpha) - \alpha$ arrives. For this sequence, the optimal cost tends to $4\alpha^2 + 2\alpha + 1$ whereas RSPT(α) pays $4\alpha^2 + 5\alpha + 1$.

This implies that $\mathcal{R}(\text{RSPT}(\alpha)) \geq \max(1/\alpha, \frac{5\alpha+1}{3\alpha+1}, \frac{10\alpha^2+4\alpha+1}{10\alpha^2+1}) \geq \frac{41+5\sqrt{57}}{31+3\sqrt{57}} =$ 1.467845. □

4 Analysis of RSPT

Outline Consider an input sequence σ. To analyze RSPT's competitive ratio on such a sequence, we will work with credits and an invariant.

Each job that arrives receives a certain amount of credit, based on its (estimated) completion time in the optimal schedule and in RSPT's schedule. We will show that each time that RSPT starts a job, we can distribute the credits of the jobs so that a certain invariant holds, using an induction. The calculations of the credits at such a time, and in particular of the estimates of the completion times in the two schedules, will be made under the assumption that no more jobs arrive later.

We will first give the invariant. Then we need to show that the invariant holds at the first time that RSPT starts a job. For the induction step, we need to show that the invariant holds again if RSPT starts a job later,

- *given* that it held after the previous start of a job;
- taking into account any jobs which arrived after that (possibly updating calculations for some jobs that had arrived before); and
- assuming no jobs arrive from now on.

Using this structure, the above-mentioned assumption that no jobs arrive after the current start of a job does not invalidate the proof. There will be one special case where the invariant does not hold again immediately. In that case, we will show the invariant is restored at some later time before the completion of σ.

Finally, we need to show that if the invariant holds at the last time that RSPT starts a job, then RSPT maintains a competitive ratio of $3/2$. We begin by making some definitions and assumptions.

Definition 1. *An* event *is the start of a job by* RSPT.

Definition 2. *An event has the property STATIC if no more jobs arrive after this event.*

At the time of an event, RSPT completes a job, interrupts a job, or is idle.

In our analysis, we will use 'global assumptions' and 'event assumptions'. We show that we can restrict our analysis to certain types of input sequences and schedules and formulate these restrictions as Global assumptions. Then, when analyzing an event (from the remaining set of input sequences), we show in several cases that it is sufficient to consider events with certain properties, and make the corresponding Event assumption. The most important one was already mentioned above:

Event assumption 1 *The current event has the property STATIC.*

The optimal off-line algorithm There can be more than one optimal schedule for a given input σ. For the analysis, we fix some optimal schedule and denote the algorithm that makes that schedule by OPT. We use this schedule in the analysis of every event. Hence, OPT takes into account jobs that have not arrived yet in making its schedule, but OPT does not change its schedule between successive events: the schedule is completely determined at time 0. OPT does not interrupt jobs, because it can simply keep the machine idle instead of starting a certain job and interrupting it later, without affecting the total completion time.

We can make the following assumption about RSPT and OPT, because the cost of OPT and RSPT for a sequence is unaffected by changing the order of jobs of the same size in their schedules.

Assumption 1. *If two or more jobs in σ have the same size, RSPT and OPT complete them in the same order.*

Definition 3. *An input sequence σ has property SMALL if, whenever RSPT is running a job of some size x from σ, only jobs smaller than x arrive. (Hence, jobs larger than x only arrive at the completion of a job, or when the machine is idle.)*

We show in the full paper [11] that if RSPT maintains a competitive ratio of $3/2$ on all the sequences that have property *SMALL*, it maintains that competitive ratio overall. Henceforth, we make the following assumption.

Assumption 2. *The input sequence σ has property SMALL.*

5 Definitions and Notations

After these preliminaries, we are ready to state our main definitions. A job J arrives at its release time $r(J)$ and has size (weight) $w(J)$. The size is the time that J needs to be run without interruptions in order to complete. For a job J_i, we will usually abbreviate $r(J_i)$ as r_i and $w(J_i)$ as w_i, and use analogous notation for jobs J', J^* etc. When RSPT is running a job J, we will be interested in J-*large* unfinished jobs, that are at least as large as J, and J-*small* unfinished jobs, that are smaller, separately. To distinguish between these sets of jobs, the unfinished large jobs will be denoted by $J, J^2, J^3, \ldots$ with sizes x, x_2, x_3 while the small jobs will be denoted by $J_1, J_2, \ldots$ with sizes $w_1, w_2, \ldots$. We let $Q(t)$ denote the queue Q of RSPT at time t.

Definition 4. *A run-interval is a half-open interval $I = (s(I), t(I)]$, where RSPT starts to run a job (denoted by $J(I)$) at time $s(I)$ and runs it continuously until exactly time $t(I)$. At time $t(I)$, $J(I)$ is either completed or interrupted. We denote the size of $J(I)$ by $x(I)$.*

Definition 5. *For a run-interval I, we denote the set of jobs that arrive during I by $ARRIVE(I) = \{J_1(I), \ldots, J_{k(I)}(I)\}$. We write $r_i(I) = r(J_i(I))$ and $w_i(I) = w(J_i(I))$ for $1 \leq i \leq k(I)$. The jobs are ordered such that $w_1(I) \leq w_2(I) \leq \ldots \leq w_{k(I)}(I)$. We denote the total size of jobs in ARRIVE(I) by $T(I)$, and write $T_i(I) = \sum_{j=1}^{i} w_j(I)$ for $1 \leq i \leq k(I)$.*

RSPT will run the jobs in $ARRIVE(I)$ in the order $J_1(I), \ldots, J_{k(I)}(I)$ (using Global assumpion 1 if necessary). We have $w_{k(I)}(I) < x(I)$ using Global assumption 2. Of course it is possible that $ARRIVE(I) = \emptyset$. In that case I ends with the completion of the job RSPT was running $(t(I) = s(I) + x(I))$.

The input sequence σ may contain jobs of size 0. Such jobs are completed instantly when they start and do not have a run-interval associated with them. Thus we can divide the entire execution of RSPT into run-intervals, completions of 0-sized jobs, and intervals where RSPT is idle. We show in the full paper [11] that all jobs in σ arrive either in a run-interval or at the end of an interval in which RSPT is idle. Also, if RSPT interrupts job $J(I)$ at time t, then $t = r_1(I)$.

Definition 6. *For the jobs in $ARRIVE(I)$, we write $r_i(I) + w_i(I) = \frac{2}{3}(s(I) + x(I)) + \tau_i(I)$ $(i = 1, \ldots, k(I))$.*

We have $\tau_i(I) > 0$ for $i = 2, \ldots, k(I)$, and $\tau_1(I) > 0$ if $J(I)$ completes at time $t(I)$, $\tau_1(I) \leq 0$ if it is interrupted at time $t(I)$.

Definition 7. *We define $f_{OPT}(I)$ as the index of the job that OPT completes first from $ARRIVE(I)$. An interruption by RSPT at time $t(I)$ is slow if OPT starts to run a job from $ARRIVE(I)$ strictly before time $t(I)$; in this case $f_{OPT}(I) > 1$ and $J_{f_{OPT}(I)}(I)$ did not cause an interruption when it arrived.*

We call such an interruption slow, because in this case it could have been better for the total completion time of RSPT if it had interrupted $J(I)$ for one of the earlier jobs in $ARRIVE(I)$ (i. e. faster); now, at time t, RSPT still has to run all the jobs in $ARRIVE(I)$, whereas OPT has already partially completed $J_{f_{OPT}(I)}(I)$. Note that whether an interruption is slow or fast depends entirely on when OPT runs the jobs in $ARRIVE(I)$. It has nothing to do with RSPT.

We now define some variables that can change over time. We will need their values at time t when we are analyzing an event at time t. They give as it were a snapshot of the current situation.

Definition 8. *If job J has arrived but is not completed at time t, $s_t(J)$ is the (next) time at which RSPT will start J, based on the jobs that have arrived until time t. For a job J that is completed at time t, $s_t(J)$ is the last time at which J was started (i.e. the time when it was started and not interrupted anymore). (For a job J that has not arrived yet at time t, $s_t(J)$ is undefined.)*

We show in the full paper that $s_t(J)$ cannot decrease between one event and the next. Hence, for a job J in $Q(t)$, $s_t(J)$ is the earliest possible time that RSPT will start to run J.

Definition 9. *A job J is interruptable at time t, if $s_t(J) < 2w(J)$ and $t \leq \frac{2}{3}(s_t(J) + w(J))$.*

I. e. a job J is interruptable if it is still possible that RSPT will interrupt J after time t (cf. Property R5).

Definition 10. $BEFORE_t(J)$ is the set of jobs that RSPT completes before $s_t(J)$ (based on the jobs that have arrived at or before time t). $b_t(J)$ is the total size of jobs in $BEFORE_t(J)$. $\ell_t(J)$ is the size of the largest job in $BEFORE_t(J)$.

Clearly, $b_t(J)$ and $\ell_t(J)$ can only increase over time, and $\ell_t(J) \leq b_t(J)$ for all times t and jobs J.

During our analysis, we will maintain an *estimate* on the starting time of each job J in the schedule of OPT, denoted by $s_t^{OPT}(J)$. We describe later how we make and update these estimates. We will maintain the following as part of our invariant (which will be defined in section 7). Denote the actual optimal completion time of a job J by OPT(J). Then at the time t of an event,

$$\sum_{J:r(J)\leq t} \text{OPT}(J) \geq \sum_{J:r(J)\leq t} (s_t^{OPT}(J) + w(J)) \qquad (2)$$

This equation implies that at the end of the sequence, $\text{OPT}(\sigma) \geq \sum_J (s_t^{OPT}(J) + w(J))$. We will use the following Lemma to calculate initial values of $s_t^{OPT}(J)$ for arriving jobs in such a way that (2) holds.

Lemma 2. *For a given time t, denote the most recent arrival time of a job by $t' \leq t$. Denote the job that OPT is running at time t' by $\Phi(t')$, and its remaining unprocessed jobs by $\Psi(t')$. The total completion time of OPT of the jobs in $\Psi(t')$ is at least the total completion time of these jobs in the schedule where those jobs are run consecutively in order of increasing size after $\Phi(t')$ is completed.*

Proof. The schedule described in the lemma is optimal in case no more jobs arrive after time t (Event assumption 1). If other jobs do arrive after time t, it is possible that another order for the jobs in $\Psi(t')$ is better overall. However, since this order is suboptimal for $\Psi(t')$, we must have that the total completion time of the jobs in $\Psi(t')$ is then not smaller. □

The fact that the optimal schedule is not known during the analysis of an event is also the reason that we check that (2) is satisfied instead of checking $\text{OPT}(J) \geq s_t^{OPT}(J) + w(J)$ for each job J separately.

Definition 11. $D_t(J) = s_t(J) - s_t^{OPT}(J)$ is the delay of job J at time t.

6 Credits

The credit of job J at time t is denoted by $K_t(J)$. A job will be assigned an initial credit at the first event after its arrival. The idea is that the credit of a job indicates how much its execution can still be postponed by RSPT without violating the competitive ratio of 3/2: if a job has δ credit, it can be postponed by δ time. At the end of each run-interval $I = (s, t]$, each job $J_i(I)$ in $ARRIVE(I)$ receives an initial credit of

$$\frac{1}{2}\left(s^{OPT}(J_i(I)) + w_i(I)\right) - D(J_i(I)) \qquad i = 1, \ldots, k(I). \qquad (3)$$

If at time t a (non-zero) interval ends in which RSPT was idle, or $t = 0$, then suppose $Q(t) = \{J_1, \ldots, J_k\}$ where $w_1 \leq \ldots \leq w_k$. The initial credit of job J_i in $Q(t)$ is then $\frac{1}{2}t + \frac{1}{2}\sum_{j=1}^{i} w(J_j)$ $(i = 1, \ldots, k(I))$.

This is a special case of (3): by Lemma 2 and Event assumption 1, OPT will run the jobs in $Q(t)$ in order of increasing size, hence $s_t^{\text{OPT}}(J_i) \geq s_t(J_i)$ for $i = 1, \ldots, k$. Therefore $D(J_i) \leq 0$ for $i = 1, \ldots, k$. Moreover, by definition of RSPT we have $s_t(J_i) = t + \sum_{j=1}^{i-1} w_i$ for $i = 1, \ldots, k$.

For the competitive ratio, it does not matter how much credit each individual job has, and we will often transfer credits between jobs as an aid in the analysis. During the analysis of events, apart from transferring credits between jobs, we will also use the following rules.

C1. If $s_t(J)$ increased by δ since the previous event, then $K(J)$ decreases by δ.
C2. If the estimate $s_t^{\text{OPT}}(J)$ increased by δ since the previous event, then $K(J)$ increases by $\frac{3}{2}\delta$.

As stated before, $s_t(J)$ cannot decrease. We will only adjust $s_t^{\text{OPT}}(J)$ in a few special cases, where we can show that (2) still holds if we increase $s_t^{\text{OPT}}(J)$. Both rules follow directly from (3): it can be seen that if $s_t(J)$ or $s_t^{\text{OPT}}(J)$ increases, J should have received more credit initially.

Theorem 1. *Suppose that after RSPT completes any input sequence σ, the total amount of credit in the jobs is nonnegative, and (2) holds. Then RSPT maintains a competitive ratio of $3/2$.*

Proof. We can ignore credit transfers between jobs, since they do not affect the total amount of credit. Then each job has at the end credit of $K(J) = \frac{1}{2}(s_t^{\text{OPT}}(J) + w(J)) - (s_t(J) - s_t^{\text{OPT}}(J))$, where we use the final (highest) value of $s_t^{\text{OPT}}(J)$ for each job J, and the actual starting time $s_t(J)$ of each job. This follows from (3) and the rules for increasing job credits mentioned above.

Thus if the total credit is nonnegative, we have $\sum_J (s(J) - s_t^{\text{OPT}}(J)) \leq \sum_J \frac{1}{2}(s_t^{\text{OPT}}(J) + w(J)) \Rightarrow \sum_J s(J) \leq \frac{3}{2} \sum_J s_t^{\text{OPT}}(J) + \frac{1}{2} \sum_J w(J) \Rightarrow \text{RSPT}(\sigma) = \sum_J (s(J) + w(J)) \leq \frac{3}{2} \sum_J (s_t^{\text{OPT}}(J) + w(J)) \leq \frac{3}{2}\text{OPT}(\sigma)$. □

Calculating the initial credit The only unknowns in (3) are $s_t^{\text{OPT}}(J_i(I))$ ($i = 1, \ldots, k(I)$). If there is an interruption at time t, Lemma 2, together with the job that OPT is running at time t, gives us a schedule for OPT that we can use to calculate $s_t^{\text{OPT}}(J_i(I))$ for all i (if OPT uses a different schedule, its overall cost is not lower, so (2) still holds). We also use the following Event assumption.

Event assumption 2 *If the run-interval I ends in a completion, all jobs in $ARRIVE(I)$ arrive no later than the time at which OPT completes $J_{f_{\text{OPT}}(I)}(I)$.*

It is known that all jobs in $ARRIVE(I)$ arrive no later than at time $t(I)$. Moreover, by the time OPT completes $J_{f_{\text{OPT}}(I)}(I)$, RSPT will not interrupt $J(I)$ anymore by Property R2. Hence Event assumption 2 does not influence RSPT's decisions or its total completion time. It is only used so that we can apply

Lemma 2 to calculate lower bounds for the completion times of OPT of these jobs. Since the optimal total cost cannot increase when release times are decreased, we have that (2) holds. Note that both after a completion and after an interruption, the schedule of OPT is not completely known even with these assumptions, because we do not know which job OPT was running at time t. Therefore we still need to consider several off-line schedules in the following analysis.

We now describe three situations in which credit is required, and try to clarify some of the intuition behind the invariant that will be defined in Section 7.

Interruptions Suppose a job J of size x is interrupted at time r_1, because job J_1 arrives, after starting at time s. Then $s < 2x$. J_1 will give away credit to J, J^2, J^3 and J^4 as follows:

s	$[0, x]$	$[0, x]$	$(x, \frac{3}{2}x]$	$(x, \frac{3}{2}x]$	$(\frac{3}{2}x, 2x)$
r_1	$[0, x]$	$(x, \frac{4}{3}x]$	$(x, \frac{3}{2}x]$	$(\frac{3}{2}x, \frac{5}{3}x]$	$(\frac{3}{2}x, 2x)$
J	$r_1 - s$	$r_1 + w_1 - s$	$r_1 + w_1 - s$	$r_1 + w_1 - s$	$r_1 + w_1 - s$
J^2	0	$r_1 - x$	$r_1 - s$	$r_1 - s$	$r_1 - s$
J^3	0	0	0	$r_1 - \frac{3}{2}x$	$r_1 - s$
J^4	0	0	0	$r_1 - \frac{3}{2}x$	$r_1 - s$
$\sum$	$r_1 - s$	$2r_1 + w_1 - (s+x)$	$2(r_1 - s) + w_1$	$4r_1 + w_1 - 2s - 3x$	$4(r_1 - s) + w_1$

We have the following properties.

INT1 The amount of lost processing time due to this interruption is $r_1 - s$. This is at most $\frac{2}{3}(s + x) - s = \frac{2}{3}x - \frac{s}{3}$, which is monotonically decreasing in s.

INT2 The size of J_1 is w_1. This is at most $\frac{2}{3}(s+x) - r_1 < \frac{2}{3}(s+x) - s = \frac{2}{3}x - \frac{s}{3}$, which is monotonically decreasing in s.

So, in the above table, J_1 appears to give away more credit if s is larger, but a) it has more (this follows from (3)); b) it needs less (we will explain this later); and c) $r_1 - s$ is smaller. From the table we can also see how much credit is still missing. For instance if $s < x$ and $x < r_1$, then J^2 receives $r_1 - x$ from J_1, but it lost $r_1 - s$ because it now starts $r_1 - s$ time later. We will therefore require that in such a case, J^2 has at least $x - s$ of credit itself, so that it still has nonnegative credit after this interruption. In general, any job that does not get all of its lost credit back according to the table above, must have the remaining credit itself. We will formalize this definition in Section 7.

Completions Suppose a job J completes at time $s + x$. We give the following property without proof.

COM1 The jobs in $ARRIVE(I)$ (where $I = (s, s + x]$) need to get at most $\frac{1}{2}(x - b_s(J))$ of credit from J.

By this ("needing" credit) we mean that the amount of credit those jobs receive initially, together with at most $\frac{1}{2}(x - b_s(J))$, is sufficient for these jobs to satisfy the conditions that we will specify in the next section.

Small jobs As long as a job J has not been completed yet, it is possible that smaller jobs than J arrive that are completed before J by RSPT. If OPT completes them after J, then $D_t(J)$ increases.

7 The Invariant: Analysis of an Event

From the previous section we see that for a job, sometimes credit is required to pay for interruptions of jobs that are run before it, and sometimes to make sure that jobs that arrive during its final run have sufficient credit (COM1). We will make sure that each job has enough credit to pay both for interruptions of jobs before it and for its own completion.

For a job J, we define the *interrupt-delay* associated with an interruption as the amount of increase of $D_t(J)$ compared to the previous event. This amount is at most $t(I) - s(I)$ at the end of a run-interval I. (It is less for a job J if $s_{OPT}(J)$ also increases).

Credit can also be required because the situation marked "Small jobs" in Section 6 occurs. The *small job-delay* of J associated with an event at time t is the total size of jobs smaller than J in $ARRIVE(I)$ that are completed before J by RSPT and after J by OPT.

When considering the credit of a job J, we will make a distinction between *interrupt-credit* $K_{INT}(J)$, which is used to pay for interrupt-delays whenever they occur, and *completion-credit* $K_{COM}(J)$, which is used to "pay for the completion" of J (see above). (We do not reserve credit for small job-delays since they will be paid for by the small jobs that cause it.) Accordingly, we now make two important definitions.

Definition 12. $N_{INT}(J,t)$ *is the maximum amount of credit that J may need to pay for (all the) interruptions of jobs that RSPT completes before it, as it is known at time t.*

I.e. this is the amount of credit needed if all the jobs before J that have arrived before time t are interrupted as often as possible. If other jobs arrive later, $N_{INT}(J,t)$ can change. For any job J that is already completed at time t by RSPT, $N_{INT}(J,t) = 0$.

Definition 13. $N_{COM}(J,t)$ *is the maximum amount of credit that J may need to pay for its own completion, as known at time t (based on arrived jobs).*

This amount can also change over time, namely if $s(J)$ increases. Again, if J is completed before time t, then $N_{COM}(J,t) = 0$.

Consider an event at time t. Suppose $Q(t) = \{J_1, \ldots, J_k\}$, where $w_1 \leq w_2 \leq \ldots \leq w_k$. We write $s_i = s_t(J_i) = t + \sum_{j=1}^{i-1} w_j$. From the table in Section 6 it can be seen that $N_{INT}(J_i, t) = \max(0, w_{i-1} - s_{i-1}) + \max(0, \frac{3}{2}w_{i-2} - s_{i-2}) + \max(0, 2w_{i-4} - t)$, where each maximum only appears if the corresponding job exists. For the third maximum in this equation, note that the total interrupt-delay of J_i caused by interruptions of the jobs $J_1, \ldots, J_{i-4}$ is at most $2w_{i-4} - t$

after time t, since RSPT starts to run J_1 at time t and does not interrupt any of the jobs $J_1, \ldots, J_{i-4}$ after time $2w_{i-4}$ by Property 4. In the full paper, we show $N_{INT}(J_i, t) \leq \max(0, w_{i-1} + \frac{1}{2}w_{i-2} + \frac{1}{2}w_{i-4} - t)$.

Using property COM1, we have $N_{COM}(J_i, t) = \max\left(0, \frac{1}{2}(w_i - b_t(J_i))\right)$. $N_{COM}(J_i, t)$ can only decrease over time (since s_i and $b_t(J_i)$ only increase). For all jobs J that have arrived at time t, we will maintain

$$K_t(J) \geq N_{COM}(J, t) + N_{INT}(J, t). \tag{4}$$

This means that each job will be able to pay for the specified parts of its interrupt-delay and for its completion. I. e. the total credit of each job will be sufficient to pay for both these things. We now define our invariant, that will hold at specific times t in the execution, and in particular when a sequence is completed:

Invariant

At time t, for all jobs that have arrived, (4) holds; furthermore, (2) holds.

Theorem 2. $\mathcal{R}(\text{RSPT}) \leq 3/2$.

Proof outline. The proof consists of a case analysis covering all possible interruptions and completions, which we have to omit. "All possible" refers to both the times at which these events occur, and the possible schedules of the off-line algorithm.

For every possible event, we will give a time at which the above invariant holds again, assuming that it held after the previous event. This will be no later than at the completion of the last job in σ. At that time, the invariant implies that all completed jobs have nonnegative credit, since for completed jobs we have $N_{COM}(J, t) = N_{INT}(J, t) = 0$. Also, (2) holds. We can then apply Theorem 1. □

Final comments The remainder of the proof consists of a large amount of "bookkeeping". In order to ensure that the invariant holds again after an event, we will often transfer credits between jobs. Also, we will use the credit that some jobs must have because the invariant was true previously, to pay for their interrupt-delay or for their completion. We need to take into account that $N_{INT}(J, t)$, $s_t^{OPT}(J)$ etc. of some jobs that arrived before or at the previous event can change as a result of the arrival of new jobs, compared to the calculations in that event (that were made under the assumption that STATIC held).

In the full paper [11], we begin by showing that (2) holds and that (4) holds for the J-large jobs besides J. After this, in the analysis of events, it is sufficient to check (or ensure) that J and the jobs in $ARRIVE(I)$ satisfy (4). We do this for all possible events.

There is only one case in which the invariant does not hold again immediately after the event. This is a slow interruption of a job J, where $s(J) \leq \frac{2}{3}w(J)$, and

OPT runs no large jobs before the jobs in $ARRIVE(I)$. If such an event occurs at some time t, we show that the invariant is restored no later than when RSPT has completed the second-smallest job in $ARRIVE(I)$. Compare this situation to the second case of Lemma 1.

For all other events, we show how to transfer credits so that each job satisfies (4) immediately after the event.

Acknowledgement

The authors wish to thank Leah Epstein for interesting discussions.

References

[1] C. Chekuri, R. Motwani, B. Natarajan, and C. Stein. Approximation techniques for average completion time scheduling. In *Proceedings of the 8th Annual ACM-SIAM Symposium on Discrete Algorithms*, pages 609–618, New York / Philadelphia, 1997. ACM / SIAM.

[2] Leah Epstein and Rob van Stee. Lower bounds for on-line single-machine scheduling. In *Proc. 26th Symp. on Mathematical Foundations of Computer Science*, volume 2136 of *Lecture Notes in Computer Science*, pages 338–350, 2001.

[3] H. Hoogeveen, C. N. Potts, and G. J. Woeginger. On-line scheduling on a single machine: maximizing the number of early jobs. *Operations Research Letters*, 27:193–197, 2000.

[4] J. A. Hoogeveen and A. P. A. Vestjens. Optimal on-line algorithms for single-machine scheduling. In W. H. Cunningham, S. T. McCormick, and M. Queyranne, editors, *Integer Programming and Combinatorial Optimization, 5th International IPCO Conference, Proceedings*, volume 1084 of *Lecture Notes in Computer Science*, pages 404–414. Springer, 1996.

[5] C. A. Phillips, C. Stein, and J. Wein. Scheduling jobs that arrive over time. In *Proceedings of the 4th Workshop on Algorithms and Data Structures (WADS'95)*, volume 955 of *Lecture Notes in Computer Science*, pages 86–97. Springer, 1995.

[6] David B. Shmoys, Joel Wein, and David P. Williamson. Scheduling parallel machines on-line. In Lyle A. McGeoch and Daniel D. Sleator, editors, *On-line Algorithms*, volume 7 of *DIMACS Series in Discrete Mathematics and Theoretical Computer Science*, pages 163–166. AMS/ACM, 1991.

[7] W. E. Smith. Various optimizers for single-stage production. *Naval Research Logistics Quarterly*, 3:59–66, 1956.

[8] L. Stougie. Unpublished manuscript, 1995.

[9] L. Stougie and A. P. A. Vestjens. Randomized on-line scheduling: How low can't you go? Unpublished manuscript.

[10] Marjan van den Akker, Han Hoogeveen, and Nodari Vakhania. Restarts can help in the on-line minimization of the maximum delivery time on a single machine. *Journal of Scheduling*, 3:333–341, 2000.

[11] R. van Stee and J. A. La Poutré. Minimizing the total completion time on a single on-line machine, using restarts. Technical Report SEN-R0211, CWI, Amsterdam, June 2002.

[12] Arjen P. A. Vestjens. *On-line Machine Scheduling*. PhD thesis, Eindhoven University of Technology, The Netherlands, 1997.

High-Level Filtering for Arrangements of Conic Arcs
(Extended Abstract)*

Ron Wein**

School of Computer Science
Tel Aviv University
wein@post.tau.ac.il

Abstract. Many computational geometry algorithms involve the construction and maintenance of planar arrangements of conic arcs. Implementing a general, robust arrangement package for conic arcs handles most practical cases of planar arrangements covered in literature. A possible approach for implementing robust geometric algorithms is to use exact algebraic number types — yet this may lead to a very slow, inefficient program. In this paper we suggest a simple technique for filtering the computations involved in the arrangement construction: when constructing an arrangement vertex, we keep track of the steps that lead to its construction and the equations we need to solve to obtain its coordinates. This construction history can be used for answering predicates very efficiently, compared to a naïve implementation with an exact number type. Furthermore, using this representation most arrangement vertices may be computed approximately at first and can be refined later on in cases of ambiguity. Since such cases are relatively rare, the resulting implementation is both efficient and robust.

1 Introduction

Given a collection $\mathcal{C}$ of curves in the plane, the *arrangement* of $\mathcal{C}$, denoted $\mathcal{A}(\mathcal{C})$ is the subdivision of the plane into cells of dimension 0 (*vertices*), 1 (*edges*) and 2 (*faces*) induced by the curves in $\mathcal{C}$. Constructing arrangements of curves in the plane is a basic problem in computational geometry. Applications relying on arrangements arise in fields such as robotics, computer vision and computer graphics. Many algorithms for constructing and maintaining arrangements under various conditions have been published. For a review see [17].

---

* Work reported in this paper has been supported in part by the IST Programme of the EU as a Shared-cost RTD (FET Open) Project under Contract No IST-2000-26473 (ECG - Effective Computational Geometry for Curves and Surfaces), by The Israel Science Foundation founded by the Israel Academy of Sciences and Humanities (Center for Geometric Computing and its Applications), and by the Hermann Minkowski – Minerva Center for Geometry at Tel Aviv University.
** This paper is part of the author's thesis, under the supervision of Prof. Dan Halperin.

The CGAL library [1] offers a very flexible framework for constructing planar arrangements for any type of curves: the library includes an arrangement template that can be instantiated with a curve type and a *traits* class that supports several predicates and operations on curves of that specific type. Having implemented the traits class and instantiated the template, the user can easily create and manipulate planar arrangements. So far, several traits classes have been implemented and are included in CGAL: for arrangements of line segment, for circles and for canonical parabolas. Since all those curve types are really special cases of conic arcs, it seems only natural to try and implement a traits class for conic arcs, supplying a unified approach rather than dealing with special cases and allowing more complicated arrangements. Conic curves, which are the simplest form of non-linear planar curves, have several properties that make them quite easy to deal with: they are always convex (see Section 2) and never self-intersecting. At the same time, most planar arrangements found in literature fall into the category of conic arc arrangements.

Typically in the literature, all atomic predicates and operations on curves are considered to take "constant time". In practice, the generalization of the traits class to handle arbitrary conic arcs leads to some non-trivial computations with algebraic numbers that may be very time-consuming, especially when exact number types are used to ensure robustness (since CGAL uses the exact computation paradigm to ensure the robustness of its algorithms). In this paper we shall review the difficulties raised by the implementation of the conic arcs' traits class and suggest a high-level filtering method that reduces the number of computations by utilizing all available geometric and algebraic information, trying to avoid unnecessary predicates and speed up costly computations.

When constructing an arrangement of curves we have to manipulate the input curves provided by the user or the application and construct new objects from them. Typically, two kinds of objects may be created: *sub-curves* and arrangement *vertices*. An input curve usually has two end-points that eventually become arrangement vertices. Each intersection point of two (or more) curves also becomes an arrangement vertex. For many practical algorithms, it is useful to compute the *vertical decomposition* of the arrangement (see Figure 1 for an example) — this enables us to perform point location queries on the arrangement in $O(\log n)$ time where n is the number of curves (see e.g. [25]). For that purpose it is more convenient to handle only x-monotone curves, so we add more vertices that split the original input curves into x-monotone sub-curves.

The key idea that we propose for speeding up the implementation is to keep the *construction history* of the new objects we create. For each arrangement vertex we create, we keep track of the conic curves that led to its construction as well as the polynomial equations that define its coordinates. We then use the construction history for quick filtering of unnecessary computations. When comparing the coordinates of such vertices we can actually use the GCD of their generating polynomials to filter out unnecessary computations. Since the GCD is usually a low degree polynomial we can construct arrangements of arbitrary conic arcs in an exact manner, without assuming general position and while keeping reasonable running times.

1.1 Related Work

The work described in this paper is partially based on CGAL's arrangement package [18] and the Conic class implemented by T. Andres [7]. Some of the notations used in this paper were taken from Andres' work.

The exact computation paradigm is a popular approach towards the implementation of robust geometric algorithms, yet it may lead to time-consuming implementations (see [26]). In order to produce algorithms that are both fast and exact, some kind of filtering must be used.

The LEDA [3], [24] and CORE [2], [19] libraries both support number types that can accurately represent algebraic expressions (leda_real in LEDA and EXPR in CORE). Both libraries use low-level arithmetic filtering to evaluate the expression values (for more details see Section 3.2; see also [26] for an extensive review). However, while arithmetic filtering may help making computations faster, high-level filtering tries to completely avoid such costly computations all together.

The MAPC library [21] implements arrangements of curves using arithmetic filtering: expressions are evaluated using the PRECISE floating-point representation with an arbitrary mantissa length. Using this technique arrangements of algebraic curves of arbitrary degrees can be constructed, but some general position assumptions are made — namely no two intersection points lie too close to one another. The technique presented in this paper takes advantage of the low degree of the conic curves to allow robust construction of any arrangement, including degeneracies of any kind.

Karasick et al. [20] were the first to use filtering for the implementation of Delaunay triangulations. Fortune and Van-Wyk [13] used filtering techniques for implementing a variety of algorithms on linear objects. Funke and Mehlhorn have implemented high-level filtering for line segments in LOOK [14]: they use lazy evaluations for the newly constructed objects (such as intersection points), such that costly computations of the exact representations of those objects are carried out only when they are needed. At present, LOOK deals only with linear objects.

Devillers et al. [11] suggest a representation of circular arcs that enables efficient arithmetic filtering on the predicates needed for constructing arrangements of arcs. An arc is represented by its supporting circle and two lines that intersect it — this allows implicit representation of intersection points. This representation leads to predicates that involve signs of resultants that can be filtered using a low-level arithmetic filter. Unfortunately, this approach cannot be easily extended for handling arbitrary conic arcs.

Geismann et al. [15] have implemented arrangements of curves with degrees up to 4 using an extension of *interval arithmetic* (see e.g. [8]) to $\mathbb{R}^2$: the event points are separated into boxes and precise computations are performed only if two (or more) boxes cannot be separated. This approach reduces most predicates and computations to a simple query regarding the position of two boxes.

The rest of the paper is organized as follows: In Section 2 we introduce the definitions and notations used in the paper. Section 3 reviews the atomic predicates and methods needed for implementing arrangements of conic arcs and discusses the selection of number types. High-level filtering for arrangements of conic arcs is presented in Section 4. Finally, experimental results are presented in Section 5.

2 Preliminaries

A planar *conic curve* C is the locus of all points (x, y) in the plane satisfying the equation $rx^2 + sy^2 + txy + ux + vy + w = 0$. A conic curve can therefore be characterized by the 6-tuple $\langle r, s, t, u, v, w \rangle$.

There are actually three types of planar conic curves. The type of a conic curve C can be determined by checking the sign of $\Delta_C = 4rs - t^2$. If $\Delta_C > 0$ then C is an *ellipse*; If $\Delta_C < 0$ then C is a *hyperbola*; If $\Delta_C = 0$ then C is a *parabola*. Notice that the special case of a curve of degree 1, i.e. $r = s = t = 0$ (yielding the line $ux + vy + w = 0$), may be considered as a degenerate parabola. Indeed it is convenient that our conic arc package supports line segments.

A finite *conic arc* $\bar{a}$ may be either one of the following:
- A full conic curve C. Since we only handle bounded arcs, C must be an ellipse.
- The 3-tuple $\langle C, p_s, p_t \rangle$ where p_s, p_t, the edge end-points, both lie on the conic boundary. In case of a conic arc of degree 2 (i.e. not a line segment), the coefficients of C actually determine whether we go from p_s to p_t in a clockwise or in a counter-clockwise direction.

In both cases, C is called the *underlying conic* of $\bar{a}$.

3 Implementing Arrangements of Conic Arcs

Arrangements of conic arcs are a natural generalization of many specialized types of planar arrangements that arise in practice: Arrangements of *line segments* are needed in numerous applications (see e.g. [4] for applications of motion-planning; also see also [6] and [16] for other applications of line segment arrangements). Arrangements of *canonical parabolas* (i.e. conic curves of the form $y = rx^2 + ux + w$) are used for solving dynamic proximity queries. Motion planning algorithms for a disc robot with polygonal obstacles involve arrangements of *arcs of circles* and *line segments* for offsetting the polygonal objects (see [12]). A new algorithm for computing the detour of polygonal curves requires the construction of an arrangement of *hyperbolic arcs* (see [5]).

While restricting ourselves to special sub-classes may yield a simpler, faster implementation, the generalization to conic arcs does not complicate the software by much — since conic objects are still fairly simple from a geometric point of view. We can therefore supply a unified approach for handling many practical cases of planar arrangements.

3.1 Arrangements in CGAL

The CGAL library offers a convenient framework for implementing planar arrangements: the user can use the arrangement template, having to instantiate it with a curve type and a traits class. The traits class must support some basic predicates and operations on the curves, such as:

- Checking whether a curve is x-monotone.
- Splitting a curve into x-monotone sub-curves.
- Checking whether a point is on, above or below a curve.
- Finding all points on a curve with a given x coordinate.
- Computing the intersection points between two curves.
- Deciding which x-monotone sub-curve is above the other at a given x coordinate.

For more details on the usage of the arrangement template in CGAL and the requirements from the traits class, see [10]. In [27] one can find a detailed overview of the steps needed to implement the conic traits class. It is important to notice that solutions of quadratic equations are involved in the implementation of most methods for conic arcs, while for the computation of intersection points one needs to solve quartic equations (i.e. finding the roots of polynomials of degree 4). At any case, the equation coefficients are obtained by applying simple arithmetic operations on the input conic coefficients.

3.2 Selecting the Number Type

The most fundamental requirement from the arrangement implementation is to be robust — that is, it should be able to run on all inputs, yielding geometrically and topologically correct results. To achieve this goal in CGAL, one needs to use an exact number type for precise computations. This is quite practical for implementing an arrangement of line segments: if we confine ourselves to segments with rational end-points (i.e. (x, y) where x, y are both rational numbers), then all vertices (including intersection points) can be expressed with rational coordinates.

Unfortunately, in the general case the intersection points of two conic curves have irrational coordinates even if all conic coefficients are rational. We therefore need to use number types that can handle algebraic numbers in an accurate manner. Since we can gain nothing from restricting ourselves to rational inputs, we actually drop this requirement and allow algebraic expressions in the input (namely in the coefficients of an input conic curve $\langle r, s, t, u, v, w \rangle$ and the coordinates of its end-points may all be algebraic numbers).

The LEDA and CORE libraries both offer a convenient representation for algebraic numbers, using more or less the same concepts: the number is represented using an expression DAG whose leaf nodes represent constant numbers and inner nodes represent operations on their child nodes (either $+, -, \times, \div$ or $\sqrt[k]{\cdot}$). A key operation on this expression DAG is determining its sign, an operation which can be efficiently implemented based on the theory of separation bounds. First, some quick calculations are carried out to obtain an initial approximation of the expression value. In case the result is far enough from 0, the sign can be safely determined. Otherwise, the expression value is iteratively refined with increasing bit-length until reaching a "safe" decision. In the worst case, we will need $\Theta(\log(\frac{1}{\varepsilon}))$ bits of precision where ε is the separation bound for the expression tree. Consequently, a naïvely written algorithm using such an exact number type may suffer from poor performance, unless it avoids unnecessary computations of expressions that may be known in advance to equal 0.

But even if we try to avoid unnecessary computations, we are still left with many computations needed to be carried out, yielding very complicated expression DAGs. In recent years separation bounds of algebraic expression have been studied and tightened (see [9] and [22] for recent results). However, those bounds may still require hundreds (or even thousands) of bits of precision when dealing with expressions resulting from conic curve manipulations, since such expressions usually have a relatively high degree (the degree of an algebraic expression is a multiplication of the degree of all algebraic numbers involved) and therefore require a very small separation bound, since the separation bound exponentially decreases with the degree (see [9]).

It is therefore clear that we should strive to work with the machine double-precision floating-point number type whenever possible. Indeed, in most cases this limited precision is sufficient, especially when there are no degeneracies in the input. Yet our algorithm should yield a topologically correct output even if the input contains degeneracies. In this case, we would like to carry out the heavy, time-consuming computations only when they are really necessary. This means that our arrangements may include objects that have been approximately computed and objects that have been exactly computed. In the next section we show how we can manage such a representation of the arrangement.

4 High-Level Filtering

4.1 Vertex Types in the Arrangement

By taking a close look at the arrangement vertices, we can easily divide them into three categories (see Figure 1), depending on the operation that constructed them. For each type of vertex we need to keep track of the conic curves that were involved to the construction of the vertex. Here are the types of vertices and the *construction history* we need to keep for each type:

1. *Original end-points* of arcs, supplied as part of the input. Since original end-points are part of the input, they have no construction history. However, for practical reasons, it is convenient to keep a pointer to the underlying curve.
2. *Vertical tangency points*, created by dividing input arcs into x-monotone sub-arcs. We shall keep a pointer to the underlying conic of the originating arc.

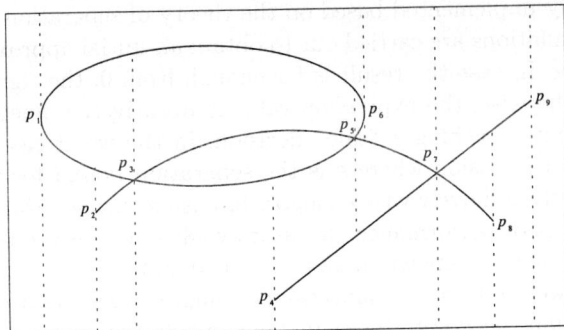

Fig. 1. Vertex types in an arrangement of conic arcs: p_2, p_4, p_8, p_9 are *original end-points*; p_1, p_6 are *vertical tangency points*; p_3, p_5, p_7 are *intersection points*. The vertical decomposition into pseudo-trapezoids is also shown

3. *Intersection points* between two arcs. In this case we need to keep pointers to the two underlying conics of the originating arcs.

A very important observation is that the coordinates of the vertical tangency points and of the intersection points can be expressed as roots of polynomials, whose coefficients are expressions involving the coefficients of the generating conics (for exact details see [27]). Given a vertex p which is a vertical tangency point or an intersection point, we shall denote the two *background polynomials* generating its x, y coordinates by $\xi_p(x)$ and $\eta_p(y)$ respectively. While we can actually store initial approximated values of the coordinates, the coefficients of the background polynomials will be represented using an exact number type.

4.2 Using the Construction History for High-Level Filtering

We can benefit from storing the construction history of the arrangement vertices in several ways. First, we can actually avoid costly evaluation of expressions whose value is zero: as explained before, if a point p lies on an arc $\bar{a}$ with an underlying conic C, then plugging p's coordinates in the equation that defines C should yield the result 0. This kind of calculation is very costly when using DAGs with separation bounds, especially if the coordinates of p are themselves non-trivial expressions. However, such zero-valued expressions usually occur when we try to check whether a point p, that was actually constructed by some operation on a conic arc whose underlying curve is C, is on the boundary of C — now if we have the construction history, we can avoid such unnecessary computations: If we wish to check whether a vertex p lies on the arc $\bar{a}$ whose underlying conic is C, we first check whether C appears in the construction history of p. Only if it does not we check whether $C(p) = 0$, and this computation should terminate quickly because it can usually be handled by the arithmetic filter.

Secondly, we can allow ourselves to work with inexact but faster number types. For example, working with double-precision floating-point numbers will be sufficient for approximating most arrangement vertices while maintaining the topological structure of the arrangement correct.

The ability to work with an approximate number type is very significant when we are coming to deal with intersection points. To find the coordinates of such points, we should find the roots of a polynomial $f(x)$ (and $f(y)$) of degree 4. A closed-form solution to this problem, known as Ferrari's formula, indeed exists, but its contains trigonometric functions and neither LEDA nor CORE can currently handle such expressions directly[1].

We therefore need to use approximate computations without associating them with exact expression DAGs only for the simple intersection points. It is possible to obtain a good approximation for the real-valued roots of $f(x)$ using Ferrari's method with double-precision floating-point numbers. It is important to notice that later on it is possible to refine each of those roots to any desired level of approximation using floating-point arithmetic with arbitrary mantissa length (the `leda_bigfloat` number type in LEDA implements such arithmetic, for example).

Lemma 1. *Given a polynomial $f(x)$ that is guaranteed not to have any multiple roots, let ζ_0 be an approximation to one of $f(x)$'s real-valued roots. Then this approximation can be refined to any desired level of precision using only simple arithmetic operations (namely $+, -, \times, \div$).*

Proof. Using Sturm sequences we can obtain a function $S_f : \mathbb{R} \to \mathbb{Z}$ such that for every interval $[a, b]$, $S_f(a) - S_f(b)$ is the number of real roots of $f(x)$ contained in the interval. The construction of Sturm sequences requires long division of polynomials and can be carried out using simple arithmetic operations. Given ζ_0, we can start from an interval $[z_l^0, z_r^0]$ that contains exactly one root using the approximate root value (that is, $S_f(z_l^0) - S_f(z_r^0) = 1$). We then refine this interval iteratively, where if we are given the segment $[z_l^k, z_r^k]$ after k iterations, we will proceed by bisecting it and selecting either $[z_l^{k+1} = z_l^k, z_r^{k+1} = \frac{1}{2}(z_l^k + z_r^k)]$ or $[z_l^{k+1} = \frac{1}{2}(z_l^k + z_r^k), z_r^{k+1} = z_r^k]$ that contains the root (again, using S_f). Thus, with every iteration we can obtain an extra bit of precision. □

For most practical cases, using approximated double-precision coordinates is sufficient, especially when we use the construction history to filter out unnecessary computations. It is necessary, however, to use exact values when comparing the x (or the y) coordinates of two points in order to ensure the correctness of the construction.

Proposition 1. *Given two vertices in an arrangement of conic arcs, it is possible to compare the x (or the y) coordinates of these vertices and to reach the*

---

[1] It is somewhat surprising, but in this case degeneracies in the input are actually easier to handle: If two curves are tangent and not only intersecting, then the polynomial $f(x)$ we need to solve to find the x coordinates of their intersection points (and also $f(y)$) has at least one multiple root. We can therefore compute $\gcd(f(x), f'(x))$, find all its roots in an exact manner (since it is a polynomial of degree 2 at most) and factor them out from $f(x)$, leaving us with a polynomial of degree 2 at most.

correct result, without the need to solve polynomials of a high degree (say more than 2).

Proof. Let us assume that we are given two points p and q and wish to compare their x coordinates. If we have approximations for the values x_p and x_q, we normally have error bounds δ_p and δ_q for those approximations. We can first check whether $|x_p - x_q| > \delta_p + \delta_q$ — in this case, the two values are well-separated from one another and we can just compare their approximated values.

In any other case, when the two x coordinates lie very close to one another, we should check the following cases:

- If both p and q are original end-points, we can just compare their coordinates — since they are part of the input, which is assumed to be correct.
- In case only one of the points (say q) is an original end-point, then the x coordinate of p is a root of the polynomial $\xi_p(x)$. We can first check whether $\xi_p(x_q) = 0$ — if so, both points have the same x coordinate.
- Similarly, if both points have construction histories we can simply check whether $\gcd(\xi_p, \xi_q) = 1$ — if so, the two x coordinates cannot be the same. If $\gcd(\xi_p, \xi_q) \neq 1$, in most cases it will be a polynomial of degree 2 at most so we can systematically find its roots. If any of the GCD roots is close enough to x_p (and x_q), we can conclude that the two points have the same x coordinate.

At any case, if we reach the decision that the x coordinates are not equal, we can iteratively refine the approximations for x_p, x_q, isolating each of the values in an increasingly refined interval until obtaining two disjoint intervals. In the very rare occasion that $\gcd(\xi_p, \xi_q)$ is of degree 3 or 4, we can use the same technique, but if we do not break the tie after a certain number of bits, we can conclude that the two point indeed have the same x coordinate.

How many bits of precision are sufficient? According to Loos [23], if $r_1, \ldots, r_n$ are the real roots of the polynomial $f(x)$ and $s_1, \ldots, s_m$ are the real roots of the polynomial $g(x)$, such that the maximal degree of $f(x), g(x)$ is d and the maximal value of all their coefficients is bounded by K, then if $\gcd(f, g) \neq 1$, the following holds:

$$\min_{\substack{1 \leq i \leq n \\ 1 \leq j \leq m}} |r_i - s_j| > K^{-d} \tag{1}$$

In our case we need to separate coordinates, that are actually roots of polynomials of degree at most 4 whose coefficients are bounded by $48M^5$, where M is a bound on the absolute value of the conic curve parameters (see [27] for details). Using the bound in Equation (1) it is clear that such roots should be at least $48^{-4}M^{-20}$ apart from one another. We therefore need at most $23 + 20\log_2 M$ bits of precision when refining the approximations of vertices in an arrangement of conic arcs. □

We can therefore ensure the correctness of our construction even without a mechanism that exactly handles root extraction from polynomials. It is worth mentioning that using the GCD of the background polynomials is a powerful technique that can be conveniently used to compare two vertices even if their coordinates can be computed exactly. If p and q are both vertical tangency points their x coordinates are solutions to quadratic equations, yet the separation bound for this comparison is about $68000^{-1}M^{-8}$ (the expression DAG has a degree of 4, since it contains two square-root operations). Using the GCD method one needs only a separation bound of about $20^{-1}M^{-6}$. The gap gets larger as the degrees of the underlying polynomials increase.

5 Experimental Results

Table 1 shows the construction times of several arrangements, using three strategies: A naïve implementation of the conic arcs' traits, an implementation of the traits that uses high-level filtering, and an implementation that uses filtering and intersection point caching. The results were obtained on a Pentium II 450 MHz machine with 528 Mb RAM.

The tng_circles_8, the rose_14 (Figure 2) and the packed_circles_41 input files contain collections of 8, 14 and 41 conic arcs respectively with many degeneracies. The mix_50 input file contains a collection of 50 random circular arcs and line segments. The tng_ellipses_9 input file (Figure 2) contains an arrangement of 9 tangent ellipses. The CGAL_logo_460 is comprised of 430 line segments and 30 circles. All those inputs have been chosen such that all the computations needed for the construction of the arrangement (including its vertical decomposition) involve only solutions of quadratic equations: the intersection points are of two circles, of a circle and a line or of two canonical ellipses. The hyper_ellip_18 input file (Figure 2) contains a collection of 16 hyperbolic arcs and two canonical ellipses, while the mix_20 input file which contains a collection of 20 random conic arcs of all types. Those arrangements contain cases where the computation of the intersection points leads to quartic equations. Since the naïve implementation is only capable of handling exact computations with the current version of LEDA, it cannot be used to construct such arrangements.

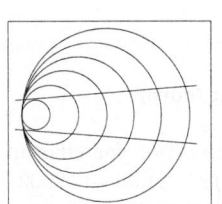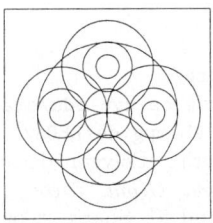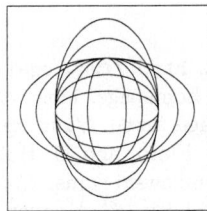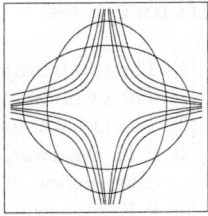

Fig. 2. Arrangements of circles and segments (from left to right): tng_circles_8; rose_14; tng_ellipses_9; hyper_ellip_18

Table 1. Construction times (in sec.) of some simple arrangements

Input file	Naïve conic arcs	Filtered conic arcs	Filtered and cached
tng_circles_8	3.19	0.9	0.88
tng_ellipses_9	14.55	4.03	0.96
rose_14	7.44	2.35	0.84
packed_circles_41	51.82	34.46	11.66
mix_50	56.01	27.11	9.55
CGAL_logo_460	638.87	86.70	38.95
hyper_ellip_18	N/A	81.68	5.52
mix_20	N/A	3.84	1.77

6 Conclusions

High-level filtering is a powerful technique that takes into consideration all available geometric knowledge in order to filter out the exact computations needed for the robust construction of planar arrangements of conic curves. The techniques used in this paper can be extended and used for constructing arrangement for algebraic curves of any degree. In this case we will encounter algebraic equations of higher degrees, yet the method of storing the background polynomials of vertex coordinates and computing GCDs of polynomials to identify degeneracies enables us using approximate coordinates while maintaining a topologically correct construction, as the GCD polynomials usually have a lower degree than the original polynomials.

Acknowledgement

I would like to thank Kurt Mehlhorn, Elmar Schömer, Stefan Funke and Nicola Wolpert for useful discussions on root separation and filtering.

References

[1] The CGAL project. http://www.cgal.org/.
[2] The CORE library homepage. http://www.cs.nyu.edu/exact/core/.
[3] The LEDA homepage. http://www.mpi-sb.mpg.de/LEDA/.
[4] P. K. Agarwal, E. Flato, and D. Halperin. Polygon decomposition for efficient construction of Minkowski sums. *Comput. Geom. Theory Appl.*, 21:39–61, 2002.
[5] P. K. Agarwal, R. Klein, C. Knauer, and M. Sharir. Computing the detour of polygonal curves. Technical Report B 02-03, Freie Universität Berlin, Fachbereich Mathematik und Informatik, 2002.
[6] Y. Aharoni, D. Halperin, I. Hanniel, S. Har-Peled, and C. Linhart. On-line zone construction in arrangements of lines in the plane. In *Proc. of the 3rd Workshop of Algorithm Engineering*, volume 1668 of *Lecture Notes Comput. Sci.*, pages 139–153. Springer-Verlag, 1999.
[7] T. Andres. Specification and implementation of a conic class for CGAL. M.Sc. thesis, Institute for Theoretical Computer Science, Swiss Federal Institute of Technology (ETH) Zurich, 8001 Zurich, Switzerland, 1999.

[8] H. Brönnimann, C. Burnikel, and S. Pion. Interval arithmetic yields efficient arithmetic filters for computational geometry. In *Proc. 14th Annu. ACM Sympos. Comput. Geom.*, pages 165–174, 1998.

[9] C. Burnikel, S. Funke, K. Mehlhorn, S. Schirra, , and S. Schmitt. A separation bound for real algebraic expressions. In *Proc. ESA 2001*, pages 254–265, 2001.

[10] *The CGAL User Manual, Release 2.3*, 2001. www.cgal.org.

[11] O. Devillers, A. Fronville, B. Mourrain, and M. Teillaud. Exact predicates for circle arcs arrangements. In *Proc. 16th Annu. ACM Sympos. Comput. Geom.*, pages 139–147, 2000.

[12] G. Elber and M.-S. Kim, editors. *Special Issue of Computer Aided Design: Offsets, Sweeps and Minkowski Sums*, volume 31. 1999.

[13] S. Fortune and C. J. van Wyk. Static analysis yields efficient exact integer arithmetic for computational geometry. *ACM Trans. Graph.*, 15(3):223–248, July 1996.

[14] S. Funke and K. Mehlhorn. LOOK: A lazy object-oriented kernel for geometric computation. In *Proc. 16th Annu. ACM Sympos. Comput. Geom.*, pages 156–165, 2000.

[15] N. Geismann, M. Hemmer, and E. Schömer. Computing a 3-dimensional cell in an arrangement of quadrics: Exactly and actually! In *Proc. 17th Annu. ACM Sympos. Comput. Geom.*, pages 264–273, 2001.

[16] M. Goldwasser. An implementation for maintaining arrangements of polygons. In *Proc. 11th Annu. ACM Sympos. Comput. Geom.*, pages C32–C33, 1995.

[17] D. Halperin. Arrangements. In J. E. Goodman and J. O'Rourke, editors, *Handbook of Discrete and Computational Geometry*, chapter 21, pages 389–412. CRC Press LLC, Boca Raton, FL, 1997.

[18] I. Hanniel. The design and implementation of planar arrangements of curves in CGAL. M.Sc. thesis, Dept. Comput. Sci., Tel Aviv University, Tel Aviv, Israel, 2000.

[19] V. Karamcheti, C. Li, I. Pechtchanski, and C. Yap. A core library for robust numeric and geometric computation. In *15th ACM Symp. on Computational Geometry, 1999*, pages 351–359, 1999.

[20] M. Karasick, D. Lieber, and L. R. Nackman. Efficient Delaunay triangulations using rational arithmetic. *ACM Trans. Graph.*, 10(1):71–91, Jan. 1991.

[21] J. Keyser, T. Culver, D. Manocha, and S. Krishnan. MAPC: a library for efficient manipulation of algebraic points and curves. In *Proc. 15th Annu. ACM Sympos. Comput. Geom.*, pages 360—369, 1999. http://www.cs.unc.edu/~geom/MAPC/.

[22] C. Li and C. Yap. A new constructive root bound for algebraic expressions. In *12th ACM-SIAM Symposium on Discrete Algorithms (SODA)*, pages 496–505, 2001.

[23] R. Loos. Computing in algebraic extensions. In B. Buchberger, G. E. Collins, R. Loos, and R. Albrecht, editors, *Computer Algebra: Symbolic and Algebraic Computation*, pages 173–187. Springer-Verlag, 1983.

[24] K. Melhorn and S. Näher. *The LEDA Platform of Combinatorial and Geometric Computing*. Cambridge University Press, 1999.

[25] K. Mulmuley. A fast planar partition algorithm, I. *J. Symbolic Comput.*, 10(3-4):253–280, 1990.

[26] S. Schirra. Robustness and precision issues in geometric computation. In J.-R. Sack and J. Urrutia, editors, *Handbook of Computational Geometry*, pages 597–632. Elsevier Science Publishers B. V. North-Holland, Amsterdam, 1999.

[27] R. Wein. High-Level Filtering for Arrangements of Conic Arcs. M.Sc. thesis, Dept. Comput. Sci., Tel Aviv University, Tel Aviv, Israel, 2002 (to appear).

An Approximation Scheme for Cake Division with a Linear Number of Cuts

Gerhard J. Woeginger

University of Twente, The Netherlands

Abstract. In the cake cutting problem, $n \geq 2$ players want to cut a cake into n pieces so that every player gets a 'fair' share of the cake by his own measure.

We prove the following result: For every $\varepsilon > 0$, there exists a cake division scheme for n players that uses at most $c_\varepsilon n$ cuts, and in which each player can enforce to get a share of at least $(1 - \varepsilon)/n$ of the cake according to his own private measure.

1 Introduction

The second paragraph of the poem *"The voice of the lobster"* by Lewis Carrol [2] gives a classical example for the unfair division of a common resource:

> *"I passed by his garden, and marked, with one eye,*
> *How the owl and the panther were sharing a pie.*
> *The panther took pie-crust, and gravy, and meat,*
> *While the owl had the dish as its share of the treat."*

Note that pie-crust, gravy, and meat might be of completely different value to the owl and to the panther. Is there any protocol which enables owl and panther to divide the food into two pieces such that both will get at least half of it by their own measure? The answer to this question is yes, and there is a fairly simple and fairly old solution due to Hugo Steinhaus [8] from 1948: The owl cuts the food into two pieces, and the panther chooses its piece out of the two. The owl is sure to get at least half the food if it cuts two equal pieces by its measure. The panther is sure to get at least half the food by its measure by choosing the better half.

In a more general and a more mathematical formulation, there is a certain resource $\mathcal{C}$ (hereinafter referred to as: *the cake*), and there are n players $1, \ldots, n$. Every player p ($1 \leq p \leq n$) has his own measure μ_p on the subsets of $\mathcal{C}$. These measures satisfy $\mu_p(X) \geq 0$ for all $X \subseteq \mathcal{C}$, and $\mu_p(X) + \mu_p(X') = \mu_p(X \cup X')$ for all disjoint subsets $X, X' \subseteq \mathcal{C}$. For every $X \subseteq \mathcal{C}$ and for every λ with $0 \leq \lambda \leq 1$, there exists a piece $X' \subseteq X$ such that $\mu_p(X') = \lambda \cdot \mu_p(X)$. The cake $\mathcal{C}$ is to be divided among the n players according to some fixed *protocol*, i.e., a step by step interactive procedure that can issue queries to the players whose answers may affect future decisions. We only consider protocols that satisfy the following properties.

- If the participants obey the protocol, then each player ends up with his piece of cake after finitely many steps.
- Each time a player is required to make a cut, he must be able to do this in complete isolation and without interaction of the other players.
- The protocol has no reliable information about the measure μ_p of player p. These measures are considered private information.

The first condition simply keeps every execution of the protocol finite. The second condition does not forbid coalitions, but it protects players from intimidation. Moreover, it eliminates any form of the moving knife procedure (Stromquist [9]). The third condition states that players cannot be trusted to reveal their true preferences. Similar and essentially equivalent conditions are stated in the papers by Woodall [11], Even & Paz [3], and Robertson & Webb [5, 6].

A *strategy* of a player is an adaptive sequence of moves consistent with the protocol. For a real number β with $0 \leq \beta \leq 1$ and some fixed protocol P, a β-*strategy* of a player is a strategy that will guarantee him at least a fraction β of the cake according to his own measure, independently from the strategies of the other $n-1$ players. (So, even if the other $n-1$ players all plot up against the nth player, the nth player in this case will still be able to get a fraction β.) A protocol is called β-*fair*, if every player has a β-strategy. A protocol for n players is called *perfectly fair*, if every player has a $\frac{1}{n}$-strategy.

Even & Paz [3] show that for $n \geq 3$ players, there does not exist a perfectly fair protocol that makes only $n-1$ cuts. Moreover, [3] describe a perfectly fair protocol for $n \geq 3$ players that uses only $n \log_2(n)$ cuts. Tighter results are known for small values of n: For $n = 2$ players, the Steinhaus protocol yields a perfectly fair protocol with a single cut. For $n = 3$ and $n = 4$ players, Even & Paz [3] present perfectly fair protocols that make at most 3 and 4 cuts, respectively. Webb [10] presents a perfectly fair protocol for $n = 5$ players with 6 cuts, and he shows that no perfectly fair protocol exists that uses only 5 cuts. For any $n \geq 2$, Robertson & Webb [6] design $1/(2n-2)$-fair protocols that make only $n-1$ cuts, and they show that this result is best possible for $n-1$ cuts. The result in [6] was rediscovered independently by Krumke et al [4]. For more information on this problem and on many other of its variants, we refer the reader to the books by Brams & Taylor [1] and by Robertson & Webb [7].

The central open problem in this area is whether there exist perfectly fair n-player protocols that only use $O(n)$ cuts. This problem was explicitly formulated by Even & Paz [3], and essentially goes back to Steinhaus [8]. The general belief is that no such protocol exists. We will not settle this problem in this paper, but we will design protocols with $O(n)$ cuts that come arbitrarily close to being $\frac{1}{n}$-fair. Our main result is as follows.

Theorem 1 *For every $\varepsilon > 0$, there exists a constant $c_\varepsilon > 0$ and a cake division scheme for n players such that*

- *each player can enforce to get a share of at least $(1-\varepsilon)/n$ of the cake, and*
- *altogether at most $c_\varepsilon n$ cuts are made.*

This seems to be the strongest possible result one can prove without settling the general question. The protocol is defined and explained in Section 2, and its fairness is analyzed in Section 3.

(S0) If there are $n \leq 2t - 1$ players,
then the cake is divided according to the Even & Paz protocol. STOP.

(S1) Each of the first $2t$ players p ($p = 1, \ldots, 2t$) makes an arbitrary cut c_p through the cake.

(S2) The first $2t$ players are divided into two groups L' and R' with $|L'| = |R'| = t$ such that for every $\ell \in L'$ and for every $r \in R'$ we have $c_\ell \leq c_r$.

(S3) Let $c^* = \max\{c_p : p \in L'\}$.
The cut c^* divides the cake $\mathcal{C}$ into a left piece $\mathcal{C}_L$ and a right piece $\mathcal{C}_R$.

(S4) Every player p in L' chooses an integer x_p with $\lceil n/2 \rceil \leq x_p \leq n$.
Every player p in R' chooses an integer x_p with $0 \leq x_p \leq \lceil n/2 \rceil$.
Every player $p \notin L' \cup R'$ chooses an integer x_p with $0 \leq x_p \leq n$.

(S5) The players are divided into two non-empty groups L and R, such that
(i) $|L| \geq t$ and $|R| \geq t$,
(ii) $x_p \geq |L|$ holds for every player $p \in L$,
(iii) $x_p \leq |L|$ holds for every player $p \in R$.

(S6) The players in L recursively share the left piece $\mathcal{C}_L$.
The players in R recursively share the right piece $\mathcal{C}_R$.

Fig. 1. The protocol $P(t)$ for n players

2 The Protocol

In this section, we define a recursive protocol $P(t)$ that is based on an integer parameter $t \geq 1$. Without loss of generality we assume that the cake $\mathcal{C}$ is the unit interval $[0, 1]$, and that all pieces generated during the execution of the protocol are subintervals of $[0, 1]$. The steps (S0)–(S6) of protocol $P(t)$ are described in Figure 1. The protocol $P(t)$ is based on a divide-and-conquer approach that is similar to that of Even & Paz [3].

Let us start with some simple remarks on $P(t)$. It is irrelevant for our arguments and for our analysis, whether the cuts in step (S1) are done in parallel or sequentially, and whether one player knows or does not know about the cuts of the other players. The same holds for the selection of the numbers x_p in step (S4). If there are two or more feasible partitions $L' \cup R'$ in step (S2), then the protocol selects an arbitrary such partition.

Next, we will discuss the exact implementation of step (S5). Let $y_1 \geq y_2 \geq \cdots \geq y_n$ be an enumeration of the integers $x_1, \ldots, x_n$. Consider the function $g(j) := y_j - j$ for $1 \leq j \leq n$. Since the t integers of players in L' are all greater or equal to $\lceil n/2 \rceil$, we have $y_t \geq \lceil n/2 \rceil \geq t$. Hence, $g(t) \geq 0$. Since the t integers of players in R' are all less or equal to $\lceil n/2 \rceil$, we have $y_{n-t+1} \leq \lceil n/2 \rceil \leq n - t$. Hence, $g(n - t + 1) < 0$. We define the splitting index s as

$$s = \min\{j : g(j) \geq 0 \text{ and } g(j+1) < 0 \text{ and } t \leq j \leq n - t\}.$$

We define the set L of players in such a way that the two multi-sets $\{y_i : 1 \leq i \leq s\}$ and $\{x_p : p \in L\}$ are identical. Moreover, we define $R = \{1, \ldots, n\} - L$. Then $|L| = s \geq t$ and $|R| = n - s \geq t$. Since $y_s \geq s$, we have $x_p \geq |L|$ for every $p \in L$. Since $y_{s+1} < s + 1$, we have $x_p \leq |L|$ for every $p \in R$. Hence, the conditions (i)–(iii) of step (S5) are indeed satisfied by these groups L and R.

In the rest of this section, we prove an upper bound on the number of cuts in the protocol $P(t)$.

Lemma 2 *If the cake is divided among n players according to protocol $P(t)$, then the players altogether make at most $2t \cdot (n-1)$ cuts.*

Proof. By induction on the number n of players. If $n \leq 2t - 1$ is small, protocol $P(t)$ becomes the Even & Paz protocol. Hence, there are at most $n \log_2(n) \leq 2t \cdot (n-1)$ cuts. For $n \geq 2t$, there are $2t$ cuts made in step (S1). Moreover, by the inductive assumption there are at most $2t(|L|-1)$ and at most $2t(|R|-1)$ cuts made in the recursion in step (S6). Altogether, this yields at most $2t(|L| + |R| - 1) = 2t \cdot (n-1)$ cuts. □

3 Proof of Fairness

In this section, we prove that the protocol $P(t)$ is $(1 - \frac{1}{t})$-fair.

Lemma 3 *Let $n \geq 2t$. Then every player p ($p = 1, \ldots, n$) can enforce that at the end of step (S4)*

$$x_p = \lceil (n-1) \cdot \mu_p(\mathcal{C}_L)/\mu_p(\mathcal{C}) \rceil.$$

Proof. The statement trivially holds for the players $p = 2t+1, \ldots, n$, since in step (S4) these players are free to choose x_p arbitrarily between 0 and n. Hence, consider a player p with $p = 1, \ldots, 2t$. We claim that a good strategy for player p is to make his cut c_p in step (S1) in such a way that $\mu_p([0, c_p]) = \mu_p(\mathcal{C}) \cdot \lceil n/2 \rceil / n$. We distinguish two cases.

In the first case, we assume that step (S2) puts player p into group L'. Then $c_p \leq c^*$ and $[0, c_p] \subseteq [0, c^*] = \mathcal{C}_L$. Hence,

$$(n-1) \cdot \frac{\mu_p(\mathcal{C}_L)}{\mu_p(\mathcal{C})} \geq (n-1) \cdot \frac{1}{\mu_p(\mathcal{C})} \cdot \mu_p(\mathcal{C}) \cdot \frac{\lceil n/2 \rceil}{n}$$

$$= (1 - \frac{1}{n})\lceil n/2 \rceil > \lceil n/2 \rceil - 1.$$

Therefore, in this case the value $\lceil (n-1) \cdot \mu_p(\mathcal{C}_L)/\mu_p(\mathcal{C}) \rceil$ is greater or equal to $\lceil n/2 \rceil$, and indeed constitutes a feasible choice for a player p from L' in step (S4).

In the second case, we assume that step (S2) puts player p into group R'. This implies $c_p \leq c^*$ and $\mathcal{C}_L = [0, c^*] \subseteq [0, c_p]$. Hence,

$$(n-1) \cdot \frac{\mu_p(\mathcal{C}_L)}{\mu_p(\mathcal{C})} \leq (n-1) \cdot \frac{1}{\mu_p(\mathcal{C})} \cdot \mu_p(\mathcal{C}) \cdot \frac{\lceil n/2 \rceil}{n}$$

$$= (1 - \frac{1}{n})\lceil n/2 \rceil \leq \lceil n/2 \rceil.$$

In this case the value $\lceil (n-1) \cdot \mu_p(\mathcal{C}_L)/\mu_p(\mathcal{C}) \rceil$ is less or equal to $\lceil n/2 \rceil$, and constitutes a feasible choice for any player from R' in step (S4). □

Lemma 4 *Let $n \geq 1$. Then every player p $(p = 1, \ldots, n)$ can enforce to get at least a fraction $\min\left\{\frac{1}{n}, \frac{t-1}{t(n-1)}\right\}$ of the cake $\mathcal{C}$.*

Proof. Player p behaves according to Lemma 3 and chooses $x_p = \lceil (n-1) \cdot \mu_p(\mathcal{C}_L)/\mu_p(\mathcal{C}) \rceil$. We prove by induction on the number n of players that this ensures him a fraction $\min\left\{\frac{1}{n}, \frac{t-1}{t(n-1)}\right\}$ of the cake $\mathcal{C}$.

For $n \leq 2t - 1$, the statement is trivial since p gets a fraction $1/n$ in step (S0). For the inductive step, we consider $n \geq 2t$ and we distinguish two cases. If in step (S6) the protocol assigns player p to the group L, then by properties (i) and (ii) from step (S5) we have $x_p \geq |L| \geq t$. By the inductive assumption, player p receives at least

$$\min\left\{\frac{1}{|L|}, \frac{t-1}{t(|L|-1)}\right\} \mu_p(\mathcal{C}_L) = \frac{t-1}{t(|L|-1)} \cdot \mu_p(\mathcal{C}_L)$$

$$\geq \frac{t-1}{t(x_p-1)} \cdot \mu_p(\mathcal{C}_L)$$

$$\geq \frac{t-1}{t} \cdot \frac{\mu_p(\mathcal{C})}{(n-1)\mu_p(\mathcal{C}_L)} \cdot \mu_p(\mathcal{C}_L)$$

$$= \frac{t-1}{t(n-1)} \cdot \mu_p(\mathcal{C}).$$

Here the first equation follows from $|L| \geq t$, and the first inequality follows from $x_p \geq |L|$. The second inequality holds since $x_p - 1 \leq (n-1)\mu_p(\mathcal{C}_L)/\mu_p(\mathcal{C})$ by the choice of x_p. This completes the first case.

In the second case, we assume that step (S6) assigns player p to the group R. Then by properties (i) and (iii) from step (S5) we have $|R| \geq t$ and $|L| \geq x_p$. Therefore, $1/|R| \geq (t-1)/t(|R|-1)$. Then by the inductive assumption, player p receives at least

$$\frac{t-1}{t(|R|-1)} \cdot \mu_p(\mathcal{C}_R) \geq \frac{t-1}{t(n-x_p-1)} \cdot \mu_p(\mathcal{C}_R)$$

$$\geq \frac{t-1}{t(n-(n-1)\mu_p(\mathcal{C}_L)/\mu_p(\mathcal{C})-1)} \cdot (\mu_p(\mathcal{C}) - \mu_p(\mathcal{C}_L))$$

$$= \frac{t-1}{t(n-1)} \cdot \mu_p(\mathcal{C}).$$

Here the first inequality follows from $x_p \leq |L| = n - |R|$. The second inequality holds since $x_p \geq (n-1)\mu_p(\mathcal{C}_L)/\mu_p(\mathcal{C})$ by the choice of x_p. This completes the second case, and also the inductive proof. □

Finally, let us prove Theorem 1. We use the protocol $P(t)$ with $t = \lceil 1/\varepsilon \rceil$. By Lemma 2, the total number of cuts is at most $2\lceil \frac{1}{\varepsilon} \rceil \cdot (n-1)$ and hence grows linearly in the number n of players. By Lemma 4, every player may enforce to get at least a fraction

$$\min\left\{\frac{1}{n}, \frac{t-1}{t(n-1)}\right\} \geq \left(1 - \frac{1}{t}\right) \cdot \frac{1}{n} \geq (1-\varepsilon) \cdot \frac{1}{n}$$

of the cake $\mathcal{C}$. This completes the proof of Theorem 1.

References

[1] S. J. BRAMS AND A. D. TAYLOR (1996). *Fair Division – From cake cutting to dispute resolution.* Cambridge University Press, Cambridge.

[2] L. CARROLL (1865). *Alice's Adventures in Wonderland.*

[3] S. EVEN AND A. PAZ (1984). A note on cake cutting. *Discrete Applied Mathematics* 7, 285–296.

[4] S. O. KRUMKE, M. LIPMANN, W. DE PAEPE, D. POENSGEN, J. RAMBAU, L. STOUGIE, AND G. J. WOEGINGER (2002). How to cut a cake almost fairly. *Proceedings of the 13th Annual ACM-SIAM Symposium on Discrete Algorithms (SODA'2002).*

[5] J. M. ROBERTSON AND W. A. WEBB (1991). Minimal number of cuts for fair division. *Ars Combinatoria* 31, 191–197.

[6] J. M. ROBERTSON AND W. A. WEBB (1995). Approximating fair division with a limited number of cuts. *Journal of Combinatorial Theory, Series A* 72, 340–344.

[7] J. M. ROBERTSON AND W. A. WEBB (1998). *Cake-cutting algorithms: Be fair if you can.* A. K. Peters Ltd.

[8] H. STEINHAUS (1948). The problem of fair division. *Econometrica* 16, 101–104.

[9] W. STROMQUIST (1980). How to cut a cake fairly. *American Mathematical Monthly* 87, 640–644.

[10] W. A. WEBB (1997). How to cut a cake fairly using a minimal number of cuts. *Discrete Applied Mathematics* 74, 183–190.

[11] D. R. WOODALL (1980). Dividing a cake fairly. *Journal of Mathematical Analysis and Applications* 78, 233–247.

A Simple Linear Time Algorithm for Finding Even Triangulations of 2-Connected Bipartite Plane Graphs

Huaming Zhang and Xin He [*]

Department of Computer Science and Engineering
SUNY at Buffalo, Amherst, NY, 14260, USA

Abstract. Recently, Hoffmann and Kriegel proved an important combinatorial theorem [4]: Every 2-connected bipartite plane graph G has a triangulation in which all vertices have even degree (it's called an *even triangulation*). Combined with a classical Whitney's Theorem, this result implies that every such a graph has a 3-colorable plane triangulation. Using this result, Hoffmann and Kriegel significantly improved the upper bounds of several *art gallery* and *prison guard* problems. A complicated $O(n^2)$ time algorithm was obtained in [4] for constructing an even triangulation of G. Hoffmann and Kriegel conjectured that there is an $O(n^{3/2})$ algorithm for solving this problem [4].
In this paper, we develop a very simple $O(n)$ time algorithm for solving this problem. Our algorithm is based on thorough study of the structure of all even triangulations of G. We also obtain a simple formula for computing the number of distinct even triangulations of G.

1 Introduction

Let $G = (V, E)$ be a 2-connected bipartite plane graph. A *triangulation* of G is a plane graph obtained from G by adding new edges into the faces of G so that all of its faces are triangles. A triangulation G' of G is called *even* if all vertices of G' have even degree. Recently, Hoffmann and Kriegel proved an important combinatorial theorem [3, 4]:

Theorem 1. *Every 2-connected bipartite plane graph has an even triangulation.*

Combined with the following classical Whitney's Theorem:

Theorem 2. *A plane triangulation is 3-colorable iff all of its vertices have even degree.*

Theorem 1 implies that every 2-connected bipartite plane graph has a 3-colorable plane triangulation. An elegant proof of Theorem 2 can be found in [7].
  In addition to its importance in Graph Theory, Hoffmann and Kriegel showed that, based on Theorem 1, the upper bounds of several *art gallery* and *prison*

---
[*] Research supported in part by NSF Grant CCR-9912418.

guard problems (a group of problems extensively studied in Computational Geometry, see [8, 9, 10]) can be significantly improved [4]. Theorem 1 was proved in [4] by showing that a linear equation system derived from the input graph G has a solution. An even triangulation of G is found by solving this linear equation system. By using the *generalized nested dissection technique* of Lipton, Rose and Tarjan [6], Hoffmann and Kriegel obtained a complicated algorithm with $O(n^2)$ runtime. This algorithm is the bottleneck of the algorithms for solving the art gallery and prison guard application problems discussed in [4].

It was conjectured in [4] that an even triangulation of G can be found in $O(n^{1.5})$ time. In this paper, we show that this problem can be solved by a very simple linear time algorithm. Thus, all algorithms for solving the art gallery and prison guard problems discussed in [4] can be uniformly improved to linear time. Our algorithm is based on newly found properties of 2-connected maximal bipartite plane graphs. We show that an even triangulation of G can be determined from the orientation of the edges of the dual graph of G. As a by-product, we also derive a simple formula for calculating the number of distinct even triangulations of G.

The rest of the paper is organized as follows. In Sect. 2, we introduce the definitions and preliminary results in [4]. In Sect. 3, we prove several theorems needed by our algorithm. In Sect. 4, we describe our algorithm.

2 Preliminaries

In this section, we give definitions and preliminary results. All definitions are standard and can be found in [1]. Let $G = (V, E)$ be a graph with $n = |V|$ vertices and $m = |E|$ edges. The *degree* of a vertex $v \in V$, denoted by $deg(v)$, is the number of edges incident to v. For a subset $V_1 \subseteq V$, $G - V_1$ denotes the graph obtained from G by deleting the vertices in V_1 and their incident edges. A vertex v of a connected graph G is called a *cut vertex* if $G - \{v\}$ is disconnected. G is *2-connected* if it has no cut vertices. $G = (V, E)$ is *bipartite* if its vertex set V can be partitioned into two subsets V_1 and V_2 such that no two vertices in V_1 are adjacent and no two vertices in V_2 are adjacent. A *k-coloring* of G is a coloring of V by k colors so that no two vertices with same color are adjacent to each other. Note that G is bipartite iff it's 2-colorable.

A *cycle* C of G is a sequence of distinct vertices $u_1, u_2, \ldots, u_k$ such that $(u_i, u_{i+1}) \in E$ for $1 \leq i < k$ and $(u_k, u_1) \in E$. We also use C to denote the set of the edges in it. If C contains k vertices, it is a *k-cycle*. A 3-cycle is also called a *triangle*. A *closed walk* is similarly defined except that it allows repeated vertices (but not repeated edges).

A *plane* graph G is a graph embedded in the plane without edge crossings (i.e. an embedded planar graph). The embedding of a plane graph G divides the plane into a number of regions. All regions are called *faces*. The unbounded region is called the *exterior face*. Other regions are called *interior faces*. The *degree* of a face is the number of edges on its boundary.

The dual graph $G^* = (V^*, E^*)$ of a plane graph G is defined as follows: For each face f of G, V^* contains a vertex v_f. For each edge e in G, G^* has an edge $e^* = (v_{f_1}, v_{f_2})$ where f_1 and f_2 are the two faces of G with e on their common boundary. e^* is called the *dual edge* of e. The mapping $e \Leftrightarrow e^*$ is a one-to-one correspondence between E and E^*.

A *diagonal* of a plane G is an edge which does not belong to E and connects two vertices of a facial cycle. G is *triangulated* if it has no diagonals. (Namely all of its facial cycles, including the exterior face, are triangles). A *triangulation* of G is obtained from G by adding a set of diagonals such that the resulting graph is plane and triangulated. An *even triangulation* is a triangulation in which all vertices have even degree.

For a 2-connected bipartite plane graph G, all of its facial cycles have even length. In order to triangulate G, we first add diagonals into the faces of G so that all facial cycles of the resulting graph are 4-cycles. This can be easily done in linear time. Thus, without loss of generality, G always denotes a 2-connected bipartite plane graph all of whose facial cycles have length 4. Such a graph will be called a 2-connected maximal bipartite plane graph (2MBP graph for short). By Euler's formula, a 2MBP graph with n vertices always has $n - 2$ faces and $m = 2n - 4$ edges. We denote the faces of G by $Q(G) = \{q_1, q_2, \ldots, q_{n-2}\}$. When G is clearly understood, we simply use Q to denote $Q(G)$.

Since G is bipartite, we can fix a 2-coloring of G with colors 0 and 1. For any face $q_i \in Q$, the set of the four vertices on the boundary of q_i is denoted by V_{q_i} and we set $Q_v = \{q_i \in Q | v \in V_{q_i}\}$. Since every facial cycle of G is a 4-cycle, every face $q_i \in Q$ has two diagonals: the diagonal joining the two 0-colored (1-colored, respectively) vertices in V_{q_i} is called the 0-diagonal (1-diagonal, respectively). Thus a triangulation of G is nothing but choosing for each face q_i either the 0-diagonal or the 1-diagonal and adding it into q_i. We associate each face $q_i \in Q$ with a $\{0,1\}$-valued variable x_i, and each triangulation T of G with a vector $\boldsymbol{x} = (x_1, x_2, \ldots, x_{n-2}) \in GF(2)^{n-2}$, where: T contains the 0-diagonal of the face q_i iff $x_i = 1$.

This defines a one-to-one mapping between the set of triangulations of G and $GF(2)^{n-2}$. We choose $GF(2)$ because most calculations in this paper are interpreted in $GF(2)$, so all additions should be understood as additions mod 2 unless otherwise specified. We use $\boldsymbol{0}$ to denote the vector $(0, \ldots, 0) \in GF(2)^{n-2}$.

Observation: If $c(v)$ is the color of a vertex $v \in V_{q_i}$, then the term $x_i + c(v)$ (mod 2) describes the increase of the degree of v after adding the diagonal of q_i which corresponds to the value x_i into the face q_i. Based on this observation, Hoffmann and Kriegel [3, 4] proved that a vector $\boldsymbol{x} = (x_i)_{1 \leq i \leq n-2} \in GF(2)^{n-2}$ represents an even triangulation of G iff $\boldsymbol{x}$ is a solution of the following linear equation system (with n equations and $n - 2$ variables) over $GF(2)$:

$$\sum_{q_i \in Q_v} x_i = deg(v) + |Q_v|c(v) \pmod{2} \quad (\forall v \in V). \tag{1}$$

In [4], Hoffmann and Kriegel showed that Equation (1) always has a solution, and hence proved Theorem 1. From now on, the term "a vector $\boldsymbol{x}$ representing an even triangulation of G" and the term "a solution vector of Equation (1)" will be used interchangeably. We will sometimes call it *a solution vector*. For a subset Q_1 of faces, if one diagonal is chosen for each face in Q_1, we will say that this subset of diagonals is a solution if there is a solution of Equation (1) such that the chosen diagonals of the faces in Q_1 appear in that solution.

To obtain all solutions of Equation (1) (i.e. all even triangulations of G), [4] introduced the concept of *straight walk*. For a 2MBP graph G, its dual graph G^* is 4-regular and connected. Consider a walk S in G^*. Since G^* is 4-regular, at every vertex of S, we have four possible choices to continue the walk: go back, turn left, go straight, or turn right. A closed walk of G^* consisting of only straight steps at every vertex is called a *straight walk* or an *S-walk*. The edge set of G^* can be uniquely partitioned into S-walks. We use $\mathcal{S}(G^*) = \{S_1, \ldots, S_k\}$ to denote this partition, where each S_i $(1 \leq i \leq k)$ is an S-walk of G^*. Each vertex of G^* (i.e. each face of G) occurs either twice on one S-walk or on two different S-walks. If a face f occurs on one S-walk twice, it is called a *1-walk face*. If f occurs on two different S-walks, it is called a *2-walk face*.

Figure 1 shows a 2MBP graph and its dual graph G^*. The edges of G are represented by solid lines. The edges of G^* are represented by dotted lines. The vertices of G^* (i.e. the faces of G) are represented by small circles. $\mathcal{S}(G^*)$ contains 3 S-walks, labeled as S_1, S_2 and S_3. The face q_1 is a 1-walk face since it occurs on S_3 twice. The face q_2 is a 2-walk face since it occurs on both S_2 and S_3.

Let S be an S-walk of G^*. Take an even triangulation of G. If we flip the diagonals of every face on S, it's easy to see that we get another even triangulation of G. This idea is formalized in [3] as follows. We associate each S-walk S with a vector $\boldsymbol{S} = (s_1, s_2, \ldots, s_{n-2}) \in GF(2)^{n-2}$ where:

$$s_j = \text{ the number of times the face } q_j \text{ occurs on } S \ . \tag{2}$$

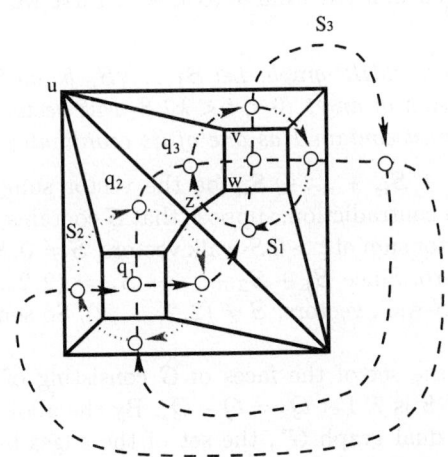

Fig. 1. A 2MBP graph G, the dual graph G^* and its S-walks

We call S the *S-walk vector of S*. In order to explicitly indicate the fact that a face q_i occurs twice on S, we set $s_i = 2$. But whenever addition is involved, the number 2 is always treated as 0. So S can and will be considered as a vector in $GF(2)^{n-2}$.

Let S be an S-walk vector. For each vector $\boldsymbol{a} \in GF(2)^{n-2}$, we define another vector $\boldsymbol{b} = (b_i)_{1 \le i \le (n-2)} = S + \boldsymbol{a} \in GF(2)^{n-2}$, where:

$$b_i = \begin{cases} a_i + 1 \pmod 2 & \text{if } s_i = 1 \\ a_i & \text{if } s_i = 0 \text{ or } 2 \end{cases} \quad (3)$$

By this definition, $S + \boldsymbol{a}$ is precisely the vector addition of S and $\boldsymbol{a}$ over $GF(2)$. The following lemma and theorem were proved in [3, 4]:

Lemma 1. *If the vector $\boldsymbol{a}$ represents an even triangulation of G and S is an S-walk of G^*, then $S + \boldsymbol{a}$ also represents an even triangulation of G.*

Theorem 3. *Let $\boldsymbol{a}$ and $\boldsymbol{b}$ be two vectors in $GF(2)^{n-2}$ representing two even triangulations of G. Then there is a collection $\{S_{i_1}, S_{i_2}, \ldots, S_{i_l}\}$ of S-walks in G^* such that*

$$\boldsymbol{b} = \boldsymbol{a} + S_{i_1} + S_{i_2} + \ldots + S_{i_l} . \quad (4)$$

3 Determine Even Triangulation Based on S-walks

The basic idea of our algorithm is as follows. First we show that if G^* has only one S-walk S, then Equation (1) has a unique solution, which can be easily determined from an orientation of S. If G^* has more than one S-walks, we show that by adding dummy diagonals into some faces of G, we can link all S-walks into a single S-walk, and the unique solution of Equation (1) for the resulting graph can be easily converted to an even triangulation of G. Throughout this section, G denotes a 2MBP graph whose vertices have been $\{0,1\}$-colored and whose faces have been indexed from 1 to $n-2$. First we prove two technical lemmas.

Lemma 2. *Let G be a 2MBP graph. Let $S_1, \ldots, S_k$ be all S-walk vectors of G. Let S be the vector sum of any t $(0 < t < k)$ S-walk vectors **before** the mod 2 operation. Then S must contain 1 as one of its coordinates.*

Proof. Let $S = S_{i_1} + S_{i_2} + \ldots + S_{i_t}$ be the vector sum before the mod 2 operation. Towards a contradiction, suppose that S contains no 1 as coordinates.

Since S is the vector sum of $t > 0$ S-walk vectors, $S \ne \boldsymbol{0}$. So some coordinates of S must be non-zero. Since $S_1 + S_2 + \ldots + S_k = (2, 2, \ldots, 2)$ and S is the vector sum of $t < k$ S-walk vectors, $S \ne (2, 2, \ldots, 2)$. So some coordinates of S must be 0.

Let $Q_1 \subset Q$ be the set of the faces of G consisting of the faces q_i where the ith coordinate of S is 2. Let $Q_2 = Q - Q_1$. By the above argument, $Q_1 \ne \emptyset$ and $Q_2 \ne \emptyset$. In the dual graph G^*, the set of the edges incident to the faces in Q_1 is exactly the set of the edges in the S-walks $S_{i_1}, S_{i_2}, \ldots, S_{i_t}$. Thus in G^*, the faces in Q_1 are not connected to the faces in Q_2. This contradicts the fact that G^* is connected.

Lemma 3. *Let G be a 2MBP graph. Let K be the solution space of the following linear equation system over $GF(2)$:*

$$\sum_{q_i \in Q_v} x_i = 0 \qquad (\forall v \in V) \ . \tag{5}$$

1. *If a is a solution vector of Equation (1), then the set of all solution vectors of Equation (1) is $\{a + S : S \in K\}$.*
2. *If $S(G^*)$ contains k S-walks, then any $k - 1$ S-walk vectors of G^* are linear independent, and form a basis for K.*

Proof. (1) Since Equation (5) is a linear equation system with $n - 2$ variables over $GF(2)$, its solution space K is a linear subspace of $GF(2)^{n-2}$. Since a is a solution vector of Equation (1) and the coefficient matrices of Equation (1) and (5) are identical, the statement 1 is clearly true (see [5]).

(2) Let $S(G^*) = \{S_1, S_2, \ldots, S_k\}$ be all S-walks of G^*. Let S_i ($1 \leq i \leq k$) be the vector corresponding to the S-walk S_i. Define $\mathcal{S} = \{S_1, \ldots, S_k\}$. Let a be a solution vector of Equation (1).

First, we show that any linear combination $S = S_{i_1} + S_{i_2} + \ldots + S_{i_l}$ of the vectors in $\mathcal{S}$ is a solution of Equation (5). By repeated applications of Lemma 1, the vector $a + S$ is also a solution of Equation (1). Thus $S = a + S - a$ is a solution vector of Equation (5).

Second, we show that every solution vector of Equation (5) is a linear combination of the vectors in $\mathcal{S}$. Theorem 3 states that every solution vector of Equation (1) can be written as $a + S$ where S is a linear combination of the vectors in $\mathcal{S}$. By statement (1), every solution vector of Equation (5) can be written as $S = a + S - a$, as to be shown.

Thus the vectors in $\mathcal{S}$ linearly span the subspace K and we can pick a basis for K from $\mathcal{S}$. Note that $\sum_{i=1}^{k} S_i = (2, 2, \ldots, 2) = 0$ over $GF(2)$. So they are linearly dependent.

Next, we show that any $k - 1$ vectors in $\mathcal{S}$ are linearly independent. By symmetry, we only need to show $S_1, S_2, \ldots, S_{k-1}$ are linearly independent.

If $k = 1$, then $S_1 = (2, 2, \ldots, 2) = 0$ over $GF(2)$ and nothing needs to be shown.

Suppose $k > 1$. Take any t vectors from $\{S_1, S_2, \ldots, S_{k-1}\}$ and consider the vector sum $S = S_{i_1} + S_{i_2} + \ldots + S_{i_t}$ **before** the mod 2 operation. By Lemma 2, S must contain 1 as its coordinates. Therefore S cannot be the zero vector 0 after the mod 2 operation. Hence, any subset of the vectors in $\{S_1, S_2, \ldots, S_{k-1}\}$ are linear independent.

Finally, since $S_1 + S_2 + \ldots + S_k = (2, 2, \ldots, 2) = 0$, $S_k = S_1 + S_2 + \ldots + S_{k-1}$ over $GF(2)$. Since any solution vector of Equation (5) can be linearly spanned by $S_1, \ldots, S_{k-1}, S_k$, it can also be linearly spanned by $S_1, \ldots, S_{k-1}$. Thus $\{S_1, \ldots, S_{k-1}\}$ form a basis of K.

The following theorem immediately follows from Lemma 3:

Theorem 4. *Let G be a 2MBP graph. If G^* has k S-walks, then G has 2^{k-1} distinct even triangulations. In particular, G has an unique even triangulation if $k=1$.*

The next lemma is crucial in the development of our algorithm.

Lemma 4. *Let G be a 2MBP graph. A face q_l of G is a 1-walk face iff the solution for x_l in Equation (1) is unique. Namely only one diagonal of q_l occurs in every even triangulation of G.*

Proof. Assume $\mathcal{S}(G^*) = \{S_1, S_2, \ldots, S_k\}$. Suppose the face q_l is a 1-walk face. Without loss of generality, we assume q_l occurs on S_k twice. Let $\boldsymbol{a}$ be a solution vector of Equation (1). By Lemma 3, any other solution vector $\boldsymbol{b}$ of Equation (1) can be written as $\boldsymbol{b} = \boldsymbol{a} + \boldsymbol{S}$, where $\boldsymbol{S}$ is a linear combination of the vectors $\boldsymbol{S}_1, \boldsymbol{S}_2, \ldots, \boldsymbol{S}_{k-1}$. Note that for any t ($1 \leq t < k$), the l'th coordinate of $\boldsymbol{S}_t$ is 0. Thus the lth coordinates of $\boldsymbol{b}$ and $\boldsymbol{a}$ are identical. Hence the solution for x_l in Equation (1) is unique.

Conversely, suppose q_l is a 2-walk face and occurs on two different S-walks S_i and S_j ($i \neq j$). The lth coordinate of $\boldsymbol{S}_i$ is 1. If a vector $\boldsymbol{a}$ is a solution of Equation (1), so is the vector $\boldsymbol{b} = \boldsymbol{a} + \boldsymbol{S}_i$. Clearly, the lth coordinates of $\boldsymbol{a}$ and $\boldsymbol{b}$ differ by 1.

An *orientation* of $\mathcal{S}(G^*)$ assigns each S-walk $S \in \mathcal{S}(G^*)$ a direction. From now on, we fix an **arbitrary** *orientation* $\mathcal{O}$ of $\mathcal{S}(G^*)$.

Definition 1. *If an S-walk S steps out of (into, respectively) a face q through an edge e on the boundary of q, then e is called an out-edge (in-edge, respectively) of face q.*

Clearly each face q of G has two out-edges and two in-edges. Note that, regardless of how the S-walks are directed in the orientation $\mathcal{O}$, the two in-edges of any face q are always incident to a common vertex on the boundary of q. Thus there are always two non-adjacent vertices on the boundary of q that are incident to both in-edges and out-edges of q. We call these two vertices the *primary vertices* and the diagonal connecting them the *primary diagonal* of q. The other diagonal is called the *secondary diagonal* of q. For example, consider the face q_3 in Fig. 1. The edge (u, v) is an out-edge of q_3. The edge (w, z) is an in-edge of q_3. The vertices u and w are incident to both in-edges and out-edges of q_3, so the diagonal (u, w) is the primary diagonal of q_3.

Note that if q is a 1-walk face that occurs on an S-walk S_i twice and if the direction of S_i is reversed, the primary diagonal of q remains unchanged. On the other hand, consider a 2-walk face q that occurs on two S-walks S_i and S_j. If both S_i and S_j reverse direction, the primary diagonal of q remains unchanged. If only one of the two S-walks reverses direction, then the primary and secondary diagonals of q swap.

Definition 2. *Let G be a 2MBP graph and q a face of G. Let $d = (v_1, v_2)$ be a diagonal of q. Add two new edges (v_1, v_0) and (v_0, v_2) into the face q, where v_0 is a new vertex. The resulting graph G' is a 2MBP graph and is called the face splitting of G on face q along diagonal d.*

Theorem 5. *Let G be a 2MBP graph and q a 1-walk face of G. Then, the primary diagonal of q is the unique diagonal that occurs in every even triangulation of G.*

Proof. Suppose the face q occurs on the S-walk S twice. We trace S following its direction in the orientation $\mathcal{O}$. S starts at q and goes out of q through an out-edge e_1 of q (see Fig. 2 (a)). It travels a section P_1 of S and comes back to q through an in-edge e_2 of q, which must be incident to e_1. (Otherwise, S ends at q and cannot pass through q twice). Let v_1 be the common vertex of e_1 and e_2. Then S goes out of q through another out-edge e_3 which is opposite to e_2. It travels through another section P_2 of S and comes back to q through another in-edge e_4 which is opposite to e_1. Let v_2 be the common vertex of e_3 and e_4. Note that $d = (v_1, v_2)$ is the primary diagonal of q. Let u_1 and u_2 be the other two vertices on the boundary of q (see Fig. 2 (a)).

We want to show that the primary diagonal d is always the diagonal of q in any even triangulation of G. Consider the graph G' obtained from G by splitting the face q along d. Call the new vertex v_0. Let q_1 and q_2 be the two subfaces of G' obtained from splitting q. We travel S, following its original direction, in the new graph G', starting at the subface q_1 through the out-edge e_1 (see Fig. 2 (b)). The S-walk travels through P_1 and comes back through e_2 and enters the other subface q_2. Then it travels through the edge (v_2, v_0) and enters q_1, completing an S-walk S_1 of G'^*. Similarly, if we start S at q_1 going through the out-edge e_3, the S-walk will go through P_2, e_4 and (v_0, v_1), completing another S-walk S_2 of G'^*.

Thus splitting the face q along its primary diagonal d also splits the S-walk S into two S-walks of G'^*. All other S-walks of G'^* are identical to the corresponding S-walks of G^*. The two subfaces q_1 and q_2 occur on both S_1 and S_2. Consider any even triangulation T' of G'. There are four cases:

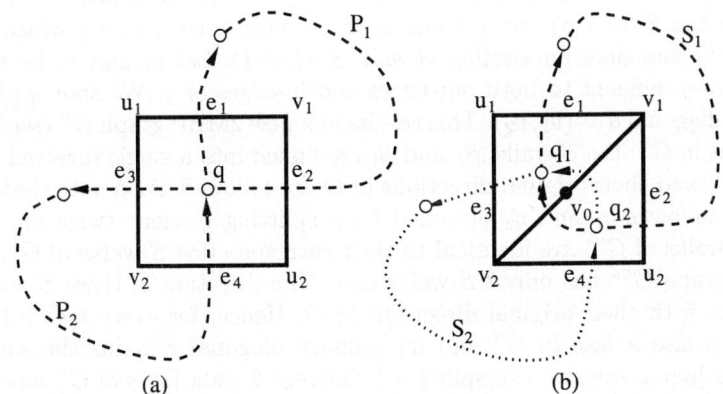

Fig. 2. (a) S passes through the face q twice; (b) S is split into two S-walks of G'^*

Case 1: The diagonal $d_1 = (v_1, v_2)$ of q_1 and the diagonal $d_2 = (v_1, v_2)$ of q_2 are in T'. For every face $p \neq q$, we select the diagonal of p that is in T'. For the face q, we chose the primary diagonal $d = (v_1, v_2)$. This gives an even triangulation of G.

Case 2. The diagonal $c_1 = (v_0, u_1)$ of q_1 and the diagonal $c_2 = (v_0, u_2)$ of q_2 are in T'. By Lemma 1, if we flip the diagonals of all faces of G' that are on the S-walk S_1, we get another even triangulation T'' of G'. Clearly, both diagonals d_1 and d_2 are in T''. This turns us back to Case 1.

Case 3: The diagonal $d_1 = (v_1, v_2)$ of q_1 and the diagonal $c_2 = (v_0, u_2)$ of q_2 are in T'. But the degree of the vertex v_0 in the resulting graph is 3. So this case is impossible.

Case 4: The diagonal $c_1 = (v_0, u_1)$ of q_1 and the diagonal $d_2 = (v_1, v_2)$ of q_2 are in T'. Similar to Case 3, this case is impossible.

Thus in every case, we can convert T' to an even triangulation of G that contains the primary diagonal of the face q. Since q is a 1-walk face, by Lemma 4, its primary diagonal occurs in every even triangulation of G.

Thus if G has only $k = 1$ S-walk, we can easily find its unique even triangulation by choosing the primary diagonal of every face. The next theorem shows that, if G has $k > 1$ S-walks, we can split some 2-walk faces of G along their primary diagonals and these split operations link the k S-walks of G into a single S-walk.

Theorem 6. *Let G be a 2MBP graph with a fixed orientation $\mathcal{O}$ of $\mathcal{S}(G^*)$. Let $\mathcal{S}(G^*) = \{S_1, S_2, \ldots, S_k\}$ be its directed S-walks. Then we can split $k - 1$ different 2-walk faces of G, each along its primary diagonal, such that the resulting graph G' has only one S-walk S, which links $S_1, S_2, \ldots, S_k$ following their original directions in the orientation $\mathcal{O}$.*

Proof. By induction on k. If $k = 1$, the theorem is trivially true.

Assume that the theorem is true for $k = t \geq 1$. Suppose G has $t + 1$ S-walks. Consider the S-walk S_1. By Lemma 2, there must exist a face q which occurs once on S_1 and once on another S-walk S_i ($i \neq 1$). Let v_1 and v_2 be the two vertices of q incident to both out-edges and in-edges of q. We split q along its primary diagonal $d = (v_1, v_2)$. This results in a new 2MBP graph G'' (see Fig. 3). Note that in G'', the S-walks S_1 and S_i are linked into a single directed S-walk S'' of G''^* and their original directions in $\mathcal{O}$ are followed. Also note that in G'', the two subfaces q_1 and q_2 obtained from splitting q occur twice on S''. All other S-walks of G''^* are identical to their corresponding S-walks of G^*.

The graph G''^* has only t S-walks now. The direction of these S-walks are consistent with their original directions in $\mathcal{O}$. Hence, for every face r ($r \neq q$) of G, r is also a face in G'' and its primary diagonal remains the same. By induction hypothesis, we can split $t - 1$ different 2-walk faces of G'' along their primary diagonals and these operations transform G'' into G' with a single S-walk, following the original directions in $\mathcal{O}$. Note that if a face q of G has been split, then in the resulting graph G'', their subfaces q_1 and q_2 become 1-walk

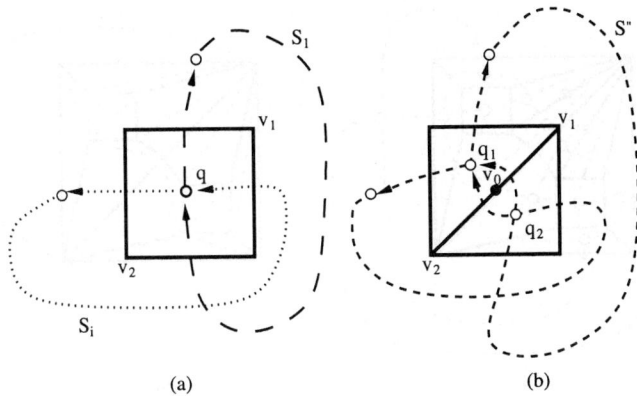

Fig. 3. (a) The S-walks S_1 and S_i pass through the face q; (b) S_1 and S_i are linked into a single directed S-walk of G'''^*

faces and hence will not be split again. Thus all the faces to be split in this process are different 2-walk faces of G. This completes the induction.

We now can prove our main theorem:

Theorem 7. *Let G be a 2MBP graph with a fixed orientation $\mathcal{O}$ of $\mathcal{S}(G^*)$. Adding the primary diagonals of all faces into G results in an even triangulation of G.*

Proof. By Theorem 6, we can split a set Q_1 consisting of $k - 1$ different 2-walk faces of G, each along their primary diagonals. This results in a 2MBP graph G' with a single S-walk, and the directions of the S-walks in the original orientation $\mathcal{O}$ of $\mathcal{S}(G^*)$ are followed. By Theorem 5, we can obtain the unique even triangulation T' of G' by choosing the primary diagonals of all faces of G'. For each face q of G that is not in Q_1, it is also a face of G'. The primary diagonal of q remains the same in G and G'. Consider a face $q \in Q_1$ of G. It is split along its primary diagonal $d = (v_1, v_2)$ into two subfaces q_1 and q_2 in the process of creating G'. Note that $d_1 = (v_1, v_2)$ is the primary diagonal of q_1 and $d_2 = (v_1, v_2)$ is the primary diagonal of q_2 in G' (see Fig. 3 (b)). Hence both d_1 and d_2 appear in T'. Now we delete the edges (v_1, v_0), (v_0, v_2) that were added into G to split q, delete d_1 and d_2, and add the diagonal $d = (v_1, v_2)$. The degree parities of v_1 and v_2 do not change. We do this for every face $q \in Q_1$. Then the even triangulation T' of G' is converted to an even triangulation of G consisting of primary diagonals of all faces of G.

Figure 4 shows the determination of the even triangulation of the graph shown in Fig. 1 by using Theorem 7. Figure 4 (a) shows the graph G' after the 2-walk faces q_2 and q_3 are split along their primary diagonals. Note that G'^* has only one S-walk whose direction is consistent with the directions of the S-walks of G^* shown in Fig. 1. Figure 4 (b) shows an even triangulation of G by adding the primary diagonals of all faces into G.

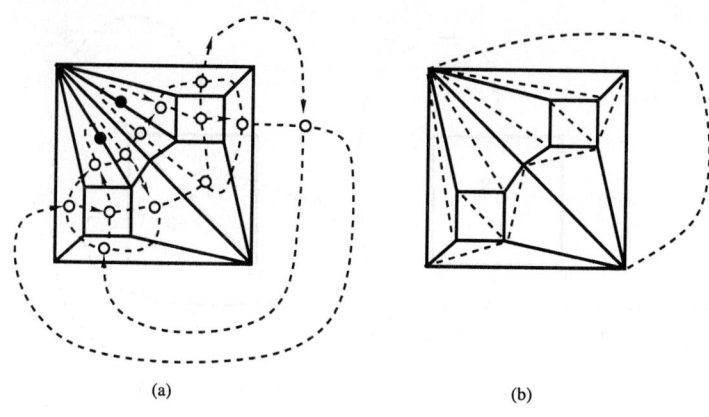

Fig. 4. Determining an even triangulation of the graph G shown in Fig. 1

Remark 1. We **do not have to** split the 2-walk face q along its primary diagonal as shown in Fig. 3. If q is split along another diagonal, the two S-walks S_1 and S_i will still be linked into a single directed S-walk, where the direction of S_1 remains the same and the direction of S_i is reversed. Continue this process $k - 1$ times, we will still get a (different) 2MBP graph G'' with a single S-walk. This will lead to a different even triangulation of G. Therefore, for each of the $k - 1$ split faces of G, we can split it along either of its two diagonals. Thus there are 2^{k-1} choices for splitting these $k - 1$ faces. Since each choice results in a different even triangulation of G, there are 2^{k-1} even triangulations of G. Again, this is precisely the number stated in Theorem 4.

4 Implementation and Analysis

Based on the discussion in Sect. 3, we obtain the following algorithm for finding even triangulations of G.

Algorithm 1: Even Triangulation
Input: A 2-connected bipartite plane graph G.
Output: An even triangulation of G, represented by a vector $(x_i)_{1 \leq i \leq n-2}$.

1. Add diagonals into G, if necessary, to make all facial cycles length 4. For simplicity, we still use G to denote the resulting graph.
2. Color the vertices of G by 0 and 1.
3. Construct the dual graph G^*.
4. Travel the edges of G^* to construct the S-walks of G^*. Each S-walk is assigned a direction when it is being traveled.
5. For each face q_i of G, identify its in-edges, out-edges and the primary diagonal. Pick the primary diagonal and assign the value to x_i accordingly.

We have:

Theorem 8. *Given a 2-connected bipartite plane graph G, Algorithm 1 finds an even triangulation of G in $O(n)$ time.*

Proof. The correctness of Algorithm 1 directly follows from Theorem 7. All steps of the algorithm can be easily implemented by using elementary graph algorithm techniques (for example see [2]), each in $O(n)$ time.

Remark 2. The $\{0,1\}$-coloring of G allows us to represents even triangulations of G by vectors $\boldsymbol{x} \in GF(2)^{n-2}$. This idea played an important role in the development of our algorithm. However, it is not essential in algorithm implementation. The even triangulation of G is obtained by choosing primary diagonals, which are completely defined by the orientation of the S-walks, and are independent from the $\{0,1\}$-coloring of G. Instead of using vector $\boldsymbol{x}$, we can specify the even triangulation of G by output the primary diagonals of the faces of G. In this case, the Step 2 of Algorithm 1 is not needed.

Remark 3. If we want to be able to generate all even triangulations of G, we only need $O(kn)$ time, where k is the number of the S-walks of G^*: It takes $O(n)$ time to find the vector $\boldsymbol{x}$ for the first solution. Writing down the S-walk vectors of G^* takes $O(k*n)$ time. All even triangulations of G can be generated from $\boldsymbol{x}$ and these S-walk vectors.

References

[1] J. A. Bondy and U. S. R. Murty, *Graph theory with applications*, North Holland, New York, 1979.
[2] T. H. Cormen, C. E. Leiserson and R. L. Rivest, *An introduction to algorithms*, McGraw-Hill, New York, 1990
[3] F. Hoffmann and K. Kriegel, A graph-coloring result and its consequences for polygon-guarding problems, Technical Report TR-B-93-08, Inst. f. Informatik, Freie Universität, Berlin, 1993.
[4] F. Hoffmann and K. Kriegel, A graph-coloring result and its consequences for polygon-guarding problems, *Siam J. Discrete Math*, Vol 9(2): 210-224, 1996.
[5] N. Jacobson, *Lectures in Abstract Algebras*, Springer-Verlag, New York, 1975.
[6] R. J. Lipton, J. Rose and R. E. Tarjan, Generalized Nested Dissection, *Siam J. Numer. Anal.*, 16: 346-358, 1979.
[7] L. Lovász, *Combinatorial Problems and Exercises*, North Holland, Amsterdam, 1979.
[8] J. O'Rourke, *Art Gallery Theorems and Algorithms*, Oxford University Press, Ne York, 1987.
[9] J. O'Rourke, Computational Geometry Column 15, *SIGACT News*, 23: 26-28, 1992.
[10] *Handbook of computational geometry*, edited by J. R. Sack and J. Urrutia, Elsevier, New York, 2000.

Author Index

Agarwal, Pankaj K. 5, 17, 29, 42, 54
Ahr, Dino 64
Althaus, Ernst 75
Arge, Lars 88

Baier, Georg 101
Barkan, Arye 114
Barrett, Chris 126
Bender, Michael A. ...139, 152, 165
Berberich, Eric 174
Berg, Mark de 187
Berman, Piotr 200
Bernstein, Joel 436
Bisset, Keith 126
Bockmayr, Alexander 75
Bodlaender, Hans L. 211
Bonis, Annalisa De 335
Bontridder, Koen M.J. De 223
Bose, Prosenjit 234
Brandes, Ulrik 247
Broersma, Hajo 211
Buchsbaum, Adam L. 257

Chen, Danny Z. 270, 284
Cohen, Edith 297
Cole, Richard 139, 152
Cook, William 1

Daneshmand, Siavash Vahdati . 795
Datar, Mayur 310, 323
Demaine, Erik D. 139, 152, 165, 348
Deshmukh, Kaustubh 361
Dessmark, Anders 374
Dey, Tamal K. 2, 387
Doerr, Benjamin 399

Efrat, Alon 411
Eigenwillig, Arno 174
Elbassioni, Khaled M. 424
Elf, Matthias 75
Enosh, Angela 436
Epstein, Leah 449, 461
Ezra, Eti 473

Fahle, Torsten 485
Farach-Colton, Martin 139, 152, 165
Ferragina, Paolo 698
Fiat, Amos 499
Fomin, Fedor V. 211

Gao, Jie 5
Gentile, Claudio 525
Goldberg, Andrew V. 361
Goodrich, Michael T. 257
Govindarajan, Sathish 17
Gudmundsson, Joachim ... 187, 234
Guibas, Leonidas J. 5
Gąsieniec, Leszek 512

Hagerup, Torben 42
Halperin, Dan 473
Hannenhalli, Sridhar 200
Har-Peled, Sariel 29
Hartline, Jason D. 361
Haus, Utz-Uwe 525
Haxell, Penny 538
He, Xin 902
Heggernes, Pinar 550
Hemmer, Michael 174
Hert, Susan 174
Hu, Xiaobo S. 270

Jacob, Riko 126
Jansen, Klaus 562
Jünger, Michael 75

Kaplan, Haim 114, 297
Kaporis, Alexis C. 574
Karavelas, Menelaos I. 586
Karlin, Anna R. 3, 361
Karpinski, Marek 200
Kasper, Thomas 75
Katz, Matthew J. 187
Kedem, Klara 436
Kirousis, Lefteris M. 574
Kliewer, Georg 845
Koberstein, Achim 845

Author Index

Kobourov, Stephen G. 411
Köhler, Ekkehard 101, 599
Köppe, Matthias 525
Kolliopoulos, Stavros G. 612
Konjevod, Goran 126
Konstantinou, Elisavet 625
Krumke, Sven O. 637

Laber, Eduardo 649
Lageweg, B.J. 223
Lalas, Efthimios G. 574
Langerman, Stefan 662
Langkau, Katharina 599
Lenstra, Jan K. 223
Levcopoulos, Christos 187
Lipmann, Maarten 674
López-Ortiz, Alejandro 348
Loryś, Krzysztof 686
Lu, X. 674
Luan, Shuang 270
Lubiw, Anna 411

Manzini, Giovanni 698
Marathe, Madhav 126
Maxová, Jana 711
Mehlhorn, Kurt 75, 174, 723
Mendel, Manor 499
Meyer, Ulrich 723
Molloy, Michael 736
Morin, Pat 662
Munro, J. Ian 348
Mustafa, Nabil H. 29
Muthukrishnan, S. 17, 323

Näher, Stefan 748
Nešetřil, Jaroslav 711
Nielsen, Benny K. 760

Ohta, Kenichiro 772
Orlin, James B. 223
Overmars, Mark H. 187

Paepe, Willem E. de 674
Pagourtzis, Aris 512
Parekh, Ojas 649
Peczarski, Marcin 785
Pelc, Andrzej 374
Polzin, Tobias 795
Potapov, Igor 512

Poutré, Han La 872
Procopiuc, Cecilia M. 54
Procopiuc, Octavian 88
Pruhs, Kirk 808
Pyatkin, Artem V. 211

Räcke, Harald 820
Rambau, Jörg 637
Rasala, April 538
Ravi, R. 649
Ray, Rahul 42
Reinelt, Gerhard 64
Rinaldi, Giovanni 525

Sadakane, Kunihiko 772
Salavatipour, Mohammad R. ...736
Schnieder, Henning 399
Schömer, Elmar 174
Schurr, Ingo 833
Seiden, Steven S. 499
Sellmann, Meinolf 845
Shachnai, Hadas 859
Sharir, Micha 42, 473
Shioura, Akiyoshi 772
Sitters, Rene A. 674
Skutella, Martin 101, 599
Smid, Michiel 42, 234, 284
Sohler, Christian 820
Stamatiou, Yiannis 625
Stappen, A. Frank van der 187
Stee, Rob van 449, 872
Steiner, George 612
Stougie, Leen 223, 674
Szabó, Tibor 833

Tamir, Tami 859
Tassa, Tamir 461
Thorup, Mikkel 4
Tokuyama, Takeshi 772
Torres, Luis M. 637

Uthaisombut, Patchrawat 808

Vaccaro, Ugo 335
Varadarajan, Kasturi R. 54
Villanger, Yngve 550
Vitter, Jeffrey Scott 88

Wang, Yusu 29
Wein, Ron 884

Weismantel, Robert 525
Welzl, Emo 42
Westermann, Matthias 820
Wilfong, Gordon 538
Winkler, Peter 538
Winter, Pawel 760
Woeginger, Gerhard J. 211, 859, 896
Wu, Xiaodong 270

Xu, Bin 284

Yu, Cedric X. 270
Yvinec, Mariette 586
Zachariasen, Martin 760
Zaroliagis, Christos 625
Zhang, Huaming 902
Zhao, Wulue 387
Zito, Jack 152
Zlotowski, Oliver 748
Zwoźniak, Grażyna 686

Lecture Notes in Computer Science

For information about Vols. 1–2387
please contact your bookseller or Springer-Verlag

Vol. 2388: S.-W. Lee, A. Verri (Eds.), Pattern Recognition with Support Vector Machines. Proceedings, 2002. XI, 420 pages. 2002.

Vol. 2389: E. Ranchhod, N.J. Mamede (Eds.), Advances in Natural Language Processing. Proceedings, 2002. XII, 275 pages. 2002. (Subseries LNAI).

Vol. 2390: D. Blostein, Y.-B. Kwon (Eds.), Graphics Recognition. Proceedings, 2001. XI, 367 pages. 2002.

Vol. 2391: L.-H. Eriksson, P.A. Lindsay (Eds.), FME 2002: Formal Methods – Getting IT Right. Proceedings, 2002. XI, 625 pages. 2002.

Vol. 2392: A. Voronkov (Ed.), Automated Deduction – CADE-18. Proceedings, 2002. XII, 534 pages. 2002. (Subseries LNAI).

Vol. 2393: U. Priss, D. Corbett, G. Angelova (Eds.), Conceptual Structures: Integration and Interfaces. Proceedings, 2002. XI, 397 pages. 2002. (Subseries LNAI).

Vol. 2394: P. Perner (Ed.), Advances in Data Mining. VII, 109 pages. 2002. (Subseries LNAI).

Vol. 2395: G. Barthe, P. Dybjer, L. Pinto, J. Saraiva (Eds.), Applied Semantics. IX, 537 pages. 2002.

Vol. 2396: T. Caelli, A. Amin, R.P.W. Duin, M. Kamel, D. de Ridder (Eds.), Structural, Syntactic, and Statistical Pattern Recognition. Proceedings, 2002. XVI, 863 pages. 2002.

Vol. 2397: G. Grahne, G. Ghelli (Eds.), Database Programming Languages. Proceedings, 2001. X, 343 pages. 2002.

Vol. 2398: K. Miesenberger, J. Klaus, W. Zagler (Eds.), Computers Helping People with Special Needs. Proceedings, 2002. XXII, 794 pages. 2002.

Vol. 2399: H. Hermanns, R. Segala (Eds.), Process Algebra and Probabilistic Methods. Proceedings, 2002. X, 215 pages. 2002.

Vol. 2400: B. Monien, R. Feldmann (Eds.), Euro-Par 2002 – Parallel Processing. Proceedings, 2002. XXIX, 993 pages. 2002.

Vol. 2401: P.J. Stuckey (Ed.), Logic Programming. Proceedings, 2002. XI, 486 pages. 2002.

Vol. 2402: W. Chang (Ed.), Advanced Internet Services and Applications. Proceedings, 2002. XI, 307 pages. 2002.

Vol. 2403: Mark d'Inverno, M. Luck, M. Fisher, C. Preist (Eds.), Foundations and Applications of Multi-Agent Systems. Proceedings, 1996-2000. X, 261 pages. 2002. (Subseries LNAI).

Vol. 2404: E. Brinksma, K.G. Larsen (Eds.), Computer Aided Verification. Proceedings, 2002. XIII, 626 pages. 2002.

Vol. 2405: B. Eaglestone, S. North, A. Poulovassilis (Eds.), Advances in Databases. Proceedings, 2002. XII, 199 pages. 2002.

Vol. 2406: C. Peters, M. Braschler, J. Gonzalo, M. Kluck (Eds.), Evaluation of Cross-Language Information Retrieval Systems. Proceedings, 2001. X, 601 pages. 2002.

Vol. 2407: A.C. Kakas, F. Sadri (Eds.), Computational Logic: Logic Programming and Beyond. Part I. XII, 678 pages. 2002. (Subseries LNAI).

Vol. 2408: A.C. Kakas, F. Sadri (Eds.), Computational Logic: Logic Programming and Beyond. Part II. XII, 628 pages. 2002. (Subseries LNAI).

Vol. 2409: D.M. Mount, C. Stein (Eds.), Algorithm Engineering and Experiments. Proceedings, 2002. VIII, 207 pages. 2002.

Vol. 2410: V.A. Carreño, C.A. Muñoz, S. Tahar (Eds.), Theorem Proving in Higher Order Logics. Proceedings, 2002. X, 349 pages. 2002.

Vol. 2412: H. Yin, N. Allinson, R. Freeman, J. Keane, S. Hubbard (Eds.), Intelligent Data Engineering and Automated Learning – IDEAL 2002. Proceedings, 2002. XV, 597 pages. 2002.

Vol. 2413: K. Kuwabara, J. Lee (Eds.), Intelligent Agents and Multi-Agent Systems. Proceedings, 2002. X, 221 pages. 2002. (Subseries LNAI).

Vol. 2414: F. Mattern, M. Naghshineh (Eds.), Pervasive Computing. Proceedings, 2002. XI, 298 pages. 2002.

Vol. 2415: J.R. Dorronsoro (Ed.), Artificial Neural Networks – ICANN 2002. Proceedings, 2002. XXVIII, 1382 pages. 2002.

Vol. 2416: S. Craw, A. Preece (Eds.), Advances in Case-Based Reasoning. Proceedings, 2002. XII, 656 pages. 2002. (Subseries LNAI).

Vol. 2417: M. Ishizuka, A. Sattar (Eds.), PRICAI 2002: Trends in Artificial Intelligence. Proceedings, 2002. XX, 623 pages. 2002. (Subseries LNAI).

Vol. 2418: D. Wells, L. Williams (Eds.), Extreme Programming and Agile Methods – XP/Agile Universe 2002. Proceedings, 2002. XII, 292 pages. 2002.

Vol. 2419: X. Meng, J. Su, Y. Wang (Eds.), Advances in Web-Age Information Management. Proceedings, 2002. XV, 446 pages. 2002.

Vol. 2420: K. Diks, W. Rytter (Eds.), Mathematical Foundations of Computer Science 2002. Proceedings, 2002. XII, 652 pages. 2002.

Vol. 2421: L. Brim, P. Jančar, M. Křetínský, A. Kučera (Eds.), CONCUR 2002 – Concurrency Theory. Proceedings, 2002. XII, 611 pages. 2002.

Vol. 2422: H. Kirchner, Ch. Ringeissen (Eds.), Algebraic Methodology and Software Technology. Proceedings, 2002. XI, 503 pages. 2002.

Vol. 2423: D. Lopresti, J. Hu, R. Kashi (Eds.), Document Analysis Systems V. Proceedings, 2002. XIII, 570 pages. 2002.

Vol. 2424: S. Flesca, G. Ianni (Eds.), Logics in Artificial Intelligence. Proceedings, 2002. XIII, 572 pages. 2002. (Subseries LNAI).

Vol. 2425: Z. Bellahsène, D. Patel, C. Rolland (Eds.), Object-Oriented Information Systems. Proceedings, 2002. XIII, 550 pages. 2002.

Vol. 2426: J.-M. Bruel, Z. Bellahsène (Eds.), Advances in Object-Oriented Information Systems.Proceedings, 2002. IX, 314 pages. 2002.

Vol. 2430: T. Elomaa, H. Mannila, H. Toivonen (Eds.), Machine Learning: ECML 2002. Proceedings, 2002. XIII, 532 pages. 2002. (Subseries LNAI).

Vol. 2431: T. Elomaa, H. Mannila, H. Toivonen (Eds.), Principles of Data Mining and Knowledge Discovery. Proceedings, 2002. XIV, 514 pages. 2002. (Subseries LNAI).

Vol. 2432: R. Bergmann, Experience Management. XXI, 393 pages. 2002. (Subseries LNAI).

Vol. 2434: S. Anderson, S. Bologna, M. Felici (Eds.), Computer Safety, Reliability and Security. Proceedings, 2002. XX, 347 pages. 2002.

Vol. 2435: Y. Manolopoulos, P. Návrat (Eds.), Advances in Databases and Information Systems. Proceedings, 2002. XIII, 415 pages. 2002.

Vol. 2436: J. Fong, C.T. Cheung, H.V. Leong, Q. Li (Eds.), Advances in Web-Based Learning. Proceedings, 2002. XIII, 434 pages. 2002.

Vol. 2438: M. Glesner, P. Zipf, M. Renovell (Eds.), Field-Programmable Logic and Applications. Proceedings, 2002. XXII, 1187 pages. 2002.

Vol. 2439: J.J. Merelo Guervós, P. Adamidis, H.-G. Beyer, J.-L. Fernández-Villacañas, H.-P. Schwefel (Eds.), Parallel Problem Solving from Nature – PPSN VII. Proceedings, 2002. XXII, 947 pages. 2002.

Vol. 2440: J.M. Haake, J.A. Pino (Eds.), Groupware: Design, Implementation and Use. Proceedings, 2002. XII, 285 pages. 2002.

Vol. 2441: Z. Hu, M. Rodríguez-Artalejo (Eds.), Functional and Logic Programming. Proceedings, 2002. X, 305 pages. 2002.

Vol. 2442: M. Yung (Ed.), Advances in Cryptology – CRYPTO 2002. Proceedings, 2002. XIV, 627 pages. 2002.

Vol. 2443: D. Scott (Ed.), Artificial Intelligence: Methodology, Systems, and Applications. Proceedings, 2002. X, 279 pages. 2002. (Subseries LNAI).

Vol. 2444: A. Buchmann, F. Casati, L. Fiege, M.-C. Hsu, M.-C. Shan (Eds.), Technologies for E-Services. Proceedings, 2002. X, 171 pages. 2002.

Vol. 2445: C. Anagnostopoulou, M. Ferrand, A. Smaill (Eds.), Music and Artificial Intelligence. Proceedings, 2002. VIII, 207 pages. 2002. (Subseries LNAI).

Vol. 2446: M. Klusch, S. Ossowski, O. Shehory (Eds.), Cooperative Information Agents VI. Proceedings, 2002. XI, 321 pages. 2002. (Subseries LNAI).

Vol. 2447: D.J. Hand, N.M. Adams, R.J. Bolton (Eds.), Pattern Detection and Discovery. Proceedings, 2002. XII, 227 pages. 2002. (Subseries LNAI).

Vol. 2448: P. Sojka, I. Kopeček, K. Pala (Eds.), Text, Speech and Dialogue. Proceedings, 2002. XII, 481 pages. 2002. (Subseries LNAI).

Vol. 2449: L. Van Gool (Ed.), Pattern Recognotion. Proceedings, 2002. XVI, 628 pages. 2002.

Vol. 2451: B. Hochet, A.J. Acosta, M.J. Bellido (Eds.), Integrated Circuit Design. Proceedings, 2002. XVI, 496 pages. 2002.

Vol. 2452: R. Guigó, D. Gusfield (Eds.), Algorithms in Bioinformatics. Proceedings, 2002. X, 554 pages. 2002.

Vol. 2453: A. Hameurlain, R. Cicchetti, R. Traunmüller (Eds.), Database and Expert Systems Applications. Proceedings, 2002. XVIII, 951 pages. 2002.

Vol. 2454: Y. Kambayashi, W. Winiwarter, M. Arikawa (Eds.), Data Warehousing and Knowledge Discovery. Proceedings, 2002. XIII, 339 pages. 2002.

Vol. 2455: K. Bauknecht, A M. Tjoa, G. Quirchmayr (Eds.), E-Commerce and Web Technologies. Proceedings, 2002. XIV, 414 pages. 2002.

Vol. 2456: R. Traunmüller, K. Lenk (Eds.), Electronic Government. Proceedings, 2002. XIII, 486 pages. 2002.

Vol. 2458: M. Agosti, C. Thanos (Eds.), Research and Advanced Technology for Digital Libraries. Proceedings, 2002. XVI, 664 pages. 2002.

Vol. 2459: M.C. Calzarossa, S. Tucci (Eds.), Performance Evaluation of Complex Systems: Techniques and Tools. Proceedings, 2002. VIII, 501 pages. 2002.

Vol. 2461: R. Möhring, R. Raman (Eds.), Algorithms – ESA 2002. Proceedings, 2002. XIV, 917 pages. 2002.

Vol. 2462: K. Jansen, S. Leonardi, V. Vazirani (Eds.), Approximation Algorithms for Combinatorial Optimization. Proceedings, 2002. VIII, 271 pages. 2002.

Vol. 2463: M. Dorigo, G. Di Caro, M. Sampels (Eds.), Ant Algorithms. Proceedings, 2002. XIII, 305 pages. 2002.

Vol. 2464: M. O'Neill, R.F.E. Sutcliffe, C. Ryan, M. Eaton, N. Griffith (Eds.), Artificial Intelligence and Cognitive Science. Proceedings, 2002. XI, 247 pages. 2002. (Subseries LNAI).

Vol. 2469: W. Damm, E.-R. Olderog (Eds.), Formal Techniques in Real-Time and Fault-Tolerant Systems. Proceedings, 2002. X, 455 pages. 2002.

Vol. 2470: P. Van Hentenryck (Ed.), Principles and Practice of Constraint Programming – CP 2002. Proceedings, 2002. XVI, 794 pages. 2002.

Vol. 2471: J. Bradfield (Ed.), Computer Science Logic. Proceedings, 2002. XII, 613 pages. 2002.

Vol. 2476: A.H.F. Laender, A.L. Oliveira (Eds.), String Processing and Information Retrieval. Proceedings, 2002. XI, 337 pages. 2002.

Vol. 2479: M. Jarke, J. Koehler, G. Lakemeyer (Eds.), KI 2002: Advances in Artificial Intelligence. Proceedings, 2002. XIII, 327 pages. (Subseries LNAI).

Vol. 2480: Y. Han, S. Tai, D. Wikarski (Eds.), Engineering and Deployment of Cooperative Information Systems. Proceedings, 2002. XIII, 564 pages. 2002.

Vol. 2483: J.D.P. Rolim, S. Vadhan (Eds.), Randomization and Approximation Techniques in Computer Science. Proceedings, 2002. VIII, 275 pages. 2002.